S0-FIV-279

Phospholipids Handbook

Phospholipids Handbook

edited by

Gregor Cevc

Technical University of Munich
Munich, Germany

Marcel Dekker, Inc. New York • Basel • Hong Kong

Library of Congress Cataloging-in-Publication Data

Phospholipids handbook / edited by Gregor Cevc.
p. cm.
Includes bibliographical references and index.
ISBN 0-8247-9050-2 (alk. paper)
1. Phospholipids--Handbooks, manuals, etc. I. Cevc, Gregor.
[DNLM: 1. Phospholipids. QU 93 P575652 1993]
QP752.P53P468 1993
574.19'247--dc20
DNLM/DLC
for Library of Congress 93-19674
CIP

The publisher offers discounts on this book when ordered in bulk quantities. For more information, write to Special Sales/Professional Marketing at the address below.

This book is printed on acid-free paper.

MARCEL DEKKER, INC.
270 Madison Avenue, New York, New York 10016

Current printing (last digit):
10 9 8 7 6 5 4 3 2 1

PRINTED IN THE UNITED STATES OF AMERICA

Preface

To date some 150,000 scientific articles refer to phospholipids. Most of these were written during the last decade, and every week several dozen new ones are added to the list. Consequently, the amount of information pertaining to phospholipids and the appreciation of this class of molecules have grown enormously over the last few years. The notion is now widely accepted that phospholipids not only represent one crucial component of all eucaryotic biological membranes but, moreover, play a subtle and frequently pivotal role in the control of many biological processes. Moreover, phospholipids are gaining importance in a variety of medical, biological, and biotechnological as well as agricultural and industrial applications.

One purpose of this handbook is therefore to catalog the current knowledge of phospholipids and their properties from many different disciplines. In doing so we have tried to overcome some of the drawbacks of the older and more specialized handbooks. This book thus attempts to encompass the whole range of topics from pure synthetic chemistry and biochemistry, through (bio)physical chemistry and structural research, to biology, immunology, and medicine. It consists of three major parts covering: (I) General properties and methods of preparation and purification (10 chapters); (II) Physical and structural properties (10 chapters); and (III) Biological and medical aspects (8 chapters). In addition, selected data characterizing the structural, thermodynamic, and some of the functional properties of phospholipids are presented in several appendixes.

We have also attempted to present the user with surveys of the main preparative and characterization techniques that are useful in phospholipid manufacturing and research. Where appropriate, convenient protocols are included to provide the reader with practical advice and recommendations. It is hoped that this will assist the ever-growing community of researchers, engineers, physicians, and students who have discovered the fascination of phospholipids to either perform fundamental studies or develop novel applications of this class of molecules.

We are confident that this comprehensive presentation of phospholipid research will make it an indispensable reference and source of inspiration for anyone interested in this class of molecules. The variety of topics covered and the didactic style, will hopefully also make it interesting to read. It is our wish that this handbook will provide the expert with an opportunity to become acquainted with complementary information, while it will offer the beginner a concise survey of the entire field of phospholipid research and a key to relevant specialized knowledge.

Gregor Cevc

Contents

Contributors

Theresa M. Allen Department of Pharmacology, University of Alberta, Edmonton, Alberta, Canada

R. Andrew Badley Immunology Section, Unilever Research, Bedford, England

Robert Bittman Department of Chemistry and Biochemistry, Queens College of The City University of New York, Flushing, New York

Alfred Blume Department of Chemistry, University of Kaiserslautern, Kaiserslautern, Germany

Hoe-Sup Byun Department of Chemistry and Biochemistry, Queens College of The City University of New York, Flushing, New York

Gregor Cevc Medical Biophysics Laboratory, Technical University of Munich, Munich, Germany

Daan J. A. Crommelin Department of Pharmaceutics, University of Utrecht, Utrecht, The Netherlands

Paul J. Davis Immunology Section, Unilever Research, Bedford, England

R. Evstigneeva Lomonosov Institute of Fine Chemical Technology, Moscow, Russia

William E. Fogler Laboratory of Experimental Immunology, Biological Response Modifiers Program, DCT, NCI–FCRDC, Frederick, Maryland

K.-H. Gober Analytical Development, Rhône-Poulenc Rorer GmbH, Köln, Germany

Mustafa Grit Department of Pharmaceutics, University of Utrecht, Utrecht, The Netherlands

B. R. Günther Process Development, Rhône-Poulenc Rorer GmbH, Köln, Germany

C. M. Gupta Institute of Microbial Technology, Chandigarh, U. T., India

Helmut Hauser Department of Biochemistry, ETH Zurich, Zurich, Switzerland

Alexander L. Klibanov Department of Pharmacology, University of Pittsburgh School of Medicine, Pittsburgh, Pennsylvania

Kenneth J. Longmuir Department of Physiology and Biophysics, College of Medicine, University of California, Irvine, California

E. M. Lünebach Analytical Development, Rhône-Poulenc Rorer GmbH, Köln, Germany

Alan D. Magid Department of Cell Biology, Duke University Medical Center, Durham, North Carolina

Thomas J. McIntosh Department of Cell Biology, Duke University Medical Center, Durham, North Carolina

Helmuth Möhwald Institute of Physical Chemistry, Johannes-Gutenberg-University of Mainz, Mainz, Germany

Yukihiro Namba Lipid Project, Nippon Fine Chemical Co., Ltd., Hyogo, Japan

S. L. Neidleman Biosource Genetics Corporation, Vacaville, California

Roger R. C. New* Cortecs Ltd., London, England

J. Wylie Nichols Department of Physiology, Emory University School of Medicine, Atlanta, Georgia

G. Repplinger Analytical Development, Rhône-Poulenc Rorer GmbH, Köln, Germany

Gerrit L. Scherphof Department of Physiological Chemistry, Groningen Institute for Drug Studies, University of Groningen, Groningen, The Netherlands

Joel M. Schnur Center for Bio/Molecular Science and Engineering, Naval Research Laboratory, Washington, D.C.

John M. Seddon Department of Chemistry, Imperial College, London, England

John R. Silvius Department of Biochemistry, McGill University, Montreal, Quebec, Canada

---

**Former affiliation*: Biocompatibles Ltd., Middlesex, England

Alok Singh Center for Bio/Molecular Science and Engineering, Naval Research Laboratory, Washington, D.C.

Herre Talsma Department of Pharmaceutics, University of Utrecht, Utrecht, The Netherlands

Suren A. Tatulian* Institute of Cytology, Academy of Sciences of the Russian Federation, St. Petersburg, Russia

D. M. Tolley Department of Chemistry, Unipath Ltd., Bedford, England

Vladimir P. Torchilin Center for Imaging and Pharmaceutical Research, Massachusetts General Hospital-East, Charlestown, Massachusetts

Anthony Watts Department of Biochemistry, Oxford University, Oxford, England

M. Wiedemann Process Development, Rhône-Poulenc Rorer GmbH, Köln, Germany

Mark A. Yorek Department of Internal Medicine, Veterans Administration Medical Center, Iowa City, Iowa

Nicolaas J. Zuidam Department of Pharmaceutics, University of Utrecht, Utrecht, The Netherlands

---

**Current affiliation*: Department of Physiology, University of Virginia, Charlottesville, Virginia

I

General and Chemical Properties

1
Structure and Nomenclature

John R. Silvius *McGill University, Montreal, Quebec, Canada*

I. INTRODUCTION

An enormous variety of phospholipid structures is found in nature, exhibiting great diversity in the structures of both the apolar and the polar moieties of the lipid molecules. While any individual lipid species may be named according to rigorous rules of organic-chemical nomenclature [1–5], a practical system of nomenclature for biologically occurring lipids must also offer reasonable brevity and simplicity to be generally useful. Moreover, even a 'purified' lipid preparation that is obtained from a biological source may be homogeneous with respect to one structural feature (e.g., the polar portions of the lipid molecules) but highly heterogeneous with respect to another (e.g., their hydrocarbon chains). A useful system of nomenclature for lipids from natural sources must allow for this fact by permitting certain structural features of the lipids in a given preparation to be described in a generic manner while other features of the structure are specified entirely.

Three general types of nomenclature for lipids will be described in this chapter. In what will be referred to as the 'fully systematic' or 'formal' system of nomenclature, all acyl and alk(en)yl residues are fully specified, using their systematic (IUPAC) designations [1], and all other structural units in the lipid molecule (polyols, monosaccharide units, amino acids, phospho- moieties, etc.) are specified individually and in full, using the nomenclature recommended in the IUPAC-IUB proposals for the Nomenclature of Lipids [4], Nomenclature of Phosphorus-Containing Compounds of Biological Importance [3] and Prenol Nomenclature [5]. In systems of nomenclature referred to below as 'trivial,' details of stereochemistry are often absent or implicit, and major portions of the molecule may be named as simple units (e.g., as a 'phosphatidyl' group) or generically. Finally, shorthand systems of nomenclature have been developed, using simple alphabetical or numerical symbols to allow concise specification of the structures of even complex lipid molecules (e.g., glycolipids) or to focus on particular details of molecular structure, such as the fatty acyl composition of a natural lipid preparation.

It is common in practice to define completely the structures of both the apolar and the polar portions of pure synthetic phospholipids, which constitute single molecular species.

By contrast, it is common to specify in detail only the polar portions of 'pure' lipid fractions from natural sources while designating the apolar portions, which are normally heterogeneous in composition, in a generic manner. Accordingly, this chapter will first describe the systems of nomenclature used to define the structures of the apolar and 'backbone' portions of lipid molecules. The following sections describe systems of nomenclature that are appropriate to describe either individual phospholipid molecular species or phospholipid fractions that are uniform in some structural features but heterogeneous in others.

II. APOLAR AND 'BACKBONE' MOIETIES OF PHOSPHOLIPIDS

Almost all biologically occurring phospholipids are constructed from two combinations of apolar and 'backbone' moieties: a glycerol (or other polyol) moiety substituted with one or two acyl or alkyl chains; or an *N*-acylated sphingoid base (i.e., a ceramide).

A. Structure and Nomenclature of Acyl/Alkylated Glycerol Moieties

While glycerol itself is a symmetrical molecule, its carbon-2 becomes a chiral center when the 1- and 3-positions are not symmetrically substituted. It is therefore useful to define a prochiral '*sn*-glycerol' (*sn* = stereospecific numbering [6]) in which the orientation of the 2-hydroxyl group and the numbering of the carbons are as shown in Fig. 1. In virtually all natural phospholipids, excepting certain archaebacterial lipids discussed below, the polar headgroup is attached to the 3-position of '*sn*-glycerol,' in which the configuration of substituents about C-2 would be designated as *R* in the (R,S)-system. The '*sn*-' convention is also applied to describe the configuration of other glycerol residues in phospholipid molecules (e.g., for the biologically occurring 1-(1',2'-diacyl-*sn*-glycero-3-phospho)-*sn*-glycerol and its 3-O-amino acyl esters [7]). Glycerol residues that are not either symmetrically or stereospecifically substituted are designated using the prefix *rac*-, as in 1-(1',2'-diacyl-*sn*-glycero-3'-phospho)-*rac*-glycerol, in which the nonacylated glycerol comprises a mixture of 1- and 3-substituted '*sn*-glycerol' residues.

1. Acyl Residues

Acyl chains are named formally using standard IUPAC rules of nomenclature for organic compounds [1], although certain features of the IUPAC system, notably the use of the (E,Z)-convention to designate the configuration about the double bond, have not received widespread use in lipid nomenclature. The systematic name of an acyl residue is derived from that of the corresponding fatty acid by replacing the suffix '-oic acid-' by '-oyl,' as in hexadecanoyl or *cis,cis*-9,12-octadecadienoyl.

Most of the fatty acids found in natural phospholipids also have one or more trivial names (Table 1; for more extensive compilations see [8,9]). To designate the name of an acyl residue using the trivial nomenclature, the suffix '-ic acid' in the common name of the parent fatty acid is replaced by the suffix '-oyl,' in analogy to the designation of acyl residues in the IUPAC system. This rule can lead to confusion in certain cases, however, notably for decanoyl chains, which it designated as 'caproyl' (from capric acid), could be mistaken for a residue of esterified hexanoic (caproic) acid.

Several common systems also exist for the shorthand designation of fatty acyl chains. The first, which is used in an IUPAC-IUB-recommended system for the shorthand

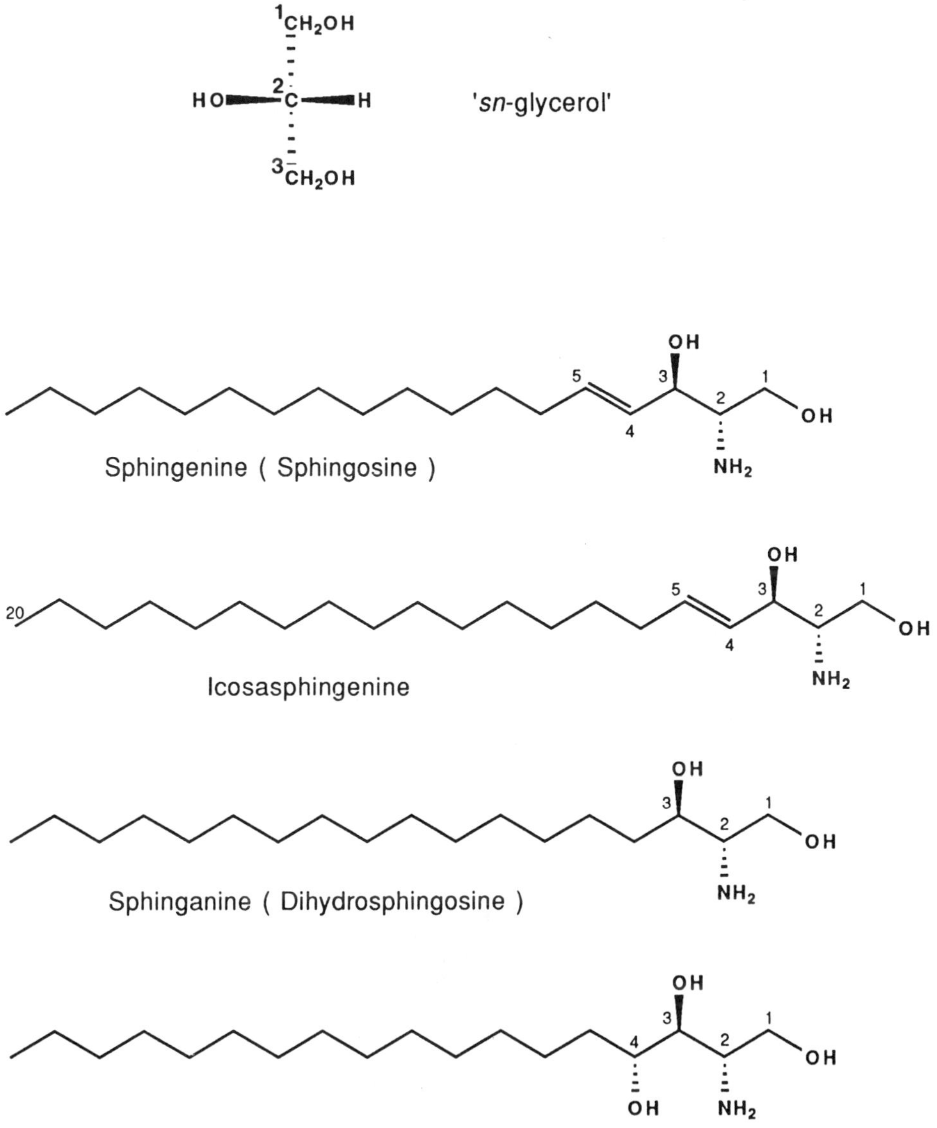

'*sn*-glycerol'

Sphingenine (Sphingosine)

Icosasphingenine

Sphinganine (Dihydrosphingosine)

4-D-Hydroxysphinganine (Phytosphingosine)

Figure 1 Structures of *sn*-glycerol and of the major sphingoid bases.

Table 1 Systematic, Trivial, and Shorthand Designations for Some Common Fatty Acyl Chains of Phospholipids

		Shorthand designations		
Systematic name	Common name	IUPAC-IUB	Δ-system	n-system
Dodecanoyl	Lauroyl	Lau	12:0	12:0
Tetradecanoyl	Myristoyl	Myr	14:0	14:0
Hexadecanoyl	Palmitoyl	Pam	16:0	16:0
Octadecanoyl	Stearoyl	Ste	18:0	18:0
Icosanoyl	Arachidoyl	Ach	20:0	20:0
cis-9-Octadecenoyl	Oleoyl	Ole	$18{:}1c\Delta^{9}$	18:1(n-9)
cis,*cis*-9,12-Octadeca dienoyl	Linoleoyl	Lin	$18{:}2cc\Delta^{9,12}$	18:2(n-6)
all-*cis*-9,12,15-Octadecatrienoyl	γ-Linolenoyl	Lnn	$18{:}3ccc\Delta^{9,12,15}$	18:3(n-3)
all-*cis*-5,8,11,14-Icosatetraenoyl	Arachidonoyl	Δ_4Ach	$20{:}4cccc\Delta^{5,8,11,14}$	20:4(n-6)

designation of lipid structures [4], consists of a set of abbreviations (Pam = palmitoyl or hexadecanoyl, Lin = linoleoyl = or *cis*,*cis*-9,12-octadecadienoyl, etc.) for common alkanoyl and alkenoyl chains. A listing of the shorthand designations of common fatty acids in this system is given in Table 1.

Other more systematic systems for the shorthand representation of fatty acyl residues are frequently used in tabulations of the fatty acyl compositions of lipid preparations. In most systems of this type, the carbon number of the fatty acid is indicated first, followed after a colon by the number of double bonds. In one such system in common use, these specifications are followed by the letters c or t to denote the configuration(s) of the double bond(s); the position(s) of the double bond(s) are indicated finally as superscripts to the symbol Δ. Using this system, a linoleoyl chain is thus designated as $18{:}2cc\Delta^{9,12}$. A related system, which is widely used for the common fatty acids found in animal lipids [9,10], designates by the symbol $(n - x)$ or $(\omega - x)$ the position of the first double bond encountered when counting from the methyl terminus of the acyl chain toward the carboxyl group (e.g., linoleic acid is designated as $18{:}2(n-6)$). The prefixes c- and t- can be added in this system to denote the configurations of double bonds, although such prefixes are often omitted for naturally occurring *cis*-unsaturated fatty acids.

2. Alkyl/alkenyl Residues

Alkyl and alkenyl hydrocarbon groups in glycerophospholipids are normally best named using formal (IUPAC) nomenclature [1,4]. The practice of deriving trivial names for alkyl groups from the trivial name of the fatty acid, replacing the suffix '-ic acid' by '-yl' (e.g., 'palmityl' for hexadecyl) can be confusing and is no longer recommended. However, the isoprenoid-derived hydrocarbon chains found in many archaebacterial lipids are rather cumbersome to specify fully using formal nomenclature, and the use of trivial ('phytanyl', 'biphytanyl') or generic names ('sesterterpanyl') has persisted in the nomenclature of these lipids (see Sec. IV). A listing of the names and structures of the common archaebacterial lipid hydrocarbon chains can be found in Ref. 11.

B. Structure and Nomenclature of Ceramide Moieties

While the prototypical 'backbone' component of a sphingolipid is sphingosine (sphingenine, (2*S*,3*R*,4*E*)-2-amino-4-octadecene-1,3-diol*; see Fig. 1), most sphingolipid preparations isolated from natural sources also contain significant proportions of other sphingoid bases that differ from sphingosine in hydrocarbon chain length, Δ^4-unsaturation, or hydroxylation at the 4-position. The structures and formal and trivial names of the common sphingoid bases are indicated in Fig. 1. For the sake of brevity, common names are widely used for the major naturally occurring sphingoid bases even in systematic systems of sphingolipid nomenclature [4]. However, it is recommended that unnatural stereoisomers of these bases be designated using the formal (IUPAC) system of nomenclature, as the trivial names of the bases imply a specific stereoconfiguration [4].

Ceramide residues, consisting of *N*-acylated sphingoid bases, are commonly named using either the trivial or the formal name of the coupled fatty acyl group when the latter can be specified. Ceramides derived from a single sphingoid base are best named as the *N*-acyl derivatives of that base, while preparations containing more than one type of base but a single type of fatty acyl residue can be named as an acyl ceramide. The IUPAC/IUB-recommended shorthand designation for a generic ceramide residue, Cer, is particularly useful in the designation of complex sphingolipids [4]. Ceramide residues bearing a single type of fatty acyl species can be abbreviated in this same scheme using the abbreviation of the fatty acid followed by Sph (for sphingosine) or Spd (for a generic sphingoid base), although this convention is not common in the literature.

III. NOMENCLATURE OF PHOSPHOLIPIDS

A. Lipids Based on a (Mono/Di)radylglycerophospho-(Monohydroxy Alcohol) Structure

As noted above, all of the glycerophosphate-based lipids found in the eubacteria and in eukaryotic organisms are formally derived from *sn*-glycerol-3-phosphate (see Fig. 1). In the fully systematic system of nomenclature, glycerophospholipids with different acyl or alkyl substituents are designated using similar conventions. However, for historical reasons different trivial nomenclatures have arisen for diacyl and alkylacyl phospholipids, which are therefore discussed separately below.

1. Diacylglycerophospholipids

Lipids of this type are named formally as phosphodiester derivatives of 1,2-diacyl-*sn*-glycero-3-phosphate, where the acyl groups are named according to the principles outlined above. Even in the formal (IUPAC/IUB) system of phospholipid nomenclature, the

*The configuration of carbon-2 of sphingosine is designated in the (R,S)-system as *S*, while that of carbon-2 of '*sn*-glycerol' is designated as *R*. In fact, however, if the structure of sphingosine is aligned with that of '*sn*-glycerol' shown in Fig. 1, with the sphingosine carbon-1 at the bottom and the carbon-3 and the remainder of the alkenyl chain at the top, the sphingosine amino substituent and the glycerol 2-hydroxyl can be seen to have the same orientation with respect to the chiral C-2 center. Since the polar headgroups of sphingolipids are attached to carbon-1 of an acylated sphingosine (or its analogues) and those of eubacterial and eukaryotic glycerophospholipids are attached to carbon-3 of the mono- or diradyl '*sn*-glycerol,' the configurations of these two classes of phospholipids about their acylated C-2 centers are in fact equivalent, in spite of their different designations in the (R,S)-system.

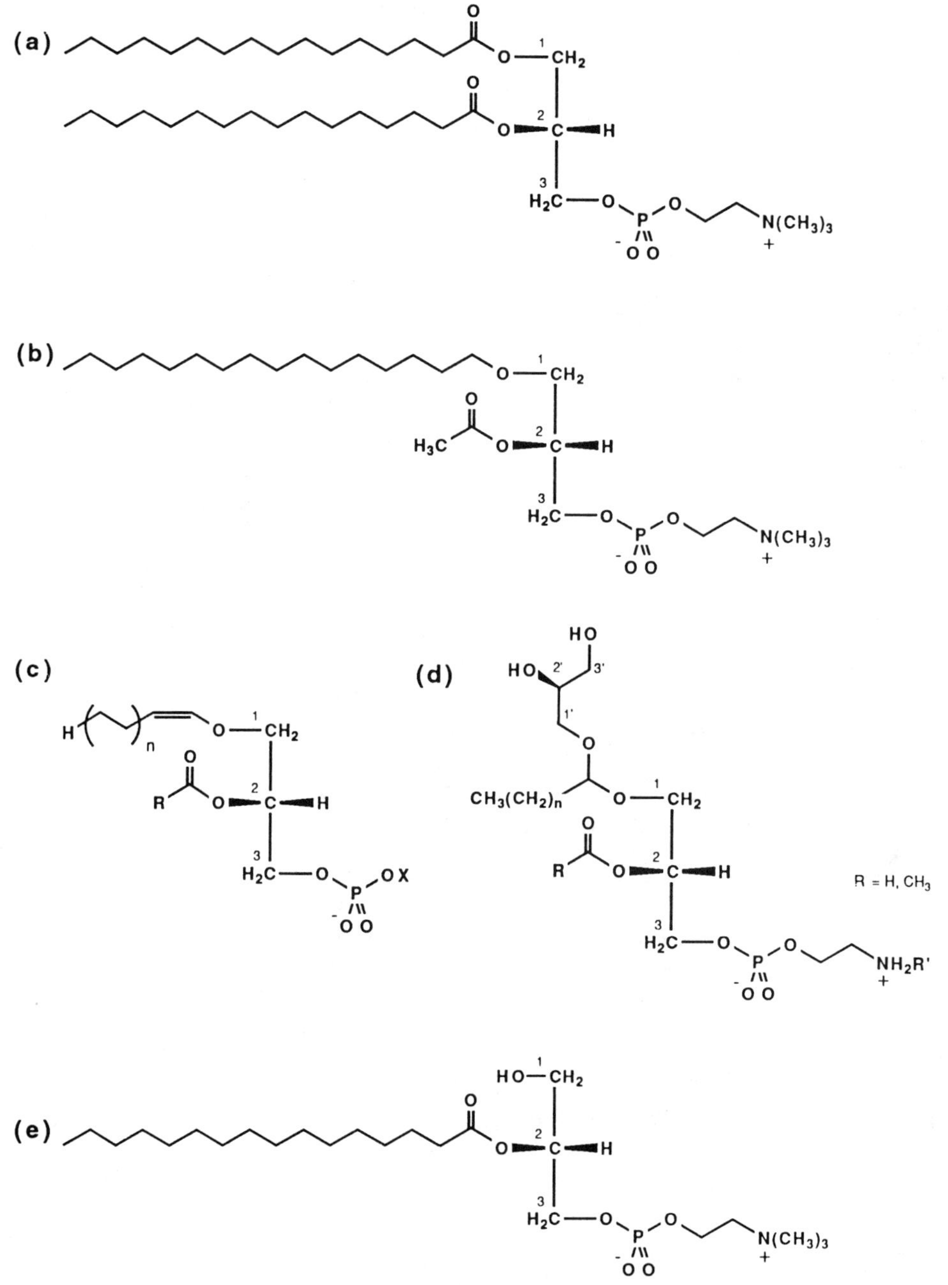

Figure 2 Structures of representative phospholipids with configurations characteristic of eukaryotic and eubacterial phospholipids. (a) 1,2-dihexadecanoyl-*sn*-glycero-3-phosphocholine or L- α-dipalmitoylphosphatidylcholine. (b) 1-hexadecyl-2-acetyl-*sn*-glycero-3-phosphocholine (platelet activating factor). (c) A 1-(alk-1'-enyl)-2-acyl-*sn*-glycero-3-phospho(alcohol), also named as an (alcohol) plasmalogen. The most common moieties X in naturally occurring phospholipids are ethanolamine and choline. (d) A 1-(1'-glyceroalkyl)-2-acyl-*sn*-glycero-3-phosphoethanolamine, found in membranes of various *Clostridium* species. (e) 2-hexadecanoyl-*sn*-glycero-3-phosphocholine, also named as 2-palmitoyl-1-*lyso*-phosphatidylcholine.

alcohols linked to the diacylglycerophospho- moiety* are normally designated by their common names (e.g., ethanolamine rather than 2-aminoethanol) unless the alcohol substituent is a rare or unnatural compound. In this system, for example, dipalmitoylphosphatidylcholine (Fig. 2a) is designated as 1,2-dihexadecanoyl-*sn*-glycero-3-phosphocholine or, less commonly in practice, 1,2-dihexadecanoyl-*sn*-glycero(3)phosphocholine [3].

Various trivial nomenclatures exist for diacylglycerophospholipids, of which the most useful (and precise) defines such compounds as derived from a phosphatidic acid, *i.e.*, a diacylglycerophosphate. Lipids in which phosphatidic acids are coupled to alcohols in phosphodiester linkage are then named as phosphatidyl alcohols (e.g., phosphatidylcholine). The term 'phosphatidyl' used without qualification generally implies a 1,2-diacyl-*sn*-glycero-3-phospho- moiety. It has been recommended [4] that this and particularly other, less common diacylglycerophospho- residues be named as 'X-*sn*-phosphatidyl' residues, where x is the position of the phospho- group on the *sn*-glycero- backbone. However, the designation '(3)-phosphatidyl(alcohol)' is seldom found in common use. While the strict definition of a phosphatidyl residue is a di*acyl*-glycerophospho- moiety, the use of the term is commonly (if incorrectly) extended to encompass mixtures of diacyl- and alkylacylglycerophospholipids in cases where the two are not resolved (e.g., in most chromatographic procedures). It should be noted that a 'phosphatidyl' residue actually comprises the unit (diacylglycerol)-O–$\overset{\text{O}}{\overset{\|}{\underset{\underset{\text{O}}{|}}{\text{P}}}}$– 'Phosphatidylethylamine,' for example, thus designates a phosphonolipid (see below) and not phosphatidylethanolamine.

Most older systems of nomenclature used for phospholipids offer no special advantages over those described above, and their use is not recommended. Terms such as 'lecithin' and 'cephalin' were originally applied to describe crude lipid fractions from particular tissues and are inappropriate for use to describe pure lipid species. The older designation of the 3-, 2- and 1-positions of '*sn*-glycerol' as the α-, β- and γ-positions, with phospholipids from eubacteria and eukaryotes defined as derivatives of 'L-α'-phosphatidic acid, should likewise be abandoned in favor of the systems described above.

As noted above, a shorthand nomenclature has been proposed for glycerophospholipids as well as other complex lipids [3,4]. The complete shorthand designation for a 1,2-diacyl-*sn*-glycero-3-phosphate-derived lipid consists of shorthand designations (see Table 1) for the acyl chains at the 1- and 2-positions (or a single abbreviation with a subscript '2' if the two chains are the same), followed successively by the abbreviation Gro and the shorthand designation of the phosphodiester-linked alcohol (see Table 2). The phosphatidyl moiety may also be denoted by the generic abbreviation Ptd. Using these conventions, dipalmitoyl phosphatidylcholine becomes Pam$_2$Gro*P*Cho (or, in a variant usage, Pam$_2$PtdCho) and phosphatidylserine (with a mixture of fatty acyl chains) becomes

*The IUPAC recommendations for nomenclature of phosphorus-containing compounds [3] suggest that phosphodiesters be designated by inserting the infix 'phospho' (as a contraction of 'phosphinico') between the names of the two coupled alcohols, and that phosphomonoesters be designated using the prefix 'phospho' (as a contraction of 'phosphono'), in cases where it is not necessary or possible to specify the other substituents on the phosphoryl residue (e.g., hydrogen). Phosphomonoester compounds may also be named as alcohol phosphates in such cases. It is recommended that the term 'phosphoryl' be reserved for cases in which all three substituents on the O=P(–O)(–O)–O center are specified. The commonly used terminologies 'glycerophosphoryl(alcohol),' 'diacylglycerophosphoryl(alcohol),' etc. are thus incorrect.

Table 2 Shorthand Designations for Common Polar Moieties found in Phospholipids

Residue	Recommended shorthand designation
(*sn*)Glycerol	(*sn*)Gro
Phosphatidyl	Ptd
Sphingosine	Sph
Sphingoid base	Spd
Ceramide	Cer
Choline	Cho
Ethanolamine	Etn
Serine	Ser
myo-Inositol	Ins
Phospho (in mono- or diester linkage)	*P*
Glucose	Glc
(*N*-Acetyl)glucosamine	GlcN(Ac)
Galactose	Gal
(*N*-Acetyl)galactosamine	GalN(Ac)
2-Aminoethylphosphono	$P\mathrm{EtNH_2}$

PtdSer, to provide two simple examples. The shorthand nomenclature is most advantageous to describe the structures of lipids with complex polar headgroups (e.g., lipids with large glycosyl moieties).

2. Alk(en)ylacyl and Dialk(en)yl Glycerophospholipids

Glycerophospholipids with one or two alkyl or alkenyl chains are named in the formal (IUPAC/IUB) system in a manner very similar to the corresponding diacyl phospholipids, with ether-linked hydrocarbon chains described as alkyl or alkenyl residues (suffix '-yl') in place of acyl residues (suffix '-oyl'). The generic term 'radyl' may be used to denote a mixture of acyl and alk(en)yl substituents at a given position of the glycero- residue in a lipid preparation with heterogeneous hydrocarbon chains. It is not necessary to apply the prefix '*O*-' to describe substituents attached to glycerol through ether linkages, since the only other possible means of alkyl substitution of glycerol, namely *C*-alkylation, would create a compound that should be named as a distinct alkanetriol. The formal name for platelet activating factor [12], for example, is thus 1-hexadecyl-2-acetyl-*sn*-glycero-3-phosphocholine (Fig. 2b).

As noted above, the trivial nomenclature of alk(en)ylacylglycerophospholipids is distinctly different from that of diacylglycerophospholipids. Historically, alk-1-enyl phospholipids (Figure 2c) were termed 'plasmalogens' because they were identified as a lipid fraction in plasma that liberated aldehyde molecules on acid treatment [13]. Terms such as 'ethanolamine plasmalogen,' 'choline plasmalogen,' etc. are still in common use to denote phospholipids containing a 1-alk-1'-enyl-2-acyl-*sn*-glycero- moiety. It has been recommended [4] that 1-alk-1'-enyl-2-acyl-*sn*-glycero-3-phosphate be named 'plasmenic acid,' by analogy to phosphatidic acid, and that plasmalogens thus be designated as plasmenyl alcohols (ethanolamine plasmalogen = plasmenylethanolamine). The older use of the term 'phosphatidal' to denote a 2-acyl-1-alk-1'-enyl-*sn*-glycero-3-phospho- residue has been criticized [14] because it is easily confused with 'phosphatidyl.' For much the same reason, the proposal to define a 2-acyl-1-alkyl-*sn*-glycero-3-phospho- residue as a 'plasmanyl' group has not received widespread usage, as this term may be easily confused with 'plasmenyl.' The terms 'plasmalogen,' 'plasmenyl,' and 'phosphatidal' are commonly used for designating alk-1-enylacyl phospholipids in a generic manner; the formal nomenclature described above is more suitable when specifying the complete structures (including those of the hydrocarbon chains) of individual molecular species.

Two interesting plasmalogen derivatives, namely the glycerol acetals of plasmenyl ethanolamine and plasmenyl *N*-methylethanolamine (Figure 2*d*) are found in *Clostridium butyricum* [15,16]. Such species are named as 1-(1'-glyceroalkyl)-2-acyl-*sn*-glycero-3-phosphoethanolamines.

The shorthand nomenclature is not commonly used to designate simple alk(en)ylacyl and dialkyl glycerophospholipids. Rules for the shorthand designation of alkyl and alkenyl residues can be found in Ref. 4.

3. Monoacyl or Monoalkyl Glycerophospholipids

No special additional rules are required to name these compounds using the formal system of nomenclature (e.g., 2-hexadecanoyl-*sn*-glycero-3-phosphocholine [Fig. 2e], 1-octadec-1'-enyl-*sn*-glycero-3-phosphoethanolamine, etc.). In the trivial nomenclature, mono-deacylated derivatives of phosphatidyl, plasmenyl or plasmanyl compounds are described using the prefix 'lyso-,' which ideally should be preceded by the position of the unsubstituted hydroxyl group on the *sn*-glycerol backbone (e.g., 2-lysophosphatidylcholine = 1-acyl-*sn*-glycero-3-phosphocholine). The term 'lysophospholipid' is a generic one that may encompass monoalkyl or monoacyl compounds or a mixture of such compounds; the term 'lysophosphatidyl-' is intended to denote specifically a monoacyl species, although in practice it is sometimes applied as well to describe unfractionated mixtures of monoacyl and monoalkyl species.

B. Phospholipids Based on a (Mono/Di)radylglycerophospho-Polyol Structure

The simplest biologically occurring compounds of this type are the 1,2-diacyl-*sn*-glycero-3-phospho derivatives of glycerol and of *myo*-inositol, known trivially as phosphatidylglycerol and phosphatidylinositol, respectively. The phosphatidylglycerols and their acylated and phosphorylated derivatives, as they occur in eubacteria and higher organisms, are based on a 1-(1',2'-diacyl-*sn*-glycero-3'-phospho)-*sn*-glycerol (or 1,2-diacyl-*sn*-glycero-3-phospho-[1'-glycerol]) structure, whose stereoconfiguration is shown in Fig. 3a.

Two other biologically important classes of polyglycerophospholipids are the 1,3-bis(1',2'-diacyl-*sn*-glycero-3'-phospho)-glycerols and their derivatives (Fig. 3b), trivially named diphosphatidyl-glycerols or cardiolipins, and the 1-(3'-acyl-*sn*-glycero-1'-phospho)-3-acyl-*sn*-glycerols (or 1-acyl-*sn*-glycero-3-phospho-[[3'-acyl]-1'-glycerols]) (Fig. 3c), trivially named as lyso-bisphosphatidic acids. The stereochemistry of the latter compounds is distinctive in that both monoacyl-*sn*-glycerol residues are linked to the central phospho- moiety via their 1-positions. This does not appear to be the case for the mono- and diacyl derivatives of phosphatidylglycerol found in certain eubacteria [17].

(a)

(b)

(c)

(d)

$X = H, OPO_3^{2-}$

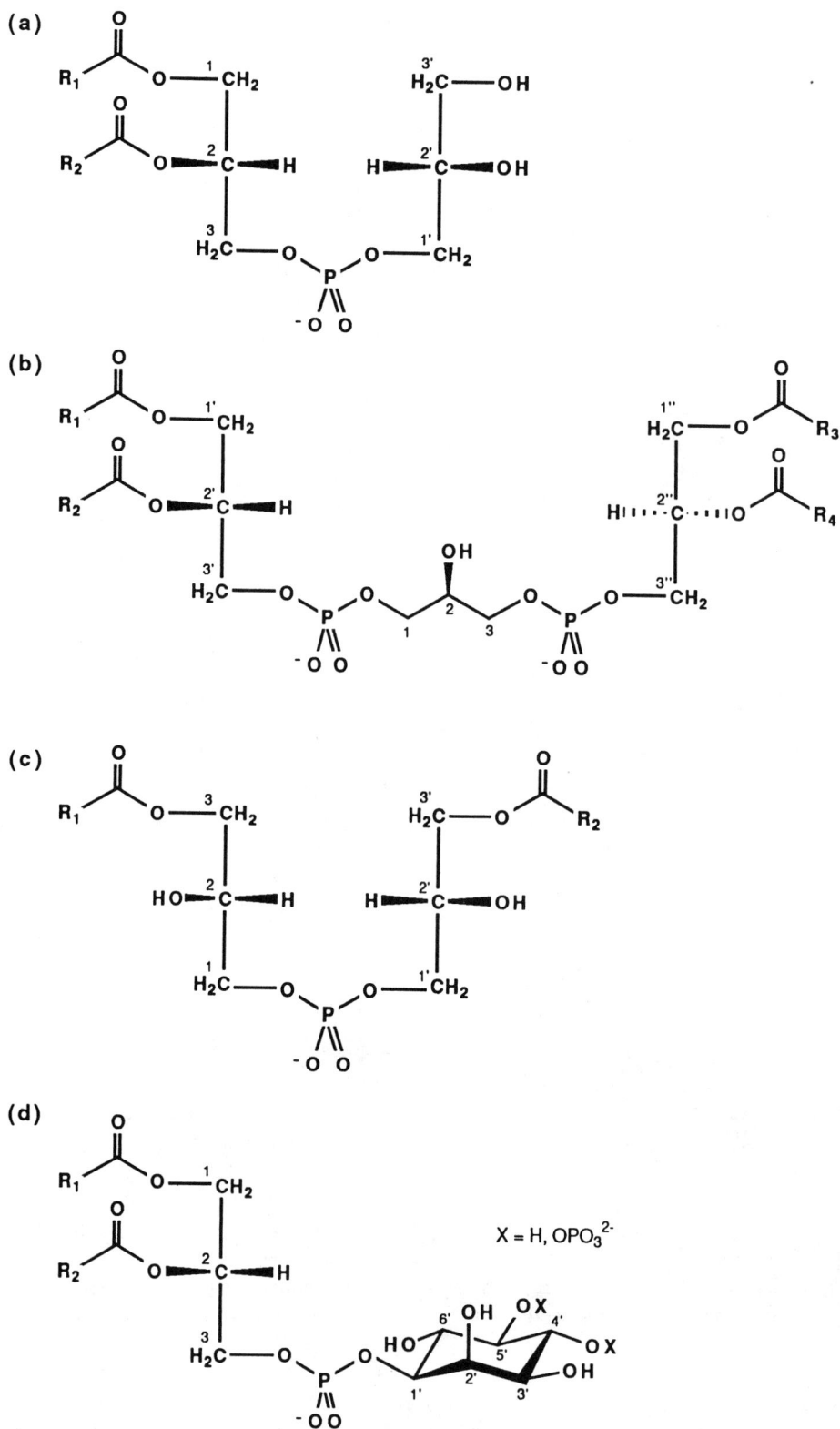

In the naturally occurring phosphatidylinositols and their phosphorylated, acylated, and glycosylated derivatives, the *myo*-inositol residue is coupled through its 1-position to the diacylphosphoglycero- moiety, as shown in Fig. 3d. The basic structure shown is named formally as a 1- (1',2'-diacyl-*sn*-glycero-3'-phospho)-L-*myo*-inositol, as a 1,2-diacyl-*sn*-glycero(3)phospho(1)-L-*myo*-inositol, or as a 1,2-diacyl-*sn*-glycero-3-phospho-(1'-*myo*-inositol). The bisphosphorylated derivative shown is most simply named as a 1-(1',2'-diacyl-*sn*-glycero-3')-phospho-L-*myo*-inositol-4,5-bis(phosphate) or as a 1,2-diacyl-*sn*-glycero(3)phospho(1)-L-*myo*-inositol-4,5-bis(phosphate). While it has recently been suggested that a revision of the inositol ring-numbering system may be useful for some classes of *myo*-inositol derivatives [18], the presently used convention is well suited for the naturally occuring inositol phospholipids. A further class of phosphatidylinositol derivatives, in which one or more glycosyl residues are attached to the inositol ring, will be discussed in Sec. C below.

Phospholipids containing multiple polyol residues are readily named by the shorthand system using the IUPAC/IUB-recommended abbreviations [3,4]; 1-(1',2'-dihexadecanoyl-*sn*-glycero-3'-phospho)-*sn*-glycerol is represented as $Pam_2GroPGro$, while phosphatidylinositol-4,5-bisphosphate is generically represented as PtdIns(4,5)P_2 and cardiolipin as Ptd_2Gro.

C. Phospholipids Based on a (Mono/Di)radylglyceroglycoside Structure

Glycoglycerolipids bearing phospho-, glycerophospho-, or mono- or diradylglycerophospho- residues can be named using the formal (IUPAC/IUB) system of nomenclature, although this is commonly done only for species bearing simple (mono- or disaccharide) glycosyl moieties. Such compounds may be named as either diradylglycerol or glycoside derivatives, in accordance with the IUPAC/IUB Rules for Carbohydrate Nomenclature [2] and the recommended rules for Nomenclature of Lipids [4], although it is more common to name these compounds as (modified) glycosyldiradylglycerols. Using these rules, for example, structure a in Fig. 4 would typically be named as 3-(6'-(1''-*sn*-glycerophospho)-α-D-glucopyranosyl)-1,2-diradyl-*sn*-glycerol or 1,2-diradyl-3-(6'-(1''-*sn*-glycerophospho)-α-D-glucopyranosyl)-*sn*-glycerol, while structure b is most clearly named as α-D-glucopyranosyl-(1-2)-(6'-(1''-*sn*-glycerophospho)-α-D-glycopyranosyl-(1-3)-1,2-diradyl-*sn*-glycerol. (Note that the latter structure in particular is best named, for clarity, with the radyl substituents listed following the complex substituent attached to the *sn*-glycerol 3-position).

←

Figure 3 Some naturally occurring phospholipids containing multiple polyol moieties. (a) A diacylphosphatidylglycerol, also named as a 1-(1',2'-diacyl-*sn*-glycero-3'-phospho)-*sn*-glycerol or as a 1,2-diacyl-*sn*-glycero-3-phospho-(1'-*sn*-glycerol). (b) A cardiolipin, formally named as a 1,3-bis(1',2'-diacyl-*sn*-glycero-3'-phospho)-glycerol. (c) A lyso-bisphosphatidic acid, formally named as a 1-(3'-acyl-*sn*-glycero-1'-phospho)-3-acyl-*sn*-glycerol or as a 1-acyl-*sn*-glycero-3-phospho-((3'-acyl)-1'-*sn*-glycerol). (d), A phosphatidylinositol (X = H for both inositol positions indicated), also named as a 1-(1',2'-diacyl-*sn*-glycero-3'-phospho)-L-*myo*-inositol, as a 1,2-diacyl-*sn*-glycero(3)phospho(1)-L-*myo*-inositol, or as a 1,2-diacyl-*sn*-glycero-3-phospho-(1'-*myo*-inositol). Formal names for the indicated phosphorylated derivatives are given in the text.

(a)

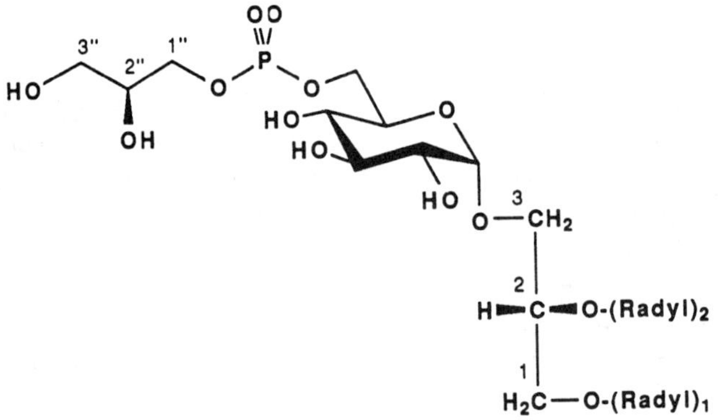

(b)

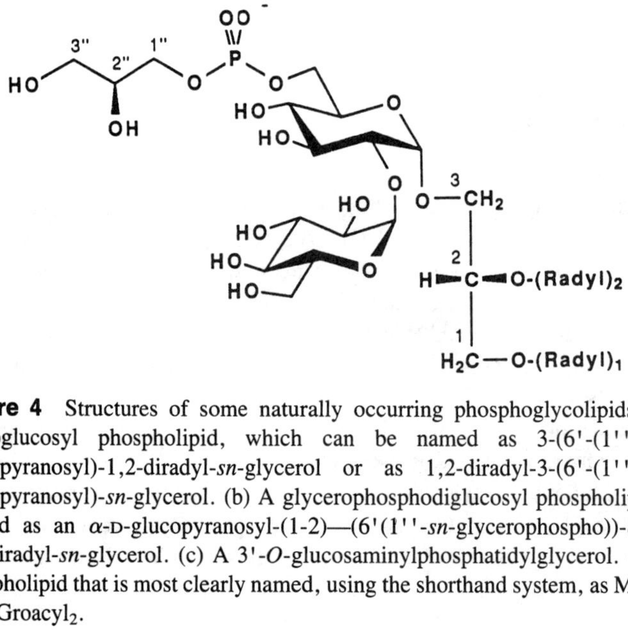

Figure 4 Structures of some naturally occurring phosphoglycolipids. (a) A glycerophosphomonoglucosyl phospholipid, which can be named as 3-(6'-(1''-*sn*-glycerophospho)-α-D-glucopyranosyl)-1,2-diradyl-*sn*-glycerol or as 1,2-diradyl-3-(6'-(1''-*sn*-glycerophospho)-α-D-glucopyranosyl)-*sn*-glycerol. (b) A glycerophosphodiglucosyl phospholipid, which is most clearly named as an α-D-glucopyranosyl-(1-2)—(6'(1''-*sn*-glycerophospho))-α-D-glucopyranosyl-(1-3)-1,2-diradyl-*sn*-glycerol. (c) A 3'-*O*-glucosaminylphosphatidylglycerol. (d) A dimannosyl-inositol phospholipid that is most clearly named, using the shorthand system, as Manα1-2(Manα1-6)Ins1-*P*-3-*sn*-Groacyl$_2$.

A number of trivial names are used to designate complex glycoglycerolipids in the literature, often incorporating succinct trivial names to denote complex oligosaccharide residues [19]. However, trivial nomenclatures are of limited usefulness to specify the structures of phosphoglycoglycerolipids and their derivatives, as it is often at best awkward in such systems to designate the points of substitution of the glycosyl moiety (which may comprise several monosaccharide residues) with phospho- and other phosphorus-containing groups.

The most generally useful system for the description of phosphoglycoglycerolipids utilizes the IUPAC/IUB-recommended shorthand nomenclature [4] to denote individual

(c)

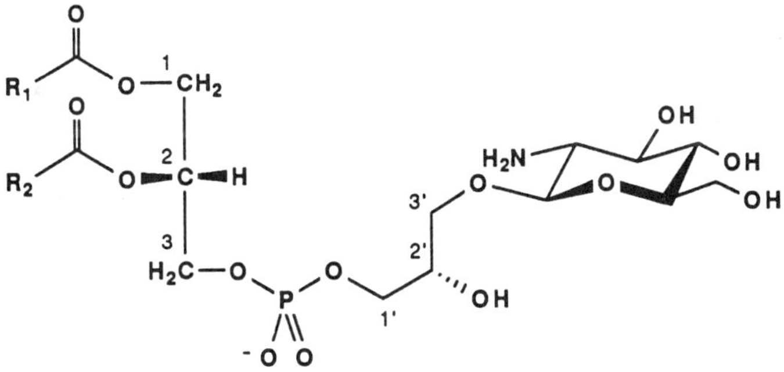

(d)

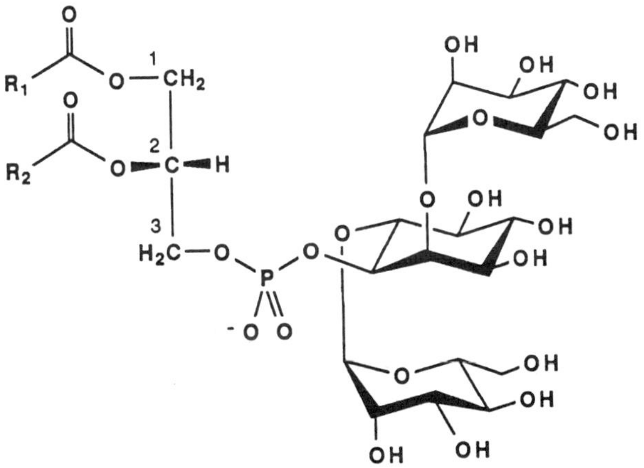

residues within even complex lipid structures. Linkages between consecutive monosaccharide (and polyol) residues in an unbranched glycosyl structure are indicated by specifying for each pair of adjacent residues the configuration of the linkage and the positions joined (e.g., Glcα1-2Glc or better, Glcα1—2Glc). In some cases, a lower case script '*f*' or '*p*' is appended to the shorthand designation of each sugar residue to indicate whether it is in the furanosyl or pyranosyl form, respectively. Additional substituents to a given sugar or polyol residue are indicated, along with the position of substitution (in parentheses, except in the case of substituents to the first-named glycosyl residue), immediately preceding the shorthand designation for that residue. The structures shown in Fig. 4a and 4b would thus be represented in this system (if radyl = acyl) as (a) 1-*sn*-Gro*P*-6Glcα1-3-Gro-1,2-acyl$_2$ and (b) Glcα1-2(1-*sn*-Gro*P*-6)Glcα1-3Gro-1,2-acyl$_2$. Examples of the shorthand nomenclature applied to more complex phosphoglycolipid structures can be found in [19].

D. Phospholipids Based on a (Mono/ Di)radylglycerophosphoglycoside Structure

Glycosylated derivatives of both phosphatidylglycerol and phosphatidylinositol are found in eubacteria [19–21]. As well, a variety of glycosylated derivatives of phosphatidylinositol have been identified in eukaryotic cells in recent years, either as free membrane lipids or as hydrophobic membrane 'anchors' for a variety of cell surface proteins [22–26]. These are most conveniently named using a shorthand nomenclature similar to that discussed in the last section [4,19], or trivially as glycosyl derivatives of phosphatidylglycerol or phosphatidylinositol. The structure shown in Figure 4c would thus be designated as GlcNβ1-3-PtdGro or as 3'-*O*-glucosaminylphosphatidylglycerol, and that in Figure 4d would most succinctly be designated as Manα1-2(Manα1-6)Insl-*P*-3-*sn*-Groacyl$_2$.

The more complex phosphatidylinositol-glycan structures found in animal cells [22–28] could in principle be named using the same conventions as those discussed above. However, it is more common for such structures to be represented using shorthand designations for the glycosyl groups but presenting the structures in two dimensions, allowing a clearer indication of the points of branching of the glycosyl residues. Structures represented in this way are of course rather inconvenient to designate within a body of text. A solution to this problem may be found when the structures of the animal PI-glycan lipids are categorized as well as those of, for example, the gangliosides, for which recommendations have been advanced to indicate complex 'core' residues using simple group designations [4].

E. Sphingosine-Containing Phospholipids

Two major classes of phosphosphingolipids are found in nature: phosphocholine derivatives of ceramides, known trivially as sphingomyelins, and glycosylated derivatives of inositol phosphoceramides, known trivially as phytoglycolipids.

Individual molecular species of sphingomylins are named in the formal (IUPAC/IUB) system of lipid nomenclature as *N*-acylsphingosylphosphorylcholines or as analogous compounds based on other sphingoid bases. When the base composition of the preparation is not homogeneous (e.g., when it is prepared using 'sphingosine' or 'sphingosinephosphorylcholine' derived from natural sphingolipids, which typically contain a mixture of sphingoid bases), the trivial term '*N*-(acyl)sphingomyelin' is best employed, as it encompasses structures containing various sphingoid bases. The shorthand nomenclature is seldom employed for sphingomyelins.

Phytoglycolipids can be named as glycosyl derivatives of ceramide phosphate, using the principles described above for glycosyl derivatives of glycerophospholipids (Sec. C). As for the latter species, the shorthand nomenclature is usually most convenient for the designation of phytoglycolipid structures.

F. Phosphonolipids

Three classes of phosphonolipids have been identified in marine invertebrates, with structures of the types shown in Fig. 5. In all cases, the phosphono- group is present in a 2-aminoethylphosphono- moiety or an *N*-methylated derivative. Species of the type shown in Fig. 5a can be named either as 1,2-diradyl-*sn*-glycero-3-(2'-aminoethyl)phosphonates or as phosphatidyl ethylamines (PtdEtNH$_2$ in the shorthand system) or their *N*-methylated derivatives. Ceramide-based species such as those shown in Fig. 5b are normally named as ceramide 2-aminoethylphosphonates or ceramide 2-methylaminoethylphosphonates, and species such as those shown in Figure 5c are named as 2-aminoethylphosphonates- or 2-methylaminoethylphosphonates.

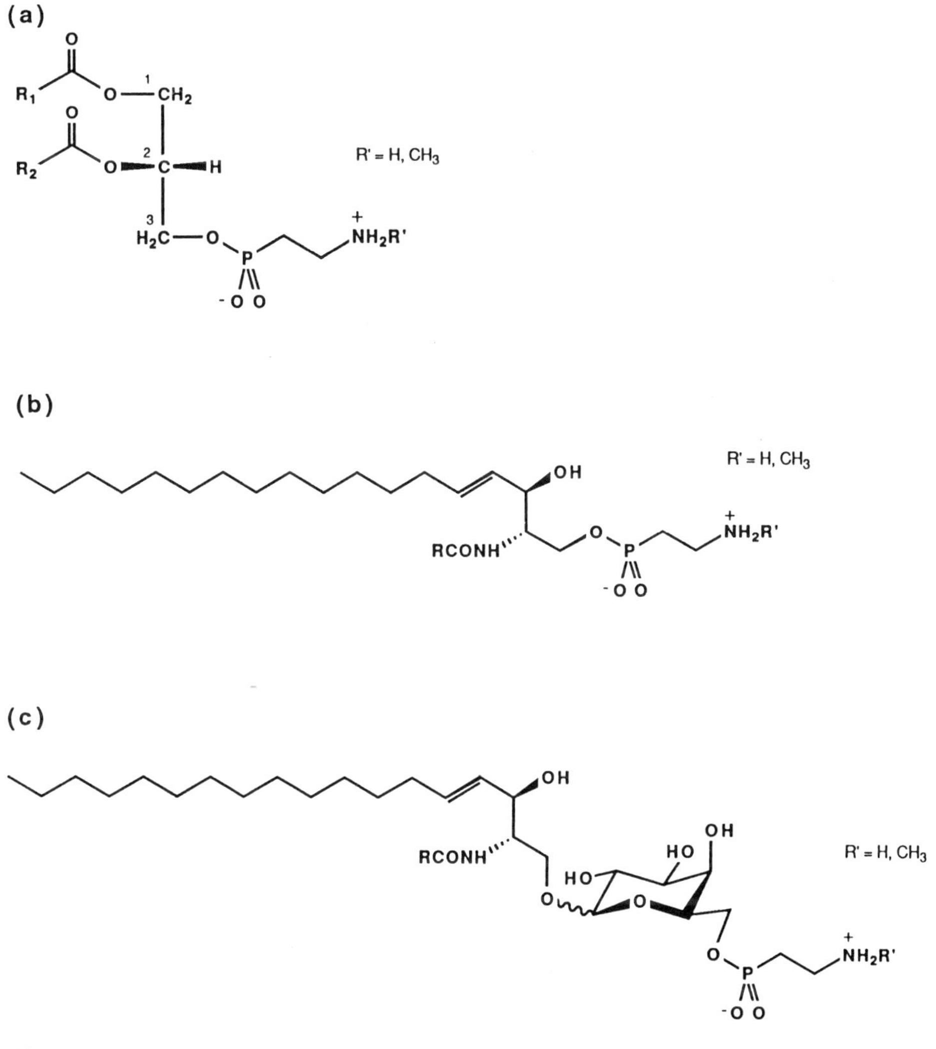

(a)

(b)

(c)

Figure 5 Some naturally occurring phosphonolipids. (a) A 1,2-diacyl-*sn*-glycero-3-(2'-aminoethyl)phosphonate (R'=H), also named as a phosphatidyl ethylamine. The *N*-methylated derivative of this species (R'=CH_3) can be named as a 1,2-diacyl-*sn*-glycero-3-(2'-methyl-aminoethyl)phosphonate or (less desirably) as a phosphatidyl *N*-methyl-ethylamine. (b) A ceramide 2-aminoethylphosphonate (R=H) or ceramide 2-methylaminoethylphosphonate (R=CH_3). (c) A 1-(6'-(2''-aminoethylphosphono)mannosyl) ceramide (R=H) or 1'-(6'-(2''-methyl-aminoethyl-phosphono)mannosyl) ceramide (R=CH_3).

IV. ARCHAEBACTERIAL PHOSPHOLIPIDS

Archaebacterial lipids exhibit several remarkable structural features that are quite different from those discussed above. First, in those archaebacterial phospholipids that are based on a glycerol backbone, the hydrophobic substituents are coupled to the 2- and 3-positions,

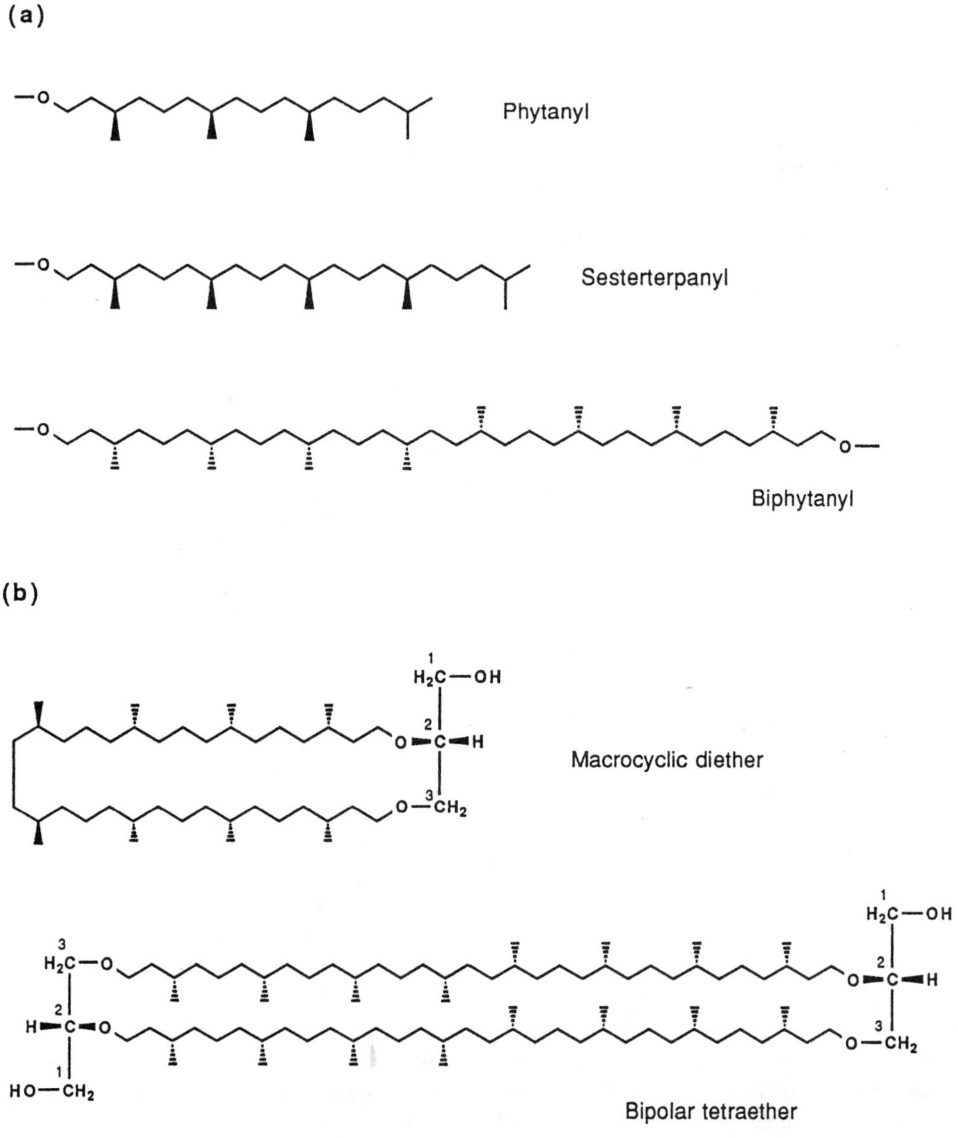

(a)

Phytanyl

Sesterterpanyl

Biphytanyl

(b)

Macrocyclic diether

Bipolar tetraether

Figure 6 Hydrocarbon chain and backbone structures of some typical archaebacterial phospholipids.

and the polar headgroup to the 1-position, of the *sn*-glycerol backbone. Tetritol and even nonitol (carditol) backbones are often found in place of glycerol in such lipids. Second, the hydrocarbon chains of these lipids comprise a series of remarkable isoprenoid-derived structures, some of which are indicated along with their commonly used designations in Figure 6a. Finally, two remarkable types of linkages between polyol backbones and alkyl chains are found in archaebacterial lipids, namely a macrocyclic diether structure in which a single biphytanyl residue is coupled in ether linkage to the 2- and 3-positions of the *sn*-glycerol residue, and bipolar tetraether structures, comprising two polyol and two alkyl residues, of the types shown in Figure 6b.

While it is common to designate simple archaebacterial lipids, such as derivatives of 2,3-diphytanyl-*sn*-glycerol, using full systematic names, more complex structures (e.g., macrocyclic diether and bipolar tetraether structures) are usually designated simply in a generic or schematic manner. The structures of a number of these unusual lipid species, and the names commonly applied to them, can be found in Ref. 11.

ACKNOWLEDGMENTS

The author wishes to thank Dr. Robert Bittman for useful comments on this manuscript. This work was supported by a Medical Research Council of Canada Scientist award to J. R. S.

APPENDIX
Molecular Weights of Phospholipids

The table presented below is divided into three parts, listing respectively the molecular weights of phospholipid acyl and alkyl chains, of the 'backbone' structures to which hydrocarbon chains and polar headgroups are attached, and of the structures found in phospholipid polar headgroups. The molecular weight of a given phospholipid is calculated by summing the molecular weights listed for these three sets of molecular constituents, e.g., the mw of 1-hexadecanoyl-2-octadecenoyl-*sn*-glycero-3-phosphocholine = 239.42 + 265.46 + 168.04 + 87.17 = 760.09.

Acyl and Alkyl Chains (RCO- and R-)		
Group	Molecular formula	Molecular weight
Hydrogen	H	1.01
Octanoyl	$C_8H_{15}O$	127.21
Nonanoyl	$C_9H_{17}O$	141.24
Decanoyl	$C_{10}H_{19}O$	155.26
Undecanoyl	$C_{11}H_{21}O$	169.29
Dodecanoyl	$C_{12}H_{23}O$	183.32
Tridecanoyl	$C_{13}H_{25}O$	197.34
Tetradecanoyl	$C_{14}H_{27}O$	211.37
Pentadecanoyl	$C_{15}H_{29}O$	225.40
Hexadecanoyl	$C_{16}H_{31}O$	239.42
Heptadecanoyl	$C_{17}H_{33}O$	253.45
Octadecanoyl	$C_{18}H_{35}O$	267.48
Nonadecanoyl	$C_{19}H_{37}O$	281.51
Icosanoyl	$C_{20}H_{39}O$	295.53
Docosanoyl	$C_{22}H_{43}O$	323.59
Tetracosanoyl	$C_{24}H_{47}O$	351.64
Hexadecenoyl	$C_{16}H_{29}O$	237.41
Octadecenoyl	$C_{18}H_{33}O$	265.46

Group	Molecular formula	Molecular weight
Acyl and Alkyl Chains (RCO- and R-) Continued		
Icosenoyl	$C_{20}H_{37}O$	293.52
Docosenoyl	$C_{22}H_{41}O$	321.57
Octadecadienoyl	$C_{18}H_{31}O$	263.45
Octadecatrienoyl	$C_{18}H_{29}O$	261.43
Icosatetraenoyl	$C_{20}H_{31}O$	287.47
Docosahexaenoyl	$C_{22}H_{31}O$	311.49
cyclo-Hexadecanoyl[a]	$C_{16}H_{29}O$	237.41
cyclo-Heptadecanoyl[a]	$C_{17}H_{31}O$	251.44
cyclo-Octadecanoyl[1]	$C_{18}H_{33}O$	265.46
cyclo-Nonadecanoyl[a]	$C_{19}H_{35}O$	279.49
cyclo-Icosanoyl[a]	$C_{20}H_{37}O$	293.52
Dodecyl	$C_{12}H_{25}$	169.33
Tetradecyl	$C_{14}H_{29}$	197.39
Hexadecyl	$C_{16}H_{33}$	225.44
Octadecyl	$C_{18}H_{37}$	253.50
Hexadecenyl	$C_{16}H_{31}$	223.43
Octadecenyl	$C_{18}H_{35}$	251.48
Hydroxyoctadecanoyl	$C_{18}H_{35}O_2$	283.48
Hydroxytetracosanoyl	$C_{24}H_{47}O_2$	367.64

Group	Molecular formula	Molecular weight
Backbone Structures		
Glycerophospho- CH_2–O– | CH–O– | CH_2–O–PO_2–O–	$C_3H_5O_6P$	168.04
Sphingosinephospho- $C_{15}H_{29}$–CH–OH | CH–NH | CH_2–O–PO_2–O–	$C_{18}H_{35}O_5NP$	376.46
Dihydrosphingosinephospho- $C_{15}H_{31}$–CH–OH | CH–NH | CH_2–O–PO_2–O–	$C_{18}H_{37}O_5NP$	378.47

Backbone Structures Continued

Group	Molecular formula	Molecular weight

Phytosphingosinephospho-
$C_{14}H_{29}CH(OH)–CH–OH$
                        |
                        CH–NH
                        |
                        $CH_2–O–PO_2–O–$
	$C_{18}H_{37}O_6NP$	394.47

Glycero- $CH_2–O–$ | CH–O– | $CH_2–O–$	$C_3H_5O_3$	89.07
Sphingos[in]yl- $C_{15}H_{29}–CH–OH$ | CH–NH | $CH_2–O–$	$C_{18}H_{35}O_2N$	297.48
Dihydrosphingos[in]yl- $C_{15}H_{31}–CH–OH$ | CH–NH | $CH_2–O–$	$C_{18}H_{37}O_2N$	299.50
Phytosphingosinephospho- $C_{14}H_{29}CH(OH)–CH–OH$ | CH–NH | $CH_2–O–$	$C_{18}H_{37}O_3N$	315.50
Bis(glycerophospho)glycerol $CH_2–O–$ | CH–O– | $CH_2–O–PO_2–O–CH_2$ | CH–OH | $CH_2–O–PO_2–O–CH_2$ | CH–O– | $CH_2–O–$	$C_9H_{16}O_{13}P_2$	394.17

Group	Molecular formula	Molecular weight
Backbone Structures Continued		
Bis-glycerophosphate CH_2–O– CH_2–O– | | CH–O– CH–O– | | CH_2–O—PO_2—O–CH_2	$C_6H_{10}O_8P$	241.12

Group	Molecular formula	Molecular weight
Polar Headgroup Moieties		
Ethanolamine	C_2H_7N	45.08
Monomethylethanolamine	C_3H_9N	59.11
Dimethylethanolamine	$C_4H_{11}N$	73.14
Choline	$C_5H_{13}N$	87.17
Serine	$C_3H_6O_2N$	88.09
Phosphonoethanolamine	$C_2H_7O_2NP$	108.06
N-Methylphosphonoethanolamine	$C_2H_9O_2NP$	122.08
Glycerol (headgroup, monosubstituted) CH_2–OH | CH–OH | CH_2–	$C_3H_7O_2$	75.09
Glycerol (headgroup, disubstituted) CH_2–O– | CH–OH | CH_2–	$C_3H_6O_2$	74.08
Glycerophosphate (headgroup, monosubstituted) CH_2–OPO_3^{-2} | CH–OH | CH_2–	$C_3H_6O_5P$	153.05
Inositol	$C_6H_{11}O_5$	163.15
Inositol phosphate	$C_6H_{10}O_8P$	241.12
Inositol bisphosphate	$C_6H_9O_{11}P_2$	319.08
Hexosyl		
Unsubstituted[b]	$C_6H_{11}O_5$	163.15
Monosubstituted[c]	$C_6H_{10}O_5$	162.14
Disubstituted[d]	$C_6H_9O_5$	161.14
Hexosamine		
Unsubstituted[b]	$C_6H_{12}O_4N$	162.17
Monosubstituted[c]	$C_6H_{11}O_4N$	161.16

Polar Headgroup Moieties Continued		
Group	Molecular formula	Molecular weight
N-Acetylhexosamine		
Unsubstituted[b]	$C_8H_{14}O_5N$	204.20
Monosubstituted[c]	$C_8H_{13}O_5N$	203.20
Phosphate ($-PO_3^{-2}$)	O_3P	78.97

Counterions	
Species	Atomic/molecular weight
H^+	1.01
Li^+	6.94
Na^+	22.99
K^+	39.10
Mg^{2+}	24.31
Ca^{2+}	40.08
NH_4^+	18.04
$N(CH_3)_4^+$	74.15

[a]Chain contains a single ring structure (e.g., cyclopropyl or cyclohexyl); the total number of carbon atoms for the chain is indicated.
[b]Refers to a glycosyl residue that is coupled to the lipid through a glycosidic linkage and in which none of the non-anomeric hydroxyl groups is substituted.
[c]Refers to a glycosyl residue that is coupled to the lipid through a glycosidic linkage and in which one of the non-anomeric hydroxyl groups is substituted (e.g., with a phospho group or another glycosyl residue).
[d]Refers to a glycosyl residue that is coupled to the lipid through a glycosidic linkage in which two of the nonanomeric hydroxyl groups are substituted (e.g., with phospho groups or other glycosyl residues).

REFERENCES

1. International Union of Biochemistry, *Nomenclature of Organic Chemistry (Sections A, B and C)*, 2nd. ed., Butterworths, London, 1966.
2. IUPAC Commission on the Nomenclature of Organic Chemistry and IUPAC-IUB Commission on Biochemical Nomenclature, *Eur. J. Biochem. 21*:455–477 (1971).
3. IUPAC-IUB Commission on Biochemical Nomenclature, *Eur. J. Biochem. 79*:1–9(1977).
4. IUPAC-IUB Commission on Biochemical Nomenclature, *Eur. J. Biochem.* 79:11–21 (1977).
5. IUPAC-IUB Commission on Biochemical Nomenclature, *Eur. J. Biochem. 167*:181–184 (1987).
6. H. Hirschmann, *J. Biol. Chem. 235*:2762–2769 (1960).
7. W. L. Koostra and P. F. Smith, *Biochemistry 8*:4794–4806 (1969).
8. P. G. Robinson, *J. Lipid Res. 23*:1251–1253 (1982).
9. D. E. Vance and J. E. Vance, *Biochemistry of Lipids and Membranes*, Benjamin/Cummings, Menlo Park, Calif, 1985, p. 190.
10. W. W. Christie, *Lipid Analysis*, Pergamon Press, Oxford, 1973, pp. 2–9.
11. M. DeRosa, A. Gambacorta, and A. Gliozzi, *Microbiol. Revs. 50*:70–80 (1986).
12. D. J. Hanahan, C. A. Demopoulos, J. Liehr, and R. N. Pinkard, *J. Biol. Chem. 255*:5514–5516 (1980).
13. R. Fuelgen and K. Voit, *Pflugers Arch. Gesamte Physiol. Menschen Tiere 206*:389–397 (1924).

14. L. A. Horrocks and M. Sharma, in *Phospholipids* (J. N. Hawthorne and G. B. Ansell, eds.), Elsevier, Amsterdam, 1982, pp. 51–93.
15. M. Matsumoto, K. Tamiya, and K. Koizumi, *J. Biochem. 69*:617–620 (1971).
16. G. K. Khuller and H. Goldfine, *Biochemistry 14*:3642–3647 (1975).
17. K. Y. Hostetler, in *Phospholipids* (J. N. Hawthorne and G. B. Ansell, eds.), Elsevier, Amsterdam, 1982, pp. 215–261.
18. IUPAC-IUB Commission on Biochemical Nomenclature, *Eur. J. Biochem. 180*:485–486 (1989).
19. I. Ishizuka and T. Yamakawa, in *Glycolipids* (H. Wiegandt, ed.), Elsevier, Amsterdam, 1985, pp. 101–197.
20. J. A. F. Op den Kamp, U. M. T. Houtsmuller, and L. L. M. van Deenen, *Biochim. Biophys. Acta 106*:438–441 (1965).
21. J. N. Hawthorne, in *Phospholipids* (J. N. Hawthorne and G. B. Ansell, eds.), Elsevier, Amsterdam, 1982, pp. 263–278.
22. M. G. Low and A. R. Saltiel, *Science 239*:268–275 (1988).
23. M. A. J. Ferguson and A. F. Williams, *Ann. Rev. Biochem. 57*:285–320 (1988).
24. M. A. J. Ferguson, S. W. Homans, R. A. Dwek, and T. W. Rademacher, *Science 239*, 753–757 (1988).
25. G. A. M. Cross, *Ann. Rev. Cell Biol. 6*:1–39 (1990).
26. M. A. J. Ferguson, *Biochem. Soc. Trans. 20*:243–256 (1992).
27. S. W. Homans, M. A. J. Ferguson, R. A. Dwek, T. W. Rademacher, R. Anand, and A. F. Williams *Nature 333*:269–272 (1988).
28. W. J. Masterson, T. L. Doering, G. W. Hart, and P. T. Englund, *Cell 56*:793–800 (1989).

2

Occurrence and Response to Environmental Stresses in Nonmammalian Organisms

S. L. Neidleman *Biosource Genetics Corporation, Vacaville, California*

I. INTRODUCTION

In preparing a chapter on the biological occurrence of phospholipids in nonmammalian organisms, a number of facts immediately influenced the approach to be taken. The first is that phospholipids are ubiquitous. The second is that within a given organism, phospholipids have multiple loci, for example, various membranes. The third is that phospholipid composition is responsive to environmental stresses. The fourth is that among these stresses are the following: temperature, pressure, pH, salt concentration, dilution rate, nutrient supply, age, light, and infection. The fifth is that the best one can hope to do in a brief chapter is to indicate the flavor and complexity of the subject and encourage the reader to think, and hopefully, experiment.

The other items to be dispensed within these introductory remarks are some of the abbreviations to be used to designate various phospholipid components in the discussion to follow: cardiolipin (CL), phosphatidic acid (PA), phosphatidyl choline (PC), phosphatidyl ethanolamine (PE), phosphatidylglycerol (PG), phosphatidylinositol (PI), phosphatidylserine (PS), phosphatidylsulfocholine (PSG), lysophosphatidylcholine (lyso-PC), lysophosphatidylethanolamine (lyso-PE), lysophosphatidylglycerol (lyso-PG), lysophosphatidylinositol (lyso-PI), phosphoryl-*N*-methylethanolamine (PME), phosphoryl-*N*, *N*-dimethylethanolamine (PDME), and sphingomyelin (SM).

II. MICROORGANISMS

More research has been accomplished in the microorganisms than in any other type of organism. The primary reason for this is the relative ease with which microorganisms may be grown and manipulated. While they clearly differ from the higher organisms, many general principles and conclusions can be reached that then can be applied to studies in other organisms.

As will be emphasized throughout this chapter, the occurrence of phospholipids in microorganisms and other organisms is ubiquitous, but the absolute composition of the phospholipid fraction is strongly influenced by many factors as noted above. Several examples of these responses to experimental variables will be noted in this section.

A. Diversity of Phospholipid Composition in Microorganisms

The relative abundance of phospholipids as a fraction of total lipids for several microorganisms is illustrated in Table 1 [1]. These data were obtained under varying experimental conditions, so at least a part of the reason for the different levels is probably due to that fact.

Among various fungi, there is a considerable variability in the components of the phospholipid fraction, illustrated in Table 2 [2]. Usually, in fungi, PC is the major phospholipid, although in some cases PE may be the most abundant, as in the case of the thermophilic fungus *Absidia ramosa*, wherein at 45°C, PE is 40% of total phospholipid and PC is 30%; while at 45 °C, PE is 67% and PE is 16% [3].

The major lipids in eubacteria are phospholipids, the most common being PG, PE, and CL. PE is often the predominant phospholipid in Gram-negative bacteria, whereas Gram-positive bacteria contain relatively more PG and CL. Archaebacteria also contain phospholipids, but in complete contrast they have largely C20 phytanyl chains (saturated isoprenoid-derived) with repeating methyl branches in ether linkage to the glycerol backbone.

In addition, in halophiles, the head groups are often more negatively charged, bearing extra phosphate or sulphate residues. The phytanyl chains of one phospholipid may be covalently linked to those of another, producing a C40 tetraether lipid that can span the membrane width. Some thermoalkaliphiles also have C25 sesterpenyl chains, and some thermoacidophiles have one to four cyclopentane rings along the alkyl chain. The various

Table 1 Phospholipid Content of Several Oleaginous Microorganisms

Microorganism	Phospholipid content (% w/w of major lipids)
Bacterium	
Arthrobacter AK19	6.7
Yeasts	
Cryptococcus terricolus	1.8
Endomycopsis vernalis	4
Lipomyces lipofer	14
L. starkeyi	15
Rhodotorula glutinis	2
Molds	
Cladosporium herbarum	7.4
Fusarium oxysporum	28
Mortierella isabellina	7
Penicillium lilacinum	6.4
Pythium irregulare	17.8
Rhizopus oryzae	11.8

Table 2 Phospholipid Composition of Several Fungi

Fungus	Relative % of phospholipid components					
	PC	PE	PS	PI	CL	PG
Mortierella isabellina	42	17	1	11	—	—
Penicillium lilacinum	21	22	5	2	5	6
Cladosporium herbarum	50	16	1	—	18	13
Pellicularia practicola	52	20	5	3	2	3
Fusarium oxysporum	25	18	16	20	2	5

C20, C25, and C40 chains give a variety of possible structures for the hydrophobic core of the membrane [4].

Among strains of a particular microorganism, there may be considerable variability in the relative amounts of the phospholipid components. This is shown in Table 3 for *Aspergillus niger* [5]. This is not to say that a reasonable estimate of phospholipid composition cannot be presented in a generalized form. For example, the range of phospholipid composition in various ascosporogenous or basidiomycetous yeasts are similar to, or at least no more variable than, the data of Table 3 for *Aspergillus niger* strains. The yeast data are given in Table 4 [2].

Another point that deserves attention is that the phospholipid composition of various membranes within a microorganism may vary, and so the phospholipid composition of the the total microorganism may or may not represent an acceptable indicator of a particular membrane. Data for membranes of *Neurospora crassa* in term of PC, PE, and PI are shown in Table 5 [6].

The methanotrophic bacteria show species variability in the composition of their phospholipids. It is emphasized that these bacteria are divided into two categories, which correspond to differences in their internal membrane organization: *Methylococcus* is in one group and *Methylosinus* and *Methylobacteria* are in the other category, for example. Data for the phospholipid composition of these bacteria are shown in Table 6 [7]. All bacteria were grown on methane as a carbon source.

Table 3 Range of Phospholipid Components in Strains of *Aspergillus niger*

Phospholipid	Relative % of total phospholipid
PC	33–54
PE	10–33
PS	7–27
CL	2–15
PG	0–6
PI	0–13
lyso-PC	0–9

Table 4 Range of Phospholipid Components in Yeast

Phospholipid	Relative % of total phospholipid
PC	35–55
PE	25–32
PI	9–22
PS	4–18
CL	1–4
PG	0–3
PA	0–10

Table 5 Phospholipid Composition of Diverse Membranes in *Neurospora crassa*

	% of total phospholipid			
Phospholipid	Vacuolar membranes	Mitochondrial membranes	Endoplasmic reticulum	Plasma membrane
PC	42	36	51	45
PE	37	43	35	39
PI	14	10	14	16

Table 6 Phospholipid Composition of *Methanotrophic Bacteria* (%)

Bacteria	PE	PC	PG	CL	PDME	PME	PS
Methylococcus capsulatus	74	8	13	5	—	—	—
Methylobacterium organophilum	57	15	—	—	24	—	1
Methylosinus trichosporium	—	11	13	—	49	21	—

B. Effects of Environment on Phospholipid Composition

The effects of environmental factors on phospholipid content and composition have been the subject of much research with microorganisms. Some individual cases involving such diverse effectors as manganese deficiency, pressure, carbon source, substrate limitation, temperature, and sodium chloride will be briefly noted to indicate even further the level of variability inherent in reporting phospholipid occurrence.

Cells of *Brevibacterium ammoniagenes* grown under conditions of manganese deficiency had a decreased phospholipid content throughout the fermentation as compared to conditions in which the manganese supply was ample. Content of PI was greatly decreased, whereas that of PG and CL were increased [8]. In the barophilic marine bacterium CNPT3, it was reported that the content of 16:1 and 18:1 fatty acids in membrane phospholipids increased with increasing pressure, whereas that of a14:1, 14:0, 16:0, and 18:0 decreased. This resembled the effects of lowering growth temperature in other bacteria [9]. Fatty acid constituents of phospholipids of *Microsporoum gypseum* grown at 20° or 27°C showed a higher degree of unsaturation when grown on glucose as compared to glycerol [10]. In *Zymomonas mobilis,* varying the concentrations of glucose or ethanol in the presence of glucose had no major effects on fatty acid composition of phospholipids; but increasing levels of each caused a decrease in PE and PG and an increase in PC and CL [11]. Table 7 shows the different effects of *N*- and *C*-limitation on the phospholipid composition of *Rhodotorula glutinis*. The major phospholipid PS is decreased under *C*-limitation, while PC is increased. An increase in dilution rate under *N*-limitation increased total phospholipid yield with a decrease occurring under *C*-limitation [12].

The most intensively studied environmental factor is temperature. Table 8 indicates some results reported for the response of phospholipid composition to this stress. The data clearly indicate that it is necessary to take information on a case by case basis, with generalizations being possible only after careful scrutiny of the experimental design of studies being compared. The proportions of the various phospholipid components of different membranes may vary in response to environmental stresses such as temperature, and this is in addition to membrane to membrane dissimilarities.

Further information on the effects of temperature is often concerned with the level of unsaturation of constituent fatty acids. Data illustrating the increase in unsaturated fatty acids with decreasing temperature in phospholipids of *Neurospora crassa* [20] and *Paecilomyces persicinus* [21] are shown in Tables 9 and 10. There are both differences and similarities, representative of species to species variation. The major similarities are the considerable decrease in 18:1 and the marked increase in 18:2 + 18:3 with lower temperature. The major difference is that in *Neurospora* the major increase is in 18:3 while with *Paecilomyces* the major increase is in 18:2.

Yet another variable that has been shown to influence phospholipid composition is sodium chloride concentration. Data for effects on *Staphylococcus aureus* are shown in Table 11 [22]. At low sodium chloride, the major phospholipid is PG, while at high sodium chloride, CL is the major component.

Table 7 Effect of *C*- and *N*-Limitation on Phospholipid Composition of *Rhodotorula glutinis* in Batch Cultures

	% by area	
Phospholipid	*C*-Limited	*N*-Limited
PS	39.9	47.9
PE	12.3	14.6
PI	10.7	12.1
lyso-PC	10.4	12.0
PC	18.7	6.9
CL	4.0	2.2

Table 8 Effect of Temperature on Phospholipid Composition

Microorganism	Temperature variations (°)	Low temperature effects	Reference
Bacillus caldotenax (membranes)	45, 65	PE ↓ PG ↑ CL↔	[13]
Bacillus stearothermophilus (membranes)	45, 65	PG ↓ CL ↑	[14]
Pseudomonas sp. (whole cells)	5, 15, 30	PE,PG,CL ↔	[15]
Yersinia enterocolitica (whole cells)	5, 25, 37	PE,PG ↔	[16]
Saccharomyces cerevisiae (whole cells)	15, 30	PC	[17]
Neurospora crassa (microsomal and mitochondrial membranes)	15, 37	PC ↓ PE ↑	[18]
Thermomyces lanuginosus (whole cells)	30, 50	PA, PI ↑ PS, PE ↔	[19]

↓ = Decrease; ↑ = Increase; ↔ = No change.

Table 9 Effect of Temperature on Fatty Acid Unsaturation in Phospholipids of *Neurospora crassa*

	Percentage of total fatty acids					
Fatty Acids	22°C	29°C	32°C	36°C	40°C	% 22°C—% 40°C
16:0	20	21	23	21	23	–3
18:0	3	4	4	3	3	0
18:1	5	12	15	24	28	–23
18:2	45	46	44	43	38	+7
18:3	27	17	14	9	8	+19

Table 10 Effect of Temperature on Fatty Acid Composition of Phospholipids of *Paecilomyces persicinus*

	Relative %		
Fatty acids	20°C	36°C	% 20°C—% 36°C
16:0	22.7	14.3	+8.4
18:0	2.2	trace	+2.2
18:1	9.6	66.9	–57.3
18:2	54.7	18:8	+35.9
18:3	10.8	trace	+10.8

Table 11 Effect of Sodium Chloride Concentration on Phospholipid Composition in *Staphylococcus aureus*

	% Sodium chloride	
Phospholipid	0.05%	10%
CL	10	50
PG	65	39
lyso-PG	12	5

III. ALGAE

In many algae, phospholipids represent a major component of the total lipid content [23]. Some representative data are shown in Table 12. The specific phospholipids detected in some microalgae are indicated in Table 13 [24]. The concentration and fatty acid composition of various phospholipids have been shown to be responsive to environmental stresses. In *Chondrus crispus,* light stimulated the labeling of PC and PG with C14-acetate [25]. Increasing % NaCl from 2.5 to 20% caused a significant increase in the proportion of PG.

IV. PLANTS

One definition given for lecithin from a commercial point of view is that it is a mixture of phospholipids, glycolipids, and some triglycerides [26]. Table 14 indicates the phospholipid composition of a number of plant lecithins, compared to those of egg and bovine brain. One message that results from scrutiny of these data is that the distribution of phospholipid components among the plant lecithins and in comparison with the non-plant material is very individualistic. This may be due to genetic differences, variations in environmental stresses, and alternative processing protocols, among other possibilities.

Table 12 Phospholipids as % of Total Lipid in Selected Algae

Alga	% Phospholipid
Chlorosarcinopsis negevensis	20
Cylindrotheca closterium	45
Cylindrotheca fusiformis	31
Dunaliella salina	17
Dunaliella primolecta	20
Fragillaria construens	26
Phaeodactylum tricornatum	47
Radiosphaera negevensis	26
Scenedesmus sp.	41
Stichococcus bacillaris	43

Table 13 Major Phospholipids in Some Microalgae

Alga	Taxonomic group	Major phospholipids
Cylindrotheca fusiformis	Bacillariophycaea	PE, PSC
Cyclotella cryptica	Bacillariophycaea	PA,PC,PE,PG,PI,CL,PME
Navicula incerta	Bacillariophycaea	PC,PE,PG
Nitzschia alba	Bacillariophycaea	PG,PI,PSC,lyso-PSC,CL
Phaeodactylum tricornutum	Bacillariophycaea	PA,PC,PE,PG,PI,PSC,CL,PME
Chlamydomonas reinhardtii	Chlorophyceae	PC,PE,PG
Chlorella vulgaris	Chlorophyceae	PC,PE,PG,PI
Dunaliella bardawil	Chlorophyceae	PC,PE,PG
Neochloris oleoabundans	Chlorophyceae	PC,PE,PG,PI,PS,CL
Volvox carteri	Chlorophyceae	PE,PG
Cryptomonas sp.	Cryptophyceae	PC,PG
Anabaena cylindrica	Cyanophyceae	PG
Chlorogloea fritschii	Cyanophyceae	PG
Cyanidium caldarium	Cyanophyceae	PC,PG,PI
Mastiglocladus laminosus	Cyanophyceae	PG
Spirulina platensis	Cyanophyceae	PG
Crypthecodinium cohnii	Dinoplyceae	PC,PE,PI,PS,lyso-PC,CL
Euglena gracilis	Euglenophyceae	PS
Nannochloropsis sp.	Eustigmatophyceae	PC,PG,PE,PI
Monodus subterraneus	Xanthophyceae	PE,PG,PI

The effects of temperature and drought as stress conditions on phospholipid content and composition in soybean oil can be seen in Table 15 [27]. High temperature in the absence of drought stress caused an increase in total phospholipid content and a specific increase in PC and PI, with a decrease in PE. Drought stress at 27°C gave similar effects to 33°C without drought. Drought stress at 33°C gave variable results. These results clearly demonstrate again the point that phospholipid content and composition are subject to stress related variability. To compare phospholipids of diverse living material is most difficult in the absence of a clear definition of the conditions under which each species has been developed.

Table 14 Phospholipid Composition of Commercial Lecithins

Phopholipid	Soy	Corn	Sunflower	Rape Seed	Egg	Bovine brain
PC	21	31	14	37	69	18
PE	22	3	24	29	24	36
PI	19	16	13	14	—	2
PA	10	9	7	—	—	2
PS	1	1	—	—	3	18
SM	—	—	—	—	1	15

Table 15 Phospholipids in Oil of Soybeans Subjected to Temperature and Drought Stress During Maturation

Temperature (°C)	Drought	Oil phospholipid	Phospholipid component (%)		
			PC	PE	PI
27	Control	1.9	29.3	66.4	4.3
	Moderate	1.8	40.8	51.4	7.7
	Severe	4.3	45.4	48.3	6.4
33	Control	5.7	42.5	44.2	13.3
	Moderate	3.6	38.8	38.1	23.0
	Severe	4.2	47.0	46.5	6.5

Further studies on the effects of temperature stress on phospholipid composition have been previously reviewed [28]. Depending upon the tissue under investigation and, presumably, the experimental protocol, nearly every possibility has been obtained: (1) no alteration of phospholipid composition upon cold-hardening at 10°C of wheat seedling roots, (2) a marked increase in PC and PE in poplar tree bark, and (3) an increase in PC and PE and a decrease in PG in rape leaves at low temperature, but with reverse trends in rape roots. Again, these data illustrate that the occurrence of phospholipids in nature is ubiquitous, but compositional definition is very much a function of environmental conditions. In a study of seasonal variations in the phospholipid content and composition of *Myriophyllum quitense*, a freshwater macrophyte, it was shown that the level of the major phospholipid PE and the total phospholipid content varied with tissue type, highest in roots and stems, lowest in flowers and fruits, and with the season, highest in June, lowest in December [29].

Another variable that may influence the phospholipid composition of biological material is storage time. In a study of cassava (*Manihot esculenta*) roots seven phospholipids were measured: PC, PE, PI, CL, PG, PS, and PA. The levels of all of these, save PG and PA, showed wide fluctuations in concentration as a function of storage time, and the relative concentrations of these components did not remain constant, so that a definitive description of phospholipid composition was on a daily basis [30].

V. PROTOZOA

The phospholipid composition of a number of protozoa has been described; among these are *Acanthamoeba castellanii*, *Entodinium caudatum*, *Tetrahymena pyriformis*, *Crithidia*, *Trypanosoma vivax*, *Entamoeba invadens*, and *Plasmodium knowlesi*. In general, the phospholipid composition of these protozoa somewhat resembles that of mammalian tissue; large amounts of PC and PE and minor amounts of PS, PI, and SM were reported to be present. For example, the distributin of phospholipid components for *E. invadens* is shown in Table 16 [31].

The complexity of the phospholipid composition of protozoa is further emphasized in the work of a number of other laboratories. It has been emphasized that the ciliate *Paramecium* has three ethanolamine sphingophospholipids and three ethanolamine sphingophosphonolipids, in addition to the ethanolamine glycerophospholipids and

Table 16 Phospholipid Composition of *E. invadens*

Phopholipid	% Total phospholipid
PC	37.2
SM	7.1
lyso-PC	2.9
PE	23.5
NCAEP	5.3
lyso-PE	1.2
PS	7.4
PI	2.7
CPI[a]	9.0
PA	2.4
Residual phospholipid	1.2

[a]Ceramide phosphorylinositol.

glycerophosphonolipids [32]. Further, two glyceryl ethers, 1-*O*-hexadecyl glycerol and 1-*O*-*cis*-octadec-11-enyl glycerol, chimyl and paramecyl alcohol, respectively, were quantified in total phospholipids as have glycerophospholipid classes for cells and cilia of the ciliated protozoon *Paramecium tetraurelia*. The ether content of 2-aminoethyl phosphonoglycerolipid was 85–90 mole %. Concentration of ethers were greatest in the ethanolamine phospholipids > phosphatidylcholines > phosphatidylserines > phosphatidylethanolamines > phosphatidylinositols. The glyceryl ether concentrations in total cellular phospholipids increased with culture age in *P. tetraurelia* cells [33].

In another study, sphingolipids were shown to make up 30 to 40 mole % of the phospholipids found in the surface membrane of *Tetrahymena pyrifomris* NT-1. Two major classes were identified: non-hydroxy fatty-acid-containing ceramide-

Table 17 Effect of Temperature on Fatty Acid Unsaturation of Phospholipids of *Tetrahymena pyriformis* NT-1()

	% of Total fatty acids		
Fatty acids	20°C	38°C	20°C—38°C
14:0	9.2	10.8	−1.6
16:0	18.0	20.0	−2.0
16:1	8.3	3.2	+5.1
18:0	1.2	2.3	−1.1
18:1	10.2	13.9	−3.7
18:2	25.6	22.9	+2.7
18:3	23.1	22.6	+0.5
19:0	1.5	1.9	−0.4
20:1	1.8	1.9	−0.1

2-aminoethylphosphonate (NCAEP) and α-hydroxy fatty-acid-containing ceramide-2-aminoethylphosphonate (HCAEP). Both classes were well represented in cells grown at 39°C. At this temperature, their principal long chain bases were *n*-hexadeca-4-sphingenine and *n*-nonadeca-4-sphingenine. The major fatty acid of NCAEP from 39°C-grown cells was palmitic acid and that of HCAEP was α-hydroxypalmitic acid. Cells grown at 15°C contained NCAEP, but only traces of HCAEP [34].

The broad effects of temperature on phospholipid composition are further illustrated in the following reports. A rise in the growth temperature of *Tetrahymena pyriformis* strain NT-1 from 15 to 34°C resulted in a marked increase in PE with a compensatory decrease in 2-aminoethylphosphonolipid. However, the level of PC remained unchanged [35]. In the same organism, the effect of temperature on unsaturation level in the phospholipids is shown in Table 17 [36]. It can be seen that the major effects on the fatty acids of the phospholipid were increases in 16:1 and 18:2 fatty acids and decreases in 16:0, 18:0, 18:1, and 19:0 fatty acids.

VI. INSECTS

Data for the phospholipid composition of cultured cells of the mosquito *Aedes aegypti* are summarized in Table 18 [37]. It can be seen that the major components are PE and PC.

In the honeybee *Apis mellifera*, the major sphingolipid has been identified as SM [38]. The long-chain bases are mainly the C_{14} homologue of sphingosine (D-*erythro*-1, 3-dihydroxy-2-amino-*trans*-4-tetradecene) and the C_{16} homologue of sphingosine (D-*erythro*-I,3-dihydroxy-2-amino-*trans*-4-hexadecene). The fatty acids are mainly normal saturated C_{20} and C_{22} fatty acids. About one tenth are 2-D-hydroxy fatty acids with similar paraffin chain distribution to the normal acids.

The adult filarial worms *Brugia pahangi* and *Brugia patei* were shown to contain PC, PE, PS, PI, PG, and CL components in labeling studies with C^{14}-glycerol-3-phosphate

Table 18 Phospholipid Composition of Cultured Cells of the Mosquito *Aedes aegypti*

Phospholipids	% of Total phospholipid
Total phospholipids	100
PA	1.1
CL	2.6
PE	46.4
PC	22.2
PS	5.5
PI	7.2
CPE[a]	10.6
CPC[b]	3.0
lyso-PC	0.1
lyso-PE	1.3

[a]= Ceramide phosphorylethanolamine.
[b]= Ceramide phosphorylcholine.

[39]. In labeling experiments with C^{14}-glycerol, the adult nematode *Haemonchus contortus* was demonstrated to synthesize PC, PE, PI, PS, SM, lyso-PC, and lyso-PE, with PC the major component.

An unusual effect of an environmental stress on phospholipid content is that of infection of the cotton leaf worm, *Spandoptera littoralis*, with *Bacillus thuringiensis*. The phospholipids were the major lipid fraction in normal and treated third instar larvae. In the normal larvae, the phospholipid was 37.5% of the total lipid, while in the treated larvae, the proportion was 61.1%, an increase of about 62% over the normal [40].

VII. GASTROPODS

The phospholipid composition of the land snail *Eobania vermiculata* was determined [41]. The major components were PC (49.2%) and PE (24.9%); with PC, 45.6% was present as the glycerylether analog, and with PE, 19.8% was the plasmalogen analog. Other phospholipids determined were NCAEP (7.5%) and DGNAEP (diglyceryl-NAEP) (6.3%) with minor amounts of PA, PI, SEA (sphingoethanolamine), and CL. This pattern of phospholipid composition of *E. vermiculata* is comparable to that given for other snails, such as the land snail *Cepaea nemoralis*, the fresh water snail *Lymnaea stagnalis*, and the marine snail *Aplysia kurodai*, as well as the slugs *Arion ater* and *Ariolimax columbianus*.

VIII. AMPHIBIANS

The phospholipid content of mature oocytes of *Bufo arenarum* and *Xenopus laevis* toads has been studied [42]. The major component was PC, followed by PE. Also detected in lesser amounts were PA, PI, PM, PS, and CL. The phospholipids present in retinal rod outer segments of the frog *Rana pipiens* have also been determined [43]. PC, PE, and PS were detected. It was noted that fatty acid composition of PC and PE were determined by a combination of factors, including rate of synthesis, rate of degradation, and selective interconversions.

IX. VARIOUS MARINE ORGANISMS

A. Jellyfish

The total lipid content found in *Pelagia noctiluca* (θ.19% of the whole-body weight) is comparable to that reported for other stinging jellyfish species. Its polar lipid components were found to represent 26.2% of the total lipids and to contain 75.7% glycerophospholipids and 24.3% sphingophosphonolipids. The overall composition of the polar lipids was CL, 9.1%; PE, 24.8% (of which 55% was glyceryl ether analog); PC, 36.3% (of which 63% was glyceryl ether analog); lyso-PE and lyso-PC, 5.0%; NCAEP, 21.0%; plus 3.3% as two minor species. Unsaturated fatty acyl groups represented about 52 and 33%, in PE and PG, respectively, while their glyceryl ether analogs were found to contain about 41 and 91% unsaturated fatty alkyl chains, respectively. In contrast, saturated fatty acyl groups with 14–16 carbon atoms were almost exclusive components (about 96%) of the major NCAEP species [44].

B. Crabs

Phospholipids constituted 70% of the total lipids of the gills of two crabs, *Cancer antennarius* and *Portunus xantusi*. PC (46–55% of the total phospholipid phosphorous) and PE (24–25%) were the principal phospholipids present. In both species 1-alkenyl glycerols were present in about 20% of the phospholipid molecules but were not detected in the neutral gill lipids. The total ether phospholipids of *C. antennarius* gills contained 62% 1-(1'-alkenyl) groups, with the remainder probably being 1-alkyl moieties [45].

C. Elasmobranchs

The phospholipids detected in the electric organ of the elasmobranch *Discopyge tschudii* were PA, PC, PE, PI, and PS. In addition, two polyphosphoinositides, PI-4-phosphate and PI-4,5 bisphosphate, were noted [46].

D. Fish

Of the total lipid composition of the orange roughy (*Hoplostethus atlanticus*), the black oreo (*Allocyttus* sp.), the small spined oreo (*Pseudocyttus maculatus*), and the sperm whale (*Physter macrocephalus*), the phospholipids were estimated to be between 0.1–1.0%]47]. In another study of the orange roughy, the distribution and fatty acid composition of the phospholipid fraction was further pursued in various tissues [48]. The data are shown in Table 19. In the roe and liver, the major fatty acids in the phospholipids were, in order, 22:6, 18:1, and 16:0.

A further demonstration of the dependence of phospholipid composition on tissue type is provided by comparing dark and white muscle of the blue mackerel (*Scomber australasicus*) as shown in Table 20 [49]. Finally, it is relevant to complicate further the consideration of phospholipid occurrence in nature by briefly commenting on the interaction between membrane type and environmental stress.

The phospholipid composition of purified outer and inner brain mitochondrial membranes from temperature-acclimated goldfish (*Carassius auratus* L.) has been reported [50]. In the outer membrane, the major components were PC and SM at both 5° and 30°C,

Table 19 Distribution of Phospholipids in Tissues of Orange Roughy (*H. atlanticus*)

Tissues	Phospholipid (mg/g lipid)
Liver	230
Swim bladder	2.7
Muscle	4.8
Skin	7.0
Roe	187
Testes	199

Table 20 Distribution of Phospholipids in Dark and White Muscle of the Blue Mackerel

	% Wt	
Phospholipid	Dark muscle	White muscle
CL	0.3	1.0
PE	1.9	4.0
PC	6.2	11.3
SM	1.5	2.5

with lesser amounts of PA, PE, PS, PI, lyso-PC, and CL at 30°C, as compared to 5°C. The only change noted was an increase in PS. In the inner membrane, the major components were PC and PE, and they were significantly reduced in concentration at 30°C. The concentrations of PA, PS, SM, and CL showed a significant increase at 30°C. Clearly, the response of each membrane differs in the adjustments it must make in phospholipid composition in response to temperature stress.

X. CONCLUSION

In considering the information presented in this chapter, there are certain unavoidable conclusions to be reached. Phospholipids are ubiquitous in their occurrence in nature. The variety of phospholipid components is impressive. The distribution of phospholipids, even within a single cell, is complex, occurring in a number of distinct membranes, for example. Within each membrane, the phospholipid components may show considerable individuality, depending upon the environment and function of the particular membrane. Beyond these internal environmental influences, there are the external stresses such as temperature, pH, salt concentration, light, and pressure, which cause alterations in phospholipid composition as the cell adjusts its chemistry to deal with and survive harsh environments. Phospholipids are an integral part of the biological machinery and are continuously fine-tuned to the advantage of living matter.

REFERENCES

1. C. Ratledge, in *Biotechnology*, Vol. 4 (H. Pape and H.-J. Rehm, eds.), Ch. 7, VCH, New York, 1986, pp. 185–213.
2. C. Ratledge, *Adv. Behavioral Sci. 33*:17–35 (1986).
3. K. S. Raju, R. Maheshwari, and P. S. Sastry, *Lipids 11*:741–746 (1976).
4. N. J. Russell, *Biochem. Soc. Trans. 11*:333–335 (1983).
5. O. Suzuki, T. Yamashima, and T. Yokochi, *Yakagaku 30*:854–862 (1981).
6. B. J. Bowman, C. E. Borgeson, and E. J. Bowman, *Exp. Mycol. 11*:197–205 (1987).
7. H. Goldfine, *J. Lipid Res. 25*:1501–1507 (1984).
8. M. Thaler and H. Diekmann, *Eur. J. Appl. Microbiol. 6*:379–387 (1979).
9. E. F. De Long and A. A. Yayanos, *Science 228*:1101–1103 (1985).

10. H. K. Jindal, V. S. Bansai, C. Kasinathan, S. Larroya, and G. K. Khuller, *Experentia 39*:151–153 (1983).
11. V. C. Carey and L. O. Ingram, *J. Bacteriol. 154*:1291–1300 (1983).
12. S. H. Yoon and J. S. Rhee, *J. Am. Oil Chem. Soc. 60*:1281–1286 (1983).
13. Y. Hasegawa, N. Kawada, and Y. Nosoh, *Arch. Microbiol. 126*:103–108 (1980).
14. J. Reizer, N. Grossowicz, and T. Barenholz, *Biochim. Biophys. Acta. 815*:268–280 (1985).
15. M. Wada, N. Fukunaga, and S. Sasaki, *Plant Cell Physiol. 28*:1209–1217 (1987).
16. H. Tsuchiya, M. Sato, N. Kanematsu, M. Kato, Y. Hoshino, N. Takagi, and I. Namikawa, *Lett. Appl. Microbiol. 5*:15–18 (1987).
17. K. Hunter and A. Rose, *Biochim. Biophys. Acta. 260*:639–653 (1972).
18. Y. Ohno, I. Yano, and M. Masui, *J. Biochem. 85*:413–431 (1979).
19. A. K. Rajasekaran and R. Maheswari, *Indian J. Exp. Biol. 28*:134–137 (1990).
20. J. D. Vogt and S. Brody, *Biochim. Biophys. Acta. 835*:176–182 (1985).
21. R. M. Parmegiani and M. A. Pisano, *Dev. Ind. Microbiol. 15*:318–323 (1974).
22. G. A. Thompson, Jr., in *Cellular Acclimitisation to Environmental Change* (A. R. Cossins and P. Sheterline, eds.), Cambridge University Press, Cambridge, 1983, pp. 33–53.
23. A. H. Scragg and R. R. Leathers, in *Single Cell Oil* (R. S. Moreton, ed.), Longman Scientific and Technical, New York, 1988, pp. 71–98.
24. A. Cobelas and J. Z. Lechado, *Grasas Y Aceites. 40*:118–145 (1989).
25. J. L. Harwood, T. P. Pettitt, and A. L. Jones, *Annual Proc. Phytochem. Soc. 28*:49–67 (1988).
26. M. Schneider, in *World Conference on Emerging Technologies in the Fats and Oils Industry* (A. R. Baldwin, ed.), American Oil Chemists Society, 1986, pp. 160–164.
27. D. L. Dornbos, R. E. Mullen, and E. G. Hammond, *J. Am. Oil Chem. Soc. 66*:1371–1373 (1989).
28. M. N. Chirstiansen, *Recent Adv. Phytochem. 18*:177–195 (1984).
29. R. J. Pollero, *Aquatic Bot. 26*:103–111 (1986).
30. F. Lalaguna and M. Agudo, *Phytochem. 28*:2059–2062 (1989).
31. H. H. D. M. van Vliet, J. A. F. Op den Kamp, and L. L. M. van Deenen, *Arch. Biochem. Biophys. 171*:55–64 (1975).
32. D. F. Matesic and E. S. Kaneshiro, *Biochem. J. 222*:229–233 (1984).
33. E. S. Kaneshiro, K. B. Meyer, and D. E. Rhoads, *J. Protozool. 34*:357–361 (1987).
34. K. Kaya, C. S. Ramesha, and G. A. Thompson, Jr., *J. Lipid Res. 25*:68–74 (1984).
35. S. Umeki, H. Maruyama, and Y. Nozawa, *Biochim. Biophys. Acta. 752*:30–37 (1983).
36. J. G. Connolly, I. D. Brown, A. G. Lee, and G. A. Kerkut, *Comp. Biochem. Physiol. 81A*:287–292 (1985).
37. A. Luukkonen, M. Brummer-Korvenkontio, and O. Renkonen, *Biochim. Biophys. Acta. 326*:256–261 (1973).
38. S.-G. Karlander, K.-A. Karlsson, H. Leffler, A. Lilja, B. E. Samuelsson, and G. O. Steen, *Biochim. Biophys. Acta. 270*:117–131 (1972).
39. A. K. Srivastava, R. D. Walter, and J. J. Jaffe, *Int. J. Parasitol. 17*:1321–1328 (1987).
40. I. Z. Boctor and H. S. Salama, *J. Invertebrate Pathol. 41*:381–384 (1983).
41. H. J. Stavrakakis, S. K. Mastronicolis, and V. M. Kapoulas, *Z. Naturforsch. 44C*:597–608 (1989).
42. T. S. Alonso, I. C. Bonini de Romanelli, and A. M. Pechen de D'Angelo, *Comp. Biochem. Physiol. 86B*:167–171 (1987).
43. K. Louie, R. D. Wiegand, and R. E. Anderson, *Biochemistry 27*:9014–9020 (1988).
44. I. C. Nakhel, S. K. Mastronicolis, and S. Miniadis-Meimaroglou, *Biochim Biophys. Acta. 985*:300–307 (1988).
45. S. Chapelle, J. L. Hakanson, J.C. Nevenzel, and A. A. Benson, *Lipids 22*:76–79 (1987).
46. N. P. Rotstein, H. R. Arias, M. I. Aveldano, and F. Barrantes, *J. Neurochem. 49*:1341–1347 (1987).

47. D. H. Buisson, D. R. Body, G. J. Dougherty, L. Eyres, and P. Vlieg, *J. AM. Oil Chem. Soc. 59*:390–395 (1982).
48. M. R. Grigor, C. R. Thomas, P. D. Jones, and D. H. Buisson, *Lipids 18*:585–588 (1983).
49. D. R. Body and P. Vlieg, *J. Food Sci. 54*:569–572 (1989).
50. M. C. J. Chang and B. I. Roots, *J. Therm. Biol. 13*:61–66 (1988).

3

Isolation and Analysis of Phospholipids and Phospholipid Mixtures

K.- H. Gober, B. R. Günther, E. M. Lünebach, G. Repplinger, and M. Wiedemann *Rhône-Poulenc Rorer GmbH, Köln, Germany*

The phospholipids are essential components of all vegetable and animal cells. This chapter provides a methodological overview of the various methods of isolating and analysing phospholipids. Only soybean and egg phospholipids and their partially or completely hydrogenated derivatives have as yet acquired great industrial importance. Hence the methods described here have been primarily developed for soybean phospholipids, but in the majority of cases they can be applied without modification to phospholipids from other sources.

The following abbreviations have come to be accepted in the literature for the most common ester phospholipids: phosphatidylcholine PC, phosphatidylethanolamine PE, phosphatidylinositol PI, phosphatidic acid PA.

I. ISOLATING NATURAL PHOSPHOLIPIDS

The classes of phospholipids discussed in this section encompass the most important phospholipids of the ester phospholipids, ether phospholipids, sphingosine phospholipids, and phosphonophospholipids [1]. Each membrane type exhibits a characteristic phospholipid distribution [2–5].

The distribution of the fatty acids among the various classes of phospholipids is by no means uniform. Thus in the ester phospholipids of the PC, PE, PI, and PA classes the saturated fatty acids are primarily to be found in the *sn*-1 position and the unsaturated fatty acids primarily in the *sn*-2 position [6]. The individual phospholipid classes PC, PA, etc. can be subdivided into different molecular species with defined fatty acid distributions [2, 7–11]. This fact must be taken into account when it is desired to isolate a whole phospholipid class by means of a preparative separation method.

The choice of a suitable method for the isolation of the phospholipids depends basically on the type of raw material available, the amounts required, and the apparatus available. In this connection preparative thin-layer chromatography has acquired practical importance as a simple method of isolation.

In general, however, resort is usually made to the classical three-step method of isolation:

1. Extraction of the phospholipids (lipids) from the cell mass
2. Fractionation (enrichment) of the desired phospholipids
3. Isolation (purification) of the phospholipids by chromatography

During all preparative work it must be remembered that the unsaturated acyl residues of the phospholipid molecules are subject to autocatalytic oxidation in the presence of air, even at room temperature. Work with biological or other unsaturated phospholipids should, therefore, be preferably carried out in an atmosphere of protective gas (argon, nitrogen) [12, 13].

A. General View of Preparative Methods

The vegetable phospholipids, and here, in particular, the soybean phospholipids, are of the greatest economic importance at present [1]. So some preparative methods for the isolation of soybean phospholipids have been included in the methodological section. The three basic techniques of isolation and enrichment of phospholipids are dealt with below.

1. Extraction of the Phospholipids from the Cell Mass

The best known process carried out on an industrial scale is the extraction of edible oils from plant seed. The vegetable phospholipids are extracted at the same time [1].

For laboratory extractions it is generally practicable to carry out the extraction with a mixture of chloroform and methanol. This extracts the lipids and the phospholipids [14]. This extraction mixture can be used to extract all classes of phospholipids (ester phospholipids, ether phospholipids, sphingosine phospholipids, and phosphonophospholipids) from the cell mass [2, 15]. The solvent systems hexane/isopropanol and tetrahydrofuran/water, to name but two, are also suitable. A complex mixture comprising phospholipids, glycolipids, sterols, carbohydrates, and neutral lipids remains after the solvents have been evaporated. This mixture can be separated in a fractionation step.

2. Fractionation of Phospholipids (Solvent Extraction)

The complex lipid extracts can be fractionated by very different methods. In de-oiling, the neutral lipids are extracted with acetone or acetone-containing mixtures [1].

Recently there have also been descriptions of de-oiling processes where the neutral lipids are extracted with supercritical gases, for example carbon dioxide. Carbohydrates are generally removed by alcoholic/aqueous extraction. Enzymatic methods are also known [1]. A process for the removal of neutral lipids, sterols, and carbohydrates is described in the methodological section (I.B.2).

The individual classes of ester phospholipids PC, PE, PI, PA, etc. have different solubilities in alcohols. This fact has led to the development of many industrial processes for the enrichment of phosphatidylcholine, using alcoholic fractionation for example [1, 16–19]. The method of alcoholic fractionation of soybean phospholipids described in the methodological section represents one of many alternatives (I.B.1).

Here the enrichment of PC is described, on the one hand, and of PA, PI, and PE on the other. The choice of a suitable fractionation method depends, among other things, on the quantity of lipid extract available. The individual substance groups are frequently isolated directly by suitable column chromatographic separations. Some typical column chromatographic methods are described below.

3. Isolation of Phospholipids (Column Chromatography)

Preparative high-pressure liquid chromatography is a very effective but technically complicated method of isolating phospholipids. The soybean phospholipids can be separated on silica gel using hexane/2-propanol/water as eluent [20] or with acetonitrile/methanol/phosphoric acid as eluent [21]. A method of chromatography on silica gel using a gradient of chloroform/methanol as eluent has been described for the isolation of egg phospholipids [22]. Furthermore, it is also possible to isolate these phospholipids by DEAE silica gel chromatography [23].

The phosphonolipids can also be isolated in pure form by silica gel chromatography using methanol and chloroform as eluent [24, 25].

The chromatographic procedures described so far are perfectly suitable for the isolation of small quantities of high purity pohospholipids or classes of phospholipids on a laboratory scale. A simple laboratory chromatography method for the isolation of individual phospholipids without involving great technical complications is described in the methodological section (I.B.3). Here the phospholipids are separated on silica gel using a chloroform/methanol gradient.

At the present time only soybean phosphatidylcholine is of any importance from the point of view of applications technology. Soybean, sunflower, rape, and egg phosphatidylcholines are readily isolated in larger quantities by chromatography on aluminum oxide or silica gel using ethanol as eluent. But it is necessary to fractionate the phospholipids before both processes. A very simple method allowing the preparation of large quantities of soybean phosphatidylcholine without previous solvent fractionation is described in the methodological section (I.B.4). Here the original lipid extract is fractionated on silica gel using hexane/2-propanol/water as eluent. This method allows the isolation of phosphatidylcholine of more than 85% purity in one step, without any special technical complication.

B. Preparative Methods

1. Fractionation of Soybean Phospholipids by Alcoholic Extraction

References

H. Pardun, *Die Pflanzenlecithine*, *Verlag f. Chem. Industrie H. Ziolkowsky*, *Augsburg*, 1988; A. J. Dijkstra, J. de Kock, Process for fractionating phosphatide mixtures, *Eur. Pat. Appl. 0 372 327 A 2*, 1990; P. Reichling, Pure natural choline diglyceride phosphates from soybean phosphatides, *Ger. Offen. 1.905. 253*, 1970; N. V. Unilever, Phosphatides extraction, *Neth. Appl. 68 13.248*, 1969; H. Liebing and J. Lau, Process for the recovery of lecithin fractions, *Fette, Seifen, Anstrichm. 78*:123–127 (1976).

Materials: Crude soybean lecithin, methanol, 96% ethanol.

Apparatus: Three-necked flask with stirrer and reflux condenser, rotary evaporator.

Method

Two modifications with different aims are described. Modification 1 is aimed at producing a product with as high a PC enrichment in the extract as possible and as high a yield of PC as possible. The aim in modification 2 is to produce a phospholipid mixture with as high a concentration of PE, PA, and PI as possible.

Modification 1

One part by weight crude soybean lecithin is extracted at room temperature with three parts by volume methanol. After stirring for 1 h the insoluble components are allowed to sediment. The clear, methanolic supernatant is decanted off and evaporated to dryness in the rotary evaporator.

Extraction yield	20–30% of theory
PC assay	35–45%
PC yield	55–65% of theory

Modification 2

One part by weight crude soybean lecithin is extracted three times exhaustively for 1 h with 3 parts by volume 96% ethanol at 75°C. The hot ethanolic supernatant is carefully separated off after each extraction. The remaining ethanol-insoluble residue is evaporated to dryness on the rotary evaporator.

	PA	PI	PE
Assay	22–28%	20–24%	20–24%
Yield % of theory	⩾ 80%	⩾ 60%	⩾ 30%

Remarks

- The ethanolic extraction is carried out on the industrial scale. The extracts find application in the foodstuff, diet and cosmetic fields. H. Pardun has produced an excellent review [1].
- The phospholipid content and the phospholipid yields of the extract (alcohol-soluble fractions) and extraction residues (alcohol-insoluble fractions) depend on the phospholipid composition of the lecithin.
- The following parameters have a great effect on the composition of the extract:

Extraction temperature
Temperature of the extract phase (sedimentation temperature)
Ratio of lecithin to alcohol
Countercurrent or cocurrent extraction
Number of Extraction steps
Water content of the alcohol

These parameters must be adjusted to suit particular problems.

2. Fractionation of Soybean Lecithin (Neutral Lipids, Carbohydrates, and Phospholipids)

References

H. Pardun, *Die Pflanzenlecithine, Verlag f. Chem. Industrie H. Ziolkowsky, Augsburg, 1988*; *US Pat. Appl. 3.798.246*, 1972; *US Pat. Appl. 2.727.046*, 1951.

Materials: Crude vegetable lecithins (soybean, sunflower, rape) dissolved in hexane, silica gel, 80% aqueous methanol.

Apparatus: Three-necked flask with stirrer, rotary evaporator, separating funnel, laboratory equipment for thin-layer chromatography.

Method

Suspend 300 g silica gel in 900 ml hexane. A solution of 150 g crude lecithin in 450 ml hexane is added with stirring. Stir for 30 minutes at room temperature and then separate off the silica gel. The neutral lipids, sterols, and sterol derivatives remain with the silica gel. The hexane filtrate is concentrated to a final volume of 300 ml and then extracted successively with four 150 ml portions of 80% aqueous methanol. The phospholipids remain in the hexane phase, while the carbohydrates are predominantly extracted into the aqueous methanol phase. The remaining hexane phase is evaporated to dryness. The phospholipid yield from the crude lecithin is 40–50% of the theory. Traces of saccharides and neutral lipids are detectable in the final product. The phospholipid fraction is composed of at least 95% phospholipids.

Remarks

- Neutral lipids, sterols, and sterol derivatives can be extracted from loaded silica gel using ethanol. After the ethanol has been distilled off, this fraction can be separated by treatment with acetone into a neutral lipid fraction (acetone-soluble) and a sterol/sterol derivative fraction (acetone-insoluble).
- The proper ratio of silica gel to crude lecithin depends on the neutral lipid content of the crude lecithin. A thin-layer chromatographic check is to be recommended in any event.
- Instead of stirring with silica gel it is also possible to carry out silica gel column chromatography. The silica gel should be activated.

3. Isolation of Pure Phospholipids

References

P. Van der Meeren, J. Vanderdeclen, M. Huys, and L. Baert, Optimization of the column loadability for the preparative HPLC separation of soybean phospholipids, *J. Am. Oil Chem. Soc. 67*:815–820 (1990); W. J. Hurst, R. A. Martin, Jr., and R. M. Sheeley, The preparative HPLC isolation and identification of phospholipids from soy lecithin, *J. Liq. Chromatogr. 9*:2969–2976 (1986); R. S. Fager, S. Shapiro and B. J. Litman, A large-scale purification of phosphatidylethanolamine, lysophosphatidylethanolamine and phosphatidylcholine by high performance liquid chromatography: A partial resolution of molecular species, *J. Lipid Res. 18*:704–709 (1977); Un Hoi Do and Shi-Lung Lo, Separation of phospholipids using a diethylaminoethylsilica gel column and thin-layer chromatography, *J. Chromatogr. 381*:233–240 (1986); M. C. Moschidis, Phosphonolipids, *Prog. Lipid Res. 23:*223–246 (1985); M. C. Moschidis, Silicic acid column chromatography of phosphonolipids, *J. Chromatogr. 435*:508–12 (1988); W. S. M. Geurts van Kessel, M. Tieman, and R. A. Demel, Purification of phospholipids by preparative high-pressure liquid chromatography, *Lipids 16*:58–63 (1981).

Materials: Phospholipid mixtures, silica gel (Mallinkrodt CC-4) activated overnight at 110°C, chloroform, methanol.

Apparatus: Chromatography columns, rotary evaporator, laboratory equipment for thin-layer chromatography.

Method

The activated silica gel is suspended in chloroform and packed into the chromatographic column. The loading ratio is 1 g phospholipid to 50 g silica gel. The phospholipid mixture is dissolved in chloroform (10 mg/ml); the solution is applied to the column. The elution is carried out with a chloroform/methanol mixture containing increasing proportions of methanol. Each step requires 1- to 5-fold the column volume in eluent. The elution volume per step depends on the phospholipid composition.

Elution series

%Methanol in chloroform	Lipid
0	neutral lipids
2	PA
2–4	PG
5–10	fatty acids
20	PS, PE
30	PI
50–60	PC
70	sphingomyelin
> 80	lysophospholipids

Notes

- The eluate fractions must be controlled by thin-layer chromatography.
- This frequently described standard method is suited to the preparation of pure phospholipids on the mg scale.
- If the degree of purity is not adequate an additional chromatography step is necessary.

4. Isolation of Vegetable and Animal Phosphatidylcholines

References

Ger. Appl. 3.445.949; *Ger. Appl. 3.445.950.*

Materials: 200 g silica gel, solvent and eluent mixtures, hexane or light petroleum ether, 2-propanol, water 1:1:0.175 (v/v/v), 450 ml of a solution of 110 g crude lecithin (soybean, sunflower, rape) or 90 g extract of egg yolk powder in the solvent mixture described.

Apparatus: Glass column with dropping funnel, rotary evaporator, laboratory equipment for thin-layer chromatography.

Method

Apply 450 ml of the lecithin solution to the silica gel-filled chromatographic column at room temperature and elute with the same solvent mixture. The complete eluate is divided into two fractions. The fractions are evaporated to dryness and all residues are analysed.

The first fraction (0–1.6 L eluate) contains neutral lipids, glycolipids and the phospholipids PE, *N*-acyl-PE, PA, and PI. The second fraction (1.8–4.4 L eluate) contains phosphatidylcholine that is almost free from accompanying phospholipids. The PC content is ⩾85%; the PC yield ⩾80% of theory.

Remarks

- This procedure is suitable for the isolation of large quantities of phosphatidylcholines with fatty acid compositions typical of soybean, sunflower, rape, and egg.
- Insoluble components of the lecithin, in particular of sunflower and rape lecithins, should be filtered off first.

- On account of autoxidation all steps should be carried out, as far as possible, under a blanket of nitrogen or argon.
- The fractionation of the eluates should be checked by thin-layer chromatography (silica gel plates, chloroform/methanol/water/25% ammonia = 65:30:4:2 (v:v:v:v)).
- Traces of glycolipids and foreign phospholipids can be removed from the final product by dissolution in ethanol and then filtration through a little aluminum oxide.

II. ANALYSIS OF PHOSPHOLIPIDS AND PHOSPHOLIPID MIXTURES

The samples for investigation are usually natural products with a particular fatty acid pattern. This means that the individual phospholipids (phosphatidylcholine, phosphatidylethanolamine, phosphatidylinositol, etc.) are not single pure substances in the chemical sense but mixtures of homologous compounds whose composition shows certain natural variations. This influences the behavior and the properties of the phospholipids and has to be taken into account in all work and investigations.

The section that follows contains a selection of analytical methods. Methods that have not previously been published are described in detail. Methods already known from the literature, that are employed in modified form, have been described briefly with details of the changes made and are accompanied by the relevant references. In all other cases only the relevant literature is cited.

A. Method Overview with Brief Commentary

1. Identification

The methods described in II.B.1 are suitable for the identification of individual phospholipids. Depending on the mode of detection it is also possible to distinguish between different groups of phospholipids.

2. Quantification

Phospholipids in general: The method given in II.B.5 (phosphorus determination after TLC separation) is suitable for the quantitative determination of all phospholipids if adequate separation has been achieved and the molecular weight is known. The time-consuming nature of the method is a disadvantage.

Soybean phospholipids: All the methods mentioned can be employed. The soybean phospholipid concerned is required as a standard substance in the case of methods II.B.2 through II.B.4 and II.C.1 through II.C.3. The methods recommended are II.B.2 to II.B.4 (TLC), II.C.2 and II.C.3 (HPLC).

Phospholipids in general with unsaturated fatty acids: What was said for soybean phospholipids applies here too with the exception that the standard substances must correspond to the phospholipids investigated in their fatty acid composition. The methods recommended are II.B.4 (TLC), II.C.2, and II.C.3 (HPLC).

Phospholipids with saturated or hydrogenated fatty acids: All methods are applicable apart from II.B.2, II.C.2, and II.C.3. In the case of the methods under II.B.3, II.B.4, and II.C.1, the standard substances and the corresponding components of the sample should agree as far as possible in their fatty acid compositions. The methods recommended are

II.B.5 (phosphorus determination after TLC separation) or II.C.1 (HPLC with light scattering detection).

Choline-containing phospholipids: Method II.D.1 involves choline determination after enzymatic cleavage followed by conversion of the choline content to phosphatidylcholine. Other choline-containing components falsify the value. The fatty acid composition must be known approximately on account of the necessary molecular weight.

Acetone-insoluble components: Method II.D.3 allows the determination of the acetone-insoluble components of a mixture of phospholipids. On account of the low specificity it can only be used to yield a broad orientation concerning the total phospholipid content.

Other components of phospholipid mixtures: Suitable methods for the determination of these components must be chosen from the references given uner II.E.

B. Thin-Layer Chromatographic Methods

1. Identification of Phospholipids

Principle of operation

The phospholipid mixture is applied to a thin-layer plate. The plate is developed. The separated components are detected using the reagent of choice. Identification is made on the basis of the Rf value or by comparison with standard substances applied at the same time.

References

DFG—Einheitsmethoden—Abteilung F—Fettbegleitstoffe F-I 6 (68); J. C. Touchstone, S. S. Levin, M. F. Dobbins, and P. J. Carter, *HRC & CC. 4*, 423–424, Differentiation of saturated and unsaturated phospholipids on TLC (1981); I. Rustenbeck and S. Lenzen, *J. Planar Chrom. 2*:470–472 (1989), Densitometric quantification of minor phospholipids via improvement of the scanning mode for cupric sulfate staining; J. C. Dittmer and R. L. Lester, *J. Lipid Res. 5*:126–127 (1964), A simple, specific spray for the detection of phospholipids on TLC; C. Touchstone and J. G. Alvarez, *J. Chrom. 429*:359–371 (1988) [*TLC/HPLC*], Phospholipids in amniotic fluid with special reference to the lecithin/spingomyelin ratio.

Materials: Precoated silica gel 60 plates (Merck) 200 × 200 mm^2.

Mobile phase: 130 ml chloroform p.a., 60 ml methanol p.a., 8 ml dist. water, 4 ml 25% ammonia p.a.

Detection reagent: (freshly prepared in the following order) 0.5 ml anisaldehyde, 50 ml glacial acetic acid p.a., 1 ml conc. sulphuric acid p.a.

Equipment

Development chamber for 200 × 200 mm^2 plates; drying cupboard.

Procedure

Prepare a 5% solution of the sample in chloroform. Apply 1–5 μl of this solution to a TLC plate.

Develop the plate with the above mentioned mobile phase (running distance ca. 120 mm).

After drying, the plate is sprayed with the detection reagent and heated to 105°C in the drying cupboard for ca. 20 min. After this time all components are clearly visible on a pale pink background (see Fig. 1).

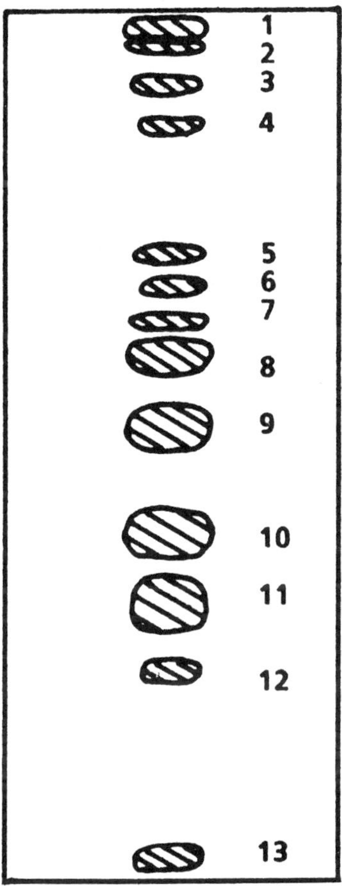

Figure 1 Thin-layer chromatographic separation of a crude soybean phosphatide. 1, triglycerides; 2, cholesterol; 3, sterol glycoside esters; 4, *N*-acylphosphatidylcholine; 5, sterol glycosides; 6, unknown; 7, free fatty acids; 8, phosphatidylethanolamine; 9, phosphatidylcholine; 10, phosphatidylinositol; 11, phosphatidic acid; 12, lysophosphatidylcholine; 13, starting point.

Recommendations

The mobile phase cited in the literature has been adapted to yield better separations. The following detection reagents can also be employed: phosphate reagent according to Dittmer and Lester; ninhydrin spray reagent 0.1% (Merck) for the detection of amines; copper salt solution (acidified with phosphoric acid) as ignition reagent.

2. Fluorescence Stimulation by Sulphuryl Chloride

Principle of operation

The phospholipids are separated by thin-layer chromatography, exposed to sulphuryl chloride vapors, and heated. The fluorescence that is produced is measured.

Materials: (3-*sn*-phosphatidyl) choline from soybeans, reference substance with exactly defined (3-*sn*-phosphatidyl) choline assay, crude phosphatide (qualitative comparison sample), 26% ammonia solution p.a., chloroform p.a., methanol p.a., dist. water, sulphuryl chloride synthesis grade, hexane p.a., paraffin viscous, precoated silica gel 60 HPTLC plates (100 × 200 mm^2) Merck.

Equipment

Drying cupboard, development chamber for 220 × 100 mm^2 plates, 50.00 ml volumetric flask, Camag Nanomat or Camag Probenautomat, hot-air dryer, Camag densitometer with evaluation, 1 ml ampoules, deep-freeze container (liquid nitrogen).

Procedure

Preparation of thin-layer plates: Predevelop the thin-layer plates in a mixture of equal volumes chloroform and methanol and activate overnight in the drying cupboard at ca. 60°C.

Reference substance solution: Weigh out exactly so much reference substance into a 50.00 ml volumetric flask that the solution contains exactly 120 mg (3-*sn*-

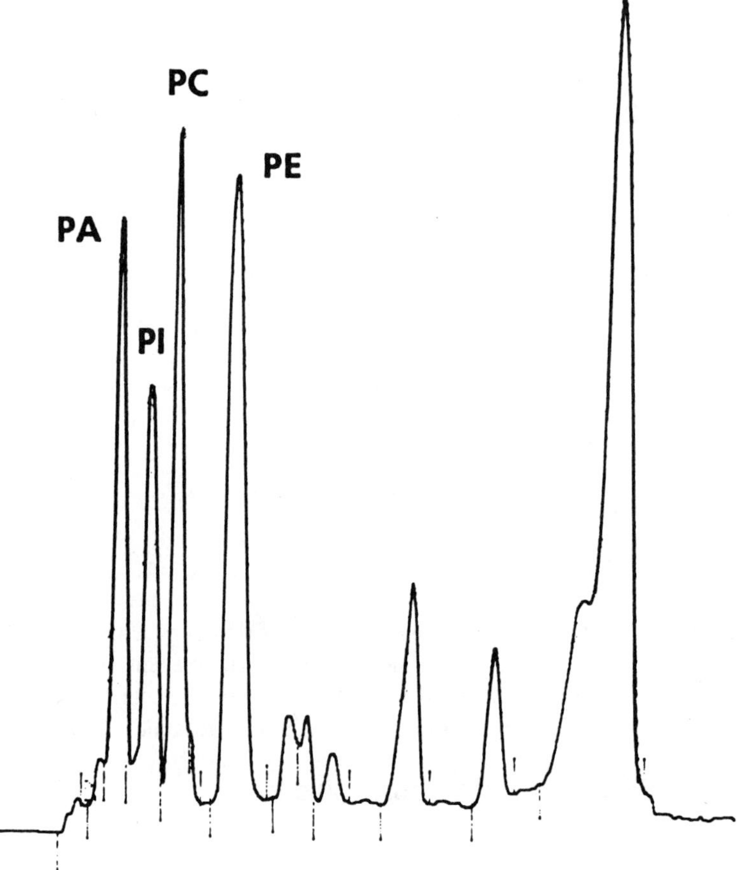

Figure 2 Scan of the fluorescence of a thin-layer chromatographic separation of a crude soybean phosphatide.

phosphatidyl)choline in 50.00 ml. Immediately after preparation fill the solution into 1 ml ampoules and store in liquid nitrogen. Ampoules removed from the deep-freeze container must be used immediately.

Qualitative comparison solution: Dissolve ca. 650 mg crude phosphatide in 50.00 ml chloroform.

Sample solution: Weigh out and dissolve in chloroform sufficient sample solution (amount weighed out = *E* [mg]) that the solution contains ca. 120 mg phosphatidylcholine in 50.00 ml.

Chromatography: Apply both the sample solution and the reference solution three times as spots to the prepared thin-layer plate. Application volume: 100 nl per spot. Also apply 1 × 100 nl qualitative comparison solution. The development is carried out at ca. 21°C in chloroform/methanol/water/ammonia (65/30/4/2 v/v) (with chamber saturation, running distance 50 mm).

The developed plate is dried in a stream of air for ca. 10 min, then exposed to sulphuryl chloride [7] for ca. 15 min, and then heated to 110°C in the drying cupboard for 10 min. The plate must be protected as much as possible from light until it is scanned. After cooling, the plate is dipped briefly into a mixture of equal volumes of hexane [8] and paraffin oil [9] (improvement of fluorescence). The hexane is evaporated in a stream of air. The densitometric evaluation follows by recording the fluorescence over 400 nm (edge filter 400 nm) with an excitation wavelength of 366 nm (see Fig. 2).

The fluorescences of the individual spots are measured. The results for the reference spots are averaged and compared with the mean calculated for the sample spots. The assignment of the spots is carried out by comparison with the spots produced by the qualitative comparison solution (see below for Rf values).

Calculation

$$\frac{120 \times Ai \times 100}{Ar \times E} = \%\ \text{phospholipid (i) in the sample}$$

E = amount of sample weighed out in mg
Ar = area of reference spot
Ai = area of spot of component (i)

Rf values of individual phospholipids:

Components	Rf value
Lysophosphatidylcholine	0.17
Phosphatidic acid	0.12
Phosphatidylinositol	0.18
Phosphatidylcholine	0.24
Phosphatidylethanolamine	0.37
(*N*-acyl-phosphatidyl)ethanolamine	0.60

Recommendations

This method can only be used to determine soybean phosphatidylcholine exactly. Should reference substances be available for all the substances of interest, then these can also be applied in defined quantities, and the amounts of the individual components can be calculated with respect to the particular reference substance.

Since the fluorescence measured for determination is affected by the fatty acid distribution, all reference substances should possess as nearly as possible the same fatty acid distribution as the corresponding components in the samples.

The fluorescence produced fades with time so the measurement should be carried out as quickly as possible.

It is advisable to check the thin-layer plate for faulty areas before use.

If the sample is not soluble in chloroform a mixture of chloroform and methanol can be employed.

Caveats

The method has only been tested for phospholipids from soybeans; it cannot be used for phospholipids containing only saturated fatty acids.

3. Fluorescence Excitation by NBD-Dihexadecylamine

Principle of operation

The sample under investigation is applied to a thin-layer plate. After the plate has been developed in chloroform, methanol, ethanol, triethylamine, 0.25% aqueous potassium chloride solution, 25% ammonium hydroxide (40/20/20/35/6/1.5 v/v) with the addition of 0.02% 4-(*NN*-dihexadecyl)amino-7-nitrobenz-2-oxa-1,3-diazole (NBD-dihexadecylamine), fluorescent spots are formed whose fluorescence is measured.

References

L. Colarow, *J. Planar Chrom. 2*:19–23 (1989). Quantitation of phospholipid classes on thin-layer plates with a fluorescence reagent in the mobile phase.

Recommendations

The following mobile phase has also been employed successfully under the conditions described in the literature: 65 ml chloroform/30 ml methanol/4 ml water/2 ml ammonia solution (25%).

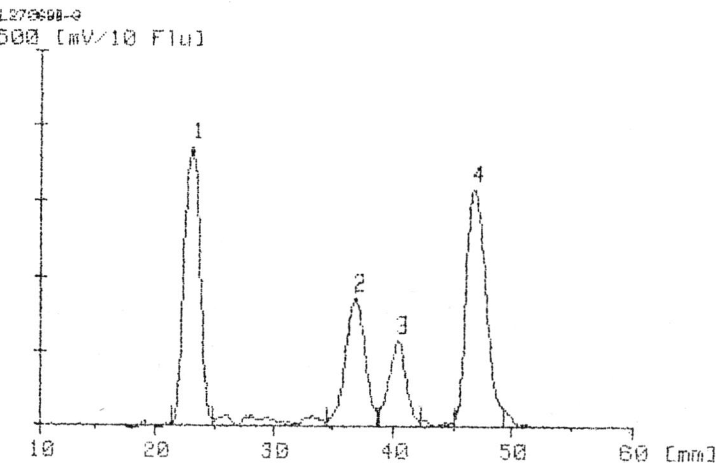

Figure 3 Scan of the fluorescence of a thin-layer chromatographic separation of PC (1), PA (2), PI (3), and PE (4).

It must be checked, from case to case, how far this method can be applied with other mobile phases.

Should reference substances be available for all components of interest, then defined quantities of these can also be applied to the plate and the calculation of the amounts of the individual components referred to the reference substance for that particular component.

According to the reference given above it is not necessary in most cases to use a conversion factor for differing phospholipids from the same source. On the other hand, it is possible to calculate a conversion factor from the number of double bonds for products with differing fatty acid distributions (from different sources). A flat chamber is recommended for development to minimize the consumption of solvent and hence of NBD-dihexadecylamine. NBD-hexadecylamine is sensitive to light and must be stored below 0°C. Figure 3 illustrates the scan of a phospholipid mixture after development in a flat chamber using the mobile phase reported in the literature.

It is advisable to check the thin-layer plate for faulty areas before use.

Caveats

The developed plates must not be immersed in liquid paraffin. It has not been investigated whether other dipping agents might be suitable for improving the fluorescence.

Only phosphatidylcholine can be determined exactly by the method as described. The other components are reported after calculation as phosphatidylcholine.

4. Heating with Copper Salts

The phospholipids are separated by thin-layer chromatography. After development the plates are dipped into an acidic copper salt solution, dried, and heated. The absorptions of the components, that become visible as dark spots, are evaluated densitometrically.

References

J. C. Touchstone, S. S. Levin, M. F. Dobbins, and P. J. Carter, *HRC & CC. 4*:423–424 (1981), Differentiation of saturated and unsaturated phospholipids on TLC; I. Rustenbeck and S. Lenzen, *J. Planar Chrom. 2*:470–472 (1989), Densitometric quantification of minor phospholipids via improvement of the scanning mode for cupric sulfate staining.

Procedure

The investigation can be carried out as described by Rustenbeck and Lenzen.
In a modification of the mobile phase described in the literature, it is also possible to develop at ca. 21°C using chloroform/methanol/water/ammonia (65/30/4/2 v/v) (with chamber saturation, distance run 50 mm).

The assignment of the spots is carried out by comparison with the spots produced by the qualitative comparison solution.

For Rf values of the individual phospholipids (in the mobile phase mentioned above) see Sec. II.B.2.

Recommendations

The developed plate should be dried thoroughly before it is dipped in the copper salt solution.

Should reference substances be available for all the components of interest, then these can also be applied in defined quantities, and the amounts of the individual components can be calculated with respect to the particular reference substance.

It is advisable to check the thin-layer plate for faulty areas before use.

Caveats

Only phosphatidylcholine can be determined exactly by the method as described. The other components are reported after calculation as phosphatidylcholine.

It can also be used for phospholipids from very different sources if the reference substances also come from the same source (i.e., they have, as near as possible, the same fatty acid distribution), since the depth of color of the spots produced on heating is also dependent on the fatty acid composition.

According to the literature (Touchstone, et al.) it is also possible to detect phospholipids containing saturated fatty acids by dipping in a solution of copper sulphate. Dipping in copper acetate solution only leads to the formation of visible spots on heating for phospholipids with unsaturated fatty acids. This makes it possible to distinguish compounds with only saturated fatty acids from compounds containing unsaturated fatty acids.

5. Phosphorus Determination after Separation

Principle of operation

The phospholipids are separated by thin-layer chromatography. The quantitative evaluation is carried out by scratching the zones concerned from the TLC plate, digesting and determining phosphorus by the method of Rouser.

References

J. C. Dittmer and R. L. Lester, *J. Lipid Res.* *5*:126–127 (1964). A simple, specific spray for the detection of phospholipids on TLC; G. Rouser, A. N. Slatokos, and S. Fleischer, *Lipids* *1*:85–86 (1966), Quantitative analysis of phospholipids by TLC and phosphorus analysis of spots; G. Rouser, G. Simon, and G. Kritchevsky, *Lipids* *4*:599–606 (1969), Species variations in phospholipid class distributions of organs: I. Kidney, liver and spleen; G. Rouser, S. Fleischer, and A. Yamamoto, *Lipids* *5*:494–96 (1970), Two dimensional TLC separation of polar lipids and determination of phospholipids by phosphorus analysis of spots; J. A. Singleton and H. E. Pattee, *J. Am. Chem. Soc*:873–875 (1981), Computation of conversion factors to determine the phospholipid content of peanut oil.

Materials: Chloroform p.a.; methanol p.a.; ninhydrin spray reagent 0.1%, Merck; phosphate reagent according to Dittmer and Lester: Place 16.0 g molybdenum trioxide in a 1 L three-necked flask and suspend in 100 ml water with continuous stirring. Add 300 ml sulphuric acid dropwise [15] and boil the solution under reflux for ca. 45 min. Then allow to cool (solution 1). Place 200 ml of this solution in a 500 ml round-bottomed flask and add 0.7 g powdered molybdenum. Boil under reflux for 15 min and cool (solution 2). Carefully add solution 1 and solution 2 to 200 ml water. Make the mixture up to 1000 ml; perchloric acid p.a.; ammonium molybdate p.a. (solution 0.3% in water); ascorbic acid p.a. (solution 10% in water); 8-anilino-1-naphthalene sulphonic acid (solution 0.01% in water); dist. water; precoated silica gel 60 F 254 plates (50×200 mm^2) Merck; precoated silica gel 60 F 254 plates (100×200 mm^2) Merck; crude soybean phosphatide (solution: 500 mg in 50 ml chloroform); ammonia solution 25% p.a.; disodium phenylphosphate dihydrate p.a. (purity 100%, 12.19% P); sulphuric acid p.a.

Equipment

Volumetric flask, 50.00 ml; Camag Linomat II; development chamber for 200×200 mm^2 plates; Camag laboratory spray made of glass; UV lamp; drying cupboard; test tubes

100 × 14 mm; heating block; heating bath (100°C); centrifuge; cuvettes 1 cm path length; photometer; Eppendorf pipettes/microlitre syringes.

Procedure

Ca. 100 mg sample, accurately weighed, is dissolved in chloroform to 50.00 ml. If the solution is not clear the sample must be dissolved in a mixture of equal volumes of chloroform and methanol (sample solution).

A volume of 0.100 ml of this solution is applied as a streak (ca. 30 mm) to each of 3 precoated silica gel plates (analysis plates). For the purpose of identifying the individual components the sample solution and the qualitative comparison solution are also applied side by side on a precoated silica gel plate (amounts and method of application as in the analysis plates) (comparison plate).

All the plates plus an additional blank precoated silica gel plate (blank plate) are developed for a running distance of ca. 150 mm with chloroform/methanol/water/ammonia solution (65/30/4/2 v/v) without chamber saturation.

After development and drying the comparison plate is sprayed with ninhydrin and heated to 110°C for a few minutes.

Phosphatidylethanolamine (PE) (Rf ca. 0.4) exibits a red color and below it lyso-PE (Rf ca. 0.2).

The comparison plate is then sprayed with phosphate reagent. A blue coloration indicates phosphorus-containing components (accompanying nonphosphorus-containing lipids become visible on heating to ca. 140°C).

After development and drying, the analysis plates are sprayed with 8-anilino-1-naphthalene sulphonic acid solution. The separated components are marked under the UV lamp.

The comparison plate is used as a guide to scrape off individually all phosphorus-containing components into labeled test tubes to which 0.5 ml perchloric acid is then added. Approximately the same quantity of silica gel is scraped off the blank plate and treated like all other components (blank 1).

A parallel three-fold determination is made of the total phosphorus. For this purpose three 0.050 ml portions of sample solution are placed in test tubes and treated with 0.5 ml portions of perchloric acid, and these test tubes are treated like all the others. One test tube contains only the reagents (blank 2) and is treated like all the others.

All test tubes are placed in a heating block and heated to 180°C for 2 h.

After cooling, 4.00 ml ammonium molybdate solution is added to each, followed by 0.50 ml ascorbic acid solution.

The samples are well mixed, heated on the heating bath at 100°C for 6 min, and cooled for exactly 1 hour. Then they are centrifuged at 6000 rpm for ca. 10 min to remove the silica gel. The absorbance (Ab) of the clear, blue solution is measured in a 1 cm cuvette at 820 nm against the corresponding blank, individual components against blank 1 and the total phosphorus against blank 2.

The extinction coefficient for calculation of the phosphorus content from the measured absorbance is determined using disodium phenylphosphate dihydrate, employing the same technique as that used for the determination of the total phosphorus.

Calculation

$$\frac{Ab \times EC \times MW \times 100}{M} = \% \text{ by weight phospholipid}$$

Ab = measured absorbance of the sample
EC = extinction coefficient per μmol phosphorus
MW = molecular weight of the corresponding phospholipid
M = amount of sample applied (μg)-M = amount of sample used (μg) for total phosphorus

Molecular weights and Rf values of phospholipids in soybeans

Phospholipid		M (g/mol)	Rf value
N-acyl-phosphatidylethanolamine	(*N*-acyl-PE)	993	0.82
Phosphatidylethanolamine	(PE)	734	0.36
Phosphatidylcholine	(PC)	782	0.27
Lysophosphatidylethanolamine	(LPE)	474	0.21
Phosphatidylinositol	(PI)	852	0.14
Phosphatidic acid	(PA)	689	0.08
Lysophosphatidylcholine	(LPC)	515	0.06
Glycerophosphorylcholine	(GPC)	257	0.00

Recommendations

The molecular weights are calculated for typical fatty acid distribution as in soybean phosphatidylcholine.

The parallel determination of total phosphorus serves as a control of whether all phosphorus-containing components have been included on the TLC plate.

The start zone of the TLC plate should be scraped off, even if no phosphorus-containing components can be detected there, and worked up accordingly.

This method is an absolute one and is suitable for checking reference substances.

A two-dimensional thin-layer chromatographic method based on the same principles is described in AOCS—Methods Ja 7-86.

Caveats

Note! All glass apparatus employed must be absolutely free from phosphate residues.

C. HPLC Methods

1. Determination of Phospholipids with Gradient Separation and Evaporative Light-Scattering Detection

Literature

W. W. Christie, Separation of lipid classes by high performance liquid chromatography with the "mass detector," *J. Chromatogra. 361*:396–399 (1986); W. W. Christie, Rapid separation and quantification of lipid classes by high performance liquid chromatography and mass (light-scattering) detection, *J. Lipid Res. 26*:507 (1985); P. van der Meeren, J. Vanderdeelen, M. Huys, and L. Baert, Simple and rapid method for high-performance liquid chromatographic separation and quantification of soybean phospholipids, *J. Chromatogr. 447*:436 (1988); J. Becart, C. Chevalier, and J. P. Biesse, Quantitative analysis of phospholipids by HPLC with light scattering evaporating detector: Application to raw materials for cosmetic use. *J. High Resolution Chromatogr. 13*:126 (1990).

Principle of operation: Assay of phospholipids in (soybean) phospholipid samples with HPLC separation on silica gel and gradient elution, detection with an evaporative light-scattering detector.

Materials: chloroform (HPLC quality), *n*-hexane (HPLC quality), tetrahydrofurane (HPLC quality), 2-propanol (HPLC quality), water (HPLC quality), serine buffer (0.5 mM serine adjusted on pH 7.5 with ethylamine).

mobile phase solvent A: *n*-hexane/tetrahydrofurane 99/1 (v/v)
solvent B: chloroform/2-propanol 1/4 (v/v)
solvent C: serine buffer/2-propanol 1/1 (v/v)

stationary phase: Spherisorb Si 3μ, 100 · 4,6 mm

standards: phospholipids (PE, PA, PI, PC, LPC) isolated and purified from natural sources, used for determination of phospholipids in phospholipid fractions; these fractions were used as working standards.

Equipment

HPLC system with ternary gradient system, autosampler, evaporative light-scattering detector (i.e., ACS 750/14), evaluation unit with the possibility to make nonlinear calibration.

Procedure

A ternary gradient elution system is used for the separation of phospholipids with 2 ml/min flow rate.

time (min)	A (%)	B (%)	C (%)
0.0	100	0	0
1.0	100	0	0
5.0	80	20	0
5.1	42	52	6
20.0	32	52	16
25.0	32	52	16
25.1	32	70	0
30.0	100	0	0
35.0	100	0	0

The calibration curve is constructed from three points. The calibration range has to correspond with the concentration of the phospholipids in the sample, so that the peak responses of different phospholipids in the sample are within the calibration ranges of the standards. In most cases, the curve is not linear. The sample concentration is 40 mg/ml.

The reproducibility of the method is better than 2.5% for PC and better than 1% for other phospholipids. The lowest detectable quantity depends on the peak shape of the eluting phospholipids; some typical values are PE, 1 μg; LPC, 1 μg; PA, 6 μg; PI, 2 μg.

Recommendations

Gradient elution permits a complete separation of the main phospholipids. The response of the evaporative light-scattering detector is independent of the fatty acid composition of the phospholipids.

Caveats

Calibration must be repeated before each analysis.

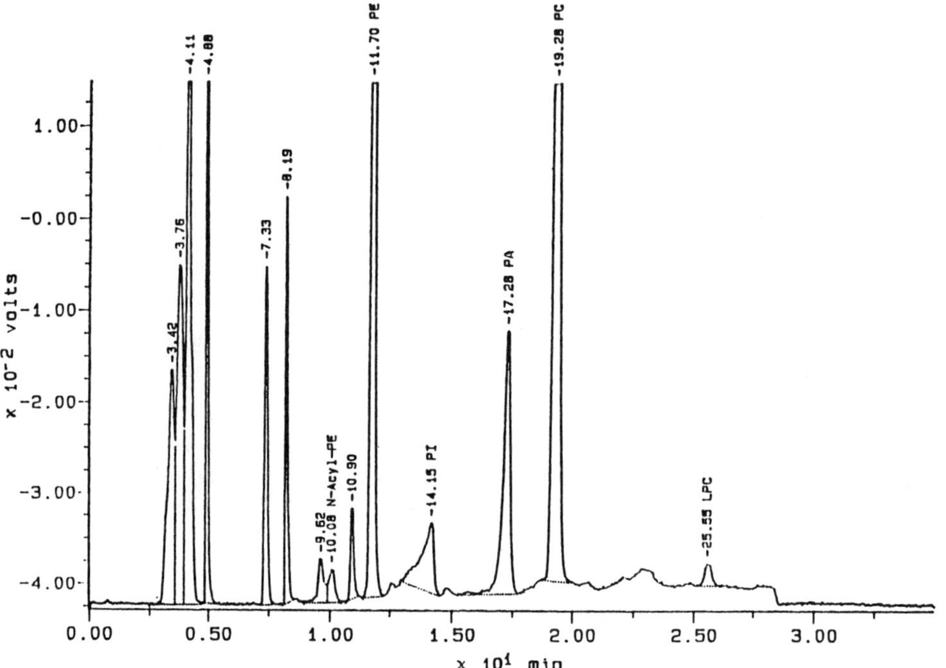

Figure 4 Chromatogram of crude lecithin with light-scattering detection.

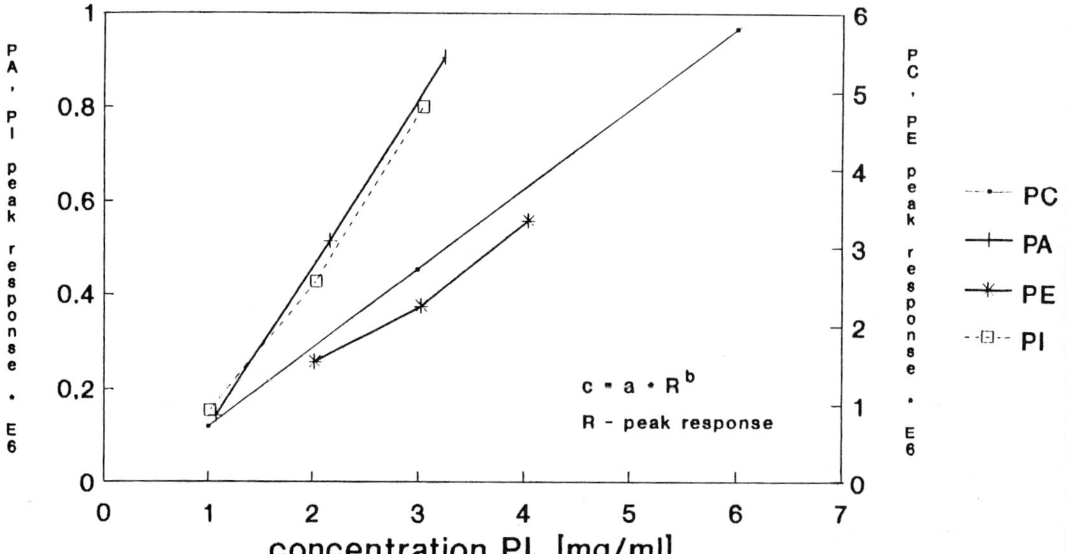

Figure 5 Calibration curve.

2. Determination of Phosphatidylcholine and Lysophosphatidylcholine, UV Detection after Isocratic Separation

Principle of operation

The assay of (3-*sn*-phosphatidyl)choline and lysophosphatidylcholine is carried out by HPLC on silica gel phase according to the "external standard" method. A separation condition has been worked out that allows rapid analysis in routine operation. UV detection is employed for detection.

Reference

Conference volume, Königsteiner Chromatographie Tage (1982) 197–218, A. Nasner reviewed numerous HPLC methods for the analysis of lecithins.

Materials/Equipment: *n*-Hexane; 2-propanol; water; methanol; *tert*-butyl methyl ether (all solvents of HPLC grade); (3-*sn*-phosphatidyl)choline from soybeans, standard substance, laboratory sample with precisely defined (3-*sn*-phosphatidyl)choline assay; lysophosphatidylcholine from soybeans, standard substance, laboratory sample with precisely defined lysophosphatidylcholine assay; HPLC facility, e.g., Merck-Hitachi, with D2500 Chromato-integrator, L6200 pump, T6300 column thermostat, L4000 UV detector; Millex HV 0.45 μm filter unit; prepacked HPLC column Kromasil Si 5μ, length 125 mm, i.d. 4 mm.

Procedure

Standard solution: Ca. 100 mg (3-*sn*-phosphatidyl)choline and ca. 30 mg lysophosphatidylcholine are accurately weighed into a 100 ml volumetric flask that is filled up to the mark with methanol at 20°C. A sample of this solution is passed through a Millex HV filter and used for analysis.

Sample solution: The amount of sample whose PC content corresponds to the concentration of the standard solution is accurately weighed into a 100 ml volumetric flask, dissolved in 10 ml *tert*-butyl methyl ether and methanol and made up to the mark with methanol at 20°C. A sample is passed through a Millex HV filter and used for analysis.

Instrumental conditions: Flow 1 ml/min, oven temperature 40°C, injection volume 10 μl, detection UV 205 nm.

The separation is carried out isocratically with the following solvent mixture: 67% 2-propanol, 17% *n*-hexane, 16% water (v/v). The solvent mixture is premixed and degassed in the ultrasonic bath.

The quantitative determination is carried out according to the external standard method. For this purpose exactly the same quantities of sample and standard solution are injected successively onto the separation column. The calculation involves comparison of the relationships between the peak areas of the standard and of the sample with the amounts weighed out.

$$\%\ \text{phospholipid} = \frac{As \times E \times f}{Ast \times S}$$

As = area of peak for sample solution
Ast = area of peak for standard solution
E = amount of standard weighed out in mg
S = amount of sample weighed out in mg
f = assay (purity) of standard

Typical retention times are (3-*sn*-phosphatidyl)choline, 3.8 min; lysophosphatidylcholine, 7.8 min. Typical chromatograms of standard and sample solutions are illustrated in Fig. 6.

Caveats
See section II.B.2.

3. Assay of Phosphatidylcholine, Phosphatidylethanolamine, Phosphatidylinositol, Lysophosphatidylcholine, UV Detection after Isocratic Separation

Principle of operation

The assay of (3-*sn*-phosphatidyl)choline, phosphatidylethanolamine, lysophosphatidylcholine, and phosphatidylinositol in crude phosphatides is carried out by HPLC on a silica gel phase using the external standard method. The parameters quoted in the literature have been adapted to our own requirements.

Reference

W. J. Hurst and R. A. Martin, *J. Am. Chem. Soc. 61/9*:1462–1463 (1984), The analysis of phospholipids in soy lecithin by HPLC.

Materials/equipment: Acetonitrile; methanol; *tert*-butyl methyl ether (all solvents HPLC grade); ortho-phosphoric acid 85% p.a.; (3-*sn*-phosphatidyl)choline from soybeans, stan-

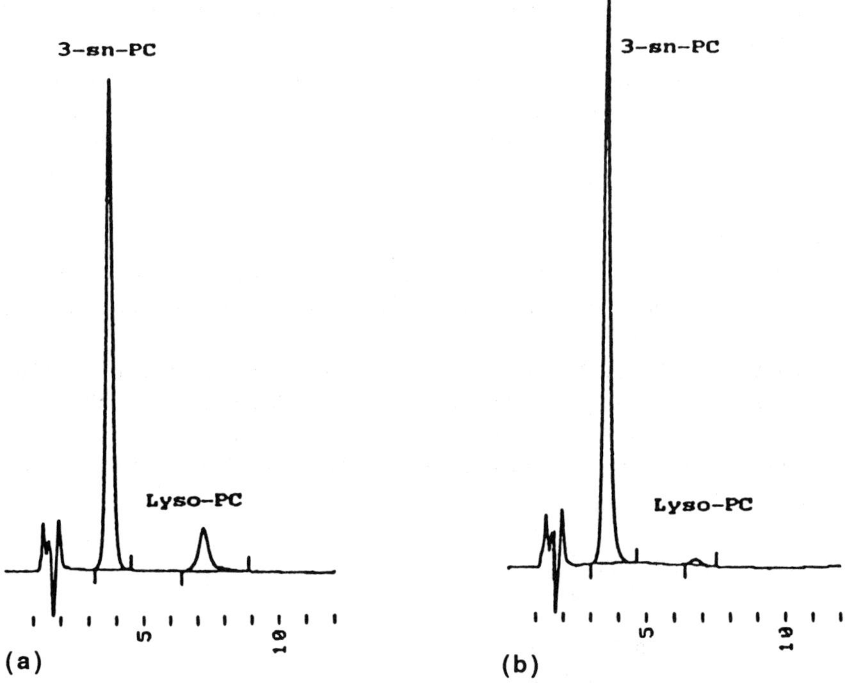

Figure 6 Chromatogram of the HPLC separation of PC and lyso-PC, standard solution (a) and sample solution (b).

dard substance, laboratory sample with precisely defined (3-*sn*-phosphatidyl)choline assay; phosphatidylethanolamine, Sigma, P 1050, 10 mg/ml; phosphatidylinositol, Sigma, P 5766, 10 mg/ml; lysophosphatidylcholine, standard substance, laboratory sample with precisely defined assay; HPLC facility, e.g., Merck-Hitachi, with D2500 Chromato-integrator, L6200 pump, T6300 column thermostat, L4000 UV detector; prepacked HPLC column μPorasilTM (Waters) length 300 mm, i.d. 3.9 mm; Millex HV 0.45 μm filter unit.

Procedure

Standard solutions: Standard 1: (3-*sn*-phosphatidyl)choline and lysophasphatidylcholine. Ca. 100 mg (3-*sn*-phosphatidyl)choline and ca. 30 mg lysophosphatidylcholine are accurately weighed into a 100 ml volumetric flask that is filled up to the mark with methanol at 20°C. A sample of this solution is passed through a Millex HV filter and used for analysis. Standard 2: Phosphatidylethanolamine and phosphatidylinositol; the ready-to-use solutions delivered are diluted 1:10 in methanol and used in the analysis (see Fig. 7).

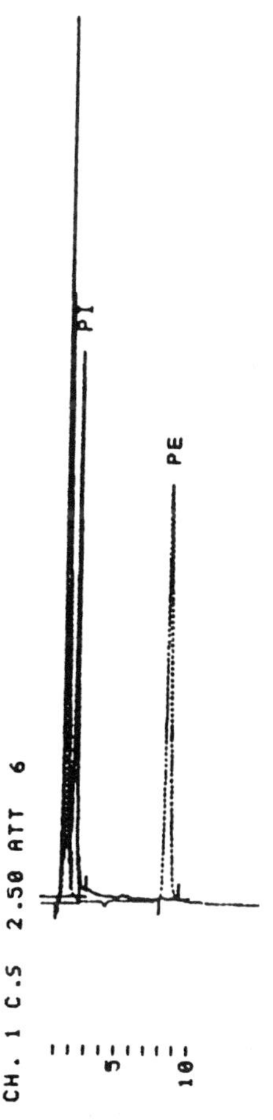

Figure 7 Chromatogram of the HPLC separation of PI and PE (standard solution).

Sample solution: The amount of sample whose PC content corresponds to the concentration of the standard solution is accurately weighed into a 100 ml volumetric flask, dissolved in 10 ml *tert*-butyl methyl ether and methanol and made up to the mark with methanol at 20°C. A sample is passed through a Millex HV filter and used for analysis (see Fig. 8).

Instrumental conditions: Flow 2 ml/min, oven temperature 35°C, injection volume 20 μl, detection 205 nm.

The separation is carried out isocratically with the following solvent mixture: 780 ml acetonitrile, 20 ml methanol, 10 ml 85% phosphoric acid. The solvent mixture is premixed and degassed in the ultrasonic bath.

The quantitative determination is carried out according to the external standard method. For this purpose exactly the same quantities of sample and standard solution are injected successively onto the separation column. The calculation involves comparison of the relationships between the peak areas of the standard and of the sample with the amounts weighed out.

The phospholipid content is calculated according to the equation in section II.c.2.

Typical retention times are PI, 2.4 min; PE, 9.3 min; PC, 14.2 min; LPC, 26.5 min.

Caveats

See section II.B.2.

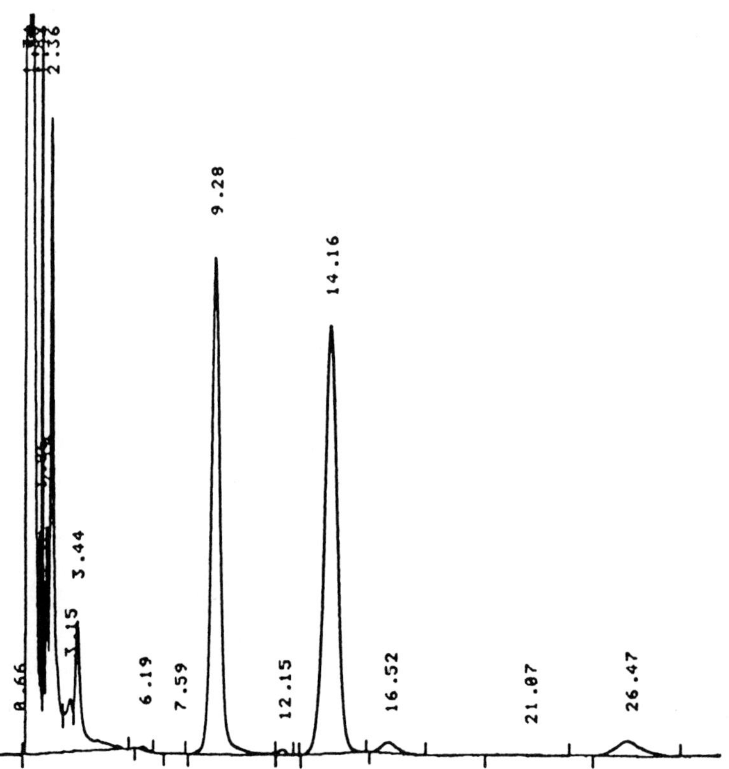

Figure 8 Chromatogram of the HPLC separation of a phospholipid mixture (PI, PE, PC, and lyso-PC).

D. Other Methods

1. Enzymatic Cleavage and Choline Determination

Principle of operation

The phosphatidylcholine in the sample under investigation is enzymatically cleaved. The free choline is determined.

References

V. Blaton, M. De Buyzere, J. Spincemaille, and B. Declercq, *Clin. Chem. 29/5*:806–809 (1983), Enzymatic assay for phosphatidylcholine and sphingomyelin in serum; M. Takayama, S. Itoh, T. Nagasaki, and I. Tanimizu, *Clin. Chim. Acta 79*:93–98 (1977), A new enzymatic method for determination of serum choline-containing phospholipids.

Procedure

The determination can be carried out according to the method reported in the literature. The method involves enzymatic cleavage of choline and the reaction of this with cholin oxidase yielding H_20_2. The H_20_2 reacts enzymatically with phenol and aminoantipyrine to yield a colored compound.

Recommendations

The method can be carried out rapidly without great difficulty.

Caveats

The phosphatidylcholine is calculated from the measured choline content. Hence no distinction is made between lysophosphatidylcholine, phosphatidylcholine, glycerophosphatidylcholine, and other choline-containing compounds and free choline. Therefore the method should not be used for strongly contaminated compounds (heated, stored, etc.).

2. NMR Spectroscopy

R. Murari, M. M. A. Abd El-Rahman, Y. Wedmid, S. Parthasarathy, and W. J. Baumann, *J. Org. Chem. 47*:2158–2163 (1982), Carbon-13 NMR spectroscopy of phospholipids in solution. Spectral and stereochemical assignments based on ^{13}C—^{31}P and ^{13}C—^{14}N couplings; P. Meneses and Th. Glonek, *J. Lipid Res. 29*:679–689 (198), High resolution ^{31}P NMR of extracted phospholipids.

3. Determination of Acetone-Insoluble Lipids

Principle of operation

The sample is dispersed in cold acetone and centrifuged. The insoluble portion is dried and weighed.

Procedure

The determination can be carried out according to one of the following methods: Deutsche Gesellschaft für Fettwissenschaft Einheitsmethoden (DGF-Einheitsmethoden) Abteilung F—Fettbegleitstoffe, F—I 5 (68); Official and Tentative Methods of the American Oil Chemists' Society, A.O.C.S. Official method Ja 4-46 (revised 1982).

E. Determination of Other Components in Phospholipids

1. Glycolipids

Sterol glycosides and sterol glycoside esters: The TLC methods discussed in connection with the identification of phospholipids can be used to identify these components.

Quantification can also be carried out if suitable reference substances are available (see list of references that follows).

References

J. Kesselmeier and E. Heinz, *Anal. Biochem. 144/2*:319–328 (1985), Separation and quantitation of molecular species from plant lipids by HPLC; Y. Kushi, C. Rokukawa, and S. Handa, *Anal. Biochem. 175*:167–176 (1988), Direct analysis of glycolipids on TLC plates by matrix assisted secondary ion mass spectrometry: Application for glycolipid storage disorders; D. Marion, R. Douillard, and G. Gandemer, *Rev. France. Corps Gras 35*:229–234 (1988), Rapid and isocratic separation of plant phospho- and glycoglycerolipids by HPLC; C. G. Walker, *Cereal Chem. 65/5*:433–435 (1988), Determination of flour glycolipids as their benzoyl derivatives by HPLC with UV detection; H. Kadowaki, K. E. Rys-Skora, and R. S. Koff, *J. Lipid Res. 30*:616–627 (1989), Separation of derivatized glycosphinogolipids into individual molecular species by HPLC; R. A. Moreau, P. T. Asman, and Helen A. Norman, *Phytochem. 29/8*:2461–2466 (1990), Analysis of major classes of plant lipids by HPLC with flame ionization detection; M. Ranny, J. Sedlacek, and C. Michalec, *J. Planar Chrom. 4*:15–18 (1991), Resolution of phospholipids on chromarods impregnated with salts of some divalent metals.

2. Sphingolipids

Methods suitable for the determination of sphongolipids are described in the references listed below.

References

R. L. Briand, S. Herold, and K. G. Blass, *J. Chrom. 223*:277–284 (1981), HPLC determination of the lecithin/sphingomyelin ratio in amniotic fluid; L. M. Brown, C. G. Duck-Chong, and W. J. Hensley, *Clin. Chem. 28/2*:344–348 (1982), Improved procedure for lecithin/sphingomyelin ratio in amniotic fluid reduces false predictions of lung immaturity [DC]; K. Tokuno, et al., *Rinsho Kensa 27/3*:322–325 (1983), Studies on an analytical method for lecithin/sphingomyelin ratio in amniotic fluid using the IATROSCAN TH 10 and the HELENA L/S ratio kit; J. C. Touchstone and J. G. Alvarez, *J. Chrom. 429*:359–371 (1988), Phospholipids in amniotic fluid with special reference to the lecithin/sphingomyelin ratio [DC/HPLC]; H. Kadowaki, K. E. Rys-Skora, and R. S. Koff, *J. Lipid Res. 30*:616–627 (1989), Separation of derivatized glycosphingolipids into individual molecular species by HPLC.

3. Ether Lipids

The ether lipids can be determined using the methods described in the following references.

References

N. Totani, Fette, Seifen, Anstrichmittel 84/2:70–73 (1982), Neue Methoden zur quantitativen Analyse von Etherlipiden; I. Rustenbeck and S. Lenzen, *J. Chromatogr. 525*:85–91 (1990), Quantitation of hexadecylphosphocholine by HPTLC with densitometry; M. L. Blank, E. A. Cress, V. Fitzgerald, and F. Snyder, *J. Chromatogr. 508*:382–385 (1990), TLC and HPTLC separation of glycerolipid subclasses as bonzoates; derivatives of the ether and ester analogs of phosphatidylcholine, phosphatidylethanolamine, and platelet activating factor.

4. Phosphonolipids

R. Hori and Y. Nozawa, Phosphonolipids, in *Phospholipids,* Elsevier Biomedical Press, 1982, 95–128.

F. Parameters Used in Commercial Phospholipid Specifications

The other parameters normally included in commercial specifications, such as peroxide value,* acid value, iodine value, iodine color value, oil content, viscosity, toluene-insoluble, etc. should be determined according to the generally applicable methods such as those described in Official and Tentative Methods of the American Oil Chemists' Society (A.O.C.S. Official Methods) or in Deutsche Gesellschaft für Fettwissenschaft Einheitsmethoden (DGF Einheitsmethoden) Abteilung F—Fettbegleitstoffe.

REFERENCES

1. H. Pardun, *Die Pflanzenlecithine.* Verlag f. Chem. Industrie H. Ziolkowsky, Augsburg, 1988.
2. H. Takamura, H. Kasai, H. Arita, and M. Kito, Phospholipid molecular species in human umbilical artery and vein endothelial cells, *J. Lipid Res. 31*:709–717 (1990).
3. J. L. Weihrauch and Y. S. Son, The phospholipid content of foods, *J. Am. Oil Chem. Soc. 60*:1971–1978 (1983).
4. J. A. Singleton and H. E. Pattee, Computation of conversion factors to determine the phospholipid content in peanut oil, *J. Am. Oil Chem. Soc. 58*:873–875 (1981).
5. G. W. Chapman, Jr., A conversion factor to determine phospholipid content in soybean and sunflower crude oils, *J. Am. Oil Chem. Soc. 57*:299–302 (1980).
6. Y. Nagender Rao, R. B. N. Prasad, and S. Venkob Rao, Positional distribution of fatty acids in oilseed phosphatidylcholines and phosphatidylethanolamines, *Fat Sci. Technol. 91*:482–484 (1989).
7. G. M. Patton, J. M. Fasulo, and S. J. Robins, Separation of phospholipids and individual molecular species of phospholipids by high performance liquid chromatograph, *J. Lipid Res. 23*:190–196 (1982).
8. A. Cantafora, M. Cardelli, and R. Masella, Separation and determination of molecular species of phosphatidylcholine in biological samples by high performance liquid chromatography, *J. Chromatogr. 507*:339–349 (1990).
9. A. Rastegar, A. Pelletier, G. Duportail, L. Freysz, and C. Leray, Sensitive analysis of phospholipid molecular species by high performance liquid chromatography using fluorescent naproxen derivatives of diacylglycerols, *J. Chromatogr. 518*:157–165 (1990).
10. T. Rezanka and M. Podojil, Preparative separation of algal polar lipids and of individual molecular species by high performance liquid chromatography and their identification by gas chromatography-mass spectrometry, *J. Chromatogr. 364*:397–408 (1989).
11. H. Rabe, G. Reichmann, Y. Nakagawa, B. Rüstow, and D. Kunze, Separation of alkylacyl and diacyl glycerophospholipids and their molecular species as naphthylurethanes by high performance liquid chromatography, *J. Chromatogr. 493*:353–360 (1989).
12. F. J. G. M. Van Kuuk and E. A. Dratz, Detection of phospholipid peroxides in biological samples, *J. Free Rad. Biol. Med. 3*:349–354 (1987).
13. C. G. Crawford, R. D. Plattner, D. J. Sessa, and J. J. Rackis, Separation of oxidized and unoxidized molecular species of phosphatidylcholine by high pressure liquid chromatography, *Lipids 15*:91–94 (1980).

*The endpoint must be determined potentiometrically in the case of highly purified phosphatidylcholines.

14. J. Folch, N. Lees, and S. G. H. Sloane, A simple method for the isolation and purification of total lipids from animal tissue, *J. Biol. Chem. 226*:497–509 (1957).
15. T. Curstedt, Chromatographic analysis, isolation, and characterization of ether lipids, in *Ether Lipids* (H. K. Mangold and F. Paltauf, eds.), Academic Press, New York, 1983.
16. A. J. Dijkstra, J. de Kock, Process for fractionating phosphatide mixtures, *Eur. Pat. Appl. 0 372 327 A 2* (1990).
17. P. Reichling, Pure natural choline diglyceride phosphates from soybean phosphatides, *Ger. Offen. 1.905.253* (1970).
18. N. V. Unilever, Phosphatides extraction, *Neth. Appl. 68 13.248* (1969).
19. H. Liebing and J. Lau, Process for the recovery of lecithin fractions, *Fette, Seifen, Anstrichm. 78*:123–127 (1976).
20. P. Van der Meeren, J. Vanderdeelen, M. Huys, and L. Baert, Optimizatin of the column loadability for the preparative HPLC separation of soybean phospholipids, *J. Am. Oil Chem. Soc. 67*:815–820 (1990).
21. W. J. Hurst, R. A. Martin, Jr., and R. M. Sheeley, The preparative HPLC isolation and identification of phospholipids from soy lecithin, *J. Liq. Chromatogr. 9*:2969–2976 (1986).
22. R. S. Fager, S. Shapiro, and B. J. Litman, A large-scale purification of phosphatidylethanolamine, lysophosphatidylethanolamine, and phosphatidylcholine by high performance liquid chromatography: a partial resolution of molecular species, *J. Lipid Res. 18*:704–709 (1977).
23. Un Hoi Do and Shi-Lung Lo, Separation of phospholipids using a diethylaminoethylsilica gel column and thin-layer chromatography, *J. Chromatogr. 381*:233–240 (1986).
24. M.C. Moschidis, Phosphonolipids, *Prog. Lipid Res. 23*:223–246 (1985).
25. M. C. Moschidis, Silicic acid column chromatography of phosphonolipids, *J. Chromatogr. 435*:508–12 (1988).

4
Phospholipid Biosynthesis

Kenneth J. Longmuir *College of Medicine, University of California, Irvine, California*

I. INTRODUCTION

This chapter presents an overview of our current understanding of the pathways of glycerophospholipid biosynthesis in cells. It begins with a presentation of fatty acid biosynthetic pathways, with particular emphasis on fatty acid modification activities (desaturation, chain elongation, and chain shortening). These activities regulate the composition of the pool of fatty acids available for glycerophospholipid biosynthesis and hence strongly influence the assortment of glycerophospholipid molecular species that are eventually produced. Next, the biosynthesis of phosphatidic acid is discussed, as this substance is the common precursor of all glycerophospholipids. The diacylglycerol pathway is then presented, which in animal cells leads to the formation of the nitrogen-containing phospholipids and triacylglycerol. This is followed by a discussion of the CDPdiacylglycerol pathways, which are utilized by all cell types (prokaryotes, yeast, animal cells, plant cells) in various ways. Next are presented the ether lipid and sphingolipid biosynthetic pathways, which are found primarily in animal cells. The chapter concludes with a discussion of fatty acid remodeling of phospholipids, a mechanism by which cells enrich certain lipid classes with polyunsaturated fatty acids.

The lipid biosynthetic pathways have been an active area of biochemical investigation over the past several decades. It is not possible in this chapter to cite the many thousands of papers that have been published that collectively have established our current understanding of how glycerophospholipids are formed in a wide variety of cell types. In each section, recent review articles have been cited to which the reader can refer in order to obtain references to the many research papers that have contributed to our current knowledge of the subject.

II. FATTY ACID BIOSYNTHESIS AND FATTY ACID MODIFICATION

A. Fatty Acid Biosynthesis

Fatty acids are supplied to cells either by *de novo* fatty acid biosynthesis or by dietary uptake. Dietary uptake is of quantitative significance principally in animal cells. Plants do not have extracellular lipid transport and must rely upon intracellular fatty acid biosynthesis for all cellular fatty acid.

Fatty acid biosynthesis in most organisms has been thoroughly and recently reviewed, including in *E. coli* [1–4], yeast [5] and other microorganisms [6], animal cells [1,7,8], and plants [9–13]. Fatty acid biosynthesis is catalyzed by the sequential action of acetyl-CoA carboxylases and by fatty acid synthetase complexes, which produce, in most organisms, primarily palmitic acid (16:0). In animal cells and in yeast, these complexes are found in the cytosol and consist of multifunctional polypeptides that do not physically separate into components capable of carrying out an individual step of fatty acid biosynthesis. In contrast, in *E. coli* and in plants, the individual steps are catalyzed by separable polypeptides. Fatty acid biosynthesis in plants takes place in the plastid.

Certain polyunsaturated free fatty acids can be metabolized by the lipoxygenases (in animal cells and in plants) and by prostaglandin H synthase (in animal cells). Otherwise, lipid metabolism using fatty acid substrates requires esterified fatty acids. In *E. coli* and in the plastid of plants, the fatty acid remains esterified to acyl carrier protein (ACP). Fatty acids can also diffuse out of the plastid of plant cells as free fatty acids. Once in the cytosol, they combine with coenzyme A to form the fatty acyl-coenzyme A thioester (fatty acyl-CoA). In animal cells, free fatty acid is released from the acyl carrier protein of the fatty acid synthetase complex; then it combines with coenzyme A to form fatty acyl-CoA. In animal cells, fatty acyl-coenzyme A ligases are present in microsomes, mitochondria, and peroxisomes. Ligase enzymes differ primarily in their chain-length dependence. A separate CoA ligase highly selective for arachidonic acid has also been demonstrated in a variety of animal cells (review [14]).

Fatty acids for phospholipid biosynthesis can be modified in four ways:

1. Desaturation
2. Chain elongation
3. Chain shortening
4. Reduction to the fatty alcohol (which provides substrate for either lipid biosynthesis; see Sec. VI)

B. Fatty Acid Desaturation

In animal cells, fatty acid desaturation is an oxygen-dependent process that can introduce a double bond at positions nine, six, five, and four from the carboxyl end of a fatty acyl-CoA thioester (reviews [15–18]). These reactions are catalyzed by separate desaturase enzymes, termed the Δ^9, Δ^6, Δ^5, and Δ^4 desaturases. Since animal cells do not possess desaturase activity that will introduce a double bond beyond position 9, fatty acids with cis double bonds beyond position 9 are obtained by a combination of dietary uptake and subsequent elongation/desaturation cycles. Linoleic and linolenic acid are the two principal fatty acids supplied by dietary uptake and are called essential fatty acids. They are required for synthesis of polyunsaturated fatty acids and, in animal cells, for prostaglandin formation.

Desaturase enzymes in animal cells are membrane bound and are located in the microsomal fraction of the cell. The Δ^9 desaturase system (also called the stearoyl-CoA desaturase) is the most thoroughly studied and requires three enzyme activities: (1) NADH-cytochrome b_5 reductase, (2) cytochrome b_5, and (3) the terminal desaturase enzyme, often call the cyanide-sensitive protein, as desaturation is inhibited by cyanide but not by carbon monoxide. NADH is the preferred source of electrons, which provide reducing power to perform the oxygen-dependent introduction of the cis double bond.

$$-CH_2-CH_2- + NADH + H^+ + O_2 \rightarrow -CH{=}CH- + NAD^+ + 2H_2O$$

Plants also carry out the oxygen-dependent introduction of cis double bonds to produce mono and polyunsaturated fatty acids (reviews [9,13,19]). Significant differences exist between the desaturation pathways of plant and animal tissues. The first desaturation step in plants is the Δ^9 desaturation of stearoyl-ACP (18:0) to form oleoyl-ACP (18:1). This enzyme is located in the plastid, is the only known soluble desaturase, and does not desaturate 16:0-ACP. NADPH is the most likely source of electrons, and reduced ferrodoxin is the immediate electron donor. This desaturation system is quite efficient, as little 18:0 fatty acid is found in plants. The 16:0-ACP and 18:1-ACP produced in the plastid can be used directly for lipid biosynthesis in the plastid by what is referred to as the "prokaryotic pathway." Otherwise, fatty acid is hydrolyzed from the ACP, diffuses into the cytosol, and is esterified to form acyl-CoA to participate in the "eukaryotic pathway" of lipid biosynthesis on the endoplasmic reticulum of plant cells.

Subsequent desaturation steps in plants are not as well characterized and appear to be quite complex, as there are at least eight genes that control desaturase activities in leaves [9]. Outside the plastid on the endoplasmic reticulum, the second desaturation step is the Δ^{12} desaturation of oleic acid (18:1 *cis*-9) to produce linoleic acid (18:2 *cis*-9,12); the third step is the Δ^{15} desaturation of linoleic acid to produce linolenic acid (18:2 *cis*-9,12,15). These desaturation steps are catalyzed by enzymes that use as substrate a fatty acid esterified to glycerolipid, not acyl-CoA or acyl-ACP. The primary substrate for Δ^{12} desaturation appears to be oleic acid esterified to phosphatidylcholine. Desaturation of linoleic acid to linolenic acid in the endoplasmic reticulum may also occur when the 18:2 fatty acid is present on phosphatidylcholine.

The plastid has fatty acid desaturation activities that can convert 18:1 to 18:2 and 18:3. Many species of plants (the "16:3" plants) also convert substantial amounts of 16:0 to 16:1, 16:2, and 16:3. These desaturation steps all utilize as substrates fatty acids esterified to glycerolipids. In the plastid, quantitatively the most important glycerolipids that are substrates for polyunsaturated fatty acid desaturation are the monogalactosyldiacylglycerols and digalactosyldiacylglycerols, the major neutral lipids of plant tissues.

In prokaryotic cells such as *E. coli*, *cis*-vaccenic acid (18:1 *cis*-11) is the principal monoenoic fatty acid. *E. coli* do not form polyunsaturated fatty acid. These organisms utilize an oxygen-independent pathway for fatty acid desaturation (reviews [2,4,6]). The branchpoint between saturated and unsaturated fatty acid biosynthesis occurs during fatty acid biosynthesis at the ten-carbon stage. For saturated fatty acid biosynthesis, the usual *trans*-α,β-double bond is formed to give *trans*-α,β-decenoyl-ACP. This double bond is subsequently reduced and several more rounds of fatty acid biosynthesis take place to give saturated fatty acids. For unsaturated biosynthesis, a cis double bond is introduced by a specific β-hydroxydecanoyl-ACP dehydrase to form *cis*-β,γ-decenoyl-ACP. This cis desaturation step yields a double bond that is not reduced. Instead, three or four more cycles of fatty acid biosynthesis follow to form palmitoleic acid (16:1 *cis*-9) or *cis*-vaccenic acid (18:1 *cis*-11).

C. Fatty Acid Chain Elongation

Chain elongation is a process whereby fatty acids, obtained either from endogenous fatty acid biosynthesis or by dietary uptake, are lengthened in two-carbon increments from the carboxylic acid end (review [18]). The chain elongation cycle takes place in four steps: (1) the initial condensation of fatty acyl-CoA with malonyl-CoA to form β-keto acyl-CoA,

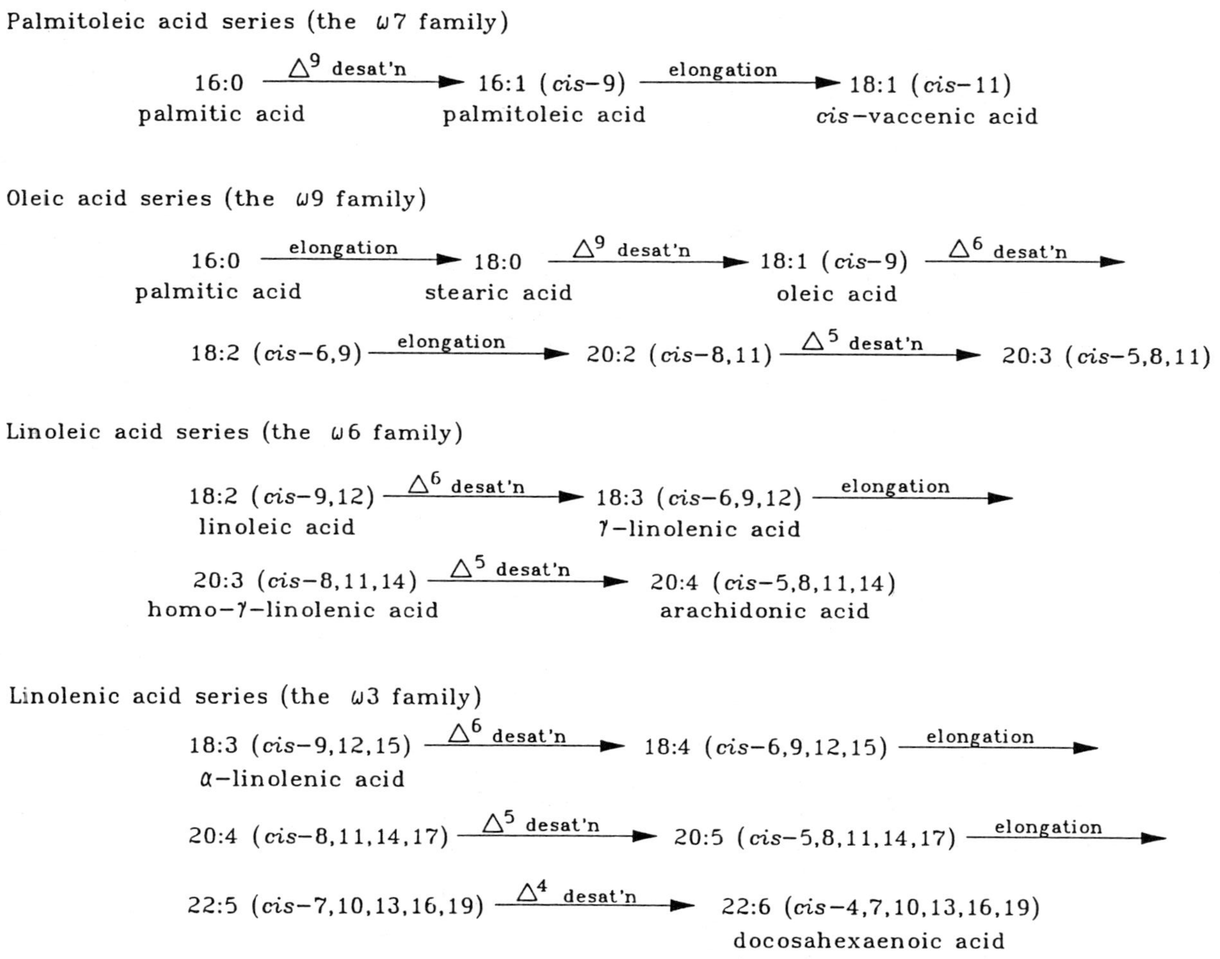

Palmitoleic acid series (the ω7 family)

16:0 (palmitic acid) —Δ^9 desat'n→ 16:1 (*cis*–9) (palmitoleic acid) —elongation→ 18:1 (*cis*–11) (*cis*–vaccenic acid)

Oleic acid series (the ω9 family)

16:0 (palmitic acid) —elongation→ 18:0 (stearic acid) —Δ^9 desat'n→ 18:1 (*cis*–9) (oleic acid) —Δ^6 desat'n→

18:2 (*cis*–6,9) —elongation→ 20:2 (*cis*–8,11) —Δ^5 desat'n→ 20:3 (*cis*–5,8,11)

Linoleic acid series (the ω6 family)

18:2 (*cis*–9,12) (linoleic acid) —Δ^6 desat'n→ 18:3 (*cis*–6,9,12) (γ–linolenic acid) —elongation→

20:3 (*cis*–8,11,14) (homo–γ–linolenic acid) —Δ^5 desat'n→ 20:4 (*cis*–5,8,11,14) (arachidonic acid)

Linolenic acid series (the ω3 family)

18:3 (*cis*–9,12,15) (α–linolenic acid) —Δ^6 desat'n→ 18:4 (*cis*–6,9,12,15) —elongation→

20:4 (*cis*–8,11,14,17) —Δ^5 desat'n→ 20:5 (*cis*–5,8,11,14,17) —elongation→

22:5 (*cis*–7,10,13,16,19) —Δ^4 desat'n→ 22:6 (*cis*–4,7,10,13,16,19) (docosahexaenoic acid)

Figure 1 Pathways of fatty acid desaturation and chain elongation in animal cells.

(2) NAPDH-dependent reduction of the β-keto compound to the secondary alcohol (β-hydroxy acyl-CoA), (3) dehydration of the secondary alcohol to form a *trans*-α,β-unsaturated fatty acid (2-*trans*-enoyl-CoA), and (4) NADPH-dependent reduction of the double bond to give the elongated fatty acid. In animal cells, most elongation takes place in microsomal compartments. The initial condensation reaction is rate-limiting, and there are several condensation enzymes that differ according to their preferences for fatty acyl-CoAs with different chain lengths and degrees of unsaturation.

There is also a mitochondrial chain elongation system in animal cells, the role of which is not well understood. The difference between it and the microsomal elongation system is that in mitochondria, the acyl-CoA chain condenses with acetyl-CoA instead of malonyl-CoA.

Chain elongation systems that use fatty acids esterified to acyl carrier proteins have been described in micoorganisms [6]. In plants, elongation of palmitic acid (16:0) to stearic acid (18:0) occurs during fatty acid biosynthesis where the fatty acid is coupled to acyl carrier protein in the plastid. Longer chain fatty acids ($>C_{18}$) are formed in plants by membrane-associated elongation systems outside the plastid, where fatty acyl-CoA is the substrate.

In mammalian tissue and particularly in liver cells, desaturation and elongation usually follow an ordered sequence [17]:

1. When palmitoyl-CoA (16:0) is the substrate, elongation to stearoly-CoA (18:0) is normally the first step.
2. For a saturated fatty acid, the first double bond is inserted at position 9 by the Δ^9 desaturase.
3. Subsequent double bonds are inserted between the carboxyl group and the double bond nearest the carboxyl group.
4. For polyunsaturated fatty acids, desaturation alternates with elongation. Methylene ($-CH_2-$) interruption between double bonds is always maintained.

These general constraints on fatty acid modification tend to limit the number of possible reactions carried out in mammalian tissue to a few principal sequences outlined in Fig. 1. Because elongation of palmitic acid usually occurs before desaturation, products of the palmitoleic acid pathway are not found in large amounts in mammalian tissue. Further metabolism of oleic acid to other products of the oleic acid series normally does not take place but will occur as the result of a dietary deficiency of essential fatty acids. The presence of the final product of the series, 5,8,11-icosatrienoic acid [20:3 (5,8,11)] is diagnostic of an essential fatty acid deficiency and is detectable before the clinical manifestations of the disease appear. Essential fatty acid deficiencies are also characterized by elevated levels of palmitoleic and oleic acids.

D. Fatty Acid Chain Shortening

Fatty acid chain-shortening is a peroxisomal β-oxidation function similar to β-oxidation in mitochondria, as it shortens fatty acids (always in the form of fatty acyl-CoA thioesters) by two carbons per cycle (review [20]). Peroxisomal chain-shortening appears to have two roles: First, further chain elongation of fatty acids to products longer than those shown in Fig. 1 is minimized. For example, when rats are fed 22:4 (7,10,13,16) (the chain-elongation product of arachidonic acid), this fatty acid is preferentially shortened to

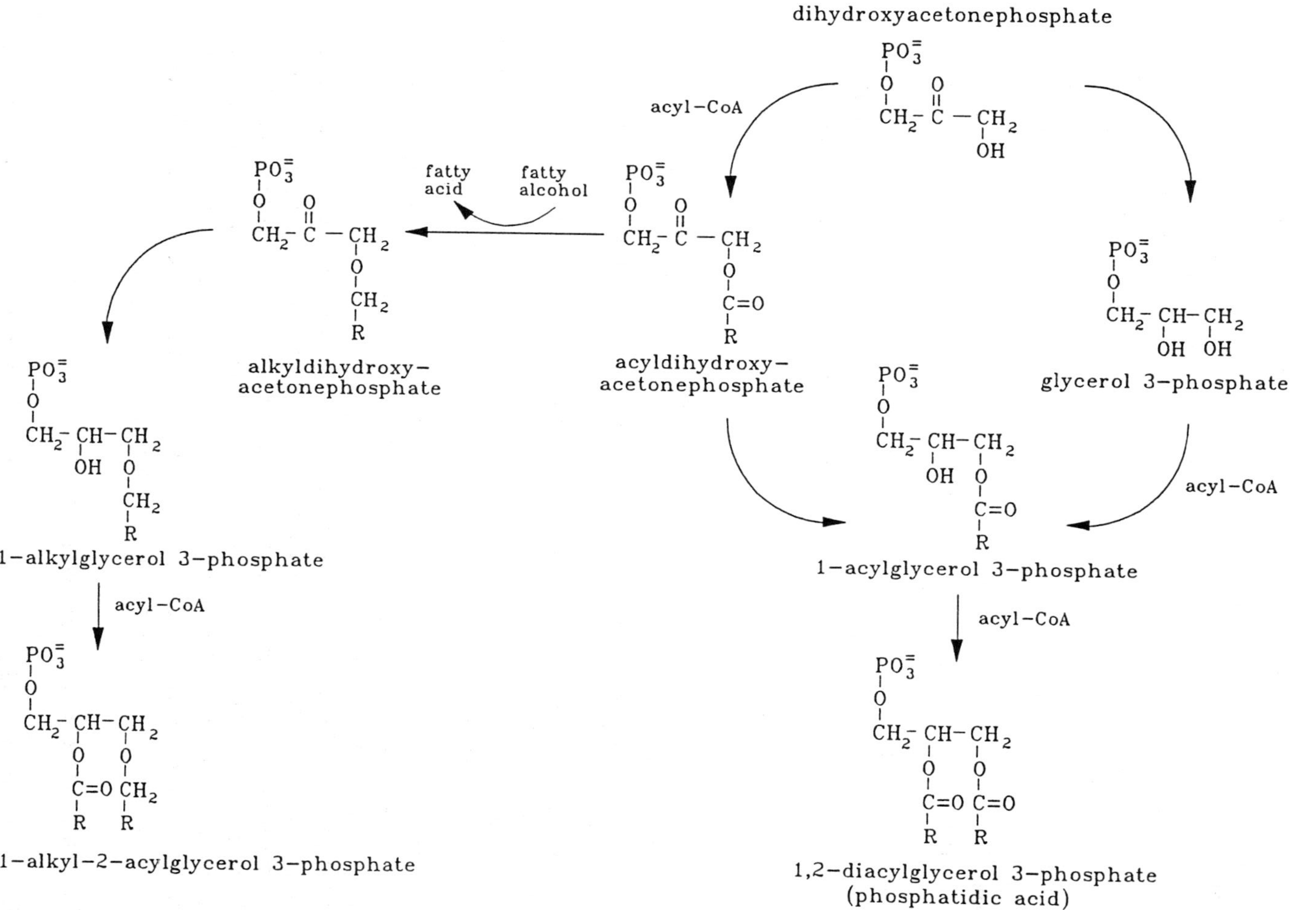

Figure 2 Biosynthesis of diacyl and alkyl-acyl phosphatidic acid.

arachidonic acid 20:4 (5,8,11,14) and then incorporated into liver lipids [20,21]. Hence chain shortening activity has considerable influence on the composition of the pool of fatty acids available for lipid biosynthesis. Second, mitochondria do not efficiently carry out the β-oxidation of fatty acids longer than about 22 carbons. These fatty acids (called very-long-chain fatty acids) must be shortened in the peroxisomes. Genetic deficiencies in one or more enzymes involved in peroxisomal β-oxidation result in the accumulation of very-long-chain fatty acids in human tissues (reviews [22,23]). The important feature of peroxisomal β-oxidation of fatty acids is that it is always incomplete. Complete β-oxidation to acetyl-CoAs then takes place in the mitochondria. As reviewed by Osmundsen et al. [20], the extent of chain shortening of an individual fatty acyl-CoA is somewhat ambiguous and differs depending upon experimental conditions. For palmitoyl-CoA, two or three β-oxidation cycles are frequently observed.

III. BIOSYNTHESIS OF PHOSPHATIDIC ACID

Phosphatidic acid is synthesized by three pathways: (1) by stepwise acylation of *sn*-glycerol 3-phosphate (all cells: prokaryotes, yeast, animal cells, plant cells), (2) by acylation of dihydroxyacetonephosphate, followed by reduction to 1-acyl-*sn*-glycerol 3-phosphate, followed by a second acylation (yeast and animal cells), and (3) by phosphorylation of diacylglycerol (all cells). Reviews: [3,4,24–30]. The biosynthetic pathways are illustrated in Figure 2.

A. Acylation of *sn*-Glycerol 3-Phosphate

sn-Glycerol 3-phosphate is obtained from the NADH-dependent reduction of dihydroxyacetonephosphate, and from the ATP-dependent phosphorylation of glycerol by glycerol kinase. This latter source appears to be of physiologic importance primarily in mammalian liver. Also, *E. coli* mutants that cannot reduce dihydroxyacetonephosphate require exogenous glycerol for growth and form *sn*-glycerol 3-phosphate via glycerol kinase [2].

The enzyme glycerophosphate acyltransferase acylates *sn*-glycerol 3-phosphate at position *sn*-1 to produce 1-acyl-*sn*-glycerol 3-phosphate. In animal cells, the activated fatty acid substrate is in the form of fatty acyl-coenzyme A thioester. In prokaryotes such as *E. coli*, the substrate is in the form of fatty acyl-ACP thioester (ACP = acyl carrier protein) that arises from *de novo* fatty acid biosynthesis. The *E. coli* enzyme can use fatty acyl-CoA as well, and there is evidence to suggest that exogenously supplied fatty acids are converted to acyl-CoA thioesters by *E. coli* for acylation reactions [30]. However, there are examples of other bacteria (e.g., *Rhodopseudomonas sphaeroides*) that rely upon acyl-ACP exclusively [30]. In plants, fatty acid in the plastid remains in the form of fatty acyl-ACP following fatty acid biosynthesis, and it is used in the prokaryotic pathway of lipid biosynthesis in the plastid. Outside the plastid, the fatty acid is in the form of fatty acyl-CoA, which is used for the eukaryotic pathway of lipid biosynthesis on the endoplasmic reticulum [13].

In animal cells, glycerophosphate acyltransferase is found both in microsomal and (outer) mitochondrial membranes ([31], reviews [24–26,28]). The microsomal and mitochondrial activities appear to arise from separate enzymes on the basis of a number of kinetic observations, including heat stability, pH optima, sensitivity to detergents, proteases, and organic solvents. The microsomal activity is sensitive to *N*-ethylmaleimide, whereas the mitochondrial activity is not. The mitochondrial enzyme has lower apparent *K*m's for both acyl-CoA and *sn*-glycerol 3-phosphate. The microsomal enzyme uses both *sn*-glycerol 3-phosphate and dihydroxyacetonephosphate as acyl acceptors, and these two substrates are mutually competitive. The mitochondrial enzyme does not accept dihydroxyacetonephosphate, nor is dihydroxyacetonephosphate a competitive inhibitor. In mammalian liver, glycerophosphate acyltransferase activity is distributed evenly between mitochondrial and microsomal compartments. In other tissues, such as adipose, heart, kidney, brain, and adrenal, and in some cultured cells, the microsomal activity is several times that found in mitochondria [31,32].

B. Acylation of Dihydroxyacetonephosphate

In animal cells and in yeast, an alternate pathway for the formation of phosphatidic acid begins with the acylation of dihydroxyacetonephosphate to produce 1-acyldihydroxyacetonephosphate. The enzyme, dihydroxyacetonephosphate acyltransferase, is found in the microsomal and peroxisomal compartments of the cells. In the microsomal fraction, it is evident, as noted above, that acylation of both *sn*-glycerol 3-phosphate and dihydroxyacetonephosphate are dual functions of a single enzyme (reviews [24,26]. This conclusion is based on kinetic analyses that find that the two microsomal enzyme activities are equally sensitive to pH, acyl chain length, heating, proteases, detergents, and inhibitors [33,34]. During differentiation of the 3T3-L1 preadipocyte, the glycerophosphate and dihydroxyacetonephosphate acyltransferase activities increase in a coordinate fashion [35]. A *S. cerevisiae* mutation in the structural gene for *sn*-glycerol 3-phosphate acyltransferase is also defective in dihydroxyacetonephosphate acyltransferase activity [36].

The dihydroxyacetonephosphate acyltransferase in peroxisomes serves to provide substrate for the committed step of ether lipid biosynthesis (reviews [24,26,37]). Based upon sensitivity to *N*-ethylmaleimide, heating, proteases, and pH, it appears that the enzymes in the microsomal and peroxisomal fractions are different. Also, the peroxisomal enzyme does not utilize (and is not inhibited by) *sn*-glycerol 3-phosphate.

1-Acyldihydroxyacetonephosphate not utilized for ether lipid biosynthesis is reduced to 1-acyl-*sn*-glycerol 3-phosphate by acyldihydroxyacetonephosphate reductase (reviews [25,26]). The reductase is found both in peroxisomal and microsomal fractions and shows a preference for NADPH over NADH. It is not clear whether the activities in the microsomal and peroxisomal compartments are from different enzymes.

C. Acylation of 1-Acyl-*sn*-Glycerol 3-Phosphate

The second fatty acid chain is placed on position *sn*-2 of 1-acyl-*sn*-glycerol 3-phosphate to form phosphatidic acid (reviews [24,26]). In yeast, animal cells, and on the endoplasmic reticulum of plant cells, fatty acyl-CoA is the activated fatty acid substrate, whereas in prokaryotes and in the plastid of plants, fatty acyl-ACP serves as the donor. The enzyme (1-acylglycerophosphate acyltransferase) in animal cells is found in microsomal membranes primarily. Its presence in mitochondria is unclear, and it does not appear to be located in peroxisomes. In *E. coli*, both the glycerophosphate and 1-acylglycerophosphate acyltransferase activities appear to be localized on the cytoplasmic face of the inner membrane.

It is well understood that esterification of dihydroxyacetonephosphate is the obligatory route for the biosynthesis of the alkyl and alkenyl lipids (the ether lipids). In contrast, the relative contributions from the glycerophosphate acyltransferase pathway and the dihydroxyacetonephosphate pathway for (diacyl) phosphatidic acid biosynthesis are not exactly known. Nor is there detailed information about how the relative contributions change under different physiologic conditions. The many reasons for this uncertainty have been reviewed [24], and only one example of the experimental strategies is given here. In one procedure, cells or tissues are incubated with [^{14}C]-glycerol plus [2-^{3}H]glycerol. Products of the acylation of *sn*-glycerol 3-phosphate contain both ^{14}C and ^{3}H labels, whereas the ^{3}H label is lost upon oxidation of *sn*-glycerol 3-phosphate to dihydroxyacetonephosphate. After the kinetic isotope effect (which discriminates against tritium) is taken into account, several investigators have concluded that both pathways provide significant (and in some cases nearly equal) amounts of phosphatidic acid for subsequent glycerophospholipid biosynthesis [38–40].

D. Diacylglycerol Kinase

The enzyme diacylglycerol kinase catalyzes the ATP-dependent phosphorylation of *sn*-1,2-diacylglycerol to form phosphatidic acid. In mammalian cells, current research interests are focused on the role of diacylglycerol kinase in regulating the cellular levels of diacylglycerol, which in turn regulates the diacylglycerol stimulation of protein kinase C. These cell regulation aspects of diacylglycerol kinase have been recently reviewed [41,42]. To summarize, diacylglycerol kinase enzyme activities are both cytosolic and membrane-bound, and there are several enzymes that have different enzymological and immunological properties. There is evidence that diacylglycerol kinase activity translocates between soluble and membrane bound forms. Also, the enzyme can be phosphorylated by protein kinases, and several investigators have suggested that the phosphorylation of the enzyme assists in the cytosolic to membrane-bound translocation. Diacylglycerol kinase will phosphorylate *sn*-2-monoacylglycerol; hence the enzyme competes with monoacylglycerol lipase, reducing release of free fatty acid.

IV. THE DIACYLGLYCEROL PATHWAY

A. Formation of Diacylglycerol from Phosphatidic Acid

Hydrolysis of phosphatidic acid (also called phosphatidate or 1,2-diacyl-*sn*-glycerol 3-phosphate) to *sn*-1,2-diacylglycerol is, in animal cells and in the endoplasmic reticulum of plants, the committed step for the *de novo* synthesis of the nitrogen-containing phospholipids and triacylglycerol. This pathway, called the diacylglycerol pathway (Figure 3), is of secondary importance in yeast, and it is not found in most prokaryotes such as *E. coli*. The enzyme phosphatidate phosphatase (also called phosphatidate phosphohydrolase) is at an important branch point of glycerophospholipid biosynthesis. In animal cells, phosphatidate hydrolysis leads to the formation of the nitrogen-containing glycerophospholipids (phosphatidylcholine, phosphatidylethanolamine, and phosphatidylserine) and triglycerol. Alternatively, phosphatidiate can be used to form CDP-diacylglycerol (Sec. V). The properties of the phosphatase have been reviewed by various authors [25,28,43,44], recently and most thoroughly in the two-volume work edited by Brindley [45].

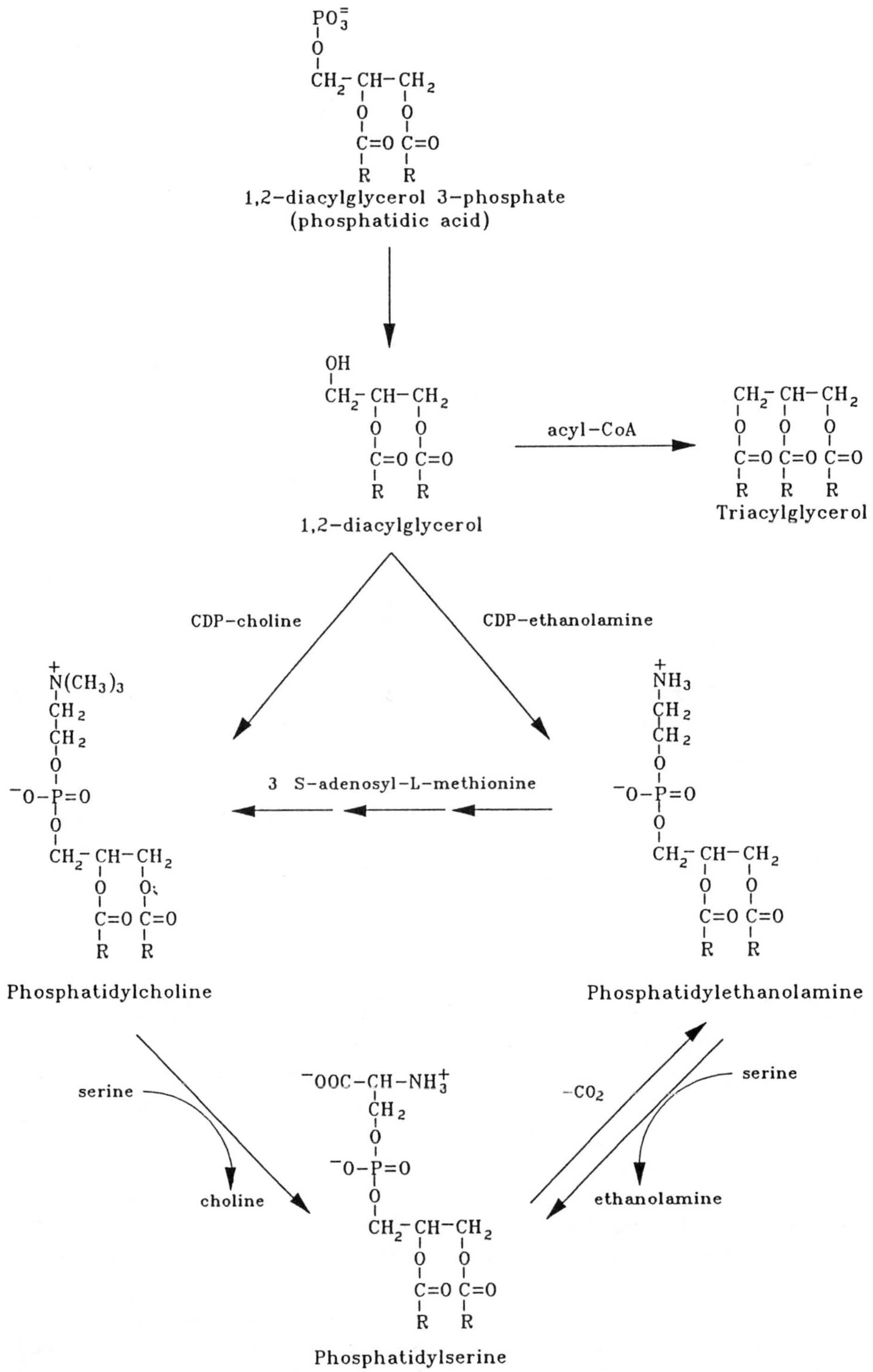

1,2-diacylglycerol 3-phosphate
(phosphatidic acid)

1,2-diacylglycerol

acyl-CoA

Triacylglycerol

CDP-choline

CDP-ethanolamine

3 S-adenosyl-L-methionine

Phosphatidylcholine

Phosphatidylethanolamine

serine

choline

$-CO_2$

serine

ethanolamine

Phosphatidylserine

Figure 3 The diacylglycerol pathway.

Phosphatidate phosphatase activity is specific for phosphatidate and not other glycerophospholipids (such as phosphatidylcholine), but it will also hydrolyze acyl-*sn*-glycerol 3-phosphate as well as phosphate esters of fatty alcohols [44]. Phosphatidate phosphatase activity is distinct from other phosphatases such as acid phosphatase, alkaline phosphatase, and phospholipase C.

Phosphatidate phosphatase activity has been reported in virtually every subcellular compartment of the cell, including the cytosol. The reason for the widespread appearance of the enzyme is puzzling, since diacylglycerol, the product of the reaction, is used primarily by microsomal enzymes for further lipid biosynthesis. Apparent phosphatidate phosphatase activity can be the result of other hydrolytic activities. As pointed out by Brindley [24], the choice of radiolabeled substrate used to assay phosphatidate phosphatase should be made with an understanding of other hydrolytic enzyme activities present in the subcellular fraction under study. In liver, deacylation of phosphatidate by phospholipases of the A type leads to the formation of water-soluble *sn*-glycerol 3-phosphate, and further hydrolysis by acid or alkaline phosphatases releases inorganic phosphate. In the presence of these activities, assay of phosphatidate phosphatase using ^{32}P-labeled phosphatidate substrate can lead to erroneously high measurements. In contrast, hydrolysis of fatty acids from the diacylglycerol product is low in liver tissue, hence phosphatidate phosphatase activity can be measured as the production of diacylglycerol from either glycerol-labeled or fatty acid-labeled phosphatidate. In adipose tissue a different situation exists, as there is little phospholipase hydrolysis of phosphatidate but significant diacylglycerol lipase activity. Hence assay of phosphatidate phosphatase by measuring the appearance of water-soluble ^{32}P from the ^{32}P-labeled substrate is recommended. An alternative substrate, 1-O-hexadecyl-rac-[2–^{3}H]glycerol 3-phosphate, was developed for the assay of activity in amniotic fluid. It contains an ether-linked hydrocarbon chain, hence avoids complications due to hydrolysis of acyl linkages in either the phosphatidic acid substrate or the diacylglycerol product [46].

A mixed micelle of phosphatidate and phosphatidylcholine serves as an effective form of the substrate for enzyme assay, provided steps have been taken to remove any residual calcium from the phosphatidate. Considerable effort has been directed toward understanding the magnesium requirement of the enzyme, and much of the literature on the enzyme refers to both magnesium-independent and magnesium-dependent activities. In general, the soluble enzyme activities appear to be highly magnesium-dependent and the particulate activities less so. However, questions always remain about residual magnesium present in the substrate, reagents, and subcellular fractions used in the assay. Brindley and coworkers have addressed this question with respect to the enzyme from rat liver by carefully treating the phosphatidate with ion-exchange resin and sonicating the substrate in EGTA + EDTA [47]. Both microsomal and cytosolic fractions show little phosphatase activity until Mg^{2+} is added. The magnesium dependency may not represent an enzyme cofactor requirement in the usual sense but instead may involve the presentation of substrate as a phosphatidate-magnesium chelate, or the ion may affect the packing arrangement of the phosphatidate and other lipids in the bilayer or micelle.

Phosphatidate phosphatase is termed an ambiquitous enzyme, as it has been shown to translocate between cytosolic and particulate fractions, with stimulation of activity occurring upon association with membranes. When microsomal [47] or mitochondrial fractions [48] are incubated with the soluble enzyme fraction in the presence of oleate, the soluble phosphatase activity becomes associated with the membrane fraction. In liver, the enzyme activity is regulated by diet, several types of drugs, ethanol, and glucocorticoids.

B. Biosynthesis of Phosphatidylcholine and Phosphatidylethanolamine from Diacylglycerol

1. Choline Phosphate and Ethanolamine Phosphate

Choline and ethanolamine are phosphorylated to form choline phosphate and ethanolamine phosphate, respectively. These reactions are catalyzed by the enzymes choline kinase and ethanolamine kinase (reviews [43,49]). Both enzyme activities are exclusively cytosolic and require ATP and Mg^{2+}. It has been a point of controversy whether the phosphorylations of choline and ethanolamine are catalyzed by the same or different enzymes (review [43]). Early studies indicated that the two activities could be separated in vitro. However, two recent reports of highly-purified kinase from rat liver [50] and rat kidney [51] suggest that a single protein can catalyze both reactions.

2. CDPcholine and CDPethanolamine

Choline phosphate and ethanolamine phosphate combine with CTP to form CDPcholine and CDPethanolamine. The enzymes are referred to by various names but are usually called choline-phosphate cytidylyltransferase and ethanolamine-phosphate cytidylyltransferase (reviews [28,43,52–54]). These two enzyme activities appear to be functions of separate proteins, as fractions enriched in choline-phosphate cytidylyltransferase activity do not contain ethanolamine-phosphate cytidylyltransferase activity, and vice-versa. The choline-phosphate cytidylyltransferase exists in two states: a soluble, cytosolic form and an aggregated, particulate form usually associated with microsomal membranes. It is the particulate form that has the higher specific activity. The soluble form is generally considered inactive, and in order to assay enzyme present in the cytosolic fractions, lipid is added to the assay medium. In contrast, ethanolamine-phosphate cytidylyltransferase activity is exclusively cytosolic.

The formation of CDPethanolamine and CDPcholine appear to be the rate-limiting steps for the synthesis of phosphatidylethanolamine and phosphatidylcholine. Typical of a rate-limiting process, the cellular pool sizes of the products CDPcholine and CDPethanolamine are usually much smaller than the pools of choline phosphate and ethanolamine phosphate. Because of its probable role as the principal regulatory enzyme of the most abundant phospholipid of eukaryotic cells, the choline-phosphate cytidylyltransferase has drawn considerable interest, with particular attention given to factors that promote the translocation of the enzyme between cytosolic and Golgi/endoplasmic reticulum/nuclear membrane locations (reviews [43,52–54]). Since the cytosolic enzyme shows little activity when assayed in the absence of lipid, the cytosolic enzyme in vivo is presumed to have little activity but to become active upon association with cell membranes. Current data indicate that this association with membranes can be induced by (1) dephosphorylation of the enzyme and (2) association with fatty acid [43,53]. There is evidence that phosphorylation of the enzyme (promoting the less active, soluble form) is mediated by both cyclic-AMP-dependent and protein kinase C-dependent pathways. Treatment of HeLa cells with fatty acids causes an increase in choline-phosphate cytidylyltransferase activity in microsomes, with a corresponding decrease in enzyme in the cytosol. Other studies have confirmed the fatty acid-induced increase in microsomal activity, but the corresponding decrease in cytosolic enzyme is not as well substantiated.

3. Phosphatidylcholine and Phosphatidylethanolamine

The final step in phosphatidylcholine or phosphatidylethanolamine biosynthesis from the diacylglycerol pathway is the combination of CDPcholine or CDPethanolamine with diacylglycerol. These reactions are catalyzed by the enzymes cholinephosphotransferase and ethanolaminephosphotransferase. Both enzymes are magnesium-dependent and appear to be exclusively microsomal. The two reactions appear to be functions of different enzymes (reviews [42,43,52]). They differ in their selectivity for various molecular species of diacylglycerol and their sensitivities to detergents, phospholipase, protease, and metal ion. A Chinese hamster ovary cell line has been isolated that is defective in ethanolaminephosphotransferase but with normal cholinephosphotransferase activity. Yeast mutants have been isolated defective in cholinephosphotransferase but not ethanolaminephosphotransferase. In in vitro assays, both enzymes appear to operate near their equilibrium points, and the reverse reactions are easily demonstrated. The physiologic significance of this finding is not clear.

While most attention has been focused on the choline-phosphate and ethanolamine-phosphate cytidylyltransferases as the most important enzyme activities that regulate phosphatidylcholine and phosphatidylethanolamine formation, the supply of microsomal diacylglycerol for the phosphotransferase reactions has been implicated as a limiting factor. Also, other systems have been described where changes in choline kinase activity regulate the level of phosphatidylcholine (review [43]).

C. Biosynthesis of Phosphatidylcholine by Methylation of Phosphatidylethanolamine

Phosphatidylethanolamine can be successively methylated with *S*-adenosyl-L-methionine to form first monomethylphosphatidylethanolamine, then dimethylphosphatidylethanolamine, then phosphatidylcholine (reviews [42,49,55]). In animal tissues this pathway appears to be of significance only in liver. Methyltransferase activity is present in other tissues, but these activities are greatly reduced in comparison to the activity present in the liver. Hence most phosphatidylcholine is formed by the transfer of CDPcholine to diacylglycerol, and choline deficiency in animal cells and tissues, including the liver, severely reduces the proportion of choline-containing lipids. Paradoxically, methylation of phosphatidylethanolamine appears essential for life, as it is the only known biosynthetic pathway for choline. In yeast, methylation of phosphatidylethanolamine is the principal route of formation of phosphatidylcholine, as the use of CDP-choline for phosphatidylcholine biosynthesis become quantitatively important only in mutations blocking methyltransferase activity. Most bacteria do not contain phosphatidylcholine. In those that do (such as *Rhodopseudomonas spheroides*) it appears that methyltransferase is the only pathway for formation of phosphatidylcholine. In all organisms, the methyl donor is always *S*-adenosyl-L-methionine.

All three methylation steps are described as catalyzed by phosphatidylethanolamine-*N*-methyltransferase activity. In rat liver, Vance and coworkers isolated an 18 kD protein that catalyzes all three steps of the methylation process [55]. The first methylation is rate limiting. Methyltransferase activity appears to be regulated by the substrates, phosphatidylethanolamine and *S*-adenosyl-L-methionine, and one of the products, *S*-adenosylhomocysteine. It has not been demonstrated that phosphatidylcholine can regulate the activity of the enzyme. The enzyme isolated from rat liver can be phosphorylated by the catalytic subunit of cAMP-dependent protein kinase, but a role of phosphorylation for the regulation of enzyme activity in vivo has not been established. The enzyme is localized primarily to rough and smooth endoplasmic reticulum, with a small amount found in the Golgi apparatus.

In yeast, genetic studies point to the existence of two methyltransferase enzymes, one that catalyzes the conversion of phosphatidylethanolamine to monomethylphosphatidylethanolamine, and a second enzyme able to catalyze all three methylation steps [42,55].

D. Biosynthesis of Phosphatidylserine by Base Exchange with Phosphatidylcholine or Phosphatidylethanolamine

In higher eukaroytes, the primary route for phosphatidylserine biosynthesis is by substitution ("base exchange") of the phosphatidylcholine or phosphatidylethanolamine headgroup with serine (reviews [25,27,56]). (In bacteria and yeast, phosphatidylserine is formed principally, if not exclusively, from CDP-diacylglycerol (Sec. VI).) Earlier studies on the phosphatidylserine base-exchange pathway indicated that phosphatidylethanolamine was the principal substrate for phosphatidylserine biosynthesis. However, recent studies of tissue culture cells have demonstrated that phosphatidylcholine can serve as the principal substrate for the serine base exchange reaction (review [27]). This is because tissue culture media are normally devoid of ethanolamine, except for a small amount from the serum supplement. Genetic studies with Chinese hamster ovary cells defective in the biosynthesis of phosphatidylserine suggest that two enzymes catalyze base exchange among the nitrogen-containing glycerophospholipids: one that can exchange serine with the choline and ethanolamine headgroups of phosphatidylcholine and phosphatidylethanolamine, and a second enzyme specific for phosphatidylethanolamine as the phosphatidyl donor. These base exchange activities (phosphatidylcholine or phosphatidylethanolamine:serine *O*-phosphatidyltransferase) are found in microsomal parts of the cell.

E. Biosynthesis of Phosphatidylethanolamine by Decarboxylation of Phosphatidylserine

Phosphatidylserine is converted to phosphatidylethanolamine by the enzyme phosphatidylserine decarboxylase (reviews [25,27,30,43,49,54]). This enzyme activity is found in prokaryotes and in yeast, where it is the primary route of synthesis of phosphatidylethanolamine. In mammalian tissue culture cells, decarboxylation of phosphatidylserine becomes the principal pathway for the formation of phosphatidylethanolamine under culture conditions where ethanolamine is limited. In animal cells, the enzyme is localized to the inner mitochondrial membrane.

F. Biosynthesis of Triacylglycerol

In addition to its important role as the precursor of the nitrogen-containing glycerophospholipids in animal cells, diacylglycerol is also acylated at position *sn*-3 to form triacylglycerol (reviews [29,43]). The enzyme that catalyzes this reaction, diacylglycerol acyltransferase, is microsomal, and it is a different enzyme from the acyltransferases discussed above (glycerol 3-phosphate acyltransferase, 1-acylglycerol 3-phosphate acyltransferase, dihydroxyacetonephosphate acyltransferase, and 1-monoacylglycerol acyltransferase). The enzyme appears to utilize a broad range of both saturated and unsaturated fatty-acyl-CoAs and a wide range of molecular species of diacylglycerol. The activity of the diacylglycerol acyltransferase can be modulated in vivo by diet, glucocorticoids, and various drugs. Some evidence has been provided that the enzyme can be reversibly phosphorylated (phosphorylated form less active).

V. THE CYTIDINE DIPHOSPHATE DIACYLGLYCEROL PATHWAY

A. Overview

Cytidine diphosphate diacylglycerol, also called CDPdiacylglycerol, CDPdiglyceride, or CMPphosphatidate, is a liponucleotide of great importance as a precursor of glycerophospholipids. In prokaryotes such as *E. coli*, CDPdiacylglycerol is utilized, in one branch of the pathway, for the biosynthesis of phosphatidylserine and phosphatidylethanolamine. A second branch of the pathway leads to the biosynthesis of phosphatidylglycerol and diphosphatidylglycerol (cardiolipin).

In yeast, the CDPdiacylglycerol pathway is the primary route for synthesis of glycerophospholipids, although yeast can also form phosphatidylcholine and phospatidylethanolamine via the diacylglycerol pathway (Sec. IV). Glycerophospholipid biosynthesis using CDPdiacylglycerol follows three routes in yeast: (1) CDPdiacylglycerol is converted to phosphatidylserine, which is decarboxylated to form phosphatidylethanolamine, which in turn can be methylated to form a phosphatidylcholine; (2) CDPdiacylglycerol can be converted to phosphatidylglycerol, which can then be used for cardiolipin biosynthesis; (3) CDPdiacylglycerol can be converted directly to phosphatidylinositol.

In animal cells, formation of phosphatidylserine from CDPdiacylglycerol does not occur. Instead, the nitrogen-containing lipids are formed via the diacylglycerol pathway (Sec. IV). CDPdiacylglycerol in animal cells is used for the formation of the anionic phospholipids phosphatidylinositol, phosphatidylglycerol, cardiolipin, and bis(monoacylglycero)phosphate. The three branches of the CDPdiacylglycerol pathway are illustrated in Figure 4. Further metabolism of phosphatidylglycerol is illustrated in Figure 5.

B. Biosynthesis of CDPdiacylglycerol

CDPdiacylglycerol is formed by the combination of CTP and phosphatidate, catalyzed by the enzyme phosphatidate cytidylyltransferase (reviews [25,27,28,30,54,57]). The enzyme has a divalent cation requirement best met by magnesium. The enzyme is membrane-associated in *E. coli*. In yeast and animal cells, it is found in both microsomes and mitochondria. The microsomal fraction appears always to have the greater specific activity.

In animal cells, the enzymes phosphatidate phosphatase and phosphatidate cytidylyltransferase reside at important committed steps in the biosynthesis of glycerophospholipids. Formation of diacylglycerol leads to the biosynthesis of triacylglycerol and the nitrogen-containing lipids phosphatidylcholine, phosphatidylethanolamine, and phosphatidylserine. Formation of CDPdiacylglycerol leads to the biosynthesis of the acidic phospholipids phosphatidylinositol, phosphatidylglycerol, cardiolipin, and bis(monoacylglycero)phosphate. The majority of phosphatidic acid formed *de novo* appears committed to the diacylglycerol pathway, as in most eukaryotic cells less than 10% of total cell glycerolipids are products of the CDPdiacylglycerol pathway. However, determining the relative fluxes of phosphatidic acid that flow into the two pathways is complicated. The supply of diacylglycerol for glycerophospholipid biosynthesis in cells is influenced not only by phosphatidate phosphatase activity but also by phospholipase activity, sphing-

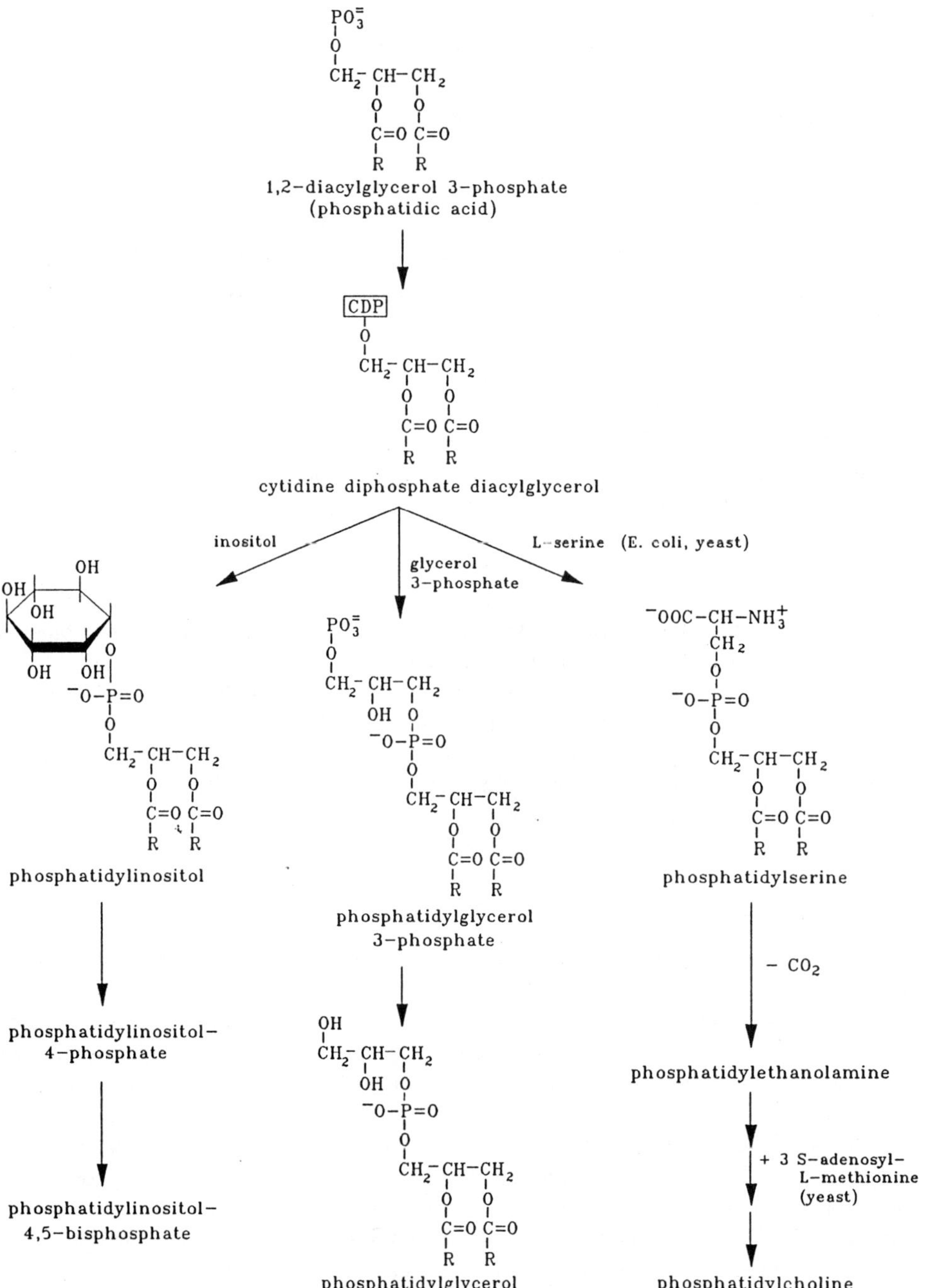

Figure 4 The CDP-diacylglycerol pathway.

omyelin biosynthetic activity (Sec. VII), and diacylglycerol kinase activity. Phosphatidylinositols, the major product of the CDPdiacylglycerol pathway in animal cells, undergo turnover more rapidly than other end-products of glycerolipid metabolism, usually via phospholipase C to produce diacylglycerol. Cells also contain pyrophosphatases that hydrolyze CDPdiacylglycerol.

The relative proportions of phosphatidic acid committed to the two pathways has been studied in microsomes in vitro by generating phosphatidic acid by phospholipase D treatment, then allowing diacylglycerol or CDPdiacylglycerol biosynthesis to take place [58]. Over a wide range of microsomal phosphatidic acid concentrations, an approximate 3:1 ratio of diacylglycerol to CDPdiacylglycerol is found. However, it should be noted that to assay the maximal phosphatidate cytidylyltransferase activity in vitro, nonphysiologic amounts of magnesium ion (40 mM) are used. As pointed out by others [59], the relative flux of phosphatidic acid into the two pathways is in fact a function of Mg^{2+} ion concentration in in vitro experiments. Most investigators conclude that the enzyme phosphatidate cytidylyltransferase operates in vivo well below the maximal activity obtainable with higher concentrations of magnesium ion.

The flow of phosphatidic acid into the CDPdiacylglycerol pathway can be enhanced by introducing a variety of amphiphilic cations into assay systems of intact cells or isolated microsomes [60]. These compounds decrease the rate of biosynthesis of products of the diacylglycerol pathway and increase the biosynthesis of the acidic phospholipids that are products of the CDPdiacylglycerol pathway. They appear to inhibit phosphatidate phosphatase activity and at the same time directly stimulate the activity of phosphatidate cytidylyltransferase at physiologic concentrations of magnesium ion [45, 59].

C. Biosynthesis of Phosphatidylinositol from CDPdiacylglycerol

1. Biosynthesis

In yeast and in animal cells, phosphatidylinositol is formed by the condensation of CDPdiacylglycerol with myoinositol, catalyzed by the enzyme phosphatidylinositol synthase (reviews [25,28,30,57]). The enzyme has a high K_m for inositol; hence the cellular levels of inositol may be important in regulating the rate of phosphatidylinositol biosynthesis. Previously it was thought that the enzyme is localized exclusively to microsomal membranes, specifically the endoplasmic reticulum. Recent evidence has been presented indicating the existence of a separate phosphatidylinositol synthase in the plasma membrane of animal cells [61,62]. Microsomal fractions of animal cells also exhibit a Mg^{2+}-dependent enzyme activity that catalyzes the exchange of free myoinositol into phosphatidylinositol ([63,64], reviews [25,28,54]). The activity is enzymatically distinct from phosphatidylinositol synthase.

2. Phosphorylation

In yeast and in animal cells, phosphatidylinositol can be sequentially phosphorylated to form phosphatidylinositol 4-phosphate, then phosphatidylinositol-4,5-bisphosphate [27,30,57,65]. The enzymes are termed phosphatidylinositol kinase and phosphatidylinositol 4-phosphate kinase. These enzymes are ATP- and Mg^{2+}-dependent. Other phosphatidylinositol kinases will phosphorylate phosphatidylinositol to form phosphatidylinositol 3-phosphate. Multiple forms of phosphatidylinositol kinase appear to be present in most eukaryotic cells, and kinase activities can be found in both soluble and membrane-bound fractions.

D. Biosynthesis of Phosphatidylglycerol from CDPdiacylglycerol

The biosynthesis of phospatidylglycerol from CDPdiacylglycerol takes place in two steps: (1) combination of CDPdiacylglycerol with *sn*-glycerol 3-phosphate to form phosphatidylglycerol 3-phosphate; (2) hydrolysis of the phosphatidylglycerol 3-phosphate to form phosphatidylglycerol (reviews [30,49,54]). The glycerol head group is therefore attached to the phosphate at position *sn*-1. This biosynthetic pathway from CDPdiacylglycerol is essentially the same in prokaryotes, yeast, animal cells, and plants. In yeast and in animal cells, both enzyme activities are localized to the inner mitochondrial membrane.

In yeast and in animal cells, phosphatidylinositol biosynthesis is a microsomal function and phosphatidylglycerol biosynthesis is a mitochondrial function. Since the phosphatidate cytidylyltransferase activity is both microsomal and mitochondrial, it is tempting to suggest that separate pools of CDPdiacylglycerol exist for phosphatidylinositol and phosphatidylglycerol. However, a variety of experimental systems have suggested that the enzymes of phosphatidylinositol and phosphatidylglycerol biosynthesis compete for common pools of CDPdiacylglycerol (review [54]).

E. Biosynthesis of Phosphatidylserine and other Nitrogen-Containing Phospholipids (Prokaryotes and Yeast)

The first step in the biosynthesis of the nitrogen-containing lipids from CDPdiacylglycerol is the formation of phosphatidylserine. CDP-diacylglycerol:L-serine phosphatidyltransferase transfers L-serine to the phosphatidate moiety of CDPdiacylglycerol. This enzyme is found in prokaryotes and in yeast (reviews [30,42,57]). It is not found in animal cells and in plants. The enzyme is membrane-associated in prokaryotes. In yeast, the single enzyme is targeted to both microsomal and mitochondrial compartments [42].

Phosphatidylethanolamine and phosphatidylcholine are formed from phosphatidylserine by pathways discussed above. First (in prokaryotes and yeast), phosphatidylserine is decarboxylated to phosphatidylethanolamine by phosphatidylserine decarboxylase, which in yeast is located in the inner mitochondrial membrane. In yeast, and in some prokaryotes [30], phosphatidylethanolamine can be methylated by one or more microsomal methyltranserases, leading to phosphatidylcholine.

F. Metabolism of Phosphatidylglycerol

1. Biosynthesis of Cardiolipin

Cardiolipin is synthesized in two ways (reviews [30,49,66]). In *E. coli*, two molecules of phosphatidylglycerol are condensed to give cardiolipin (diphosphatidylglycerol) and free glycerol. In contrast, in higher eukaryotes, cardiolipin is synthesized by the combination of phosphatidylglycerol with CDPdiacylglycerol (Figure 5). In eukaryotes, the enzyme (cardiolipin synthase) is located exclusively in the inner mitochondrial membrane. The partially purified enzyme from rat liver has an unusual selectivity for metal ion of the order $Co^{2+} > Mn^{2+} > Mg^{2+}$. In animal tissues, cardiolipin has an unusual fatty acid composition, as it contains mostly (>70%) linoleic acid. However, the cardiolipin synthase does not show great selectivity for phosphatidylglycerols or CDPdiacylglycerols enriched in linoleic acid, suggesting that fatty acid remodeling occurs after the initial biosynthesis of the molecule.

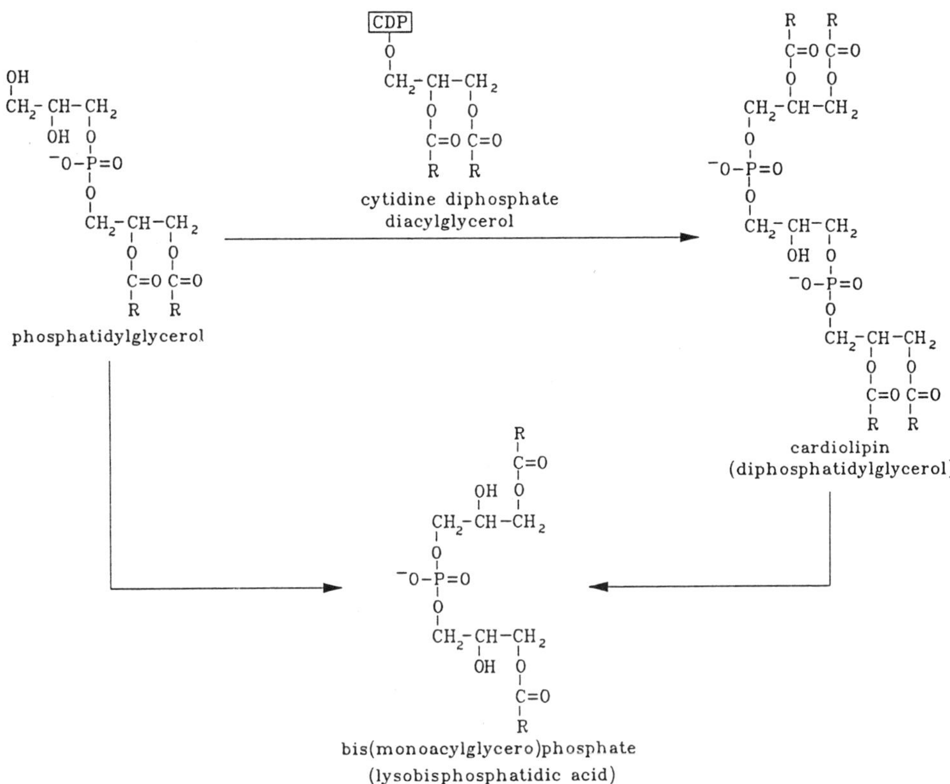

Figure 5 Lipid biosynthesis using phosphatidylglycerol.

2. Biosynthesis of Bis(monoacylglycero)phosphate

Bis(monoacylglycero)phosphate, also called lysobisphosphatidic acid, is an unusual lipid found in the lysosomes of animal cells ([67], review [54]). A distinctive feature of the glycerophospholipid is its unique stereoconfiguration. It has a backbone of *sn*-1-glycerophospho-*sn*-1'-glycerol rather than the usual *sn*-3-glycerophosphate backbone found in other glycerophospholipids. Both cardiolipin and phosphatidylglycerol can be converted to bis(monoacylglycerol)phosphate. The conversion of phosphatidylglycerol appears to require deacylation and reacylation steps, rather than migration of a fatty acid from the glycerol backbone to the glycerol headgroup [67]. Acyl-CoA is not the fatty acid donor. Instead, fatty acids are apparently obtained from other glycerophospholipids. The enzymatic steps involved in the rearrangement of the glycerol backbone stereochemistry have not been elucidated.

VI. BIOSYNTHESIS OF ETHER LIPIDS

Two types of ether lipids are common in cells. First, the alkyl glycerol ether lipids have a long chain alcohol (instead of a fatty acid) attached to the glycerol backbone by an ether bond, usually at the *sn*-1 position. The second class of ether lipids has a cis double bond between carbons 1' and 2' of the hydrocarbon chain attached by the ether linkage. These are called 1-(1'-alkenyl)-2-acylglycerolipids, often referred to by their common name, plasmalogens. Ether lipids are found in animal cells and in some anaerobic bacteria, but not in plants.

The substrate for the committed step of ether lipid biosynthesis is 1-acyl-dihydroxyacetonephosphate (reviews [54,68–71]). The alkyl ether bond is formed by an enzymatic reaction where a long-chain alcohol replaces the acyl group, to give 1-alkyldihydroxyacetonephosphate and free fatty acid. While the exact enzyme mechanism is still under investigation, it has been shown that the oxygen of the ether bond originates from the alcohol and not from the acyl chain.

The enzyme, alkyldihydroxyacetonephosphate synthase, appears to be exclusively peroxisomal. The enzyme shows little selectivity toward long-chain fatty alcohols. Instead, the control of chain specificity probably resides with the oxidoreductase that reduces fatty acyl-CoA thioesters to the corresponding alcohols [54]., In rat brain, this enzyme effectively utilizes palmitoyl-, stearoyl-, and oleoyl-CoAs as substrates, and these alkyl chains are the predominant ones found in ether lipids in mammalian tissues. Shorter acyl-CoAs and polyunsaturated acyl-CoAs are not effective substrates. (Polyunsaturated hydrocarbon chains are not found in ether linkages.)

Once formed, the 1-alkyldihydroxyacetonephosphate is reduced to 1-alkyl-*sn*-glycerol 3-phosphate by an NADPH-dependent oxidoreductase that is found both in peroxisomal and microsomal compartments. The 1-alkyl-*sn*-glycerol 3-phosphate is acylated at position *sn*-2 to form the 1-alkyl analog of phosphatidic acid. This acylation step takes place outside the peroxisomes. The ether lipid can then be metabolized along the same pathways discussed above for the diacylglycerolipids to form the alkyl ether analogs of diacylphosphatidylcholine, diacylphosphatidylethanolamine, and so forth.

The 1-(1'-alkenyl)-2-acyl-*sn*-glycero-3-phosphoethanolamines (the ethanolamine plasmalogens) are obtained by insertion of a cis double bond at the 1' position of the alkyl chain of 1-alkyl-2-acyl-*sn*-glycero-3-phosphoethanolamine. The reaction is catalyzed by a microsomal Δ^1 desaturase enzyme complex similar to the desaturase systems for fatty acyl-CoAs. Desaturation of the 1-alkyl chain has been shown only for the ethanolamine ether lipids. Hence it is likely that choline, serine, and inositol plasmalogens are all derived fron ethanolamine plasmalogen. Various pathways to form these other classes of plasmalogens can be followed [69]. For example, phospholipase C can hydrolyze the phosphoethanolamine linkage to form the 1-(1'-alkenyl) analog of diacylglycerol. This can be used as a substrate for the variety of products of the diacylglycerol pathway. Phosphorylation of the ether analog of diacylglycerol can form the ether analog of phosphatidic acid, which can then flow into the CDP-diacylglycerol pathway. Serine plasmalogen can be formed by base exchange with ethanolamine plasmalogen. Ethanolamine plasmalogen can be methylated to form choline plasmalogen.

VII. BIOSYNTHESIS OF SPHINGOLIPIDS

A. Ceramide

Sphingomyelin biosynthetic pathways have been recently reviewed [72] and are illustrated in Figure 6. The initial step is the condensation of serine with acyl-CoA to form 3-ketosphinganine by the enzyme serine palmitoyltransferase (also called 3-keto-

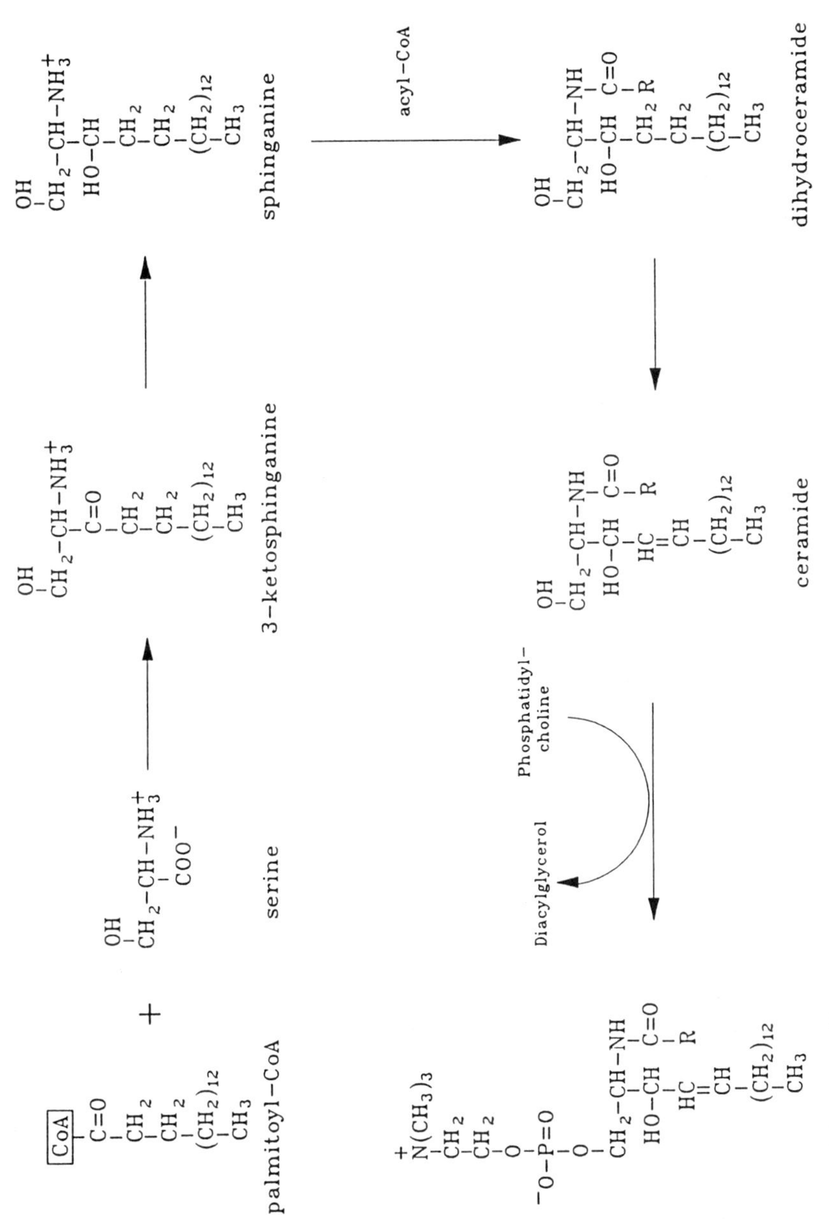

Figure 6 Biosynthesis of sphingomyelin.

sphinganine synthase). This pyridoxal 5'-phosphate-dependent enzyme exhibits a high level of selectivity for palmitoyl-CoA, resulting in the formation of, primarily, the 18-carbon sphinganine base (16 carbons from palmitic acid, 2 from serine). The 3-keto moiety is then rapidly reduced by a microsomal NADPH-dependent reductase to form sphinganine, as 3-ketosphinganine is not detected in intact cells.

The order of the next two steps in ceramide synthesis is not clear, but it appears to proceed by the following sequence: First, the amine is *N*-acylated with fatty acyl-CoA to form dihydroceramide. A CoA-independent acylation pathway has also been reported. The dihydroceramide is oxidized to ceramide by introduction of the 4,5-double bond on the long chain base, catalyzed by a presumed dihydroceramide dehydrogenase that has not been characterized.

B. Sphingomyelin

According to several recent studies, most sphingomyelin is formed by the transfer of phosphocholine from phosphatidylcholine to ceramide, catalyzed by the enzyme phosphatidylcholine:ceramide cholinephosphotransferase [73–75]. (Transer of the phosphocholine group from CDP-choline appears to be an alternate pathway, but it is believed to make a minor contribution to sphingomyelin biosynthesis.) The enzyme activity colocalizes with membranes enriched in 5'-nucleotidase activity, i.e., plasma membrane and Golgi. Studies with fluorescently labeled ceramide, which allow for a comparison of subcellular location (viewed by fluorescence microscopy) with metabolism, indicate that most of the sphingomyelin formation from ceramide occurs in the Golgi apparatus, prior to transport to the plasma membrane.

It should be noted that considerable sphingomyelin can be delivered to a cell by circulating low-density and high-density lipoproteins. The relationships between *de novo* synthesis and uptake from exogenous lipoprotein have not been established.

C. Sphingosine

Sphingomyelin is converted to back to ceramide by the action of sphingomyelinase. Both a neutral sphingomyelinase (apparently on the extracellular surface of the plasma membrane) and an acidic (lysosomal) sphingomyelinase have been identified. Ceramide is hydrolyzed to sphingosine by the removal of the amide-linked fatty acid. Three types of ceramidases (*N*-acylsphingosine deacylase) have been reported: a lysosomal enzyme with an acid pH optimum, an enzyme with a neutral pH optimum, and one with an alkaline pH optimum. Sphingosine is metabolically cleared by phosphorylation by an ATP-dependent kinase, which forms sphingosine 1-phosphate. The product is cleaved by sphingosine 1-phosphate lyase to produce ethanolamine phosphate and trans-2-hexadecanal.

VIII. PHOSPHOLIPID FATTY ACID REMODELING

Glycerophospholipids contain mostly saturated fatty acids at position *sn*-1 and unsaturated fatty acids at position *sn*-2. It appears that at least some of this specificity is introduced into the glycerol backbone at the level of phosphatidic acid biosynthesis. However, after the end-products of lipid metabolism are formed, they can be remodeled to form glycerolipids with fatty acid profiles enriched in polyunsaturated fatty acids and more characteristic of an individual phospholipid class (reviews [14,76].

In vitro studies of acylation of *sn*-glycerol 3-phosphate, most often using mammalian liver, indicate that the glycerophosphate acyltransferase (which acylates *sn*-glycerol 3-phosphate at position *sn*-1) preferentially utilizes saturated fatty acyl-CoA thioesters. In contrast, the acylglycerophosphate acyltransferases show preferences for unsaturated fatty acids when acylating 1-acyl-*sn*-glycerol 3-phosphate at position *sn*-2. A fatty acid profile of primarily saturated fatty acid at position *sn*-1 and unsaturated fatty acid (principally monoenoic and dienoic fatty acid) at position *sn*-2 is indeed what is found for phosphatidic acid in liver tissue [77].

Glycerophospholipid products of phosphatidic acid can then be remodeled by a variety of mechanisms: (1) A deacylation-reacylation cycle, where a fatty acid is removed from a glycerophospholipid by phospholipase A_1 or phospholipase A_2, then reacylated with fatty acyl-CoA; (2) CoA-independent or CoA-dependent transacylation, where a diacylglycerophospholipid serves as fatty acid donor, and a lysophospholipid (1-acyl, 1-alkyl, or 1-(1'-alkenyl)) serves as fatty acid acceptor; (3) A lysophospholipid-lysophospholipid transacylase. Reacylation activities, when measured with free fatty acid as substrate, are ATP and Mg^{2+} dependent. Transacylase activities are ATP and Mg^{2+} independent.

Fatty acid remodeling of phospholipids in mammalian cells has been recently reviewed [14], with particular emphasis on the role of remodeling in the cellular control of the arachidonic acid enrichment of various phospholipids, and with emphasis on the regulation of arachidonic acid metabolism to bioactive lipids. It is found that fatty acid remodeling pathways vary considerably among different cell types in activity, fatty acid specificity, and donor and acceptor specificity. In platelets and inflammatory cells (e.g., neutrophils and macrophages), deacylation-reacylation appears to be the major pathway for enrichment of diacylglycerophospholipids with arachidonic acid and other polyunsaturated fatty acids. There is evidence that there are separate acyl-CoA:lysophosphatide acyltransferases for nitrogen-containing lysophospholipids (lysophosphatidylcholine and lysophosphatidylethanolamine) and the acidic lysophospholipids such as lysophosphatidylinositol. In contrast to the reacylation of acyl-linked lysophospholipids, reacylation of alkyl- or 1-(1'-alkenyl)-linked lysophospholipids with acyl-CoA is low. Instead, there is compelling evidence that in platelets and inflammatory cells, transacylase activity is the primary pathway by which ether lipids become enriched with polyunsaturated fatty acids. In these reactions, diacylglycerophospholipids serve as fatty acid donors, and 1-alkyl-*sn*-glycero-3-phosphorylcholine and 1-(1'-alkenyl)-*sn*-glycero-3-phosphoethanolamine are the principal acceptors. The CoA-independent transacylase pathway appears to be quantitatively more important for enrichment of ether lipids with arachidonic acid in inflammatory cells, as the CoA-dependent transacylase activity is not as selective for arachidonic acid.

In brain, deacylation-reacylation activities and transacylase activities have been demonstrated. Brain tissue also appears capable of introducing sizable levels of polyunsaturated fatty acids into phospholipids during *de novo* biosynthesis. In contrast, in liver, deacylation-reacylation appears to be the primary mechanism for enriching glycerophospholipids with arachidonic and other polyunsaturated fatty acids. In liver, CoA-dependent transacylation activity is also present, while CoA-independent transacylation activity is low.

For a review of the enzymes of phospholipid biosynthesis, see Table 1.

Table 1 Enzymes of Phospholipid Biosynthesis. Enzymes are Listed for Which the Nomenclature Committee of the International Union of Biochemistry Have Assigned Reference Numbers. (Refs. 78–80.)

Recommended name	Systematic name	Other names	E.C. number	Reaction catalyzed
	Formation of Phosphatidic Acid			
Glycerol kinase	ATP:glycerol 3-phosphotransferase		E.C. 2.7.1.30	ATP + glycerol → ADP + *sn*-glycerol 3-phosphate
Glycerol-3-phosphate dehydrogenase (NAD^+)	*sn*-Glycerol-3-phosphate:NAD^+ 2-oxidoreductase		E.C. 1.1.1.8	*sn*-Glycerol 3-phosphate + NAD^+ → glycerone phosphate (dihydroxyacetonephosphate) + NADH
Glycerol-3-phosphate dehydrogenase ($NAD(P)^+$)	*sn*-Glycerol-3-phosphate:$NAD(P)^+$ 2-oxidoreductase		E.C. 1.1.1.94	*sn*-Glycerol 3-phosphate + $NAD(P)^+$ → glycerone phosphate (dihydroxyacetonephosphate) + NAD(P)H
Long-chain-acid-CoA ligase	Acid:CoA ligase (AMP-forming)	Acyl-CoA synthetase Fatty acid thiokinase (long chain) Acyl-activating enzyme	E.C. 6.2.1.3	ATP + long-chain carboxylic acid + CoA → AMP + pyrophosphate + acyl-CoA
Long-chain-fatty-acid-acyl-carrier-protein ligase	Long-chain-fatty-acid:protein ligase (AMP-forming)	Acyl-[acyl-carrier-protein] synthetase	E.C. 6.2.1.20	ATP + long-chain carboxylic acid + ACP → AMP + pyrophosphate + acyl-ACP
Arachidonate-CoA ligase	Arachidonate:CoA ligase (AMP-forming)	Arachidonoyl-CoA synthetase	E.C. 6.2.1.15	ATP + arachidonate + CoA → AMP + pyrophosphate + arachidonoyl-CoA
Glycerol-3-phosphate acyltransferase	Acyl-CoA:*sn*-glycerol-3-phosphate *O*-acyltransferase		E.C. 2.3.1.15	Acyl-CoA + *sn*-glycerol 3-phosphate → CoA + 1-acyl-*sn*-glycerol 3-phosphate
Glycerone-phosphate acyltransferase	Acyl-CoA:glycerone-phosphate *O*-acyltransferase	Dihydroxyacetonephosphate acyltransferase	E.C. 2.3.1.42	Acyl-CoA + glycerone phosphate → CoA + acylglycerone phosphate
1-Acylglycerol-3-phosphate acyltransferase	Acyl-CoA:1-acyl-*sn*-glycerol-3-phosphate *O*-acyltransferase		E.C. 2.3.1.51	Acyl-CoA + 1-acyl-*sn*-glycerol 3-phosphate → CoA + 1,2-diacyl-*sn*-glycerol 3-phosphate

Acylglycerone-phosphate reductase	1-Palmitoylglycerol-3-phosphate:NADP$^+$ oxidoreductase	Palmitoyldihydroxyacetonephosphate reductase	E.C. 1.1.1.101	1-Palmitoylglycerol 3-phosphate + NADP$^+$ → 1-palmitoylglycerone phosphate + NADPH (also acts on alkylglycerone-3-phosphate and alkylglycerol 3-phosphate)
Long-chain-alcohol dehydrogenase	Long-chain-alcohol:NAD$^+$ oxidoreductase		E.C. 1.1.1.192	Long-chain alcohol + 2 NAD$^+$ + H_2O → a long-chain acid anion + 2 NADH
Alkylglycerone-phosphate synthase	1-Acyl-glycerone-3-phosphate: long-chain-alcohol *O*-3-phospho-2-oxopropanyl-transferase	Alkyldihydroxyacetonephosphate synthase	E.C. 2.5.1.26	1-Acyl-glycerone 3-phosphate + a long-chain alcohol → 1-alkyl-glycerone 3-phosphate + a long-chain acid anion
Diacylglycerol kinase	ATP:1,2-diacylglycerol 3-phosphotransferase	Diglyceride kinase	E.C. 2.7.1.107	ATP + 1,2-diacylglycerol → ADP + 1,2-diacyl-*sn*-glycerol 3-phosphate

Diacylglycerol Pathway

Recommended name	Systematic name	Other names	E.C. number	Reaction catalyzed
Phosphatidate phosphatase	3-*sn*-Phosphatidate phosphohydrolase	Phosphatidate phosphohydrolase Phosphatidic acid phosphohydrolase	E.C. 3.1.3.4	3-*sn*-phosphatidate + H_2O → 1,2-diacyl-*sn*-glycerol + orthophosphate
Choline kinase	ATP:choline phosphotransferase		E.C. 2.7.1.32	ATP + choline → ADP + *O*-phosphocholine
Ethanolamine kinase	ATP:ethanolamine *O*-phosphotransferase		E.C. 2.7.1.82	ATP + ethanolamine → ADP + *O*-phosphoethanolamine
Choline-phosphate cytidylyltransferase	CTP:choline-phosphate cytidylyltransferase	CDPcholine synthetase Phosphorylethanolamine transferase	E.C. 2.7.7.15	CTP + choline phosphate → pyrophosphate + CDPcholine
Ethanolamine-phosphate cytidylyltransferase	CTP:ethanolamine-phosphate cytidylyltransferase	CDPethanolamine synthetase Phosphorylethanolamine transferase	E.C. 2.7.7.14	CTP + ethanolamine phosphate → pyrophosphate + CDPethanolamine

Table 1 Continued

Diacylglycerol Pathway				
Recommended name	Systematic name	Other names	E.C. number	Reaction catalyzed
Cholinephosphotransferase	CDPcholine:1,2-diacylglycerol cholinephosphotransferase	Phosphorylcholine-glyceride transferase	E.C. 2.7.8.2	CDPcholine + 1,2-diacylglycerol → CMP + phosphatidylcholine
Ethanolaminephosphotransferase	CDPethanolamine: 1,2-diacylglycerol ethanolaminephosphotransferase		E.C. 2.7.8.1	CDPethanolamine + 1,2-diacylglycerol → CMP + phosphatidylethanolamine
Diacylglycerol acyltransferase	Acyl-CoA:1,2-diacylglycerol *O*-acyltransferase	Diglyceride acyltransferase	E.C. 2.3.1.20	Acyl-CoA + 1,2-diacylglycerol → CoA + triacylglycerol
Phosphatidylethanolamine methyltransferase	*S*-Adenosyl-L-methionine:phosphatidylethanolamine *N*-methyltransferase		E.C. 2.1.1.17	*S*-Adenosyl-L-methionine + phosphatidylethanolamine → *S*-adenosyl-L-homocysteine + phosphatidyl-*N*-methylethanolamine
Phosphatidyl-*N*-methylethanolamine methyltransferase	*S*-Adenosyl-L-methionine:phosphatidyl-*N*-methylethanolamine *N*-methyltransferase		E.C. 2.1.1.71	*S*-Adenosyl-L-methionine + phosphatidyl-*N*-methylethanolamine → *S*-adenosyl-L-homocysteine + phosphatidyl-*N*-dimethylethanolamine
1-Acylglycerolphosphocholine acyltransferase	Acyl-CoA:1-acyl-*sn*-glycerol-3-phosphocholine *O*-acyltransferase	Lysolecithin acyltransferase	E.C. 2.3.1.23	Acyl-CoA + 1-acyl-*sn*-glycero-3-phosphocholine → CoA + 1,2-diacyl-*sn*-glycero-3-phosphocholine
1-Alkylglycerophosphocholine acyltransferase	Acyl-CoA:1-alkyl-*sn*-glycero-3-phosphocholine *O*-acyltransferase		E.C. 2.3.1.63	Acyl-CoA + 1-alkyl-*sn*-glycero-3-phosphocholine → CoA + 1-alkyl-2-acyl-*sn*-glycero-3-phosphocholine
Phosphatidylserine decarboxylase	Phosphatidyl-L-serine carboxy-lyase		E.C. 4.1.1.65	Phosphatidyl-L-serine → phosphatidylethanolamine + CO_2

CDP-Diacylglycerol Pathway				
Recommended name	Systematic name	Other names	E.C. number	Reaction catalyzed
Phosphatidate cytidylyltransferase	CTP:phosphatidate cytidylyltransferase	CDPdiglyceride cytidylyltransferase	E.C. 2.7.7.41	CTP + phosphatidate → pyrophosphate + CDPdiacylglycerol
CDPacylglycerol arachidonyltransferase	Arachidonyl-CoA:CDPacylglycerol *O*-arachidonyltransferase		E.C. 2.3.1.70	Arachidonyl-CoA + CDPacylglycerol → CoA + CDPdiacylglycerol
CDPdiacylglycerol—inositol 3-phosphatidyltransferase	CDPdiacylglycerol:myo-inositol 3-phosphatidyltransferase	CDPdiglyceride—inositol phosphatidyltransferase Phosphatidylinositol synthase	E.C. 2.7.8.11	CDPdiacylglycerol + myo-inositol → CMP + phosphatidyl-1D-myo-inositol
1-Phosphatidylinositol kinase	ATP:1-phosphatidyl-1D-myo-inositol 4-phosphotransferase		E.C. 2.7.1.67	ATP + 1-phosphatidyl-1D-myo-inositol → ADP + 1-phosphatidyl-1D-myo-inositol 4-phosphate
1-Phosphatidylinositol-4-phosphate kinase	ATP:1-phosphatidyl-1D-myo-inositol-4-phosphate 5-phosphotransferase	Diphosphoinositide kinase	E.C. 2.7.1.68	ATP + 1-phosphatidyl-1D-myo-inositol 4-phosphate → ADP + 1-phosphatidyl-1D-myo-inositol 4,5-bisphosphate
CDPdiacylglycerol—glycerol-3-phosphate 3-phosphatidyltransferase	CDPdiacylglycerol:*sn*-glycerol-3-phosphate 3-phosphatidyltransferase	Glycerophosphate phosphatidyltransferase 3-Phosphatidyl-1′-glycerol-3′-phosphate synthase	E.C. 2.7.8.5	CDPdiacylglycerol + *sn*-glycerol 3-phosphate → CMP + 3-(3-*sn*-phosphatidyl)-*sn*-glycerol 1-phosphate
Phosphatidylglycerophosphatase	Phosphatidylglycerophosphate phosphohydrolase		E.C. 3.1.3.27	Phosphatidylglycerophosphate + H_2O → phosphatidylglycerol + orthophosphate
CDPdiacylglycerol-serine *O*-phosphatidyltransferase	CDPdiacylglycerol:L-serine *O*-phosphatidyltransferase	CDPdiglyceride—serine *O*-phosphatidyltransferase Phosphatidylserine synthase	E.C. 2.7.8.8	CDPdiacylglycerol + L-serine → CMP + *O* -*sn*-phosphatidyl-L-serine

	Sphingomyelin Biosynthesis	E.C. number	Reaction catalyzed
Serine palmitoyltransferase	Palmitoyl-CoA:L-serine C-palmitoyltransferase (decarboxylating)	E.C. 2.3.1.50	Palmitoyl-CoA + L-serine → CoA + 3-dehydro-*O*-sphinganine + CO_2

REFERENCES

1. A. W. Alberts and M. D. Greenspan, Animal and bacterial fatty acid synthetase: Structure, function and regulation, *Fatty Acid Metabolism and Its Regulation* (S. Numa, ed.), Elsevier, Amsterdam, 1984, pp. 29–58.
2. C. R. H. Raetz, Molecular genetics of membrane phospholipid synthesis, *Ann. Rev. Genetics 20*:253–295 (1986).
3. T. V. Boom and J. E. Cronan, Jr., Genetics and regulation of bacterial lipid metabolism. *Ann. Rev. Microbiol. 43*:317–343 (1989).
4. C. O. Rock and J. E. Cronan, Jr., Lipid metabolism in procaryotes, in *Biochemistry of Lipids and Membranes* (D. E. Vance and J. E. Vance, eds.), Benjamin/Cummings, Menlo Park, Calif., 1985, pp. 73–115.
5. E. Schweizer, Genetics of fatty acid biosynthesis in yeast, in *Fatty Acid Metabolism and Its Regulation* (S. Numa, ed.), Elsevier, Amsterdam, 1984, pp. 59–84.
6. E. Schweizer, Biosynthesis of fatty acids and related compounds, *Microbial Lipids, Vol. 2* (C. Ratledge and S. G. Wilkinson, eds.), Academic Press, San Diego, 1989, pp. 3–50.
7. S. J. Wakil, Fatty acid synthase, a proficient multifunctional enzyme, *Biochemistry 28*:4523–4530 (1989).
8. A. G. Goodridge, Fatty acid synthesis in eucaryotes, in *Biochemistry of Lipids and Membranes* (D. E. Vance and J. E. Vance, eds.), Benjamin/Cummings, Menlo Park, Calif., 1985, pp. 143–180.
9. C. Somerville and J. Browse, Plant lipids: Metabolism, mutants, and membranes, *Science 252*:80–87 (1991).
10. P. K. Stumpf, The biosynthesis of saturated fatty acids, in *The Biochemistry of Plants: A Comprehensive Treatise,* Vol. 9 (P. K. Stumpf, ed.), Academic Press, Orlando, Fla., 1987, pp. 121–136.
11. J. B. Ohlrogge, Biochemistry of plant acyl carrier proteins, in *The Biochemistry of Plants: A Comprehensive Treatise,* Vol. 9 (P. K. Stumpf, ed.), Academic Press, Orlando, Fla, 1987, pp. 137–158.
12. P. K. Stumpf, Fatty acid biosynthesis in higher plants, in *Fatty Acid Metabolism and Its Regulation* (S. Numa, ed.), Elsevier, Amsterdam, 1984, pp. 155–180.
13. J. B. Ohlrogge, J. Browse, and C. R. Somerville, The genetics of plant lipids, *Biochim. Biophys. Acta 1082*:1–26 (1991).
14. J. I. S. MacDonald and H. Sprecher, Phospholipid fatty acid remodeling in mammalian cells, *Biochim. Biophys. Acta 1084*:105–121 (1991).
15. R. Jeffcoat and A. T. James, The regulation of desaturation and elongation of fatty acids in mammals, *Fatty Acid Metabolism and Its Regulation* (S. Numa, ed.), Elsevier, Amsterdam, 1984, pp. 85–112.
16. P. W. Holloway, Fatty acid desaturation, *The Enzymes 16*:63–83 (1983).
17. R. Jeffcoat, The biosynthesis of unsaturated fatty acids and its control in mammalian liver, *Essays Biochem. 15*:1–36 (1979).
18. H. W. Cook, Fatty acid desaturation and chain elongation in eucaryotes, in *Biochemistry of Lipids and Membranes* (D. E. Vance and J. E. Vance, eds.), Benjamin/Cummings, Menlo Park, Calif., 1985, pp. 181–212.
19. J. G. Jaworski, Biosynthesis of monoenoic and polyenoic fatty acids, in *The Biochemistry of Plants: A Comprehensive Treatise,* Vol. 9 (P. K. Stumpf, ed.), Academic Press, Orlando, Fla., 1987, pp. 159–174.
20. H. Osmundsen, J. Bremer, J. I. Pedersen, Metabolic aspects of peroxisomal beta-oxidation, *Biochim. Biophys. Acta 1085*:141–158 (1991).
21. H. Sprecher, Biochemistry of essential fatty acids, *Prog. Lipid Res. 20*:13–22 (1981).
22. A. Poulos, Lipid metabolism in Zellweger's syndrome, *Prog. Lipid Res. 28*:35–51 (1989).

23. R. J. A. Wanders, C. W. T. van Roermund, R. B. H. Schutgens, P. G. Barth, H. S. A. Heymans, H. van den Bosch, and J. M. Tager, The inborn errors of peroxisomal beta-oxidation: A review, *J. Inherit. Metab. Dis. 13*:4–36 (1990).
24. D. N. Brindley, General introduction, *Phosphatidate Phosphohydrolase,* vol 1 (D. N. Brindley, ed.), CRC press, Boca Raton, Fla., pp. 2–19 (1988).
25. J. D. Esko and C. R. H. Raetz, Synthesis of phospholipids in animal cells, *The Enzymes 16*:207–253 (1983).
26. R. M. Bell and R. A. Coleman, Enzymes of triacylglycerol formation in mammals, *The Enzymes 16*:87–111 (1983).
27. W. R. Bishop and R. M. Bell, Assembly of phospholipids into cellular membranes: Biosynthesis, transmembrane movement and intracellular translocation, *Ann. Rev. Cell Biol. 4*:579–610 (1988).
28. R. M. Bell and R. A. Coleman, Enzymes of glycerolipid synthesis in eukaryotes, *Ann. Rev. Biochem. 49*:459–487 (1980).
29. D. N. Brindley, Metabolism of triacylglycerols, in *Biochemistry of Lipids and Membranes* (D. E. Vance and J. E. Vance, eds.), Benjamin/Cummings, Menlo Park, Calif., 1985, pp. 213–241.
30. R. A. Pieringer, Biosynthesis of non-terpenoid lipids, in *Microbial lipids,* Vol. 2 (C. Ratledge and S. G. Wilkinson, eds.), Academic Press, San Diego, Calif., 1989, pp. 51–114.
31. D. Haldar, W.-W. Tso, and M. E. Pullman, The acylation of *sn*-glycerol 3-phosphate in mammalian organs and Ehrlich ascites tumor cells, *J. Biol. Chem. 254*: 4502–4509 (1979).
32. S. M. Fitzpatrick, G. Sorresso, and D. Haldar, Acyl CoA:*sn*-glycerol-3-phosphate *O*-acyltransferase in rat brain microsomes, *J. Neurochem. 39*:286–289 (1982).
33. D. M. Schlossman and R. M. Bell, Triacylglycerol synthesis in isolated fat cells. Evidence that the *sn*-glycerol 3-phosphate and dihydroxyacetone phosphate acyltransferases are dual catalytic functions of a single microsomal enzyme, *J. Biol. Chem. 251*:5738–5744 (1976).
34. D. M. Schlossman and R. M. Bell, Micosomal *sn*-glycerol 3-phosphate acyltransferase activities from liver and other tissues. Evidence for a single enzyme catalysing both reactions, *Archiv. Biochem. Biophys. 182*:732–742 (1977).
35. R. A. Coleman and R. M. Bell, Selective changes in enzymes of *sn*-glycerol 3-phosphate and dihydroxyacetonephosphate pathways of triacylglycerol biosynthesis during differentiation of 3T3-L1 preadipocytes, *J. Biol. Chem. 255*:7681–7687 (1980).
36. T. S. Tillman and R. M. Bell, Mutants of *Saccharomyces cerevisiae* defective in *sn*-glycerol 3-phosphate acyltransferase, *J. Biol. Chem. 261*:9144–9149 (1986).
37. A. K. Hajra, S. Horie, and K. O. Webber, The role of peroxisomes in glycerol ether lipid metabolism, in *Biological Membranes: Abberations in Membrane Structure and Function,* Alan R. Liss, New York, 1988, pp. 99–116.
38. R. Manning and D. N. Brindley, Tritium isotope effects in the measurement of the glycerol phosphate and dihydroxyacetone phosphate pathways of glycerolipid biosynthesis in rat liver, *Biochem. J. 130*:1003–1012 (1972).
39. R. Rognstad, D. G. Clark, and J. Katz, Pathways of glyceride glycerol synthesis, *Biochem. J. 140*:249–251 (1974).
40. R. J. Mason, Importance of the acyl dihydroxyacetone phosphate pathway in the synthesis of phosphatidylglycerol and phosphatidylcholine in alveolar type II cells, *J. Biol. Chem. 253*:6650–6653 (1978).
41. H. Kanoh, K. Yamada, and F. Sakane, Diacylglycerol kinase: A key modulator of signal transduction? *Trends Biochem. Sci. 15*:47–50 (1990).
42. R. H. Hjelmstad and R. M. Bell, Molecular insights into enzymes of membrane bilayer assembly, *Biochemistry 30*:1731–1740 (1991).
43. L. B. M. Tijburg, M. J. H. Geelen, and L. M. G. van Golde, Regulation of the biosynthesis of triacylglycerol, phosphatidylcholine, and phosphatidylethanolamine in the liver, *Biochim. Biophys. Acta 1004*:1–19 (1989).

44. J. E. Bleasdale and J. M. Johnston, Phosphatidic acid production and utilization, in *Lung Development: Biological and Clinical Perspectives* (P. M. Farrell, ed.), Academic Press, New York, 1982, pp. 259–294.
45. D. N. Brindley, *Phosphatidate Phosphohydrolase,* CRC Press, Boca Raton, Fla., 1988.
46. J. E. Bleasdale, C. S. Davis, A. K. Hajra, and B. W. Agranoff, A rapid sensitive assay for phosphatidate phosphohydrolase, *Anal. Biochem. 87*:19–27 (1978).
47. A. Martin, P. Hales, and D. N. Brindley, A Rapid assay for measuring the activity and the Mg^{2+} and Ca^{2+} requirements of phosphatidate phosphohydrolase in cytosolic and microsomal fractions of rat liver, *Biochem. J. 245*:347–355 (1987).
48. M. Freeman and E. H. Mangiapane, Translocation to rat liver mitochondria of phosphatidate phosphohydrolase, *Biochem. J. 263*:589–595 (1989).
49. D. E. Vance, Phospholipid metabolism in eucaryotes, in *Biochemistry of Lipids and Membranes* (D. E. Vance and J. E. Vance, eds.), Benjamin/Cummings, Menlo Park, Calif., 1985, pp. 242–270.
50. T. J. Porter and C. Kent, Purification and characterization of choline/ethanolamine kinase from rat liver, *J. Biol. Chem. 265*:414–422 (1990).
51. K. Ishidate, K. Furusawa, and Y. Nakazawa, Complete co-purification of choline kinase and ethanolamine kinase from rat kidney and immunological evidence for both kinase activities residing on the same enzyme protein(s) in rat tissues, *Biochim. Biophys. Acta 836*:119–124 (1985).
52. G. M. Hatch, O. Karmin, and P. C. Choy, Regulation of phosphatidylcholine metabolism in mammalian hearts, *Biochem. Cell Biol. 67*:67–77 (1989).
53. C. Kent, G. M. Carman, M. W. Spence, and W. Dowhan, Regulation of eukaryotic phospholipid metabolism, *FASEB Journal 5*:2258–2266 (1991).
54. K. J. Longmuir, Biosynthesis and distribution of lipids, *Curr. Topics Membr. Transport 29*:129–174 (1987).
55. D. E. Vance and N. D. Ridgway, The methylation of phosphatidylethanolamine, *Prog. Lipid Res. 27*:61–79 (1988).
56. J. Baranska, Biosynthesis and transport of phosphatidylserine in the cell, *Adv. Lipid Res. 19*:163–184 (1982).
57. G. M. Carman and S. A. Henry, Phospholipid biosynthesis in yeast, *Ann. Rev. Biochem. 58*:635–669 (1989).
58. G. P. H. van Heusden and H. van den Bosch, The influence of exogenous and of membrane-bound phosphatidate concentration on the activity of CTP:phosphatidate cytidylyltransferase and phosphatidate phosphohydrolase, *Eur. J. Biochem. 84*:405–412 (1978).
59. R. G. Sturton and D. N. Brindley, Factors controlling the activities of phosphatidate phosphohydrolase and phosphatidate cytidylyltransferase, *Biochem. J. 162*:25–32 (1977).
60. D. Allan and R. H. Michell, Enhanced synthesis de novo of phosphatidylinositol in lymphocytes treated with cationic amphiphilic drugs, *Biochem. J. 148*:471–478 (1975).
61. A. Imai and M. C. Gershengorn, Independent phosphatidyl inositol synthesis in pituitary plasma membrane and endoplasmic reticulum, *Nature 325*:726–728 (1987).
62. A. Imai and M. C. Gershengorn, Regulation by phosphatidylinositol of rat pituitary plasma membrane and endoplasmic reticulum phosphatidylinositol synthase activities, *J. Biol. Chem. 262*:6457–6459 (1987).
63. T. Takenawa and K. Egawa, Phosphatidyl inositol:myo-inositol exchange enzyme from rat liver: Partial purification and characterization, *Archiv. Biochem. Biophys. 202*:601–607 (1980).
64. J. E. Bleasdale and P. Wallis, Phosphatidylinositol-inositol exchange in rabbit lung, *Biochim. Biophys. Acta 664*:428–440 (1981).
65. H. Kanoh, Y. Banno, M. Hirata, and Y. Nozawa, Partial purification and characterization of phosphatidylinositol kinases from human platelets, *Biochim. Biophys. Acta 1046*:120–126 (1990).

66. P. V. Ioannou and B. T. Golding, Cardiolipins: Their chemistry and biochemistry, *Prog. Lipid Res. 17*:279–318 (1979).
67. S. J. Huterer and J. R. Wherrett, Formation of bis(monoacylglycero)phosphate by a macrophage transacylase, *Biochim. Biophys. Acta 1001*:68–75 (1989).
68. H. K. Mangold and N. Weber, Biosynthesis and biotransformation of ether lipids, *Lipids 22*:789–799 (1987).
69. F. Paltauf, Biosynthesis of 1-O(1'-alkenyl)glycerolipids (plasmalogens), in *Ether Lipids: Biochemical and Biomedical Aspects* (H. K. Mangold and F. Paltauf, eds.), Academic Press, New York, 1983, pp. 107–128.
70. A. K. Hajra, S. Horie, and K. O. Webber, The role of peroxisomes in glycerol ether lipid metabolism, in *Progress in clinical and biological research,* Vol. 282 (M. L. Karnovsky, A Leaf, and L. C. Bolis, eds.), Alan R. Liss, New York, 1987, pp. 99–116.
71. F. Snyder, Metabolism of platelet activating factor and related ether lipids: Enzymatic pathways, subcellular sites, regulation, and membrane processing, in *Progress in Clinical and Biological Research,* Vol. 282 (M. L. Karnovsky, A. Leaf, and L. C. Bolis, eds.), Alan R. Liss, New York, 1988, pp. 57–72.
72. A. H. Merrill, Jr., and D. D. Jones, An update of the enzymology and regulation of sphingomyelin metabolism, *Biochim. Biophys. Acta 1044*:1–12 (1990).
73. W.-D. Marggraf, F. A. Anderer, and J. N. Kanfer, The formation of sphingomyelin from phosphatidylcholine in plasma membrane preparations from mouse fibroblasts, *Biochim. Biophys. Acta 664*:61–73 (1981).
74. D. R. Voelker and E. P. Kennedy, Cellular and enzymic synthesis of sphingomyelin, *Biochemistry 21*:2753–2759 (1982).
75. J. T. Bernert and M. D. Ullman, Biosynthesis of sphingomyelin from erythro-ceramides and phosphatidylcholine by a microsomal cholinephosphotransferase, *Biochim. Biophys. Acta 666*:99–109 (1981).
76. B. J. Holub and A. Kuksis, Metabolism of molecular species of diacylglycerophospholipids, *Adv. Lipid Res. 16*:1–125 (1978).
77. F. Possmayer, G. L. Scherphof, T. M. A. R. Dubbelman, and L. M. G. van Golde, Positional specificity of saturated and unsaturated fatty acids in phosphatidic acid from rat liver, *Biochim. Biophys. Acta 176*:95–110 (1969).
78. Nomenclature Committee of the International Union of Biochemistry, *Enzyme Nomenclature,* Academic Press, Orlando, Fla., 1984.
79. Nomenclature Committee of the International Union of Biochemistry, Enzyme nomenclature. Recommendations 1984. Supplement 1: Corrections and additions, *Eur. J. Biochem. 157*:1–26 (1986).
80. Nomenclature Committee of the International Union of Biochemistry, Enzyme nomenclature. Recommendations 1984. Supplement 2: Corrections and additions, *Eur. J. Biochem. 179*:489–533 (1989).

5

Chemical Preparation of Sphingosine and Sphingolipids: A Review of Enantioselective Syntheses

Hoe-Sup Byun and Robert Bittman *Queens College of The City University of New York, Flushing, New York*

I. STRUCTURE AND NOMENCLATURE OF SPHINGOSINE

The generic term sphingosine is sometimes used to designate the naturally occurring long-chain sphingoid bases that are present in sphingolipids such as sphingomyelin, gangliosides, and cerebrosides. This term includes aliphatic 2-amino-1,3-diols without regard to variations involving chain length, stereoisomerism, and presence or absence of an olefinic double bond. In this review the term sphingosine is used to designate a long-chain aliphatic 2-amino-1,3-diol with a Δ^4-*trans* double bond; this compound has also been referred to in the literature as *trans*-4-sphingenine. Although the length of the aliphatic chain varies in nature, the 18-carbon sphingosine is the predominant long-chain base found in most mammalian sphingolipids [1] and has also been found in HL-60 cells [2], neutrophils [3], and liver [4]. Unless otherwise specified, sphingosine refers to the 18-carbon species in this review. Since sphingosine has two asymmetric carbon atoms (C-2 and C-3), four stereoisomers can exist. The structures of the four stereoisomers of sphingosine (D-*erythro*, or 2*S*, 3*R*; L-*erythro*, or 2*R*, 3*S*; D-*threo*, or 2*R*, 3*R*; and L-*threo*, or 2*S*, 3*S*) are shown in Fig. 1. Only the D-*erythro* form is thought to occur naturally [1]; therefore many chemical synthetic studies emphasize the preparation of this form.

II. INTRODUCTION AND SCOPE OF THIS REVIEW

In addition to their function as providing the backbone of sphingolipids, sphingosine and other long-chain bases having at least sixteen carbons and an amine group are potent inhibitors of protein kinase C in vitro and of cellular processes mediated by this Ca^{2+}- and phospholipid-dependent enzyme (see Refs. 5 and 6 for recent reviews). The metabolism of sphingomyelin and possible involvement of sphingolipids in cell function have been reviewed recently [7]. As a result of the recent discovery that sphingosine and lysosphingolipids are inhibitors of protein kinase C and thus may be possible regulators of intercellular signal transduction mechanisms, there has been a renewed interest in the preparation of chemically well-defined sphingosines and other sphingolipids of sphingomyelin metabolism for use in studies involving control of signal transduction pathways.

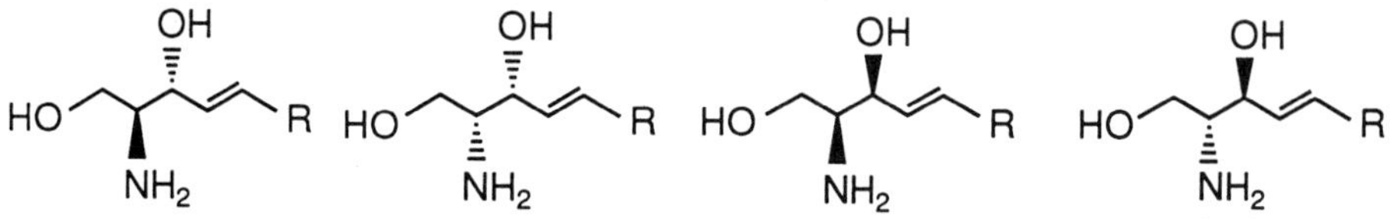

(1a) (1b) (1c) (1d)

Figure 1 (a) 2*S*, 3*R*; (b) 2*R*, 3*R*; (c) 2*S*, 3*S*; (d) 2*R*, 3*S*.

Many chiral syntheses of the D- and L-*erythro* and *threo* isomers of sphingosine (1,3-dihydroxy-2-amino-*trans*-4-octadecene, (1a-d)), the most widely occurring of the sphingolipid bases, have been developed in recent years. This review summarizes the enantioselective total syntheses of sphingosine that have appeared over the last two decades. Also included in this review is a discussion of the practical methods (including laboratory directions and spectral data) of preparing the key intermediates from which one may synthesize either *threo* or *erythro* sphingosine, ceramides, glycosphingolipids, and sphingomyelins. We place emphasis in this review on reactions that offer a great deal of synthetic versatility in producing sphingosine derivatives with the desired stereochemistry of naturally occurring sphingolipids, i.e., D-*erythro*-sphingosine. This review does not include syntheses of the saturated form of sphingosine, sphinganine (dihydrosphingosine), or of hydroxylated derivatives such as 4-D-hydroxysphinganine (phytosphingosine). *O*-Substituted derivatives of sphingosine are also not discussed, with the exception of 3-*O*-benzoylsphingosine, which has been used in early syntheses as an intermediate in the preparation of D- and L-*erythro*- and *threo*-sphingosines.

The first total chemical syntheses of sphingosine and sphingomyelin were reported by Shapiro et al. [8–10]. This classical method involves a Knoevenagel condensation of malonic acid with myristaldehyde, giving *trans*-2-hexadecenoic acid. The latter was converted into 3-*O*-benzoyl-DL-sphingosine in a lengthy procedure. 3-*O*-Benzoyl-DL-sphingosine was resolved into the D- and L-isomers by treatment of the benzoylsphingosine sulfuric acid salt with optically active tartaric acid and recrystallization of the diastereomeric salt from ethanol. 3-*O*-Benzoyl-D-ceramide was prepared by *N*-acylation and then converted to sphingomyelin by reaction with 2-chloroethylphosphoryl dichloride, followed by treatment with trimethylamine. The synthesis of 3-*O*-benzoyl-D-ceramide was repeated by Baer and Sarma with modifications of several experimental procedures [11]. The classical syntheses of sphingomyelin and synthetic routes to phosphonoglycero- and phosphonosphingolipids were reviewed by Rosenthal [12]. Another modification of the resolution of DL-sphingolipid intermediates involves treatment of ethyl DL-*erythro*-2-acetamino-3-hydroxy-4-(*E*)-octadecenoate with (L)-(+)-acetylmandeloyl chloride, followed by separation of the mandelate diastereomers by column or thin-layer chromatography [13].

This review does not include a detailed discussion of the methods for preparing racemic sphingosine. Some new methods that have been developed to prepare DL-*erythro*-sphingosine include the following nonchirospecific syntheses: (1) condensation of 2-nitroethanol with 2-hexadecynal catalyzed by potassium carbonate [14,15] or with (*E*)-2-hexadecenal in the presence of triethylamine [16], (2) reaction of 1-hexadecene and 2-nitroethyl tetrahydropyranyl ether in the presence of phenyl isocyanate [17], (3) aldol reaction of tris(trimethylsilyl)glycine with (*E*)-2-hexadecenal, then reduction of the aldol adduct [18], (4) intramolecular Diels-Alder reaction of (*E*,*Z*)-2,4-octadienyl-*N*-sulfinylcarbamate [19], and (5) iodocyclization of 1-trichloroacetimido-(2*E*,4*E*)-octadecadiene [20].

III. SYNTHESIS OF SPHINGOSINE FROM SERINE DERIVATIVES

Derivatives of serine, whose chiral center corresponds to that at C-2 of sphingosine, are important precursors of sphingosine. All four stereoisomeric sphingosines can be prepared by a single synthetic strategy, depending only on the choice of D- or L-serine as the starting material. Moreover, the adjacent chiral center at C-3 of sphingosine may be induced from the chirality of the C-2 center. Four types of reactions have been reported starting from D- or L-serine derivatives: (1) reaction of serinal with *trans*-vinylalane, (2) reaction of serinal with lithium alkyne, (3) reaction of serinal with a formylation agent followed by Wittig olefination, and (4) reaction of serine amide with lithium alkyne. Recently, a fifth method was reported, i.e., reaction of a Schiff base ester derived from L-serine with i-$Bu_2AlH \cdot i$-Bu_3Al at $-78°C$, followed by (*E*)-1-lithiopentadecene, giving 1c with very high stereoselectivity [116].

A. From Serinal Derivatives

1. By Reaction with *trans*-Vinylalane [(*E*)-*n*-$C_{13}H_{27}C{=}CHAl(Bu\text{-}i)_2$]

Newman communicated the synthesis of sphingosine starting from serine (2) as shown in Scheme 1 [21]. The amine and hydroxy groups were protected by using phthalimido and acetoxy groups by the reaction of the amine with *N*-carbethoxyphthalimide in the presence of sodium carbonate followed by acetylation with acetic anhydride. The protected serine 3 was converted to the corresponding serinal 4 in two steps: acid chloride formation with thionyl chloride in benzene and catalytic hydrogenation of the acid chloride. During the hydrogenation partial racemization did not occur, but only 50% of the desired

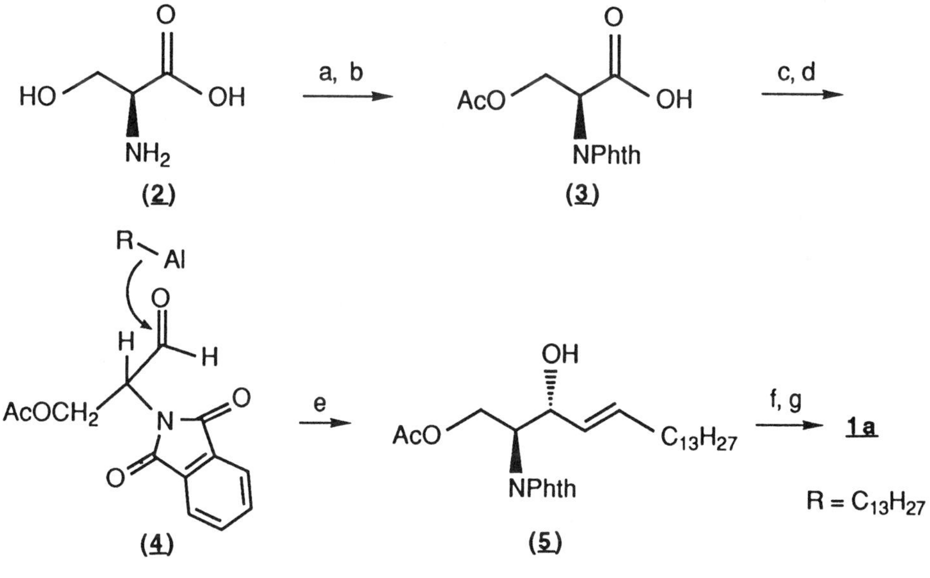

Scheme 1 Reaction of *trans*-vinylalane with serinal 4. (a) $PhthCO_2Et/Na_2CO_3$, (b) Ac_2O, (c) $SOCl_2/C_6H_6$, (d) H_2/Pd-C, (e) (*E*)-*n*-$C_{13}H_{27}CH{=}CHAl(i\text{-}Bu)_2$, (f) $MeOH/H^+$, (g) NH_2NH_2/EtOH.

aldehyde was obtained as judged from the NMR spectrum (comparison of intensity ratios of the aldehyde proton and the phthaloyl and -CH_2CH- protons). Addition of a solution of aldehyde 4 in benzene-hexane (2 : 1) to a solution of *trans*-pentadecenyldiisobutylalane (*trans*-vinylalane) in hexane at 5–10°C gave D-*erythro*-1-*O*-acetyl-*N*-phthaloylsphingosine (5) as an oil in 13.3% yield, along with impure *threo* isomer (3.5% yield). The *erythro* selectivity of the reaction was explained by preferential attack on conformation 4 by the organometallic reagent from the less hindered side. The synthesis of 1a was completed by acid-catalyzed methanolysis and hydrazinolysis to remove the acetoxy and the phthaloyl groups, respectively.

Another reaction that involves *trans*-vinylalane is the addition to 4-formyl-2-phenyl-Δ^2-oxazoline (7) as shown in Scheme 2 [22]. The oxazoline was obtained by reaction of serine methyl ester hydrochloride with benziminoethyl ether. The ester group of the oxazoline was reduced by the addition of diisobutylaluminum hydride (DIBAL) at –70°C. Extraction followed by concentration under high vacuum gave the unstable aldehyde 7 as a yellow oil; the reaction was shown by TLC to be 90% complete in aldehyde with a trace of unreacted methyl ester. Dropwise addition of the crude aldehyde to the *trans*-vinylalane prepared from 1-pentadecyne and DIBAL resulted in a 51% yield of a 1 : 1 mixture of *erythro* and *threo* sphingosine oxazoline (8a and b), which were separable by silica gel chromatography by eluting with ether-petroleum ether (10 : 7). In order to avoid

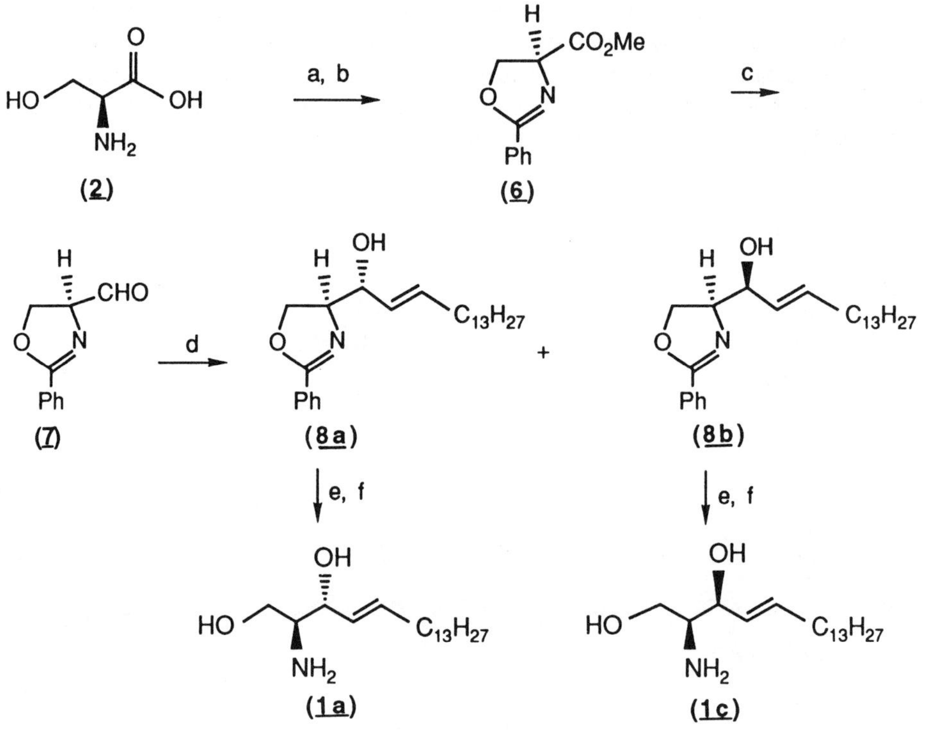

Scheme 2 Reaction of *trans*-vinylalane with oxazoline aldehyde 7. (a) HCl/MeOH, (b) HN=C(Ph)OEt/CH_2Cl_2, (c) DIBAL/toluene-hexane (3:1), (d) (*E*)-*n*-$C_{13}H_{27}CH$=CHAl-(*i*-Bu)$_2$/ether-toluene (3:1), (e) HCl/THF, (f) NaOH.

the isolation of unstable aldehyde 7, the oxazoline ester 6 was reduced with DIBAL in ether at –77°C, and *trans*-vinylalane was added to the reduction mixture in situ [23,24]. Each oxazoline derivative was converted to the known *N,O,O*-triacetylsphingosine by the following sequence of reactions: (1) *N*-acetyl-1-*O*-benzoylsphingosine formation by hydrolysis of the oxazoline ring with 2 N HCl followed by treatment with *p*-nitrophenyl acetate in pyridine, and (2) debenzoylation with 1 N NaOH in methanol followed by acetylation with acetic anhydride/pyridine. Similarly, sphinga-4,8-dienine was synthesized [25]. Recently, L-*threo*-3-fluorosphingosine was prepared by the reaction of oxazoline 8b with 2-chloro-1,1,2-trifluorotriethylamine followed by hydrolysis. In addition, the D-*erythro* 1-fluoro analog of 1a was prepared by using (*S*)-3-fluoroalanine instead of L-serine as the starting material [25a].

As shown in Scheme 3, the addition reaction of *trans*-vinylalane to oxazolidine aldehyde 11 gave a 2:1 mixture of the *threo* and *erythro* adducts 13a and 13b [26]. This diastereofacial selectivity was explained by chelation control in the transition state shown in Scheme 3 and is based on the observation of other Lewis acid catalyzed additions to aldehyde 11 [27]. Hydrolytic deprotection of the 2:1 adducts with 1 N HCl afforded a 2:1 mixture of *threo* and *erythro* sphingosine in 60% combined yield from the aldehyde. A *threo*-enriched adduct (7:1) was obtained by chromatography on silica gel (hexane/ethyl acetate 12:1); two recrystallizations from hexane-methylene chloride gave pure L-*threo*-sphingosine (1c), which was characterized as the *N,O,O*-triacetylsphingosine.

2. By Reaction with Lithium Alkyne

Four different research groups developed the synthesis of sphingosine by the addition reaction of lithium pentadecyne to a protected, configurationally stable serine aldehyde as shown in Scheme 4.1 [26,28–30]. The preparation of oxazolidine ester 10 from L-serine

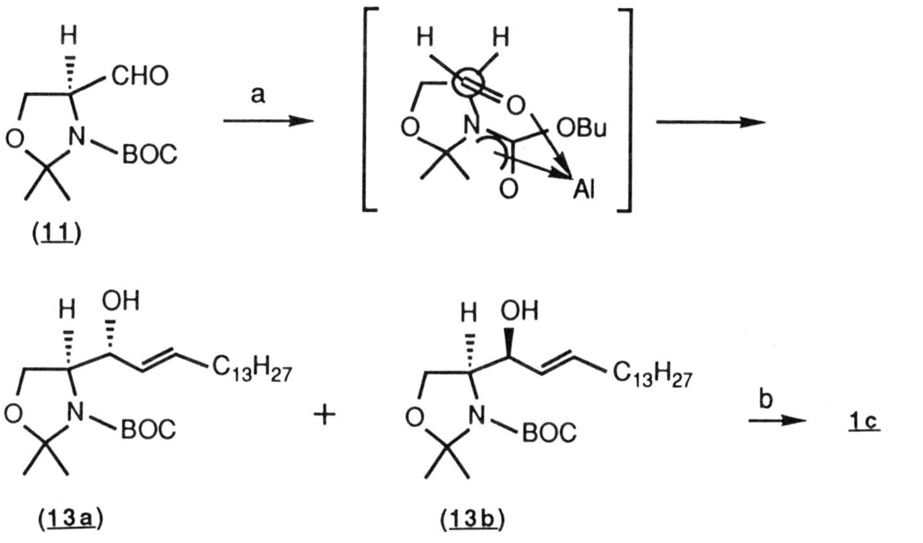

Scheme 3 Reaction of *trans*-vinylalane with oxazolidine 11. (a) (*E*)-*n*-$C_{13}H_{27}CH{=}CHAl(i\text{-}Bu)_2$/toluene, (b) HCl/THF.

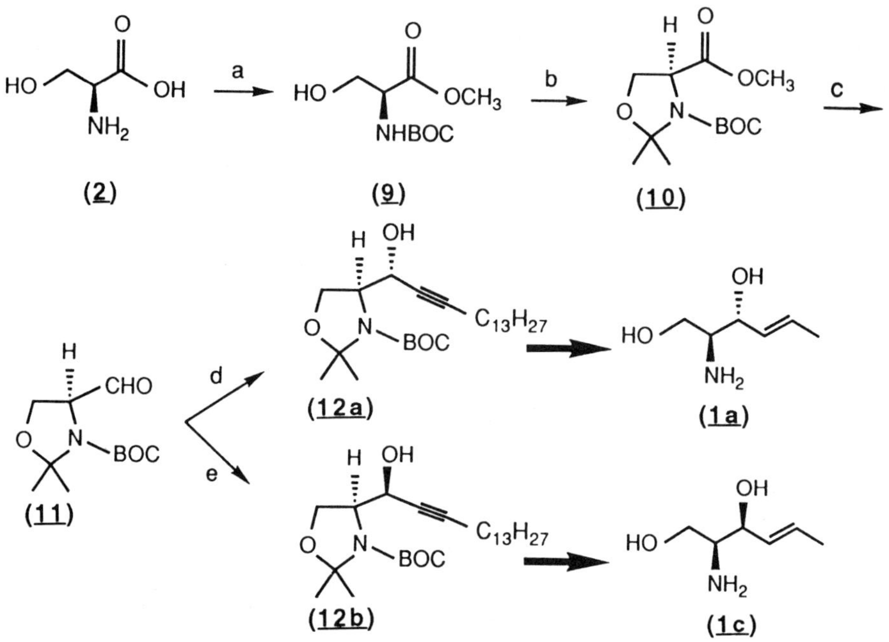

Scheme 4.1 Reaction of lithium pentadecyne with oxazolidine aldehyde 11. (a) i. $(BOC)_2O$/NaOH, ii. CH_2N_2/Et_2O, (b) $Me_2C(OMe)_2/C_6H_6$, (c) DIBAL/toluene, (d) lithium pentadecyne/THF/HMPA, (e) lithium pentadecyne/$ZnBr_2/Et_2O$.

(2) generally involves three steps: (1) protection of the amine with di-*tert*-butyl dicarbonate [$(BOC)_2O$] above pH 10, (2) esterification of the crude *N*-BOC-L-serine with diazomethane, and (3) 2,2-dimethyloxazolidine formation with 2,2-dimethoxypropane under acid-catalyzed conditions. Reduction of the oxazolidine ester 10 with DIBAL led to oxazolidine aldehyde 11, which was purified by vacuum distillation. The purified aldehyde was contaminated with 5% of the starting ester. The same sequence of preparation was carried out with D-serine to give the antipode of 11 [31].

The addition of lithium pentadecyne to the oxazolidine aldehyde 11 can produce either the *erythro* or the *threo* stereochemistry, depending on the reaction conditions used. *Erythro* selectivity was increased remarkably by the addition of hexamethylphosphoramide (HMPA), a cation-complexing solvent. High *threo* selectivity was achieved with chelation conditions such as the presence of anhydrous zinc bromide in ether solvent. Both adducts were formed with 95% diastereoselectivity, as determined by ^{1}H-NMR spectroscopy, without any racemization taking place at C-2 during the reaction [28]. The diastereoselectivity of the condensation reaction may be explained by the Felkin-Anh open-chain model [32,33] for asymmetric induction in which the nucleophile (pentadecynyl anion) attacks preferentially at the *re* face of aldehyde 11 to give *erythro* 12a. In contrast, in the presence of a Lewis acid such as zinc bromide, the diastereofacial selectivity is reversed, because of chelation control in the transition state, resulting in attack of the nucleophile at the *si* face to give *threo* 12b.

As shown in Scheme 4.1, addition of the oxazolidine aldehyde 11 to the alkynyllithium (prepared from pentadecyne and *n*-butyllithium) in the presence of HMPA at

–78°C yielded a 20:1 *erythro:threo* ratio of oxazolidine propargylic alcohols 12a:12b in 71% yield [28]. For the synthesis of the *threo* isomer 12b, reaction of the aldehyde with the alkynyllithium was carried out in the presence of zinc bromide in ether. A 20:1 ratio of *threo* to *erythro* adduct was obtained in 84% yield. The same reaction without HMPA produced the *erythro:threo* oxazolidine propargylic alcohol 12a,b in an 8:1 [26] or 9:1 [29] ratio, from which the pure *erythro* adduct was isolated in good yield after flash chromatography on silica gel. The *erythro* propargylic alcohol could be isomerized to the corresponding *threo* isomer 12b in 70% yield by using the Mitsunobu reaction (triphenylphosphine/diethyl azodicarboxylate/benzoic acid) [29].

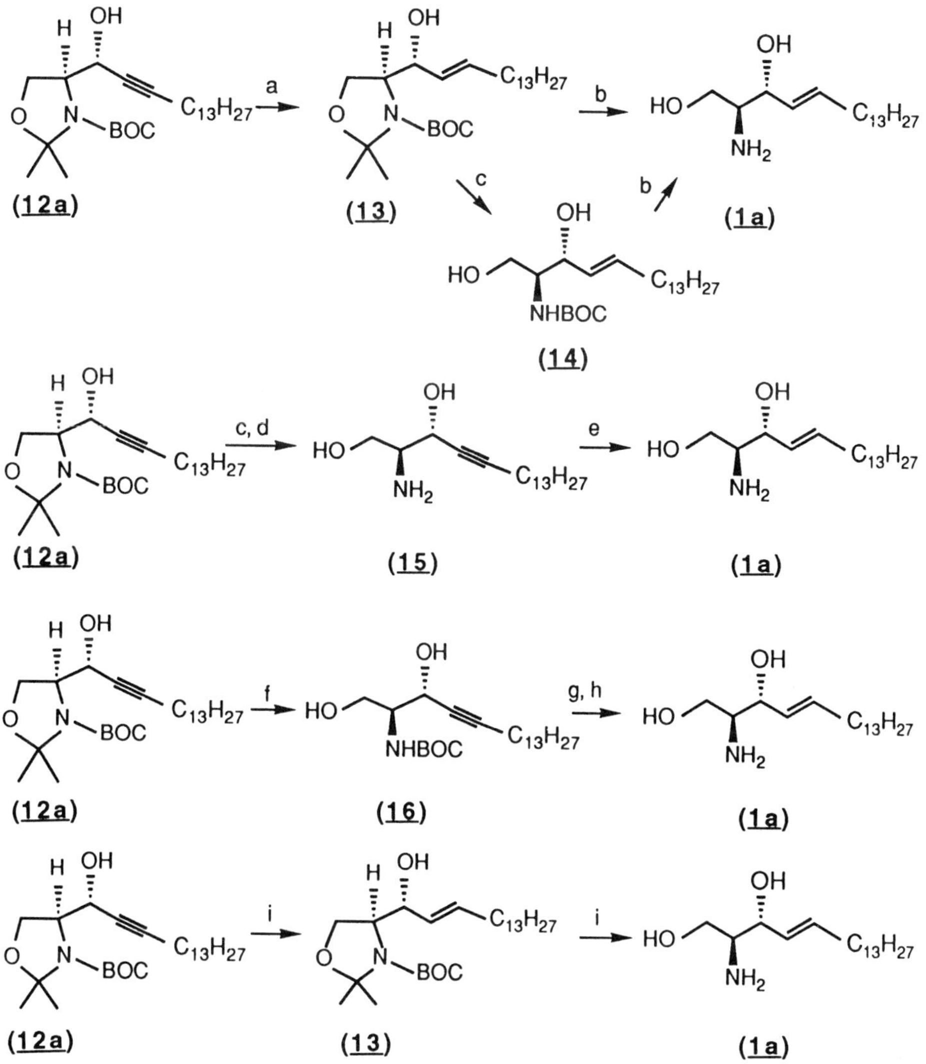

(12a) (13) (1a) (14)

(12a) (15) (1a)

(12a) (16) (1a)

(12a) (13) (1a)

Scheme 4.2 Conversion of propargylic alcohol 12a to sphingosine. (a) Li/NH_3 (liq.), (b) HCl/MeOH, (c) *p*-TsOH/MeOH, (d) HCl/EtOAc, (e) Na/NH_3 (liq.)/THF or $LiAlH_4$/DME, (f) Amberlyst 15/MeOH, (g) Red-Al/Et_2O, (h) 1 N HCl/dioxane, (i) Li/$EtNH_2$.

The oxazolidine propargylic alcohols 12 have been converted into sphingosines by four different routes (Scheme 4.2). First, dissolving metal reduction of the oxazolidine propargylic alcohol with lithium/liquid ammonia for 15 h provided oxazolidine allylic alcohol 13 [30]. The allylic alcohol was treated with a catalytic amount of *p*-toluenesulfonic acid in methanol to give D-*erythro*-*N*-BOC-sphingosine (14) in low (37%) yield; removal of the BOC group with methanolic HCl gave D-*erythro*-sphingosine (1a). One-step treatment of the oxazolidine allylic alcohol 13 with HCl also gave D-*erythro*-sphingosine in 34% yield. Second, 2-amino-propargylic-1,3-diol 15 was obtained by methanolysis of oxazolidine propargylic alcohol 12 followed by treatment with HCl in ethyl acetate [29]. Exposure of the 2-amino-propargylic-1,3-diol 15 to a refluxing solution of excess lithium in liquid ammonia/tetrahydrofuran (THF) for 7 h produced a 9 : 1 mixture of sphingosine and starting amino-propargylic diol. This procedure could not be applied on a large scale because the reduction requires a long reaction time (>24 h) and it is difficult to maintain a constant concentration of ammonia in the reaction mixture over prolonged time. Alternatively, lithium aluminum hydride reduction of 2-amino-propargylic-1,3-diol 15 in refluxing dimethoxyethane gave sphingosine (1a) in 70% yield on scales up to 5 g. Third, selective cleavage of the acetal moiety from the oxazolidine propargylic alcohol 12 with Amberlyst 15 resin produced *N*-BOC-propargylic-1,3-diols 16 in about 75% yield for either the *erythro* or the *threo* isomer [28]. The *trans* double bond was introduced into *erythro*- or *threo*-*N*-BOC-sphingosine by reduction of the triple bond of 16 with sodium bis(2-methoxyethoxy)aluminum hydride (Red-Al) in ether. A *cis* double bond was formed by partial hydrogenation of *N*-BOC-propargylic-1,3-diol 16 over Lindlar catalyst. Fourth, treatment of the propargylic alcohol 12a with lithium/ethylamine for 1 h at –78°C resulted in nearly quantitative formation of oxazolidine allylic alcohol 13a, which was contaminated with a small amount of sphingosine 1a [26]. The latter apparently arose from fragmentation of the

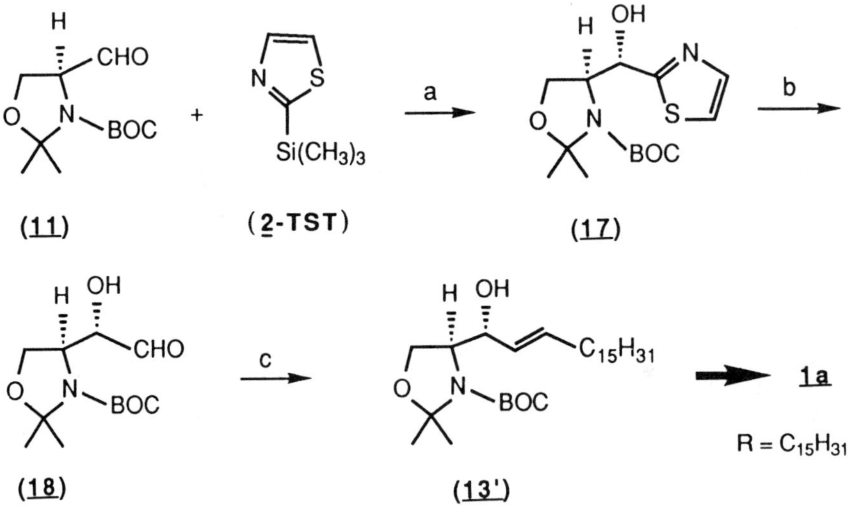

H CHO O N BOC (11) + N S Si(CH$_3$)$_3$ (2-TST) a → H OH N S O N BOC (17) b →

H OH CHO O N BOC (18) c → H OH C$_{15}$H$_{31}$ O N BOC (13') → 1a R = C$_{15}$H$_{31}$

Scheme 5 Reaction of 2-TST with oxazolidine aldehyde 11. (a) i. 2-TST/CH_2Cl_2, ii. *n*-Bu_4NF, (b) i. MeI (10 equiv)/CH_3CN, ii. $NaBH_4$ (2 equiv), iii. $HgCl_2$ (1.2 equiv), (c) *n*-$C_{16}H_{33}PPh_3Br$/PhLi/LiBr.

N-BOC-oxazoline system under the strongly reducing conditions. The one-pot reaction from the oxazolidine aldehyde 11 to sphingosine 1a proceeded in 65–68% overall yield.

3. By Reaction with 2-(Trimethylsilyl)thiazole

Dondoni et al. reported the synthesis of sphingosine via one-carbon elongation of the oxazolidine aldehyde 11 by formylation followed by Wittig olefination as shown in Scheme 5 [34,35]. The reaction of oxazolidine aldehyde 11 with 2-(trimethylsilyl)thiazole for 20 h in methylene chloride, followed by desilylation, gave the anti addition adduct, glycitol 17, in 85% yield and with 92% diastereoselectivity [34]. The relative stereochemistry of the glycitol was determined from the coupling constants between the newly formed methyne proton (CHOH) and the adjacent CH*N*BOC (anti isomer $J = 2.8$ Hz; syn isomer $J = 8.7$ Hz), and by inference from the x-ray structure of the one-carbon higher homologue prepared by coupling of hydroxy aldehyde 18 with 2-(trimethylsilyl)thiazole. The major anti glycitol could be isomerized into the minor syn glycitol by the sequence of potassium permanganate oxidation and L-Selectride reduction in 95% diastereoselectivity. The mechanism of anti stereoselectivity in this addition reaction is the same as that in the reaction with oxazolidine aldehyde and alkynyllithium discussed earlier. The hydroxy aldehyde 18 was obtained in three steps in 65% yield: (1) *N*-thiazolium iodide formation with 10-fold excess of methyl iodide; (2) thiazolidine formation by reduction of the thiazolium iodide with sodium borohydride (2 equiv) in methanol; (3) hydrolysis of the thiazolidine ring with mercuric chloride (1.2 equiv) in acetonitrile-water (4:1). The hydroxy aldehyde was used in a Wittig reaction using *n*-hexadecyltriphenylphosphorane. The latter was generated in situ from the reaction of *n*-hexadecyltriphenylphosphonium bromide with phenyllithium in the presence of excess lithium bromide. An excess of lithium bromide is required to achieve *trans* selectivity. Unfortunately, the yield of alkene 13' was low (31%), possibly because of poor solubility of the betaine/lithium bromide adduct in THF. The *trans* selectivity of the Wittig reaction in the presence of excess lithium salt will be discussed later. From the Wittig adduct, D-*erythro*-triacetyl-C_{20}-sphingosine was prepared from 13' by removal of the protecting groups with trifluoroacetic acid followed by treatment with acetic anhydride/pyridine.

B. From Serine Amide Derivatives

1. Reaction with Lithium Alkyne

Boutin and Rapoport demonstrated two chirospecific syntheses of sphingosine from L-serine (2) via an α'-amino-α,β-ynone system [36]. One method involves introduction of the sphingosine alkyl chain with lithium pentadecyne and diastereoselective reduction of the keto group in the α'-amino-α,β-ynone system. This route produced D-*erythro*-sphingosine (1a) in five steps from *N*-benzylcarbomethoxy (CBZ)-serine in 22% overall yield, as shown in Scheme 6. A different route to sphingosine 1a (not shown in a scheme) involves the coupling reaction of lithium acetylene with serine oxazolidide 19, and reduction of the resulting *N*-CBZ-amino-1-hydroxy-4-pentyn-3-one and alkylation of the terminal acetylene group with 1-iodotridecane. This route is not discussed here because the serine isoxazolidide 19 containing an unprotected hydroxy group did not undergo the coupling reaction with lithium acetylene at all. Even the coupling reaction of *O*-*tert*-butyldimethylsilyl-*N*-CBZ-serine oxazolidide was not reproducible. In addition,

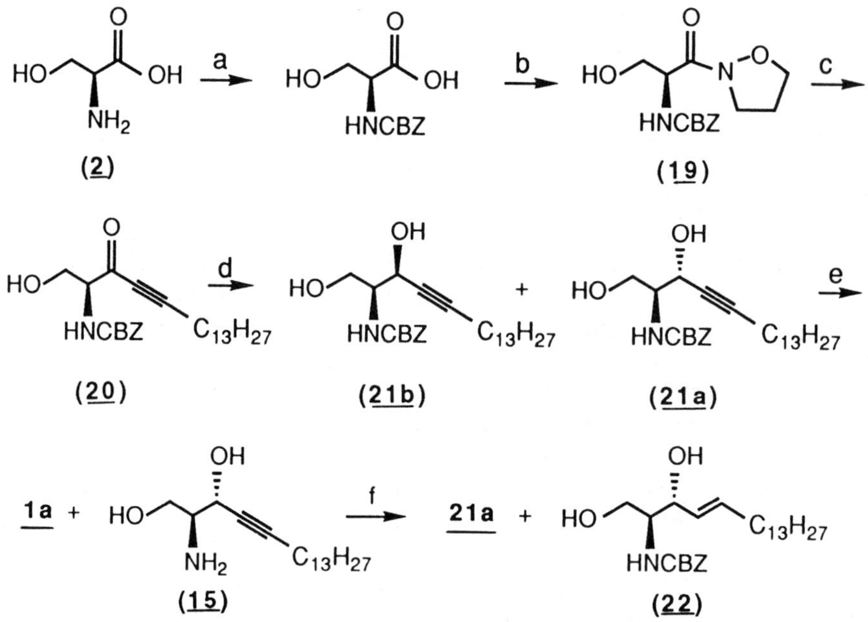

Scheme 6 Reaction of lithium pentadecyne with serine isoxazolidide 19. (a) BnOCOCl/ $NaHCO_3$, (b)*i*-BuOCOCl/*N*-methylmorpholine/isoxazolidine, (c) lithium pentadecyne, (d) $NaBH_4$, (e) Li/NH_3 (liq.), (f) BnOCOCl/Et_3N.

protection of the hydroxy groups in *N*-CBZ-amino-4-pentyne-1,3-diols, obtained by reduction of the acetylenic ketone, was necessary prior to introduction of the alkyl chain to the terminal acetylene, and the yield of the cross-coupling reaction was also low (36% yield).

The starting *N*-CBZ-L-serine isoxazolidide 19 was prepared in 76% yield by coupling of L-CBZ-serine with isoxazolidine in the presence of *N*-methylmorpholine and isobutyl chloroformate. Reaction of isoxazolidide 19 with *n*-pentadecyllithium produced *N*-CBZ-ynone 20, which could be separated in partially racemic form by flash chromatography on silica gel, eluting with hexane-ethyl acetate (2:1). The ynone 20 was reduced in diastereoselective fashion with sodium borohydride, giving a 4:1 *erythro*:*threo* mixture and 92% chemical yield. To avoid racemization during chromatographic separation, the crude ynone was reduced directly to give the desired *N*-CBZ-propargylic diol 21a in 53% overall yield from the isoxazolidide 19. The diastereomers of *N*-CBZ-propargylic diol 21 can be separated either by preparative HPLC or by column chromatography on sodium borate impregnated silica gel. The factor that potentially controls this diastereoselectivity is chelation of the free hydroxy or carbamate group in ynone 20 [37]. The reduction with tetramethylammonium triacetoxyborohydride [38] gave 21a, which was purified by crystallization in isooctane [39]. Finally, *N*-CBZ-propargylic diol 21 was treated with lithium (25 equiv) in liquid ammonia/THF followed by resubjecting the crude product to the dissolving metal reduction, giving a mixture of sphingosine 1a and 2-amino-1,3-dihydroxy-4-octadecyne (15); about 25% of the latter alkyne was found to persist in the mixture [36]. This mixture was treated with benzyl chloroformate and triethylamine to lead to a mixture of D-*erythro*-*N*-CBZ-sphingosine (22) and starting *N*-CBZ-propargylic diol 21a. Purification by chromatography gave pure *N*-CBZ-sphingosine (22) in 50% yield and recovered *N*-CBZ-propargylic diol 21a in 21% yield. Pure sphingosine (1a) was obtained by removing the carbobenzyloxy group with lithium in liquid ammonia. In addition to sphingosine, D-*erythro*-*N*-CBZ-*cis*- and D-*erythro*-*N*-CBZ-dihydrosphingosine were synthesized.

IV. SYNTHESIS OF SPHINGOSINE FROM CARBOHYDRATES

Carbohydrates have been important chiral starting materials for the preparation of sphingosine. Generally, two steps are involved in this method of preparation. One is removal of the "unnecessary" carbon units from the carbohydrate by oxidative cleavage of a glycol, generating an aldehyde that is used in Wittig olefination to give the long hydrocarbon chain of sphingosine. The other step is introduction of the amino group by activation of a sugar hydroxy group, followed by nucleophilic substitution with azide ion.

A. Amino Sugars as Starting Materials

Reist and Christie developed a synthesis of sphingosine starting from 3-amino-3-deoxy-1,2:5,6-di-*O*-isopropylidene-α-allofuranose (23), as shown in Scheme 7 [40]. After the primary amine was protected by the reaction of 23 with alkyl (methyl, ethyl, or benzyl) chloroformate and pyridine, one isopropylidene protecting group was hydrolyzed selectively in 75% aqueous acetic acid at 60°C. The oxidative cleavage of the resulting diol

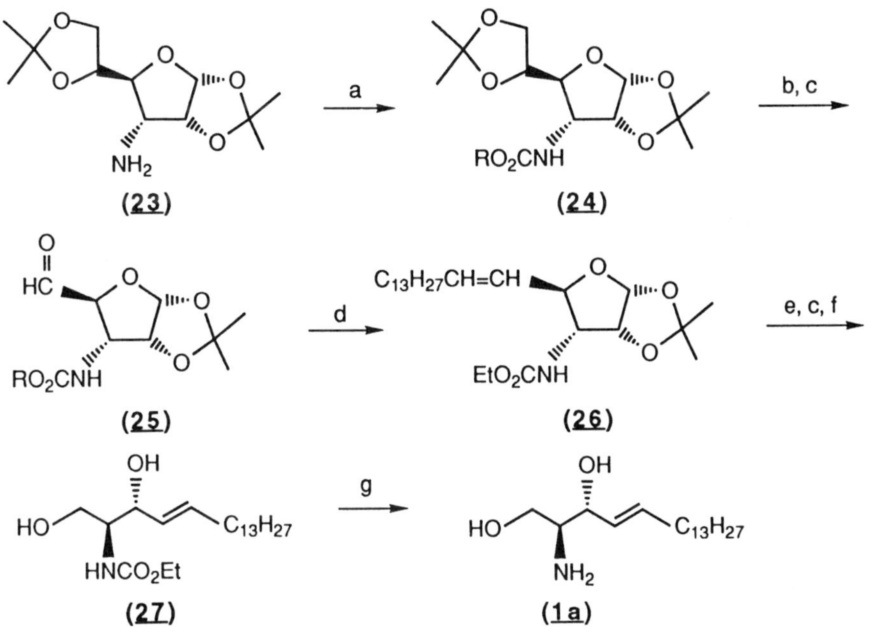

NH2 (23) —a→ RO2CNH (24) —b, c→

HC(=O) ... RO2CNH (25) —d→ $C_{13}H_{27}CH{=}CH$ ... EtO_2CNH (26) —e, c, f→

HO ... OH ... $C_{13}H_{27}$ ... $HNCO_2Et$ (27) —g→ HO ... OH ... $C_{13}H_{27}$ ... NH_2 (1a)

Scheme 7 Conversion of aminofuranose 23 to sphingosine. (a) ROCOCl/Py, (b) 75% CH_3CO_2H, (c) $NaIO_4$, (d) n-$C_{14}H_{29}PPh_3Br$/PhLi/LiBr, (e) 80% CH_3CO_2H, (f) $NaBH_4$, (g) $Ba(OH)_2$/dioxane.

with sodium metaperiodate generated the sugar aldehyde 25. Wittig reaction between *n*-tetradecyltriphenylphosphorane and an aldehyde led to the *cis* or *trans* adducts 26, depending on the reaction conditions. In the presence of excess lithium salt, the transformation of the zwitterionic betaine to the olefin is strongly hindered, allowing sufficient time for equilibrium to be established between the *erythro*-betaine and the thermodynamically favored *threo*-betaine [41]. During the Wittig reaction with an excess of lithium salt, the *N*-protected benzyl or methyl carbamate group in 26 was replaced with an ethyl group [40]. This transesterification reaction can be rationalized by considering that the carbamate group undergoes attack by ethoxide ion produced from the reaction of solvent ether with excess phenyllithium. The remaining isopropylidene group in 26 was removed by refluxing in 80% acetic acid and oxidative cleavage, followed by sodium borohydride reduction, furnishing *N*-ethoxycarbonylsphingosine (27) in 19% overall yield based on the *N*-protected sugar 24. The ethoxycarbonyl group in 27 was removed by using a number of different reaction conditions such as potassium hydroxide in methanol,

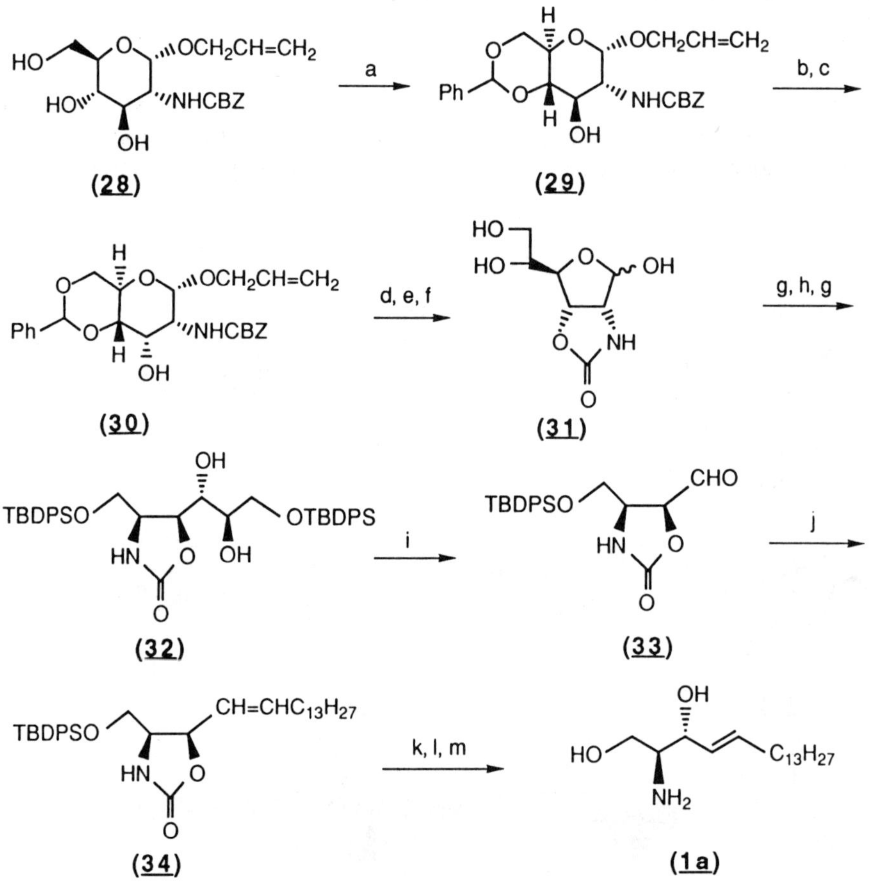

Scheme 8 Conversion of aminopyranose 28 to sphingosine. (a) $PhCHO/ZnCl_2$, (b) $(COCl)_2$/DMSO/Et_3N, (c) $NaBH_4$/MeOH/DMF, (d) NaH/DMF, (e) $(Ph_3P)_3RhCl$/DABCO, (f) HCl/MeOH, (g) *t*-$BuPh_2SiCl$/imidazole/DMF, (h) $NaBH_4$/MeOH, (i) $NaIO_4$/MeOH, (j) *n*-$C_{14}H_{29}PPh_3Br$/PhLi/THF, (k) 1-methyltetrazol-5-yl disulfide/AIBN/C_6H_6, (l) *n*-Bu_4NF/THF, (m) NaOH/EtOH.

anhydrous hydrogen fluoride, or sulfuric acid, but only a trace amount of sphingosine was obtained. Only freshly prepared aqueous barium hydroxide was found to be a satisfactory reagent, giving sphingosine in 60% yield based on ethoxycarbonylsphingosine.

As shown in Scheme 8, D-*erythro*-sphingosine was synthesized in 13 steps [42] starting from allyl *N*-benzyloxycarbonylamino-2-deoxy-α-D-glucopyranoside (28) [43]. The 4,6-diol of α-glucoside 28 was protected as the benzylidene acetal by reaction with benzaldehyde and zinc chloride. Inversion of the unprotected alcohol was carried out by Swern oxidation followed by sodium borohydride reduction [44]. The hydroxy group of 30 was protected as a cyclic carbamate by treatment with an equimolar amount of sodium hydride [42]. Isomerization of the allyl group by $(Ph_3P)_3RhCl$ in the presence of 1,4-diazabicyclo[2.2.2]octane (DABCO) followed by hydrolysis provided 2-amino-2-*N*, 3-*O*-carbonyl-2-deoxy-α,β-D-allofuranose (31). The sequence of protection of the 6-hydroxy group as the *tert*-butyldiphenylsilyl ether, reduction of the hemiacetal with sodium borohydride, and reprotection of the 1-hydroxy function gave the key intermediate, D-allitol 32. Oxidative cleavage of the allitol with sodium metaperiodate led to aldehyde 33, which was subjected to Wittig olefination. Chromatographic separation provided *trans*- and *cis*-alkene 34a and 34b in 6.6 and 36.7% yields, respectively. The *cis*-alkene was isomerized to the *trans* isomer by treatment with 1-methyltetrazol-5-yl disulfide in refluxing benzene in the presence of 2,2'-azobisisobutyronitrile (AIBN), giving a 10:1 *trans*:*cis* mixture, from which the pure *trans* isomer was isolated in 63% yield. The stereochemistry of the alkene was determined by ^{13}C-NMR data of the allylic carbon atoms, C-3 and C-6, in the sphingosine. (The chemical shifts of C-3 and C-6 are at δ 80.4 and 32.5 ppm for *trans*-alkene 34a and 75.2 and 28.0 ppm for *cis*-alkene 34b.) Finally, removal of all of the protecting groups led to sphingosine 1a.

B. Pyranose and Furanose as Starting Materials

D-*erythro*-Sphingosine was prepared in 10 steps in 20–30% overall yield starting from 1,2:5,6-di-*O*-isopropylidene-α-D-glucofuranose 35 as shown in Scheme 9 [45,46]. The Wittig reaction of aldehyde 36 (which was obtained from the glucofuranose) with *n*-tetradecyltriphenylphosphorane in THF afforded a mixture of *trans*- and *cis*-olefins 37a and 37b in 42 and 48% yields, respectively [45]. Photochemical rearrangement of the isolated *cis*-olefin in 1,4-dioxane/cyclohexane (1:19) in the presence of diphenyldisulfide led to a 14.5:1 *trans*:*cis* mixture of alkene 37 [47,48]. The two-step transformation (the Wittig reaction and photochemical isomerization) from the aldehyde 36 to *trans*-alkene 37a was achieved in 87% yield [45]. After the hydroxy group at C-3 was activated with mesylate, one carbon in the Wittig adduct was removed by oxidative cleavage, and the aldehyde was reduced to yield diol 38, which was protected with a 1-ethoxyethyl group. The 1,3-bis(1-*O*-ethoxyethyl)sphingosine 39 was prepared by azide displacement followed by either reduction of the azido group with sodium borohydride or hydrogenation using Lindlar catalyst [49].

As shown in Scheme 10, D-*erythro*-sphingosine (1a) was prepared from the 2,4-di-*O*-protected-D-threoses 42 and 43 [50–55]. The threose was obtained from either D-galactose or D-xylose in a two-step procedure: (1) partial acetalization of galactose 41 with benzaldehyde and zinc chloride [56] or partial ketalization of xylose 40 with 2,2-dimethoxypropane [57], and (2) oxidative cleavage of the partially protected sugar with sodium metaperiodate. The Wittig reaction of the 2,4-*O*-isopropylidene-D-threose 42 with the phosphorane generated by *n*-hexadecyltriphenylphosphonium bromide and

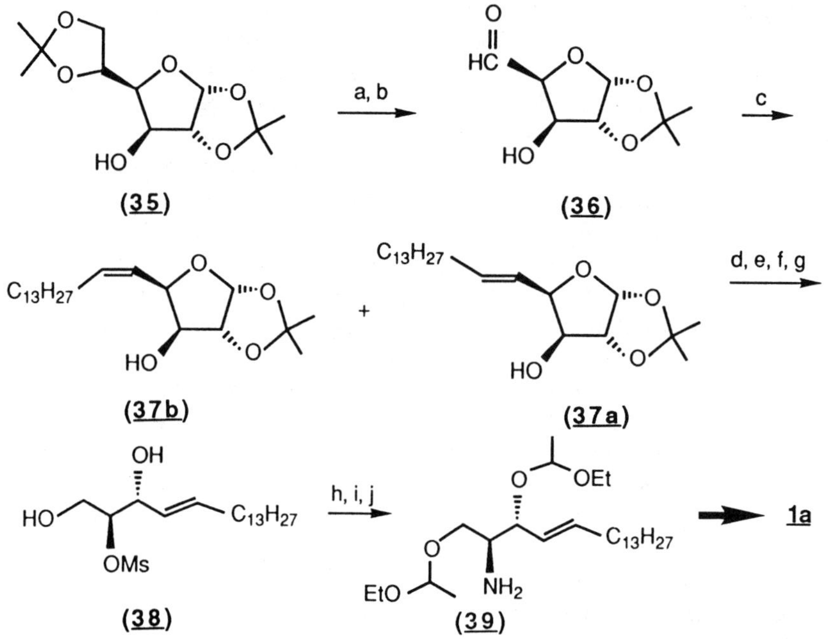

Scheme 9 Conversion of glucofuranose 35 to sphingosine. (a) 75% CH_3CO_2H, (b) $NaIO_4$, (c) n-$C_{14}H_{29}PPh_3Br$/n-BuLi/THF, (d) MsCl/Py, (e) 80% CH_3CO_2H, (f) $NaIO_4$/EtOH, (g) $NaBH_4$/EtOH, (h) PPTS/ethyl vinyl ether, (i) NaN_3/DMF, (j) $NaBH_4$/i-PrOH.

potassium *tert*-butoxide gave a mixture of *cis*- and *trans*-alkenes 44a in 40% and 35% yield, respectively [50]. The hydroxy group in the Wittig adduct was activated with methanesulfonyl chloride and pyridine, and the substitution reaction was carried out with sodium azide in *N,N*-dimethylformamide (DMF). When the substitution reaction was tried with trifluoromethanesulfonyl (triflate) derivatives, two major elimination by-products were obtained. The Wittig reaction of 2,4-*O*-benzylidene-D-threose (43) with the phosphorane in the presence of an excess of lithium bromide affords practically exclusively the *trans*-olefin 44b [52,58]. Activation by triflate followed by azide substitution gave 2-azido-1,3-*O*-benzylidene-4-*trans*-octadecene (45) [52]. The azido group was reduced selectively to an amino group by using one of the following reduction procedures: hydrogen sulfide [52], sodium borohydride in 2-propanol [59], triphenylphosphine-water [54], or hydrogenation in the presence of Lindlar catalyst [49]. In the experimental section of this chapter, we give laboratory directions for the reduction of azide 45 to the corresponding amine by using triphenylphosphine in tetrahydrofuran-water at room temperature. This method has several advantages compared with the other reducing methods that have been used to reduce selectively azides in the presence of double bonds and carbonyl groups, such as ease of handling and workup, mild reaction conditions, and high yield.

As shown in Scheme 11, D-*erythro*-sphingosine (1a) was synthesized by a ring-opening reaction of 2,3-dihydrofuran 53, which was obtained from D-(+)-mannose (46) [60]. After the extra carbon from the protected mannose 47 was removed by oxidative cleavage, the 2,3-dihydrofuran was prepared in a long sequence of reactions including reductive elimination. Bromination of 52 followed by dehydrobromination gave 4-

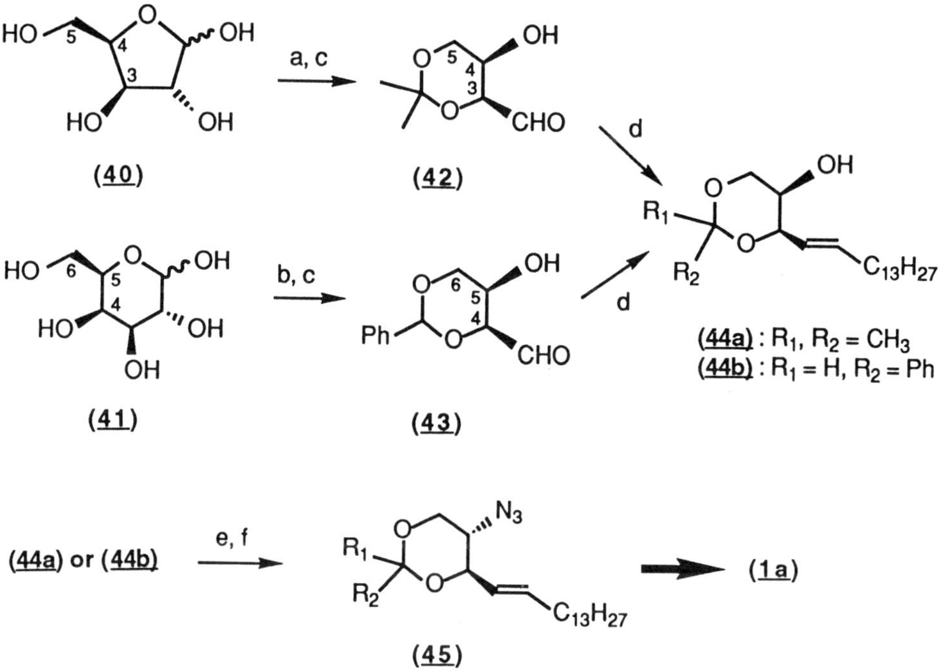

Scheme 10 Conversion of threose 42 or 43 to sphingosine. (a) $(CH_3)_2C(OMe)_2$/*p*-TsOH/DMF, (b) $PhCHO/ZnCl_2$, (c) $NaIO_4$, (d) *n*-$C_{14}H_{29}PPh_3Br$/PhLi/LiBr/THF, (e) MsCl/Py or Tf_2O/Py, (f) NaN_3/DMF.

bromo-3-*O*-methoxymethyl (MOM)-2-*O*-MOM-methyl-2,3-dihydrofuran (53a) as the key intermediate. Ring opening of dihydrofuran 53 with *n*-butyllithium provided acetylene 54. The lithium salt of the acetylene was alkylated with 1-bromotridecane in THF/HMPA. The amino group was introduced by nucleophilic substitution using lithium azide in DMF followed by reduction. Similarly, L-*threo*-sphingosine (1c) was synthesized starting from D-(+)-ribono-1,4-lactone 58, as shown in Scheme 12. The D-(+)-ribonolactone 58 was prepared from D-(–)-arabinose in a four-step sequence of reactions [61].

C. From Miscellaneous Sugar Derivatives

D-*erythro*-Sphingosine was prepared from chlorofumaric acid as shown in Scheme 13 [62]. Chlorofumaric acid (60) was converted to L-*threo*-chloromalic acid (61) in 99.5% enantiomeric purity by hydration of the carbon-carbon double bond catalyzed by fumarase. Malate 61 was reduced to chlorotriol 62 with borane in THF. The vicinal diol of the triol was protected with an isopropylidene group and then treated with benzyl isocyanate. The intramolecular displacement of chloride in the acyclic carbamate 64 by treatment with potassium *tert*-butoxide led to the cyclic carbamate 65 in 48% yield. After the isopropylidene group was hydrolyzed, the primary hydroxy group was blocked as a *tert*-butyldimethylsilyl ether. The remaining secondary hydroxy group was *O*-benzylated to yield the fully protected aminotriol 66, which was contaminated with about 30% of the rearranged 1-*O*-benzyl-2-*O*-*tert*-butyldimethylsilyl isomer; the latter

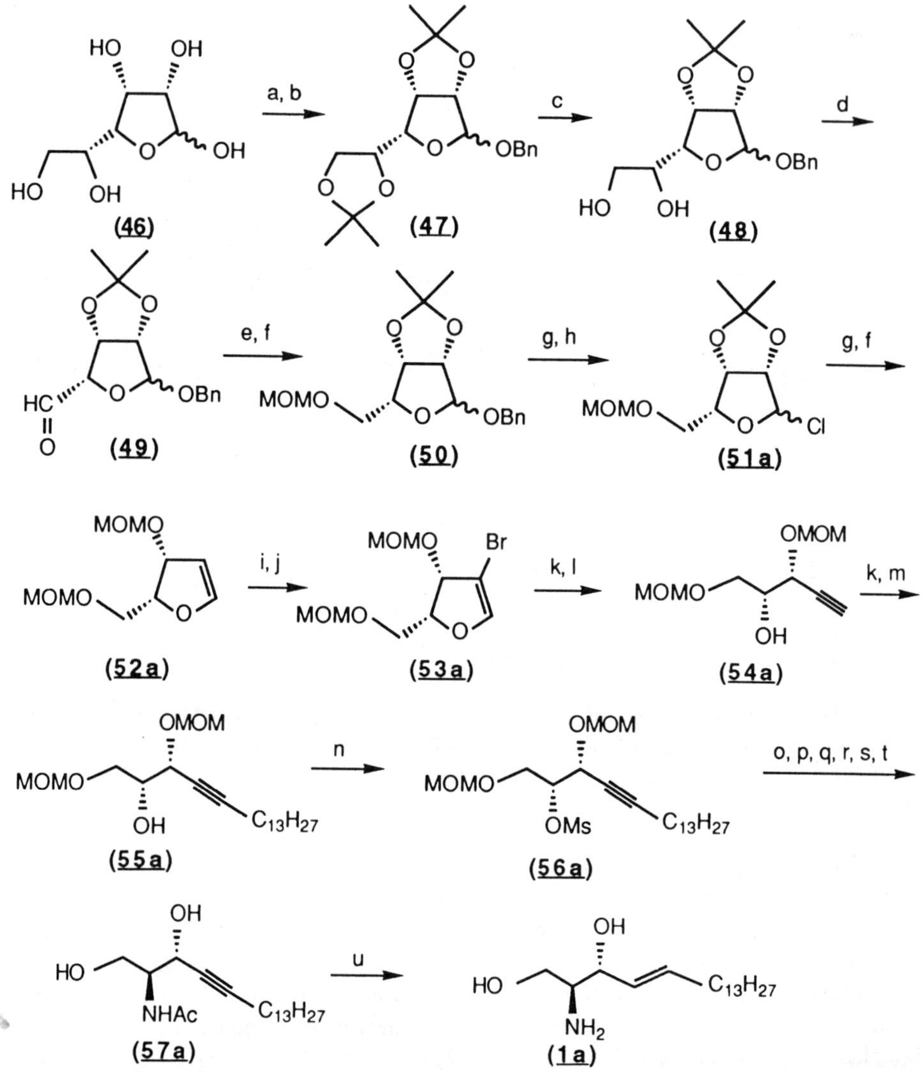

Scheme 11 Conversion of mannose 46 to sphingosine. (a) $(CH_3)_2CO/H^+$, (b) NaH/BnCl/DMF, (c) HCl/MeOH, (d) $NaIO_4$/MeOH, (e) $NaBH_4$/EtOH, (f) $ClCH_2OCH_3/(i\text{-Pr})_2EtN$, (g) Li/NH_3 (liq.), (h) CCl_4/PPh_3, (i) Br_2/CCl_4, (j) DBU/THF, (k) *n*-BuLi/THF, (l) H_2O, (m) $n\text{-}C_{13}H_{27}Br$/HMPA, (n) $MsCl/(i\text{-Pr})_2EtN$, (o) LiN_3/DMF, (p) Ph_3P, (q) $Ac_2O/(i\text{-Pr})_2EtN$, (r) H_2O/100°C, (s) TMSBr, (t) $NaHCO_3/H_2O$, (u) Na/*n*-BuOH.

results from silyl group migration under the benzylation conditions. Desilylation of the major product followed by Swern oxidation afforded the aldehyde 68, which was stable to chromatography on silica gel but epimerized when exposed to acid during workup. Reaction of aldehyde 68 with lithiated phenyl *n*-tetradecyl sulfone followed by acetylation yielded the acetoxysulfone 69 as a mixture of diastereomers. Subjection of the acetoxysulfone to Na-Hg reduction gave the protected sphingosine 70 in 28% overall yield based on the aldehyde. Finally, debenzylation under Birch conditions followed by hydrolysis of the cyclic carbamate group gave sphingosine (1a).

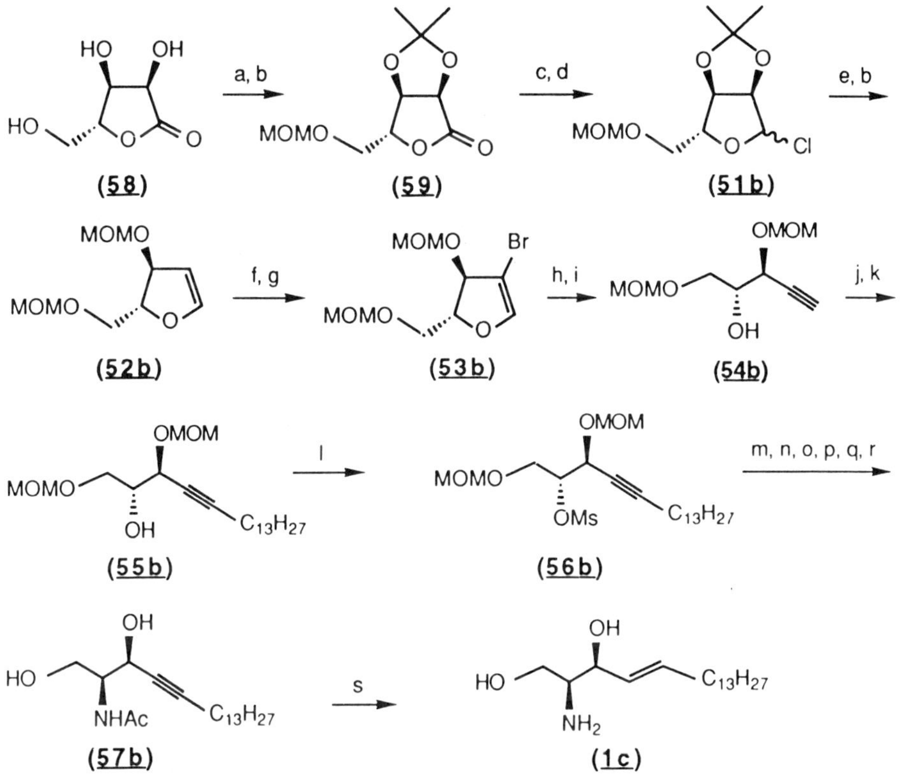

Scheme 12 Conversion of ribonolactone 58 to sphingosine. (a) $(CH_3)_2CO/(CH_3)_2C(OMe)_2/H^+$, (b) $ClCH_2OCH_3/(i\text{-}Pr)_2EtN$, (c) $DIBAL/Et_2O$, (d) CCl_4/PPh_3, (e) Li/NH_3 (liq.), (f) Br_2/CCl_4, (g) DBU/THF, (h) *s*-BuLi/THF, (i) H_2O, (j) *n*-BuLi/THF, (k) $n\text{-}C_{13}H_{27}Br/HMPA$, (l) MsCl/$(i\text{-}Pr)_2EtN$, (m) LiN_3/DMF, (n) Ph_3P/toluene, (o) $Ac_2O/(i\text{-}Pr)_2EtN$, (p) H_2O, (q) TMSBr, (r) $NaHCO_3/H_2O$, (s) Na/*n*-BuOH.

Sphingosine (1a) was synthesized starting from methyl 2,3-*O*-isopropylidene-D-glycerate (71) as shown in Scheme 14 [63]. The reaction of the protected methyl glycerate 71 with lithiated dimethyl methylphosphonate produced the ketophosphonate 72 as a colorless oil in 97% yield. The Wittig-Horner reaction of the ketophosphonate with *n*-tetradecanal in the presence of cesium carbonate in 2-propanol produced the enone 73 in 85% yield. Diastereoselective reduction of the enone with L-Selectride gave a 80% yield of the desired *threo* alcohol 74 together with 8% of the *erythro* isomer. Deprotection of the isopropylidene group followed by reprotection of 75 with a benzylidene group led to 2-hydroxy-1,3-*O*-benzylidene-octadecyne (44b) in 49% yield. Finally, introduction of the amino group was achieved by the following three steps: (1) mesylation (MsCl, Et_3N, CH_2Cl_2), (2) reaction with sodium azide, and (3) reduction of the azido group (Ph_3P, THF-H_2O, 7:1) [64].

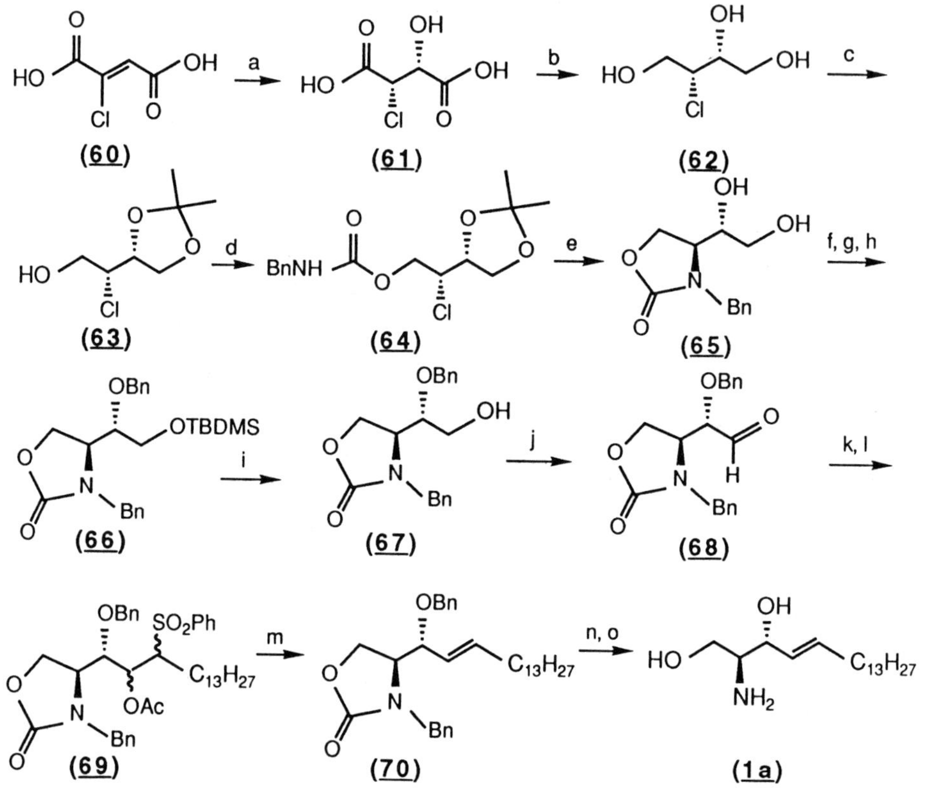

Scheme 13 Conversion of L-*threo*-chloromalic acid (61) to sphingosine. (a) fumarase, (b) BH_3/THF, (c) $(CH_3)_2CO/(CH_3)_2C(OMe)_2$/*p*-TsOH, (d) BnNCO/(*i*-Pr)$_2$EtN/C_6H_6, (e) *t*-BuOK/THF/DMF, (f) 1 N HCl/THF, (g) TBDMSCl/Et_3N/DMAP/Py, (h) NaH/BnBr/*n*-Bu_4NI/THF, (i) *n*-Bu_4NF/THF, (j) $(COCl)_2$/DMSO/Et_3N/CH_2Cl_2, (k) *n*-$C_{14}H_{29}SO_2Ph$/*n*-BuLi/THF, (l) Ac_2O, (m) Na-Hg/MeOH/EtOAc, (n) Na/NH_3 (liq.), (o) 1 N NaOH/EtOH.

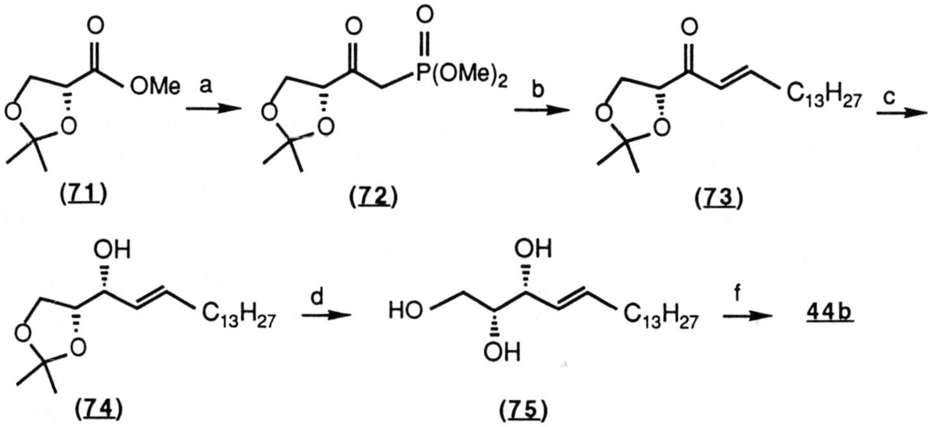

Scheme 14 Horner-Wittig reaction of glyceroylmethylphosphonate 72 with tetradecanal. (a) $CH_3PO(OMe)_2$/*n*-BuLi/THF, (b) *n*-$C_{13}H_{27}CHO/Cs_2CO_3$/*i*-PrOH, (c) L-Selectride/THF, (d) H^+, (f) $PhCH(OMe)_2/H^+$.

V. SYNTHESIS OF SPHINGOSINE BY ALDOL REACTIONS

The diastereoselective aldol reaction is a powerful method for constructing the C-2 and C-3 chiral centers of sphingosine in one step. An aldol reaction of methyl α-isocyanoacetate (77) with aldehydes in the presence of 1 mol % of a chiral (aminoalkyl)-ferrocenylphosphine-gold (I) complex gave aldol adducts in good yield [65–68]. As shown in Scheme 15, the gold-catalyzed asymmetric aldol reaction of isocyanoacetate 77 with *trans*-2-hexadecenal (78) gave optically active oxazoline derivative 79, which contained *trans* and *cis* isomers in a ratio of 89 : 11 [69]. The enantiomeric purities of these isomers were determined to be 93% and 20%, respectively, as judged by ^1H NMR spectroscopy using Eu(dcm) shift reagent. Treatment of the isolated major *trans* oxazoline derivative with concentrated HCl gave β-hydroxy-α-amino acid methyl ester hydrochloride 80 in quantitative yield. Reduction of the amino acid ester with lithium aluminum hydride led to L-*threo*-sphingosine (1c). After protection of the amino and primary hydroxy groups as acetates, the *threo* isomer was isomerized to D-*erythro*-sphingosine (1a) via the Mitsunobu reaction.

Another aldol strategy for the synthesis of sphingosine was based on the chiral oxazolidinone derivatives developed by Evans et al. [70–72] and Pridgen et al. [73,74]. As shown in Scheme 16, the boron enolate of bromoacetyl oxazolidinone derivative 81 was condensed with *trans*-2-hexadecenal to afford aldol adduct 82 in 72% yield [75]. Nucleophilic substitution of the corresponding bromide with sodium azide in dimethyl sulfoxide led to azide 83 in 92% yield with complete inversion of stereochemistry at the reaction center. 2-Azido-3-*O*-*tert*-butyldimethylsilylsphingosine (85) was obtained by protection of the allylic hydroxyl group as the *tert*-butyldimethylsilyl ether followed by lithium borohydride reduction in THF. The synthesis of 1a was completed by reaction of

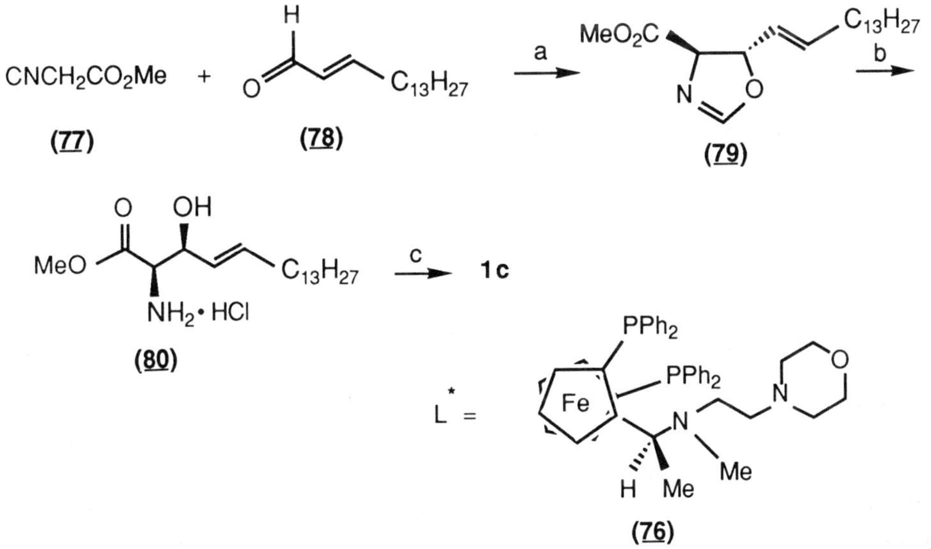

$CNCH_2CO_2Me$ + (77) (78) $\xrightarrow{a}$ (79) $\xrightarrow{b}$ (80) $\xrightarrow{c}$ **1c**

MeO, O, OH, $C_{13}H_{27}$, $NH_2 \cdot HCl$, MeO_2C, N, O, H, L^* =, Fe, PPh_2, N, Me, H, Me, (76)

Scheme 15 Aldol reaction of α-isocyanoacetate 77 catalyzed by a chiral ligand 76-gold complex. (a) Au(I)/L*(76)/CH_2Cl_2, (b) HCl/MeOH, (c) $LiAlH_4$/THF.

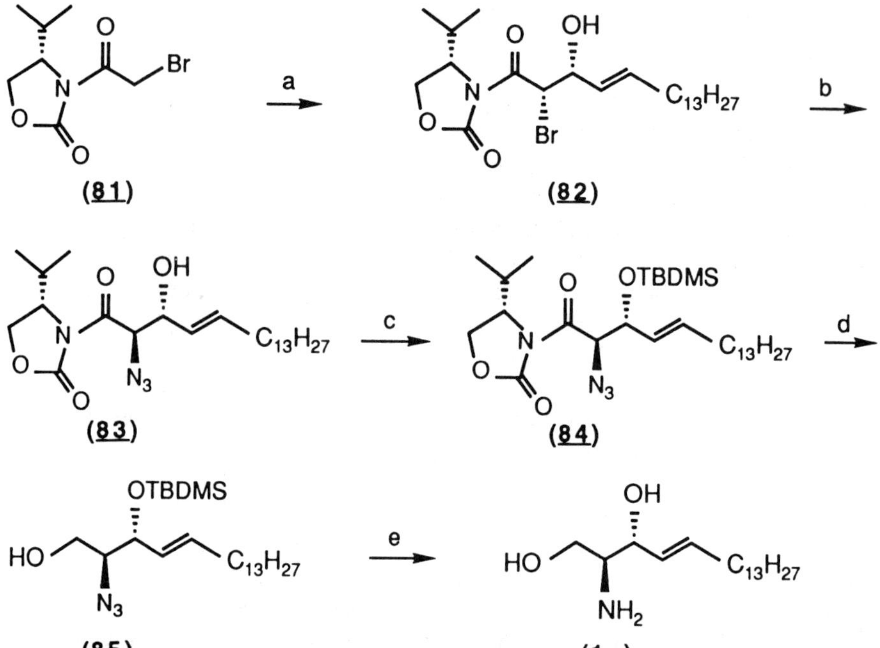

Scheme 16 Aldol reaction of boron enolate of oxazolidinone 81 with (*E*)-2-hexadecenal. (a) $(n\text{-Bu})_2BOTf/Et_3N/(E)$-2-hexadecenal/$Et_2O/H_2O_2$/MeOH, (b) NaN_3/DMSO, (c) TBDMSOTf/2,6-lutidine/CH_2Cl_2, (d) $LiBH_4$/THF, (e) i. $n\text{-}Bu_4NF$/THF, ii. $HS(CH_2)_3SH/Et_3N$/MeOH.

85 in the sequence of (1) desilylation, (2) azide reduction with propylenedithiol in the presence of triethylamine [76a], and (3) acetylation, which furnished the triacetate of product 1a.

An aldol reaction of (*E*)-hexadec-2-enal with the lithiated bislactim ether of cyclo-(L-Val-Gly) gave D-*erythro*-sphingosine (1a) in 21% overall yield [76b].

VI. SYNTHESIS OF SPHINGOSINE BY SHARPLESS EPOXIDATION

Sphingosine has been prepared by the asymmetric epoxidation of a prochiral allyl alcohol in conjunction with selective epoxide cleavage with a nitrogen nucleophile [77,78]. Sharpless epoxidation of 2,4-octadecadiene initially gave a single product, which subsequently decomposed to several products; therefore enynol 89 was chosen as a starting material [79]. As shown in Scheme 17, the enynol was prepared in four steps: (1) reaction of epichlorohydrin 86 and lithium acetylide, (2) protection of the resulting alcohol as the tetrahydropyranyl (THP) ether, (3) alkylation of the terminal acetylene with 1-bromotridecane in THF/HMPA, and (4) deprotection of the THP group [80]. The Sharpless epoxidation of the enynol in the presence of $Ti(OBu\text{-}t)_4$ [81] and D-(−)-diethyl tartrate [(−)-DET] as catalyst gave α,β-epoxy alcohol 90 in 86% yield and 98% enantiomeric excess. Because of the low solubility of the enynol under the reaction conditions, the epoxidation reaction should be controlled either by addition of co-solvent

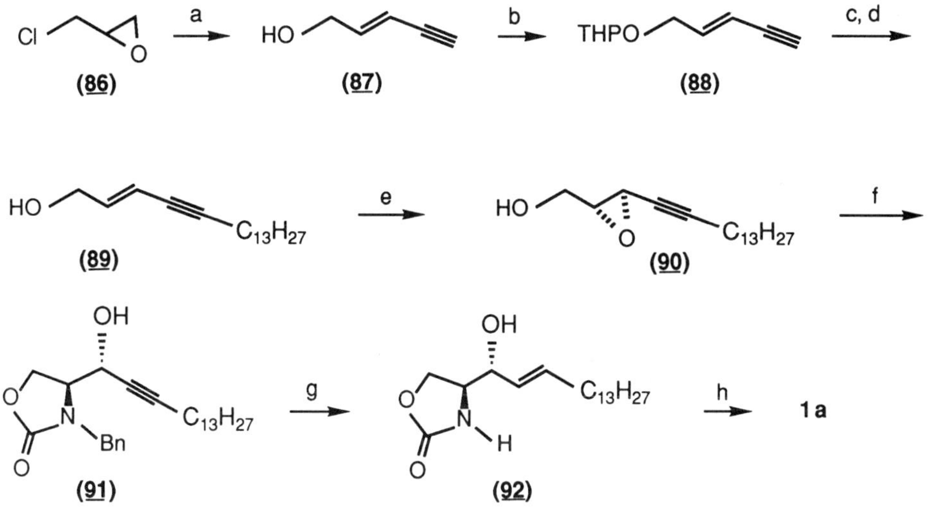

Scheme 17 Sharpless epoxidation of enynol 89. (a) Sodium acetylide, (b) DHP/*p*-TsOH, (c) *n*-BuLi/THF/HMPA/*n*-$C_{13}H_{27}Br$, (d) MeOH/THF/*p*-TsOH, (e) Ti(OBu-*t*)$_4$/(–)-DET/*t*-BuOOH/CH_2Cl_2, (f) BnNCO/NaH/THF, (g) Li/$EtNH_2$/*t*-BuOH, (h) 2N NaOH/EtOH.

(2,3-dimethyl-2-butene and methylene chloride 1:1) or by very slow addition of the enynol 89 to the mixturc of the tartrate and the titanium catalyst in methylene chloride; thus crystallization of the enynol is prevented. To the homogeneous reaction mixture *tert*-butyl hydroperoxide was added. The regioselective intramolecular ring opening of an acyclic carbamate, which was prepared from the epoxy alcohol 90 and benzyl isocyanate, led to the cyclic carbamate 91 [82]. *N*-Debenzylation and partial reduction of the triple bond of the cyclic carbamate 92 was achieved in one step by using lithium in ethylamine. Finally, base-catalyzed hydrolysis of the oxazolidinone ring in 92 was achieved in nearly quantitative yield. Similarly, dihydrosphingosine [82] and phytosphingosine [83] were also prepared.

Another approach to the synthesis of the four diastereomers of sphingosine involves Sharpless epoxidation of *cis*- and *trans*-2-butene-1,4-diol, followed by a regioselective ring-opening reaction of the resulting chiral epoxide with azide ion, as shown in Scheme 18.1 [84]. After monoprotection of the hydroxy function in diol 93 as a propionic ester, a sequence of oxidation of the alcohol with pyridinium chlorochromate and reduction with sodium borohydride was used to give 4-hydroxy-*trans*-2-butenyl propionate (95) in 33% overall yield. The *m*-methoxytrityl (*m*-MTr) ethers 96 and 97 were prepared from the propionates by tritylation followed by alkaline hydrolysis. Sharpless epoxidation of the *trans*-trityl alcohol with either (–)-DET or (+)-DET gave (2*R*,3*R*)- and (2*S*,3*S*)-epoxy alcohols 98a,b in 79% and 84% yields, respectively. Similarly, (2*R*,3*S*)- and (2*S*,3*R*)-epoxy alcohols 99a,b were synthesized from *cis*-2-butenyl propionate in 75% and 73% yields, respectively. The enantiomeric excess exceeded 93%, as judged from NMR analysis of the corresponding epoxy Mosher esters. The ring-opening reaction of (2*S*,3*R*)-epoxy alcohol 98a by Sharpless procedures [85] with Ti(*O*-*i*-Pr)$_2$(N_3)$_2$ gave a mixture of 1,2- and 1,3-diols 100a,b in a ratio of 14:1 and in 83% yield (Scheme 18.2). The major diol was converted to hydroxy-amide 101 in 69% yield by use of the following sequence

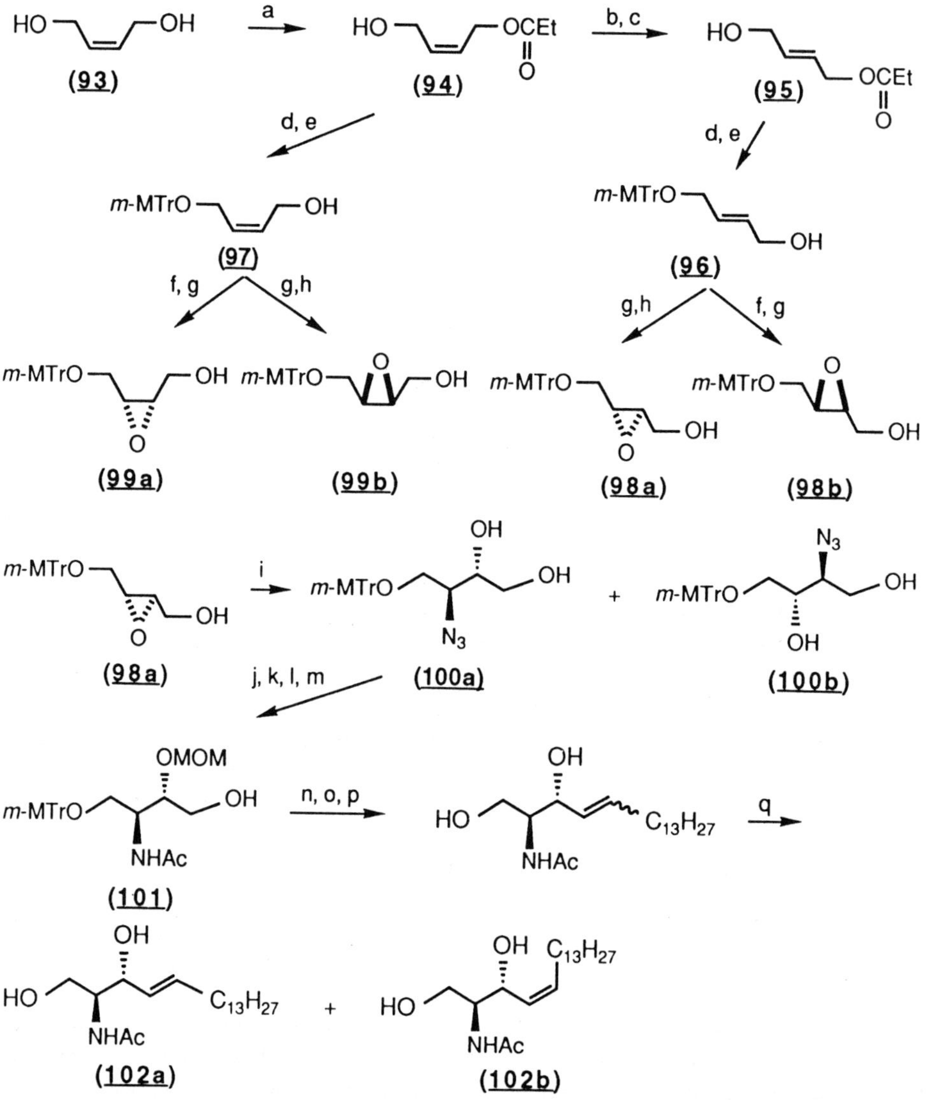

HO–CH₂CH=CHCH₂–OH (93) →a HO–CH₂CH=CHCH₂–OC(O)Et (94) →b, c HO–CH₂CH=CHCH₂–OC(O)Et (95)

(94) →d, e *m*-MTrO–CH₂CH=CHCH₂–OH (97); (95) →d, e *m*-MTrO–CH₂CH=CHCH₂–OH (96)

(97) →f, g (99a); (97) →g,h (99b); (96) →g,h (98a); (96) →f, g (98b)

(98a) →i (100a) + (100b)

(100a) →j, k, l, m (101) →n, o, p →q (102a) + (102b)

Scheme 18.1 Sharpless epoxidation of tritylbuten-1,4-diols 96 and 97 and their conversion to sphingosine. (a) $(EtCO)_2O/(CH_3)_2CO$, (b) PCC/CH_2Cl_2, (c) $NaBH_4$, (d)*m*-MTrCl/Py, (e) 1% KOH/MeOH, (f) (+)-DET, (g) *t*-BuOOH, (h) (–)-DET, (i) $Ti(OPr\text{-}i)_2(N_3)_2/C_6H_6$, (j) C_6H_5COCl/Et_3N, (k) $ClCH_2OCH_3/(i\text{-}Pr)_2EtN$, (l) $LiAlH_4$, (m) $Ac_2O/MeOH$, (n) $(COCl)_2/DMSO/Et_3N/CH_2Cl_2$, (o) $n\text{-}C_{14}H_{29}PPh_3Br/BuLi$, (p) 9% HCl/MeOH, (q) hν/PhSSPh.

of reactions: (1) benzoylation with benzoyl chloride and triethylamine, (2) methoxymethylation with chloromethyl methyl ether and diisopropylethylamine, (3) reduction with lithium aluminum hydride, and (4) acetylation with acetic anhydride in methanol. Swern oxidation of hydroxy-amide 101, then Wittig reaction of the resulting aldehyde with *n*-tetradecyltriphenylphosphonium bromide and *n*-butyllithium in THF, and acidic hydrolysis gave a *cis* and *trans* mixture of *N*-acetylsphingosine (102) in 73% yield.

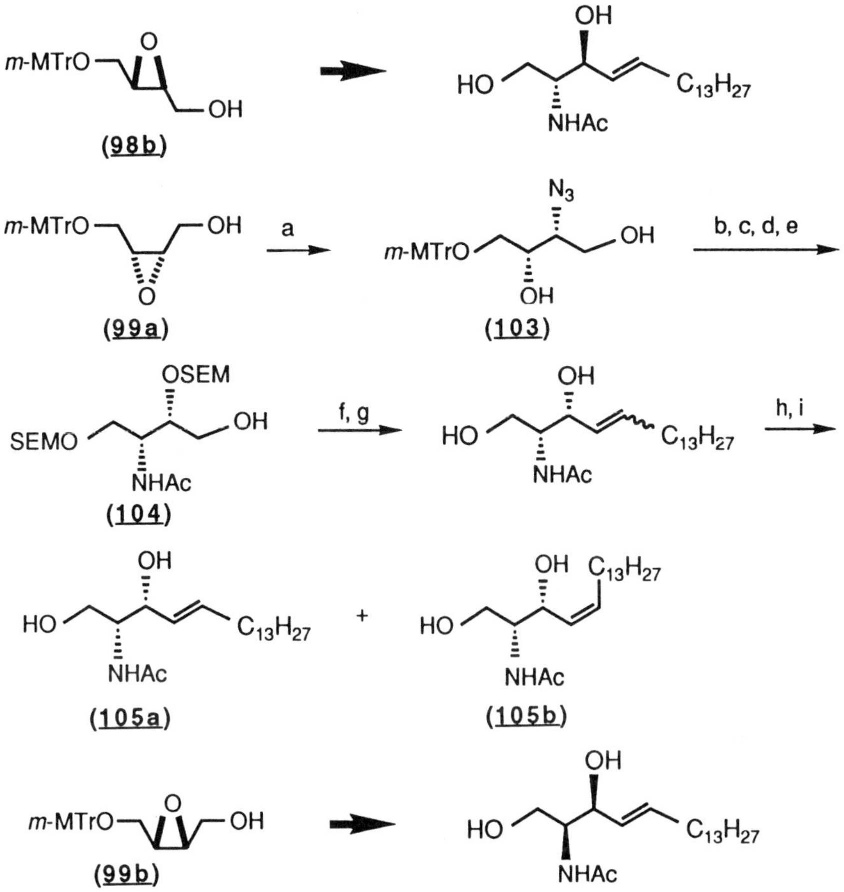

(98b) (99a) (103) (104) (105a) (105b) (99b)

Scheme 18.2 Conversion of epoxides to sphingosine. (a) $Ti(OPr\text{-}i)_2(N_3)_2$,/$C_6H_6$, (b) SEMCl/ $(i\text{-}Pr)_2EtN$, (c) $LiAlH_4$, (d) Ac_2O/Py, (e) 1% HCl/MeOH, (f) $(COCl)_2$/DMSO/Et_3N/CH_2Cl_2, (g) $n\text{-}C_{14}H_{29}PPh_3Br$/$n$-BuLi, (h) 9% HCl/MeOH, (i) hν/PhSSPh.

After photoisomerization in the presence of diphenyldisulfide in cyclohexane-dioxane (4:1) and acetylation, triacetyl-D-*erythro*-sphingosine was isolated by HPLC. Similarly, L-*erythro*-sphingosine (1d) was synthesized starting from (2*S*,3*S*)-epoxy alcohol 98b. D- and L-*threo*-Sphingosines (1b and 1c) were synthesized from (2*R*,3*S*)-epoxy alcohol 99a and (2*S*,3*R*)-epoxy alcohol 99b by a similar manner, except that protection of the hydroxy group was via the trimethylsilylethoxymethyl (SEM) ether.

VII. SYNTHESES OF CERAMIDES, GLYCOSPHINGOLIPIDS, AND SPHINGOMYELIN

As mentioned in the Introduction, aspects of the chemical syntheses of these compounds are reviewed in less detail than those used to prepare sphingosine (1). Selective *N*-acylation of sphingosine with either *p*-nitrophenyl alkanoate or *N*-succinimidyl alkanoate gave ceramides in good yield. For example, treatment of 1-*O*-benzoylsphingosine hydrochloride (formed by hydrolysis of oxazoline 8 in aqueous HCl-THF) with *p*-nitrophenyl oleate in pyridine followed by hydrolysis of the benzoate in sodium methoxide/methanol gave *N*-oleoylsphingosine in 71% overall yield [22]. Reduction of azide 45 to the amine with triphenylphosphine and *N*-acylation with *p*-nitrophenyl palmitate in pyridine gave *N*-palmitoyl-1,3-*O*-benzylidenesphingosine in 95% yield [54]. The latter was hydrolyzed with 2 N HCl in THF to give the ceramide in 72% yield. Base-catalyzed hydrolysis of oxazolidinone 92 followed by selective *N*-acylation with *N*-succinimidyl stearate gave *N*-stearoylsphingosine in 88% yield [80].

Prior protection of the allylic hydroxy group (C-3) of ceramide is required for many reactions taking place at the C-1 position. Ceramides have been protected at the C-3 position as benzoyl [54] and acetoxy esters [86] and *O*-silyl [22,75] and *O*-tetrahydropyranyl ethers [87]. For example, *N*-palmitoyl-3-*O*-benzoyl-D-*erythro*-sphingosine was prepared from the ceramide in 58% overall yield by the following sequence of reactions: (1) blocking of the primary hydroxy group with *tert*-butyldiphenylsilyl (TBDPS) chloride and imidazole in DMF, (2) benzoylation of the allylic hydroxy group with benzoyl chloride in pyridine, and (3) desilylation with tetra-*n*-butylammonium fluoride in THF [54]. Examples of migration of acyl protecting groups from C-3 to C-1 are known to occur during protection-deprotection sequences. The C-1 hydroxy group of *N*-acetyldihydrosphingosine was protected as the *O*-trityl ether, and the 3-hydroxy group was then protected by either acetylation or benzoylation [86]. Detritylation of *N*-acetyl-3-*O*-benzoyl-1-*O*-trityldihydrosphingosine in refluxing 90% acetic acid did not result in benzoyl migration from C-3 to C-1, but detritylation of the analogous 3-*O*-acetyl compound was accompanied by migration of the acetate group. However, when the aldehyde of *N*-acetyl-3-*O*-benzoyldihydrosphingosine was reduced with NaB^3H_4, a mixture of 3-*O*-benzoyl- and 1-*O*-benzoyl-*N*-[1-^{3}H]-acetyldihydrosphingosine was obtained [86].

A series of sphingomyelin analogs was synthesized in a route that involves 1-*O*-(*tert*-butyldiphenylsilyl)-2-*N*-acyl-*erythro*-sphingosine [87a,b]. The 3 position was alkylated in order to prepare 3-*O*-alkyl-sphingomyelins [87b]. To prepare sphingomyelins with various *N*-acyl chains, the 3 position was converted to the 3-tetrahydropyranyloxy ether [87a]. Desilylation and insertion of the phosphocholine group complete the reaction sequence (see below).

Scheme 19 shows some syntheses of glycosphingolipids. *N*-Palmitoyl-3-*O*-benzoyl-D-*erythro*-sphingosine (107) was condensed with α-D-tetra-*O*-acetylgalactopyranosyl bromide (106a) to prepare glycolipid 108a [54]. The Koenigs-Knorr reaction of bromide 106a and ceramide 107 catalyzed by mercuric cyanide in nitromethane gave a mixture of three products. The mixture was treated with trimethylsilyl triflate in methylene chloride in the presence of 4A molecular sieves, giving 1-*O*-β-D-tetraacetylgalactosylceramide 108a in 42% yield from the ceramide. The same reaction in the presence of stannous triflate, 1,1,3,3-tetramethylurea as a base, and 4A molecular sieves gave galactosylceramide 108a in 47% yield. Azide 85, in which the azide group serves as a masked equivalent to a primary amine, is a coupling partner to form lysoglycosphingolipid precursors 108b,c on reaction with suitable carbohydrate donors. The reaction of tetra-*O*-pivaloyl-*O*-α-D-glucosyl trichloroacetimidate (106b) with 3-*O*-benzoyl-2-azidosphingosine (85a) in the presence of boron trifluoride etherate as a catalyst gave glucoside 108b in 94% yield [88]. Protected 1-fluorosugars 106c have also been used in coupling reactions with 3-*O*-*tert*-butyldimethylsilyl-2-azidosphingosine (85b) in the presence of silver triflate-stannous chloride, leading to lysoglycosphingolipid precursor 108c in good yield [75,89]. Reduction of the azide and *N*-acylation gave the

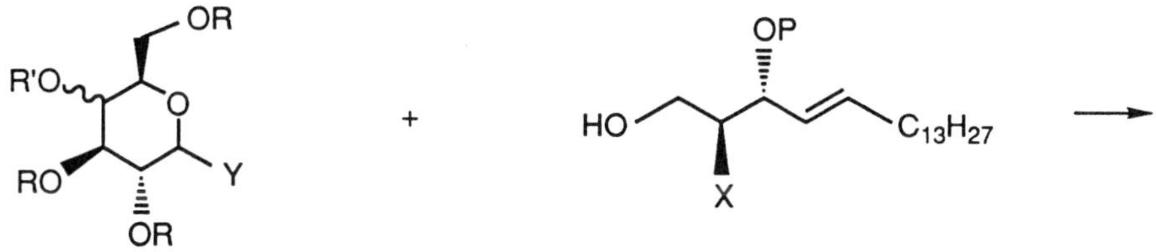

(106a) Y = Br, R and R' = CH_3CO
(106b) Y = $CCl_3C(NH)O$, R and R' = $(CH_3)_3CCO$
(106c) Y = F, R = $(CH_3)_3CCO$, R' = Carbohydrate

(107) X = $NHCOC_{15}H_{31}$, P = Bz
(85a) X = N_3, P = Bz
(85b) X = N_3, P = TBDMS

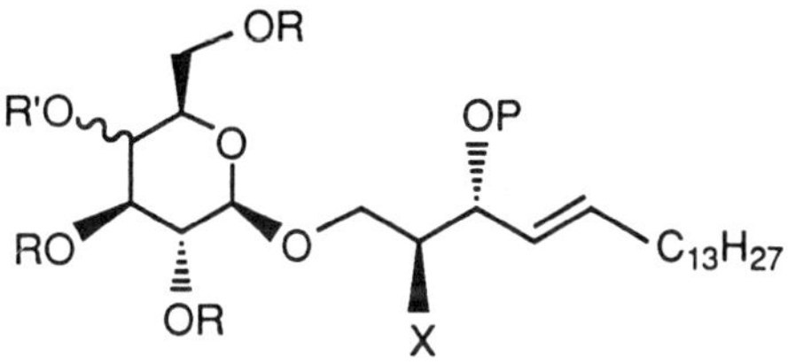

(108a) R and R' = CH_3CO, X = $NHCOC_{15}H_{31}$, P = Bz
(108b) R and R' = $(CH_3)_3CCO$, X = N_3, P = Bz
(108c) R = $(CH_3)_3CCO$, R' = Carbohydrate, X = N_3, P = TBDMS

Scheme 19 Synthesis of glycosphingolipids 108 from protected ceramides 106 or azidosphingosines 85.

long-chain amide, completing the synthesis of the glycosphingolipids. Many other preparations of glycosphingolipids have been outlined in a recent compilation [90a].

3-*O*-Acetyl-*N*-acylsphingosine underwent the Koenigs-Knorr reaction with acetobromoglucose and mercuric cyanide in benzene-nitromethane (90°C, 2 h), giving the crude glucosphingolipid in 61% yield [90b].

Semisynthetic methods for the preparation of sphingomyelins having chemically defined amide chains have been reported. The mixed *N*-acyl chains present in bovine-brain and other natural sphingomyelins were hydrolyzed by using butanolic HCl [91] or methanolic HCl [92] to give sphingosylphosphocholine (lysosphingomyelin), which was *N*-acylated by using fatty acids activated as *O*-acyl isoureas (carbodiimide method) [93,94], *N*-hydroxysuccinimide esters of fatty acids [95], *p*-nitrophenyl alkanonates [96], and acyl imidazolidides (1,1'-carbonyldiimidazole method) [97]. The latter method has the disadvantage that 3-*O*-acylation accompanies the *N*-acylation reaction; the 3-hydroxy group is regenerated by mild alkaline hydrolysis [94]. The method involving acylation of sphingosylphosphorylcholine using nitrophenyl fatty acid derivatives requires prior preparation of the *p*-nitrophenyl ester, mediated by 1,3-dicyclohexylcarbodiimide and 4-(dimethylamino)pyridine [98], although many *p*-nitrophenyl fatty acid esters are now available commercially. Sphingosine-1-phosphate was prepared from sphingosylphosphocholine in about 70% yield by using phospholipase D from *Streptomyces chromofuscus* [99].

The use of natural sphingomyelins as precursors of chemically defined sphingomyelins suffers from two disadvantages. First, acid-catalyzed hydrolysis of sphingomyelin gives rise to partial loss of the stereochemical integrity at the allylic (C-3) position. ^{13}C-NMR studies showed that the extent of inversion at C-3 of sphingosylphosphocholine is greater when sphingomyelin is hydrolyzed by using 1-butanol at 95°C for 90 min than by using methanol at 70°C for 20 h [100]. Second, natural sphingolipids contain several different long-chain (sphingoid) bases. Sphingomyelinase treatment of bovine-brain sphingomyelin, followed by derivatization of the ceramides with 4-biphenylcarbonyl chloride, showed that C_{18}-sphingosine is the predominant sphingoid base (~90% [100]; 82% [101a]), but significant quantities of other molecular species are present, such as C_{20}-sphingosine and C_{18}-dihydrosphingosine.

Alternatively, sphingosine may be isolated by the following sequence of steps: (1) alkaline hydrolysis of brain cerebrosides, (2) sulfuric acid catalyzed hydrolysis of the

Table 1 Melting Points and Optical Rotation Data of Chemically Synthesized Sphingosines and *N,O,O*-Triacetylsphingosines

Compound	mp (°C)	[α]	(Conc, solvent, temp)	Ref.
(1a)	72–75	−0.58°	(1.67, $CHCl_3$,—)	26
	72–75	−0.78°	(2.02, $CHCl_3$,—)	26
	72–75	−1.3°	(3.5, $CHCl_3$, 21°C)	36
	79–82	−2.5°	(6, $CHCl_3$, 22°C)	52
	81–82	−2.8°	(—, $CHCl_3$, 24°C)	84
(1b)	84–85	+2.8°	(—, $CHCl_3$, 24°C)	84
(1c)	86–87	−2.65°	(1.13, $CHCl_3$,—)	26
	84–85	−2.7°	(—, $CHCl_3$, 24°C)	84
(1d)	81–82	+2.8°	(—, $CHCl_3$, 24°C)	84
(1a)-tri-Ac	—	−10.6	(0.3, $CHCl_3$,—)	21
	103.5–104.5	−12.9°	(— $CHCl_3$, 24°C)	22
	102.5–103.5	−13.0°	(1.08, $CHCl_3$—)	26
	104.5–105	−12.9°	(1.0, $CHCl_3$, 25°C)	28
	99–101	−13°	(0.5, $CHCl_3$, 24°C)	40
	103.5–104.5	−12.6°	(0.75, $CHCl_3$, 25°C)	42
	103.5–104	−13.3°	(1.4, $CHCl_3$, 25°C)	62
	103	−12.2°	(1.0, $CHCl_3$, 24°C)	69
	101–102	−12.1°	(1, $CHCl_3$, 25°C)	80
	101–102	−11.4°	(—, $CHCl_3$, 24°C)	84
	104	−22.3°	(2, CH_3CO_2H, 21°C)	53
	104.6–106	−21.6°	(0.88, CH_3CO_2H, 22°C)	55
(1b)-tri-Ac	41–42	−8.9°	(—, $CHCl_3$, 24°C)	84
	44.5–45.5	+10.4°	(1.0, $CHCl_3$, 25°C)	28
(1c)-tri-Ac	43.0–43.5	+8.43°	(—, $CHCl_3$, 24°C)	22
	42–44	+7.02°	(2.05, $CHCl_3$,—)	26
	41–42	+8.5°	(—, $CHCl_3$, 24°C)	84
	—	+8.78°	(1.2, $CHCl_3$, 24°C)	69
	—	+8.2°	(2.2, $CHCl_3$, —)	116
(1d)-tri-Ac	101–102	+12.1°	(—, $CHCl_3$, 24°C)	84
C_{20}-(1a)-tri-Ac	103–105	−22.5°	(1.07, CH_3CO_2H, 23°C)	34
	106–107	−22.7°	(0.5, CH_3CO_2H, 25°C)	42

resulting galactosylceramide (psychosine), and (3) silica gel column chromatography [101b]. About 400 mg of sphingosine was obtained from 2 g of brain cerebrosides. It was reported that no *threo*-sphingosine was detected by TLC. However, the sphingosine obtained was not characterized as its triacetate derivative, which would confirm the stereochemistry (see Table 1).

Thus a total chemical synthesis of sphingomyelin from one of the stereoisomers of sphingosine prepared by one of the enantioselective methods described earlier is required to produce sphingomyelin in high optical purity. For example, oxazoline sphingosine 8 (see Scheme 2) was hydrolyzed with aqueous HCl, and the resulting 1-*O*-benzoyl-D-*erythro*-sphingosine was acylated with *p*-nitrophenyl stearate to give 1-*O*-benzoyl-*N*-stearoyl-D-*erythro*-sphingosine [23]. The allylic hydroxy group of the latter ceramide was protected as the *tert*-butyldiphenylsilyl ether, and the 1-*O*-benzoyl group was removed by base-catalyzed hydrolysis to give 3-*O*-*tert*-butyldiphenylsilyl-*N*-stearoyl-D-*erythro*-sphingosine (109). Scheme 20 outlines the synthesis of *N*-stearoyl-D-*erythro*-sphingomyelin (110) from 3-*O*-*tert*-butyldiphenylsilyl-protected ceramide 109 by phosphitylation using chloro-*N*,*N*-diisopropylmethoxyphosphine in the presence of triethylamine. Reaction of the phosphoramidite with choline tosylate in the presence of excess 1 *H*-tetrazole, then oxidation of the phosphite triester with *tert*-butyl hydroperoxide, demethylation of the methyl phosphate with trimethylamine, and finally desilylation using tetra-*n*-butylammonium fluoride gave *N*-stearoyl-D-*erythro*-sphingomyelin (110) in 70% overall yield.

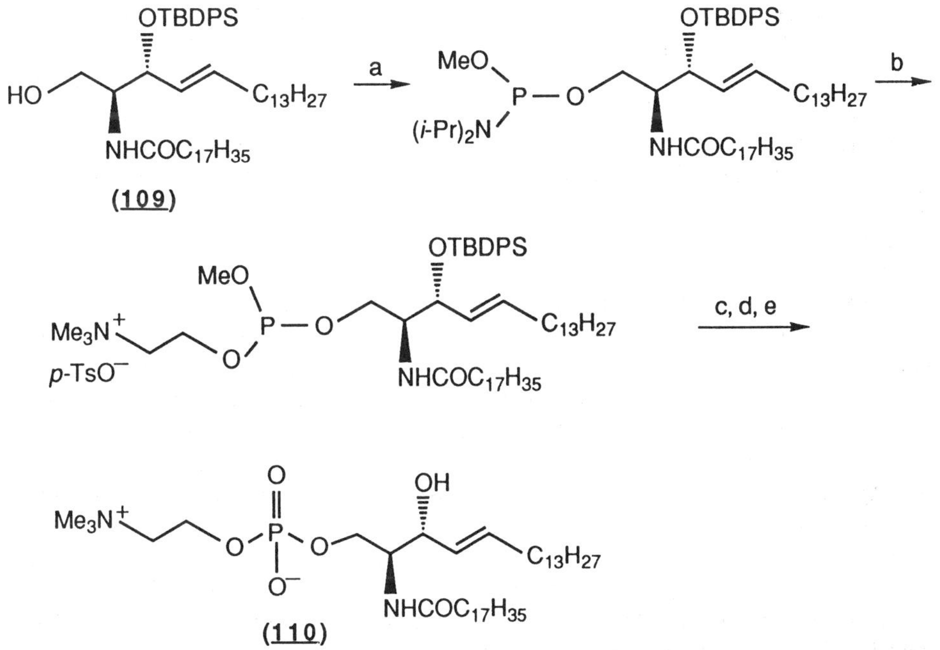

OTBDPS
HO — — $C_{13}H_{27}$ —a→ MeO / $(i\text{-}Pr)_2N$ P—O — — $C_{13}H_{27}$ (OTBDPS, $NHCOC_{17}H_{35}$) —b→
$NHCOC_{17}H_{35}$
(109)

Me_3N^+ p-TsO$^-$ — O — P(MeO) — O — — $C_{13}H_{27}$ (OTBDPS, $NHCOC_{17}H_{35}$) —c, d, e→

Me_3N^+ — O — P(=O)(O$^-$) — O — — $C_{13}H_{27}$ (OH, $NHCOC_{17}H_{35}$)
(110)

Scheme 20 Synthesis of sphingomyelin from ceramide via phosphitylation. (a) $i\text{-}Pr_2N(OMe)PCl/Et_3N/CHCl_3$, (b) $Me_3N^+(CH_2)_2OH$ p-TsO$^-$/tetrazole/MeCN/THF, (c) t-BuOOH/THF, (d) Me_3N/toluene, (e) $n\text{-}Bu_4NF$/THF.

A similar phosphitylation approach was used to convert 3-*O*-tetrahydropyranyl-protected ceramides into sphingomyelins with a *N*-(α-hydroxypalmitoyl) chain [87a]. Sphingomyelins have also been synthesized by phosphorylation of 3-*O*-tetrahydropyranyl-protected ceramides, 3-*O*-alkylceramides, or 3-deoxyceramides using 2-chloro-2-oxo-1,3,2-dioxaphospholane or 2-bromoethylphosphoric acid dichloride, then treatment with trimethylamine and deprotection of the THP group [87a,b]. The overall yields from the various ceramide derivatives to the sphingomyelin analogs via the phosphorylation approach ranged from 36 to 61% [87b].

VIII. EXPERIMENTAL PROCEDURES

See Refs. 102 and 103 for methods of drying the solvents used in the chemical syntheses described below.

(S)-4-(Carbomethoxy)-2-phenyl-Δ^2-oxazoline (6) [22]. A stirred biphasic solution of serine methyl ester hydrochloride (10.9 g, 70 mmol) in water (7 mL) and benziminoethyl ether [104,105] (15.6 g, 105 mmol) in methylene chloride (45 mL) was stirred for 16 h at room temperature. To the mixture sufficient water was added to dissolve the precipitated ammonium chloride; then the product was extracted with methylene chloride. The organic layer was dried over $MgSO_4$ and concentrated using a rotary evaporator. The product was purified by vacuum distillation to give 11.3 g of 6 (79% yield) as a colorless oil; bp 109–110°C (0.01 torr) [lit. [106] 110–112°C (0.02 torr); IR (neat) 1740, 1640 cm^{-1}; ^1H-NMR ($CDCl_3$) δ (ppm) 7.96–7.99 (m, 2H), 7.36–7.49 (m, 3H), 4.95 (dd, 1H, J = 10.7, 8.2 Hz), 4.68 (dd, 1H, J = 8.4, 8.5 Hz), 4.57 (dd, 1H, J = 10.7, 8.8 Hz), 3.79 (s, 3H).

(S)-4-Formyl-2-phenyl-Δ^2-oxazoline (7) [22]. To a stirred solution of the oxazoline ester 6 (3.1 g, 15 mmol) in dry hexane-toluene (120 mL, 1:3) at –77°C a solution of diisobutylaluminum hydride (DIBAL, 15 mmol) in toluene was added dropwise at a rate such that the temperature did not exceed –70°C. After the reaction mixture was stirred for an additional 3 h, methanol was added slowly, maintaining the temperature at –70°C for an additional 30 min. Ethyl acetate (10 mL) and saturated aqueous sodium potassium tartrate solution (40 mL) were added, and the mixture was allowed to warm to room temperature. The reaction mixture was transferred to a separatory flask and extracted with ethyl acetate and saturated tartrate solution. The combined ethyl acetate extracts were dried over $MgSO_4$, filtered, and evaporated by rotary evaporation followed by high vacuum to give a yellow oil (2.95 g). The aldehyde content was 90%, with a trace of unreacted methyl ester 6 shown by TLC (R_f, 0.22 for the aldehyde, R_f 0.60 for the ester, eluting solvent: $CHCl_3$-CH_3OH, 95:5). The unstable aldehyde 7 was stored in benzene at 20°C prior to use; IR (neat) 1720, 1640 cm^{-1}.

(4S,1'R)-(8a) *and (4S,1'S)-4-(1'*-Hydroxy-2'-*(E)-hexadecenyl-2-phenyl-Δ^2-oxazoline* (8b) [22]. To a solution of 1-pentadecyne (6.26 g, 30 mmol) in dry hexane (40 mL) a solution of DIBAL (30 mmol) in toluene was added. The mixture was heated at 60°C for 2 h under nitrogen [26,107,108]. The reaction was monitored by TLC (R_f 0.66 for pentadecyne, R_f 0.92 for *trans*-vinylalane, elution with 100% hexane). To this *trans*-vinylalane solution the aldehyde 7 (5.3 g, 30 mmol) in ether-toluene (12 mL, 3:1) was added at 5–10°C, and then the reaction mixture was warmed to room temperature and stirred for 1 h. The reaction mixture was worked up by the ethyl acetate-tartrate extraction procedure described for 7. The *erythro* and *threo* products were separated by column chromatography on silica gel, eluting with ether-petroleum ether (10:7).

Erythro isomer (8a). 3.0 g (26% yield); mp 89–90°C; R_f 0.39 (ether-petroleum ether 10:7); IR (KBr) 1640 cm^{-1}; ^{1}H-NMR ($CDCl_3$) δ (ppm) 7.91 (m, 2H), 7.35–7.50 (m, 3H), 5.81 (dt, 1H, J = 15.4, 7.1 Hz), 5.44 (dd, 1H, J = 15.4, 5.9 Hz), 4.52 (m, 1H), 4.39 (m, 3H), 2.42 (br s, 1H), 1.62 (m, 2H), 1.26 (s, 22H), 0.88 (t, 3H, J = 6.6 Hz).

Threo isomer (8b). 2.9 g (25% yield); mp 70–70.5°C; R_f 0.23 (ether-petroleum ether 10:7); IR (KBr) 1640 cm^{-1}; ^{1}H-NMR ($CDCl_3$) δ (ppm) 7.91 (m, 2H), 7.36–7.53 (m, 3H), 5.80 (dt, 1H, J = 15.4, 7.1 Hz), 5.40 (dd, 1H, J = 15.3, 6.0 Hz), 4.50 (m, 1H), 4.37 (m, 3H), 2.21 (br s, 1H), 1.62 (m, 2H), 1.26 (s, 22H), 0.88 (t, 3H, J = 6.7 Hz).

N-(tert-Butoxycarbonyl)-L-serine, methyl ester (9) [31]. To a stirred solution of $(BOC)_2O$ (78.4 g, 395 mmol) in dioxane (280 mL) was added a solution of L-serine (31.73 g, 302 mmol) in 1 N NaOH (620 mL) at 0°C. After stirring for 30 min at 5°C and 3.5 h at room temperature, the mixture was concentrated to half of its original volume by rotary evaporation at 35°C, cooled in an ice bath, and acidified to pH 2–3 by addition of 1 N $KHSO_4$. The product was extracted with ethyl acetate. The organic phase was dried over $MgSO_4$, filtered, and concentrated to give *N*-BOC-L-serine as a colorless sticky foam. This product was dissolved in ether (600 mL) and treated with ten 50-mL aliquots if a 0.6 M of etheral diazomethane [109] at 0°C. After excess diazomethane was destroyed by addition of acetic acid, the reaction mixture was washed with half-saturated $NaHCO_3$ solution and brine solution, dried over $MgSO_4$, and then concentrated to give 60.1 g of 9 (90% yield) as a colorless, sticky foam. This product was used without further purification. R_f 0.38 (hexane-EtOAc 1:1); IR (neat) 3400, 1720 (br) cm^{-1}; ^{1}H-NMR (C_6D_6, 17°C) δ (ppm) 5.6 (m, 1H), 4.4 (m, 1H), 3.76 (dd, 1H, J = 11, 4 Hz), 3.66 (dd, 1H, J = 11, 4 Hz), 3.26 (s, 3H), 2.5 (br s, 1H), 1.41 (s, 9H).

3-tert-Butyloxycarbonyl (S)-4-carbomethoxy-2,2-dimethyl-3,4-oxazolidine (10) [31]. A solution of methyl ester 9 (48.5 g, 221 mmol), $Me_2C(OMe)_2$ (55 mL, 450 mmol), and *p*-TsOH·H_2O (0.593 g, 3.12 mmol) in benzene (770 mL) was refluxed for 30 min and then distilled until 660 mL of distillate had been collected. Additional $Me_2C(OMe)_2$ (14 mL, 110 mol) and fresh benzene (310 mL) were added and the procedure was repeated, collecting 250 mL of distillate. The reaction mixture was diluted with ether (600 mL) and washed with saturated $NaHCO_3$ solution and brine. The organic layer was dried over $MgSO_4$, filtered, and concentrated to give the crude product as an oil. The product was purified by vacuum distillation through a 10-cm Vigreaux column to give 40.3 g (70% yield) as a very pale yellow oil, which has 95% purity; bp 101–102°C (2 torr); R_f 0.78 (EtOAc-hexane 1:1); $[\alpha]_D$ −46.7° (*c* 1.30, $CHCl_3$). The identical procedure starting from D-serine gave the antipode of 10 in 80% yield; $[\alpha]_D$ + 53° (*c* 1.30, $CHCl_3$). On further purification of distillation and column chromatography, the product had a maximum rotation of +57°. IR (neat) 1760, 1704 cm^{-1}; ^{1}H-NMR (C_6D_6, 75°C) δ (ppm) 4.26 (m, 1H), 3.81 (dd, 1H, J = 8.5, 3.5 Hz), 3.75 (dd, 1H, J = 8.5, 8.1 Hz), 3.35 (s, 3H), 1.81 (br s, 3H), 1.53 (br s, 3H), 1.41 (s, 9H). (At ambient temperature, the NMR spectrum was more complicated, indicating that the oxazolidine system undergoes a dynamic equilibrium.)

1-tert-Butyloxycarbonyl (S)-4-formyl-2,2-dimethyl-3-oxazolidine (11) [31]. To a stirred solution of oxazolidine ester 10 (40.2 g, 155 mmol) in dry toluene (300 mL) at −78°C under nitrogen a DIBAL solution (263 mmol) in toluene was added slowly via cannula so as to keep the temperature of the reaction mixture below −65°C. The reaction mixture was stirred for an additional 2 h at −78°C, then quenched by slow addition of methanol (60 mL). The resulting white emulsion was slowly poured into ice-cold 1 N HCl solution (1 L) with swirling over 15 min and then extracted with ethyl acetate (3 × 100 mL). The combined organic layers were washed with brine solution, dried over $MgSO_4$, filtered, and concentrated under reduced pressure to give a crude product as a colorless oil, which was purified by vacuum distillation through a 10-cm Vigreaux column to give 26.86 g of 11 (76% yield); bp 83–88°C (1.0–1.4 torr); $[\alpha]_D$ −91.7° (*c* 1.34, $CHCl_3$). The identical procedure starting from D-serine gave the antipode in 85% yield; $[\alpha]_D$ +95° (*c* 1.34, $CHCl_3$). The distilled product was purified further by column chromatography to give a product with a maximum rotation of +105°. R_f 0.33 (hexane-EtOAc 4:1); IR (neat 1735, 1705 cm^{-1}; ^{1}H-NMR (C_6D_6, 60°C) δ (ppm) 9.34 (br s, 1H), 3.90 (m, 1H), 3.65 (dd, 1H, J = 8.7, 2.9 Hz), 3.52 (dd, 1H, J = 8.7, 8.3 Hz), 1.59 (br s, 3H), 1.40 (br s, 3H), 1.34 (s, 9H).

1-tert-Butyloxycarbonyl (4S,1'R)-2,2-dimethyl-4-(1'-hydroxy-2'-hexadecynyl)-3-oxazolidine (12a). Method A. [28]. To a solution of 1-pentadecyne (28.72 g, 138 mmol) in dry THF (750 mL) *n*-butyllithium (126 mmol) was added dropwise at −20°C, and the mixture was stirred for an additional 2 h at −20°C. After 37 mL of HMPA (200 mmol) was added, a solution of aldehyde 11 (24.32 g, 106 mmol) in dry THF (60 mL) was added and the mixture was stirred for 1 h at −78°C. The reaction mixture was allowed to warm to −20°C within 2 h and was then quenched by the addition of saturated NH_4Cl (1.2 L). After removal of the volatile solvents under vacuum, the residue was diluted with water (600 mL) and extracted with ether (3 × 500 mL). The organic layer was washed with 0.5 N HCl and brine solution, dried, and evaporated under reduced pressure. Filtration through silica gel, first using petroleum ether as the solvent to recover excess 1-pentadecyne, followed by elution with petroleum ether-EtOAc (6:1) gave 32.9 g of a 20:1 mixture of 12a/12b (71% yield) as a colorless oil, which was purified by column chromatography (petroleum ether-EtOAc 6:1); $[\alpha]_D^{25}$ −40.1° (*c* 1.0, $CHCl_3$); ^{1}H-NMR (CD_3SOCD_3) δ (ppm) 5.45 (d, 1H, J = 6.5 Hz), 4.41–4.50 (m, 1H), 3.71–4.06 (m, 2H), 2.06–2.21 (m, 2H), 1.11–1.50 (m, 37H), 0.85 (t, 3H, J = 7.5 Hz).

Method B [26]. To a solution of 1-pentadecyne (3.49 g, 16.75 mmol) in dry THF (150 mL) was added *n*-butyllithium (14.3 mmol), and the mixture was stirred for 30 min at −23°C under nitrogen. To the mixture a solution of aldehyde 11 (2.82 g, 12.31 mmol) in dry THF (75 mL) was added, and the mixture was stirred for 1.5 h at −23°C. Extractive workup gave a crude 8:1 mixture of 12a/12b, which was purified by column chromatography on silica gel, eluting with hexane-EtOAc (9:1, 74% yield); $[\alpha]_D$ −39.7° (*c* 1.41, $CHCl_3$); R_f 0.43 for 12a, and 0.39 for 12b (hexane-EtOAc 4:1); IR (neat) 3440, 2220, 1700 cm^{-1}; ^{1}H-NMR (C_6D_6, 60°C) δ (ppm) 4.70 (m, 1H), 3.65–4.20 (m, 3H), 2.06 (pseudo t, 2H), 1.70 (br s, 3H), 1.48 (br s, 3H), 1.38 (s, 9H), 1.30 (br s, 22H), 0.91 (pseudo t, 3H), 0.45 (br s, 1H).

1-tert-Butyloxycarbonyl (4S,1'S)-2,2-dimethyl-4-(1'-hydroxy-2'-hexadecynyl)-3-oxazolidine (12b) [28]. To a solution of 1-pentadecyne (36.12 g, 173 mmol) in dry ether (900 mL), *n*-butyllithium (160 mmol) was added dropwise at −20°C. After the mixture was stirred for an additional 1 h at −20°C, zinc bromide (42.0 g, 186 mmol) was added to the suspension, and the mixture was stirred for 1 h at 0°C and for 1 h at room temperature. After dropwise addition of a solution of aldehyde 11 (30.57 g, 133 mmol) in ether (185 mL) at −78°C, the reaction mixture was stirred overnight at room temperature and then quenched by the addition of saturated NH_4Cl solution (600 mL) at −20°C. The product was extracted with ether and washed with brine solution, dried, filtered, and concentrated. Filtration through silica gel, as described above, afforded 48.8 g of a 20:1 mixture of 12b/12a (84% yield). An analytical sample was purified by column chromatography, eluting with petroleum ether-EtOAc (6:1); $[\alpha]_D^{25}$ −32.4° (*c* 1.3, $CHCl_3$); ^{1}H-NMR

(CD_3SOCD_3) δ (ppm) 5.51 (d, 1H, J = 5.5 Hz), 5.54 (d, 1H, J = 5.5 Hz), 4.56–4.70 (m, 1H), 3.88–4.06 (m, 2H), 3.73–3.85 (m, 1H), 2.13 (t, 2H, J = 6 Hz), 1.10–1.52 (m, 37H), 0.85 (t, 3H, J = 7.5 Hz).

1-tert-Butyloxycarbonyl (4S, 1'R)-2,2-dimethyl-4-(1'-hydroxy-2'-(E)-hexadecenyl)-3-oxazolidine (13a). *Method A* [30]. Laboratory directions for a large-scale synthesis of 13a are as follows. To liquid ammonia (4.8 L) lithium metal (60 g, 8.6 mol) was added portionwise at –60°C. To this blue solution propargylic alcohol 12a (75 g, 171 mmol) in dry THF (6 L) was added slowly. After refluxing for 15 h, the reaction was quenched by the addition of solid NH_4Cl (2.3 kg), and then the solvent was allowed to evaporate. The residue was treated with H_2O-THF (500 mL, 1:1) and the product was extracted with ether, dried, and concentrated to give 70 g of oxazolidine 13a (93% yield); $[\alpha]_D^{20}$ –25.2° (c 0.215, $CHCl_3$); IR (film) 3424, 2926, 2854, 1701, 1670, 1390, 1366, 1258, 1176 cm^{-1}; ^{1}H-NMR ($CDCl_3$) δ (ppm) 5.74 (dt, 1H, J = 15.5, 6.7 Hz), 5.43 (dd, 1H, J = 15.5, 6.3 Hz), 3.65–4.4 (m, 4H), 2.04 (m, 2H), 1.15–1.65 (m, 37H), 0.89 (t, 3H, J = 7 Hz).

Method B [26]. The reaction mixture containing 12a prepared by method B was added directly –78°C to a blue solution of lithium metal (0.451 g, 65 mmol) in $EtNH_2$ (75 mL) via cannula under nitrogen. After stirring for 1 h at –78°C, the reaction mixture was quenched at –78°C by the addition of solid NH_4Cl (8.8 g). The ethylamine was allowed to evaporate at room temperature overnight followed by removal of residual solvent on a rotary evaporator. Water and ether were added to the resulting residue, and the product was extracted with ether. The combined ether layers were washed with brine, dried over Na_2SO_4, filtered, and concentrated to give 2.66 g of a waxy solid, which was contaminated with sphingosine. Pure 13a was obtained as a colorless oil after column chromatography on silica gel, eluting with 12:1 hexane-EtOAc; $[\alpha]_D$ –28° (c 0.65, $CHCl_3$); R_f 0.37 (hexane-EtOAc 4:1); IR (neat) 3400, 1700 cm^{-1}; ^{1}H-NMR (400 MHz, C_6D_6, 60°C) δ (ppm) 5.79 (dt, 1H, J = 15.3, 6.6 Hz), 5.55 (dd, 1H, J = 15.3, 5.3 Hz), 4.30 (br s, 1H), 3.97 (br s, 1H), 3.79 (br s, 1H), 3.66 (dd, 1H, J = 8.8, 6.8 Hz), 2.03 (q, 1H, J = 7.1 Hz), 1.64 (br s, 3H), 1.45 (br s, 3H), 1.39 (br s, 9H), 1.13 (br s, 22H), 0.90 (t, 3H, J = 6.6 Hz), 0.43 (br s, 1H). The above dissolving-metal reduction was carried out for 4 h at –78°C and the same extractive workup was used to prepare crude spingosine directly. Trituration of the crude material with cold pentane (10 mL) followed by recrystallization from 1:1 hexane-EtOAc afforded 1.339 g (68% yield from the aldehyde 11) of pure sphingosine (1a) as a white solid; mp 72–75°C; $[\alpha]_D$ –0.78° (c 2.02, $CHCl_3$); R_f 0.62 (n-BuOH-H_2O-AcOH 4:1:1); IR (KBr) 3300, 1579 cm^{-1}; ^{1}H-NMR (400 MHz, $CDCl_3$) δ (ppm) 5.75 (dt, 1H, J = 15.4, 7.5 Hz), 5.46 (dd, 1H, J = 15.5, 6.9 Hz), 4.04 (t, 1H, J = 6.2 Hz), 3.67 (dd, 1H, J = 10.9, 4.5 Hz), 3.60 (dd, 1H, J = 10.7, 5.9 Hz), 2.87 (q, 1H, J = 5.1 Hz), 2.04 (q, 2H, J = 7.0 Hz), 1.95 (br s, 4H), 1.24 (br s, 22H), 0.86 (t, 3H, J = 6.7 Hz).

1-tert-Butyloxycarbonyl (4S,1'S)-2,2-dimethyl-4-(1'-hydroxy-2'-hexadecenyl)-3-oxazolidine (13b) [26]. To a solution of *trans*-vinylalane prepared from 1-pentadecyne (1.0 g, 4.82 mmol) by the method described for 8a,b a solution of aldehyde 11 (0.850 g, 3.71 mmol) in dry toluene (3.3 mL) was added at –78°C. The resulting suspension was stirred for 2 h, during which time the mixture was allowed to warm to –60°C, whereupon a colorless solution formed. The mixture was poured into ice water, acidified with 1 N HCl to pH 1, and extracted with ether. The combined extracts were washed with brine, dried over $MgSO_4$, filtered, and concentrated to give 1.86 g of crude product as a colorless oil, which contained a 1:2 mixture of 13a,b. A 13b-enriched sample (a:b =

1:7) was obtained by column chromatography on silica gel, eluting with hexane-EtOAc (12:1); $[\alpha]_D$ –39° (*c* 0.25, $CHCl_3$); R_f 0.44 (hexane-EtOAc 3:1); IR (neat) 3470, 1700, 1670 cm^{-1}; ^{1}H-NMR (400 MHz, C_6D_6, 60°C) δ (ppm) 5.70 (dt, 1H, J = 15.6, 6.8 Hz), 5.52 (dd, 1H, J = 15.5, 7.1 Hz), 4.40 (t, 1H, J = 7.2 Hz), 3.95 (pseudo t, 1H, J = 6.3 Hz), 3.89 (br d, 1H, J = 9.1 Hz), 3.67 (dd, 1H, J = 6.4, 4.9 Hz), 1.99 (q, 2H, J = 6.8 Hz), 1.65 (br s, 3H), 1.46 (br s, 3H), 1.39 (br s, 9H), 1.31 (br s, 22H), 0.91 (t, 3H, J = 6.9 Hz), 0.55 (br s, 1H).

N-tert-Butyloxycarbamyl (2S,3R)-1,3-dihydroxy-4-(E)-octadecyne (16) [28]. To a solution of 12a (26.37 g, 60 mmol) in methanol (600 mL) Amberlyst 15 (31 g) was added. The heterogenous mixture was stirred for 41 h at room temperature, then filtered through a Celite pad. Evaporation under reduced pressure gave a residue that was purified by filtration on silica gel with hexane-EtOAc (1:1) to give 17.4 g of 16 as a waxy solid (72% yield); mp 43–44°C; $[\alpha]_D^{25}$ –8.5° (*c* 1.0, $CHCl_3$); ^{1}H-NMR (CD_3SOCD_3) δ (ppm) 6.18 (d, 1H, J = 8 Hz), 5.31 (d, 1H, J = 6 Hz), 4.50 (t, 1H, J = 5 Hz), 4.15–4.23 (m, 1H), 3.36–3.53 (m, 2H), 2.15 (t, 2H, J = 6 Hz), 1.15–1.48 (m, 31H), 0.85 (t, 3H, J = 7.5 Hz). Similarly, the (2*S*, 3*S*)-isomer 16 was prepared from 12b in 75% yield; $[\alpha]_D^{25}$ –14.0° (*c* 0.5, $CHCl_3$); ^{1}H-NMR (CD_3SOCD_3) δ (ppm) 6.15 (d, 1H, J = 8 Hz), 5.18 (d, 1H, J = 7 Hz), 4.60 (t, 1H, J = 5 Hz), 4.25–4.36 (m, 1H), 3.26–3.50 (m, 2H), 2.15 (t, 2H, J = 6 Hz), 1.13–1.50 (m, 31H), 0.85 (t, 3H, J = 7.5 Hz).

N-BOC-Sphingosine (14) [28]. To an ether solution of Red-Al (6.92 mmol in toluene) a solution of 16 (5.0 g, 12.6 mmol) in dry ether (20 mL) was added dropwise at 0°C. The reaction mixture was stirred for 24 h at room temperature, then quenched by the addition of methanol (9 mL) at 0°C. After the mixture was diluted with ether (100 mL), a saturated solution of potassium sodium tartrate (100 mL) was added, and the biphasic mixture was stirred vigorously at room temperature for 3 h. The organic layer was separated and the product was extracted with ether. The combined ether extracts were washed with a saturated solution of potassium sodium tartrate, brine, dried, and concentrated to give a crude product. Purification by chromatography (hexane/EtOAc 1:1) and crystallization with hexane gave 3.26 g of 14 (64.9% yield); mp 64–65°C; $[\alpha]_D^{25}$ –1.4° (*c* 1.1, $CHCl_3$); ^{1}H-NMR (CD_3SOCD_3) δ (ppm) 6.20 (d, 1H, J = 8.5 Hz), 5.53 (dt, 1H, J = 15, 6.5 Hz), 5.38 (dd, 1H, J = 15, 6.5 Hz), 4.76 (d, 1H, J = 5 Hz), 4.40 (t, 1H, J = 5.5 Hz), 3.83 (td, 1H, J = 6.5, 5 Hz), 3.35–3.51 (m, 2H), 3.21–3.35 (m, 1H), 1.75–2.0 (m, 2H), 1.21–1.42 (m, 31H), 0.85 (t, 3H, J = 7.5 Hz).

2-(Trimethylsilyl)thiazole [110]. *Method A*. To a solution of *n*-butyllithium (51 mmol) in ether (50 mL) was added a solution of thiazole (4.25 g, 50 mmol) in ether (50 mL) over 30 min at –78°C. After the mixture had stirred for 30 min, a solution of trimethylsilyl chloride (5.45 g, 50 mmol) in ether (50 mL) was added. The reaction mixture was stirred for 1 h and then washed with saturated solution of $NaHCO_3$. The organic layer was dried over anhydrous $NaSO_4$. The solvent was removed and the product 2-TST was purified by vacuum distillation; yield, 7.3 g (93%); bp 58–60°C (16 torr).

Method B [110]. Similarly, 2-TST was prepared in 95% yield from a reaction mixture of bromothiazole (20 g, 122 mmol) in ether (200 mL), *n*-butyllithium (130 mmol), and trimethylsilyl chloride (13.3 g, 122 mmol) in ether (200 mL); IR (film) 2950 cm^{-1}; ^{1}H-NMR ($CDCl_3$) δ (ppm) 8.01 (d, 1H, J = 3 Hz), 7.40 (d, 1H, J = 3 Hz), 0.40 (s, 9H).

(1S)-2-Amino-2-N-tert-butoxycarbonyl)-2-deoxy-2,3-N,O-isopropylidene-1-(2-thiazolyl)-L-glycitol (17) [34,111]. To a stirred solution of aldehyde 11 (0.23 g, 1.0 mmol) in dry CH_2Cl_2 (5 mL) was added dropwise a solution of 2-TST (0.23 g, 1.5 mmol) in the same solvent (3 mL) at room temperature. The mixture was stirred for 20 h, the

solvent was removed, and tetra-*n*-butylammonium fluoride (1.5 mmol in THF) was added. After the mixture had been stirred for 1 h, the solvents were evaporated under vacuum, and a saturated solution of $NaHCO_3$ was added. Extractive workup gave a crude product that was purified by column chromatography on silica gel, eluting with petroleum ether/EtOAc (4:1); 0.27 g (85% yield); mp 168–171°C; $[\alpha]_D$ –48.3° (*c* 0.87, $CHCl_3$); IR (KBr) 3200, 1700 cm^{-1}; ^{1}H-NMR ($CDCl_3$-D_2O) δ (ppm) 7.76 (d, 1H, J = 3.2 Hz), 7.30 (d, 1H, J = 3.2 Hz), 5.22 (d, 1H, J = 2.8 Hz), 3.87–4.58 (m, 3H), 1.47 (s, 15H).

*(1S)-2-Amino-2-N-(tert-butoxycarbonyl)-2-deoxy-2,3-N,O-isopropylidene-*L*-glycitol* (18) [34]. To a solution of alcohol 17 (0.44 g, 1.4 mmol) in acetonitrile (30 mL) methyl iodide (0.9 mL, 14 mmol) was added. The reaction mixture was refluxed for 12 h and the solvent was removed. The resulting *N*-thiazolium iodide was dissolved in methanol (30 mL) and treated with sodium borohydride (2 equiv) at –10°C. After 30 min the reaction was quenched by the addition of acetone (2 mL). The solvents were removed and the residue was treated with a saturated $NaHCO_3$ solution. The product was extracted with ether and dried over Na_2SO_4. Removal of the solvent under vacuum gave crude thiazolidine, which was dissolved in acetonitrile (5 mL) and added to a solution of $HgCl_2$ (1.2 equiv) in 4:1 acetonitrile-water. Extractive workup with ether gave hydroxy-aldehyde 18 in 65% yield, which was used in the Wittig reaction without further purification. IR (film) 3340, 1690 cm^{-1}; ^{1}H-NMR ($CDCl_3$) δ (ppm) 9.62 (br s, 1H).

Isoxazoline hydrochloride [112]. To a solution of potassium hydroxide (26.71 g, 480 mmol) and hydroxyurethane (50.0 g, 480 mmol) in ethanol (210 mL) was added 1,3-dibromopropane (24 mL, 0.23 mol). The resulting suspension was refluxed for 1 h, and then additional potassium hydroxide (13.35 g, 240 mmol) and 1,3-dibromopropane (12 mL, 120 mmol) were added. After the mixture was refluxed for an additional 1 h, the solvent was removed and the residue was suspended in boiling ether and filtered. The white salts were digested a second time in hot ether and filtered. The combined filtrates were dried over $NaSO_4$, filtered, and evaporated. The residue was fractionally distilled to give *N*-(ethoxycarbonyl)isoxazolidine (43.14 g, 85% yield based on 1,3-dibromopropane). ^{1}H-NMR ($CDCl_3$) δ (ppm) 4.20 (q, 2H, J = 7.0 Hz), 3.92 (t, 2H, J = 7.0 Hz), 3.65 (dd, 2H, J = 6.8, 7.2 Hz), 2.26 (quin, 2H, J = 6.8 Hz), 1.30 (t, 3H, J = 7.0 Hz). A solution of *N*-(ethoxycarbonyl)isoxazolidine (17.48 g, 120 mmol) in aqueous HCl (5.2 M, 93 mL, 480 mmol) was refluxed for 2 h. After being cooled to 20°C, the solution was washed with ether (3 × 40 mL) and then evaporated, affording crude isoxazolidine hydrochloride. Recrystallization in ethanol-ether gave 6.25 g of isoxazolidine hydrochloride (47% yield); mp 123–125°C (lit. [113] 124–125°C); ^{1}H-NMR (CD_3SOCD_3) δ (ppm) 4.40 (br s, 2H) 4.19 (t, 2H, J = 7.0 Hz), 3.49 (t, 2H, J = 7.0 Hz), 2.40 (quin, 2H, J = 7 Hz).

*N-(Benzyloxycarbonyl)-*L*-serine isoxazolidide* (19) [36]. To a biphasic solution of isoxazolidine hydrochloride (3.67 g, 33.5 mmol) in THF (50 mL) and water (2 mL) was added anhydrous potassium carbonate (9.25 g, 66.9 mmol). The mixture was stirred for 3 h. In a separate flask was placed a solution of *N*-CBZ-L-serine (6.66 g, 27.9 mmol) in THF (200 mL), and *N*-methylmorpholine (3.06 mL, 27.9 mmol) and isobutyl chloroformate (3.62 mL, 27.9 mmol) were added at –15°C under nitrogen. After 1 min the isoxazoline solution was added rapidly, and the mixture was stirred for 45 min at –15°C. The reaction was quenched with water (30 mL), the solvents were removed, and a 5% aqueous citric acid solution (100 mL) was added. The product was extracted with ethyl acetate (2 × 150 mL). The combined extracts were washed with 5% citric acid solution and saturated $NaHCO_3$, dried, and concentrated. Crystallization of the residue from

EtOAc-isooctane gave serine oxazolidide 19 as white needles; 6.25 g (76% yield); mp 134–135°C; R_f 0.17 (EtOAc); $[\alpha]_D^{21}$ –4.89° (*c* 1.78, MeOH); IR (KBr) 1720, 1630 cm^{-1}; ^{1}H-NMR ($CDCl_3$) δ (ppm) 7.33 (s, 5H), 6.06 (br d, J = 8 Hz), 5.11 (s, 2H), 4.84 (m, 1H), 4.10 (m, 1H), 3.90 (m, 4H), 3.59 (m, 1H), 2.33 (m, 2H).

(2S,3R)-2-[(Benzyloxycarbonyl)amino]-1,3-dihydroxy-4-octadecyne (21a) *and (2S,-3S)* (21b) [36]. To a 1-pentadecyne solution (4.37 g, 21 mmol) in THF (150 mL) was added *n*-butyllithium (21 mmol of a solution in hexane). The mixture was stirred for 20 min at –23°C under nitrogen, and was then transferred to a solution of *N*-CBZ-L-serine isoxazolidide (19) (1.24 g, 4.2 mmol) in THF (50 mL) via cannula. After 45 min the reaction mixture was poured into 1 M NaH_2PO_4 solution, and the product was extracted with ethyl acetate. The combined extracts were washed with 1 M NaH_2PO_4 and brine solution, dried, and concentrated. The residue was dissolved in 2-propanol (100 mL) and sodium borohydride (0.48 g, 12.6 mmol) was added. After the mixture had been stirred for 1 h at 0°C, the solvents were removed and the residue was dissolved in methanol (100 mL) and 1 N HCl (20 mL) was added dropwise. The solvent was evaporated, the residue was dissolved in ethyl acetate (100 mL), and 1 N HCl (50 mL) was added. The product was extracted with ethyl acetate, the combined extracts were washed with 1 N HCl, saturated $NaHCO_3$ solution, and brine. The diols 21a,b were separated by sodium borate impregnated silica gel. Ynone 20 was isolated in partially racemic form by flash chromatography on silica gel, eluting with hexane/EtOAc (2 : 1); yield 89%; mp 48–50°C; R_f 0.33 (hexane/EtOAc 1 : 1); IR (KBr) 2220, 1675 cm^{-1}; ^{1}H-NMR ($CDCl_3$) δ (ppm) 7.37 (s, 5H), 5.84 (br d, 1H, J = 8 Hz), 5.13 (s, 2H), 4.49 (m, 1H), 4.13 (m, 2H), 2.38 (t, 2H, J = 7.2 Hz), 1.27 (m, 22H), 0.88 (t, 3H, J = 7.1 Hz).

Sodium borate impregnated silica gel purification of 21 [36]. Silica gel (150 g, 230–400 mesh) was treated with saturated $Na_2B_2O_7 \cdot H_2O$ (13.75 g) in water (250 mL). Most of the water was evaporated, and then the silica gel was dried in an oven at 110°C for 24 h. The treated silica gel was equilibrated with the atmosphere for a minimum of 72 h. A column was poured using a slurry of 150 g of the impregnated silica gel in 400 mL of chloroform. Unreacted pentadecyne was eluted with chloroform (100 mL), 21a was eluted with 3% 2-propanol in chloroform (750 mL), and 21b was eluted with 5% 2-propanol in chloroform.

21a: crystallized from isooctane (0.946 g, 53% yield); mp 81–83°C; R_f 0.42 (hexane-EtOAc 1 : 1), 0.31 ($CHCl_3$-*i*-PrOH 95 : 5); $[\alpha]_D^{21}$ –4.29° (*c* 1.7, $CHCl_3$); IR (KBr) 2210, 1690 cm^{-1}; ^{1}H-NMR ($CDCl_3$) δ (ppm) 7.36 (m, 5H), 5.57 (m, 1H), 5.13 (s, 2H), 4.63 (m, 1H), 4.14 (m, 1H), 3.83 (m, 2H), 2.69 (br d, 1H, J = 6 Hz), 2.21 (dt, 3H, J = 7.0, 2.0 Hz), 1.50 (m, 2H), 1.26 (m, 20H), 0.88 (t, 3H, J = 6.9 Hz).

21b: crystallized from isooctane (0.187 g, 10% yield); mp 67–69°C; R_f 0.42 (hexane-EtOAc 1 : 1), 0.31 ($CHCl_3$-*i*-PrOH 95 : 5); $[\alpha]_D^{21}$ –8.88° (*c* 3.2, $CHCl_3$); IR (KBr) 2205, 1680 cm^{-1}; ^{1}H-NMR ($CDCl_3$) δ (ppm) 7.36 (m, 5H), 5.36 (br m, 1H), 5.13 (s, 2H), 4.64 (m, 1H), 3.86 (m, 3H), 2.61 (br m, 1H), 2.26 (br m, 1H), 2.18 (dt, 2H, J = 7.0, 1.8 Hz), 1.49 (m, 2H), 1.25 (m, 20H), 0.88 (t, 3H, J = 6.9 Hz).

3,5-O-Isopropylidene-D-xylofuranose [57]. To a stirred solution of D-xylose (40) (4.0 g, 26.6 mmol) in DMF (50 mL) were added *p*-toluenesulfonic acid (60 mg) and 2,2-dimethoxypropane (10 mL). The mixture was stirred for 2 to 3 h at 40–45°C and then treated with Amberlite IRA-410 (OH^-) ion-exchange resin to remove the acid. The resin was removed by filtration and washed with methanol. After concentration the residue was purified by column chromatography on silica gel. The product was obtained by elution first with benzene, which gave 0.98 g of the diisopropylidene compound, then with 100 : 1

benzene-methanol, and then with 50:1 benzene-methanol, giving 1.6 g of isopropylidene-D-xylofuranose (32% yield); $[\alpha]_D^{20}$ + 19.2° (*c* 0.5, MeOH); IR (film) 3370, 840 cm^{-1}; ^{1}H-NMR ($CDCl_3$) δ (ppm) 5.67 (d, α-H, J = 3.8 Hz), 5.2 (s, β-H, J = ~0 Hz), 3.8–4.4 (m, 5H), 3.5 (s, 1H), 1.45 (s, 6H).

2,4-O-Isopropylidene-D-threose (42) [50]. To a solution of isopropylidene xylofuranose (5 g, 26 mmol, prepared as described above) in methanol (250 mL) was added sodium metaperiodate (7.4 g, 35 mmol), and the mixture was stirred for 3 h at room temperature. The resulting precipitate was removed by filtration, and the filtrate was concentrated. The residue was purified by column chromatography on silica gel to give a 1:1 mixture of threose 42 and its formate. The mixture was used in the Wittig reaction without further purification; $[\alpha]_D$ –50° (*c* 0.6, $CHCl_3$); IR (film) 3700–3100, 1720, 850 cm^{-1}; ^{1}H-NMR ($CDCl_3$) δ (ppm) 9.54 (s, 0.5H), 8.06 and 8.16 (2s, 1H), 1.49, 1.50, 1.52, 1.56 (4s, 6H).

4,6-O-Benzylidene-D-galactose [56]. To a mixture of benzaldehyde (30 mL, 295 mmol) and D-(+)-galactose (13 g, 72 mmol) was added fused zinc chloride (10 g, 73 mmol), and the suspension was stirred. After 3 h, more benzaldehyde (30 mL) was added and stirring was continued. After 21 h the mixture became a syrup, containing a small amount of undissolved zinc chloride. Water (30 mL) was added and the mixture slowly separated into two phases on standing at 5°C. The lower aqueous phase was separated and the upper organic phase was washed with water (2 × 30 mL). The combined aqueous layers were made alkaline with 10% sodium carbonate, forming a zinc carbonate precipitate, which was removed by filtration. The filtrate was extracted with petroleum ether (bp 40–60°C) and evaporated to dryness under vacuum. The solid residue was extracted with boiling ethyl acetate (2 × 250 mL). The combined extracts were concentrated to 200 mL, and the residue was crystallized. Recrystallization from ethanol gave pure 4,6-*O*-benzylidene-D-galactose; mp 190–191°C; $[\alpha]_D^{18}$ + 118.5°C (*c* 1.0, MeOH).

2,4-O-Benzylidene-D-threose (43) [52]. To a solution of 4,6-*O*-benzylidene-D-galactose (30.0 g, 111 mmol) in phosphate buffer (pH 7.6, 1.2 L) sodium metaperiodate (55 g, 257 mmol) was added at room temperature. The pH of the reaction mixture was adjusted between 7 and 8 by the addition of 2 N sodium hydroxide solution. After stirring for 1.5 h at room temperature the reaction mixture was concentrated to dryness. The product was extracted with THF, and the combined THF extracts were dried over $MgSO_4$ and concentrated under reduced pressure to give threose 43 (20 g, 85% yield), which was used in the Wittig reaction without any further purification; R_f 0.64 (toluene-EtOAc 3:1).

(2R,3R)-1,3-O-Isopropylidene-4-(E)-octadecene-1,2,3-triol (44a) [50]. To a THF solution (100 mL) of *n*-tetradecyltriphenylphosphonium bromide (18.8 g, 34.8 mmol), which was prepared [114] by refluxing a solution of 1-bromotetradecane and triphenylphosphine in xylene for 16 h, was added a phenyllithium solution (34.8 mmol, 17.4 mL of a 2 M solution in 7:3 cyclohexane-ether). After the mixture was stirred for 30 min at room temperature under nitrogen, the resulting clear solution was cooled to –60°C, and a solution of the threose 42 (4.7 g, 29.3 mmol) in THF (19 mL) was added. After 30 min at –30°C more phenyllithium solution (27.7 mL) was added to the cream-colored precipitate. The resulting dark red solution was kept for 40 min at room temperature and then poured into ice-cold water. The product was extracted with ether, and the extracts were washed with water, dried, and evaporated. The product was purified by chromatography on silica gel with alcohol-free chloroform to give *trans* isomer 44a (3.67 g, 40%) and *cis* isomer (3.21 g, 35%), which were recrystallized from aqueous ethanol.

trans isomer: mp 44.5–45.5°C; $[\alpha]_D$ –26° (*c* 0.7, $CHCl_3$); ^{1}H-NMR ($CDCl_3$) δ (ppm) 5.80 (m, 1H), 5.6 (m, 1H), 4.36 (d, 1H, J = 6.6 Hz), 4.07 (dd, 1H, J = 1.5 Hz), 3.84 (dd, 1H, J = 12.1, 1.8 Hz), 3.36 (s, 1H), 2.06 (q, 2H, J = 6.6 Hz), 1.49 (s, 3H), 1.46 (s, 3H), 1.15–1.5 (m, 22H), 0.88 (t, 3H).

cis isomer: mp 44.5–45.5°C; $[\alpha]_D$ –3° (*c* 0.4, $CHCl_3$); ^{1}H-NMR ($CDCl_3$) δ (ppm) 5.55–5.70 (m, 2H), 4.70 (d, 1H, J = 6.2 Hz), 4.09 (dd, 1H, J = 1.5 Hz), 3.84 (dd, 1H, J = 12.1, 1.8 Hz), 3.33 (s, 1H), 2.83 (br s, 1H), 2.0–2.2 (m, 2H), 1.52 (s, 3H), 1.45 (s, 3H). 1.15–1.5 (m, 22H), 0.88 (t, 3H).

(2R,3R)-1,3-O-Benzylidene-4-(E)-octadecene-1,2,3-triol (44b). *Method A*. [52]. To a suspended solution of *n*-tetradecyltriphenylphosphonium bromide (70.0 g, 130 mmol) in toluene (1 L) was added phenyllithium at –30°C. The phenyllithium was prepared from lithium metal (6.50 g, 940 mmol) and bromobenzene (74.0 g, 470 mmol) in ether (200 mL). Then a solution of threose 43 (21.6 g, 104 mmol) in THF (150 mL) was added to the reaction mixture over a 20-min period. After 20 min the reaction was quenched by the addition of methanol (150 mL) and water (250 mL). Extraction and purification by chromatography on silica gel [elution with petroleum ether-ether (9 : 1)] gave 27.0 g of product 44b (68% yield); mp 54–55°C; $[\alpha]_D^{22}$ –3.8° (*c* 0.5, $CHCl_3$); R_f 0.64 (petroleum ether-ether 9 : 1); ^{1}H-NMR ($CDCl_3$) δ (ppm) 7.52 and 7.38 (2m, 5H), 5.87 (m, 1H), 5.65 (m, 2H), 4.42 (d, 1H, J = 6.1 Hz), 4.25 (dd, 1H, J = 1.8 Hz), 4.07 (dd, 1H, J = 12, 1.2 Hz), 3.53 (dd, 1H, J = 10.4, 1.8 Hz), 2.62 (d, 1H, J = 10.4 Hz), 2.08 (m, 2H), 1.18–1.50 (m, 22H), 0.87 (t, 3H, J = 6.4 Hz).

Method B [54]. To a stirred solution of tetradecynyltriphenylphosphonium bromide (12.85 g, 23.8 mmol) in THF (30 mL) was added a solution of potassium *tert*-butoxide (4.69 mmol, 41.8 mmol) in THF (20 mL) at 0°C under argon atmosphere. After 20 min a solution of 2,4-benzylidene-D-threose 43 (3.67 g, 17.6 mmol) in THF (30 mL) was added at 0°C. The reaction mixture was warmed to room temperature, stirred overnight, and quenched with saturated ammonium chloride solution. The product was extracted with ethyl acetate and the combined extracts were washed with brine, dried over $MgSO_4$, and concentrated. The residue was treated with hexane, and filtered. The filtrate was concentrated. The residue was purified by chromatography on silica gel, eluting with benzene-EtOAc (97 : 3) to afford 4.25 g of 44b (62% yield); mp 48.0–50.3°C. The product was shown to be a 4 : 1 mixture of *cis* and *trans* isomers by GLC (3% OV-101); IR (neat) 3400, 1660, 1100, 760, 700 cm^{-1}; ^{1}H-NMR ($CDCl_3$) δ (ppm) 7.2–7.55 (m, 5H), 5.5–5.8 (m, 3H), 4.72 (dd, 1H, J = 7, 1.8 Hz), 4.24 (dd, 1H, J = 11.3, 2.1 Hz), 4.10 (dd, 1H, J = 11.3, 1.8 Hz), 3.42 (m, 1H), 2.0–2.3 (m, 2H), 1.9 (s, 1H), 1.0–1.5 (m, 22H), 0.88 (t, 1H, J = 6.9 Hz).

cis and trans isomerization of 44b [54]. A solution of the 4:1 *cis-trans* mixture 44b (1.20 g, 3.09 mmol) and diphenyldisulfide (0.346 g, 1.58 mmol) in a mixture of cyclohexane (120 mL) and dioxane (30 mL) was irradiated with a 100-W high-pressure mercury lamp for 3 h. After removal of the solvents, the residue was purified by chromatography on silica gel, eluting with benzene-EtOAc (97:3) to afford a 1 : 10 *cis-trans* mixture; mp 50.0–52.6°C; IR (neat) 3400, 1100, 960, 760 cm^{-1}; ^{1}H-NMR ($CDCl_3$) δ (ppm) 7.2–7.55 (m, 5H), 5.96 (s, 1H), 5.84 (dt, 1H, J = 15.1, 8.0 Hz), 5.53 (dd, 1H, J = 15.1, 7.0 Hz), 4.40 (dd, 1H, J = 7.0, 1.8 Hz), 4.23 (dd, 1H, J = 11.3, 2.1 Hz), 4.06 (dd, 1H, J = 11.3, 1.8 Hz), 3.42 (m, 1H), 2.64 (br s, 1H), 1.96–2.2 (m, 2H), 1.0–1.5 (m, 22H), 0.88 (t, 3H, J = 6.9 Hz).

(2S,3R)-2-Azido-1,3-O-benzylidene-4-(E)-octadecene-1,3-diol (45). *Method A* [52]. To a mixture of 44b (10.0 g, 25.0 mmol) and pyridine (5 mL) in dry methylene chloride

(70 mL) trifluoromethanesulfonic anhydride (8.70 g, 31.0 mmol) was added slowly at –51°C. After the mixture had stirred for 15 min, DMF (250 mL) and sodium azide (7.50 g, 100 mmol) were added. The reaction mixture was stirred for 5 h at room temperature and poured into water. The product was extracted with petroleum ether, concentrated, and purified by chromatography on silica gel, eluting with petroleum ether-EtOAc (9:1) to give 8.5 g (82% yield) as a colorless oil; $[\alpha]_D^{20}$ –11.7° (*c* 3, $CHCl_3$); R_f 0.80 (petroleum ether-ether 9:1); ^{1}H-NMR ($CDCl_3$) δ (ppm) 7.48 and 7.37 (m, 5H), 5.97 (m, 1H), 5.58 (dd, 1H, J = 15.6, 7.3 Hz), 5.49 (s, 1H), 4.33 (dd, 1H, J = 11, 4.8 Hz), 4.05 (dd, 1H, J = 9.1, 7.3 Hz), 3.61 (dd, 1H, J = 11, 11 Hz), 3.48 (ddd, 1H, J = 11, 9.1, 4.8 Hz), 2.11 (m, 2H), 1.10–1.50 (m, 22H), 0.87 (t, 3H, J = 6.4 Hz).

Method B [54]. To a solution of 44b (1.05 g, 2.69 mmol) and triethylamine (1.5 mL, 10.8 mmol) in methylene chloride (13 mL) was added methanesulfonyl chloride (0.45 mL, 5.50 mmol) dropwise at 0°C. The reaction mixture was warmed to room temperature and stirring was continued for 1.5 h. The reaction mixture was poured into saturated $NaHCO_3$ solution, and the product was extracted with methylene chloride. The combined organic phase was washed with brine and concentrated. The residue was dissolved in ether, washed with brine, dried over $MgSO_4$, and evaporated to give the mesylate as a pale yellow solid (1.25 g, 99.4% yield); mp 81.5–83.0°C; IR (Nujol) 1170, 770, 710 cm^{-1}; ^{1}H-NMR ($CDCl_3$) δ (ppm) 7.2–7.9 (m, 5H), 5.96 (s, 1H), 5.88 (dt, 1H, J = 15.1, 8.0 Hz), 5.48 (dd, 1H, J = 15.1, 7.0 Hz), 4.57 (dd, 1H, J = 7, 1.8 Hz), 4.36 (dd, 1H, J = 11.3, 2.1 Hz), 4.05 (dd, 1H, J = 11.3, 1.8 Hz), 3.7–3.8 (m, 1H), 3.09 (s, 3H), 1.2–1.5 (m, 22H), 0.88 (t, 3H, J = 6.9 Hz).

To a suspension of sodium azide (1.01 g, 15.5 mmol) in dry DMF (2 mL) was added a solution of the crude mesylate (1.94 g, 4.01 mmol) in DMF (3 mL). The reaction mixture was heated at 90°C for 3 h under argon atmosphere. After removal of solvent under vacuum, the residue was partitioned between ice-water and methylene chloride. The organic layer was washed with water and brine, dried over $MgSO_4$, and concentrated. The product was purified by chromatography on silica gel, eluting with hexane-benzene (3:1) to give 0.953 g of the azide 45 (57% yield) as a pale yellow oil. IR (neat) 2100, 1090, 780, 710 cm^{-1}; ^{1}H-NMR ($CDCl_3$) δ (ppm) 7.2–7.5 (m, 5H), 5.96 (s, 1H), 5.76 (dt, 1H, J = 15.1, 8.0 Hz), 5.53 (dd, 1H, J = 15.1, 7.0 Hz), 4.1–4.42 (dt, 1H, J = 11.0, 6.9 Hz), 3.82–4.03 (ddd, 1H, J = 11.0, 6.2, 4.2 Hz), 3.52 (dd, 1H, J = 11.3, 4.2 Hz), 3.40 (dd, 1H, J = 11.3, 6.2 Hz), 1.95–2.14 (m, 2H), 1.0–1.5 (m, 22H), 0.88 (t, 3H, J = 6.9 Hz).

*(2S,3R)-1,3-O-Benzylidene-*D*-sphingosine* [54]. A mixture of 45 (0.438 g, 1.06 mL) and triphenylphosphine (0.278 g, 1.06 mmol) in THF (3.5 mL) and water (0.5 mL) was stirred overnight at room temperature. The mixture was diluted with methylene chloride and the layers were separated. The organic phase was dried and evaporated. The residue was extracted with hexane. The combined extracts were evaporated, leaving the benzylidene-sphingosine as a colorless oil (0.408 g, 99.5% yield). IR (neat) 3400, 1590, 1090, 690 cm^{-1}; ^{1}H-NMR ($CDCl_3$) δ (ppm) 7.1–7.5 (m, 5H), 5.96 (s, 1H), 5.86 (dt, 1H, J = 15.1, 8.0 Hz), 5.53 (dd, 1H, J = 15.1, 7.0 Hz), 4.24 (dd, 1H, J = 11.0, 6.9 Hz), 3.80 (ddd, 1H, J = 11.0, 6.2, 4.2 Hz), 3.0 (dd, 1H, J = 11.3, 4.2 Hz), 2.88 (dd, 1H, J = 11.3, 6.2 Hz), 2.15 (m, 2H), 1.95–2.3 (m, 2H), 1.0–1.5 (m, 22H), 0.88 (t, 3H, J = 6.9 Hz).

Phenyl n-tetradecyl sulfide [62]. A mixture of n-tetradecyl bromide (2.77 g, 10 mmol), thiophenol (1.03, 10 mmol), triethylamine (1.53 mL, 11 mmol), and tetra-*n*-butylammonium iodide (40 mg) in THF (50 mL) was stirred for 6 h. After concentration, the residue was taken up in ether, filtered, and concentrated to give a white solid, which was recrystalized from EtOAc-MeOH to yield colorless flat needles; 2.75 g (90% yield); mp 41°C; R_f 0.56 (hexane); IR (film) 2920, 2850, 1460, 1440, 730, 685 cm^{-1}; ^{1}H-NMR ($CDCl_3$) δ (ppm) 7.1–7.4 (m, 5H), 2.91 (t, 2H, J = 7 Hz), 1.64 (p, 2H, J = 7 Hz), 1.2–1.5 (m, 22H), 0.88 (t, 3H, J = 7 Hz).

Phenyl n-tetradecyl sulfone [62]. To a solution of phenyl *n*-tetradecyl sulfide (4.39 g, 14.3 mmol) in methylene chloride (50 mL) was added dropwise a solution of *m*-chloroperbenzoic acid (6.18 g, 35.8 mmol) in methylene chloride (70 mL). After 1 h, the solution was filtered, washed with 10% aqueous Na_2SO_3 solution (15 mL) and saturated $NaHCO_3$ solution, dried over Na_2SO_4, filtered, and evaporated to give 4.86 g (100% yield) of phenyl *n*-tetradecyl sulfone as a white solid. Recrystallization from methanol (4°C) yielded fine colorless needles; 92% yield; mp 43–44°C; IR 3060, 2920, 2850, 1770, 1450, 1310, 1215, 1150, 1090, 725, 690, 600, 570 cm^{-1}; ^{1}H-NMR ($CDCl_3$) δ (ppm) 7.85–7.95 (m, 2H), 7.5–7.7 (m, 3H), 3.07 (m, 2H), 1.70 (m, 2H), 1.1–1.4 (m, 22H), 0.88 (t, 3H, J = 7 Hz).

(4S,1'S)-3-Benzyl-4-[2-acetoxy-1-(benzyloxy)-3-(phenylsulfonyl)hexadecyl]-1,3-oxazolidin-2-one (69) [62]. To a stirred solution of phenyl *n*-tetradecyl sulfone (67.7 mg, 0.2 mmol) in THF (4 mL) *n*-butyllithium (2.5 M in hexane, 20 μL) was added slowly under nitrogen until a yellow color persisted at –70°C. Additional *n*-butyllithium (80 μL, 0.2 mmol) was added dropwise, resulting in a clear solution. Aldehyde 68 (30 mg, 0.09 mmol) in THF (0.5 mL) was added dropwise to the cooled sulfone solution. After 30 min, acetic anhydride (75 μL, 0.8 mmol) was added, and the reaction mixture was stirred for 1 h at –70°C. After the reaction was quenched by the addition of saturated NH_4Cl solution (0.5 mL), the product was extracted with ether. The combined organic layer was dried over $MgSO_4$, filtered, and concentrated to a colorless oil. The oil was purified by chromatography on silica gel, eluting with EtOAc-hexane (1:2.5), to afford 39.1 mg (55% yield) of 69; R_f 0.28; for the intermediate hydroxy sulfone, R_f 0.11 (EtOAc-hexane 1:3). ^{1}H-NMR ($CDCl_3$) δ (ppm) 7.1–7.5 (m, 15H), 2.8–5.3 (m, 10H), 1.86–1.89 (m, 3H), 1.0–1.5 (m, 24H), 0.88 (m, 3H).

(4S,1'R,2'E)-3-Benzyl-4-[1-(benzyloxy)hexadec-2-enyl]-1,3-oxazolidin-2-one (70) [62]. A solution of sulfone acetate 76 (627 mg, 0.85 mmol) in MeOH-EtOAc (2:1, 18 mL) was stirred with 2% Na-Hg (10 g) for 6 h. The solution was decanted from the amalgam and extracted with 5 mL of 50% saturated NH_4Cl solution. The aqueous phase was extracted with ether (50 mL). The combined organic layer was dried over $MgSO_4$, concentrated, and chromatographed on silica gel, eluting with hexane-EtOAc (5:1) to give 70 as a colorless oil (321 mg, 51% yield); R_f 0.66 (hexane-EtOAc, 2:1); $[\alpha]_D^{22}$ –48.2° (c 1.2, $CHCl_3$); IR (film) 2920, 2855, 1755, 1420, 1205, 1065 cm^{-1}; ^{1}H-NMR ($CDCl_3$) δ (ppm) 7.1–7.4 (m, 10H), 5.75 (dt, 1H, J = 15.4, 7 Hz), 5.24 (dd, 1H, J = 15.4, 8 Hz), 4.76 (d, 1H, J = 15.1 Hz), 4.55 (d, 1H, J = 11.9 Hz), 4.21 (d, 1H, J = 11.9 Hz), 4.17–4.24 (m, 2H), 3.95 (d, 1H, J = 15.1 Hz), 3.79 (dd, 1H, J = 8, 3 Hz), 3.65 (ddd, 1H, J = 9, 6, 3 Hz), 2.09 (q, 2H, J = 7 Hz), 1.15–1.4 (m, 22H), 0.88 (t, 3H, J = 7 Hz).

(4R,5S)-4-(Carbomethoxy)-5-[1'-(E)-pentadecenyl]-Δ^2-oxazoline (79) [66,69]. Methyl α-isocyanoacetate (77) (0.20 g, 5.0 mmol) was added to a mixture of bis(cyclohexyl isocyanide)gold(I) tetrafluoroborate (25.1 mg, 50 μmol), (*S*)-*N*-methyl-*N*-[2-(morpholino)ethyl]-1-[(*R*)-1',2-bis(diphenylphosphino)ferrocenyl]ethylamine (76) (37.3 mg, 51 μmol), and (*E*)-2-hexadecenal (1.19 g, 5.0 mmol) in dry methylene chloride (5 mL). The mixture was stirred for 35 h at 25°C. The reaction mixture was filtered through a

short column of Florisil (elution with CH_2Cl_2) to give a quantitative yield of the crude product, which was isolated by silica gel MPLC (hexane-EtOAc 1:1) in 89% yield; $[\alpha]_D^{20}$ –173° (*c* 1.2, THF); ^{1}H-NMR ($CDCl_3$) δ (ppm) 6.79 (d, 1H, J = 2 Hz), 5.78 (dt, 1H, J = 15, 7 Hz), 5.38 (dd, 1H, J = 15, 7 Hz), 5.00 (t, 1H, J = 7 Hz), 4.30 (dd, 1H, J = 7, 2 Hz), 3.72 (s, 3H), 1.8–2.3 (m, 2H), 1.0–1.8 (br s, 22H), 0.7–1.0 (m, 3H).

(4S,2'S,3'R)-3-[2'-Bromo-3'-hydroxy-4'-(E)-octadecenoyl]-4-isopropyl-2-oxazolidinone (82) [75]. To a stirred solution of (4*S*)-3-bromoacetyl-4-isopropyl-2-oxazolidinone (11.0 g, 44 mmol) in ether (150 mL) was added triethylamine (8.6 mL, 61.7 mmol) at –78°C. After 5 min, 11.3 mL (52.4 mmol) of freshly prepared di-*n*-butylboron triflate [115] was added. After the reaction mixture was stirred for 15 min at –78°C, allowed to warm to room temperature, and stirred for an additional 2 h, the reaction mixture was cooled to –78°C. A solution of (*E*)-2-hexadecenal (6.66 g, 27.9 mmol) in ether (130 mL) was added, and the mixture was stirred for 45 min at –78°C and at 0°C for 1.5 h. The reaction mixture was diluted with ether (250 mL) and washed with 1 N $NaHSO_4$ solution and brine. The ether layer was concentrated under vacuum, giving a residue that was dissolved in ether (150 mL). To the ether solution was added a 1:1 solution of MeOH: 30% H_2O_2 (150 mL). After the solution had stirred for 1 h at 0°C, the product was extracted with ether. The combined ether layer was washed with saturated $NaHCO_3$ solution and brine, dried over $MgSO_4$, and concentrated. The product was purified by chromatography on silica gel, eluting with petroleum ether-EtOAc (20:3) to afford pure 82 (10.21 g, 72% yield).

(2R,3R)-2,3-Epoxyoctadec-4-yn-1-ol (90) [80]. To a solution of freshly distilled Ti(O*t*-Bu)$_4$ (41.1 mL, 107.6 mmol) in dry methylene chloride (100 mL) was added (–)-DET (35 mL, 3.14 M in dry CH_2Cl_2, 110 mmol) during 15 min at –25°C. After 15 min a solution of alcohol 89 (15.0 g, 56.7 mmol) in methylene chloride (200 mL) was added at such a rate that the mixture remained homogeneous, and then *tert*-butyl hydroperoxide (34 mL, 3.79 M in toluene, 129 mmol) was added. After stirring for 4–5 h at –30°C the reaction mixture was quenched by the addition of 10% aqueous DL-tartaric acid (500 mL). The mixture was warmed to room temperature and the product was extracted with ether. The combined extracts were washed with 10% DL-tartaric acid solution and brine. The organic layer was dried over $MgSO_4$, concentrated, and dried under high vacuum to afford a yellow oil, which was purified by column chromatography on silica gel, eluting with hexane-EtOAc (4:1) to give recovered 89 (1.05 g, 6.6%) and 90 (12.62 g, 86% based on recovered starting material (89); ≥98% enantiomeric excess (ee) determined by HPLC of the Mosher ester. Two recrystallizations from hexane gave pure 90 (100% ee); mp 55–56°C; R_f 0.38 (hexane-EtOAc 2:1); $[\alpha]_D^{25}$ –2.0° (*c* 2.05, $CHCl_3$); $[\alpha]_{365}^{22}$ –41.5° (*c* 2.05, $CHCl_3$); IR (KBr) 3300, 3180, 3000, 2960, 2920, 2850, 2240, 1460, 1320, 1070, 1030, 875, 725 cm^{-1}; ^{1}H-NMR ($CDCl_3$) δ (ppm) 3.94 (ddd, 1H, J = 12.9, 4.9, 2.2 Hz; with D_2O: dd, J = 12.9, 2.2 Hz), 3.70 (ddd, 1H, J = 12.9, 7.9, 3.4 Hz; with D_2O: dd, J = 12.9, 3.4 (Hz), 3.43 (q, 1H, J = 1.7 Hz), 3.27 (ddd, 1H, J = 3.4, 2.2, 1.7 Hz), 2.20 (td, 2H, J = 7.0, 1.7 Hz) 1.55 (m, 1H), 1.26 (m, 22H), 0.88 (t, 3H, J = 6.7 Hz).

(4S,1'R)-3-Benzyl-4-(1'-hydroxyhexadec-2'-ynyl)-1,3-oxazolidin-2-one (91) [80]. To a solution of the epoxy alcohol 90 (10.0 g, 35.66 mmol) and benzyl isocyanate (5.70 g, 42.79 mmol) in dry THF (175 mL) was added sodium hydride (2.14 g, 89.15 mmol; a commercial sodium hydride suspension in mineral oil was washed with dry hexane and dried) under nitrogen. After 1 h at room temperature, the mixture was heated to 60°C for

3 h. Excess sodium hydride was destroyed by careful addition of acetic acid at 5°C. The reaction mixture was diluted with ether (300 mL) and washed with water (2 × 80 mL), saturated aqueous $NaHCO_3$ solution (1 × 80 mL), and brine (1 × 80 mL). The organic layer was dried over $MgSO_4$ and concentrated under vacuum to give crude product 91 (17.7 g), which was purified by chromatography (hexane-EtOAc, first with 4:1, then with 2:1) followed by recrystallization from hexane to afford pure product 91 (12.84 g, 87% yield); mp 51–52°C; R_f 0.21 (hexane-EtOAc 2:1); $[\alpha]_D^{25}$ –28.9° (*c* 1.0, $CHCl_3$); IR ($CHCl_3$) 3610, 3400, 3000, 2980, 2930, 2850, 2230, 1745, 1605, 1420, 1380, 1355, 1220, 1135, 1110, 1095, 1070, 1030, 970 cm^{-1}; ^{1}H-NMR (400 MHz, $CDCl_3$) δ (ppm) 7.34 (m, 5H), 4.74 (d, 1H, J = 15.3 Hz), 4.45 (ddt, 1H, J = 4.0, 3.1, 1.9 Hz; with D_2O: dd, J = 3.1, 1.9 Hz), 4.40 (dd, 1H, J = 9.2, 5.3 Hz), 4.35 (d, 1H, J = 15.3 Hz), 4.29 (t, 1H, J = 9.1 Hz), 3.74 (ddd, 1H, J = 9.1, 5.3, 3.1 Hz), 2.16 (dt, 1H, J = 7.2, 1.9 Hz), 1.89 (d, 1H, exchangeable with D_2O, J = 4.0 Hz), 1.47 (quin, 2H, J = 7.2 Hz), 1.25 (m, 20H), 0.88 (t, 3H, J = 6.8 Hz).

*N-Stearoyl-*D*-erythro-sphingomyelin* (110) and *threo* isomer [23]. To a solution of D-*erythro*-3-*O*-*tert*-butyldiphenylsilyl-*N*-stearoylsphingosine (109) (1.104 g, 1.37 mmol) and triethylamine (370 μL, 2.7 mmol) in chloroform (5 mL) was added chloro-*N*,*N*-diisopropylaminomethoxyphosphine (324 mg, 1.64 mmol) at room temperature. After 0.5 h the mixture was concentrated to dryness under vacuum, tetrazole (350 mg, 4.92 mmol) and dry choline tosylate (1.01 g, 4.0 mmol) were placed in the reaction flask. Ten milliliters of acetonitrile-THF (1:1) were added, and the mixture was stirred for 3 h at room temperature. The solvents were removed under vacuum, and the residue was dissolved in THF and added to *tert*-butyl hydroperoxide (170 mg, 80% solution in *t*-BuOH). After 2 h the reaction mixture was diluted with ethyl acetate (10 mL) and washed with triethylammonium hydrogen carbonate buffer (1 M, pH 7.5). The organic phase was concentrated and the residue was dried by evaporation with dry toluene. The residue was dissolved in toluene (10 mL) and treated with anhydrous trimethylamine (3 mL) for 12 h at room temperature. Finally, the crude product was dissolved in THF (5 mL) and treated with tetra-*n*-butylammonium fluoride (520 mg, 2 mmol) for 48 h at room temperature. The crude product was precipitated with acetone and then purified by column chromatography [elution with chloroform-methanol-water (65:35:4)]. There was obtained 700 mg of sphingomyelin (110); 70% yield from ceramide 109; R_f 0.2 (chloroform-methanol-water 65:35:4); ^{1}H-NMR (500 MHz, CD_3OD) δ (ppm) 5.702 (dtd, 1H), 5.449 (ddt, 1H), 4.27 (m, 2H), 4.106 (ddd, 1H), 4.046 (t, 1H), 3.969 (ddd, 1H) 3.935 (m, 1H), 3.625 (m, 2H), 3.21 (br s, 9H), 2.180 (m, 2H), 2.027 (br q, 2H), 1.58 (br, m, 2H) 1.378 (br m, 2H) 1.288 (br s, 48H), 0.898 (t, 6H). The L-*threo* isomer of 110 was synthesized from L-*threo*-3-*O*-*tert*-butyldiphenylsilyl-*N*-stearoylsphingosine (400 mg, 0.5 mmol) in a similar fashion; yield, 260 mg (71%); R_f 0.2 (chloroform-methanol-water 65:35:4); ^{1}H-NMR (500 MHz, CD_3OD) δ (ppm) 5.726 (dtd, 1H), 5.452 (ddt, 1H), 4.333 (ddd, 1H), 4.272 (m, 2H), 4.052 (dt, 1H), 3.991 (dt, 1H) 3.843 (dt, 1H), 3.640 (m, 2H), 3.23 (br s, 9H), 2.22 (m, 2H), 2.033 (dt, 2H), 1.599 (br, m, 2H) 1.376 (br m, 2H) 1.288 (br s, 46H), 0.898 (t, 6H).

ACKNOWLEDGMENT

The work by the authors was supported by Grant HL-16660 (to R. Bittman) from the National Institutes of Health.

C.A. Registry Numbers

D-*erythro*-sphingosine (1a), 123-78-4; L-*erythro*-sphingosine (1d), 6036-75-5; D-*threo*-sphingosine (1b), 6036-85-7; L-*threo*-sphingosine (1c), 25695-95-8; *N*-stearoyl-D-*erythro*-sphingosine, 2304-81-6; *N*-stearoyl-L-*threo*-sphingosine, 95037-06-2; *N*-stearoyl-D-*erythro*-sphingomyelin (110).

ABBREVIATIONS

Ac	acetyl
AIBN	2,2'-azobisisobutyronitrile
BOC	*tert*-butoxycarbonyl
Bn	benzyl
CBZ	carbobenzyloxy
DABCO	1,4-diazabicyclo[2.2.2]octane
DBU	1,8-diazabicyclo[5.4.0]undec-7-ene
DCC	*N,N'*-dicyclohexylcarbodiimide
DET	diethyl tartrate
DHP	dihydropyran
DMAP	4-(dimethylamino)pyridine
DME	1,2-dimethoxyethane
DMF	*N,N*-dimethylformamide
DMSO	dimethyl sulfoxide
DIBAL	diisobutylaluminum hydride
HMPA	hexamethylphosphoramide
HPLC	high-pressure liquid chromatography
MOM	methoxymethyl
Ms	methanesulfonyl
m-MTr	*m*-methoxytrityl
PCC	pyridinium chlorochromate
Phth	phthalimide
PPTS	pyridinium *p*-toluenesulfonate
Py	pyridine
Red-Al	sodium bis(2-methoxyethyl)aluminum hydride
SEM	trimethylsilylethoxymethyl
TBDMS	*tert*-butyldimethylsilyl
TBDPS	*tert*-butyldiphenylsilyl
Tf	trifluoromethanesulfonyl
THF	tetrahydrofuran
THP	tetrahydropyranyl
TMS	trimethylsilyl
TLC	thin-layer chromatography
p-Ts	*p*-toluenesulfonyl
2-TST	2-(trimethylsilyl)thiazole

REFERENCES

1. K.-A. Karlsson, *Chem. Phys. Lipids 5*:6 (1970).
2. A. H. Merrill, Jr., A. M. Sereni, V. L. Stevens, Y. A. Hannun, R. M. Bell, and J. M. Kinkade, Jr., *J. Biol. Chem. 261*:12610 (1986).

3. E. Wilson, E. Wang, R. E. Mullins, D. C. Liotta, J. D. Lambeth, and A. H. Merrill, Jr., *J. Biol. Chem. 263*:9304 (1988).
4. A. H. Merrill, Jr., E. Wang, and R. E. Mullins, *Biochemistry 27*:340 (1988).
5. A. H. Merrill, Jr., and V. L. Stevens, *Biochim. Biophys. Acta 1010*:131 (1989).
6. Y. A. Hannun, and R. M. Bell, *Science 243*:500 (1989).
7. A. H. Merrill, Jr., and D. D. Jones, *Biochim. Biophys. Acta 1044*:1 (1990).
8. D. Shapiro, and K. H. Segal, *J. Am. Chem. Soc. 76*:5894 (1954).
9. D. Shapiro, H. Segal, and H. M. Flowers, *J. Am. Chem. Soc. 80*:1194 (1985).
10. D. Shapiro, and H. M. Flowers, *J. Am. Chem. Soc. 84*:1047 (1962).
11. E. Baer, and G. R. Sarma, *Can. J. Biochem. 47*:603 (1969).
12. A. F. Rosenthal, in *Methods in Enzymology*, Vol. 35, Academic Press, New York, 1975, pp. 429–529.
13. Y. Shoyama, H. Okabe, Y. Kishimoto, and C. Costello, *J. Lipid Res. 19*:250 (1978).
14. C. A. Grob, and F. Gadient, *Helv. Chim. Acta 40*:1145 (1957).
15. K. Mori, and Y. Funaki, *Tetrahedron 41*:2369 (1985).
16. T. Hino, K. Nakakyama, M. Taniguchi, and M. Nakagawa, *J. Chem. Soc., Perkin Trans. 1*:1687 (1986).
17. W. Schwab, and V. Jäger, *Angew. Chem. Int. Ed. Engl. 20*:603 (1981).
18. R. R. Schmidt, and R. Kläger, *Angew. Chem. Int. Ed. Engl. 21*:210 (1982).
19. R. S. Garigipati, A. J. Freyer, R. R. Whittle, S. M. Weinreb, *J. Am. Chem. Soc. 106*:7861 (1984).
20. G. Cardillo, M. Orena, S. Sandri, and C. Tomasini, *Tetrahedron 42*:917 (1986).
21. H. Newman, *J. Am. Chem. Soc. 95*:4098 (1973).
22. P. Tkaczuk, and E. R. Thornton, *J. Org. Chem. 46*:4393 (1981).
23. K. S. Bruzik, *J. Chem. Soc. Perkin Trans. 1*:423 (1988).
24. K. S. Bruzik, *J. Chem. Soc., Chem. Commun.* 329 (1986).
25. K. Mori, and Y. Funaki, *Tetrahedron 41*:2379 (1985).
25a. A. P. Kozikowski, and J.-P. Wu, *Tetrahedron Lett. 31*:4309 (1990).
26. P. Garner, J. M. Park, and E. Malecki, *J. Org. Chem. 53*:4395 (1988).
27. P. Garner, and S. Ramakanth, *J. Org. Chem. 51*:2609 (1986).
28. P. Herold, *Helv. Chim. Acta 71*:354 (1988).
29. S. Nimkar, D. Menaldino, A. H. Merrill, and D. Liotta, *Tetrahedron Lett. 29*:3037 (1988).
30. H.-E. Radunz, R. M. Devant, and V. Eiermann, *Justus Liebigs Ann. Chem.* 1103 (1988).
31. P. Garner, and J. M. Park, *J. Org. Chem. 52*:2361 (1987); P. Garner, and J. M. Park, *Org. Synth. 70*:18 (1991).
32. M. Chérest, H. Felkin, and N. Prudent, *Tetrahedron Lett.* 2199 (1968).
33. N. T. Anh, *Top. Curr. Chem. 88*:144 (1980).
34. A. Dondoni, G. Fantin, M. Fogagnolo, and P. Pedrini, *J. Org. Chem. 55*:1439 (1990).
35. A. Dondoni, G. Fantin, M. Fogagnolo, and A. Medici, *J. Chem. Soc., Chem. Commun.* 10 (1988).
36. R. H. Boutin, and H. Rapoport, *J. Org. Chem. 51*:5320 (1986).
37. G. J. Karabatsos, *J. Am. Chem. Soc. 89*:1367 (1967).
38. D. A. Evans, and K. T. Chapman, *Tetrahedron Lett. 27*:5939 (1986).
39. Y. H. Lee, H. S. Byun, and R. Bittman, unpublished results.
40. E. J. Reist, and P. H. Christie, *J. Org. Chem. 35*:4127 (1970).
41. M. Schlosser, *Angew. Chem. Int. Ed. Engl. 7*:650 (1968).
42. T. Sugawara, and M. Narisada, *Carbohydr. Res. 194*:125 (1989).
43. S. Kusumoto, S. Imaoka, Y. Kambayashi, and T. Shiba, *Tetrahedron Lett. 23*:2961 (1982).
44. S. Hanessian, R. Masse, and T. Nakagawa, *Can. J. Chem. 56*:1509 (1978).
45. K. Koike, M. Numata, M. Sugimoto, Y. Nakahara, and T. Ogawa, *Carbohydr. Res 158*:113 (1986).

46. K. Koike, Y. Nakahara, and T. Ogawa, *Glycoconjugate J. 1*:107 (1984); *Chem. Abstr. 103*:104777v (1985).
47. K. H. Schulte-Elte, and G. Ohloff, *Helv. Chim. Acta 51*:548 (1968).
48. J. Rokach, R. N. Young, M. Kakushima, C.-K. Lau, R. Seguin, R. Frenette, and Y. Guindon, *Tetrahedron Lett. 22*:979 (1981).
49. E. J. Corey, K. C. Nicolaou, R. D. Balanson, and Y. Machida, *Synthesis* 590 (1975).
50. M. Kiso, A. Nakamura, Y. Tomita, and A. Hasegawa, *Carbohydr. Res. 158*:101 (1986).
51. M. Kiso, A. Nakamura, J. Nakamura, Y. Tomita, and A. Hasegawa, *J. Carbohydr. Chem. 5*:335 (1986).
52. P. Zimmermann, and R. R. Schmidt, *Justus Liebigs Ann. Chem.* 663 (1988).
53. R. R. Schmidt, and P. Zimmermann, *Tetrahedron Lett. 27*:481 (1986).
54. K. Ohashi, S. Kosai, M. Arizuka, T. Watanabe, Y. Yamagiwa, T. Kamikawa, and M. Kates, *Tetrahedron 45*:2557 (1989).
55. K. Ohashi, Y. Yamagiwa, T. Kamikawa, and M. Kates, *Tetrahedron Lett. 29*:1185 (1988).
56. E. G. Gros, and V. Deulofeu, *J. Org. Chem. 29*:3647 (1964).
57. M. Kiso, and A. Hasegawa, *Carbohydr. Res. 52*:95 (1976).
58. M. Schlosser, H. B. Tuong, and B. Schaub, *Tetrahedron Lett. 26*:311 (1985).
59. Y. Ali, and A. C. Richardson, *Carbohydr. Res. 5*:441 (1967).
60. M. Obayashi, and M. Schlosser, *Chem. Lett.* 1715 (1985).
61. K. L. Bhat, S.-Y. Chen, and M. M. Joullié, *Heterocycles 23*:691 (1985).
62. M. A. Findeis, and G. M. Whitesides, *J. Org. Chem. 52*:2838 (1987).
63. T. Yamanoi, T. Akiyama, E. Ishida, H. Abe, M. Amemiya, and T. Inazu, *Chem. Lett.* 335 (1989).
64. J. R. Falck, S. Manna, J. Viala, A. K. Siddhanta, C. A. Moustakis, and J. Capdevila, *Tetrahedron Lett. 26*:2287 (1985).
65. Y. Ito, M. Sawamura, and T. Hayashi, *J. Am. Chem. Soc. 108*:6405 (1986).
66. Y. Ito, M. Sawamura, E. Shirakawa, K. Hayashizaki, and T. Hayashi, *Tetrahedron Lett. 29*:235 (1988).
67. Y. Ito, M. Sawamura, M. Kobayashi, and T. Hayashi, *Tetrahedron Lett. 29*:6321 (1988).
68. Y. Ito, M. Sawamura, H. Hamashima, T. Emura, and T. Hayashi, *Tetrahedron Lett. 30*:4681 (1989).
69. Y. Ito, M. Sawamura, and T. Hayashi, *Tetrahedron Lett. 29*:239 (1988).
70. D. A. Evans, J. V. Nelson, E. Vogel, and T. R. Taber, *J. Am. Chem. Soc. 103*:3099 (1981).
71. D. A. Evans, E. B. Sjogren, A. E. Weber, and R. E. Conn, *Tetrahedron Lett. 28*:39 (1987).
72. D. A. Evans, and A. E. Weber, *J. Am. Chem. Soc. 109*:7151 (1987).
73. A. Abdel-Magid, L. N. Pridgen, D. S. Eggleston, and I. Lantos, *J. Am. Chem. Soc. 108*:4595 (1986).
74. A. Abdel-Magid, I. Lantos, and L. N. Prigden, *Tetrahedron Lett. 25*:3273 (1984).
75. K. C. Nicolaou, T. Caulfield, H. Kataoka, and T. Kumazawa, *J. Am. Chem. Soc. 110*:7910 (1988); Prof. K. C. Nicolaou, personal communication.
76a. H. Bayley, D. N. Standring, and J. R. Knowles, *Tetrahedron Lett.* 3633 (1978).
76b. U. Groth, U. Schöllkopf, and T. Tiller, *Tetrahedron*: 2835 (1991).
77. T. Katsuki, and K. B. Sharpless, *J. Am. Chem. Soc. 102*:5974 (1980).
78. Y. Gao, R. M. Hanson, J. M. Klunder, S. Y. Ko, H. Masamune, and K. B. Sharpless, *J. Am. Chem. Soc. 109*:5765 (1987).
79. B. Bernet, and A. Vasella, *Tetrahedron Lett. 24*:5491 (1983).
80. R. Julina, T. Herzig, B. Bernet, and A. Vasella, *Helv. Chim. Acta 69*:368 (1986).
81. R. C. Mehrota, *J. Am. Chem. Soc. 76*:2266 (1954).
82. W. R. Roush, and M. A. Adam, *J. Org. Chem. 50*:3752 (1985).
83. S. Sugiyama, M. Honda, and T. Komori, *Justus Liebigs Ann. Chem.* 619 (1988).

84. H. Shibuya, K. Kawashima, M. Ikeda, and I. Kitagawa, *Tetrahedron Lett. 30*:7205 (1989).
85. M. Caron, P. R. Carlier, and K. B. Sharpless, *J. Org. Chem. 53*:5185 (1988).
86. M. W. Crossman, and C. B. Hirschberg, *J. Lipid Res. 25*:729 (1984).
87a. L. Grönberg, Z.-S. Ruan, R. Bittman, and J. P. Slotte, *Biochemistry 30*:10746 (1991).
87b. C.-C. Kan, Z.-S. Ruan, and R. Bittman, *Biochemistry 30*:7759 (1991).
88. R. R. Schmidt, and P. Zimmermann, *Angew. Chem. Int. Ed. Engl. 25*:725 (1986).
89. K. C. Nicolaou, T. J. Caulfield, H. Katakoa, and N. A. Stylianides, *J. Am. Chem. Soc. 112*:3693 (1990).
90a. *Synform*, *Sphingolipids*, Verlag Chemie, Weinheim, 1988, pp. 262–368.
90b. K. Mori, and T. Kinsho, *Liebigs Ann. Chem.* 1309–1315 (1991).
91. H. Kaller, *Biochem. Z. 334*:451 (1961).
92. R. C. Gaver, and C. C. Sweeley, *J. Am. Oil Chem. Soc. 42*:294 (1965).
93. S. Hammarström, *J. Lipid Res. 12*:760 (1971).
94. R. Cohen, Y. Barenholz, S. Gatt, and A. Dagan, *Chem. Phys. Lipids 35*:371 (1984).
95. T. Y. Ahmad, J. T. Sparrow, and J. D. Morrisett, *J. Lipid Res. 26*:1160 (1985).
96. D. Shapiro, E. S. Rachaman, Y. Rabinsohn, and A. Diver-Haber, *Chem. Phys. Lipids 1*:183 (1967).
97. W. I. Calhoun, and G. G. Shipley, *Biochemistry 18*:1717 (1979); W. F. Boss, C. J. Kelley, and F. R. Landsberger, *Anal. Biochem. 64*:289 (1975).
98. A. Hassner, and V. Alexanian, *Tetrahedron Lett.* 4475 (1978).
99. P. P. Van Veldhoven, R. J. Fogeslong, and R. M. Bell, *J. Lipid Res. 30*:611 (1989).
100. P. K. Sripada, P. R. Maulik, J. A. Hamilton, and G. G. Shipley, *J. Lipid Res 28*:710 (1987).
101a. F. B. Jungalwala, J. E. Evans, E. Bremer, and R. H. McCluer, *J. Lipid Res. 24*:1380 (1983).
101b. N. S. Radin, *J. Lipid Res. 31*:2291 (1990).
102. A. Vogel, *Vogel's Textbook of Practical Organic Chemistry*, 4th ed., Longman, London, 1978, pp. 264–319.
103. *Organic Synthesis*, Wiley, New York, 1947, Vols. I–VI.
104. Dox, A. W., *in Organic Synthesis*, Wiley, New York, 1941, Vol. I, pp. 5–7.
105. F. C. Schaefer, and G. A. Peters, *J. Org. Chem. 26*:2778 (1961).
106. D. F. Elliot, *J. Chem. Soc.* 589 (1949).
107. H. Newman, *Tetrahedron Lett.* 4571 (1971).
108. E. Negishi, T. Takahashi, and S. Baba, *Org. Synth. 66*:60 (1987) and references cited therein.
109. F. Arndt, in *Organic Synthesis*, Wiley, New York, 1943, Vol. II, pp. 165–167.
110. A. Dondoni, G. Fantin, M. Fogagnolo, A. Medici, and P. Pedrini, *J. Org. Chem. 53*:1748 (1988).
111. A. Dondoni, G. Fantin, M. Fogagnolo, A. Medici, and P. Pedrini, *J. Org. Chem. 54*:702 (1989).
112. T. L. Cupps, R. H. Boutin, and H. Rapoport, *J. Org. Chem. 50*:3972 (1985).
113. H. King, *J. Chem. Soc.* 432 (1942).
114. Y. Le Bigot, M. Delmas, and A. Gaset, *Synth. Commun. 12*:107 (1982).
115. T. Inoue, T. Uchimaru, and T. Mukaiyama, *Chem. Lett.* 153 (1977).
116. R. Polt, M. A. Peterson, and L. DeYoung, *J. Org. Chem. 57*:5469 (1992).

6
Chemical Preparation of Glycerolipids: A Review of Recent Syntheses

Robert Bittman *Queens College of The City University of New York, Flushing, New York*

I. INTRODUCTION AND SCOPE OF THIS REVIEW

The field of chemical synthesis of phospholipids has expanded significantly in recent years in response to the need for phospholipids with chemically defined structures for a variety of biochemical and biophysical applications involving membrane-associated processes. Phospholipids have been synthesized with spectroscopic probes or photoactivatable groups in the fatty acyl chains or in the head group for physical studies of model and biological membranes. The recognition that certain lipids have important functional properties in cells in addition to their well-known structural-based properties, but are present in low natural abundance, has stimulated a renewed interest in the development of new chemical approaches to many classes of phospholipids. The semisynthetic approach of phospholipid synthesis, i.e., enzymatic cleavage followed by chemical replacement of the cleaved substituent, is generally limited to the preparation of small quantities, and may also not generate chemically defined phospholipids because the natural source may be heterogeneous at more than one site. The synthesis of unnatural stereoisomeric forms of bioactive phospholipids (e.g., 2,3-diacyl-*sn*-glycero-1-phospholipids, which have the *S* configuration at C2) has also attracted interest in order to provide analogs for analyzing the role of binding to receptors in the action of certain phospholipids on cellular processes.

The structural identification of platelet activating factor in 1979 as 1-*O*-alkyl-2-acetyl-*sn*-3-glycerophosphocholine led to a change in thinking about the role of phospholipids in nature, which was previously considered to be limited to a structural role. Many lipids are known to exert effects of a large number of cellular processes such as proliferation, metabolism, contraction, and secretion. An example of a glycerolipid that has a function in a pathway involving signal transduction is phosphatidylinositol 4,5-bisphosphate; this phospholipid is a source of the second messengers inositol 1,4,5-trisphosphate, which induces calcium release, and diacylglycerol, which is the physiological activator of the regulatory enzyme protein kinase C. The hydrolysis of diacylphosphatidylcholines and phosphatidylinositols results from the activation of phospholipases; these enzymes are activated when various extracellular signaling molecules (hormones, growth factors, neurotransmitters, immunoglobulins, etc.) bind to cell-surface receptors. For a recent review of the generation of lipid second messengers from the action of phospholipases on phospholipids, see Dennis et al. (1991).

Analogs of the naturally occurring platelet activating factor, such as 1-*O*-alkyl (hexadecyl or octadecyl)-2-*O*-methyl-*sn*-glycero-3-phosphocholine and thioether derivatives, are examples of other lipids with potent physiological properties. They have antineoplastic properties in vitro and in vivo; these compounds may exert cytotoxicity by several mechanisms, one of which is inhibition of protein kinase C. Nucleoside phospholipid conjugates have been synthesized for use as possible prodrugs of nucleoside derivatives. Phosphorothioate analogs and acylamino analogs of phospholipids have been used to study the mechanisms of various membrane-related enzymes. The properties of these compounds have generated considerable interest in the synthesis of new analogs for study of the mechanism of action and for analysis of biological activity, and for use in new chemotherapeutic approaches.

Chemical syntheses of glycerolipids of the types described above are reviewed in this chapter. Detailed examples of the experimental methods of preparation of glycerides having different structural features are presented. See Table 1 for a summary of the groups attached to the glycerol backbone that are presented in the present review. Methods of preparing both optically active and optically inactive phosphoglycerides from a variety of precursors are reviewed in this chapter. An attempt is made to include procedures that provide phospholipid structural features such as ester, ether, thioether, thioester, amide, and carbon-phosphorus linkages. New procedures for preparing chiral triglyceride analogs are also included. Laboratory procedures for the preparation of ether-linked phospholipids in which the alkyl group contains a functional group, an aromatic moiety, or an isotopic label at the end of the chain are also included. Examples of procedures for sulfur substitution at the phosphate group and for conjugation of thiosugar and nucleoside moieties to glycerolipids are presented. This chapter also includes experimental procedures for inserting the phosphocholine group into glycerol derivatives by phosphorylation and phosphitylation approaches.

The decision of which examples to include is admittedly quite arbitrary. The choices are made based on a review of new procedures or modifications of previous procedures that have appeared in the literature between the end of 1988 and the time of writing of this review (middle of 1990); several additional procedures that appeared in 1991 or early 1992 have been included, while this book was awaiting editing. The experimental procedures described in this chapter are taken from the cited articles. In some cases, references are given to modifications of the procedures. Previous review articles should be consulted for earlier approaches used to prepare phospholipids, e.g., van Deenen and De Haas (1964), Rosenthal (1975), Kates (1977), and Eibl (1980, 1984).

II. GENERAL PROCEDURES

Lipid chemists are familiar with problems that arise in chemical syntheses because of the unusual properties of lipids. Such problems include slow reactions caused by steric hindrance in long-chain compounds, solubility incompatibilities in reactions involving a long-chain compound and a polar starting material, and the need for extensive chromatographic procedures. Lipids tend to be purified frequently by chromatography rather than by crystallization, since in many instances the compounds may be oils or waxlike solids, and even when they are crystalline, recrystallization may not remove impurities completely. As in other fields of organic synthesis, lipid syntheses that include oxygen- or moisture-sensitive reactions are run in oven-dried or flame-dried glassware under argon or nitrogen atmosphere. Reactions in which polyunsaturated compounds are used are generally run in the presence of antioxidants. Solvents are generally removed using a rotary evaporator at bath temperatures of ≤40°C.

A. Drying of Solvents

Solvents used in reactions are generally dried as outlined below. The procedures have been taken from different publications. Additional information about drying solvents is available in Gordon and Ford (1972) and Vogel (1989).

Acetonitrile. Predry over type 4A molecular sieves, then dry over P_2O_5. The filtered solvent is distilled from CaH_2 (bp 81–82°C).

Benzene and toluene. Dry over sodium metal and then distill and store over sodium wire or type 5A molecular sieves. (*CAUTION*: Since benzene is a carcinogen, all procedures that use benzene must be conducted in a well-ventilated hood, and protective gloves should be worn.)

Diethyl ether. If absolute diethyl ether is not used, check for peroxide content and remove peroxides, if necessary, according to standard procedures before distillation is attempted (Gordon and Ford, 1972, p. 437). Dry over KOH and distill from lithium aluminum hydride, CaH_2, or sodium (bp 35°C), taking care that no flame or hot plate is in the vicinity.

N,N-Dimethylformamide (DMF) and dimethyl sulfoxide (DMSO). Mix with 0.2 vol of benzene and remove benzene in a rotary evaporator at 40°C, or store over KOH pellets. The solvent is filtered, vacuum distilled from calcium oxide or barium oxide, and stored over type 4A molecular sieves under nitrogen.

Ethanol-free chloroform. To remove ethanol that is added as a stabilizer (to suppress oxidation to give phosgene), shake several times with 0.5 vol of distilled water, dry the chloroform layer over $CaCl_2$, distill from P_2O_5, and store in the dark for no more than 2 days under a nitrogen atmosphere (to avoid formation of phosgene). Alternatively, shake with concentrated sulfuric acid, wash several times with water, dry over $CaCl_2$, and distill or pass through a column of activated grade I alumina (~25 g per 500 mL of chloroform) before use. Hydrocarbon-stabilized chloroform is available from several suppliers and can be used directly when alcohol-free chloroform is required.

Ethyl acetate. Distill from P_2O_5 (bp 77°C).

Hexamethylphosphoric triamide. Distill over calcium oxide.

Methylene chloride and *carbon tetrachloride*. Dry over $CaCl_2$, then distill from P_2O_5. Other methods are (1) dry over $CaCl_2$, then pass through alumina 60 active basic, grade I, Merck; (2) dry over P_2O_5 and distill from CaH_2 before use; (3) wash with concentrated sulfuric acid, then with aqueous sodium carbonate, and with water, then dry over $CaCl_2$, and distill from P_2O_5.

Pyridine. Dry over KOH pellets, reflux with barium oxide for 2 h prior to distillation, taking care to exclude moisture (bp 115°C). Store over KOH pellets.

Tetrahydrofuran (THF). Test for peroxides with aqueous potassium iodide solution/starch solution and treat if necessary to remove peroxides before distillation. One method for removal of peroxides is to shake 1 L of solvent with 20 mL of an acidified ferrous sulfate solution (60 g of ferrous sulfate dissolved in 6 mL of concentrated sulfuric acid and 110 mL of water). The THF is then predried over KOH pellets, refluxed with sodium/benzophenone ketyl (10 g of sodium and 10 g of benzophenone per 1 L of THF) until the

Table 1 Summary of Groups Attached to the Glycerol Moiety Covered in This Chapter[a]

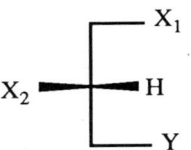

X_1	X_2	Y	References
OCOR	OH	OH	Burgos et al. (1987); Kodali (1987)
OCOR	OCOR	OH	Fröling et al. (1984)
OCOR	OCOR	SH	Snyder (1987)
OCOR	OCOR	OPC	Cubero Robles and van Berg (1969); Gupta et al. (1977); Warner and Benson (1977); Lammers et al. (1978); Patel et al. (1979); Radhakrishnan et al. (1981); Kanda and Wells (1981); Mason et al. (1981); Perly et al. (1984); Ali and Bittman (1988, 1989); Mangroo and Gerber (1988); Runquist and Helmkamp (1988); Hermetter et al. (1989); Singh (1990); Hébert and Just (1990); Parks et al. (1992)
SCOR	SCOR	OPC	Hendrickson and Hendrickson (1990); Hendrickson et al. (1983); Ward (1988); Yu et al. (1990)
OCOR	NHCOR	OPC	Dijkman et al. (1990)
NHCOR	NHCOR	OPC	Sunamoto et al. (1990)
NHCOR	OMe	OPC	Marx et al. (1988)
NHCOR	OCOR	OMe	Deveer et al. (1991)
OCOR	OR	OPC	Ali and Bittman (1990)
OCOR	OMe	NHCOR	Deveer et al. (1991)
OCOR	OCOR	OPI	Young et al. (1990); Lin and Tsai (1989); Salamonczyk and Bruzik (1990)
OCOR	OCOR	SPC	Snyder (1987); Mlotkowska and Markowska (1990); Mlotkowska (1991)

OCOR	OCOR	OPC_n	Ukawa et al. (1989); Ali and Bittman (1989)
OCOR	OCOR	OPS_s	Lindh and Stawinski (1989); Loffredo and Tsai (1990)
OCOR	OCOR	OPE	Martin and Josey (1988)
OCOR	OCOR	OPE-Me	Eibl (1978); McGuigan and Swords (1990)
OCOR	OCOR	CH=CHPC	Schwartz et al. (1988)
OR	OAc	OPC	Guivisdalsky and Bittman (1989c); Prakash et al. (1989); Hirth et al. (1983); Tsai et al. (1988)
OR	OCOR	OPC	Guivisdalsky and Bittman (1989c)
OR	SCOR	OPC	Aarsman et al. (1985); Bhatia and Hajdu (1989)
SR	NHCOR	OPC	Bhatia and Hajdu (1988); Yu et al. (1990)
OR	OR'	OPC	Guivisdalsky and Bittman (1989b); Abdelmageed et al. (1990); Yamauchi et al. (1987); Bittman and Witzke (1989); Meyer et al. (1989)
OR	OCOR	OPC	Hendrickson et al. (1987); Lai et al. (1989)
$O(CH_2)_nCO_2H$	OAc	OPC	Prahad et al. (1990)
OR	OH	OH	Johnson et al. (1989
OR	OH	OCH_2CF_3	Jain et al. (1991)
NH_2	OH	OH	Sutowardoyo and Sinou (1991)
OR	OR	CH_2PC	Deroo et al. (1976)
OR	NHCOR	$OP(CH_2)PC$	Turcotte et al. (1991)
OR	NHCOR	$(CH_2)_2PC$	Yuan et al. (1989)
OR	OR	phosphino	Vargas and Rosenthal (1983)
OCOR	OCOR	OPOPOnucl	Hostetler et al. (1990)
SR	OCOR	OPOPOnucl	Hong et al. (1990)
OR	OR	CH_2PCH_2POnucl	Vargas et al. (1984)

[a]Abbreviations: PC, phosphocholine; PE, phosphoethanolamine; PI, phosphoinositol; OPC_n, $OP(O)(O^-)(CH_2)_nNR_3^+$; OPS_s, thiophosphoserine; CH_2PC and $(CH_2)_2PC$, phosphonocholine; $OP(CH_2)PC$, phosphonocholine with C–P bond on choline side; nucl, nucleoside.

blue-purple color persists and then distilled from sodium/benzophenone ketyl before use. An alternative method is to dry over KOH and reflux over and distill from lithium aluminum hydride or CaH_2.

Triethylamine. Distill from ninhydrin under nitrogen to remove primary and secondary amines, and store over type 4A molecular sieves. Alternatively, reflux with and distill from KOH pellets (bp 88–90°C).

B. Chromatography

Flash chromatography on silica gel (230–400 mesh, ASTM) is frequently used to purify compounds. For a general reference on this technique, see Still et al. (1978). Preparative high-pressure liquid chromatography is also used in lipid synthetic work in order to reduce the time required for purification. Rapid separation of phospholipid classes has been achieved on a preparative scale using normal-phase preparative HPLC columns (e.g., Ellingson and Zimmerman, 1987), and ether-linked lipids have been purified by using tandem normal-phase preparative HPLC columns (Han and Gross, 1991). Reverse-phase HPLC has also been used with detection at 203 nm. Laser light-scattering detection of lipids has been used on an analytical scale (Lutzke and Braughler, 1990); if the effluent were split so that only a small fraction went through the nebulizer, this detection technique could be used on a preparative scale. Methods used for monitoring glycerophospholipids eluted from HPLC columns have been reviewed (Shukla, 1988; Amari et al., 1992).

Glass-backed thin-layer chromatography (TLC) plates precoated with silica gel are used to monitor reactions and to determine lipid purity. They are available from many suppliers (Analtech, Baker, Brinkmann, Merck, Whatman). Thin-layer plates used for preparative chromatography may be cleaned by prerunning the plates in chloroform-methanol (1:1), then methanol before air drying and activation in an oven at ≥110°C.

Silica gel plates coated with silver nitrate are frequently used for separations of unsaturated lipids. Although $AgNO_3$-coated plates are available commercially, they can be prepared simply by developing silica gel plates in a solution of silver nitrate (e.g., 10–20%, w/v) in acetonitrile or methanol or methanol-water (2:1), followed by air drying (in the dark, of course) and heat activation at ≥110°C. Solid-phase extraction columns packed with a bonded cation exchanger can be converted to the silver ion form and then used to separate lipids differing in the degree of unsaturation (Christie, 1989). Incidentally, solid-phase extraction columns (such as Bond Elut-aminopropyl from Analytichem International, a subsidiary of Varian Associates, Harbor City, Calif.; Bakerbond extraction columns from J. T. Baker; or similar phases from Burdick and Jackson, available from Baxter Healthcare) are used for rapid separations of neutral lipids from polar lipids and for separation of phospholipid classes (Kim and Salem, 1990).

C. Detection of Lipids on TLC Plates

Many sprays are available for detection of lipids on TLC plates. For a summary of the procedures for preparation of the sprays, see Kates (1986). For phospholipids that contain a phosphoric ester linkage, the *phosphomolybdic acid spray* (Dittmer and Lester, 1964) or a modification thereof (e.g., Ryu and MacCoss, 1979; Vaskovskii and Kostetskii, 1968) is often used. It is available under the name "phospray" from Supelco (Bellefonte, Pa.). For lipids containing a primary amino group, visualization is by spraying with 0.25% ninhydrin in acetone or acetone-lutidine (9:1). A phospholipid generally can be identified as a gel when the plate is sprayed heavily with water vapor, and the gel can be collected and the phospholipid then extracted with chloroform-methanol. [Since the silica adsorbent is partially soluble in alcohol and water, it must be removed from the extracted phospholipid. This is achieved by dissolving the lipid residue in chloroform and passing the solution through a Metricel™ 0.45-μm syringe filter several times. The chloroform-resistant Acrodisc-CR or Cameo filters are made by Gelman Sciences (Ann Arbor, MI) or Micron Separations, Inc. (Westboro, MA), respectively, and are available from major suppliers such as Fisher Scientific and Baxter Healthcare.] Many compounds may be visualized with iodine vapor. Short-wavelength UV light is often useful for compounds containing chromophores.

Other lipids are often visualized on TLC plates with various sprays that are based on the presence of long-chain hydrocarbon groups. The plates are sprayed with or dipped into the following solutions, and the lipid bands are visualized under UV or visible light.

Rhodamine 6G: a 0.001 to 0.005% aqueous solution
Cobalt chloride: 1% in acetone (Donner and Lohs, 1965)
2',7'-Dichlorofluorescein: 0.1% in methanol
Primulin: 1% in acetone-water (4:1)
Coomassie blue: 0.03% in methanol-water (1:4) (Nakamura and Handa, 1984)

Charring procedures are used frequently. The TLC plates are sprayed with various reagents (see list below), and the plates are then heated for about 5 min at 110–140°C on a hot plate in the hood.

~10 to 50% sulfuric acid in ethanol: the most general procedure for charring of organic compounds, and used routinely in many laboratories
p-Anisaldehyde-sulfuric acid (Yuan et al., 1989)
5% Ammonium molybdate (VI) in 10% sulfuric or perchloric acid or arsenomolybdic acid (Yuan et al., 1989)
10% Copper sulfate/phosphoric acid spray (Touchstone et al., 1983; Rustenbeck and Lenzen, 1990)
2% Potassium dichromate-30% aqueous sulfuric acid (Rouser et al., 1976)
0.5% $KMnO_4$ in 1 M NaOH (Yuan et al., 1987)
0.5% α-Naphthol-sulfuric acid for glycolipids

D. Catalysts Used for Coupling Reactions

The catalysts used for the coupling reactions are purified as follows:

Dicyclohexylcarbodiimide (DCC) may be recrystallized from chloroform-ether (1:1) before use and dried in a desiccator over P_2O_5, but for routine use recrystallization is usually not necessary.

4-(N,N-Dimethylamino)pyridine (DMAP) is recrystallized from chloroform-ether (1:1) Gupta et al., 1977) or from ether-hexane (Singh, 1990) before use and dried in a desiccator over P_2O_5; mp 112°C.

4-Pyrrolidinopyridine is dissolved in a minimum volume of chloroform and extracted into petroleum ether. The petroleum ether is evaporated under vacuum, the residue is redissolved in a minimum volume of chloroform, and the extraction is repeated until the yellow color disappears (Mangroo and Gerber, 1988). Alternatively, 4-pyrrolidinopyridine is recrystallized from hexane-ether (9:1) and stored under nitrogen at –20°C.

Trichloroacetonitrile is distilled before use; bp 83–84°C.

2,4,6-Triisopropylbenzenesulfonyl chloride (TPS) is recrystallized from pentane before use; mp 95–96°C.

1-(Mesitylene-2-sulfonyl)-3-nitro-1,2,4-triazole (MSNT) may be recrystallized from benzene; mp 130–132°C.

III. PREPARATION OF 1-ACYL-LINKED LIPIDS

A. 1-Acyl-*sn*-Glycerol

1. Preparation from Glycidol

Key features of the method:

Titanium-assisted nucleophilic opening of glycidol (2,3-epoxy-1-propanol, **1**) with stearic acid at 0°C gives 1-acyl-*sn*-glycerol, which serves as an intermediate for conversion to di- and triacylgylcerols and glycerophospholipids. No significant loss of ee occurs during the epoxide-opening reaction; when (*S*)-glycidol of 88% ee is used, the ee of the product 1-stearoyl-*sn*-glycerol could be improved to 96% ee by one crystallization from methylene chloride. The regioisomer (2-acyl-*sn*-glycerol) was not formed when the action was carried out at 0°C.

1-Acylglycerols have been purified by crystallization from ether solvent mixtures (diethyl ether-diisopropyl ether or light petroleum ether-diethyl ether) rather than by chromatographic methods, which may induce isomerization.

Advantage of the method: (*R*)- and (*S*)-Glycidol are available commercially in high enantiomeric purity (≥88% enantiomeric excess) from Aldrich, Fluka, Daiso Co., Osaka, and ARCO Chemical, Newtown Square, Pa. (*S*)-Glycidol is thus a convenient starting material for the synthesis of ester-linked lipids with the natural configuration at C2.

Disadvantages of the method: The yield of the titanium-assisted opening of glycidol is low (<50%), possibly because of the reactivity of glycidol toward other nucleophiles. Emulsive titanium salts may cause a problem in achieving phase separation during workup.

Literature: Burgos et al. (1987).

Related literature: The synthetic utility of ring-opening reactions of epoxy alcohols with nucleophiles is well known in the field of asymmetric synthesis (e.g., Jurczak et al., 1986; Caron and Sharpless, 1985; Hanson, 1991). In addition to the example shown in Scheme 1, underivatized glycidol has been used frequently as the starting material for the preparation of lipids containing ester groups. For example, (*S*)-glycidol was converted into optically active 1-acyl-*sn*-glycerol with use of tri-*n*-butylamine as a mild basic catalyst; *rac*-glycidol was converted into *rac*-mono-acylglycerol and into 1,2-diacylglycerols in the presence of a quaternary ammonium salt (Lok et al., 1976; Hendrickson et al., 1983; Lok et al., 1985; Zlatanos et al., 1985).

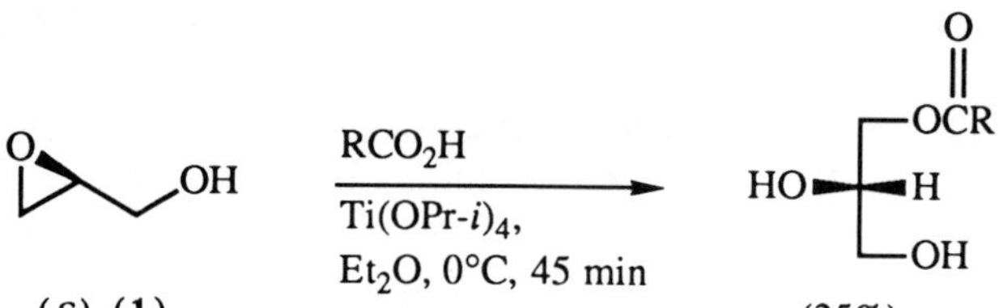

(*S*)-(**1**) —[RCO_2H; $Ti(OPr\text{-}i)_4$, Et_2O, 0°C, 45 min ($R = C_{17}H_{35}$)]→ HO–C–H ($CH_2OC(O)R$, CH_2OH) (25%)

Scheme 1

Method: To a mixture of 18.95 g (66.6 mmol) of stearic acid in 111 mL of ether is added 14.2 g (50 mmol) of titanium (IV) isopropoxide in one portion via syringe. After the solution is cooled to 0°C, 2.47 g (33.3 mmol) of (*S*)-(–)-glycidol is added dropwise, and stirring is continued for 45 min. The reaction is quenched by the addition of triethanolamine (50 mL of a 1 M solution in methylene chloride) at 0°C. The mixture is allowed to warm to rt, and ether (300 mL), water (250 mL), and 10% aqueous sodium carbonate (100 mL) are added. The mixture is stirred briefly; phase separation becomes a problem if stirring is continued for long periods. The mixture is allowed to stand without stirring at rt for 2 h, during which time hydrolysis of the titanium complex takes place; then it is filtered through a Celite pad, which is washed with ether. Two phases are obtained, the aqueous layer of the filtrate is removed, and the organic layer is washed with 10% aqueous sodium carbonate until excess stearic acid is removed (monitored by TLC). The organic layer is dried (Na_2SO_4), filtered, and evaporated under reduced pressure to give 3.0 g (25%) of 1-stearoyl-*sn*-glycerol.

Use of a levulinoyl protecting group, which is removed by using hydrazine, allows the synthesis of 1-saturated-2-unsaturated-diacylglycerols from glycidol (Fröling et al., 1984). 1-Palmitoyl-3-levulinoyl-*rac*-glycerol is prepared by the ring opening of *rac*-glycidol palmitate with levulinic acid in the presence of tetraethylammonium bromide at 105°C. After acylation with oleoyl chloride in hexane/pyridine, the levulinoyl group is removed with hydrazine in pyridine/acetic acid solution, and the excess hydrazine is decomposed by adding 2,4-pentadione. The extent of acyl migration is low (<3%) during the deprotection step, which is carried out in 3 min at rt.

2. Preparation from L-Arabinose via 2,3-*O*-Isopropylidene-*sn*-Glycerol

Key features of the method:

The isopropylidene group is removed in good yield with dimethylboron bromide at low temperature without acyl migration.

2,3-*O*-Isopropylidene-*sn*-glycerol is prepared from L-(+)-arabinose by a modification of the procedure of Kanda and Wells (1980). It was synthesized from D-mannitol in 1939 (Baer and Fisher, 1939a); see modification by Eibl (1981). It is now commercially available.

Literature: Kodali (1987).

Method: 1,1'-Diethylmercapto-L-arabinose (2) is prepared by adding L-(+)-arabinose (90 g, 0.60 mol) in portions and with stirring to zinc chloride (75 g, 0.55 mol) in 250 mL (3.35 mol) of ethanethiol. After the mixture has stirred for 5 h at rt, unreacted ethanethiol is removed by evaporation under vacuum, leaving a residue that is dissolved in a minimum volume of methanol and filtered. The product (2) is precipitated by adding ice to the filtrate, then crystallized from ethanol-water 1 : 1 at 0°C, giving 146 g (95%) of the product as a white solid, mp 127°C; $[\alpha]^{22}_D$ + 11.2° (*c* 5, MeOH).

The 4,5-isopropylidene derivative of 1,1'-diethylmercapto-L-arabinose (3) is prepared as follows. To a solution of 25.6 g (0.1 mol) of 1,1'-diethylmercapto-L-arabinose (2) in 100 mL of dry DMF at 0°C are added 2 g of Drierite ($CaCl_2$) and 200 mg of *p*-toluenesulfonic acid. 2-Methoxypropene (7.2 g, 0.1 mol) is added dropwise with stirring. Additional (2.9 g, 40 mmol) 2-methoxypropene is added after 15 min, and stirring is continued for 1 h. The reaction is quenched by adding 5 g of Na_2CO_3 and stirring for 10 min. The mixture is filtered, and the filtrate is poured onto crushed ice, yielding a precipitate that is isolated by filtration, then dried and crystallized from

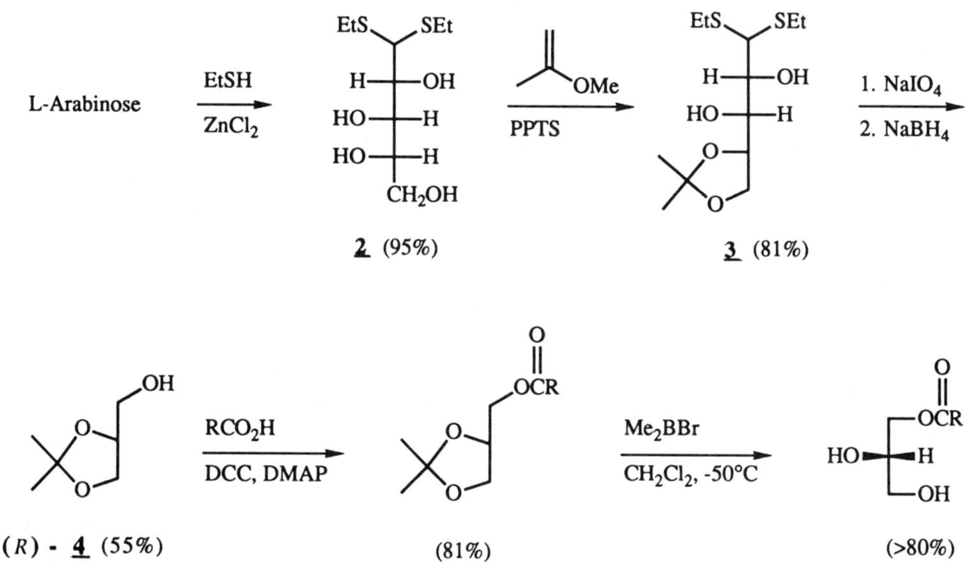

Scheme 2

diisopropyl ether at 0°C to give 24 g (81%) of the ketal 3; mp 77°C [lit. (English and Griswold, 1948) mp 75.6°C]; $[\alpha]^{22}_D$ + 7.5° (*c* 8.5, MeOH) [lit. (English and Griswold, 1948) $[\alpha]^{22}_D$ +7.6° (*c* 8.5, MeOH)].

The isopropylidene derivative 3 is cleaved to yield 2,3-*O*-isopropylidene-*sn*-glycerol [(*R*)-4] as follows. To a solution of 64.2 g (0.3 mol) of sodium periodate in 600 mL of water at 0°C is added 29.6 g (0.10 mol) of 1,1'-diethylmercapto-4,5-isopropylidene-L-arabinose (3) in portions (maintaining the temperature below 15°C) over a 20-min period with stirring. Stirring is continued for 20 min, and ethanol (1 L) is added to precipitate sodium iodate, which is removed by filtration. The pH of the filtrate is adjusted to 8.0 by adding about 40–45 mL of 10% sodium hydroxide solution. Sodium borohydride (15.2, 0.4 mol) is added to the alkaline solution, and the solution is stirred for about 40 min at rt. The reaction mixture is diluted with 250 mL of water, and the product is extracted into chloroform. The organic layer is isolated, washed with water, and dried over Na_2SO_4. Evaporation of the solvents under reduced pressure gives a residue that is distilled [bp 45°C (~0.25 mm Hg)] to yield 7.2 g (55%) of product 4; further purification by redistillation gives $[\alpha]^{22}_D$ −14.9° (neat) [lit. (Kanda and Wells, 1980) $[\alpha]^{22}_D$ −14.5° (neat); bp 94–95°C (15 mm Hg)]. The overall yield from L-arabinose is 40%.

Isopropylideneglycerol [(*R*)-4] was shown to be optically pure by chiral HPLC of the benzoyl derivatives (prepared by acylation of each enantiomer of isopropylidene-*sn*-glycerol with 1 equivalent of benzoic acid in the presence of DCC and DMAP) on a 4.6 × 250 mm Chiralcel-OB column, J. T. Baker Chemical Co., elution with hexane-*i*-PrOH, 95:5, flow rate of 0.5 mL/min, detection at 229 nm; retention times: 1,2-isopropylidene-3-benzoyl-*sn*-glycerol, 13.00 min; 1-benzoyl-2,3-isopropylidene-*sn*-glycerol, 16.51 min.

Acylation of 2,3-*O*-isopropylidene-*sn*-glycerol (4) (see Scheme 2) is carried out by adding 1.1 mmol of long-chain saturated or unsaturated fatty acid and 122 mg (1.0 mmol) of DMAP with stirring to a solution of 132 mg (1.0 mmol) of 2,3-isopropylidene-*sn*-

glycerol in 15 mL of carbon tetrachloride. A solution of 226 mg (1.1 mmol) of DCC in 10 mL of carbon tetrachloride is added at rt over a 15-min period. After the mixture has stirred for 1.5 h, the reaction is complete when monitored by TLC (elution with diisopropyl ether-hexane, 1:2). Dicyclohexyl urea is removed by filtration, the precipitate is washed with carbon tetrachloride, and the filtrate is concentrated under reduced pressure to give a residue that is purified by flash chromatography (elution with hexane and then a gradient of 1–10% diisopropyl ether in hexane); yield of 1-acyl-2,3-*O*-isopropylidene-*sn*-glycerol, 81%.

The final step in Scheme 2 is ketal cleavage of 1-acyl-2,3-*O*-isopropylidene-*sn*-glycerol by using dimethylboron bromide at –50°C. A solution of 1 mmol of 1-acyl-2,3-*O*-isopropylidene-*sn*-glycerol in methylene chloride under argon atmosphere is cooled to –50°C, and 1 mmol of a dimethylboron bromide solution in methylene chloride is added with stirring. [Me_2BBr is made under argon from BBr_3 and tetramethyltin, then distilled and stored at –20°C, as described by Guindon et al. (1984).] The progress of the reaction is monitored by TLC (chloroform-acetone-methanol, 95:4:1); after 2 h of stirring, the deketalization reaction is complete. The reaction is quenched by the slow addition of 3 mL of saturated aqueous $NaHCO_3$ solution. The mixture is warmed to room temperature and the product is extracted into chloroform. The chloroform extract is washed with water, dried (Na_2SO_4), and evaporated. The product is crystallized from ether at 0°C; yield, >80%. No evidence of 2-monoacyl-*sn*-glycerol by-product is found by HPLC on a C18 analytical column (detection at 215 nm, elution with water-*i*-PrOH-CH_3CN-THF mixtures) and by TLC on boric acid impregnated plates, indicating that acyl migration does not occur during the ketal-cleavage procedure.

It was reported recently that Me_2BBr in methylene chloride at low temperature deprotects diacylglycerol bearing benzyl, 4-methoxybenzyl, or trityl ether groups, with little or no acyl migration (Kodali and Duclos, 1992; Hébert et al., 1992).

B. Diacylphosphatidylcholines

1. Preparation from D-Mannitol and Other Natural Products

The classical approach to diacyl-PCs (Baer and Kates, 1950) is from D-mannitol, which is acetonated in the presence of anhydrous zinc chloride to give 1,2:5,6-diisopropylidene-D-mannitol (5) (see Scheme 3). The bis(acetonide)5 is treated with lead tetraacetate to give 1,2-diisopropylidene-*sn*-3-glyceraldehyde [(*R*)-2,3-*O*-isopropylideneglyceraldehyde], which is reduced (originally by catalytic hydrogenation, now with sodium borohydride or lithium aluminum hydride) to (*S*)-glycerol 1,2-acetonide; the latter is converted to *sn*-glycero-3-phosphate (Baer, 1952) and then to phosphoglycerides 6, in which X can represent a variety of head groups, such as choline, ethanolamine, serine, and glycerol (reviewed by Baer, 1965) (see Scheme 3).

Improved procedures have been developed for many of the intermediates in the synthetic route from D-mannitol to phospholipids. For example, modified methods have appeared for the preparation of 1,2:5,6-di-*O*-isopropylidene-D-mannitol (5) (Chittenden, 1980a,b; Debost et al., 1983), for cleaving D-mannitol bis(acetonide) (5) with periodate in the presence of potassium carbonate to give acid-free aldehyde (Dumont and Pfander, 1983; Häfele and Jäger, 1987), for reduction of (*R*)-glyceraldehyde acetonide to 1,2-*O*-isopropylidene-*sn*-glycerol [(*S*)-4] (Eibl, 1981) or its enantiomer (from L-ascorbic acid) to 2,3-*O*-isopropylidene-*sn*-glycerol [(*R*)-4] (Hubschwerlen, 1986; Jung and Shaw,

acetone $ZnCl_2$ → $Pb(OAc)_4$ → $NaBH_4$ →

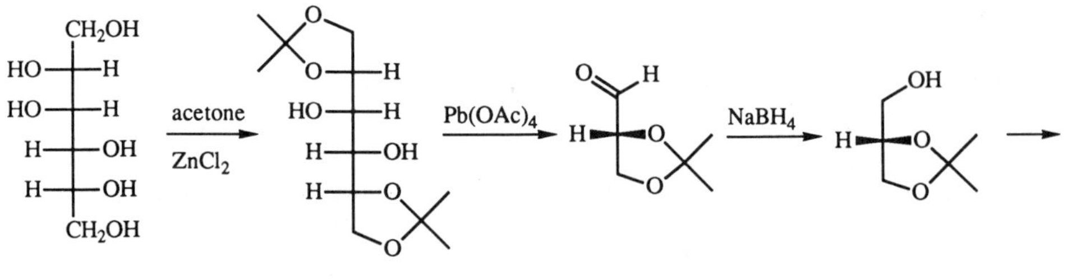

D-Mannitol 5 (*S*)-glycerol acetonide (*S*)-4

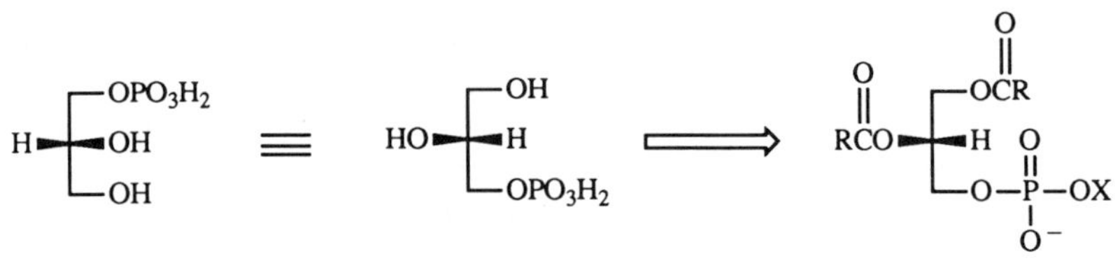

sn-glycero-3-phosphate 6

Scheme 3

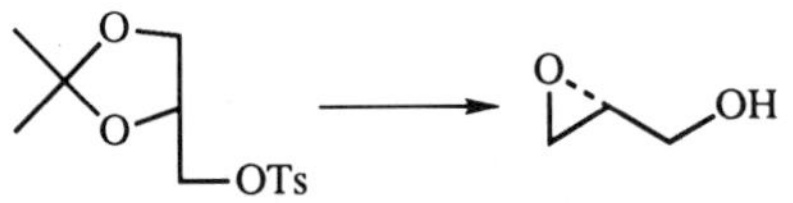

7 (*R*)-1

Scheme 3a

1980), and for use of different protecting groups in D-mannitol acetals (Peters et al., 1987) and in *sn*-glycerols (Eibl and Woolley, 1986).

The classical route described above is rarely used now. Disadvantages of the classical approach to PCs are that the procedure is lengthy, requiring nine steps to convert D-mannitol into symmetric-acid PCs (6) and fourteen steps to mixed-acid PCs, and that 1,2:5,6-di-*O*-isopropylidene-D-mannitol (5) is somewhat unstable, decomposing on storage to D-mannitol and monoisopropylidene-D-mannitol (Rosenthal, 1975). Moreover, the intermediate (*S*)-glyceraldehyde acetonide, which is obtained by lead tetraacetate or sodium periodate oxidation of the diacetonide of D-mannitol (5), may undergo epimerization in the presence of acids or bases and is therefore reduced immediately with sodium borohydride to isopropylidene-*sn*-glycerol [(*S*)-4]. The latter has been reported to undergo partial racemization during storage (Eibl, 1984; Ohno, et al., 1985) and complete racemization in the presence of a trace of acidic impurity (Baldwin et al., 1978). However, a recent report indicated that configurational stability of (*S*)-glycerol 1,2-acetonide is retained for several months if the material is worked up at pH 7.2 and if $NaHCO_3$ is added before distillation; $[\alpha]^{25}_D + 14.6°$ (neat) (Nakamura et al., 1990). This is a modification of the workup procedure of Baldwin et al. (1978), who reduced glyceraldehyde acetonide with sodium borohydride and extracted glycerol acetonide into ethyl acetate; the organic layer was washed with 5% aqueous sodium hydroxide and brine, dried (Na_2SO_4) and concentrated, and the product was distilled (bp 80–90°C/20 mm Hg).

In addition to D-mannitol and mannitol derivatives such as 1,3:4,6-di-*O*-benzylidene-D-mannitol (Peters et al., 1987), many other natural products have been used as chiral precursors of phospholipids via enantioselective synthesis of glycerol derivatives. For example, L-serine (Lok et al., 1976), L-ascorbic acid (Jung and Shaw, 1980; Mikkilineni et al., 1988), L-glyceric acid (Bhatia and Hajdu, 1988), L-arabinose (Baer and Fischer, 1939b), L-inositol (Angyal and Hoskinson, 1963), (*S*)-malic acid, and D- and L-tartaric acids (Ohno et al., 1985) have been used to give 2,3- or 1,2-*O*-isopropylidene-*sn*-glycerol (4). Enantioselective reduction of the monoesters of 1,3-dihydroxypropanone with baker's yeast gives (*S*)- and (*R*)-1,2-*O*-isopropylideneglycerol (Aragozzini et al., 1989).

Isopropylidene-*sn*-glycerol derivatives with high stereochemical stability are available commercially and have been used as phospholipid precursors. For example, 1,2-*O*-isopropylidene-*sn*-glycerol 3-tosylate (7) was used as the starting material for the preparation of 1-*O*-hexadecyl-2-thioacyl-2-deoxy-GPCs (Bhatia and Hajdu, 1989). Furthermore, the *O*-tosyl derivative of (*S*)-glycerol 1,2-acetonide (7) is converted into (*R*)-glycidol [(*R*)-1] by acid-catalyzed acetonide cleavage followed by base-catalyzed epoxidation (Baldwin et al., 1978) (see Scheme 3a). Benzyl glycidyl ether (*O*-benzyl glycidol), which can be made from glycidyl tosylate in two steps in high yield (Byun and Bittman, 1989) or from 1-*O*-tosyl-3-*O*-benzyl-*sn*-glycerol (Hirth and Barner, 1982), is another useful glycerolipid precursor.

Acylation procedures have received a great deal of attention. Symmetric-chain PCs are often prepared by acylation of *sn*-glycero-3-phosphocholine (GPC) (Cubero Robles and van Berg, 1969; Warner and Benson, 1977; Lammers et al., 1978; Kanda and Wells, 1981) and its cadmium chloride complex (GPC·$CdCl_2$) (Gupta et al., 1977; Patel et al., 1979; Radhakrishnan et al., 1981). It is recommended that GPC·$CdCl_2$ be dried at 78°C over P_2O_5. The acylating species is usually the fatty acid anhydride (Gupta et al., 1977; Patel et al., 1979; Radhakrishnan et al., 1981; Kanda and Wells, 1981) or the acyl imidazolide (Warner and Benson, 1977; Lammers et al., 1978; Mangroo and Gerber, 1988). Fatty acid anhydrides are made by stirring a solution of the fatty acid and 0.55 equivalents of DCC in carbon tetrachloride or methylene chloride at rt for 3 h (Selinger and Lapidot, 1966). Fatty acyl imidazolides are made by the reaction of fatty acids and carbonyl diimidazole at rt; addition of DMSO anion to a solution of oleoyl imidazolide and GPC in DMSO at rt results in complete conversion to dioleoyl-GPC in only 5 min (Warner and Benson, 1977).

In some acylation reactions, an excess of free fatty acid is used (e.g., 4 equivalents relative to GPC·$CdCl_2$) together with DCC (an equimolar amount relative to fatty acid) in order to generate the anhydride (Samuel et al., 1985; Runquist and Helmkamp, 1988). These methods are carried out in the presence of a molar excess (usually 2 equivalents relative to GPC·$CdCl_2$) of catalysts such as DMAP (Gupta et al., 1977) or 4-pyrrolidinopyridine (Patel et al., 1979). Recently, multiple increments of 0.1–0.15 equivalents (relative to fatty acid) were added every 4 h in order to avoid using a large excess of fatty acid (Han et al., 1992).

The acylation reaction is carried out with stirring under nitrogen in a water bath at rt, and the reaction mixture is protected from light. In the absence of catalyst, high temperatures and long reaction times are necessary to obtain a homogeneous mixture of GPC·$CdCl_2$ in the aprotic solvent used for the acylation reaction; such severe reaction conditions result in extensive acyl and phosphoryl migration reactions (Lammers et al.,

1978; Keough and Davis, 1979), and the by-products are difficult to separate from the desired products. With the use of DMF as solvent, efficient diacylation of GPC·$CdCl_2$ (>90% yield) was reported in 3 h at rt by using 2.5 equivalents of fatty acid anhydride and 2.1 equivalents of DMAP (Perly et al., 1984).

In addition to anhydrides, many other activated acylating agents have been used in lipid chemistry. Recently, enol esters [$RCO_2C(R')$=CH_2] have been used as activated acylating reagents in lipase-catalyzed stereoselective acylation reactions of glycerol derivatives in organic solvents (Wang and Wong, 1988; Wang et al., 1988). Long-chain enol carboxylates such as vinyl stearate are made by adding a fatty acid to acetylene in a ruthenium-catalyzed reaction (Mitsudo et al., 1987). Since the alcohol released is tautomerized to acetone (when R' is methyl) or acetaldehyde (when R' is hydrogen), the acylation reaction is irreversible because the carbonyl compound is volatile and does not undergo the reverse reaction. The lipase-catalyzed stearoylation of 2-*O*-benzylglycerol with vinyl stearate in diisopropyl ether was reported recently (Baba et al., 1990). 1-Stearoyl-2-*O*-benzyl-*sn*-glycerol is formed in 50% yield, which is the maximum yield of one enantiomer that can be obtained from a racemic substrate, and with high enantioselectivity (92% ee). Thus the lipase-catalyzed acylation reaction offers an alternative to the DMAP/DCC method and has the advantage that is stereoselective when two primary hydroxy groups are present. The lipase-catalyzed kinetic resolution approach to optically active compounds via esterification in organic solvents has been reviewed (Chen and Sih, 1989).

2. Preparation from 1-Acyl-2-lyso-*sn*-glycero-3-phosphocholine

Key features of the method:

Acylation of lyso-PC is carried out with a 10-fold molar excess of fatty acid anhydride in dry, alcohol-free chloroform at rt in 5 to 6 h in the presence of 1.2 equivalents of 4-pyrrolidinopyridine (or DMAP) as a catalyst. The lyso-PC and anhydride are both predried over P_2O_5 under high vacuum. Yields of mixed-acid PCs are high (>90%) and the extent of acyl migration is low (<1%).

Disadvantage of the method: An excess of fatty acid anhydride is necessary to complete the acylation reaction within 6 h at rt; reaction within 6 h appears to be important in order to avoid acyl migration. When the fatty acid is expensive, as in the use of isotopically labeled or highly unsaturated compounds, care should be taken to recover the unreacted fatty acid by applying the crude product to a silica gel column and first eluting the nonpolar fraction with chloroform. The fatty acid anhydride can then be recycled. Although the acylation conditions may be modified to use lower amounts of fatty acid

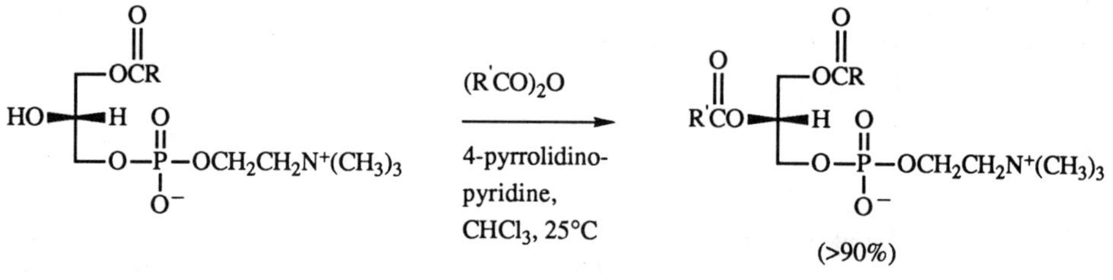

HO–C(H)(CH₂OC(O)R)–CH₂–O–P(O)(O⁻)–$OCH_2CH_2N^+(CH_3)_3$ —[$(R'CO)_2O$; 4-pyrrolidinopyridine, $CHCl_3$, 25°C]→ R'C(O)O–C(H)(CH₂OC(O)R)–CH₂–O–P(O)(O⁻)–$OCH_2CH_2N^+(CH_3)_3$ (>90%)

Scheme 4

anhydrides by using higher temperature, the extent of migration of acyl groups increases with increasing temperature. For example, with 3–3.5 equivalents of fatty acid anhydride good yields of mixed-acid PCs are obtained in 2 h at 40°C in the presence of DMAP in chloroform-pyridine 4:1. However, appreciable acyl migration takes place under these conditions.

Modifications of the method: Increasing the amount of 4-pyrrolidinopyridine has also been reported to increase the rate of acylation of 1-acyllyso-PC (solubilized as the trifluoroacetate salt in chloroform) by fatty acid anhydrides at room temperature without affecting the extent of acyl group migration (Mangroo and Gerber, 1988). The reaction proceeds to completion when two equivalents of catalyst and two equivalents of anhydride are used, and the product is purified on Sephadex LH-20 (elution with chloroform-methanol 1:1). In another modification, lyso-PC was acylated with 1.2 equivalents of fatty acid anhydride in the presence of 1.25 equivalents of DMAP in chloroform, giving mixed-acid-diacyl-PCs in 1.5 h at rt in >85% yield and with <5% acyl migration (Perly et al., 1984). The latter method was recently modified by using ultrasound to assist the reaction instead of magnetic stirring (Singh, 1990). DCC has also been included (one equivalent relative to anhydride) in the acylation of lyso-PC with fatty acid anhydride in alcohol-free, dry chloroform (Runquist and Helmkamp, 1988).

Instead of using the anhydride to acylate lyso-PC, one may use an excess of fatty acid in the presence of an excess of DMAP and DCC. The reaction may be carried out in a brown screw-cap tube under nitrogen or argon. For example, 41.5 μmol of a lyso-PC is acylated using 53.7 μmol of fatty acid, 163.9 μmol of DMAP, and 82.5 μmol of DCC in a minimal volume of methylene chloride at rt for 16 h (Szolderits et al., 1991). The reaction mixture is diluted with methylene chloride, water is added, and the organic phase is isolated and dried (Na_2SO_4). Evaporation gives a residue that is purified by silica gel chromatography (elution with gradients of chloroform/methanol); yield of diacyl-PC, ~45%. A similar reaction of lyso-PC with fatty acid in the presence of DCC and DMAP was reported recently in which the reaction mixture was sonicated for 2 h, but no yield was given (Parks et al., 1992).

Caveat: Care should be taken to maintain the reaction temperature (see Scheme 4) at 25°C by use of a water bath. At 35°C or above, the extent of acyl migration becomes appreciable. Ethanol-free (hydrocarbon-stabilized) chloroform is used.

Literature: (a) Ali and Bittman (1989); (b) Mason et al. (1981); (c) Mangroo and Gerber (1988).

Method (From Ref. a): A typical procedure is given for the preparation of 1-stearoyl-2-decanoyl-*sn*-glycero-3-phosphocholine by acylation of 1-stearoyllyso-PC. (Lyso-PC is obtained by phospholipase A_2 treatment of PC: Chakrabarti and Khorana, 1975. It is lyophilized from benzene and dried before use.) Capric anhydride (1.6 g, 4.7 mmol) and 4-pyrrolidinopyridine (85 mg, 0.56 mmol) are added to a suspension of 1-stearoyllyso-PC (250 mg, 0.47 mmol) in 35 mL of hydrocarbon-stabilized chloroform under nitrogen. After the reaction mixture is stirred under nitrogen at rt for 6 h, the solvent is removed under reduced pressure. The residue is dissolved in a minimum volume of chloroform and purified by flash chromatography (elution first with chloroform to remove excess anhydride and other nonpolar impurities, then with chloroform-methanol 9:1, and finally with chloroform-methanol 3:2). Fractions that contain the product are combined, and the solvents are evaporated. The residue is dissolved in chloroform and filtered through a 0.45-μm Metricel filter (Acrodisc-CR, Gelman Sciences) to remove suspended silica gel particles. A light yellow residue is obtained on evaporation of the solvent, which is lyophilized from about 10 mL of benzene to afford 300 mg (93%) of diacyl-PC as a white solid. If the product is still yellow, further purification by a second lyophilization from benzene or by preparative TLC may be performed.

3. Preparation from 1-*O*-Trityl-*sn*-glycero-3-phosphocholine via 1-Lyso-2-acyl-*sn*-glycero-3-phosphocholine

Key features of the method:
Mixed-chain diacyl-PCs are obtained in high yields and high isomeric purities. 1-*O*-Trityl-*sn*-glycero-3-phosphocholine (8), which is prepared from *sn*-glycero-3-phosphocholine (GPC), is used as the starting material to prepare mixed-chain diacyl-PCs in two steps.

Acylation is carried out at the available *sn*-2 position, then a one-pot detritylation-acylation reaction sequence, with 4 equivalents of boron trifluoride etherate and 2 equivalents of fatty acid anhydride in methylene chloride at 0°C, is used to give the product. The one-pot detritylation-acylation reaction avoids acyl migration from the *sn*-2 to *sn*-1 position.

Literature: Hermetter et al. (1989).

Method: The starting material (8) is prepared from *sn*-glycero-3-phosphocholine (GPC) as follows. To 10.0 g (37 mmol) of GPC suspended in 100 mL of dry DMF is added 5.0 g (35 mmol) of powdered zinc chloride. Trityl chloride (recrystallized from hexane containing a few drops of acetyl chloride) (10.6 g, 38 mmol) is added and the reaction mixture is stirred at 5°C for 10 h. The crude product (8) is precipitated by adding

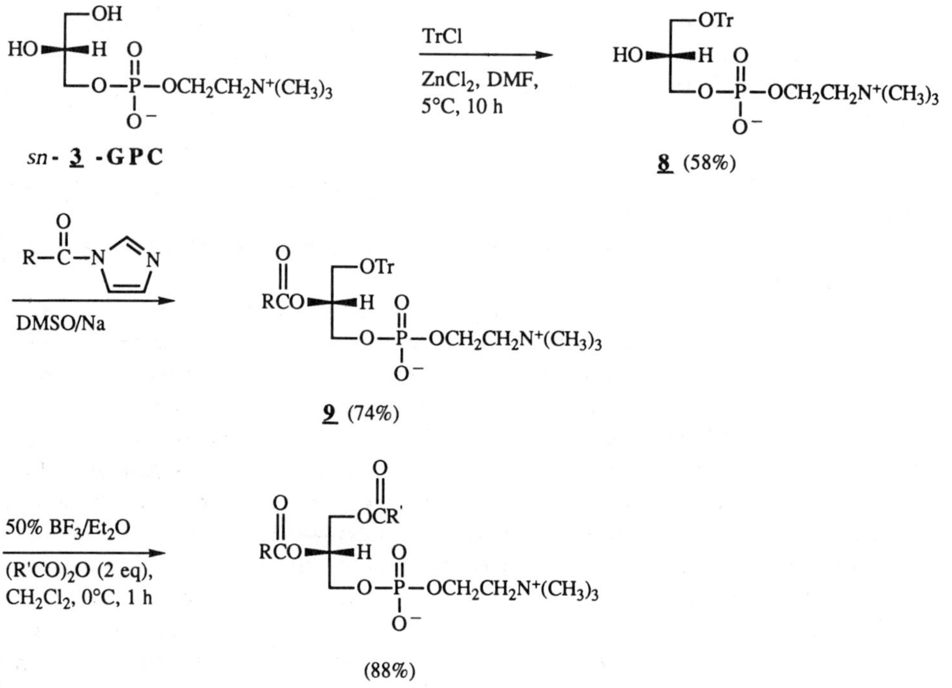

sn-3-GPC $\xrightarrow[\text{ZnCl}_2\text{, DMF, 5°C, 10 h}]{\text{TrCl}}$ 8 (58%)

$\xrightarrow[\text{DMSO/Na}]{\text{R–C(=O)–imidazole}}$ 9 (74%)

$\xrightarrow[\text{(R'CO)}_2\text{O (2 eq), CH}_2\text{Cl}_2\text{, 0°C, 1 h}]{\text{50\% BF}_3\text{/Et}_2\text{O}}$ (88%)

Scheme 5

ether (200 mL). The precipitate is collected by centrifugation and dissolved in 200 mL of chloroform-2-butanol (2:1). The chloroform/2-butanol solution is washed with 4% aqueous NH_4OH(2 × 70 mL), water (2 × 60 mL), then dried (Na_2SO_4). The product is reprecipitated by adding two volumes of ether per volume of chloroform/2-butanol solution, giving 14.6 g (77%) of crude product (8). Further purification is achieved either by trituration of the crude product with 200 mL of acetonitrile, which removes the 1,2-di-*O*-trityl-3-GPC by-product, or by chromatography on silica gel (elution with methylene chloride/methanol/25% NH_4OH 40:50:4) to provide pure 8, mp 225–229°C; R_f 0.25 (chloroform/methanol/25% NH_4OH 65:35:5); $[\alpha]^{20}_D$ −10.58° (*c* 5, chloroform/methanol 2:1); $[\alpha]^{20}_D$ −10.23° (*c* 5, methanol).

Acylation of 8 with 2 equivalents of oleoylimidazolide is carrried out as follows. A solution of 5.1 g (10 mmol) of 1-*O*-trityl-GPC (8) and 20 mmol of oleoylimidazolide [prepared from the reaction of 20 mmol of oleic acid with 22 mmol of 1,1'-carbonyldiimidazole in THF (Warner and Benson, 1977) or in DMF-benzene (Mangroo and Gerber, 1988)] in 25 mL of DMSO is added to 52 mL of DMSO to which 690 mg (30 mmol) of sodium had been added. After the mixture is stirred at rt for 30 min, 62.5 mL of 0.5 M acetic acid is added to protonate the DMSO anion. The product is extracted with 1 L of chloroform-methanol 2:1, the organic layer is washed twice with methanol-water (1:1) containing NH_4OH, then evaporated under reduced pressure. The residue is dried azeotropically with toluene. Chromatography on silica gel [elution with chloroform/methanol (the latter containing 0.5 vol % of 25% NH_4OH), first with 9:1, then with increasing proportions of methanol to 1:1] removes the residual free fatty acid and gives 5.9 g (74%) of pure 1-*O*-trityl-2-oleoyl-*sn*-glycero-3-phosphocholine (9).

The latter is subjected to detritylation with BF_3 and acylation with 2 equivalents of palmitic anhydride as follows. To the solution of 2.51 g (3.25 mmol) of 9 and 3.36 g (6.5 mmol) of palmitic anhydride in 80 mL of methylene chloride is added 3.5 mL (13 mmol) of boron trifluoride etherate (50% BF_3). The reaction mixture is stirred at 0°C for 1 h. The reaction is quenched by adding 10 g of sodium bicarbonate in 20 mL of water, then stirring for 10 min. Methanol (40 mL) is added, the salt is removed by filtration, and the organic phase is washed with methanol/water (1:1) (3 × 25 mL). Removal of the solvents under reduced pressure gives a residue that is purified by chromatography on silica gel (elution with a gradient of chloroform/methanol) to give 2.2 g (88%) of pure mixed-chain diacyl-PC. Phospholipase A_2 digestion, followed by isolation of the released fatty acids and GC analysis of the methyl esters, indicated that <3% acyl migration had occurred during the reaction.

4. Preparation from 3-Arenesulfonate Derivatives of Glycidol

Key features of the method:

This is a simple method for the preparation of symmetric-chain saturated *sn*-3 and the enantiomeric *sn*-1 glycerophospholipids. It is a simpler route to *sn*-1 phosphoglycerides [(*S*)-12] than via diacylation of 1-*O*-trityl-*sn*-glycerol (Virtanen et al., 1980).

The starting materials [(*R*)- and (*S*)-glycidyl tosylate, 10] are stable, crystalline compounds available commercially (from Aldrich Chemical Co., Fluka, Daiso Co., Ltd., Osaka) in high optical purity. [They are made by asymmetric epoxidation of allyl alcohol and in situ derivatization (Gao et al., 1987) or by treatment of commercially available (*R*)- or (*S*)-glycidol (~88–91% ee) (prepared by the catalytic Sharpless asymmetric epoxidation of allyl alcohol using titanium (IV) isopropoxide, cumene hydroperoxide, diisopropyl tartrate, and 3Å molecular sieves (Hanson and Sharpless, 1986) with an arenesulfonyl chloride and triethylamine in toluene or methylene chloride at –10°C (Klunder et al., 1989).] The ee of the starting materials (10) can be improved by multiple recrystallizations from ethanol (Klunder et al., 1989).

Yields of isolated 1,2-diacyl-*sn*-glycero-3-phosphocholines (12) are good (>70%). Enantiomeric excesses are high, since the reaction proceeds without loss of chiral purity. If highly optically pure diacyl-PCs (12) are not required, the number of steps can be shortened by omitting the tosylate-to-iodide conversion step.

The route may be modified to vary the substituent at the 3 position, e.g., a different phosphate-containing head group can be substituted for phosphocholine, or an acyl chain can be inserted, giving a chiral triacylglycerol having three identical acyl groups or the same acyl group at the *sn*-1 and *sn*-2 positions and a different one at the *sn*-3 position.

Outline of the reactions:

Lewis acid catalyzed opening of (*R*)-(–)-glycidyl tosylate [(*R*)-10] with fatty acid anhydrides (1.5 equivalents) in the presence of boron trifluoride etherate as catalyst in refluxing methylene chloride for 2 h under nitrogen gives good yields of 1,2-diacyl-*sn*-glycero-3-*p*-toluenesulfonate (11). The arenesulfonate group serves as a "protecting group" at the *sn*-3 position, blocking acyl migration from the *sn*-2 position. The tosylate group can be converted directly into a phosphate ester functionality in good yield by refluxing the tosylate with silver diphenyl phosphate in xylene; however, partial racemization occurs at refluxing xylene temperature. Therefore, chiral phosphocholines (12) are

OTs

(*R*) - 10

$(RCO)_2O$ (1.5 eq)

$BF_3{\cdot}Et_2O$, CH_2Cl_2, 40°C, 2 h

O || RCO — H, OCR, OTs

11 (76%)

1. NaI, acetone, reflux

2. $AgOP(O)(OPh)_2$, benzene, reflux, 4 h

O || RCO — H, OCR, $OP(O)(OPh)_2$

(37% overall)

H_2, PtO_2

cyclohexane-HOAc, 4 h

O || RCO — H, OCR, OPO_3H_2

$HOCH_2CH_2N^+(CH_3)_3\ OTs^-$

Cl_3CCN, Py, 50°C, 48 h

O || RCO — H, OCR, $O{-}P(O)(O^-){-}OCH_2CH_2N^+(CH_3)_3$

(*R*)- 12

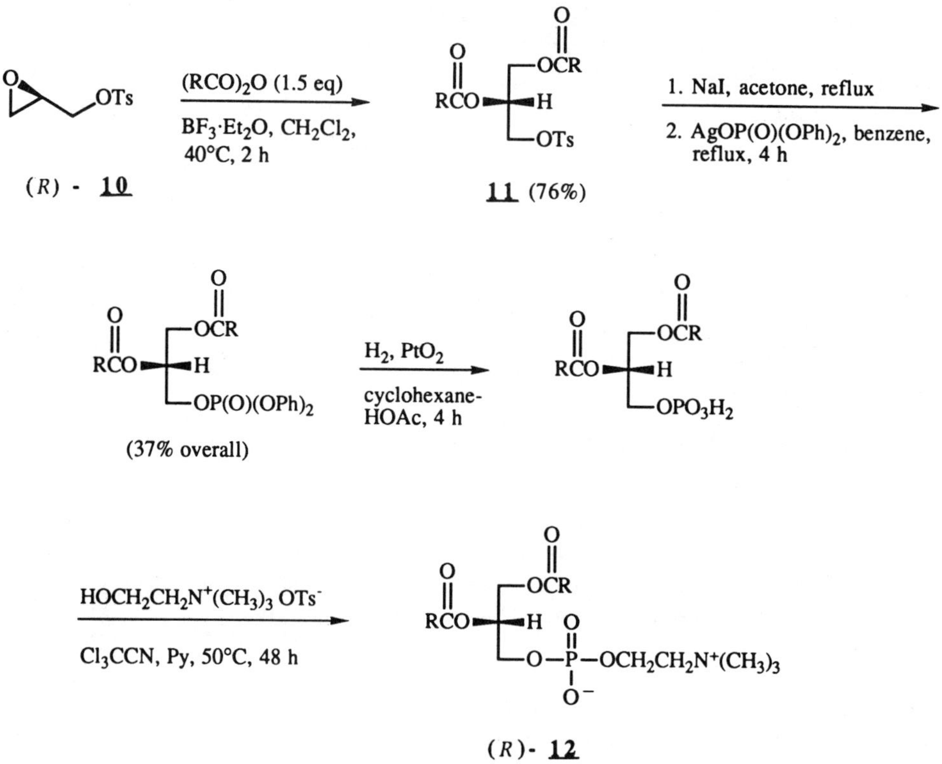

Scheme 6

prepared by first converting the tosylate 11 into the corresponding iodide by refluxing with excess sodium iodide in acetone; the iodide reacts with silver diphenyl phosphate in refluxing benzene. The diphenyl ester must be converted into the phosphatidic acid, which is then converted to diacylphosphocholine [(*R*)-12] by using standard procedures. The preparation of 2,3-diacyl-*sn*-glycero-1-phosphocholines [(*S*)-12], which have the unnatural configuration of glycerolipids derived from *sn*-glycerol-3-phosphate, can also be achieved conveniently from (*S*)-(+)-glycidyl tosylate [(*S*)-10] by the same method.

Disadvantages of the method: The chiral glycidyl derivatives 10 currently available are expensive, making large-scale syntheses of glycerolipids impractical. If unsaturated acyl PCs are to be prepared, the scheme must be redesigned, since the catalytic hydrogenolysis step cannot be used.

Literature: Ali and Bittman (1988).

Method: Preparation of distearoyl-PC (12, R = $C_{17}H_{35}$): Stearic anhydride (crystallized from hexane before use; 0.75 to 1.0 mmol) is placed in a dry two-neck 100-mL flask fitted with a condenser under nitrogen atmosphere. A solution of 114.1 mg (0.5 mmol) of (*R*)-(–)-glycidyl tosylate (10) in 20 mL of dry methylene chloride is added, followed by 4–5 drops (about 0.4 mmol) of freshly distilled $BF_3 \cdot Et_2O$. The mixture is refluxed with stirring for 2 h, then cooled to rt. Ether (30 mL) is added, the organic layer is washed with 10 mL of 5% aqueous sodium bicarbonate solution, 10 mL of brine, then dried (Na_2SO_4), filtered, and evaporated under reduced pressure. The residue is purified by flash chromatography to give 295 mg (76%) of 1,2-distearoyl-*sn*-glycerol 3-*p*-toluenesulfonate (11). The latter (200 mg, 0.26 mmol) is converted into the iodide by refluxing overnight with 250 mg (1.7 mmol) of sodium iodide in 10 mL of dry acetone under nitrogen. Acetone is removed by using a rotary evaporator. Ether (25 mL) is added to the residue, and the mixture is filtered through a Celite pad, which is washed with ether. Removal of the ether leaves a yellow residue that is purified by recrystallization from methanol-petroleum ether 9:1. The iodide (90 mg, 0.12 mmol) is dried in a desiccator and then dissolved in 10 mL of refluxing benzene in a flask protected from light with aluminum foil. Silver diphenyl phosphate (175 mg, 0.35 mmol) is added, and the reaction mixture is refluxed for 4 h, then cooled to rt, and filtered through a sintered-glass funnel packed with Celite. The Celite is washed with chloroform. Evaporation of the filtrate under vacuum leaves a residue that is dissolved in hexanes/ethyl acetate (95:5) and purified by flash chromatography (elution with the same solvent), yielding 60 mg (59%) of the diphenyl phosphate ester. Catalytic hydrogenolysis of the phenyl ester is carried out by reducing 80 mg of platinum oxide (Adams' catalyst) suspended in 10 mL of glacial acetic acid under hydrogen atmosphere with stirring until black granules of platinum black appear (about 1 h). A solution of 60 mg (0.070 mmol) of the diphenyl phosphate ester in 10 mL of cyclohexane-glacial acid (1:1) is injected rapidly into the flask through a rubber septum, and the mixture is stirred for 3 h at rt. The mixture is filtered through a sintered-glass funnel packed with Celite, which is then washed with chloroform. Evaporation of the filtrate under vacuum gives a solid that is dissolved in a minimum volume of chloroform. The phosphatidic acid is obtained by precipitation with 10 mL of cold (–20°C) acetonitrile and is then precipitated a second time, dried, and used in the next step without further purification.

The choline group is incorporated as follows. The phosphatidic acid (36 mg, 0.050 mmol) is dissolved by heating in 5–10 mL of dry pyridine under nitrogen at 50°C for 30 min. Freshly dried (in a desiccator over P_2O_5 or in a drying pistol over refluxing benzene for several days) choline tosylate (140 mg, 0.50 mmol, recrystallized from dry acetone before drying) and 2 mL of trichloroacetonitrile are added, and the reaction mixture is stirred for 48 h at 50°C, during which time the mixture becomes brown. [Alternatively, phosphatidic acid reacts with the tetraphenylborate salt of choline in the presence of 2,4,6-triisopropylbenzenesulfonyl chloride as the condensing agent (Harbison and Griffin, 1984).] The solvent is removed under vacuum. To ensure removal of pyridine, the residue is dissolved in 10–25 mL of chloroform/methanol (1:1), and the solvents are evaporated; this procedure is repeated two times. The residue is dissolved in 20 mL of THF/water (9:1), and the solution is applied to a mixed-bed ion-exchange column (20 g of Amberlite MB-3) that has been preequilibrated with the same solvent mixture. Elution with ~250 mL of THF/water 9:1, followed by evaporation of the solvents under reduced pressure and azeotropic removal of water with 2-propanol (3 × 25 mL), gives the crude product as a tan solid; negatively and positively charged species are retained on the mixed-bed resin. The desired phosphocholine 12 is then purified by flash chromatography on silica gel (elution with chloroform/methanol, 9:1 then 3:2, or with chloroform/methanol/water, 65:25:4) to give 32 mg (55%) of the product 12 that contains silica gel. To remove the silica gel, the product is dissolved in chloroform, and the solution is filtered three times through a 0.45-μm Metricel filter.

Caveats: The specific rotation of the starting material 10 should be measured before use; if the [α] value is low, the starting material is recrystallized, dried, and its [α] remeasured.

5. Thioester-PCs: Preparation of 1-Thioester-2-ester and 1,2-Dithioester-PC from Trityloxy Derivative of Glycidol

Key features of the method:

(*R*)- and (*S*)-Trityloxyglycidol (13) are prepared by tritylation of glycidol. The nucleophilic ring-opening reaction of (*R*)-trityloxyglycidol with an excess of thioacid (without use of a catalyst) gives 1-thioacyl-3-*O*-trityl-*sn*-glycerol (14) as an intermediate, which is acylated, detritylated under conditions that avoid acyl migration, and converted to 1-thioester-2-ester-PC (15). Ring opening of the starting material 13 with methyl xanthate gives a 1,2-trithiocarbonate derivative of glycerol (16), which is converted into 1,2-bisthioacyl-PC (17) (Hendrickson et al., 1983; Yu et al., 1990).

A thioether group can be incorporated at the *sn*-1 position of 18 by using 1.2 equiv of a long-chain thiol under base-catalyzed conditions (a catalytic amount of *n*-butyllithium in THF) to open the epoxide 13.

Literature: Hendrickson and Hendrickson (1990); Hendrickson et al. (1983); Ward (1988).

Advantage of the method: The trityl ether of glycidol (13) is a versatile C_3 synthon for the synthesis of a variety of glycerolipids. The reaction of 13 with hexadecyl mercaptan to yield 18 was reported to proceed with >98% ee, as judged by ^{1}H NMR analysis of the Mosher ester derived from the ring-opened intermediate. This value may be too high, however, since the Mosher ester was purified by chromatography prior to NMR analysis. [It is advisable to *avoid* purification of the Mosher esters for ee determination, and simply remove polar impurities by passing the crude Mosher ester through a short plug of silica gel. An example of the workup of a Mosher ester of a glycerol derivative is the following. The reaction mixture [(*R*)-or (*S*)-MTPA chloride and the glycerol derivative] is quenched by adding 3-(dimethylamino)propylamine, the solvent is removed under reduced pressure, and the residue is dissolved in spectral-grade hexane-

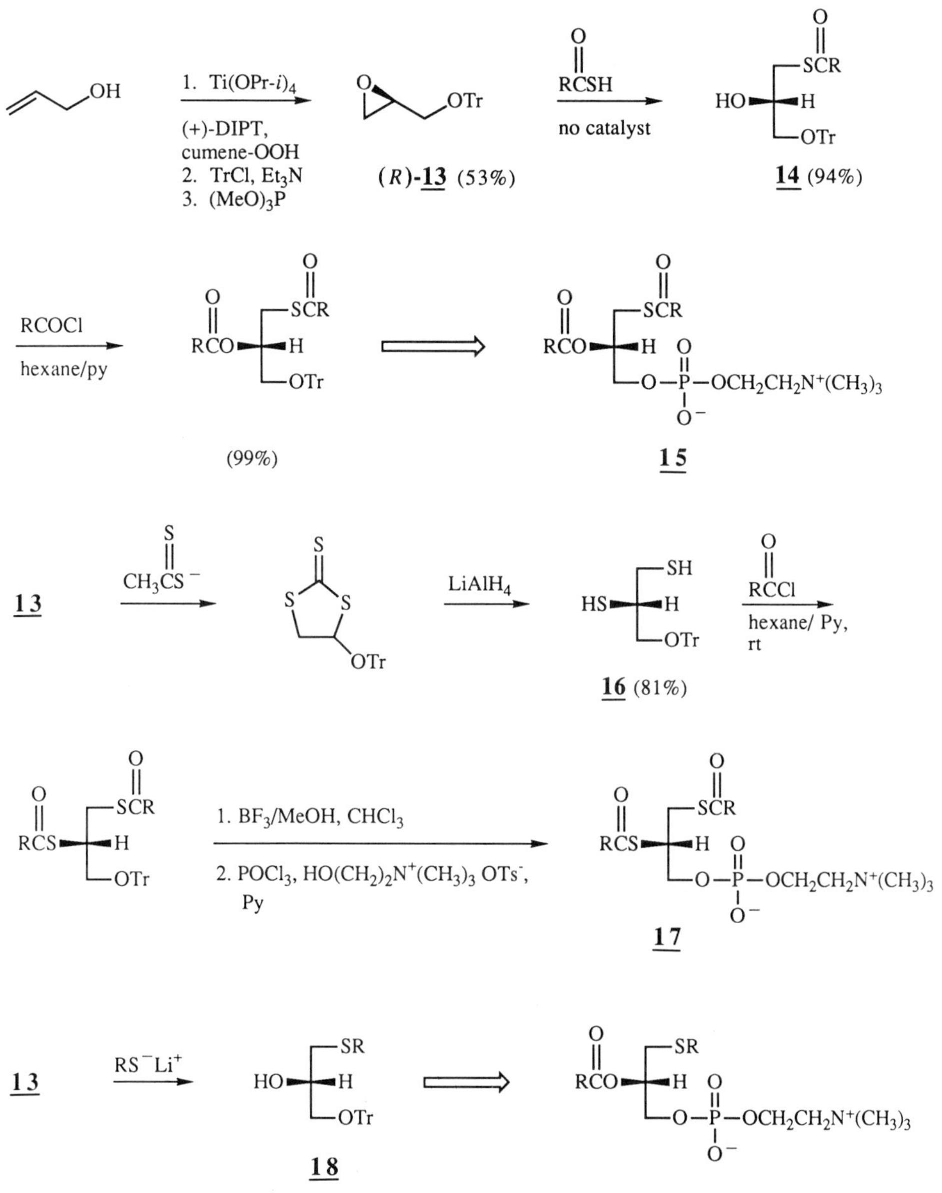

1. $Ti(OPr\text{-}i)_4$
(+)-DIPT, cumene-OOH
2. TrCl, Et_3N
3. $(MeO)_3P$

(*R*)-13 (53%)

RCSH, no catalyst

14 (94%)

RCOCl, hexane/py

(99%)

15

13

CH_3CS^-

$LiAlH_4$

16 (81%)

RCCl, hexane/ Py, rt

1. BF_3/MeOH, $CHCl_3$
2. $POCl_3$, $HO(CH_2)_2N^+(CH_3)_3$ OTs^-, Py

17

13

RS^-Li^+

18

Scheme 7

EtOAc. The solution is passed through a sintered-glass funnel packed with silica gel in order to remove polar impurities to give the crude Mosher ester after evaporation of the solvents; Ali and Bittman (1990).]

Also, use of a second method of ee determination of the derived Mosher esters is advisable when it is critical to have a reliable estimate of the optical purity; for methods of ee analysis of glycerol derivatives, see Table 2. For analysis by HPLC on a chiral stationary phase, see Gao et al. (1987); Guivisdalsky and Bittman (1989a); for analysis by capillary gas chromatography, see Hirth and Walther (1985).

Disadvantages of the method: Since the (*R*)- and (*S*)-trityloxyglycidols (13) are not currently available commercially, they must be prepared from allyl alcohol or from commercially available (*R*)- and (*S*)-glycidol. The % ee values of the (*R*)- and (*S*)-trityloxyglycidols (13) are not established. *O*-Benzyl glycidol, which is available commercially in both enantiomeric forms, may also be useful for the preparation of 15 and 17.

Method: (*R*)-Glycidyl trityl ether (13) is prepared in 53% yield (mp 99–100°C; [α] not reported) by catalytic asymmetric epoxidation of allyl alcohol and in situ tritylation. Trityl chloride (1 equiv) and triethylamine (1.2 equiv) are added at –30°C to (*R*)-glycidol that is prepared from allyl alcohol and treated carefully with trimethyl phosphite (for reduction of excess hydroperoxide) at the same temperature. The reaction mixture is stirred overnight at 2°C, then filtered through a Celite pad. The filtrate is washed with 10% aqueous tartaric acid (2 × 30 mL), saturated aqueous sodium bicarbonate (2 × 30 mL), and brine (2 × 30 mL). The organic layer is dried (Na_2SO_4), filtered, and concentrated, leaving an oil that is dissolved in hexane-ether (94 : 6) and purified by silica gel chromatography (elution with hexane-ether, 94 : 6 then 88 : 12).

The epoxide opening with thiodecanoic acid is carried out as follows. To a solution of 1.58 g (5.0 mmol) of (*R*)-trityloxyglycidol (13) in 40 mL of hexane-toluene 1 : 1 is added 1.94 g (10.3 mmol) of thiodecanoic acid (prepared from decanoyl chloride and sodium sulfide in pyridine) dropwise. Water is added to the mixture, and the organic layer is washed with methanol/0.5 M NH_4OH (3 : 1) and twice with methanol/water (3 : 1). The organic phase is dried (Na_2SO_4), and the solvents are removed under vacuum to give 3.2 g

Table 2 Methods for Determination of the Optical Purity of Glycerol Derivatives[a]

Method	Derivative	References
Capillary gas chromatography	Chiral Mosher ester	Hirth and Walther (1985)
Chiral stationary phase HPLC	Chiral Mosher ester Benzoyl	Guivisdalsky and Bittman (1989a,c) Kodali (1987)
Nuclear magnetic resonance	Chiral Mosher ester	Guivisdalsky and Bittman (1989a,b); Ali and Bittman (1990); Johnson et al. (1989); Abdelmageed et al. (1990); Hendrickson and Hendrickson (1990)
Nuclear magnetic resonance	Chiral shift reagent	Johnson et al. (1989); Mikkilineni et al. (1988)

[a]Since optical rotations are not considered to be reliable estimates of chiral purity, the assessment of enantiomeric purity of glycerol derivatives is made by a variety of other methods. The most common methods are chiral stationary phase HPLC and NMR analysis of diastereomers obtained by reaction of the glycerol derivative with a chiral reagent. In these methods it is important to achieve conditions that result in baseline separations of the respective peaks of the diastereomers. It is advisable to prepare the same derivative from each glycerol enantiomer, or from one glycerol enantiomer and the racemic mixture of the glycerol compound. The Mosher [(*R*)- or (*S*)-α-methoxy-α-(trifluoromethyl)phenylacetic acid] ester is the most common derivative made from the glycerol compound. For ^{1}H-NMR analysis, spectra are generally recorded at high field (400 or 500 MHz) in order to separate the signals from the enantiotopic H_a and H_b protons at C-1 of the (*R*)- or (*S*)-Mosher esters of the glycerol derivatives; see Ali and Bittman (1990).

of crude product 14. Purification by chromatography on silica gel (elution with hexane then hexane-ether 98 : 2) gives 2.37 g (94%) of product 14. Acylation with 1.5 equivalents of acyl chloride in hexane-pyridine at 2°C for 10 min and at rt for 30 min, followed by workup, gives 1-thioacyl-2-acyl-3-*O*-trityl-*sn*-glycerol in 99% yield. The synthesis of 15 is completed by detritylation with the boron trifluoride/methanol complex (Hermetter and Paltauf, 1981) and reaction of the alcohol with phosphorus oxychloride and choline tosylate.

For the preparation of dithioester-PCs (17), 3-*O*-trityl-1,2-dideoxy-1,2-(thiocarbonyldithio)-*sn*-glycerol is reduced to the mercaptan 16 with lithium aluminum hydride (Yu et al., 1990). The mercaptan 16 is dissolved in acetone and purified by silica gel chromatography (elution with hexane-acetone 50 : 1); the column is packed with silica gel that has been deoxygenated under vacuum and stored under argon. To a solution of mercaptan 16 in hexane-pyridine is added slowly a solution of acyl chloride in hexane. After the reaction mixture is stirred at rt for 7 h, benzene is added, and the organic phase is washed sequentially with water, 0.5 M ammonium hydroxide in methanol-water 3 : 1, and methanol-water 1 : 1. Evaporation of the solvents gives an oil that is detritylated by adding boron trifluoride-methanol to a chloroform solution of the dithioester, followed by stirring at rt for 2 h.

The detritylated compound is used in the phosphorylation reaction without purification in order to minimize the possibility of acyl migration and hydrolysis of the thioester linkages. The phosphorylation reaction is carried out by adding a solution of freshly distilled phosphorus oxychloride in chloroform to a solution of 1,2-dithioacyl-*sn*-glycerol in chloroform containing pyridine. After the mixture is stirred at 50°C for 1 h, choline tosylate and pyridine are added at rt. The residue obtained after workup is dissolved in chloroform-methanol-water 65 : 25 : 4 and passed through a column of Rexyn I-300, eluting with the same solvent system. Evaporation of the eluate gives a residue that is purified by silica gel chromatography, giving 17.

C. 1-Acyl-2-*O*-alkyl-*sn*-glycero-3-phosphocholines

1. Preparation from 1-*O*-Benzyl-*sn*-glycerol 3-*O*-(*p*-Toluenesulfonate)

Outline of the method:

The regio- and stereospecific ring opening of the epoxide ring of (*R*)- and (*S*)-glycidyl arenesulfonates (10) with benzyl alcohol (2 equivalents) in the presence of $BF_3 \cdot Et_2O$ gives the chiral C_3-synthon 1-*O*-benzyl-*sn*-glycerol 3-*O*-arenesulfonate (or its enantiomer) (19). The sequence of alkylation with alkyl triflate in the presence of an excess of hindered base (2,6-di-*tert*-butyl-4-methylpyridine), debenzylation, and acylation gives 1-acyl-2-*O*-alkyl-*sn*-glycerol 3-arenesulfonate. Replacement of the arenesulfonate group by phosphocholine completes the synthesis. The optical purities of the PCs made by this method are very high (93–96% ee), as shown by high-field ^1H NMR spectroscopy of a Mosher ester derivative.

Literature: Ali and Bittman (1990).

Method: Two equivalents of benzyl alcohol and a catalytic amount (4 drops, ~0.40 mmol) of BF_3 etherate are added to a solution of ~4 mmol of glycidyl tosylate (10) or glycidyl 3-nitrobenzenesulfonate in 40 mL of methylene chloride. The reaction mixture is stirred under nitrogen at rt for 2 h. Ether (50 mL) is added, the organic phase is washed with 5% aqueous sodium bicarbonate-saturated sodium chloride, then dried (Na_2SO_4).

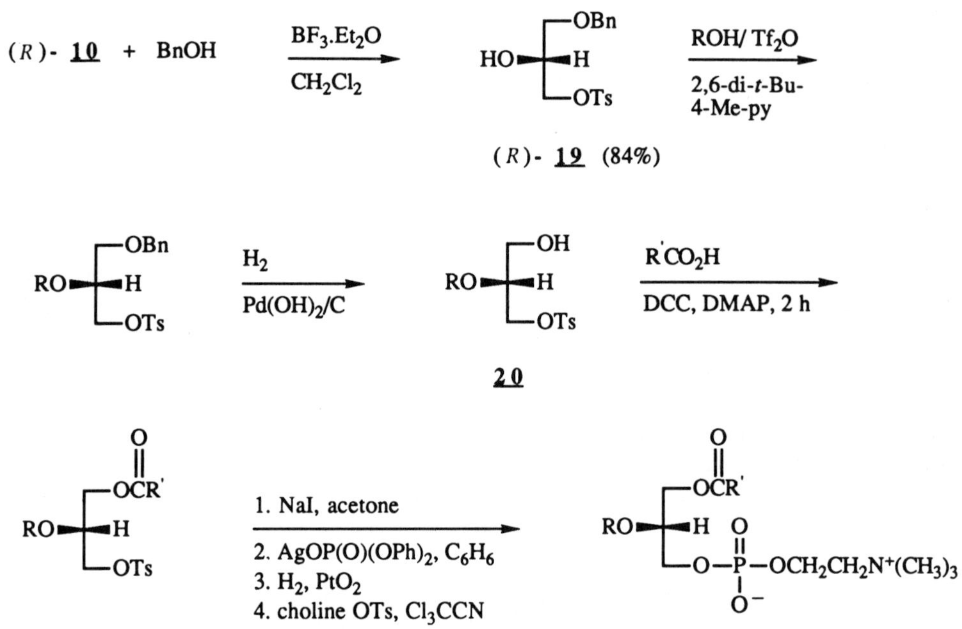

(*R*)- 10 + BnOH —$BF_3.Et_2O$ / CH_2Cl_2→ (*R*)- 19 (84%) —ROH/ Tf_2O / 2,6-di-*t*-Bu-4-Me-py→

—H_2 / $Pd(OH)_2/C$→ 20 —$R'CO_2H$ / DCC, DMAP, 2 h→

1. NaI, acetone
2. $AgOP(O)(OPh)_2$, C_6H_6
3. H_2, PtO_2
4. choline OTs, Cl_3CCN

Scheme 8

The solvent is removed under reduced pressure, leaving a residue that is dissolved in ethyl acetate-light petroleum ether 15:85; yield, ~80%; $[\alpha]^{25}{}_D$ –6.70° (*c* 7.20, benzene). Alkylation of 19 at the available *sn*-2 position is carried out by using two equivalents of long-chain alcohol and 4 equivalents of hindered base. For example, a solution of 68 mg (0.28 mmol) of 1-hexadecanol and 115 mg (0.56 mmol) of 2,6-di-*tert*-butyl-4-methylpyridine in 10 mL of methylene chloride is cooled to –78°C. The solution is flushed with nitrogen several times, and triflic anhydride (71 mg, 0.42 mmol) is added. After the mixture is allowed to warm to rt and stirred for 1.5 h, 50 mg (0.14 mmol) or tosylate 19 is added to the solution. The mixture is refluxed for 3 days under nitrogen. The precipitate of triflic acid is removed by filtration, and the filtrate is evaporated to leave an oil that is purified by flash chromatography with ethyl acetate-light petroleum ether 5:95; yield, 37 mg (43%); $[\alpha]^{25}{}_D$ –2.16° (*c* 1.57, chloroform). The benzyl group is converted to toluene by catalytic hydrogenolysis, giving alcohol 20, which is acylated with a fatty acid in the presence of DCC and DMAP. The arenesulfonate group is replaced by iodide, and the iodo compound is converted to the diphenyl phosphate ester as in Scheme 6. The latter is converted to the 1-acyl-2-*O*-alkylphosphocholine by standard procedures.

D. Acylaminophosphatidylcholines

1. 1,2-Diacylamido-1,2-dideoxy-PC: Preparation from 2,3-Diaminopropionic Acid or from L-Asparagine

Outline of the method:

1,2-Diacylamido-1,2-dideoxy-PC is prepared via 1,2-diacylamido-3-propanol (21). The latter is prepared in synthetic schemes that start either with 2,3-dibromopropionic

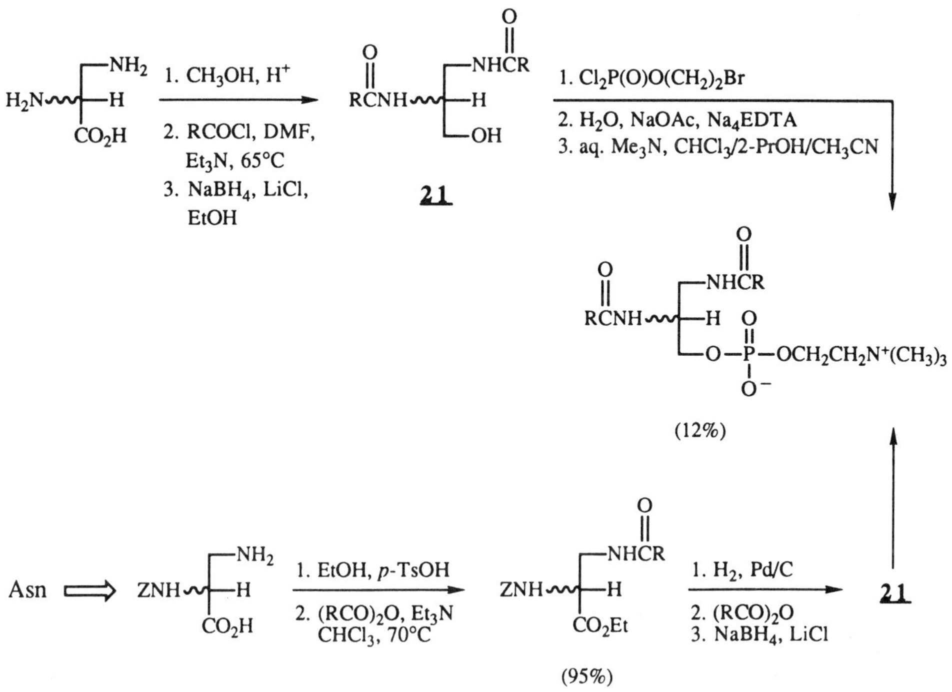

1. CH_3OH, H^+
2. RCOCl, DMF, Et_3N, 65°C
3. $NaBH_4$, LiCl, EtOH

21

1. $Cl_2P(O)O(CH_2)_2Br$
2. H_2O, NaOAc, Na_4EDTA
3. aq. Me_3N, $CHCl_3$/2-PrOH/CH_3CN

(12%)

Asn ⟹ ZNH

1. EtOH, p-TsOH
2. $(RCO)_2O$, Et_3N $CHCl_3$, 70°C

(95%)

1. H_2, Pd/C
2. $(RCO)_2O$
3. $NaBH_4$, LiCl

21

Scheme 9

acid or with L-asparagine (Asn). *rac*-2,3-Diaminopropionic acid is obtained by heating 2,3-dibromopropionic acid with ammonium hydroxide in a sealed tube at 100°C, then converted to the methyl ester. The ester is acylated and reduced to the alcohol 21 with sodium borohydride in the presence of lithium chloride. Alcohol 21 is made in the asparagine route in a scheme that involves converting N^2-benzyloxycarbonyl-L-asparagine into N^2-benzyloxycarbonyl-2,3-diaminopropionic acid. The latter is converted to *N,N*-diacylamidopropionate, which is esterified and reduced to alcohol 21. The phosphocholine group is inserted via (2-bromoethyl)phosphodichloridate, hydrolysis, and trimethylamination.

Literature: Sunamoto et al. (1990).

2. 1-Acyl-2-acylamino-2-deoxy-PC: Preparation from 2-Aminopropanol

Outline of the method:

Monoacylation of *rac*-2-aminopropanediol (22) is carried out by adding a long-chain acyl chloride in THF to an aqueous solution of excess 2-aminopropanediol in the presence of triethylamine. After the reaction mixture is heated at 40°C overnight, the solvents are removed under vacuum and the residue is dissolved in chloroform. The chloroform layer is washed with 5% HCl and water, and the crude 2-acylaminopropanediol (23) is crystallized from chloroform; yield, 75%. Acylation of one of the hydroxy groups with 1.04 equivalents of palmitoyl chloride in chloroform containing 2.4 equivalents of

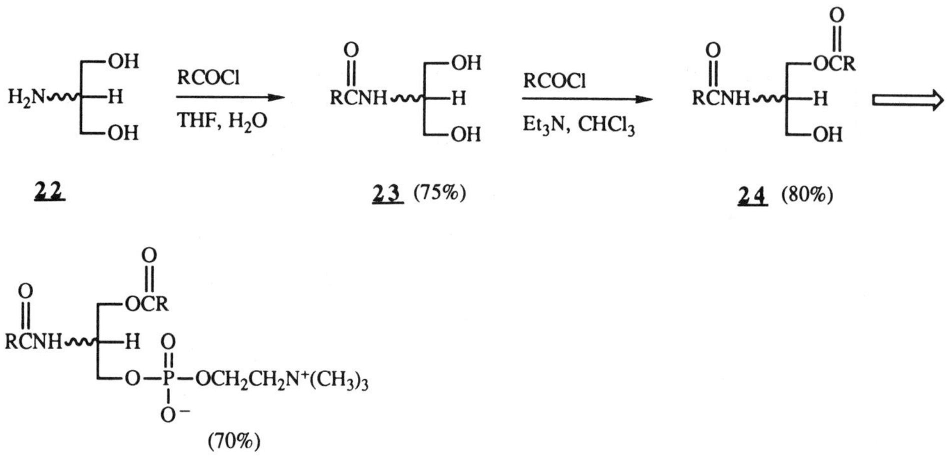

Scheme 10

triethylamine gives *rac*-1-palmitoyl-2-acylaminopropanediol. The chloroform solution is washed with 5% HCl and water, the solvent is evaporated, and the residue is crystallized from chloroform-hexane; yield of 24, 80%. Phosphorylation of 24 with 2-chloro-2-oxo-1,3,2-dioxaphospholane in benzene containing triethylamine, followed by amination with anhydrous trimethylamine, gives the phosphocholine product; yield, 70% after purification by silica gel chromatography.

Literature: Dijkman et al. (1990).

3. 1-Acylamino-1-deoxy-2-*O*-methyl-PC: Preparation from 3-Amino-1,2-propanediol

Outline of the method:

rac-3-Amino-1,2-propanediol is acylated with 1 equivalent of palmitoyl chloride in pyridine-DMF, giving *rac*-3-hexadecanamido-1,2-propanediol (25) in 91% yield after

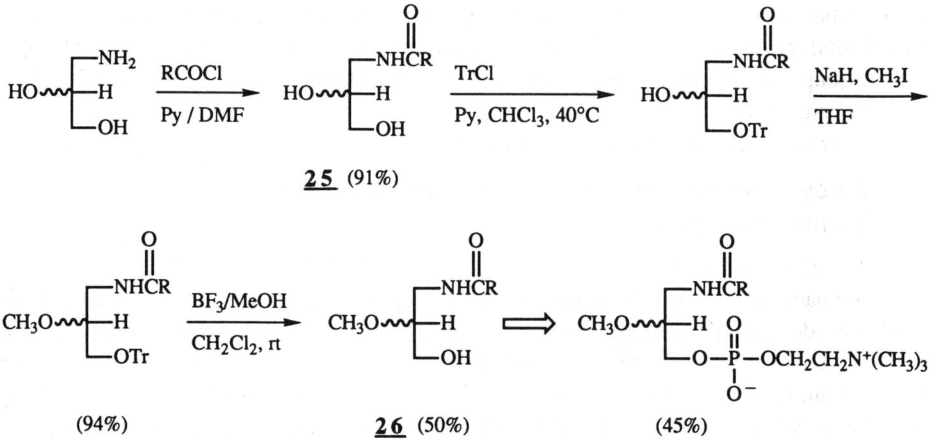

Scheme 11

recrystallization from ethanol and 2-propanol. After tritylation of 25 at the primary hydroxy group, alkylation of the secondary hydroxy group is carried out by using sodium hydride in THF, followed by addition of methyl iodide. Detritylation with boron trifluoride/methanol complex in methylene chloride gives *rac*-3-palmitoylamido-2-methoxy-1-propanol (26), which is purified by precipitation from hot petroleum ether and chromatography on silica gel (elution with hexane-ethyl acetate 2:1). The phosphocholine group is introduced into 26 as described above for 1-acyl-2-acylamino-2-deoxy-PC; yield, 45% after purification by chromatography on silica gel (elution with chloroform-methanol-ammonium hydroxide (75:25:5) and precipitation from cold acetone. The ether amido-PC inhibits protein kinase C and leukemic cell growth.

Literature: Marx et al. (1988).

E. A Chiral "Triglyceride" with One Ester Linkage: 1-Acyl-2-*O*-methyl-*sn*-3-(acylaminodeoxy)glycerol

Outline of the method:

O-Benzyl glycidol [prepared from (*R*)-glycidyl tosylate [(*R*)-10] by the procedure of Byun and Bittman (1989)] is opened with aqueous NH_4OH, giving (*R*)-1-*O*-benzyl-3-amino-3-deoxyglycerol (27) in quantitative yield. Acylation selectively at nitrogen without accompanying *O*-acylation is carried out by using 0.9 equivalent of decanoyl chloride. After the amide is purified by recrystallization from ether/hexane 1:1, the methyl ether is prepared by using 3 equivalents of sodium hydride and 3 equivalents of methyl iodide.

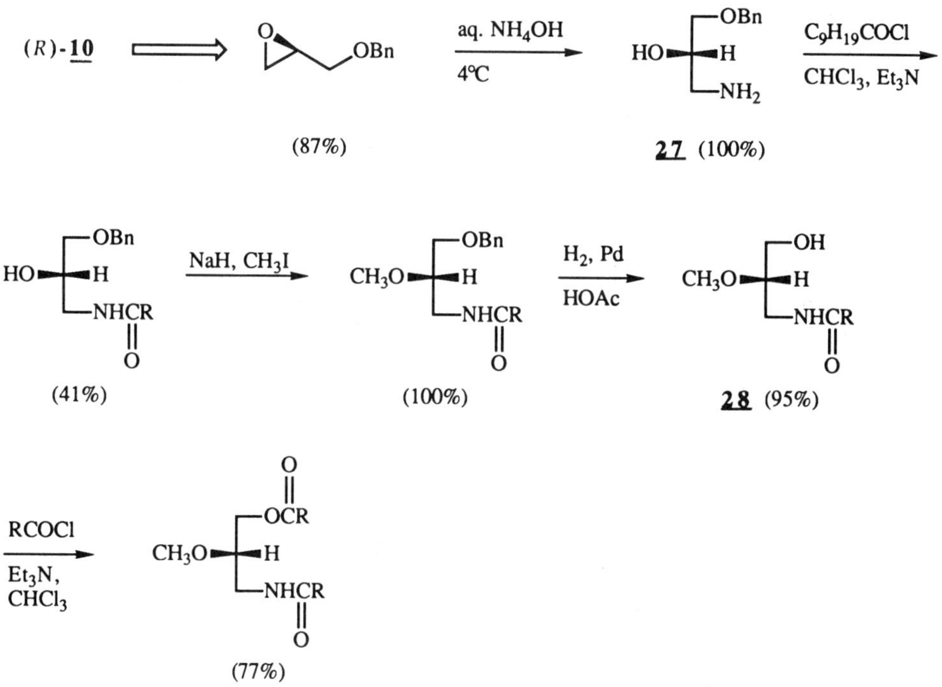
(R)-10
O
OBn
aq. NH4OH
4°C
(87%)
OBn
HO
H
NH2
27 (100%)
C9H19COCl
CHCl3, Et3N
OBn
HO
H
NHCR
O
(41%)
NaH, CH3I
OBn
CH3O
H
NHCR
O
(100%)
H2, Pd
HOAc
OH
CH3O
H
NHCR
O
28 (95%)
RCOCl
Et3N,
CHCl3
O
OCR
CH3O
H
NHCR
O
(77%)

Scheme 12

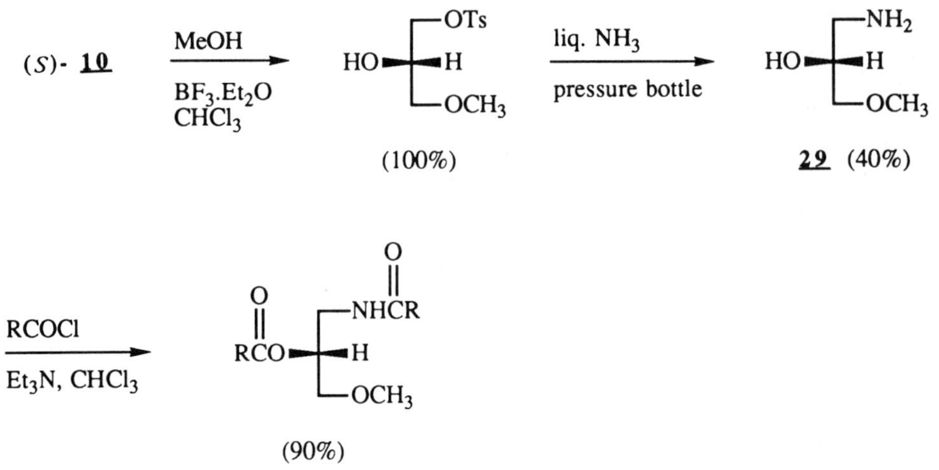

(S)- 10 → MeOH / $BF_3.Et_2O$, $CHCl_3$ → HO–C(H)(–OTs)(–OCH_3) (100%) → liq. NH_3 / pressure bottle → HO–C(H)(–NH_2)(–OCH_3) 29 (40%)

→ RCOCl / Et_3N, $CHCl_3$ → RC(=O)O–C(H)(–NHC(=O)R)(–OCH_3) (90%)

Scheme 13

After debenzylation (H_2/Pd in acetic acid), the glycerol derivative 28 is purified by silica gel chromatography (elution with ether/hexane 1 : 1). Acylation of 28 gives 1-acyl-2-*O*-methyl-*sn*-3-(acylaminodeoxy)glycerol; $[\alpha]^{20}{}_D$ $+7.2°$ (neat). The enantiomer is made from (*S*)-glycidyl tosylate [(*S*) -10] by the same procedure. The chiral triglycerides having only one ester bond are used to analyze the stereo- and regiospecificity of *Pseudomonas glumae* lipase.

The hydrolyzable ester bond is placed at the *sn*-2 position by first opening (*S*)-glycidyl tosylate [(*S*)-10] with methanol using $BF_3 \cdot Et_2O$ as the catalyst (Guivisdalsky and Bittman, 1989b), displacing the tosylate with liquid ammonia in a pressure bottle (Sowden and Fischer, 1942), followed by simultaneous diacylation of the amino and hydroxy groups of 29. This sequence of reactions is outlined in Scheme 13.

Literature: Deveer et al. (1991).

Amino-linked glycerol derivatives can also be prepared by ring opening of (*R*)-10 with trimethylsilyl azide using catalytic amounts of Ti(OPr-*i*)$_4$ in THF or Al(OPr-*i*)$_3$ in methylene chloride, as outlined in Scheme 14. The azide group of 30 is reduced readily to the amino group.

Literature: Sutowardoyo and Sinou (1991).

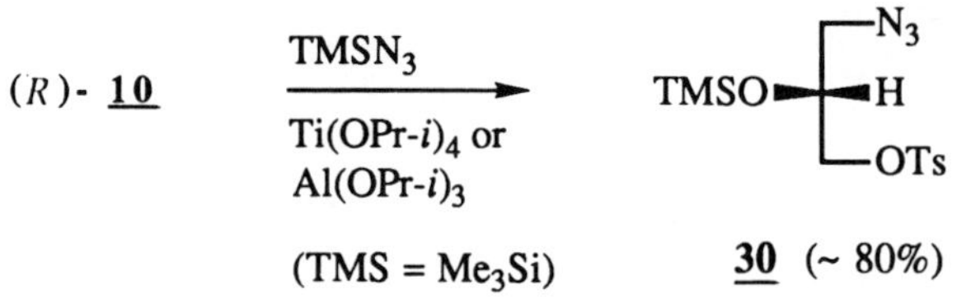

(R)- 10 → $TMSN_3$ / Ti(OPr-*i*)$_4$ or Al(OPr-*i*)$_3$ (TMS = Me_3Si) → TMSO–C(H)(–N_3)(–OTs) 30 (~ 80%)

Scheme 14

IV. PREPARATION OF 1-ETHER-LINKED LIPIDS

A. 1-*O*-Alkyl-*sn*-glycerol

1-*O*-Alkyl-*sn*-glycerol is prepared in moderate yield (<60%) by heating glycidol with $Ti(OR)_4$ (where R is the long-chain alkyl group) without solvent at 70–75°C. About 5% of the 2-*O*-alkyl isomer is obtained as a by-product, as shown in Scheme 15.

Literature: Johnson et al. (1989).

The $BF_3 \cdot Et_2O$ catalyzed opening of glycidyl arenesulfonates with alcohols (Guivisdalsky and Bittman, 1989c) (see Scheme 16) is a better method for preparing ether-linked glycerolipids than the titanium-mediated opening of the parent glycidol, since it is regio- and stereospecific. In addition, the enantiomeric purity of glycidyl derivatives commercially available at this time is higher than that of glycidol.

B. Ether/Ester-PCs (Including Platelet-Activating Factor)

At physiological (nanomolar) concentrations, platelet-activating factor (PAF), 1-*O*-alkyl-2-acetyl-*sn*-glycero-3-phosphocholine, plays many roles in biological and pathological processes involving various tissues and cell types. For a recent review, see Braquet (1991). New methods of preparing PAF and PAF analogs are presented here.

1. Platelet-Activating Factor and Long-Chain Esters at *sn*-2: Preparation of Ether/Ester-PCs from Arenesulfonate Derivatives of Glycidol

Key features of the method:

Glycidyl arenesulfonate 10 are opened at C-3 with excellent regio- and stereospecificity with a long-chain saturated or unsaturated alcohol in methylene chloride in the presence of $BF_3 \cdot Et_2O$ as catalyst.

The ring-opened arenesulfonate is protected as the *O*-benzyl ether 31 by using basic conditions that proved to be so mild that epoxide formation via arenesulfonate displacement did not take place.

1-*O*-Alkyl-2-*O*-benzyl-*sn*-glycerol (32) is prepared by arenesulfonate displacement using cesium acetate, followed by lithium aluminum hydride reduction of the acetoxy group; alternatively, the acetoxy group can be converted into a hydroxy group via hydrolysis. The phosphocholine group is incorporated, the benzyl group is removed, and the resulting 1-ether-2-lyso-PC is acylated to give chiral ether-ester-PC.

Either the natural (*R* configuration) or unnatural enantiomer (*S* configuration) of PC can be synthesized, depending on the choice of (*R*)- or (S)-10 as the starting material. The ring opening of 10 is stereospecific, as shown by analysis of the diastereomeric (*R*)-Mosher esters of the ring-opened intermediate (1-*O*-alkyl-*sn*-glycerol 3-tosylate) by HPLC on a chiral stationary phase column and high-field ^1H-NMR spectroscopy (Guivisdalsky and Bittman, 1989a,b,c).

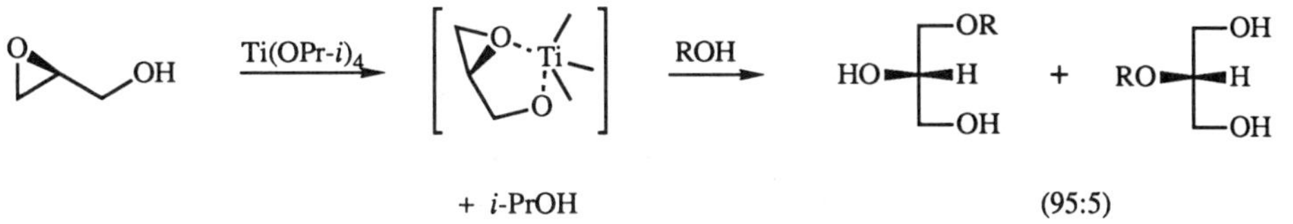

$Ti(OPr\text{-}i)_4$ → ROH → (95:5)

+ *i*-PrOH

Scheme 15

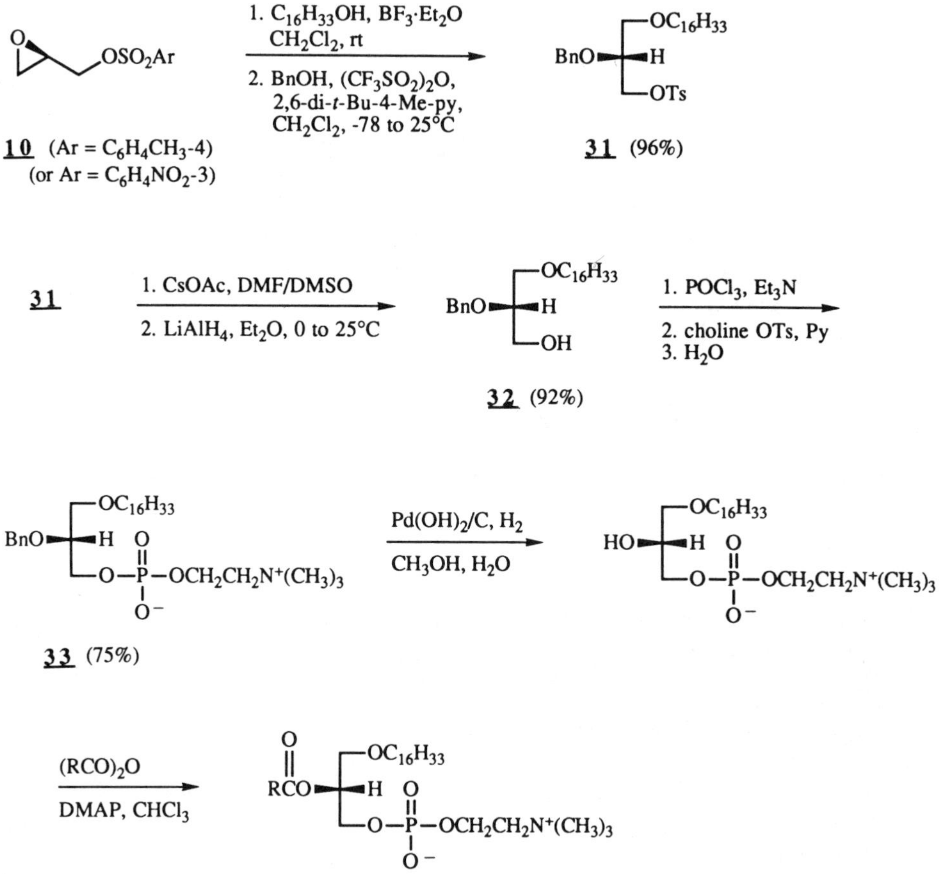

10 (Ar = $C_6H_4CH_3$-4) (or Ar = $C_6H_4NO_2$-3)

1. $C_{16}H_{33}OH$, $BF_3 \cdot Et_2O$, CH_2Cl_2, rt
2. BnOH, $(CF_3SO_2)_2O$, 2,6-di-*t*-Bu-4-Me-py, CH_2Cl_2, -78 to 25°C

31 (96%)

31 → 1. CsOAc, DMF/DMSO; 2. $LiAlH_4$, Et_2O, 0 to 25°C

32 (92%)

1. $POCl_3$, Et_3N
2. choline OTs, Py
3. H_2O

33 (75%)

$Pd(OH)_2/C$, H_2; CH_3OH, H_2O

$(RCO)_2O$; DMAP, $CHCl_3$

Scheme 16

Literature: Guivisdalsky and Bittman (1989c).

Advantages of the method: Use of BF_3 as the catalyst has several advantages over use of the stoichiometric titanium (IV) isopropoxide procedure. The nucleophilic species liberated in Ti(OPr-*i*)$_4$-mediated reactions may compete with the long-chain alcohol in the ring-opening reaction, whereas the nonnucleophilic molecule ether is displaced when BF_3 etherate is used. Workup is much easier with BF_3, since titanium-mediated reactions generally require the use of sulfuric acid or sodium hydroxide with prolonged stirring to achieve phase separation. In contrast, workup of small-scale BF_3 etherate mediated reactions involves merely evaporation of the solvent on a rotary evaporator and purification of the residue by flash chromatography; in large-scale reactions, 10% aqueous sodium bicarbonate solution is added, followed by extraction of the glyceryl ether with dichloromethane.

Method for preparation of 1-O-alkyl-2-acetyl- and 1-O-alkyl-2-palmitoyl-sn-glycero-3-phosphocholines: To a mixture of 0.40 mmol of (*R*)-(–)-glycidyl 3-nitrobenzenesulfonate [recrystallized twice from ethanol before use, then dried; $[\alpha]^{25}{}_D$–23.3° (*c* 2.14, $CHCl_3$)] and 0.40 mmol of long-chain alcohol in 3 mL of dichloromethane is added 4 drops (~5 mol %) of a 10% stock solution of boron trifluoride etherate in dichloromethane (prepared by diluting 1 volume of freshly distilled $BF_3 \cdot Et_2O$ with 9 volumes of CH_2Cl_2). The mixture is stirred at rt for 18 h under nitrogen. The solvent is removed under reduced pressure, leaving a residue that is purified by flash chromatography (elution with 20% ethyl acetate-hexanes); ee >99% by analysis of the (*R*)-(+)-MTPA ester on a Pirkle type IA column, eluted with hexanes/2-propanol 87.5 : 12.5, flow rate 0.5 mL/min. A similar procedure is used with (*R*)-(–)-glycidyl tosylate as the starting material, to give >94% ee of the ring-opened product (analysis by chiral HPLC, elution with hexanes-2-propanol 9 : 1); when 1-*O*-hexadecyl-*sn*-glycerol 3-tosylate is recrystallized three times from ether-hexanes, the chiral purity is raised to >97% ee.

The benzyl ether is formed by reaction with benzyl trifluoromethanesulfonate (triflate) as follows. To a dry, 50-mL round-bottom flask equipped with a Claisen head and a nitrogen-filled balloon are added 5 mL of dry dichloromethane and 168 μL (1.0 mmol) of trifluoromethanesulfonic acid anhydride. A solution of 104 μL (1.0 mmol) of benzyl alcohol and 205 mg (1.0 mmol) of 2,6-di-*tert*-butyl-4-methylpyridine in 2 mL of dry dichloromethane is added dropwise over a 5-min period. The reaction mixture is stirred at –78°C for 15 min. A solution of 0.5 mmol of 1-*O*-alkyl-*sn*-glycerol 3-tosylate and 268 mg (1.3 mmol) of 2,6-di-*tert*-butyl-4-methylpyridine in 2 mL of dry dichloromethane is added dropwise over a 5-min period. After the mixture has stirred at –78°C for 30 min, it is warmed to rt, and stirring is continued until 1-*O*-alkyl-*sn*-glycerol 3-tosylate has disappeared (about 4 h), as monitored by TLC (developed with hexanes/ethyl acetate 3 : 1). The excess of benzyl triflate is destroyed by slowly adding 167 μL (2.07 mmol) of pyridine. The reaction mixture is then diluted with dichloromethane (30 mL) and washed with water (3 × 10 mL). The organic layer is dried (Na_2SO_4), and the solvents are removed under reduced pressure. The residue is purified by flash chromatography (eluted with hexanes-ethyl acetate 8 : 1) to give the pure *O*-benzyl product 31 as a colorless oil.

The arenesulfonate group is removed by acetate displacement followed by lithium aluminum hydride reduction as follows. The *sn*-3-tosylate group is displaced by adding 2 equivalents of cesium acetate in 5 mL of dry DMSO-DMF (4 : 1) to 1 equivalent of tosylate 31. After the mixture has stirred for 36 h (but only a few hours are needed when the 3-nitrobenzenesulfonate is used instead of the tosylate group), water (30 mL) is added and the product is extracted with ether (3 × 50 mL). The organic phase is dried (Na_2SO_4), and the solvents are concentrated under reduced pressure to a volume of about 10 mL. The solution is cooled in an ice bath, and 40 mg (1.0 mmol) of lithium aluminum hydride is added. The mixture is stirred for 30 min at 0°C and then at rt for 2 h. Water is added, and the aluminum salts are removed by filtration. The product is extracted with 60 mL of chloroform. Evaporation of the solvents gives the *sn*-3 alcohol 32 in 92% yield. Alternatively, the acetate intermediate can be hydrolyzed with aqueous carbonate solution to afford alcohol 32.

The phosphocholine group is introduced by stirring with 1.25 equivalents of phosphorus oxychloride and 1.25 equivalents of triethylamine in alcohol-free chloroform at –10°C for 30 min, then at rt for 30 min; then 1.5 equivalents of choline tosylate and 0.5 mL of pyridine are added, and the mixture is stirred for 16 h. Water (0.2 mL) is added, and the mixture is stirred for 30 min. The solvents are removed under reduced pressure, and 30 mL of dichloromethane-toluene 1 : 1 is added to the residue. The mixture is filtered, and the filtrate is evaporated to leave a residue that is dissolved in THF-water 9 : 1 and passed through an Amberlite column two times (elution with THF-water 9 : 1). The eluate is evaporated and the residue is purified by flash chromatography (elution with chloroform-methanol-water 65:35:4) to give 1-*O*-alkyl-2-*O*-benzyl-*sn*-glycero-3-phosphocholine (33) as a white solid in 75% yield.

The phosphocholine moiety has also been incorporated into 1-*O*-hexadecyl-2-*O*-benzyl-*sn*-glycerol (32) in 70% yield by the reaction with (2-bromoethyl)phosphodichloridate ($Cl_2P(O)OCH_2CH_2Br$), then hydrolysis of the P-Cl bond with 0.1 M KCl, and amination with trimethylamine in the presence of silver acetate (Ohno et al., 1983).

Catalytic hydrogenolysis of 0.24 mmol of 33 is carried out using 100 mg of 20% palladium hydroxide on carbon (Pearlman's catalyst) in 9 mL of methanol and 1 mL of water, with stirring under hydrogen atmosphere for 24 h. The mixture is filtered through Celite, the Celite is washed with methanol, and the filtrate is evaporated, leaving a residue that is dried by azeotropic removal of water using 2-propanol.

The lyso-PC is acylated by stirring with 5 equivalents of fatty acid anhydride and 1 equivalent of DMAP in alcohol-free chloroform under nitrogen for 24 h. The solvents are removed under reduced pressure, leaving a residue that is purified by flash chromatography (elution with chloroform, then chloroform-methanol 9:1, then chloroform-methanol 3:2). A chloroform solution of the product is passed through a 0.45-μm Metricel filter to remove suspended silica gel. Finally, the product 1-*O*-alkyl-2-acyl-PC is further purified by lyophilization from benzene. When acetic anhydride is used in the acylation reaction, PAF is the product; $[\alpha]^{25}_{D}$ −3.39° (*c* 0.53, $CHCl_3$-CH_3OH, 1:1) (Guivisdalsky and Bittman, 1989c). When palmitic acid is used in the acylation reaction, 1-*O*-alkyl-2-palmitoyl-PC is obtained in 98% yield.

Glycerol acetonide *S*-4 has been converted to (*R*)-PAF in a route that involves introduction of a 3-isoxazolyloxy moiety at the *sn*-2 position of 3-*O*-alkyl-1-*O*-trityl-*sn*-glycerol (Nakamura et al., 1990). This reaction occurs with inversion, since a Mitsunobu reaction is used with 3-hydroxyisoxazole, triphenylphosphine, and dimethyl azodicarboxylate. This is an alternate route to chiral PAF that does not involve 2-*O*-benzyl derivatives.

2. A Photoreactive, Radioiodinated PAF Derivative [1-*O*-(4'-Azido-2'-hydroxy-3-iodobenzamido)-undecyl Ether of PAF]: Preparation from Isopropylidene-*rac*-glycerol

Key features of the method:

A phthalimidoundecyl group is incorporated into PAF via alkylation of isopropylidene-*rac*-glycerol (4) with $Br(CH_2)_{11}OMs$ (1.5 equiv, 3 equiv NaH, DMSO/THF (5.5:4.5), 0°C, 2 h), then displacement of the mesyl group by potassium phthalimide in DMSO (95°C, 2 h), hydrolysis, tritylation, acetylation, and incorporation of the phosphocholine group.

The phthalimido group is converted to a photoreactive arylazide (Bette-Bobillo et al., 1985; Chau et al., 1989) for use as a photoaffinity probe of PAF binding sites in membranes.

Disadvantage of the method: About 5% of the undesired isomer 1-*O*-(11-phthalimidoundecyl)-3-acetyl-*rac*-glycero-2-phosphocholine (39b) is formed as a by-product. Its retention time on a preparative HPLC column is very similar to that of the desired isomer (39a). Therefore it is advisable to use a synthetic scheme in which the acetyl group is introduced by acetylation of lyso-PAF; 1-*O*-alkyl-2-*O*-benzyl-*sn*-glycerol (32) is commonly used as an intermediate in PAF synthesis (see Scheme 16). After this compound is converted to phosphocholine 33, the benzyl group is removed by hydrogenolysis to afford lyso-PAF.

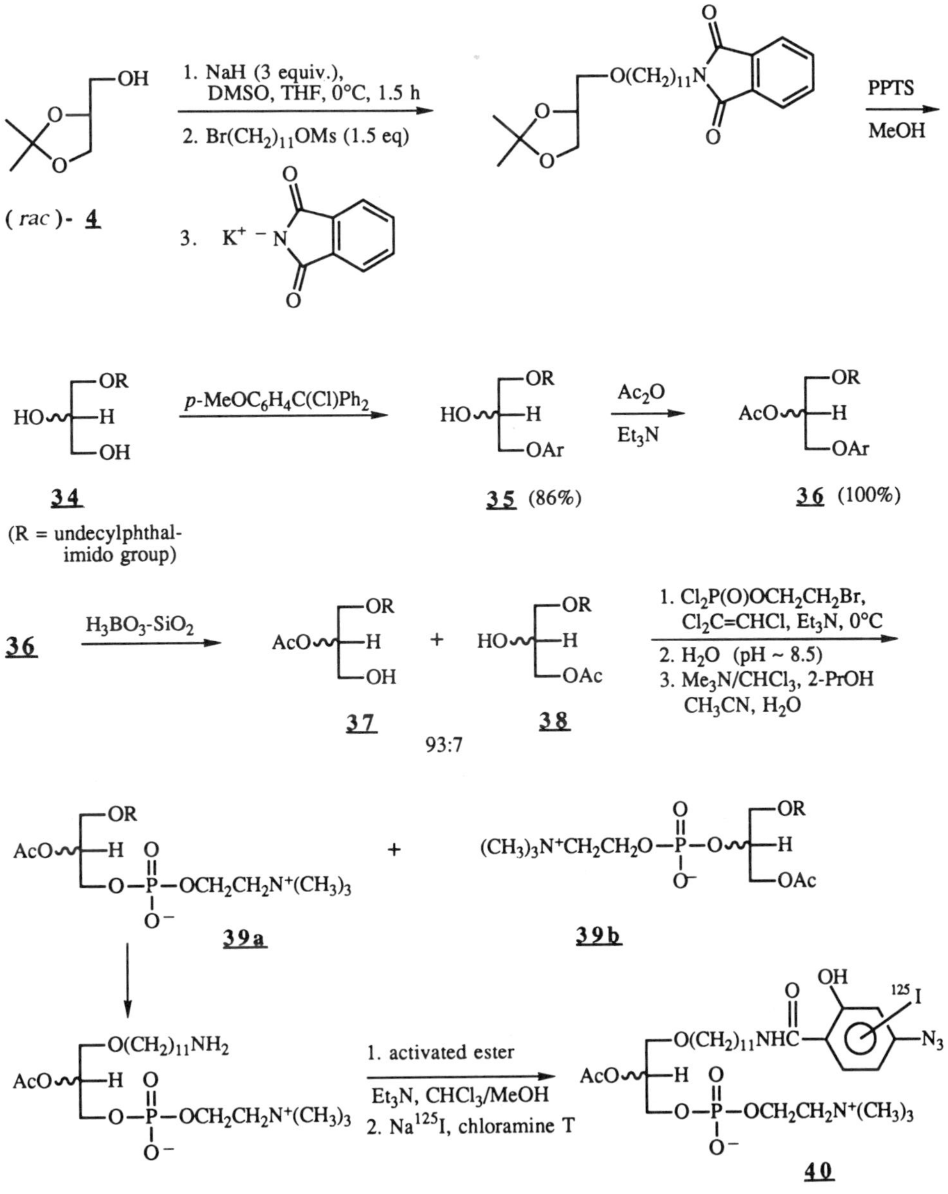

Scheme 17

Literature: Tsai et al. (1988).

Method: To a solution of 1.22 g (3.1 mmol) of diol 34 and 0.86 mL (6.2 mmol) of dry triethylamine in 5 mL of methylene chloride at 0°C under nitrogen is added dropwise a solution of 1.53 g (4.9 mmol) of (4-methoxyphenyl)diphenylmethyl chloride in 5 mL of methylene chloride over a 30-min period. The mixture is stirred at 0°C for 30 min, then diluted with ether (100 mL) and water (50 mL). The organic layer is isolated, washed with brine (50 mL), dried ($MgSO_4$), and concentrated under reduced pressure to give a yellow oil. Chromatography on silica gel (elution with ethyl acetate/hexane 1:1) gives 1.76 g

(86%) of 1-*O*-(4-methoxyphenyldiphenyl)methyl-3-*O*-(11-phthalimidoundecyl)-*rac*-glycerol (35) as a yellow oil. The alcohol 35 is acetylated as follows. To a solution of 2.15 g (3.2 mmol) of alcohol 35, 0.91 mL (6.5 mmol) of dry triethylamine, and 8 mg (66 μmol) of DMAP in 10 mL of methylene chloride under nitrogen is added 0.49 mL (5.2 mmol) of acetic anhydride in one portion. After the solution has stirred at rt for 3 h, ether (100 mL) and water (50 mL) are added. The organic layer is isolated, washed with brine (50 mL), dried ($MgSO_4$), and concentrated under reduced pressure to give 2.32 g (100%) of acetate 36 as a yellow oil. The oil is used in the next step without purification in order to reduce the opportunity of acyl migration.

Detritylation of 36 is achieved by using boric acid impregnated silica gel (Buchnea, 1974) as follows. Silica gel (10 g) is mixed thoroughly with a hot solution of 6.3 g of boric acid in 34 mL of water. The mixture is filtered by suction, and the silica gel is dried at 100°C for 24 h. Crude acetate 36 (76 mg, 0.11 mmol) is detritylated on a column of the boric acid treated silica gel (2 g) (elution with ethyl acetate/hexane (15:85)). Fractions that contain unreacted 36 are recycled by concentrating the fractions, reloading on the treated silica gel, and eluting again with ethyl acetate/hexane (15:85). The detritylated product 37/38 is eluted with gradients of ethyl acetate (first 3:7, then 1:1). Fractions that contain alcohol 37,38 are combined and concentrated to give an orange solid that is purified by chromatography on boric acid treated silica gel (elution with ethyl acetate/hexane, 3:7, then 1:1, then 4:6). The fractions containing the product are combined, washed with water to remove boric acid, dried ($MgSO_4$), and concentrated under reduced pressure to give 38.5 mg (84% from alcohol 34) as a pale yellow oil. The alcohol is a mixture of two regioisomers 37 and 38 in a 93:7 ratio: ^{1}H-NMR (300 MHz, $CDCl_3$): major isomer (37): δ 2.06 (s, 3H, $COCH_3$); minor isomer (38): δ 2.04 (s, 3H, $COCH_3$).

The alcohol (37,38) is reacted with (2-bromoethyl)phosphodichloridate, then hydrolyzed and aminated as follows. To a solution of 27 μL (0.2 mmol) of (2-bromoethyl)phosphodichloridate (Eibl and Nicksh, 1978) in 200 μL of trichloroethylene cooled to 0°C under nitrogen is added 25 μL (0.18 mmol) of dry triethylamine in one portion. The mixture of alcohols 37, 38 (34.5 mg, 0.080 mmol) is added slowly and the reaction mixture is stirred at 0°C for 35 min, then filtered and concentrated under vacuum to give a residue that is dissolved in 0.67 mL of THF. Then 0.67 mL of a 0.5 M sodium acetate solution, pH 8.5, and 40 μL of a 0.5 M EDTA solution, pH 10.5, are added. The mixture is stirred for 1 h and then extracted with 24 mL of chloroform/methanol 9:1 to hydrolyze the remaining P-Cl bond. The organic layer is dried ($MgSO_4$) and concentrated under vacuum to give 60.9 mg of a residue. To the residue are added 0.26 mL of chloroform, 0.45 mL of 2-propanol, and 0.45 mL of acetonitrile, followed by 0.62 mL of a 45% aqueous trimethylamine solution. After the mixture has stirred for 4 h at 60°C, the product 39 is extracted with 25 mL of a 9:1 mixture of chloroform/methanol. The organic layer is dried ($MgSO_4$) and concentrated under vacuum to give 57.3 mg of a residue that is purified by chromatography on silica gel (elution with methylene chloride/methanol/water, 65:15:1, then 65:35:5); yield, 31.4 mg (66%) of product 39 as a colorless oil. HPLC (Nucleosil column, 4.7 $\times$ 25 cm, elution with acetonitrile/methanol/85% phosphoric acid, 130:5:1.5) analysis indicates that 5% of the undesired regioisomer, 1-*O*-(11-phthalimidoundecyl)-3-acetyl-*rac*-glycero-2-phosphocholine (39b), is present. The latter is removed from 39a by preparative HPLC, but the difference in retention times of the two regioisomers is only one minute under the conditions used.

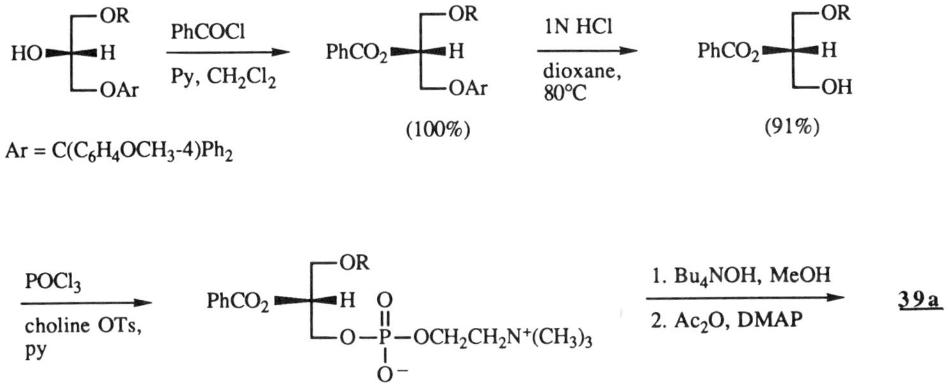

HO–C(H)(–OR)(–OAr) → [PhCOCl; Py, CH_2Cl_2] → $PhCO_2$–C(H)(–OR)(–OAr) (100%) → [1N HCl; dioxane, 80°C] → $PhCO_2$–C(H)(–OR)(–OH) (91%)

Ar = $C(C_6H_4OCH_3\text{-}4)Ph_2$

→ [$POCl_3$; choline OTs, py] → $PhCO_2$–C(H)(–OR)–O–P(=O)(O⁻)–$OCH_2CH_2N^+(CH_3)_3$ → [1. Bu_4NOH, MeOH; 2. Ac_2O, DMAP] → 39a

Scheme 18

In contrast, no regioisomer by-product was formed using the reaction sequence (see Scheme 18) of benzoylation of 1-*O*-alkyl-3-*O*-(*p*-methoxyphenyldiphenyl)methyl-*sn*-glycerol followed by detritylation with 2 N HCl in dioxane at 80°C (purification by silica gel chromatography using elution with toluene-ethyl acetate 95 : 5 , toluene-ethyl acetate-pyridine 94 : 5 : 1, then with ether-pyridine 99 : 1), phosphocholine insertion (phosphorus oxychloride, choline tosylate), debenzoylation (tetra-*n*-butylammonium hydroxide in methanol), and acetylation (acetic anhydride, DMAP). Therefore the route to 39a shown in Scheme 18 is superior.

Literature: Hirth et al. (1983).

The hydrolysis of the phthalimido group and coupling of the resulting amino group to the *N*-hydroxysuccinimidoyl ester of the 4-azido-2-hydroxybenzoic acid are carried out as described previously (Bette-Bobillo et al., 1985). Radioiodination with $Na^{125}I$ in the

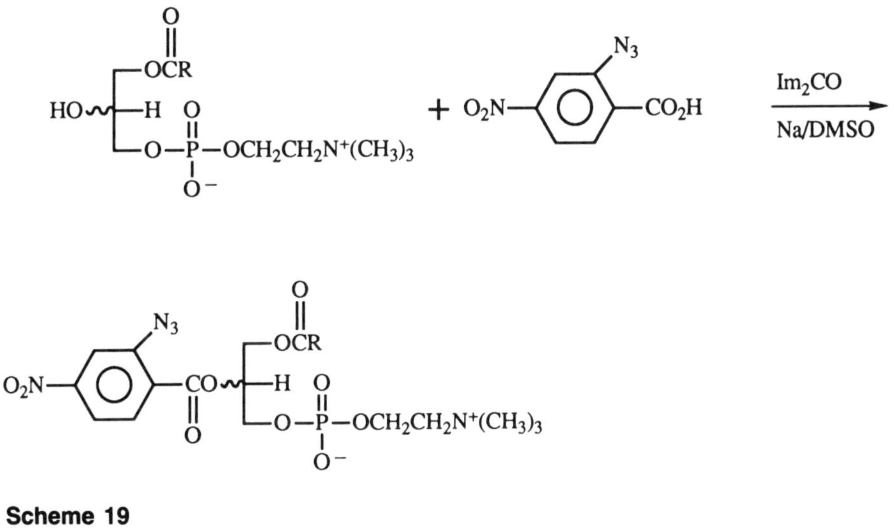

HO–C(H)(–OC(=O)R)–O–P(=O)(O⁻)–$OCH_2CH_2N^+(CH_3)_3$ + O_2N–$C_6H_3(N_3)$–CO_2H → [Im_2CO; Na/DMSO]

O_2N–$C_6H_3(N_3)$–C(=O)O–C(H)(–OC(=O)R)–O–P(=O)(O⁻)–$OCH_2CH_2N^+(CH_3)_3$

Scheme 19

presence of chloramine T, followed by extraction of the ^{125}I-product with chloroform and purification by HPLC, gives 40.

Literature: Chau et al. (1989).

A diacylphospholipid containing a photoactivable group at the *sn*-2 position has been prepared by coupling of the imidazolide of a 2-azido-4-nitrobenzoic acid to myristoyllyso-PC, as shown in Scheme 19 (Pradhan et al., 1989). Purification of the phospholipid is by silica gel chromatography using chloroform/methanol 9:1.

3. A 1-*O*-(ω-Carboxy)alkyl Derivative of PAF [1-*O*-(15-Carboxypentadecyl)-2-acetyl-*sn*-glycero-3-phosphocholine]: Preparation from 2,3-*O*-Isopropylidene-*sn*-glycerol

Key features of the method:

The ω-methyl group of the hexadecyloxy group at the *sn*-1 position of PAF is replaced by a carboxyl group. The chain introduced into the *sn*-1 position can be tritiated, providing a radiolabeled PAF analog that can be conjugated to proteins and used as a potential hapten.

Literature: Prahad et al. (1990).

Method: Sodium hydride (1.64 g of a 60% suspension in oil, 41.0 mmol, washed with pentane) is placed in a two-neck 250-mL flask fitted with a condenser and rubber septum under nitrogen atmosphere. After toluene (11 mL) is added, a solution of 4.07 g (11.7 mmol) of 16-mesyloxy-(*E*)-9-hexadecenoic acid (41) in 35 mL of toluene is added dropwise with stirring. After 5 min, a solution of 2.32 g (17.6 mmol) of 2,3-*O*-isopropylidene-*sn*-glycerol [(*R*)-4] in 5 mL of toluene is added dropwise. The mixture is refluxed for 24 h, then cooled to 0°C and acidified with 3 N HCl. The acetonide is extracted with ethyl acetate. The organic phase is dried (Na_2SO_4) and concentrated, leaving a residue that is dissolved in 150 mL of methanol and hydrolyzed with 3 mL of

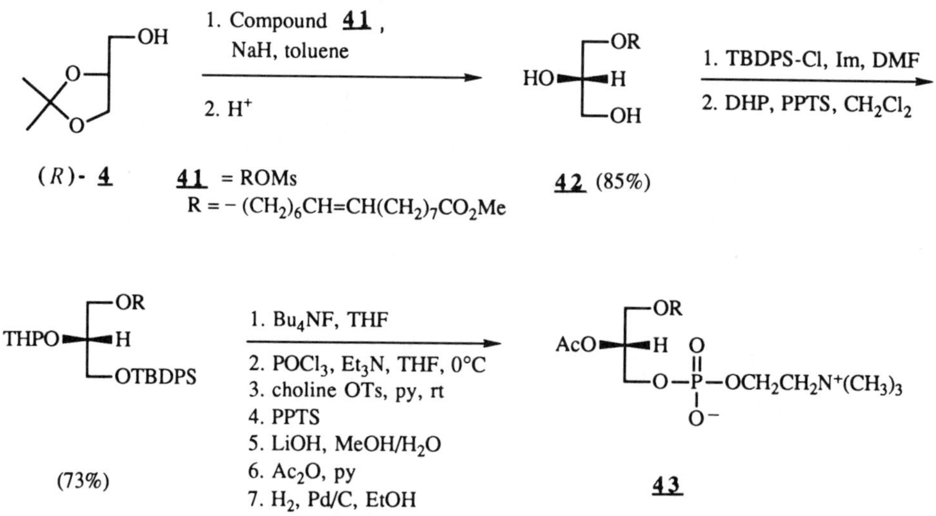

(*R*)-4 → [1. Compound 41, NaH, toluene; 2. H^+] → 42 (85%) → [1. TBDPS-Cl, Im, DMF; 2. DHP, PPTS, CH_2Cl_2]

41 = ROMs
R = –$(CH_2)_6CH{=}CH(CH_2)_7CO_2Me$

(73%) → [1. Bu_4NF, THF; 2. $POCl_3$, Et_3N, THF, 0°C; 3. choline OTs, py, rt; 4. PPTS; 5. LiOH, MeOH/H_2O; 6. Ac_2O, py; 7. H_2, Pd/C, EtOH] → 43

Scheme 20

concentrated sulfuric acid. The mixture is refluxed for 2 h, then cooled in an ice bath, neutralized with sodium bicarbonate, and concentrated. The residue is treated with brine and extracted with ethyl acetate. The organic layer is dried (Na_2SO_4) and concentrated, giving a residue that is purified by flash chromatography (elution with petroleum ether-ethyl acetate 3:7, then ethyl acetate); yield of diol 42, 3.55 g (85%). The conversion of 1-*O*-(15-carbomethoxypentadec-7(*E*)-enyl)-*sn*-glycerol (42) to PAF analog 43 is completed by the reaction sequence of silylation of the *sn*-3 hydroxy group with *tert*-butyldiphenylsilyl chloride (TBDPS-Cl) in the presence of imidazole in DMF, then protection of the *sn*-2 hydroxy group as the 2-*O*-tetrahydropyranyloxy derivative, desilylation, phosphocholine insertion (using phosphorus oxychloride, then choline tosylate), deprotection of the THP group, hydrolysis of the methyl ester with lithium hydroxide in aqueous methanol at rt, acetylation (acetic anhydride in pyridine), and catalytic reduction of the double bond (H_2, 10% Pd-C, EtOH).

4. A Spin-Labeled Derivative of PAF: Preparation from 2,3-*O*-Isopropylidene-*sn*-glycerol [(*R*)-4]

Key features of the method:

2,3-*O*-Isopropylidene-*sn*-glycerol [(*R*)-4] is alkylated with 1-bromooctadecane, deketalized, and tritylated. The spin label is incorporated into the *sn*-2 position of 1-*O*-octadecyl-3-*O*-trityl-*sn*-glycerol by acylation using 4-doxylpentanoic acid in the presence of DCC. After detritylation using BF_3-methanol complex, the phosphocholine moiety is introduced by using (2-bromoethyl)phosphochloridate, followed by hydrolysis and trimethylamination.

Literature: Lai et al. (1989). For the preparation of a diester-PC containing the doxyl group in an ester linkage at the *sn*-2 position, see Joseph and Lai (1988).

The nitroxide spin label is also introduced into the *sn*-2 position in an ether linkage by a reaction sequence (see Scheme 22) that involves displacement of 2,5,5-trimethyl-1,3-dioxane-2-propylmethane sulfonate by the alkoxide of 1-*O*-octadecyl-3-*O*-trityl-*rac*-glycerol. *rac*-1-*O*-Octadecyl-2-*O*-[(2,5,5-trimethyl-1,3-dioxane)propyl]-3-*O*-glycerol is converted into the diether-PC as shown in Scheme 22.

Literature: Joseph et al. (1991).

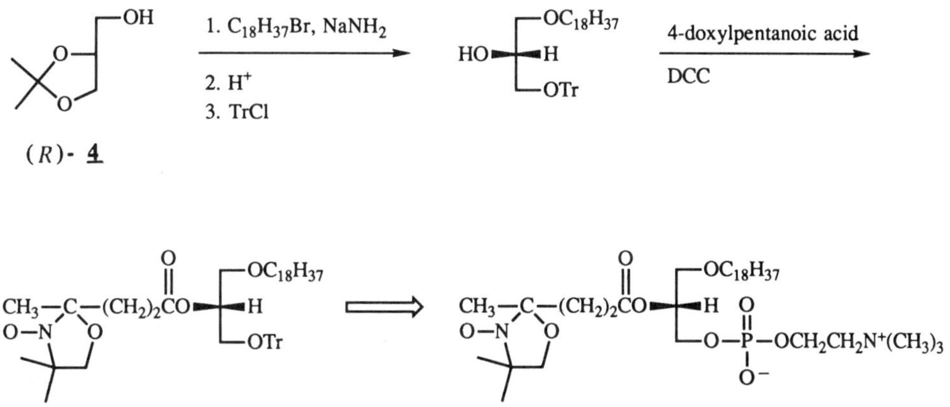

(*R*)-4 — 1. $C_{18}H_{37}Br$, $NaNH_2$; 2. H^+; 3. TrCl → HO–C(H)($CH_2OC_{18}H_{37}$)(CH_2OTr) — 4-doxylpentanoic acid, DCC → CH_3–C(doxyl)–$(CH_2)_2$CO–O–C(H)($CH_2OC_{18}H_{37}$)(CH_2OTr) ⟹ CH_3–C(doxyl)–$(CH_2)_2$CO–O–C(H)($CH_2OC_{18}H_{37}$)(CH_2–O–P(O)(O$^-$)–$OCH_2CH_2N^+(CH_3)_3$)

Scheme 21

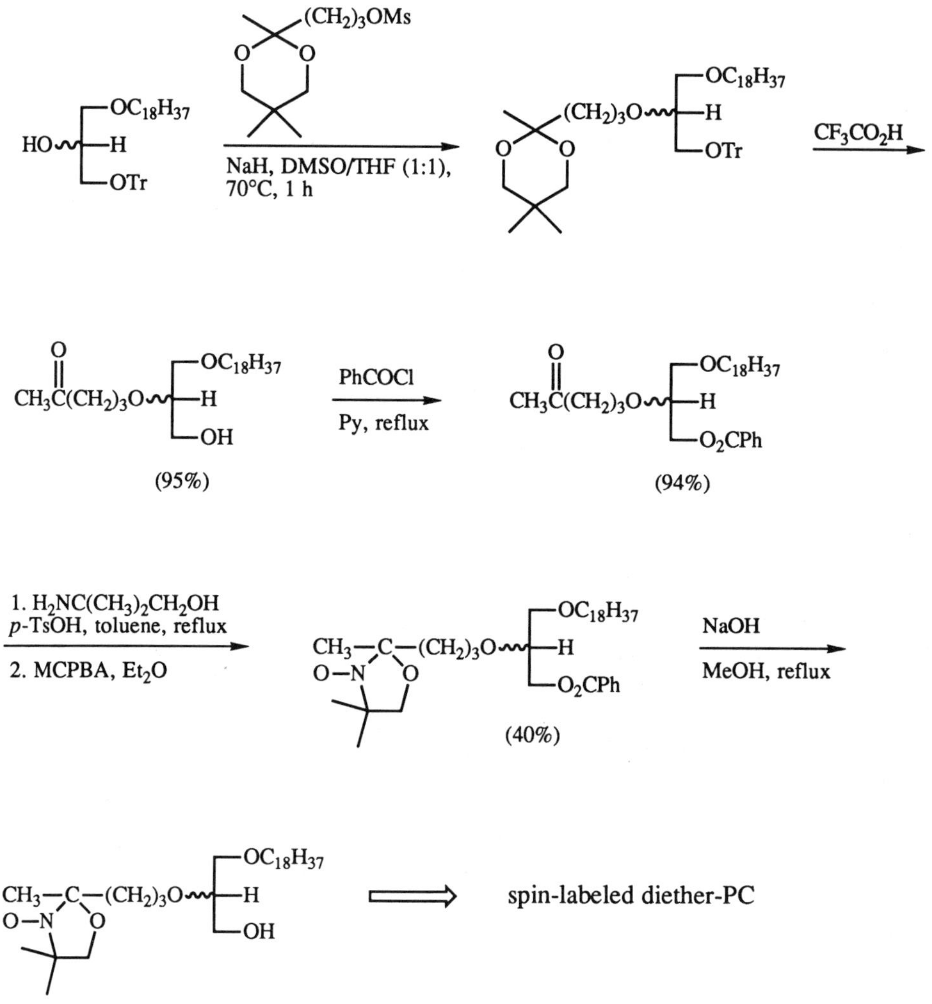

HO–CH(OC$_{18}$H$_{37}$)(OTr) → [2,2,5,5-tetramethyl-1,3-dioxane-(CH$_2$)$_3$OMs; NaH, DMSO/THF (1:1), 70°C, 1 h] → dioxane-(CH$_2$)$_3$O–glycerol(OC$_{18}$H$_{37}$)(H)(OTr) → [CF_3CO_2H]

$CH_3C(=O)(CH_2)_3O$–glycerol(OC$_{18}$H$_{37}$)(H)(OH) (95%) → [PhCOCl; Py, reflux] → $CH_3C(=O)(CH_2)_3O$–glycerol(OC$_{18}$H$_{37}$)(H)(O_2CPh) (94%)

→ [1. $H_2NC(CH_3)_2CH_2OH$, p-TsOH, toluene, reflux; 2. MCPBA, Et_2O] → CH_3–C(oxazolidine N–O)–$(CH_2)_3O$–glycerol(OC$_{18}$H$_{37}$)(H)(O_2CPh) (40%) → [NaOH; MeOH, reflux]

CH_3–C(oxazolidine N–O)–$(CH_2)_3O$–glycerol(OC$_{18}$H$_{37}$)(H)(OH) ⟹ spin-labeled diether-PC

Scheme 22

5. An Aryl-Terminated Ether Derivative of PAF: Preparation from 2,3-*O*-Isopropylidene-*sn*-glycerol

Key features of the method:

12-(2-Naphthyl)-dodec-11-en-1-ol tetrahydropyranyl ether (44) is prepared by a Wittig reaction and then converted to naphthylvinyl-PC (46), an *sn*-1 ether-linked PC. Compound 46 is used as a substrate in an assay for phospholipase A_2 using HPLC with fluorescence detection to separate and quantify 46 and lyso-naphthylvinyl-PC on a normal-phase silica gel column (elution with hexane-*i*-PrOH-H_2O 6:8:1.6).

Literature: Hendrickson et al. (1987).

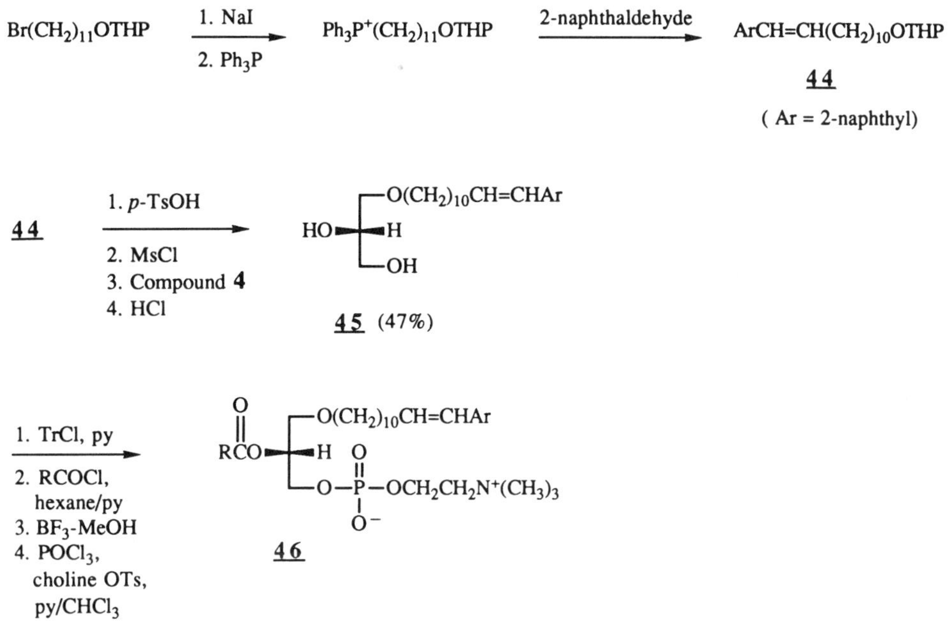

$Br(CH_2)_{11}OTHP$ —(1. NaI; 2. Ph_3P)→ $Ph_3P^+(CH_2)_{11}OTHP$ —(2-naphthaldehyde)→ $ArCH{=}CH(CH_2)_{10}OTHP$

44

(Ar = 2-naphthyl)

44 —(1. *p*-TsOH; 2. MsCl; 3. Compound **4**; 4. HCl)→

CH_2—$O(CH_2)_{10}CH{=}CHAr$; HO—C—H; CH_2—OH

45 (47%)

—(1. TrCl, py; 2. RCOCl, hexane/py; 3. BF_3-MeOH; 4. $POCl_3$, choline OTs, py/$CHCl_3$)→

CH_2—$O(CH_2)_{10}CH{=}CHAr$; RC(=O)O—C—H; CH_2—O—P(=O)(O^-)—$OCH_2CH_2N^+(CH_3)_3$

46

Scheme 23

Method: 11-Bromoundecyl tetrahydropyranyl ether is first converted to the iodo compound. The phosphonium salt is prepared (see Bergelson, 1980), dried by evaporation with benzene, and converted to the ylide by addition of THF and *n*-BuLi under nitrogen. The solution becomes orange, indicating the formation of the ylide. After the solution is stirred for 15 min, 2-naphthaldehyde is added, and stirring under nitrogen is continued for 2 h. The reaction is quenched with water. Dichloromethane is added, and the organic phase is extracted with ether and washed with water three times. The organic layer is dried over $MgSO_4$ and concentrated to give an oil that contains crystals of triphenylphosphine oxide. The oil is decanted and the crystals are washed twice with cold hexane. The crude oil is purified by silica gel chromatography, giving a pure yellow oil on elution with hexane-EtOAc 95:5. The THP group is removed by treatment with *p*-TsOH in ethanol at 60°C, giving 12-(2-naphthyl)-dodec-11-en-1-ol, which is crystallized from hexane and then is converted to the mesylate with mesyl chloride in pyridine-hexane.

The mesylate is crystallized from hexane-acetone and used to alkylate 2,3-*O*-isopropylidene-*sn*-glycerol (4) as follows. 2,3-*O*-Isopropylidene-*sn*-glycerol (1.4 g, 10.71 mmol) is added slowly at rt to a solution of 0.6 mg (25 mmol) of sodium hydride in 35 mL of benzene. 12-(2-Naphthyl)-dodec-11-enyl mesylate (3.75 g, 9.65 mmol) is added, and the mixture is refluxed for 4 h. After the mixture is cooled in an ice bath, methanol is added to destroy the excess NaH. The mixture is neutralized with acetic acid in methanol and diluted with 40 mL of benzene. The organic phase is separated, washed with methanol-water (1:1, 2 × 40 mL), dried over $MgSO_4$, and evaporated to dryness. The acetonide is dissolved in 5 mL of methanol and hydrolyzed by adding 20 mL of 10% conc. HCl in methanol and heating on a steam bath for 20 min. The mixture is concentrated to about one-third of its volume under vacuum, diluted with chloroform (50 mL), and washed twice with methanol-water 1:1. The organic phase is isolated, evaporated, and applied to a column of silica gel (30 g). The product is eluted with mixtures of acetone in hexane (first with 10% acetone in hexane, then 20% and 30%), affording 1.75 g (47%) of diol 45 as a pale yellow solid. Tritylation with trityl chloride in pyridine, then acylation with 1.5 equiv of acyl chloride in hexane-pyridine follow standard procedures. Detritylation with BF_3/methanol complex as described by Hendrickson et al. (1983) gives crude 1-*O*-alkyl-2-acyl-*sn*-glycerol, which is purified by silica gel chromatography using elution with ether-hexane. The alkyl/acyl-glycerol is converted to the phosphocholine 46 by adding the alcohol to phosphorus oxychloride in chloroform-pyridine at 0°C, followed by stirring at rt for 1 h. Choline tosylate in pyridine is added, and the mixture is stirred overnight at rt. Purification by column chromatography on silica gel (elution with chloroform-methanol 3:1, then with chloroform-methanol-conc. NH_4OH-water mixtures (65:25:0.5:0.5, 65:25:1:1, and 65:35:2:2), affords the pure product; the yield of the phosphorylation/choline tosylate reaction is 46%.

6. An ω-CD_3-Derivative of PAF [1-*O*-(16'-$^2H_3$)Hexadecyl-2-acetyl-PC]: Preparation from Isopropylidene-*rac*-glycerol

Outline of the method:

The ω-methyl group of PAF is deuterated to prepare a trideuterated sample for use as an internal standard for quantitative analyses based on mass spectrometry methods. The

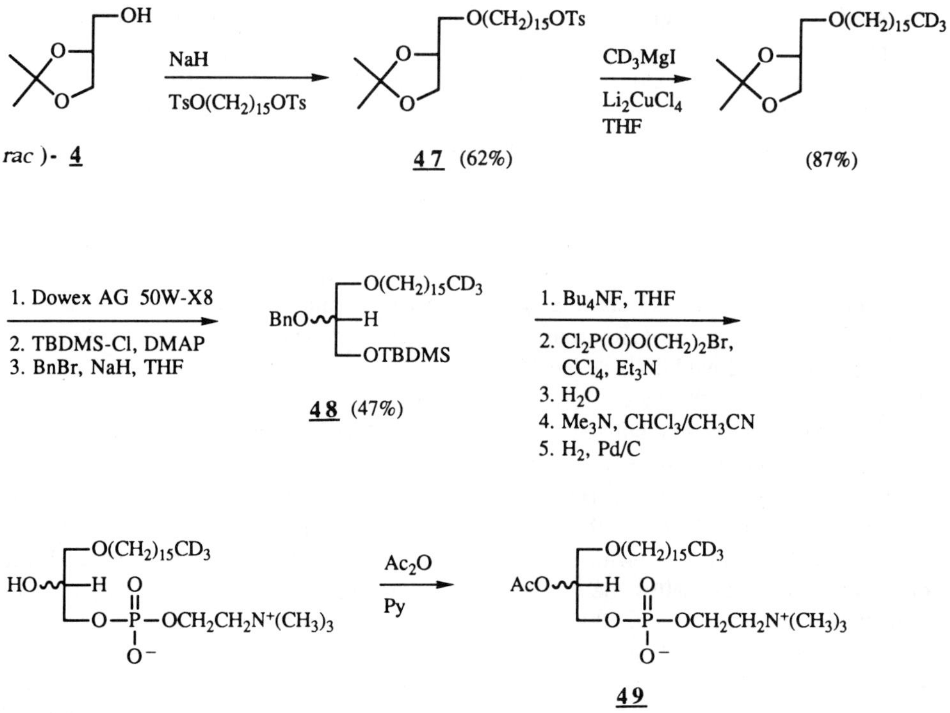

(*rac*)-4 —[NaH, $TsO(CH_2)_{15}OTs$]→ 47 (62%), $-O(CH_2)_{15}OTs$ —[CD_3MgI, Li_2CuCl_4, THF]→ $-O(CH_2)_{15}CD_3$ (87%)

—[1. Dowex AG 50W-X8; 2. TBDMS-Cl, DMAP; 3. BnBr, NaH, THF]→ 48 (47%) (BnO, H, $O(CH_2)_{15}CD_3$, OTBDMS) —[1. Bu_4NF, THF; 2. $Cl_2P(O)O(CH_2)_2Br$, CCl_4, Et_3N; 3. H_2O; 4. Me_3N, $CHCl_3/CH_3CN$; 5. H_2, Pd/C]→

HO, H, $O(CH_2)_{15}CD_3$, $O-P(O)(O^-)-OCH_2CH_2N^+(CH_3)_3$ —[Ac_2O, Py]→ AcO, H, $O(CH_2)_{15}CD_3$, $O-P(O)(O^-)-OCH_2CH_2N^+(CH_3)_3$ 49

Scheme 24

synthesis involves use of the 1,15-bis(tosylate) of pentadecane-1,15-diol to alkylate glycerol 2,3-acetonide [(*rac*)-4] in DMF in the presence of sodium hydride, giving 1-*O*-15'-*O*-*p*-toluenesulfonylpentadecyl-*rac*-glycerol 2,3-acetonide (47) in 62% yield. A solution of [$^2H_3$]methylmagnesium iodide is added to a solution of the tosyl derivative in THF containing a catalytic amount of Li_2CuCl_4, affording 1-*O*-16'-[$^2H_3$]hexadecyl-*rac*-glycerol 2,3-acetonide; the latter is hydrolyzed by stirring with Dowex resin AG 50W-X8 in methanol. Silylation of the primary hydroxy group with *tert*-butyldimethylsilyl chloride (TBDMS-Cl) in methylene chloride in the presence of DMAP at 0°C, then benzylation of the available 2-hydroxy group with benzyl bromide using sodium hydride in THF at –5°C at rt, gives 1-*O*-alkyl-2-*O*-benzyl-3-*O*-(*tert*-butyldimethylsilyl)-*rac*-glycerol (48). The synthesis of the deuterated PAF 49 is completed by the sequence of desilylation, phosphocholine insertion, catalytic hydrogenolysis, and acetylation. Note that debenzylation is carried out after the desilylation-phosphorylation sequence, since (as mentioned earlier) acetate migration from the *sn*-2 to *sn*-1 position would contaminate the product if the sequence had been reversed.

Literature: Prakash et al. (1989).

C. Diether-PCs

A great deal of interest has been directed to phospholipids bearing two ether-linked chains in recent years, since they have cytotoxic properties (for reviews, see Weltzien and Munder, 1983; Berdel et al., 1985). The prototype of this class of alkyllysophospholipid agents is 1-*O*-octadecyl-2-*O*-methyl-*rac*-glycero-3-phosphocholine (ET-18-OCH_3). These analogs of PAF have a broad range of biological and pharmacological effects, including inhibition of malignant cell growth, activation of neutrophils and macrophages, alteration of phospholipid metabolism, and interference with membrane function. The antineoplastic and growth-inhibitory effects of these agents may arise by several mechanisms, including inhibition of membrane-associated enzymes such as protein kinase C (e.g., Shoji et al., 1988) and Na,K-adenosine triphosphatase (Zheng et al., 1990). Analogs of ET-18-OCH_3 have been synthesized for use in studies of the sites of actions of the ether lipids and for evaluation as new anticancer drugs.

1. Preparation from Glycidyl Derivatives

Key features of the method:

1,3-Disubstituted derivatives of glycerol are prepared in one step by BF_3-catalyzed ring opening of glycidyl arenesulfonates (10) or glycidyl *tert*-butyldiphenylsilyl ethers with long-chain alcohols in high ee and high regiospecificity.

A variety of alkyl groups can be introduced at the *sn*-2 position by alkylation of the ring-opened intermediate.

Since the number of protection-deprotection reactions is low, this route represents an efficient approach to both of the enantiomers of cytotoxic alkyllysophospholipids having a 2-*O*-methyl group.

Literature: Guivisdalsky and Bittman (1989b).

Method: Acid-catalyzed epoxy ring opening of (*R*)-glycidyl tosylate (10) with BF_3 etherate and 1-hexadecanol (see Scheme 16) gives 1-*O*-hexadecyl-*sn*-glycerol 3-*O*-*p*-toluenesulfonate. 1-*O*-Hexadecyl-2-*O*-methyl-3-*O*-(*p*-tolylsulfonyl)-*sn*-glycerol (50) is made by methylation of 1-*O*-hexadecyl-*sn*-glycerol 3-*O*-*p*-toluenesulfonate using diazomethane in the presence of silica gel as follows. To a suspension of 235 mg (0.5

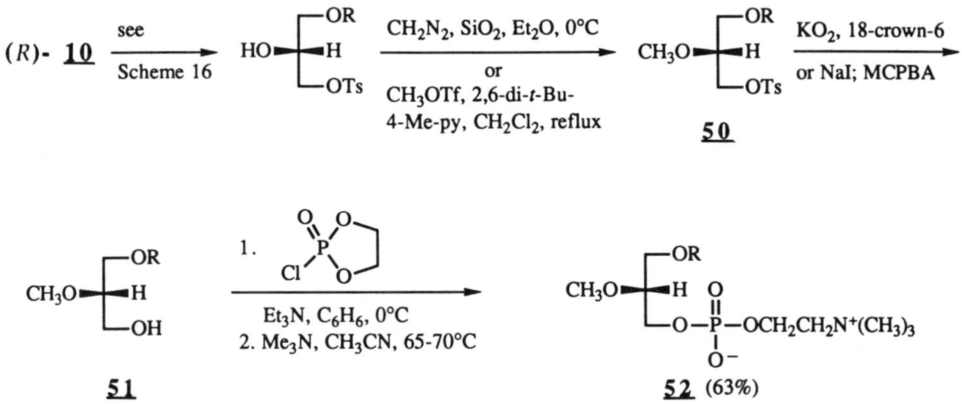

(*R*)- 10 $\xrightarrow{\text{see Scheme 16}}$ HO–[–OR, H, –OTs] $\xrightarrow[\text{or } CH_3OTf,\ 2,6\text{-di-}t\text{-Bu-4-Me-py},\ CH_2Cl_2,\ \text{reflux}]{CH_2N_2,\ SiO_2,\ Et_2O,\ 0°C}$ CH_3O–[–OR, H, –OTs] (50) $\xrightarrow[\text{or NaI; MCPBA}]{KO_2,\ 18\text{-crown-6}}$

CH_3O–[–OR, H, –OH] (51) $\xrightarrow[\text{2. } Me_3N,\ CH_3CN,\ 65\text{-}70°C]{\text{1. 2-chloro-2-oxo-1,3,2-dioxaphospholane},\ Et_3N,\ C_6H_6,\ 0°C}$ CH_3O–[–OR, H, –O–P(=O)(O^-)–$OCH_2CH_2N^+(CH_3)_3$] 52 (63%)

Scheme 25

mmol) of 1-*O*-hexadecyl-*sn*-glycerol 3-*O*-*p*-toluenesulfonate in 15 mL of ether and 1.2 g (50 wt equiv) of silica gel (Baker, 60–200 mesh, dried overnight at 120°C) is added a solution of diazomethane (20 mol equiv) in ether at 0°C. The mixture is stirred for at least an additional 6 h at 0°C, filtered, and washed with ether. The solvents are evaporated under vacuum, and the residue is purified by flash chromatography (elution with hexane-ethyl acetate 8:1), giving 155 mg (64%) of the *O*-methyl ether 50. A higher yield (90%) is obtained by carrying out the methylation using methyl triflate in the presence of the hindered 2, 6-di-*tert*-butyl-4 methylpyridine in refluxing methylene chloride. In this method, 141 mg (0.3 mmol) of 1-*O*-hexadecyl-*sn*-glycerol 3-*O*-*p*-toluenesulfonate and 616 mg (3.0 mmol) of 2,6-di-*tert*-butyl-4-methylpyridine in 3 mL of methylene chloride is treated with 340 μL (3.0 mmol) of methyl triflate. The mixture is refluxed for 16 h under nitrogen, and then the solvents are removed under vacuum. To the residue are added 50 mL of ethyl acetate and 30 mL of 2 N HCl. The organic phase is separated and washed again with 30 mL of 2 N HCl. (The excess of the hindered base can be recovered by neutralizing the combined aqueous phase, neutralizing with 20% aqueous NaOH, and extracting 2,6-di-*tert*-butyl-4-methylpyridine into methylene chloride.) The ethyl acetate phase is washed with ether, saturated aqueous sodium bicarbonate, and water, and then dried (Na_2SO_4). The solvents are evaporated to give a residue that is purified by flash chromatography (elution with hexane-ethyl acetate 8:1). Tosylate 50 (70 mg, 0.15 mmol) is converted into alcohol 51 by reaction with potassium superoxide (42 mg, 0.6 mmol) in the presence of 18-crown-6 (159 mg, 0.6 mmol) in 3 mL of DMSO/DMF/1,2-dimethoxyethane (1:1:1) at rt under nitrogen. The alcohol is extracted with ether and purified by flash chromatography (elution with hexane-ethyl acetate 4:1). Alternatively, detosylation of 50 is carried out via conversion to the iodide, followed by MCPBA treatment, or via the acetate, which is reduced or hydrolyzed.

The alcohol 51 is phosphorylated by adding a solution of 94 mg (0.30 mmol) of alcohol and 50 μL (0.36 mmol) of triethylamine in 3 mL of benzene to a solution of 51.3 mg (0.36 mmol) of 2-chloro-2-oxo-1,3,2-dioxaphospholane in 3 mL of benzene at 0°C. After the mixture has stirred for 2 h at rt, TLC (hexane-ethyl acetate 4:1) indicates that the starting alcohol has disappeared. Triethylamine hydrochloride is removed by filtration and the solvent is evaporated under vacuum, giving 128 mg (100%) of the desired cyclic phosphate triester, which is used in the amination step without purification. The cyclic phosphate triester (128 mg, 0.3 mmol) is transferred to a glass pressure tube (Ace Glass, Vineland, N.J.) with 4 mL of acetonitrile. The mixture is cooled to –78°, and about 0.5 mL of trimethylamine is allowed to condense into the solution. The tube is closed and heated in an oil bath at 65–70°C for 36 h. The crude product phosphocholine 52 is precipitated when the mixture is cooled to –20°C for 3 h. Purification by flash chromatography (elution with chloroform-methanol-water 65:25:4) gives 94 mg (63%) of product 52; $[\alpha]^{25}_D$ –5.41° (*c* 0.95, chloroform-methanol 1:1).

Scheme 26 shows the use of the *tert*-butyldiphenylsilyl ether of glycidol (53) as a starting material for 52. Ring opening with hexadecanol at 4°C in the presence of $BF_3 \cdot Et_2O$ gives a mixture of the desired regioisomer 54a (C_3 attack) and the undesired isomer 54b (C_2 attack) in a ratio of 9:1 (Guivisdalsky and Bittman, 1989a). In contrast, attack of 1-hexadecanol on the tosylate derivative of glycidol (10) at room temperature gives the desired regioisomer exclusively. The *tert*-butyldiphenylsilyl derivative 54a can be alkylated by using strongly basic conditions (sodium hydride/methyl iodide in benzene), whereas alkylation of the arenesulfonate intermediate requires mild basic conditions in which the arenesulfonate group is not displaced by the *sn*-2 hydroxy group. Desilylation with *n*-Bu_4NF in THF gives the alcohol 51, which is converted to the phosphocholine as shown in Scheme 25.

2. Preparation from 2,3-*O*-Isopropylidene-*sn*-glycerol via 1-*O*-Alkyl-2,3-*O*-benzylidene-*sn*-glycerol

Outline of the method:

1-*O*-Alkyl-2,3-*O*-isopropylidene-*sn*-glycerol is transketalized to the 2,3-*O*-benzylidene derivative 55, which is reduced to 3-*O*-benzyl-1-*O*-hexadecyl-*sn*-glycerol (56) as the major regioisomer. The latter is alkylated in refluxing xylene with a long-chain alkyl bromide in the presence of excess potassium *tert*-butoxide and a phase transfer catalyst. After catalytic hydrogenolysis, phosphorylation and amination steps are carried out, giving the diether-PC.

Literature: Abdelmageed et al. (1990).

Method: 2,3-*O*-Isopropylidene-*sn*-glycerol [(*R*)-4] is alkylated with 1-bromohexadecane in refluxing toluene in 4 h. The reaction mixture is cooled, the excess

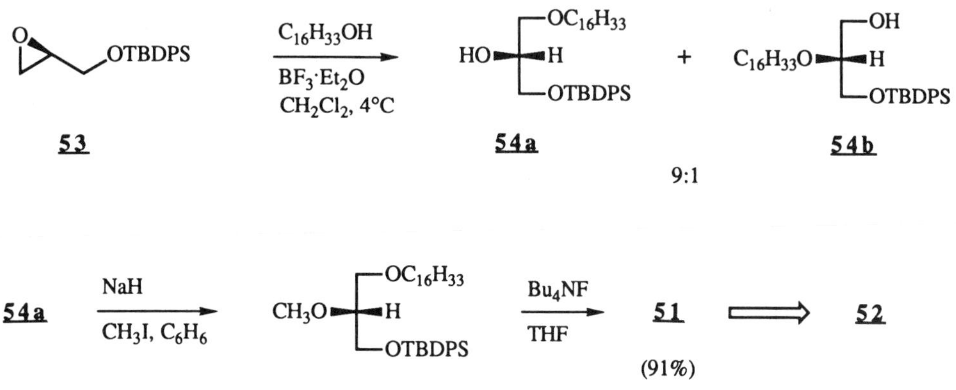

53 (glycidyl OTBDPS) —$C_{16}H_{33}OH$, $BF_3 \cdot Et_2O$, CH_2Cl_2, 4°C→ 54a (HO–C–H; $OC_{16}H_{33}$, OTBDPS) + 54b ($C_{16}H_{33}O$–C–H; OH, OTBDPS)

9:1

54a —NaH, CH_3I, C_6H_6→ CH_3O–C–H ($OC_{16}H_{33}$, OTBDPS) —Bu_4NF, THF→ 51 (91%) ⟹ 52

Scheme 26

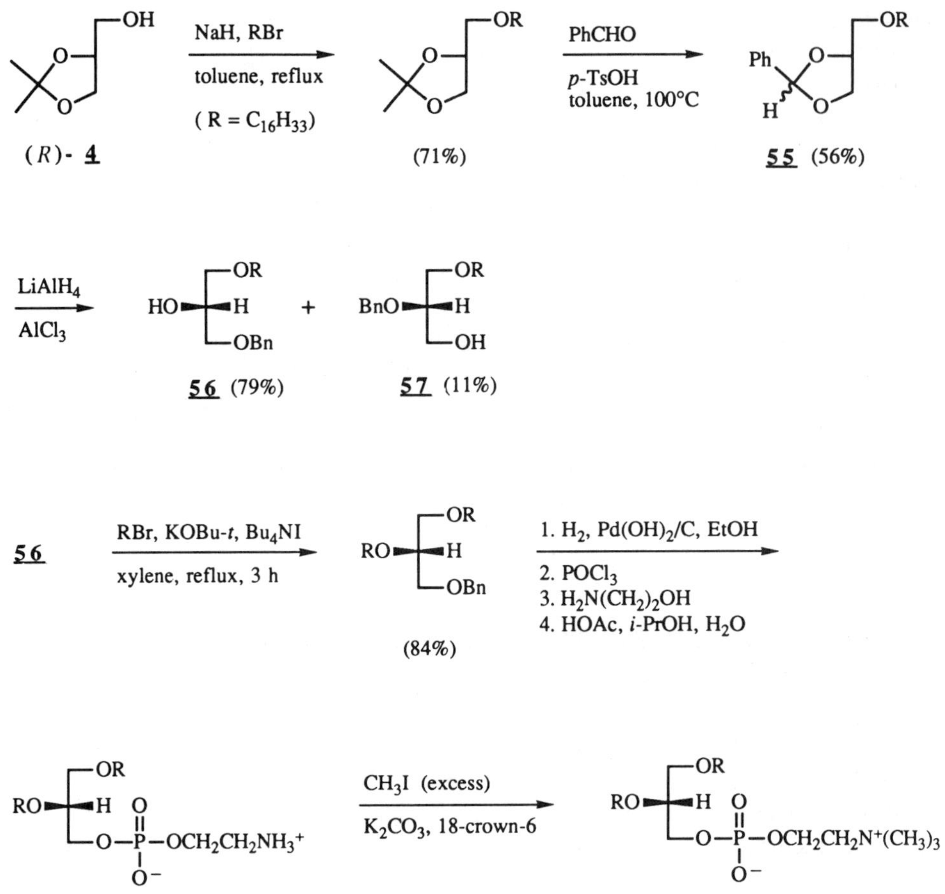

NaH, RBr / toluene, reflux (R = $C_{16}H_{33}$) — PhCHO / p-TsOH toluene, 100°C

(R)-4 (71%) 55 (56%)

$LiAlH_4$ / $AlCl_3$

56 (79%) + 57 (11%)

56 — RBr, KOBu-t, Bu_4NI / xylene, reflux, 3 h — (84%) — 1. H_2, $Pd(OH)_2/C$, EtOH 2. $POCl_3$ 3. $H_2N(CH_2)_2OH$ 4. HOAc, i-PrOH, H_2O

CH_3I (excess) / K_2CO_3, 18-crown-6

$-O-P(O)(O^-)-OCH_2CH_2NH_3^+$ → $-O-P(O)(O^-)-OCH_2CH_2N^+(CH_3)_3$

Scheme 27

sodium hydride is quenched with methanol, and the product is extracted with ether. The ether layer is dried (Na_2SO_4), the ether is evaporated, and the residue is purified by flash chromatography (elution with petroleum ether to remove 1-bromohexane, then with chloroform-petroleum ether 1:1); yield, 71%. Transketalization is carried out in the presence of benzaldehyde and *p*-toluenesulfonic acid at 100°C in toluene; acetone is removed by distillation. The benzylidene ring of 55 is reduced with lithium aluminum hydride-aluminum chloride in ether, giving 3-*O*-benzyl-1-*O*-hexadecyl- and 2-*O*-benzyl-1-*O*-hexadecyl-*sn*-glycerol (56 and 57) in a 84:16 ratio. The alkylation of 3-*O*-benzyl-1-*O*-hexadecyl-*sn*-glycerol (56) with 1-bromohexadecane is carried out by refluxing in xylene for 3 h in the presence of 4 equivalents of potassium *tert*-butoxide and 0.1 equivalent of tetra-*n*-butylammonium iodide. Catalytic hydrogenolysis (20% $Pd(OH)_2/C$ in ethanol) gives the alcohol, which is converted into the phosphocholine.

Scheme 28 shows another route from 1,2-*O*-isopropylidene-*sn*-glycerol [(*S*)-4] to a diether-PC (1,2-di-*O*-octadecyl-*sn*-glycero-3-phosphocholine). The reaction sequence involves the protection of the *sn*-3-hydroxy group of 4 with a 2-methoxyethoxymethyl (MEM) or 2-(trimethylsilyl)ethoxymethyl (SEM) group, followed by deketalization, heating the protected glycerol with octadecyl tosylate at 190–200°C in the absence of solvent for 30 min, demasking of the *sn*-3 hydroxy group with titanium tetrachloride, phosphorylation with (2-bromoethyl)phosphochloridate, and amination with trimethylamine (Yamauchi et al., 1987). The MEM and SEM acid-labile protecting groups offer advantages compared with the commonly used benzyl group, which is removed by catalytic hydrogenolysis and is thus incompatible with compounds containing unsaturation and is impractical in large-scale preparations. (The related methoxymethyl, MOM, group has also been applied to protection of the *sn*-2-hydroxy group: Guivisdalsky and Bittman, 1989c.)

Literature: Yamauchi et al. (1987).

Method: To a solution of 3.5 g (26.5 mmol) of 1,2-*O*-isopropylidene-*sn*-glycerol [(*S*)-4] in 200 mL of THF is added 1.4 g (35 mmol) of sodium hydride (60% in oil). The mixture is stirred at rt for 30 min, chilled to 0°C, and 3.6 mL (31.5 mmol) of 2-methoxyethoxymethyl chloride (MEM-Cl) is added dropwise. After stirring for 2 h at 0°C, the solution is concentrated under vacuum, the residue is extracted with 300 mL of chloroform, and the organic layer is concentrated to give 4.8 g (82%) of 1,2-*O*-isopropylidene-3-*O*-(2-methoxyethoxymethyl)-*sn*-glycerol; bp 92–94°C (2 mm Hg); $[\alpha]^{25}_{D}$ + 11.3° (neat).

Deketalization is achieved by adding 1.5 g (7.9 mmol) of *p*-toluenesulfonic acid monohydrate to a solution of 4.7 g (21 mmol) of 1,2-*O*-isopropylidene-3-*O*-(MEM)-*sn*-glycerol in 55 mL of methanol-water (10:1), followed by stirring for 2 h at rt. The solution is neutralized with sodium bicarbonate. Evaporation of the solvents gives a residue that is extracted with chloroform. The chloroform solution is dried over Na_2SO_4 and evaporated to give 3.4 g (89%) of 3-*O*-(2-methoxyethoxymethyl)-*sn*-glycerol; bp 120–125°C (1 mm Hg); $[\alpha]^{25}_{D}$ −1.58° (*c* 9.2, CH_3OH).

Dialkylation is carried out by first preparing the dialkoxide and then adding excess octadecyl tosylate as follows. A mixture of 1.5 g (8.3 mmol) of 3-*O*-(MEM)-*sn*-glycerol and 700 mg (18 mmol) of sodium hydride (60% in oil) in 100 mL of THF is stirred at rt for 30 min. Octadecyl tosylate (8.5 g, 20.0 mmol) is added, and the mixture is heated to remove the solvent by distillation. The residue is heated in oil bath at 190–200°C for 30

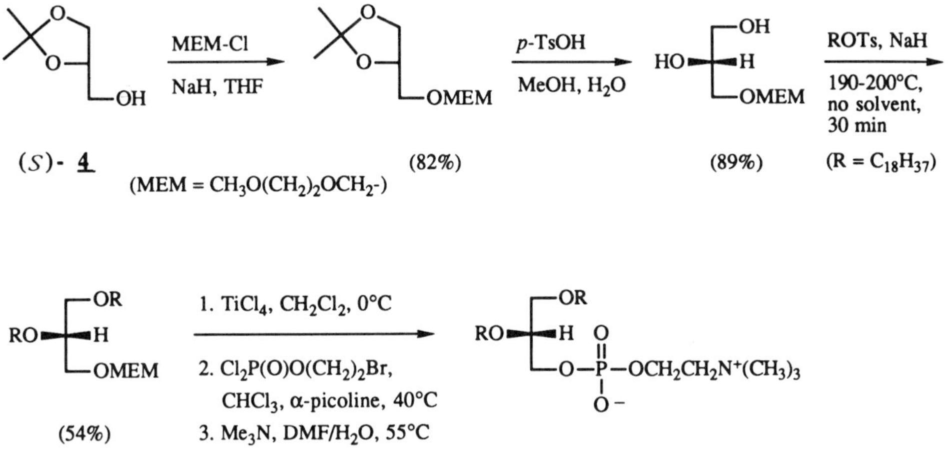

(*S*)-4 — MEM-Cl; NaH, THF → (82%) — *p*-TsOH; MeOH, H_2O → (89%) — ROTs, NaH; 190-200°C, no solvent, 30 min (R = $C_{18}H_{37}$)

(MEM = $CH_3O(CH_2)_2OCH_2$-)

(54%) — 1. $TiCl_4$, CH_2Cl_2, 0°C; 2. $Cl_2P(O)O(CH_2)_2Br$, $CHCl_3$, α-picoline, 40°C; 3. Me_3N, DMF/H_2O, 55°C → $-O-P(O)(O^-)-OCH_2CH_2N^+(CH_3)_3$

Scheme 28

min. The mixture is cooled to rt and diluted with chloroform, washed with water, then concentrated to give a residue that is dissolved in 150 mL of ethanol. A solid forms after the solution is stored at about 10°C overnight. The solid is collected by filtration and purified by chromatography on silica gel (elution with hexane-ethyl acetate 10:1), followed by recrystallization from acetone to give 3.1 g (54%) of 1,2-di-*O*-octadecyl-3-*O*-MEM-*sn*-glycerol; mp 32–33°C; $[\alpha]^{22}_D$ −2.2° (*c* 2.68, $CHCl_3$). Titanium tetrachloride (0.4 mL, 3.64 mmol) is added to an ice-cold solution of 0.74 g (1.1 mmol) of 1,2-di-*O*-octadecyl-3-*O*-MEM-*sn*-glycerol in dichloromethane. After stirring for 1 h, the solution is neutralized with 1 M sodium bicarbonate. The product 1,2-di-*O*-octadecyl-*sn*-glycerol is extracted with chloroform, the chloroform solution is washed with water, then dried (Na_2SO_4) and concentrated. Purification of the residue by silica gel chromatography (elution with hexane-ethyl acetate 8:1) gives 0.39 g (61%) of di-*O*-octadecyl-*sn*-glycerol; mp 57–58°C $[\alpha]^{25}_D$ −7.47° (*c* 1.6, $CHCl_3$).

The phosphocholine moiety is then inserted as follows. A mixture of 0.5 g (0.83 mmol) of di-*O*-octadecyl-*sn*-glycerol, 0.31 g (1.3 mmol) of (2-bromoethyl)phosphodichloridate, and 2 mL of α-picoline in 30 mL of chloroform is stirred at 40°C overnight. Amination is carried out with 24 mL of trimethylamine in 6.25 M aqueous DMF at 55°C overnight. The solution is concentrated to give a residue that is suspended in 30 mL of 90% aqueous methanol and stirred with 1.1 g (6.6 mmol) of silver acetate for 1 h. The precipitated silver bromide is removed by filtration, and the filtrate is concentrated and applied to a silica gel column. Elution with chloroform-methanol-water (65:35:5) and recrystallization from ethyl acetate gives 0.26 g (42%) of the product 1,2-di-*O*-octadecyl-*sn*-glycero-3-phosphocholine; mp 205°C; $[\alpha]^{25}_D$ +2.1° (*c* 0.5, $CHCl_3$-CH_3OH, 2:1); R_f 0.30 ($CHCl_3$-CH_3OH-conc. NH_4OH, 65:35:5).

3. Preparation of *C*2-Methyl-1-2,di-*O*-alkyl-PCs from Methylallyl Alcohol

Outline of the method:

Dialkylation of *C*2-methyl-1-*O*-tetrahydropyranyl-*rac*-glycerol is accomplished in the presence of potassium hydroxide in refluxing toluene, with removal of water under a

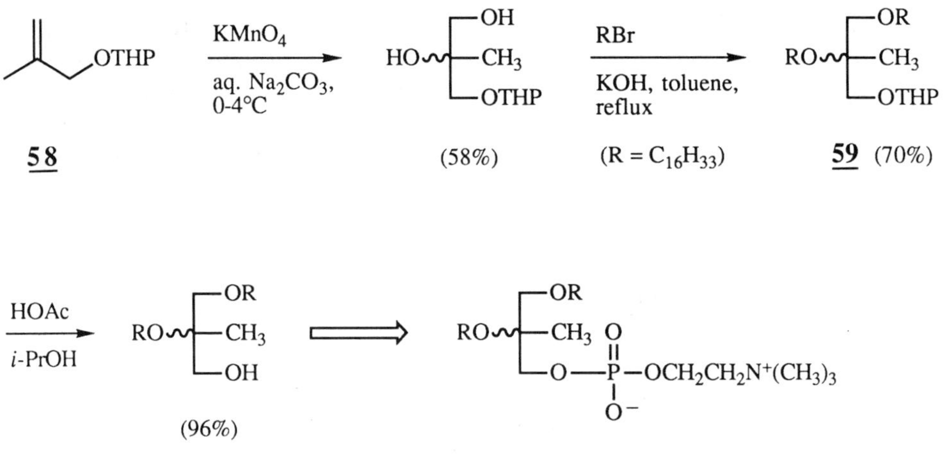

OTHP

58

$KMnO_4$

aq. Na_2CO_3, 0-4°C

HO — OH, CH_3, OTHP

(58%)

RBr

KOH, toluene, reflux

($R = C_{16}H_{33}$)

RO — OR, CH_3, OTHP

59 (70%)

HOAc

i-PrOH

RO — OR, CH_3, OH

(96%)

RO — OR, CH_3, O—P(=O)(O⁻)—$OCH_2CH_2N^+(CH_3)_3$

Scheme 29

Dean-Stark water trap. The $KMnO_4$-mediated dihydroxylation process affords the racemic diol. However, asymmetric catalytic dihydroxylation may be achieved by using osmium tetroxide and a chiral amine ligand (Tomioka et al., 1987; Jacobsen et al., 1989), suggesting that optically active *C*2-methyl-phospholipids could be prepared via chiral dihydroxylation of methylallyl alcohol.

The synthesis of a conformationally restricted PC analog having a *C*2-alkyl group is completed by replacement of the THP group by the phosphocholine group.

Literature: Bittman and Witzke (1989).

Method: To a solution of 0.2 g (2 mmol) of sodium carbonate in 60 mL of water is added 6.4 g (40 mmol) of *C*2-methylallyl THP ether (58) while the temperature is maintained at 2°C. A solution of 6.4 g (40 mmol) of potassium permanganate in 120 mL of water is added dropwise over a 45-min period with vigorous stirring at 2–4°C. The reaction mixture is cooled at 0°C for 2 h, then heated in a boiling water bath for 40 min to coagulate the precipitate, which is removed by filtration, pressed well, and washed with 20 mL of water. The filtrate is cooled to 0°C and treated with potassium carbonate (0.8 g/mL). The product is extracted with ether (2 × 40 mL), and the combined ether extract is dried (K_2CO_3). Evaporation of the THP ether gives a viscous residue that is vacuum distilled using a short-path, wide-diameter distillation apparatus; yield, 4.4 g (58%); bp 98–104°C (1.5–2.0 mm Hg).

The dialkylation is performed as follows. A mixture of 1.95 g (10.2 mmol) of *C*2-methyl-1-*O*-THP and 3.7 g (57 mmol) of 86% powdered potassium hydroxide in 90 mL of toluene is refluxed with stirring under a Dean-Stark trap for 45 min. 1-Bromohexadecane (6.5 g, 21.3 mmol) is added and the mixture is refluxed for 4 h. An additional portion (5.8 g, 19.0 mmol) of 1-bromohexadecane is then added, and refluxing is continued for 30 h. Since TLC indicates that the reaction is still incomplete, more potassium hydroxide (3.7 g, 57 mmol) and 1-bromohexadecane (6.0 g, 19.6 mmol) are added, and the mixture is refluxed with stirring for 50 h. The mixture is cooled, diluted with 100 mL of ether, and washed with water (2 × 100 mL). The combined aqueous washing is extracted with 50 mL of ether, which is washed with 30 mL of water. The organic layer is dried (K_2CO_3), the solvents are removed under vacuum, and the residue is purified by chromatography on silica gel (elution with 1,2-dichloroethane-ethyl acetate 98:2); yield of 59, 4.6 g (70%) as a colorless oil that solidifies below 15°C. Deprotection of the THP group is carried out by refluxing a mixture of 1.02 g (1.60 mmol) of 59 in 5 mL of 1-propanol, 4 mL of glacial acetic acid, and 1 mL of water for 4 h. After the mixture is cooled, 1 mL of water is added. The mixture is allowed to stand overnight at 5°C, giving a precipitate that is collected by filtration, washed with cold 50% aqueous methanol, and dried under vacuum to give 0.85 g (96%) of the product as white crystals, mp 37.0–38.5°C after recrystallization from 95% ethanol. The phosphocholine group is introduced by standard procedures.

4. An Aryl-Terminated Iodinated *sn*-1-Ether/*sn*-2-*O*-methyl-PC: Preparation from 1-*O*-Benzyl-3-*O*-trityl-*rac*-glycerol

Key features of the method:

The (*m*-iodophenyl)dodecyl group is incorporated into the phospholipid in a synthetic scheme starting with a Wittig reaction involving (11-carbethoxyundecyl)triphenylphosphonium bromide and *m*-nitrobenzaldehyde. 1-*O*-Benzyl-3-*O*-trityl-*rac*-glycerol is alkylated with methyl iodide and sodamide in dioxane or with methyl iodide and sodium hydride in THF. After detritylation in refluxing 80% acetic

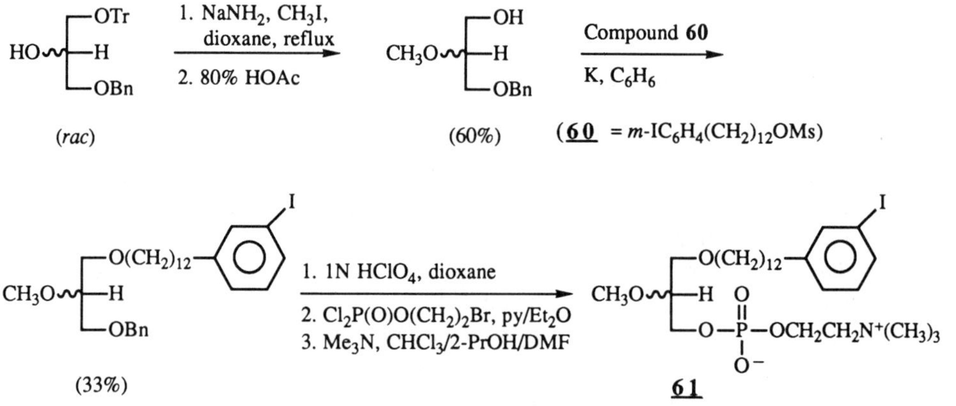

HO∿ –OTr –H –OBn (*rac*) → 1. $NaNH_2$, CH_3I, dioxane, reflux; 2. 80% HOAc → CH_3O∿ –OH –H –OBn (60%) → Compound **60**; K, C_6H_6 (**60** = *m*-$IC_6H_4(CH_2)_{12}OMs$)

CH_3O∿ –$O(CH_2)_{12}$–(*m*-iodophenyl) –H –OBn (33%) → 1. 1N $HClO_4$, dioxane; 2. $Cl_2P(O)O(CH_2)_2Br$, py/Et_2O; 3. Me_3N, $CHCl_3$/2-PrOH/DMF → CH_3O∿ –$O(CH_2)_{12}$–(*m*-iodophenyl) –H –$O-P(O)(O^-)-OCH_2CH_2N^+(CH_3)_3$ **61**

Scheme 30

acid, alkylation with 12-(*m*-iodophenyl)dodecyl mesylate 60 in benzene in the presence of potassium metal gives *rac*-1-*O*-[12-(*m*-iodophenyl)dodecyl]-2-*O*-methyl-3-*O*-benzylglycerol in 33% yield. Debenzylation by heating with 1 N perchloric acid in dioxane, followed by reaction with 2-(bromoethyl)phosphodichloridate in ether containing pyridine, then amination with trimethylamine in chloroform/2-propanol/dimethylformamide complete the synthesis of PC 61. Radioiodination of the phospholipid by isotope exchange with aqueous $Na^{125}I$ gives radioiodinated 61, which is used in tissue distribution studies in tumor-bearing rats.

Literature: Meyer et al. (1989).

Method: 12-(*m*-Iodophenyl)dodecanol is prepared by reduction of 12-(*m*-iodophenyl)dodecanoic acid with diborane in THF. To 12-(*m*-iodophenyl)dodecanol (5.88 g, 15.2 mmol) in 30 mL of pyridine in a three-neck 100-mL round-bottom flask equipped with a reflux condenser and nitrogen inlet at 0°C is added dropwise freshly distilled methanesulfonyl chloride (2.0 mL, 26.0 mmol). The mixture is stirred at rt for several hours, then poured into ice-cold water, and the resulting precipitate is filtered. The solid is dissolved in ether and the solution is extracted with water, 1 N HCl, and water. The ether layer is dried ($MgSO_4$), filtered, and concentrated under vacuum, affording a residue that is recrystallized from hexane-ethyl acetate; mp 35.5–37.0°C; yield of mesylate 60, 71%. Alkylation of *rac*-1-*O*-benzyl-3-*O*-tritylglycerol with methyl iodide is carried out by adding 616 mg (15.0 mmol) of sodium amide to a two-neck 50-mL round-bottom flask containing a solution of 5.30 g (12.5 mmol) of the protected glycerol in 30 mL of 1,4-dioxane. The reaction mixture is heated at reflux for 1 h. Methyl iodide (3.55 g, 25.0 mmol) is added dropwise, and refluxing is continued overnight. The mixture is cooled and diluted with ether and water. The ether layer is washed with 1 N HCl, water, saturated $NaHCO_3$, water, and brine, dried ($MgSO_4$), and filtered. Removal of the solvent under vacuum gives a yellow oil that solidifies on standing. The crude solid is recrystallized with hexane; yield, 79%. Detritylation in refluxing 80% acetic acid, then workup by neutralization with 10% potassium hydroxide and extraction with ether, gives 1-*O*-benzyl-2-*O*-methyl-*rac*-glycerol in 60% overall yield from 1-*O*-benzyl-3-*O*-trityl-*rac*-glycerol.

Alkylation with mesylate 60 is carried out as follows. Benzene (5 mL) and potassium metal (74.8 mg, 1.91 mmol) are added to a three-neck 50-mL round-bottom flask equipped with a reflux condenser and nitrogen inlet. The mixture is heated at reflux to melt the potassium metal, and a solution of 423 mg (2.15 mmol) of 1-*O*-benzyl-2-*O*-methyl-*rac*-glycerol in 5 mL of benzene is added dropwise via a syringe. After the reaction is heated at reflux for 1 h, a solution of 868 mg (1.86 mmol) of mesylate 60 in 4 mL of benzene is added and heating is continued overnight. The reaction is cooled to 0°C and water is added slowly. The organic layer is separated, washed with water, 1 N H_2SO_4, and water, dried ($MgSO_4$), and filtered. Removal of the solvent under vacuum gives the crude product, which is purified by silica gel chromatography (elution with hexane-ethyl acetate 8:1); yield, 33%. Debenzylation and phosphocholine insertion complete the synthesis of PC 61.

D. 1,3-Di-*O*-alkyl-*sn*-glycerols: Preparation from Glycidyl Derivatives

Outline of the method:

The 1-*O*-alkyl group is introduced by $BF_3{\cdot}Et_2O$-mediated ring opening of recrystallized (*R*)- or (*S*)-glycidyl 3-nitrobenzenesulfonate, the C_2 hydroxyl group is protected as its methoxymethyl ether (see Scheme III in Guivisdalsky and Bittman, 1989c), and the 3-nitrobenzenesulfonate group is displaced by refluxing in trifluoroethanol in the presence of NaOH.

Literature: Jain et al. (1991).

Method: The ring-opening reaction of (*R*)- or (*S*)-glycidyl 3-nitrobenzenesulfonate (recrystallized three times from ethanol, then dried thoroughly) with 1.2 equivalents of hexadecanol is carried out in dichloromethane at rt under nitrogen as described above (see Scheme 16). The ring-opened product is purified by silica gel chromatography (elution with hexane/EtOAc 4:1). The ring-opened product is added to an excess of dimethoxymethane and phosphorus pentoxide in chloroform at rt, and the mixture is stirred under nitrogen for 24 h. After the mixture is cooled to 0°C, 10% aqueous sodium carbonate is added dropwise to destroy the excess of phosphorus pentoxide. The product is extracted with chloroform, and the organic layer is dried (K_2CO_3) and evaporated.

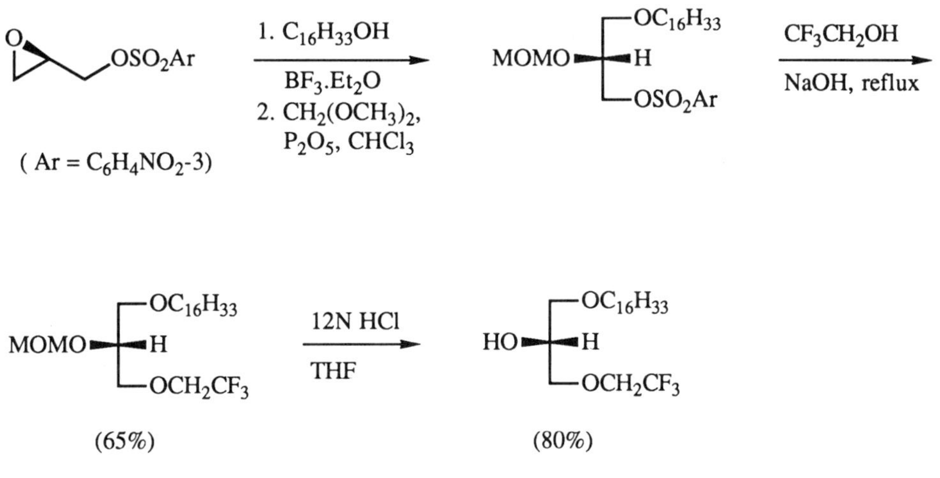

O, OSO_2Ar

($Ar = C_6H_4NO_2$-3)

1. $C_{16}H_{33}OH$, $BF_3.Et_2O$
2. $CH_2(OCH_3)_2$, P_2O_5, $CHCl_3$

MOMO—H; $OC_{16}H_{33}$; OSO_2Ar

CF_3CH_2OH, NaOH, reflux

MOMO—H; $OC_{16}H_{33}$; OCH_2CF_3

(65%)

12N HCl, THF

HO—H; $OC_{16}H_{33}$; OCH_2CF_3

(80%)

Scheme 31

A solution of 1-*O*-hexadecyl-2-*O*-(methoxymethyl)-3-*O*-((*m*-nitrophenyl)sulfonyl)-*sn*-glycerol in trifluoroethanol is refluxed in the presence of NaOH for 3 h. The mixture is cooled to rt, poured into water, and extracted with ether. The ether is dried ($MgSO_4$) and evaporated, leaving a residue that is purified by chromatography on silica gel (elution with hexane/EtOAc 9:1). Deprotection of the methoxymethyl group is then carried out by treatment with 12 N HCl in THF. The product is extracted with ether, the ether layer is dried ($MgSO_4$) and evaporated, and the residue is purified by chromatography on silica gel (elution with hexane/EtOAc 4:1).

V. THIO-CONTAINING PHOSPHOLIPIDS (SEE ALSO SEC. III.B.5)

A. 1-Ether-2-thioester-PC: Preparation from 1,2-*O*-Isopropylidene-*sn*-glycerol 3-*O*-(*p*-Toluenesulfonate)

Key features of the method:

1,2-Isopropylidene-*sn*-glycerol 3-tosylate, a commercially available configurationally stable glycerol precursor, is used to prepare 2-thioacyl-PC. The thioacetyl group is introduced into the glycerol intermediate by S_N2 displacement of a *p*-nitrobenzenesulfonate group with potassium thioacetate in acetonitrile. Displacement does not take place when the bulky trityl group is present at the *sn*-3 position; however, in a related system, displacement of an *sn*-2 tosylate group by potassium thioacetate in DMF did proceed in good yield without removal of the methoxytrityl protecting group. Migration of the thioacetyl group to the *sn*-3 position is minimized by use of Sephadex LH-20 instead of silica gel in the column chromatography step.

Literature: Bhatia and Hajdu (1989); Aarsman et al. (1985).

Method: A solution of 9.68 g (40 mmol) of 1-hexadecanol in 100 mL of THF is added dropwise to a cold (0°C) suspension of sodium hydride (44 mmol, 1.76 g of a 60% dispersion in mineral oil, washed with three 40-mL portions of petroleum ether under nitrogen). The mixture is stirred at rt for 15 min, then heated at 60°C for 1.5 h, and cooled to 0°C. A solution of 11.44 g (40 mmol) of 1,2-isopropylidene-*sn*-glycerol 3-tosylate (62) in 100 mL of THF is added dropwise to the suspension, and the mixture is heated at reflux for 36 h. The mixture is diluted with water (100 mL), and most of the solvents are evaporated under vacuum, giving a residue that is extracted with ether (3 × 100 mL). The combined ether layer is washed with brine, dried ($MgSO_4$), and evaporated, leaving 12.0 g (84%) of 3-*O*-hexadecyl-1,2-*O*-isopropylidene-*sn*-glycerol as an oil. The latter is dissolved in 40 mL of chloroform and deketalized by bubbling dry hydrogen chloride gas through a methanolic solution of the acetonide with stirring at rt for 1 h. The resulting 3-*O*-hexadecyl-*sn*-glycerol is obtained by evaporating the solvents under vacuum, drying the solid over potassium hydroxide under vacuum, and crystallization from chloroform-petroleum ether (bp 30–60°C); yield, 8.2 g (65%); mp 65°C; $[\alpha]^{23}{}_D$ −1.44° (*c* 1.32, chloroform-methanol, 1:4). Tritylation is carried out by suspending 4.9 g (15.5 mmol) of 63 in 70 mL of toluene containing 2.12 g (21 mmol) of triethylamine, then heating at 65°C for 60 h. The solid (Et_3NH^+ Cl^-) is removed by filtration and washed with 20 mL of benzene. The solvent is evaporated from the filtrate to leave a brown oil that is purified by chromatography on neutral alumina (elution with chloroform-petroleum ether 3:1). The fractions containing the product are combined and concentrated. The residue is diluted with 100 mL of petroleum ether (bp 30–60°C) to precipitate triphenylmethyl carbinol,

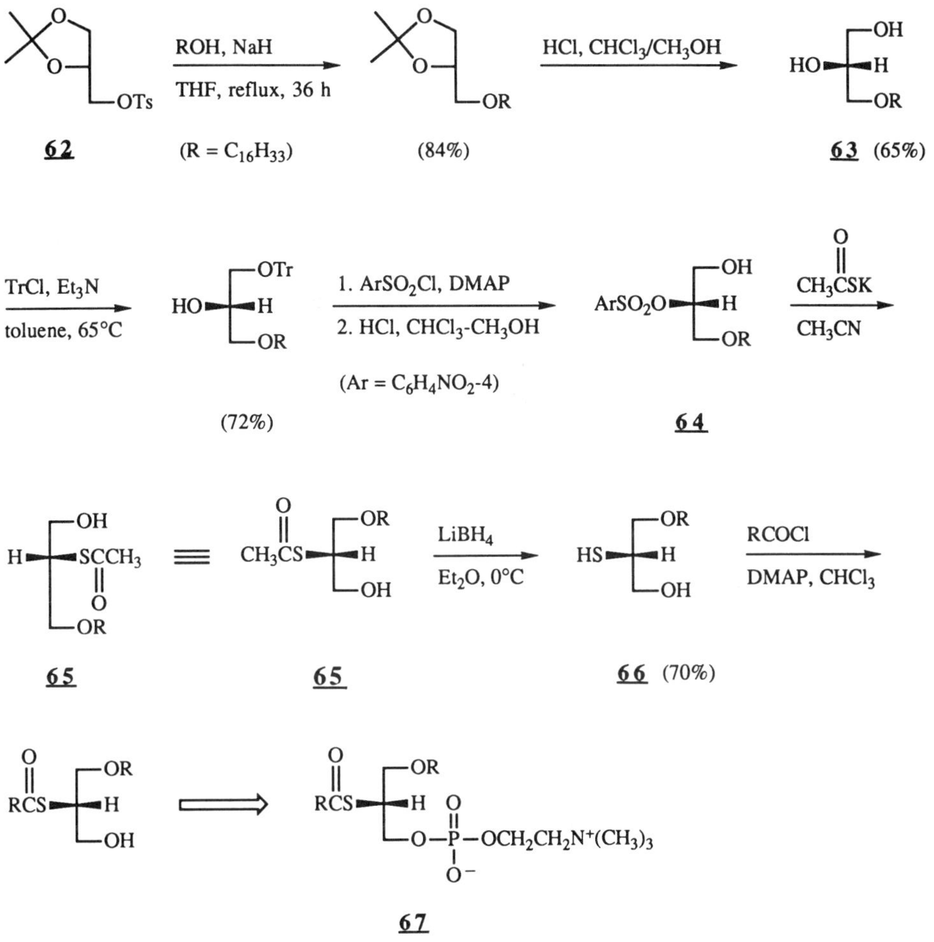

Scheme 32

which is removed by filtration. The trityl ether is crystallized from petroleum ether at –20°C; yield, 6.2 g (72%; mp 49°C; $[\alpha]^{23}{}_D$ –3.22° (*c* 1.55, chloroform-methanol, 1:4).

Reaction of the trityl ether (4.2 g, 7.53 mmol) in 100 mL of chloroform with an excess (2.1 g, 9.46 mmol) of *p*-nitrobenzenesulfonyl chloride in the presence of 1.22 g (10 mmol) of DMAP at rt for 36 h gives the sulfonate ester in 84% yield after workup (addition of water, extraction with chloroform, chromatography on silica gel). The *p*-nitrobenzenesulfonate (3.6 g, 4.84 mmol) is dissolved in 30 mL of chloroform and added to 100 mL of chloroform-methanol (1:1) through which dry hydrogen chloride has been bubbled for 40 min. The solution is stirred at rt for 1 h, the solvents are removed, and the residue is dried over potassium hydroxide pellets under vacuum. Chromatography on silica gel (elution with chloroform) gives 2.2 g (76%) of the alcohol 64. A suspension of 1.8 g (3.6 mmol) of 64 in 35 mL of dry acetonitrile is stirred with 0.5 g (4.39 mmol) of potassium thioacetate at rt for 6 h. The precipitated salt is removed by filtration, the solvent is removed, and thioacetate 65 is purified by chromatography on Sephadex

LH-20 (elution with chloroform/methanol 1 : 1); yield, 1.15 g (85%). To obtain products having a long-chain thioacyl group at the *sn*-2 position, thioacetate 65 (1.15 g, 3.07 mmol) is dissolved in 50 mL of ether, cooled to 0°C, and reduced by using lithium borohydride (75 mg, 3.4 mmol) with stirring at 0°C for 30 min and at rt for 1.5 h. After excess borohydride is decomposed by adding 20 mL of 0.5 M HCl, the product is extracted with ether (3 × 50 mL), dried ($MgSO_4$), and evaporated to give thiol 66, which is acylated without further purification and without delay. Alternatively, deacetylation of 65 is carried out with sodium borohydride in absolute ethanol at 40°C; to complete the reaction, 10% potassium hydroxide in absolute ethanol is added. Solutions used for washing of the thiol product are deoxygenated, and extractions are done under nitrogen.

The thiol 66 (0.7 g, 2.1 mmol) is then acylated with an acyl chloride in the presence of DMAP in chloroform under nitrogen, giving 1-ether-2-thioester-*sn*-glycerol. The latter is purified by Sephadex LH-20 chromatography (elution with chloroform-methanol 1 : 1) and then lyophilized from benzene. The dry alcohol is phosphorylated with 2-(bromoethyl)phosphodichloridate in the presence of triethylamine in chloroform, then aminated with aqueous trimethylamine to give PC 67. Alternatively, the phosphocholine group is incorporated by reaction of the alcohol with 2-chloro-2-oxo-1,3,2-dioxaphospholane in benzene containing triethylamine; the solvent is removed under vacuum to give the phosphate triester, which is suspended in acetonitrile, transferred to a pressure bottle, cooled to –80°C, then aminated by adding an excess of dry, liquefied trimethylamine. The mixture is stirred in the pressure bottle and heated in an oil bath at 65°C for 24 h. The mixture is cooled, the product is collected by filtration and purified by chromatography on silica gel (elution with chloroform-methanol-water 65 : 25 : 4), then lyophilized from benzene.

B. 1-Thioether-2-amide-phospholipids: Preparation from D-Serine Methyl Ester

Outline of the method:

The amino group of D-serine methyl ester is acylated with palmitoyl chloride in chloroform containing triethylamine and DMAP. Tosylation of the serine hydroxy group, followed by displacement of the tosyl group by the sodium salt of hexadecyl mercaptan, gives *S*-hexadecyl-*N*-palmitoyl-D-cysteine methyl ester. Reduction of the ester with lithium borohydride in methanol gives 1-hexadecylthio-2-palmitoylamido-1,2-dideoxy-*sn*-glycerol, which is converted into the phosphoethanolamine derivative 68 by reaction with phthalimidoethylphosphoryl dichloride, followed by removal of the phthalimido group by using hydrazine in ethanol (see Scheme 33).

Literature: Yu et al. (1990); Bhatia and Hajdu (1988).

C. From *rac*-3-Thioglycerol

Key features of the method:

A glycerophosphate oxygen is replaced by a sulfur atom to give a substrate for phospholipase C; the thioglyceride product containing the P-S-C bond can be used in a continuous spectrophotometric assay. Thiol-functionalized diacylglycerols are also useful for coupling of *S*-activated lipids to peptides and proteins (Moroder et al., 1990).

Literature: Snyder (1987).

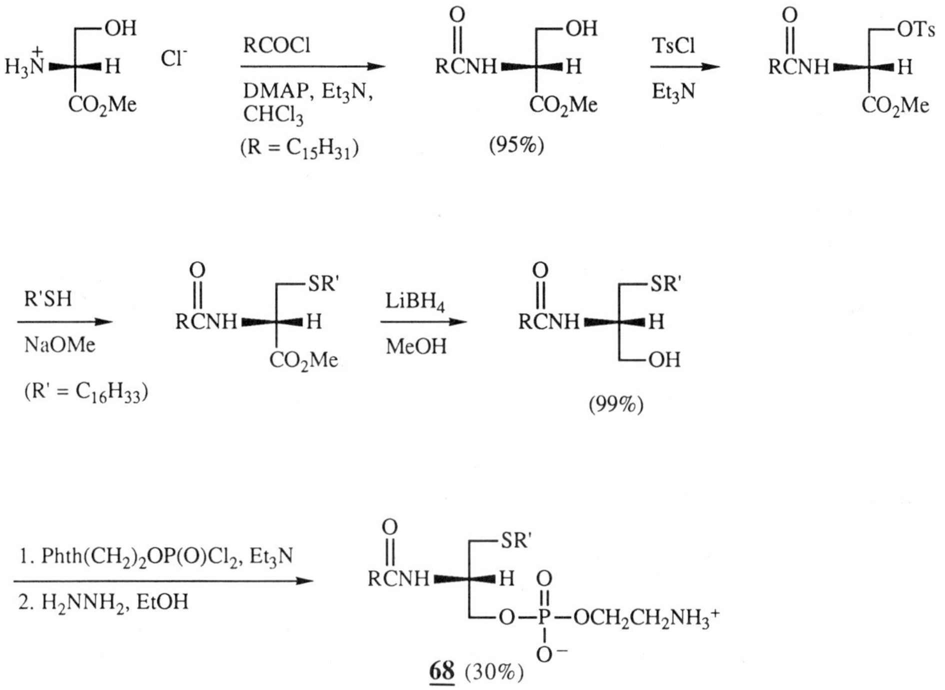

Scheme 33

Method: *rac*-3-Mercapto-1,2-propanediol (thioglycerol) (69) is oxidized to the disulfide with 30% H_2O_2 at 30–40°C. The disulfide (30 mmol) is dissolved in ethanol-free chloroform (200 mL) containing 180 mmol of dry pyridine. Acyl chloride (150 mmol) is added with stirring, and the mixture is stirred overnight at rt. Workup gives crude *rac*-3,3'-dithiobis(1,2-diacylpropanediol), which is purified by chromatography on basic alumina (Cox et al., 1979). The disulfide (15 mmol) is reduced to the diacylthiol 70 with

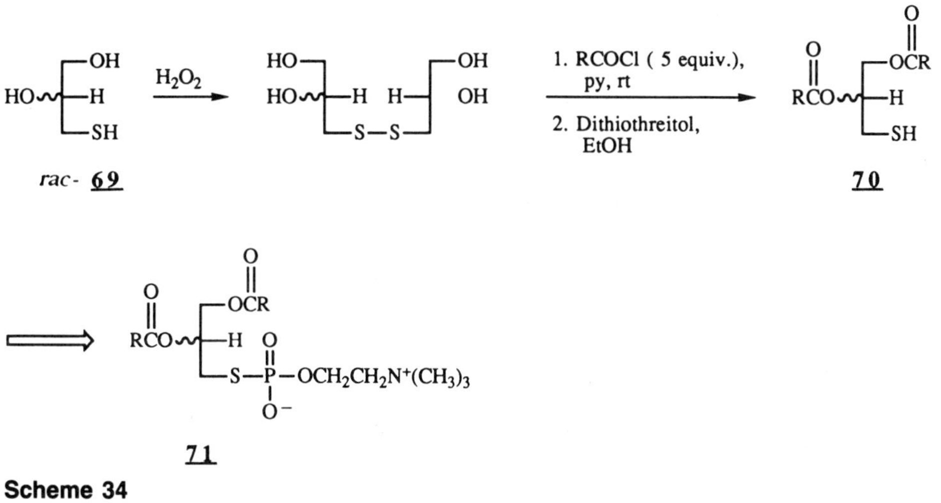

Scheme 34

dithiothreitol (30 mmol) in ethanol (300 mL) containing 0.6 mL of 29% aqueous NH_4OH; IR of thiol (2570 cm^{-1}). *rac*-2,3-Diacyl-1-thiophosphatidylcholine (71) is obtained by using standard procedures, i.e., by using 2-(bromoethyl)phosphodichloridate (Snyder, 1987) or 2-chloro-2-oxo-1,3,2-dioxaphospholane (Mlotkowska and Markowska, 1990) followed by reaction with trimethylamine (see Scheme 34).

Alternative Method: *rac*-2,3-Diacyloxypropanethiol (70) is also prepared from *rac*-3-mercapto-1,2-propanediol (69) as shown in Scheme 35. The thiol function is protected by using 1-(*tert*-butylthiohydrazine)-1,2-dicarboxymorpholide as a *tert*-butylthio donor (Moroder et al., 1990). Symmetrical diacylthioglycerols 70 are prepared in high yields by acylation of the *S*-protected derivative 72 with 3 equivalents of fatty acid in THF in the presence of 3 equivalents of DCC and 0.1 equivalent of DMAP at rt. The dicyclohexylurea is removed by filtration, the solvent is removed, and the residue is purified by column chromatography on silica gel (elution with hexane/*t*-BuOMe/AcOH, 88:10:2).

For the preparation of mixed-acid analogs of 71, the primary alcohol group of 72 is protected as the trityl ether by using 1.2 equivalents of trityl chloride in toluene in the presence of pyridine at 60°C (see Scheme 35). After esterification of the secondary hydroxyl group is carried out by using fatty acid and DCC in the presence of DMAP, detritylation is accomplished without substantial acyl migration by using 10 equivalents of anhydrous zinc bromide in methylene chloride containing methanol. The undesired 1-acyl isomer is removed by silica gel chromatography (elution with hexane/*t*-BuOMe, 2:1). The primary hydroxyl group of 72 is acylated, the residue is purified by chromatography (elution with hexane/*t*-BuOMe, 8:1), and the thiol-protecting group is removed by reduction with tri-*n*-butylphosphine in argon-saturated trifluoroethanol/*t*-BuOMe solution

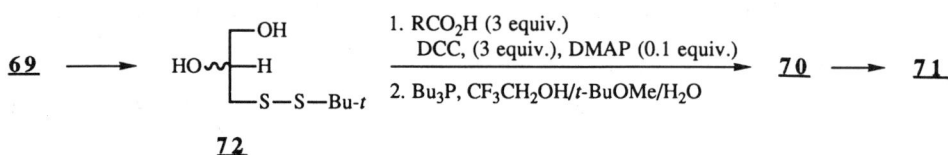

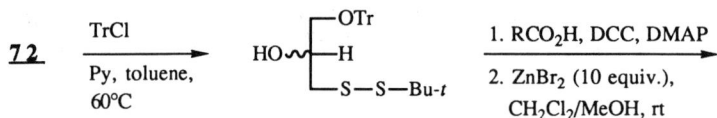

69 ⟶ HO∿[—OH, —H, —S—S—Bu-*t*] (72) — 1. RCO_2H (3 equiv.), DCC, (3 equiv.), DMAP (0.1 equiv.); 2. Bu_3P, CF_3CH_2OH/*t*-BuOMe/H_2O ⟶ 70 ⟶ 71

72 — TrCl; Py, toluene, 60°C ⟶ HO∿[—OTr, —H, —S—S—Bu-*t*] — 1. RCO_2H, DCC, DMAP; 2. $ZnBr_2$ (10 equiv.), CH_2Cl_2/MeOH, rt ⟶

RC(=O)O∿[—OH, —H, —S—S—Bu-*t*] — 1. $R'CO_2H$, DCC, DMAP; 2. Bu_3P, CF_3CH_2OH/*t*-BuOMe/H_2O ⟶ RC(=O)O∿[—OC(=O)R, —H, —SH]

Scheme 35

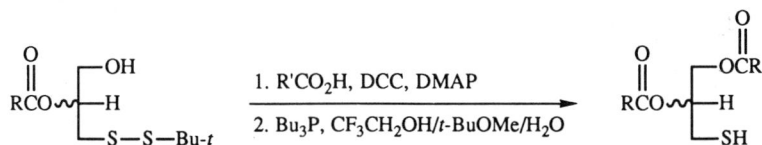

in the presence of water. Under these conditions desulfurization and *O*-to-*S* acyl migration do not occur.

Thiol 70 is phosphorylated using 2-(bromoethyl)phosphodichloridate and converted into the phosphocholine as described previously (Cox et al., 1979; Aarsman et al., 1976). The crude thio-PC is dried, dissolved in methanol/chloroform 4:1, and purified by flash chromatography, eluting with the same solvent mixture. The eluted fractions are monitored by TLC (chloroform/methanol/H_2O, 60:40:10); overall yield from diacylthioglycerol 70 to product 71, 30–40%.

D. From *rac*-bis(2,3-Diacyloxypropyl) Disulfide (73)

Another route to *rac*-2,3-diacyl-1-thio-PC 71 begins with bis[2,3-di(stearoyloxy)propyl] sulfide (73), which is converted into the corresponding sulfenyl chloride 74 with an excess of sulfuryl chloride in benzene at 5°C (see Scheme 36).

Literature: Mlotkowska (1991).

Method: The crude sulfenyl chloride 74 is added dropwise to a solution of 2,4-dichlorophenyl dimethyl phosphite in benzene at 5°C. Direct halide displacement (Arbuzov reaction) gives the *O*-aryl-*O*-methylthiophosphate in 86% yield. Demethylation is achieved by using trimethylamine. The coupling of the choline group is carried out at rt with choline 2-mesitylenesulfonate in the presence of an excess of 2-mesitylenesulfonyl chloride and *N*-methylimidazole as activating agents. Chromatography on silica gel (elution with chloroform, then chloroform/methanol 1:1) gives the *O*-arylthiophosphocholine in 72% yield. The 2,4-dichlorophenyl group is removed with *sym*-4-nitrobenzaldoxime and *N,N,N',N'*-tetramethylguanidine in THF, completing the synthesis of 71.

Sulfenyl chloride 74 is also converted to thiophosphocholine 71 by an Arbuzov reaction with the cyclic phosphite shown at the bottom of Scheme 36, followed by reaction with trimethylamine in benzene-acetonitrile at 55–60°C (Mlotkowska and Markowska, 1988).

VI. GLYCEROGLUCOPYRANOSIDES AND GLYCERO-1-THIO-GLUCOPYRANOSIDES

A. Ether-Linked Glycolipids and Thioglycolipids

Outline of the method:

1-*O*-Hexadecyl-2-*O*-methyl-3-*O*-(*p*-tolylsulfonyl)-*sn*-glycerol (75a) reacts with 1-thio-β-D-glucose tetra-*O*-acetate in the presence of 1,8-diazabicylco[5.4.0]undec-7-ene (DBU) to give a 3:1 mixture of the α and β anomers of 1-thioglycopyranosides 76a and 76b (see Scheme 37). The anomers are separated by flash chromatography; the R_f values in hexane-ethyl acetate are 76a, 0.38; 76b, 0.27. The tetra-*O*-acetates are hydrolyzed quantitatively by using methanolic barium oxide at rt for 24 h, followed by neutralization by addition of an ion-exchange resin (Dowex 50W-X8, H^+ form). The resin is removed by filtration through Celite, and the filtrate is dried over sodium bicarbonate to remove traces of acid. The filtrate is concentrated under vacuum, affording the product as white crystals.

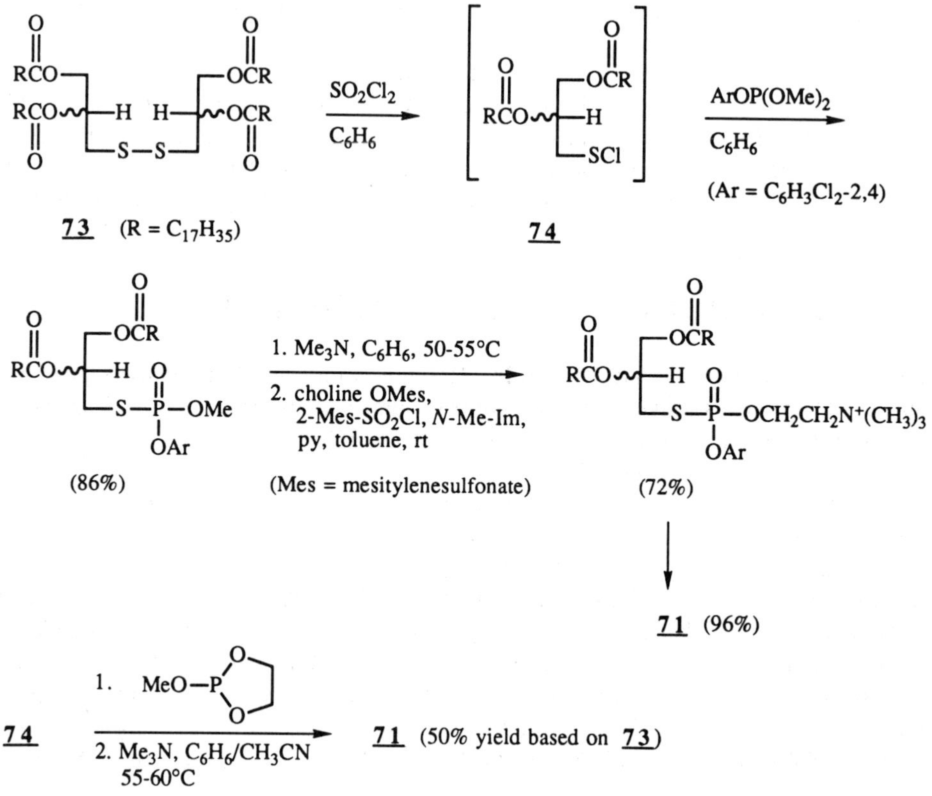

$$\mathbf{73}\ (R = C_{17}H_{35}) \xrightarrow[C_6H_6]{SO_2Cl_2} [\mathbf{74}] \xrightarrow[C_6H_6]{ArOP(OMe)_2} (Ar = C_6H_3Cl_2\text{-}2,4)$$

(86%)

1. Me_3N, C_6H_6, 50-55°C
2. choline OMes, 2-Mes-SO_2Cl, *N*-Me-Im, py, toluene, rt

(Mes = mesitylenesulfonate)

(72%) → 71 (96%)

74 → 1. MeO–P(cyclic) 2. Me_3N, C_6H_6/CH_3CN 55-60°C → 71 (50% yield based on 73)

Scheme 36

1-*O*-Hexadecyl-2-*O*-methyl-*sn*-glycerol (75b) reacts with acetobromoglucose in the presence of mercuric cyanide in benzene-nitromethane 1:1 (the Helferich modification of the Königs-Knorr procedure) to give 1-*O*-hexadecyl-2-*O*-methyl-*sn*-glycero-3-*O*-(β-D-glucopyranosyl 2,3,4,6-tetraacetate) (77) in 85% yield. The tetra-*O*-acetate 77 is hydrolyzed quantitatively to the target β-glucoside with methanolic sodium hydroxide. The α-linked thioglycolipid and the β-linked 3-*O*-D-glucopyranosyl-lipid that bear a long-chain *O*-alkyl group at the *sn*-1 position and a *O*-methyl group at the *sn*-2 position of glycerol have selective cytotoxic activity against leukemic cells. In recent years many new methods of glycosylation involving various sugar derivatives and activating reagents have been used to prepare biologically interesting glycoconjugates (e.g., Schmidt, 1989).

Literature: Guivisdalsky et al. (1990).

B. Ester-Linked Glycolipids

Outline of the method:

1,2-Di-*O*-(but-2-enyl)-*sn*-glycerol is coupled to 2,3,4,6-tetra-*O*-benzyl-α-D-glucopyranosyl bromide in methylene chloride/DMF (10:1) in the presence of *N*,*N*-diisopropylethylamine, tetraethylammonium bromide, and 4Å molecular sieves, giving

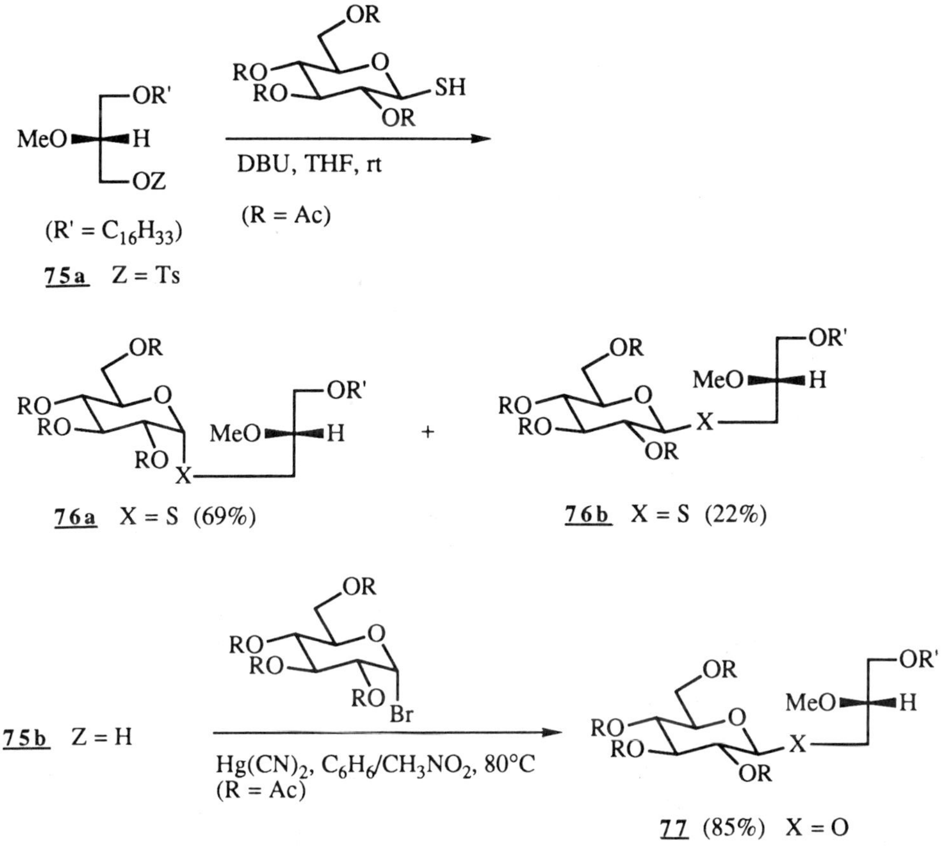

(R' = $C_{16}H_{33}$)

75a Z = Ts

DBU, THF, rt (R = Ac)

76a X = S (69%)

76b X = S (22%)

75b Z = H

$Hg(CN)_2$, C_6H_6/CH_3NO_2, 80°C (R = Ac)

77 (85%) X = O

Scheme 37

3-*O*-(2,3,4,6-tetra-α-D-glucopyranosyl)-*sn*-glycerol (78) in 65–70% yield after column chromatography on silica gel (elution with hexane/ether). The butenyl groups are removed by using potassium *tert*-butoxide in DMSO, giving 3-*O*-(2,3,4,6-tetra-*O*-benzyl-α-D-glucopyranosyl)-*sn*-glycerol. The latter is acylated at the *sn*-1 and *sn*-2 positions, then debenzylated with 10% Pd/C (catalyst was pretreated with hydrogen in ethanol to remove acidic materials) in methanol/ethyl acetate/acetic acid (3:3:1).

Literature: Mannock et al. (1990).

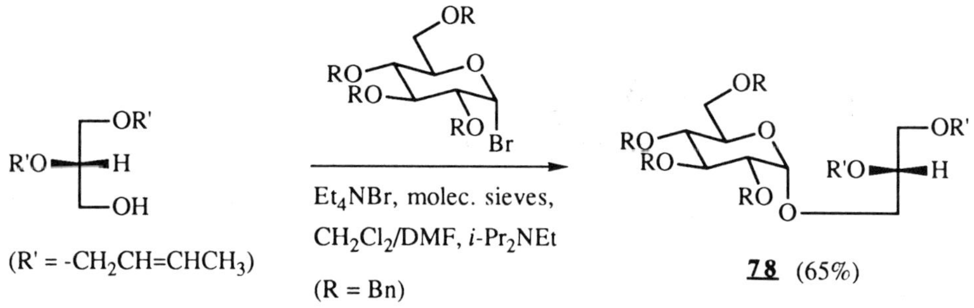
OR
RO
RO
O
RO
Br
OR'
R'O
H
OH
(R' = -CH2CH=CHCH3)
Et4NBr, molec. sieves,
CH2Cl2/DMF, i-Pr2NEt
(R = Bn)
OR
O
RO
RO
RO
O
OR'
R'O
H
78 (65%)

Scheme 38

VII. METHODS FOR PHOSPHORYLATION OF GLYCEROL DERIVATIVES

The classical method for introducing the phosphocholine group into diacyl- or di-*O*-alkyl-*sn*-glycerol is by the reaction sequence of phosphorylation with (2-bromoethyl)phosphodichloridate, hydrolysis, and trimethylamination (Hirt and Berchtold, 1958) (see Scheme 39a). The phosphorylating agent is prepared by adding 2-bromoethanol dropwise to a solution of excess phosphorus oxychloride (distill immediately before use; bp 105–106°C) in trichloroethylene; a stream of dry nitrogen is passed through the solution to remove hydrogen chloride as it is formed. After the reaction mixture is stirred overnight under nitrogen at rt, toluene is added and the volatiles (including unreacted phosphorus oxychloride) are evaporated under reduced pressure at 40°C (Eibl and Nicksch, 1978). Although most investigators use the phosphodichloridate directly without further purification, it may be purified by vacuum distillation (bp 89° at 1.5 mm Hg) and then stored in sealed ampoules at low temperature; distillation is advisable if the phosphodichloridate to be used was not freshly prepared. In a typical

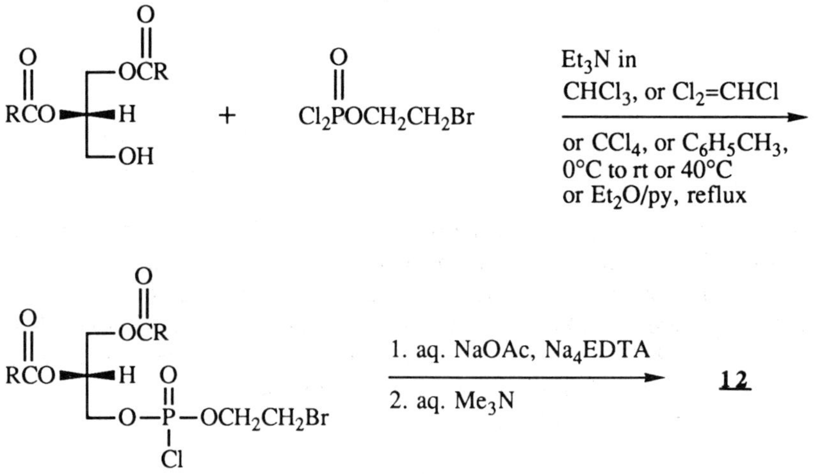
O
OCR
O
RCO
H
OH
+
O
Cl2POCH2CH2Br
Et3N in
CHCl3, or Cl2=CHCl
or CCl4, or C6H5CH3,
0°C to rt or 40°C
or Et2O/py, reflux
O
OCR
O
RCO
H
O
O–P–OCH2CH2Br
Cl
1. aq. NaOAc, Na4EDTA
2. aq. Me3N
12

Scheme 39a

phosphorylation procedure, a solution of 5 mmol of the disubstituted glycerol in about 15 mL of chloroform (or trichloroethylene, carbon tetrachloride, benzene, or toluene) is added dropwise to a solution of 4.9 g (20 mmol) of (2-bromoethyl)phosphodichloridate and 4.3 g (42 mmol) of triethylamine in chloroform at 0°C (Aarsman et al., 1976).

In an alternative procedure, the glycerol derivative is allowed to react with 3 equivalents of (2-bromoethyl)phosphodichloridate in refluxing ether containing pyridine; ether was substituted for chloroform in order to avoid formation of chlorinated by-products (Hansen et al., 1982). The precipitate of triethylamine hydrochloride or pyridine hydrochloride is removed by filtration, the solvents are evaporated under vacuum, and the phosphorylchloride is hydrolyzed to the phosphate diester by dissolving the residue in a minimum volume of THF and adding an equal volume of 0.5 M sodium acetate solution, pH 8.5, and 0.06 vol of 0.5 M Na_4EDTA solution, pH 10.5. After the mixture (pH~8) is stirred for 1 h, the bromoethyl phosphate ester is extracted with chloroform-methanol 9:1, and the organic layer is dried ($MgSO_4$), filtered, and concentrated, giving a residue that is dissolved in chloroform-2-propanol-acetonitrile (3:5:5) and allowed to react with an excess of aqueous trimethylamine. [The displacement of the bromide may also be carried out in acetone (or carbon tetrachloride) with anhydrous trimethylamine in a pressure bottle at 60–75°C (Hansen et al., 1982).] After the amination reaction is complete, the volatiles are removed, and the residue is dissolved in chloroform-methanol-water 5:4:1 and passed through an ion-exchange column (Amberlite or Rexyn) to remove salts; the eluate is evaporated, and the product is further purified by silica gel chromatography. In small-scale work, many investigators omit the ion-exchange column chromatography step. The product can be dissolved in a minimum volume of chloroform and precipitated with 10 volumes of cold acetonitrile or acetone, then lyophilized from benzene. The phosphocholine head group can be exchanged to phosphoethanolamine or other head groups via transphosphatidylation catalyzed by phospholipase D (Cox et al., 1979).

Poor yields have been obtained when (2-bromoethyl)phosphochloridate is used to phosphorylate sterically hindered glycerols (Bonsen et al., 1972; Chandrakumar and Hajdu, 1981). 2-Chloro-2-oxo-1,3,2-dioxaphospholane (Nguyen Thanh Thuong and Chabrier, 1974), which is prepared by bubbling oxygen through a solution of ethylene chlorophosphite in benzene followed by vacuum distillation (Edmundson, 1962), is more

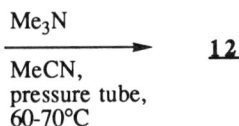

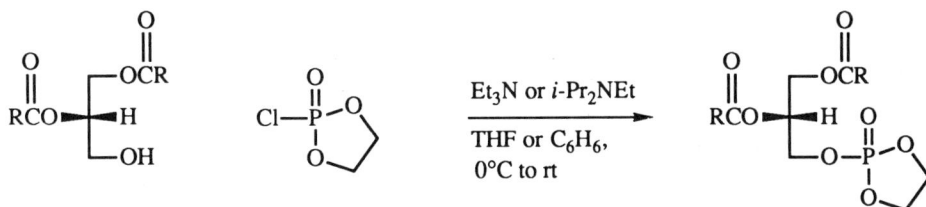

Scheme 39b

reactive than (2-bromoethyl)phosphochloridate with sterically hindered glycerol derivatives (Chandrakumar and Hajdu, 1982; Dijkman et al., 1990). In a typical procedure, the alcohol to be phosphorylated is dried azeotropically with benzene or under vacuum over P_2O_5, dissolved in THF or benzene, and stirred at 0°C; then 1.2–1.5 equivalents of triethylamine are added, followed by dropwise addition 1.2 equivalents of 2-chloro-2-oxo-1,3,2-dioxaphospholane (e.g., Magolda and Galbraith, 1989) (see Scheme 39b). When the reaction is complete, as monitored by TLC, the mixture is diluted with cold tetrahydrofuran, $MgSO_4$ is added with stirring, and the mixture is filtered and evaporated. The residue (cyclic phospholane) is dried azeotropically with benzene, then dissolved in chloroform or acetonitrile, and transferred into a glass pressure tube equipped with a teflon screw cap (available from Ace Glass, Vineland, N.J.). If chloroform is used, it is evaporated by using a stream of nitrogen, and acetonitrile is added. A large excess of anhydrous trimethylamine is added to the pressure tube; this may be carried out by condensing the desired volume of trimethylamine by bubbling the gas through a chilled (–78°C) tube. The pressure tube is closed after trimethylamine is added, and the mixture is heated on an oil bath at 60–70°C with stirring for about 24 to 36 h. The mixture is cooled, the pressure tube is opened, the volatiles are evaporated, and the residue is purified by silica gel chromatography. Yields of PCs (based on the alcohol) are generally in the range of 50–65%.

Literature: Chandrakumar and Hajdu (1982, 1983); Agarwal et al. (1984); Guivisdalsky and Bittman (1989b).

Scheme 39b has been modified (Miyazaki et al., 1989) by using a higher boiling solvent and a hindered base, resulting in a better yield than that obtained with 2-chloro-2-oxo-1,3,2-dioxaphospholane in benzene containing triethylamine. A solution of 2.39 mmol of alcohol and 0.83 mL (4.76 mmol) of *N,N*-diisopropylethylamine in 17 mL of 1,2-dichloroethane is added in one portion with stirring to a solution of 678 mg (4.76 mmol) of 2-chloro-2-oxo-1,3,2-dioxaphospholane in 3 mL of 1,2-dichloroethane. The mixture is stirred at 80°C for 24 h, cooled, and concentrated to dryness under nitrogen. The phosphate triester residue is dissolved in 6 mL of acetonitrile, mixed with a solution of 7.00 g (118 mmol) of trimethylamine in 10 mL of acetonitrile in a pressure bottle, and heated on an oil bath at 70–80°C for 65 h. The mixture is cooled, the solvent is evaporated, and the residue is chromatographed on silica gel (elution with methylene chloride-methanol-water 65 : 35 : 5); yield, 64%.

The cyclic phosphotriester ester is readily converted to polar head groups other than phosphocholine. Hydrolysis instead of amination of the phosphate triester is carried out in 50% aqueous acetic acid at rt, giving the phosphoglycol [$ROP(O)(O^-)OCH_2CH_2OH$] in 65–70% yield (Dijkman et al., 1990). [The latter compound has also been made by the reaction of phosphatidic acid dichloride with ethylene glycol, followed by ring opening with 20% aqueous acetic acid in 2-propanol (Eibl, 1978).] Reaction with diethylamine instead of triethylamine gives phosphatidylethanol-*N,N*-dimethylamine (Bhatia and Hajdu, 1988). The yield of the ring-opening reaction with nucleophiles is improved by addition of 2 equivalents of trimethylsilyl triflate to the solution of the cyclic phosphotriester in dichloromethane (Gadek, 1989).

Modifications in the distance between the phosphate and trimethylammonium group in the phosphocholine moiety are made by the reaction of diester-, diether, or ether/ester-glycerols with phosphorus oxychloride and either $HO(CH_2)_nN^+Me_3$ (<u>79</u>) or $Br(CH_2)_nOP(O)Cl_2$ (see Scheme 39c). [In fact, reaction of the glycerol derivative with phosphorus oxychloride and choline tosylate gives PCs (Witzke and Bittman, 1986).]

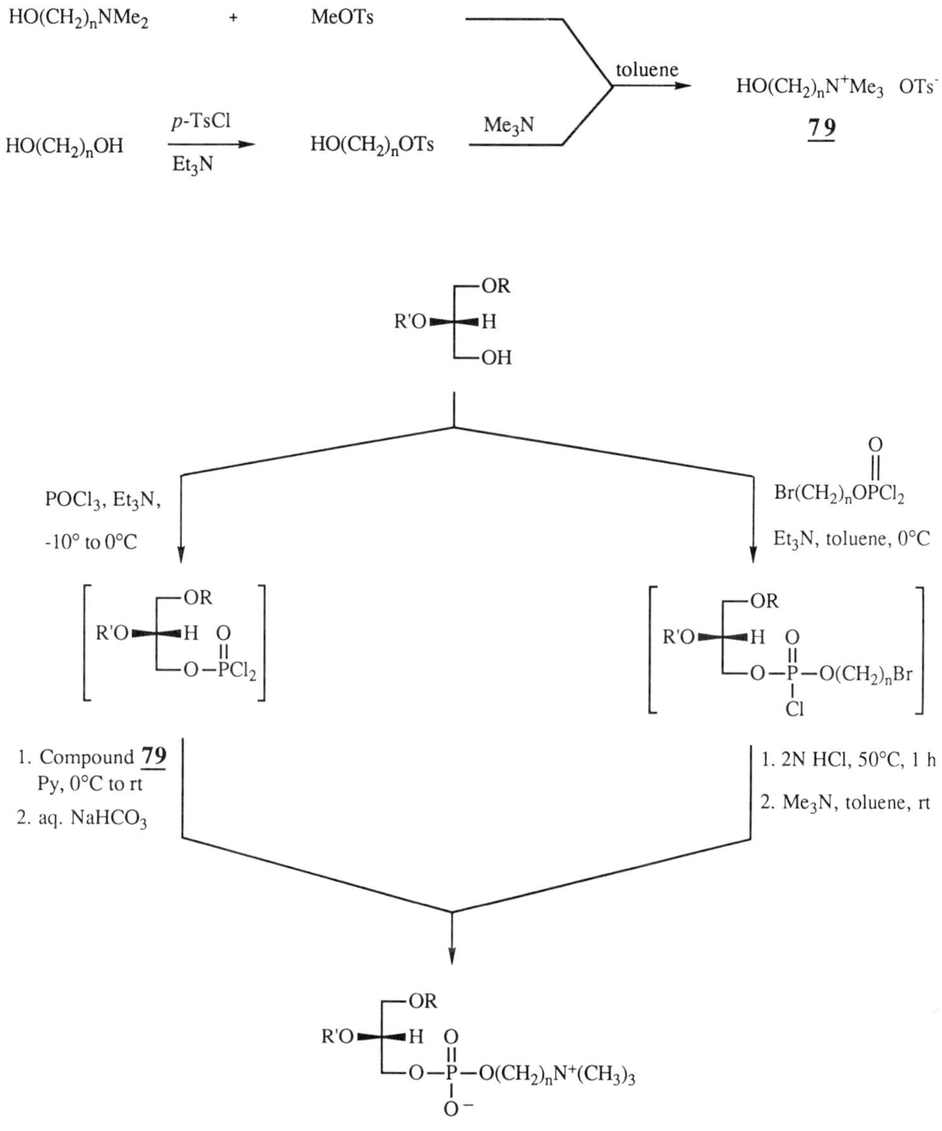

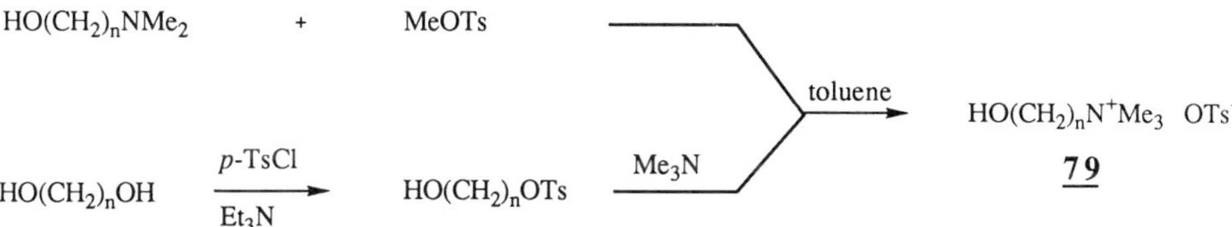

R'O–C(H)(–OR)(–OH)

$POCl_3$, Et_3N,
-10° to 0°C

[R'O–C(H)(–OR)–O–P(O)Cl_2]

1. Compound **79**
Py, 0°C to rt
2. aq. $NaHCO_3$

$Br(CH_2)_nOP(O)Cl_2$
Et_3N, toluene, 0°C

[R'O–C(H)(–OR)–O–P(O)(Cl)–$O(CH_2)_nBr$]

1. 2N HCl, 50°C, 1 h
2. Me_3N, toluene, rt

R'O–C(H)(–OR)–O–P(O)(O^-)–$O(CH_2)_nN^+(CH_3)_3$

80

Scheme 39c

Phosphocholine analogs with head groups containing additional methylene groups between the phosphorus and nitrogen functions or with *N*-alkyl groups larger than *N*-methyl have been used in a wide variety of biophysical studies (see references in Ali and Bittman, 1989). Such compounds have also been used to show that the P-N distance is important in analogs of *rac*-ET-18-OCH_3-PC for activation of platelets and for high antitumor efficacy (Ukawa et al., 1989).

ω-Hydroxyalkyltrimethylammonium tosylate [$HO(CH_2)_nN^+Me_3\ OTs^-$] (79) may be prepared by adding a solution of methyl *p*-toluenesulfonate in toluene to a solution of ω-dimethylamino-1-alkanol in toluene at rt with stirring over a period of 30 min to overnight (TLC is used to monitor disappearance of starting alkanol) (Ukawa et al., 1989) (Scheme 39c). The crystals are collected by filtration, washed with toluene and ether, and dried under vacuum.

An alternative method that is more difficult for the preparation of 79 is from α,ω-diols (Scheme 39c). A mixture of triethylamine (1.4 equiv), diol (2 equiv), and *p*-toluenesulfonyl chloride (1 equiv) is stirred overnight at room temperature. The residue obtained after evaporation of the reaction mixture to dryness is dissolved in dichloromethane, washed with water, 1 N HCl, and water. The organic layer is separated, dried ($MgSO_4$), and concentrated under vacuum. The residue is purified by chromatography on silica gel, eluting with dichloromethane-methanol 96:4, giving the monotosylate of the α,ω-diol [$HO(CH_2)_nOTs$]. The tosylate is then allowed to react with trimethylamine (18% in toluene) at rt over several days.

Phosphocholine analogs 80, which have modifications in the P-N distance, i.e., $-OP(O)(O^-)(CH_2)_nN^+Me_3$, where $n = 2 - 10$, are prepared by the reaction of the glycerol derivative with ω-hydroxyalkyltrimethylammonium tosylate (Ukawa et al., 1989). A solution of the glycerol derivative in ethanol-free chloroform is added dropwise to a mixture of phosphorus oxychloride and 5 equivalents of triethylamine in ethanol-free chloroform at 0°C (or preferably at –10°C). The reaction mixture is stirred for 1 h at rt, then cooled to 0°C. To this mixture, a solution of 79 (1.4 equiv) in pyridine is added at 0°C. The mixture is stirred for several days at rt, then treated with saturated $NaHCO_3$ solution and evaporated to dryness. The residue is extracted with chloroform-toluene 1:1. The solvents are removed, leaving a residue that is purified by silica gel chromatography, eluting with chloroform-methanol-water (65:25:4).

Another method for the preparation of 80 involves the addition of a solution of the glycerol derivative in toluene to a mixture of 1.5 equivalents of ω-bromoalkylphosphodichloridate [$Br(CH_2)_nOP(O)Cl_2$] in toluene at 0°C with stirring (Ukawa et al., 1989). After the mixture is allowed to react at rt for several hours, 2 N HCl is added and the mixture is heated at 50°C for 1 h. The bromoethylphospholipid intermediate is extracted with ether. The ether layer is washed with water, dried with $MgSO_4$, and evaporated under vacuum, to give a residue that is dissolved in 20% (w/v) trimethylamine/toluene. After reaction at rt for 3 days, the solvents are removed and the residue is purified by chromatography on silica gel (elution with chloroform-methanol-water 65:25:4).

The coupling of *N,N,N*-trialkylammonium alkanols (as their *p*-toluenesulfonate or tetraphenylborate salt) to phosphatidic acid is achieved by the following method (Scheme 40), affording PC analog 81 (Isaacson et al., 1979; Ali and Bittman, 1989). About 50 mg of phosphatidic acid is dissolved in 5–10 mL of pyridine by heating in a round-bottom flask under nitrogen with stirring at 50°C for 30 min. Ten equivalents of the desired choline analog (as the *p*-toluenesulfonate or tetraphenylborate salt; previously dried over P_2O_5) and 2–3 mL of trichloroacetonitrile are added. The reaction mixture is stirred under nitrogen at 50°C for 2 days. The solvent is removed under vacuum, leaving a brown residue. To assure removal of pyridine, 20 mL of chloroform-methanol 1:1 is added to the residue, and the solvents are evaporated. The crude product 81 is dissolved in 20 mL of THF-water 9:1 and applied to a column of Amberlite MB-3 (20 g) that has been equilibrated in the same solvent system. After elution with about 400 mL of THF-water

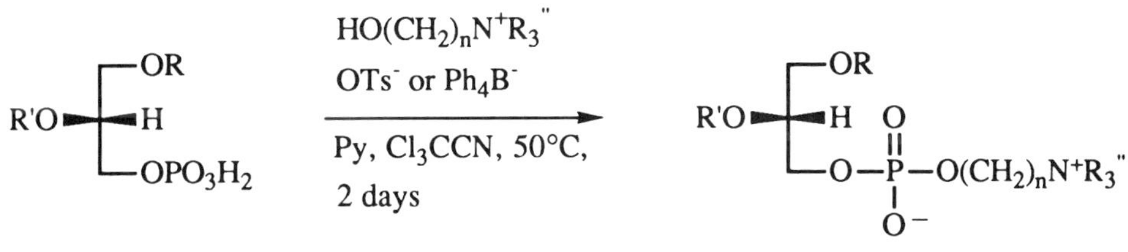

$$R'O \blacktriangleright\!\!\begin{bmatrix} -OR \\ \blacktriangleleft H \\ -OPO_3H_2 \end{bmatrix} \xrightarrow[\text{Py, } Cl_3CCN,\ 50^\circ C,\ 2\text{ days}]{HO(CH_2)_nN^+R_3''\ \ OTs^- \text{ or } Ph_4B^-} R'O \blacktriangleright\!\!\begin{bmatrix} -OR \\ \blacktriangleleft H \\ -O-P(=O)(O^-)-O(CH_2)_nN^+R_3'' \end{bmatrix}$$

81

Scheme 40

9:1, the eluate is evaporated; water is removed by azeotropic distillation with 2-propanol. The tan residue is dissolved in a minimum volume of chloroform and purified by flash chromatography (elution first with chloroform, then with chloroform-methanol, 9:1 and 3:2). Yields vary with the extent of steric bulk in the choline analog; typical yields range from 30 to 60%.

Dimethyl phosphochloridate is another useful phosphorylating agent (Bittman et al., 1984). The methyl esters are removed by transsilylation followed by hydrolysis.

Bis(2,2,2-trichloroethyl)chlorophosphate and diaryl chlorophosphate are other useful phosphorylating agents that allow facile removal of protecting phosphate esters functionalities.

VIII. Phosphitylation of Glycerol Derivatives

Reactions of hydroxyl groups in lipids with a trivalent phosphorus reagent (phosphite or H-phosphonate) are reviewed in this section. The name H-phosphonate is used to emphasize the presence of a P-H bond, whereas the term phosphite indicates a tricoordinated species derived from the phosphite form of phosphonic acid: $(HO)_2P(O)H \rightleftharpoons (HO)_3P$

A. Via H-Phosphonates

Outline of the method:

Scheme 41 shows the preparation of diacylglycero-H-phosphonate 82 from the reaction of 1,2-diacyl-*sn*-glycerol with PCl_3/imidazole in toluene. The H-phosphonate intermediate 82 reacts with hydroxy groups when activated with a condensing agent, which is typically an acyl chloride. Phosphorochloridates and arenesulfonyl chlorides have also been used to promote the condensation reaction. The coupling reaction is carried out in pyridine. In a typical procedure for the preparation of different phospholipid classes, an excess of anhydrous hydroxylic component [choline tosylate for PC synthesis, *N*-(*tert*-butoxycarbonyl)ethanolamine for phosphatidylethanolamine (PE) synthesis, or *N*-(*tert*-butoxycarbonyl)-L-serine for phosphatidylserine (PS) synthesis] and an excess of condensing agent (e.g., pivaloyl chloride or trichloroacetonitrile) are added to a solution of diacylglycero-H-phosphonate 82 in pyridine. The reaction is complete after stirring for 5–10 min at rt and is quenched by adding triethylammonium bicarbonate or aqueous pyridine. The crude H-phosphonate is extracted with chloroform and purified by chromatography on a silica gel column. Oxidation of the H-phosphonate with 2 equiv-

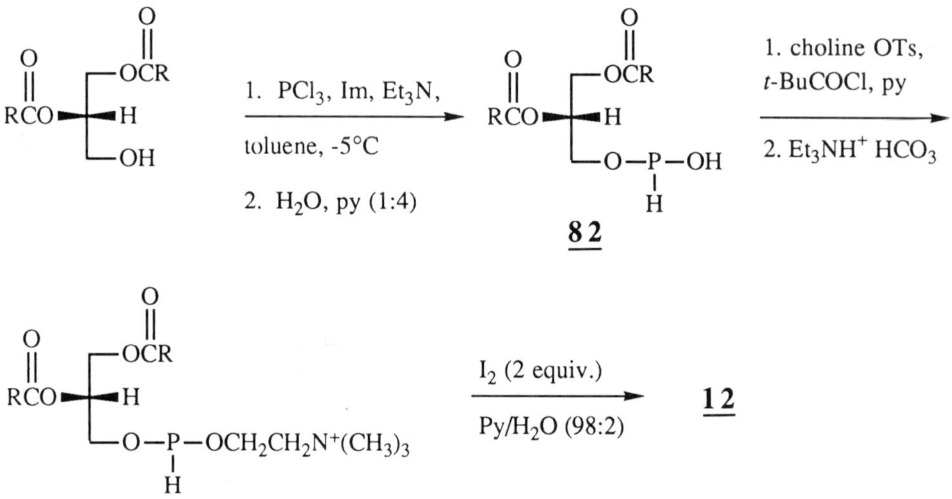
O
OCR
O
RCO
H
OH
1. PCl3, Im, Et3N,
toluene, -5°C
2. H2O, py (1:4)
O
OCR
O
RCO
H
O—P—OH
H
82
1. choline OTs,
t-BuCOCl, py
2. Et3NH+ HCO3
O
OCR
O
RCO
H
O—P—OCH2CH2N+(CH3)3
H
I2 (2 equiv.)
Py/H2O (98:2)
12

Scheme 41

alents of iodine in pyridine-water 98:2 at rt for 5 min gives saturated diacyl-PCs (12). Oxidation under mild conditions with MCPBA, *tert*-butyl hydroperoxide, or hydrogen peroxide instead of iodine in aqueous pyridine also converts the H-phosphonate into the phosphate ester, i.e., $(RO)_2P(O)(H)R' \rightarrow (RO)_2P(O)(O^-)R'$. Use of sulfur as the oxidant affords phosphorothioate analogs, $(RO)_2P(O)(S^-)R'$. Selenization has been used to introduce the Se bond. Deprotection of the *N-tert*-butoxycarbonyl group with trifluoroacetic acid and perchloric acid in dichloromethane is necessary to obtain PE and PS.

Literature: Lindh and Stawiński (1989).

B. Via Phosphoroamidites

Phosphitylation by use of P(III) amide chlorides, which are more reactive than P(V) compounds, has been used extensively in coupling reactions in oligonucleotide synthesis (Atkinson and Smith, 1984; Froehler et al., 1988; Goodchild, 1990). These coupling procedures have recently been applied to phospholipid synthesis (Bruzik et al., 1986; Lemmen et al., 1990). In a typical example, a phosphoramidite is formed by treating a solution of a diprotected glycerol in chloroform containing 2 equivalents of triethylamine at 0°C with 1.2 equivalents of *N,N*-diisopropylmethylphosphoramidic chloride $[ClP(OCH_3)N(Pr\text{-}i)_2]$. After TLC monitoring of the reaction mixture (e.g., 5 min) indicates that no starting alcohol remains, the solvents and excess triethylamine are removed under vacuum. Excess 1*H*-tetrazole (4 equivalents) and choline tosylate (2–3 equivalents) or other hydroxy component are added to the phosphoramidite. (Tetrazole serves as an acid catalyst, and it forms a tetrazolide phosphinate intermediate, which reacts rapidly with hydroxy groups.) The coupling reaction is carried out in acetonitrile-THF at rt. After the coupling reaction is completed, the solvents are evaporated, toluene is added, and the phosphite ester is oxidized with an excess of *tert*-butyl hydroperoxide,

MCPBA, or hydrogen peroxide. The phosphite triester intermediate has been formed by using a variety of other chlorophosphoramidites, $ClP(OR)(NR')_2$.

Phosphite oxidation is also achieved by using an excess of iodine in pyridine-water or an iodine-based oxidant in nonaqueous solvent (Fourrey and Varenne, 1985). The reaction mixture is washed with triethylammonium bicarbonate buffer (pH 7.0), the organic phase is concentrated, and the residue is dried thoroughly by repeated evaporation with benzene. Demethylation with trimethylamine is carried in toluene in a pressure bottle at rt for about 10 h. The product is purified by flash chromatography (elution with chloroform/methanol/water mixtures). An example is given below in which the oxidation step is replaced by sulfurization (reaction with excess sulfur in toluene), giving chiral phosphorothioate analogs of phospholipids in which the nonbridging oxygen atom in the phosphate group is replaced by a sulfur atom (Bruzik, 1988). In this example, a protected serine derivative is used as the hydroxy compound. Diastereomers have been synthesized and used to examine the stereoselectivity of the hydrolysis reactions catalyzed by phospholipases.

A long-chain alcohol is converted into an alkyl *N*-methylphosphocholine by the phosphitylation/oxidation sequence shown in Scheme 42. In this example, nitrogen tetroxide is used as the oxidant (McGuigan and Swords, 1990).

An alternative synthesis of phospholipids that have an *N*-methylphosphocholine head group involves reaction of phosphatidic acid dichloride with *N*-methylethanolamine to give a cyclic oxazaphospholane intermediate, which undergoes acid-catalyzed hydrolysis (Eibl, 1978).

The dialkyl phosphoramidite $Br(CH_2)_2OP[O(CH_2)_2CN]N(Pr\text{-}i)_2$ has been used in an inverse phosphite triester approach to convert diacylglycerols into phosphatidylcholines (Hébert and Just, 1990). The dialkyl phosphoramidite is made by the reaction of 2-bromoethanol with 2-cyanoethyl *N*,*N*-diisopropylchlorophosphoramidite in methylene chloride in the presence of triethylamine. A solution of diacylglycerol is added at rt to the dialkyl phosphoramidite in the presence of 1*H*-tetrazole, forming the trialkyl phosphite $RCO_2CH_2CH(OCOR)CH_2OP(OCH_2CH_2Br)(OCH_2CH_2CN)$. The latter is purified by silica gel chromatography and oxidized to the phosphate triester. Removal of the cyanoethyl group and displacement of bromide ion with trimethylamine is carried out in acetonitrile solution in a pressure bottle at 65°C.

C. Via Other Phosphitylating Agents

Methyl dichlorophosphite ($MeOPCl_2$) and phenyl dichlorophosphite ($PhOPCl_2$) in the presence of 3 equivalents of *N*,*N*-diisopropylethylamine have been used to convert glycerol derivatives into phospholipids via coupling of two alcohols to the phosphorus.

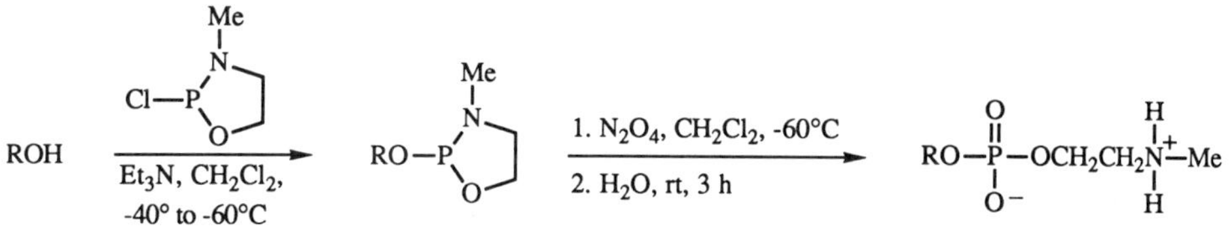

ROH → (Cl–P(cyclic oxazaphospholane, N-Me); Et_3N, CH_2Cl_2, -40° to -60°C) → RO–P(cyclic oxazaphospholane, N-Me) → (1. N_2O_4, CH_2Cl_2, -60°C; 2. H_2O, rt, 3 h) → $RO\text{–}P(=O)(O^-)\text{–}OCH_2CH_2N^+H_2\text{–}Me$

Scheme 42

Selective phosphitylation of the primary alcohol is achieved as follows. In the first step, a solution of the less hindered alcohol ($HOCH_2CH_2NHBoc$) in THF is added to a solution of 1.2 equivalents of $ROPCl_2$ and $(i\text{-}Pr)_2NEt$ in THF at –78°C. Then a solution of 1-acyl-*sn*-glycerol in a minimum volume of THF is added slowly. After the mixture is stirred for 2 h at –78°C, it is allowed to warm to rt and then stirred for 1 h. The solvent and excess *N,N*-diisopropylethylamine are evaporated, and the phosphite is purified by flash chromatography (elution with hexane/ethyl acetate). The oxidation, acylation, Arbuzov, and hydrolysis reactions are carried out using standard procedures.

Outline of the method:

Phosphatidylethanolamine is prepared by coupling of 1-acyl-*sn*-glycerol and *N*-Boc-ethanolamine to methyl dichlorophosphite in the presence of *N,N*-diisopropylethylamine, followed by oxidation with hydrogen peroxide. The secondary hydroxy group is esterified by using fatty acid in the presence of DCC and DMAP. Demethylation is carried out with sodium iodide in refluxing methyl ethyl ketone, and the *N*-Boc group is removed with trifluoroacetic acid.

Literature: Martin and Josey (1988).

1. Phosphorothioate of Phosphatidylserine via Phosphitylation

Key features of the method:

The amino group of serine is protected as an *N*-trityl group, and the carboxyl group is protected as the methyl ester. 1,2-Dipalmitoyl-*sn*-glycerol is phosphitylated to give a phosphoramidite intermediate, which is condensed with the protected serine to form a phospite triester adduct. The phosphite triester is sulfurized to give the phosphorothioate triester 83 in 70% overall yield from 1,2-dipalmitoyl-*sn*-glycerol. Deprotection of the amino and carboxyl groups of serine by acid hydrolysis, followed by demethylation with trimethylamine, gives the product as a mixture of R_p and S_p diastereomers (see Scheme 44). The R_p isomer is hydrolyzed to the corresponding lyso-PsS by bee venom phospholipase A_2. The other isomer (S_p)-DPPsS, is left unreacted. In related work, the R_p stereoisomer of platelet activating factor is prepared via phospholipase A_2 catalyzed

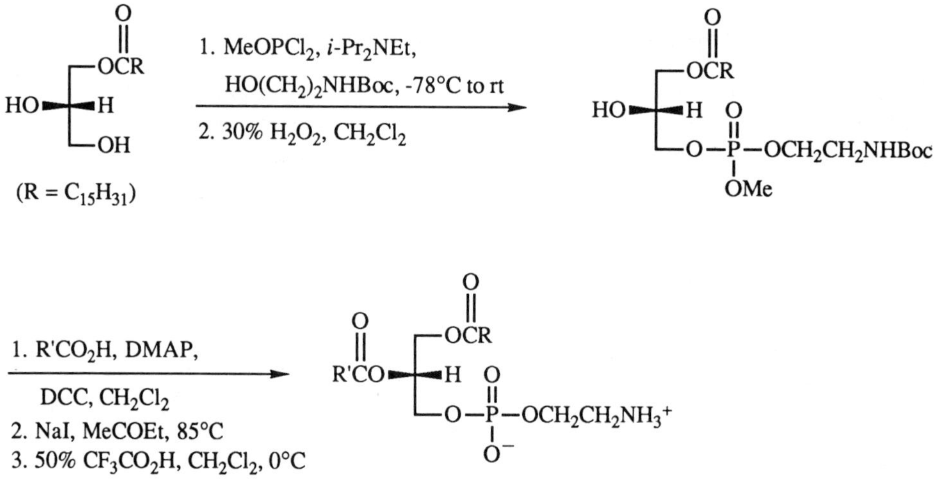

HO–C(H)(CH₂OC(O)R)(CH₂OH) $(R = C_{15}H_{31})$

1. $MeOPCl_2$, $i\text{-}Pr_2NEt$, $HO(CH_2)_2NHBoc$, -78°C to rt
2. 30% H_2O_2, CH_2Cl_2

→ HO–C(H)(CH₂OC(O)R)(CH₂O–P(O)(OMe)–OCH_2CH_2NHBoc)

1. $R'CO_2H$, DMAP, DCC, CH_2Cl_2
2. NaI, MeCOEt, 85°C
3. 50% CF_3CO_2H, CH_2Cl_2, 0°C

→ R'C(O)O–C(H)(CH₂OC(O)R)(CH₂O–P(O)(O^-)–$OCH_2CH_2NH_3^+$)

Scheme 43

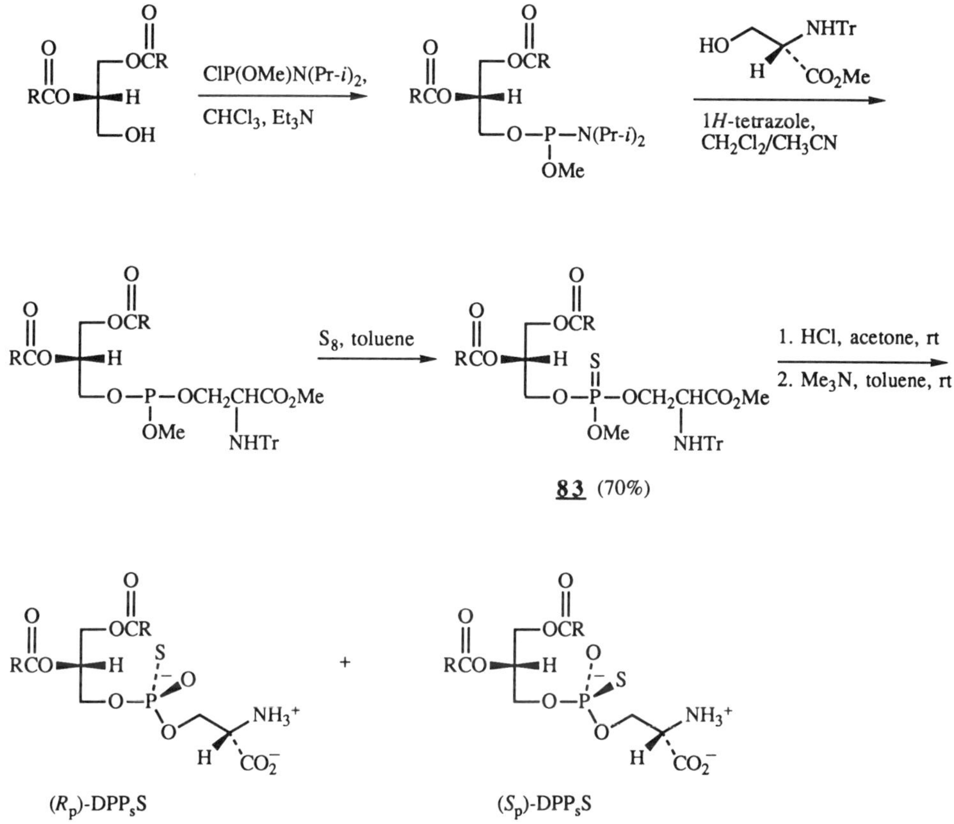

$ClP(OMe)N(Pr\text{-}i)_2$, $CHCl_3$, Et_3N

1*H*-tetrazole, CH_2Cl_2/CH_3CN

S_8, toluene

83 (70%)

1. HCl, acetone, rt
2. Me_3N, toluene, rt

(R_p)-DPP$_s$S + (S_p)-DPP$_s$S

Scheme 44

hydrolysis of (R_p+S_p)-1-*O*-hexadecyl-2-palmitoyl-*sn*-glycero-3-thiophosphocholine, followed by acetylation of (R_p)-lysoPsC (Rosario-Jansen et al., 1988).

Literature: Loffredo and Tsai (1990).

2. Phosphorothioate of PAF via Phosphitylation

Key features of the method:

The thiophosphocholine group is introduced into 1-*O*-hexadecyl-2-palmitoyl-*sn*-glycerol by phosphitylation with *N*,*N*-diisopropylmethylphosphoramidic chloride, then coupling to choline tosylate, and sulfurization. Bee venom phospholipase A_2 is used to hydrolyze the R_P stereoisomer in the mixture of (R_P+ S_P)-1-*O*-hexadecyl-2-palmitoyl-*sn*-glycero-3-thio-PC. The unreacted S_P isomer is isolated and hydrolyzed chemically with tetra-*n*-butylammonium hydroxide. Each isomer of lyso-3-thio-PC is acetylated with acetic anhydride in the presence of DMAP (see Scheme 45).

Literature: Rosario-Jansen et al. (1988).

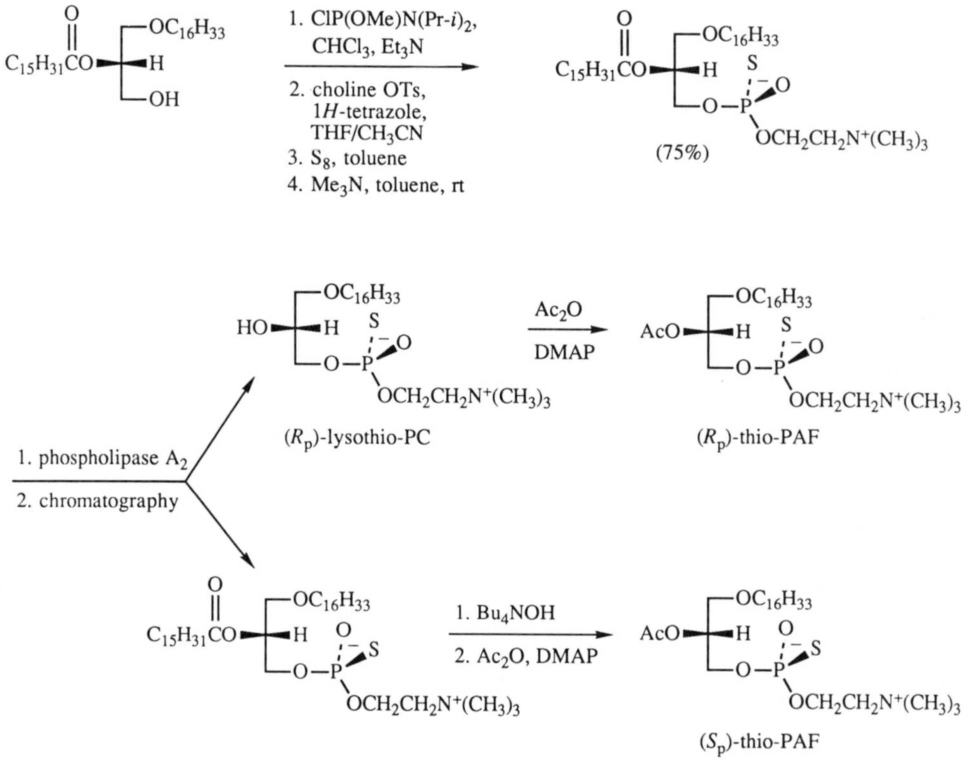

Scheme 45

3. 1,2-Dipalmitoyl-*sn*-glycero-3-thiophospho-1′-inositol (See Sec. IX.B.)

IX. PHOSPHATIDYLINOSITOLS

The phosphoinositide cell-signaling system is brought about by secondary messengers generated by the receptor-controlled hydroylsis of phosphatidylinositol 4,5-bisphosphate (for a review, see Berridge and Irvine, 1989). Syntheses of inositol phosphates on the phosphatidylinositol (PI) pathway as well as analogs of the metabolic products of the PI cycle have been achieved. These compounds are useful as probes of the mechanisms involved in transmembrane signaling and in the regulation of various cellular processes.

A. Preparation via Coupling of Phosphatidic Acid with Protected Inositols. From 1,2-Dipalmitoyl-*sn*-3-glyceryl Phenyl Phosphate

Outline of the method:

Phosphatidylinositols have two chiral centers; the inositol ring may be substituted at the D-1 or L-1 (D-3) position, and the configuration at C2 of the glycerol moiety may be (*S*) or (*R*). The four synthetic stereoisomers of 1,2-dipalmitoyl-PI are prepared by

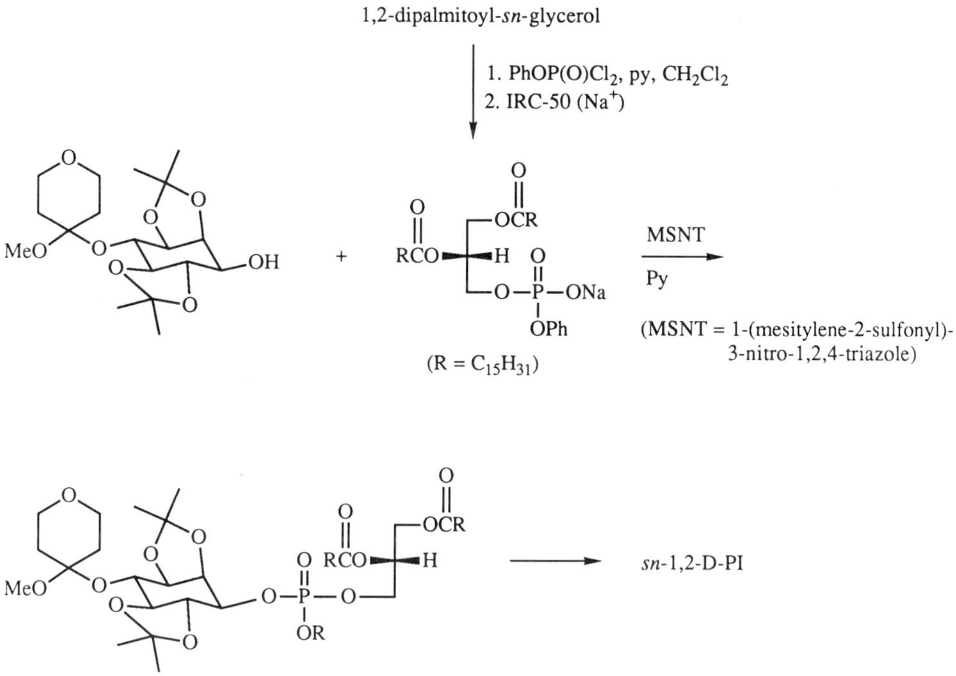

1,2-dipalmitoyl-*sn*-glycerol

1. $PhOP(O)Cl_2$, py, CH_2Cl_2
2. IRC-50 (Na^+)

($R = C_{15}H_{31}$)

MSNT / Py

(MSNT = 1-(mesitylene-2-sulfonyl)-3-nitro-1,2,4-triazole)

84 (R = Ph)

sn-1,2-D-PI

Scheme 46

coupling the sodium salts of 1,2-dipalmitoyl-*sn*-3-glyceryl phenyl phosphate (Scheme 46) and its enantiomer, 2,3-dipalmitoyl-*sn*-1-glyceryl phenyl phosphate, with each enantiomer of pentaprotected *myo*-inositol. The sodium salt of 1,2-diacyl-*sn*-3-glyceryl phenyl phosphate is prepared by the reaction of 1,2-diacyl-*sn*-3-glycerol with phenyl dichlorophosphate in pyridine/methylene chloride, followed by ion-exchange chromatography using Amberlite IRC-50 (sodium form). The phosphatidic acid is coupled with the pentaprotected *myo*-inositol in pyridine using 1-(mesitylene-2-sulfonyl)-3-nitro-1,2,4-triazole (MSNT) as the coupling agent, affording the phenyl ester of PI in 60–75% yield. Catalytic hydrogenolysis with PtO_2 in ethanol and conversion to the sodium salt (Amberlite IRC-50, sodium form, elution with aqueous ethanol) gives PI. The PI stereoisomers have been used to study the stereoselectivity of partially purified PI 4-kinases from human erythrocyte membranes. The chirality of the *myo*-inositol ring is critical for efficient phosphorylation of PI to PI-4-phosphate, i.e., the inositol ring must be linked to the glyceryl moiety through the D-1 position of *myo*-inositol. However, the configuration at the glycerol C2 does not affect the erythrocyte PI-4-kinase-catalyzed phosphorylation reaction, since *sn*-1,2-D-PI and *sn*-2,3-D-PI react at similar rates. A placental soluble PI-4-kinase did show a preference for the *sn*-1,2-diacyl configuration of the glyceryl moiety.

Literature: Young et al. (1990); Macphee et al. (1992).

It was also found recently that the configuration at C2 did not affect the activity of PI-selective phospholipase C (Bruzik et al., 1992).

B. Preparation via Coupling of Inositol Phosphate with 1,2-Di-*O*-alkyl (or acyl)-*sn*-glycerol. From the Pyridinium Salt of 2,3,4,5,6-Penta-*O*-benzyl-*myo*-inositol-1-phosphate

Outline of the method:

rac-1-*O*-Octadecyl-2-*O*-methylglycerol was coupled with *rac*-2,3,4,5,6-penta-*O*-benzyl-*myo*-inositol-1-phosphate by a modification of a published procedure (Hermetter et al., 1982). The protected inositol-1-phosphate reacts with the diether glycerol in pyridine in the presence of 2,4,6-triisopropylbenzenesulfonyl chloride. At high concentrations, the inositol derivative 85 has growth-inhibitory properties against HL-60 cells and BG3 ovarian adenocarcinoma cells, and it also inhibits diacylglycerol-stimulated protein kinase C activity in an in vitro assay system.

Literature: Ishaq et al. (1989); Noseda et al. (1987).

C. By Action of PI Synthase on Short-Chain *ara*-CDP-diacyl-*sn*-glycerol

Outline of the method:

ara-CDP-1,2-dihexanoyl-*sn*-3-glycerol (85) is prepared by coupling of phosphatidic acid (55 mg, 0.14 mmol) with *ara*-cytidine-5'-monophosphomorpholidate 4-morpholine *N,N*'-dicyclohexylcarboxamidinium salt (120 mg, 0.17 mmol) under nitrogen. The

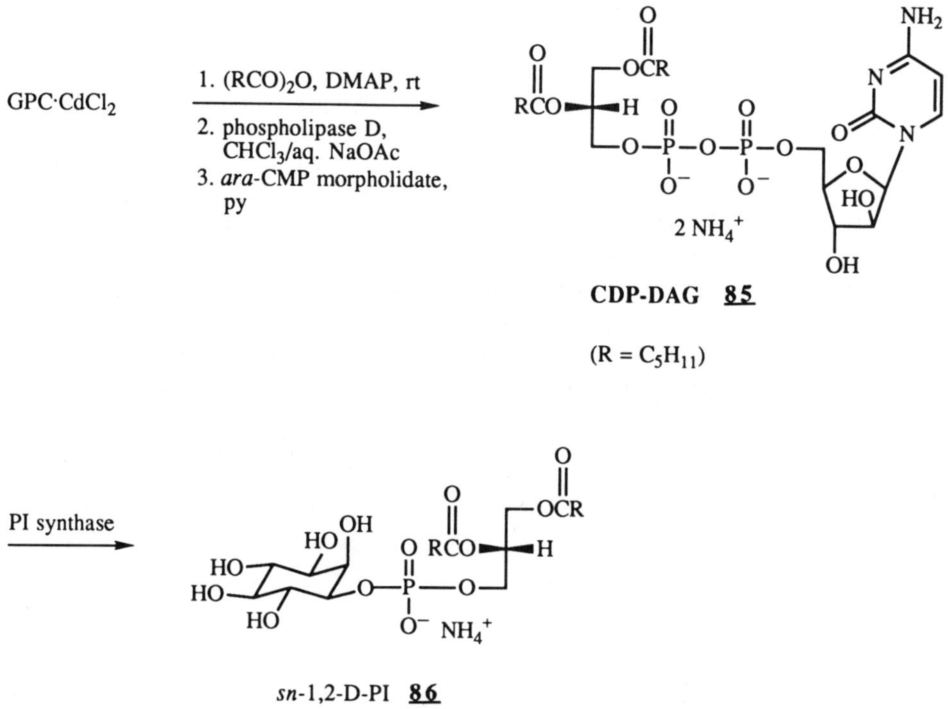

$GPC \cdot CdCl_2$ → 1. $(RCO)_2O$, DMAP, rt; 2. phospholipase D, $CHCl_3$/aq. NaOAc; 3. *ara*-CMP morpholidate, py

CDP-DAG 85

($R = C_5H_{11}$)

PI synthase →

sn-1,2-D-PI 86

Scheme 47

pyridine is evaporated under reduced pressure by addition and reevaporation of toluene several times. The residue is treated with 10 mL of a solution of 10% HOAc in $CHCl_3$:MeOH:H_2O (2:3:1). During a 1-h period of stirring all of the solid residue dissolves; 10 mL of water is then added, giving a clear solution that is extracted into 200 mL of $CHCl_3$:MeOH (2:1). Evaporation of the solvents gives a pale-yellow residue that is purified by medium-pressure liquid chromatography on a Cellulose DE-52 (OAc^-) column (1.2 × 35 cm) equilibrated in $CHCl_3$:MeOH:H_2O (2:3:1). The column is first eluted with the same solvent system (200 mL) and then with 600–800 mL of $CHCl_3$:MeOH:H_2O (2:3:1) solution containing ~0.1% (w/v) of ammonium acetate. The last fractions (~700–800 mL) are pooled and concentrated. The product 85 (R_f 0.40, $CHCl_3$:MeOH:H_2O; 6:4:1) is obtained as the di-NH_4^+ salt, which is suspended in 2 mL of benzene and lyophilized to give a white powder; yield, 38 mg (36%); mp 200–204°C (dec.).

The conversion of CDP-DAG to PI 86 is carried out using PI synthase (*S. cerevisiae*) immobilized on an *ara*-CDP-1,2-dipalmitoylglycerol-agarose resin. The reaction mixture contains [2-^{3}H]*myo*-inositol (1.5–3 mM) and CDP-DAG (1–2 mM) in 50 mM Tris buffer, pH 8.0, containing 3 mM $MnCl_2$, 10 mM 2-mercaptoethanol, and 0.04% (v/v) Triton X-100. After the reaction is complete, the resin is removed by centrifugation and the filtrate is lyophilized. The crude product is extracted with ethanol/ether 3:1. The solvents are evaporated, and the residue is dissolved in 5% aqueous acetonitrile containing 0.2 M ammonium acetate, pH 6.0. The product is purified by reverse-phase HPLC.

The short-chain PIs serve as substrates for the PI-specific phospholipase C-δ_1 isozyme from bovine brain cytosol at PI concentrations below the critical micelle concentration. Although Ca^{2+} is required to produce hydrolysis, the Ca^{2+} requirement is not related to membrane adsorption to or penetration into the membrane surface.

Literature: Rebecchi et al. (1993).

D. Preparation Using Phosphite Chemistry

Phosphitylation of 1,2-dipalmitoyl-*sn*-glycerol with bis(*N,N*-diisopropyl) benzyl phosphoramidite, $[(i\text{-}Pr)_2N]_2POBn$, followed by coupling of a pentaprotected derivative of D-*myo*-inositol with 1*H*-tetrazole as the catalyst leads to DPPI. Use of *trans*-1-propenyl protecting groups at the 4 and 5 positions of 86 allows selective deprotection by hydrolysis at these positions; PI-4,5-bisphosphate is obtained by phosphitylation of the 4 and 5 hydroxy groups, followed by oxidation and hydrogenolysis of the benzyl groups.

Outline of the method:

First, the bifunctional phosphitylating reagent *N,N*-diisopropyl benzyl phosphite reacts with 1,2-dipalmitoyl-*sn*-glycerol in dichloromethane/acetonitrile in the presence of 1*H*-tetrazole to give the phosphoramidite 87 in 92% yield. The latter is coupled to the pentaprotected inositol in the presence of 1*H*-tetrazole to afford a phosphite triester, which is oxidized to the phosphate triester with *tert*-butyl hydroperoxide in situ at 0°C. Acid-catalyzed hydrolysis of the propenyl groups in the inositol moiety followed by phosphitylation at the 4 and 5 positions with *N,N*-diisopropyl dibenzylphosphoramidite, then oxidation, gives the two phosphite triesters. The benzyl-protected derivative is purified by silica gel column chromatography, and catalytic hydrogenolysis gives dipalmitoyl-PI-4,5-bisphosphate in good yield (see Scheme 48).

Literature: Dreef et al. (1988).

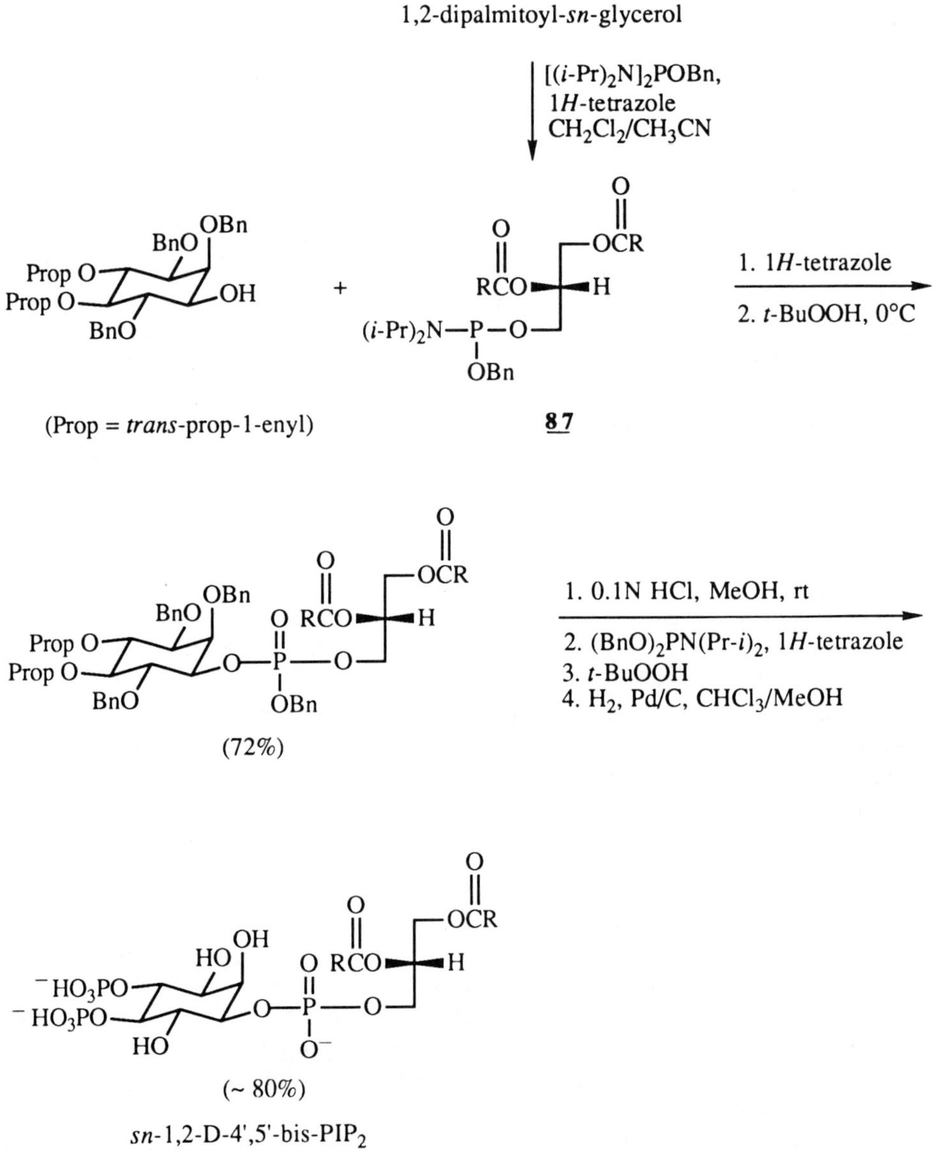

1,2-dipalmitoyl-*sn*-glycerol

$[(i\text{-Pr})_2N]_2POBn$,
1*H*-tetrazole
CH_2Cl_2/CH_3CN

(Prop = *trans*-prop-1-enyl)

87

1. 1*H*-tetrazole
2. *t*-BuOOH, 0°C

(72%)

1. 0.1N HCl, MeOH, rt
2. $(BnO)_2PN(Pr\text{-}i)_2$, 1*H*-tetrazole
3. *t*-BuOOH
4. H_2, Pd/C, $CHCl_3$/MeOH

(~ 80%)

sn-1,2-D-4',5'-bis-PIP_2

Scheme 48

1,2-Dipalmitoyl-*sn*-glycero-3-thio-1'-phosphoinositol (DPPsI) (88) is prepared as shown in Scheme 49.

Outline of the method:

The R_p and S_p isomers of 1,2-dipalmitoyl-*sn*-glycero-3-thiophosphoinositol (DPPsI) (88) are synthesized via a similar phosphitylation coupling reaction. Methoxymethyl-bis(*O*-cyclohexylidene)-protected inositol is converted to the phosphoramidite 89 by reaction with *N*,*N*-diisopropylmethylphosphoramidic chloride, $ClP(OMe)N(Pr\text{-}i)_2$. The latter is used without purification in a tetrazole-mediated coupling reaction with 1,2-

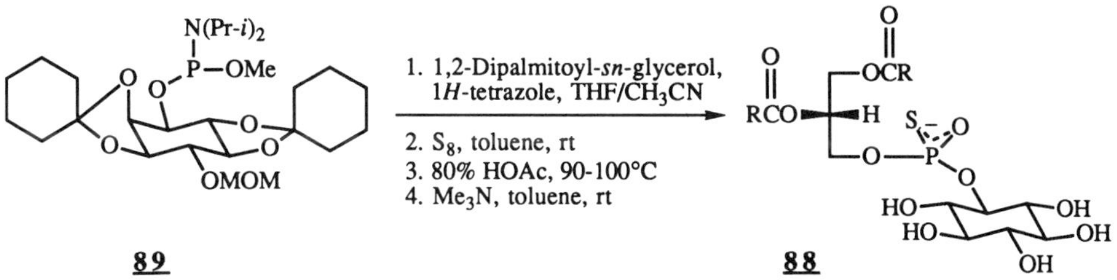

89

1. 1,2-Dipalmitoyl-*sn*-glycerol, 1*H*-tetrazole, THF/CH_3CN
2. S_8, toluene, rt
3. 80% HOAc, 90-100°C
4. Me_3N, toluene, rt

88

Scheme 49

dipalmitoyl-*sn*-glycerol. The preparation of 88 is completed by reaction with excess sulfur in toluene, acid hydrolysis of the cyclohexylidene ketals and methoxymethyl ether, and demethylation of the phosphorothioate with trimethylamine. The diastereomers are of interest for testing the stereoselectivity of PI-specific phospholipase C and for use as possible antimetabolites in blocking steps in inositol phosphate metabolism.

Literature: Lin and Tsai (1989).

In a similar approach, the phosphoramidite of 1,2-dipalmitoyl-*sn*-glycerol is condensed with D-(–)-2,3,4,5,6-pentabenzyl-*myo*-inositol in THF-acetonitrile in the presence of tetrazole, then sulfurized with elemental sulfur to give a mixture of diastereomeric phosphorothioates that are separated by silica gel chromatography (elution with carbon tetrachloride-acetone (100:1). The separated diastereomers are then deprotected (demethylation of the phosphorothioate with trimethylamine and debenzylation of the protected inositol hydroxy functions with BF_3 etherate in ethyl mercaptan) to give (R_p)- and (S_P)-DPPsI. Both reports found that the stereoisomer of DPPsI that is hydrolyzed by phospholipase C from *Bacillus cereus* has the same relative configuration as the stereoisomer of DPPsC that is hydrolyzed by bee venom phospholipase A_2.

Literature: Salamonczyk and Bruzik (1990).

Short-chain diacylglycero-H-phosphonates were recently coupled to the hydroxy group of a pentaprotected inositol in good yield (Garigapati and Roberts, 1993).

X. PREPARATION OF PHOSPHONOCHOLINES AND PHOSPHINOCHOLINES

Isosteric or nonisosteric replacement of the phosphocholine functionality with phosphonocholine gives analogs that are not susceptible to hydrolysis by phospholipase C (if the carbon-phosphorus bond is on glycerol side) or by phospholipase D (if the carbon-phosphorus bond is on head-group side) or by both phospholipases C and D (if each P–O–C bond is replaced by P–C–C bond, i.e., as a phosphinocholine).

A. Nonisosteric Phosphonocholines

Phosphonate diesters have been prepared by the Arbuzov reaction of an alkyl halide with a trialkyl phosphite, $P(OR)_3$, at 150–165°C. Tris(trimethylsilyl) phosphite, $(Me_3SiO)_3P$, may be used instead of a trialkyl phosphite. Tris(trimethylsilyl) phosphite is prepared in good yield by silylation of phosphorous acid (obtained from the trisodium salt of phosphonoformic acid) using chlorotrimethylsilane (Sekine et al., 1982) or by silylation of the tris(trimethylammonium) salt of phosphorous acid (Deroo et al., 1976). On Arbuzov reaction with diacyloxypropyl iodide, trimethylsilyl dialkyl phosphites are formed at about 125°C (Deroo et al., 1976). Trimethylsilyl iodide and excess tris-

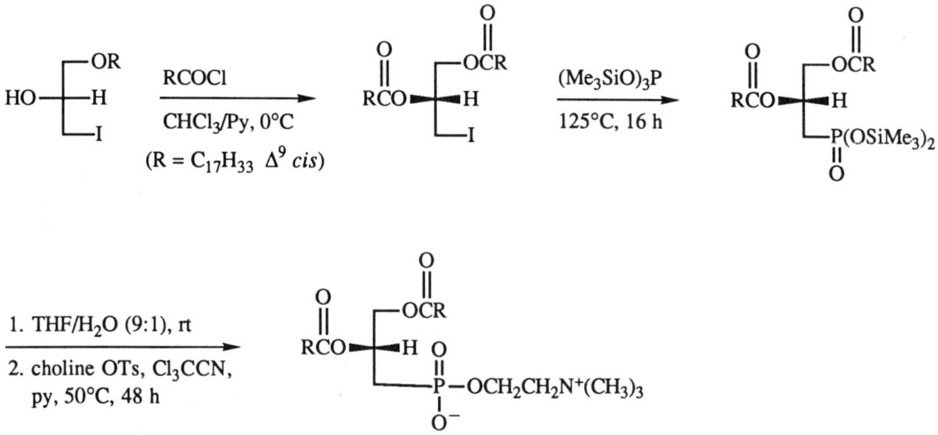

Scheme 50

(trimethylsilyl) phosphite are removed by vacuum distillation. The iodohydrin diacylates are easily prepared by acylation of glycerol α-iodohydrin. Hydrolysis of the phosphonate silyl-diesters in aqueous THF is facile, giving the corresponding phosphonic acid. The latter is converted to the phosphonocholine by reaction with an excess of choline tosylate in the presence of trichloroacetonitrile in pyridine at 50°C (Deroo et al., 1976) (see Scheme 50).

B. Isosteric Phosphonocholines

Phosphonocholines in which the C–P bond is on the glycerol side are prepared in isosteric form by using a 1,2-dihydroxybutyl moiety in place of glyceryl moiety as the backbone; thus, the CH_2 group replaces the missing oxygen atom. The lipid contains a four-carbon rather than a three-carbon backbone as in the example shown in Scheme 50. The four-carbon backbone makes the phosphonolipid isosteric with the phosphate moiety in phosphatidylcholines. Many of the synthetic schemes have been reviewed recently (Engel, 1992).

1. 1-*myo*-Inosityl 10-Carboxy-1-decylphosphonate: Preparation by the Arbuzov Reaction of Benzyl 11-Bromoundecanoate with Tris(trimethylsilyl)phosphite

Method: A mixture of 1.08 g (3.04 mmol) of benzyl 11-bromoundecanoate and 5.20 g (17.4 mmol) of tris(trimethylsilyl)phosphite is heated with stirring at 135°C for 25 h under argon. The excess phosphite is removed by distillation under reduced pressure (0.05 mm, temperature <100°C). The residue is dissolved in 8 mL of THF, water (2 mL) is added, and the mixture is stirred at rt for 36 h. The solvents are removed under vacuum and the residue is extracted into ether (10 mL). The ether is evaporated, and traces of water are removed by addition of ethanol, followed by evaporation. Crystallation from chloroform/hexane gives 1.07 g (99%) of phosphonic acid 90. The coupling of 90 to pentabenzylinositol in the presence of trichloroacetonitrile is carried out as follows. To a solution of pentabenzylinositol (80 mg, 0.13 mmol) and phosphonic acid 90 (50 mg, 0.14 mmol) in 0.5 mL of dry pyridine is added 0.58 g (4.0 mmol) of trichloroacetonitrile.

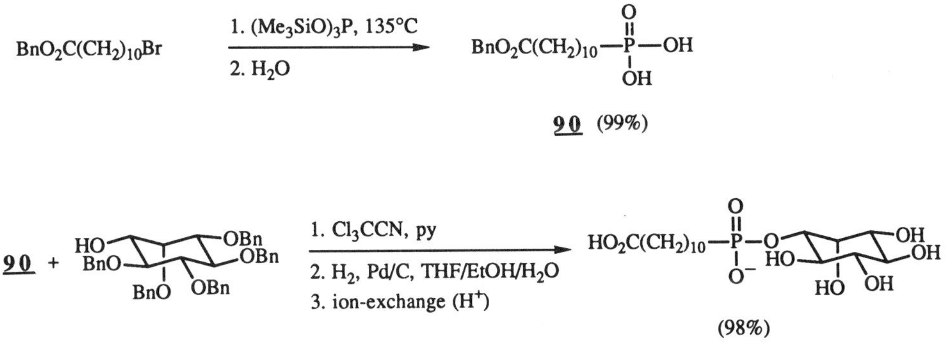

$BnO_2C(CH_2)_{10}Br$ → 1. $(Me_3SiO)_3P$, 135°C; 2. H_2O → $BnO_2C(CH_2)_{10}$—P(=O)(OH)—OH

90 (99%)

90 + (HO, BnO, OBn, OBn, BnO, OBn inositol) → 1. Cl_3CCN, py; 2. H_2, Pd/C, THF/EtOH/H_2O; 3. ion-exchange (H^+) → $HO_2C(CH_2)_{10}$—P(=O)(O^-)—O—(HO, OH, OH, HO, OH inositol)

(98%)

Scheme 51

The mixture is heated with stirring at 60°C for 56 h. The mixture is cooled to rt, then evaporated to dryness, with addition of toluene to remove traces of pyridine. The residue is purified by chromatography on silica gel (elution with chloroform/methanol 98:2). Deprotection of the benzyl groups is carried out by dissolving the phosphonate in 4.6 mL of THF/water/ethanol (3:0.8:0.8) and hydrogenolysis with 50 mg of 30% Pd-C at rt for 24 h. The catalyst is removed by centrifugation, and the solvents are evaporated. The residue is dissolved in 5 mL of ethanol/water (1:3), and the solution is passed through a cation-exchange resin (H^+ form) (elution with ethanol/water 1:3). The solvents are evaporated and the solid is dried; yield, 98%.

Literature: Shashidhar et al. (1990).

2. Diester Isosteric Phosphonocholines from Isopropylidene-*sn*-glyceraldehyde

Method: The vinylic phosphonate diisopropyl ester 91 is prepared by the Wadsworth-Emmons modification of the Wittig reaction. To a solution of 17.94 g (57 mmol) of tetraisopropyl methylenediphosphonate in 200 mL of heptane under nitrogen is added 38.5 mL (52 mmol) of a 1.5 M solution of *n*-butyllithium in hexane. The mixture is stirred for 2 h, then cooled to 0°C, and a solution of 6.78 g (52 mmol) of 1,2-isopropylidene-*sn*-glyceraldehyde in 25 mL of heptane is added. The reaction mixture is heated at reflux for 2 h, then cooled to rt and quenched by the addition of 500 mL of water. The organic phase is separated, and the aqueous layer is washed with heptane (2 × 100 mL). The combined organic layer is dried ($MgSO_4$), filtered, and concentrated under reduced pressure, giving a residue that is vacuum distilled (bp 100–115°C/0.025 mm Hg); yield of 91, 14.57 g (96%); $[\alpha]^{25}_D$ +15.48°(*c* 9.3, ethanol). Acid hydrolysis of the acetonide functionality gives the diol, which is treated with acyl chloride (3 equiv) in chloroform containing 3 equivalents of pyridine at 0°C. After the mixture is stirred at rt for 48 h it is diluted with ether, and the ether layer is washed with 0.1 N sulfuric acid, ice water, 10% sodium thiosulfate, and water. The organic layer is dried (Na_2SO_4) and evaporated, giving a residue that is purified by chromatography on silica gel (elution with chloroform) and then recrystallized from methanol.

The cleavage of the phosphonodiester linkages is accomplished by using trimethylsilyl bromide followed by aqueous workup and extraction of the phosphonic acid product

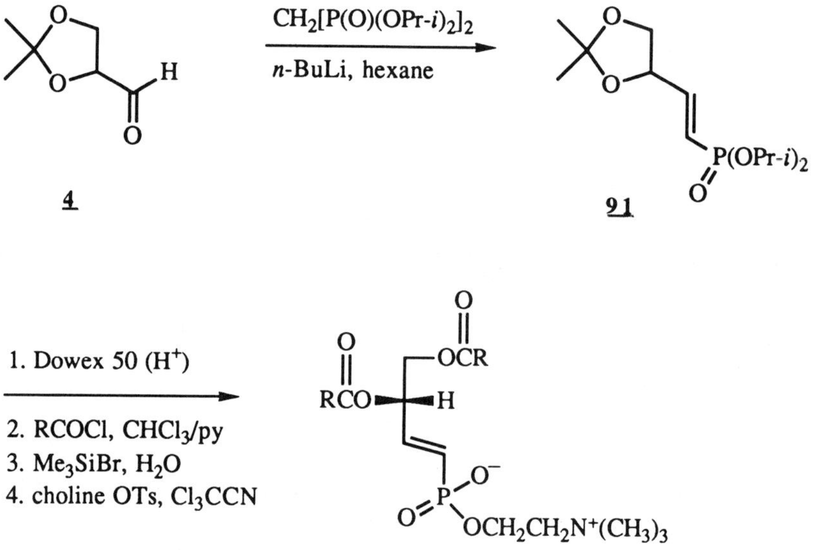

$CH_2[P(O)(OPr\text{-}i)_2]_2$

n-BuLi, hexane

4 → 91

1. Dowex 50 (H^+)
2. RCOCl, $CHCl_3$/py
3. Me_3SiBr, H_2O
4. choline OTs, Cl_3CCN

Scheme 52

into ether. The ether layer is dried (Na_2SO_4) and evaporated to give a residue that is recrystallized from methanol. The choline group is coupled to the phosphonic acid by heating a mixture of phosphonic acid (1.3 g, 2.0 mmol), choline tosylate (3.3 g, 12.0 mmol), and trichloroacetonitrile (10.2 g, 71 mmol) in 37 mL of pyridine at 50°C for 48 h with stirring. The crude product is then precipitated by addition of cold acetonitrile, and the solid is purified by recrystallization from acetone to give 1.2 g (80%) of the phosphonocholine. Instead of recrystallization from acetone, the crude product could be dissolved in THF/water 9:1 and passed through an Amberlite MB-3 column, eluting with THF/water 9:1, leaving salts on the column.

Literature: Schwartz et al. (1988); Lalinde et al. (1983).

3. Diester Isosteric Phosphonocholines from 3,4-Dihydroxy-1-bromobutane

Method: To 1.37 g (2.9 mmol) of 3,4-dicapryloxy-1-bromobutane (obtained by diacylation of *rac*-3,4-dihydroxy-1-bromobutane) under a nitrogen atmosphere is added 7.68 g (29 mmol) of tris(trimethylsilyl)phosphite (Deroo et al., 1976). The solution is heated at 165–170°C for 16 h. The volatile materials are removed under vacuum, giving a residue that is dissolved in 25 mL of THF/water (9:1) and heated at reflux for 2 h. Evaporation of the solvents gives a residue, to which 100 mL of chloroform is added. The chloroform-soluble portion is decanted, chloroform is removed, and the residue is purified by chromatography on silica gel (elution with chloroform, then with gradients of chloroform/methanol up to 25:75). The fractions that contain the product are combined, the solvent is removed, and the product is dissolved in chloroform and passed through a Metricel filter to remove the suspended silica gel. Evaporation of the solvent gives 240 mg (18%) of the phosphonic acid.

Literature: Schwartz et al. (1988).

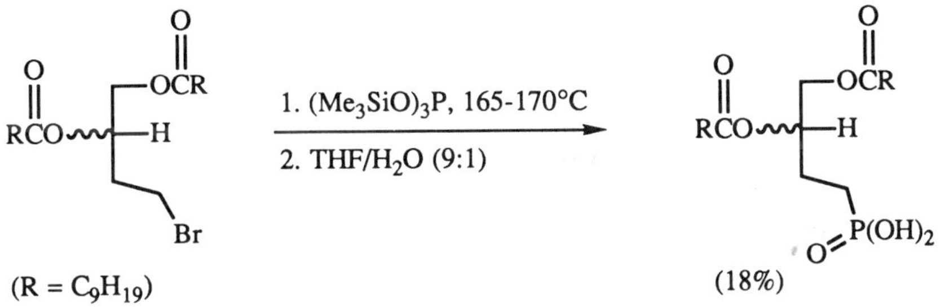

(R = C_9H_{19})

1. $(Me_3SiO)_3P$, 165-170°C
2. THF/H_2O (9:1)

(18%)

Scheme 53

C. 1-Ether-2-acylaminophosphonocholines

1. C–P Bond on the Glycerol Side. [4-(Octadecyloxy)-3(*S*)-Hexadecanoylamino)but-1-yl]phosphonocholine

Key features of the method:

The C–P bond is introduced by displacement of tosylate from tosylamino tosylate 92 by the lithium salt of diethyl methanephosphonate. 1-*O*-Alkyl-2-*N*-acylamino-deoxyphosphonocholine (93) is prepared from the *N*-tosylphosphonic acid.

Literature: Yuan et al. (1989).

Method: The simultaneous tosylation of the amino and alcohol functionalities of 1-*O*-octadecyl-2-aminodeoxyglycerol (Chandrakumar and Hajdu, 1983) gives 92 in 77% yield. The *O*-tosyl group is displaced from 92 as follows. A solution of 1.94 mL (13.3 mmol) of diethyl methanephosphonate in 7 mL of THF under argon is cooled to –78°C, and 6.2 mL (13.3 mmol) of *n*-butyllithium in hexane is added dropwise. After the solution has stirred for 30 min at –78°C, a solution of 2.16 (3.32 mmol) of tosylamino

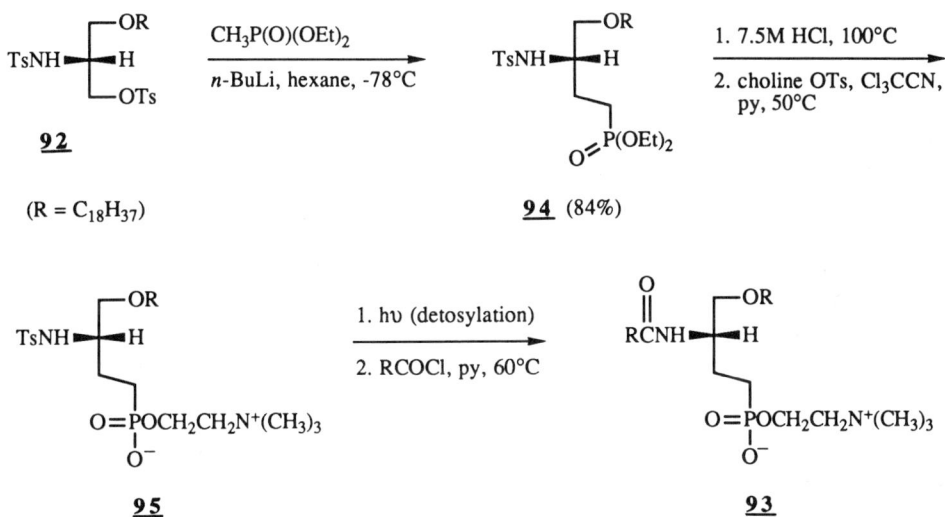

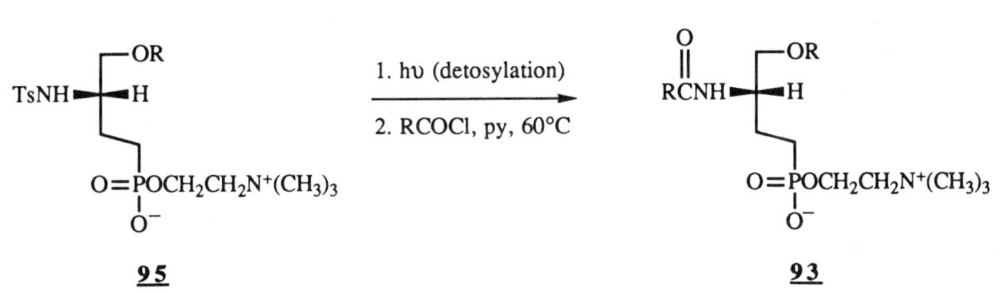

92 (R = $C_{18}H_{37}$)

$CH_3P(O)(OEt)_2$, *n*-BuLi, hexane, -78°C

94 (84%)

1. 7.5M HCl, 100°C
2. choline OTs, Cl_3CCN, py, 50°C

95

1. hυ (detosylation)
2. RCOCl, py, 60°C

93

Scheme 54

tosylate 92 in 5 mL of THF is injected rapidly, and stirring is continued for 3 h at –20°C. The reaction is quenched by adding 1 N HCl, the product is extracted with ether, and the organic layer is dried ($MgSO_4$) and concentrated. The residue is purified by flash chromatography (elution with chloroform, then chloroform-methanol 99 : 1), then lyophilized from benzene; yield of diethyl phosphonate 94, 1.76 g (84%). The ester is hydrolyzed by heating in 7.5 M HCl at 100°C with vigorous stirring overnight, giving the corresponding phosphonic acid in 87% yield after lyophilization from benzene. Pyridine (35 mL) is added to 1.0 g (1.74 mmol) of phosphonic acid 94 and 2.87 g (10.44 mmol) of choline tosylate. The mixture is warmed to 50°C, and 6.28 mL (62.63 mmol) of trichloroacetonitrile is added dropwise. After the solution has stirred at 50°C for 48 h, pyridine is evaporated under vacuum, leaving a residue that is dissolved in chloroform-methanol 1 : 1 and passed through a mixed-bed ion-exchange column (Rexyn I-300) and eluted with the same solvent. The eluate is evaporated to dryness under vacuum, then purified further by flash chromatography (elution with chloroform-methanol-water 65 : 25 : 4); yield of tosylamino phosphonocholine 95, 0.91 g (79%).

The *N*-tosyl group is removed from 95 by photolysis in ether-methanol 8 : 1 with use of a 550-W mercury lamp (1 h). After the solvents are evaporated, the residue is applied to an ion-exchange column (Bio-Rad AG50W-X4, H^+ form) equilibrated with chloroform-methanol-water 5 :5 : 1. The column is washed with 12 column volumes of the same solvent to remove neutral and negatively charged impurities. The amine is eluted with chloroform-methanol-concentrated ammonium hydroxide 5 : 5 : 1. The fractions that contain the product are concentrated to give the amine (167 mg, 43%) as a yellow powder. The amine (115 mg, 0.227 mmol) is suspended in 3 mL of pyridine at 60°C, and 156 mg (0.568 mmol) of palmitoyl chloride is added dropwise. After the mixture is stirred overnight at 60°C, pyridine is removed under vacuum, leaving a residue that is dissolved in chloroform-methanol 1 : 1. The solution is passed through a Rexyn I-300 column and eluted with chloroform-methanol 1 : 1. The eluate is evaporated to dryness, giving a residue that is purified by flash chromatography (elution with chloroform-methanol-water 65 : 25 : 4); yield of 93, 51 mg, 30%) (see Scheme 54).

2. C–P Bond on the Choline Side. 1-*O*-Hexadecyl-2-hexadecanoylamino-*sn*-3-glycerophosphonocholine

Key features of the method:

The C–P bond is introduced on the choline side by phosphonylation of (*S*)-2-*N*-(palmitoylamido)-3-hexadecyloxy-1-propanol with (3-bromopropyl)phosphonochloridic acid in ethanol-free chloroform in the presence of triethylamine. Amination of the resulting bromophosphonate with aqueous trimethylamine in 2-propanol-DMF-$CHCl_3$-H_2O (25 : 25 : 15 : 35) at 50°C gives the phosphonocholine.

Method: L-Serine is converted to *O-tert*-butyl-L-serine *tert*-butyl ester (96). The amino group is protected by reaction with *p*-anisyl chlorodiphenylmethane in chloroform in the presence of triethylamine and 3Å molecular sieves. The serine derivative is purified by chromatography on alumina (elution with petroleum ether), the ester group is reduced, and the resulting alcohol is alkylated. Detritylation of the amino group and deprotection of the *O-tert*-butyl ether are carried out simultaneously by passing hydrogen chloride gas through a chloroform solution at 0°C. The amino group is acylated selectively by using palmitoyl chloride in chloroform in the presence of DMAP at rt, and the product is obtained after phosphonylation and amination as outlined above. It is interesting that rearrangement to an oxazoline (97) does not take place (see Scheme 55).

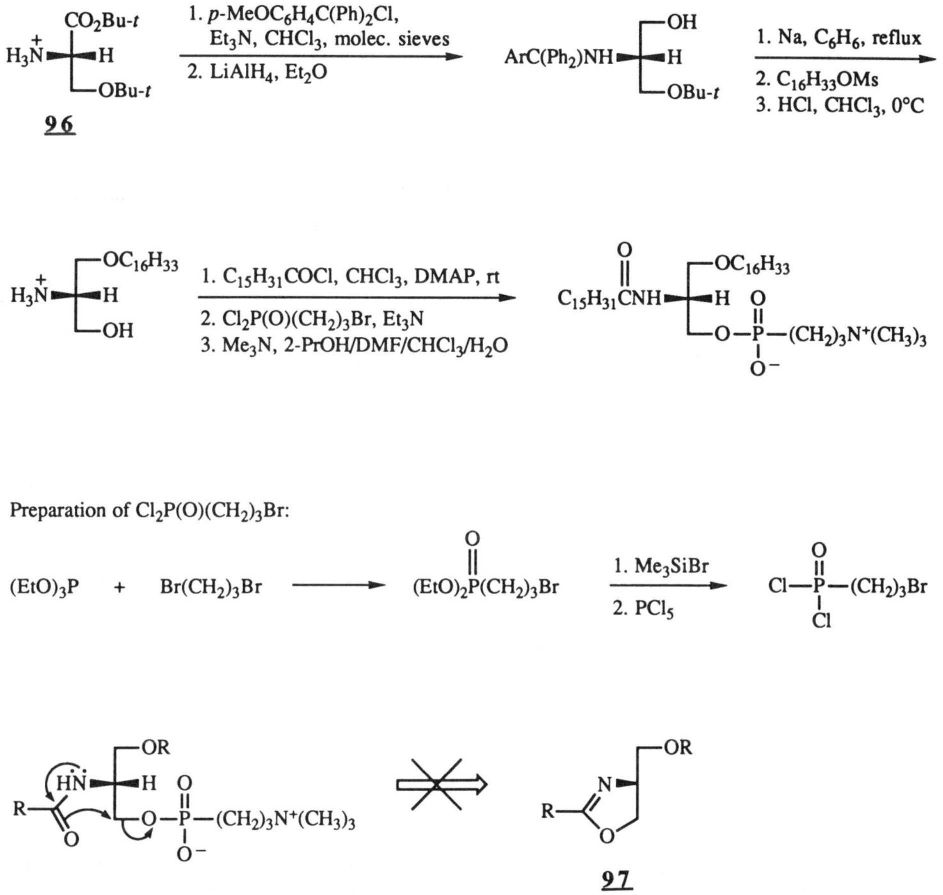

Scheme 55

Phosphonocholines having the C–P bond on the choline side are readily prepared by reaction of the glycerol derivative with (3-bromopropyl)phosphonochloridic acid ($Cl_2P(O)CH_2CH_2CH_2Br$). The latter phosphonylating agent is prepared by Arbuzov reaction of 1,3-dibromopropane with triethyl phosphite (Baer and Rao, 1966), followed by silylation of the resulting diethyl phosphite ester, $(EtO)_2P(O)CH_2CH_2CH_2Br$, with trimethylsilyl bromide (TMSBr), and treatment of $(TMSO)_2P(O)CH_2CH_2CH_2Br$ with phosphorus pentachloride or with excess oxalyl chloride and catalytic DMF in methylene chloride (Bhongle et al., 1987).

Literature: Turcotte et al. (1991).

D. Phosphinocholines

Diether phosphinocholines are synthesized in a long route from 1,2:5,6-*O*-isopropylidene-D-mannitol via 3,4-dialkoxy-1-butene (Rosenthal et al., 1972). The chloromethyl derivative 98 is the key reagent used to form the phosphinate (C–P–C–P) linkage via a Wittig reaction followed by an Arbuzov reaction. (*R*)-Glyceraldehyde

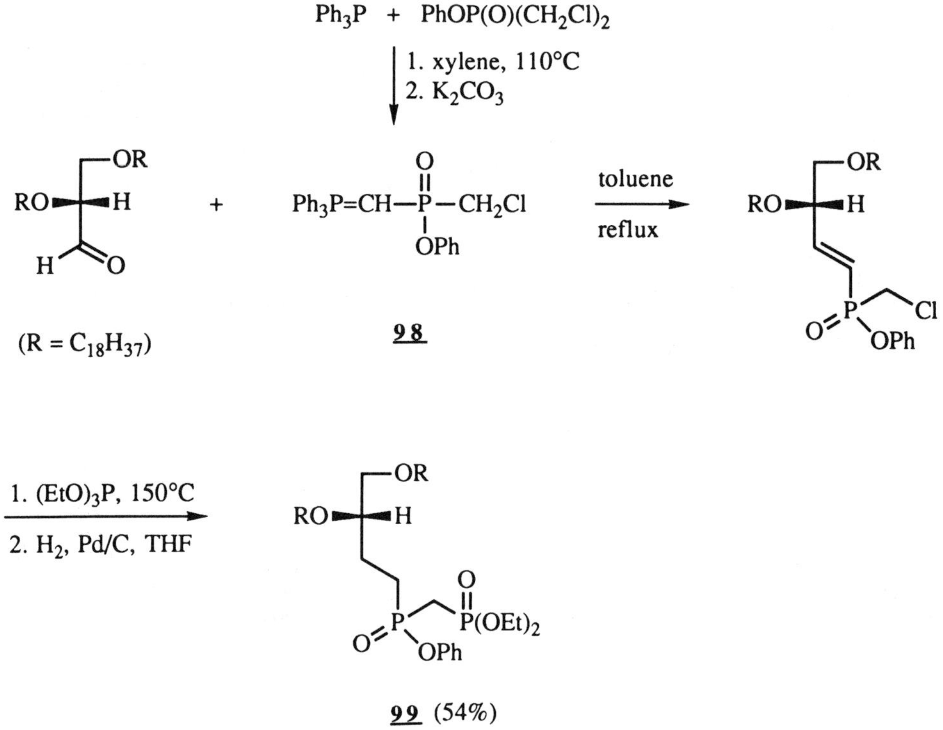

Ph_3P + $PhOP(O)(CH_2Cl)_2$

1. xylene, 110°C
2. K_2CO_3

$(R = C_{18}H_{37})$ + $Ph_3P{=}CH{-}P(O)(OPh){-}CH_2Cl$ (98) → toluene, reflux

1. $(EtO)_3P$, 150°C
2. H_2, Pd/C, THF

99 (54%)

Scheme 56

2,3-dioctadecyl ether reacts with 98 to give the Wittig product. Arbuzov reaction of the chloromethyl phosphinate intermediate with triethyl phosphite, followed by catalytic hydrogenation of the olefin, gives phosphonate-phosphinate 99 in 54% yield.

Literature: Vargas and Rosenthal (1983).

XI. PREPARATION OF LIPONUCLEOTIDE PRODRUGS

A. Azidothymidine and Dideoxynucleoside Conjugates of Phosphatidic Acid

Outline of the method:

Inhibitors of the reverse transcriptase of human immunodeficiency virus (HIV) are coupled to phosphatidic acid in the presence of 2,4,6-triisopropylbenzenesulfonyl (TPS) chloride or via the nucleoside-monophosphate morpholidate (see Scheme 57). Phospholipids having azidothymidine or a dideoxynucleoside (2',2'-dideoxycytidine or 3'-deoxythymidine) as the polar head group are incorporated into liposomes prepared from PC, phosphatidylglycerol, and cholesterol for delivery to macrophages. The liponucleotides are active in vitro as antiretroviral agents in HIV-infected cell lines, where cellular lipases and phosphodiesterases degrade the liponucleotide conjugate to release the nucleoside inhibitor as the free 5'-hydroxy or 5'-monophosphate form.

Literature: Hostetler et al. (1990); van Wijk et al. (1992).

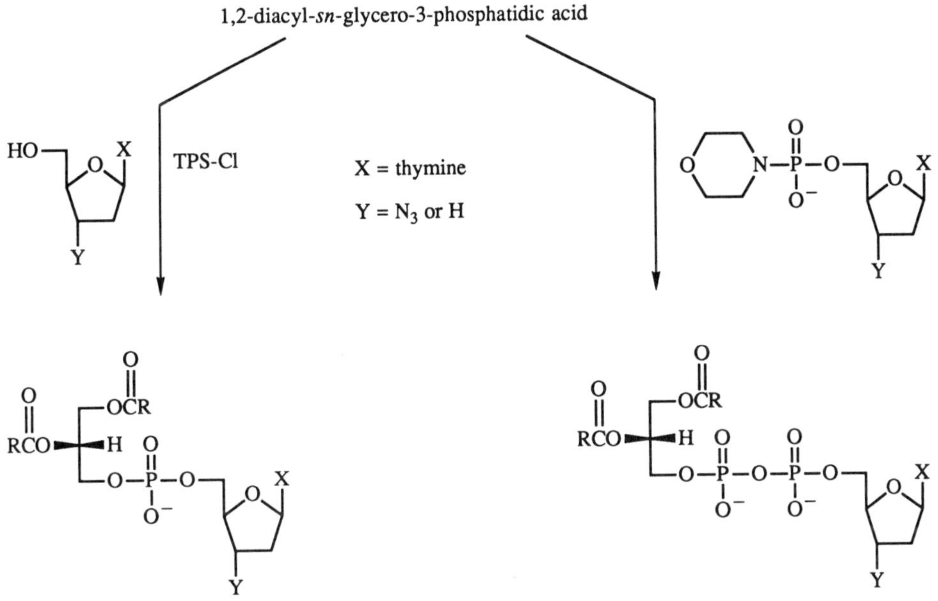

1,2-diacyl-*sn*-glycero-3-phosphatidic acid

TPS-Cl

X = thymine

Y = N_3 or H

Scheme 57

B. *ara*-Cytosine and *ara*-Cytosine 5'-Diphosphate Thioether Lipid Conjugates

Outline of the method:

Cytotoxic conjugates of thioether lipids with 1-β-D-arabinofuranosylcytosine (*ara*-C) and with cytidine are prepared and tested for effectiveness as antitumor agents in mice bearing implanted lymphoid leukemia. After enzymatic action, two cytotoxic agents, *ara*-C and 1-*S*-alkyllysophospholipid, are released from the conjugate. The coupling of 1-*S*-octadecyl-2-palmitoyl-*rac*-glycero-3-phosphatidic acid to *ara*-CMP morpholidate in pyridine, followed by purification by ion-exchange chromatography [(diethylamino)ethyl]cellulose-52 (acetate) and Amberlite CG-50 (Na^+) columns gives *ara*-CDP-*rac*-1-*S*-octadecyl-2-palmitoyl-1-thioglycerol (100) in 38% yield. The phosphatidic acid is prepared by the reaction of 1-*S*-octadecyl-2-palmitoyl-*rac*-glycerol with phosphorus oxychloride in hexane containing triethylamine, followed by hydrolysis and crystallization from cold hexane, then recrystallization from ether at rt. 1-*S*-Octadecyl-2-palmitoyl-*rac*-glycerol is prepared by acylation of 1-*S*-octadecyl-3-*O*-(*tert*-butyldimethylsilyl)-*rac*-glycerol with palmitoyl chloride in toluene containing pyridine. On desilylation with tetra-*n*-butylammonium fluoride in THF, acyl migration occurs. The regioisomers are separated by dissolving them in boiling 95% ethanol; the undesired 1-*S*-octadecyl-3-palmitoyl-*rac*-glycerol precipitates on cooling to rt and is removed by filtration, and the filtrate contains mostly 1-*S*-octadecyl-2-palmitoyl-*rac*-glycerol, which crystallizes at 0–3°C (see Scheme 58). [Significant acyl migration may be avoided by protecting the primary hydroxy group as a benzyl ether rather than as the silyl ether, since rearrangement can take place during fluoride ion induced cleavage of the Si–O bond (Dodd et al., 1975).] Debenzylation by catalytic hydrogenolysis, followed by silica gel

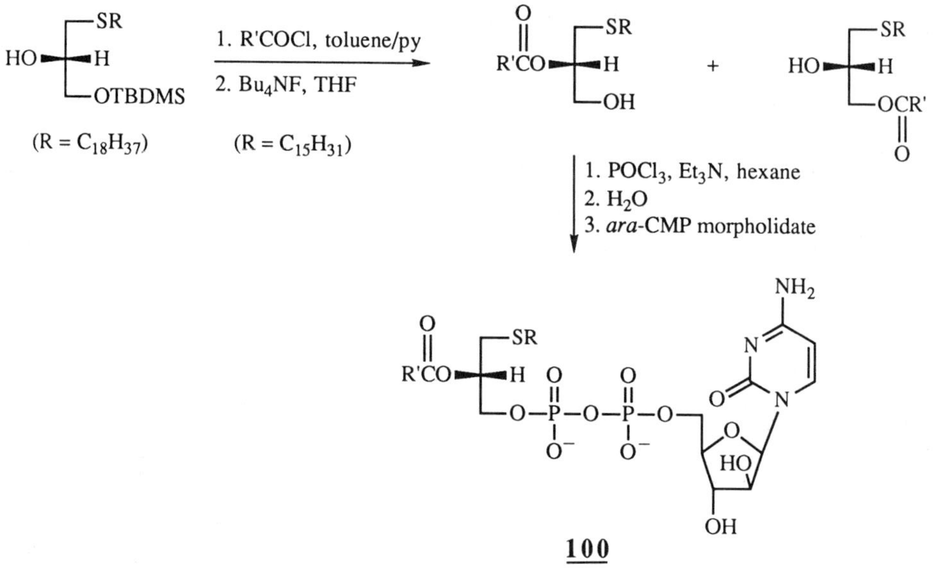

Scheme 58

chromatography on short columns (elution with hexane-ether) or crystallization, may avoid extensive isomerization to the undesired 1,3-diacylglycerol derivatives.

Literature: Hong et al. (1990).

C. Phosphonylmethyl Derivatives of Cytosine

1. Cytidine (*S*)-3,4-Dioctadecoxybutylphosphinato-(methylenephosphonate)

Key features of the method:

A di-*O*-alkylglycerol is linked covalently to a nucleoside via a phosphonylphosphinyl (glycerol-$CH_2P(O)(O^-)CH_2P(O)(O^-)O$-nucleoside) bridge, in which the bridging oxygen atom of the diphosphate is replaced by a carbon. The liponucleotide bearing phosphinate and phosphonate linkages, which is not a possible phosphatidyl donor, inhibits platelet phosphatidylinositol synthetase. The synthesis involves coupling of phosphinatomethylenephosphonic acid 101 with *N*-phenoxyacetyl-2',3'-isopropylidene-cytidine 102, followed by deprotection of the *N*-phenoxyacetyl group with trifluoroacetic acid and hydrolysis of the phenyl phosphate ester with cesium fluoride in methanol (see Scheme 59).

Literature: Vargas et al. (1984).

2. (*S*)-1-[3-Hydroxy-2-(phosphonylmethoxy)propyl]cytosine

Key features of the method:

The side chain of the phosphonate derivatives, which have antiviral activity against herpes simplex virus and human cytomegalovirus, is prepared in a synthetic route starting with 2,3-*O*-isopropylidene-*sn*-glycerol [(*R*)-4]. Benzylation with 1.8 equivalents of benzyl bromide in the presence of 0.02 equivalents of benzyltriethylammonium bromide as phase transfer catalyst in 10 N aqueous sodium hydroxide at 90–95°C gives 1-*O*-benzyl-

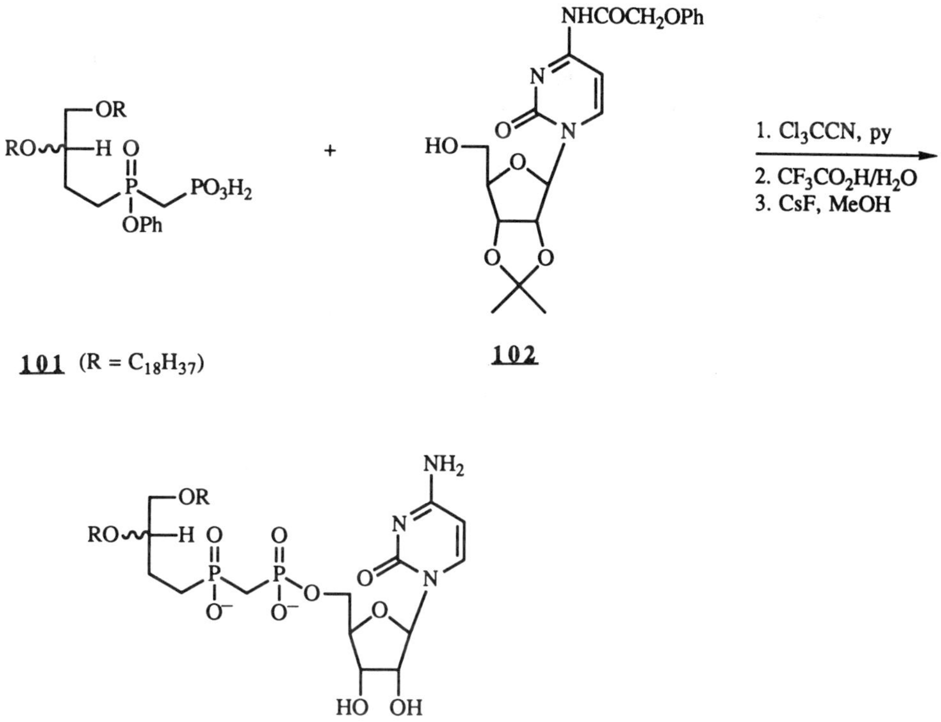

1. Cl_3CCN, py
2. CF_3CO_2H/H_2O
3. CsF, MeOH

101 ($R = C_{18}H_{37}$)

102

Scheme 59

2,3-*O*-isopropylidene-*sn*-glycerol (see Scheme 60). The acetonide is cleaved in 1.5 N sulfuric acid solution at 90°C. The water-soluble 1-*O*-benzyl-*sn*-glycerol is obtained by first removing dibenzyl ether and benzyl alcohol by washing with petroleum ether, then adjusting the pH of the aqueous phase to 10–12 with sodium hydroxide, and extracting the diol with ethyl acetate. Removal of ethyl acetate under vacuum affords crude 1-*O*-benzyl-*sn*-glycerol as a yellow-orange oil, and distillation gives the product as a pale yellow oil, bp 132–135°C (0.15 mm Hg); $[\alpha]^{22}_{D}$ −5.85° (neat). The primary hydroxy group of the diol is protected as the monomethoxytrityl ether by tritylation with (*p*-methoxyphenyl)diphenylmethyl chloride in the presence of DMAP and triethylamine in dichloromethane. The phosphomethyl ether group is introduced into bis-protected glycerol derivative 103 by alkylation with diethyl [(tosyloxy)methyl]phosphonate. The alkylation is carried out by adding a solution of 103 in THF to sodium hydride suspended in THF under argon, followed by stirring at room temperature for 30 min and then at reflux for 5 h; the mixture is cooled to 0°C and transferred to a solution of diethyl [(tosyloxy)methyl]phosphonate in THF at 0°C, followed by stirring with an overhead mechanical stirrer at 0°C for 1 h and at room temperature for 14 h. Phosphonate 104 is extracted with ethyl acetate, and the organic layer is washed with aqueous ammonium chloride and brine, dried (Na_2SO_4), and concentrated; yield of 104, 55%, after purification by silica gel chromatography (elution with hexane/ethyl acetate mixtures). Detritylation with either 80% aqueous acetic acid on a steam bath or with Amberlyst-15 ion-exchange resin gives the primary alcohol in ~90%

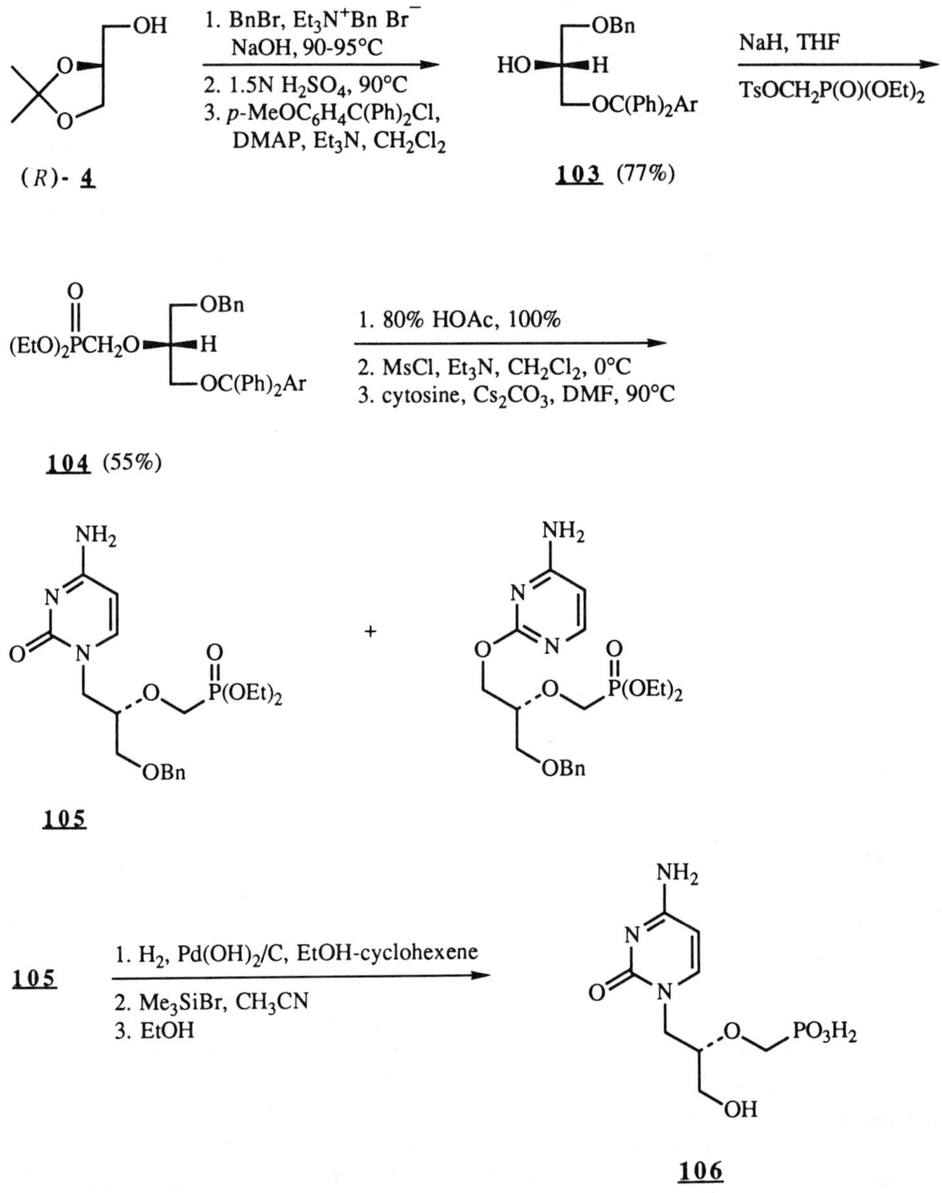

(R)- 4

1. BnBr, $Et_3N^+Bn\ Br^-$, NaOH, 90-95°C
2. 1.5N H_2SO_4, 90°C
3. p-$MeOC_6H_4C(Ph)_2Cl$, DMAP, Et_3N, CH_2Cl_2

103 (77%)

NaH, THF
$TsOCH_2P(O)(OEt)_2$

104 (55%)

1. 80% HOAc, 100%
2. MsCl, Et_3N, CH_2Cl_2, 0°C
3. cytosine, Cs_2CO_3, DMF, 90°C

105 +

105

1. H_2, $Pd(OH)_2/C$, EtOH-cyclohexene
2. Me_3SiBr, CH_3CN
3. EtOH

106

Scheme 60

yield. The alcohol is converted to the mesylate derivative using 1.2 equivalents of methanesulfonyl chloride in the presence of 2 equivalents of triethylamine in methylene chloride. The mesylate is coupled to cytosine at 90°C in dimethylformamide containing cesium carbonate. The desired *N*-alkylated product 105 is isolated and purified by column chromatography on silica gel (elution with methylene chloride-methanol mixtures); a 3 : 1 mixture of *N*- and *O*-alkylated isomers is obtained. Debenzylation by transfer hydrogenation with 20% $Pd(OH)_2$–C in cyclohexene-ethanol gives 106, together with about 5–15% of the corresponding dihydrouracil derivative; the latter arises from reduction of the cytosine ring. The debenzylated cytosine product is treated with trimethylsilyl bromide in acetonitrile. The pure product is obtained by precipitation from ethanol and lyophilization from water. The nine-step route from (*R*)-2,3-*O*-isopropylidene-*sn*-glycerol (4) to 106 proceeds in 18% overall yield on a multigram scale.

Literature: Bronson et al. (1989).

XII. PROTECTION AND DEPROTECTION PROCEDURES

Many of the synthetic schemes show the use of a variety of protecting groups in lipid chemistry. The protecting groups used in the schemes are listed below.

Benzyl. Schemes 8, 12, 16, 24, 27, 30, 51, and 60. In addition to the methods for debenzylation cited in the text (catalytic hydrogenolysis and dimethylboron bromide), $BF_3 \cdot Et_2O$-ethanethiol in carbon tetrachloride has been used (Yamauchi et al., 1989; Salamonczyk and Bruzik, 1990).
Benzoyl. Schemes 18 and 22.
Benzylidene. Scheme 27. See also Ohno et al. (1983).
O-Butenyl. Scheme 38.
tert-Butoxy. Scheme 55.
tert-Butyldimethylsilyl. Schemes 24 and 58.
tert-Butyldiphenylsilyl. Schemes 20 and 26.
Methoxyethoxymethyl. Scheme 28.
Tetrahydropyranyl. Schemes 23 and 29.
Trimethylsilyl. Scheme 14.
(2-Trimethylsilylethoxy)methyl. Scheme 28.
Trityl or monomethyltrityl (*p*-methoxyphenylchlorodiphenylmethane). Schemes 5, 7, 11, 18, 21, 22, 23, 30, 32, 35, 44, 55, and 60. Note that the schemes show a variety of acid-catalyzed conditions for detritylation, i.e., *p*-toluenesulfonic acid in methanol/THF 1:1 (Wissner et al., 1984); conc. HCl in 95% ethanol (Witzke and Bittman, 1985); HCl gas in chloroform-methanol 1:1 (Bhatia and Hajdu, 1988); boric acid impregnated silica gel (Buchnea, 1974); $BF_3 \cdot 2MeOH$ complex in methylene chloride at 0°C (Hermetter and Paltauf, 1981); glacial acetic acid in 1-propanol/water 5:1 (Bittman et al., 1987); zinc bromide in methylene chloride/methanol (10:1) (Kohli et al., 1980; Moroder et al., 1990).

Other protecting groups discussed in the text are: allyl (Hébert and Just, 1990); levulinoyl (Fröling et al., 1984) and 2,2,2-trichloroethoxycarbonyl (Hasegawa et al., 1987).

ACKNOWLEDGMENTS

The work cited in this chapter from the author's laboratory was supported in part by N.I.H. Grant HL-16660. I thank Catherine Vilcheze for skillful preparation of the synthetic schemes.

ABBREVIATIONS

Ac	acetyl
ara	1-β-D-arabinofuranosyl
ee	enantiomeric excess
Bn	benzyl
Boc	*tert*-butoxycarbonyl
Bz	benzoyl
CDP	cytosine 5'-diphosphate
DAG	diacylglycerol
DBU	1,8-diazabicylco[5.4.0]undec-7-ene
DCC	dicyclohexylcarbodiimide
DIPT	diisopropyl tartrate
DMAP	4-(*N*,*N*-dimethylamino)pyridine
DMF	*N*,*N*-dimethylformamide
DMSO	dimethyl sulfoxide
doxyl	4,4-dimethyloxazolidinyl-*N*-oxy
EDTA	ethylenediaminetetraacetate
MCPBA	*m*-chloroperbenzoic acid
MEM	methoxyethoxymethyl
Mes	mesityl
MOM	methoxymethyl
Ms	methanesulfonyl
MSNT	mesitylenesulfonyl-3-nitro-1,2,4-triazole
PAF	platelet-activating factor
PC	phosphatidylcholine
PE	phosphatidylethanolamine
PI	phosphatidylinositol
PPTS	pyridinium *p*-toluenesulfonate
PS	phosphatidylserine
Py	pyridine
rt	room temperature
SEM	(2-trimethylsilylethoxy)methyl
TBDMS	*tert*-butyldimethylsilyl
TBDPS	*tert*-butyldiphenylsilyl
Tf	trifluoromethylsulfonyl
THF	tetrahydrofuran
THP	tetrahydropyranyl
TMS	trimethylsilyl
Ts	*p*-toluenesulfonyl
TsOH	*p*-toluenesulfonic acid
Z	carbobenzyloxy

REFERENCES

Aarsman, A. J., L. L. M. van Deenen, and H. van den Bosch (1976), *Bioorg. Chem.* *5*:241–253.

Aarsman, A. J., C. F. P. Roosenboom, G. A. van der Marel, B. Shadid, J. H. van Boom, and H. van den Bosch (1985), *Chem. Phys. Lipids 36*:229–242.

Abdelmageed, O. H., R. I. Duclos, Jr., E. Abushanab, and A. Makriyannis (1990), *Chem. Phys. Lipids 54*:49–59.

Agarwal, K., A. Bali, and C. M. Gupta (1984), *Chem. Phys. Lipids 36*:169–177.
Ali, S., and R. Bittman (1988), *J. Org. Chem. 53*:5547–5549.
Ali, S., and R. Bittman (1989), *Chem. Phys. Lipids 50*:11–21.
Ali, S., and R. Bittman (1990), *Biochem. Cell Biol. 68*:360–365.
Amari, J. V., P. R. Brown, and J. G. Turcotte (1992), *American Laboratory 24*(2):23–29; *24*(3):26–36.
Angyal, S. J., and R. Hoskinson (1963), *Methods Carbohydr. Chem. 2*:87–89.
Aragozzini, F., E. Maconi, D. Potenza, and C. Scolastico (1989), *Synthesis* 225–227.
Atkinson, T., and M. Smith (1984), in *Oligonucleotide Synthesis—A Practical Approach* (M. J. Gait, ed.), IRL, Oxford.
Baba, N., K. Yoneda, S. Tahara, J. Iwasa, T. Kaneko, and M. Matsuo (1990), *J. Chem. Soc., Chem. Commun.* 1281–1282.
Baer, E. (1952), in *Biochemical Preparations*, Vol 2 (E. G. Ball, ed.), Wiley, New York, pp. 31–38.
Baer, E. (1965), *J. Am. Oil Chem. Soc. 42*:257–266.
Baer, E., and H. O. L. Fisher (1939a), *J. Am. Chem. Soc. 61*:761–765.
Baer, E., and H. O. L. Fisher (1939b), *J. Biol. Chem. 128*:463–473.
Baer, E., and M. Kates (1950), *J. Am. Chem. Soc. 72*:942–949.
Baer, E., and K. V. J. Rao (1966), *Can. J. Biochem. 45*:317–325.
Baldwin, J. J., A. W. Raal, K. Menster, B. H. Arison, and D. E. McClure (1978), *J. Org. Chem. 43*:4876–4878.
Berdel, W. E., R. Andreesen, and P. G. Munder (1985), in *Phospholipids and Cellular Regulation* (J. F. Kuo, ed.), CRC Press, Boca Raton, Fla, Vol. 2, pp. 41–73.
Bergelson, L. D. (1980), *Lipid Biochemical Preparations*, Elsevier/North Holland, Amsterdam, p. 84.
Berridge, M. J., and R. F. Irvine (1989), *Nature 341*:197–205.
Bette-Bobillo, P., A. Bienvenue, C. Broquet, and L. Maurin (1985), *Chem. Phys. Lipids 37*:215–226.
Bhatia, S. K., and J. Hajdu (1988), *J. Org. Chem. 53*:5034–5039.
Bhatia, S. K., and J. Hajdu (1988), *Tetrahedron Lett. 29*:31–34.
Bhatia, S. K., and J. Hajdu (1989), *Synthesis* 16–20.
Bhongle, N. N., R. H. Notter, and J. G. Turcotte (1987), *Syn. Commun. 17*:1071–1076.
Bittman, R., A. F. Rosenthal, and L. A. Vargas (1984), *Chem. Phys. Lipids 34*:201–205.
Bittman, R., and N. M. Witzke (1989), *Chem. Phys. Lipids 50*:99–103.
Bittman, R., N. M. Witzke, T-C. Lee, M. L. Blank, and F. Snyder (1987), *J. Lipid Res. 28*:733–738.
Bonsen, P. P. M., G. H. De Haas, W. A. Pieterson, and L. L. M. van Deenen (1972), *Biochim. Biophys. Acta 270*:364–382.
Braquet, P. G. (1991), *Handbook of PAF and PAF Antagonists,* CRC Press, Boca Raton, Fla.
Bronson, J. J., I. Ghazzouli, M. J. M. Hitchcock, R. R. Webb II, and J. C. Martin (1989), *J. Med. Chem. 32*:1457–1463.
Bruzik, K. S. (1988), *J. Chem. Soc., Perkin Trans 1*:423–431.
Bruzik, K. S., A. M. Morocho, D.-Y. Jhon, S. G. Rhee, and M.-D. Tsai, (1992), *Biochemistry 31*:5183–5193.
Bruzik, K. S., G. Salamónczyk, and W. J. Stec (1986), *J. Org. Chem. 51*:2368–2370.
Buchnea, D. (1974), *Lipids 9*:55–57.
Burgos, C. E., D. E. Ayer, and R. A. Johnson (1987), *J. Org. Chem. 52*:4973–4977.
Byun, H.-S., and R. Bittman (1989), *Tetrahedron Lett. 30*:2751–2754.
Caron, M., and K. B. Sharpless (1985), *J. Org. Chem. 50*:1557–1560.
Chakrabarti, P., and H. G. Khorana (1975), *Biochemistry 14*:5021–5033.
Chandrakumar, N. S., and J. Hajdu (1981), *Tetrahedron Lett. 22*:2949–2952.
Chandrakumar, N. S., and J. Hajdu (1982), *J. Org. Chem. 47*:2144–2147.

Chandrakumar, N. S., and J. Hajdu (1983), *J. Org. Chem. 48*:1197–1202.
Chau, L.-Y., Y.-M. Tsai, and J.-R. Cheng (1989), *Biochem. Biophys. Res. Commun. 161*:1070–1076.
Chen, C-S., and C. J. Sih (1989), *Angew. Chem., Intl. Ed. Engl. 28*:695–707.
Chittenden, C. J. F. (1980a), *Carbohydr. Res. 84*:350–352.
Chittenden, C. J. F. (1980b), *Carbohydr. Res. 87*:219–226.
Christie, W. W. (1989). *J. Lipid Res. 30*:1471–1473.
Cox, J. W., W. R. Snyder, and L. A. Horrocks (1979), *Chem. Phys. Lipids 25*:369–380.
Cubero Robles, E., and D. van Berg (1969), *Biochim. Biophys. Acta 187*:520–529.
Debost, J.-L., J. Gelas, and D. Horton (1983), *J. Org. Chem. 48*:1381–1382.
Dennis, E. A., S. G. Rhee, M. M. Billah, and Y. A. Hannun (1991). *FASEB J. 5*:2068–2077.
Deroo, P. W., A. F. Rosenthal, Y. A. Isaacson, L. A. Vargas, and R. Bittman (1976), *Chem. Phys. Lipids 16*:60–70.
Deveer, A. M. Th. J., R. Dijkman, M. Leuveling-Tjeenk, L. van den Berg, S. Ransac, M. Batenburg, M. Egmond, H. M. Verheij, and G. H. De Haas (1991), *Biochemistry 30*:10034–10042.
Dijkman, R., N. Dekker, and G. H. De Haas (1990), *Biochim. Biophys. Acta 1043*:67–74.
Dittmer, J. C., and R. L. Lester (1964), *J. Lipid Res. 5*:126–127.
Dodd, G. H., B. T. Golding, and P. V. Ioannou (1975), *J. Chem. Soc., Chem. Commun.* 249–250.
Donner, R., and Kh. Lohs (1965), *J. Chromatogr. 17*:349–354.
Dreef, C. E., C. J. J. Elie, P. Hoogerhout, G. A. van der Marel, and J. H. van Boom (1988), *Tetrahedron Lett. 29*:6513–6516.
Dumont, R., and H. Pfander (1983), *Helv. Chim. Acta 66*:814–823.
Edmundson, R. S. (1962), *Chem. Ind. (London)* 1828 [*C. A. 58*: 5497d (1963)].
Eibl, H. (1978), *Proc. Natl. Acad. Sci. 75*:4074–4077.
Eibl, H. (1980), *Chem. Phys. Lipids 26*:405–429.
Eibl, H. (1981), *Chem. Phys. Lipids 28*:1–5.
Eibl, H. (1984), *Angew. Chem., Intl. Ed. Engl. 23*:257–271.
Eibl, H., and A. Nicksh (1978), *Chem. Phys. Lipids 22*:1–8.
Eibl, H., and P. Woolley (1986), *Chem. Phys. Lipids 41*:53–63.
Ellingson, J. S., and R. L. Zimmerman (1987), *J. Lipid Res. 28*:1016–1018.
Engel, R. (1992), in *Handbook of Organophosphorus Chemistry*, Marcel Dekker, New York, pp. 559–600.
English, J., and P. H. Griswold (1948), *J. Am. Chem. Soc. 70*:1390–1392.
Fourrey, J.-L., and J. Varenne (1985), *Tetrahedron Lett. 26*:1217–1220.
Froehler, B., P. Ng, and M. Matteucci (1988), *Nucleic Acids Res. 16*:4831–4839.
Fröling, A., H. J. J. Pabon, and J. P. Ward (1984), *Chem. Phys. Lipids 36*:29–38.
Gadek, T. R. (1989), *Tetrahedron Lett. 30*:915–918.
Gao, Y., R. M. Hanson, J. M. Klunder, S. Y. Ko, H. Masamune, and K. B. Sharpless (1987), *J. Am. Chem. Soc. 109*:5765–5780.
Garigapati, V. R., and M. F. Roberts (1993), *Tetrahedron Lett. 34*:769–772.
Goodchild, J. (1990), *Bioconj. Chem. 1*:165–187.
Gordon, A. J., and R. A. Ford (1972), *The Chemist's Companion*, Wiley, New York.
Guindon, Y., C. Yoakim, and H. E. Morton (1984), *J. Org. Chem. 49*:3912–3920.
Guivisdalsky, P. N., and R. Bittman (1989a), *J. Am. Chem. Soc. 111*:3077–3079.
Guivisdalsky, P. N., and R. Bittman (1989b), *J. Org. Chem. 54*:4637–4642.
Guivisdalsky, P. N., and R. Bittman (1989c), *J. Org. Chem. 54*:4643–4648.
Guivisdalsky, P. N., R. Bittman, Z. Smith, M. L. Blank, F. Snyder, S. Howard, and H. Salari (1990), *J. Med. Chem. 33*:2614–2621.
Gupta, C. M., R. Radhakrishnan, and H. G. Khorana (1977), *Proc. Natl. Acad. Sci. 74*:4315–4319.
Häfele, B., and V. Jäger (1987), *Liebigs Ann. Chem.* 85–87.
Han, X., and R. W. Gross (1991), *Biochim. Biophys. Acta 1063*:129–136.

Han, X., L. A. Zupan, S. L. Hazen, and R. W. Gross (1992), *Analyt. Biochem. 200*:119–124.
Hansen, W. J., R. Murari, Y. Wedmid, and W. J. Baumann (1982), *Lipids 17*:453–459.
Hanson, R. M. (1991), *Chem. Rev. 91*:437–475.
Hanson, R. M., and K. B. Sharpless (1986), *J. Org. Chem. 51*:1922–1925.
Harbison, G. S., and R. G. Griffin (1984), *J. Lipid Res. 25*:1140–1142.
Hasegawa, E., K. Eshima, Y. Matsushita, H. Nishide, and E. Tsuchida (1987), *Synthesis* 60–62.
Hébert, N., A. Beck, R. B. Lennox, and G. Just (1992), *J. Org. Chem. 57*:1777–1783.
Hébert, N., and G. Just (1990), *J. Chem. Soc., Chem. Commun.* 1497–1498.
Hendrickson, H. S., and E. K. Hendrickson (1990), *Chem. Phys. Lipids 53*:115–120.
Hendrickson, H. S., E. K. Hendrickson, and R. H. Dybvig (1983), *J. Lipid Res. 24*:1532–1537.
Hendrickson, H. S., E. K. Hendrickson, and T. J. Rustad (1987), *J. Lipid Res. 28*:864–872.
Hermetter, A., and F. Paltauf (1981), *Chem. Phys. Lipids 29*:191–195.
Hermetter, A., F. Paltauf, and H. Hauser (1982), *Chem. Phys. Lipids 30*:35–45.
Hermetter, A., H. Stütz, R. Franzmair, and F. Paltauf (1989), *Chem. Phys. Lipids 50*:57–62.
Hirt, R., and R. Berchtold (1958), *Pharm. Acta Helv. 33*:349–356.
Hirth, G., and R. Barner (1982), *Helv. Chim. Acta 65*:1059–1084.
Hirth, G., H. Saroka, W. Bannwarth, and R. Barner (1983), *Helv. Chim. Acta 66*:1210–1240.
Hirth, G., and W. Walther (1985), *Helv. Chim Acta 68*:1863–1871.
Hong, C. I., A. J. Kirisits, A. Nechaev, D. J. Buchheit, and C. R. West (1990), *J Med. Chem. 33*:1380–1386.
Hostetler, K. Y., L. M. Stuhmiller, H. B. M. Lenting, H. van den Bosch, and D. D. Richman (1990), *J. Biol. Chem. 265*:6112–6117.
Hubschwerlen, C. (1986), *Synthesis* 962–964.
Isaacson, Y. A., P. W. Deroo, A. F. Rosenthal, R. Bittman, J. O. McIntyre, H.-G. Bock, P. Gazzotti, and S. Fleischer (1979), *J. Biol. Chem. 254*:117–126.
Ishaq, K. S., M. Capobianco, C. Piantadosi, A. Noseda, L. W. Daniel, and E. J. Modest (1989), *Pharm. Res. 6*:216–224.
Jacobsen, E. N., I. Marko, M. B. France, J. S. Svendsen, and K. B. Sharpless (1989), *J. Am. Chem. Soc. 111*:737–739.
Jain, M. K., W. Tao, J. Rogers, C. Arenson, H. Eibl, and B-Z. Yu (1991), *Biochemistry 30*:10256–10268.
Johnson, R. A., C. E. Burgos, and E. G. Nidy (1989), *Chem. Phys. Lipids 50*:119–126.
Joseph, J., and C.-S. Lai (1988), *J. Lipid Res. 29*:1101–1104.
Joseph, J., C. C.-Y. Shih, and C.-S. Lai (1991), *Chem. Phys. Lipids 58*:19–26.
Jung, M. E., and T. J. Shaw (1980), *J. Am. Chem. Soc. 102*:6304–6311.
Jurczak, J., S. Pikul, and T. Bauer (1986), *Tetrahedron 42*:447–488.
Kanda, P., and M. A. Wells (1980), *J. Lipid Res. 21*:257–258.
Kanda, P., and M. A. Wells (1981), *J. Lipid Res. 22*:877–879.
Kates, M. (1977), *Methods Membr. Biol. 8*:219–290.
Kates, M. (1986), *Techniques of Lipidology: Isolation, Analysis, and Identification of Lipids,* 2nd ed., Elsevier, pp. 239–251.
Keough, K. M. W., and P. J. Davis (1979), *Biochemistry 18*:1453–1459.
Kim, H.-Y., and N. Salem, Jr. (1990), *J. Lipid Res. 31*:2285–2289.
Klunder, J. M., T. Onami, and K. B. Sharpless (1989), *J. Org. Chem. 54*:1295–1304.
Kodali, D. R. (1987), *J. Lipid Res. 28*:464–469.
Kodali, D. R., and R. I. Duclos, Jr. (1992), *Chem. Phys. Lipids 61*:169–173.
Kohli, V., H. Blöcker, and H. Köster (1980), *Tetrahedron Lett. 21*:2683–2686.
Lai, C.-S., J. Joseph, and C. C.-Y. Shih (1989), *Biochem. Biophys. Res. Commun. 160*:1189–1195.
Lalinde, N., B. E. Tropp, and R. Engel (1983), *Tetrahedron 39*:2369–2372.
Lammers, J. G., Th. J. Liefkens, J. Buo, and J. van der Meir (1978), *Chem. Phys. Lipids 22*:293–305.
Lemmen, P., K. M. Buchweitz, and R. Stumpf (1990), *Chem. Phys. Lipids 53*:65–75.

Lin, G. L., and M.-D. Tsai (1989), *J. Am. Chem. Soc. 111*:3099–3101.
Lindh, I., and J. Stawiński (1989), *J. Org. Chem. 54*:1338–1342.
Loffredo, W. M., and M.-D. Tsai (1990), *Bioorg. Chem. 18*:78–84.
Lok, C. M., J. P. Ward, and D. A. van Dorp (1976), *Chem. Phys. Lipids 16*:115–122.
Lok, C. M., A. P. J. Mank, and J. P. Ward (1985), *Chem. Phys. Lipids 36*:329–334.
Lutzke, B. S., and J. M. Braughler (1990), *J. Lipid Res. 31*:2127–2130.
Macphee, C. H., A. N. Carter, F. Ruiz-Larrea, J. G. Ward, R. C. Young, and C. P. Downes (1992), *J. Biol. Chem. 267*:11137–11143.
Magolda, R. L., and W. Galbraith (1989), *J. Cell Biochem. 40*:371–386.
Mangroo, D., and G. E. Gerber (1988), *Chem. Phys. Lipids 48*:99–108.
Mannock, D. A., R. N. A. H. Lewis, and R. N. McElhaney (1990), *Chem. Phys. Lipids 55*:309–321.
Martin, S. F., and J. A. Josey (1988), *Tetrahedron Lett. 29*:3631–3634.
Marx, M. H., C. Piantadosi, A. Noseda, L. W. Daniel, and E. J. Modest (1988), *J. Med. Chem. 31*:858–863.
Mason, J. T., A. V. Broccoli, and C-H. Huang (1981), *Analyt. Biochem. 113*:96–101.
McGuigan, C., and B. Swords (1990), *J. Chem. Soc., Perkin Trans. 1*:783–787.
Meyer, K. L., S. W. Schwendner, and R. E. Counsell (1989), *J. Med. Chem. 32*:2142–2147.
Mikkilineni, A. B., P. Kumar, and E. Abushanab (1988), *J. Org. Chem. 53*:6005–6009.
Mitsudo, T., Y. Hori, Y. Yamakawa, and Y. Watanabe (1987), *J. Org. Chem. 52*:2230–2239.
Miyazaki, H., N. Ohkawa, N. Nakamura, T. Ito, T. Sada, T. Oshima, and H. Koike (1989), *Chem. Pharm. Bull. 37*:2379–2390.
Mlotkowska, B. (1991), *Liebigs Ann. Chem.* 1361–1362.
Mlotkowska, B., and A. Markowska (1988), *Liebigs Ann. Chem.* 191–193.
Mlotkowska, B., and A. Markowska (1990), *Liebigs Ann. Chem.* 923–925.
Moroder, L., H.-J. Musiol, and G. Sigmüller (1990), *Synthesis* 889–892.
Nakamura, K., and Handa, S. (1984), *Analyt. Biochem. 142*: 406–410.
Nakamura, N., H. Miyazaki, N. Ohkawa, T. Oshima, and H. Koike (1990), *Tetrahedron Lett. 31*:699–702.
Nguyen Thanh Thuong, and P. Chabrier (1974), *Bull. Soc. Chim. Fr.* 667–671.
Noseda, A., M. E. Berens, C. Piantadosi, and E. J. Modest (1987), *Lipids 22*:878–883.
Ohno, M., K. Fujita, H. Nakai, S. Kobayashi, M. Yamahita, K. Inoue, and S. Nojima (1983), in *Platelet-Activating Factor INSERM Symposium No. 23* (J. Benveniste and B. Arnoux, eds.), Elsevier, pp. 9–20.
Ohno, M., K. Fujita, H. Nakai, S. Kobayashi, K. Inoue, and S. Nojima (1985), *Chem. Pharm. Bull. 33*:572–582.
Paltauf, F., in *Ether Lipids: Biochemical and Biomedical Aspects* (H. K. Mangold and F. Paltauf, eds.), Academic Press, New York, 1983, pp. 49–84.
Parks, J. S., T. Y. Thuren, and J. D. Schmitt (1992), *J. Lipid Res. 33*:879–887.
Patel, K. M., J. D. Morrisett, and J. T. Sparrow (1979), *J. Lipid Res. 20*:674–677.
Perly, B. E. J. Dufourc, and H. C. Jarrell (1984), *J. Labelled Compd. Radiopharm. 21*:1–13.
Peters, U., W. Bankova, and P. Welzel (1987), *Tetrahedron 16*:3803–3816.
Pradhan, D., P. Williamson, and R. A. Schlegel (1989), *Biochemistry 28*:6943–6949.
Prahad, M., J. C. Tomesch, and J. R. Wareing (1990), *Chem. Phys. Lipids 53*:121–126.
Prakash, C., S. Saleh, D. F. Taber, and I. A. Blair (1989), *Lipids 24*:786–792.
Radhakrishnan, R., R. J. Robson, Y. Takagaki, and H. G. Khorana (1981), *Methods Enzymol. 72*:408–433.
Rebecchi, M. J., R. Eberhardt, T. Delaney, S. Ali, and R. Bittman (1993), *J. Biol. Chem. 268*:1735–1741.
Rosario-Jansen, T., R.-T. Jiang, M.-D. Tsai, and D. J. Hanahan (1988), *Biochemistry 27*:4619–4624.
Rosenthal, A. F. (1975), *Methods Enzymol. 35B*:429–529.
Rosenthal, A. F., L. Vargas, and S. C. Han (1972), *Biochim. Biophys. Acta 260*:369–379.

Rouser, G., G. Kritchevsky, and A. Yamamoto (1976), in *Lipid Chromatographic Analyses*, 2nd ed., Marcel Dekker, New York, Vol. 3, pp. 713–776.

Runquist, E. A., and G. M. Helmkamp, Jr. (1988), *Biochim. Biophys. Acta 940*:10–20.

Rustenbeck, I., and Lenzen, S. (1990), *J. Chromatog. 525*: 85–91.

Ryu, E. K., and M. MacCross (1979), *J. Lipid Res. 20*:561–563.

Salamonczyk, G. M., and K. S. Bruzik (1990), *Tetrahedron Lett. 31*:2015–2016.

Samuel, N. K. P., M. Singh, K. Yamaguchi, and S. L. Regen (1985), *J. Am. Chem. Soc. 107*:42–47.

Schmidt, R. R. (1989), *Pure and Appl. Chem. 61*:1257–1270.

Schwartz, P. W., B. E. Tropp, and R. Engel (1988), *Chem. Phys. Lipids 48*:1–7.

Sekine, M., H. Mori, and T. Hata (1982), *Bull. Chem. Soc. Jpn. 55*:239–242.

Selinger, Z., and Y. Lapidot (1966), *J. Lipid Res. 7*:174–175.

Shashidhar, M. S., J. F. W. Keana, J. J. Volwerk, and O. H. Griffith (1990), *Chem. Phys. Lipids 53*:103–113.

Shoji, M., R. L. Raynor, W. E. Berdel, W. R. Vogler, and J. F. Kuo (1988), *Cancer Res. 48*:6669–6673.

Shukla, V. K. S. (1988), *Prog. Lipid Res. 27*:5–38.

Singh, A. (1990), *J. Lipid Res. 31*:1522–1525.

Snyder, W. R. (1987), *J. Lipid Res. 28*:949–954.

Sowden, J. C., and O. L. Fischer (1942), *J. Am. Chem. Soc. 64*:1291–1294.

Still, W. C., M. Kahn, and A. Mitra (1978). *J. Org. Chem. 43*:2923–2925.

Sunamoto, J., M. Goto, K. Iwamoto, H. Kondo, and T. Sato (1990), *Biochim. Biophys. Acta 1024*:209–219.

Sutowardoyo, K. I., and D. Sinou (1991), *Tetrahedron Asymmetry 2*:437–444.

Szolderits, G., G. Daum, F. Paltauf, and A. Hermetter (1991), *Biochim Biophys. Acta 1063*:197–202.

Tomioka, K, M. Nakajima, and K. Koga (1987), *J. Am. Chem. Soc. 109*:6213–6215.

Touchstone, J. C., S. S. Levine, M. F. Dobbins, L. Matthews, P. C. Beers, and S. G. Gabbe (1983), *Clin. Chem. 29*:1951–1954.

Tsai, Y.-M., S.-L. Jiang, H.-J. Chen, L.-Y. Chau, and J.-T. Lin (1988), *J. Chinese Chem. Soc. 35*:429–435.

Turcotte, J. G., W. H. Lin, N. C. Motola, P. E. Pivarnik, N. N. Bhongle, H. R. Heyman, S. S. Shirali, Z. Lu, and R. H. Notter (1991), *Chem. Phys. Lipids 58*:81–95.

Ukawa, K., E. Imamiya, H. Yamamoto, K. Mizuno, A. Tasaka, Z.-I. Terashita, T. Okutani, H. Nomura, T. Kasukabe, M. Hozumi, I. Kudo, and K. Inoue (1989), *Chem. Pharm. Bull. 37*:1249–1255.

van Deenen, L. L. M., and G. H. De Haas (1964), *Adv. Lipid Res. 2*:167–234.

van Wijk, G. M. T., K. Y. Hostetler, C. N. S. P. Suurmeijer, and H. van den Bosch (1992), *Biochim. Biophys. Acta 1165*:45–52.

Vargas, L. A., L. X. Miao, and A. F. Rosenthal (1984), *Biochim. Biophys. Acta 796*:123–128.

Vargas, L. A., and A. F. Rosenthal (1983), *J. Org. Chem. 48*:4775–4776.

Vaskovskii, V. E., and E. Ya. Kostetskii (1968), *J. Lipid Res. 9*:396.

Virtanen, J. A., J. R. Brotherus, O. Renkonen, and M. Kates (1980), *Chem. Phys. Lipids 27*:185–190.

Vogel, A. I. (1989), *Textbook of Practical Organic Chemistry*, fifth ed. (B. S. Furnis, A. J. Hannaford, P. W. G. Smith, and A. R. Tatchell, eds.), Wiley, New York.

Wang, Y.-F., J. J. Lalonde, M. Momongan, D. E. Bergbreiter, and C.-H. Wong (1988), *J. Am. Chem. Soc. 110*:7200–7205.

Wang, Y.-F., and C.-H. Wong (1988), *J. Org. Chem. 53*:3129–3130.

Ward, J. P. (1988), *Chem. Phys. Lipids 47*:217–224.

Warner, T. G., and A. A. Benson (1977), *J. Lipid Res. 18*:548–552.

Weltzien, H. U., and P. G. Munder (1983), in *Ether Lipids. Biochemical and Biomedical Aspects* (H. K. Mangold and F. Paltauf, eds.), Academic Press, New York, pp. 277–308.

Wissner, A., P-E. Sum, R. E. Schaub, C. A. Kohler, and B. M. Goldstein (1984), *J. Med Chem. 27*:1174–1181.

Witzke, N. W., and R. Bittman (1985), *J. Lipid Res.* *26*:623–628.
Witzke, N. W., and R. Bittman (1986), *J. Lipid Res.* *27*:344–351.
Yamauchi, K., M. Hihara, and M. Kinoshita (1987), *Bull. Chem. Soc. Jpn.* *60*:2169–2172.
Yamauchi, K., A. Moriya, and M. Kinoshita (1989), *Biochim. Biophys. Acta* *1002*:151–160.
Young, R. C., C. P. Downes, D. S. Eggleston, M. Jones, C. H. Macphee, K. K. Rana, and J. G. Ward (1990), *J. Med. Chem.* *33*:641–646.
Yu, L., R. A. Deems, J. Hajdu, and E. A. Dennis (1990), *J. Biol. Chem.* *265*:2657–2664.
Yuan, W., R. J. Berman, and M. H. Gelb (1987), *J. Am. Chem. Soc.* *109*:8071–8081.
Yuan, W., K. Fearon, and M. H. Gelb (1989), *J. Org. Chem.* *54*:906–910.
Zheng, B., K. Oishi, M. Shoji, H. Eibl, W. E. Berdel, J. Hajdu, W. R. Vogler, and J. F. Kuo (1990), *Cancer Res.* *50*:3025–3031.
Zlatanos, S. N., A. N. Sagredos, and V. P. Papageourgiou (1985), *J. Am. Oil Chem. Soc.* *62*:1575–1577.

7
Polymerizable Phospholipids

Alok Singh and Joel M. Schnur *Center for Bio/Molecular Science and Engineering, Naval Research Laboratory, Washington, D.C.*

I. INTRODUCTION

During the past 40 years a large number of papers have been published on synthetic and natural phospholipids [1–5]. It is clearly reflected from the reviews that a large number of those papers have focused on the potential for using microstructures derived from these lipids for applications from encapsulation, controlled release, biosensors, and enzyme immobilization to functional protein reincorporation. The usefulness of polymerizable lipids is reflected by a number of patents describing the preparation of polymerized vesicles [6,7–10], encapsulation of materials for use in food, cosmetics, and pharmaceuticals [7,11–14], controlled release of bioactive materials [15], magnetic vesicles [16], enhanced oil recovery [17], enzyme encapsulation [12] or immobilization [18], and applications of novel tubule microstructures [19]. The term vesicles is used in the text to describe spherical or ellipsoidal, single or multicompartmented, closed bilayer structures, regardless of their chemical composition. These structures have been used as models for biological membranes [1–3,20]. In a large number of these areas proof of principle experiments have been successfully demonstrated. Yet few if any of these are currently in commercial use. One reason for this is the lack of stability of lipid-based microstructures.

This stability problem has been understood since the beginning stages of research on liposomes. Strategies for improved stability have included the incorporation of proteins, sugars, and cholesterol [21,22]. Indeed, considerable progress has been reported in the use of sugar to stabilize both red blood cells and an artificial oxygen-carrying fluid made by encapsulating hemoglobin into vesicles [23,24]. These incorporation strategies have had only limited success in some specific applications. They also suffer because they add another component to the system, a component that may well profoundly perturb the system that the incorporated substance was to stabilize. Two strategies seem to have the potential for wide application: the use of long-chain phospholipids, which form highly robust structures (see Sec. III), and the modification of the phospholipids to enable them to form stabilized microstructures via polymerization [25–28]. However, the latter approach can also lead to complications. The modification of the lipid by the incorporation of a polymerizable moiety can inhibit the formation of the microstructure that was supposed to be stabilized [29]. On the other hand, these structural modifications can also lead to new structures that also are of technological importance [30,31].

This chapter will review a number of the approaches being used to stabilize microstructures formed from polymerizable phospholipids, the properties of the lipids, structures formed from such polymerizable lipids, and the potential technological applications of these approaches. The following section will elucidate the various possibilities for monomer placement in lipid molecules, which will be followed by a section on the properties of microstructures derived from polymerizable phospholipids. We will conclude with a description of the technological applications of these lipids and try to provide a prospectus for their future technological development.

II. THE ARCHITECTURE OF A POLYMERIZABLE PHOSPHOLIPID

Since the first appearance of reports [25–28] describing the concept of polymerized vesicles, polymerizable lipids have attracted considerable attention from different disciplines. As a result, a good number of reports on the design and synthesis of polymerizable phospholipids have appeared in the literature. Figure 1 illustrates that there are four different sites available for the attachment of polymerizable moieties in a phospholipid

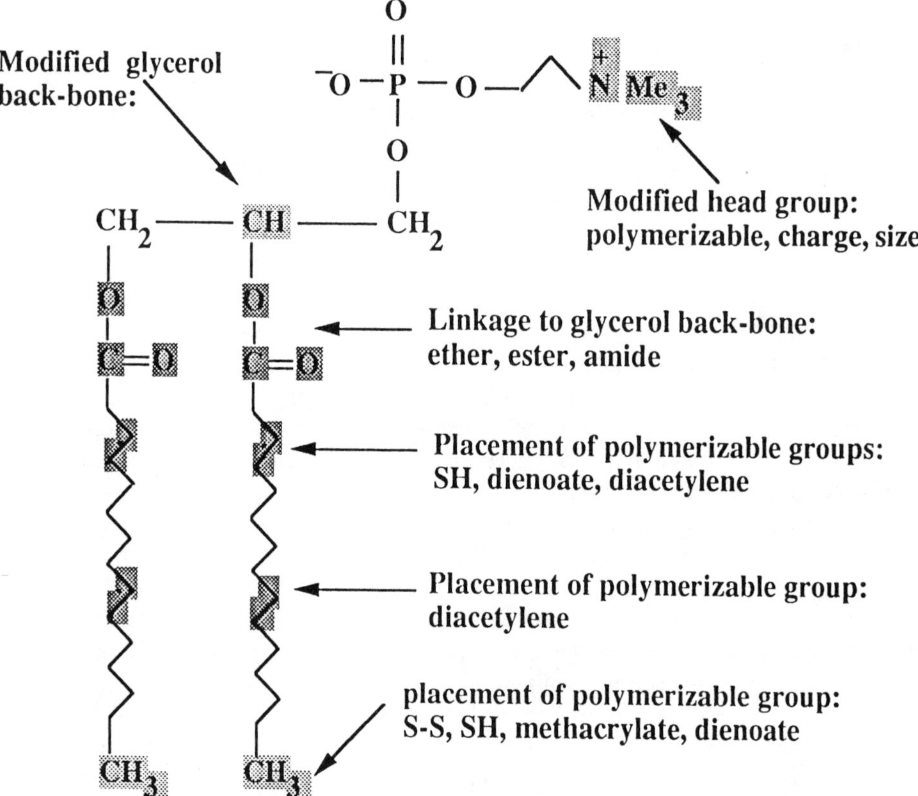

Figure 1 Possible sites for the placement of polymerizable moieties and possible modifications in a phospholipid.

molecule. The head group region provides one site. Other sites are in the acyl region at the terminus, in the mid-chain, and near the carboxylic end. As the field has progressed, the focus on polymerizable lipids has shifted from stabilization of the microstructures to specific applications based on polymerization (e.g., induced phase separation, biodegradable carriers, morphology transformations, controlled release, etc.). Most of these subjects are briefly addressed in the reviews that have appeared on the subject during the last decade [5,20,32–38].

Polymerizable moieties and their placement in the lipid molecule are likely to influence the physical properties of the vesicles or other microstructures formed from the polymerizable lipids. The polymerization in the hydrophilic region of the molecule may change the properties of the head groups [39], while the polymerization at the chain terminus may cause inter-layer cross-linking. Mid-region cross-linking also can perturb the bilayer structure [29]. As our knowledge about the effects of molecular geometry upon microstructure formation increases, it is thus likely that the properties of biologically derived membranes can be modified to meet a specific need by carefully selecting the polymerizable groups and their placement in phospholipid molecules.

To date a number of polymerizable moieties have been incorporated in phospholipid molecules. A comprehensive list of resulting lipids is compiled in Appendix 1.

III. THE ARCHITECTURE OF A POLYMERIZABLE PHOSPHOLIPID

A. Linkages and Placement of Polymerizable Moieties

The polymerizable moieties incorporated in the phospholipids are vinyl, methacrylate, diacetylene, styryl, acetylene, and dienoyl (sorbate). There are two other classes of compounds that have also been used for achieving the cross-linking in the vesicles; the sulfhydryl and disulfides in the acyl region and the substituted amino acids in the head group region of the phospholipids. Both of these constrain the neighboring molecule by condensation reaction in order to provide overall stabilization.

Methacrylate lipids. The methacryloyl group has been attached to the phospholipids via an ester linkage. This is the only functionality that has been linked to both hydrophilic [39–44] and lipophilic [45–56] sites. Acyl chains bearing polymerizable groups have been linked to the glycerol backbone in position 1, 2 or both. Another feature of the methacryloyl moiety is that it can be polymerized in bilayer membranes employing either radical initiator azobisisobutyronitrile (AIBN) or by photo initiation employing ultraviolet (UV) irradiation [25,45,46].

Thiol and disulfide lipids. Thiols are known to undergo the thiol-disulfide redox cycle reversibly. In biological systems the protein fragments are held together via a disulfide linkage that on reduction produces fragments containing sulfhydryl groups (–SH). This reaction cycle has been used in the attachment of antibodies to solid surfaces [57]. The idea has been extended to lipid molecules to achieve polymerization and depolymerization in the phospholipid vesicles. Due to the intermolecular reactivity between –SH groups of phospholipids dispersed in water, the resulting vesicles have been polymerized or depolymerized on demand. Placement of the –SH group has been reported both near the head group and at the end of the fatty acyl chain [58–61]. Placement of the –SH group near the head group has been claimed to give the highest capture volume and slowest release rate of encapsulant [59].

The sulfhydryl strategy worked successfully in vesicle stabilization by producing linear polymers. The scheme has then been extended to cyclic disulfide moiety (lipoyloxy, $-CH[S]-CH_2-CH_2[S]$) to achieve rugged cross-linked polymers. The lipoyloxy groups were placed at the terminus of the fatty acyl chain. Ring opening polymerization has been achieved either by ultraviolet irradiation at 333 nm or by using a catalytic amount of dithiothreitol (DTT). Due to the presence of two sulphur atoms per chain, extensive cross-linking has been achieved [62,63] leading to stable vesicle preparation.

Dienoate lipids. Dienoate or sorbate functionalities have been incorporated in the phospholipids only in the acyl chain region. The placement of this group was accomplished at two positions, at the chain terminus and near the head group. Due to the nature of this group (conjugated diene), a mid-chain placement is also possible, but to date no such lipid has been reported. This could be attributed to the synthetic reasons. The selective hydrogenation of the isomeric diacetylenic lipids may provide a simple and efficient route to the dienyl lipids.

Dienoate groups near the glycerol backbone are linked via an ester linkage employing 2,4-alkadienoic acid. Both racemic [39,64–68] and optically active [69–83] phospholipids bearing such groups have been synthesized. The dienoate moiety has also been placed at the terminus of the acyl chain via an ether linkage employing 2,4-hexadienol [84] or ester linkage by making use of sorbic acid [71,85]. A variety of phospholipid molecules bearing this chromophore has been reported (see Appendix 1).

The dienoate chromophore, which strongly absorbs near 257 nm due to the carbonyl group conjugated to the diene, has been polymerized by ultraviolet irradiation [69–71]. Polymerization of this group in lipids has also been achieved using hydrophilic and lipophilic free-radical initiators, azobis(2-amidinopropane) dihydrochloride (AAPD), and azobis isobutyronitrile (AIBN), respectively [73,74,80].

Styryl lipids. In order to obtain improved stability of the lipid vesicles, the styryl groups have been incorporated. Styrene has been linked to the acyl chain through a ketone [86–92], ether [86], and amide [86] group. The placement of the styryl group is accomplished in one chain linked to the *sn*-2 position of the glycerol backbone. Lipids containing styryl group in both chains did not form vesicles [87]. The second acyl chain in these lipids was usually linked via an ether linkage following a synthetic route that seldom gave pure enantiomeric lipid. Polymerized vesicles from these lipids were stable enough to withstand freeze-drying and redispersion in organic solvents without losing structural integrity [92]. Polymerization of the styryl moiety in the vesicles was achieved by ultraviolet irradiation only.

Diacetylenic lipids. The incorporation of diacetylenes into lipid bilayers was prompted by the fact that diacetylenes polymerize in the solid state when the molecules are packed in an ordered manner leading to a colored conjugated polymer backbone made of alternate single, double, and triple bonds [93,94]. Phospholipid molecules exhibit an ordered packing of the hydrocarbon chain (i.e., the chains adopt an all trans conformation) when dispersed in aqueous medium below their chain melting transition temperature. This common property led to the synthesis of diacetylenic phospholipids for the construction of polymer-stabilized vesicles where the degree of polymerization could be monitored by recording visible spectrum. Placement of the diacetylene moiety was achieved only in the acyl chains. Polymerization has been achieved using UV irradiation except in those few instances where gamma radiation has been employed [52]. For the purpose of clarity in referring to a diacetylenic phosphatidylcholine, we shall use the $DC_{m,n}PC$ terminology, where m is the number of methylene units between diacetylene and carboxylate group and n refers to the number of methylenes on the terminal methyl segment.

Lipid based tubules. Phospholipids containing the diacetylenic moiety are among the most studied polymerizable lipids. These lipids have been studied from two distinct points of view. In one instance the lipids have been used to construct vesicles [26–28,37,39,42,49,53,70,93–101] and in the other to form Langmuir or self-assembled monolayers [93] at the gas-water interface. While studying the former, Yager and Schoen [29] at the Naval Research Laboratory (NRL) discovered very unusual 0.5 micron diameter hollow cylinders that they called "tubules." This discovery is an exciting aspect of studying the polymerizable lipid systems from the theoretical as well as the practical point of view [30,102]. Helical microstructures and hollow microcylinders have already attracted the attention of experimentalists [103–105] and theoreticians [106–107] alike. A variety of synthetic [30,108–112], mechanistic [30,113–120], fabrication-oriented [121–124], thermotropic [123–125], spectroscopic [126–131], and application related [19,132–140] studies have contributed to the progress on tubule-related topics.

Tubules research is discussed in separate sections covering the physical characterizations, properties, and potential technological applications, in order to deal adequately with this complex and interesting topic.

Heterobifunctional lipids. Because of inefficient polymerization of diacetylene in lipid tubules, other polymerizable moieties and approaches [30] have been introduced to stabilize these structures. Two polymerizable groups, olefinic and methacrylate, have been placed at the terminus of acyl chains of the diacetylenic lipid [52]. In the case of methacrylate, extensive polymerization has been achieved after UV irradiation, but the tubule structures were disrupted by polymerization. An olefinic moiety has been polymerized by gamma radiation, resulting in more stable microstructures.

Miscellaneous polymerizable lipids. In an intriguing system the polymerizable moiety itaconate ($-OCO-CH_2-C(=CH_2)-COO-$) is linked to the glycerol backbone to construct giant stable vesicles [67]. Polymerization of the itaconate moiety in vesicles is achieved by photoinitiation by UV light. Usual radical polymerization of itaconate is slow, but in assembled structures it is reported to be very rapid. The strategy is technologically attractive since the polymer backbone is closer to the glycerol backbone and thus could be hydrolyzed by a variety of water soluble or dispersible reagents. However, this system has not been tested for hydrolysis, nor has the extent of polymerization been determined.

The polymerizable isocyano moiety [141,142] may also constitute a potentially useful system because of the rigid helical configuration with four repeating units per helical turn in its polymer. The cross-linking has been reported to proceed parallel to the vesicle surface. This system is the first one in which some transbilayer cross-linking has been reported. Heretofore it has been used only in ammonium surfactants, and it now awaits its turn to be used in phospholipids.

In an interesting approach to cross-link the neighboring lipid or protein molecules, photocleavable carbene precursors such as omega-(2)-diazo-3,3,3,-trifluoropropionyloxy ($-OCOCN_2CF_3$) [143,144], diazirinophenoxy ($-O-C_6H_4-CHN_2$), 2-nitro-4-azidophenoxy ($-O-C_6H_4(NO_2)N_3$), *m*-azidophenoxy [145], alpha, beta-unsaturated keto ($-CH{=}CH-CO-$) [145,146], and aliphatic and aromatic azido groups were linked to the acyl chain of phospholipid [143,145,147].

Nitrene generated from azido group (aromatic or aliphatic) did not work satisfactorily in forming cross-links by insertion in C–H bonds of neighboring molecules. We believe that the latter type of system may be of use in the stabilization of the self-assembled microstructures, however, while not necessarily creating a high molecular weight cross-linked polymer.

In a recent report, an α,β unsaturated keto group has been used as a polymerizable moiety to stabilize the vesicles [146] by photopolymerization. Unlike other polymerizable vesicles, the morphological integrity of the vesicles prepared from this phospholipid could not be preserved during photopolymerization, and closed tubular structures have been obtained.

B. Modifications Near the Head Groups

Phospholipids containing a polymerizable moiety in the polar head group region have not been pursued as actively as their acyl counterpart. The phospholipid analogs bearing polymerizable moieties, usually methacrylate, usually lack a phosphate or choline group [39,42,43]. The true phosphocholine analogs equipped with the methacrylate moiety have been reported by Kusumi et al. [40] and Ohkatsu et al. [148]. In the latter case, the lipid was meant to produce flexible membranes mimicking the skin. Polymerization by UV irradiation at 60°C produced a polymer of 98,000 molecular weight, as determined by gel permeation chromatography using polystyrene molecular weight standards.

In a closely related system, diacetylenic phospholipids containing a modified polar head group with hydroxyalkanol have been synthesized and studied [138,149,150]. Both polymerization and morphology of the resulting microstructures were found to be dependent on pH, cations, and anions present in the dispersion medium.

C. Ionic Interaction of Polymerizable Moieties

In order to create a novel polymer network in an organized system the counterions of the polar surfactants can be exchanged with a polymerizable moiety. Polymerization of these counterions then creates a thin protective shield on the vesicles. This type of system has been reviewed from the point of view of its physical properties and applications [20]. Most of the experiments have been carried out on the phosphate or ammonium surfactants. Thus polymer-encased vesicles were prepared from dioctdecyldimethylammonium methacrylate [151–153] by replacing chloride ion with methacrylate ion, by incorporating 4-vinylpyridine to sonicated aqueous dispersion of dicetyl phosphate [154], and by incorporating the styrene sulfonate moiety as counterion to glutamate amphiphile containing ammonium head group [155]. Formation of vesicles from 1,2-dimyristoylphosphatidic acid containing ammonium methacrylate or diethyldimethacrylate dimethylammonium counterion [156] constitutes the only example of ionic interaction between phospholipid and polymerizable moiety.

In general, polymerization of monomer counterions in organized bilayers has been achieved by one of the following techniques:

1. UV irradiation [151–153,155,156] (methacrylate ions)
2. Addition of protic acid after the salt formation [154] of vinylpyridine with anionic surfactant
3. Addition of photo initiator (EtO–C(S)–S–CH_2CO–(bis undecyl)–L–glutamate) [155] (styrene sulfonate)

Water-soluble radical initiators initiate [156] polymerization of cholinemethacrylate counterion efficiently, while in the case of methacrylate counterion, the same initiators are completely ineffective 53]. Systems have been used to improve the stability of vesicles and exert control over leak rate of the encapsulant by controlling the degree of polymerization.

The mechanical stability of polymerized vesicles has been demonstrated by Ohno et al. [89]. This group prepared giant vesicles by mixing polymerizable lipid, 1-(9-(*p*-vinylbenzoyl)-nonanoyl)-2-O-octadecyl-*rac*-glycero-3-phosphocholine with cholesterol. The vesicle mixture being photopolymerized, freeze-dried, and washed with chloroform to produce mechanically tough, skeletonized frame structures of the polymerized vesicles. This system might be useful for the construction of controlled release carriers where a variety of nonpolymerizable surfactants could be used in the fine-tuning of the release mechanism.

IV. ROLE OF POLYMERIZABLE PHOSPHOLIPIDS IN THE ADVANCEMENT OF TECHNOLOGY

Studies on conventional vesicles under carefully controlled laboratory conditions have demonstrated the technological usefulness of these closed lipid bilayer structures. The major drawback of conventional vesicles has been their lack of stability against chemical, biochemical, and mechanical stress. As we have described earlier, the major reason behind the development of polymerizable lipids was to produce stabilized vesicles for practical applications. Reviewing the increasing number of articles published each year, it is apparent that polymerized vesicles not only have proven worthy of the efforts but also have extended the list of applications. Now, their usefulness may apply from the biomedical field [54–56,75,91] to the microelectronic [136,157]. In the following paragraphs we shall discuss the role of polymerizable phospholipids in conventional and novel technological applications. Novel applications have been mainly due to the polymerizable lipids, e.g., formation of tubules, corking and decorking of vesicles, and mixed lipid systems consisting of monomer and polymer domains.

A. Stabilized Vesicles

Incorporation of polymerizable moieties in a phospholipid or any other surfactant (anionic or cationic) molecule may result in the stabilization of the self-assembled assemblies, vesicles, etc., while retaining the morphological features intact. However, the criteria for determining the stability of polymerized vesicles varies from one research group to another. The stabilized vesicles should demonstrate at least the following properties:

1. Prolonged shelf-life
2. Physical and mechanical stability
3. Chemical resistance to biological or other nearby fluids

In a number of research articles on polymerized vesicles, decreased permeability and resistance to precipitation [27,42,58,59,61,94,99,101], stability against ethanol and surfactants [27,42,45,46,48,58,59,96], mechanical stability during freeze-drying and redispersion steps [48,50], lipase resistance, biocompatibility [47,48,56,158], etc., have been reported. Obviously, the improved stability and its implications have given rise to numerous patents issued on the subjects, such as preparation of stabilized vesicles [6,159], mono and multilayers [160–162], utilization of polymerizable phospholipids in antigen determination [164], enzyme immobilization [18], preparation of biocompatible surfaces [160,161,165], and contact lenses [159,166].

This section will review the results pertaining to the stability of the polymerized vesicles irrespective of the nature of the polymerizable moieties, their placement in the molecule, and the polymerization technique.

Vesicles from diacetylenic phospholipids ($DC_{m,n}PC$, and 1-acyl, 2-alkadiynoyl PC) have been initially studied by Chapman, Ringsdorf, and O'Brien. Chapman's criterion for the stability of vesicles has been the absence of precipitated vesicles in the polymerized dispersion. Polymerized vesicles containing polystyryl backbone have been reported to have the same particle size for a month [86,87]. Diacetylenic PC (**38,48,49**) (see App. 1) dispersions, however, from both polymerized and unpolymerized vesicles, have precipitated within 1 h at 50°C, but corresponding polymerized vesicles have an improved stability at lower temperatures. For example, at 4°C the precipitation first became apparent after 7 days [99] and later after a month [101]. Polymerized vesicles from diacetylenic lipids were less thrombogenic than normal vesicles and were suggested to be useful for coating the blood containing artificial devices [158]. Similar stability and compatibility have also been reported for polymerized methacrylate-derived lipid assemblies [47,48].

Another approach for testing vesicle stability was to study the resistance to known perturbants such as ethanol [27,45,46,48,96] or detergents [42,48,58,59,96]. In a typical experiment, both polymerized and polymerizable assemblies are diluted with ethanol and their turbidities are measured. The turbidity of polymerized vesicles remains unchanged up to 25% ethanol concentration. The spherical structure remains unchanged even after the precipitation by salt (KC1), which osmotically shocks the vesicles [27]. Vesicles from diacetylenic lipids became slightly turbid upon polymerization but remain in stable suspension [96]. The stability of these vesicles is dependent on the size of the vesicles, however. Large vesicles (~10 μm in diameter) are much less stable than small ones (~1 μm). Thus vesicles from $DC_{8,7}PC$ are stable before and after polymerization, while vesicles prepared from $DC_{8,12}PC$ decompose rapidly.

Mechanical stability of the vesicles has been tested by following techniques: (1) polymerized vesicles are subjected to a heating and cooling cycle; (2) detergent is added to release the vesicle content, which is usually a fluorescent dye [42] or a radioactive marker [48]; stability against detergent may also be measured by turbidimetry; and (3) mild sonication [51]. As the field matures, the conditions for testing vesicle stability become harsher. Thus vesicles containing cross-linked disulfide polymer backbone were tested for their stability in the presence of SDS and heating up to 60°C [62].

B. Encapsulation and Controlled Release

Polymerizable lipids provide means for developing controlled-release systems in the area of microencapsulation, drug delivery, catalysts, sensors, micro repair of surfaces (biological and nonbiological), etc. Perhaps this is the reason why so many reports and patents on polymerizable phospholipids deal with entrapment and permeability loss from the corresponding lipid vesicles. Stability is directly associated with the polymerization efficiency irrespective of the type of monomer. It should be emphasized here that stability can occur from physical interactions as well as chemical cross-linking. Thus extensive polymerization is not necessarily a technological necessity. The nature of the monomer and its placement in the lipid molecule also have an impact on the properties of polymerized vesicles produced. The following paragraphs summarize the role of polymerized vesicles in the area of controlled release.

The entrapment and permeability rates in the vesicles from phospholipids containing methacrylate monomer at *sn*-1 (**2**), *sn*-2, 2 (**3**), or both positions (**1**) have been studied [45,46,51]. Vesicles from lipid **3** retained 53% of the encapsulant after 8 h as compared to 16% retention in the case of unpolymerized dispersion. The vesicles from **2** retained 50 and 54% marker after 8 h before and after polymerization, respectively. An 80:20 copolymer of either lipids with **1** has retained of most of the marker even after 24 h. The vesicles from **1** are tested for biomedical applications [48] by subjecting the entrapped polymerized vesicles to lyophilization, redispersion, incubation, and ultrasound agitation cycle. Brief ultrasound agitation caused 29% DPPC to release from polymerized vesicles as compared to 85% release from the nonpolymerized vesicles. Vesicles from lipid **1** have modest effect on platelet aggregation and may provide innocuous nonthrombogenic coatings for the biomedical devices in contact of blood [47].

Polymer liposomes inhibited antifungal growth in vitro but did not protect mammalian cells against amphotericin B induced toxicity. The lipid provided a stable, reproducible, pharmaceutically acceptable liposome preparation, however [54,55].

Stabilized vesicles for biomedical applications have been constructed from biodegradable polymers starting with thiol-containing phosphatidylcholines [58,59]. The thiol moiety, placed in the acyl chain near the head group or at the chain terminus [59], was polymerized either by UV irradiation at 254 nm or by treating with H_2O_2 at 40°C. Polymerized vesicles retained higher percentages of marker than nonpolymerized vesicles, but the amount released after 2 and 4 h was the same in both cases. The results indicate that polymerized vesicles have a larger captured volume.

Highly cross-linked vesicles were prepared from dilopoyl PC (**13**) [62,63] by ring-opening polymerization of the 1,2 dithiolane (lipoic) moiety; this was achieved by raising the pH of the medium to 8.5 and adding 10 mol% dithiothreitol (DTT). The polymerized vesicles were reported to be very leaky to glucose but able to retain sucrose.

In 1-palmitoyl 2-(12-lipoyloxy)dodecanoyl-*sn*-glycero-3-phosphocholine (**14**) vesicles, the palmitoyl group significantly increased the lipid packing density [63] and decreased the permeability to markers already in the nonpolymerized vesicles. The mechanism of decreased permeability in the case of mixed PC (**13** and **14**) is attributed to the polymer boundaries produced in membrane, which serve as primary avenues for the release of marker. Thus the permeability rate decreases as the ratio of **13** to **14** increases by decreasing the effective number of lipid domains and consequently of polymer boundaries.

Biocompatibility studies on **13** revealed that this lipid is suitable for in vivo use in the contexts of drug delivery systems or biomaterial development [56]. Vesicles from this lipid did not affect fibrin clot formation or platelet aggregation in either polymerized or nonpolymerized form.

In an interesting approach, the hydrophilic core of a vesicle has been polymerized to yield microparticles encapsulated by vesicles [66].

The self-quenching fluorescent dye carboxyfluorescein (CF) [96] makes a nonradioactive but sensitive alternative marker to study the leak kinetics of lipid vesicles. The vesicles from di-2,4-octadecadienoyl pcho (**15**) released all the trapped CF in ~50 h at 20°C (above T_c), polymerization making them completely tight; even the addition of 30% ethanol caused only 15% release. Diacetylenic lipid, $DC_{8,12}PC$, retains CF in vesicles for a longer period. However, upon cooling to ~0°C, necessary for topochemical polymerization, all the trapped markers were released in seconds. This was probably due to the transformation of vesicles to tubules [29].

Leaver et al. [100] have studied the glycerol permeability in the vesicles derived from diacetylenic lipid, $DC_{8,6;8,9 \text{ and } 8 11}PC$ (**41,42,43**) before and after polymerization. They have obtained anomalous results that are as yet not understood. Interesting results have been reported by Sonnenschein and Weiss [118] during their fluorescence quenching studies to probe the microenvironment of vesicles from diacetylenic lipid, $DC_{8,9}PC$. In their studies, the fluorescent probes embedded in the lipid bilayers became more available to water molecules below the lipid chain melting transition temperature than above.

O'Brien et al [69,70,167] have reported an inefficient polymerization in diacetylenic lipid bilayers and efficient polymerization of the vesicles from di-2,4-hexadecadienoyl PC, **16**, (95+%). About 20% glucose was released rapidly from the latter type of vesicles, but it took a week for the rest of the glucose to leak. The results were attributed to two populations of lipids in polymerized vesicles.

Preparation of unilamellar vesicles (size ranging 35–85 nm) and their polymerization have been reported for 1,3-di-2,4 (dodeca-, tetradeca-, hexadeca-, and octdecadienoyl)-*rac*-glycero-2-phosphocholine (**18**) [75]. The shorter chain (C = 12) PC gave the largest size (85 nm) vesicles.

C. Oxygen Carriers

Diacetylenic phospholipids, $DC_{8,9}$ (**42**), $_{8,11}$ (**43**) or 1-acyl, 2-$_{8,9 \text{ or } 8,11}$ (**48,49**), have been used in the construction of vesicles filled with human hemoglobin (Hb) to make hemosomes [168,169]. The UV polymerized hemosomes remained monodispersed up to 8 h at room temperature; for prolonged storage they were kept at 4°C, however. Hemosomes were capable of reversibly binding the dissolved gas; the polymerized membrane, being gas-permeable, could be suitable for the development of surrogate erythrocytes.

The second approach has been reported by Tsuchida et al. [72,76,78,81,88,90,91], where an oxygen-supplying medium, 5,10,15,20-tetra ($\alpha,\alpha,\alpha,\alpha$-O-(2',2'-dimethyl-20'-(2''-trimethylammonioethyl) phosphonatoxyeicosanamido) phenyl) porphinatoiron(II), has been embedded in bilayers of polymerizable vesicles to bind oxygen reversibly under physiological conditions. The polymerizable lipids in these studies were dienoyl and styryl PC.

The technique of embedding lipid-heme(s) in polymerized bilayers for making artificial blood has been patented [14]. In addition to this technique, patents have also been issued on the techniques of hemoglobin encapsulation in polymerized vesicles constructed from dienate [13] and eleostearate [11] PC. The latter example utilizes the technique of redispersion of polymerized and freeze-dried vesicles in saline containing hemoglobin to prepare Hb-containing stable vesicles. The ultimate success of these approaches for a replacement for natural oxygen-carrying systems is yet to be seen.

D. Surface Modification, Enzyme Immobilization, and Biosensors

Polymerized lipids have been successfully used in schemes leading to covalent enzyme immobilization [50,77,82] or incorporation into the lipid bilayers [83,133–137]. The enzymes normally are covalently attached to a surface via an amino or carboxylic group. One approach [50] is to polymerize a head group modified methacrylate phosphatidylcholine (**5**) in vesicle form. Such polymerized vesicles are coupled with the amino group of chymotrypsin. The immobilized enzyme retained 15% of its original activity. Similarly, phospholipase A_2 [82], trypsin, and soybean trypsin inhibitor (77) have been immobilized on vesicle surfaces containing polybutadiene backbone.

Polymerized matrices provide a chemically and mechanically stable environment to membrane proteins. Bacteriorhodopsin is a widely studied protein for reincorporation in polymerizable lipids. This protein is inserted into lipid bilayers by cosonication followed by polymerization [135,170]. A net inward proton flow in the presence of visible light proved the stability and vectorial orientation of the protein in lipid bilayers. The procedure was improved by inserting rhodopsin directly into the prepolymerized layers from a mixture of **15** and DOPC [83], and $DC_{8,9}PC$ (**42**) and DNPC [133]. In the latter case, the presence of short chain lipids enhanced the polymerization efficiency of diacetylenes in bilayers. In order to combine the mechanical stability of polymerizable lipids and the flexibility of conventional lipids, asymmetric monolayers were deposited on glass patch electrodes to incorporate ion channels in the membranes [134,136,137]. Thus, alamethicin has been incorporated in the asolectin/$Dc_{8,9}PC$ bilayer [137], or asolectin-cholesterol/$DC_{8,9}PC$ bilayers, [136] and acetylcholine receptor in the $DC_{8,9}PC$/DOPE:DOPS:DOPC:Cholesterol bilayers [134]. The techniques for enzyme immobilization on porous vesicles prepared from dienoate PC [18] and enzyme encapsulation in vesicles from styryl PC [12] have been patented for application.

E. Molecular Recognition

Recognition phenomena have been mimicked by preparing model cells prepared from polymerizable phospholipids. Recognition and association either occurred in monolayers on the air-water interface [20] or else in the inhibition of enzyme activity [77] in vesicles and lysis of the target cells [82]. In the former case, polymerized vesicles were prepared from **15** and 1-acyl$_{C\text{-}16\text{-}18}$,2-octadecyl-2,4-dienoyl PE (**17b**), the trypsin and soybean trypsin inhibitor (STI) having being immobilized separately. Addition of STI vesicles to trypsin-bound vesicles inhibited the hydrolysis of *N*-tosyl-L-lysine methyl ester hydrochloride. The second scheme involves two vesicle populations, one consisting of **15** and **17b**, to which biotin and phospholipase A_2 were covalently attached, the other having been made from **15**, **17b**, and DMPC; the surface was decorated with avidin [82]. The latter vesicle population was filled with carboxy fluorescein. Upon mixing the two vesicle populations, a rapid increase in fluorescence was observed, signifying recognition, association, and lysis of the target cells.

F. Tubules and Technological Opportunities

Tubules are open ended, hollow cylindrical microstructures with an average diameter of 0.5 μm. The length of the tubules is process-dependent [121,122,124] and may be as long as 600 μm. The morphology of the tubules itself, control over the dimensions, and ease of fabrication of these tubules have led to a number of applications [19]. Earlier studies did not consider open-ended tubules for encapsulation or controlled-release purposes. Studies on tubules prepared in the presence of fluorescent lipophiles concluded that the tubule structures were extremely rigid and tightly packed [117]. Recently it was demonstrated, however, that material can be encapsulated in these microstructures [132]. Successful uniform coating of the tubules with a variety of metals [157], moreover, has already led to their electronic applications. Successful encapsulation into these metal structures has made the tubules attractive candidates for controlled-release applications.

Price et al. [139], for example, have utilized copper-coated tubules to achieve sustained release for over one year in the ocean environment. A slurry of tetracycline and polymer binder was incorporated into the 0.4 μm inner core of the copper-coated tubules by capillary action. The binder was then cured and the tubules were washed. Incorporation of such tubules into a paint is useful for antifouling applications.

The release mechanism is not well understood. It is thought to derive from a combination of capillarity effects, surface tension effects, diffusivity of the paint matrix, and the corrosion parameters of the copper coating. This approach offers a number of potential applications where prolonged release in adverse conditions is required.

Tubules can be oriented in an external magnetic field [171]. Tubule-based composites can exhibit dielectric constants well over 30 at 4% weight loading in the 10 GHz regime [172]. Gold-coated tubules also have been used for high-power microwave cathode applications [173].

G. Mixed Lipid Systems and Technical Applications

Mixed lipid systems, which we shall describe here, are mostly composed of polymerizable and nonpolymerizable phospholipids. We shall discuss the physical properties of these systems and their technical implications. Three topics will be highlighted; *the formation of rippled phases, the tuning of marker release, and the enhancement in polymerization.*

1. Phase Separated Lipid Bilayers

Bueschl et al. [94] have reported that polymerizable diacetylenic lipid, $DC_{8,12}PC$, mixed with a nonpolymerizable lipid, DLPC, can form monolayers and vesicles containing "islands" of nonpolymerizable lipids. The idea of using these islands for the incorporation of proteins for the selective opening of the polymerized vesicles was proven by selectively hydrolyzing the polymerized islands on a monolayer by the action of phospholipase A_2 ("corkscrews"). Mixed lipids (dimethyl-bis[2-octadeca-2,4-dienoyloxyethyl] ammonium-bromide and DMPC) have produced ripples on the surface of vesicles due to the lateral diffusion and phase separation as a result of polymerization [174]. Similar observations have been made for the mixtures of DPPC and **15** as evidenced by the appearance of two transition temperatures (18 and 41°C for **15** and DPPC, respectively) [175].

Hybrid vesicles, prepared by mixing nonpolymerizable lipids consisting of noncholine head groups, might be useful for the construction of controlled-release systems. A mixture of a diacetylenic lipid, $DC_{m,n}PC$(m = 8; n = 7, 8, and 9) and a phospholipid containing methacrylate moiety on the head group was shown to be stable and flexible due to the mosaiclike arrangement of the cross-linked and monomeric domains [42]. The latter could serve as matrices for the incorporation of ligands that could be recognized by target tissues or participate in a release mechanism triggered by pH, temperature, or ions [42,68,85,176]. Thus **15** was mixed with egg phosphatidylethanolamine (PE), dioleoyl PE or DOPC, and photopolymerized [85]. The polymerization of **15** in vesicles caused PE to phase separate, triggering its transformation into nonlamellar phase resulting in an increase in membrane permeability. Similarly, stimuli-responsive release from polymerized vesicles was achieved by mixing **15** with DMPC or DPPC, DPPE, and bovine brain phosphatidylserine (PS) [176]. In the case of **15** and DPPC/DMPC vesicles, the content was rapidly released rate at the chain melting transition temperatures (*Tm*) of the nonpolymerizable lipids, i.e., at 37 and 19°C, respectively. In the vesicles from a mixture of **15** with DPPE, marker release was attained by changing the pH from 11 to 7.5. Vesicles with PS gave larger release upon addition of Ca^{2+} ions. In a different approach, polymerizable lipid bilayers from *rac*-DODPC (**15**) or *rac*-1-stearoyl 2-(2,4-octadecadienoyl PC (**20**) were corked in the pores of nylon capsules [68]. Upon polymerization, the lipid cork became physically stable, but the porous defects decreased the barrier effect and NaCl permeation increased.

2. Short Chain Lipid Spacer

Short chain phospholipids with acyl chain lengths similar to the *m* segment of diacetylenic PC have been reported to improve the rate and efficiency of polymerization [108,177,178]. This mixed lipid system differs from others reported in the literature in its requirement for the specific chain length of the nondiacetylenic lipid. Most probably this is because the short chain lipid must fit well with the diacetylenic lipid in a manner suitable for topotactic polymerization. Indeed, preliminary results suggest that the mixture of these two lipids does not phase separate, as shown by DSC [108].

For polymerization, matching the acyl chain lengths is crucial. Lipids with an acyl chain shorter or longer than *m* segments did not enhance polymerization [177].

This system appears to have technological importance; one application has been explored by Ahl et al. [133], who have shown that after polymerization the nonpolymerizable spacer lipid can be successfully replaced by membrane protein bacteriorhodopsin.

V. PHYSICO-CHEMICAL PROPERTIES

A. Polymerization and Polymer Characterization

Polymerization of monomer lipids in bilayer microstructures has not been studied extensively yet. The following is a brief review of the published data. However, no reports on nonphospholipid detergents are included.

Methacrylate PC. Photopolymerization of the phospholipid **2** in vesicles has been found to be slower than that of lipid **3**, which has a methacrylate moiety in the acyl chain at the *sn*-2 position [45]. There is no apparent reason reported for this behavior. The course of polymerization has been routinely monitored by both TLC and NMR. NMR revealed the disappearance of the vinyl methyl protons caused by polymerization. Typically, one hour of UV irradiation is needed to accomplish complete polymerization [45,46]. There is no report available on the degree of polymerization of methacrylate PC. However, the number-average degree of polymerization has been reported for a methacrylate ammonium surfactant: 600 for the radical-catalyzed polymer and 135 for UV polymerized vesicles [179].

Methacrylate moieties placed on the head group of phospholipids or glycerides have been extensively studied in monolayers [39,41,44]. Polymerization of these lipids in vesicles has been achieved both by UV irradiation [42] and by heating in the presence of the radical initiator 2,2'-azoisobutyronitrile, AIBN, at 60°C for 20 h. Monolayers have been polymerized by 254 nm UV irradiation for 1 h.

Polymers in these systems have not been well characterized. There are two reports in the literature where the methyl group on choline has been replaced with the polymerizable moieties acrylate [148] and methacrylate [45]. The former lipid produced a photopolymer of 98,000 as determined by gel permeation chromatography, using chloroform as eluant and polystyrene as standard. In the latter case, complete polymerization was observed by TLC, but no polymer analysis was made.

Thiol and disulfide PC. The chemistry of the sulfhydryl or thiol group is simple and straightforward. Upon oxidation thiol dimerizes to disulfide. In two chain amphiphiles, such as phosphatidylcholine, the dimerization of thiols leads to a polymer whose highest molecular weight is dependent on the vesicle size.

Polymerized vesicles from a lipid equipped with thiol moiety, either at the terminus or near the head group, have been prepared by exposing lipid dispersions to UV light or by heating with H_2O_2 at 40°C [51,58,59]. Hydrogen peroxide oxidation has been reported to be pH-sensitive, the thiol reacting at pH 7.0 and pH 8.5, being 55% and 95%, respectively [58]. Depolymerization was achieved by reacting the polymerized vesicles with tri-*n*-butylphosphine.

Polymerization was monitored by TLC. Based on the thiol content in lipid vesicles from **9**, **10**, and **11**, the extent of the polymerization was reported to be 17, 25, and 20 lipids per polymer chain, respectively [59]. Lipid **13** has been polymerized at pH 8.5 by treating the vesicle dispersions with DTT at 27°C for 4 h [61–63]. Polymerization was followed by TLC and also by UV spectroscopy, which is based on absorption at 333 nm characteristic of the five-membered ring cyclic disulfide. In the absence of DTT the dispersion polymerizes at 23°C in 72 h and at 50°C in 6 h. Extensively cross-linked polymer is insoluble in chloroform or chloroform/methanol (1:1).

Styryl PC. Styryl moieties were polymerized by ultraviolet irradiation at 50°C for 1 to 4 h. The course of polymerization was monitored by measuring UV absorption at 265 nm (due to the vinyl group), and by ^{13}C NMR spectra [88,90,91]. A complete disappearance of the 265 nm peak and signals at δ 116.8 and 136.1 ppm indicated the completion of polymerization. The styryl lipid has been reported to polymerize more rapidly in the presence of porphinato iron than in pure form [90]. The latter has been reported to function as a photosensitizer accelerating the polymerization, even though iron functions as a retardant due to its reductive nature. The sturdy nature of the polymer made from lipid **32** has been demonstrated by making channels in giant polymerized vesicles with the aid of conventional lipids [89]. Polymerization of lipid **32** indicates that this process may proceed intermolecularly via *sn*-1 chain. The resulting polymer has been analyzed after acetolysis or methanolysis of the lipids. The number-average degree of polymerization of the methanolyzed polymer has been reported to be 400 [92].

Lipids **31** and **33** do not form vesicles. Polymerization in these and other (**32** and **34**) lipids caused the 265 nm peak shift to 251 nm. [86,87].

There is only one report in the literature on the polymerization of the styryl groups in phospholipid vesicles by the free radical initiators 2,2' azobis-[2-(imidazolin-2-yl)propane]dihydrochloride(VA-44) and 2,2'-azobis(4-methoxy-2,4-dimethyl) valeronitrile(V-70) [180]. The polymerization extent is then temperature-dependent; it takes 6 h at 45°C as compared to 3.5 h at 60°C to achieve complete polymerization.

Dienoate PC. Vesicles prepared from dienoate PC have been polymerized by UV irradiation and radical initiated thermal polymerization. Photopolymerization was achieved by exposing vesicle dispersions to 254 nm UV light at 50°C for 1 to 2 h. The course of polymerization was monitored by observing the disappearance of the diene peak at 255–260 nm [39,55,69–72,75,84,85,96]. In most of the reports, the disappearance of dienoate absorption has been identified with complete polymerization. The proof that oligomers or polymers rather than dimers were created was the insolubility of freeze-dried vesicles in methylene chloride [96]. The degree of polymerization of dienoate in vesicles has been estimated by gel permeation chromatography to be 26 [81]. The polymer, obtained after acetolysis of the vesicles, was dissolved in chloroform and tetrahydrofuran, and the solution was checked by GPC. The average molecular weight of UV polymerized molecules in vesicles was around 5000 [74].

Radical initiated polymerization of the 2,4-dienoate groups in phospholipid has been studied in detail by Ohno et al. [73,74,80] and Takeoka et al. [181]. In their studies, the unequivalent chemical environment faced by the diene group in phospholipid vesicles has been demonstrated. When the polymerization in dienoate PC, **15**, was initiated either by lipophilic radical initiators, AIBN, or the hydrophilic initiator azobis(2-amidinopropane) dihydrochloride (AAPD), about 50% conversion was achieved; this did not change upon increasing the reaction time [73,74], but chain fluidization was advantageous [80]. This effect was attributed to the fact that the diene group on the 2-acyl chain is exposed to the aqueous phase, while diene on the *sn*-1 chain does not come in contact with the aqueous environment. AAPD has been reported to initiate polymerization of vesicles prepolymerized with AIBN, but simultaneous initiation caused higher polymer conversion.

The polymer from either AIBN or AAPD is partially soluble in ethanol, methanol, or chloroform; polymer obtained from both AIBN and AAPD radical initiation, however, is insoluble in these solvents [74].

Potassium peroxodisulfate has been found to elicit polymerization of lipids in vesicles prepolymerized by either AIBN or AAPD. This effect has been attributed to the smaller size of OH radicals which, presumably, can approach and attack the diene group in the 1-acyl chain.

Diacetylenic PC. Solid state polymerization of diacetylenes initiated by the absorption of ultraviolet light has been well studied in nonphospholipid systems [93,94]. The reaction is topotactic, its efficiency thus depending on the correct alignment of the monomers. Consequently, the polymerization is particularly efficient in the gel state, monolayers, multilayers, and vesicles. Diacetylene stabilizes the phospholipid vesicles without appreciably affecting their size [27,28]. After polymerization, the colorless dispersion turns red, due to the production of alternating single, double, and triple bonds [97,98]. The production of colored dispersion also produces a well-defined visible spectrum exhibiting absorption maxima near 485, 525, and 625 nm. This is the reason why most of the polymer characterizations reported in the literature are based on the interpretation of the visible part of spectrum [27,28,42,70,96,97,168,169,182–184].

Polymerization is usually affected by the method of vesicle preparation, temperature, and the time and intensity of irradiation. It is found that highly colored polymers are formed after about 50% cross-linking.

Vesicles prepared from 1-stearoyl,2-$C_{8,9 \text{ or } 8,13}$PC (**49**) or 1-palmitoyl,2-$C_{8,9 \text{ or } 8,13}$PC do not polymerize well, as observed by recording absorbance in the visible range, but mixed phospholipids made from lyso egg PC polymerize efficiently [99]. Polymerized vesicles from diacetylenic ethanolamine and phosphatidic acid [96] and phosphatidylcholines show thermochromism [26,182].

Sonication produces smaller vesicles whose photoreactivity is low [70,96]. Probably, lipid chain order is lost as the size of the vesicles decreases. In diacetylenic lipids, polymerization is incomplete, and detailed molecular weight determinations remain to be done. Addition of phospholipids containing a chain length equal to the upper segment above the diacetylene in the phospholipid, by some unexplained mechanism, improves polymerization, however [108,177].

In an interesting experiment, diacetylenic lipids were incorporated into membrane phospholipids by growing a bacterium in diacetylenic substrate. Polymerization of diacetylene in the *Acholeplasma laidlawaii* cells produced red polymer, while the cultures of *Bacillus cereus* yielded a metallic blue polymer [169].

B. Thermotropic Phase Behavior of Lipids in Vesicles

Temperature and pressure affect the structure and phase behavior of lipid membranes. Phase transition temperatures of polymerizable lipids have been found to be lipid-dependent [185]. Lipids **9**,**10**, containing a thiol group [59] at the chain terminus, did not exhibit any phase transition, despite the presence of longer segments with 11 and 16 carbon atoms per chain in their acyl moiety. Lipid **11**, however, which is equipped with an -SH group next to the carboxylate with the same number of carbon atoms, exhibits a transition at 22°C. This remains unchanged even after polymerization.

Similar results have been observed for lipids with methacryloyl moiety in the acyl region [45,46]. The transition temperature of **3** was reported to be 11°C, broadened upon polymerization [8–10°C]. When the methacrylate is incorporated in the head group region, the phase transition temperatures increase, however, upon polymerization [40,41].

The transition temperatures in the dienoate-phospholipid dispersions could not be measured after polymerization. A technique has been introduced to measure the phase transition temperatures in the dienoate lipids as well as other lipids, based on the temperature-dependent interaction of diene groups of the lipid in vesicles [181]. The absorption maximum of the vesicles in the liquid crystalline state is 256.8 nm, which shifts to 241.8 nm in the gel state due to the interactions between the diene groups. This shift has been used for determining the phase transition temperature of partially polymerized dienoate lipids. The phase transition temperature of other lipids can be measured spectrophotometrically using the dienoate lipid (5 mol %) as an internal probe. The effect is more pronounced in lipids containing the dienoyl moiety at the chain terminus [71]. In the gel state, a hyposchromic shift from 257 nm to 242 nm, with a diminished extinction coefficient, has been attributed to the aggregation of the terminal sorbyl (dienoyl) group. Above the chain-melting phase transition temperature, the diene absorption shifts back to 257 nm with doubling of the extinction coefficient. No phase transition temperature has been reported on styryl polymerizable phospholipids.

Diacetylenic lipids have shown an interesting thermal phase behavior. The transition temperature (T_m) of diacetylenes in an aqueous dispersion broadens and then disappears as polymerization proceeds. This indicates that polymerization suppresses the fluid-to-gel transition of $DC_{8,7}PC$ [42]. Unlike isomeric unsaturated phosphocholines [186], the phase transition temperatures of isomeric diacetylenic lipids with fixed acyl chain lengths do not depend on diacetylene position [110].

C. Lipid Tubules

A large number of tubule-forming diacetylenic lipids have been synthesized and their physical properties characterized [52,110,111,140,177,182,183] in an attempt to understand the molecular basis of tubule formation. Raman, FT-IR, microscopic, x-ray, and molecular modelling techniques have been used to study tubule formation. Helices have also been studied as possible precursors for tubules [103,121]. Formation of right handed helices from L-$DC_{8,9}PC$ [115,122] suggests that chirality [118] may be important in the formation of tubules. But other factors may also be involved, as seen from the formation of both left- and right-handed helical structures and tubules from the DL-$DC_{8,9}PC$ [113] and by formation of tubule-like microstructures from an achiral diacetylenic surfactant [112]. Continuous transfer of lipid bilayers from vesicles by a rolling-up process may be one of these factors [114,116].

Acyl chains in tubules are highly ordered in the gel state [115,129,130], but longitudinal acoustic modes (LAM) probably exist in the diacetylenic bilayers [115,126,128]. These LAMs are not typically observed in lipid samples. Diacetylenic PCs show intense LAMs in the lower segment (n) of the diacetylenic chain, indicating that the upper segment of the diacetylene is vibrationally decoupled from the lower segment (n) [128].

Rhodes et al. [131,187] and Lando et al. [188] have studied the diacetylenic PC, **42**, by x-ray diffraction and electron-diffraction techniques, respectively. While it is not possible to study oriented tubules, some pertinent conclusions can be drawn from these studies. It is clear that there is bilayer structure in the tubules and that lipids are oriented at some angle (probably about 30°) relative to the layer plane. Prost (personal communication) has speculated that tubules form via a hexatic phase. Before the sophisticated theory of Prost can be verified, high resolution x-ray studies must be performed. In particular, it is vital to ascertain the nature of the phase from which the tubules are formed.

Other Head-Groups and Tubules

To date there is no report of phospholipids other than diacetylenic lipids that form tubules as discussed above. Nonphospholipid amphiphiles derived from glutamic acid [103,105,189] or galactonamides [104], however, have produced helices and tubulelike structures. Recently, a diacetylenic phospholipid in which the choline group is replaced by hydroxyethanol has been reported to produce tubules [138,150,193]. The resultant microstructure morphologies were both pH and ion dependent. Detailed studies on this type of system may provide deeper understanding of the formation of microstructures.

APPENDIX 1: POLYMERIZABLE PHOSPHOLIPIDS

PHOSPHOLIPID STRUCTURE		COMMENTS	REFERENCES
METHACRYLATES			
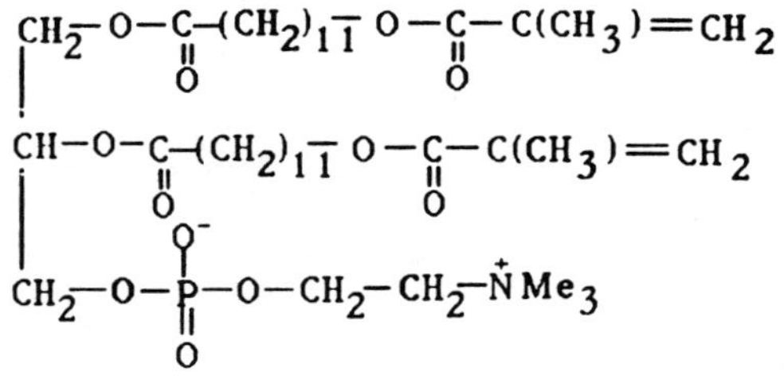 $CH_2-O-C(=O)-(CH_2)_{11}-O-C(=O)-C(CH_3)=CH_2$ $CH-O-C(=O)-(CH_2)_{11}-O-C(=O)-C(CH_3)=CH_2$ $CH_2-O-P(=O)(O^-)-O-CH_2-CH_2-\overset{+}{N}Me_3$ **1**	1.2 bis(12-(methacryloy) decanoyl)-*sn*-glycero-3-phosphocholine	UV, 254 nm, 1 h	Regen et al. [45,46]
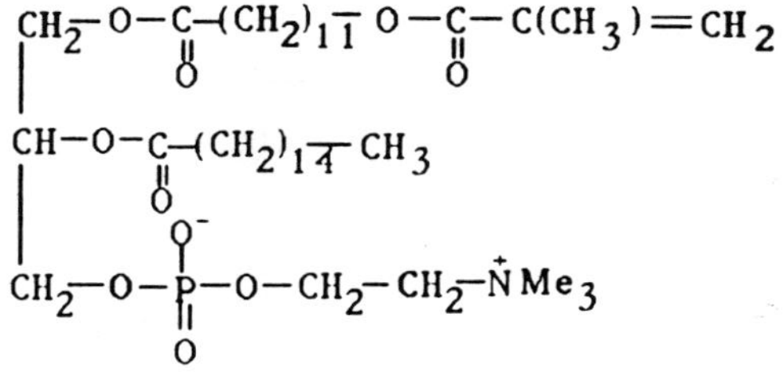 $CH_2-O-C(=O)-(CH_2)_{11}-O-C(=O)-C(CH_3)=CH_2$ $CH-O-C(=O)-(CH_2)_{14}-CH_3$ $CH_2-O-P(=O)(O^-)-O-CH_2-CH_2-\overset{+}{N}Me_3$ **2**	1-(12-methacryloyloxy) dodecanoyl)-2-palmitoyl-*sn*-glycero-3-phosphocholine	UV 254 nm, 1 h	Regen et al. [45]

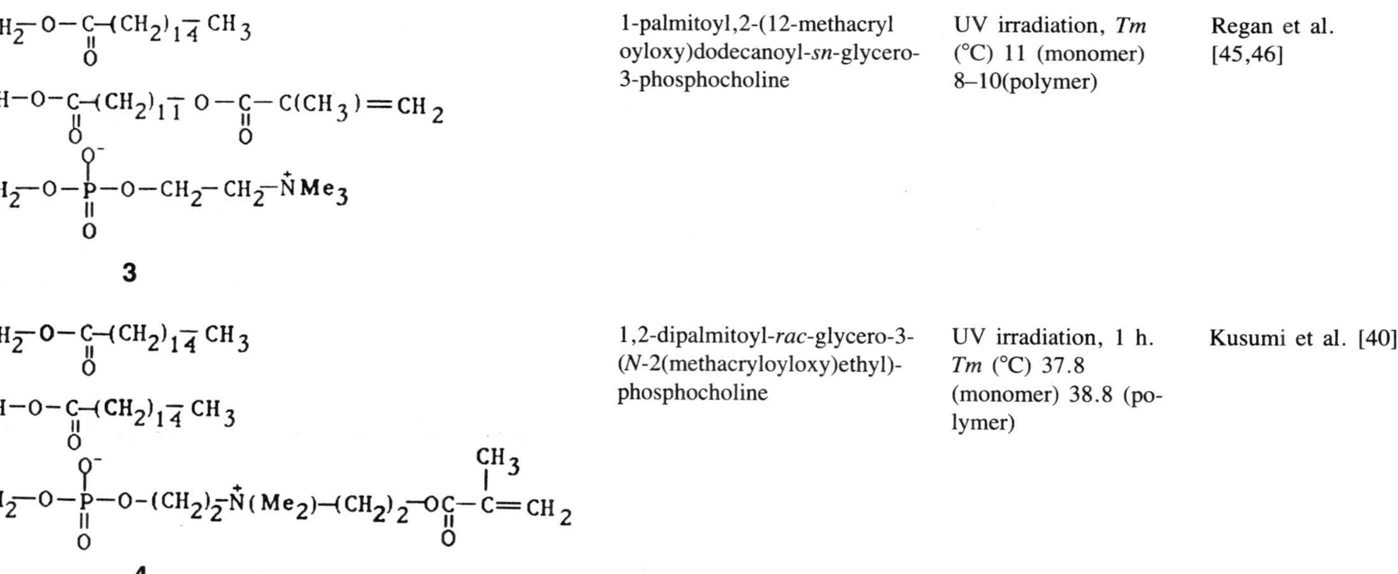

3	1-palmitoyl,2-(12-methacryl oyloxy)dodecanoyl-*sn*-glycero-3-phosphocholine	UV irradiation, *Tm* (°C) 11 (monomer) 8–10(polymer)	Regan et al. [45,46]
4	1,2-dipalmitoyl-*rac*-glycero-3-(*N*-2(methacryloyloxy)ethyl)-phosphocholine	UV irradiation, 1 h. *Tm* (°C) 37.8 (monomer) 38.8 (polymer)	Kusumi et al. [40]

APPENDIX 1: *(Continued)*

Structure	Name	Conditions	Reference
$CH_2-O-C(=O)-(CH_2)_{11}-O-C(=O)-C(CH_3)=CH_2$ $CH-O-C(=O)-(CH_2)_{11}-O-C(=O)-C(CH_3)=CH_2$ $CH_2-O-P(=O)(O^-)-O-(CH_2)_2-\overset{+}{N}(Me_2)-(CH_2)_2-CH(OEt)_2$ **5**	1,2 bis(12-(methacryloyloxy)dodecanoyl)-*rac*-glycero-3-(*N*-3.3 di-ethoxypropyl)-phosphocholine	254 nm, 1h	Regen et al. [50,51]
$CH_2-O-C(=O)-(CH_2)_{14}-CH_3$ $CH-O-C(=O)-(CH_2)_{14}-CH_3$ $CH_2-\overset{+}{N}(Me_2)-(CH_2CH_2O)_3-OC(=O)-C(CH_3)=CH_2$ I^- **6***	2,3 bis-(hexadecanoyloxy)propyl-9-methacryloyl-3,6,9-trioxanonyl dimethylammonium iodide	Photopolymerization, radical initiator UV irradiation, 1 h. *Tm* (°C) 42(monomer) 49(polymer)	Elbert et al. [41]

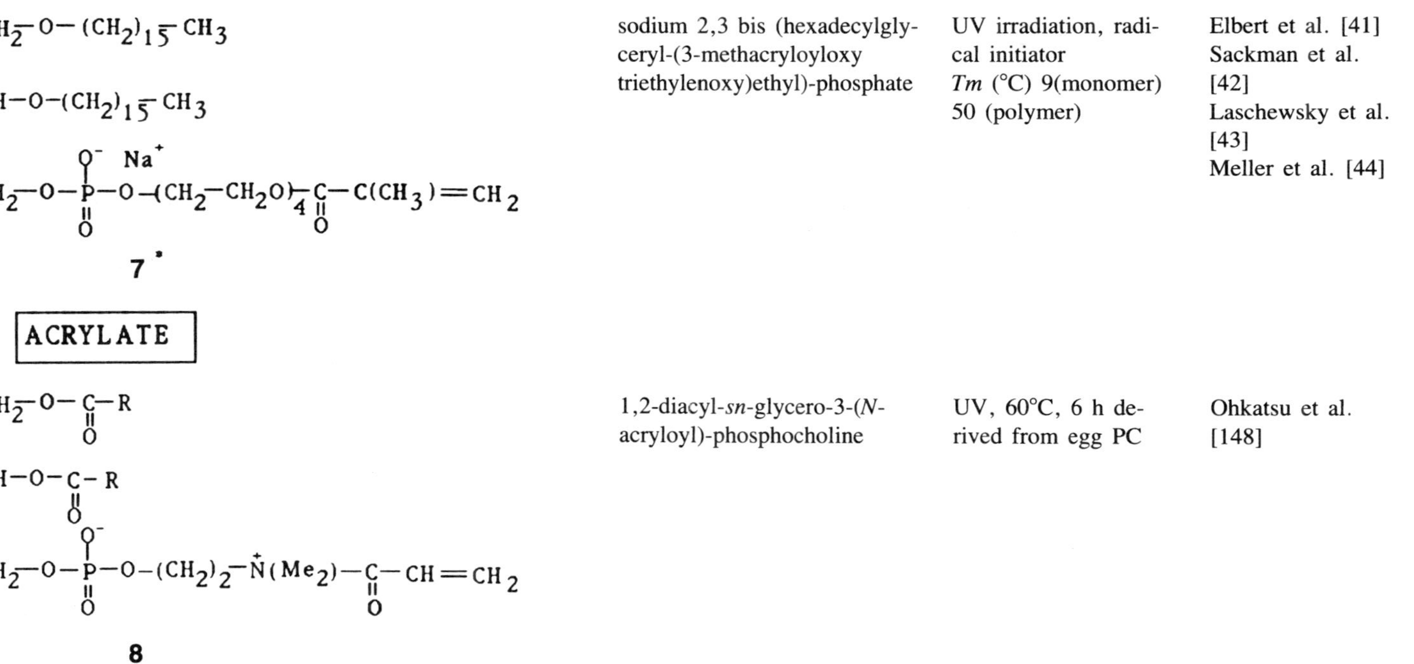

7	sodium 2,3 bis (hexadecylglyceryl-(3-methacryloyloxy triethylenoxy)ethyl)-phosphate	UV irradiation, radical initiator *Tm* (°C) 9(monomer) 50 (polymer)	Elbert et al. [41] Sackman et al. [42] Laschewsky et al. [43] Meller et al. [44]
8	1,2-diacyl-*sn*-glycero-3-(*N*-acryloyl)-phosphocholine	UV, 60°C, 6 h derived from egg PC	Ohkatsu et al. [148]

APPENDIX 1: *(Continued)*

THIOLS AND DISULFIDES

Structure	Name	Conditions	Reference
$CH_2-O-C(=O)-(CH_2)_{15}-SH$ $CH-O-C(=O)-(CH_2)_{15}-SH$ $CH_2-O-P(=O)(O^-)-O-CH_2-CH_2-\overset{+}{N}Me_3$ **9**	1,2 bis(16-mercaptohexadecanoyl)-*sn*-glycero-3-phosphocholine	30% H_2O_2, 40°C, 3h no *Tm* observed	Regen et al. [59]
$CH_2-O-C(=O)-(CH_2)_{10}-SH$ $CH-O-C(=O)-(CH_2)_{10}-SH$ $CH_2-O-P(=O)(O^-)-O-CH_2-CH_2-\overset{+}{N}Me_3$ **10**	1,2 bis(11-mercaptoundecanoyl)-*sn*-glycero-3-phosphocholine	UV 254 nm, 0.5–1 h 30% H_2O_2, 40°C, 3h no *Tm* observed	Regen et al. [58] Regen et al. [59]

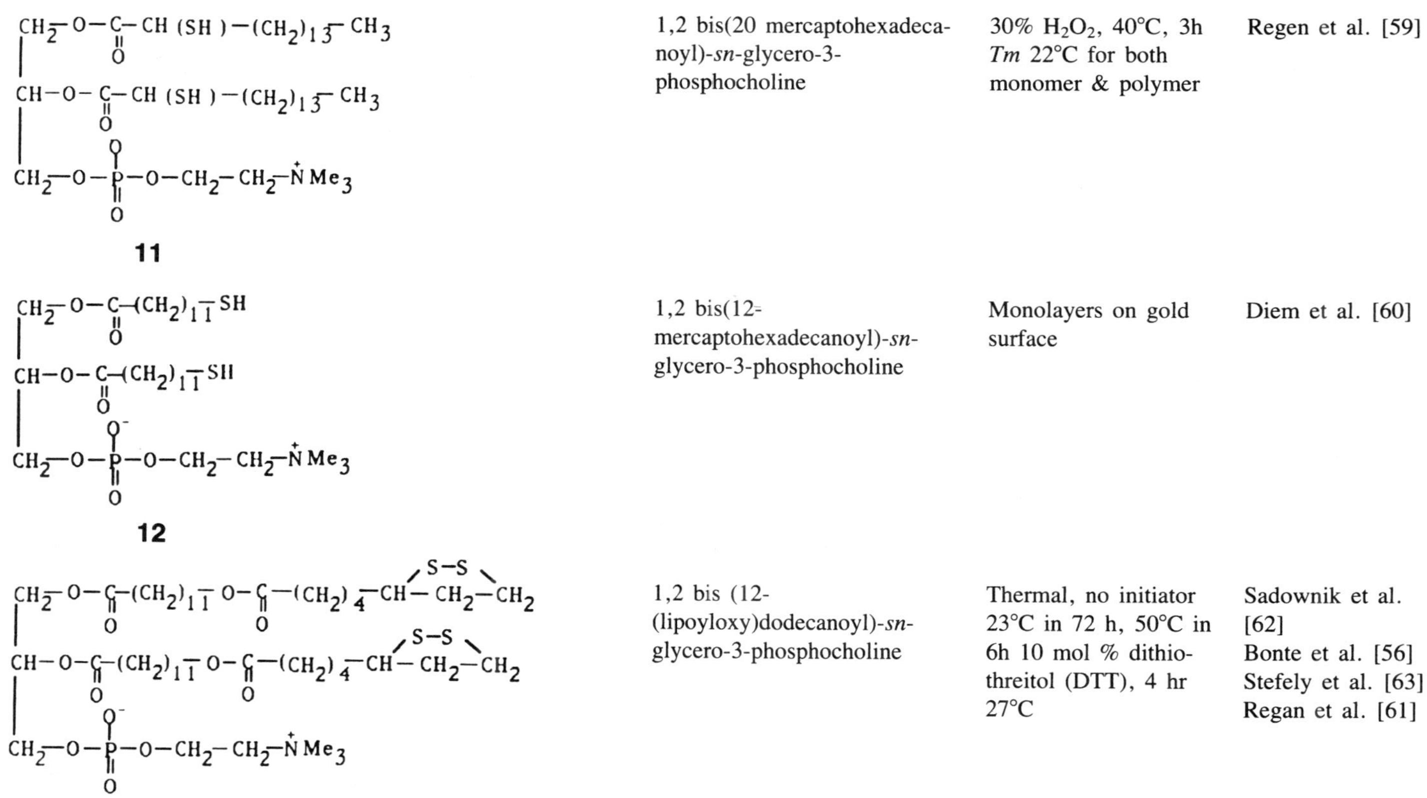

1,2 bis(20 mercaptohexadecanoyl)-*sn*-glycero-3-phosphocholine	30% H_2O_2, 40°C, 3h *Tm* 22°C for both monomer & polymer	Regen et al. [59]
1,2 bis(12-mercaptohexadecanoyl)-*sn*-glycero-3-phosphocholine	Monolayers on gold surface	Diem et al. [60]
1,2 bis (12-(lipoyloxy)dodecanoyl)-*sn*-glycero-3-phosphocholine	Thermal, no initiator 23°C in 72 h, 50°C in 6h 10 mol % dithiothreitol (DTT), 4 hr 27°C	Sadownik et al. [62] Bonte et al. [56] Stefely et al. [63] Regan et al. [61]

APPENDIX 1: *(Continued)*

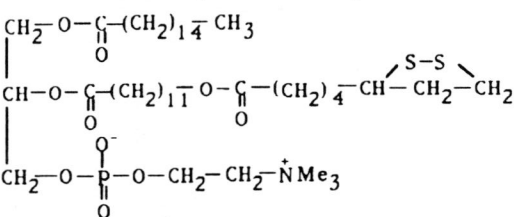

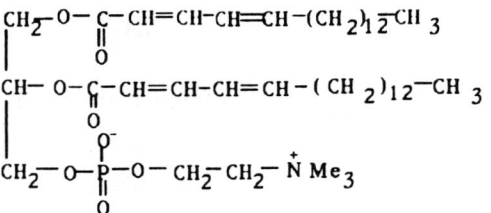

Structure	Name	Polymerization / properties	References
$CH_2-O-C(=O)-(CH_2)_{14}-CH_3$ $CH-O-C(=O)-(CH_2)_{11}-O-C(=O)-(CH_2)_4-CH(-S-S-)-CH_2-CH_2$ $CH_2-O-P(=O)(O^-)-O-CH_2-CH_2-\overset{+}{N}Me_3$ **14**	1-palmitoyl, 2-(12-(lipoyloxy)dodecca-noyl)-*sn*-glycero-3-phosphocholine	DTT	Stefely et al. [63]
DIENOATES			
$CH_2-O-C(=O)-CH=CH-CH=CH-(CH_2)_{12}-CH_3$ $CH-O-C(=O)-CH=CH-CH=CH-(CH_2)_{12}-CH_3$ $CH_2-O-P(=O)(O^-)-O-CH_2-CH_2-\overset{+}{N}Me_3$ **15**	1,2 bis (2,4-octadecadienoyl)-*sn*-glycero-3-phosphocholine	Radical initiator UV irradiation Tm = 18°C M.P. 220°C [24]	Pure enantiomer [73,74,76,77,80,82, 175,180] Racemic [14,39,65,66,72, 77,79,82,96,166]
$CH_2-O-C(=O)-CH=CH-CH=CH-(CH_2)_{10}-CH_3$ $CH-O-C(=O)-CH=CH-CH=CH-(CH_2)_{10}-CH_3$ $CH_2-O-P(=O)(O^-)-O-CH_2-CH_2-\overset{+}{N}Me_3$ **16**	1,2 bis (2,4-hexadecadienoyl)-*sn*-glycero-3-phosphocholine	Enantiomer UV Irradiation	O'brien et al. [70] Dorn et al. [60]

Structure	Name	Notes	Reference
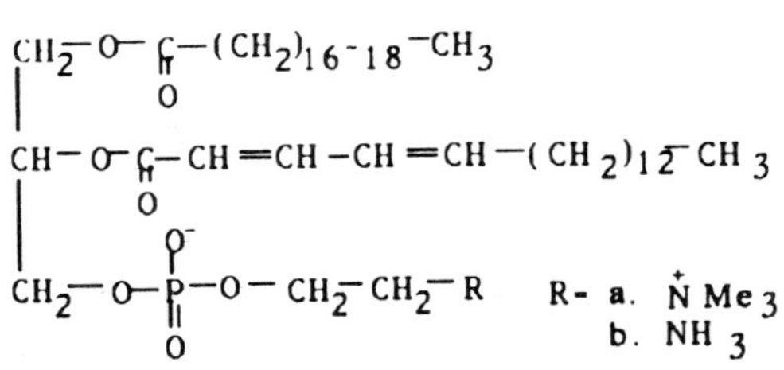 $CH_2-O-C(=O)-(CH_2)_{16-18}-CH_3$ $CH-O-C(=O)-CH=CH-CH=CH-(CH_2)_{12}-CH_3$ $CH_2-O-P(=O)(O^-)-O-CH_2-CH_2-R$ R- a. $\overset{+}{N}Me_3$ b. NH_3 **17**	1-acyl,2-(2,4-octadecadienoyl)-*sn*-glycero-3-phosphocholine a. choline, b. ethanolamine	Radical initiator, AAPD, followed by UV exposure for 10 h egg PC derived	Kitano et al. [77,82]
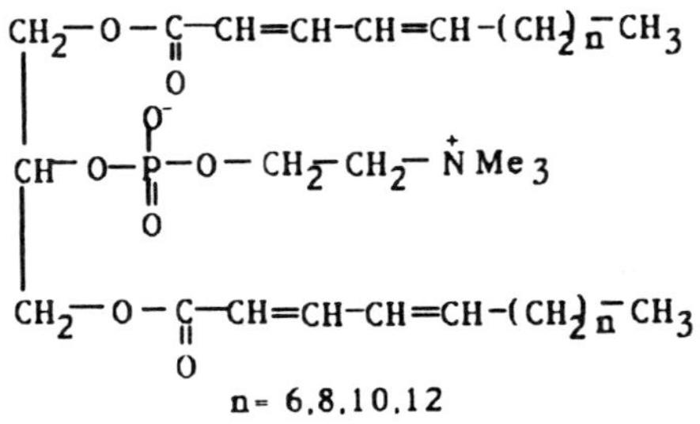 $CH_2-O-C(=O)-CH=CH-CH=CH-(CH_2)_n-CH_3$ $CH-O-P(=O)(O^-)-O-CH_2-CH_2-\overset{+}{N}Me_3$ $CH_2-O-C(=O)-CH=CH-CH=CH-(CH_2)_n-CH_3$ n= 6,8,10,12 **18**	1,3 dialkana-*trans*,*trans*-2,4-dienoyl-glycero-2-phosphocholine	*Tm* (*n*=12) 19.5°C	Hasegawa et al. [75]
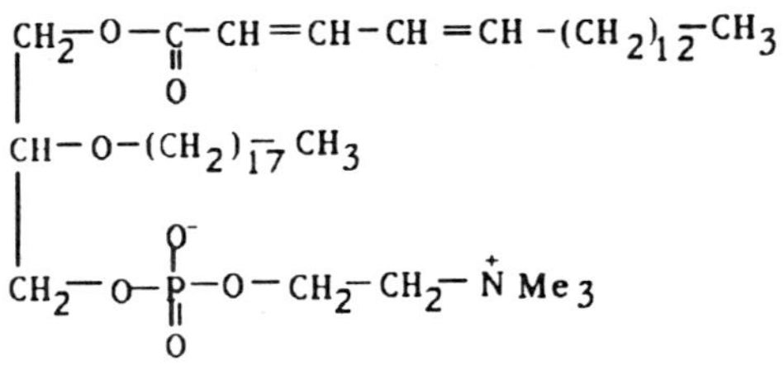 $CH_2-O-C(=O)-CH=CH-CH=CH-(CH_2)_{12}-CH_3$ $CH-O-(CH_2)_{17}CH_3$ $CH_2-O-P(=O)(O^-)-O-CH_2-CH_2-\overset{+}{N}Me_3$ **19**	1-alkana-*trans*-2,4-dienoyl,2-octadecyloxy-rac-glycero-3-phosphocholine	Racemic *Tm* 40.3°C	Hasegawa et al. [75]

APPENDIX 1: *(Continued)*

Structure	Name	Remarks	Reference
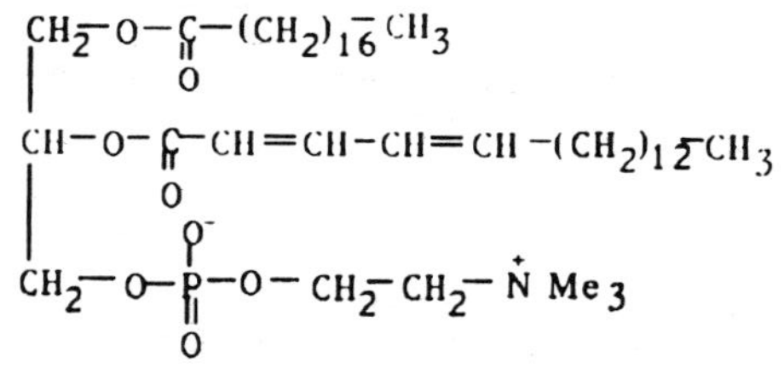 $CH_2-O-C(=O)-(CH_2)_{16}-CH_3$ $CH-O-C(=O)-CH=CH-CH=CH-(CH_2)_{12}-CH_3$ $CH_2-O-P(=O)(O^-)-O-CH_2-CH_2-\overset{+}{N}Me_3$ **20**	1-octadecanoyl, 2-(2,4-octadecadienoyl)-*rac*-glycero-3-phosphocholine	Racemic	Okahata et al. [68]
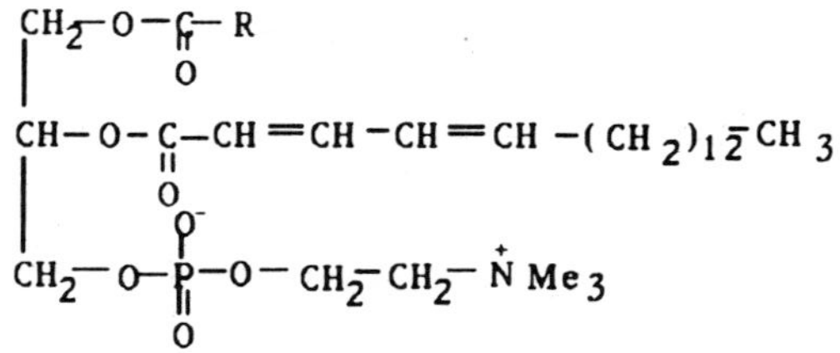 $CH_2-O-C(=O)-R$ $CH-O-C(=O)-CH=CH-CH=CH-(CH_2)_{12}-CH_3$ $CH_2-O-P(=O)(O^-)-O-CH_2-CH_2-\overset{+}{N}Me_3$ **21**	1-acyl, 2-(2,4-octadecadienoyl)-*sn*-glycero-3-phosphocholine	Egg-derived enantiomer UV irradiation	Tyminski et al. [71]
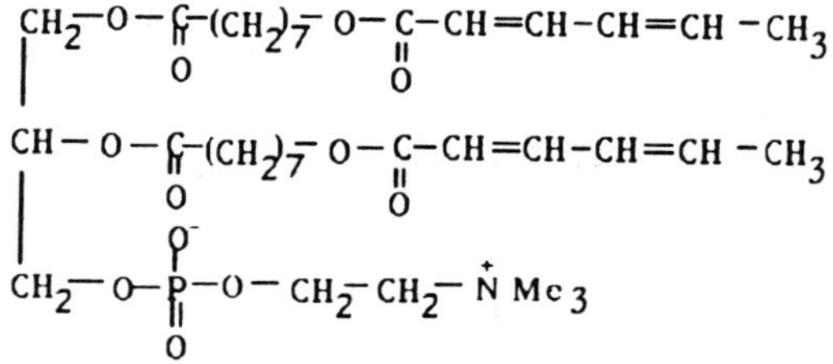 $CH_2-O-C(=O)-(CH_2)_7-O-C(=O)-CH=CH-CH=CH-CH_3$ $CH-O-C(=O)-(CH_2)_7-O-C(=O)-CH=CH-CH=CH-CH_3$ $CH_2-O-P(=O)(O^-)-O-CH_2-CH_2-\overset{+}{N}Me_3$ **22**	1,2 bis (8-(2,4-hexadienoyloxy)octadecanoyl)-*sn*-glycero-3-phosphocholine	UV Irradiation UV spectrum changes with solvent medium *Tm* 11°C	Tyminski et al. [71]

Structure	Name	Properties	Reference
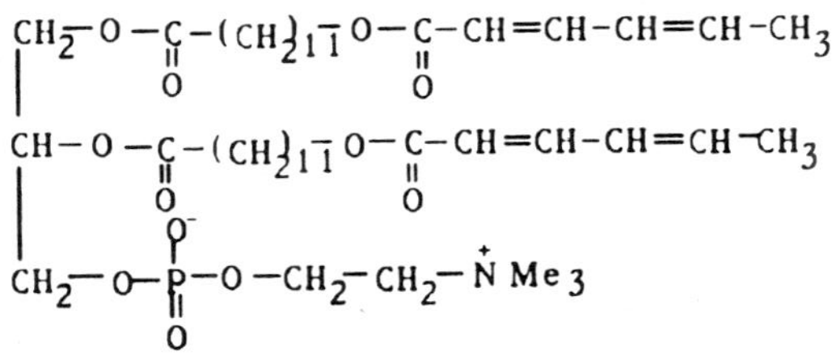 $CH_2-O-C(=O)-(CH_2)_{11}-O-C(=O)-CH=CH-CH=CH-CH_3$ $CH-O-C(=O)-(CH_2)_{11}-O-C(=O)-CH=CH-CH=CH-CH_3$ $CH_2-O-P(=O)(O^-)-O-CH_2-CH_2-\overset{+}{N}Me_3$ **23**	1,2 bis (12-(2,4-hexadienoyloxy)dodecanoyl)-*sn*-glycero-3-phosphocholine	*Tm* 33°C	Tyminski et al. [71]
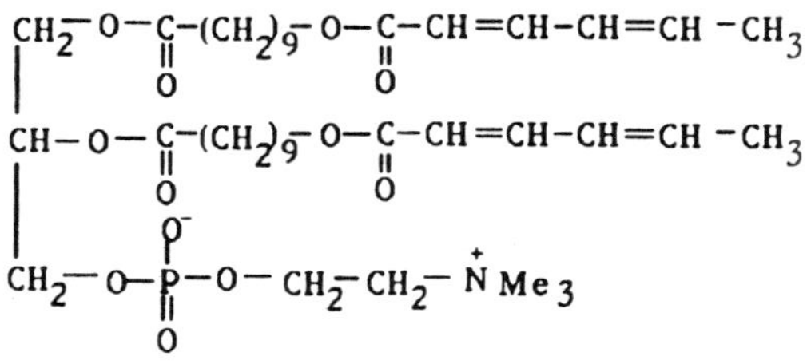 $CH_2-O-C(=O)-(CH_2)_9-O-C(=O)-CH=CH-CH=CH-CH_3$ $CH-O-C(=O)-(CH_2)_9-O-C(=O)-CH=CH-CH=CH-CH_3$ $CH_2-O-P(=O)(O^-)-O-CH_2-CH_2-\overset{+}{N}Me_3$ **24**	1,2 bis (10-(2,4-hexadienoyloxy)decanoyl)-*sn*-glycero-3-phosphocholine	254 nm UV	Frankel et al. [85]
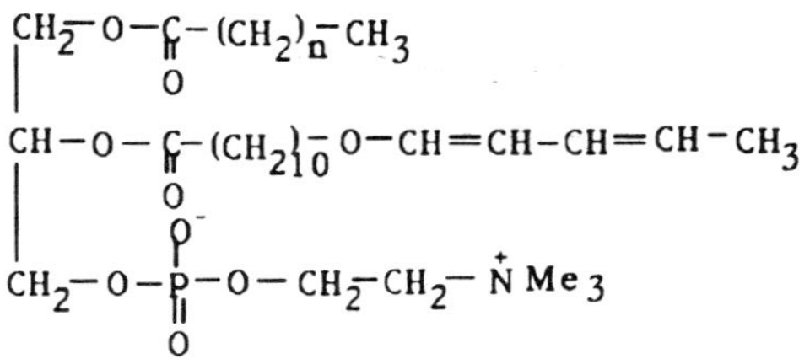 $CH_2-O-C(=O)-(CH_2)_n-CH_3$ $CH-O-C(=O)-(CH_2)_{10}-O-CH=CH-CH=CH-CH_3$ $CH_2-O-P(=O)(O^-)-O-CH_2-CH_2-\overset{+}{N}Me_3$ n= 12,14 **25**	1-acyl-2(11-(2,4-hexadienoyloxy)undecanoyl)-*sn*-glycero-3-phosphocholine	*n* *Tc*°C 12 18.0 14 19.5 UV irradiation	Hasegawa et al. [84]

APPENDIX 1: *(Continued)*

Structure	Name	Data	Reference
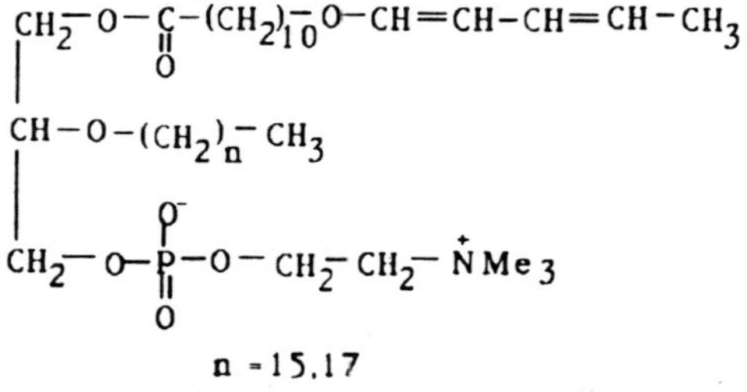 $CH_2-O-C(=O)-(CH_2)_{10}-O-CH=CH-CH=CH-CH_3$ $CH-O-(CH_2)_n-CH_3$ $CH_2-O-P(=O)(O^-)-O-CH_2-CH_2-\overset{+}{N}Me_3$ n = 15,17 **26**	1-(11-(2,4-hexadiennyloxy)undecanoyl)-2-*O*-alkyl-*rac*-glycero-3-phoaphocholine	*n* *Tc*°C 13 25.9 15 31.7 17 54.3	Hasegawa et al. [84]
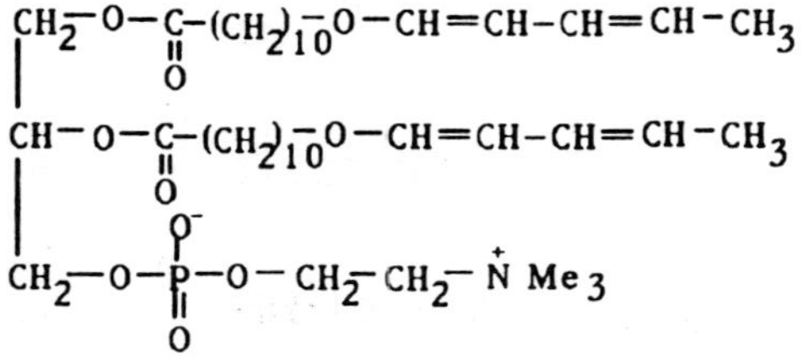 $CH_2-O-C(=O)-(CH_2)_{10}-O-CH=CH-CH=CH-CH_3$ $CH-O-C(=O)-(CH_2)_{10}-O-CH=CH-CH=CH-CH_3$ $CH_2-O-P(=O)(O^-)-O-CH_2-CH_2-\overset{+}{N}Me_3$ **27**	1,2-bis(11-(2,4-hexadienyloxy)-2-*O*-alkyl-*rac*-glycero-3-phosphocholine	*Tc* 28.5°C	Hasegawa et al. [84]

Structure	Name	Polymerization / application	Reference
$CH_2-O-\underset{\underset{O}{\parallel}}{C}-(CH_2)_7-(CH{=}CH)_3-(CH_2)_3-CH_3$ $CH-O-\underset{\underset{O}{\parallel}}{C}-(CH_2)_7-(CH{=}CH)_3-(CH_2)_3-CH_3$ $CH_2-O-\overset{O^-}{\underset{\underset{O}{\parallel}}{P}}-O-CH_2-CH_2-\overset{+}{N}Me_3$ **28**	1,2 dieleostearoyl-*sn*-glycero-3-phosphocholine	Encapsulation of hemoglobin and radiation-sensitive coatings	Suzuki et al [11], Yoshioka et al. [165], Noguchi [7]
VINYLBENZENES			
$CH_2-O-\underset{\underset{O}{\parallel}}{C}-(CH_2)_8-NH\underset{\underset{O}{\parallel}}{C}-C_6H_4-CH{=}CH_2$ $CH-O-(CH_2)_n-CH_3$ $CH_2-O-\overset{O^-}{\underset{\underset{O}{\parallel}}{P}}-O-CH_2-CH_2-\overset{+}{N}Me_3$ $n = 13, 15, 17, 19$ **29**	1-*N*-(12-(4-vinylbenzamidododecnoyl-2-*O*-alkyl-*rac*-glycero-3-phosphocholine	UV irradiation racemic	Hasegawa et al. [9,86,163]

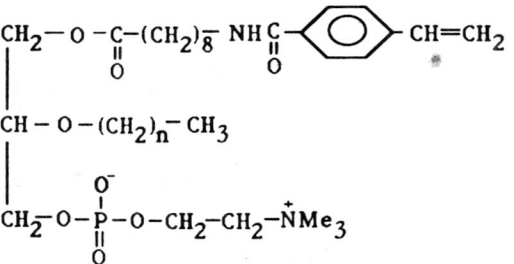

APPENDIX 1: *(Continued)*

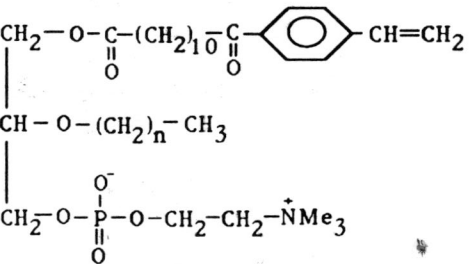

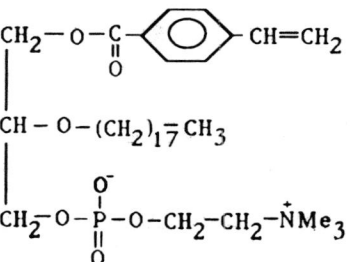

Structure	Name	Properties	Reference
$CH_2-O-C(=O)-(CH_2)_{10}-C(=O)-C_6H_4-CH=CH_2$ $CH-O-(CH_2)_n-CH_3$ $CH_2-O-P(=O)(O^-)-O-CH_2-CH_2-\overset{+}{N}Me_3$ n = 13, 15, 17, 19 **30**	1-(4-vinylbenzoyl)undecanoyl-2-*O*-alkyl-*rac*-glycero-3-phosphocholine	Racemic *n* *Tc*°C 15 33.1 17 41.2 19 41.6	Hasegawa [86]
$CH_2-O-C(=O)-C_6H_4-CH=CH_2$ $CH-O-(CH_2)_{17}-CH_3$ $CH_2-O-P(=O)(O^-)-O-CH_2-CH_2-\overset{+}{N}Me_3$ **31**	1-(4-vinylbenzoyl)-2-*O*-octadecyl-*rac*-glycero-3-phosphocholine	Racemic, no vesicles	Hasegawa et al. [87]

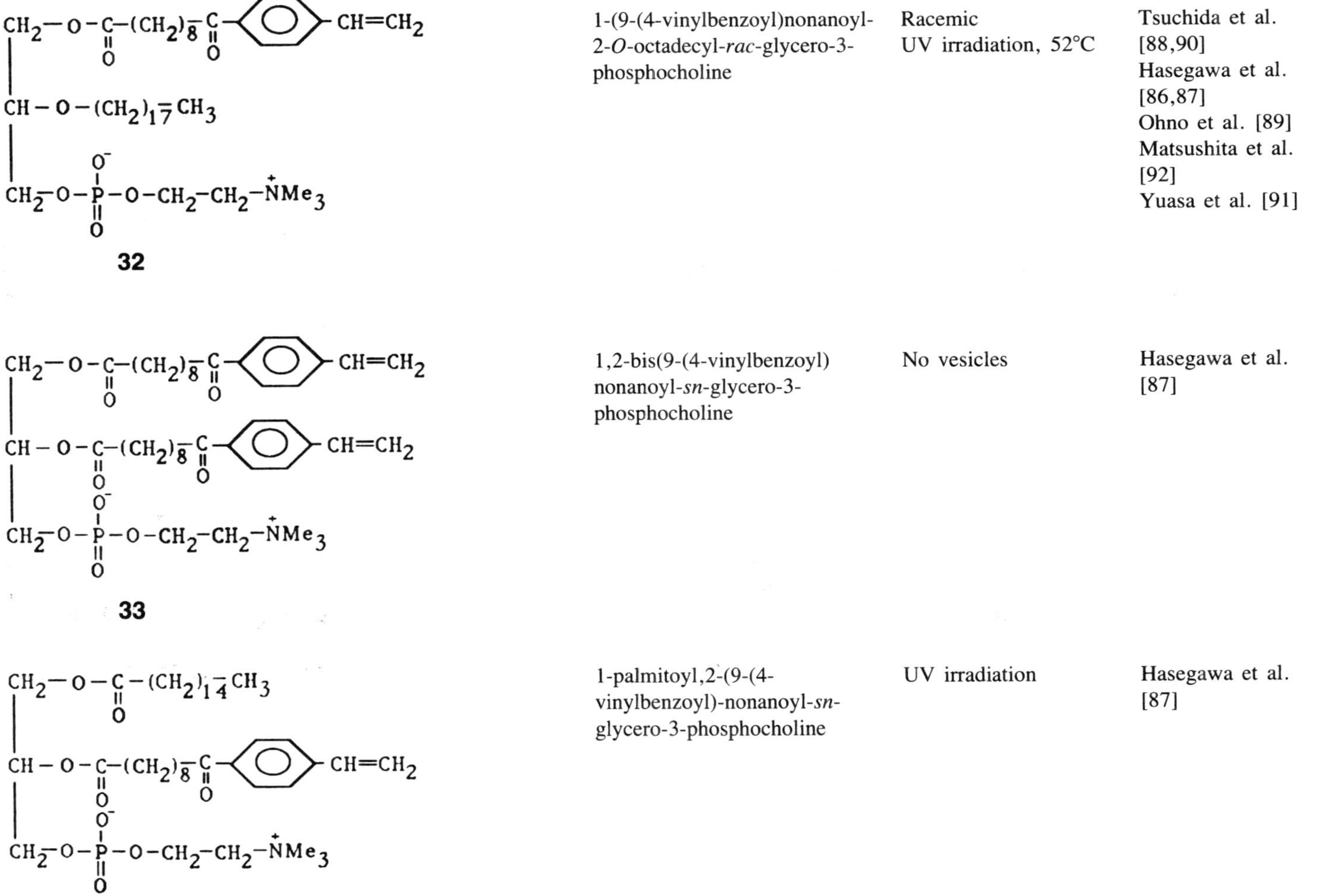

32	1-(9-(4-vinylbenzoyl)nonanoyl-2-*O*-octadecyl-*rac*-glycero-3-phosphocholine	Racemic UV irradiation, 52°C	Tsuchida et al. [88,90] Hasegawa et al. [86,87] Ohno et al. [89] Matsushita et al. [92] Yuasa et al. [91]
33	1,2-bis(9-(4-vinylbenzoyl) nonanoyl-*sn*-glycero-3-phosphocholine	No vesicles	Hasegawa et al. [87]
34	1-palmitoyl,2-(9-(4-vinylbenzoyl)-nonanoyl-*sn*-glycero-3-phosphocholine	UV irradiation	Hasegawa et al. [87]

APPENDIX 1: *(Continued)*

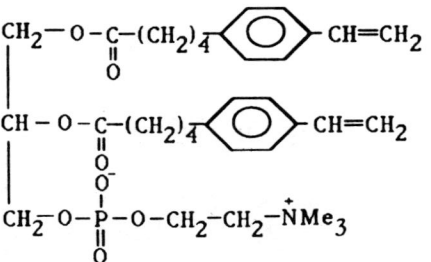

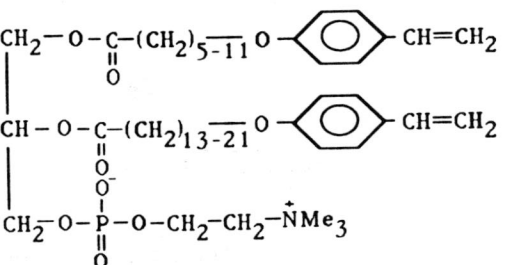

Structure	Name	Polymerization	Reference
$CH_2-O-C(=O)-(CH_2)_4-C_6H_4-CH{=}CH_2$ $CH-O-C(=O)-(CH_2)_4-C_6H_4-CH{=}CH_2$ $CH_2-O-P(=O)(O^-)-O-CH_2-CH_2-\overset{+}{N}Me_3$ **35**	1,2 bis (5-(4-vinylbenzyl)pentanoyl)-*sn*-glycero-3-phosphocholine	Radical initiator	Yamaguchi et al. [180]
$CH_2-O-C(=O)-(CH_2)_{5-11}-O-C_6H_4-CH{=}CH_2$ $CH-O-C(=O)-(CH_2)_{13-21}-O-C_6H_4-CH{=}CH_2$ $CH_2-O-P(=O)(O^-)-O-CH_2-CH_2-\overset{+}{N}Me_3$ **36**	1-*O*-(11-(4-vinylphenoxy)alkanoyl)-2-*O*-alkyl-*rac*-glycero-3-phosphocholine		Tsuchida et al. [12] Hasegawa et al. [86]

DIACETYLENES

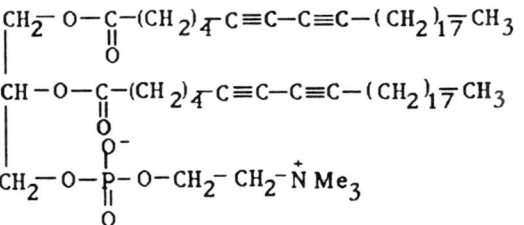

Structure	Name	Properties	Reference
$CH_2-O-C(=O)-(CH_2)_4-C\equiv C-C\equiv C-(CH_2)_{17}-CH_3$ $CH-O-C(=O)-(CH_2)_4-C\equiv C-C\equiv C-(CH_2)_{17}-CH_3$ $CH_2-O-P(=O)(O^-)-O-CH_2-CH_2-\overset{+}{N}Me_3$ **37**	1,2 bis(heptacosa-6,8-diynoyl)-*sn*-glycero-3-phosphocholine	*Tm* 64.8°C	Rudolph et al. [129]
$CH_2-O-C(=O)-(CH_2)_5-C\equiv C-C\equiv C-(CH_2)_n-CH_3$ $CH-O-C(=O)-(CH_2)_5-C\equiv C-C\equiv C-(CH_2)_n-CH_3$ $CH_2-O-P(=O)(O^-)-O-CH_2-CH_2-\overset{+}{N}Me_3$ **38**	1,2 bis (alka-7,9-diynoyl)-*sn*-glycero-3-phosphocholine	*n* *Tm*°C 13 39 16 55.4	Singh et al. [110,111]

APPENDIX 1: *(Continued)*

Structure	Name	Data	References
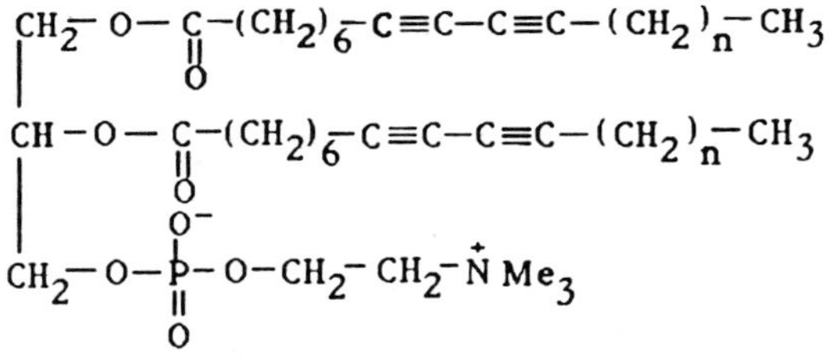 $CH_2-O-C(=O)-(CH_2)_6-C\equiv C-C\equiv C-(CH_2)_n-CH_3$ $CH-O-C(=O)-(CH_2)_6-C\equiv C-C\equiv C-(CH_2)_n-CH_3$ $CH_2-O-P(=O)(O^-)-O-CH_2-CH_2-\overset{+}{N}Me_3$ **39**	1,2 bis (alka-8,10-diynoyl)-*sn*-glycero-3-phosphocholine	*n* *Tm*°C 13 46.6 15 58.9	Singh et al. [110,111] Rudolph et al. [129]
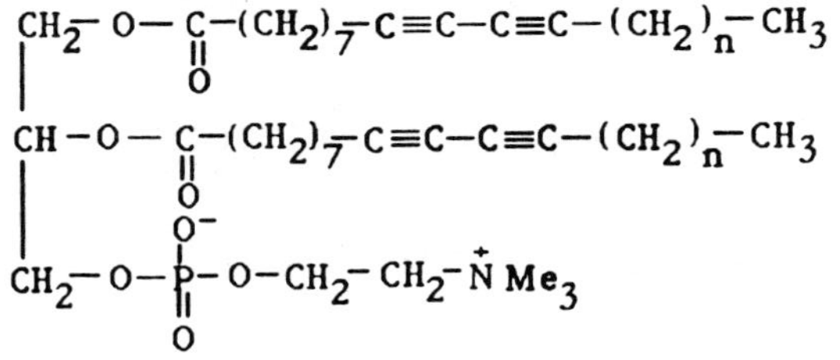 $CH_2-O-C(=O)-(CH_2)_7-C\equiv C-C\equiv C-(CH_2)_n-CH_3$ $CH-O-C(=O)-(CH_2)_7-C\equiv C-C\equiv C-(CH_2)_n-CH_3$ $CH_2-O-P(=O)(O^-)-O-CH_2-CH_2-\overset{+}{N}Me_3$ **40**	1,2 bis (alka-9,11-diynoyl)-*sn*-glycero-3-phosphocholine	*n* *Im*°C 9 26 14 54.9	Singh et al. [110,111] Rudolph et al. [129]
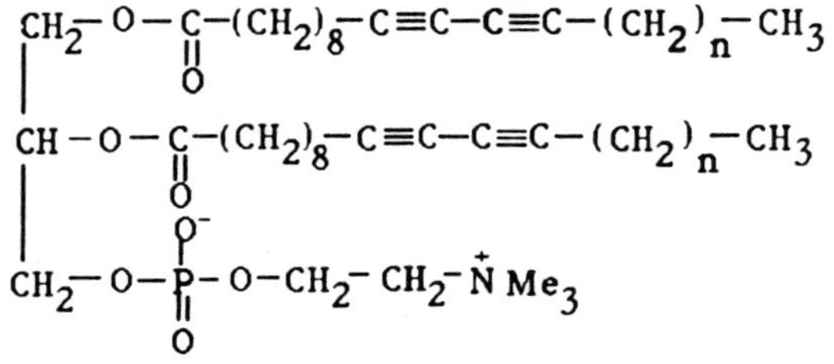 $CH_2-O-C(=O)-(CH_2)_8-C\equiv C-C\equiv C-(CH_2)_n-CH_3$ $CH-O-C(=O)-(CH_2)_8-C\equiv C-C\equiv C-(CH_2)_n-CH_3$ $CH_2-O-P(=O)(O^-)-O-CH_2-CH_2-\overset{+}{N}Me_3$ **41**	1,2 bis (alka-10,12-diynoyl)-*sn*-glycero-3-phosphocholine	*n* *Im*°C 3 — 6 22 7 28	Faltynowicz et al. [53] Leaver et al. [100] Singh et al. [111] Sackman et al. [42] Regen et al. [49]

Structure	Name	Properties	References
$CH_2-O-C(=O)-(CH_2)_8-C\equiv C-C\equiv C-(CH_2)_9-CH_3$ $CH-O-C(=O)-(CH_2)_8-C\equiv C-C\equiv C-(CH_2)_9-CH_3$ $CH_2-O-P(=O)(O^-)-O-CH_2-CH_2-\overset{+}{N}Me_3$ **42**	1,2 bis (tricosa-10,12-diynoyl)-*sn*-glycero-3-phosphocholine	*Tm*°C 37 38 39 42	10, 16, 28–30, 42, 70, 98, 100, 108, 109, 113–116, 121–135, 159, 169, 184
$CH_2-O-C(=O)-(CH_2)_8-C\equiv C-C\equiv C-(CH_2)_n-CH_3$ $CH-O-C(=O)-(CH_2)_8-C\equiv C-C\equiv C-(CH_2)_n-CH_3$ $CH_2-O-P(=O)(O^-)-O-CH_2-CH_2-\overset{+}{N}Me_3$ **43**	1,2 bis (alka-10,12-diynoyl)-*sn*-glycero-3-phosphocholine	*n* *Tm*°C 11 48 12 52 13 61.6	Leaver et al. [100] Hub et al. [27] Hupfer et al. [64,96] Bueschi et al. [95] Nishide et al. [72] Singh et al. [110] Johnston et al. [99] Sheridan [128]
$CH_2-O-C(=O)-(CH_2)_9-C\equiv C-C\equiv C-(CH_2)_{12}-CH_3$ $CH-O-C(=O)-(CH_2)_9-C\equiv C-C\equiv C-(CH_2)_{12}-CH_3$ $CH_2-O-P(=O)(O^-)-O-CH_2-CH_2-\overset{+}{N}Me_3$ **44**	1,2 bis (heptacosa-10,12-diynoyl)-*sn*-glycero-3-phosphocholine	*Tm* 55.9°C	Singh et al. [110] Rudolph et al. [129]

APPENDIX 1: *(Continued)*

Structure	Name	Properties	References
$\begin{array}{l} CH_2-O-C(=O)-(CH_2)_{10}-C\equiv C-C\equiv C-(CH_2)_n-CH_3 \\ CH-O-C(=O)-(CH_2)_{10}-C\equiv C-C\equiv C-(CH_2)_n-CH_3 \\ CH_2-O-P(=O)(O^-)-O-CH_2-CH_2-\overset{+}{N}Me_3 \end{array}$ **45**	1,2 bis (alka-12,14-diynoyl)-*sn*-glycero-3-phosphocholine	*n* *Tm*°C 9 42 11 60.7 13 64.3	Singh et al. [110,182] Schnur et al. [183]
$\begin{array}{l} CH_2-O-C(=O)-(CH_2)_{11}-C\equiv C-C\equiv C-(CH_2)_{10}-CH_3 \\ CH-O-C(=O)-(CH_2)_{11}-C\equiv C-C\equiv C-(CH_2)_{10}-CH_3 \\ CH_2-O-P(=O)(O^-)-O-CH_2-CH_2-\overset{+}{N}Me_3 \end{array}$ **46**	1,2 bis(heptacosa-13,15-diynoyl)-*sn*-glycero-3-phosphocholine	*Tm* 57.4°C	Singh et al. [110] Rudolph et al. [129] Georger et al. [122]
$\begin{array}{l} CH_2-O-C(=O)-(CH_2)_{15}-C\equiv C-C\equiv C-(CH_2)_6-CH_3 \\ CH-O-C(=O)-(CH_2)_{15}-C\equiv C-C\equiv C-(CH_2)_6-CH_3 \\ CH_2-O-P(=O)(O^-)-O-CH_2-CH_2-\overset{+}{N}Me_3 \end{array}$ **47**	1,2 bis (heptacosa-17,19-diynoyl)-*sn*-glycero-3-phosphocholine	*Tm* 61.8°C	Rudolph et al. [129]

Structure	Name	Properties	Reference
$CH_2-O-C(=O)-R$ $CH-O-C(=O)-(CH_2)_8-C\equiv C-C\equiv C-(CH_2)_n-CH_3$ $CH_2-O-P(=O)(O^-)-O-CH_2-CH_2-\overset{+}{N}Me_3$ **48**	1-acyl,2-(alka-10,12-diynoyl)-*sn*-glycero-3-phosphocholine	Egg-derived *n* *Tm*°C 9 20 13 33 Both polymerize well	Hayward et al. [168,169]
$CH_2-O-C(=O)-(CH_2)_{16}CH_3$ $CH-O-C(=O)-(CH_2)_8-C\equiv C-C\equiv C-(CH_2)_n-CH_3$ $CH_2-O-P(=O)(O^-)-O-CH_2-CH_2-\overset{+}{N}Me_3$ **49**	1-stearoyl,2-(alka-10,12-diynoyl)-*sn*-glycero-3-phosphocholine	*n* *Tm*°C 9 23 13 — Do not polymerize well	Hayward et al. [168,169] Johnston et al. [99]
$CH_2-O-C(=O)-(CH_2)_8-C\equiv C-C\equiv C-(CH_2)_n-CH_3$ $CH-O-C(=O)-(CH_2)_{10}C\equiv C-C\equiv C-(CH_2)_7-CH_3$ $CH_2-O-P(=O)(O^-)-O-CH_2-CH_2-\overset{+}{N}Me_3$ **50**	1-alka-10,12-diynoyl, 2-tricosa-12,14-diynoyl-*sn*-glycero-3-phosphocholine	$n=7$, no *Tm* observed $n=9$, *Tm* –°C Inefficient polymerization	Singh et al. [111]

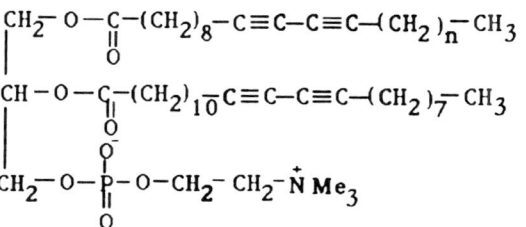

APPENDIX 1: *(Continued)*

Structure	Name	Remarks	Reference
$CH_2-O-C(=O)-(CH_2)_7-CH_3$ $CH-O-C(=O)-(CH_2)_8-C\equiv C-C\equiv C-(CH_2)_9-CH_3$ $CH_2-O-P(=O)(O^-)-O-CH_2-CH_2-\overset{+}{N}Me_3$ **51**	1-nonanoyl,2-(tricosa-10,12-diynoyl)*sn*-glycero-3-phosphocholine	Inefficient polymerization	
$CH_2-O-C(=O)-(CH_2)_8-C\equiv C-C\equiv C-(CH_2)_9-CH_3$ $CH-O-C(=O)-(CH_2)_7-CH_3$ $CH_2-O-P(=O)(O^-)-O-CH_2-CH_2-\overset{+}{N}Me_3$ **52**	1-(tricosa-10,12-diynoyl),2-nonanoyl-*sn*-glycero-3-phosphocholine	Inefficient polym.	Singh et al. [177] Rhodes and Singh [178]
$CH_2-O-C(=O)-(CH_2)_8-C\equiv C-C\equiv C-(CH_2)_6-C\equiv CH$ $CH-O-C(=O)-(CH_2)_8-C\equiv C-C\equiv C-(CH_2)_6-C\equiv CH$ $CH_2-O-P(=O)(O^-)-O-CH_2-CH_2-\overset{+}{N}Me_3$ **53**	1,2 bis (hencosa-10,12-diyn-20-ynoyl)-*sn*-glycero-3-phosphocholine	Polymerize well *Tm* 24°C	Singh, A. (unpublished)

Structure	Name	Properties	Reference
$CH_2-O-C(=O)-(CH_2)_8-C\equiv C-C\equiv C-(CH_2)_8-CH=CH_2$ $CH-O-C(=O)-(CH_2)_8-C\equiv C-C\equiv C-(CH_2)_8-CH=CH_2$ $CH_2-O-P(=O)(O^-)-O-CH_2-CH_2-\overset{+}{N}Me_3$ **54**	1,2 bis (tricosa-10,12-diyn-22-enoyl)-*sn*-glycero-3-phosphocholine	*Tm* 27.9°C Stabilizes tubules	Singh et al. [52]
$CH_2-O-C(=O)-(CH_2)_8-C\equiv C-C\equiv C-(CH_2)_{11}-O-C(=O)-C(CH_3)=CH_2$ $CH-O-C(=O)-(CH_2)_8-C\equiv C-C\equiv C-(CH_2)_{11}-O-C(=O)-C(CH_3)=CH_2$ $CH_2-O-P(=O)(O^-)-O-CH_2-CH_2-\overset{+}{N}Me_3$ **55**	1,2 bis (24-methacryoyloxy(tetracosa-10,12-diynoyl)-*sn*-glycero-3-phosphocholine	*Tm* 27.9°C Tubules disrupts after polymerization	Singh et al. [52]
$CH_2-O-C(=O)-(CH_2)_8-C\equiv C-C\equiv C-(CH_2)_{12}-CH_3$ $CH-O-C(=O)-(CH_2)_8-C\equiv C-C\equiv C-(CH_2)_{12}-CH_3$ $CH_2-O-P(=O)(O^-)-O-CH_2-CH_2-\overset{+}{N}H_3$ **56**	1,2 bis (hexacosa-10,12-diynoyl)-*rac*-glycero-3-phosphoethanolamine	*Tm* 48.5°C	Hupfer et al. [96]

APPENDIX 1: *(Continued)*

Structure	Name	Conditions	Reference
$CH_2-O-C(=O)-(CH_2)_{16}-CH_3$ $CH-O-C(=O)-(CH_2)_8-C\equiv C-C\equiv C-(CH_2)_{12}-CH_3$ $CH_2-O-P(=O)(O^-)-OH$ **57**	1-stearoyl,2-(hexacosa-10,12-diynoyl)-*rac*-glycero-3-phosphate	*Tm* 36°C	Hupfer et al. [96]
MISCELLANEOUS			
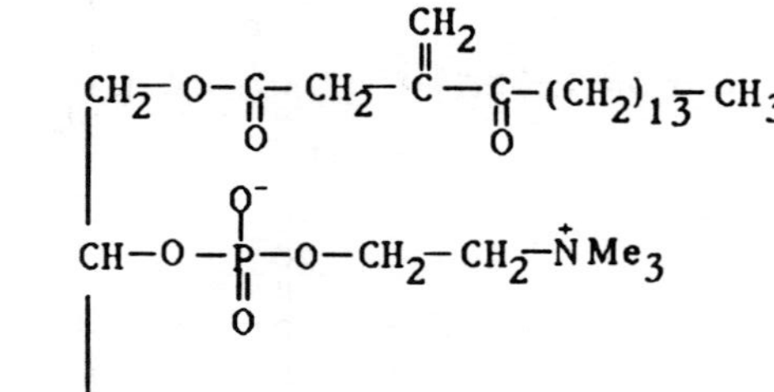 n= 12,16 **58**	1-acyl-3-(tetradecyloxycrbonylmethyl)acryloyl)-glycero-2-phosphocholine	UV, 254 nm, 1 h.	Takane et al. [67]

Structure	Name	Remarks	Reference
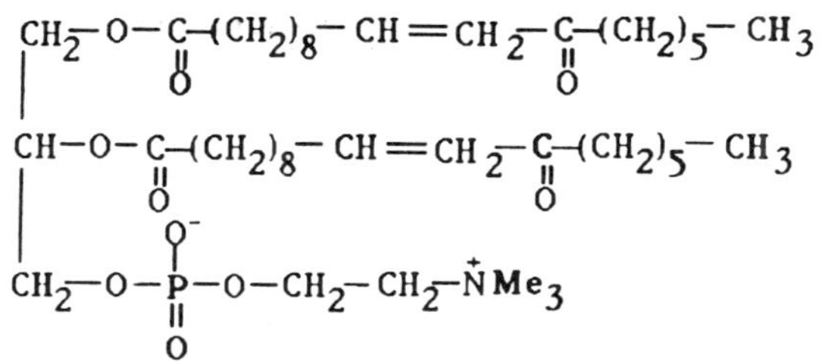 $CH_2-O-C(=O)-(CH_2)_8-CH=CH_2-C(=O)-(CH_2)_5-CH_3$ $CH-O-C(=O)-(CH_2)_8-CH=CH_2-C(=O)-(CH_2)_5-CH_3$ $CH_2-O-P(=O)(O^-)-O-CH_2-CH_2-\overset{+}{N}Me_3$ **59**	1,2 di(12-oxooctadec-*trans*-10-enoyl)-*sn*-glycero-3-phosphocholine	Tm 30–40°C UV polymerization	Molotkovskii et al. [146] Gupta et al. [145,147]
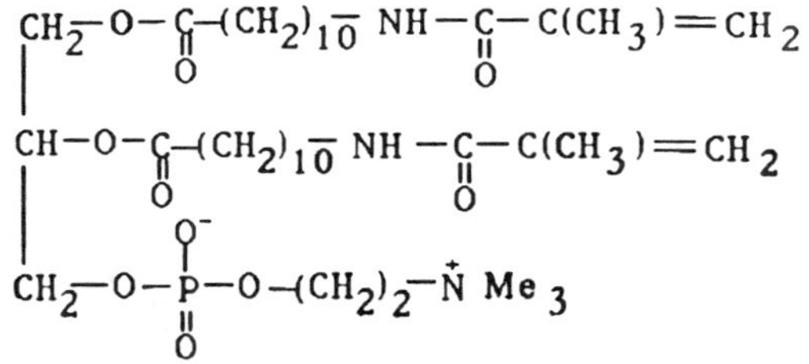 $CH_2-O-C(=O)-(CH_2)_{10}-NH-C(=O)-C(CH_3)=CH_2$ $CH-O-C(=O)-(CH_2)_{10}-NH-C(=O)-C(CH_3)=CH_2$ $CH_2-O-P(=O)(O^-)-O-(CH_2)_2-\overset{+}{N}Me_3$ **60**	1,2-di-(11-methacryloyloxylaminoundecanoyl)-*sn*-glycero-3-phosphocholine	UV polym. vesicles Close tubular structure on polymerization	Molotkovskii et al. [146]
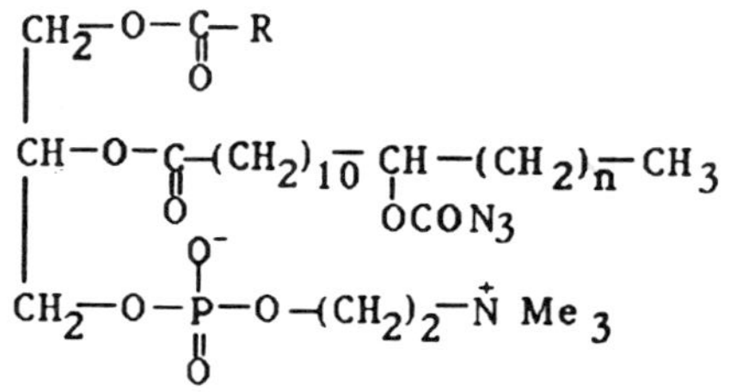 $CH_2-O-C(=O)-R$ $CH-O-C(=O)-(CH_2)_{10}-CH(OCON_3)-(CH_2)_n-CH_3$ $CH_2-O-P(=O)(O^-)-O-(CH_2)_2-\overset{+}{N}Me_3$ **61**	1-acyl-2-(12-azidoformyloxystearoyl)-*sn*-glycero-3-phosphocholine	Derived from lyso egg PC	Tsirenina et al. [144]

APPENDIX 1: ***(Continued)***

Structure	Name	Source	Reference
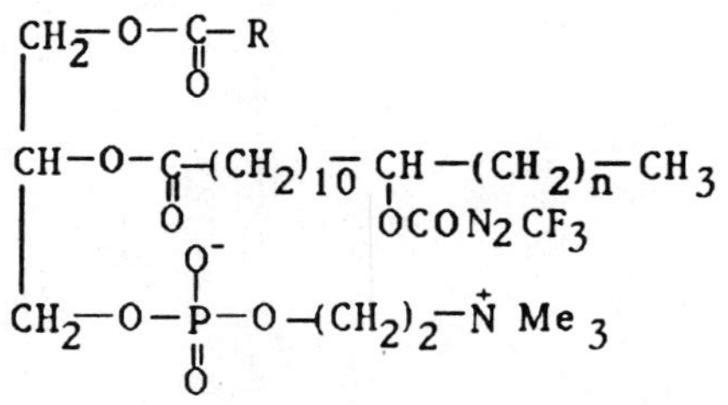 $CH_2{-}O{-}C({=}O){-}R$ / $CH{-}O{-}C({=}O){-}(CH_2)_{10}{-}CH(OCON_2CF_3){-}(CH_2)_n{-}CH_3$ / $CH_2{-}O{-}P({=}O)(O^-){-}O{-}(CH_2)_2{-}\overset{+}{N}Me_3$ **62**	1-acyl-2(12-(2-diazo-3.3.3-trifluoropropionyloxy)stearoyl)-*sn*-glycero-3-phosphocholine	Derived from lyso egg PC	Tsirenina et al. [144] Gupta et al. [145,147]

*included because of general relevance to the topics in this chapter

APPENDIX 2: SYNTHETIC TECHNIQUES

Strategies for phospholipid synthesis have been the subject of several reports and reviews [8,145,190]. The most widely used procedure for the synthesis of phosphatidylcholines has been reported by Gupta et al. [145]. In the procedure, *N,N*-dimethylaminopyridine (DMAP) base has been used to facilitate efficient acylation of sn-glycero-3-phosphocholine (GPC). In polymerizable phospholipids, conventional fatty acids have been replaced by the acids bearing polymerizable moieties. Considering the sensitive nature of the monomers, a time-efficient synthesis is desired. The problem has been solved by replacing magnetic stirring with ultrasound [109]. In this method, the acylation time has been reduced to hours from day(s), the yield of the products has been improved, and the side reactions are minimized. In a typical reaction, 1 mol vacuum dried GPC or GPC·$CdCl_2$ complex is reacted with acid anhydride (3 mol) and DMAP (3 mol) in freshly distilled chloroform. The volume of choroform was kept to a minimum: just enough to facilitate stirring. Agitation is achieved by a common laboratory ultrasonic cleaner. The course of the reaction is maintained by thin layer chromatography on silica gel employing the standard lipid solvent system: chloroform, methanol, and water in 65:25:4 ratio. The pure lipid was obtained by passing the reaction mixture through mixed bed ion exchange resin (to remove ionic impurities) followed by column chromatography and finally by acetone precipitation. Lipids are usually stored at or below –20°C, well protected from laboratory light. Purity of the samples should be checked routinely by both thin layer chromatography and high pressure liquid chromatography.

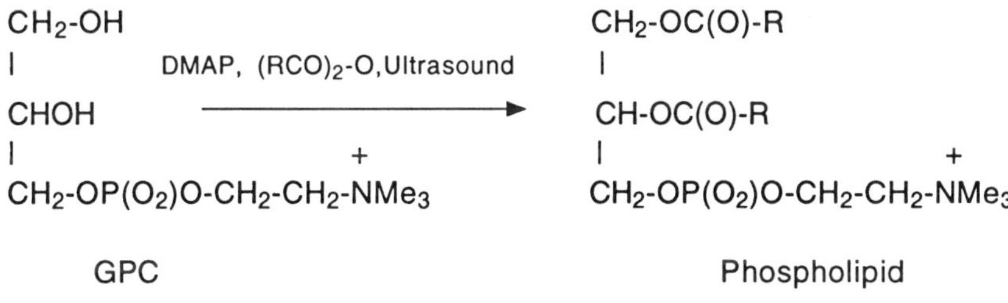

Scheme 1

A. Synthesis of Polymerizable Precursors

GPC acylation to produce phospholipid is essentially the same for each lipid preparation. In the following paragraphs, we shall describe the synthesis of the carboxylic acids bearing the polymerizable functionalities only: methacrylate, dienoate, diacetylene, thiols and disulfides, and styryl-containing acids have been reported.

B. Synthesis of Acids Containing Methacrylate Moiety

The synthesis of 12-(methacryloyloxy)-dodecanoic acid [45,46] can be considered as representative for the introduction of a methacryloyl moiety in an acid. The reaction is carried out in tetrahydrofuran (THF) at 0°C. 12-Hydroxydodecanoic acid is reacted with an equimolar amount of methacryloyl chloride in the presence of 1.5 mol eq. pyridine.

The reaction was complete after stirring at room temperature for 10 h. Removal of THF and extraction of the acid with ether affords typically the acid in 53% yield.

$$HOOC\text{-}(CH_2)_{11}\text{-}OH \xrightarrow[\text{PYRIDINE}]{ClOC\text{-}C(CH_3)=CH_2} HOOC\text{-}(CH_2)_{11}\text{-}OOC\text{-}C(CH_3)=CH_2$$

Scheme 2

Precautions

1. The methacrylate acid should be protected from heat and light.
2. Methacrylate acid undergoes a side reaction when dicyclohexyl carbodiimide (DCC) is employed for anhydride formation.

The acid is converted into its anhydride by first reacting the acid in THF with equimolar ethylchloroformate in the presence of triethylamine at –20°C. To the resulting unsymmetrical anhydride additional mol of acid and triethylamine in THF are added at low temperature. Additional stirring at room temperature and usual work-up affords the product in 94% yield. The formation of anhydride can be confirmed by observing the bands at 1730 and 1800 cm^{-1} in the infrared spectrum.

C. Synthesis of Thiol-Containing Acids

1. Thiol Moiety in the Carboxylic Group [59]

The same strategy is used in the synthesis of acids containing a thiol group next to the acid residue or at the terminus. We report here the synthesis of 2-(ethyldithio)hexadecanoic acid. 2-Mercaptohexadecanoic acid (6 mmol) prepared from 2-bromohexadecanoic acid is dissolved in chloroform and reacted with ethyl ethanethiosulfinate (8 mmol) in the presence of triethylamine (5 mmol) by stirring at room temperature for 24 h. The resulting disulfide is obtained in 72% yield, following the purification step on a silica column.

$$CH_3\text{-}(CH_2)_{13}\text{-}CH(Br)\text{-}COOH \longrightarrow CH_3\text{-}(CH_2)_{13}\text{-}CH(SH)\text{-}COOH \xrightarrow{EtSS(O)Et}$$

$$CH_3\text{-}(CH_2)_{13}\text{-}CH(S\text{-}SEt)\text{-}COOH \xrightarrow{Bu_3P} \text{deprotected acid}$$

Scheme 3

Owing to its high reactivity, the protected mercapto acid, in this case, was used in the synthesis of phospholipid. In a typical deprotection reaction tri-n-butyl phosphine (0.45 mmol) was added to dithio phospholipid dissolved in 2 mL 50% ethanol. The reaction mixture was stirred in the dark, at room temperature, for 11 h. Solvent removal followed by column chromatography on silica gel gives thiol lipid in 76% yield.

2. Cyclic Disulfide Incorporated Carboxylic Acid [62]

The synthesis of cyclic disulfide containing phospholipid is performed starting with a hydroxy-protected 12-hydroxy dodecanoic acid, which is first reacted with GPC to yield the hydroxy terminated phospholipid, then with lipoic anhydride to produce lipid **11**.

12-Hydroxydodecanoic acid (10 mmol) was suspended in dry THF (20 mL) and reacted with dihydropyran (16.5 mmol) in the presence of p-toluenesulfonic acid monohydrate (0.105 mmol) for two h at room temperature. Work-up and purification by flash chromatography (chloroform) affords 12-(tetrahydropyranyloxy) dodecanoic acid in 80% yield. This acid (3.15 mmol) is reacted with GPC·$CdCl_2$ (1.8 mmol) in the presence of dicyclohexylcarbodiimide (8 mmol) in methylene chloride. After 40 h of stirring at room temperature (in the dark), the solvent is removed and the residue is treated with a resin AG MP50 in 95% methanol to deprotect the hyroxyl group. The phospholipid, 1,2-bis(12-hydroxydodecanoyl)-sn-glycero-3-phosphocholine, is obtained in 48% yield.

The free hydroxy groups present at the acyl chain terminus of phospholipid (0.06 mmol) are reacted with a 2 mL 0.15 M methylene chloride solution of DL-1,2-dithiolane-3 pentanoic (lipoic) acid anhydride in the presence of DMAP (0.13 mmol) to give the final lipid in 90% yield.

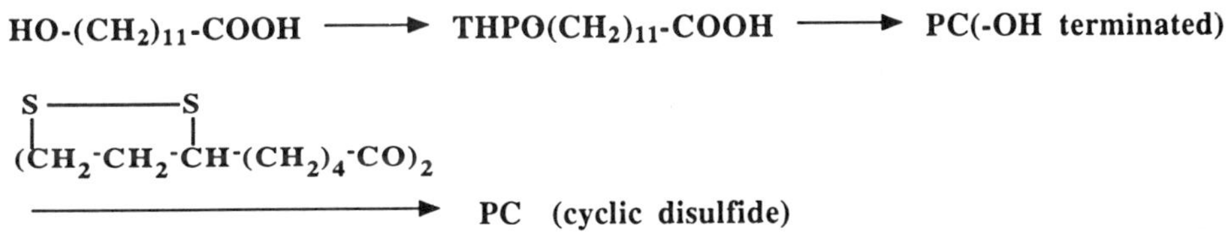

HO-$(CH_2)_{11}$-COOH ⟶ THPO$(CH_2)_{11}$-COOH ⟶ PC(-OH terminated)

S———S (ring bond linking the first CH_2 and the CH)
$(CH_2\text{-}CH_2\text{-}CH\text{-}(CH_2)_4\text{-}CO)_2$
⟶ PC (cyclic disulfide)

Scheme 4

D. Synthesis of Diacetylenic Acids [140]

Isomeric diacetylenic acids are prepared by coupling a w-alkynoic acid with an iodoalkyne. Both alkyne and alkynoic acids have been prepared starting from their bromo analog. The bromo compounds are reacted with slight excess (1.2 meq) of lithium acetylide ethylenediamine complex employing dimethylsulfoxide (DMSO) as a reaction medium. The amount of DMSO employed is kept minimum, enough to make a slurry. The temperature is maintained at 4°C during the addition of the bromo compound to the lithium acetylide complex. Reaction is quenched by addition of water at 8–10°C till gas evolution ceases, followed by the addition of 10% HCl and hexane extraction. The corresponding alkyne is obtained in high yields (alkynoic acids ~60% and alkyne >90%). Alkynes were then converted to iodoalkyne by reacting with ethylmagnesium bromide and iodine [91].

Heterocoupling between acetylenic acid and iodoalkyne is achieved by the slow addition of iodoalkyne, (10 mmol) dissolved in a methanol-ethyl ether (1:1) solution, to the solution of potassium salt of alkynoic acid (10 mmol) in water. The later also contains

0.25 mol eq. Cu_2Cl_2 dissolved in (10 mL) aqueous ethylamine and 0.5 wt % of hydroxylamine hydrochloride ($NH_2OH{\cdot}HCl$) crystals. During the addition of iodoalkyne, the original yellow color of the reaction mixture turns blue; the original color is recovered by addition of 10% aqueous $NH_2OH{\cdot}HCl$ solution. The process is repeated until all iodoalkyne is reacted.

Finally, the reaction mixture is acidified with dilute sulfuric acid and extracted with ether. The acid is purified by column chromatography on silica gel followed by crystallization from hexane. The pure acid is light-sensitive and must be stored at room temperature in the dark under inert atmosphere.

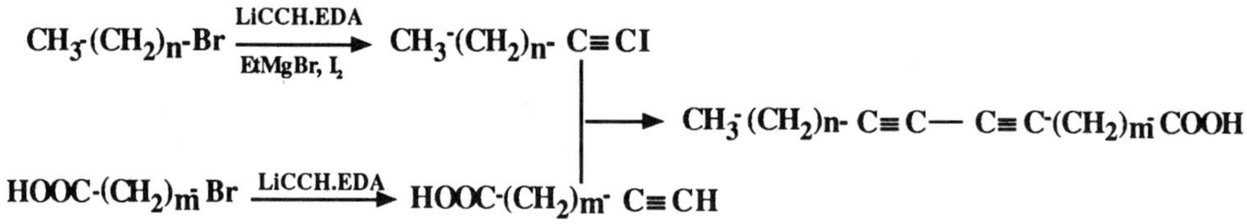

$$CH_3\text{-}(CH_2)_n\text{-}Br \xrightarrow[\text{EtMgBr, } I_2]{\text{LiCCH.EDA}} CH_3\text{-}(CH_2)_n\text{-}C{\equiv}CI$$

$$HOOC\text{-}(CH_2)_m\text{-}Br \xrightarrow{\text{LiCCH.EDA}} HOOC\text{-}(CH_2)_m\text{-}C{\equiv}CH$$

$$\longrightarrow CH_3\text{-}(CH_2)_n\text{-}C{\equiv}C\text{—}C{\equiv}C\text{-}(CH_2)_m\text{-}COOH$$

Scheme 5

E. Synthesis of Dienoates

There have been two types of dienoic acids reported in the literature: 2,4 dienoic acids and the acids containing 2,4 hexadienoyl group at the terminus. The terminal groups have been linked to the carboxylic acid terminus via both ester and ether bonds.

1. Alkyl-2,4-Dienyl Groups Near the Carboxylic Group [69,166,192]

Two different approaches can be employed in the synthesis of the title compound. In the first approach [192] (Scheme 6), octadecadienal ($m=12$) was first prepared as follows. The 1-methoxybut-1-en-3-yne (0.51 mol) was reacted with equimolar amounts of ethyl magnesium bromide suspended in 300 mL of THF at 40°C. Tetradecanal (0.45 mol) in THF is added to this reagent followed by refluxing for 30 min. The contents are then cooled to room temperature and 18 mL ethanol is added. Twenty minutes later 0.5 mol solid liithium aluminum hydride (LAH) is added in small portions. The mixture is stirred for two h, left overnight unstirred, and quenched successively with ethyl acetate, water, and 4N sulfuric acid. Alkadienal is contained in organic phase. Removal of solvent provides 46% alkadienal.

The aldehyde (0.05 mol, in ethanol) is then oxidized with silver oxide (prepared by reacting 0.05 mol silver nitrate dissolved in 100 mL water, with 0.08 mol sodium hydroxide dissolved in 20 mL water) suspended in sodium hydroxide solution (0.26 mol in 150 mL water). The reaction mixture is first heated at 50–60°C for 20 min, then stirred over a period of three h, acidified, and extracted with ether. Removal of ether affords crystal of the dienoic acid in 29% yield.

$$MeO\text{-}CH{=}CH\text{-}C{\equiv}C\text{-}MgBr \xrightarrow[LiAlH_4]{CH_3\text{-}(CH_2)_m\text{-}CHO} CH_3\text{-}(CH_2)_m\text{-}CH{=}CH\text{-}CH{=}CH\text{-}CHO$$

$$\xrightarrow[EtOH,\ 50\text{-}60^{\circ}C]{Ag_2O/NaOH} CH_3\text{-}(CH_2)_m\text{-}CH{=}CH\text{-}CH{=}CH\text{-}COOH$$

Scheme 6

In the second approach [69], triethyl, 4-phosphonocrotonate (12.5 g) is stirred with a suspension of sodium hydride (2.4 g) in 100 mL THF. To the mixture, after stirring at room temperature for 20 min, 9.2 g dodecanecarboxaldehyde is slowly added. The resulting mixture is further stirred at 0°C and then at room temperature; subsequently it is quenched with water and extracted by ether. Removal of the ether gives 10.5 g 2,4 hexadecadienoate, which upon gentle refluxing with methanolic potassium hydroxide affords hexadeca-2,4-dienoic acid.

$$CH_3\text{-}(CH_2)_{10}\text{-}CHO + (C_2H_5)_2\text{-}P(O)\text{-}CH_2\text{-}CH{=}CH\text{-}COOC_2H_5 \longrightarrow$$

$$CH_3\text{-}(CH_2)_{10}\text{-}CH{=}CH\text{-}CH{=}CH\text{-}COOC_2H_5 \xrightarrow{KOH/EtOH}$$

$$CH_3\text{-}(CH_2)_{10}\text{-}CH{=}CH\text{-}CH{=}CH\text{-}COOH$$

Scheme 7

2. Dienyl Group at the Terminus of Carboxylic Acid

The dienyl group can attached to the carboxylic acid terminus by ester or ether linkage. The ester linkage [69] is prepared by reacting sorbyl chloride to one of the hydroxyl groups of appropriate diols. The second hydroxyl group is oxidized to carboxylic acid by pyridinium chlorochromate.

$$HO\text{-}(CH_2)m\text{-}OH \xrightarrow{CH_3\text{-}CH{=}CH\text{-}CH{=}CH\text{-}COCl} CH_3\text{-}CH{=}CH\text{-}CH{=}CH\text{-}COO\text{-}(CH_2)m\text{-}OH$$

$$\xrightarrow{\text{Pyridinium chlorochromate}} CH_3\text{-}CH{=}CH\text{-}CH{=}CH\text{-}COO\text{-}(CH_2)m-1\text{-}COOH$$

Scheme 8

The ether linkage is produced as follows [84]. Sorbyl alcohol (0.51 mol) is stirred with equimolar amounts of sodium hydride in 700 mL dimethyl formamide (DMF), first at ice-water temperature (1 h), then at room temperature (1 h), and finally at 50°C (4 h). Methyl-11-bromoundecanoate is added to the ice-cooled solution of this mixture and heated to 60°C after the addition. After 6 h of heating, the reaction is cooled and quenched with water to be then extracted with hexane. Solvent removal yields a mixture of the methyl esters, which are converted to acid by stirring with methanolic 2N sodium hydroxide for a day. The reaction mixture is then acidified and the precipitated acid collected, washed, and recrystallized with petroleum ether to afford the desired compound in 36% yield.

$$CH_3\cdot CH{=}CH\cdot CH{=}CH\cdot CH_2OH \xrightarrow{NaH/DMF} CH_3\cdot CH{=}CH\cdot CH{=}CH\cdot CH_2ONa$$

$$\xrightarrow{MeOOC\cdot(CH_2)_{10}\cdot Br} CH_3\cdot CH{=}CH\cdot CH{=}CH\cdot CH_2O\cdot(CH_2)_{10}\cdot COOMe$$

$$\xrightarrow{2N\ NaOH/MeOH} CH_3\cdot CH{=}CH\cdot CH{=}CH\cdot CH_2O\cdot(CH_2)_{10}\cdot COOH$$

Scheme 9

F. Styryl Group Incorporated Carboxylic Acid

In addition to direct linkage, the styryl moiety can be attached to the acid-chain via amide, carbonyl, and ether linkage. Each of the acids was synthesized as follows.

5-(4-Vinylphenyl)pentanoic acid is prepared [180] by treating 4-(4-bromobutyl)styrene (25 mmol) with activated Mg (49 mmol) suspended in 50 mL THF at 0°C. After 5 h of stirring, the resulting Grignard is added dropwise to 100 g of dry ice and the mixture is stirred for 1 h. The reaction mixture is acidified and extracted with diethyl ether to yield styryl-incorporated acid in 56% yield.

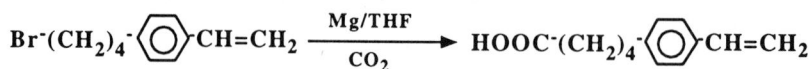

$$Br\cdot(CH_2)_4\cdot C_6H_4\cdot CH{=}CH_2 \xrightarrow[CO_2]{Mg/THF} HOOC\cdot(CH_2)_4\cdot C_6H_4\cdot CH{=}CH_2$$

Scheme 10

Methyl 11-aminoundecanoate and 4-vinyl benzoyl chloride are reacted in methylene chloride in the presence of triethylamine to give methyl, 11-(4-vinylnenzamido)-undecanoate. The latter, upon base hydrolysis with sodium hydroxide-ethanol produced the corresponding acid [86].

The synthetic scheme for attaching the styryl moiety to the acid vie ether linkage has been reported as follows [12,86]. Para hydroxystyrene (19 g) dissolved in 300 mL of dry

DMF is added at 0°C to 7.2 g sodium hydride (washed with dry hexane) suspended in 300 mL dry DMF. To the resulting sodium phenoxide, 48.6 g 11-bromoundecanoate in DMF is added dropwise. After 90 minutes of stirring, the reaction is quenched by the addition of methanol. The product, methyl ester, was purified by column chromatography (55% yield). The dioxane solution of methyl ester is hydrolized to acid by stirring with 2N sodium hydroxide in ethanol in 15 h.

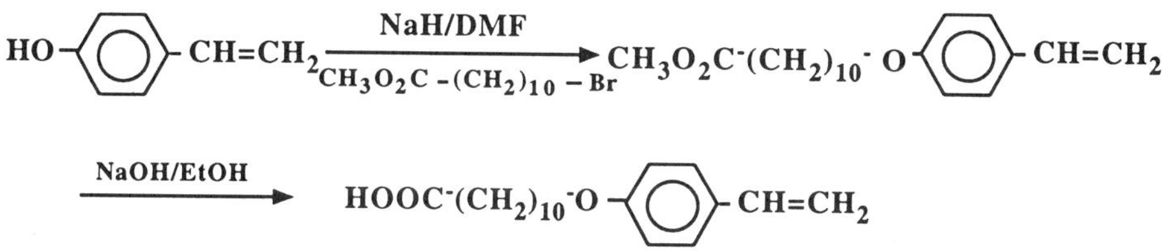

$$HO-C_6H_4-CH=CH_2 \xrightarrow[CH_3O_2C-(CH_2)_{10}-Br]{NaH/DMF} CH_3O_2C-(CH_2)_{10}-O-C_6H_4-CH=CH_2$$

$$\xrightarrow{NaOH/EtOH} HOOC-(CH_2)_{10}-O-C_6H_4-CH=CH_2$$

Scheme 11

The synthesis of 1-(9-(*p*)-vinylbenzoyl)nonanoic acid has been described as follows [86,87,92]. The reaction of 2-bromo ethylbenzene (335 mmol) with 402 mol ethyl, (9-chloroformyl) nonanoate in nitrobenzene, in the presence of anhydrous aluminum trichloride, gives 840 g 4-(2-bromoethyl)-phenyl, ethyl decanedioate in 49% yield. The product is purified on a column of silica gel and subjected to dehydrobromination, hydrolyzed reaction by refluxing with 750 g potassium hydroxide dissolved in 150 mL ethanol for 3 h. Neutralization with concentrated hydrochloric acid, extraction, and subsequent recrystallization with toluene/petroleum ether affords the corresponding acid in 46% yield.

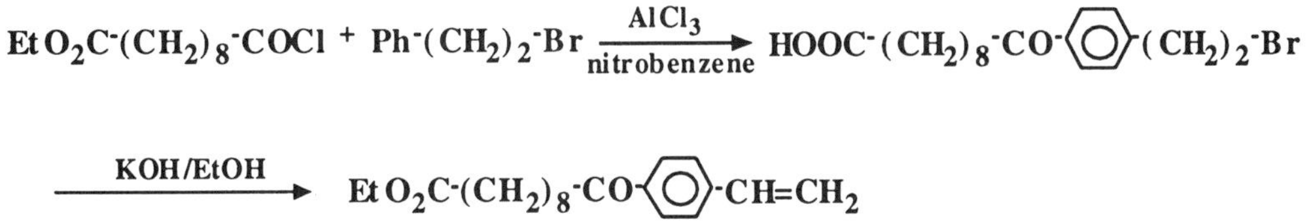

$$EtO_2C-(CH_2)_8-COCl + Ph-(CH_2)_2-Br \xrightarrow[nitrobenzene]{AlCl_3} HOOC-(CH_2)_8-CO-C_6H_4-(CH_2)_2-Br$$

$$\xrightarrow{KOH/EtOH} EtO_2C-(CH_2)_8-CO-C_6H_4-CH=CH_2$$

Scheme 12

REFERENCES

1. A. D. Bangham *Progress Biophys. Mol. Biol. 18*:31–95 (1968). Membrane models with phospholipids.
2. C. G. Knight, ed., *Liposomes: From Physical Structures to Therapeutic Applications*, Elsevier/North-Holland Biomedical Press, New York, 1981.
3. A. D. Bangham, M. W. Hill, and N. G. A. Miller *Methods Membr. Biol. 1*:1–67 (1974). Preparation and use of liposome as models for biological membranes.

4. J. H. Fendler and P. Tundo *Acc. Chem. Res. 17*:1–16 (1984). Polymerized surfactant aggregates: Characterization and utilization.
5. H. Ringsdorf, B. Schlarb, and J. Venzmer, *Angew. Chem. Int. Ed. Engl. 27*:113–158 (1988). Molecular architecture and function of polymeric oriented systems: Models for the study of organization, surface recognition and dynamics of biomembranes.
6. Y. Nagata, A. Akimoto, Y. Muneda, A. Miyamoto, and F. Schichino. *Jpn. Kokai Tokkyo Koho JP 62 081,394*, 14 April 1987, 10 pp. Preparation of mixed acid polymerizable phospholipid derivatives.
7. Y. Noguchi and O. Nakachi, *Jpn. Kokai Tokkyo Koho JP 61129190*, June 17, 1986, 6 pp. Polymerizable glycerophospholipids.
8. H. Eibl and A. Nicksh, *Ger. Offen. DE 3010185*, September 24, 1981, 35 pp. Polymerizable phospholipids.
9. N. Hasegawa, K. Ejima, Y. Matsushita, and H. Tsuchida, *Jpn. Kokai Tokkyo Koho JP 61000091*, January 6, 1986, 10 pp. Polymerizable phosphatidylcholines.
10. T. Nakaya, M. Yasuzawa, and M. Imoto, *Jpn. Kokai Tokkyo Koho JP 6120591*, September 11, 1986, 7 pp. Polymerizable phospholipids.
11. K. Suzuki, and H. Yoshioka, *Eur. Pat. Appl. EP 186211*, July 2, 1986, 22 pp. Polymerizable liposome-forming lipid and its use.
12. H. Tsuchida, K. Ejima, Y. Matsushita, and H. Tsuchida, *Jpn. Kokai Tokkyo Koho JP 61129192*, June 17, 1986, 8 pp. Polymerizable phospholipids.
13. H. Ono, T. Takahashi, and H. Tsuchida, *Jpn, Kokai Tokkyo Koho JP 63232841*, September 28, 1988, 6 pp. Manufacture of liposomes containing polymerized phospholipids.
14. K. Ejima, E. Hasegawa, Y. Matsushita, H. Nishide, and H. Tsuchida, *Jpn. Kokai Tokkyo Koho JP 62292762*, December 19, 1987, 7 pp. Preparation of polymerizable omega-imidazole-1-ylalkadienoates and their use for preparation of polymerized liposome-embedded heme-imidazole complexes.
15. T. Shigehara, M. Takane, and H. Tsuchida, *Jpn. Kokai Tokkyo Koho JP 60214794*, October 28, 1985, 8 pp. Polymerizable phosphatidylcholines.
16. E. L. Chang, *U. S. Pat. Appl. US 714711*, August 16, 1985, 17 pp. Magnetically localizable polymerized lipid vesicles containing pharmaceuticals, and a method of releasing them.
17. K. D. Schmitt, *U. S. Patent 4582137*, April 15, 1986, 6 pp. Polymerizable surfactants for permeability control in water flooding.
18. H. Ono, K. Ukaji, and H. Tsuchida, *Jpn. Kokai Tokkyo Koho JP 62, 104,844*, May 15, 1987, 6 pp. Porous liposomes for enzyme immobilization.
19. P. E. Schoen, P. Yager, and J. M. Schnur, *U. S. Pat. Appl. US 852596*, October 24, 1986, 24 pp. Lipid tubules.
20. C. Tanford, *Physics of Amphiphiles: Micelles, Vesicles and Microemulsions*, (V. Degiorgio and M. Corti, eds.), North-Holland Physics Publishing, Amsterdam, 1985, pp. 547–554.
21. L. M. Crowe, J. H. Crowe, A. R. Rudolph, C. Womersley, and L. Apple, *Arch. Biochem. Biophys. 242*:240–247 (1985). Preservation of freeze-dried liposomes by trehalose.
22. R. Brederhorst, F. S. Ligler, A. W. Kusterbeck, E. L. Chang, B. P. Gaber, and C. W. Vogel, *Biochemistry 25*:5693–5698 (1986). Effect of covalent attachment of immunoglobulin fragments on liposomal integrity.
23. M. C. Farmer, and B. P. Gaber, *Methods Enzymol. 149*:184–200 (1987). Liposome-encapsulated hemoglobin as an artificial oxygen-carrying system.
24. A. S. Rudolph, *Cryobiology 25*:277–284 (1988). The freeze-dried preservation of liposome encapsulated hemoglobin: A potential blood substitute.
25. S. L. Regen, B. Czech, and A. Singh *J. Amer. Chem. Soc. 102*:6638–6640 (1980). Polymerized vesicles.
26. D. S. Johnston, S. Sanghera, M. Pons, and D. Chapman, *Biochim. Biophys. Acta 602*:57–69 (1980), Phospholipid polymers: Synthesis and spectral characteristics.
27. H. Hub, B. Hupfer, H. Koch, and H. Ringsdorf *Angew. Chem. Int. Ed. Engl. 19*:938–940 (1980). Polymerizable phospholipid analogues. New stable biomembrane and cell models.

28. D. F. O'Brien, R. T. Klingbiel, and T. H. Whitesides, *J. Polym. Sci., Polym. Lett. Ed.* *19*:95–101 (1981). The polymerization of lipid diacetylenes in bimolecular-layer membranes.
29. P. Yager, and P. E. Schoen, *Mol. Cryst. Liq. Cryst.* *106*:371–1381 (1984). Formation of tubules by a polymerizable surfactant.
30. J. M. Schnur, R. Price, P. Schoen, P. Yager, J. Calvert, J. Georger, and A. Singh, *Thin Solid Films* *152*:181–206 (1987). Lipid based tubule microstructures.
31. A. S. Rudolph, J. M. Calvert, P. E. Schoen, and J. M. Schnur, Technological development of lipid based tubule microstructures, in *Biotechnological Applications of Lipid Microstructures* (B. P. Gaber, J. M. Schnur, and D. Chapman, eds.), Plenum Press, New York, 1988, p. 305.
32. J. H. Fendler, *Acc. Chem. Res.* *13*:7–13 (1980). Surfactant vesicles as membrane mimetic agents: Characterization and utilization.
33. L. Gross, H. Ringsdorf, and H. Schupp, *Angew. Chem. Int. Ed. Engl.* *20*:305–325 (1981). Polymeric antitumor agents on a molecular and on a cellular level.
34. B. Hupfer, and H. Ringsdorf. Polymeric monolayers and liposomes as model of biomembranes and cells, in *Polymer Science Overview* (G. A. Stahl, ed.), ACS Symposium Series, Washington, D.C., 1981, p. 209.
35. J. H. Fendler, *Science* *223*:890–894 (1984). Polymerized sufactant vesicles: Novel membrane mimetic agents.
36. S. L. Regen, *Polymer News* *10*:68–73 (1984). Polymerized vesicles.
37. D. S. Johnston, and D. Chapman, Polymerized liposomes and vesicles, in *Liposome Technology* (G. Gregoriadis, ed.), CRC Press, Boca Raton, Fla., 1984, Vol. 1, p. 123.
38. J. H. Fuhrhop, and J. Mathieu, *Angew. Chem. Int. Ed. Engl.* *23*:100 (1984). Routes to functional membranes without protein.
39. A. Akimoto, K. Dorn, L. Gros, H. Ringsdorf, and H. Schupp, *Angew. Chem. Int. Ed. Engl.* *20*:90–91 (1981). Polymer model membranes.
40. A. Kusumi, M. Singh, D. A. Tirrell, G. Oehme, A. Singh, N. K. P. Samuel, J. S. Hyde, and S. L. Regen, *J. Amer. Chem. Soc.* *105*:2975–2980 (1983). Dynamic and structural properties of polymerized phosphatidylcholine vesicle membranes.
41. R. Elbert, A. Laschewsky, and H. Ringsdorf, *J. Amer. Chem. Soc.* *107*:4134–4141 (1985). Hydrophilic spacer groups in polymerizable lipids: Formation of biomembrane models from bulk polymerized lipids.
42. E. Sackmann, P. Eggl, C. Fahn, H. Bader, H. Ringsdorf, and M. Schollmeier, *Ber. Bunsenges. Phys. Chem.* *89*:1198–1208 (1985). Compound membranes of linearly polymerized and cross-linked macrolids with phospholipids: Preparation, microstructure and applications.
43. A. Laschewsky, H. Ringsdorf, G. Schmidt, and J. Schneider, *J. Amer. Chem. Soc.* *109*:788–796 (1987). Self-organization of polymeric lipids with hydrophilic spacers in side groups and main chain: Investigation in monolayers and multilayers.
44. P. Meller, R. Peters, and H. Ringsdorf, *Colloid. Polym. Sci.* *267*:97–107 (1989). Microstructure and lateral diffusion in monlayers of polymerizable amphiphile.
45. S. L. Regan, A. Singh, G. Oehme, and M. Singh, *J. Amer. Chem. Soc.* *104*:791–795 (1982). Polymerized phosphatidylcholine vesicles: Synthesis and characterization.
46. S. L. Regen, A. Singh, G. Oehme, and M. Singh, *Biochem. Biophys. Res. Commun.* *101*:131–136 (1981). Polymerized phosphatidylcholine vesicles. Stabilized and controllable time-release carriers.
47. R. L. Juliano, M. J. Hsu, D. Peterson, S. L. Regen, and A. Singh, *Exp. Cell. Res.* *146*:422–427 (1983). Interaction of conventional or photopolymerized liposomes with platelets in vitro.
48. R. L. Juliano, S. L. Regen, M. Singh, M. J. Hsu, and A. Singh, *Biotechnology* *1*:882–885 (1983). Stability properties of photopolymerized liposomes.

49. S. L. Regen, P. Kirszensztejn, A. Singh, *Macromolecules 16*:335–338 (1983). Polymer-supported membranes. A new approach for modifying polymer surfaces.
50. S. L. Regen, M. Singh, and N. K. P. Samuel, *Biochem. Biophys. Res. Commun. 119*:646–651 (1984). Functionalized polymeric liposomes. Efficient immobilization of alpha chymotrypsin.
51. S. L. Regen, *Ann. N.Y. Acad. Sci. 446*:296–307 (1985). Polymerized phosphatidylcholine vesicles as drug carriers.
52. A. Singh, R. Price, P. E. Schoen, P. Yager, and J. M. Schnur, *Polymer Preprints 27*(2):393–394. Tubule formation by heterobifunctional polymerizable lipids: Synthesis and characterization.
53. Z. Foltynowicz, K. Yamaguchi, B. Czajka, and S. L. Regen, *Macromolecules 18*:1394–1401 (1985). Modification of low-density polyethylene film using polymerizable surfactants.
54. R. Mehta, M. J. Hus, R. L. Juliano, H. J. Krause, and S. L. Regen *J. Pharm. Sci. 75*:579–581 (1986). Polymerized phospholipid vesicles containing amphotericin B: Evaluation of toxic and antifungal activities in vitro.
55. H. J. Krause, R. L. Juliano, and S. Regan, *J. Pharm. Sci. 76*:1–5 (1987). In vivo behavior of polymerized lipid vesicles.
56. F. Bonte, M. J. Hsu, A. Papp, K. Wu, S. L. Regen, and R. L. Juliano, *Biochim. Biophys. Acta 900*:1–9 (1987). Interaction of polymerizable phosphatidylcholine with blood components: Relevance to biocompatibility.
57. J. F. Martin, W. J. Hubbell, and D. Papahadjopoulos, *Biochemistry 20*:4229–4238 (1981). Immunospecific targeting of liposomes to cells: A novel and efficient method for covalent attachment of Fab fragments via disulfide bonds.
58. S. L. Regen, K. Yamaguchi, N. K. P. Samuel, and M. Singh, *J. Amer. Chem. Soc. 105*:6354–6355 (1983). Polymerized-depolymerized vesicles. A reversible phosphatidylcholine-based membrane.
59. N. K. P. Samuel, M. Singh, K. Yamaguchi, and S. L. Regen, *J. Amer. Chem. Soc. 107*:42–47 (1985). Polymerized-depolymerized vesicles. Reversible thiol-disulfide-based phosphatidylcholine membranes.
60. T. Diem, B. Czajka, B. Weber, and S. L. Regen, *J. Amer. Chem. Soc. 108*:6094–6095 (1986). Spontaneous assembly of phospholipid monolayers via adsorption onto gold.
61. S. L. Regen, *NATO ASI Series, Ser. C. 215*:317–324 (1987). Polymerized vesicles.
62. A. Sadownik, J. Stefely, and S. L. Regen, *J. Amer. Chem. Soc. 108*:7789–7791 (1986). Polymerized liposomes formed under extremely mild conditions.
63. J. Stefely, M. A. Markowitz, and S. L. Regen, *J. Amer. Chem. Soc. 110*:7463–7469 (1988). Permeability characteristics of lipid bilayers from lipoic acid derived phosphatidylcholines: Comparison of monomeric, cross-linked and non-cross-linked polymerized membranes.
64. B. Hupfer, H. Ringsdorf, and H. Schupp *Makromol. Chem. 182*:247–253 (1981). Polymeric phospholipid monolayers.
65. H. Bueschl, H. Ringsdorf, and U. Zimmermann, *FEBS Letters 150*:38–42 (1982). Electric field-induced fusion of large liposomes from natural and polymerized phosphoplipids.
66. V. P. Torchilin, A. L. Klibanov, N. N. Ivanov, H. Ringsdorf, and B. Schlarb, *Makromol. Chem., Rapid Commun. 8*:457–460 (1987). Polymerization of liposome-encapsulated hydrophilic monomers.
67. M. Takane, K. Shigehara, and E. Tsuchida, *Makromol. Chem. 187*:853–856 (1986). Polymerized phospholipids and their polymeric liposomes.
68. Y. Okahata, K. Ariga, and T. Seki, *J. Amer. Chem. Soc. 110*:2496–2500 (1988). Polymerized lipid-corked capsule membranes. Polymerization at different positions of corking lipid bilayers on the capsule and effect of polymerization on permeation behavior.
69. K. Dorn, R. T. Klingbiel, D. P. Specht, P. N. Tyminski, H. Ringsdorf, and D. F. O'Brien, *J. Amer. Chem. Soc. 106*:1627–1633 (1984). Permeability characteristics of polymeric bilayer membranes from methacryloyl and butadiene lipids.

70. D. F. O'Brien, R. T. Klingbiel, D. P. Specht, and P. N. Tyminski, *Ann. N.Y. Acad. Sci. 446*:282–295 (1985). Preparation and characterization of polymerized liposomes.
71. P. N. Tyminski, I. S. Ponticello, and D. F. O'Brien, *J. Amer. Chem. Soc. 109*:6451–6452 (1987). Polymerizable dienoyl lipids as spectroscopic bilayer membrane probes.
72. H. Nishide, M. Yuasa, Y. Hashimoto, and E. Tsuchida, *Macromolecules 20*:459–461 (1987). Amphiphilic and polymerizable porphyrins and their copolymerization with phospholipid: Oriented fixation of porphyrins in a bilayer membrane.
73. H. Ohno, S. Takeoka, and E. Tsuchida, *Bull. Chem. Soc. Jpn. 60*:2945–2951 (1987). Unequivalent chemical environment of diene groups in 1- and 2-acyl chains of polymerizable lipids analyzed by radical polymerization.
74. H. Ohno, Y. Ogata, and E. Tsuchida, *Macromolecules 20*:929–933 (1987). Polymerization of liposomes composed of diene containing lipids by uv and radical initiators: Evidence for the different chemical environment of diene groups on 1- and 2-acyl chains.
75. E. Hasegawa, K. Eshima, Y. Matsushita, H. Ohno, and E. Tsuchida, *Polymer Bull.* (Berlin)*18*:65–71 (1987). Characterization of polymerized vesicles derived from polymerizable 1,3-diglycero-2-phosphocholines: Formation of large unilamellar vesicles by ultrasonication.
76. M. Yuasa, H. Nishide, E. Tsuchida, and A. Yamagishi *J. Phys. Chem. 92*:2987–2990 (1988). Oriented fixation of synthetic heme complexes in phospholipid bilayer membranes: Electro optic measurement.
77. H. Kitano, N. Kato, Tanaka, and N. Ise, *Biochem. Biophys. Acta 942*:131–138 (1988). Mutual recognition between polymerized liposomes: Enzyme and enzyme inhibitor system.
78. E. Tsuchida, H. Nishide, M. Yuasa, T. Babe, and M. Fukuzumi, *Macromolecules 22*:66–72 (1989). Synthesis of polymerizable amphiphiles (porphinato) irons and their copolymers with polymerizable phosphlipid.
79. E. Wang, E. Shouji, H. Ohno, and E. Tsuchida, *Polym. Bull.* (Berlin) *21*:195–202 (1989). Fluorescence behavior of porhpinato zinc derivative in the molecular assembly of polymerized lipid.
80. H. Ohno, S. Takeoka, H. Iwai, and E. Tsuchida, *Macromolecules 22*:61–66 (1989). Effect of phase transition on photosensitized radical polymerization of direne-containing lipids as liposomes.
81. E. Tsuchida, H. Nishide, M. Yuasa, E. Hasegawa, K. Eshima, and Y. Matsushita, *Macromolecules 22*:2103–2107 (1989). Polymerizable liposome/lipid-heme as an oxygen transporter under physiological conditions.
82. H. Kitano, N. Kato, and N. Ise, *J. Amer. Chem. Soc. 111*:6809–6813 (1989). Mutual recognition between polymerized liposomes: Macrophage model system by polymerized liposomes.
83. P. N. Tyminski, L. H. Latimer, and D. F. O'Brien, *Biochemistry 27*:2696–2705 (1988). Reconstitution of rhodopsin and cGMP cascade in polymerized bilayer membranes.
84. E. Hasegawa, Y. Matsushita, K. Eshima, H. Ohno, and E. Tsuchida, *Polym. Bull.* (Berlin) *15*:397–403 (1986). Polymerizable glycerophosphocholines containing terminal 2,4-hexadienoyloxy groups and their polymerized vesicles.
85. D. A. Frankel, H. Lamparski, U. Liman, and D. F. O'Brien, *J. Amer. Chem. Soc. 111*:9262–9263 (1989). Photoinduced destabilization of bilayer vesicles.
86. E. Hasegawa, K. Eshima, Y. Matsushita, H. Nishide, and E. Tsuchida, *Polym. Bull. 14*:31–38 (1985). Synthesis of polymerizable glycerophosphocholines and their polymerizable vesicles.
87. E. K. Hasegawa, K. Matsushita, and K. Eshima, *Makromol. Chem. Rapid Commun. 5*:779–784 (1984). Synthesis of novel styrene groups containing glycerophosphocholines and their polymerization as liposomes.
88. E. Tsuchida, E. Hasegawa, Y. Matsushita, K. Eshima, M. Yuasa, and H. Nishide, *Chemistry Lett.* 969–972 (1985). Polymerized liposomes as the carrier of heme. A physically stable oxygen carrier under physiological conditions.

89. J. Ohno, S. Takeoda, and E. Tsuchida, *Polym. Bull. 14*:487–490 (1985). Skeletonized hybrid liposomes.
90. E. Tsuchida, H. Nishide, and M. Yuasa, *J. Macromol. Sci.-Chem. A24*:333–341 (1987). Oxygen-binding profile of the iron porphyrin complex embedded in polymerized liposome: Effect of the polymerized liposomes on the stability and the oxygen-binding ability of the iron porphyrin complex.
91. M. Yuawa, E. Hasegawa, H. Nishide, and E. Tsuchida, *J. Macromol. Sci.-Chem. A24*:661–668 (1987). Oxygen-exchange reaction between artificial lung device: The heme embedded in polymerized lipo liposome as an artificial oxygen carrier.
92. Y. Matsushita, E. Hasegawa, and K. Eshima, *Makromol. Chem., Rapid Commun. 8*:1–6 (1987). Synthesis and properties of polymerizable phospholipids. 5. Molecular weight of polymeric liposomes.
93. B. Tieke, G. Lieser, and G. Wegner *J. Polym. Sci., Polym. Chem. Ed. 17*: 1631–1644 (1979). Polymerization of diacetylene in multilayers.
94. R. H. Baughman, *J. Appl. Phys. 43*:4362–4370 (1972). Solid state polymerization of diacetylenes.
95. R. Bueschl, B. Hupfer, and H. Ringsdorf *Makromol. Chem., Rapid Commun. 3*:588–596 (1982). Mixed monolayers and liposomes from natural and polymerizable lipids.
96. B. Hupfer, H. Ringsdorf, and H. Schupp, *Chem. Phys. Lipids 33*:355–374 (1983). Liposome from polymerizable phospholipids.
97. M. Pons, D. S. Johnston, and D. Chapman *J. Polym. Sci.: Chem. Ed. 20*:513–520 (1982). A study of the spectra of diacetylenic phospholipid polymers in solvents and dispersions.
98. M. Pons, C. Villaverde, and D. Chapman, *Biochim. Biophys. Acta 730*:306–312 (1983). A ^{13}C-NMR study of 10,12-tricosadiynoic acid and the corresponding phospholipid and phospholipid polymer.
99. D. S. Johnston, L. R. McLean, M. A. Whittman, A. D. Clark, and D. Chapman, *Biochemistry 22*:3194–3202 (1983). Spectra and physical properties of liposomes and monolayers of polymerizable phosphlipids containing diacetylene groups in one or both acyl chains.
100. J. Leaver, A. Alonso, A. A. Durrani, and D. Chapman, *Biochim. Biophys. Acta 732*:210–218 (1983). The physical properties and photopolymerization of diacetylene-containing phospholipid liposomes.
101. D. S. Johnston, and D. Chapman, Polymerized liposomes and vesicles, in *Liposome Technology* (G. Gregoriadis, ed.) *1*:123–129, CRC Press, Boca Raton, Fla., 1984.
102. Science News Report. *Science* (U.S.A.) *247*:1410 (1990). Physicists tackle theory, tubes, and temperature.
103. N. Nakashima, S. Asakuma, and T. Kunitake, *J. Amer. Chem. Soc. 107*:509–510 (1985). Optical microscopic study of helical superstructures of chiral bilayer membranes.
104. J.-H. Fuhrhop, P. Scneider, E. Boekema, and W. Helfrich, *J. Amer. Chem. Soc. 110*:2861–2867 (1988). Lipid bilayer fibers from diastereomeric and enantiomeric *N*-octylaldonamides.
105. K. Yamada, H. Ihara, T. Ide, T, Fukumoto, and C. Hirayama, *Chem. Lett.* 1713–1716 (1984). Formation of helical super structure from single-walled bilayers by amphiphiles with oligo-L-glutamic head group.
106. W. Helfrich, *J. Chem. Phys. 85*:1085–1987 (1986). Helical bilayer structures due to spontaneous torsion of the edges.
107. P. G. De Gennes, *C. R. Acad. Sci. Paris 304*:259–263 (1987). Surface and interphase physics. Electrostatic buckling of chiral lipid bilayers.
108. A. Singh, B. Herendeen, B. Gaber, and J. P. Sheridan, *Polym. Mat. Sci. Eng. 56*:283–285 (1987). Polymerization behavior of diacetylenic phosphatidylcholine vesicles in the presence of analogous acetylenic phosphatidylcholine.
109. A. Singh, *J. Lipid Res. 31*:1522–1525 (1990). An efficient synthesis of phosphatidylcholines.

110. A. Singh, B. P. Singh, B. P. Gaber, B. Herendeen, R. Price, T. G. Burke, P. E. Schoen, J. M. Schnur, and Y. Yager, in *Surfactants in Solution* (K. L. Mittal, ed.), Plenum Press, New York, *8*:467–476 (1989). Synthesis and characterization of positional isomers of 1,2 bis heptacosadiynoyl phosphatidylcholine.
111. A. Singh, and J. M. Schnur, *Polymer Preprints 26*(2):184–185 (1985). Polymerized diacetylenic phosphatidylcholine vesicles: Synthesis and characterization.
112. A. Singh, P. E. Schoen, and J. M. Schnur *J. Chem. Soc., Chem. Commun.* (18):1222–1223 (1988). Self-assembled microstructures from a polymerizable ammonium surfactant: di-(hexacosa-12,14-diynyl) dimethylammonium bromide.
113. A. Singh, T. G. Burke, J. M. Calvert, J. H. Georger, B. Herendeen, R. R. Price, P. E. Schoen, and P. Yager, *Chem. Phys. Lipids 47*:135–1348 (1988). Lateral phase separation based on chirality in a polymerizable lipid and its influence on formation of tubular microstructures.
114. P. Yager, R. R. Price, J. M. Schnur, P. E. Schoen, A. Singh, and D. G. Rhodes, *Chem. Phys. Lipids 46*:171–179 (1988). The mechanism of formation of lipid tubules from liposomes.
115. P. Schoen, P. Yager, J. P. Sheridan, R. Price, J. M. Schnur, A. Singh, D. G. Rhodes, and S. L. Blechner, *Mol. Cryst. Liq. Cryst. 153*:357–363 (1987). Order in diacetylenic microstructures.
116. P. Yager, P. E. Schoen, J. H. Georger, R. R. Price, and A. Singh, *Biophys. J. 49*:320a (1986). Two mechanisms for forming novel tubular microstructures from polymerizable lipids.
117. A. L. Plant, D. M. Benson, and G. L. Trusty, *Biophys. J. 57*:925–933 (1990). Probing the structure of diacetylenic phospholipid tubules with fluorescent lipophiles.
118. M. F. Sonnenschein, and R. G. Weiss, *Photochem. Photobiol.* (1990). Depth profiles for permeation of water and oxygen in vesicles and tubule phases by fluorescence quenching studies.
119. R. Treanor, and M. D. Pace, *Biochim. Biophys. Acta 1046*:1–11 (1990). Microstructure, order and fluidity of 1,2 bis (tricosa-10,12-diynoyl)-*sn*-glycero-3-phosphocholine ($DC_{8,9}PC$), a polymerizable lipid, by ESR and NMR.
120. J. Chappell, and P. Yager, *Chem. Phys. Lipids 58*:253–258 (1991). A model for crystalline order within helical and tubular structures of chiral bilayers.
121. P. Yager, P. E. Schoen, C. Davies, R. Price, and A. Singh, *Biophys. J. 48*:899–906 (1985). Structure of lipid tubules formed from a polymerizable lecithin.
122. J. H. Georger, A. Singh, R. R. Price, J. M. Schnur, P. Yager, and P. E. Schoen, *J. Amer. Chem. Soc. 109*:6169–6175 (1987). Helical and tubular microstructures formed by polymerizable phosphatidylcholines.
123. A. S. Rudolph, J. M. Calvert, M. E. Ayers, and Joel M. Schnur, *J. Amer. Chem. Soc. 111*:8516–8517 (1989). Water-free self-assembly of phosphlipid tubules.
124. J. M. Schnur, R. Price, P. Yager, P. Schoen, J. H. Georger, and A. Singh, U.S. Patent 4,877,501, Oct. 31, 1989. Process for fabrication of lipid microstructures.
125. T. G. Burke, A. S. Rudolph, R. R. Price, J. P. Sheridan, A. W. Dalziel, A. Singh, and P. E. Schoen, *Chem. Phys. Lipids 48*:215–230 (1988). Differential scanning calorimetric study of the thermotropic phase behavior of a polymerizable, tubule-forming lipid.
126. A. S. Rudolph, P. E. Schoen, M. Nagumo, F. Behroozi, T. G. Burke, M. E. Ayers, A. Singh, and R. Treanor, *SPIE 1057*, 57–63 (1989). Spectroscopic studies of tubule-forming polymerizable lecithins.
127. P. E. Schoen, and P. Yager *J. Polym. Sci, Polym. Phys. Ed. 23*:2203–2216 (1985). Spectroscopic studies of polymerized surfactants: 1,2-bis(10, 12-tricosadiynoyl)-*sn*-glycero-3-phosphocholine.
128. J. S. Sheridan, *NRL Memorandum Report 5975*, Naval Research Laboratory, Washington, D.C., 1988. Conformational order in lipid tubules formed from a diacetylenic lecithin: A Raman spectroscopic study.

129. A. S. Rudolph, B. P. Singh, A. Singh, and T. G. Burke, *Biochim. Biophys. Acta 943*:454–462 (1988). Phase characteristic of positional isomers of 1,2 bis heptacosadiynoyl-*sn*-glycero-3-phosphocholines.
130. A. S. Rudolph, and T. G. Burke, *Biochim. Biophys. Acta 902*:349–359 (1987). A Fourier-transform infrared spectroscopic study of the polymorphic phase behavior of 1,2-bis(tricosa-10,12-diynoyl)-*sn*-glycero-3-phosphocholine, 1 polymerizable lipid which forms novel microstructures.
131. D. G. Rhodes, S. L. Blechner, P. Yager, and P. E. Schoen, *Chem. Phys. Lipids 49*:39–47 (1988). Structure of polymerizable lipid bilayers. I. 1,2-Bis (10,12-tricosadiynoyl)-*sn*-glycero-3-phosphocholine, a tubule-forming phosphatidylcholine.
132. T. G. Burke, A. Singh, and P. Yager, *Ann. N.Y. Acad. Sci. 507*:330–332 (1987). Entrapment of 6-carboxyfluorescein within cylindrical phospholipid microstructures.
133. P. L. Ahl, R. Price, J. Schumuda, B. P. Gaber, and A. Singh, *Biochim. Biophys. Acta 1028*:141–153 (1990). Insertion of bacteriorhodopsin into polymerized diacetylenic phosphatidylcholine bilayers.
134. A. W. Dalziel, J. Georger, R. R. Price, A. Singh, and P. Yager, Progress report on the fabrication of an acetylcholine receptor-based biosensor, in *Membrane Proteins* (Steven C. Goheen, ed.), Bio-Rad Laboratories, 1986, pp. 643–673.
135. P. Yager, *Biosensors 2*:363–373 (1986). Functional reconstitution of a membrane protein in a diacetylenic polymerizable lecithin.
136. F. S. Ligler, T. L. Fare, K. D. Seib, J. W. Smuda, A. Singh, M. E. Ayers, A. Dalziel, and P. Yager, *Med. Inst. 22*:247–256 (1988). Fabrication of key components of receptor based biosensor.
137. T. L. Fare, A. Singh, K. D. Seib, J. W. Smuda, P. L. Ahl, F. S. Ligler, and J. M. Schnur, Incorporation of ion channels in polymerized membranes and fabrication of a biosensor, in *Molecular Electronics* (Felix T. Hong, ed.), Plenum Press, New York, 1989, pp. 305–315.
138. M. Markowitz and A. Singh, *Langmuir 7*(2):16–18 (1991). Self-assembling properties of 1,2 diacyl-*sn*-glycero-3-phosphohydroxyethanol: A headgroup modified diacetylenic phospholipid.
139. R. R. Price and M. Patchan, *J. Microencapsulation 8*:301–306 (1991). Controlled release from cylindrical microstructures.
140. J. M. Schnur and A. Singh, *U. S. Patent 4867,917*, Sept. 19, 1989. A general method for the synthesis of diacetylenic acids.
141. M. C. Cleji, M. F. M. Rocks, and R. J. M. Nolte, *Polymer Prepr. 28*(2):432–433 (1987). Effect of polymerization on properties of vesicles derived from isocyano surfactant.
142. M. F. M. Roks, R. S. Dezentje, V. E. M. Kaats-Richters, W. Drenth, J. Verkleij, and R. J. M. Nolte, *Macromolecules 20*:920–929 (1987). Synthesis and characterization of vesicles stabilized by polymerization of isocyano functions.
143. C. M. Gupta, C. C. Costello, and H. G. Khorana, *Proc. Natl. Acad. Sci. U.S.A. 76*:3139–3143 (1979). Site of intermolecular crosslinking of fatty acyl chains in phospholipids carrying a photoactivable carbene precursor.
144. M. L. Tsirenina, T. N. Simonova, N. A. Koltovaya, E. F. Golubeva, and A. N. Ushakov, *Soviet J. Bioorg. Chem. 7*:671–679 (1981). A study of lipid-lipid and lipid-protein interactions in membranes using phospholipids containing photoreactive groupings. Synthesis of new photoreactive phosphatidylcholines.
145. C. M. Gupta, R. Radhakrishnan, and H. G. Khorana, *Proc. Natl. Acad. Sci. U.S.A. 74*:4315–4319 (1977). Glycerophospholipid synthesis: Improved general method and new analogues containing photoactivable groups.
146. Y. G. Molotkovskii, A. A. Dergousov, and L. D. Bergel'son, *Soviet J. Biorg. Chem. 14*:849–856 (1988). New type of polymerizable phosphatidylcholines: Synthesis and properties.

147. C. M. Gupta, R. Radhakrishnan, G. E. Gerber, W. L. Olsen, S. L. Quay, and H. G. Khorana, *Proc. Natl. Acad. Sci. U.S.A. 76*:2595–2599 (1979). Intermolecular crosslinking of fatty acyl chains in phospholipids: Use of photocleavable carbene precursors.
148. Y. M. Ohkatsu, M. Yokotsu, and T. Kusano, *Makromol. Chem. 189*:775–760 (1988) Synthesis and polymerization of macromonomer derived from phosphatidylcholine.
149. A. Singh, S. Marchywka, and B. P. Gaber, *Polym. Mat. Sci. Eng. 61*:931–935 (1989). Polymerization behavior of aqueous dispersions of diacetylenic and short chain phospholipid mixtures.
150. A. Singh, and S. Marchywka, *Polym. Mat. Sci. Eng. 61*:675–678 (1989). Synthesis and characterization of headgroup modified 1,3 diacetylenic phospholipids.
151. S. L. Regen, J. S. Shin, and K. Yamaguchi, *J. Amer. Chem. Soc. 106*:2446–2447 (1984). Polymer-encased vesicles.
152. S. L. Regen, J. S. Shin, J. F. Hainfeld, and J. S. Wall, *J. Amer. Chem. Soc. 106*:5756–5757 (1984). Ghost vesicles.
153. H. Fukuda, T. Diem, J. Stefely, F. J. Kezdey, and S. L. Regen, *J. Amer. Chem. Soc. 108*:2321–2327 (1986). Polymer-encased vesicles derived from dioctadecyldimethylammonium methacrylate.
154. K. V. Aliev, H. Ringsdorf, B. Schlarb, and K. H. Leister, *Makromol. Chem., Rapid Commun. 5*:345–352 (1984). Liposome in net: Spontaneous polymerization of 4-vinylpyridine on acidic liposomal surfaces.
155. N. Higashi, T. Adachi, and M. Niwa, *J. Chem. Soc., Chem. Commun*:1573–1575 (1988). Molecular weight control of photopolymerization at an oriented bilayer surface using phase separation of flurocarbon- and hydrocarbon-amphiphiles.
156. H. Ringsdorf, and B. Schlarb, *Polymer Prepr. 27*(2):195–196 (1986). Liposomes in a net from lipids with ionically or covalently bound polymerizable headgroups.
157. J. M. Schnur, P. Yager, R. Price, J. M. Calvert, P. E. Schoen, and J. H. Georger, *U.S. Patent #4,911,981*, March 27, 1990. Metal clad lipid microstructures.
158. B. Hall, R. le R. Bird, and D. Chapman, *Angew, Makromol. Chem. 166/167*:169–178 (1989). Phosphlipid polymers and new haemocompatible materials.
159. D. Chapman, *Eur. Pat. Appl. EP 32622*, July 29, 1981, 38 pp. Polymerizable phospholipids and polymers, their use in coating substrates and forming liposomes and the resulting coated substrates and liposome compositions.
160. S. L. Regen, *Eur. Pat. Appl. EP 153133*, 28 Aug 1985, 27 pp. Assembling multilayers of polymerizable surfactant on a surface of a solid material.
161. S. L. Regen, *US Patent #4,560,599*, December 24, 1985, 16 pp. Assembling multilayers of polymerizable surfactant on a surface of a solid material.
162. K. Dorn, E. V. Patton, R. T. Klingbiel, D. F. O'Brien, and H. Ringsdorf, *Makromol. Chem. Rapid Commun. 4*:513–517 (1983). Molecular weight of polymers from methacryloyl lipids in bilayer membranes.
163. E. Hasegawa, K. Ejima, Y. Matsushita, and H. Tsuchida, *Kokai Tokkyo Koho JP 61178996*, August 11, 1986, 8 pp. Polymerizable phospholipids.
164. K. Hirotake, S. Kobayashi, H. Matsumura, H. Yokoyama, M. Aizama, and Y. Katayama, *Jpn. Kokai Tokkyo Koho JP 63,274,870*, Nov. 11, 1988, 8 pp. Preparation of stabilized liposomes or micelles for use in agglutination tests.
165. H. Yoshioka, and Suzuki, K. *Eur. Pat. Appl. EP 245,799*, November 19, 1987, 35 pp. Radiation-sensitive polymerizable coating material containing eleostearate residues for testing surfaces to make them biocompatible.
166. Toyo Soda Mfg. Co., *Jpn. Kokai Tokkyo Koho JP 60067489*, April 17, 1985, 5 pp. Manufacture of polymerizable phospholipids.
167. D. F. O'Brien, *Polymer. Prepr. 28*(2):438–439 (1987). Permeability of polymerized vesicles.

168. J. A. Hayward, M. L. Daniel, N. Lawrence, R. S. Sanford, D. S. Johnston, and D. Chapman, *FEBS Letters 187*:261–266 (1985). Polymerized liposomes as stable oxygen-carriers.
169. J. A. Hayward, D. S. Johnston, and D. Chapman *Ann. N.Y. Acad. Sci. 446*:267–281 (1985). Polymeric phospholipids as new biomaterials.
170. R. Pabst, H. Ringsdorf, H. Koch, and K. Dose, *FEBS Lett. 154*:5–9 (1983). Light-driven proton transport of bacteriorhodopsin incorporated into a long-term stable liposome of a polymerizable sulfolipid.
171. C. Rosenblatt, P. Yager, and P. E. Schoen, *Biophys. J. 52*:295–301 (1987). Orientation of lipid tubules by a magnetic field.
172. F. Behroozi, M. Orman, W. Stockton, J. Calvert, F. Rochford, and P. Schoen, *J. Appl. Phys. 68*:3688–3693 (1990): Interaction of metallized tubules with electromagnetic radiations.
173. D. A. Kirkpatrick, P. E. Schoen, W. B. Stockton, R. Price, S. Baral, K. Brian, J. M. Schnur, M. Levinson, and B. M. Ditchek *IEEE Transactions on Plasma Science 19*:749–756 (1989). Measurements of vacuum field emission from bio-molecular and semiconductor-metal eutectic composite microstructures.
174. H. Gaub, E. Sackmann, R. Bueschl, and H. Ringsdorf, *Biophys. J. 45*:725–731 (1984). Lateral diffusion and phase separation in two-dimensional solutions of polymerized butadiene lipid in dimyristoylphophatidylcholine bilayers.
175. N. Seki, E. Tsuchida, K. Ukaji, T. Sekiya, and Y. Nozawa, *Polymer Bull. 13*: 489–492 (1985). Phase separation of polymerized lipids in hybrid liposomes.
176. S. Takeoda, H. Ohno, N. Hayashi, and E. Tsuchida, *J. Controlled Release 9*:177–186 (1989). Control of release of encapsulated molecules from polymerized mixed liposomes induced by physical or chemical stimuli.
177. A. Singh, and B. P. Gaber, *Applied Bioactive Polymeric Materials* (C. G. Gebelein, C. E. Carraher, Jr., and V. R. Foster, eds.), Plenum Press, New York, 1988, 239–249. Influence of short chain lipid spacers on the properties of diacetylenic phosphatidylcholine bilayers.
178. D. G. Rhodes, and A. Singh, *Chem. Phys. Lipids 59*:215–224 (1991). Structure of polymerizable lipid bilayers IV: Mixture of long chain diacrtylenic and short chain saturated phosphatidylcholine and analogous asymmetric isomers.
179. D. Bolikal, and S. L. Regen, *Macromolecules 17*:1287 (1984). Degree of polymerization of vesicle membrane.
180. K. Yamaguchi, S. Watanabe, and S. Nakahama, *Makromol. Chem. 190*:1195–1205 (1989). Emulsion polymerization of styrene using phospholipids as emulsifier. Immobilization of phospholipids on the latex surface.
181. S. Takeoda, H. Sakai, L. Wang, H. Ohno, and E. Tsuchida, *Polymer J. 21*:641–648 (1989). Study of the phase transition behavior of polymerized liposomes through the interaction of diene-groups in their acyl chains.
182. A. Singh, R. B. Thompson, and J. M. Schnur, *J. Amer. Chem. Soc. 108*:2785–2787 (1986). Reversible thermochromism in photopolymerized phosphatidylcholine vesicles.
183. J. M. Schnur, A. Singh, *Polymer Preprints 26*:186–187 (1985). Reversible thermochromism in photopolymerized phosphatidylcholine vesicles.
184. E. Lopez, D. F. O'Brien, and T. H. Whitesides, *J. Amer. Chem. Soc. 104*:305–307 (1982). Structural effects on the photopolymerization of bilayer membranes.
185. A. Blume, *Chem. Phys. Lipids 57*:253–273 (1991). Phase transitions of polymerizable phospholipids.
186. J. R. Silvius, B. D. Read, and R. N. McElhanney, *Biochim. Biophys. Acta 555*:175 (1979). Thermotropic phase transitions of phosphatidylcholines with odd-numbered *n*-acyl chains.
187. S. L. Blechner, W. Morris, P. E. Schoen, P. Yager, A. Singh, and D. G. Rhodes, *Chem. Phys. Lipids 58*:41–54 (1991). Structure of polymerizable lipid bilayers. II. Two heptacosadiynoyl phosphotidylcholine isomers.

188. J. B. Lando and R. V. Sudiwala, *Chem. Matr.* 2:594–599 (1990). Structural investigations of Langmuir-Blodgette films and tubules of 1,2 bis (10,12-tricosadiynoyl)-*sn*-glycero-3-phosphocholine ($DC_{8,9}PC$) using electron diffraction techniques.
189. D. Frankel and D. F. O'Brien, *J. Amer. Chem. Soc.* *113*:7436–7437 (1991). Supramolecular assemblies of diacetylenic lipids.
190. H. Eibl, *Angew. Chem. Int. Ed. Engl.* *23*:257–271 (1984). Phospholipids as functional constituents of biomembranes.
191. T. H. Vaughn, and J. A. Nieuwland, *J. Amer. Chem. Soc.* *55*:3456–3458 (1933). The direct iodination of monosubstituted acetylenes.
192. H. Ringsdorf and H. Schupp, *J. Macromol. Sci. Chem.* *A15*:1015–1026 (1981). Polymerization of substituted butadienes at the gas-water interface.
193. M. A. Markowitz, J. M. Schnur, and A. Singh, *Chem. Phys. Lipids* *62*:193–204 (1992). The influence of the polar headgroups of acidic diacetylenic phospholipids on tubule formation, microstructure morphology and Langmuir film behavior.

8
Coupling and Labeling of Phospholipids

Vladimir P. Torchilin *Massachusetts General Hospital-East, Charlestown, Massachusetts*

Alexander L. Klibanov *University of Pittsburgh School of Medicine, Pittsburgh, Pennsylvania*

I. INTRODUCTION

Chemical modification of phospholipid molecules allows us to increase substantially the possibilities of their application for research, clinical, and industrial purposes, especially in the field of applied bioengineering and biotechnology. The aim of this chapter is to inform the reader about the possibilities provided by chemically modified phospholipids in bioengineering. Simple procedures for the coupling of phospholipids to proteins, peptides, carbohydrates, etc. will be provided. Application of phospholipids as "anchors" for the attachment of proteins and other natural and synthetic molecules to liposomes and biomembranes will be briefly discussed as well as the labeling of liposomes with modified phospholipids.

We focus on the chemical modification of the polar region of the phospholipid molecule (in fact, the most widely used reaction nowadays is acylation or alkylation of the primary amino group of phosphatidylethanolamine). The more traditional chemical modification of the hydrophobic, nonpolar region of the phospholipid molecule will be mentioned only briefly [1,2]. The phospholipid's fatty acids have been replaced with derivatives carrying specific labels, e.g., fluorescent or photoaffinity ones. But chain attachment or replacement is dealt with in the chapter on the chemical synthesis of phospholipids.

II. REASONS FOR THE ATTACHMENT OF A PHOSPHOLIPID TO A CERTAIN COMPOUND

Attachment of a hydrophobic anchor to a given compound in order to link the latter to a cell surface, liposome membrane or a lipid monolayer.

Attachment of peptides, hormones, carbohydrates, or proteins/polymers to a liposome. This may help to enhance antibody response against an appropriate compound upon in vivo injection.

Use of certain phospholipid derivatives for the specific delivery of liposomes to specific cells or tissues or the protection of liposomes from such cells or tissues (e.g., from recognition and uptake by RES cells).

Construction of a liposome-based immunoassay system, immobilizing either antibody or antigen on the surface of liposomes.
Attachment of labels or label carriers to a phospholipid for monitoring the fate of resulting liposomes in a body or cell.
Construction of inactive, nontoxic phospholipid derivatives of a given drug, which will be activated by the intracellular enzymes within a target cell.
Preparation of positively charged phospholipid derivatives to infect cells with foreign genetic material.

III. COUPLING OF WATER-SOLUBLE COMPOUNDS TO THE SURFACE OF PREFORMED LIPOSOMES

Historically, the first attempt to modify a phospholipid headgroup in an aqueous environment was associated with a question: Does the incorporation of phosphatidylethanolamine into a lipid membrane influence the reactivity of the PtdEtn amino group? [3]. It was shown that the primary amino group of a PtdEtn molecule incorporated into a liposome can react with trinitrobenzene sulfonic acid or formaldehyde. The continuation of research in this field has led to the development of protocols for the coupling of certain ligands to the preformed lipid membranes containing PtdEtn.

It was shown, for example, that carboxylic groups of immunoglobulins can be activated by water-soluble carbodiimide; activated protein then can be bound to PtdEtn-containing liposomes. Some of the antibody carboxylic groups form an amide bond with a PtdEtn amino group; the coupling yield of the reaction is relatively low [4]; however, increase in the antibody concentration results in significant inactivation of antibody and in the formation of by-products: cross-linked protein molecules as well as cross-linked liposome aggregates both appear [5].

Surface phospholipid headgroups of the preformed liposomes can also be activated with homo-cross-linking agents, like glutaric aldehyde or dimethyl suberimidate. Experimentally, preformed liposomes were treated with a cross-linking agent. Then the excess of the latter was rapidly removed by a gel filtration/dialysis, and ligand was immediately added. This technique allowed attachment of up to $1–2 \times 10^{-4}$ mol protein/mol liposomal lipid. Antibodies [6], enzymes, such as chymotrypsin [7] and lysozyme [8], transferrin [9], peptides [10], and *p*-aminophenyl-derivatives of certain carbohydrates [11] have already been attached successfully to liposomes by this method. The main advantage of the latter approach is that it is simple; its disadvantage is that by-reactions occur, such as hydrolysis of the active imidoester group [7] or cross-linking of PtdEtn molecules in the liposomal membrane [12]. Cross-linking and aggregation of liposomes also can take place. It is evident that the described reactions seem to be far from optimal in terms of the coupling yield and the degree of liposome coating with a ligand (i.e., the quantity of ligand molecules attached to one liposome). One can increase the degree of liposome coating by using a large excess of the ligand, though this causes the coupling yield to drop significantly.

To improve the coupling protocol, phospholipids can be premodified, for example with the aid of heterobifunctional cross-linking reagents (e.g., in organic solvent); purification prior to the attachment of a ligand is also useful (see Sec. VI for details). This approach, in some instances, increases the coupling effectiveness and narrows the range of by-products. Chemical modification of a PtdEtn primary amino group with a heterobifunctional reagent was initially carried out by Leserman et al. [13]. The popular *N*-

succinimidyl-3(2-pyridyldithio)propionate (SPDP) reagent was used for the synthesis of a PtdEtn derivative, which was further used for the coupling to SH-containing proteins after lipid purification (see Sec. VI and Worksheet 1 for the details of PDP-PtdEtn preparation).

Some proteins, or protein fragments (e.g., Fab' fragments of immunoglobulin molecules), carry reactive thiol, SH groups; on other proteins, such groups can be created by additional chemical modification, by the modification of protein primary amino groups with SPDP, for example [13]. Then starting dithiopyridine residues in such procedures are reduced with dithiothreitol, thiopyridine is released, and the SH group is generated. To avoid cleavage of the natural disulfide bonds in a protein, mild acidic pH is often the choice. The release of thiopyridine can be assessed by the increase of optical density at 343 nm (molar extinction = 8080).

Another possibility is to rely on the reaction of iminothiolane with protein amino groups (as in Ref. 14). This reagent is mild; a thiol group is generated in a single-step reaction and does not require additional chemical modification. Moreover, the positive charge on the modified primary amino group is retained; this preserves the electrostatic balance in a protein globule and also its active conformation. (In case of SPDP modification, the positive charge on the amino group is lost.)

Most interesting is the application of free thiol groups on immunoglobulin Fab' fragments. It is believed that these SH groups are located far from the antigen-binding sites, enabling the phospholipid-bound antibody fragments to retain their avidity toward antigens.

The main advantages of this procedure are its simplicity and the possibility of controlling the progress of the reaction as well as the quality of reagents prior to performing the final step by the described measurement of thiopyridine release. The main disadvantages are

1. Undesired reactions between the free thiol groups with the formation of disulfide bonds. To minimize this reaction, a nitrogen or argon inert atmosphere can be used. However, it was reported that even under an argon atmosphere the antibody Fab' fragments can reassociate and form disulfide bonds [15]. Freshly prepared thiol-containing ligands, therefore, should be used for the coupling to corresponding reactive phospholipids (see also Worksheet 8).
2. Disulfide bonds between a ligand and a phospholipid may be unstable under physiological conditions; Fab' fragments have been reported [15], for example, to dissociate from liposomes during prolonged incubations with serum.

To solve this problem, a different coupling chemistry can be applied, that is, the thiol groups on a ligand (protein) are reacted with the maleimide-carrying phospholipid molecules (see Fig. 1). This approach was proposed by Martin and Papahadjopoulos [16] and is now one of the most widely used in research and practical applications. Different commercially available maleimide reagents can be used for the preparation of maleimide-carrying phospholipids in a simple single-step procedure (see Worksheet 1).

Various high and low molecular weight compounds have been attached to liposomes by using pyridyldithiopropionyl-PtdEtn or maleimide reagents. These include carbohydrates [17], coenzymes [18], fragments of immunoglobulin M [19], Fab' fragments of IgG antibodies [15,16,20], protein A and other proteins with chemically introduced free thiol groups [13,21], carboxylic polymers (to provide pH-sensitive release of liposome contents [22]), etc. Coupling yields were usually close to 10–20% with regard to the

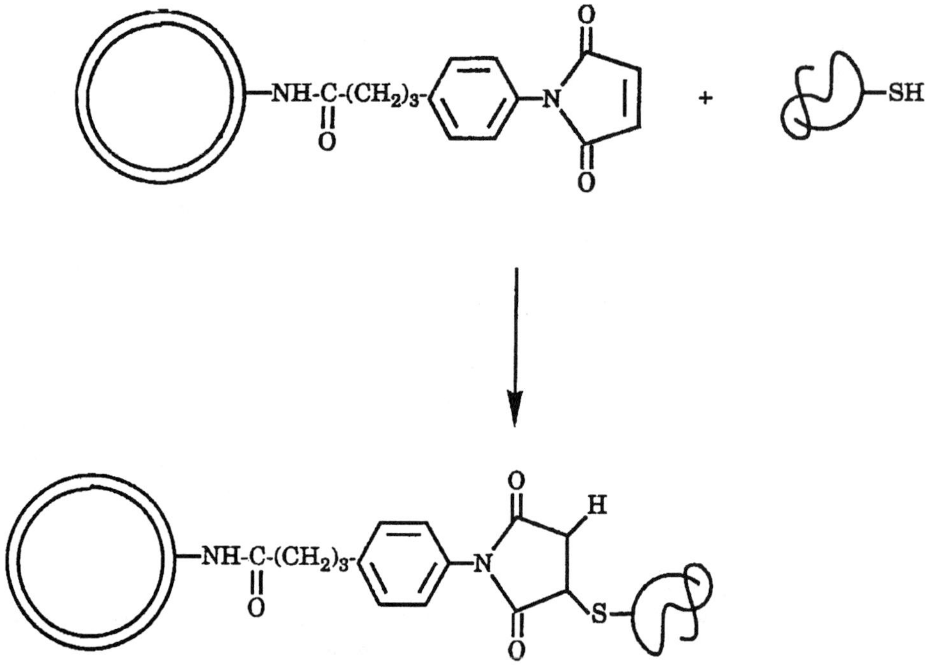

Figure 1 Covalent coupling of thiol-carrying ligand with maleimidophenyl-butyryl-bearing liposomes. (From Ref. 16, © American Society of Biol. Chemists., 1982.)

initially added ligand. Precautions have to be taken to avoid the increase in liposome permeability after protein attachment [23], especially if the relative concentration of the reactive phospholipid exceeds 2.5% mol. Another problem is the poor storage stability of maleimide reagents; certain reagents (such as those carrying cyclohexyl ring) are considered more stable than others, however [24].

Some proteins carry covalently attached carbohydrate residues. These can be easily oxidated to yield aldehyde groups that can react with aminophospholipids (e.g., PtdEtn), with the formation of a Schiff base [25]. To avoid homo-cross-linking, the primary protein amino groups must be preblocked. To avoid unnecessary protein modification, specially synthesized lipid-hydrazine complex can be used [26]. This shifts the reaction equilibrium away from the formation of a Schiff base from primary amino-groups and aldehyde groups and prevents undesirable homo-cross-linking. Hydrazine derivatives react with the aldehyde groups even at mild acidic pH, but reaction takes several hours.

Amide bonds between ligands and phospholipids may offer several advantages:

1. Simplicity and short procedure time
2. Relatively high yields
3. Availability and low price of the reagents (important for practical applications)

Initially water-soluble carbodiimide has been used for the generation of proteo-lipid conjugates [4]. This produced a nondefined mixture of aggregates and by-products, however [5]. To avoid such undesired by-reactions, phospholipids can be premodified and activated prior to the addition of protein or some other ligand. For this purpose, carboxyacyl derivatives of PtdEtn proves useful [27,28,29]. Initially [27], the synthesis as well as the activation of the terminal carboxylic group of phospholipid were performed in an organic solvent (see Sec. VI). *N*-succinyl-PtdEtn was prepared by the reaction of PtdEtn with succinic anhydride in the presence of triethylamine (see Fig. 2 and Worksheet 2).

Multilamellar liposomes were prepared by the adding of DNP-lysine containing aqueous buffer with vigorous stirring to dry lipid film, containing preactivated lipid [28]. *N*-hydroxysuccinimide ester of the phospholipid (Fig. 3) was covalently attached to the ligand primary amino group. However, hydrolysis was substantial, and most of the ligand was lost.

Later, liposomes containing carboxyl-bearing derivatives of PtdEtn were used for the attachment of different ligands (29). These liposomes can be prepared by various techniques and activated with water-soluble carbodiimide directly prior to ligand addition. Mouse IgG could be successfully attached to these liposomes with high yield [29] as well as peptides and low molecular weight aminophenyl derivatives of carbohydrates [30,31] (see Worksheet 6).

To perform the coupling procedure, carbodiimide is added to liposome dispersion at mild acidic pH; a few minutes later, amino group containing ligand (e.g., protein, peptide, or amino sugar) is added at mild alkaline pH. In this way, homo-cross-linking between the carboxylic groups and amino groups on a protein is prevented. The addition of *N*-hydroxysulfosuccinimide to the reaction medium also can increase the coupling yield [32] (Fig. 4).

The main advantages of this approach are the relatively high yield of ligand attachment to liposomes and the easy separation of unbound ligands. For the latter, one relies on the fact that liposome size and density differ from those of free ligands, and one uses gel filtration, sedimentation, or flotation. However, the presence of unreacted phospholipids may modify the liposome behavior in vivo.

The disadvantage of this approach is that some of the reactive phospholipids may react with substances incorporated into a lipid membrane or inside the liposome. This constrains the choice of lipids and encapsulated compounds. For example, the PtdEtn-based liposomes, used for the construction of pH-sensitive systems, should be avoided, but ganglioside-containing vesicles can be used.

The presence of a lipid membrane creates steric hindrances for the attachment of

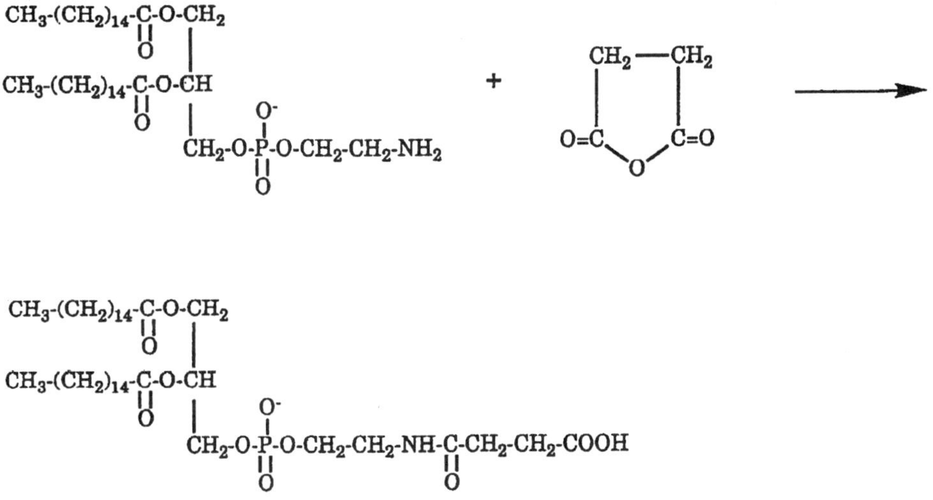

CH_3-$(CH_2)_{14}$-C(=O)-O-CH_2
CH_3-$(CH_2)_{14}$-C(=O)-O-CH
CH_2-O-P(=O)(O^-)-O-CH_2-CH_2-NH_2

+ CH_2—CH_2 (O=C–O–C=O ring, succinic anhydride) ⟶

CH_3-$(CH_2)_{14}$-C(=O)-O-CH_2
CH_3-$(CH_2)_{14}$-C(=O)-O-CH
CH_2-O-P(=O)(O^-)-O-CH_2-CH_2-NH-C(=O)-CH_2-CH_2-COOH

Figure 2 Synthesis of *N*-succinylphosphatidylethanolamine.

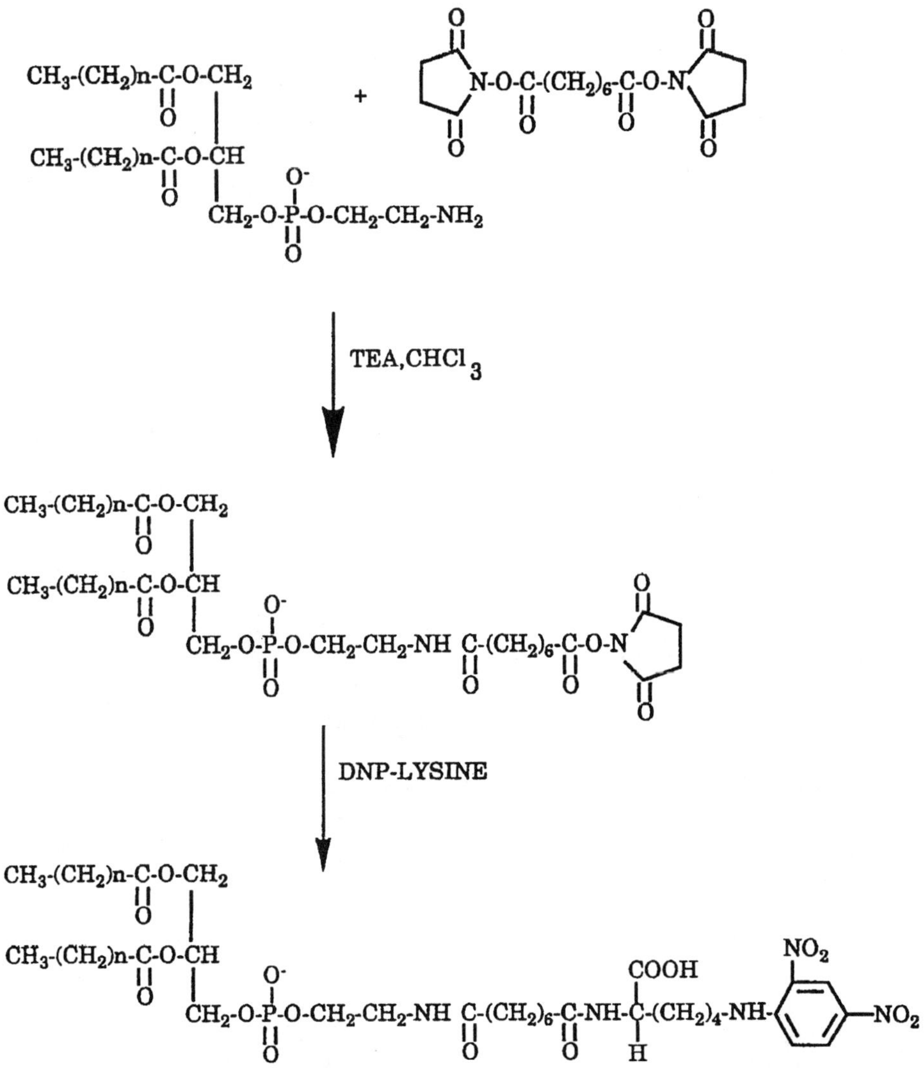

TEA, $CHCl_3$

DNP-LYSINE

Figure 3 Synthesis of NHS-suberylphosphatidylethanolamine and its reaction with DNP-lysine. (From Ref. 28, © Elsevier Science Publishers B. V., 1984.)

ligands to the reactive groups, especially in the case of high molecular weight ligands, e.g., proteins. This lowers the yield of the coupling reaction and results in ligand loss.

IV. CONJUGATION OF LIGANDS TO NONLIPOSOMAL PHOSPHOLIPIDS

The following steps are necessary for such conjugation:

1. Preparation of a chemically modified phospholipid
2. Activation, if necessary

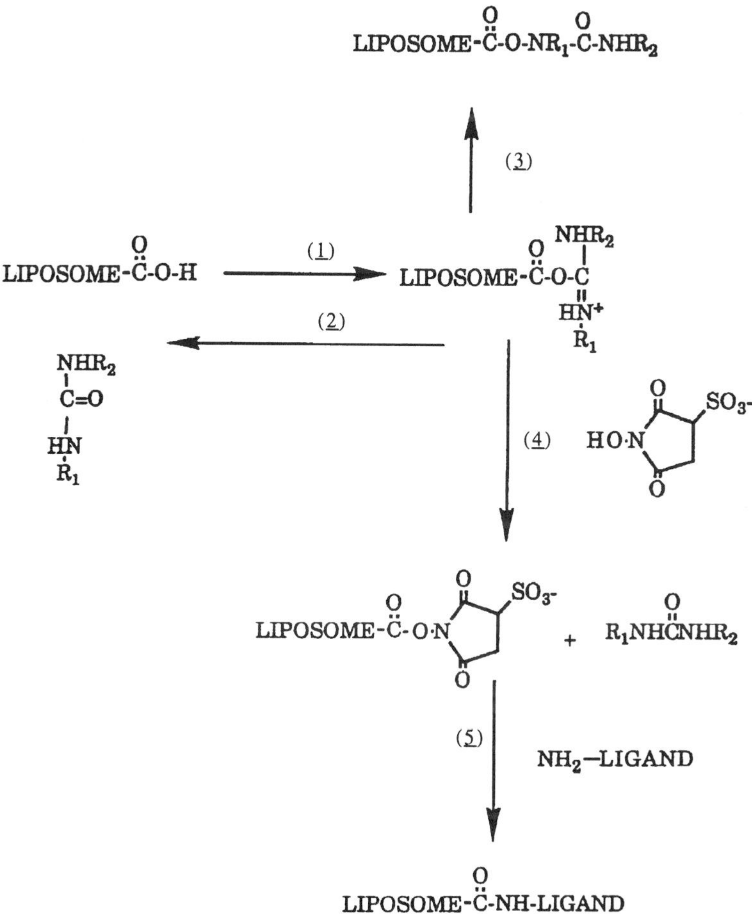

Figure 4 The carbodiimide activation of liposomes carrying carboxylic groups: (1) modification of the carboxylic group with the formation of *O*-acylisourea; (2) the hydrolysis of *O*-acylisourea; (3) the isomerisation of *O*-acylisourea with the formation of *N*-acylurea; (4) the formation of *N*-hydroxysulfosuccinimide ester in the presence of *N*-hydroxysulfosuccinimide; (5) reaction of *N*-hydroxysulfosuccinimide ester with amino-ligand. (From Ref. 32, © FEBS, 1988.)

3. Coupling of a ligand (in aqueous buffer, in detergent solution, or in organic solvent)
4. Purification, if necessary
5. Incorporation of the conjugate into a lipid membrane or a cell

The main advantage of this strategy is that it permits an optimization of each individual step. In general, it also ensures high yields of coupling and incorporation. This approach, moreover, offers the possibility of separation of the unreacted components, which could alter the system behavior in an undesirable way (unbound lipids, for example, may cause cell damage or a modification of the liposome properties).

Basic principles of the phospholipid attachment to proteins were proposed by Karush [33] more than a decade ago. Since then, the general strategy of coupling has remained basically the same, but some methodological simplifications have been introduced.

Initially, Bence-Jones lambda-chain protein was used as a model for the covalent attachment to the specially synthesized phospholipid, *N*-(*N*-iodoacetyl-*N*-dansyllysyl)-phosphatidylethanolamine. Bence-Jones protein carries a single S–S bond, which can be reduced to an SH-group (one per polypeptide chain). This SH-group is then reacted with a lipid iodoacetyl moiety in detergent, initially, sodium dodecyl sulfate (SDS) solution.

Later, Fab' fragments of the antilactose antibody were successfully modified with the same phospholipid reagent. To avoid protein denaturation by SDS, the latter was replaced by a nonionic detergent Nonidet P-40, however [34].

An important point is that the Fab' fragment of the antibody carries a single SH group sterically separated from the antigen-binding site, so that the attachment of a protein can be performed in a controlled, properly oriented manner.

During the phospholipid-protein reaction, the unreacted phospholipid as well as the unreacted protein and detergent can be removed easily from the system; chromatography or electrophoresis can be used for this purpose [33,34]. Chromatography of the reaction mixture on a DEAE-Sephadex ion exchange column, for example, allows (1) the removal of Nonidet P-40 detergent while protein and phospholipid remain on the column; (2) elution of the unmodified protein upon increasing the ionic strength of eluent; (3) elution of phospholipid-protein conjugates by the addition of cholate to the reaction mixture.

A product with protein-to-phospholipid 1:1 molar ratio was obtained, intrinsic affinity of the antibody fragment to the antigen being decreased only slightly.

Phospholipid conjugates with Bence-Jones monomer protein are water-soluble even in the absence of detergents [33]. They exist in the form of protein-phospholipid "supermicelles," which permit an irreversible, spontaneous incorporation of the modified protein into phospholipid vesicles (with yields over 30%) after prolonged incubation. Likewise, the phospholipid-modified proteins bind to the plasma membrane of the mammalian cell. Up to 10^5 protein molecules per erythrocyte ghost are typically bound.

The synthesis of *N*-(*N*-iodoacetyl-*N*-dansyllysyl)-PtdEtn is a complicated multistep procedure. To perform a simpler, single-step coupling of the sulfhydryl-containing ligands to phospholipids (e.g., in a detergent solution), maleimide-carrying derivatives of PtdEtn (MPB-PtdEtn, etc.) are successfully used. Mannino et al. [35], for example, were able to attach maleimide phospholipids to the SH-containing peptides in the controlled, oriented way. These peptide phospholipids were then used to enhance the antipeptide T and B immune response.

The maleimide-derived structures have the disadvantage of not being completely natural; effects of these groups in vivo are as yet unclear. For biological applications, simple proteo-phospholipid conjugates are therefore recommended. The simplest method for obtaining such conjugates is to oxidize the inositol ring of phosphatidyl inositol by sodium periodate (see Worksheet 5); the resulting intermediate is then reacted with the protein amino group in an aqueous solution [36].

However, this procedure is time-consuming, and phosphatidyl inositol is not always easily available. To improve this, different PtdEtn derivatives are often used for the coupling of ligands via amide bond formation. Jansons and Mallett [37], for example, have mixed protein solution with an aqueous dispersion of phosphatidylethanolamine and added water-soluble carbodiimide as a cross-linking agent. To prevent massive homo-cross-linking of the protein, its primary amino groups were protected by citraconic anhydride treatment (in some cases this procedure may cause protein inactivation). During the coupling process several reactions take place; to narrow the range of products obtained, phospholipid derivatives should be prepared first, then activated with carbodiimide, and finally mixed with a protein solution under conditions that do not favor protein cross-linking.

A phospholipid derivative of carboxylic acid can be easily prepared by the reaction of PtdEtn with cyclic anhydride or dicarboxylic acid (e.g., succinic [27] or glutaric anhydride [38]; see Worksheet 2). These compounds can be dispersed by sonication in a mildly acidic aqueous medium containing dimethylsulfoxide [38] or a detergent (e.g., octyl glucoside) [39]. Immediately after that, water-soluble carbodiimide is added [38] (or carbodiimide/*N*-hydroxysulfosuccinimide for improved yield [39]). After 5–10 min the phospholipid active ester formed is mixed with the solution of protein in mildly alkaline buffer. (At alkaline pH, carbodiimide is unable to activate the carboxylic groups of the protein and thus does not cause homo-cross-linking of the protein molecules.)

This procedure is easy; *N*-glutaryl PtdEtn is stable during storage and activated with carbodiimide within several minutes. Less activated phospholipid is required, for the direct coupling to a protein, than when phospholipids are incorporated into a liposomal membrane.

Chemical modification of the phospholipid molecules with water-soluble ligands becomes even simpler when preactivated *N*-hydroxysuccinimide esters of *N*-carboxyacyl-PtdEtn are used. These compounds can be synthesized in organic solvent and purified; they are relatively stable during storage (Worksheet 3) [28,40,41]. The active compound is added to the aqueous solution of a protein in the presence of detergent [40] or in the reversed micellar system in the presence of organic solvent [41].

The former approach resembles the use of *N*-hydroxysuccinimide ester of palmitic acid [42] and is thus well known. It may require lengthy purification procedures, such as gel filtration in detergent [40]. The latter approach has the advantage of reduced hydrolysis due to the low water content in the reaction system. It requires initial protein concentration to be rather high, over 30 mg/mL, but it allows an easier purification by the protein precipitation with cold acetone, which disrupts the reverse micelles; finally, protein can be additionally purified by the ammonium sulfate precipitation.

The question of solubility of the protein-lipid conjugate in aqueous buffers creates some controversy. Karush [33] has reported the preparation of protein-phospholipid (1:1) conjugates, which formed water-soluble "micelles" with $n = 20$. After controversial reports by other authors [43], McConnell has dissolved the antibody-phospholipid and peptide-phospholipid complexes in 0.5% deoxycholate [40].

In our hands [41], the water solubility of modified immunoglobulins is lower (<0.6 mg/mL) than that of native protein. Phospholipid-protein complexes with molar ratio higher than 1:1 are insoluble. The soluble micelles of the phospholipid-protein complex can be formed only when only one phospholipid residue is attached per protein molecule. Randomly situated multiple phospholipid residues cause the formation of large aggregates with subsequent protein precipitation. The presence of unreacted phospholipid seems to favor the formation of such aggregates.

The solubility question is important for the application of phospholipid-protein conjugates. Karush has successfully incorporated "soluble" conjugates into liposomal membrane during coincubation. With the detergent present [39], liposomes have to be prepared by the dialysis method, however, which may lower the entrapment of the low-molecular weight solutes inside liposomes.

The incubation of phospholipid-protein conjugate with erythrocyte ghosts results in the enhanced conjugate binding to the cells [33]. The incorporation of the detergent-solubilized phospholipid-protein complexes into the plasma membrane of living cells requires up to 100-fold dilution to avoid cell lysis, however [40].

V. CHEMICAL MODIFICATION OF PHOSPHOLIPIDS IN NONAQUEOUS MEDIA

Phospholipid molecules often need to be attached to low molecular weight compounds or to organic polymers that are not inactivated upon contact with organic solvents. Organic solvents in such situations can be used as reaction media. This increases the coupling yield and decreases the danger of side reactions, such as hydrolysis of the active esters or anhydrides.

The general strategy for such synthesis includes the preparation of a PtdEtn solution in an organic solvent (usually chloroform and/or methanol supplemented with pyridine or triethylamine) and the subsequent addition of a nucleophilic agent capable of reacting with the PtdEtn amino groups (usually *N*-hydroxy-succinimide ester of the ligand or some heterobifunctional reagent (see Worksheet 1). This technique was used to prepare hapten-modified phospholipids for use in liposome-based immunoassays [44] and for the preparation of target-sensitive liposomes [45].

Chemical modification of the primary amino group of the PtdEtn with the anhydride of carboxylic acid in an organic solvent (Worksheet 2) allows the preparation of reactive phospholipids: (1) *N*-succinyl PtdEtn molecules are pH sensitive; this makes them suitable for the controlled release of the liposomal contents from *N*-succinyl PtdEtn/PtdEtn-containing liposomes at mildly acidic pH [46]; (2) *N*-succinyl-, *N*-glutaryl-, *N*-suberyl-, and other PtdEtn- derivatives can be converted to active esters and used for the phospholipid attachment to proteins [29], haptens, and carbohydrates.

N-hydroxysuccinimide ester of *N*-suberyl-PtdEtn or *N*-succinyl-PtdEtn can be synthesized by the reaction of *N*-hydroxysuccinimide with the carboxyl-bearing phospholipids in the presence of dicyclohexyl carbodiimide [27,41]. The simplified single-step synthesis of related compounds is based on the interaction of PtdEtn with a large excess of the bifunctional cross-linking NHS-eester reagents [28,40]. These derivatives are stable during storage and suitable for the conjugation of various ligands carrying primary amino groups both in organic solvents and in the aqueous phase. Phospholipids activated with NHS esters can be used for several months; this eliminates the need for the repeated activation of phospholipid. Another possible application of the *N*-hydroxysuccinimide esters (of *N*'-suberyl type) is their use as the reagent for the automated peptide synthesis during preparation of peptide-phospholipid conjugates; the latter are now extensively investigated as a basis for potent vaccines [35,47,48]. (The deblocked terminal alpha amino group of the peptide is easily forced to form an amide bond attaching phospholipids to the *N*-terminus of a peptide, in a way similar to the use of *N*-hydroxysuccinimide esters of protected amino acids.)

Another important group of compounds suitable for phospholipid-ligand coupling in organic solvents includes PtdEtn modified with sulfhydryl-reactive heterobifunctional reagents (see Fig. 5 and Worksheet 1). The moieties carrying maleimide, dithiopyridyl, iodoacetyl, or bromacetyl residues in this approach can be coupled to the phospholipid headgroups. The resulting phospholipids can react with the free thiol groups of various

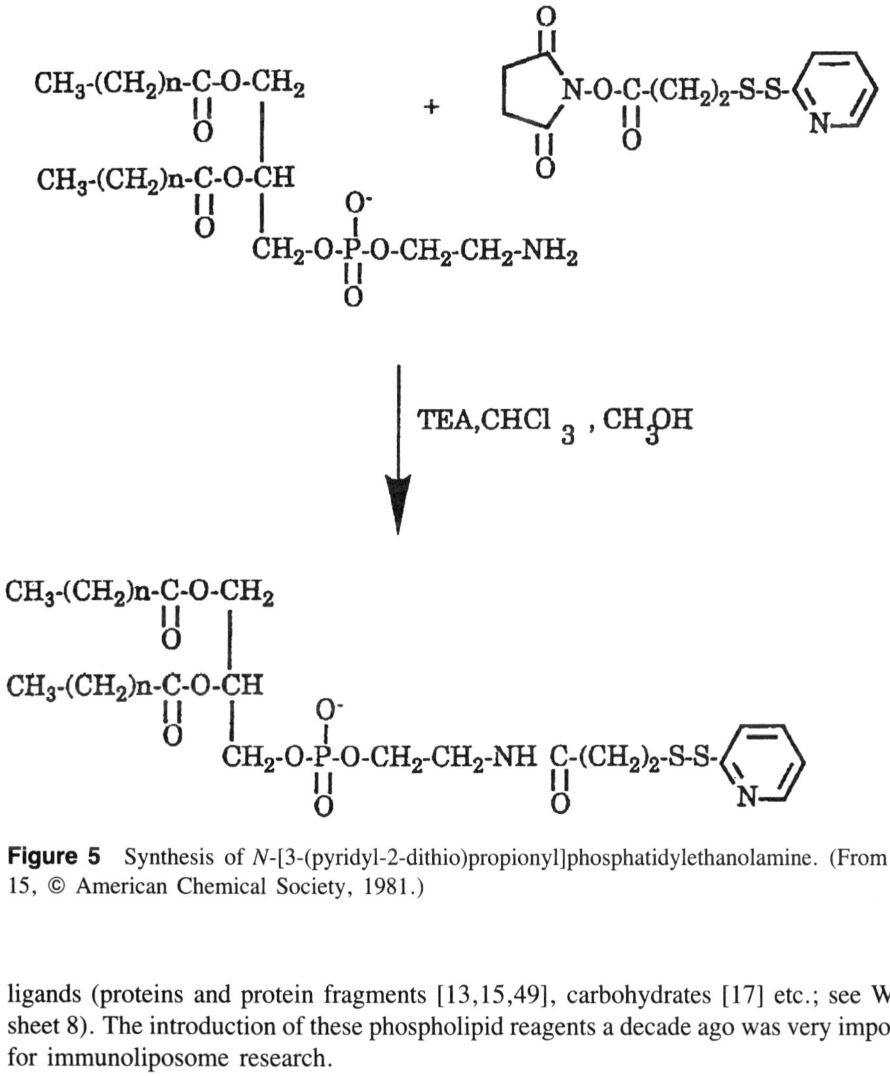

$CH_3\text{-}(CH_2)n\text{-}C(=O)\text{-}O\text{-}CH_2$

$CH_3\text{-}(CH_2)n\text{-}C(=O)\text{-}O\text{-}CH$

$CH_2\text{-}O\text{-}P(=O)(O^-)\text{-}O\text{-}CH_2\text{-}CH_2\text{-}NH_2$

+

$N\text{-}O\text{-}C(=O)\text{-}(CH_2)_2\text{-}S\text{-}S\text{-}$ (pyridyl)

TEA, $CHCl_3$, CH_3OH

$CH_3\text{-}(CH_2)n\text{-}C(=O)\text{-}O\text{-}CH_2$

$CH_3\text{-}(CH_2)n\text{-}C(=O)\text{-}O\text{-}CH$

$CH_2\text{-}O\text{-}P(=O)(O^-)\text{-}O\text{-}CH_2\text{-}CH_2\text{-}NH\ C(=O)\text{-}(CH_2)_2\text{-}S\text{-}S\text{-}$ (pyridyl)

Figure 5 Synthesis of *N*-[3-(pyridyl-2-dithio)propionyl]phosphatidylethanolamine. (From Ref. 15, © American Chemical Society, 1981.)

ligands (proteins and protein fragments [13,15,49], carbohydrates [17] etc.; see Worksheet 8). The introduction of these phospholipid reagents a decade ago was very important for immunoliposome research.

The primary amino group of PtdEtn in the presence of triethylamine can be modified easily with the *N*-hydroxysuccinimide ester of polyethylene glycol succinate, a commercially available reactive polymer, in an organic solvent (see Fig. 6 and Worksheet 11). Liposomes carrying polyethylene glycol-PtdEtn stay in circulation substantially longer than regular phosphatidylcholine/cholesterol liposomes of the same size ($t_{1/2}$ = 5 hours for the liposomes of 0.2 μm size) [50].

Negatively charged natural polymers, DNA or RNA can be bound electrostatically to the positively charged phospholipids [51]. Positively charged phospholipids ("lipopolyamine," "lipospermine") that are required for this purpose can be prepared by condensing the carboxyl derivative of the BOC-protected spermine to the primary amino group of PtdEtn. The resulting lipopolyamine can coat plasmids with a lipid sheet (the resulting structures are not liposomes!) and allows successful infection of mammalian cells with foreign genetic material.

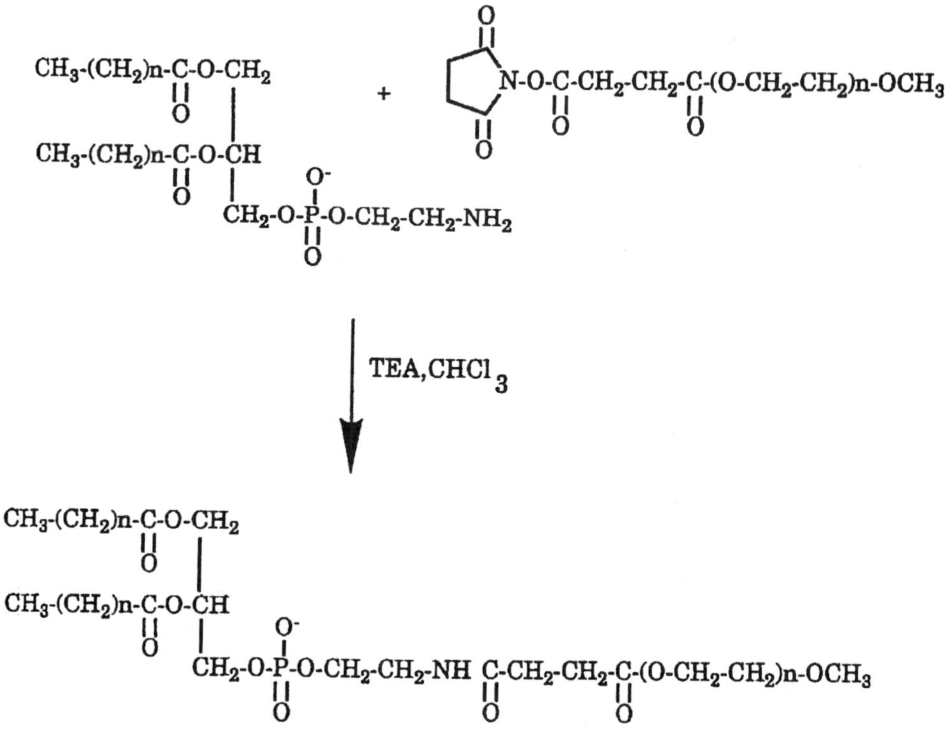

CH_3-(CH_2)n-C-O-CH_2

CH_3-(CH_2)n-C-O-CH

CH_2-O-P-O-CH_2-CH_2-NH_2

+

N-O-C-CH_2-CH_2-C-(O-CH_2-CH_2)n-OCH_3

TEA,$CHCl_3$

CH_3-(CH_2)n-C-O-CH_2

CH_3-(CH_2)n-C-O-CH

CH_2-O-P-O-CH_2-CH_2-NH C-CH_2-CH_2-C-(O-CH_2-CH_2)n-OCH_3

Figure 6 Synthesis of polyethylene glycol-phosphatidylethanolamine.

VI. LABELING OF PHOSPHOLIPIDS

Labeled phospholipids are needed for studies of localization in the organism and in the cells.

The coupling of fluorescent, radioactive, affinity and photoaffinity labels to phospholipids is performed in accordance with the general rules described in previous sections. *N*-hydroxysuccinimide esters or anhydrides of the carboxyl derivatives of the label, for example, are easily attached to the primary amino group of PtdEtn. Reaction can be performed in organic solvent containing traces of triethylamine or pyridine.

Biotinylated phospholipids (see Worksheet 1 for synthesis and Worksheet 4 for the purification procedure) have been used, for example, to monitor the intracellular fate of the biotin-labeled lipids and liposomes by means of electron microscopy; ferritin-avidin was used as a staining agent [52]. Due to the high affinity of avidin for biotin ($K_D = 10^{-15}$ M), it is possible to localize the biotin-labeled phospholipids indirectly with any kind of avidin-attached label. For instance, Trubetskoy et al. [53] have been able to assess the amount of biotin-carrying liposomes on the cell surface with the help of labeled avidin. The endocytosis rate was estimated by the use of dual labels: biotin-PtdEtn provided information about the number of liposomes attached to the cell surface; the radio-labeled lipid revealed the extent of total cell-liposome association (external and internal).

Direct incorporation of radio-labeled compounds into phospholipid molecules is well developed. A variety of beta-emitting ^{14}C- and ^{3}H-radio-labeled phospholipids is thus available commercially. However, experimental procedures with gamma-emitting radioisotopes are much easier, and these isotopes are affordable. In the latter case, a very high specific activity (over 10 μCi/μg lipid) can be easily achieved. To bind gamma-radioisotopes, such as radioiodine (e.g., ^{125}I) to a phospholipid molecule, phenol derivatives can be used, which easily incorporate preoxidized radioiodine.

p-Hydroxybenzimidate (Wood's reagent) or *N*-hydroxysuccinimide ester of *p*-hydroxyphenylpropionic acid (Bolton-Hunter reagent) can be covalently conjugated to PtdEtn (see Worksheet 1 for synthesis and purification). Compounds, prelabeled with ^{125}I, thus can be conjugated with PtdEtn. Alternatively, "cold" phospholipid derivatives can be synthesized and consequently labeled using labeling procedures that are otherwise used for protein labeling: chloramine T [54] or Iodogen procedures (see Worksheet 9) are examples.

Metabolizable radioiodine can be successfully used to trace liposomes in vivo [54]. For particular studies, however, ^{111}In-radioisotope is a better choice. This is because this label is not metabolized or excreted rapidly after internalization. Phospholipid derivatives of DTPA, chelator with high affinity to ^{111}In, ^{113m}In, or ^{99m}Tc [55] (see Worksheet 10), can therefore be used to attach these radiolabels to phospholipids. Phospholipid-DTPA derivatives also bind paramagnetic metal labels, used for NMR tomography studies, such as Gd^{3+}. This may allow the imaging of certain disorders of the spleen and liver, e.g., liver metastasis of certain tumors.

Phospholipids with attached photoaffinity labels can be used to determine the disposition of certain membrane proteins and enzymes in natural and model membranes. Azidosalicylic acid-derived photoaffinity labels can also be radio-labeled easily with radioiodine. Photoaffinity residues can be attached either to the fatty acid chain region [56] or to the polar headgroup of phospholipids, thus allowing distinction between the membrane-embedded proteins and proteins attached to the surface of the membrane.

VII. PRODRUGS

Recently, chemically modified inactive phospholipid derivatives of various pharmacologically active compounds have been introduced [57]. These compounds after injection undergo metabolic conversion with the formation of active compounds, presumably at the target site.

The application of phospholipid prodrugs incorporated into phospholipid bilayers (as compared to the incorporation of hydrophilic compounds inside lipid vesicles) brings several benefits [58]:

1. The efficiency of the prodrug incorporation is high.
2. Prodrugs do not leak from the liposome into the aqueous phase.
3. The drug is protected against metabolic degradation.
4. Long-lasting therapeutic drug levels can be achieved.

The application of phospholipid-derived prodrugs may improve the specificity of their delivery to the target cells, where active ingredients are liberated metabolically [59]. This may allow the resistance of a certain type of tumor cells to a given drug to be overcome, as an alternative pathway of drug delivery into the cell is used.

Prodrug-phospholipid conjugates usually involve anticancer and antiviral agents. Methods for their synthesis are similar to those described in Secs. IV and V. Methotrexate, for example, has been conjugated to Myr_2PtdEtn by the carbodiimide-mediated coupling of the methotrexate carboxylic group [60]. Several distinct products with different cytotoxic action have been isolated; the target enzyme is always dihydrofolate reductase, however [61].

Further phospholipid prodrugs are derivatives of arabinofuranosyl-cytosine [62], acyclovir [63], and other nitrogenous bases. Recently, Hostetler et al. [59] have also synthesized the conjugates of phosphatidic acid with azidothymidine and 3'-deoxythymidine, the promising compounds for the AIDS therapy. Monophosphates were obtained by condensing the phosphatidic acid phosphate group with a hydroxyl group of the azidothymidine molecule in the presence of triisopropylbenzenesulfonylchloride. This synthetic method had been known for a long time [64,65], but it was not widely used until recently. The approach opens a wide range of possibilities for the synthesis not only of prodrugs but also of other phospholipid derivatives.

WORKSHEETS

Worksheet 1. Chemical Modification of the Headgroup of Phosphatidylethanolamine with *N*-Hydroxysuccinimide Reagents

Principle of operation: Acylation of the primary amino group of PtdEtn with *N*-hydroxysuccinimide ester of carboxylic acid reagent in organic media in the presence of triethylamine.

Literature: L. D. Leserman, J. Barbet, F. Kourilsky, and J. N. Weinstein. *Nature 288*:602 (1980). F. J. Martin, W. L. Hubbell, and D. Papahadjopoulos, *Biochemistry 20*:4229 (1981). F. J. Martin, and D. Papahadjopoulos, *J. Biol. Chem. 257*:286 (1982). B. Wolff and G. Gregoriadis. *Biochim. Biophys. Acta 802*:259 (1984). E. A. Bayer, B. Rivnay, and E. Skutelsky. *Biochim. Biophys. Acta 550*:464 (1979).

Materials: To modify 10 μmol phospholipid, use 15–30 μmol of *N*-hydroxysuccinimide ester reagent (e.g., *N*-hydroxysuccinimide ester of biotin, SPDP, or SMPB) in 0.5–3 mL of chloroform or chloroform-methanol mixture and 23–60 μmol of triethylamine, >99% pure.

Equipment: Vortex mixer, microscale reaction vessel or tube, micropipettes, analytical balance.

Procedure: Dissolve PtdEtn in chloroform or chloroform/methanol 2:1 mixture. Add 1.5- to 3-fold molar excess of *N*-hydroxysuccinimide ester reagent (e.g., biotin-NHS, or biotin-amidocaproyl-NHS, or Bolton-Hunter reagent) in chloroform, methanol, or dimethylformamide. Add 1.5-to 2-fold molar excess of triethylamine over *N*-hydroxysuccinimide ester reagent and leave overnight at room temperature.

Check the completion of the reaction by thin-layer chromatography with ninhydrin staining of primary amino group of PtdEtn. Usually the unreacted PtdEtn ninhydrin-positive spot completely disappears.

Purification: purify conjugated phospholipid derivative from unreacted *N*-hydroxysuccinimide ester, TEA, etc. by preparative TLC, column chromatography or HPLC, or simply by dialysis (see Worksheet 4).

Recommendations: Reactions proceed faster at increased concentrations of reactants, so if solubility allows keep the reaction volume as small as possible.

Some lipids, like Pam_2PtdEtn, do not require an oxygen-free atmosphere. Nevertheless, most of the natural phosphatidylethanolamines contain polyunsaturated fatty acids; therefore reaction at room temperature should be performed under argon or at least under nitrogen.

Caveats: PtdEtn of natural origin or with unsaturated fatty acid residues is usually soluble in chloroform; Pam_2PtdEtn is soluble (10 mg/mL) in chloroform-methanol mixtures (2:1 v/v) or upon moderate heating (37°C).

Predissolve the *N*-hydroxysuccinimide ester reagent in methanol or dimethylformamide, if it is not freely soluble in chloroform, and mix it with chloroform solution of PtdEtn.

Use chloroform and methanol containing a minimal amount of water (analytical grade, <0.01% H_2O), or dry them over a 3A-molecular sieve.

Use triethylamine free of primary amino group contaminants (i.e., the highest grade available). If TEA is contaminated, the coupling yield may drop significantly. Purification of TEA may be achieved by its refluxing over ninhydrin with subsequent distillation at atmospheric pressure.

Check the *N*-hydroxysuccinimide ester reagent on TLC; in some instances it may be partially hydrolyzed. In this case increase the input ratio of the reagent, as well as the amount of triethylamine.

Worksheet 2. Synthesis of Carboxyacyl Derivatives of Phosphatidylethanolamine

Principle of operation: Cyclic anhydride or mixed anhydride is used to acylate the primary amino group of PtdEtn in organic media in the presence of triethylamine.

Literature: V. T. Kung and C. T. Redemann, *Biochim. Biophys. Acta 862*:435 (1986). V. Weissig, J. Lasch, A. L. Klibanov, and V. P. Torchilin. *FEBS Lett. 202*:86 (1986). A. N. Lukyanov, A. L. Klibanov, A. V. Kabanov, V. P. Torchilin, A. V. Levashov, and K. Martinek, *Bioorg. Khim. 14*:670 (1988). S. C. Kinsky, J. E. Loader, and A. L. Benson. *J. Immunol. Methods 65*:295 (1983).

Materials: To modify 10 μmol phospholipid, use 15–60 μmol of glutaric anhydride (or other anhydride) and 20–70 μmol of triethylamine (>99% purity). Use up to 1 mL of water-free chloroform (analytical grade).

Equipment: Vortex mixer, microscale reaction vessel or tube, micropipettes, analytical balance.

Procedure: Dissolve PtdEtn in chloroform at 5 to 20 mg/mL, add corresponding anhydride (powder or solution in chloroform or methylene chloride) and triethylamine and leave overnight at room temperature. Completion of the reaction can be monitored by thin-layer chromatography; the ninhydrin-positive PtdEtn spot disappears from TLC pattern.

Purification: See Worksheet 4.

Recommendations: See Worksheet 1.

Caveats: Use water-free solvents and pure triethylamine (see Worksheet 1).

Worksheet 3. Preparation of *N*'-Hydroxysuccinimide Ester of *N*-Carboxyacyl Derivatives of Phosphatidylethanolamine

Principle of operation: Dicyclohexylcarbodiimide is used to activate the free carboxyl group of the *N*-carboxyacyl derivative of PtdEtn and convert it to *N*'-hydroxysuccinimide ester.

Literature: S. C. Kinsky, J. E. Loader, and A. L. Benson, *J. Immunol. Methods* *65*:295 (1983). S. C. Kinsky, K. Hashimoto, J. E. Loader, and A. L. Benson, *Biochim. Biophys. Acta 769*:543 (1984). A. N. Lukyanov, A. L. Klibanov, A. V. Kabanov, V. P. Torchilin, A. V. Levashov, and K. Martinek, *Bioorg. Khim. 14*:670 (1988). N. L. Thompson, A. A. Brian, and H. M. McConnell, *Biochim. Biophys. Acta 772*:10 (1984).

Materials: To modify 10 μmol phospholipid, use 11–25 μmol of dicyclohexyl carbodiimide and 10–20 μmol of *N*-hydroxysuccinimide (excess of the modifying reagent varies for different publications). Coupling of PtdEtn with dicarboxylic acid is described in Worksheet 2.

Equipment: Vortex mixer, microscale reaction vessel or tube, micropipettes, analytical balance.

Procedure 1: Dissolve *N*-glutaryl-PtdEtn or *N*-succinyl-PtdEtn in chloroform (5–10 mg/mL), add dicyclohexylcarbodiimide and *N*-hydroxysuccinimide, mix vigorously, and leave overnight at room temperature. Cool reaction mixture to 4°C, and 1 h later filter the dicyclohexyl urea formed.

Purification 1: Thin-layer chromatography can be used to separate *N*-hydroxysuccinimide ester of *N*'-carboxyacyl-PtdEtn from unreacted phospholipid and other contaminants. Silica gel 60 or Unisil was used, and the plate was developed with chloroform:methanol:water (70:30:5). However, water present in the chromatography developing system may cause hydrolysis of the desired *N*-hydroxysuccinimide ester. For derivatives based on $Pam_2PtdEtn$, precipitation with cold acetone of the Mg^{2+} salt of the lipid could be used (35 mg product in 0.3 mL chloroform was precipitated by the addition of 4 mL acetone containing 50 μL 10% $MgCl_2$ in ethanol).

Procedure 2: Alternatively, PtdEtn (1 mg/mL in chloroform) can be modified directly with 10–25-fold molar excess of bis-*N*-hydroxysuccinimide ester homo-cross-linking reagent, e.g., disuccinimidyl suberate in chloroform or chloroform/acetone in the presence of triethylamine for 30 min.

Purification 2: Excess of the unreacted cross-linking agent can be removed by thin-layer chromatography on silica in acetone-water (95:5) and chloroform:methanol:water (95:25:3).

Worksheet 4. Purification of *N*-Acyl Derivatives of Phosphatidylethanolamine

Principle of operation: Water-soluble contaminants are extracted by the aqueous phase, whereas phospholipid-bound compounds arc retained by organic phase.

Literature: L. D. Leserman, J. Barbet, F. Kourilsky, and J. N. Weinstein, *Nature 288*:602 (1980). V. T. Kung and C. T. Redemann. *Biochim. Biophys. Acta 862*:435. (1986). S. C. Kinsky, J. E. Loader, and A. L. Benson. *J. Immunol. Methods 65*:295 (1983). A. N. Lukyanov, A. L. Klibanov, A. V. Kabanov, V. P. Torchilin, A. V. Levashov, and K. Martinek, *Bioorg. Khim. 14*:670 (1988).

Materials: For dialysis or extraction of phospholipid derivative, use 0.05 M sodium acetate buffer pH 4.5, 0.01 M borate buffer pH 8, and deionized distilled water. A stream of nitrogen can be used to remove organic solvent.

Equipment: Magnetic stirrer or preparative centrifuge + Vortex mixer, dialysis tubing, rotary evaporator or lyophilizer, analytical balance.

Purification procedures: PtdEtn derivative in chloroform or chloroform/methanol is brought in contact with water phase.

1. Extraction procedure: 1 mL of chloroform phase containing 1–10 mg of the phospholipid derivative is added to 10 mL of acidic aqueous buffer in a glass centrifuge tube and vortexed. Phases are separated by centrifugation (e.g., 5000 g, 5 min) and the aqueous phase discarded. Then thc procedure is repeated with alkaline aqueous buffer and finally with deionized water to remove salts. Organic solvent is removed by the stream of nitrogen or evaporated under vacuum; product is weighed and redissolved in chloroform for storage in the freezer.

2. Dialysis procedure: 1 mL of chloroform phase containing modified phospholipid, triethylamine, and unreacted acylating agent is put in the dialysis tubing and dialyzed against 1 L of aqueous phase with several subsequent changes of the buffer: 50 mM sodium acetate pH 4.5, sodium borate pH 8, and two changes of deionized water (12 h for each dialysis step). After completion of the dialysis procedure the organic solvent as well as aqueous phase inside the dialysis bag is collected and organic solvent removed by rotary evaporation or under the stream of nitrogen. Water is removed by freeze-drying. Dry powder of phospholipid is weighed and redissolved in chloroform for storage in the freezer.

Recommendation: The extraction/centrifugation technique is more rapid; the advantage of the dialysis technique is in the much smaller number of manipulations required. Losses of phospholipid are very small in the dialysis procedure; during centrifugation part of the phospholipid obviously stays in the aqueous phase (upper phase after centrifugation is cloudy). Dialyzed product contains fewer impurities than the product purified by extraction/centrifugation.

Caveats: For the purification of phospholipids, preparative thin layer chromatography or column chromatography is widely used. In this particular case, i.e., the removal of water-soluble polar compound of low molecular weight from the nonpolar hydrophobic derivative, the application of adsorption chromatography appears much more time-and effort-consuming, especially for laboratories specializing in immunology and cell biology. Silica gel adsorption chromatography may be successfully used; though, preparation may be contaminated with silicic acid dissolved in methanol used for elution; if the methanol content of the eluent media is low, a substantial fraction of the phospholipid may stay bound to silica gel. This may lower the yield of the product.

Worksheet 5. Chemical Activation of the Polar Headgroup of Phosphatidylinositol: Oxidative Conversion into Aldehyde Residues; Coupling of Activated Phosphatidylinositol to Proteins

Principle of operation: A dispersion of phosphatidylinositol in mildly acidic buffer is treated with sodium periodate. Unreacted periodate is reduced with excess ethylene glycol; low molecular weight contaminants are removed by gel filtration. The resulting aldehyde-carrying lipid reacts with protein amino groups, and the Schiff base formed can be converted to a stable secondary amine bond by borohydride reduction.

Literature: V. P. Torchilin, A. L. Klibanov, V. N. Smirnov. *FEBS Lett 138*:117 (1982). E. Claassen and N. van Rooijen. *Prep. Biochem. 13*:167 (1983).

Materials: Yeast phosphatidyl inositol was used (BDH, England, supplied under methanol). To modify 10 μmol phospholipid, use 100 μmol of sodium periodate. Acetate buffer pH 5.6 was used as the reaction medium in the lipid activation step. Ethylene glycol was used to block the unreacted periodate. Sephadex G-50M was used for gel filtration in the centrifuge. 0.05 M borate or carbonate buffer pH 8.5–9.2 was used for protein coupling to aldehyde-phosphatidylinositol. Sodium borohydride or sodium cyanoborohydride was used for the Schiff base reduction.

Equipment: Magnetic stirrer, ultrasound disintegrator, microscale reaction vessel or tube, micropipettes, analytical balance, preparative centrifuge, 5-mL polypropylene column with porous frit for gel filtration in the centrifuge.

Procedure: Disperse phosphatidylinositol (1 mM) in 50 mM acetate buffer pH 5.6 by sonication to produce an optically clear solution. Add solution of sodium periodate in water (10 μM final concentration); perform oxidation at 4°C in dark for 5–10 h. Add a twofold molar excess of ethylene glycol, put the sample in the Sephadex G-50 minicolumn, which was previously centrifuged 3 min at 1000 g, and perform gel filtration in the centrifuge. After application of the sample of oxidized phospholipids to the column, it is centrifuged in the bucket rotor for the second time. Running conditions (time and centrifugal speed) after sample application to the column should be similar to the running conditions for the first run.

Activated phospholipid is eluted from the column during centrifugation; low molecular weight contaminants are retained by the gel. Product can be stored in liquid nitrogen for several months.

For the preparation of protein-phospholipid conjugate, add oxidized phosphatidyl inositol solution to the protein solution (1 mL $5 \cdot 10^{-5}$ M protein + 0.25 mL 1 mM activated PI) in 50 mM carbonate or borate buffer pH 8.5–9.2, incubate overnight, add 1 mg of sodium borohydride, and incubate for additional 12–15 h.

Purification: Remove low molecular weight contaminants by gel filtration or dialysis. If necessary, unbound protein can be removed from protein-phospholipid micelles by gel filtration; unreacted phospholipid can be removed either by gel filtration in detergent or by ion exchange or affinity chromatography.

Recommendations: Sodium borohydride is a stronger reducing agent than sodium cyanoborohydride. It may cause some inactivation of protein due to the reduction of disulfide bounds.

Avoid foaming of the reaction mixture during the borohydride reduction procedure, keeping temperature low and pH high.

Caveats: Natural phosphatidylinositol contains a large fraction of polyunsaturated fatty acids; to avoid lipid peroxidation, perform sonication under nitrogen or argon gas.

Carefully remove ammonium, glycine, or tris from protein solution. Submillimolar concentrations of these compounds can block protein coupling to phospholipid.

Worksheet 6. Chemical Coupling of Proteins, Peptides and Other Ligands Containing Primary Amino Groups to Carboxyacyl Phosphatidylethanolamine-Containing Liposomes

Principle of operation: Liposomes, carrying a carboxyl derivative of phospholipid, are activated by water-soluble carbodiimide in mild acid media, and this activated derivative reacts with primary amino groups of protein ligand in mild alkaline media.

Literature: V. T. Kung and C. T. Redemann, *Biochim. Biophys. Acta 862*:435 (1986). G. Gregoriadis, V. Weissig, L. Tan, and Q. Xiao. *Biochem. Soc. Trans. 17*:128 (1989). V. Weissig, J. Lasch, and G. Gregoriadis, *Biochim. Biophys. Acta 1003*:54 (1989). A. A. Bogdanov, A. L. Klibanov, and V. P. Torchilin, *FEBS Lett. 231*:381 (1988).

Materials: For liposome preparation use 1–10 mg/mL phosphatidylcholine (e.g., from egg yolk) + 0.5–5 mg/mL cholesterol, with 5–10 mol % *N*-glutaryl-PtdEtn or *N*-suberyl-PtdEtn (see Worksheet 2 for preparation). 1–15 mg/mL 1-ethyl-3(3-dimethyl-aminopropyl)carbodiimide is used. 10 mM *N*-hydroxysulfosuccinimide (Pierce Chem. Co.) can be used to enhance the coupling efficiency. Activation of liposomes with carbodiimide is performed at pH 4.5–5.5, in the presence of 50 mM MES buffer. Coupling of protein to activated liposomes is performed in 0.1 M borate buffer pH 7.5–8.5 (in various publications, different concentrations of reagents and coupling reagents are used).

Equipment: Ultrasound disintegrator, rotary evaporator, microscale reaction tube, micropipettes, analytical balance, preparative ultracentrifuge.

Procedure: Prepare liposomes by dehydration-rehydration, reverse-phase evaporation, or sonication, from phosphatidylcholine, cholesterol, and *N*-glutaryl-PtdEtn (5–10 mol % of the latter). Total phospholipid concentration in liposome dispersion should be approx. 5–10 mM in % mM MES-saline, pH 4.5–5.5.

To 200 μl of liposomes in MES-saline, 20 μL 1-ethyl-3(3-dimethylaminopropyl)carbodiimide (0.25 M in water) was added. 20 μL of *N*-hydroxysulfosuccinimide (0.1 M in water) can be added to improve the coupling yield. 5–10 min after addition of carbodiimide, 200 μl of 1 mg/mL protein solution is added in isotonic borate buffer, and pH is shifted to slightly alkaline (7.5–8.5). Reaction mixture is incubated over 3 h, and unreacted ligand is removed by gel filtration on Sepharose CL-6B or by ultracentrifugation.

Recommendations: Instead of protein, peptide, aminophenyl derivative of sugar, or other amino-containing ligand can be used. In this case, usage of the large excess of ligand (up to 18 mM ligand) allows one to utilize for coupling up to 50% of the *N*-glutaryl-PtdEtn incorporated in liposome membrane.

Caveats: Avoid phosphate, glycine, and tris buffer for coupling of ligands to liposomes via carbodiimide.

Worksheet 7. Chemical Modification of the Proteins by Carboxyacyl Derivatives of Phosphatidylethanolamine

Principle of operation: The carboxyl group of a phospholipid derivative is converted to active ester by carbodiimide and hydroxysuccinimide reagents; the active ester reacts with the primary amino group of the protein with the formation of an amide bond.

Literature: V. Weissig, J. Lasch, A. L. Klibanov, and V. P. Torchilin. *FEBS Lett.* *202*:86 (1986). E. Holmberg, K. Maruyama, D. C. Litzinger, S. Wright, M. Davis, G. W. Kabalka, S. J. Kennel, and L. Huang, *Biochem. Biophys. Res. Commun.* *165*:1272 (1989). N. L. Thompson, A. A. Brian, and M. McConnell, *Biochim. Biophys. Acta* *772*:10 (1984). A. N. Lukyanov, A. L. Klibanov, A. V. Kabanov, V. P. Torchilin, A. V. Levashov, and K. Martinek, *Bioorg. Khim.* *14*:670 (1988).

Materials: To modify 1.5 mg IgG, use 50–150 μg of carboxyacyl derivatives of PtdEtn (for synthesis see Worksheets 2 and 3). Octyl glucoside can be used to solubilize the phospholipid. 1-ethyl-3(3-dimethylaminopropyl) carbodiimide is used for coupling.

Equipment: Vortex mixer, ultrasound disintegrator, microscale reaction vessel or tube, micropipettes.

Procedure 1: Solubilize 1 mg/mL *N*-glutaryl-PtdEtn (see Worksheet 2 for preparation) in 10 mM MES buffer pH 4–5, 150 mM NaCl, 5 mg/mL octyl glucoside. To 200 μl of the *N*-glutaryl PtdEtn solution add 40 μl 0.25 M 1-ethyl-3(3-dimethylaminopropyl)carbodiimide in water and 40 μl 0.1 M *N*-hydroxysulfosuccinimide in MES buffer and incubate for 5 min. Add 200 μl of the protein solution in 0.1 M HEPES or borate buffer pH 7.5–8.5 and incubate for 8 h at 4°C with stirring or for a shorter period of time at 20°C.

Procedure 2: Prepare a lipid film of 50 μg *N*-glutaryl-PtdEtn *N*'-hydroxysuccinimide ester on the bottom of the glass tube. Add 0.5 mL 3 mg/mL IgG in 0.05 M $NaHCO_3$, 0.5% deoxycholic acid pH 8.3 and allow reaction to proceed for 1–2 h at room temperature.

Purification: Conjugate can be purified from unbound phospholipid by gel filtration on Sephadex G-75 in 0.5% deoxycholate or octyl glucoside.

Recommendations: For coupling avoid buffers containing primary amine. Different proteins possess different amounts of primary amino groups, and the yield of the reaction may vary. Usually, 1–2 phospholipid residues are attached per protein globule.

Caveats: It is more difficult to disperse in aqueous buffers derivatives of phospholipid based on Pam_2PtdEtn, so Ole_2PtdEtn-based derivatives can be recommended.

Worksheet 8. Coupling of Ligands to Phospholipids Premodified with Sulfhydryl Reagents

Principle of operation: Presynthesized phospholipid derivatives of heterobifunctional reagents are covalently attached to thiol-containing ligands.

Literature: L. D. Leserman, J. Barbet, F. Kourilsky, and J. N. Weinstein. *Nature 288*:602 (1980). F. J. Martin, W. L. Hubbell, and D. Papahadjopoulos, *Biochemistry 20*:4229 (1981). F. J. Martin and D. Papahadjopoulos, *J. Biol. Chem. 257*:286 (1982). B. Wolff and G. Gregoriadis, *Biochim. Biophys. Acta 802*:259 (1984). J. T. P. Derksen and G. L. Scherphof. *Biochim. Biophys. Acta 814*:151 (1985). D. Sinha and F. Karush, *Biochem. Biophys. Res. Commun. 90*:554 (1979). D. Sinha and F. Karush. *Biochim. Biophys. Acta 684*:187 (1982).

Materials: To modify 10 μmol ligand, use 15–30 μmol of reactive phospholipid. Nonidet P-40, octyl glucoside, or SDS are used to solubilize phospholipids. Dithiothreitol or mercaptoethanol is used for reduction of disulfide bonds.

Equipment: Vortex mixer, microscale reaction vessel or tube, micropipettes, analytical balance.

Procedure: PtdEtn is iodoacetylated, modified with SPDP or SMPB as described in Worksheet 1. Phospholipid is dispersed in aqueous buffer media, pH 7–8, containing detergent (Nonidet P-40 0.2%, sodium dodecyl sulfate 0.2%, or octyl glucoside 1%).

The thiol group is generated on the ligand by reduction of its natural disulfide bonds or disulfide bonds resulting from premodification of ligand. Excess of the reducing agent is removed from the ligand by gel filtration immediately prior to coupling; and ligand is transferred to mild acidic media (pH 3.5–6), containing 1 mM EDTA, in oxygen-free atmosphere (N_2 or Ar). Thiolated ligand is mixed with reactive phospholipid solution and incubated at pH 7–8 for 3–8 h in oxygen-free atmosphere. Phospholipid concentration is in the range of 0.1–5 mg/mL.

Purification: Depending on the molecular weight of the ligand, conjugate can be purified from excess ligand by gel filtration or dialysis. Protein conjugates can be purified by ion-exchange chromatography on DEAE-Sephadex; conjugate is eluted by 0.72 M NaCl in the presence of detergent.

Recommendations: Ligands, premodified with SPDP, contain disulfide bonds, which can be reduced by mercaptoethanol or dithiothreitol to thiol group.

Free thiol groups may form disulfide bonds even in an oxygen-poor environment. In order to avoid this adverse reaction, pH is kept acidic, and EDTA is added to chelate heavy metal ions, which catalyse adverse reactions.

Avoid large excess (over tenfold molar excess of the disulfide bond to be split) of reducing agents (mercaptoethanol or dithiothreitol), because it may be difficult to remove them completely from the ligand.

Alternatively, ligand primary amino group can be modified with succinimidylacetylthioacetate, and thiol group generated later by addition of hydroxylamine.

Caveats: Not only thiol-reactive phospholipid derivatives may be used to modify thiol-containing ligands in aqueous detergent solutions. They may be preincorporated in liposomes and reacted with ligands later.

Worksheet 9. Synthesis and Radioiodine Labeling of Phenol-Based Derivatives of Phosphatidylethanolamine

Principle of operation: Phenol derivative is attached to the polar head group of PtdEtn; radioiodination of the "cold" phospholipid phenol ring is performed consequently.

Literature: P. M. Abra, H. Schreier, and F. C. Szoka. *Res. Commun. Chem. Pathol. Pharm. 37*:199 (1982). A. J. Schroit. *Biochemistry 21*:5323 (1982).

Materials: To modify 10 μmol phospholipid, use 15–30 μmol of hydroxyphenylpropionic acid *N*-hydroxysuccinimide ester, or *p*-hydroxybenzimidate in chloroform or chloroform-methanol in the presence of 20–35 μmol triethylamine. Preparation of PtdEtn derivatives of phenol (PPE) is performed essentially as in Worksheet 1. Purification of PPE is performed as in Worksheet 4.

For labeling of 10–100 μg phospholipid with 0.1–1 mCi ^{125}I, use 10–50 μg of Chloramine T or Iodogen.

Equipment: Vortex mixer, microscale reaction vessel or tube, micropipettes, analytical balance, ultrasound disintegrator, thin-layer chromatography equipment.

Procedure 1: Labeling of PPE with ^{125}I via Chloramine T. 10 μg PPE in 200 μL chloroform in a conical centrifuge tube is added to 180 μL of 0.05 M borate buffer pH 8, followed by addition of 10 μL 2 mM KI, $Na^{125}I$ and 10 μL of 4 mM Chloramine T. After 30 min stirring, the reaction is quenched with 10 μL 40 mM sodium bisulfite and 10 μL 1 M KI. Product is initially purified by sixfold extraction with 2 mL phosphate-buffered saline.

Procedure 2: Labeling of PPE with ^{125}I via Iodogen. 0.02–0.1 mg PPE is dispersed in 0.1–0.5 mL 0.1 M phosphate buffer pH 8 and sonicated to obtain an optically transparent solution. Dry 20–50 μg Iodogen from 200–500 μL chloroform in a microcentrifuge tube. Add PPE solution to Iodogen-containing tube; immediately add $Na^{125}I$ (carrier free) 0.1–1 mCi and incubate at room temperature with occasional slow rocking for 30 min. Carefully remove PPE solution from Iodogen-containing tube, dialyze against saline and several changes of distilled water, lyophilize and redissolve in chloroform.

Purification: Final purification of radiolabeled PPE derivatives can be performed by thin-layer chromatography on silicic acid.

Worksheet 10. Synthesis, Purification and Labeling with ^{111}In of the DTPA Derivative of Phospholipid

Principle of operation: The primary amino group of phosphatidylethanolamine is acylated with cyclic anhydride of chelator DTPA, which tightly binds In^{3+}, Fe^{3+}, Gd^{3+}.

Literature: C. W. Grant, S. Karlik, and E. Florio. *Magn. Res. Med. 11*:236 (1989).

Materials: To modify 100 μmol dipalmitoyl phosphatidylethanolamine, use 1000 μmol of the cyclic dianhydride of diethylene triamine pentaacetic acid. As a solvent, pyridine is used (dried over CaH_2).

Equipment: reaction vessel with condenser and heating mantle, analytical balance.

Procedure: Dissolve PtdEtn in pyridine at 10 mg/mL upon heating. Separately reflux DTPA cyclic dianhydride in pyridine at 50 mg/mL until it dissolves. Add dianhydride to PtdEtn via the reaction vessel sidearm and reflux it for an additional 70 min. Add H_2O (0.25 mL per 1 mL of mixture) to open the unreacted cyclic anhydride ring and reflux for additional 70 min. Evaporate reaction media to dryness.

Purification: Thin layer chromatography on silica gel can be used in 65:25:4:1 $CHCl_3$:CH_3OH:H_2O:formic acid system. Alternatively, dialysis procedure can be used (see Worksheet 4).

Labeling with metal isotope: Prepare liposomes incorporating purified DTPA-PtdEtn in 0.15 M NaCl solution. Add metal isotope, e.g., aliquots of $GdCl_3$ in saline or $^{111}InCl_3$ as solution in 0.1 M citrate buffer pH 6. Unbound metal can be removed from liposomes by gel filtration or liposome sedimentation.

Alternatively, 1 μL of radiolabeled $^{111}InCl_3$ in 0.01 M HCl can be added to 50 μL of DTPA-PtdEtn solution in chloroform (0.1–1 mg/mL) and 0.5 μL triethylamine added 5 min later to compensate the acidic media formed. Radiolabeled lipid can be later incorporated in liposomes.

Recommendations: In order to avoid the formation of insoluble hydroxides and salts of transition metals in aqueous media, avoid alkaline pH and phosphate and borate buffers. Citrate complexes of these metals can be initially formed to avoid the problem.

Caveats: Cyclic dianhydride of DTPA possesses two reactive moieties, and it may serve as a cross-linking agent, covalently linking two phospholipid molecules. This adverse reaction should be avoided, by using high DTPA-PtdEtn initial molar ratio, and/or prehydrolyzing part of the cyclic anhydride residues of PtdEtn by water.

Worksheet 11. Synthesis and Purification of Polyethyleneglycol-Phosphatidylethanolamine

Principle of operation: The primary amino group of PtdEtn is acylated in organic solvent in the presence of triethylamine by monomethoxypolyethyleneglycol succinate *N*-hydroxysuccinimide ester.

Literature: A. L. Klibanov, K. Maruyama, V. P. Torchilin, and L. Huang. *FEBS Lett. 268*:235 (1990).

Materials: To modify 10 μmol phospholipid, use 30 μmol of monomethoxypolyethyleneglycol succinimidyl succinate (M = 5000, Sigma Chem. Co.) in the presence of 35 μmol triethylamine in chloroform. Bio-Gel A1.5M is used for product purification.

Equipment: Vortex mixer, microscale reaction vessel or tube, micropipettes, analytical balance, fraction collector, and column for gel filtration.

Procedure: An aliquot of PEG-OSu in $CHCl_3$ (142 mg/mL) was added to a solution of Ole_2PtdEtn in $CHCl_3$ (25 mg/mL), followed by addition of triethylamine. The final molar ratio of these components in the reaction mixture was 3:1:3.5 (PEG-OSU: PtdEtn:triethylamine). The reaction mixture was incubated overnight at room temperature, and $CHCl_3$ was evaporated with a stream of nitrogen gas. Full conversion of the PtdEtn primary amino group was confirmed by thin-layer chromatography, using a $CHCl_3$:CH_3OH:H_2O solvent system (65:30:4) and silica gel 60-coated glass plates.

Purification: The reaction mixture after $CHCl_3$ evaporation was redissolved in saline (0.9% NaCl). Unreacted PEG-OSu is rapidly hydrolyzed in aqueous media. The resulting mixture in saline was applied to a Bio-Gel A1.5M column (1.5 × 30 cm) preequilibrated with saline. PEG-succinate is retained by the column, whereas PEG-PtdEtn forms micelles that are excluded by the column gel. PEG-PtdEtn peak fractions were pooled, dialyzed against water, and lyophilized.

Recommendations: Polyethyleneglycol active ester is hygroscopic and may be partially hydrolyzed. Use a large excess of reagent for complete blocking of the primary amino group of PtdEtn. Use pure triethylamine, free from contaminating primary and secondary amines.

REFERENCES

1. L. D. Bergelson, *Lipid Biochemical Preparations*, Elsivier/North-Holland Biomedical Press, New York, 1980.
2. R. P. Haugland, ed., *Handbook of Fluorescent Probes and Research Chemicals*, Molecular Probes, Inc., Eugene, Oregon, 1989, pp. 134–138.
3. D. Papahadjopoulos and L. Weiss, Amino groups at the surfaces of phospholipid vesicles, *Biochem. Biophys. Acta 183*:417 (1969).
4. J. K. Dunnick, I. R. McDougall, S. Aragon, M. L. Goris, and T. P. Kriss, Vesicle interactions with polyaminoacids and antibody: In vitro and in vivo studies, *J. Nucl. Med. 16*:483 (1975).
5. H. Endoh, Y. Suzuki, and Y. Hashimoto, Antibody coating of liposomes with 1-ethyl-3-(3-dimethylaminopropyl)carbodiimide and the effect on target specificity, *J. Immunol. Methods 44*:79 (1981).
6. V. P. Torchilin, B. A. Khaw, V. N. Smirnov, and E. Haber, Preservation of antimyosin antibody activity after covalent coupling to liposomes, *Biochem. Biophys. Res. Commun. 89*:1114 (1979).
7. V. P. Torchilin, V. S. Goldmacher, and V. N. Smirnov, Comparative studies on covalent and noncovalent immobilization of protein molecules on the surface of liposomes. *Biochem. Biophys. Res. Commun. 85*:983 (1978).
8. T. Arvinte, P. Wahl, and C. Nicolau, Low pH fusion of mouse liver nuclei with liposomes bearing covalently bound lysozyme, *Biochim. Biophys. Acta 899*:143 (1987).
9. J. C. Stavridis, G. Dcliconstantinos, M. C. Psallidopoulos, N. A. Armenaka, D. J. Hadjimin, and J. Hadjimin, Construction of transferrin-coated liposomes for in vivo transport of exogenous DNA to bone-marrow erythroblasts in rabbits. *Exp. Cell Res. 164*:568 (1986).
10. A. R. Neurath, S. B. Kent, and N. Strick, Antibodies to hepatitis-B surface-antigen (HBSAG) elicited by immunization with a synthetic peptide covalently linked to liposomes, *J. Gen. Virol. 65*:1009 (1984).
11. P. Ghosh and B. K. Bachhawat, Grafting of different glycosides on the surface of liposomes and its effect on the tissue distribution of ^{125}I-labeled globulin encapsulated in liposomes, *Biochim. Biophys. Acta 632*:562 (1980).
12. M. R. Roth, R. B. Avery, and R. Welti, Cross-linking of phosphatidylethanolamine neighbors with dimethylsuberimidate is sensitive to the lipid phase, *Biochim. Biophys. Acta 986*:217 (1989).
13. L. D. Leserman, J. Barbet, F. Kourilsky, and J. N. Weinstein, Targeting to cells of fluorescent liposomes covalently coupled with monoclonal antibody or protein A, *Nature 288*:602 (1980).
14. J. Lasch, G. Niedermann, A. A. Bogdanov, and V. P. Torchilin, Thiolation of preformed liposomes with iminothiolane, *FEBS Lett. 214*:13 (1987).
15. F. J. Martin, W. L. Hubbell, and D. Papahadjopoulos, Immunospecific targeting of liposomes to cells: A novel and efficient method for covalent attachment of F_{ab}' fragments via disulfide bonds, *Biochemistry 20*:4229 (1981).
16. F. J. Martin and D. Papahadjopoulos, Irreversible coupling of immunoglobulin fragments to preformed vesicles. *J. Biol. Chem. 257*:286 (1982).
17. G. D. Muller and F. Schuber, Neo-mannosylated liposomes: Synthesis and interaction with mouse Kupffer cells and resident peritoneal macrophages, *Biochim. Biophys. Acta 986*:97 (1989).
18. J. Salord and F. Schuber, In vitro drug delivery mediated by ecto-NAD^+-glycohydrolase ligand-targeted liposomes, *Biochim. Biophys. Acta 971*:197 (1988).
19. Y. Hashimoto, M. Sugawara, and H. Endoh, Coating of liposomes with subunits of monoclonal IgM antibody and targeting of the liposomes. *J. Immunol. Methods 62*:155 (1983).
20. M. Egger, S. P. Heyn, and H. E. Gaub, Two-dimensional recognition pattern of lipid-anchored F_{ab}' fragments, *Biophys. J. 57*:669. (1990).

21. K. K. Matthay, A. M. Abai, S. Gobb, K. Hong, D. Papahadjopoulos, and R. M. Straubinger, Role of ligands in antibody-directed endocytosis of liposomes by human T leukemia cells, *Cancer Res. 49*:4879 (1989).
22. M. Maeda, A. Kumano, and D. A. Tirell, H^+-Induced release of contents of phosphatidylcholine vesicles bearing surface-bound polyelectrolyte chains, *J. Am. Chem. Soc. 110*:7455 (1988).
23. R. Bredehorst, F. S. Ligler, A. W. Kusterbeck, E. L. Chang, B. P. Gaber, and C.-W. Vogel, Effect of covalent attachment of immunoglobulin fragments on liposomal integrity, *Biochemistry 25*:5693 (1986).
24. S. Yoshitake, M. Imagawa, E. Ishikawa, Y. Niitzu, I. Urushiza, M. Nishiura, R. Kanazawa, H. Kurosaki, S. Tachiban, and N. Nakazawa, Mild and efficient conjugation of rabbit F_{ab}' and horseradishperoxidase using a malamide compound and its use for enzyme immunoassay, *J. Biochem. 92*:1413 (1982).
25. T. D. Heath, D. Robertson, M. S. C. Birbeck, and A. J. S. Davies, Covalent attachment of horseradish peroxidase to the outer surface of liposomes, *Biochem. Biophys. Acta 599*:42 (1980).
26. M.-M. Chua, S.-T. Fan, and F. Karush, Attachment of immunoglobulin to liposomal membrane via protein carbohydrate, *Biochim. Biophys. Acta 800*:291 (1984).
27. S. C. Kinsky, J. E. Loader, and A. L. Benson, An alternative procedure for the preparation of immunogenic liposomal model membranes, *J. Immunol. Methods 65*:295 (1983).
28. S. C. Kinsky, K. Hashimoto, J. E. Loader, and A. L. Benson, Synthesis of *N*-hydroxysuccinimide esters of phosphatidylethanolamine and some properties of liposomes containing these derivatives, *Biochim. Biophys. Acta 769*:543 (1984).
29. V. T. Kung and C. T. Redemann, Synthesis of carboxyacil derivatives of phosphatidylethanolamine and use as an efficient method for conjugation of protein to liposomes, *Biochim. Biophys. Acta 862*:435 (1986).
30. G. Gregoriadis, V. Weissig, L. Tan, and Q. Xiao, A novel method for the covalent coupling of sugars to liposomes, *Biochem. Soc. Trans. 17*:128 (1989).
31. V. Weissig, J. Lasch, and G. Gregoriadis, Covalent coupling of sugars to liposomes, *Biochim. Biophys. Acta 1003*:54 (1989).
32. A. A. Bogdanov, A. L. Klibanov, and V. P. Torchilin, Protein immobilization on the surface of liposomes via carbodiimide activation in the presence of *N*-hydroxysulfosuccinimide, *FEBS Lett 231*:381 (1988).
33. D. Sinha and F. Karush, Attachment to membranes of exogenous immunoglobulin conjugated to a hydrophobic anchor, *Biochem. Biophys. Res. Commun. 90*:554 (1979).
34. D. Sinha and F. Karush, Specific reactivity of lipid vesicles conjugated with oriented anti-lactose antibody fragments, *Biochim. Biophys. Acta 684*:187 (1982).
35. G. Goodman-Snitkoff and R. J. Mannino, Identification of helper T cell epitopes from GP 160 of HIV through the use of peptide-phospholipid conjugates, *J. Cell Biochem. Suppl. 13A*:275 (1989).
36. V. P. Torchilin, A. L. Klibanov, and V. N. Smirnov, Phosphatidylinositol may serve as the hydrophobic anchor for immobilization of proteins on liposome surface, *FEBS Lett. 138*:117 (1982).
37. V. K. Jansons and P. L. Mallett, Targeted liposomes: A method for preparation and analysis, *Anal. Biochem. 111*:54 (1981).
38. V. Weissig, J. Lasch, A. L. Klibanov, and V. P. Torchilin, A new hydrophobic anchor for the attachment of proteins to liposomal membranes, *FEBS Lett. 202*:86. (1986).
39. E. Holmberg, K. Maruyama, D. C. Litzinger, S. Wright, M. Davis, G. W. Kabalka, S. J. Kennel, and L. Huang, Highly efficient immunoliposomes prepared with a method which is compatible with various lipid compositions, *Biochem. Biophys. Res. Commun. 165*:1272 (1989).
40. N. L. Thompson, A. A. Brian, and M. McConnell, Covalent linkage of a synthetic peptide to a fluorescent phospholipid and its incorporation into supported phospholipid monolayers, *Biochim. Biophys. Acta 772*:10 (1984).

41. A. N. Lukyanov, A. L. Klibanov, A. V. Kabanov, V. P. Torchilin, A. V. and K. Martinek, Phospholipid covalent binding with proteins in system of reverse micelles, *Bioorg. Khim, 14*:670 (1988).
42. A. Huang, L. Huang, and S. J. Kennell, Monoclonal antibody covalently coupled with fatty acid, *J. Biol. Chem. 255*:8015 (1980).
43. A. Huang, Y. S. Tsao, S. J. Kennell, and L. Huang, Characterization of antibody covalently bound to liposomes. *Biochim. Biophys, Acta 716*:140 (1982).
44. T. Masaki, O. Noriko, R. Yasuda, and H. Okada, Assay of complement activity in human serum using large unilamellar liposomes, *J. Immunol. Meth. 123*:19 (1989).
45. R. J. Ho and L. Huang, Interactions of antigen-sensitized liposomes with immobilized antibody: A homogeneous solid-phase immunoliposome assay, *J. Immunology 134*:4035 (1985).
46. R. Nayar, I. J. Fidder, and A. Y. Schroit, Potential applications of pH-sensitive liposomes as drug delivery systems, in *Liposomes as Drug Carriers* (G. Gregoriadis, ed.), John Wiley, New York, 1988, p. 771.
47. K. Deres, H. Schild, K.-H. Weismuller, G. Jung, and H.-G. Rammensee, In vivo priming of virus-specific cytotoxic T lymphocytes with synthetic lipopeptide vaccine, *Nature 342*:561 (1989).
48. P. R. Dal Moute and F. C. Szoka, Antigen presentation by B cells and macrophages of cytochrome c and its antigenic fragments when conjugated to the surface of liposomes, *Vaccine 7*:401 (1989).
49. B. Wolff and G. Gregoriadis, The use of monoclonal anti-Thy_1IgG_1 for the targeting of liposomes to AKR-A cells in vitro and in vivo, *Biochim. Biophys. Acta 802*:259 (1984).
50. A. L. Klibanov, K. Maruyama, V. P. Torchilin, and L. Huang, Amphipathic polyethyleneglycols effectively prolong the circulation time of liposomes, *FEBS Lett. 268*:235 (1990).
51. J.-P. Behr, B. Demeneix, J.-P. Loeffler, and J. Perez-Mutul, Efficient gene transfer into mammalian primary endocrine cells with lipopolyamine-coated DNA, *Proc. Natl. Acad. Sci. USA 86*:6982 (1989).
52. E. A. Bayer, B. Rivnay, and E. Skutelsky, Mode of liposome-cell interactions biotin-conjugated lipids as ultrastructural probes, *Biochim. Biophys. Acta 550*:464 (1979).
53. V. S. Trubetskoy, E. V. Dormeneva, V. P. Tsibulskiy, V. S. Repin, and V. P. Torchilin, Use of enzyme label for quantitative evaluation of liposome adhesion on cell-surface-studies with J774 macrophage monolayers, *Analyt. Biochem. 172*:185 (1988).
54. P. M. Abra, H. Schreier, and F. C. Szoka, The use of a new radioactive-iodine labelled lipid marker to follow in vivo disposition of liposomes: Comparison with an encapsulated aqueous space marker, *Res. Commun. Chem. Pathol. Pharm. 37*:199 (1982).
55. C. W. Grant, S. Karlik, and E. Florio, A liposomal MRI contrast agent: phosphatidylethanilamine-DTPA, *Magn. Res. Med. 11*:236 (1989).
56. A. J. Schroit, J. Madsen, and A. E. Ruoho, Radioiodinated photoactivatable phosphatidylcholine and phosphatidylserine transfer properties and differential photoreactive interaction with human erythrocyte membrane proteins, *Biochemistry 26*:1812 (1987).
57. J. G. Furcotte, S. P. Srivastava, W. A. Merezak, B. A. Rizkalla, F. Louzon, and T. P. Wung, Cytotoxic liponucleotide analogs, *Biochim. Biophys. Acta 619*:604 (1980).
58. W. Rubas, A. Supersaxo, H. G. Weder, H. R. Hartmann, H. Hengartner, H. Schott, and R. Schwendener, Treatment of Murine L1210 lymphoid leukemia and melanoma b16 with lipophilic cytosine arabinoside prodrugs incorporated into unilamellar liposomes, *Int. J. Cancer 37*:149 (1986).
59. K. Y. Hostetler, L. M. Stuhmiller, H. B. M. Lenting, H. Bosch, and D. D. Richman, Synthesis and antiretroviral activity of phospholipid analogs of azidothymidine and other antiviral nucleosides, *J. Biol. Chem. 265*:6112 (1990).
60. K. Hashimoto, J. E. Loader, and S. C. Kinsky, Synthesis and characterization of methotrexate-dimyristoylphosphatidylethanolamine derivatives and the glycerophosphorylethanolamine analogs. *Biochim. Biophys. Acta 816*:163 (1985).

61. S. C. Kinsky, K. Hashimoto, J. E. Loader, M. S. Knight, and D. Y. Fernandes, Effect of liposomes sensitized with MTX-DMPE on cells that are resistant to methotrexate, *Biochim. Biophys. Acta 885*:129 (1986).
62. C. I. Hong, S. H. An, D. J. Buchheit, A. Nechev, A. J. Kirisits, C. R. West, E. K. Ryu, and M. Maccoss, 1-beta-D-arabinofuranosylcytosine-phospholipid conjugates as prodrugs of 1-beta-D-arabinofuranosylcytosine, *Cancer Drug Deliv. 1*:181 (1984).
63. C. J. Welch, A. Larsson, A. C. Ericson, B. Oberg, R. Datema, and J. Chattopadhyaya, The chemical synthesis and antiviral properties of an acyclovir-phospholipid conjugate, *Acta Chem. Scand. Ser. B Org. Chem. Biochem. 39*:47 (1985).
64. R. Aneja, J. S. Chadha, and A. P. Davies, A general synthesis of glycerophospholipids, *Biochim. Biophys. Acta 218*:102 (1970).
65. R. Aneja and J. S. Chadha, A total synthesis of phosphatidylcholines, *Biochim. Biophys. Acta 248*:455 (1971).

9
Chemical Stability

R. Evstigneeva *Lomonosov Institute of Fine Chemical Technology, Moscow, Russia*

I. CHEMICAL PROPERTIES OF PHOSPHOLIPIDS

Phospholipids are functionally active components of biological membranes. The wonderful properties of living membranes—their structural stability together with the dynamic equilibrium of membranous components involved in metabolic processes—are based on the chemical structure of phospholipids. There are more than a hundred variations of lipid molecules in a biological membrane, which differ one from another by the fatty acid composition and the character of the substituents X located at the polar head of the molecule. The chemical bonds in phospholipids are of three types only, C–C bonds, ester bonds, and phosphoester bonds. Phospholipids can be regarded as asymmetric phosphoric acid diesters [1].

$$\mathrm{RO-\overset{\overset{\displaystyle O}{\|}}{\underset{\underset{\displaystyle O^-}{|}}{P}}-OX}$$

These chemical features determine the main properties of phospholipids and their stability. Phospholipids easily undergo hydrolytic splitting in acidic and alkaline media. Only at pH 7 are phospholipids stable enough, as under these conditions the ester bond hydrolysis does not proceed to any significant degree [2].

Hydrolytic splitting of lipids is extensively used for structure determinations and for isolation of certain lipidic components such as fatty acids.

Alkaline hydrolysis of glycerophospholipids carrying ester bonds leads initially to fatty acids and glycerophosphates. The latter are apt to undergo further hydrolysis, because the phosphoester bond linking up the hydrophilic component X to the phospholipid moiety is not stable enough under alkaline conditions and splits up to give a cyclic phosphate. When the cycle opens up, it gives a (1 : 1) mixture of 2- and 3-glycerophosphates.

$$\begin{array}{l} {}^{1}CH_2OCOR \\ | \\ RCOO-{}^{2}CH \quad O \\ | \quad \| \\ {}^{3}CH_2O-P-OX \\ \quad | \\ \quad O^- \end{array} \xrightarrow[-2RCOOH]{OH^-} \begin{array}{l} CH_2OH \\ | \\ HO-CH \quad O \\ | \quad \| \\ CH_2O-P-OX \\ \quad | \\ \quad O^- \end{array} \xrightarrow{-XOH}$$

$$\begin{array}{l} O \quad O-CH(CH_2OH) \\ \;\backslash\backslash / \quad | \\ \;\;P \quad\; | \\ \;/ \backslash \quad | \\ {}^-O \quad O-CH_2 \end{array} \longrightarrow \begin{array}{l} CH_2OH \\ | \\ HO-CH \quad O \\ | \quad \| \\ CH_2O-P-O^- \\ \quad | \\ \quad O^- \end{array} + \begin{array}{l} CH_2OH \quad O \\ | \quad \| \\ CH-O-P-O^- \\ | \quad | \\ CH_2OH \quad O^- \end{array}$$

In an acidic medium the equilibrium between 2- and 3-phosphates is shifted to 3-phosphates. In an alkaline medium the hydrolysis at C-2 can be accompanied by configuration inversion.

In the group of glycerophospholipids carrying an ether bond, the less stable compounds are the plasmalogens containing an acid-labile cis-vinyl ether bond.

$$\begin{array}{l} \quad\quad H \quad H \\ \quad\quad | \quad | \\ CH_2-OC=C-R \\ | \\ R^1COO-CH \quad O \\ | \quad \| \\ CH_2O-P-OR^2 \\ \quad | \\ \quad O^- \end{array} \xrightarrow{H^+} \begin{array}{l} CH_2-OH \\ | \\ R^1COO-CH \quad O \\ | \quad \| \\ CH_2O-P-OR^2 \\ \quad | \\ \quad O^- \end{array} + \begin{array}{l} \quad O \\ \quad // \\ RCH_2C \\ \quad \backslash \\ \quad H \end{array}$$

Sphingolipids undergo acidic and alkaline hydrolysis to give sphingosine bases and acids.

$$\begin{array}{l} R-CH-CH-CH_2OH \\ \quad | \quad\; | \\ \quad OH \;\; NHCOR^1 \end{array} \xrightarrow{H^+ \text{ or } OH^-} \begin{array}{l} R-CH-CH-CH_2OH \\ \quad | \quad\; | \\ \quad OH \;\; NH_2 \end{array} + R^1-COOH$$

Isolation of sphingosine bases which chemically are aliphatic amino alcohols can be successfully achieved by acidic hydrolysis. Usually the hydrolysis is carried out by a water-methanolic 1 N solution of HCl; under these conditions, the undesirable side reactions are minimal.

The phosphoester bond is also apt to split under the action of chemical agents. Thus the treatment of phospholipids with acetic anhydride induces phosphoester bond cleavage from the hydrophobic component side with subsequent acetylation of the diglyceride formed. The acetylation can be accompanied by acyl group migration.

Diazomethane contrariwise cleaves the phosphoester bond with the polar component X to give a dimethyl ester of the phosphatidic acid. No bond inversion takes place, and the configuration of the ester is retained.

The treatment of diacyl phospholipids with lithium aluminum hydride leads to the splitting of phosphoester and ester bonds and the reduction of acidic residues to give alcohols.

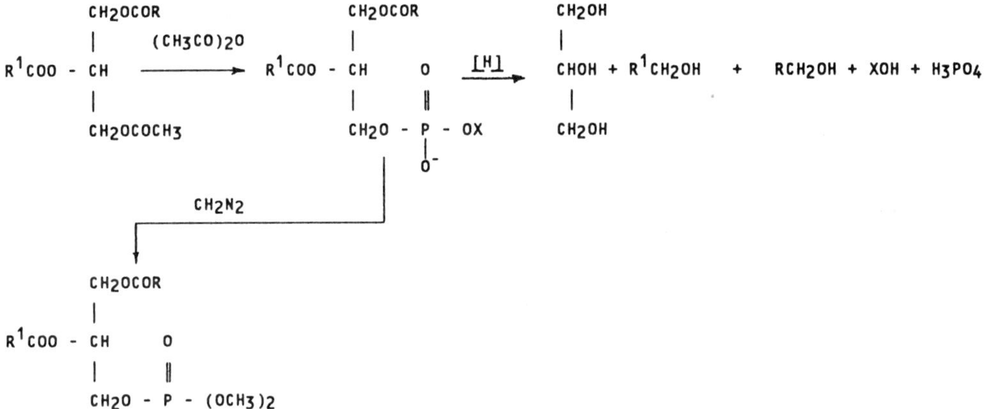

Ether phospholipids treated with lithium aluminum hydride undergo partial splitting, their ether bond remaining unruptured.

```
              CH2OR                           CH2OR
              |                  LiAlH4       |
R1COO  -  CH          O    ----------->       CHOH  +  R1CH2OH  +  XOH  +  H3PO4
              |          ‖                    |
              CH2O  -  P  -  OX               CH2OH
                         |
                         O-
                                  R= - (CH2)nCH3      or      - CH = CHR2
```

II. ENZYMATIC DIGESTION

Under physiological conditions, the ester bonds in phospholipids are easily digested by relevant enzymes [3]. There are four types of phospholipases that react with glycerophospholipids with high regio-and stereoselectivity. These are phospholipases A_1, A_2, C, and D.

```
                       PLA1  O
                          ↓  ‖
        O        CH2- O — C - R
        ‖        |
R1- C - O - CH           O PLD
         ↑       |           ‖ ↓
        PLA2     CH2O — P —— OX
                          ↑ |
                        PLC O-
```

Enzymatic hydrolysis using these four phospholipases is often employed in scientific research for structural and synthetic studies.

Phospholipase A_2 hydrolyzes the ester bond at C–2 in the natural phospholipids such as lecithins, cephalins, phosphatidylserine, phosphatidylglycerol and its amino acid derivatives, phosphatidylmyoinositol, and plasmalogens. The structures of the hydrophilic and hydrophobic moieties do not seem to have any effect on the enzyme specificity, but they markedly affect the hydrolysis rate.

Phospholipase A_2 is highly stereospecific, digesting the 3-*sn*-glyerophosphate derivatives only; apparently the nature of the fatty acid residues does not impair its specificity but affects the reaction rate. Optimal pH for phospholipase A_2 is 7.2, the enzyme being activated by Ca^{2+} ions. Optimum pH for phospholipase A_1 is 4.2.

Pancreatic lipase (glyceryl ester hydrolase EC 3.1.13) hydrolyzes ester bonds but is not so sterospecific as phospholipases A_1 and A_2.

Substitution of 1-alkenylester group for the fatty acid residue in plasmalogens does not affect the specificity of the lipolytic enzymes but considerably inhibits the hydrolysis rate. Thus plasmalogens split more slowly than do acyl type phospholipids.

III. ENZYMATIC OXYGENATION

The stability of phospholipids depends strongly on the nature of the constituent acyl and alkyl residues. Phospholipids carrying unsaturated bonds tend to change their structure under the action of the air oxygen; they are especially prone to undergo nonenzymatic reactions of autooxidation and photooxygenation, resulting in hydroperoxide formation with concomitant acyl migration and stereomutation of the double bonds. The hydroperoxides formed are subject to further transformations either by splitting to molecules of lower molecular mass or contrariwise by dimerization to compounds of higher molecular mass.

The oxidation results initially in allyl hydroperoxides, the double bond in which is prone to migrate or undergo stereomutation [4]. For example, a simple substrate, methyl oleate, gives hydroperoxides of four types:

```
                         - CH2 - CH = CH - CH2-
                                    Cis
                                     |  - H+
- CH2 - ĊH - CH = CH -   <-----------+----------->   - CH = CH - ĊH - CH2 -
  |                                                                    |
  |       - CH2 - CH = CH - ĊH -          - ĊH - CH = CH -CH2 -        |
  |                  |     OOH              OOH     |                  |
  |                  v      |                |      v                  |
  |       - CH2- CH = CH - CH -           - CH - CH = CH - CH2-        |
  |       cis/trans (26%)                  cis/trans (26%)             |
  v         OOH                                    OOH                 v
            |                                       |
- CH2 - CH - CH = CH -                    - CH = CH - CH - CH2 -
      trans    (23%)                            trans   (24%)
```

Compounds containing several double bonds give a much more complicated oxidation picture; there additionally appear dihydroperoxides, hydroperoxy peroxides, and hydroperoxy diperoxides.

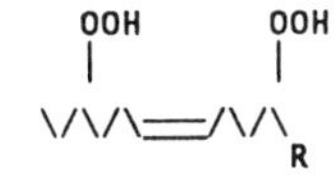

```
OOH       OOH
 |         |
\/\/\==/\/\
           R
```

dihydroperoxide

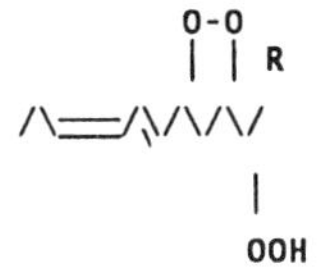

```
      O-O
      | | R
/\==/\/\/\/
         |
        OOH
```

dihydroperoxy peroxide

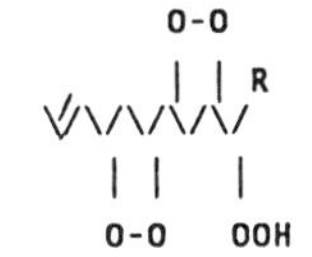

```
         O-O
         | | R
\/\/\/\/\/\/
  | |   |
  O-O  OOH
```

hydroperoxy diperoxide

R = $(CH_2)_7COOMe$

Autooxidation is a radical chain process involving chain initiation, propagation, and termination reactions.

$$RH \longrightarrow R^{\cdot}$$

$$R^{\cdot} + O_2 \longrightarrow ROO^{\cdot}$$

$$ROO^{\cdot} + RH \longrightarrow ROOH + R^{\cdot}$$

$$R^{\cdot} + R^{\cdot} \longrightarrow R - R$$

The mechanism of the initiation step is not very well understood; however, initiation is known to be accelerated by metals, such as copper and iron, and also by temperature increase. Heating and metals catalyze peroxide decomposition and favor the formation of radicals initiating chain reactions.

The autooxidation reaction sequence can be controlled via prooxidant and antioxidant substances. Prooxidants favor chain propagation, while antioxidants induce chain breakdown. Antioxidant activity is enhanced by synergic compounds that chelate metals and thus inhibit the metal-catalyzed chain initiation. Such synergists are ascorbic acid, citric acid, and ethylenediamine tetraacetic acid.

Natural tocopherols and sterically hindered synthetic phenols also act as antioxidants.

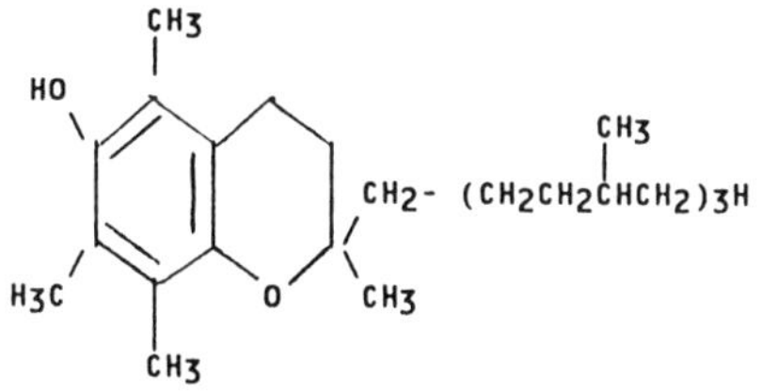

```
      CH3
HO                      CH3
                        |
          CH2- (CH2CH2CHCH2)3H

H3C        O   CH3
      CH3
```

α - tocopherol

```
                 OH
(H3C)3 - C             C - (CH3)3


                 CH3
```

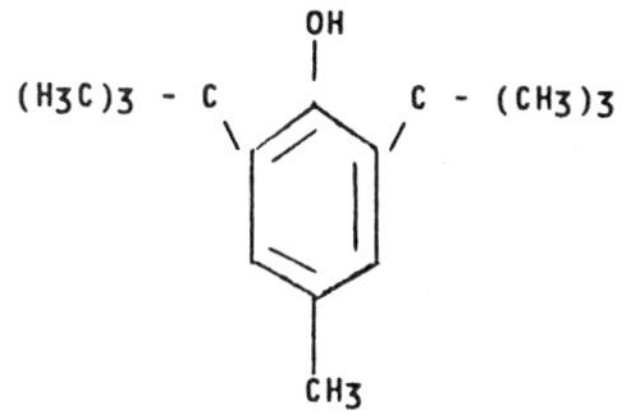

ionol

The oxidations of monoene, diene, and triene compounds seem to proceed by actually the same reaction route, whereas oxidation and photooxygenation pathways apparently differ one from another; this can be concluded from their reaction products, which are similar but not identical.

Photooxygenation involves the interaction between the double bond and the singlet oxygen produced from the triplet oxygen after illumination in the presence of such sensitizers (sens) as chlorophyll and certain dyes (Bengal rose, methylene blue).

$$sens + h\nu \longrightarrow {}^1sens \longrightarrow {}^3sens \xrightarrow{O_2} {}^1O_2$$

This reaction is not a chain process and has no induction period, so that it is not inhibited by antioxidants. Inhibition of photooxygenation can be achieved by singlet oxygen quenchers, for instance carotene. The interaction of unsaturated compounds with singlet oxygen is accompanied by double bond migration and stereomutation.

```
R     R1                              R1
 \==/        +   O = O   ------>  R/\\__/
                                        \
                                         O
                                        /
                                   H - O
```

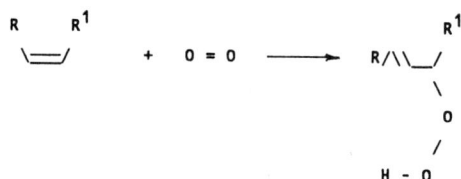

IV. EPOXIDATION OF UNSATURATED MOIETIES

Epoxidation of unsaturated compounds is a reaction well known in organic chemistry. In oxidative metabolism of phospholipids epoxidation also plays an important role.

```
                                   O
                                  / \
- CH = CH -   -------->     - HC  -  CH -
```

Epoxides can be prepared either by chemical procedures or by enzymic methods. Usually epoxidation is performed by treatment with peroxy acids: peroxy formic, peroxy acetic, peroxy trifluoroacetic, perbenzoic, and others. Peroxy acids in their turn are prepared by action of H_2O_2 on free acids, anhydrides, and chloroanhydrides.

$$\left.\begin{array}{l} RCOOH \\ (RCO)_2O \\ RCOCl \end{array}\right\} \xrightarrow{H_2O_2} RCO_3H$$

Since epoxidation is an isothermic reaction, one should not use high concentrations of peroxy acids for preparing acid-labile epoxides.

The mechanism of epoxidation is not fully clarified; the most probable is the following mechanism proposed by Bartlett:

```
\ /        H   O              \ /
 C         |   ||              C.   .H ··· O
 ||    +   O   C     ------>   | ·O·  |    :|      ------>
 C          \ / \              C·   ·O —·— C ——— R
/ \          O   R            / \

            \ /
             C           HO
              \           |
------>       | O   +  O = C
              /           \
             C             R
            / \
```

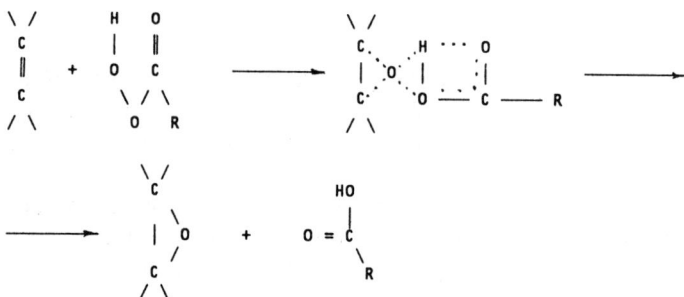

Compounds with cis double bonds give rise to cis epoxy compounds, while compounds with trans double bonds afford trans epoxy products. Oleic and elaidic acids give cis epoxystearic and trans epoxystearic acids, respectively.

Under acidic conditions, epoxides are easily hydrolized to give diols and monoesters.

```
        O
       / \            H2O
- HC  - CH -      ----------->    - CH(OH) - CH(OH) -
      |                                   ^
      | AcOH                              |
      v                       H2O         |
   - CH(OAc) - CH(OH)     ----------------
```

The reaction is stereospecific and is accompanied by configuration inversion, so that cis epoxides result in threo diols and trans epoxides in erythro diols.

```
   O                                  OH
\ / \ /                               |
 C - C        ----------->    - CH - CH -
/     \                         |
  cis                           OH
                               threo

   O
\ / \ /
 C - C        ----------->    - CH - CH -
/     \                         |     |
 trans                          OH    OH
                                erythro
```

Epoxidation in vivo proceeds as an enzymatic reaction; enzymatic epoxidations can also be reproduced in vitro. Epoxy compounds are the major metabolites of unsaturated fatty acids.

V. ENZYMATIC OXYGENATION

Splitting off of an acyl moiety from the C-2 atom of glycerol gives rise to a number of acids, some of which are unsaturated. The following eicosapolyenoic acids were obtained in this process: arachidonic (5Z,8Z,11Z,14Z-eicosatetraenoic), dihomo-γ-linolenic (8Z,11Z,14Z-eicosatrienoic), and tymnodonic (5Z, 8Z, 11Z, 14Z, 17Z-eicosapentaenoic) acids; these acids undergo enzymatic oxygenation to give a number of biologically active metabolites: hydroperoxy acids (HPETE), hydroxy acids (HETE), prostaglandins (PG), thromboxanes (TX), levuglandins (LG), leukotrienes (LT), lipoxins (LX), and hepoxylins (HX) (Scheme 1).

At present, two pathways for enzymatic transformation of polyunsaturated acids are comparatively well investigated. These are the cyclogenase pathway resulting in prostaglandins, thromboxanes, prostacyclin, and levuglandins, and the lipoxygenase pathway leading to acyclic metabolites, leukotrienes, and a number of hydroxy acids and lipoxins.

Enzymatic oxygenation gives two types of oxygen-containing metabolites: the peroxy acids and the expoxides. Their further transformations in the organism lead to oxy compounds. Oxygenation reactions involve a stereoselective removal of the pro-R or pro-S hydrogen from the methylene grouping; simultaneously, the Z double bond isomerizes to give an E double bond, thus forming a conjugated system and a radical.

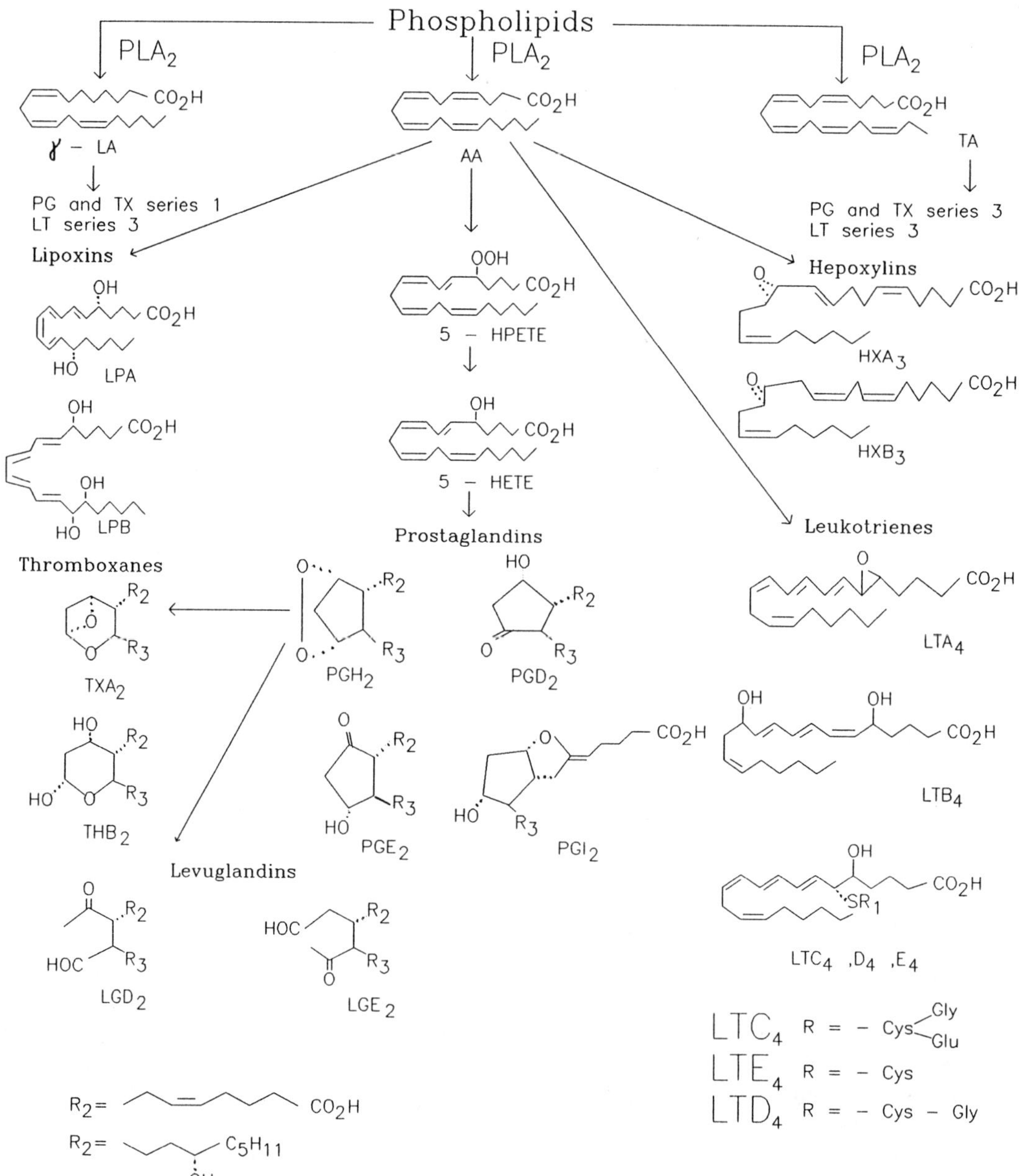

Scheme 1 Enzymatic transformations of eicosapolyenoic acids.

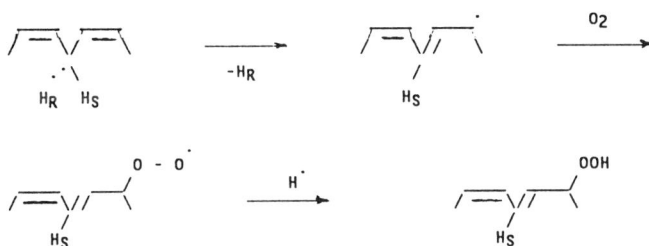

The mechanism of an asymmetric site formation was established using a 5-lipoxygenation reaction [5]. The reaction proceeds via an intermediate composed of a conjugated diene complexed with Fe(III), which further transforms into an Fe(III)-hydroperoxide of the same configuration. After a hydrolytic splitting, the iron-hydroperoxide complex gives a hydroperoxy grouping.

Monohydroperoxy acids (HPETE) actually are the common intermediates in the biosynthesis of metabolites for the cascade of arachidonic acid as well as for that of other polyenoic acids.

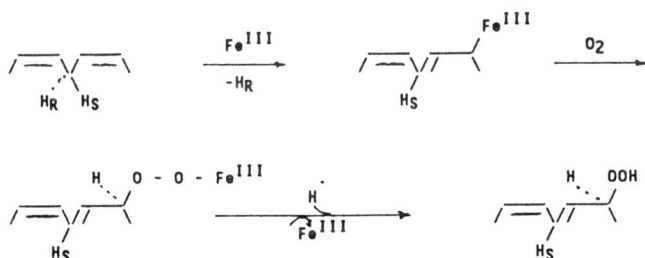

There is another very important pathway in the phosphatidylinositol metabolism. Inositol lipids are split by phospholipase C to be further involved in the phosphatidylinositol turnover taking part in the transmittance of information signals. When an extracellular signal reaches a relevant receptor on the biological membrane, phosphatidylinositol decomposes. The process is accompanied by protein kinase C activation and mobilization of calcium ions (Scheme 2) [6].

Phospholipase C splits the inositol triphosphate derivative. The resulting 1,2-diacyl glyceride (DG) activates protein kinase C; the inositol triphosphate (IP_3) formed further serves to generate phosphatidylinositol.

Arachidonic acid cascade and the phosphatidylinositol cycle are the major metabolic pathways based on degradation of phospholipid molecules. These processes take part in cell bioregulation and are involved in the signaling system. Inositol lipids give an example of the structural peculiarity of phospholipids that are constituent parts of the molecular ensemble, ensuring the stability of biological membranes and the dynamism of certain constituents that supply metabolites for bioregulatory processes.

Unsaturated compounds participate in other transformations that result from further degradation processes or are produced by oxidative condensation; however, they are not considered here, as they are not very important for phospholipids, whereas for lipids these reactions have been discussed in earlier publications.

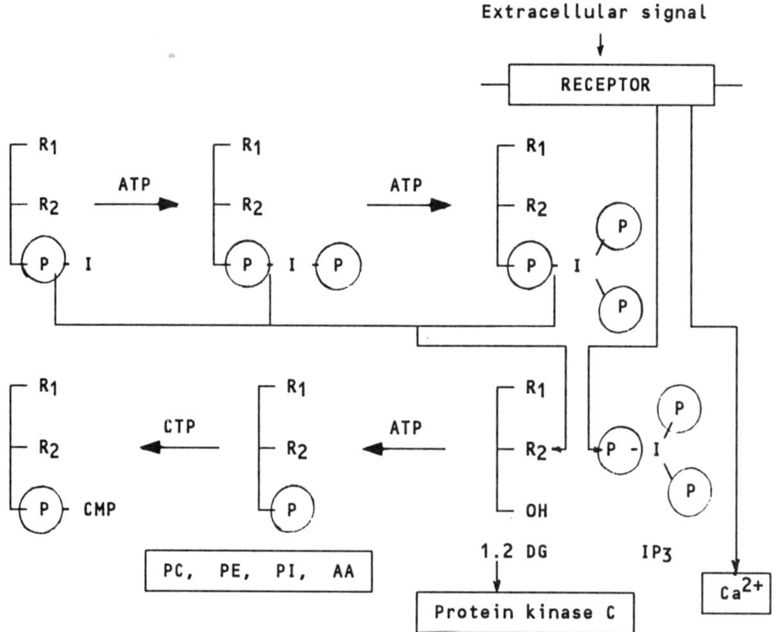

Scheme 2 Inositol lipid metabolism. ATP, adenosine triphosphate; CMP, citidine monophosphate; PC, phosphatidylcholine; PE, phosphatidylethanolamine; PI, phosphatidylinositol; AA, arachidonic acid.

VI. STORAGE OF LIPID PREPARATIONS

Since lipids are prone to decomposition by oxidation, they must be stored at low temperature in the dark and should be protected from air oxygen. Very convenient is to store them in Dewar reservoirs filled with dry ice; the atmosphere therein is inert and the temperature is low.

As air oxygen cannot be completely excluded during operations, less stable preparations, such as unsaturated lipids, are generally supplemented with antioxidants (0.05% of the sample weight) [7].

Lipid decomposition is catalyzed by the glass walls of the reaction vessel, so that lipids are better stored as solutions. The choice of solvent depends on the nature of the lipid; neutral lipids are usually stored as hexane solutions, whereas phosphatidylcholines are kept in benzene and other lipids in (9:1) mixtures of water saturated chloroform and methanol. One must remember that methanol, as well as other alcohols, can cause lipid esterification, though on the other hand, alcohols as free radical acceptors are capable of inhibiting the autooxidation of lipids.

Before use, the purity of lipid preparations should be checked by TLC, GLC, or HPLC. A more detailed description of some of the problems associated with the storage of lipid suspensions in water is given in the following chapter.

REFERENCES

1. N. A. Preobrazhenskii and R. P. Evstigneeva, eds., *The Chemistry of Biologically Active Natural Compounds*, Khimiya, Moscow, 1976.

2. R. P. Evstigneeva, E. N. Zvonkova, G. A. Serebrennikova, and V. I. Schvets, *Lipid Chemistry*, Khimiya, Moscow, 1983.
3. D. Chapman, *Introduction to Lipids*, McGraw-Hill, London, 1969.
4. F. D. Gunstone, J. L. Harwood, and F. B. Padley, eds, *The Lipid Handbook*, Chapman and Hall, New York. 1986.
5. E. J. Corey, S. W. Wright, and S. P. T. Matsuda, Stereochemistry and mechanism of the biosynthesis of leukotriene A_4 from 5(S) hydroperoxy-6(E), 8, 11, 14 (Z)-eicosatetraenoic acid. Evidence for an organoiron intermediate. *J. Am. Chem. Soc. 111*:1452 (1989).
6. Y. Nishizuka, Protein kinases in signal transduction. *TIBS*:163 (1984).
7. L. D. Bergelson, E. V. Dyatlovitskaya, Yu. G. Molotkovskii, S. G. Batrakov, L. I. Barsukov, and N. V. Prokazova, *Preparative Biochemistry of Lipids*, Nauka, Moscow, 1981.

10
Physical Stability on Long-Term Storage

Daan J. A. Crommelin, Herre Talsma, Mustafa Grit, and Nicolaas J. Zuidam *University of Utrecht, Utrecht, The Netherlands*

I. INTRODUCTION

Vesicles based on phospholipid bilayers dispersed in aqueous dispersions are presently under investigation for several quite different reasons. For instance, in biochemistry these vesicles are considered as model systems better to understand biochemical processes; in the pharmaceutical sciences their potential to act as carriers for diagnostic agents and pharmaceuticals is studied. In biochemical studies, liposome stability is evaluated over periods of hours, days, or weeks. Liposomes designed as drug carriers, however, require a different time-frame for stability assessment: years. This chapter deals with the state of the art concerning the long term stability—shelf life—of pharmaceutical formulations of (phospho)lipid vesicles.

The first parenterally applied formulation (Ambisome™, a liposomal amphotericin formulation) for the therapy of disseminated fungal infections frequently occurring in immunosuppressed patients (e.g., after chemotherapy or after infection with HIV virus) was launched on the market (in Ireland) in 1990. It showed both a higher intrinsic therapeutic activity and a reduced toxicity compared to the original product (Crossley, 1990). Other drug-containing liposome formulations are in the process of clinical testing. Extensive clinical work has been done with MTP-PE, muramyltripeptide-phosphatidylethanolamine, liposomes, and doxorubicin liposomes (Zonneveld and Crommelin, 1988; Gabizon, 1989a,b, Schumann et al., 1989; Urba et al., 1990) and amphotericin formulations developed by other groups (e.g., Mufson et al., 1990).

The stability of the present and future liposomal products should preferably meet the standards of conventional pharmaceutical products: a shelf life of one year is considered to be an absolute minimum. Both chemical and physical stability aspects are involved. The chemical instability mainly concerns hydrolysis of the ester bonds in phospholipids and oxidation of their unsaturated acyl chains, if present. Chapter 9 in Part I of this handbook provides information about the background of chemical degradation processes of phospholipids. The present chapter contains a short introduction to the pH- and temperature-dependency of the hydrolysis reaction kinetics of aqueous liposomal phosphatidylcholine dispersions. Further information can be found in extensive reviews on hydrolytic and oxidation reactions of phospholipids (e.g., Konings, 1984; Grit et al., 1992a). However, the major issue dealt with in this chapter is the *physical* stability of liposomes. In the literature on the physical stability of liposomes, attention has been focused so far on two processes affecting the quality, and therefore the acceptability, of liposomes. First, the encapsulated drug can leak from the vesicles into the extraliposomal compartment. Secondly, liposomes can aggregate and/or fuse, forming larger units. Both processes (leakage and aggregation/fusion) change the disposition of the compound in vivo and thereby can affect the therapeutic index of the compound involved. Apart from these two points of concern, other physical parameters may change on storage as well. For instance, hydrolysis of phospholipids causes the formation of fatty acids (FA) and lysophospholipids. The presence of these compounds can affect the physical properties of the phospholipid bilayer considerably. Changes in the characteristics of liposome dispersions are only acceptable when they stay within certain limits; as an acceptability criterion one may choose that they neither affect the disposition of the compound, as compared to the disposition of the "freshly prepared" liposome, nor have a negative effect on the safety of liposomes. Unfortunately, detailed studies evaluating drug disposition for "freshly prepared" liposomes and "stored" liposomes are not available in the literature.

Several approaches have been developed to ensure the physical stability of lipid vesicles on storage; as mentioned above, high retention and prevention of aggregation or fusion over periods of many months are the major goals to achieve.

Two categories of approaches for stabilization can be discerned and will be discussed separately in subsequent sections:

1. In the case of storage of aqueous liposome dispersions, the structure of the bilayer can be adjusted to induce optimum stability.
2. The aqueous liposomal dispersions can be (freeze)-dried.

Different concepts to solve the problem of liposome instability are to formulate lipid systems or liposome dispersions in such a way that they: (a) produce liposomes, or (b) are loaded with drug, *in situ*, 'at the bedside', respectively, following the proliposome concept (a) or the remote-loading approach (b). As these last mentioned options (a and b) are closely related to the topic discussed in this chapter, attention will be paid to proliposomes and remote loading techniques in Sec. V.

II. STABILIZATION THROUGH SELECTION OF BILAYER COMPONENTS

The choice of bilayer components strongly affects the permeability of the bilayer for encapsulated compounds. Availability of pure compounds, toxicological considerations, and price tend to limit the choice of phospholipids to the family of the phosphatidylcholines and—if required for stability reasons—phosphatidylglycerols. The selection of compounds with long and saturated alkyl chains (DSPC, distearoylphosphatidylcholine and DPPC, dipalmitoylphosphatidylcholine), or saturated hydrogenated soybean or egg phosphatidylcholine tends to produce bilayers with low permeabilities for small nonbilayer interacting compounds; carboxyfluorescein (CF) and sugars are often used as model compounds for this class of molecules.

Alternatively, cholesterol is often added to leaky bilayers to reduce leakage rates (e.g., Scherphof et al., 1984; New, 1990). An example of the effect of bilayer composition on permeability is given in Fig. 1.

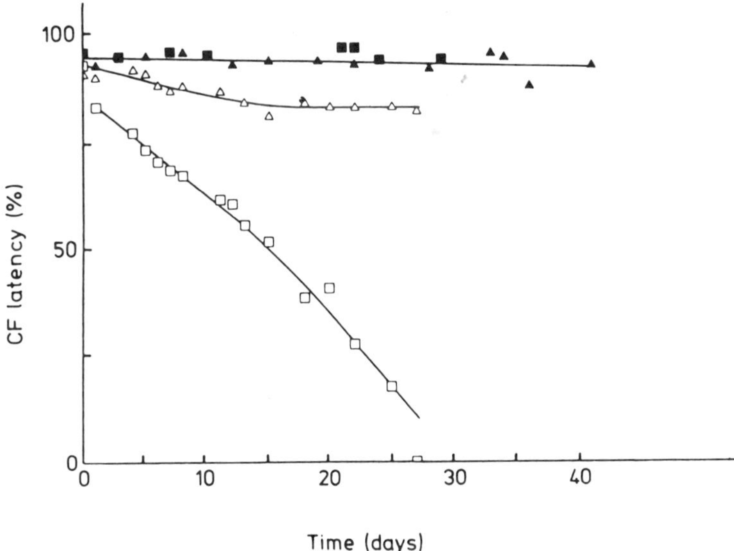

Figure 1 CF retention (latency) in percent for reverse phase evaporation vesicles (REV) stored at 4–6°C. The aqueous phase consisted of isoosmotic sodium chloride/0.01 M tris solutions, pH 7.4. □: PC/phosphatidylserine (PS) 9/1; △: PC/PS/cholesterol 10/1/4; ■: DSPC/dipalmitoylphosphatidylglycerol (DPPG) 10/1; ▲: DSPC/DPPG/cholesterol 10/1/5. (From Crommelin and Van Bommel, 1984)

The choice of bilayer constituents can influence the vesicle capacity for incorporating lipophilic compounds (e.g., Ma et al., 1991). Extruding liposomes to reduce their size and to narrow down their particle size distribution tends to decrease the solubilization potential of liposomes (personal observation). In such cases, leakage of lipophilic compounds from liposomes on storage is limited, because of their tendency to partition into the lipid bilayer. Problems can occur, however, when the temperature at which the compound is solubilized is higher than the storage temperature. Post-preparation precipitation of the compound is then often observed even at relatively late times (Van Bloois et al., 1987). It may take several days or even weeks before a lipid suspension will have reached equilibrium.

Charged amphipatic compounds, such as the cytostatic drug doxorubicin, which is positively charged at acid pH, take an intermediate position as far as leakage is concerned. In particular, in negatively charged liposomes such compounds interact strongly with the bilayer. The encapsulation efficiency is higher than expected on the basis of the encapsulated water volume; doxorubicin, for example, leaks out of the vesicles only slowly, even if bilayers have a rather high permeability for water soluble compounds (Storm et al., 1989; Amselem et al., 1990).

A special situation occurs when liposomes are stored at the temperature where the phospholipid undergoes its phase transition (Yatvin et al., 1978; Magin and Weinstein, 1984): vesicles show a dramatic increase in permeability at this temperature.

In conclusion, for hydrophilic, low molecular weight and nonbilayer interacting compounds, selection of proper bilayer constituents is critical, if a high retention in liposomes over long periods of time is aimed for. For compounds that strongly interact with the bilayer, however, leakage is not a major problem.

III. HYDROLYSIS KINETICS IN AQUEOUS LIPOSOME DISPERSIONS: pH AND TEMPERATURE EFFECTS

Phospholipids can degrade through oxidation and hydrolysis processes. The oxidation processes, as discussed in Chap. 9 of Part I, can be minimized by the addition of antioxidants such as α-tocopherol and BHT (butylhydroxytoluol) and the use of proper manufacturing conditions for dispersions as, for example, the reduction of oxygen pressure by flushing with nitrogen or argon. Moreover, if possible, phospholipids should be selected with saturated acyl chains or with acyl chains with low degrees of unsaturation.

In phospholipids such as phosphatidylcholine (PC), four ester bonds can be discerned. The two fatty acid ester bonds are the most labile bonds and are hydrolyzed first. If one fatty acid is left, lyso-PC is formed, which can dramatically change the physicochemical characteristics of the lipid bilayer (Grit and Crommelin, 1992c). At low levels of degradation, lyso-PC and the hydrolysed free fatty acid chain cause a reduction of the bilayer permeability. For partially hydrogenated PC and egg phosphatidylglycerol (PG) bilayers, an increase in permeability was only observed when over 10% of the PC was hydrolyzed.

Temperature and pH strongly influence hydrolysis kinetics. Grit et al. (1989 and 1992a,b) studied the effect of these parameters for soybean phosphatidylcholine, hydrogenated soybean phosphatidylcholine, partially saturated egg phosphatidylcholine, and PG. For all lipid dispersions studied, the disappearance of PC in time could be well described by (pseudo) first-order kinetics. A typical example of the pH-dependency of hydrogenated soybean PC hydrolysis kinetics (first-order rate constant k_{obs}) is shown in Fig. 2.

Arrhenius plots were drawn (k_{obs} versus $1/T$) in the temperature range between 30 and 82°C (Fig. 3). Below and above the transition temperature, Arrhenius kinetics applied. Around the transition temperature, pH-dependent discontinuities could be observed. For hydrogenated soybean PC dispersions (pH 4), the k_{obs} value at 6°C obtained by extrapola-

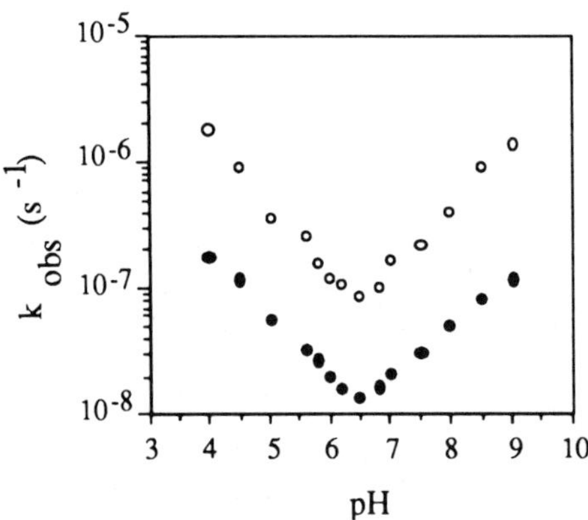

Figure 2 The effect of pH on the hydrolysis kinetics of hydrogenated soybean PC in aqueous dispersions. k_{obs} is the (pseudo) first-order rate constant of PC disappearance. Buffers were acetate (pH 4.0–5.0), citrate (pH 5.6–6.5), Hepes (pH 6.8–7.5), and tris (pH 8.0–9.0). Buffer concentration = 0.05 M; each point represents the mean of at least two separate determinations. ●: 40°C; ○: 70°C. (From Grit et al., 1922b.)

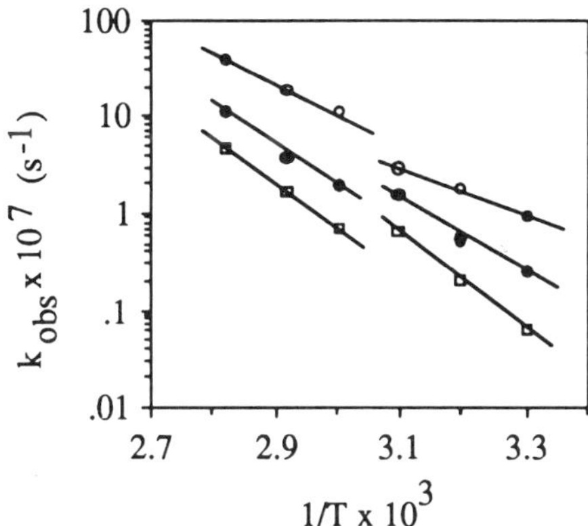

Figure 3 The effect of the temperature ($1/T$) on the hydrolysis of hydrogenated soybean PC (Arrhenius plot) for different pH values. k_{obs} is the (pseudo) first-order rate constant of PC disappearance. The lines were calculated by linear regression analysis. Each point represents the mean of at least two separate determinations. ○: pH 4.0; ●: pH 5.0; □: pH 7.0. (From Grit et al., 1922b.)

tion of the Arrhenius plot below the transition temperature and the k_{obs} of dispersions that were actually stored at 6°C were in reasonably good agreement.

Apart from strong pH and temperature influences, minor catalytic effects of the buffer species involved were observed as well.

In conclusion, if aqueous liposome dispersions have to be stored for prolonged periods of time, pH and temperature conditions have to be carefully considered.

IV. STABILIZATION THROUGH (FREEZE)-DRYING

If a compound requires a highly condensed bilayer with a low permeability, it is advisable to use saturated and long acyl chains for bilayer core formation. Such choice can interfere, however, with the preferred bilayer constitution for the in vivo application (Chap. 2 of Part III). For example, if in vivo a rather fast release is required, the optimum choice for bilayer constituents for prolonged shelf life and in vivo release characteristics may be mutually exclusive. Then (freeze)-drying is a logical alternative to achieve the required long term stability on the shelf and the fast release in vivo.

Several groups have studied freezing, drying (e.g., Hauser and Strauss, 1987), or freeze-drying of liposomes. Most groups have focused on freezing or freeze-drying as a means of (cryo)preservation. Therefore, only cryopreservation will be discussed in this review. In spite of many efforts, questions remain open concerning the mechanism(s) of cryopreservation and the optimum experimental conditions to be selected (e.g., Özer et al., 1988). The major factors affecting the stability in a freeze-drying and rehydration cycle are listed in Table 1.

Of critical importance for the success of any freeze-drying of phospholipid suspensions is the interaction of the encapsulated substance with the bilayer. For molecules that strongly interact with the bilayer (e.g., amphotericin), freeze-drying generally does not present a problem, if proper cryoprotectants (see below) are present in the system. Retention and particle size remain then essentially unchanged upon rehydration. However, for hydrophilic compounds that do not interact intimately with a phospholipid bilayer,

Table 1 Factors Influencing the Physical Stability of Liposomes in a Freeze-Drying/Rehydration Cycle

Bilayer structure
Interaction compound/bilayer
Liposome size
Presence and type of cryoprotectant
Process variables
freezing/drying time
freezing/drying temperature(s)
cooling/drying rate
rehydration conditions

such as CF, the retention after a freezing/thawing or freeze-drying/rehydration cycle tends to vary and to depend on the exact experimental conditions. Cryoprotectants play an important role in the physical stabilization of liposomes during freezing, drying, or freeze-drying. The effect of a cryoprotectant (and other experimental variables such as freezing temperature and liposome size) on CF retention after a freezing/thawing cycle is shown in Fig. 4. From these results one can see that under optimal conditions 100% CF retention (no leakage) is possible after a full freezing/thawing cycle.

The results obtained during the freeze-drying process of CF containing liposomes (average size 0.28 μm and 0.13 μm) are shown in Fig. 5. In this comparison, retention upon thawing is initially high for the small-size liposomes (0.13 μm), but this retention upon rehydration drops with the removal of water and finally reaches a value of around 25% for the 'dry' product.

In the absence of cryoprotectants, aggregation and fusion often occur during a freezing/thawing cycle (cf. Fig. 6). Although many compounds with very different chemical natures (e.g., saccharides, proteins and amino acids, electrolytes and (poly)alcohols) show some cryoprotective activity, most attention is now focused on the group of saccharides and some polyalcohols (glycerol). Discussions about the optimal cryoprotectant (or mixture of cryoprotectants) are ongoing. Progress will certainly be quicker when the mechanism(s) of cryoprotection are fully understood. For glycerol, the following possible mechanisms of cryoprotection were proposed by Fransen et al. (1986): (1) a direct interaction of glycerol with the phospholipid bilayers, (2) reduction of the rate of ice crystallization and ice-crystal growth, and (3) lowering of the eutectic temperature.

Other authors have mentioned the possibilities that certain disaccharides would bind through hydrogen bonds to the polar head groups (the phosphate group being involved) of the dried bilayer; this would change the physical state of the bilayer and prevent direct interbilayer contact (reviewed by Crowe et al., 1987). Vesicles could also be embedded and immobilized in a rigid external phase (Harrigan et al., 1990). Crowe et al. (1987) have discussed the experimental evidence for bilayer-cryoprotectant interactions collected with different experimental techniques.

V. IN SITU PREPARATION OF LIPOSOMES

An entirely different approach is either (1) to form new liposomes by hydration of lipids "at the bedside," or (2) to load preexisting "empty" liposomes just prior to use. These concepts will be dealt with separately in the following paragraphs.

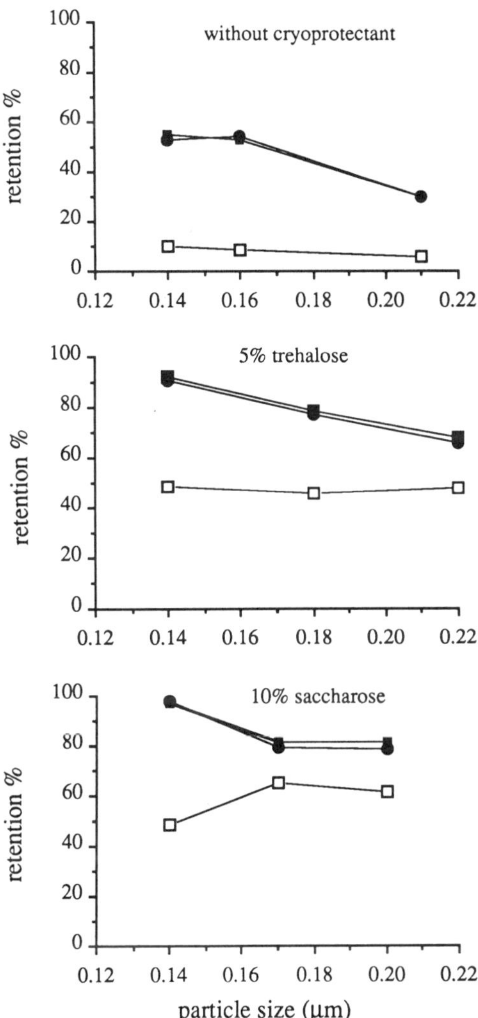

Figure 4 An example demonstrating the critical importance of freezing temperature, liposome size, and presence of cryprotectants on the retention of carboxyfluorescein (CF). Liposome dispersions (hydrogenated soy bean phosphatidylcholine/dicetylphosphate; molar ratio 10:1) were prepared as described (Fransen et al., 1986) with slight modifications: 5 mM CF was encapsulated (for details: Talsma et al., 1991b). Cooling rate: 10°C/min. Storage time at –25 (□), –50 (●), and –75 (■) °C: 45 min. After storage the dispersions were immediately heated up to 20°C. (From Talsma, 1991b.)

Freeze-drying of lipids from a suitable organic solvent (e.g., butanol) in order to form liposomes for parenteral administration "at the bedside" was described in the 1970s (ICI, 1978). These liposomes were designed for intraarticular administration of corticosteroids in the treatment of rheumatoid arthritis (Dingle et al., 1978). One of the problems encountered at that stage was the required high temperature at which liposome formation had to take place (Shaw et al., 1976). Later, a similar concept was used with the MTP-PE containing liposomes developed for nonspecific immunotherapy by macrophage activation. Here, 1-palmitoyl-2-oleyl-*sn*-glycero-3-phosphocholine (POPC) and 1,2 dioleyl-*sn*-glycero-3-phospho-L-serine (DOPS) in a molar ratio of 7:3 were used to form liposomes

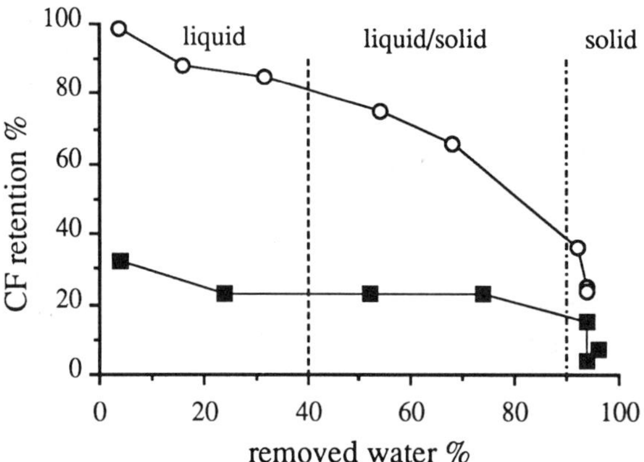

Figure 5 Retention of encapsulated CF after thawing or rehydration as a function of the percentage of water removed by freeze-drying. Liposome dispersions (hydrogenated soy bean phosphatidylcholine/dicetylphosphate; molar ratio 10 : 1) contained 10% m/m saccharose in/outside the vesicles. The liposomes were prepared as described by Fransen et al. (1986) with slight modifications. 5 mM CF (in 10 mM tris buffer, pH 7.4) was encapsulated. Particle size: ■: 0.28 μm; ○: 0.13 μm. Freezing temperature: boiling liquid nitrogen; primary drying phase: 2–14 h, dP (Pa) 10, temperature –40°C; secondary drying phase: temperature was raised from –40 to 20°C at a rate of 10°C/h. The vials were sequentially stoppered during the freeze-drying process. The samples with less than 40% water removed—the vials closed in the early stage of the process—still contained liquid material upon thawing after the cycle was finished. When between 40% and 90% water was removed, both solid and fluid material was present. When over 90% water was removed, the system was solid. (From Talsma, 1991a.)

by shaking at room temperature (Van Hoogevest and Fankhauser, 1989). In their article, Van Hoogevest and Fankhauser have discussed the physical and chemical properties of the MTP-PE vesicles that were generated by hydration of the sterile freeze-dried cake of lipids. The mass average diameter (measured with a Coulter Counter) of liposomes in a number of dispersions taken from two different batches ranged between 2.0 and 3.5 μm. Liposomes larger than 20 μm in size were very rare. The lipids and MTP-PE were

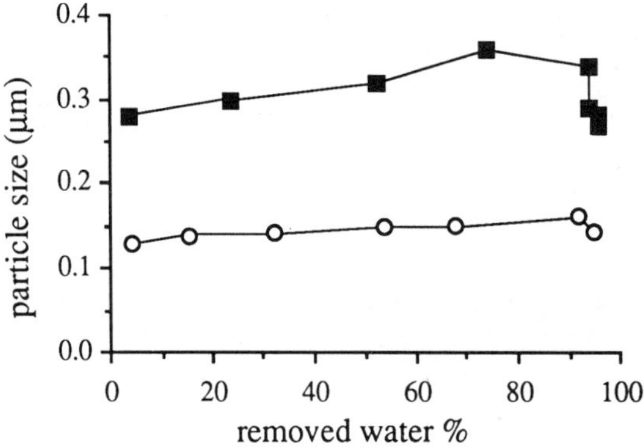

Figure 6 Mean particle size of the dispersion after thawing or rehydration as a function of the percentage of removed water by freeze-drying (for experimental conditions, see Fig. 5). Particle size: ■: 28 μm; ○: 0.13 μm. (From Talsma, 1991a.)

chemically stable in freeze-dried form; there was no indication of a liposome-size-dependent, inhomogeneous distribution of MTP-PE. This product meets all (pharmaceutical) technological requirements and is currently being evaluated in phase II clinical trials in the U.S.A.

Payne and coworkers (1986a,b) developed the so-called proliposome concept for "bedside" preparation of liposomes. Here, the lipids are casted, finely divided, from an organic solvent on a fine powder, such as sodium chloride and sorbitol. A certain degree of control over the particle size upon hydration of the lipid films is possible by selection of the carrier, the nature of the lipid, and the lipid film thickness. Dependent on whether DMPC/DMPG or egg PC/ergosterol were used, the proliposomes were formed by hydration and shaking at 37°C or room temperature, respectively. The liposomes described by Payne and coworkers all contained the antifungal drug amphotericin B.

A different approach to avoid leakage of drugs during storage is to load preexisting "empty" liposomes with the desired drug or radiolabel just before administration of the vesicles to the patient. Three different ways to reach this goal have been described: ^{111}In-containing liposomes were designed for imaging of certain tumors. The ^{111}In is loaded just before administration in stable and small DSPC/cholesterol liposomes with nitrilotriacetate as a strong ^{111}In binding chelator in the internal aqueous phase. After adding the $^{111}InCl_3$ solution to the liposomes, the mixture is heated to 65°C for 15 min to facilitate the transport of ^{111}In across the DSPC/cholesterol bilayer by the ionophore A23187. The ^{111}In loading efficiency is over 90%.

A second way of loading liposomes with a drug "at the bedside" is based on the establishment and maintenance of a pH gradient between the internal (within the liposome) and external water phase (Nichols and Deamer, 1976; Mayer et al., 1990a,b). As the cytostatic doxorubicin is often used to demonstrate the potential of this approach, this drug will be used to explain the principles. Basic drugs, such as doxorubicin (pK_a 8.15) will partition and accumulate in the vesicle interior if the pH of the intravesicular compartment is low (e.g., pH 4) compared to the external phase (pH 7.4). A prerequisite for efficient loading is that the nonionized species of the compound are able to pass through the bilayers to reach equilibrium, i.e., equal concentrations of the nonionized species at both sides of the membrane. In order to reach this equilibrium within an acceptable time frame, elevated temperatures are usually used for bilayers that are in the gel state at room temperature. Buffer species should be selected that cannot pass through the bilayers under the chosen pH conditions. Citric acid is a common choice.

In practice, the following protocol can be used: empty liposomes are prepared and kept in a medium with a low pH until the loading procedures starts. If the low pH causes unacceptable hydrolysis reactions (cf. Sec. III above), these empty liposomes can be stored in freeze-dried form in the presence of a cryoprotectant (see above). For loading, a doxorubicin-containing solution with a high pH (e.g., pH 7.4) is added, while the external pH is considerably higher (close to neutral) than the internal pH. If the protocol (including a short incubation at elevated temperatures) is properly performed, over 90% of the doxorubicin is entrapped in the liposomes, which are then ready for use.

An elegant procedure to establish a pH gradient over the liposomal membrane (low pH inside) was introduced by Cohen et al. (1991). Ammonium sulfate solutions are used for the preparation of the empty liposomes. Upon dilution (with a doxorubicin-containing solution with no or a low concentration of ammonium sulfate), a pH gradient is established; the internal pH drops because of the diffusion of (uncharged) ammonia through the membrane out of the vesicle. In the case of doxorubicin, over 90% of this cytostatic could be encapsulated. At present, the applicability of the above-described pH gradient loading techniques is under investigation. As preparation of these liposome formulations "at the bedside" requires a number of technical manipulations, the pharmaceutical and medical staff should be well informed and trained in order to avoid improper loading of the liposomes.

VI. CONCLUSIONS

In this chapter different approaches for formulating liposomes for pharmaceutical use with a long shelf life have been discussed. It is clear that for lipophilic, bilayer interacting compounds, such as amphotericin B and MTP-PE, stable formulations can be obtained by freeze-drying and in-situ preparation techniques. The "at the bedside" preparation of new liposomes by hydration of lipids can only be used if rather large liposomes and broad particle size distributions of the generated liposomes are acceptable for administration. Liposome dispersions with a narrow size distribution, that are freeze-dried in the presence of proper cryoprotectants, will form dispersions with a particle size distribution similar to that of the original liposome dispersion. Therefore this technique is preferred if small-size liposomes with a narrow size distribution are required.

For low molecular weight hydrophilic compounds such as CF, (freeze)-drying techniques have failed so far to preserve acceptable retention levels upon rehydration. The only way to formulate stable liposome dispersions for this class of compounds is to select a rigid bilayer with a low permeability and store these liposomes as aqueous dispersions. pH and temperature control is critical to avoid high hydrolysis rates.

During the last decade considerable progress has been made in understanding the mechanisms of cryoprotection. Nevertheless, solutions for problems are still found through trial and error rather than through utilizing basic insights into the events taking place during freezing/thawing or (freeze)-drying/rehydration. This lack of knowledge hampers further progress. Basic work in the field of cryoprotection should therefore have a high priority and be stimulated.

If the hydrophilic low molecular weight compounds cannot be encapsulated in liposomes with a rigid bilayer and a low permeability, e.g., because of the required disposition in vivo, then a shelf life problem exists. The described techniques to load these compounds just before administration, "at the bedside," have limitations. Only those compounds with the proper pK_a or ligand binding affinity and a certain capacity to pass through membranes are potential candidates. Besides, these molecules must have a certain capacity to pass through bilayers in order to accumulate in the liposome core. Finally, one should keep the protocol to prepare these liposomes "at the bedside" simple, to avoid acceptance problems by the pharmaceutical and medical staff. As these "remote loading" techniques are relatively new developments, further progress can be expected in this field.

With hydrophilic low molecular weight compounds one can expect shelf life problems and low encapsulation efficiencies. Therefore, pharmaceutically, a lipophilic analog of the drug in question is preferred. The use of MTP-PE instead of the highly water soluble, hydrophilic and poorly encapsulated MDP (muramyl dipeptide) is an example of this approach to switch to lipophilic membrane interacting drugs in the case of liposomal delivery.

WORKSHEETS

Worksheet 1. Freeze-Drying of Liposomes

Principle of operation: Stabilization of liposomes in cryoprotectant-containing solution by sublimation of crystallized water present in the dispersion.

Literature: H. Talsma, thesis, University of Utrecht, 1991a; Y. Özer, H. Talsma, D. J. A. Crommelin, A. A. Hincal, *Acta Pharm. Technol. 34*:129 (1988).

Materials: To freeze-dry 1–2 ml of 30–50 μmol/mL of (e.g., Phospholipon 100H/DCP (10 : 1)) liposomes, one requires a cryoprotectant containing liposome dispersion (e.g., 5–10% m/m trehalose or saccharose), boiling liquid nitrogen or, alternatively, an acetone/dry ice mixture, freeze-drying vials.

Equipment: A temperature- and pressure-controlled freeze-dryer (e.g., a Leybold-Heraeus GT4).

Procedure: For freezing of the dispersions see Worksheet 2. The primary drying phase should take place at a temperature below the collapse temperature of the cryoprotectant mixture, e.g. –40°C for saccharose and –25°C for trehalose, and last until ≤3% pressure rise is found after closing the drying chamber for 1 min. The primary drying time is strongly dependent on the type and geometry of the vials and the amount of dispersion that should be freeze-dried. For 0.25–2 mL the required time would be between 5 and 25 h. For the secondary drying phase the temperature should be raised by 10°C/h until 20°C and kept at this temperature for 2–6 h. The vials should be stoppered under a nitrogen atmosphere. A typical residual moisture content of 1–2% mass of the total mass of the dried cake is found after freeze-drying of the dispersion. Rehydration of the freeze-dried dispersions is achieved by adding distilled water at room temperature.

Recommendations: The optimum temperature for the primary drying phase (below the collapse temperature) can be detected either by differential scanning calorimetry or by a freezing analyzer. For maximum stability of the dispersions, a residual moisture content of 1–2% is necessary.

Caveats: The freeze-drying of lipophilic compounds that associate with the lipid bilayer has been successfully performed following the above protocol. Both retention and average particle size did not change after hydration. However, freeze-drying of dispersions containing encapsulated water-soluble, nonbilayer interacting compounds ('CF-class') with acceptable retention levels after rehydration is hard to achieve. Here the marker retention is typically liposome-size-dependent. For 5 mM encapsulated CF, a retention of 25% was found for a dispersion with a particle size of 0.13 μm after a complete freeze-drying cycle, while the retention of 0.28 μm vesicles was only 7%.

Worksheet 2. Stabilization of Liposomal Dispersions by Freezing

Principle of operation: Preservation of the physical stability of liposomes in a cryoprotectant-containing solution by crystallization of water present in the dispersion.

Literature: H. Talsma, M. J. Van Steenbergen, and D. J. A. Crommelin, *Int. J. Pharm. 77*:119 (1991b); J. Kristiansen and A. Hvidt, *Cryo. Lett. 11*:137 (1990); Y. Özer, H. Talsma, D. J. A. Crommelin, and A. A. Hincal, *Acta Pharm. Technol. 34*:129 (1988).

Materials: To stabilize 30–50 μmol/mL of (e.g., Phospholipon 100H/DCP (10 : 1)) liposomes, one requires a cryoprotectant-containing liposome dispersion (e.g., 5–10% m/m trehalose, or saccharose), boiling liquid nitrogen or, alternatively, an acetone/dry ice mixture.

Equipment: A storage device for the frozen dispersions, with a storage temperature below the eutectic temperature of the cryoprotectant dispersion, a waterbath of 20°C for the rapid thawing of the dispersions.

Procedure: Small quantities of the dispersion (0.25–2 mL) in glass vials should be immersed directly in boiling liquid nitrogen or an acetone/dry ice mixture and kept there for 5–10 min. Afterwards, the vials can be stored in a device with a constant temperature below the eutectic temperature. To thaw the dispersions, the vials should be immersed in a water bath of 20°C and kept there for 5 min. After thawing, the particle size is not significantly affected.

Recommendations: The nucleation of the internal volume of small liposomes is prevented during cooling. This protection is cooling-time- and rate-dependent. If (internal) crystallization of liposomes occurs below the homogeneous nucleation temperature (at temperatures below –40°C), almost complete retention of encapsulated low molecular weight, water-soluble, nonbilayer interacting compounds (carboxyfluorescein: "CF-class") is found. This is generally the case for high cooling rates and small liposome sizes (<0.15 μm). For large liposomes, a particle-size-dependent retention of these compounds is found as (partly) heterogeneous internal nucleation during cooling occurs. To check whether homogeneous nucleation is the main crystallization route, the solution can be frozen in a differential scanning calorimeter (DSC). The fraction of internal volume crystallizing at the homogeneous nucleation temperature can be estimated by measuring the heat-flow below –40°C.

Caveats: The higher the storage temperature the higher the chance of recrystallization of the internal volume; this may induce leakage of encapsulated compounds upon thawing of the dispersion.

The cooling rate should be fast enough. Larger samples are generally more difficult to cool down rapidly, because of the large crystallization enthalpy of water in the dispersion and the unfavourable surface/volume ratio.

Retention of low molecular weight, water-soluble, nonbilayer interaction compounds (e.g., CF) in the internal cavity of the liposomes requires the careful selection of experimental conditions (cooling rate, cooling temperature, bilayer constituents, storage temperature, cryoprotectant). With bilayer interacting compounds a retention of around 100% is readily reached.

REFERENCES

S. Amselem, A. Gabizon, and Y. Barenholz, Optimization and upscaling of doxorubicin-containing liposomes for clinical use, *J. Pharm. Sci. 79*:1045–1052 (1990).

R. Cohen, G. Haran, L. K. Bar, and Y. Barenholz, Ammonium ion gradients in liposomes: A method to obtain efficient entrapment and controlled release of amphipatic molecules, 15th Int. Congress of Biochemistry, 1991, p. 97.

D. J. A. Crommelin and E. M. G. Van Bommel, Stability of liposomes on storage: Freeze dried, frozen or as an aqueous dispersion. *Pharm. Res. 1*:159–163 (1984).

R. J. Crossley, Industrial development of liposomes at Vestar, in *Liposomes in Drug Delivery: 21 Years On* (G. Gregoriadis, H. Patel, and A. T. Florence, eds.), London, 1990.

J. H. Crowe, L. M. Crowe, J. F. Carpenter, and C. A. Wistrom, Stabilization of dry phospholipid bilayers and proteins by sugars, *Biochem. J. 242*: 1–10 (1987).

J. T. Dingle, J. L. Gordon, B. L. Hazleman, C. G. Knight, D. P. Page Thomas, N. C. Phillips, I. H. Shaw, F. J. T. Fildes, J. E. Oliver, G. Jones, E. H. Turner, and J. S. Lowe, Novel treatment for joint inflammation, *Nature 271*:372–373 (1978).

G. J. Fransen, P. J. M. Salemink, and D. J. A. Crommelin, Critical parameters in freezing of liposomes, *Int. J. Pharm. 33*:27–35 (1986).

A. Gabizon, Liposomes as a drug delivery system in cancer chemotherapy, in *Drug Carrier Systems: Horizons in Biochemistry and Biophysics*, Vol. *9* (F. H. Roerdink and A. M. Kroon, eds.), John Wiley, Chichester, 1989a, pp. 185–211.

A. Gabizon, A. Sulkes, T. Peretz, S. Druckmann, D. Goren, S. Amselen, and Y. Barenholz, Liposome-associated doxorubicin: Preclinical pharmacology and exploratory clinical phase, in *Liposomes in the Therapy of Infectious Diseases and Cancer* (G. Lopez-Berestein, and I. J. Fidler, eds.), Alan R. Liss, New York, 1989b, pp. 391–402.

M. Grit, J. H. de Smidt, A. Struijke, and D. J. A. Crommelin, Hydrolysis of phosphatidylcholine in aqueous liposome dispersions, *Int. J. Pharm. 50*:1–6 (1989).

M. Grit, N. J. Zuidam, and D. J. A. Crommelin, Analysis and hydrolysis kinetics of phospholipids in aqueous dispersions, in *Liposome Technology*, 2nd edition (G. Gregoriadis, ed.), CRC Press, Boca Raton, Fla., 1992a.

M. Grit, W. J. M. Underberg, and D. J. A. Crommelin, Hydrolysis of saturated soybean phosphatidylcholine in aqueous liposome dispersions, *J. Pharm. Sci.*: 1992b (in press).

M. Grit and D. J. A. Crommelin, The effect of aging on the physical stability of liposome dispersions, *Chem. Phys. Lipids 62*:113–122 (1992c).

P. R. Harrigan, T. D. Madden, and P. R. Cullis, Protection of liposomes during dehydration or freezing, *Chem. Phys. Lipids 52*:139–149 (1990).

H. Hauser and G. Strauss, Stabilization of smaller unilamellar phospholipid vesicles during spray-drying, *Biochim. and Biophys. Acta 897*:331–334 (1987).

Imperial Chemical Industries Ltd., Belgian Patent 866.697, 1978.

A. W. T. Konings, Lipid peroxidation in liposomes, in *Liposome Technology*, Vol. I (G. Gregoriadis, ed.), CRC Press, Boca Raton, Fla., 1984, pp. 139–162.

L. Ma, C. Ramachandran, and N. D. Weiner, Partitioning of an homologous series of alkyl *p*-aminobenzoates into multilamellar liposomes: Effect of liposome composition, *Int. J. Pharm. 70*:209–218 (1991).

R. L. Magin and J. N. Weinstein, The design and characterization of temperature-sensitive liposomes, in *Liposome Technology*, Vol. III (G. Gregoriadis, ed.), CRC Press, Boca Raton, Fla., 1984, pp. 137–155.

L. D. Mayer, M. B. Bally, and P. R. Cullis, Strategies for optimizing liposomal doxorubicin, *J. Liposome Res. 1*:463–480 (1990).

L. D. Mayer, L. C. L. Tai, M. B. Bally, G. N. Mitilenes, R. S. Ginsberg, and P. R. Cullis, Characterization of liposomal systems containing doxorubicin entrapped in response to pH gradients, *Biochim. Biophys. Acta 1025*:143–151 (1990).

D. Mufson, L. S. S. Guo, and R. M. Fielding, Amphotericin B-cholesterol-sulfate complex (ABCD™): Stability and tissue distribution of a novel dosage form, *Proc. Intern. Symp. Control. Rel. Bioact. Mater. 17*:81–82 (1990).

R. R. C. New, *Liposomes: A Practical Approach*, IRL Press at Oxford Univ. Press, Oxford, 1990.

J. W. Nichols and D. W. Deamer, Cathecholamine uptake and concentration by liposomes maintaining pH gradients, *Biochim. Biophys. Acta 455*:269–271 (1976).

Y. Özer, H. Talsma, D. J. A. Crommelin, and A. Hincal, Influence of freezing and freeze-drying on the stability of liposomes dispersed in aqueous media, *Acta Pharm. Technol. 34*:129–139 (1988).

N. I. Payne, P. Timmins, C. V. Ambrose, M. Ward, and F. Ridgway, Proliposomes: A novel solution to an old problem, *J. Pharm. Sci. 75*:325–329 (1986a).

N. I. Payne, I. Browning, and C. A. Hynes, Characterization of proliposomes, *J. Pharm. Sci. 75*:330–333 (1986b).

G. Scherphof, J. Damen, and J. Wilschut, Interactions of liposomes with plasma proteins, in *Liposome Technology*, Vol. III (G. Gregoriadis, ed.), CRC Press, Boca Raton, Fla., 1984, pp. 205–224.

G. Schumann, P. Van Hoogevest, P. Fankhauser, A. Probst, A. Peil, M. Court, J.-C. Schaffner, M. Fischer, T. Skripsky, and P. Graepel, Comparison of free and liposomal MTP-PE: Pharmacological, toxicological and pharmacokinetic aspects, in *Liposomes in the Therapy of Infectious Diseases and Cancer* (G. Lopez-Berestein and I. J. Fidler, eds.), Alan R. Liss, New York, 1989, pp. 191-203.

I. H. Shaw, C. G. Knight, and J. T. Dingle, Liposomal retention of a modified anti-inflammatory steroid, *Biochem. J. 158*:473–476 (1976).

G. Storm, L. Van Bloois, P. A. Steerenberg, E. Van Etten, G. De Groot, and D. J. A. Crommelin, Liposome encapsulation of doxorubicin: Pharmaceutical and therapeutic aspects, *J. Control. Rel. 9*:215–229 (1989a).

H. Talsma, Preparation, Characterization and Stabilization of Liposomes, thesis, University of Utrecht, 1991a.

H. Talsma, N. J. von Steenbergen, and D. J. A. Crommelin, The cryopreservation of liposomes: 3. Almost complete retention of a water-soluble marker in small liposomes in a cryoprotectant containing dispersion after a freezing/thawing cycle, *Int. J. Pharm. 77*:119–126 (1991b).

W. J. Urba, L. C. Hartmann, D. L. Longo, R. G. Steis, J. W. Smith II, I. Kedar, S. Creekmore, M. Sznol, K. Conlon, W. C. Kopp, C. Huber, M. Herold, W. G. Alvord, S. Snow, and J. W. Clark, Phase I and immuno-modulatory study of a muramylpeptide, muramyltripeptide phosphatidylethanolamine, *Cancer Res. 50*:2979–2986 (1990).

L. Van Bloois, D. D. Dekker, and D. J. A. Crommelin, Solubilization of lipophilic drugs by amphiphiles: Improvement of the apparent solubility of almitrine bismesylate by liposomes, mixed micelles and O/W emulsions, *Acta Pharm. Technol. 33*:136–139 (1987).

P. Van Hoogevest and P. Fankhauser, An industrial liposomal dosage form for muramyl-tripeptide-phosphatidylethanolamine (MTP-PE), in *Liposomes in the Therapy of Infectious Diseases and Cancer* (G. Lopez-Berestein and I. J. Fidler, eds.), Alan R. Liss, New York, 1989, pp. 453–466.

M. B. Yatvin, J. N. Weinstein, W. H. Dennis, and R. Blumenthal, Design of liposomes for enhanced local release of drugs by hyperthermia, *Science 202*:1290–1293 (1978).

G. M. Zonneveld and D. J. A. Crommelin, Liposomes: parenteral administration to man, in *Liposomes as Drug Carriers: Recent Trends and Progress* (G. Gregoriadis, ed.), John Wiley, Chichester, 1988, pp. 795–817.

II
Physical and Structural Properties

11
Physical Characterization

Gregor Cevc *Technical University of Munich, Munich, Germany*

John M. Seddon *Imperial College, London, England*

I. TECHNIQUES

A variety of techniques can be used to investigate the physical properties of phospholipids and their aggregates; every year a few new or improved ones are introduced. For example, depending on whether structural, thermodynamic, or dynamic information is required, a number of diffraction techniques and microscopy, calorimetry, or different spectroscopic methods, respectively, can be used. In the first part of this chapter, consequently, the basic features of certain standard techniques for phospholipid research will briefly be reviewed; in the second part, different physical characteristics of phospholipid systems and some guidelines for the best choice of corresponding experimental methods will be given.

A. Diffraction Techniques

1. X-Ray and Neutron Diffraction

X-ray diffraction remains the principal technique for studying single crystal phospholipid samples [1], although electron diffraction has also been usefully employed for some lipid systems [2]. One of the disadvantages of the latter method is that it requires a vacuum and hence is unsuitable for studying hydrated phases. For these hydrated structures, x-ray diffraction, either using conventional sealed tube or rotating anode generators, or synchrotron sources, is by far the most important technique. However, neutron diffraction has certain extremely important advantages, namely the possibility of contrast variation (see later), the ability to detect hydrogen atom positions, and the possibility of studies of dynamics, employing quasi-elastic or inelastic scattering.

X-rays are a form of electromagnetic radiation, with wavelengths typically in the region of 1 Å (0.1 nm or 1×10^{-10} m). They are scattered predominantly by the atomic electrons in a sample. The scattering power or atomic scattering factor f of a given atom is thus simply proportional to the number of electrons it has, i.e., to its atomic number Z. Because the electrons around a given atom occupy a volume whose dimensions are comparable to the wavelength of the x-rays, the amplitude of the scattered wave is angle-dependent, being strongest in the forward direction (along the incident beam direction) and falling off with angle. The atomic scattering factors for all atoms as functions of the scattering angle 2θ are tabulated in [3]. However, for computational purposes, it is more convenient to use analytic approximate expressions, also given in the same reference.

Neutrons are neutral nuclear particles of mass $m = 1.675 \times 10^{-27}$ kg. A neutron travelling in a beam with a speed of v has a momentum $p = mv$ and a kinetic energy $E = p^2/2m$. From the de Broglie relation, this is equivalent to a wave of wavelength $\lambda = h/p$, where h is Planck's constant. Typically, the wavelengths used for diffraction experiments are in the range of 1 ··· 12 Å (0.1 ··· 1.2 nm). Because neutrons are scattered from the nuclei of atoms, which are much smaller than the neutron wavelength, the scattering from an atom is isotropic (unlike x-rays). The scattering power of atoms shows no discernible pattern with atomic number, with certain light atoms such as hydrogen scattering quite strongly. Furthermore, different isotopes of a given atom (e.g., ^{1}H and ^{2}H [deuterium]) may have drastically different scattering powers. This feature is utilized in the technique of contrast variation, where by isotopic substitution of specific chemical groups, their scattering power for neutrons may be systematically altered, whilst making only a very minor perturbation to the system.

The fundamental problem with any diffraction technique is that the experimentally measured quantity is the scattered intensity $I(\theta, \phi)$, where θ and ϕ are the polar and azimuthal angles of the scattered beam with respect to the incident beam (see Fig. 1). To deduce the structure one requires the scattered *amplitude* $F(\theta, \phi)$, however. The intensity is the modulus squared of the amplitude, i.e., $I = FF^*$, where F^* is the complex conjugate of F. This means that the phases of the scattered waves are lost and must be deduced in some way (such as isomorphous replacement, direct methods, etc.) in order to obtain the structure from the diffraction pattern. Once this has been done for the whole diffraction pattern, then the electron density distribution (map) may be calculated by Fourier transformation of the scattered amplitudes (for neutrons it is the neutron scattering length density profile that is derived). However, the usual methods of phasing are not suitable for labile liquid crystalline structures, which invariably give rise to only a few Bragg reflections. One method that has been widely employed for phasing diffraction patterns from lamellar phases is to swell the structure by varying the water content [4]. This method has the disadvantage that it relies on the structure of the bilayer remaining unchanged during the swelling, which is not in general the case. Another approach is to use neutron diffraction in conjunction with solvent contrast variation, varying the ratio of H_2O to D_2O in the water [5]. For more complex structures such as cubic phases, the only method that has so far been successfully applied is a pattern recognition approach, whereby all of the possible phase combinations are screened according to certain crystallographic criteria, and the most probable is then selected [6,10].

A great deal of valuable information on the structure of the lyotropic mesophases can also be deduced from the diffraction patterns simply by analyzing the positions of the peaks.

(a) Scattering geometry. The incident and scattered x-ray or neutron beams are described by their wave vectors $\mathbf{k}_i$, whose moduli are equal and given by $|\mathbf{k}| = 2\pi/\lambda$, where λ is the wavelength of the radiation. The geometry of the diffraction experiment is shown in Fig. 1. The Ewald sphere, which defines all the possible scattered wave vectors $\mathbf{k}_s$, is of radius $2\pi/\lambda$ and is centred on the sample.

It is convenient to express the direction of the scattered beam in terms of the scattering vector $\mathbf{Q}$, where $\mathbf{Q} = \mathbf{k}_s - \mathbf{k}_i$ and has modulus $Q \equiv |\mathbf{Q}| = 4\pi \sin\theta/\lambda$, 2θ being

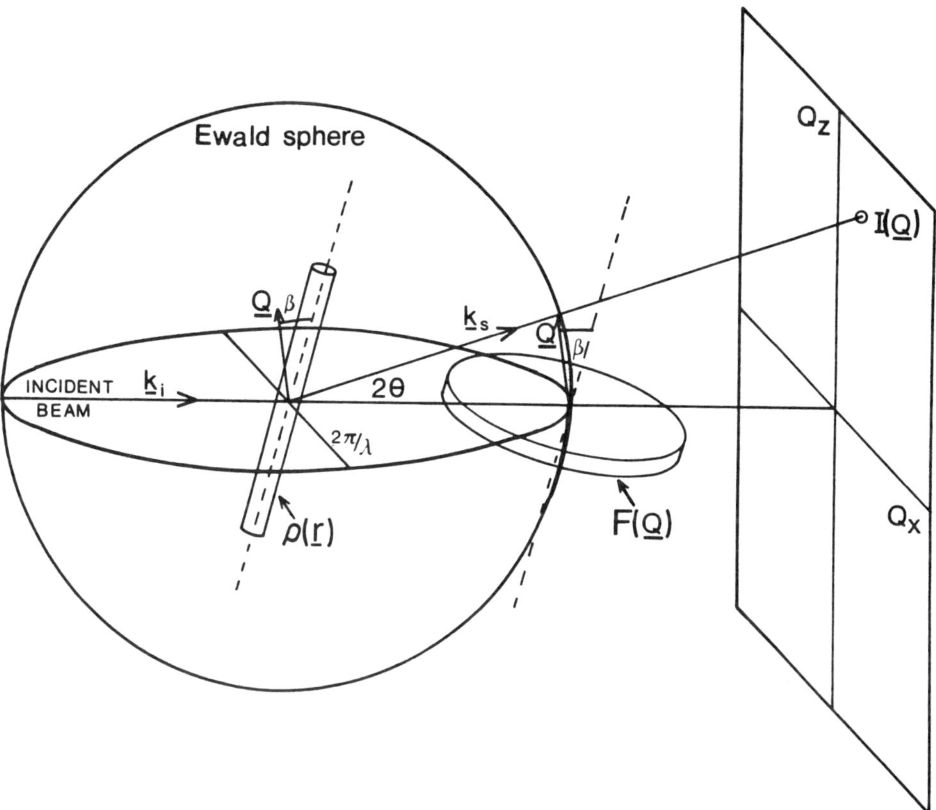

Figure 1 Geometry of the diffraction experiment. (From Ref. 11.)

the angle between the incident and scattered beams. The function $F(\mathbf{Q})$ is the amplitude of the wave of scattering vector $\mathbf{Q}$, and it is the Fourier transform of the sample structure $\rho(\mathbf{r})$ (electron density or neutron scattering length density distribution). The transform $F(\mathbf{Q})$ is centered on the surface of the Ewald sphere at the point where the undiffracted beam emerges. The orientation of the sample, defined by the density function $\rho(\mathbf{r})$, and its transform $F(\mathbf{Q})$ are coupled together: rotating the sample about its center has the effect of rotating the transform about its center by the same amount. The intensity scattered in a direction defined by some particular value of $\mathbf{Q}$ is given by

$$I(\mathbf{Q}) = | F(\mathbf{Q}) |^2 \tag{1}$$

Thermal disorder does not affect the sharpness of the Bragg peaks, but it decreases their intensities progressively towards higher diffraction angles. In calculations, this may be allowed for by scaling the structure factors by an effective Debye-Waller (temperature) factor:

$$\exp(-\alpha Q^2) \tag{2}$$

Effects due to factors such as surface roughness of the water/lipid cylinders may be absorbed into this factor.

When the lipid aggregates are ordered onto a lattice, then the intensity is zero everywhere except for certain discrete scattering angles (Q values), which are given by Bragg's law

$$d_{hkl} = \frac{\lambda}{2 \sin \theta_{hkl}} = \frac{2\pi}{Q_{hkl}} \tag{3}$$

d_{hkl} is the spacing between the crystallographic planes defined by the Miller indices h, k, l.

The lattice may be one, two, or three dimensional; for lyotropic liquid crystals, examples would be a lamellar, a hexagonal, and a cubit phase, respectively. Schematic examples of these types of lattice are sketched in Fig. 2. Also shown in this picture are the corresponding reciprocal lattices (the Fourier transforms of the real space lattices), which in conjunction with the Ewald sphere construction (cf. Fig. 1) illustrate the form of the diffraction pattern for an arbitrary sample geometry. In order to observe a particular Bragg peak, the sample (and hence its reciprocal lattice) must be oriented so as to bring the reciprocal lattice point into the diffracting condition (i.e., touching the surface of the Ewald sphere). The vector $\mathbf{k}_s$ then gives the direction of the diffracted beam. Note that for an actual phase, consisting of structural elements at each lattice point (e.g., a single bilayer for the lamellar phase), the diffraction patterns shown in Fig. 2 are modified (multiplied) by the form factor (Fourier transform) of the basic structural unit.

Owing to the short-range liquid-like disorder within (truly) liquid-crystalline mesophases such as L_α and H_{II}, the Bragg peaks are largely confined to the low-angle region—cf. Eq. (2)—and the scattered intensities are insensitive to the precise atomic configurations. For calculations of scattered intensity, it is, therefore, appropriate to follow the approach adopted in solution small-angle scattering, whereby the particles are modelled by idealized regions or shells of uniform scattering density. For a particle of a defined shape and dimensions, it is then possible to calculate the scattered amplitude $F(\mathbf{Q})$ (the form factor). In general, one must also take into account the orientational distribution of the particles, and the positional or orientational correlations between particles (interparticle interference terms) to calculate the scattering. This may be difficult or impossible to calculate. However, in the case of the transitionally ordered mesophases, such as the H_{II} phase, inasmuch as they can be modelled as lattices of simple structure elements, the problem is simpler (see, for example, [11]).

(b) X-ray apparatus, sample preparation, and sample environment. For x-ray diffraction, intense synchroton sources are now available, which allow time-resolved (msec) diffraction experiments to be carried out [12,13]. The recent techniques of x-ray [14,15] and neutron reflectivity [16] allow structural studies of liquid interfaces to be carried out. For example, it is now possible to study single phospholipid monolayers adsorbed at the air-water interface of a Langmuir-Adams trough using either x-ray [17] or neutron [18] reflectivity. Furthermore, processes such as the formation of multilayers at the air-water interface can also be studied by this technique [15].

Instruments for reflectivity or neutron diffraction experiments will not be described here. Details of these may be obtained from the relevant literature, or from the central facilities themselves (Institut Laue-Langevin, Grenoble; ISIS, Rutherford Appleton Laboratory).

Here we will only discuss laboratory based x-ray diffraction equipment, which is well adapted to routine identification and structural study of lipid phases. The study of

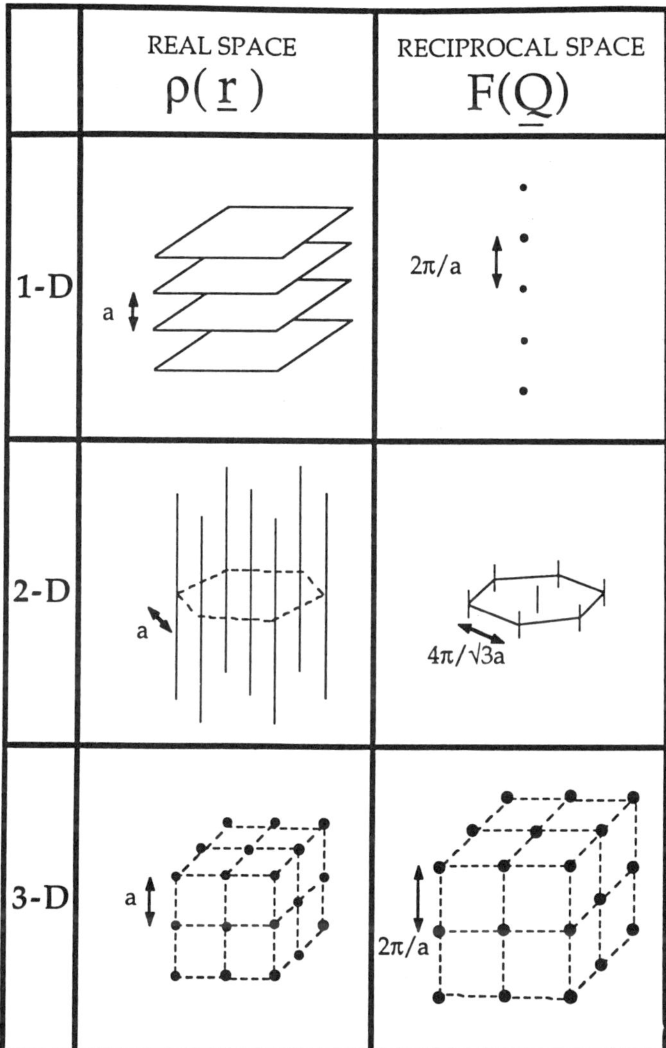

Figure 2 Lattices and their reciprocal lattices.

lyotropic phases of lipids by diffraction requires a number of special features in the apparatus. First, the radiation should be as nearly monochromatic as possible. This may be achieved using a suitable crystal monochromator (such as quartz or germanium). Second, it is necessary to be able to cover both the low-angle (d-spacings $\leq$ 10 Å), and the wide-angle regions (d-spacings $\geq$ 10 Å). The former requirement, particularly when the d-spacings exceed 100 Å, means that there should be very little parasitic scatter at low angles. In order to achieve this, in addition to careful camera design and setting, it is advisable to operate the camera in vacuo, or with a helium (low Z) atmosphere, to eliminate air scattering. The samples may be mounted in a suitably machined brass block, whose temperature is regulated within 1 °C by electrical heating, fluid circulation, or Peltier thermoelectric elements. If greater temperature accuracy is required, then a multistage oven with suitable windows (e.g., thin mica, mylar, or polyimide films) must be used.

The most versatile system meeting these requirements for unaligned samples is the Guinier camera, available from a number of manufacturers. The disadvantage of this camera is that it operates with a line focus; consequently, it is not (very) suitable for the studies of aligned lipid systems. For oriented samples, the optimal system may rely on a 100 μm point focus from a rotating anode x-ray generator, focussed either with Franks double mirrors or with toroid optics. The former optics is extremely clean at low angles, allowing d-spacings in excess of 800 Å to be measured. It has the disadvantage, however, of giving a relatively low intensity. Toroid optics gives a very intense beam, particularly important for detecting wide-angle scattering, but is restricted to d-spacings less than approximately 80 Å.

For detection of x-ray diffraction patterns, possible systems range from x-ray film, scintillation counters, single- or multiwire proportional counters, charged-coupled devices (CCDs), TV-based detectors, and storage phosphor plates. The relative merits of these various devices have been discussed in [19].

Most diffractograms from lipid samples to date have been stored on x-ray films, which have the advantage of low cost, high spatial resolution (approx. 25 μm), uniform response across its whole area, and no counting-rate limitations. Films, moreover, provide a compact and permanent record. The disadvantages of film detection arc the largc intrinsic background noise (which leads to relatively long exposure times), the limited dynamic range ($\geq 10^3$), and the need to process and optically scan the films to extract the data.

Electronic detectors have the advantage that the positional and intensity data are directly available in digitized form for rapid readout and display. This allows experiments to be carried out 'interactively'. For aligned samples, the use of a two-dimensional (area or 2D-) detector is highly advantageous, as it allows an entire section of the diffraction pattern to be recorded simultaneously.

(c) Sample preparation and mounting. For routine phase identification and measurement of *d*-spacings it is advisable in the first instance to use unaligned samples. This method has four main advantages. First, there is the simplicity of sample preparation and mounting on the x-ray camera. Second, the sample temperature and composition may be accurately regulated. Third, by slowly ramping the sample temperature whilst sweeping the film behind a masking slit, it is possible to record the diffraction pattern as a continuous function of temperature (this facility is enormously useful for routine study of both lyotropic and thermotropic liquid crystals). Fourth, it is straightforward to obtain accurate measurements of peak positions and intensities (for line focus systems, correction for the line geometry is necessary).

However, unaligned samples also have a number of important disadvantages. First, only a one-dimensional, spherical average of the diffraction pattern is measured. Thus all information about the relative orientations of the various diffracted peaks, and hence the direct information on the symmetry and orientation of the lattice, is lost. For complex structures, it may be difficult or impossible to index these diffraction patterns. Second, the presence of systematic absences, crucial for space group determination, is much less obvious than in a single crystal pattern. Third, for cubic phases in particular, not only will all (*hkl*) permutations (largest multiplicity of 48 when $h \neq k \neq l$) appear as a single Bragg reflection, but so will reflections, such as (221) and (300), which share the same *d*-spacing. Fourth, for samples, such as cubic phases, that have a strong tendency to form relatively large 'single crystal' domains, the powder patterns from a line focus system such as a Guinier camera may be uninterpretable, due to orientation effects. This problem may, at least partially, be overcome by slow rotation of the sample in the x-ray beam, which helps to produce a uniform statistical sampling of all sample orientations. Finally, analysis of structured diffuse scattering, due to effects such as defect formation, lowered dimensionality, and anisotropic short-range or quasi-long-range ordering, is not feasible unless the sample is aligned.

Various possibilities for the preparation of nonaligned and oriented samples are discussed in Sec. C.

2. Light Scattering and Optical Diffraction

When light shines on a small molecule, the electric field vector corresponding to this light induces an oscillating dipole moment in the irradiated volume which, in response, emits electromagnetic waves (see, e.g., [20]). The intensity of the resulting scattered light is proportional to the second power of the molecular polarizability and contains information about the molecular size and shape. However, visible light is not suitable for the observation of single phospholipid molecules, owing to its relatively long wavelength ($350 \leq \lambda \leq 750$ nm). However, such light is convenient for the investigation of phospholipid aggregates, such as micelles, monolayers at the air-water interface, or vesicles in a suspension [21] or near an interface.

In the case of elastic light scattering, the wavelengths of the incident and diffracted light beams remain the same, independent of the size of the scattering particle. (Quasi-) Rayleigh scattering is more common in phospholipid research than Mie scattering, owing to the relatively small phospholipid aggregate size: $r_v \ll \lambda/2$, where r_v is the scattering particle size; for Mie scattering $r_v \geq \lambda$.

For a nonabsorbing, light scattering vesicle suspension, the turbidity $\tau = -\ln(I/I_0)$, or the relative absorbance, $A = \tau/2.303$, are both functions of the integral of the particle scattering function $P(\theta)$ over the scattering angle:

$$Q = \frac{3}{8} \int_0^{\pi} P(\theta)(1 + \cos^2 \theta) \sin \theta \, d\theta$$

so that

$$\tau = -\ln\left(1 + \frac{16\pi R_\theta Q}{3}\right) \tag{4}$$

The Rayleigh ratio in the corresponding limit is defined as

$$R_\theta = \frac{2\pi^2 n_0^2 (dn/dc)^2}{N_A \lambda^4} c \equiv Kc$$

where $M_v \propto R_v^3$ is the effective mass of each scattering vesicle. The parameter K is a function of the refractive index of the medium, n_0, the particle concentration, $c = M_v N/N_A$, and the change of the effective refractive index with particle concentration [22]; N_A is Avogadro's number, and N is the particle number density.

The interference of light emerging from the different scattering centers on a phospholipid aggregate causes the diffracted light to have a pronounced angular dependence. Thus for maximum experimental accuracy the concentration and angular dependences of the scattered light intensity should both be studied and then extrapolated to zero (c, $\theta_{\text{scatter}} \rightarrow 0$). This is normally done in so-called Zimm diagrams and yields directly the effective particle mass. The Zimm procedure, moreover, yields a value for the radius of gyration of the scattering particle. From this radius the particle nonsphericity can be assessed from

$$\frac{R_\theta}{Kc}\bigg|_{c=0} = \left\{\left[1 + \left(16\pi^2 R_G^2 \sin^2\left(\frac{\theta}{2}\right) 3\lambda^2\right)\right] \frac{1}{M_v + 2Bc}\right\}^{-1}$$

R_G is the corresponding radius of gyration. The virial coefficient B is a measure of the interparticle interaction. Rayleigh limit is obtained when the value of the term inside the square brackets is 1.

Eq. 4 relates the turbidity and vesicle radius changes; with independent radius information for at least one vesicle suspension, this result can even be used to calculate approximate vesicle sizes from the light absorption data. In general, however, it is difficult or impossible to obtain reliable estimates of phospholipid vesicle or micelle size from static light experiments unless cumbersome calibration procedures are carried out. Simple light scattering experiments (turbidimetry) are thus often used to follow changes in vesicle size and other scattering properties semiquantitatively as functions of time or external variables (cf. Fig. 13). To get more quantitative information, dynamic, quasi-elastic light scattering (photon correlation spectroscopy, PCS) should be used (for a review see [23], for the basic formulae see [24], and for recent developments see [25]).

The spectrum of quasi-elastically scattered light is different from that of the original light, owing to the (translational) mobility of the scattering particles and the Doppler effect; Fig. 3 illustrates this schematically. But also the half-width $\Delta\nu$ of the scattered monochromatic light beam in frequency terms is always wider than in the incident beam by a diffusion-constant-dependent amount

$$\Delta\nu = D\frac{8\pi}{\lambda^2} \sin^2\left(\frac{\theta}{2}\right) \equiv \frac{4kT}{\lambda^2} \frac{\sin^2(\theta/2)}{3\eta r_v}$$

The second expression follows from the first upon the substitution of the Stokes-Einstein relation for the diffusion constant as a function of the medium viscosity η.

This equation can be used for the determination of phospholipid aggregate mobility and thus for studies of particle size distribution. The initial part of the correlation function (I^2 *vs.* t) is normally linear and gives diffusion time; the terminal curvature is a measure of the sample polydispersity.

A dynamical light scattering experiment is typically performed by directing a beam of monochromatic light onto a suspension of phospholipid vesicles and then determining the correlation function of the scattered light. If required, this can be carried out as a function of the scattering angle θ. With most of the liposomes suspensions small scattering angles typically used, to minimize the effects of diffusion broadening from the Brownian motion. The phospholipid concentration in such experiments is typically between 0.1 and 10 mM, depending on the incident light intensity and the average lipid vesicle size. In surface quasi-elastic light scattering experiments also the phase behavior of phospholipid monolayers can be investigated [26].

In elastic Raman scattering the wavelength of the incident light is chosen so as to excite molecular vibrations in the irradiated phospholipid molecules; from the wavelengths of Raman resonances, conclusions about the type and state of lipid molecules

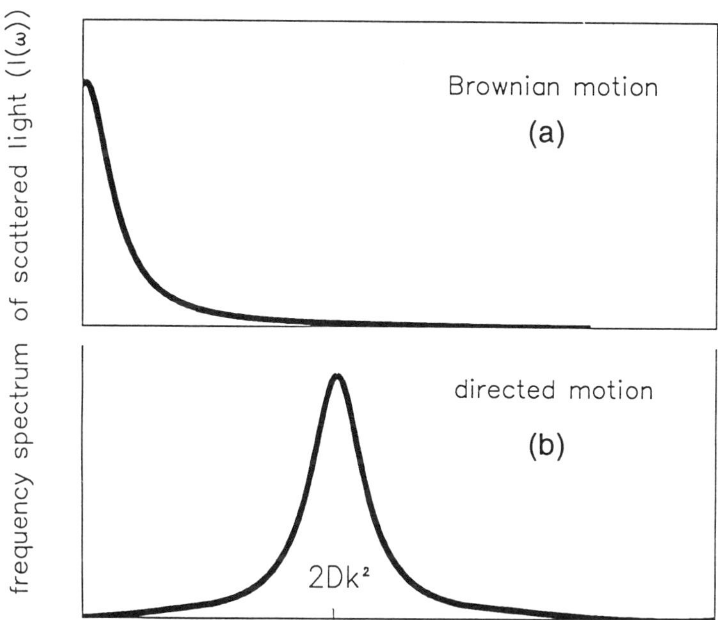

Figure 3 Dynamic light scattering: peak corresponding to the monochromatic light (with a frequency ω) scattered by (a) nondrifting vesicles and (b) unidirectionally moving vesicles. In the former case, only a thermally broadened peak with original frequency is observed; in the latter case, the peak is shifted, the shift being proportional to the vesicle velocity and thus the diffusion constant D.

can be drawn. Raman spectroscopy thus belongs in the world of scattering as well as spectroscopic techniques and provides information similar to that obtained from infrared spectroscopy (see further discussion). This is the chief reason why the importance of Raman spectroscopy in phospholipid research is continuously decreasing, at least since the advent of Fourier transform infrared spectrometers, which now make it feasible to study lipid samples in the presence of water.

(a) Practical aspects. When performing light scattering experiments with phospholipid suspensions, it should be kept in mind that the scattered light intensity increases with the sixth power of the vesicle radius. This makes such experiments extremely sensitive to contamination by dust and other large particles. Use of clean rooms—or at least of carefully filtered solvents and meticulously washed cuvettes—is therefore recommended; multiple, independent measurements (preferably with a dilution series to check for the effects of multiple scattering) are obligatory. Strong light sources, such as powerful lasers, moreover, may be necessary for the investigation of very small vesicles or micelles. For a 5-mW He-Ne laser, for example, the fading of the scattering intensity sets a limit to the measurable vesicle size at ~50 nm.

The main advantage of static light scattering experiments over other types of diffraction is that they are extremely fast and cheap to perform; accurate turbidimetric measurements are feasible with nearly all fluorimeters (90° scattering) or UV vis spectrometers (forward or 0° scattering). The theoretical description of such experiments is rather complicated, however. This causes data analysis to be rather opaque and subject to misinterpretation. A simple trick, which may improve the accuracy of vesicle-size estimations based on standard turbidimetry data, is to compare the light scattering intensity at two different wavelengths (e.g. 350 and 450 nm) as a function of the variable of interest.

The chief advantage of the dynamic light scattering method is that it is very versatile and, potentially, highly accurate; in combination with Doppler velocitometry it can be used, for example, for electrophoretic vesicle or micelle mobility studies [27]. The price for this is the high cost and complexity of the equipment required (for example, Autosizer III (Malvern Ind., UK) is priced above 50 k$); complex program packages, moreover, are needed for data evaluation. Multimodal approximations (e.g., Contin, written by Provencher [28]), in the majority of cases, should be given preference over cumulant analysis [29], which is only valid for vesicle populations with low polydispersity.

When the angle between the incident and scattered light beams gets close to or greater than the Brewster angle (in water ~49°), the dominant process becomes light reflection. A geometrical arrangement in which the incident angle is near 50°, therefore, is the basis for total internal reflection fluorescence (see further discussion) and ellipsometric measurements.

B. Ellipsometry

Ellipsometry is an optical method that exploits the fact that the optical properties of a flat surface cause the polarization of the reflected beam to differ from that of the incident beam (for a review see [30].) In a typical zero-point ellipsometric experiment, for example (Fig. 4), the incident beam of light is polarized elliptically, by means of a polarizer (P) and a compensator (C, typically a $\lambda/4$ plate), in such a way that after the light reflection at the sample surface (S) the secondary beam becomes linearly polarized; this is checked by means of an analyzer (A); the reflected beam intensity, for appropriate setting, must finally be zero.

Theoretical background for ellipsometric data analysis is given in [30]. By such analysis hydrocarbon thickness and refractive indices can be calculated from the values of two pairs of angular positions ((A_2, P_2) and (A_4, P_4) at $C = -\pi/4$) at which the reflected beam intensity has a zero (or minimum) value. For the compensator in the form of a $\lambda/4$ plate, the corresponding angle values, moreover, should fulfill the conditions $A_2 = -A_4$ and $P_2 + \pi/2 = P_4$. Deviations from these identities are indicative of experimental error and/or sample surface anisotropy.

It is often difficult to determine the absolute value of each of the two characteristic ellipsometric angles $\Delta \equiv -2P_2 - \pi/2 = -2P_4 + \pi/2$ and $\psi = A_2 = -A_4$ with sufficient accuracy. In phospholipid research this difficulty can be overcome, by and large, by using relative values in which the characteristic water-surface data Δ_0 and ψ_0 are subtracted from the data obtained with lipid layers.

For thin, nonadsorbing films, only one such relative ellipsometric angle value, $\delta\Delta = \Delta - \Delta_0$, varies if the structural and morphological properties of the studied phospholipid layer are changed, while $\delta\psi = \psi - \psi_0$ remains zero. Thus all the information about film thickness and refractive indices is included in the former angle [31]. In order to extract information pertaining to molecular lipid properties from this angle, suitable models must be used. The most common ones are based on the evaluation of the variation of Δ with molecular area and certain structural input from space-filling models.

Ellipsometry can also be used to infer the overall orientation of hydrocarbon chains in

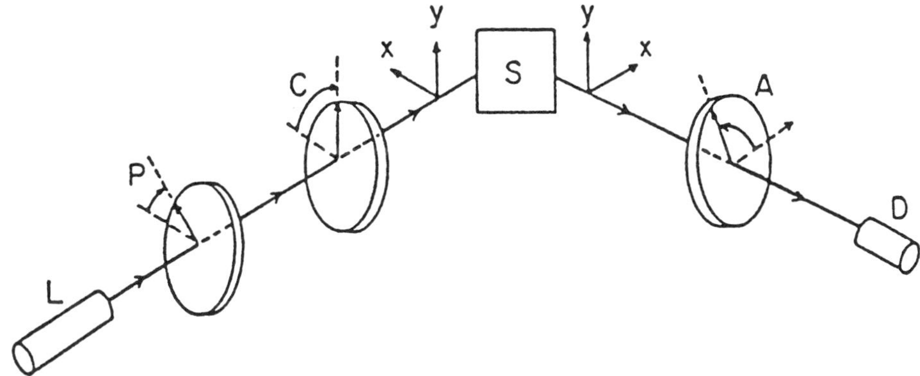

Figure 4 Schematic representation of a zero-point ellipsometer for measuring lipid layer thickness and birefringence. L: light source; P: polarizer; C: compensator; S: sample; A: analyzer; D: detector.

the studied molecules based on the refractive index anisotropy. As the lipid (bi)layer becomes more and more vertically oriented, $n_{\parallel}$ increases at the expense of $n_{\perp}$. Negative birefringence is indicative of the presence of conformational excitations in the hydrocarbon chains (kinks). During the chain melting phase transition of phosphatidylcholines, for example, a fivefold increase in the chain birefringence is observed [31].

(a) Practical aspects. Apart from the high cost of the equipment, the difficulty of ensuring good contrast for ellipsometry is the most severe obstacle to wider use of this technique in phospholipid research. On the one hand, most standard ellipsometers work in the vertical mode and thus can only be used for supported lipid (bi)layers. Horizontal ellipsometers, on the other hand, are sensitive to the presence of phospholipid molecules or vesicles in the subphase (that lower the optical contrast) and thus can only be used for samples at concentrations below ~1 μM lipid. (In the majority of studies to date, simple monolayers at the air-water interface were used.)

In most diffraction experiments, monochromatic exciting light is used and the angular or time dependence of the scattered light is observed. In spectroscopic measurements, however, the energy of some suitable form of electromagnetic field, in principle at least, is changed over the range of interest in order to test the molecular—and sometimes the macromolecular—properties of the sample.

C. Spectroscopic Techniques

An electromagnetic field of the right frequency causes molecular or atomic excitations in the investigated system (see [32] for a series of reviews). Resonant absorption of energy thus may provide information about the energy and conformation states of the phospholipid molecules.

1. Magnetic Resonances

In the magnetic resonance technique, which is a special form of absorption spectroscopy, in an external magnetic field transitions between the Zeeman energy levels of the atomic nuclei or lone-pair electrons are excited, giving rise to *n*uclear *m*agnetic *r*esonance) (NMR) and *e*lectron *s*pin *r*esonance (ESR), respectively (for an introductory review see

[33,34]. The resonant frequency ν is related to the excitation energy as $\Delta E = h\nu$, where h is Planck's constant and thus typically falls in the range of 5–30 GHz for ESR and 50–500 MHz for NMR, depending on the strength of the magnetic field causing the energy-level splitting. In standard magnetic resonance experiments of the early days, the frequency of the exciting electric field was kept constant whilst the magnetic field strength was swept across the resonance region. In modern NMR experiments, however, the response of a sample to high-frequency pulses are measured. (For better resolution ESR spectra are habitually represented in derivative form.)

The use of magnetic resonance in phospholipid research is discussed in detail in Chaps. 13 and 20 of this part. In the following, therefore, only the most important general aspects of NMR and ESR spectroscopy are briefly recapitulated.

(a) Electron spin resonance. To study phospholipid-ion interactions, paramagnetic (most often transition metal, such as Cu^{2+}, Fe^{3+}, or Mn^{2+}) ions can be observed directly by means of ESR. Phospholipid molecules normally do not contain any lone-pair electrons, however, with the exceptions of some lipid degradation intermediates. (To stabilize and study the very short-lived free radicals appearing during lipid oxidation it may be necessary to use spin traps.)

Normal phospholipid molecules, consequently, must be spin labelled in order to become visible to ESR spectrometers. This is frequently done by the introduction of nitroxide groups into the lipid headgroup or chain regions, some relevant resulting molecules being shown in Fig. 5. The resulting stable organic radicals arc chcmical derivatives of either oxazolidin or piperidine (pyrrolidin) rings. The ESR spectrum of such groups in the isotropic case thus contains three lines at least (for nitrogen, $I = 1$, see Fig. 6). The separation between the outer pair of lines, which is proportional to the hyperfine splitting constant A, is sensitive to the local polarity near the spin label. The ESR line width is a rough measure of the probe's mobility.

An anisotropic ESR signal contains at least two pairs of outer lines with two different hyperfine splitting constants ($A_\parallel$, $A_\perp$); their difference is proportional to the order parameter of the labelled groups: $S = (A_\parallel - A_\perp)/(A_{zz} - A_{xx})$, where A_{zz} and A_{xx} are the corresponding maximal and minimal A values (cf. Fig. 6). Thus by using a series of phospholipid derivatives with spin labels located at different positions, order parameter profiles can be mapped. In the simplest approximation [34], the average order parameter for spin labelled phospholipids in a bilayer is given by

$$S = 1.6221 \, \frac{A_\parallel - A_\perp}{A_\parallel + 2A_\perp}$$

Molecular or segmental mobility, which is fast on the ESR time scale, tends to average out the effects of the local anisotropy. Detailed analysis of the ESR line shapes thus may reveal dynamical information about the spin label and chain mobility, but not before several simplifying assumptions are made about the relative significance of various processes that may affect the ESR line shape, such as motional line narrowing and exchange broadening [35]. At low spin label concentrations ($\leq 2\%$), the former prevails; at high concentrations ($5 \cdots 50\%$), the latter becomes more dominant.

In the simplest approximation, the rotational correlation time can be estimated from an empirical relation between the width of the central ESR line, ΔB_0, and the ratio of the high field and middle line intensities (h_{-1}, h_0, respectively): $\tau_R = 6.5 \times 10^{-10} \times \Delta B_0[(h_0/h_{-1})^{1/2} - 1]$. From the value of τ_R, moreover, the local viscosity can be deduced by using Stokes law.

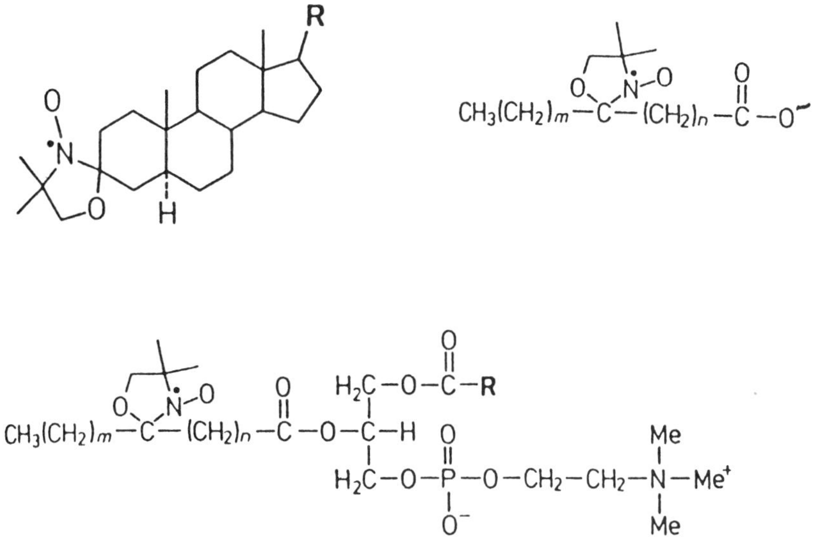

Figure 5 Some frequently used spin labelled lipids. Cholestane spin labels (top left) mimic the behavior and properties of the cholesterol derivatives; fatty acids with a nitroxide group at a fixed position along the chain (top right) can be used to map the transbilayer profiles, for example; but they may have a different distribution in the phospholipid bilayers than spin labeled phospholipids (bottom). Spin labels can also be attached to the headgroup of the latter.

Electron spin resonance is useful for monitoring the rotational or wobbling motions in the time-window around 10^{-9} s. To extend this range towards longer relaxation times, up to 10^{-3} s, saturation-transfer ESR (STESR) can be used. This highly sophisticated (and time consuming) ESR technique is based on the relaxation of the saturation of different parts of the ESR spectrum owing to the spin diffusion mediated transfer of this saturation (for a review see [36,37]).

The main advantages of ESR are its short characteristic time scale (i.e., high frequency domain), its sensitivity, and the possibility of using opaque and highly concentrated samples. Its chief problem is that spin labels, which are rather bulky, may perturb the investigated system. This handicap is normally greater for the ordered, more highly packed phases. To get data that are quantitatively reliable it is thus mandatory to study samples with spin labels at different dilutions and then extrapolate to $c_{SL} \rightarrow 0$. It is clear that even than direct information is obtained only for the labelled entity itself.

The lateral diffusion coefficient of the spin labelled phospholipid molecules [34] can be calculated by first extracting the lipid exchange frequency from the ESR line width by computer simulations. As a function of this frequency and total spin label concentration, the coefficient of lateral diffusion, in one particular model [38], is given by $D_L = 6 \times 10^{-16} \ \nu_{ex}/c_{SL}$, the numerical value accounting for the geometrical factors.

To obtain information about the transbilayer diffusion of phospholipid molecules, the headgroup labelled molecules on one side of a membrane are reduced by ascorbic acid. From the resulting temporal dependency of the signal, the diffusion coefficient of transverse motion (flip-flop) is then calculated [39].

ESR is also useful for studying the polarity profiles of phospholipid bilayers [40].

(b) Nuclear magnetic resonance. The basic principle of the (now obsolete) continuous

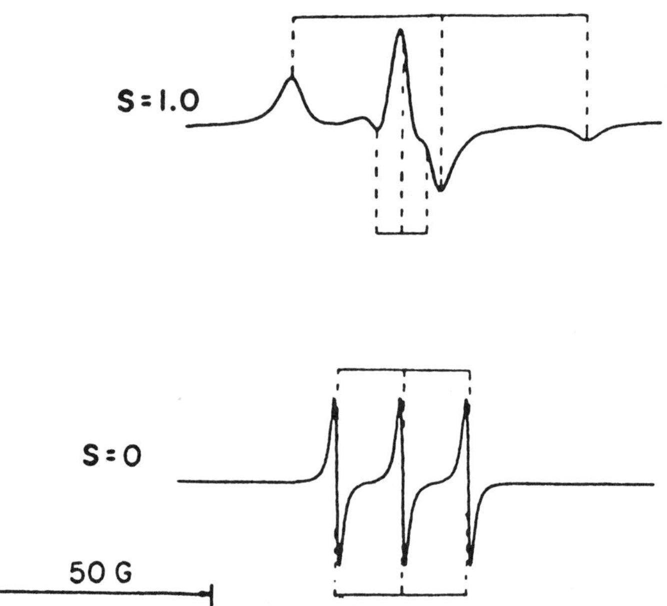

Figure 6 Typical electron spin resonance spectra of spin labels in randomly oriented samples with order parameter $S = 1$ (top) and $S = 0$ (bottom). $A_{\parallel}$ and $A_{\perp}$ are given by the hyperfine splitting (outer and inner pairs of dotted lines, respectively).

wave NMR is similar to that of ESR: upon sweeping the magnetic field, in which the sample irradiated by an electromagnetic field is held, resonant absorption peaks are observed [33]. Their absolute positions are determined by the chemical shift of each individual atom in the given position. Interactions of this atom with its neighbors (spin-spin coupling) furthermore may cause each resonance line to split into a series of lines (multiplet). Such splitting is analogous to the hyperfine splitting in ESR spectra [33]. In a single experiment, only one type of nucleus can thus be investigated.

Nowdays, essentially all NMR experiments are conducted in the Fourier transform mode with the use of pulsed techniques. For details on this technique, see Chap. 20 in this book.

The recovery of the macroscopic magnetization to equilibrium after a high-frequency pulse is characterized by the two characteristic relaxation times T_1 and T_2. These depend on the spin-lattice and spin-spin interactions, respectively. To get the relaxation time values, pulsed techniques are normally used: T_2 can be determined by the spin-echo method; to obtain T_1 values, the inversion-relaxation and the saturation method are most often used. From the knowledge of such characteristic time constants one can finally deduce the corresponding correlation times of the underlying interactions.

Depending on the type of nucleus observed, spin-lattice relaxation times on the order of up to 100 s in the fluid phase, and on the order of hours or even days for the solid specimen, can be measured. With the aid of appropriate simulation models, typical correlation times for various molecular motions of phospholipids (τ) can be deduced from such relaxation times (see Chap. 13 for more details).

Most commonly, deuterium (^{2}H) and phosphorus-31 (^{31}P), less frequently carbon-13

(^{13}C), and only occasionally oxygen-17 (^{17}O) or nitrogen-15 (^{15}N) are used to study phospholipid molecules by means of nuclear magnetic resonance. Normally, chemical shift anisotropies of those nuclei with spin ½, δ, are determined as functions of all relevant system parameters (such as temperature, system composition, pressure, etc.).

In principle, the δ patterns should provide enough information for the $S = 1/2$ systems for a precise conformational analysis of the investigated molecules to be possible. To date, however, structural investigations similar to those that are becoming nearly routine in the field of protein structure studies are not possible with phospholipid suspensions. This is a consequence of the prohibitively large size of phospholipid aggregates, which prevents an efficient line narrowing by motional averaging and thus spoils the experimental resolution. It can be anticipated, however, that with new sample preparation and measuring techniques this situation will improve; magic-angle spinning measurements, in which all anisotropic tensor interactions are averaged out, are already indicative of this [41,42].

Broad lines caused by the anisotropic interactions between the nuclear spins in phospholipid molecules hamper the use of high-resolution techniques; but they may provide information about the molecular order and the mechanism as well as the rates of molecular reorientations, provided that data analysis involves appropriate line shape simulation motional models [43,44]. Oriented lipid samples also contribute to clarifying the experimental analysis [45,46].

The fast rotation of the phospholipid molecules around their long axes in the fluid phase causes an averaging of the nonaxial symmetric component of the interaction tensor and thus reduces the NMR spectra of such phospholipids to a sharp, axially symmetric powder pattern that is interpretable in terms of an effective order parameter [33]. For each carbon-deuterium bond the average value of this parameter is

$$S_{CD} = \frac{4}{3}\frac{h}{e^2qQ}\,\Delta\nu_Q = 3.3 \times 10^{-3}\,M_1$$

where $\Delta\nu_Q$ is the observed quadrupole splitting, $e^2qQ/h \simeq 169Hz$ is the quadrupole constant of the C–D bonds, and M_1 in the alternative expression is the first moment of the ^2H–NMR line shape.

The time domain of standard NMR measurements is $10^{-6} \cdots 10^{-4}$ s; from NMR relaxation data motional correlation times in the range of $10^{-12} \cdots 10^{-1}$ s can thus be determined. This makes NMR measurements particularly suitable for studies of segmental molecular motions, such as lipid headgroup or terminal CH_3-group rotations [48]. Wobbling motions and collective fluctuations are also measurable.

NMR measurements are nonperturbing and, in the axially symmetric powder-pattern limit, relatively easy to analyze. Their handicap is, however, that good, modern instruments suitable for multinuclear measurements are extremely costly; furthermore, material requirements for a typical NMR experiment with less abundant isotopes may be quite high (50–250 mg phospholipid); for proton or deuterium NMR measurements, however, less than 5 mg of material are normally sufficient.

Conventional high-resolution NMR is only suitable for studying small, fluid-phase lipid vesicles or lipid micelles for which rapid particle tumbling and molecular diffusion cause an isotropic averaging of the NMR lines. For investigations of ordered lipid phases, macroscopically oriented lipid samples or magic-angle spinning should be used. (To prepare such samples, techniques similar to those described in the last section of this chapter can be used.) In every case, the interpretation of such data requires extensive computer simulation and modelling.

2. Dielectric Measurements

Owing to their low polarizability, hydrocarbon chains contribute only relatively little to the dielectric dispersion of phospholipid samples. But the electric dipole moments of the phospholipid headgroup and/or backbone region provide a naturally occuring, sensitive probe for dynamical studies of phospholipid molecules. Furthermore, dielectric measurements in the high-frequency region are suitable for investigations of the lipid-associated water.

Dielectric relaxation measurements in the 0.05–50 GHz region, for example, distinguish between the lipid bound and free water [49], the increasing viscosity of the former with decreasing lipid hydration being reflected in the lower values of the corresponding characteristic frequencies [50]. Dispersion in the 10–50 MHz region is diagnostic of the phospholipid headgroup motion in the plane of the bilayer [51]. Pure nonionic phospholipids, such as phosphatidylcholines, do not show appreciable dielectric dispersion below 10 MHz, but this frequency region is sensitive to the bulk ion concentration, vesicle coagulation and fusion [53]. Contamination with ionic amphiphiles, therefore, may affect the low-frequency dispersion of phospholipid suspensions dramatically.

For dielectric dispersion experiments with phospholipid-water mixtures, Boonton bridges combined with a coaxial cutoff cell at intermediate frequencies and suitable microwave bridges beyond 4 GHz have proven useful. Owing to the relatively large sample size and mass (50–200 mg phospholipid), vigorous mechanical mixing prior to the experiments (by ultrasound, for example) is often necessary. Samples thus prepared should be aged to avoid the deleterious effects of membrane defects. Moreover, corrections for electrode polarization should be made for measurements below about 1 MHz [53].

3. Ultrasound

The frequency range below ~1MHz is the domain of ultrasonic measurements. The ultrasonic velocity and absorption in a sample per unit wavelength are described for a single relaxation process as $v^2 = (v_0^2 + (v_\infty^2 - v_0^2))/(1 + \omega^{-2}\tau^{-2})$ and $\alpha\lambda = \pi\{v_\infty^2 - v_0^2)/[v^2(\omega^{-1}\tau^{-1} + \omega\tau)] + B\omega/v^2\}$, where v, v_0, and v_∞ are the ultrasonic velocities at the angular frequencies of ω, zero, and infinity, respectively. τ is the relaxation time [54].

The second term in the above expression describes classical absorption of sound due to shear and bulk viscosity as well as thermal conductivity. For very high or low frequencies, this is the only term contributing to the ultrasonic dispersion. Since the enhanced isothermal compressibility near a critical point causes a sharp decrease in v_0, however, the ultrasonic velocity at intermediate frequencies also shows the anomalous dip at T_c. The resulting attenuation is greatest for $\omega\tau \sim 1$.

When the volume concentration, c_L, is much smaller than unity, the magnitude of the sharp change in the ultrasonic velocity Δv of the liposome suspension is proportional to the volume change of the phospholipid bilayer:

$$\frac{\Delta V}{V} \simeq \frac{\Delta v_m}{v_m} c_L$$

where v and v_m are the ultrasonic velocities of the liposome suspension and the phospholipid bilayer, respectively, and Δv_m is the velocity change [54].

In a representative experiment, an ultrasonic pulse is excited in the measuring cell by a pulse generator and reflected many times in the cell until it is completely attenuated. After a certain number of echos, a new excitation is triggered. By parallel experiments with a sample-filled and reference cell, excess ultrasonic velocity Δv can thus be determined [55]. It is very convenient to combine ultrasonic with densitometric measurements.

4. Light Spectroscopy

Light of certain wavelength can be partly absorbed in phospholipid molecules and used for the induction of electronic excitations in this material. In the majority of cases, this causes transitions from the ground into the first excited state, which is sensitive to the material properties and thus suitable for the phospholipid investigations (for review see [52]).

The probability of light-induced transitions is proportional to the square of the transition dipole moment, that is, to the difference between the squares of the dipole moments of the scattering entity in the ground and the excited state. (This quantity is traditionally also called the dipole strength and is directly proportional to the molar extinction coefficient ϵ_{max}.)

(a) Absorption spectroscopy. More specifically speaking, the transition probability depends on the relative positions and energy differences of the electronic molecular states (molecular orbitals) to be excited. In phospholipid research, only the $\pi \rightarrow \pi^*$ and $n \rightarrow \pi^*$ transitions are of some significance.

Solvent molecules or other polar entities, which also possess large dipole moments, may affect the energy levels of the light-absorbing molecules and thus influence the transition probabilities. On the one hand, this gives rise to the solvatochromic effect, that is, to the shift of absorption maxima in surroundings of different polarity. (Transitions between two π orbitals normally show a red solvatochromic shift; for the $n \rightarrow \pi^*$ transitions the reverse is true.) On the other hand, the addition of appropriate solvents may increase the extinction coefficient as well.

(b) Ultraviolet spectroscopy. Native phospholipids with unsaturated chains undergo $\pi \rightarrow \pi^*$ transitions and absorb light in the UV region (typically around 185 nm for simple double bonds and around 230 for two adjacent double bonds); each further double bond, as a rule, results in a redshift of the absorption maximum of ~30 nm. The corresponding extinction coefficients in benzene are of the order of 2.5×10^4 for mono- and dienes, around 3 and 7.5×10^4 for the conjugated tri- and tetraenes, and beyond 10^5 for polyunsaturated fatty acids. In water, smaller extinction coefficients are measured. The main field of application of UV spectroscopy in phospholipid research is thus to study hydrocarbon unsaturation and oxidation [56]. Further characteristic absorption bands relevant in phospholipid research are given in Table 1.

Suspensions or solutions of monounsaturated phospholipids with ≤ mmol/L are suited for the standard measurements. By recording spectra derivatives, the sensitivity may be improved considerably, however [57]. Polyunsaturated fatty acids absorb light more strongly than monounsaturated ones; hexaenoic (parinaric) fatty acids are even useful as fluorescent membrane markers.

(c) Fluorescence spectroscopy. Fluroescence emission is observed after an electron that has been brought into an energetically excited state by the absorption of a photon of suitable wavelength (within ~10^{-15} s) spontaneously returns to a lower energy state and thus is deactivated (for a review see [58]). This normally occurs through a $\pi^* \rightarrow \pi$ transition between the states with antiparallel (singlet, $S = 0$) or parallel spin orientation

Table 1 Absorption Bands in the UV Region for Some Chromophores Found on Phospholipid Molecules

Group	CH	C–C	C=C	C=O ($\pi \rightarrow \pi^*$)	C= ($n \rightarrow \pi^*$)	–N–
Wavelength (nm)	122	130	185	187	273	194

(triplet, $S = 1$). The corresponding energy change, and thus the wavelength of the emitted light, depends on the electronic interactions and, as a rule, is lower for the $S = 1$ state. Transitions between the first excited and the ground singlet states give rise to a measurable fluorescence signal; this normally occurs on the time scale of $10^{-9} \cdots 10^{-7}$ s.

With only a few exceptions, nearly all lipid molecules need to be labelled with fluorescent groups in order to become useful for fluorescence measurements with visible light excitation (see further discussion). Such fluorescent labels, as a rule, contain conjugated double bonds, to be excitable even with low energy photons. Many are commercially available from Molecular Probes (Eugene, Oreg.), for example. Selections of fluorescently labelled phospholipids are also offered by Avanti Polar Lipids (Birmingham, Ala.) and several other lipid suppliers. A selected list of the fluorescently labelled phospholipids is given in Fig. 7. In addition to these phospholipids, the fluorescent anionic pyrenedodecanoic acid, 9-antracenedodecanoic acid, 5-(*N*-octadecanoyl)aminofluorescein, the cationic 3,3'-dioctadecylthiacarbocyanine perchlorate (DiS-C_{18}(3)), octadecylrhodamine B (R18), 1-pyrenemethyltrimethylammonium, and the nonionic 1,6-diphenyl-1,3,5-hexatriene (DPH), *N*-phenyl-1-napthylamine have also proven useful in phospholipid research.

Two factors are particularly important for the choice of suitable fluorescent labels: quantum yield and the (relative) position of the absorption and emission maxima. For fluorescence polarization measurements, the relative position of the transition dipole moment may also play some role [60].

The quantum yield of a given fluorophore is given by the ratio of the number of photons emitted to the number absorbed. This quantity, in the first approximation, is identical to the fluorescence transition rate divided by the sum of the fluorescence, inner conversion, intercombination, and quenching rates.

(d) Quenching. The quantum yield is independent of the excitation wavelength but is strongly sensitive to quenching by other light-absorbing molecules. Dynamic and collision-dependent or static and complexation-dependent processes may both be involved in this. The effects of dynamic quenching, which is active over separations ≤ 10 nm, in the simplest approximation, are described by

$$\frac{I_{F0}}{I_F} = Kc_Q\tau_F + 1$$

c_Q is the quencher concentration, τ_F is the lifetime of the excited state, and K (the quenching constant) is identical to the ratio of the quenching rate to the quencher concentration. In the case of static quenching one has analogously

$$\frac{I_{F0}}{I_F} = k_{FQ}c_Q + 1$$

where k_{FQ} is now the fluorophore-quencher association constant.

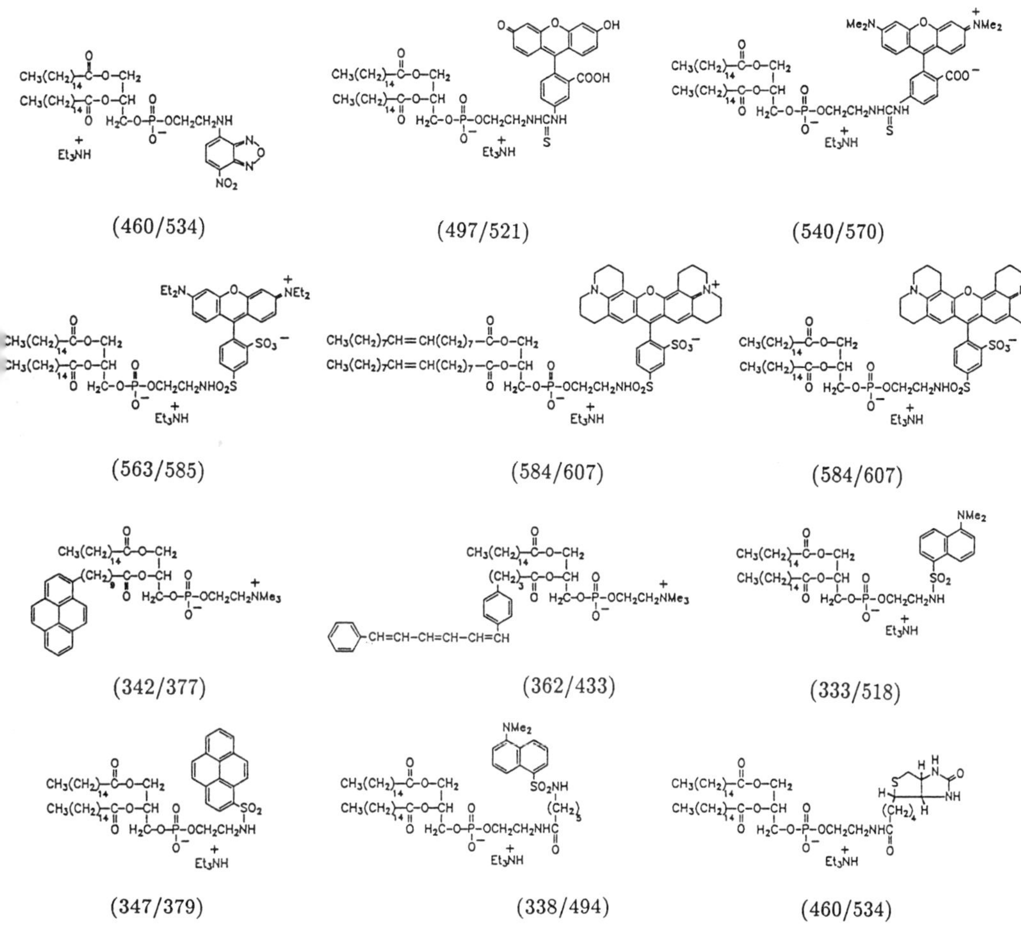

(460/534) (497/521) (540/570)

(563/585) (584/607) (584/607)

(342/377) (362/433) (333/518)

(347/379) (338/494) (460/534)

Figure 7 Some fluorescently labelled phospholipids. Numbers in parentheses give excitation and emission wavelengths (nm) in methanol, respectively (modified from ref. 59).

Fluorescence quenching is a basis for a variety of assays common in phospholipid research. Vesicle encapsulation and leakage rates, for example, are often tested with carboxyfluorescein, a fluorescent marker with self-quenching in the millimolar region [61]. To assess the efficiency of mixing of phospholipid vesicle interiors, terbium-dipicolinic acid [62] or carboxyfluorescein dilution [63] assay have proven useful. All these methods depend on the (self) quenching of a fluorescence signal.

When quenchers are fluorescent themselves, the efficiency of energy transfer (Förster transfer) between donor and acceptor molecules determines the fluorescence of the latter. For an efficient energy transfer of this type, the emission band of the donor and the excitation band of the acceptor molecules must overlap at least partly, and the average separation between both types of molecules, in general, should be smaller than 5–10 nm. (The intensity of fluorescence signal decreases with the inverse six-power of this separation.) This provides a means for measuring the separation R between such molecules from the energy-transfer efficiency

$$R = \frac{R_0}{(E_{FT} - 1)^{1/6}}$$

where $1 \leq R_0/nm \leq 5$ is a separation constant, characteristic for each given donor-acceptor pair, and E_{FT} is proportional to the ratio of the fluorescence intensity in the presence and absence of acceptor molecules: $E_{FT} = 1 - I_{F,D\text{-}A}/I_{F,D}$.

Förster energy transfer is the rationale behind several membrane fusion assays, such as the NDP/Rhodamine test [64]. (For more assays see [65].)

In special cases, donor and acceptor molecules are similar. In fact, Förster energy transfer is effective for essentially all common fluorescent labels, often in the millimolar concentration range. When fluorescently labelled lipids are incorporated into a phospholipid aggregate, the intensity of their fluorescence therefore first increases, then saturates, and finally decreases with increasing label concentration; for diacyl phospholipids the saturation limit is normally near 5 mol %.

To minimize the danger of label self-quenching—and in order not to perturb the investigated phospholipid system too much—in most fluorimetric experiments with lipid aggregates, fluorescent probe concentrations below 2.5 mol % and preferably lower than 0.25 mol % are used.

(e) 'Solvatochromism'. The position of the excitation and emission maxima is primarily a function of the molecular structure of the fluorescing entity. Some characteristic excitation and emission wavelengths of labels most often used in phospholipid research are given in Fig. 7. However, the local surrounding of a fluorophore may change the corresponding spectrum characteristics. In some cases, the emission maximum is shifted by more than 20 nm if the polarity of the surrounding medium is changed, for example [66]. This solvatochromic effect is caused by a spillover of energy from the excited electron to adjacent molecular segments or to proximal solvent molecules. Apolarity of the surroundings often results in a blue shift; highly polar environments, conversely, cause the fluorescence maximum to shift upwards in wavelength, in the majority of cases. Unfortunately, exceptions to this rule are too numerous to permit reliable generalizations. For trustworthy conclusions to be drawn from the shifts of fluorescence maximum, meticulous calibration procedures, therefore, are necessary.

This fact notwithstanding, the solvatochromic effect has been successfully used to assay the local polarity near or in lipid bilayer membranes [67] as well as to monitor the changes in this polarity induced by temperature variations, interactions with solvent molecules, or macromolecular adsorption [68].

(f) Fluorescence polarization. Fluorescence light induced by a beam of polarized light may itself be polarized. This latter polarization depends on the average orientation of the light-emitting transition dipole moment and thus is sensitive to the order and mobility of the fluorophore group. Fluorophores embedded into a phospholipid membrane, which is a quasi-two-dimensional and thus highly anisotropic system, therefore offer a means for studying the dynamics and average orientation of the hydrocarbon chains (for a review see [60]). The fluorescence anisotropy

$$r_\infty = \frac{I_{F,\parallel} - I_{F,\perp}}{I_{F,\parallel} + 2I_{F,\perp}}$$

is thus a measure of the order parameter of the fluorescent probes in a lipid bilayer, $S = \sqrt{r_\infty/r_0}$, where r_0 is the zero-time anisotropy value measured in dynamic experiments ([69], see also Chap. 13). If the motional correlation time is much shorter than the lifetime of the excited state, $r_0 = 0.4$ and the order parameter value can be obtained directly from the static fluorescence anisotropy value. In the reverse case, static measurements can only yield a value for r_0, depolarization then being entirely absent.

This shows that static fluorescence polarization measurements can yield no direct information on the membrane "fluidity," in contrast to a rather common belief. To get such information, time-resolved experiments should be performed [69]. Only in convenient frequency-domain windows, static red-edge effect measurements [70] may provide some auxilliary information on the fluorophore dynamics in the lipid bilayers. In such experiments, excitation by low-energy quanta leads to selective excitation of only a small part of the molecular states (dark area in Fig. 8) and shifts the fluorescence spectrum to longer wavelengths. On relaxation, there may be a decrease of the mean excited-state energy level, which then eliminates the excitation-wavelength dependence.

(g) Fluorescence recovery after photobleaching (FRAP). FRAP is a technique with which the lateral mobility of the (labelled) phospholipid molecules is measured directly. In a typical FRAP experiment, a spot or some characteristic pattern on the layer of

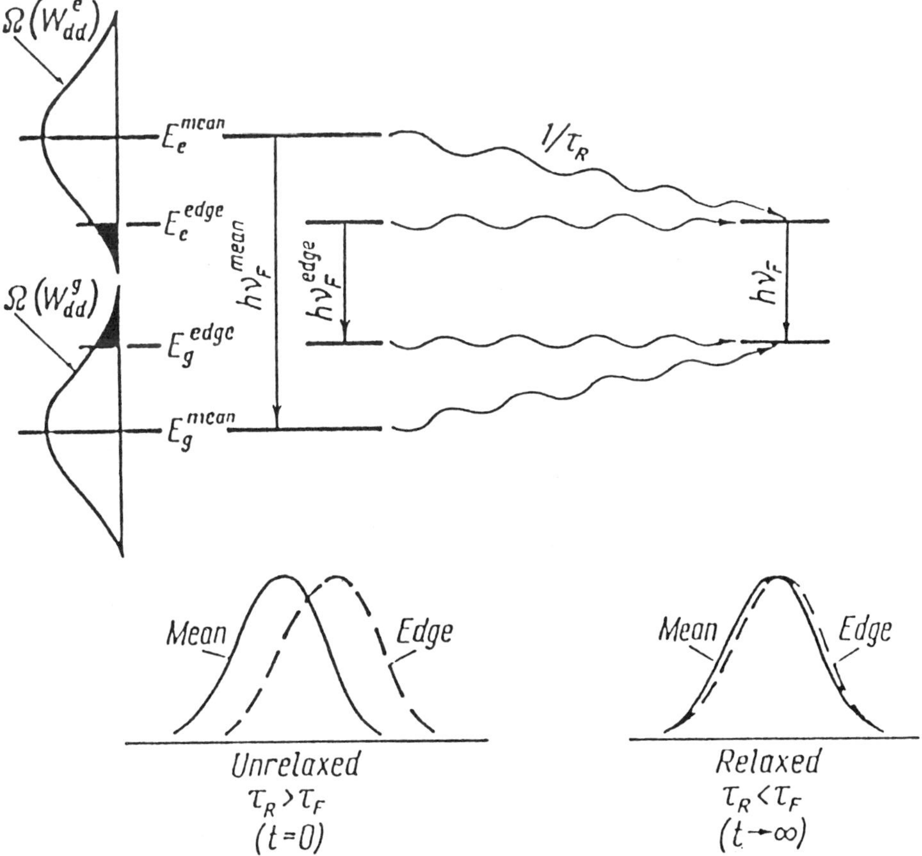

Figure 8 Scheme of a fluorescence measurement which exploits the "red-edge" effect to yield dynamical information from the static data. (From Ref. 70.)

phospholipid molecules is illuminated with a very strong pulse of light. This results in local fluorophore bleaching and in a transient fluorescence decrease. Subsequent inflow of the unbleached fluorophores into the bleached region is accompanied by a signal recovery; this is proportional to the lipid lateral diffusion coefficient [166].

In a special form of FRAP experiment, the so-called TIR-FRAP, the sample is illuminated from underneath under the conditions of total internal reflection. This minimizes the signal intensity from the bulk and improves the surface sensitivity of the experiment. Solvent layers as thin as 50 nm can be studied in this fashion, which makes the TIR-FRAP technique particularly valuable for surface adsorption studies [71].

(h) Infrared spectroscopy. In the infrared wavelength region ($2.5 \leq \lambda/\mu m \leq 250$), the photon energy is so low that only molecular vibrations and rotations can still be excited (for a review see [72]). (In IR spectroscopy it is customary to give the wavelength in terms of the so-called wave number, defined as: $\tilde{\nu}$ [cm^{-1}] = $10^4/\lambda$ [μm].) In spite of this, the basic features of infrared spectroscopy are the same as for other spectroscopic techniques, the band intensity, for example, being proportional to the transition dipole moment. A detailed description of the use of IR techniques for membrane research is given in [72,73]. A representative spectrum of dipalmitoylphosphatidylcholine is shown in Fig. 9.

To study the hydrocarbon chains in phospholipid molecules the region between 800 and 1200 cm^{-1} is usually inspected. This may reveal variations in the C–H and C–C vibrations as a function of temperature and other external variables; CH_2 vibrations are particularly strong and instructive in this respect. To study the properties of phospholipid headgroups and lipid-water interactions, chiefly the C=O streching vibrations, at 1700–1725, and P–O vibrations, at 815–825 cm^{-1}, are frequently studied. Both are sensitive to hydrogen bond formation in the headgroup region [47, 74], the band frequencies normally being shifted downwards with an increase in such binding. A more complete list of the characteristic vibrations observed in IR spectroscopy of phospholipids is given in Table 2.

As in the case of NMR studies, labelling of specific carbon atoms with deuterium, which shifts the resonance positions to lower wave numbers, can be used for easier identification and better experimental sensitivity [75]. Use of polarized light may also bring additional advantages [76].

In phospholipid-protein interaction studies, amide II and, in particular, amide I bands

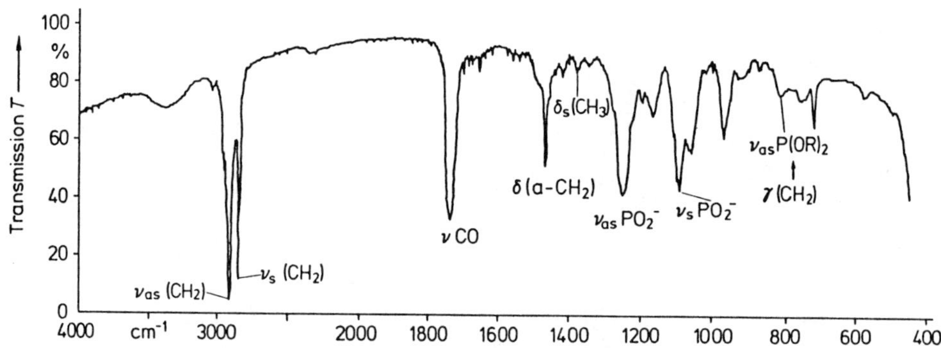

Figure 9 Attenuated total reflection infrared spectrum of dipalmitoylphosphatidylcholine at 22°C. (From Ref. 72.)

Table 2 Positions of Some Characteristic Vibrations of Phospholipid Molecules in the IR Spectrum (in Wavenumber Units)

Group	Vibration										
	Stretch			Double-stretch		Wag	Scissor.			Rock	Twist
	Sim.		Asym.	Sim.	Asym.		Sym.		Asim.		
CH_3			3030							819	
$CH_2 - CH_3$	2860–2890		2950–2970				1370–1380		1430–1470	1170	
$-CH_2-$	2840–2865		2880			1180–1345		1445–1485		720–1000	1296
C–C (chain)		1135									
C–C (all *trans*)		1063									
C–C (*sn*-1 chain, $k = 1$)		1104									
C–C (*sn*-2 chain, $k = 1$)		1098									
C=O		1700–1750									
COO–	1040			1300–1420	1550–1610						
H–O		3500–3560 2500–2700						1210–1320			
P=O (in PO_2^-)	1085–1110		1220–1260								
–P–O (in $P-(OR)_2^-$)		760	815–825								
C–C (head)		1012									
$-CH_2-$ (head)											1224
$R - NH_3^+$	3020		3200				1570–1620				
$N-CH_3$	900–930	711	950–970				1470–1480				
CH_3 on $N - (CH_3)_3$		3091	3020–3030				1470–1480			950	
H_2O			1200–1400				3200–3600				

near 1550 and 1650 cm^{-1}, respectively, are the most frequently analyzed. In such experiments, and in certain other FT-IR studies of phospholipids, ordinary water in the majority of cases is replaced by deuterium oxide (heavy water) in order to suppress the very strong signals in the region between 1200 and 1400 cm^{-1}. (Note that pH and pD differ by some 0.35 units; to compensate for this, pH calibrations should be performed for D_2O or D_2O/H_2O mixtures.)

(i) Instrumental considerations and sample preparation. IR spectroscopy has been revived by the introduction of the Fourier transform approach. This has made background and water-signal subtraction possible, has improved instrumental stability, and has provided a much higher spectral resolution; FT-IR spectrometers also require less time for data accumulation than classical IR spectrometers.

During a typical FT-IR measurement a sample is irradiated with polychromatic light (chiefly from a very hot rare-earth oxide electrode) with known spectral characteristics; the decay of the sample signal is then measured, over 1 ms, say, in 500–2000 consecutive scans. Finally the averaged transmitted signal is analyzed by means of Fourier transformation, and the desired information pertaining to the absorption in the sample is extracted by numerical routines: if no absorption has occured, the Fourier transform of the interferogram is identical to the original spectrum; otherwise, absorption bands characteristic of the investigated molecules are detected. The position and the intensity (integrated area) of these bands provide information about the physicochemical state of the irradiated molecules and its changes with changing system properties, such as temperature or sample composition.

FT-IR measurements with unoriented phospholipid samples are normally performed on suspensions of small or middle-sized vesicles (to prevent too strong light scattering). It is advisable to degas each sample prior to an experiment, especially for high-temperature experiments. If D_2O is used instead of H_2O, water vapor should also be meticulously eliminated from the sample housing. This may require several hours of purging.

For standard measurements with phospholipids, instruments with a resolution on the order of 0.05 wave numbers are sufficiently good for most applications. Higher resolutions (offered by Perkin-Elmer model 1850, Nicolet 170SX, Bruker IFS 113V, for example) is seldom required for simple phospholipid research. An ATR facility, however (*vide infra*) is often extremely valuable. As a caveat it should be noted that different software packages, claiming to perform the same type of analysis, may yield highly disparate results, especially in the higher-derivative mode.

In the *a*ttenuated *t*otal *r*eflection (ATR) mode of IR spectroscopy, a sample is applied on a solid substrate (most often in the form of a trapezoidal germanium plate) and IR light is then guided into this substrate from the side. Under conditions of total reflection, the light beam irradiates only a very thin layer of material near the substrate surface, which absorbs light at the resonance frequencies. The light beam that leaves the plate thus carries the desired information and can be analyzed in a standard manner.

ATR-IR is particularly suitable for the investigation of highly opaque materials that cannot be studied in the ordinary transmission mode. In combination with polarized light, ATR-IR also permits conclusions to be drawn about the average orientation and flexibility of the phospholipid molecules in the oriented sample near the plate surface. A special variant of this technique, in which the ambient conditions (e.g., humidity) are varied periodically [77], facilitates the identification of segments on the investigated molecules that are affected by the corresponding system variations.

D. Electron and Atomic Force Microscopy

In contrast to the situation with proteins, phospholipids have not yet been investigated by means of electron microscopy (EM) at molecular resolution. This is owing to the smallness as well as to the low electron density of the phospholipid molecules. Lack of highly developed sample preparation techniques (e.g., for the generation of highly ordered semimacroscopic specimens) also may have played some role. In spite of this, electron microscopy has been a valuable tool for studies of phospholipid aggregate structure, in some cases with a resolution better than 3 nm.

Originally, the lipidic structures in a sample were visualized by different negative and positive staining techniques [78]. Then it was realized that the resulting phospholipid stain complexes may be in a form quite different from that of the unstained specimen [79]. At present, freeze-fracture and freeze-etching techniques, in combination with some suitable shadowing procedures, are thus the most widely used EM methods for phospholipid research.

Sample preparation for such techniques, in general, consists of two steps. First, a specimen to be inspected is rapidly frozen to about –200°C within 10 msec or longer. Then the frozen sample is fractured with a sharp knife in vacuum. The fracture plane, in the majority of cases, falls in the weakest regions, that is, in the middle of a membrane. (An additional freeze-etching step may be included to expose the partly hydrated surfaces somewhat better.) Ultimately, an ultrathin layer of metal, usually platinum, is evaporated at some fixed angle onto the probe surface. This now forms an electron-dense replica that can be inspected in the ordinary transmission EM mode. Such a replica has a high contrast and may reveal even relatively small details of the original surface with dimensions of the order of appr. 2–3 nm—but it is only a replica.

Lipid bilayer replicas have been investigated with transmission [80] as well as with scanning tunneling microscopy (STM) [81]. However, a potentially more interesting approach than STM is to use atomic force microscopy (AFM), as this permits direct visualization of the lipid samples in situ [82]. The AFM method is particularly suitable to monitor periodic lipid structures, potentially with a resolution greater than 1 nm [83]. Its main disadvantage is that structural identifications are difficult and that undesired conformational or phase changes may result from the proximity of a scanning tip during the AFM measurements.

The recently introduced technique of cryoelectron microscopy, in many respects, is the most powerful version of EM available to date for routine phospholipid research [88]. For observation with this technique, lipidic structures are embedded in an ultrathin layer of amorphous ice. This, in combination with the use of energy filters, improves the material contrast considerably and permits experimental resolution, in the best case, to exceed 3 nm. The normal resolution of cryo-EM for phospholipid suspensions is on the order of 10 nm, however. An example of the visualization of lamellar and nonlamellar phospholipid phases by means of cryo-EM is given in Fig. 10.

Suspensions to be investigated by transmission electron microscopy should contain around 1 weight % of lipid. For the scanning techniques, lipid layers on a solid substrate are required. This puts severe constraints on the sample preparation and makes studies of symmetrical specimens, for example, difficult or impossible.

While preparing phospholipid samples for any sort of EM investigation, care must be taken to avoid osmotic stress, as this is prone to alter significantly the morphology of lipid

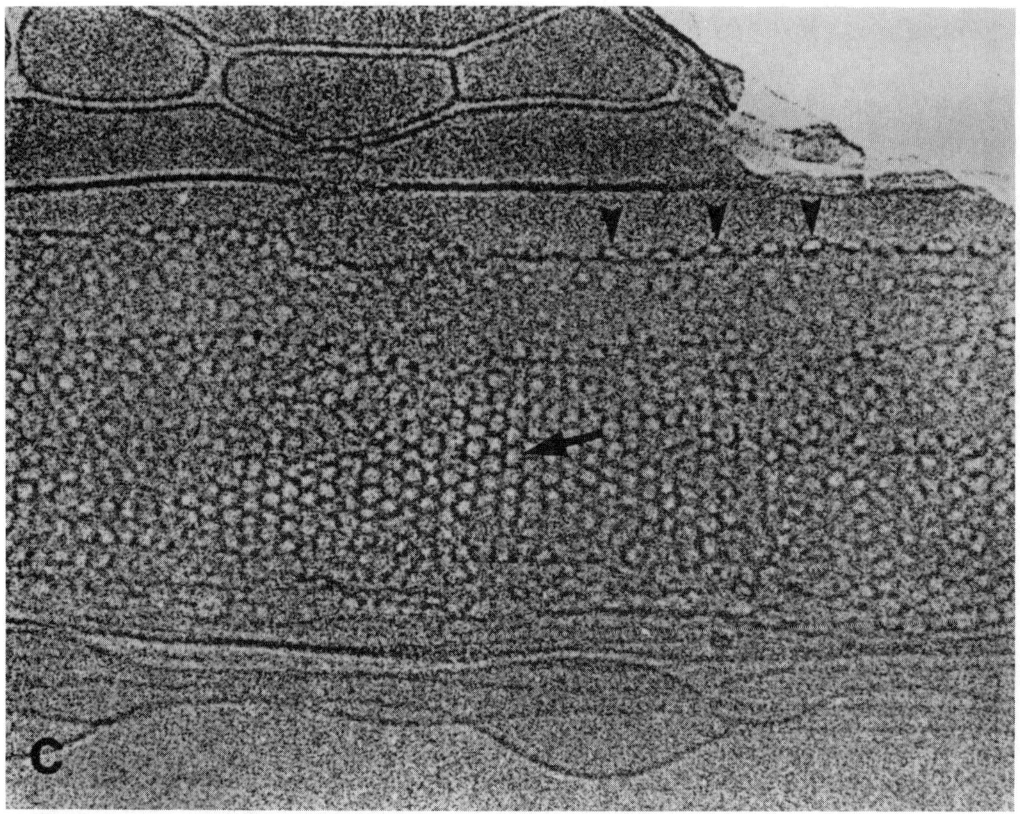

Figure 10 Cryoelectron microscopy picture of an dioleoylphosphatidylethanolamine/dioleoylphosphatidylcholine/cholesterol (3 : 1 : 1) mixture in water showing inverted hexagonal structures (arrowhead), isotropic, cubic structures (arrow), and lipid bilayers. (From Ref. 88.)

vesicles. Temperature gradients caused by insufficiently rapid cryofixation are another potential source of artifacts. In every case, a sufficiently high number of pictures should be taken to ensure good statistics; in the analysis of the results obtained with cryo-EM, moreover, the danger of nonuniform vesicle distribution in the ultrathin water matrix, and thus a bias towards underestimation of the vesicle size, should be kept in mind.

E. Optical (Polarizing) Microscopy

Polarizing microscopy, in principle, offers a simple and rapid technique for qualitatively mapping out lipid/water phase diagrams [89]. Different liquid crystalline phases have different optical properties [90,91]. For example, lamellar phases normally have a positive birefringence (refractive index maximum along the optic axis, i.e., along the layer normal); hexagonal phases have negative birefringence (refractive index minimum along the optic axis, i.e., along the cylinders); and cubic phases are optically isotropic. When viewed between crossed polarizers, the former two phases will thus transmit light (unless the optic axis of the phase is macroscopically aligned along either of the two polarization directions, or normal to them both), whereas cubic phases will not.

The sign of the birefringence may be determined using a retardation plate. However, the lamellar and hexagonal phases have rather characteristic optical 'textures' that are quite different from each other, and so these phases can usually be easily distinguished by inspection. Cubic phases may be differentiated from isotropic solutions by their much higher viscosity, which can be assessed simply by gently pressing on the sample.

In the water penetration technique, a thin layer ($\sim$100 μm) of dry lipid is sandwiched between two glass plates (usually a microscope slide below and a cover slip above). Water or buffer is then added to the edge of the plates and is sucked in around the lipid sample by capillary action. As water diffuses into the lipid, a lateral concentration gradient is set across the sample. In favorable cases, one can then observe a horizontal slice across the entire binary phase diagram, hence determining the phase sequence, although without locating the phase boundaries accurately. Varying the temperature then allows the entire phase diagram to be mapped out. Although this approach is immensely useful for simple surfactant systems, it is less successful with phospholipids, as with the latter the characteristic textures do not develop so readily (this is probably due to the greater stiffness of phospholipid layers).

A similar approach is also useful for measuring water diffusion rates in phospholipid multibilayer systems with the aid of the x-ray diffraction technique [92].

F. Calorimetry

Nearly all structural transformations and binding reactions are accompanied by an evolution or absorption of heat. The determination of this heat change, calorimetry, therefore offers an attractive and sensitive means for the investigation of changing phospholipid properties (for a concise review see [93]).

Most frequently, differential scanning calorimetry (DSC) is used for this purpose [94]. The basic idea was proposed in 1963 and consists of measuring a difference between the thermal powers released or absorbed in two calorimetric chambers—one for the sample and one for the reference. Both measuring chambers are surrounded by thermal screens and heated with a constant power. This power input causes the temperature in the measuring cells to rise continuously in the absence of excessive heat evolution or absorption by the sample; in this situation, the thermal difference between the two chambers has a zero value. As soon as the physicochemical state of the sample, and thus the heat balance of the sample, starts to change anomalously, a temperature gradient is established, however, between the sample and the reference chambers. Most instruments compensate electronically for the resulting difference in thermal powers, the signal proportional to the compensation current being a measure of the total heat absorbed or created within the sample. After suitable calibration or numerical processing, this signal is expressed in energy units.

A widely used DSC instrument suitable for phospholipid research is the DSC-7 (Perkin-Elmer, Norwalk, Conn.). In its current version this instrument is useful for relatively rapid measurements with small probe volumes (up to 50 μL) containing highly concentrated (5–30%) lipids. With reasonable sensitivity, the DSC-7 allows measurements to be performed at temperatures between ≤ -20°C (with Intracooler) and more than 150°C.

Most of the modern DSC instruments, such as the MC-2 (MicroCal, Northampton, Mass.) or the DASM-4 (Russian Academy of Sciences, Moscow), work in the adiabatic mode and have a much higher sensitivity; they use an extra feedback loop to maintain the

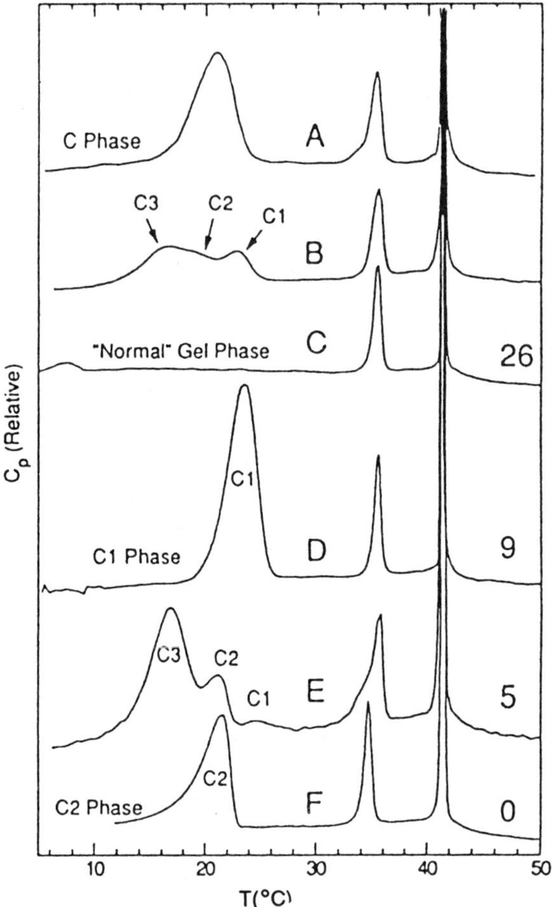

Figure 11 Differential scanning calorigram of dipalmitoylphosphatidylcholine in water. As the sample temperature is scanned, heat is being absorbed or released (for the endothermic or exothermic transitions, respectively). The resulting peak intensity is proportional to the enthalpy change in the system. (From Ref. 185.)

temperature of the adiabatic shield at the same temperature as that of the sample cell. Typical sensitivity of these instruments is of the order of 5×10^{-5} J/degree at a scanning rate of 1 degree per minute. (Higher scanning rates may slightly improve this sensitivity at the cost of potentially bringing the system out of equilibrium.) This implies that for the detection of the chain melting phase transition of common phospholipids with 16 carbon atoms per chain, more than 1 nmol of material (or at least 1 μg) are needed; for the identification of less energetic phase transitions the minimum required quantities are correspondingly larger (for the pretransition or bilayer-to-nonbilayer phase transition 10 μg are needed). Fig. 11 gives an example of the scanning calorigram of dipalmitoylphosphatidylcholine in water.

In related, but far less common, adiabatic titration experiments, the change of heat is measured as a function of the amount of titrant added for a binding reaction. Again, the working principle is based on a power feedback compensation between the sample cell, in which the titration is performed, and a reference cell that provides a differential thermal standard. The enthalpy of this reaction, the so-called van't Hoff enthalpy ΔH_{vH}, is proportional to the temperature variation of the binding constant K:

$$\Delta H_{vH} = RT^2 \left(\frac{d \ln K}{dT}\right)_p \tag{5}$$

Depending on the magnitude of the reaction heat, binding constants of the order of 10^2 to 10^8 M^{-1} can be determined with adiabatic titration instruments. With suitable equipment (such as OMEGA by MicroCal) available, corresponding measurements are rather straightforward to perform, for monomolecular or micellar dispersions, at least. Titration of phospholipid suspensions, however, brings the danger of molecular aggregation: in phospholipid aggregates, such as vesicles, an appreciable proportion of the molecules studied is confined to the "inner" membrane halves, which are not readily accessible to the titrating substances. Kinetic and molecular trapping problems in such situations may become prohibitive experimental obstacles.

The free energy of any system undergoing a first-order phase transition at T_t does not change. From the measured transition enthalpy ΔH_t, the transition entropy can thus be calculated as

$$\Delta S_t = \frac{\Delta H_t}{T_t}$$

From this entropy value and the pressure gradient of the transition temperature, the apparent volume change at the phase transition is obtained from the Clausius-Clapeyron equation [93]:

$$\Delta V_t = \Delta S_t \frac{dT_t}{dp}$$

Furthermore, from the ratio of the transition enthalpy and the van't Hoff enthalpy at the center of the phase transition (that is, calculated for $K = 0.5$ and $T = T_t$), the size of the "cooperative unit" $\Delta H_{vH}/\Delta H_t$ can be estimated.

Heat, by definition, is an unspecific quantity. Therefore the correspondence between the thermal change in a sample detected by a calorimeter and the underlying molecular process causing it must be established separately based on some supplementary information. Most frequently, various diffractometric or spectroscopic data are used for this purpose; polarizing microscopy is also useful.

Phospholipid specimens for standard calorimetric measurements are typically prepared by weighing the lipids into a calorimetric vessel and subsequently adding water of buffer (in the case of DSC-7 or similar instruments). For use in adiabatic calorimeters, appropriate amounts of lipids are suspended in a degassed aqueous subphase, 0.5–1 mg of lipid in a volume of 1.3–1.5 mL typically being used. Multibilayer samples normally have a narrower chain melting phase transition than suspensions of unilamellar, sonicated vesicles, the latter also often showing no pretransition. (Both these effects probably are consequences of the elastic stress in lipid bilayers [95] rather than being caused by a diminished cooperativity unit [96] or kinetic effects [93].)

Convenient scanning rates for studying the chain melting phase transition of common phospholipids are 30–60 degrees/h; when the lipid pretransition or bilayer-to-nonbilayer phase transition are the center of interest, scanning rates on the order of 30 degrees/h or less are recommended. In general, all transitions between highly ordered or structured phases should be scanned more slowly than those between fluid-phase transitions, owing to the difficulties of equilibration and solution transport in the former case. (The lipid subtransition [rotator phase transition in the chains region], for example, for many phospholipids is likely to occur on the time scale of many hours and up to weeks, in the presence of excess water.) Less strongly hydrated samples normally attain their equilibrium more rapidly, showing that restricted water diffusion may be a decisive factor in determining the rate of phospholipid thermal equilibration.

Except for samples that have been aged under controlled conditions, the first scan may be nonrepresentative and thus should be excluded from the data analysis. This also is largely owing to the difficulty of achieving a homogeneous water and solute distribution in the multilamellar lipid samples used for calorimetric experiments.

G. Mechanical and Elastic Measurements

When exposed to a mechanical stress, phospholipid bilayers change their morphological properties (for a theoretical review see [97]). Most frequently their area and/or surface curvature under stress become different; after an unequal expansion of the two bilayer halves, moreover, local curvature anisotropies may appear, such as isolated bilayer protrusions or invaginations, in the case of positive and negative area difference, respectively [98].

The magnitude of these effects depends largely on the (isothermal) area compressibility modulus K_A of each lipid bilayer, which for phosphatidylcholines is on the order of 0.15 and ≥ 0.9 N m^{-1} in the fluid and gel phases, respectively [99]; the apparent value in the undulated phase is even smaller, $\sim$0.05 N m^{-1}. In the gel phase, this modulus is prone to reflect chiefly the strains in the bilayer interior [100]; in the fluid phase, surface electrostatics may dominate [101].

The mechanical properties of phospholipid bilayers depend only little on the bulk modulus K_B of membrane compressibility, owing to the low volume compressibility of the lipid matrix. The latter is of the order of $\leq 3 \times 10^9$ N m^{-2} and thus, on the appropriate scale (K_A/d_b, where d_b is the bilayer thickness), it is by two orders of magnitude greater than the bilayer area compressibility modulus [102]. The bilayer volume compressibility may thus often be neglected. If so, the thickness elastic modulus is simply related to the area elastic modulus by $K_t = K_A/d_b$. Likewise, the elastic curvature modulus of a lipid bilayer can be estimated from the area elastic modulus as $B = K_A d_b^2/2$ [103], and vice versa.

The area elastic modulus of phospholipid bilayers can be measured in many ways. The most direct and easy to interpret, but experimentally very tricky, are micropipette aspiration measurements [103]. In such measurements, a small section of the surface of a giant ($\geq$ 10 μm) unilamellar vesicle is sucked into a pipette, which then acts as a "manometer," the length of the vesicle segment inside the pipette being directly proportional to the applied pressure. To obtain some dynamical information, micropipette suction experiments can be complemented with a spectral analysis of the surface undulations (flicker measurements) [104], which requires complicated instrumental equipment, however. A key method for elucidating the dynamic mechanical properties of phospholipid systems is the measurement of ultrasonic wave propagation [105]. Ultrasonic velocity is a measure of the lateral area compressibility of the bilayers, while the attenuation is related to the viscosity. This is particularly clear in laser-induced phonon spectroscopy [106].

The simplest method for obtaining the compressibility modulus is to analyze the microscopically observed wavelike fluctuations of very elongated ($\geq 10\mu m$) tube-like lipid aggregates [107].

Relatively simple osmotic swelling experiments, introduced by Katchalsky and extensively exploited by Haines and colleagues (see [101], for example) also can be performed relatively simply and at low cost to determine the value of K_A with moderate precision. In such experiments, the increase in the average size of a uniform population of lipid vesicles is studied as a function of the osmotic stress imposed on these vesicles upon diluting the suspension. From the resulting area *vs.* pressure curves, the value of K_A is obtained by a numerical fitting procedure.

H. Electric Measurements

All phospholipids carry atomic, local excess charges on their surface polar residues. (These could be determined by a combination of neutron and x-ray diffraction data and appropriate theoretical analysis, but to date such an analysis has only been performed for the phosphorylethanolamine "headgroup" [108,109].) Moreover, some lipids may possess a net, normally negative, charge on the phosphate or carboxylic groups that can be detected directly by means of acid-base titration, for example (see Chap 14 for a discussion of lipid ionization). Still other lipids carry one negative and one positive charge and are thus zwitterionic. Electrical measurements, therefore, can be used to gain information about the molecular and supramolecular state of individual phospholipid molecules and their aggregates.

1. Coulombic Surface Potential

Techniques for probing the electrostatic, coulombic potential and other electric properties of phospholipid bilayers may be direct or indirect, depending on whether they rely on using probes. In principle, direct methods are superior, as they are nonperturbing. However, in practice, indirect techniques frequently offer a much higher time-resolution and are sometimes more sensitive. Often they are also more convenient and less costly to use.

Direct measurements are chiefly based on the determination of the mobility of phospholipid headgroups [110] or whole lipid vesicles (vesicle electrophoresis, zeta-potential measurements) (see [111] for a general discussion and [112] for a review article biased towards phospholipid research) in an intrinsic local or external field, respectively. The former method requires high-frequency dielectric bridges or an NMR spectrometer and is thus very costly to perform. The latter only needs an optical microscope equipped with a suitable measuring cell (e.g., Mark II) or a laser Doppler velocimeter (such as Zetasizer IV, Malvern Ind., Malvern, Worcestershire), especially the former setup being relatively cheap.

Due to its simplicity, vesicle electrophoresis is perhaps the most commonly used method for the determination of phospholipid "surface charge density" and surface "zeta-potential." In a typical vesicle electrophoresis experiment, the movement of one lipid vesicle or the flow of a (monodisperse) population of lipid vesicles is studied under the microscope or in the measuring cuvette of a dynamic light scattering instrument, respectively, with a homogeneous external electric field in both cases being applied along the measuring cell. The resulting speed u of vesicle movement, or the vesicle mobility value, is then directly measured. Both increase with the charge density on the surface of each lipid vesicle; they are constant, however, during each individual experiment, since the electrical force and the resistance of the medium with viscosity η in the steady state are balanced.

The zeta potential (ζ-potential) is defined as the electrostatic potential at the plane of shear between the membrane-associated and the stationary part of the double layer, the latter not being dragged through the measuring cell by the moving vesicle(s). In an improved coulombic approximation [113], suitable for many practical applications, this zeta potential ζ is related to the vesicle speed by the relation

$$u = \frac{\epsilon\epsilon_0}{\eta}\,\zeta\left\{\left[(1 + R_{\mathrm{el}}^{-1}) + 4R_{el}\ \ln\cosh\left(\frac{Ze_0\zeta}{4kT}\right)\right]\left[\frac{Ze_0\zeta}{4kT}\ (2 + R_{el}^{-1})\right]^{-1}\right\}$$

where $R_{el} = (\lambda_D/r_v)\exp(e_0\,|\,\psi_x\,|\,/kT)$, λ_D is the Debye length, and the other symbols have their standard meanings. For large vesicles or high salt concentrations, the magnitude of the expression in the braces, which accounts for surface conductivity effects, is approximately unity. This limit is thus normally used for the analysis of the data obtained with microscopy measurements [112].

The main advantage of zeta-potential measurements by means of microscopy is that they require relatively cheap equipment; surface conductivity and size-polydispersity corrections for the individual large vesicles used in such measurements, moreover, are unimportant, but the time requirements for the experiments are unfavorable. The chief benefit of experiments with laser velocimetry equipment is that these are very rapid and statistically significant; their main disadvantage is that the position and state of the measuring capillary may affect the final results [150]. In both cases, experiments are difficult to perform in electrolytes with a salt concentration higher than 0.1 molar owing to heating problems caused by the ohmic currents. Careful calibration (preferably with monodisperse, 60 nm and up to 150 nm large phosphatidylglycerol or phosphatidylserine vesicles) in this and other cases are absolutely essential. In laser experiments all results that show a multipeak distribution for a homogeneous suspension should be discarded.

Another possibility of determining the electrostatic properties of phospholipids is to study the spatial variation of the "native ion" concentration near the lipid bilayer surface [102,114], or its direct and indirect consequences on the transbilayer mass flow. The electrostatic potential at the ion binding plane near a bilayer surface is given, in the first approximation, by the result

$$c(\text{surface}) = c(\text{bulk})\ \exp\left[\frac{Ze_0\psi(\text{surface})}{kT}\right] \tag{6}$$

In general, however, it is better to use rigorous equations that also take into account noncoulombic interactions [115].

To detect the lipid-associated ions, NMR [116] and ESR [117], radioactive decay [118] and ion fluorescence upon optical [119] or x-ray excitation [120] may be employed. Under suitable experimental conditions, most of these methods can distinguish ions near the phospholipid molecules from those in the bulk and thus provide data that upon electrostatic analysis give information about electrical properties of phospholipid molecules and their aggregates.

Detection of small, easily visible surrogate ions and displacement studies bridge the gap between direct and probe measurements of phospholipid electrostatics. For example, the binding of the trivalent fluorescent ion Tb^{3+} to bilayers can be used to study the receptor sites for Ca^{2+} on the phospholipid membranes [119]; electron paramagnetic resonance of manganese is convenient to follow the substitution of this ion near the surface of phospholipid bilayers by other divalent and monovalent ions [117]; certain ion isotopes can be used for the corresponding neutron experiments [121], etc.

By far the most common indirect approach to studying phospholipid ionization is to measure the fluorescence-marker binding. In a typical such experiment, the fluorescence spectrum and intensity of a given marker are measured as a function of the marker and salt concentration, and the surface potential is calculated by means of Eq. 6. Fig. 12 illustrates the results of one such measurement.

Owing to their dependence on the phospholipid charge and ionization state, many structural or functional phospholipid characteristics such as lipid phase transitions are also indicative of certain electrostatic properties of phospholipid bilayers [122,123,124]. Quantitatively correct conclusions are difficult to extract from the data pertaining to such characteristics, however, owing to the interdependence between surface electrostatics and hydration effects, the latter often being dominant [129,115].

A relatively detailed account of the analysis of electrostatic potential measurements is given in [129].

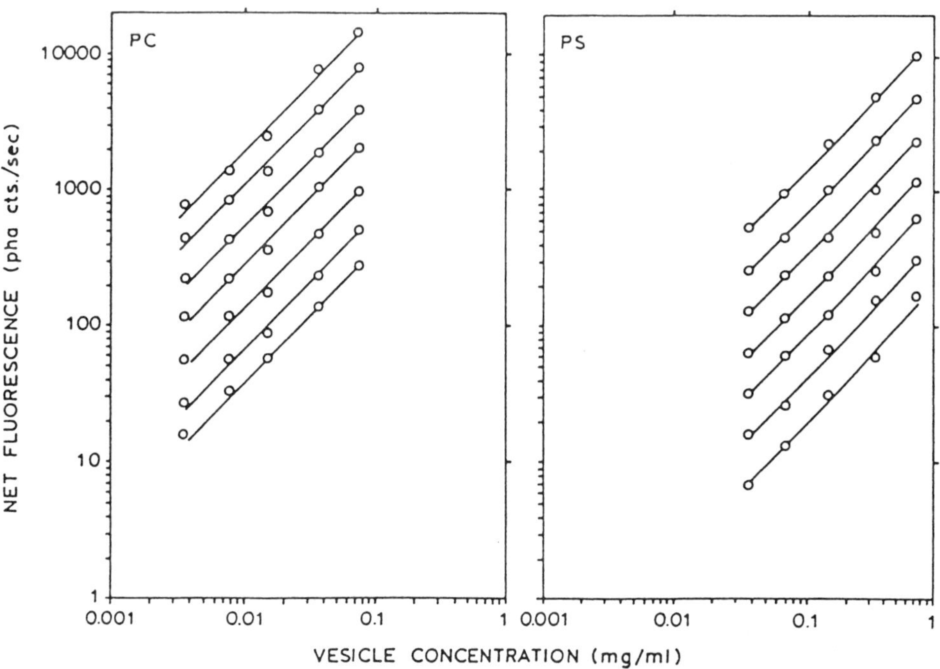

Figure 12 Surface potential measurement by means of fluorescent label adsorption. Intensity of the light emitted by charged fluorescent labels adsorbed at the phospholipid-electrolyte interface is proportional to the electrostatic bilayer potential, which is essentially zero for phosphatidylcholine (PC) and strongly negative for phosphatidylserine (PS). This potential, therefore, can be deduced from the fluorescence intensity after appropriate calibration measurements. (From Ref. 172.)

2. Dipolar Surface Potential

In addition to the coulombic bilayer surface potential, which is only significantly different from zero for charged phospholipids, all lipid membranes also exhibit another, so-called dipolar, surface potential. To measure this potential the ionizing gap, the static capacitor, or the "vibrating plate" methods have proven useful [126], the last approach being the most popular of these to date. It consists of bringing one condensor electrode to a separation of some 0.05–1.5 mm away from the lipid bilayer surface and then varying the electrode position periodically; the other (Ag/AgCl) electrode is immersed in water under the lipid layer or is in contact with the conducting lipid support. Plate vibration results in capacity variations $C(t)$, and in an electric current, $I = \Delta V\, dC(t)/dt$. The 'dipolar' surface potential is then identified with the compensation potential that must be applied to the "condensor plates" to minimize the signal in a loop (containing a lock-in amplifier).

The dipolar surface potential is often, but without any strict justification, expressed in terms of the surface dipole moment density, $\mu = \Delta V\, \epsilon_0 \epsilon_r A/\cos\,\theta$, where ϵ_r is the relative dielectric permittivity, A is the molecular area, and θ is the angle between the hypothetical dipole and the lipid (bi)layer surface. In reality, the apparent dipole moment is a sum of contributions from the hydrocarbon termini (10–30%), from the acyl-chain carbonyls (approximately 30–50%) [127], and from the lipid-associated water ($\leq$50%), at least [128,129]. To identify the magnitude of each individual contribution to this sum, extensive measurements with a variety of cunningly chosen lipid derivatives, therefore, is always necessary.

For common phospholipids, the measured values of the dipolar surface potential typically range between 150 and 600 mV and are very sensitive to a shift of the reference voltage and to lipid layer heterogeneities. An even more important source of error is the accumulation of surface active impurities near the air-water interface. Dipolar surface potential measurements, consequently, are quite suitable for phospholipid purity tests.

The same is true for lipid lateral conductivity measurements. These have repeatedly been claimed to be a fairly straightforward assay for the existence of interfacial hydrogen bond networks. Whilst it is most likely that such networks exist, and the experimental setup needed for carying out such measurements is easily constructed [130], utmost prudence is required in the data analysis and conclusions: surface-tension changes and counterion accumulation effects are prone to cause effects much larger than those that are actually intended to be studied. Several publications have failed to realize this.

II. PROPERTIES

Detailed descriptions of various experimental and instrumental aspects as well as physicochemical lipid data are given in other chapters and in the appendices. This section, therefore, only summarizes the most important techniques suitable for investigations of individual properties of phospholipid molecules or phospholipid aggregates, with some general recommendations and caveats, where appropriate.

A. Molecular Structure

The structures of phospholipids and their aggregates in water can be studied either by direct imaging methods or indirectly by scattering techniques. The former methods encompass to some extent electron and, in particular, atomic force [83,82] or scanning tunneling microscopy [81]. Amongst the scattering techniques, single crystal x-ray diffraction is most rewarding (for an early publication see [131]; for more recent references see [132]), but this approach is hampered by the difficulty of obtaining suitable single crystal specimens; perhaps the use of surface aligned phospholipid bilayers [15] will provide a partial solution to this problem. Complementary information for (partly) hydrated systems can also be obtained by neutron diffraction on selectively deuterated phospholipids [133,134]. When phospholipid aggregates are small enough to ensure efficient anisotropic averaging, simple [135] or multidimensional high-resolution NMR may also prove useful [136,137].

B. Molecular Packing

To study phospholipid packing properties, a much lower resolution is needed than in the case of phospholipid structure studies. X-ray [138,139] and neutron diffraction [133] on macroscopically oriented or powderlike samples are thus well suited for this purpose. Transmission or reflection mode are both useful, but for the latter only special radiation sources, such as the synchrotron, are brilliant enough to ensure good resolution.

The reflections of relatively large angles are diagnostic of the hydrocarbon packing; the positions of peaks in the angular region of 22°, for example, define the interchain separation perpendicular to the long chain axis. The small-angle reflections (below 3°, by and large) provide information about the lamellar or nonlamellar packing density. Detailed phospholipid and hydrocarbon packing assignment, to date, is only possible on the basis of diffractometric data; with some experience, IR spectroscopy is also useful for this purpose [140].

The molecular area in a lipid bilayer is not directly accessible to simple diffractometry. To determine this quantity a series of x-ray or neutron diffraction experiments is performed for samples containing increasing amounts of water. The cross-section of each lipid is then calculated from the known limiting value of the lamellar repeat distances and relative volume data [141].

The average bilayer thickness can also be obtained from such experiments [141]. Alternatively, ellipsometric [31] and bilayer capacitance [142] measurements may be used to get information on the hydrocarbon core thickness. If the hydrocarbon order parameter value is known, this thickness can also be calculated as $d_b = 2l_h(\bar{S}_{CD} + 0.5)$, where the maximum chain length is: $l_h \approx 0.175(n_C - 1)$ or $l_h \approx 0.263(n_C - 1)$ below and above the chain melting transition temperature, respectively. For phospholipids with tilted chains, these estimates must be corrected by a factor of $\cos \Phi_{tilt}$.

The transbilayer phosphate-phosphate separation, $d_{p\text{-}p}$, can be deduced directly from the calculated electron density profiles of lipid multibilayer stacks [138,143]. (Recently, new algorithms have been elaborated for this purpose [6] that are also useful with single experiment data.) The difference $d_b - d_{p\cdot p}$ provides an estimate for the headgroup region thickness. Unfortunately, the systematic error in most such determinations is relatively large, often on the order of 0.5 nm. This, on the one hand, is a consequence of the low resolution of the available data; on the other hand, the calculated values rely on a number of *ad hoc* assumptions, such as the postulate that lipid and water in a multibilayer stack are completely separated.

For the gel phase, the hydrocarbon tilt angle Φ_{tilt} can be calculated from the diffractometric data obtained for powderlike systems from the known bilayer thickness and molecular area [141]. The proviso for this is that lipid density (and thus molecular volume) is known. For a simpler estimate of the hydrocarbon tilt angle, the measured molecular area is divided by the molecular area of a phospholipid molecule with fully extended chains perpendicular to the bilayer surface: $\Phi_{\text{tilt}} \simeq \arccos(A/A_o)$, where $0.38 \le A_o/\text{nm}^2 \le 0.42$. The approximate relation arccos (d_b/l_h) is similarly reliable. More direct and exact information on the hydrocarbon tilt angle are obtained from oriented sample measurements [144].

Moreover, spin and fluorescence label probes can be used to monitor the tilt between a preferential probe axis and the bilayer normal, as well as its variations. Phospholipid molecules spin labelled in the upper part of the hydrocarbon chains, chiefly C-5 nitroxide spin labels, or rodlike fluorescence labels, such as diphenylhexatriene, are most often used for this purpose. (When performing experiments with such probes, care must be taken not to misinterpret the data, which are strongly sensitive to motion averaging on a very rapid time scale.)

C. Supramolecular Packing and Polymorphism

The interbilayer water layer thickness, d_w, in the simplest approximation, is given by the difference of bilayer repeat distance and bilayer thickness values. A more reliable estimate for this quantity is obained by measuring the corresponding phosphate-phosphate separation across the aqueous subphase from the electron density profiles. It should be realized, however, that any such d_w value is an average value at best. One reason for this is that water, with a variable probability, can protrude deep in the hydrocarbon region [145,146] and that lipid headgroups can also penetrate into the aqueous subphase. The other reason is the deviation of the surface geometry from planarity [147,148]. Defect textures in the lipid bilayers are most easily observed by means of EM [149].

Electron [150] or scanning-tunneling microscopy provide the simplest means for studying static bilayer surface undulations [80,81]. But it is also possible to reconstruct the corresponding surface patterns from the high-resolution x-ray diffraction data [147], measurable by synchrotron radiation.

These techniques, moreover, can be used to study quantitatively the long-range packing properties of phospholipid assemblies. X-ray diffraction is perhaps the most reliable and easily accessible technique for the determination of interlayer separation, but it should be used with prudence; this is especially true for the samples containing nonbilayer "cubic" phases [10]. For more qualitative studies of phosphilipid polymorphism, ^{31}P NMR, fluorescence spectroscopy, and polarizing microscopy are also often used. Whilst the latter can distinguish clearly between hexagonal and cubic nonbilayer phases, the former two methods may be misleading if they are not flanked by supplementary data.

To get information about the lipid vesicle size, electron microscopy and dynamic light scattering are chiefly used; liposomes greater than 300–500 nm can be visualized with good laboratory microscopes using Nomarski (phase contrast) optics. Among the newer techniques that can be used for this purpose are capillary hydrodynamic and field flow fractionation as well as disc centrifuge photosedimentometry or image analysis [25]. With suitable calibration, gel-exclusion chromatography (for example on Sepharose 4B or Sephacryl S100 columns [24] or plates [151] and even simple turbidimetry become suitable for the determination of phosphilipid vesicle size. Fig. 13 gives an example of this latter method.

In [24] the results of dynamic light scattering experiments (with various evaluation options), of gel filtration, and of electron microscopy are compared.

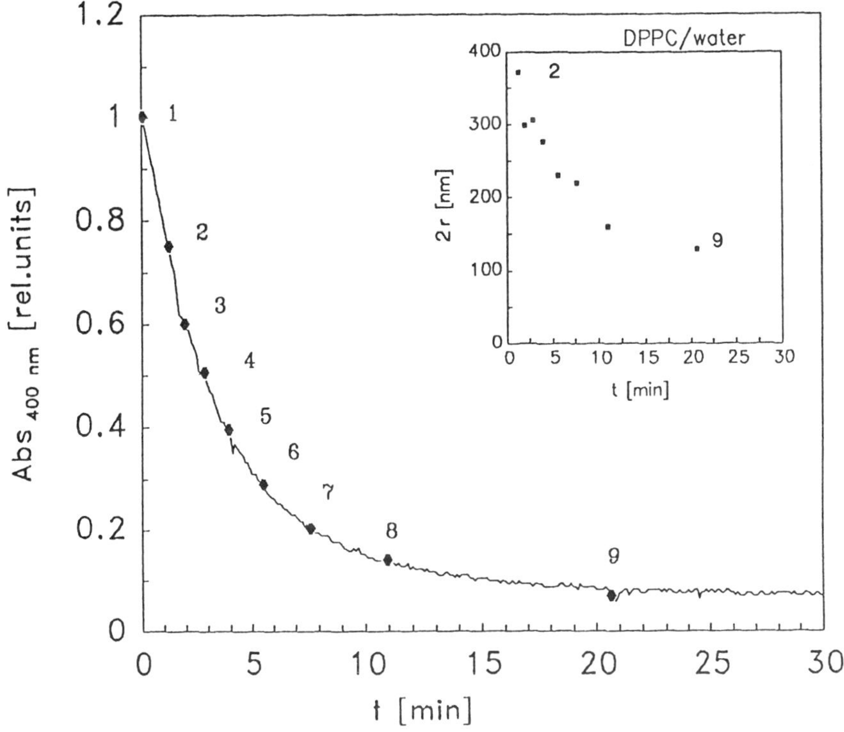

Figure 13 Interdependence between the absorption in (optical density of) a dipalmitoylphosphatidylcholine suspension and the corresponding vesicle diameter (insert), as determined by means of dynamic light scattering as a function of sonication time.

Macroscopic sample appearance is a very crude, but often useful, indicator of the phospholipid phase state. Formation of dense precipitates from the phospholipid suspensions suggests that lipids are probably in the ordered state. If the precipitate is moderately hydrated (this can be decided on the basis of the wet and dry weights after high-speed centrifugation, equilibration under humid atmosphere, and final drying), phospholipids are probably in the gel, L_β or L'_β phase; very low water contents (below two water molecules per lipid) is diagnostic of the formation of lipid crystals (L_c or L'_c). Optical transparency of an unfragmented suspension signifies lipid solubilization or formation of inverted cubic phases, the later being also highly viscous and sticky. Lipid stickiness and the presence of large, macroscopically quasi-continuous aggregates in excess water point to the existence of inverted nonbilayer phases. Cubic nonlamellar phases normally have a higher electrical conductivity than the inverted-hexagonal phases; the former also have a characteristic appearance under the polarizing microscope (cf. Sec. I.E).

To assess the macroscopic phospholipid packing density, volumetry with a double resonator (available, for example, from A. Parr, Warmminster, Penn.) is a good method of choice. The main disadvantage of this technique is that it requires a lot of material; it is also highly sensitive to the presence of bubbles or cavities and is not suitable for measurements at extreme temperatures. If only small amounts of phospholipid are avilable and/or if data over a wide temperature region are needed, the buoyancy method can be used [7]. This is based on determining the composition of a D_2O/H_2O mixture in which the phospholipid vesicles upon centrifugation neither sink nor float, the phospholipid density then being equal to the effective density of this mixture.

1. Interfacial Solvent Properties

To determine water activity near the surface of phospholipid bilayers, solute-induced phase transition shifts can be analyzed [9]. Direct measurements of the water activity coefficient [8] provide an average value for the interbilayer space. To establish the corresponding water activity profile hydration force data as a function of interbilayer separation may be analysed [175]. Dynamic information pertaining to the phospholipid-associated water can be obtained from NMR proton relaxation data, for example [154].

Effective water viscosity near phospholipid bilayer surfaces can be deduced from the dielectric relaxation data [50]. Corresponding water density data may be misleading because phospholipid headgroups and the first water molecules of hydration are intercalated and form an inseparable interphase. From computer simulations it has been concluded, however, that only the first one or two bound water layers are anomalously densely packed [153].

D. Thermodynamic Properties

For a complete phospholipid phase characterization, calorimetric measurements should be combined with diffractometry and, if possible, volumetry. At least two consecutive scans should always be performed to check whether or not some of the observed transitions are between two or more metastable states. Scan rates ideally should not exceed 1 degree per minute for chain melting phase transition studies; transformations between ordered phases normally require even slower scanning rates. In the diffractometric experiments, care should be taken to avoid or minimize gradients and/or coexistence of phases.

Any method that provides direct information about the state of the hydrocarbon chains, and to a lesser extent of the lipid headgroups, can reveal the phospholipid transition temperature, but not the corresponding transition enthalpy and entropy; x-ray diffraction, nuclear magnetic resonance, infrared (or Raman) spectroscopy, and dielectric measurements are among the best established such methods. Less direct methods include fluorescence spectroscopy with water soluble [152], lipid soluble [122], or lipid-bound labels [155], electron spin resonance spectroscopy [156], and turbidimetry [22]. For the latter, essentially any vis-spectrometer or even simple photodiode can be used, which makes this method particularly suitable.

The common disadvantage of all optical methods is that they are sensitive to lipid aggregation and flocculation; both cause concentration gradients within a probe and may also give rise to an anomalous scattering intensity that is superimposed on other signals. A general difficulty with label-dependent methods is that they yield transition temperature results that are sensitive to the label concentration. In fluorescence measurements, additional problems arise from the fact that the phase-transition effects may be superimposed on, or modified by, relatively trivial polarity and viscosity effects.

E. Mechanical Properties

The most direct method for studying the mechanical properties of phospholipid aggregates is elastometry. This can be performed by the micropipette technique [103], by measuring bilayer curvature as a function of external stress [107,101], or by analyzing surface waves on lipid vesicles [104] or planar films [157]. Elastometric spectroscopy, moreover, is possible with ultrasonic measurements [54,158]. The osmotic-stress technique is the only one of these methods that can be performed relatively easily and rapidly [101], especially if instead of dynamic light scattering simple turbidimetry in combination with appropriate calibration experiments is used.

Information on the bulk phospholipid compressibility modulus is normally obtained from ultrasonic dispersion [55] or straightforward volumetric [159,160] experiments. The latter technique is also used when knowledge about thermal volume expansivity is required. The area compression modulus K_A is normally obtained from ultrasonic or elastometric experiments; in the absence of better data, monolayer compressibility data are also useful [102]. Because phospholipid bulk compressibility is two orders of magnitude lower than lateral compressibility, the value of K_A can also be deduced from the known bilayer thickness compressibility. Bilayer thickness expansivity is given by $\alpha_T \equiv d_b^{-1}(dd_b/dT) \simeq [C_p(T)/V(T)T](dT_t/dp)$, where $C_p(T)$ is the isobaric heat capacity of the system, measurable in calorimetric experiments [161]. Bilayer area expansivity is best determined directly from the temperature dependence of the radius of the relaxed unilamellar phospholipid vesicles [162].

F. Kinetic and Dynamic Properties

Kinetic data normally require time-resolved measurements (see Chap. 13), except when certain time-average values are analyzed with suitable theoretical models, the experimental results in the latter case being sensitive to the choice of model. In general, rapid changes call for the use of rapid techniques: to monitor hydrocarbon chain melting at the level of single molecules, which occurs on the time scale of nanoseconds, fluorescence and electron spin resonance spectroscopy are useful; to observe cluster growth and long-range changes in extended phospholipid (multi)bilayers occurring on the time-scale of microseconds and up to several hundred milliseconds, nuclear magnetic resonance, ultrasonic dispersion, or even time-resolved x-ray diffraction with a synchrotron light source may be used [13].

To follow the kinetics of defect annealing, permeability studies [163], vesicle size increase [164], or partitioning and spectroscopic studies [165] are suitable, for fast ($\leq$ 1 h), intermediate ($\leq$ 1 day), and slow ($\leq$ 1 month) processes, respectively.

For lateral mobility studies, ESR [34], fluorescence recovery after bleaching [166], and NMR [167] are often used. Transbilayer flip-flop rates are determinable by ESR [39] or headgroup labeling [168] methods, among others. To study phospholipid transfer across an aqueous subphase fluorescence quenching [169], radioactive label transfer [170], or calorimetry, in the case of mixed vesicle systems [171], are particularly useful.

For a detailed discussion of various possibilities for phospholipid dynamics studies, Chap. 13 should be consulted.

G. Electrical Properties

To calculate the surface charge density in a lipid layer, the molecular ionization state must be known. The most direct method for determining this is an acid-base titration; essentially any method that is sensitive to the changing phospholipid properties can be used for this purpose, but it should be realized that in most cases the resulting change is not a linear function of the lipid headgroup ionization. This makes data interpretation difficult, especially when several titration regions overlap. Ionization studies with direct visualization of the titrated group (by means of ^{31}P, ^{17}O, ^{13}C, or ^{15}N NMR, or by FT-IR of the –OH, P–OH, or –NH groups), consequently, are less prone to misinterpretation than less direct titration of the chain melting phase transition shift, partitioning, or molecular mobility measurements. From the known degree of ionization, α, and area per lipid, A_L, the area per charge and thus a precise value for the surface charge density is obtained: $\sigma_{el} = \alpha e_o/A_L$.

Surface polarity (hydrophilicity) is often nearly as important as surface charge density. The difficulty with this quantity is that it varies strongly throughout the interfacial region, values corresponding to $\epsilon \sim 70$, ~ 30, and ~ 5 having been reported for the outer, central, and deep interfacial region, respectively. The dielectric constant of the bilayer interior is believed to be ~ 2.5. Different methods for the determination of the phospholipid "surface polarity" (such as the use of solvatochromic, electron spin resonance, or dielectric probes) may thus yield widely different effective values. The measured values, moreover, are sensitive to variations in the overall bilayer properties. To map the dielectric profile throughout a phospholipid bilayer properly, a series of independent experiments normally should be made, often with several probes to test different regions. For hydrocarbon polarity (dielectric constant) studies, ellipsometry can be used.

The surface potential of phospholipid vesicles is often assayed "directly" by electrophoretic mobility measurements [111,112] or indirectly by studying the adsorption of various ions [116,117,119] or charged labels [152] to the lipid bilayer surface. In both cases, the potential value has to be calculated from the measured data using more or less plausible assumptions [129,172]. Therefore results from adsorption studies in particular are seldom correct in absolute terms; but they are qualitatively (and relatively) reliable, by and large, provided that changes in the surface charge density by the label or ion adsorption in the data analysis have been accounted for.

H. Interactions with Nonlipid Molecules

Most phospholipid properties are sensitive to the presence of nonphospholipid molecules. Molecular adsorption onto phospholipids can thus be studied by any technique that is sensitive to some easily measurable phospholipid property. For optimal reliability, the most "linear" and perturbation-free method, rather than the most sensitive method, should be chosen. (The linearity of a chosen method should be assessed in a series of calibration experiments with well characterized and known reference phospholipids.)

Adsorption of fluorescent molecules, for example, can be monitored directly, preferably by means of the total internal reflection technique, on lipid monolayers [174] or, potentially, on surface-bound lipid bilayers [15]. With neutron diffraction the adsorption of isotopically labelled substances [173], and with x-ray diffraction [121] the adsorption of electron-dense materials, can also be followed directly. If the molecules of interest are not easily detectable, such bound molecules can be exchanged by others that are more easily visualized.

High adsorbent concentrations often cause major changes in phospholipid system properties (transition temperature, packing density, segmental mobility, etc.). This may be an obstacle to the measurements, but it may also provide a basis for an adsorption assay. In any case, care should be taken to avoid sample contamination with multilamellar or multivesicular systems, which inevitably cause the apparent association or binding constants to be underestimated; bilayer aggregation may have similar effects.

Conversely, lipid domain boundaries and/or membrane defects bring the danger of an overestimation of such constants, as they may act as high-affinity binding sites. This is especially true for organic or partly hydrophobic molecules (e.g., peptides or proteins).

Furthermore, kinetic aspects of molecular adsorption to the phospholipid layers should never, *a priori*, be forgotten. The on-off rates for small ions adsorbing to lipid bilayers are on the order of nanoseconds; larger organic ions are $10^3 \ldots 10^6$ times less rapid in this respect. It is nearly generally true that the larger is an adsorbing or adsorbed polar molecule the longer it will take before it reaches an equilibrium with the bulk molecules. Large water-soluble molecules (such as serum albumin) may require many hours, if not days, for this.

I. Interactions Between Phospholipid Membranes

The most direct method for measuring the force between two lipid bilayers is the micropipette technique. This is only applicable in the weak-interaction range, however. In order to extend such measurements to smaller interfacial separations, "osmotic" [175] and direct force [176] measurements must be used. In the former, the average separation between lipid bilayers in a multibilayer stack exposed to an external osmotic (or hydrostatic) stress or to an atmosphere of variable humidity is measured; this is a very straightforward approach, which only requires access to x-ray small-angle scattering equipment and a few polymer solutions. In the latter type of force measurement, lipid bilayers are deposited onto two crossed mica cylinders and then pushed towards each other or away from each other. Simultaneously, the force and separation are measured directly by means of a cantilever and by interferomctry, respectively. In principle, this allows very high resolution and, moreover, makes direct measurements of interbilayer attraction possible, but experiments with freely suspended bilayers are not possible. This latter method also has the disadvantage of being extremely sensitive to particulate impurities and is not suitable for experiments at very high or low temperatures.

Vesicle aggregation studies are far less direct than interfacial force measurements and at best yield semiquantitative results [177,178].

III. GENERAL CONSIDERATIONS

There are many possibilities for making phospholipid samples, the best choice depending largely on the question of interest. For lipid packing and structural investigations, multilamellar systems are normally used. Lipid multilamellae, or at least oligolamellae, are also well suited for calorimetric measurements, as they minimize the danger of elastic stress. In the simplest case they are prepared simply by mixing dry lipids with the appropriate amount of water and subsequent equilibration. If the lipid is hydrated inside a narrow test vial, care must be taken (by repeated centrifugation through a constriction, incubation, and/or temperature cycling) to achieve a homogeneous hydration of the samples (in many cases, this should be apparent from the diffraction patterns or other pertinent sample properties). Note that to be able to prepare low hydration samples, whose water content is known with an accuracy of better than 1%, requires careful use of a six- or seven-figure weighing balance. To keep lipid hydration during the measurements constant, care should be taken to minimize the void volume of the sample container; suitably formed teflon plugs may be used for this purpose.

NMR and optical techniques necessitate the use of small vesicles, by and large, as do essentially all transport studies. To obtain such vesicles, multilamellar systems are fragmented by ultrasound, extrusion through fine pores, and french-press or high-pressure homogenization. Physical suspension properties are little sensitive to the detailed conditions of fragmentation, but chemical sample integrity may deteriorate rapidly, especially after ultrasonication or pressure homogenization. (This is largely due to the lipid chains' oxidation and may be suppressed by using inert [nitrogen or argon] atomosphere.)

Moreover, care should be exercised when preparing small unilamellar vesicles not to overstress the bilayers. With phosphatidylcholines, for example, vesicles as small as 20 nm in diameter can be prepared; these are prone to relax into the less curved and thus less energetic state, with diameters in excess of 100 nm, upon equilibration of a highly concentrated lipid suspension for a day or two. Net charges on the lipid bilayers may slow down this process; system equilibrium may then be incomplete even after weeks of incubation. Decreasing the lipid hydrophilicity also tends to increase the equilibrium vesicle size, as does the reduction of the water content in the system. Phosphatidylethanolamines, for example, and phospholipid preparations with more than ~ 30% of lipid, do not form stable gel phase vesicles at all.

Essentially stress-free unilamellar vesicles can be prepared by dialysis [85,86,87] or different flotation techniques [99,98,84].

For diffraction, there are two basic methods for preparing unaligned lipid samples. First, they may be mounted in a suitable capillary. For x-ray diffraction experiments, this capillary is made of thin-walled (10 μm) glass or quartz tubes of approximately 1 mm diameter, which are then carefully sealed by flame and/or epoxy adhesive. If the lipid is first prepared and hydrated in a vial, then for viscous samples it may be impossible to transfer them to the fragile capillaries without breaking them. In this case it is preferable to clamp the samples between thin mica windows with a 1 mm teflon spacer in metal holders made for the purpose. Although in principle the water content can be determined after the experiment, it is nonetheless more difficult with this method than with capillaries to be sure that the water content has not changed during the experiment.

Both methods of sample mounting can be used up to quite high temperature (> 150°C), because in a sealed container water will not boil, unless the pressure buildup has burst the sample tube. Of course, the sample purity must be checked by TLC after the experiment to ensure that significant degradation of the lipid has not occurred.

A number of methods have been developed for preparing macroscopically aligned samples. For example, the lipid may be deposited on a suitable substrate by careful evaporation from an organic solvent, by evaporation of a liposomal disperson, or by dipping the substrate repeatedly through a lipid monolayer on a Langmuir trough, to produce a Langmuir-Blodgett film [179]. The advantages of the latter technique are that a controlled number of layers may be deposited, the bilayers may be made asymmetric, and a high degree of alignment (low mosaic spread) of the layers is possible. The substrate may be any of a number of suitable materials such as glass, aluminum, silicon, mica, or Mylar, and it may be flat or cylindrically curved. For fluid lamellar phases, alignment may also be promoted by shearing a thin sample (< 100 μm) between two parallel surfaces. In any case, only the properties of the first 8–12 lipid layers are affected by the substrate.

It is also possible to produce free standing oriented lipid films by carefully scraping a hydrated sample across a hole (~1 cm) in a solid plate [180]. The advantage of this latter technique is that the structure of the film, and thus its diffraction pattern, is free from effects due to the substrate (except of course at the edges). Furthermore, extremely low mosaic spreads can be achieved with this method.

For hexagonal phases (particularly of type I), it is sometimes possible to achieve structural alignment by shearing the sample in narrow (< 100 μm) capillaries; alternatively, partial alignment may be obtained by direct deposition on a flat substrate, or by starting with an aligned lamellar phase and changing the conditions (temperature, composition, etc.) to bring the sample finally into the hexagonal phase.

For cubic phases, where it would be highly desirable to have monodomain samples for structural study, it is more difficult to get a good alignment. In some favorable cases, such as the polyoxyethylene surfactant $C_{12}EO_6$, the sample will spontaneously grow into a single monodomain upon slowly traversing a phase boundary [181]. In most instances, however, it will be necessary to rely on surface effects, or the presence of an adjacent aligned lamellar phase, to be able to achieve even a partial alignment of these phases.

A major problem associated with the common methods for producing aligned samples is that it is extremely difficult to regulate the water content of the samples during the experiments. This is normally done by controlling the relative humidity of the atmosphere around the sample, either by equilibrating with various saturated salt solutions [182] or preferably by direct regulation [183]. However, even with this control, some problems remain. First, many lipids are found not to hydrate fully in an atmosphere of 100% relative humidity. The reason for this striking and reproducible effect has not yet been satisfactorily explained. Second, metastable or phase separation effects are often enhanced and may be difficult to control, particularly given that sample reequilibration times are quite slow when changes to the humidity are made. Third, precise temperature regulation of the sample is more difficult.

The number of lipid bilayers in a multilamellar system can be easily determined in the negative staining and cryoelectron microscopic pictures by simple counting. A thorough analysis of the x-ray diffraction data also provides semiquantitative estimates of this system's characteristics [184] as well as does freeze-fracture electron microscopy.

REFERENCES

1. H. Hauser, I Pascher, I. H. Pearson and S. Sundell, Preferred conformation and molecular packing of phosphatidylethanolamine and phosphatidycholine, *Biochim. Biophys Acta 650*:21–51 (1981).
2. D. L. Dorset, Electron diffraction structure analysis of phospholipids, *J. Electron Microsc. Tech. 7*:35–46 (1987).
3. J. A. Ibers and W. C. Hamilton, eds., *International Tables for Crystallography* Vol. IV, Kynoch Press, Birmingham, 1974.
4. N. P. Franks and Y. K. Levine, Low-angle X-ray diffraction, In *Membrane Spectroscopy* (E. Grell, ed.), Springer Verlag, Berlin, 1981, pp. 437–487.
5. G. Zaccai and B. Jacrot, Small angle neutron scattering, *Ann. Rev. Biophys. Bioeng. 12*:139–157 (1983).
6. P. Mariani, V. Luzzati and H. Delacroix, Cubic phases of lipid containing systems: Structure analysis and biological implications, *J. Mol. Biol. 204*:165–189 (1988).
7. J. M. Seddon, G. Cevc, R. D. Kaye and D. Marsh, X-ray diffraction study of the polymorphism of hydrated diacyl- and dialkylphosphatidylethanolamines, *Biochemistry 23*:2634–2644 (1988).
8. K. Hammond, I. G. Lyle and M. N. Jones, Vesicle-vesicle interaction and forces between bilayers in phospholipid systems incorporating phosphatidylinositol, *Coll. Surf. 237*:241–257 (1987).

9. G. Cevc, Polymorphism of bilayer membranes in the ordered phase and the molecular origin of lipid pretransition and rippled lamellae, *Biochim. Biophys. Acta 1062*:59–69 (1991).
10. V. Luzatti, A. Vargas, A. Gulik, P. Mariani, J. M. Seddon, and E. Rivas, Lipid polymorphism: A correction. The structure of the cubic phase of extinction symbol Fd—consists of two types of disjoined reverse micelles embedded in a 3D hydrocarbon matrix, *Biochemistry 31*:279–285 (1992).
11. J. M. Seddon, Structure of the inverted hexagonal (H_{II}) phase and non-lamellar phase transitions of lipids, *Biochim. Biophys. Acta 1031*:1–69 (1990).
12. M. Caffrey, The study of lipid phase transition kinetics by time-resolved x-ray diffraction. *Ann. Rev. Biophys. Biophys. Chem. 18*:159–186 (1989).
13. P. Laggner and M. Kriechbaum, Phospholipid phase transitions: Kinetics and structural mechanisms, *Chem. Phys. Lipids 57*:121–145 (1991).
14. P. Pershan, Liquid crystal surfaces, *J. Physique* (*Colloq.*) *50* (*C7*):1–20 (1989).
15. G. Cevc, W. Frenzl, and L. Sigl, Surface-induced x-ray reflection visualization of membrane orientation and fusion into multibilayers, *Science 249*:1161–1163 (1990).
16. J. B. Hayter, R. R. Highfield, B. J. Pullman, R. K. Thomas, A. I. McMullen, and J. Penfold, Critical reflection of neutrons: A new technique for investigating interfacial phenomena, *J. Chem. Soc. Faraday Trans I 77*: 1437–1448 (1981).
17. C. A. Helm, H. Möhwald, K. Kjaer, and J. Als-Nielsen, Phospholipid monolayer density distribution perpendicular to the water surface. A synchrotron x-ray reflectivity study, *Europhys. Lett. 4*:697 (1987).
18. T. M. Bayerl, R. K. Thomas, J. Penfold, A. Rennie, and E. Sackmann, *Biophys. J. 57*:1095 (1990).
19. U. W. Arndt, X-ray position sensitive detectors, *J. Appl. Cryst.* 19:145–163 (1986).
20. B. J. Berne and W. R. Pecora, in *Dynamic Light Scattering*, John Wiley, New York, 1976, Chap 10.
21. M. Warner, Theory of light scattering from vesicles, *Colloid Polym. Sci. 261*:508–519 (1983).
22. N. O. Petersen, S. I. Chan, The effects of the thermal prephase transition and salts on the coagulation and flocculation of phosphatidylcholine bilayer vesicles, *Biochim. Biophys. Acta 509*:111–128 (1978).
23. V. A. Bloomfield, and T. K. Lim, Quasielastic light scattering, *Meth. Enzym. 48*:415–494 (1978).
24. P. Schurtenberger and H. Hauser, Characterization of the size distribution of unilamellar vesicles by gel filtration, quasi-elastic light scattering, and electron microscopy. *Biochim. Biophys. Acta 778*:470–480 (1984).
25. T. Provder, ed., *Particle size distribution*, ACS Symposium series, Vol. 472, ACS, Washington, D.C., 1991.
26. B. B. Sauer, Y. L. Chen, G. Zografi, and H. Yu, *Langmuir 4*:111–127 (1988).
27. E. E. Uzgiris, Laser Doppler methods in electrophoresis, *Progr. Surface Sci. 10*:56–164 (1978).
28. S. W. Provencher, Contin: A programme for the evaluation of particle size from the correlated light spectroscopy data, *Comput. Phys. Commun. 227*:213–227 (1982).
29. D. E. Koppel, *J. Chem. Phys. 57*:4814–4820 (1972).
30. R. M. A. Azzam and N. M. Bashara, *Ellipsometry and Polarized Light*. North-Holland, Amsterdam, 2nd ed., 1986.
31. D. Ducharme, J.-J. Max, C. Sallese, and R. M. Leblanc, Ellipsometric study of the physical states of phosphatidylcholines at the air-water interface. *J. Phys. Chem. 94*:1925–1932 (1990).
32. E. Grell, ed., *Membrane Spectroscopy*, Springer Verlag, Berlin, 1981.
33. J. Seelig, Deuterium magnetic resonance: Theory and applications to lipid bilayers, *Quart. Rev. Biophys. 10*:353–418 (1977).

34. D. Marsh, Electron spin resonance: Spin labels, in *Membrane Spectroscopy* (E. Grell, ed.), Springer Verlag, Berlin, 1982, pp. 51–142.
35. J.-H. Sachse, M. D. King, and D. Marsh, ESR determination of lipid translational diffusion coefficients at low spin-label concentrations in biological membranes, using exchange broadening, exchange narrowing, and dipole-dipole interactions, *J, Magn. Reson. 71*:385–404 (1987).
36. D. D. Thomas, L. R. Dalton, and J. S. Hyde, Rotational diffusion studied by passage saturation transfer electron paramagnetic resonance, *J. Chem. Phys. 65*:3007–318 (1976).
37. D. D. Thomas, *Enzymes of Biological Membranes* (A. Martonosi, ed.), Plenum Press, New York, Vol. 1., (1985), pp. 287ff.
38. H.-J. Gala and W. Hartmann, Excimer forming lipids in membrane research, *Chem. Phys. Lipids 27*:199–213 (1980).
39. R. D. Kornberg and H. M. McConnell, Inside-outside transition of phospholipids in vesicle membranes, *Biochemistry 10*:1111–1120 (1971).
40. O. H. Griffith, P. J. Dehlinger, and S. P. Van, Shape of the hydrophobic barrier of phospholipid bilayers (Evidence for water penetration in biological membranes), *J. Membrane Biol. 15*:159–192 (1974).
41. E. Oldfield, J. L. Bowers, and J. F. Forbes, High-resolution proton and carbon-13 NMR of membranes: Why sonicate? *Biochemistry 26*:6919–6923 (1987).
42. K. S. Bruzik, B. Sobon, and G. M. Salamonszyk, *Biochim. Biophys. Acta 1023*:143 ff. (1990).
43. R. J. Whittebort, E. T. Olejniczak, and R. G. Griffin, Analysis of deuterium nuclear magnetic resonance line shapes in anisotropic media, *J. Chem. Phys. 86*:5411–5418 (1987).
44. C. Mayer, K. Müller, K. Wcisz, and G. Kothe, Deuterium NMR relaxation studies of phospholipid membranes, *Liq. Crystals* 3:797–806 (1988).
45. C. Mayer, G. Gröbner, K. Müller, K. Weisz, and G. Kothe, Orientation-dependent deuteron spin-lattice relaxation times in bilayer membranes: Characterization of the overall lipid motions, *Chem. Phys. Letters 165*:155–161 (1990).
46. W. Hübbner and A. Blume, 2H-NMR spectroscopy of oriented phospholipid bilayers in the gel phase, *J. Phys. Chem. 94*:7726–7730 (1990).
47. J. Grdadolnik, J. Kidrič, D. Hadži, *Chem. Phys. Lipids* 59:57–68 (1991).
48. H. H. Füldner, Characterization of a third phase transition in multilammelar dipalmitoyllecithin liposomes, *Biochemistry 20*:5707–5710 (1981).
49. U. Kaatze and K. Lautscham, High-frequency ultrasonic absorption spectroscopy on aqueous suspensions of phospholipid bilayer vesicles, *Biophys. Chem. 32*:153–167 (1989).
50. G. Nimtz and W. Weiss, Relaxation time and viscosity of water near hydrophilic surfaces, *Z. Phys. B—Condensed Matter 67*:483–487 (1987).
51. J. C. W. Shepherd and G. Büldt, Zwitterionic dipoles as a dielectric probe for investigating head group mobility in phospholipid membranes, *Biochim. Biophys. Acta 514*:83–94 (1978).
52. C. N. Banwell, *Fundamentals of Molecular Spectroscopy*, McGraw-Hill, Maidenhead, 1983.
53. V. Uhlendorf, Fatty acid contamination and dielectric relaxation in phospholipid vesicle suspensions, *Biophys. Chem. 20*:261–273 (1984).
54. S. Mitaku, A. Ikegami, and A. Sakanishi, Ultrasonic studies of lipid bilayer phase transition in synthetic phosphatidylcholine liposomes, *Biophys. Chem. 8*:295–304 (1978).
55. S. Mitaku and K. Okano, Ultrasonic measurements of two-component lipid bilayer suspensions, *Biophys. Chem. 14*:147–158 (1981).
56. J. M. C. Gutteridge and B. Halliwell, The measurement and mechanism of lipid peroxidation in biological systems, *TIBS 15*:129–135 (1990).

57. F. P. Corongiu and A. Milia, An improved and simple method for determining diene conjugation in autooxidized polyunsaturated fatty acids, *Chem.-Biol. Interactions 44*:289–297 (1983).
58. L. M. Loew, ed., *Spectroscopic Membrane Probes*, Volumes 1,2,3, CRC Press, Boca Raton, Fla., 1988.
59. R. P. Haugland, *Handbook of Fluorescent Probes and Research Chemicals,* Molecular Probes, Eugene, Oreg., 1989.
60. L. B.-A. Johansson and G. Lindblom, Orientation and mobility of molecules in membranes studied by polarized light spectroscopy, *Quart. Rev. Biophys. 13*:63–118 (1980).
61. J. Ruiz and A. Alonso, Surfactant-induced release of liposomal contents. A survey of methods and results, *Biochim. Biophys. Acta 937*:127–134 (1988).
62. J. Wilschut and D. Papahdjopoulos, Ca^{2+}-induced fusion of phospholipid vesicles monitored by mixing of aqueous contents, *Nature 281*:690–692 (1979).
63. F. C. Szoka, K. Jacobson, and D. Papahdjopoulos, The use of aqueous space markers to determine the mechanism of interaction between phospholipid vesicles and cells, *Biochim. Biophys. Acta 551*:295–303 (1979).
64. D. K. Struck, D. Hoekstra, and R. E. Pagano, Use of resonance energy transfer to monitor membrane fusion, *Biochemistry 20*:4093–4099 (1981).
65. J. Wilschut, Membrane fusion: Lipid vesicles as a model system, *Chem. Phys. Lipids 40*:145–166 (1986).
66. C. Reichardt, In *Solvent Effects in Organic Chemistry*, VCH, Weinheim, 2nd ed., 1988.
67. C. J. Drummond, F. Grieser, and T. W. Healy, A single spectroscopic probe for the determination of both the interfacial solvent properties and electrostatic surface potential of model lipid membranes, *Faraday Discuss. Chem. Soc. 81*:95–106 (1986).
68. M. B. Lay, C. J. Drummond, P. J. Thistlethwaite, and F. Grieser, ET(30) as a probe for the interfacial microenvironment of water-in-oil microemulsions, *J. Colloid. Interf. Sci. 128*:602–604 (1989).
69. F. Jähnig, Structural order of lipids and proteins in membranes: Evaluation of fluorescence anisotropy data, *Proc. Natl. Acad. Sci. USA 76*:6361–6365 (1979).
70. A. A. Demchenko, Site-selective excitation: A new dimension in protein and membrane spectroscopy, *TIBS 13*:374–376 (1988).
71. R. M. Zimmerman, C. F. Schmidt, and N. H. E. Gaub, Absolute quantities and equilibrium kinetics and macromolecular adsorption measured by fluorescence photobleaching in total internal reflection, *J. Colloid Int. Sci. 139*:268–280 (1990).
72. U. P. Fringeli and H. H. Günthard, Infrared membrane spectroscopy, in *Membrane Spectroscopy* (E. Grell, ed.), Springer Verlag, Berlin, 1981.
73. D. G. Casal and H. H. Mantsch, Polymorphic phase behaviour of phospholipid membranes studied by infrared spectroscopy, *Biochim. Biophys. Acta 779*:381–401 (1983).
74. K. Leberle, I. Kempf, and G. Zundel, An intramolecular hydrogen bond with large proton polarizability within the head group of phosphatidylserine: An infrared investigation, *Biophys. J. 55*:637–648 (1989).
75. D. G. Casal, H. L. Cameron, H. H. Boulanger, Y. Mantsch and I. C. P. Smith, The thermotropic behavior of dipalmitoyl phosphatidylcholine bilayers: A fourier transform infrared study of specifically labeled lipids, *Biophys. J. 35*:1–16 (1981).
76. H. Vogel, F. Jähnig, V. Hoffmann, and J. Stuempel, The orientation of melittin in lipid membranes: A polarized infrared spectroscopy study, *Biochim. Biophys. Acta 733*:201–209 (1983).
77. U. P. Fringeli and H. H. Günthard, Hydration sites of egg phosphatidylcholine determined by means of modulated excitation infrared spectroscopy, *Biochim. Biophys. Acta 450*:101–106 (1976).
78. J. R. Harris, A negative staining study of natural and synthetic L-(alpha)-lysophosphatidylcholine micelles, macromolecular aggregates and crystals. *Micron & Microscopica Acta 17*:289–305 (1986).

79. P. Fromherz and D. Rüppel, Lipid vesicle formation: The transition from open disks to closed shells, *FEBS Letters 179*:155–161 (1985).
80. R. Krbecek, C. Gebhardt, H. Gruler, and E. Sackmann, Three dimensional microscopic surface profiles of membranes reconstructed from freeze etching electron micrographs, *Biochim. Biophys. Acta 554*:1–22 (1979).
81. J. A. N. Zasadzinski, J. Schnier, V. Gurley, V. Elings, and P. K. Hansma, Scanning tunneling microscopy of freeze-fracture replicas of biomembranes, *Science 239*:1013–1015 (1988).
82. A. L. Weisenhorn, M. Egger, F. Ohnesorge, A. C. Gould, S.-P. Heyn, H. G. Hansma, R. L. Sinsheimer, H. E. Gaub, and P. K. Hansma, Molecular resolution images of Langmuir-Blodgett films and DNA by atomic force microscopy, *Langmuir 7*:8–12 (1991).
83. M. Egger, S. P. Heyn, and H. E. Gaub, Two-dimensional recognition pattern of lipid-anchored Fab' fragments, *Biophys. J. 57*:669–673 (1990).
84. D. S. Dimitrov, and M. I. Angelova, Lipid swelling and liposome formation mediated by electric fields, *Bioelectrochem. Bioenerg. 19*:323–331 (1988).
85. M. H. W. Milsmann, R. A. Schwendener, and H.-G. Weder, The preparation of large single bilayer liposomes by a fast and controlled dialysis, *Biochim. Biophys. Acta 512*:147–155 (1978).
86. D. Schwarz, D. Zirwer, K. Gast, H. W. Meyer, and U. Lachmann, Preparation and properties of large octylglucoside dialysis/adsorption liposomes, *Biomed. Biochim. Acta 47*:609–621 (1989).
87. H. Alpes, K. Allmann, H. Plattner, J. Reichert, R. Riek, and S. Schulz, Formation of large unilamellar vesicles using alkyl maltoside detergents. *Biochim. Biophys. Acta 862*:294–302 (1986).
88. P. M. Frederik, K. N. J. Burger, M. C. A. Stuart, and A. J. Verkleij, Lipid polymorphism as observed by cryo-electron microscopy. *Biochim. Biophys. Acta 1062*:133–141 (1991).
89. N. H. Hartshorne, Optical properties of liquid crystals, in *Liquid Crystals and Plastic Crystals* (G. W. Gray and P. E. Winsor, eds.), Ellis Horwood, Chicester, (1974).
90. D. Demus and L. Richter, *Textures of Liquid Crystals*, Verlag Chemie, Weinheim, 1978.
91. G. W. Gray and J. W. Goodby, *Smectic Liquid Crystals*, Leonard Hill, Glasgow, 1984.
92. H. C. Gerritsen and M. Caffrey, Water transport in lyotropic liquid crystals and lipid-water systems: mutual diffusion coefficient determination, *J. Phys. Chem 94*:944–948.
93. R. L. Biltonen, A statistical therodynamic view of cooperative structural changes in phospholipid bilayer membranes: Their potential role in biological function. *J. Chem. Thermodynamics 22*:1–19 (1990).
94. M. L. Johnson, W. van Osdol, and R. L. Biltonen, *Meth. Enzym. 130*:534–551 (1986).
95. B. R. Lentz, R. L. Biltonen, and E. Freire, Fluorescence and calorimetric studies of phase transitions in phophatidylcholine multilayers: Kinetics of the pretransition, *Biochemistry 17*:4475–4482 (1978).
96. B. Grünewald, S. Stankowski, and A. Blume, Curvature influence on the cooperativity and the phase transition enthalpy of lecithin vesicles, *FEBS Letters 102*:227–229 (1970).
97. A. G. Petrov, and I. Bivas, Elastic and flexoelectric aspects of out-of-plane fluctuations in biological and model membranes, *Progr. Surf. Sci. 16*:389–511 (1984).
98. J. Käs and E. Sackmann, Shape transitions and shape stability of giant phospholipid vesicles in pure water induced by area-to-volume changes, *Biophys. J. 60*:825–844 (1991).
99. D. Needham, T. J. McIntosh, and E. Evans, Thermomechanical and transition properties of dimyristoylphosphatidylcholine/cholesterol bilayers, *Biochemistry 27*:4668–4673 (1988).
100. S. Ljunggren and J. K. Eriksson, Comments on the origin of the curvature elasticity of vesicle bilayers, *J. Colloid Interf. Sci. 107*:138–145 (1985).
101. T. H. Haines, W. Li, M. Green, and H. Z. Cummins, The elasticity of uniform, unilamellar vesicles of acidic phospholipids during osmotic swelling is dominated by the ionic strength of the media, *Biochemistry 26*:5439–5447 (1987).

102. G. Cevc and D. Marsh, *Phospholipid Bilayers. Physical Principles and Models*, John Wiley, New York, 1987.
103. E. A. Evans and R. M. Hochmut, in *Current Topics in Membranes and Transport*, Vol. 10, Academic Press, New York, 1978, pp. 1–64.
104. H. T. Duwe, J. Käs, and E. Sackmann, Bending elastic moduli of lipid bilayers: Modulation by solute, *J. de Phys. 51*:945–962 (1990).
105. J. P. Le Pesant, L. Powers, and P. S. Pershan, Brillouin light scattering measurement of the elastic properties of aligned multilamellar lipid samples, *Proc. Natl. Acad. Sci. USA 75*:1792–1795 (1978).
106. G. Eyring and M. D. Fayer, A laser-induced ultrasonic probe of the mechanical properties of aligned lipid multibilayers, *Biophys. J. 47*:37–42 (1985).
107. E. M. Servuss, W. Harbich, and W. Helfrich, Measurement of the curvature-elastic modulus of egg-lecithin bilayers, *Biochim. Biophys. Acta 436*:900–903 (1976).
108. S. Swaminathan and B. M. Craven, Electrostatic properties of phosphorylethanolamine at 123K from crystal diffraction, *Acta Cryst. B40*:511–518 (1984).
109. H.-P. Weber, McMullan, S. Swaminathan, and B. M. Craven, The structure and thermal motion of phosphorylethanolamine at 122K from neutron diffraction, *Acta Cryst. B40*:506–511 (1984).
110. J. Seelig, P. M. MacDonald, P. G. Scherer, Phospholipid head groups as sensors of electric charge in membranes, *Biochemistry 26*:7535–7541 (1987).
111. R. J. Hunter, *Zeta Potential in Colloid Science. Principles and Applications*, Academic Press, London, 1981.
112. D. Cafiso, A. McLaughlin, S. McLaughlin, and A. Winiski, Measuring electrostatic potential adjacent to membranes, *Meth. Enzym. 171*:342–364 (1989).
113. E. M. Egovora, A. S. Dukhin, and I. E. Svetlova, Effect of pH, surface conductivity and double layer polarization on zeta-potential, obtained from electrophoretic measurements on liposomes, *Biochim. Biophys. Acta 1104*:102–110 (1992).
114. G. Cevc and D. Marsh, Properties of the electrical double layer near the interface between a charged bilayer membrane and electrolyte solution, *J. Phys. Chem. 87*:376–382 (1983).
115. G. Cevc, The molecular mechanism of interaction between monovalent ions and polar surfaces, such as lipid bilayer membranes, *Chem. Phys. Letters 170*:283–288 (1990).
116. F. M. Raushel and J. J. Villafranca, A multinuclear nuclear magnetic resonance study of the monovalent-divalent cation sites of pyruvate kinase, *Biochemistry 19*:5481–5485 (1980).
117. J. S. Puskin and M. T. Coene, Na^+ and H^+ dependent Mn^{2+} binding to phospatidylserine vesicles as a test of the Gouy-Chapman-Stern theory, *J. Membrane Biol. 52*:69–74 (1980).
118. R. Fleming, R. H. Guy, and J. Hadgraft, Interfacial transfer kinetics of 22Na across a synthetic phospholipid protein membrane, *J. Colloid Interface Sci. 94*:54–59 (1983).
119. H. N. Halladay and M. Petersheim, Optical properties of Tb^{3+}-phospholipid complexes and their relation to structure, *Biochemistry 27*:2120–2126 (1988).
120. J. M. Bloch, W. B. Yun, X. Yang, M. Ramanathan, P. A. Montano, and C. Copasso, Adsorption of counterions to a stearate monolayer spread at the water-air interface: A synchrotron X-ray study, *Phys. Rev. Letters 61*:2941–2944 (1988).
121. M. P. Hentschel, M. Mischel, R. C. Oberthur, and G. Büldt, Direct observation of the ion distribution between charged lipid membranes, *FEBS Letters 193*:236–238 (1985).
122. H. Träuble, M. Teubner, P. Wooley, and H. Eibl, Electrostatic interactions at charged lipid membranes. I. Effects of pH and univalent cations on membrane structure, *Biophys. Chem. 4*:319–342 (1976).
123. G. Cevc, and D. Marsh, Titration of the phase transition of phsophatidylserine bilayer membranes. Effects of pH, surface electrostatics, ion binding, head-group hydration, *Biochemistry 20*:4955–4965 (1981).
124. G. Cevc, How membrane chain melting properties are regulated by the polar surface of the lipid bilayer, *Biochemistry 26*:6305–6310 (1987).

125. G. Cevc, Isothermal lipid phase transitions, *Chem. Phys. Lipids 57*:293–307 (1991).
126. J. L. Venselaar, A. J. Kruger, L. M. Verbakel, and J. A. Poulis, The static capacitor method of measuring the effective dipole moment of surfactant molecules. *J. Colloid Interface Sci. 70*:149–152 (1979).
127. F. Paltauf, H. Hauser, and M. C. Phillips, Monolayer characteristics of some 1,2-diacyl, 1-alkyl-2-acyl, and 1,2-dialkyl phospholipids at the air-water interface. *Biochim. Biophys. Acta 249*:539–547 (1971).
128. J. M. Smaby and H. L. Brockman, Surface dipole moments of lipids at the argon-water interface, *Biophys. J. 58*:195–204 (1990).
129. G. Cevc, Membrane electrostatics, *Biochim. Biophys. Acta. Reviews on Membranes*:1031-3, 311–382 (1990).
130. I. Sakurai and Y. Kawamura, Lateral electrical conductivity along a phosphatidylcholine monolayer, *Biochim. Biophys. Acta 904*:405–409 (1987).
131. P. B. Hitchcock, R. Mason, K. M. Thomas, and G. G. Shipley, Structural chemistry of 1,2-dilauroyl-DL-phosphatidylethanolamine: Molecular conformation and intermolecular packing of phospholipids, *Proc. Natl. Acad. Sci. USA 71*:3036–3040 (1974).
132. I. Pascher and S. Sundell, Membrane lipids: prefered conformational states and their interplay. The crystal structure of dilauroylphosphatidyl-*N,N*-dimethylethanolamine, *Biochim. Biophys. Acta 855*:68–78 (1986).
133. G. Büldt, H. U. Gally, J. Sellig, and G. Zaccai, Neutron diffraction studies on phosphatidylcholine model membranes. I. Head group conformation, *J. Mol. Bio. 134*:637–691 (1979).
134. G. Zaccai, G. Büldt, A. Seelig, and J. Seelig, Neutron diffraction studies on phosphatidylcholine model membranes. II. Chain conformation and segmental order, *J. Mol. Biol. 134*:693–706 (1979).
135. H. Hauser, W. Guyer, and M. Spiess, The polar group conformation of a lysophosphatidylcholine analogue in solution. A high-resolution nuclear magnetic resonance study, *J. Mol. Biol. 137*:265–282 (1980).
136. J. Forbes, C. Husted, and E. Oldfield, High-field, high-resolution proton 'magic-angle' sample spinning nuclear magnetic resonance spectroscopic studies of gel and liquid crystalline lipid bilayers and the effects of cholesterol, *J. Am. Chem. Soc. 110*:1059–1065 (1988).
137. C. W. B. Lee and R. G. Griffin, Two dimensional 1H/13C heteronuclear chemical shift correlation spectroscopy of lipid bilayers, *Biophys. J. 55*:355–358 (1989).
138. A. Tardieu, V. Luzzati, and F. C. Reman, Structure and polymorphism of the hydrocarbon chains of lipids, A study of lecithin-water phases, *J. Mol. Biol. 75*:711–733 (1973).
139. M. C. Wiener, R. M. Suter, and J. F. Nagle, Structure of the fully hydrated gel phase of dipalmitoylphosphatidylcholine, *Biophys. J. 55*:315–325 (1989).
140. D. G. Casal, H. L., Cameron, E. F. Gudgin, and H. H. Mantsch, The gel phase of dipalmitoyl phosphatidylcholine. An infrared characterization of the acyl chain packing, *Biochim. Biophys. Acta 596*:463–467 (1980).
141. M. J. Janiak, D. M. Small, and G. G. J. Shipley, Temperature and compositional dependence of the structure of hydrated dimyristoyl lecithin, *J. Biol. Chem. 254*:6068–6078 (1979).
142. M. Stelzle and E. Sackmann, Sensitive detection of protein adsorption to supported lipid bilayers by frequency dependent capacitance measurements and microelectrophoresis, *Biochim. Biophys. Acta 981*:135–142 (1989).
143. J. F. Nagle and M. C. Wiener, Relations for lipid bilayers. Connection of electron density profiles to other structural quantities, *Biophys. J. 55*:309–313 (1989).
144. R. Hosemann, M. Hentschel, and W. Helfrich, Direct x-ray study of the molecular tilt in dipalmitoyl lecithin bilayers, *Z. Naturforschung 35a*:643–644 (1980).
145. H. L. Casal, On the water content of micelles: Infrared spectroscopic studies, *J. Am. Chem. Soc. 110*:5203–5205 (1988).
146. J. M. Smaby, A. Hermetter, P. C. Schmidt, F. Paltauf, and H. L. Brockman, Packing of ether and ester phospholipids in monolayers. Evidence for hydrogen bonded water at the sn-1 acyl group of phosphatidylcholine, *Biochemistry 22*:5808–5813 (1983).

147. D. C. Wack and W. W. Webb, Synchrotron x-ray study of the modulated lamellar phase P-beta in the lecithin water system, *Phys. Rev. A 40*:2712–2730 (1989).
148. S. Simon and T. D. McIntosh, Surface ripples cause the large fluid spaces between gel phase bilayers containing small amounts of cholesterol, *Biochim. Biophys. Acta 1064*:69–74 (1991).
149. E. Sackmann, D. Rueppel, and C. Gebhardt, Defect structure and texture of isolated bilayers of phospholipids and phopholipid mixtures, in *Springer Series in Chemical Physics 11* (W. Helfrich and G. Heppke, eds), Springer Verlag, Berlin, 1980, pp. 309–326.
150. G. Cevc, Electrostatic characterization of liposomes, *Chem. Phys. Lipids:* in press (1993).
151. A J. B. M. van Renswoude, R. Blumenthal, and J. N. Weinstein, Thin-layer chromatography with agarose gels—A quick, simple method for evaluating liposome size, *Biochim. Biophys. Acta 595*:151–156 (1980).
152. R. Gibrat, C. Romieu, and C. Grignon, A procedure for estimating the surface potential of charged or neutral membranes with 8-anilino-1-naphthalenesulphonate probe. Adequacy of the Gouy-Chapman model, *Biochim. Biophys. Acta 736*:196–202 (1983).
153. H. Frischleder, Monte Carlo and molecular dynamics simulation of the structure and dynamical properties of molecules in phospholipid/water lamellar systems, in *Sixth School on Biophysics of Membrane Transport,* Poland, 1981.
154. A. Llor and P. Rigny, Some tentative models of molecular motion applied to water in small reversed micelles, *J. Am. Chem. Soc. 108*:7533–7541 (1986).
155. J. Teissie, Fluorescence temperature jump relaxations of dansyl-phosphatidylethanolamine in aqueous dispersions of dipalmitoylphosphatidylcholine during the gel to liquid-crystal transition, *Biochim. Biophys. Acta 555*:553–557 (1979).
156. E. J. Luna and H. M. McConnell, Lateral phase separations in binary mixtures of phospholipids having different charges and different crystalline structures, *Biochim. Biophys. Acta 470*:303–316 (1977).
157. J. Daillant, L. Bosio, J. J. Benattar, and J. Meunier, Capillary waves and bending elasticity of monolayers on water studied by X-ray reflectivity as a function of surface pressure, *Europhys. Letters 8*:453–458 (1989).
158. R. C. Gamble and P. R. Schimmel, Nanosecond relaxation processes of phospholipid bilayers in the transition zone. *Proc. Natl. Acad. Sci. USA 75*:3011–3014 (1978).
159. D. L. Melchior, F. J. Scavitto, and J. M. Steim, Dilatometry of dipalmitoyl-lecithin-cholesterol bilayers, *Biochemistry 19*:4828–4834 (1980).
160. D. A. Wilkinson and J. F. Nagle, Dilatometry and calorimetry of saturated phosphatidylethanolamine dispersions, *Biochemistry 20*:187–192 (1981).
161. I. Hatta, K. Suzuki, and S. Imaizumi, Pseudo-critical heat capacity of single lipid bilayers, *J. Phys. Soc. Jpn. 52*:2790–2797 (1983).
162. A. Milon, J. Ricka, S.-T., Sun, T. Tanaka, Y. Nakatani, and G. Ourisson, Precise determination of the hydrodynamic radius of phospholipid vesicles near the phase transition, *Biochim. Biophys. Acta 777*:331–333 (1984).
163. R. Lawaczeck, M. Kainosho, J.-L., Girardet, and S. I Chan, Effects of structural defects in sonicated phospholipid vesicles on fusion and ion permeability, *Nature 256*:584–586 (1975).
164. J. Suurkuusk, Y. Barenholz, and T. E. Thompson, Calorimetric and fluorescent probe study of the gel-liquid crystalline phase transition in small, single-lamellar dipalmitoylphosphatidylcholine vesicles, *Biochemistry 15*:1393–1401 (1976).
165. E. Meirovitch and J. H. Freed, Effects of defect annealing on concentration-dependent ESR spectra from hydrated dimyristoylphosphatidylcholine, *J. Phys. Chem. 85*:1617–1620 (1981).
166. J. L. Rubenstein, B. A. Smith, and H. M. McConnell, Lateral diffusion in binary mixtures of cholesterol and phosphatidylcholines, *Proc. Natl. Acad. Sci. USA 76*:15–18 (1979).
167. R. W. Fisher and T. L. James, Lateral diffusion of the phospholipid molecule in dipalmitoylphosphatidylcholine bilayers: An investigation using nuclear spin-lattice relaxation in the rotating frame; studies of lateral diffusion in PC, *Biochemistry 17*:1177–1183 (1978).

168. R. D. Koynova and B. G. Tenchov, Effect of ion concentration on phosphatidylethanolamine distribution in mixed vesicles, *Biochim. Biophys. Acta 727*:351–356 (1983).
169. M. A. Roseman and T. E. Thompson, Mechanism of the spontaneous transfer of phospholipids between bilayers, *Biochemistry 19*:439–444 (1980).
170. M. D. Cuyper and M. Joniau, Spontaneous intervesicular transfer of anionic phospholipids differing in the nature of their polar headgroup, *Biochim. Biophys. Acta 814*:374–380 (1985).
171. T. M. Bayerl, C. F. Schmidt, and E. Sackmann, Kinetics of symmetric and asymmetric phospholipid transfer between small sonicated vesicles studied by high-sensitivity differential scanning calorimetry, NMR, electron microscopy and dynamic light scattering, *Biochemistry 27*:6078–6085 (1988).
172. M. Eisenberg, T. Gresalfi, T. Riccio, and S. McLaughlin, Adsorption of monovalent cations to bilayer membranes containing negative phospholipids, *Biochemistry 18*(23):5213–5223 (1979).
173. L. Herbette, C. A. Napolitano, and R. V. McDaniel, Direct determination of the calcium profile structure for dipalmitoyllecithin multilayers using diffraction, *Biophys. J. 46*:677–685 (1984).
174. C. F. Schmidt, R. M. Zimmerman, and N. H. E. Gaub, Multilayer adsorption of resorcin on the hydrophobic substrate, *Biophys. J. 57*:577–588 (1990).
175. R. P. Rand, Interacting phospholipid bilayers: Measured forces and induced structural changes, *Ann. Rev. Biophys. 10*:277–314 (1981).
176. J. Marra and J. Israelachvili, Direct measurements of forces between phosphatidylcholine and phosphatidylethanolamine bilayers in aqueous electrolye solutions, *Biochemistry 24*:4608–4618 (1985).
177. S. Ohki, N. Duezguenes, and K. Leonards, Phospholipid vesicle aggregation: Effect of monovalent and divalent ions, *Biochemistry 21*:2127–2129 (1982).
178. G. Cevc, J. M. Seddon, R. Hartung, and W. Eggert, Properties of phosphatidylcholine-fatty acid bilayer membranes I. Effects of protonation, salt, temperature, and chain length on the colloidal and phase behavior, *Biochim. Biophys. Acta 940*:219–240 (1988).
179. G. G. Roberts, An applied science perspective of Langmuir-Blodgett films, *Adv. Phys. 34*:475–512 (1985).
180. G. S. Smith, C. R. Safinya, D. Roux, and N. A. Clark, X-ray study of freely suspended films of a multilamellar lipid system, *Mol. Cryst. Liq. Cryst. 144*:235–255 (1987).
181. Y. Rançon and J. Charvolin, Epitaxial relationships during phase transformations in a lyotropic liquid crystal, *J. Phys. Chem. 92*:2646–2651 (1988).
182. F. E. M. O'Brien, The control of humidity by saturated salt solutions, *J. Sci. Instrum. 25*:73–76 (1948).
183. S. M. Gruner, Controlled humidity gas circulators, *Rev. Sci. Instrum. 52*:134–136 (1981).
184. H. Jousma, H. Talsma, F. Spies, J. G. H. Joosten, H. E. Junginger, and D. J. A. Crommelin, The influence of extrusion of multilamellar vesicles through polycarbonate membranes on particle size, particle size distribution and number of bilayers, *Int. J. Pharmac. 35*:263–274 (1987).
185. C. P. Yang, M. C. Wiener, and J. F. Nagle, New phases of DPPC/water mixtures, *Biochim. Biophys. Acta 945*:101–104 (1988).

12

Lipid Polymorphism: Structure and Stability of Lyotropic Mesophases of Phospholipids

John M. Seddon *Imperial College, London, England*

Gregor Cevc *Technical University of Munich, Munich, Germany*

I. INTRODUCTION

The first part of this chapter (Sec. II) is concerned with the identification of the mesophases adopted by phospholipids and the determination of their structures (the experimental techniques that are used to derive this information have been covered in the previous chapter). The second part (Sec. III) deals with the forms of the phase diagrams and the transitions between different phases, and the factors that control their relative stabilities. These phases range from lamellar crystals exhibiting full three-dimensional translational and orientational ordering, liquid crystalline phases exhibiting varying degrees of order, to micellar solutions in which little or no ordering occurs. The aim is to give a qualitative yet relatively complete overview of the structures of the known phases formed by amphiphilic molecules. In addition, the various factors that control the phase stability will be described qualitatively. This chapter is in no way intended as a comprehensive review of this vast subject; it is merely intended to serve as an introduction to the area.

A number of reviews of the structures of lipid phases are available [1–10], and these should be consulted for further details. A recent issue of *Chemistry and Physics of Lipids* [11] was devoted to the subject of lipid polymorphism, and an issue of *Journal de Physique* [12] to geometry and interfaces, relating mainly to lyotropic liquid crystals. Recent books have dealt with phospholipid bilayers [13] and the physics of amphiphilic layers [14]; theoretical approaches to membrane conformations have been recently reviewed [15, 16].

II. STRUCTURE OF LYOTROPIC MESOPHASES OF LIPIDS

Amphiphiles have a unique ability to form structures that combine long range periodicity in one, two, or three dimensions with short-range disorder. This feature is unusual, as long range order is normally propagated through short range order. Luzzati and coworkers carried out the first detailed structural studies of the many phases formed by anhydrous and hydrated amphiphiles [1], and even today much of our knowledge of lyotropic phase structures [17] stems from the work of this research group.

A. Phase Identification

Diffraction methods, in particular x-ray scattering, are the most reliable way of carrying out lipid phase identification. However, many other technqiues have been used, although most of them can under certain circumstances lead to incorrect assignments. Freeze-fracture electron microscopy is an attractive approach, although it is very difficult to ensure that the freezing procedure does not lead to artifacts. The best way of controlling this is by carrying out x-ray diffraction experiments on the sample, before and after freezing it [18]. Even with this precaution, the resolution is not quite high enough to solve the structure unambiguously.

Polarizing microscopy can provide a useful and rapid way of measuring phase transition temperatures, of tentatively identifying phases, and of gaining a qualitative view of phase sequences as a function of water content. The latter information is gained by carrying out penetration experiments, whereby a water concentration gradient is established across a thin sample sandwiched between two microscope slides. However, in practice the technique works best with simple surfactants, where the characteristic optical textures develop readily, and is less successful with phospholipids.

Various spectroscopic techniques (NMR, ESR, etc.) have been used extensively by certain authors for identification of lipid phases. Although it is inherently dangerous to infer the long range organization of a phase from such techniques, which measure single molecule properties, applied with caution they may usefully complement diffraction studies.

The characterization of lipid mesophases by diffraction is based firstly on symmetry [1]. There are two basic regions of the diffraction pattern that are used to identify the phase. The low angle region identifies the symmetry and long range organization of the phase, whereas the wide angle region gives information on the molecular packing, or short range organization of the phase.

Aggregation of amphiphilic molecules into micelles leads to diffuse scattering in the small angle region. If the positions of the aggregates are completely independent of each other (e.g., very dilute micellar solutions), then the observed scattering is simply the sum of the scattering from each individual aggregate. For a monodisperse solution (i.e., each aggregate being identical), the form of this scattering curve can be calculated from a model structure for the aggregate. With increasing concentration of the amphiphile, the micelles will start to interact (for example through long range electrostatic forces, or shorter range hydration forces) and hence influence each others' mutual positions. This causes the diffuse scattering to be modulated by an interference term arising from the nonuniform radial distribution function of the aggregates. The stronger the interactions, the greater the extent of correlation in the mutual positions of the aggregates, and the more sharply peaked the scattering becomes. Note that the aggregate size and shape may either stay essentially constant or change radically with increasing concentration, depending on the particular lipid, its chain length, or the nature of any associated counterions. In any event, at some point the repulsive interactions are strong enough to lead to an ordering of aggregates onto a lattice, forming one of the translationally ordered liquid crystalline phases. At this transition, the aggregates may simply become ordered on the lattice, or they may simultaneously radically change their size and shape. In any event, the signature for the observation of a translationally ordered mesophase is the appearance of one or more sharp (Bragg) peaks in the low angle region.

The long range translational ordering of the lipid-water aggregates (bilayers, cylinders, micelles, etc.) onto one-, two-, or three-dimensional lattices gives rise to Bragg reflections whose reciprocal spacings ($s_{hkl} = 1/d_{hkl}$) are in characteristic ratios, for example:

Lamellar: $s_1 = l/d$ (Ratios 1, 2, 3, 4 . . .)

Hexagonal: $s_{hk} = 2(h^2 + k^2 - hk)^{1/2}/\sqrt{3}\,a$ (1, $\sqrt{3}$, 2, $\sqrt{7}$, 3. $\sqrt{12}$, $\sqrt{13}$ . . .)

Cubic: $s_{hkl} = (h^2 + k^2 + l^2)^{1/2}/a$ (1, $\sqrt{2}$, $\sqrt{3}$, 2, $\sqrt{5}$, $\sqrt{6}$, $\sqrt{8}$, 3 . . .)

Schematic illustrations of the unaligned ("powder") diffraction patterns from these phases are shown in Fig. 1. The intensities (moduli-squared of the scattered amplitudes) of the various Bragg peaks are determined by the distribution of matter (electron density) in the

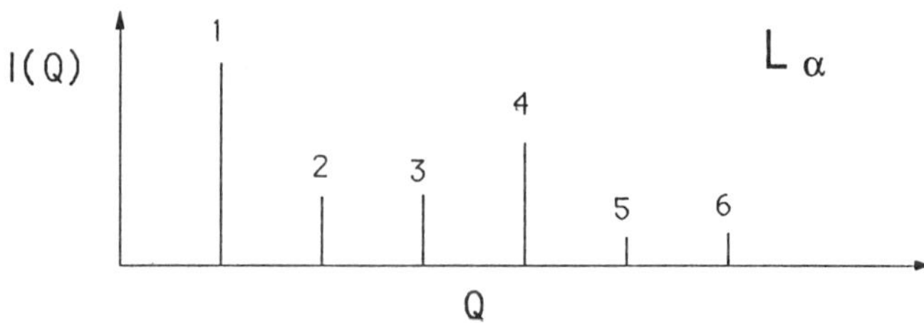

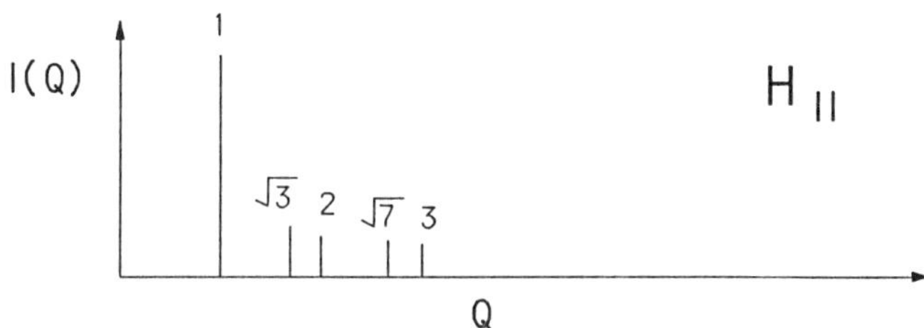

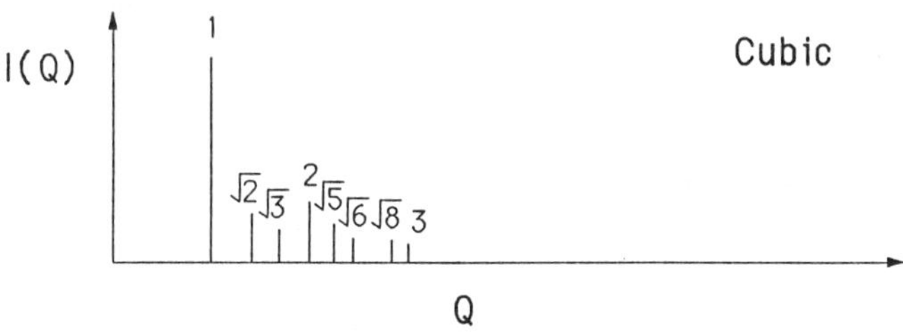

Figure 1 Schematic low-angle diffraction patterns of lamellar, hexagonal, and cubic phases $Q_{hkl} = 2\pi s_{hkl}$.

unit cell. The scattered amplitudes are essentially given by the Fourier transform of this electron density distribution. Not all of the reflections may be observed for a particular sample: some may be absent for additional symmetry reasons. For example, a body-centered cubic phase only gives reflections for which $h + k + l = 2n$ (i.e., even); additional symmetry elements within a particular crystallographic space group may give rise to other systematic absences. Furthermore, a symmetry-allowed reflection may nonetheless have zero intensity because the unit cell Fourier transform happens to pass through zero at that particular diffraction angle. It is thus of paramount importance in structural studies of lipid mesophases to ensure that the symmetry of the phase is correctly identified. Unfortunately, this is often not so trivial as it appears, because usually only a few diffraction orders are detected, due to the large thermal disorder inherent in liquid crystalline phases. There are many examples in the literature where incorrect assignments appear to have been made. Note that if the $\sqrt{3}$ and $\sqrt{7}$ reflections of a hexagonal phase are too weak to be observed, the pattern might be incorrectly indexed as lamellar; similarly, a cubic phase whose $\sqrt{2}$ (110) reflection was weak or missing could be mistaken for a hexagonal phase, if only the first three reflections were detected (it is by no means unusual for so few reflections to be observed). One way to guard against such errors is to explore around a particular point in the phase diagram, checking that the diffraction data index consistently as the temperature and composition are altered within the single phase region. In addition, other techniques such as polarizing microscopy and NMR can be invaluable for corroborating the phase identification.

A major problem with the study of lipid phases is the difficulty of ensuring that the sample is at equilibrium. In part this may be due to the very slow rate at which a phase comes to equilibrium. However, a further problem is that the phase itself may be metastable, reverting to more stable forms over a time scale that can span seconds to months. For example, both gel and fluid lamellar phases of phosphatidylethanolamines are metastable within certain temperature ranges and will spontaneously convert to lamellar crystals on incubation [19].

B. Topology Determination

Apart from the lamellar L_α phase, all of the lyotropic liquid-crystalline phases may potentially occur either as type I (normal topology, oil-in-water), or as type II (inverse topology, water-in-oil) structures. It is essential to determine which type one is dealing with, but this is by no means trivial. In order to establish the topology of a lipid phase, a number of approaches may be employed [1,9].

1. There is a strong rule, obeyed for almost every lipid system, that the area S always increases or stays constant, but never decreases, with increasing water content. This rule is essentially simply an expression of the fact that the favorable interaction between water and the amphiphile polar headgroup normally causes a lateral expansion of the headgroups. If the interfacial area per molecule S is plotted versus water concentration, it often stays rather constant for a type I phase but increases dramatically for a type II structure. This may be understood qualitatively in the following way: a lattice of lipid rods in a water matrix can (in principle) swell at constant surface area (as can the lamellar phases); however an inverse phase such as H_{II} must necessarily undergo an increase in interfacial area as increasing amounts of water penetrate the rods comprising the polar groups. When data are plotted assuming the incorrect topology, the wrong dependence on water content is usually observed.

2. Observation of a fluid nonlamellar phase in coexistence with an excess aqueous phase is by itself strong evidence in favor of an inverted type II structure; type I phases of lipid systems almost invariably break up into micellar solutions beyond a certain limiting water content.
3. If a phase of unknown topology is adjacent in the phase diagram to a nonlamellar phase of known topology, then it almost certainly has the same topology. For example, bicontinuous cubic phases occurring between L_α and H_{II} may be safely inferred to be inverse.
3. The value of interfacial area S, measured at the (effective) water-lipid interface, is normally lower for an inverse topology phase than for an adjacent L_α phase, when the transition is driven either by varying the composition or by varying the temperature. On the other hand, normal topology phases normally have larger values of S than L_α.
4. Analysis of the intensities of the Bragg peaks with those calculated from models of the structure, for a range of water contents, can provide unambiguous evidence for one or the other topology.
5. In principle it should be possible to probe phase topology by employing neutron diffraction contrast variation techniques, although as yet this has been little employed in structural studies of lipid polymorphism.

C. Phase Dimensions

A great deal of useful structural information can be deduced simply from the positions of the diffraction lines, in conjunction with the chemical parameters such as the lipid molecular weight M_L, the water and lipid partial volumes v_w and v_L, and the lipid weight concentration c_L (lipid/(lipid + water)). The partial specific volumes may be measured by using either a neutral bouyancy technique or an oscillating tube method. The volume concentration (fraction) of lipid is then

$$\phi_L = \left[1 + \frac{v_W}{v_L}\left(\frac{1-c}{c}\right)\right]^{-1} \tag{1}$$

and the volume concentration of water is $\phi_W = 1 - \phi_L$.

For lamellar phases, the bilayer and water layer thicknesses are related to the measured d spacings by

$$d_L = d\phi_L \quad \text{and} \quad d_w = d\phi_W \tag{2}$$

The interfacial area per molecule is given by

$$S = \frac{2Mv_L}{d_L N_A} \tag{3}$$

For the gel phases, the lattice parameter of the hexagonal chain packing is given from the wide angle d spacing by

$$a = \frac{2}{\sqrt{3}} d_{\text{chain}} \tag{4}$$

and the cross-sectional area per chain by

$$S_{\text{chain}} = \frac{2}{\sqrt{3}} d^2_{\text{chain}} \tag{5}$$

The angle of the tilt of the (stiff) chains to the bilayer normal is then given (for a diacyl lipid) by

$$\theta_t = \cos^{-1}\left(\frac{2S_{\text{chain}}}{S}\right) \tag{6}$$

For the H_{II} phase (see Fig. 7), the lattice spacing is related to the d spacing (the (10) Bragg peak) by

$$a = \frac{2}{\sqrt{3}} d \tag{7}$$

and the diameter of the water cylinders is given by

$$d_w = \left[\frac{2\sqrt{3}}{\pi}(1 - \phi_L)a^2\right]^{1/2} \tag{8}$$

The area per molecule at the lipid-water interface is given by

$$S = \frac{2\pi d_w M v_L}{\sqrt{3}\, a^2 \phi_L N_A} \tag{9}$$

For more complex phase structures, estimates of the aggregate dimensions and areas can similarly be derived, although they will vary in different regions of the unit cell of the phase.

Note that for the lamellar phases, the interfacial area per molecule is unambiguous and is constant across the layer. However, for all nonlamellar phases, it varies with position in the phase, and it is necessary to define the position, usually either at the polar-nonpolar interface or at the lipid-water interface. Both of these interfaces, however, are not sharply defined (there is a partial interpenetration of the polar lipid headgroups with some of the water), and so a degree of ambiguity remains.

The widely used operational approach proposed by Luzzati assumes that

1. The aggregates consist of simple, uniform shapes, e.g., flat bilayers and circular cylinders in the lamellar and hexagonal phases, respectively.
2. All of the water is segregated from the lipid within each phase.
3. The densities of the components within the phase are equal to their bulk values at the same temperature and pressure.

These assumptions are clearly only approximations to the real situation, but the approach has nevertheless proved to be extremely useful.

D. Lyotropic Phase Structures

1. Nomenclature

As yet there is no universally accepted nomenclature for lyotropic mesophases, although the most comprehensive and widely used system is that proposed by Luzzati [1], and this will be used throughout this chapter. The lattice type is denoted by a capital letter, L for lamellar, H for hexagonal, and Q for cubic. Subscripts I and II are used to denote normal (oil in water) or reversed (water in oil) type phases. A Greek subscript is used to denote the chain conformation: β for ordered gellike, α for liquidlike, $\alpha\beta$ for coexisting gellike and liquidlike regions, and δ for a helically coiled chain conformation. In addition, the subscript c is frequently used to denote a crystalline packing.

A list of the well-established lyotropic mesophases is given in Table 1.

2. Solidlike Lamellar Phases

(a) 3-D lamellar crystals. One or more crystalline L_c phases are formed at low temperatures and/or hydrations by essentially all phospholipids. These phases exhibit both long and short range order in three dimensions and are therefore true crystals. They are

Table 1 Principal Lyotropic Mesophases

Solidlike lamellar phases		
Type	Name	Phase structure
3-D	L_c	3-D crystal
2-D	L_c^{2D}	2-D crystal
	$P_{\beta'}$	Rippled gel
	P_δ	Ordered ribbon phase
	B	Ordered ribbon phase?
1-D	L_β	Untilted gel
	$L_{\beta'}$	Tilted gel
	$L_{\beta I}$	Interdigitated gel
	$L_{\alpha\beta}$	Partial gel
Fluid phases		
Type	Name	Phase structrure
1-D	L_α	Fluid lamellar
2-D	H	Hexagonal
	H^c	Complex hexagonal
	R	Rectangular
	M	Oblique
3-D	Q	Cubic
	T	Tetragonal
	R	Rhombohedral
	O	Orthorhombic

sometimes anhydrous, but they may also contain a number of water molecules of cocrystallization. The primary technique for investigating the molecular conformation and intermolecular interactions in the solid state is x-ray crystallography. Although experimental difficulties are encountered in obtaining sufficiently good single crystals of phospholipids, a number of structures have been solved.

The structures of the lamellar crystalline phases of dilauroyl phosphatidylethanolamine (DLPE) [20], dilauroyl dimethylphosphatidylethanolamine (2M-DLPE) [21], and dimyristoyl phosphatidylcholine (DMPC) [22] are shown in Fig. 2. The structures show that methylation of the phospholipid terminal ammonium group has a profound effect on the molecular packing. All three lipids pack into bilayers, but for DLPE a tight network of headgroup-headgroup hydrogen bonds is formed, with the headgroups parallel to the plane of the layer, whereas DMPC headgroups (lacking any donor groups) interact via bridging water molecules. Surprisingly, the intermediate dimethylated compound 2M-DLPE exhibits headgroup interdigitation between adjacent layers, with no cocrystallised water.

For DLPE the lateral packing requirements of the two hydrocarbon chains and the headgroup are nearly matched, and so there is little chain tilt. On the other hand, the phosphorylcholine group of DMPC requires a headgroup area of 47–54 Å^2, which is considerably larger than the cross-section of two crystalline hydrocarbon chains. To accommodate both polar and nonpolar regions, the bilayer packing has to be adjusted. This is achieved by a tilting of the chains relative to the bilayer normal. For 2M-DLPE the headgroup interdigitation leads to a cross-sectional area per molecule of 45.2 Å^2, and the hydrogen chains tilt at 33° to accommodate this increased molecular area.

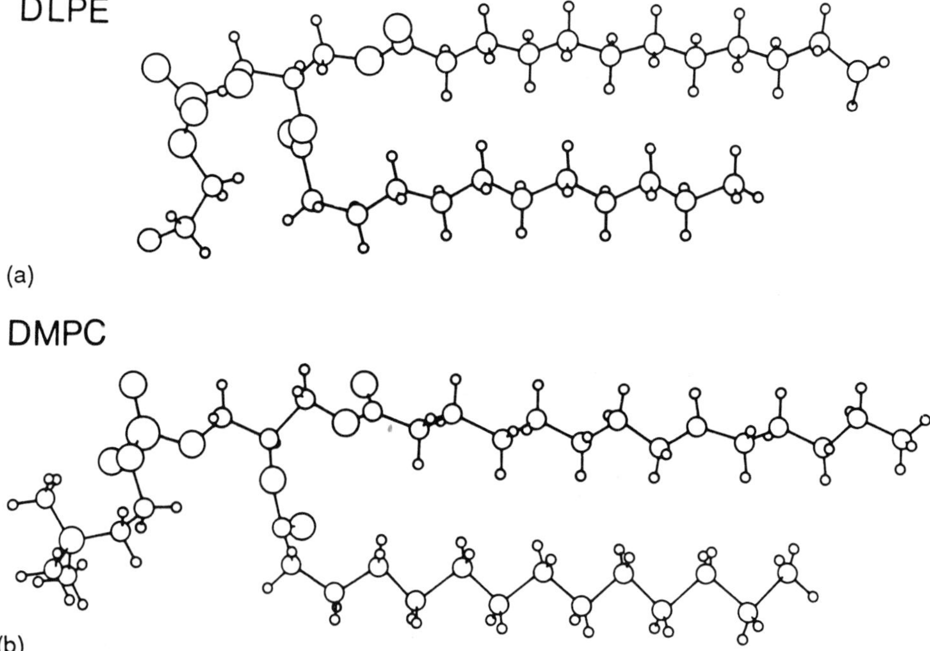

Figure 2 Crystal structures of phospholipids: (a) DLPE · acetic acid; (b) DMPC dihydrate; (c) *N,N*-dimethyl DLPE. (From Refs. 243,21.)

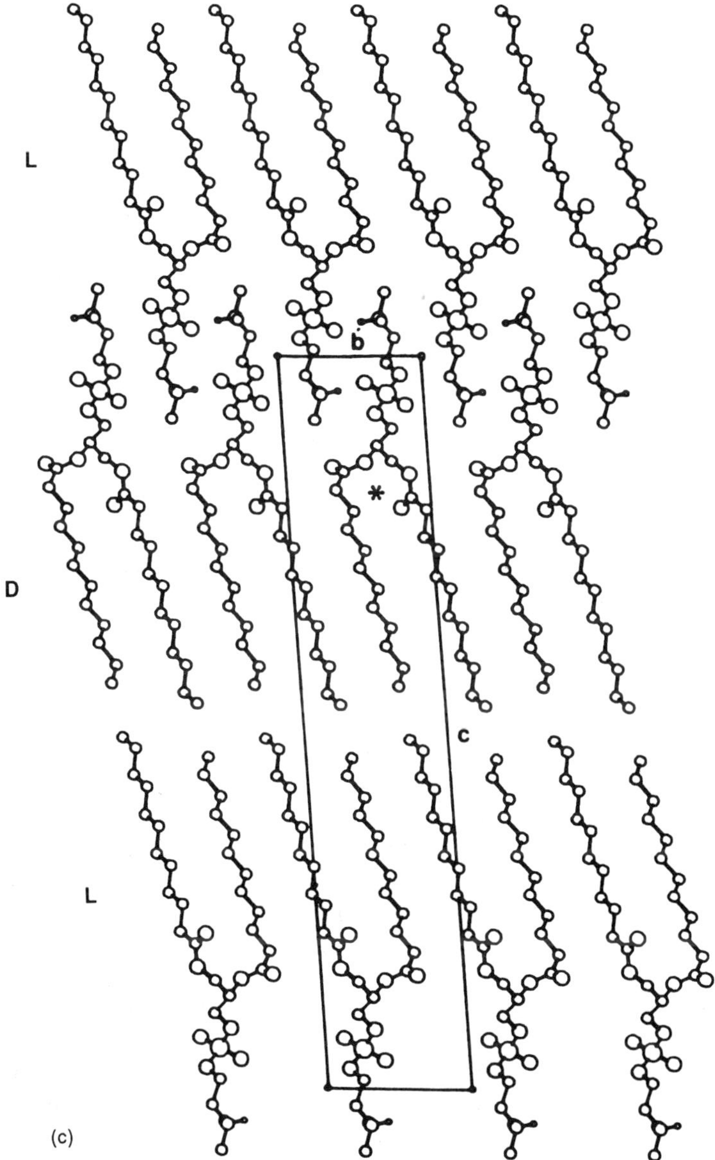

(b) 2-D lamellar crystals. Recently the sub-gel, a more ordered phase existing below the normal gel phase, has been the subject of much interest. It consists of crystalline sheets that are not regularly stacked and has been termed a two-dimensional crystal [23]. For a charged lipid such as phosphatidylglycerol (PG), the crystalline bilayers can be swollen apart by electrostatic repulsion [23].

(c) 2-D modulated ordered lamellar phases. A number of phases exist that are partially disordered, and that exhibit translational ordering in two dimensions. They are all based upon modulated lamellar structures.

The $P_{\beta'}$ ripple phase, shown in Fig. 3a, occurs below the L_α phase with temperature and has been observed in phosphatidylcholines [24] and phosphatidylglycerol at neutral pH [25], and in phosphatidylethanolamine [26] and phosphatidic acid [27] at high pH.

The lattice is usually oblique (2-D space group p2, No. 2), and the phase contains rippled lamellae. The chains are essentially in the tilted gel-like β' conformation. The phase P_δ (2-D space group cmm, No. 9) is found in the anhydrous side of PC phase diagrams [24]. In this phase, ribbon-like strips of bilayer are packed onto a 2-D centered rectangular lattice. The hydrocarbon chains have the unusual helically coiled δ conformation. For anhydrous saturated phosphatidylethanolamines, a phase (denoted B in [28] occurs between the crystalline and inverse hexagonal phases. Its structure has not yet been elucidated but is likely to bear some similarity to the P_δ phase.

(d) 1-D ordered lamellar phases. In the 1-D ordered lamellar phases, the molecules are arranged into bilayers, which stack into a multilayer with layers of water between each bilayer. The bilayers are equidistant and parallel to each other but exhibit little or no lateral correlation with adjacent layers. The thickness of the water layer depends on factors such as the water content, temperature, headgroup size, polarity, and charge. Within the hydrocarbon region the chain packing can be gellike (β), partially gellike ($\alpha\beta$), or helically coiled (δ).

In the gel phases the chains are stiff, essentially fully extended and packed onto a two-dimensional hexagonal lattice and undergo hindered rotation (on a timescale of 100 nsec) about their long axis. The hydrocarbon chains can be either parallel to the bilayer normal (L_β) (Fig. 3b), tilted ($L_{\beta'}$) (Fig. 3c) or interdigitated ($L_{\beta I}$) [29] (Fig. 3d). The diffraction patterns of the L_β and $L_{\beta I}$ phases show a single fairly sharp peak in the wide angle region at a spacing close to 4.2 Å, which arises from the 2-D hexagonal chain lattice. This gives a value characteristically close to 21 Å^2 for the cross-sectional area per chain. No diffraction arising from the headgroups (which cannot lie on the same lattice as the chains for a diacyl lipid) is normally observed, which implies that they are disordered. This is consistent with dielectric relaxation measurements, which indicate correlation times in the gel phase of a few nanoseconds for headgroup rotation about the P–O bond to the glycerol backbone [30]. If the headgroup packing requirement is close to 42 Å^2, or twice the cross-sectional area of a gel-phase chain, then the untilted L_β phase results. However, if it is larger, then the tilted $L_{\beta'}$ phase forms to accommodate the packing mismatch. If the mismatch is extreme, then the interdigitated $L_{\beta I}$ phase may be adopted, which gives an available area close to 84 Å^2 for each headgroup.

The L_δ phase has been observed in dry phosphatidylcholines [24,31]. Again the molecules are in lamellae, but now the hydrocarbon chains are coiled into helices and are arranged on a two-dimensional square lattice. The polar headgroups are oriented perpendicular to the bilayer and are interdigitated with the opposite monolayer. The headgroups are also on a square lattice the length of which is the diagonal of the square lattice of chains.

3. Fluid Phases

(a) 1-D fluid phases (lamellar). Above a certain temperature characteristic of each system or under the combined influence of temperature and water, the chains undergo a transition from the β or β' conformation to a liquidlike conformation (L_α) (Fig. 4) in which rapid lateral diffusion occurs. There is typically an expansion of 15–30% in the interfacial area per molecule. In certain lyotropic systems, it is possible to swell the lamellar phase to extremely large spacings, by the addition of either water or oil [32]. The swelling may be driven either by electrostatic repulsion (if the layers are charged) or by thermal undulations (if the layers are quite flexible). In the case of oil swelling, the bilayer separates into two monolayers.

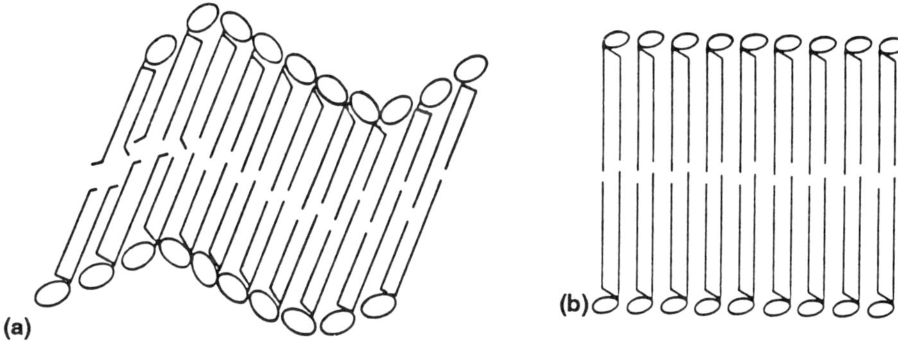

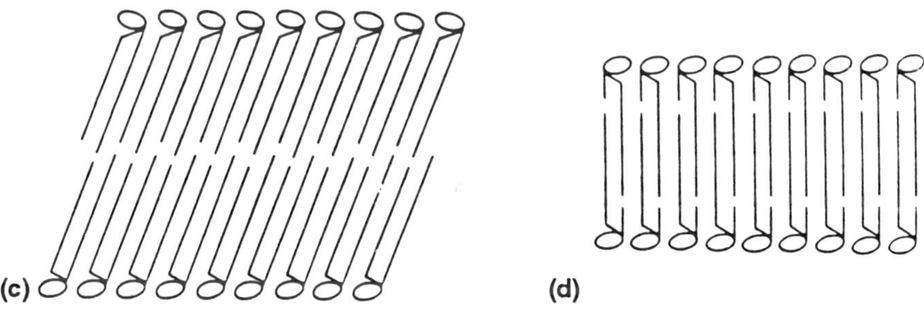

Figure 3 Gel phase structures: (a) $P_{\beta'}$; (b) L_β; (c) $L_{\beta'}$; (d) $L_{\beta I}$.

(b) 2-D fluid phases. The best established of the two-dimensional fluid phases are the normal and reverse topology hexagonal phases H_I and H_{II} (2-D space group p6m, No. 17) shown in Fig. 5. The chains are fluid; either they are contained within the cylinders, type I or normal hexagonal (H_I), or water is within the cylinders and the fluid hydrocarbon chains fill the spaces between the rods (H_{II}). The H_I phase is very common in simple surfactant systems, but it tends not to be formed by diacyl phospholipids. However, lysophospholipids frequently adopt this phase. The H_{II} phase is very common in phospholipids such as PE, having small weakly hydrated headgroups, and having attractive headgroup-headgroup interactions [9]. Although the vast majority of reported hexagonal phases are based on aggregates having a single curved lipid layer (monolayer), a more complex type, denoted H^c, has been found in certain systems, whose structure appears to be based on a hexagonal packing of cylinders formed by curved lipid bilayers [1].

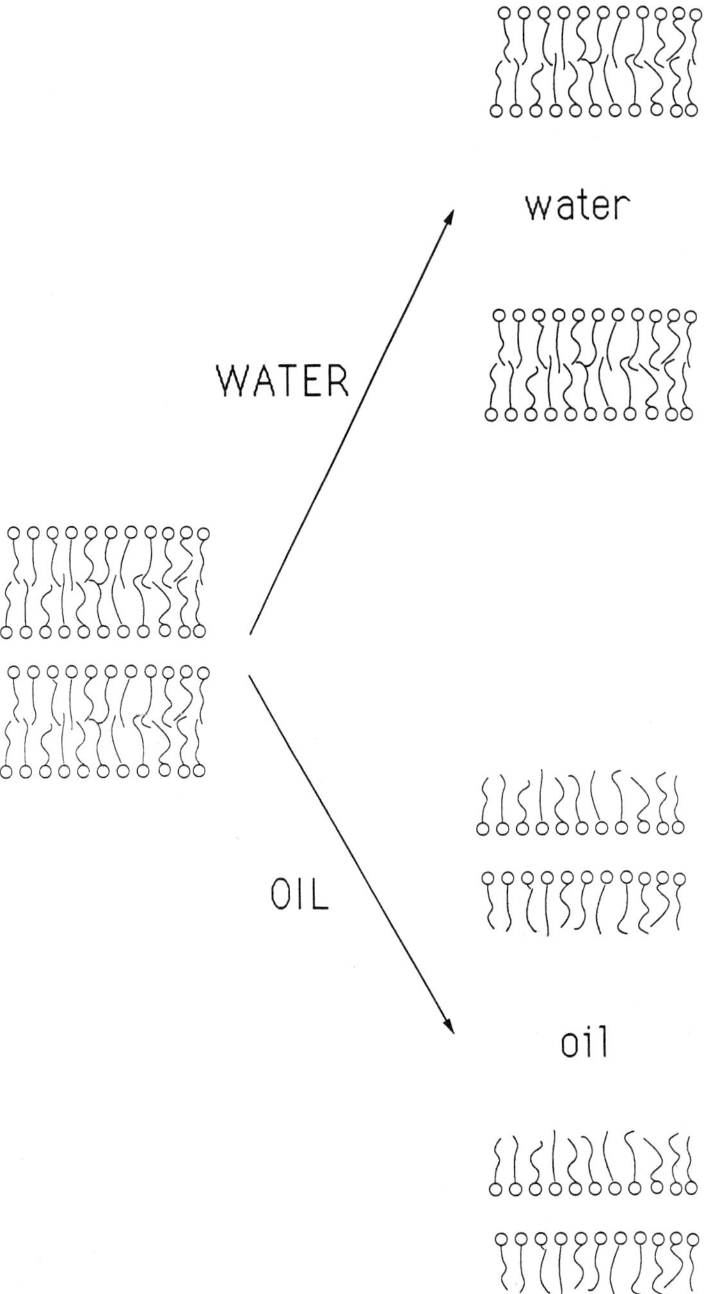

Figure 4 The fluid lamellar L_α phase, and its water and oil swollen versions.

(a)

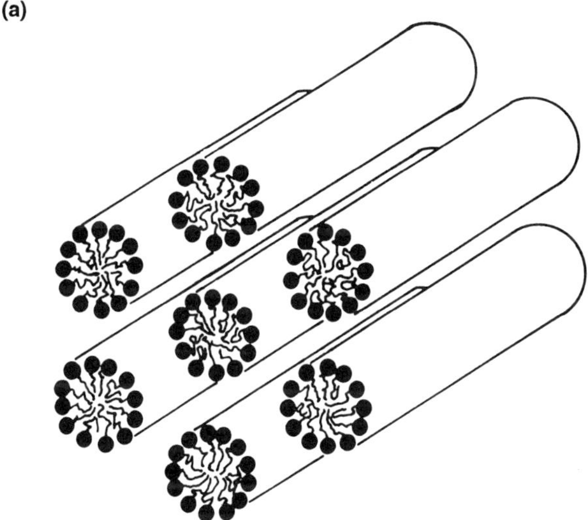

(b)

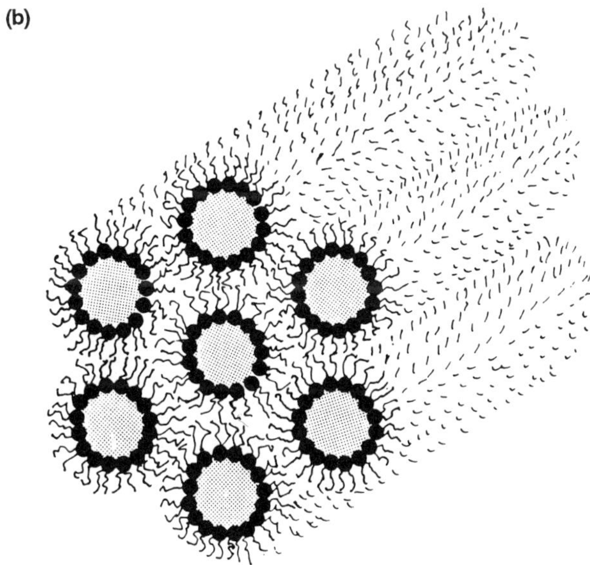

Figure 5 The normal and inverse hexagonal phases: (a) H_I; (b) H_{II}. (From Ref. 9.)

Phases similar to the H_I phase but having rectangular or oblique symmetry have been observed in certain surfactant systems such as sodium dodecyl sulphate (SDS) and water [33].

(c) 3-D fluid phases. In the 3-D fluid phases, a complex pattern of Bragg peaks is observed in the low angle region (see Fig. 1 for a schematic pattern of a cubic phase), along with a single diffuse peak in the wide angle region close to 4.6 Å, characteristic of

fluid hydrocarbon chains. The phase may be either of normal topology (oil in water) or inverse (water in oil). To characterize these phases it is necessary first to identify the phase symmetry and then to try to identify the specific crystallographic space group to which the phase belongs. This is done by comparing the characteristic absences (unobserved Bragg peaks) with those tabulated in the International Tables for Crystallography [34]. The vast majority of three-dimensional fluid phases so far detected are of cubic symmetry, although rhombohedral, tetragonal, and orthorhombic phases of inverse topology have been detected in a few lipid systems at low hydrations [1].

The rhombohedral phase (space group $R\overline{3}m$, No. 166) contains rod-like elements connected three by three in planar two-dimensional hexagonal arrays that are then stacked onto a three-dimensional rhombohedral lattice (Fig. 6a). In the tetragonal phase (space group I422, No. 97) the structure is formed from rods linked four by four into planar two-dimensional square arrays that are then stacked onto a regular three-dimensional body centered tetragonal lattice (Figure 6b). In the body-centered orthorhombic phase (space groups mmm, No. 47, or 222, No. 16) observed in certain anhydrous soap systems [1], the structure appears to consist of planar disks of polar region embedded in a hydrocarbon matrix.

Cubic phases are quite easily detected, since they are optically isotropic, yet unlike isotropic or micellar solutions, they are very viscous. So far five different cubic phases have been reliably identified in lipid-water systems, with a further one that appears to occur in certain lipid-protein mixtures at low hydration. These are listed in Table 2, along with their space group number and cubic aspect (this is, strictly speaking, what is identified from the pattern of systematic absenses). Cubic phases appear to fall into two distinct families, one based on periodic minimal surfaces (bicontinuous) and the other on complex packings of discrete lipid aggregates (micellar). In both cases the phases may be either of normal or of inverse topology. However, apart from the cubic phase Ia3d (Q230), for a given space group the topology seems to be only one type or the other. It is not yet clear whether this apparent asymmetry is merely accidental. It is also not clear how many other cubic phases remain to be discovered. The lattice parameters observed so far for cubic phases fall into the range of 80–250 Å. There are some theoretical grounds for believing that the latter figure may be close to an upper stability limit, but this remains to be established.

The first cubic phase structure to be determined was Ia3d [35]. The structural elements are rods (water or lipid, depending on the topology) connected coplanarly three by three in two unconnected but mutually interwoven chiral networks (Fig. 7a). Note, however, that the cubic phase itself is centrosymmetric, and this remains true even when the phospholipid is chiral [36]. The structure of Pn3m was identified by Tardieu [37] and independently confirmed eleven years later by Longley and McIntosh [38]. As for Ia3d, there are two networks interwoven yet unconnected channels, but this time the aqueous channels are connected tetrahedrally four by four, forming a double diamond lattice (Fig. 7b). In the structure of Im3m, the aqueous channels are orthogonal and are connected six by six (Fig. 7c).

We now realize that these bicontinuous cubic phases can be usefully described in terms of periodic minimal surfaces—mathematical surfaces of zero mean curvature where the Gaussian curvature is everywhere negative or zero [39–44,12]. In this differential geometric description, the inverse topology cubic phases Ia3d, Pn3m, and Im3m are formed by draping a continuous lipid bilayer onto the gyroid, F-minimal, and P-minimal surfaces, respectively.

The structure of the cubic phase Pm3n [45] has been the subject of much controversy over the last two decades. It is now agreed that the structure [46] consists of a cubic packing of two types of micelle, one quasi-spherical (two per unit cell) and the other slightly asymmetric (six per unit cell), as shown in Fig. 8a. However, it is not yet fully established whether the asymmetric micelles are disklike [47,48] or rodlike, with rotational disorder around one of the short axes [49,50].

The face-centered cubic phase Fd3m has been observed in a variety of hydrated binary lipid systems, such as diglyceride-phosphatidylcholine mixtures [51]. The structure (Fig. 8b) has recently been solved [52]. It is the first well-established example of a cubic phase structure based solely on a packing of inverse micelles. As in the normal topology micellar cubic phase Pm3n, there are two types of micelle per unit cell. However, in Fd3m both types of micelle are quasi-spherical but of different sizes. There are eight larger and sixteen smaller inverse micelles per unit cell. It is fascinating that this structure appears to form only when more than one lipid component is present, and with widely differing hydrophilicities. The reason is probably that this permits an asymmetric lipid composition between the two types of inverse micelle, allowing them to have different sizes.

The cubic phase $P4_332$ has so far only been observed in one ternary lipid-protein-water system. The proposed structure is derived from that of Ia3d [17]. One of the networks of rods present in Ia3d is preserved in $P4_332$, the other being replaced by quasi-spherical inverse micelles (in which the protein is located) centered on every second three-rod junction. The fascinating feature of this cubic phase is that it is chiral.

(d) Solution phases. In this brief overview of the phase structures that may be formed by phospholipids, there was insufficient space to discuss the translationally disordered solution phases such as micellar solutions, microemulsions, or so-called L_3 (sponge) phases.

Micelles can be formed by short chain phospholipids (typically C_6 or C_8) or by lysophospholipids in water [1]. They may have a variety of shapes, such as spheres, rods, discs. For anisotropic aggregates (for certain concentration ranges), there is a possibility of forming nematic phases, where the aggregates are not packed on a lattice but are preferentially aligned along a particular direction in space. If the molecules are strongly chiral, this direction spirals round in a helical path, forming a chiral nematic (cholesteric) phase. Inverse micellar solutions can be formed by hydrated phospholipids in the presence of certain organic solvents such as benzene or cholorobenzene [53]. Such systems are of great interest because they can entrap catalytically active enzymes [54,55]. Furthermore, at specific, lower water contents (typically 2–8 waters per lipid) phosphatidylcholines in certain organic solvents such as alkanes form stiff, nonbirefringent gels [56] whose structure seems to consist of entangled flexible inverse cylindrical micelles [57]. Inverse micelles can also be observed for certain hydrated mixtures of phospholipids with less amphiphilic lipids such as diglycerides, in the absence of organic solvents.

Microemulsions are transparent yet microstructured solutions formed by surfactant-oil-water mixtures [58]. For certain compositions, the structure is based on discrete micelles or inverse micelles. However, when the amounts of water and oil are similar, bicontinuous structures may form. The microstructure is believed to be based upon thermally disordered minimal surfaces [39,40]. In some surfactant systems, stiff microemulsion gels have been found, which have recently been shown to have a cubic phase structure, although one that appears to be based on an interfacial *monolayer* rather than a bilayer as in the phospholipid bicontinuous cubic phases [44,59,60]. It is, however,

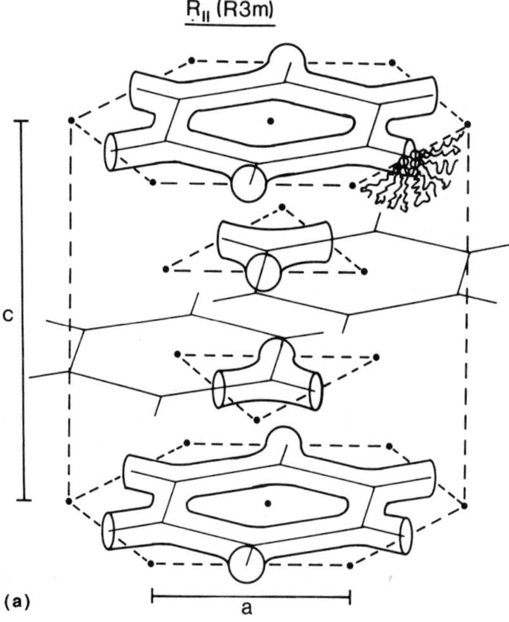

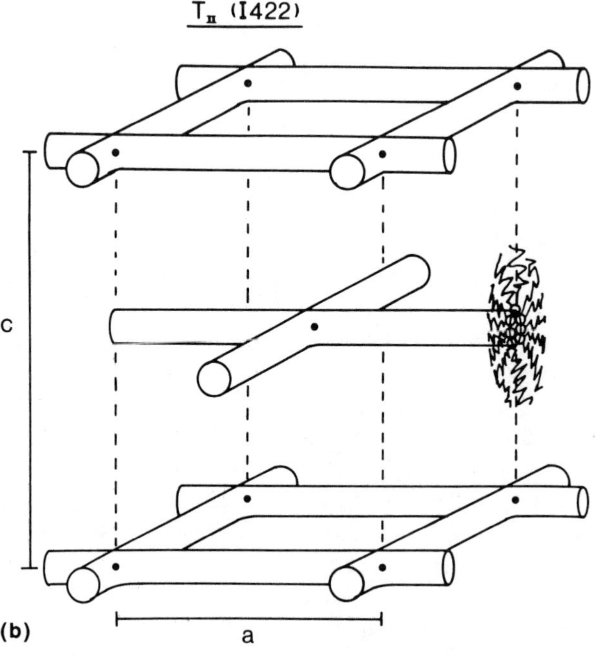

Figure 6 Non-cubic 3-D fluid phases: (a) inverse rhombohedral; (b) tetragonal. The examples shown are both of inverse topology. (From Ref. 9.)

Table 2 Established Lyotropic Cubic Phases

Space group	Number	Cubic aspect	Topology	Structure
$P4_332$	212	Q3	I	?
Pm3n	223	Q5	N	Micellar
Pn3m	224	Q4	I	Bicontinuous
Fd3m	227	Q15	I	Inv. micellar
Im3m	229	Q8	I	Bicontinuous
Ia3d	230	Q12	N + I	Bicontinuous

The phase topology is denoted N (normal) and I (inverse). Further details of cubic space groups are given in [34].

possible that such monolayer cubic phases could be formed by phospholipids in the presence of organic solvents.

In certain surfactant systems, very highly swollen lamellar phases are formed, which may transform upon dilution to a so-called L_3 or sponge phase [61,62]. The structure of this phase appears to be essentially a disordered version of the bicontinuous cubic phases: the interface is highly flexible, and thermal excitations break down the long range order of the network of channels. Such a phase has not yet been found in any phospholipid-containing system but might occur in the presence of cosurfactants such as pentanol, which should drastically lower the rigidity of the lipid bilayer, thereby enhancing thermal fluctuations.

III. LYOTROPIC PHASE TRANSITIONS

A. Lyotropic Phase Diagrams

For binary lipid-water systems, phase transitions may be induced by varying either the temperature or the water content. A hypothetical binary lipid-water phase diagram, where the transitions are driven predominantly by the latter, is shown in Fig. 9. This shows the various possible fluid phases in their "natural" sequence, in terms of the average mean curvature of the polar-nonpolar interface [63,40,42,9]. Although ionic surfactant phase diagrams often show a striking similarity to parts of the hypothetical diagram, phospholipid phase diagrams invariably show a strong dependence also on temperature (see Fig. 10).

A very useful collection of lyotropic binary and ternary phase diagrams has been presented by Ekwall [64]. Reviews of phase diagrams of lipid mixtures have been given [65,66,13]. Binary phospholipid phase diagrams are usually either of the lipid-water or of the lipid-lipid type in the presence of excess water (note that this is strictly speaking a ternary system). For various binary lipid mixtures, a range of types of phase diagram may be observed, from perfect mixing to eutectic, peritectic, or monotectic behavior. In general, deviations from ideality become stronger when the lipids differ strongly in chain length or headgroup type. A compendium of lipid phase diagrams has been presented [67], and databases of transition temperatures and enthalpies, and of phase diagrams, are currently being assembled [68]. A number of reviews of lamellar and nonlamellar phases and lipid phase transitions has appeared [69–73,5,74,6–9,36,75,76,10].

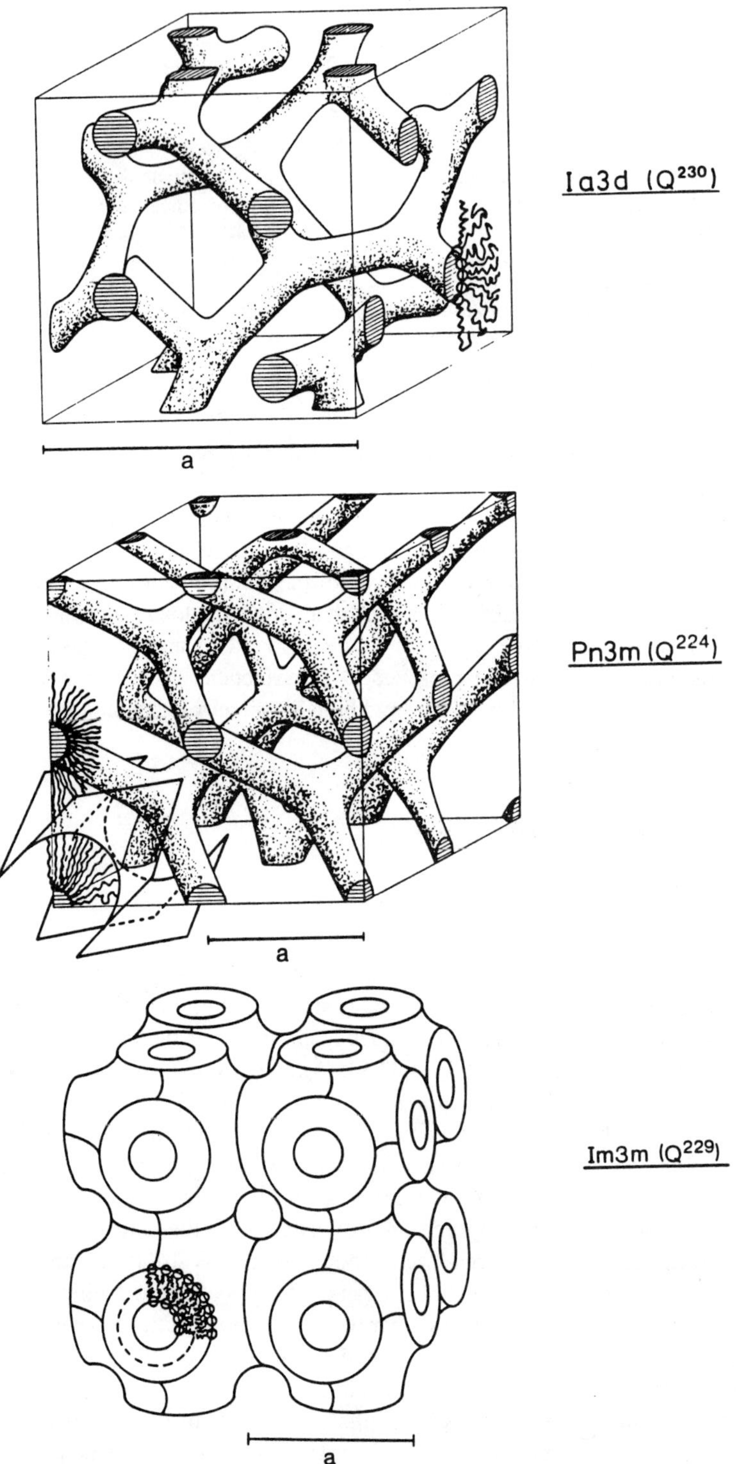

Figure 7 The bicontinuous cubic phases Ia3d, Pn3m, and Im3m. The examples shown are all of inverse topology. (From Ref. 9.)

B. Types of Transition

Phase transitions between translationally ordered lyotropic phases break down into three main types:

1. Solid-solid transitions. These always involve lamellar phases. Examples are crystal-crystal, crystal-gel, gel-gel (e.g.. $L_{\beta'}$-$P_{\beta'}$). Furthermore, it appears that various of the gel phases can undergo transitions to interdigitated versions.
2. Chain melting transitions. The lower temperature phase is invariably lamellar, whereas the higher temperature phase, which is at least partially fluid, need not be lamellar, and need not even be liquid crystalline (e.g., it could be a micellar solution).
3. Fluid phase transitions. Both of the phases are fluid; the transition involves a change of symmetry and/or topology. Examples are lamellar-hexagonal, lamellar-cubic, cubic-cubic.

This breakdown is not however a hard and fast one, since gel phases may be partially fluid, and certain phases (such as P_{δ}) may have both fluid and solidlike regions. Furthermore, one can in principle have transitions from isotropic solution phases (micellar solutions or microemulsions) to any of the translationally ordered phases.

Phase transitions can be induced by changing any of the intensive thermodynamic variables such as temperature, pressure, etc., but they may also be induced isothermally by changes in hydration, pH, salt concentration, etc. Generally speaking, the sensitivity of a transition to a given perturbation tends to be proportional to the free energy shift induced by the perturbation, divided by the transition entropy [77,13].

C. Lipid Polymorphism

1. Crystalline and Gel Lamellar Phases

Dry amphiphiles and hydrated lipids at relatively low temperatures typically form densely packed crystalline structures, as described in previous sections. Hydrocarbon chains in such structures often form an orthorhombic hybrid chain-subcell lattice [78–81]. With increasing temperature, these lipid crystals become energetically unfavorable owing to the thermal, initially chiefly rotational, chain excitations. Consequently, at some subtransition temperature $T = T_s$ the two-dimensional lipid crystal normally transforms into a more expanded lipid gel phase of the β' or, more frequently, β type [24,82], with tilted or untilted chains, respectively. Different crystal forms of similar lipid types may also have different subtransition temperatures [83]. Increasing hydration (and sometimes also increasing salt concentration) may lower somewhat the phospholipid subtransition temperature. Lipid chains in the gel state are packed on a (skewed) hexagonal lattice and are moderately tilted ($L_{\beta'}$) or untilted (L_b) [78]. The properties of the phospholipid chains in such phases resemble those of simple long-chain hydrocarbons, such as alkanes, above the transition to the so-called rotator phase.

2. Undulated Phases

Heating lipids above the subtransition temperature enhances the torsional oscillations of the hydrocarbon chains until these turn into essentially unhindered long-axis rotation [84,85].[1] This often occurs in a cooperative manner at a pretransition temperature $T = T_P$.

[1] Crystalline chain order is lost at the latest at the chain melting phase transition temperature.

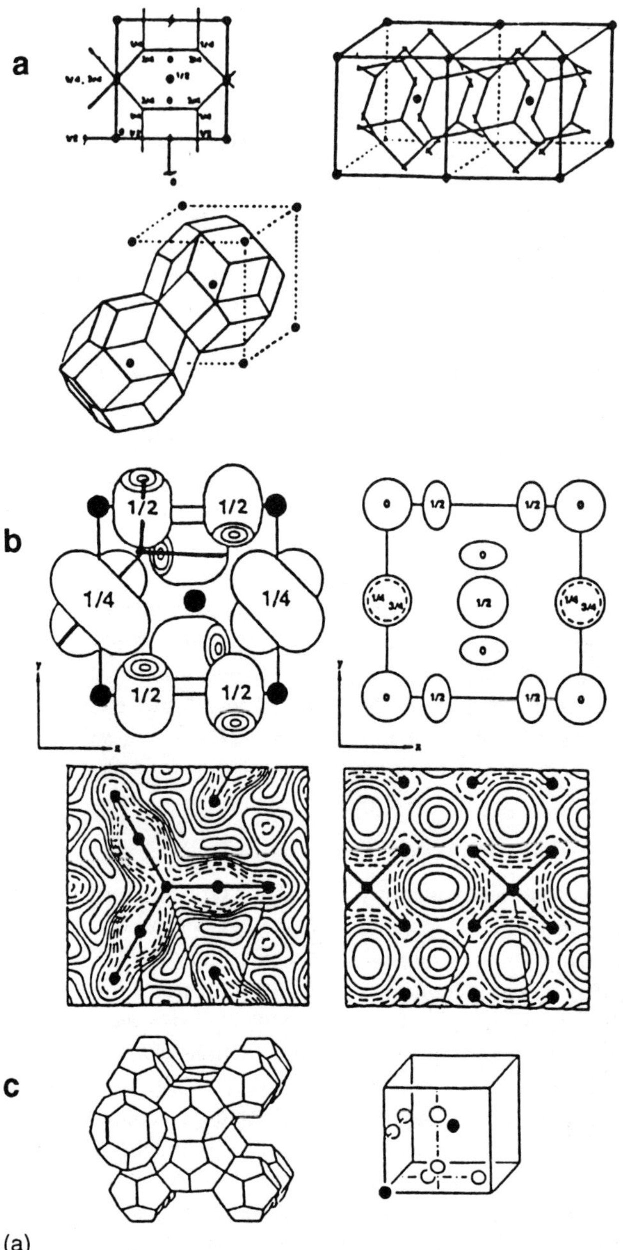

Figure 8 The discontinuous cubic phases. (a) Pm3n (micellar); (b) Fd3m (inverse micellar). (From Refs. 8 and 186.)

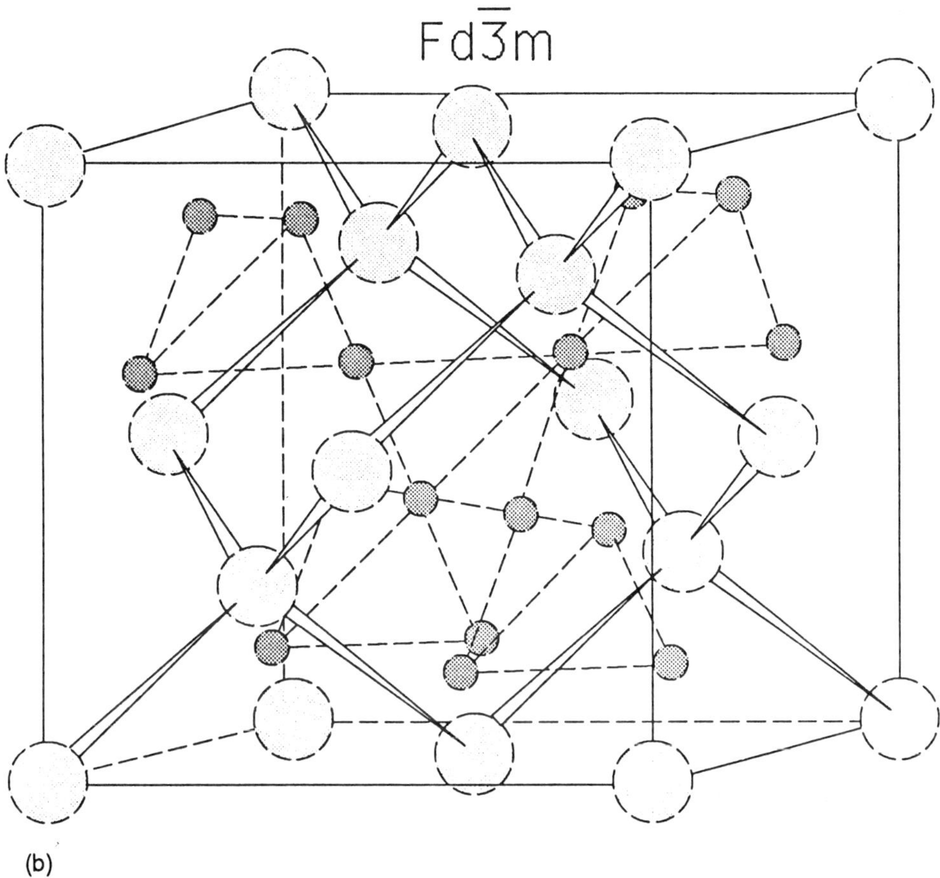

(b)

The lipid headgroup mobility as well as the interfacial area per molecule increase substantially at this point (see also Chap. 13 of this book). The area per phosphatidylcholine headgroup, for example, below and above the pretransition temperature, is believed to be near 0.525 nm^2 and 0.65 nm^2, respectively [86,87], the concomitant change in the chain area being only a few percent.

It is very probable that during a phospholipid pretransition individual chains mutually shift along their long axes to stay in close contact. This may explain why unsaturated phospholipids do not form undulated bilayers; it is also the most likely reason why the bilayer surfaces at $T = T_p$ break up into a series of periodic, asymmetric, quasi-lamellar bilayer segments that give rise to the appearance of the surface undulations or ripples characteristic of the $P_{\beta'}$ or P_β-phase [88].

One, and probably the main, reason for the rippling (and pretransition) of lipid bilayers below the chain melting phase transition is the tendency of the polar lipid headgroups to achieve a certain degree of fluidity and solvation whilst the hydrocarbon chains remain ordered [89]. Phospholipids, consequently, undergo a pretransition only if the polar lipid headgroups are sufficiently hydrophilic and hydrated and if the inter-chain packing is sufficiently weak.

At least 12 ± 2 water molecules must be associated with each phospholipid headgroup for a bilayer undulation to be feasible. The closer the system hydration is to

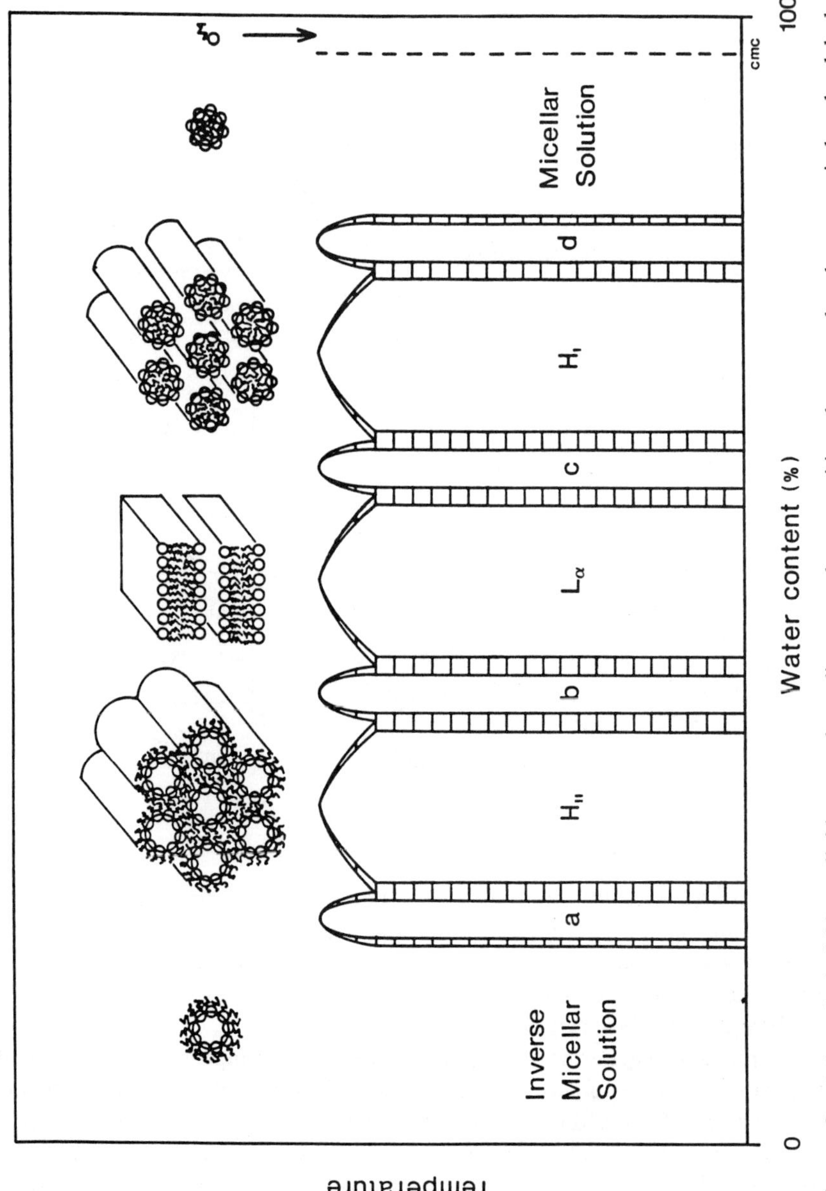

Figure 9 A hypothetical binary lipid-water phase diagram, where transitions between the phases are induced solely by varying the water content. (From Ref. 9.)

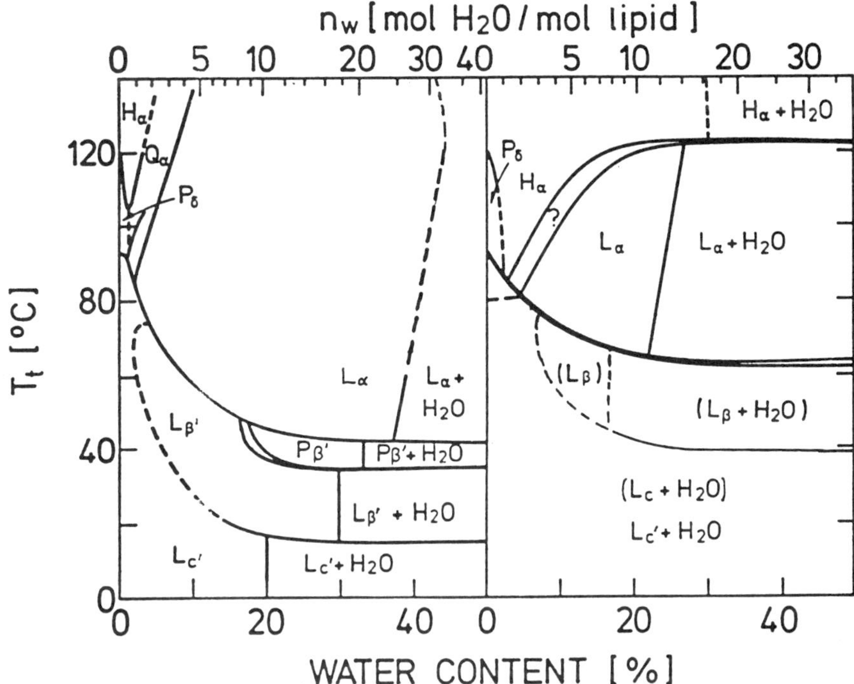

Figure 10 Partial phase diagram of dipalmitoylphosphatidylcholine (DPPC) and dipalmitoylphosphatidylethanolamine (DPPE) in water. DPPE has a less hydrophilic headgroup than DPPC. This causes regions of the fluid-lamellar phase for the former lipid to be narrower. For a similar reason, the gel phase is metastable, the undulated gel phase is not observed, and the inverted hexagonal phase is very prominent for DPPE. (From Ref. 13.)

this limiting value, the longer is the repeat distance for the surface undulations and the less stable is the undulated bilayer phase. The minimum required amount of lipid-bound water is itself essentially chain-length independent and unaffected by the method of hydration variation; physical dehydration, hydrational competition between the lipid molecules and dissolved substances, or decreasing the lipid headgroup polarity all increase the pretransition temperature similarly on the appropriate scale. However, even in an excess of water, bilayers can only undulate as long as each of the two identical lipid chains contains between 12 ± 1 and 22 ± 1 carbon atoms; for shorter chains, lipid crystallization prevents surface rippling; for phospholipids with very long chains, the headgroup effects become too small to cause such rippling. The width of the P_β phase region, as a rule, increases with the lipid polarity [89].

The pretransition temperature can be scaled in terms of the hydration-induced chain melting phase transition shift or of the lipid surface hydrophilicity. The minimum lipid hydrophilicity and the critical amount of lipid-bound water also can be related to the lipid chain melting transition temperature; the latter must not exceed some maximum, chain-length dependent value if a pretransition is to exist [89].

3. Fluid Lamellar Phases

In the low temperature gel phases, the hydrocarbon chains are in a conformationally well-ordered state, essentially in the all-trans configuration; their extension is thus close to maximum under such conditions (see, e.g., [79,24]). At higher temperatures, however, this chain order is lost, owing to conformational chain excitations. A cooperative chain melting (order-disorder, gel-to-fluid) phase transition, at $T = T_m$, results from this (cf. Fig. 10).

From the biological point of view, the lamellar fluid phase L_α is particularly crucial. Living organisms try to keep their membrane bilayers close to a fluid-phase optimum [90], often at some distance from bilayer-to-nonbilayer phase transitions [7], either by means of chemical chain modifications [90,91] or by synthesizing appropriate soluble molecules [92]. This pertains only to long observation times and large membrane areas, however. Transient and/or regional deviations from a fluid lamellar phase are not infrequent. Small patches of crystalline phase, or at least ordered-chain domains, may form in the regions where two adjacent membranes are closely apposed [93]; they are also prone to appear in areas covered by a protein or glycocalix coat [94,95]. Structures deviating from standard bilayer, such as locally nonbilayer formations, furthermore, may appear in the vicinity of membrane proteins [70], defects [96], or membrane fusion [97] and disruption sites.

4. Fluid Nonlamellar Phases

The structural properties of nonlamellar phospholipid phases are described in previous sections. In the following, only one possible rationale for the formation of such phases is given [98]; an alternative approach is to consider a spontaneous surface curvature as the driving force for nonlamellar phase formation [99,100].

Nonbilayer phospholipid structures should begin to appear precisely at the point at which the repulsive lateral pressure of the hydrocarbon chains exceeds the corresponding pressure that arises from the interfacial region, including solvation layers, and thus gives rise to a positive spontaneous curvature value. The former pressure increases with the disorder and the amount of material in the hydrocarbon membrane interior. The latter pressure increases with the lipid headgroup size and hydration and normally decreases with increasing salt content or osmotic pressure in the system.

Any phospholipid or system modification that increases the repulsion between lipid headgroups, such as an increase in lipid hydrophilicity, therefore lowers the tendency for the formation of nonlamellar fluid-phase structures. Any change that enhances the interchain repulsion, conversely, makes the generation of nonbilayer structures, more probable. Consequently, during the course of any monotonic system variation that affects the lateral pressures in the interfacial or hydrocarbon region, a lower limit to the bilayer-to-nonbilayer phase transition temperature is reached at which the bilayer-to-nonbilayer phase transition temperature coalesces with the bilayer chain melting phase transition temperature. Lamellar, bilayer membranes kept below this critical temperature are stable indefinitely; above this temperature; only nonlamellar phases can exist [98].

D. Chain Effects

1. Chain Length

The chief determinant of the phospholipid phase behavior is the type of hydrocarbon chains (the main reason for this is the chain dependence of the melting entropy).

Increasing chain length thus causes all transition temperatures at which the interfacial packing density decreases to become higher; the more polar is a given lipid, the stronger is this length dependence. This pertains to the lipid sub- or rotator phase transition, to the pretransition, as well as to the lipid chain melting transition. Conversely, the temperature of such transitions, at which the interfacial packing density increases, normally, is lowered by hydrocarbon chain elongation; transitions into nonbilayer (cubic, hexagonal) inverse phases provide examples of this.

The transition enthalpy inevitably increases with the lipid chain length in an approximately linear manner (see Appendix B). The same is true for the transition entropy. Both these quantities decrease with lipid headgroup hydration [101] and are also sensitive to the lipid packing properties. The published results for ΔH_c and ΔS_c, therefore, are quite variable. For dipalmitoylphosphatidylcholine, for example, values between 21 and 39 kJ/mol have been reported.

To calculate transition temperatures for different chain lengths, one can use a simple polynomial expression of the form

$$T_c(n) = T_c(\infty)\left(1 + \frac{n_c}{n} + \frac{n_h}{n^2} + \cdots\right) \tag{10}$$

where c = s, p, m, (i.e., sub-, pre-, or main transition). The term n_c corresponds approximately to the length of the shortest segment for which a first-order phase transition is possible; the term n_h allows phenomenologically for headgroup and other end effects [102]. Suitable values of n_c and n_h are given in Appendix B, at the end of this volume. The parameter $T_c(\infty)$ refers to the corresponding transition temperature of a hypothetical lipid with infinitely long chains; for phospholipid chain melting transitions its value is approximately 414K.

The chain-length dependence of the chain melting phase transition of all common disaturated phospholipids is illustrated in Fig. 11.

2. Chain Unsaturation and Branching

Chain unsaturation and branching effectively decouples the parts of the hydrocarbon chain on either side of the double bond or the side chain. Corresponding chain modifications thus reduce the length of parallel, strongly interacting chain segments and thus mimic the consequences of chain shortening [103]. The corresponding effects of double bonds in a *trans* configuration are weaker by approximately 50% than the effects of *cis* double bonds in the same location.

When a chain contains a *cis* unsaturated bond, the variable n in Eq. (10) must be replaced by its effective chain length n_{eff}, equal to the length of whichever chain segment on either side of the double bond is the longer (when the longer segment is nearer the headgroup, the value n_{eff} must be reduced by one methylene group). If both chain segments separated by double bonds have similar lengths, numerical corrections are necessary for an accurate prediction of the experimental results [103].

Phospholipid chain melting temperatures T_m thus increase with the length of the longest ordered and aligned segment on each chain; for the diacylenoylphospholipids (and probably also for the dialkylenoyl compounds) this transition temperature, therefore, first decreases and then increases as the double-bond position moves from the upper to the lower half of each hydrocarbon chain; phospholipids with double bonds halfway between the chain terminus and the glycerol backbone, consequently, have the lowest possible

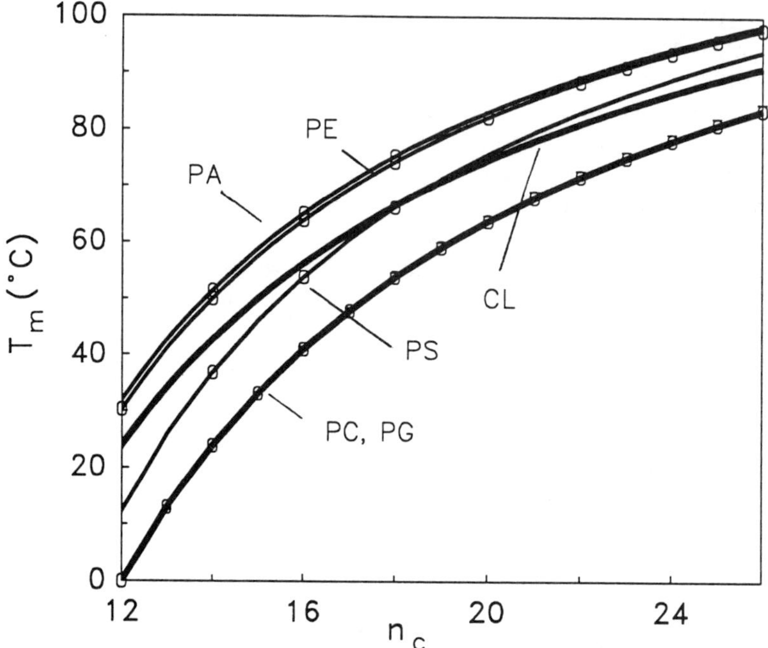

Figure 11 Effect of phospholipid head group and acyl chain length on the chain melting phase transition temperature of lipid bilayers in water or 0.1 M buffer (pH 7). Curves were calculated by means of Eq. 10, and dots correspond to measured T_m values.

chain melting phase transition temperature for a given chain length [104]. The chain melting phase transition enthalpy also shows a minimum near $n_{eff} \approx n_c/2$ [104]. Fig. 12 illustrates this for lipids with different chain lengths. Results of chain branching are similar [105].

Increasing the number of double bonds per chain causes only a small further decrease of T_m [106]; the relative significance of the chemical chains modification always decreases with the total chain length, however. The data presented in Appendix B corroborate these conclusions.

For lipid mixtures with identical headgroups and unequal hydrocarbon chains, the temperature of the gel-to-fluid phase transition, in the simplest approximation, is given by the arithmetic mean of individual phase transition temperatures calculated for the corresponding phospholipids with symmetric chains, except when the chain lengths differ by more than three or four methylene groups [103].

3. Chain Asymmetry

Transition temperature data for many phosphatidylcholines with two different acyl chains have been published by Mattai et al. [107] and by the group of Huang [108,109]. These data show that with a growing discrepancy in the lengths of the chains in the *sn*-1 and *sn*-2 positions, Δn, the total inter-chain attraction diminishes. The effective phospholipid chain length with increasing chain asymmetry also becomes shorter, at least as long as lipids form bilayers. The phospholipid chain packing density in the intercalated terminal hydrocarbon regions and the chain melting transition entropy both decrease. The chain melting transition temperature is thus shifted downwards, and is given approximately by

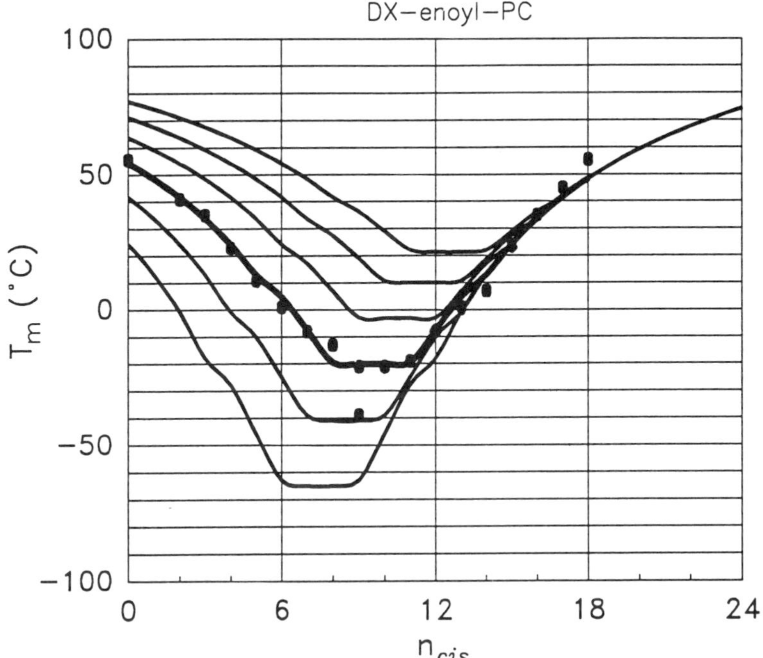

Figure 12 Chain melting phase transition temperature of symmetric, unsaturated dialkenoylphosphatidylcholines as a function of the hydrocarbon length and position of the individual double bonds; dots give experimental points. n_c = 22 . . . 12, from top to bottom respectively. Reprinted with permission from [103].

$T_m(n_c\Delta n) = T_m(n_c) - \Delta n\ (T_m(n_c) - T_m(n_c - 1))$, where $T_m(n_c)$ is calculated according to Eq. 10 [103]. This is illustrated by Fig. 13.

Phospholipids with very asymmetric chains ultimately tend to form interdigitated gel phase membranes, provided that their headgroups are sufficiently hydrated [107–109]. Corresponding fluid phase lipids can have partly interdigitated chains above the transition temperature [110]. A transition into an interdigitated state is normally accompanied by an appreciable increase in the chain melting phase transition enthalpy by approximately a factor of two [109] and may cause phase separation in the mixed lipid systems [111].

A method for the calculation of the chain melting phase transition temperature of lipid having arbitrary asymmetric chains has been described by Cevc [103]; an alternative method applicable to saturated asymmetric phospholipids has been introduced by Huang [112]. To estimate the chain melting phase transition temperatures of phospholipids with asymmetric as well as unsaturated chains, the combined effects of effective chain length modulation must be taken into account [103].

4. Chain Attachment

The chain melting phase transition temperatures of dialkylphospholipids are higher by some 1–5° than the T_m values of corresponding diacylphospholipids; for phosphatidylcholines with n_c = 16 the values are 43.5 and 42°C, respectively. Interestingly, the transition temperature of hexadecylpalmitoylphosphatidylcholine is even lower (40°C). The most

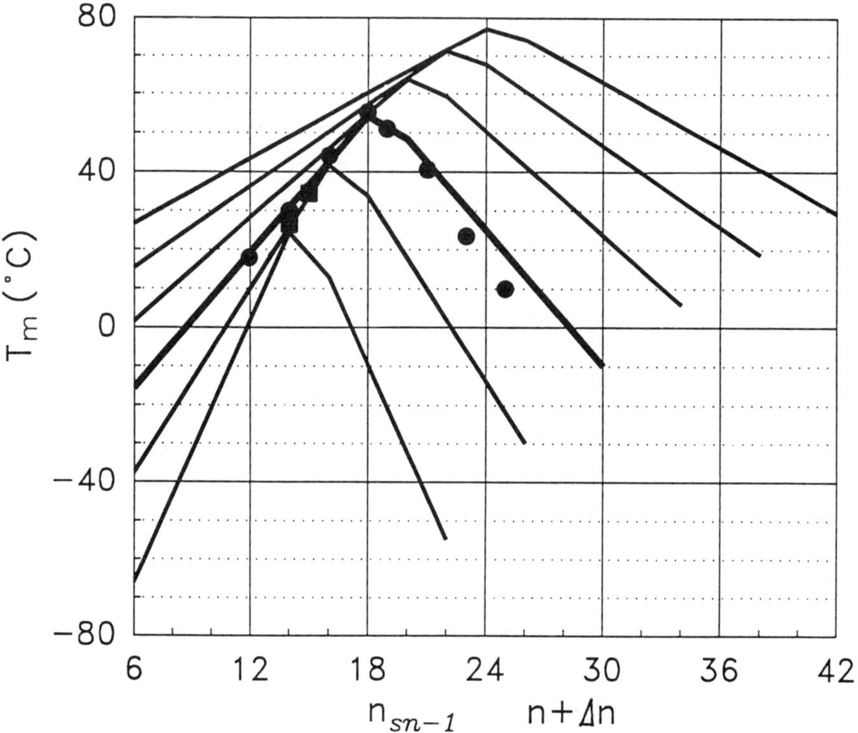

Figure 13 Effect of the hydrocarbon length and chain asymmetry on the chain melting phase transition temperature of phosphatidylcholine membranes in water. Curves were calculated according to the rule given in the text, and dots correspond to experimental data. n is the length of the longer chain, Δn the chain length difference, i.e. results are represented as if they were mirrored around the $n_{sn-1} = n$ position. Reprinted with permission from [103].

probable reasons for these discrepancies, which decrease with the hydrocarbon length, are the different hydration and different packing properties of these lipids in pure and mixed gel phases [113–115]; phosphatidylcholines form interdigitated low temperature phases whenever chains in the *sn*-1 positions are of the alkyl type [116,117]. Chains in the *sn*-2 positions seem to be inefficient in this respect (V. Gordeliy, personal communication). Ether bonds, moreover, increase the lipid sensitivity to water-soluble inducers of chain interdigitation [118].

Replacement of chain ester with ether bonds in the glycerol region favors the formation of inverse nonbilayer phases, as has been shown in detail by Seddon et al. [119,28]. Replacement of the phospholipid diacylglycerol group with the sphingosine-fatty acid conjugate as the hydrocarbon part of the lipid has similar thermodynamic consequences, probably owing to the possibility of this latter phospholipid type of forming direct hydrogen bonds between the proximal hydroxyl and esterified amino groups on the lipid chains.

E. Headgroup Effects

Lipids with different headgroups differ in their sensitivity to the solvent as well as to interlipid-bonds effects. Chemical structure [102,120] and stereoisomerism [121,122] both play some role in this, especially when the headgroup region is poorly hydrated and/or densely packed.

1. Lipid Polarity

Apart from the hydrocarbon chains, the lipid polarity is the most decisive factor for the phospholipid polymorphism; other factors, such as the ionic or zwitterionic character, or the size of lipid headgroups, play a much smaller role.

In general, the thermodynamic significance of each phospholipid headgroup and its interactions with neighbouring headgroups and solute molecules increases with the relative hydrophilicity of the lipid polar residues and with the degree of the hydrocarbon unsaturation (Figs. 11 and 14a). Unsaturated or short chain phosphatidylcholine, phosphatidylglycerol, and cardiolipin, which belong to the most polar phospholipids, consequently, are more sensitive to headgroup effects than are the saturated or long chain phospholipids or the less polar phosphatidylserine, phosphatidylethanolamine, or phosphatidic acid; sphingophospholipids take an intermediate position. Phospholipids at high pH, which as a rule are at their maximum polarity, also respond more strongly to headgroup variations than do lipids at low pH. This is in part due to lipid charge effects and, more importantly, to lipid hydrophilicity variations (see also the following sections).

2. Deprotonation and Methylation

Charged phospholipids are always more sensitive to salt effects than are zwitterionic or nonionic lipids, but the effect of the net lipid charge may be overwhelmed by the consequences of surface (de)protonation. Phospholipid deprotonation always increases lipid polarity and thus increases lipid sensitivity to headgroup effects, irrespective of whether or not it finally leads to a charged membrane state. The method used to achieve lipid headgroup deprotonation is also unimportant; increasing the pH value, replacement of lipid-bound protons by other (larger) ions or by chemically bound methyl groups, or even breaking interlipid hydrogen bonds all cause qualitatively similar thermodynamic effects. All such manipulations, for example, lower the chain melting phase transition temperature. The chain melting transition temperatures of nearly all common phospholipids, consequently, steadily decrease with increasing pH of the suspending solution, independently of the resulting state of lipid ionization [102,123,120]. Acidification, conversely, normally causes the chain metling phase transition temperature of all common phospholipids to increase [102,120].

3. Headgroup Size

The importance of headgroup size for lipid bilayer stability and polymorphism is less important than is often believed (unless bound water molecules and other solutes are mechanistically included in the size evaluation). This factor, consequently, should only be considered directly when the lipid headgroup size variations are crucial for the interlipid bonding patterns. Dipalmitolyphosphatidylcholine analogs with minimal direct interheadgroup interactions and phosphate-ammonium group separations ranging between n_c = 2 and 12 all melt between 40 and 44°C, for example [124]. In contrast to this, the chain melting phase transition temperature of various phosphatidylalkanolamines decreases appreciably with increasing headgroup length; the T_m decrease is highly nonlinear, however, indicative of the significance of direct headgroup-headgroup interaction effects [125,120]. Some further examples are given in Appendix B at the end of this volume.

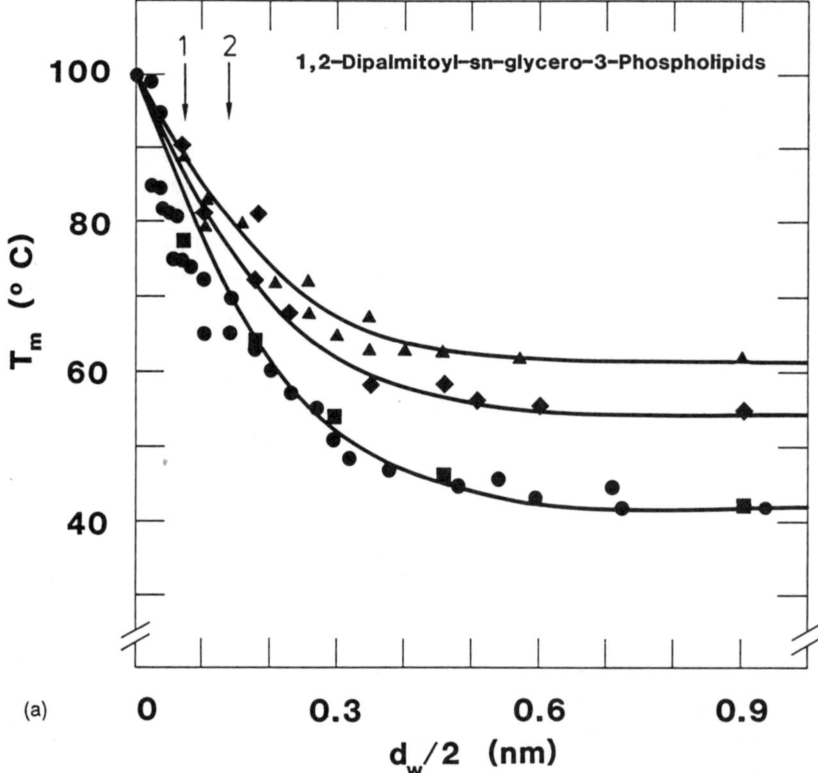

Figure 14 Effect of phospholipid (a) (from Ref. 120) and chain desaturation (b) on the hydration decrease of the chain melting phase transition temperature of various phospholipids (modified from Ref. 106). More hydrophilic and unsaturated hydrocarbons become fluid at relatively lower temperatures. Headgroup polarity and chain unsaturation, moreover, enhance the lipid sensitivity to system variations, such as changing hydration. (Curves are drawn solely to guide the eye.) Interlamellar water layer thickness d_w is approximately proportional to the bulk water concentration c_w. In (a): Phosphatidylcholine (●), phosphatidylglycerol·Na$^+$ (□), phosphatidylserine·Na$^+$ (◇), phosphatidic acid·Na$^+$ (∇), phosphatidylethanolamine (Δ). (From Ref. 120).

F. Solution Effects

The hydrocarbon chains of different dry lipids melt at nearly identical transition temperatures, as long as their effective lengths are appoximately the same [120] (cf. Fig. 14a). It should be noted that chain tilt shortens, and chain interdigitation increases, this length. Dry saturated phospholipids undergo conformational chain isomerization at quite high temperatures, typically above 70°C, most frequently around 100°C. Chain melting temperatures for the corresponding anhydrous unsaturated phospholipids are lower, below room temperature, by and large [5,106] (cf. Fig. 14b).

1. Solvent

Lipid hydration, as a rule, lowers the temperature of chain fluidization [126,127]. Initially, this lowering is quantized [126,128], indicative of distinct lipid hydration states, and it is approximately linearly proportional to the number of bound water molecules. The

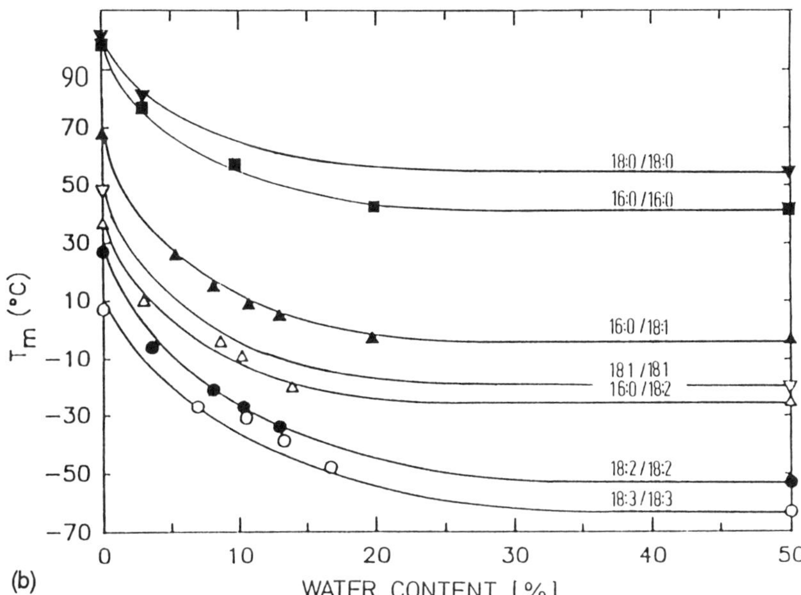

(b)

resulting total solvent-induced shift can exceed 50° in the case of diacylphosphatidylcholines. The lipid chain melting phase transition, therefore, acts as a thermodynamic osmometer: the chain melting phase transition shift is directly proportional to the change of the logarithm of the bulk water activity coefficient [123], the "osmometer sensitivity" decreasing with the effective hydrocarbon chain length (Fig. 15).

The presence of water also lowers the transition temperatures for all other phase changes that are accompanied by a decrease in the headgroup packing density at T_c (cf. Fig. 10). The lipid subtransition, pretransition, and chain melting transition exemplify this, the first and the third transition types being more strongly, and the second less strongly, hydration dependent, owing to the differential transition entropy values. In contrast to this, the presence of water increases all such phase transition temperatures that are accompanied with an interfacial packing density increase upon heating; lipid lamellar-to-non-lamellar phase transitions fall into this category. (See [127] and Chap. 13 for more detailed discussions.)

2. Solutes

In every system, some solvent molecules must be shared between the phospholipid and any other polar or amphiphilic molecules present. Increasing the bulk concentration of polar solutes, consequently, lowers the effective number of water molecules available for binding to each lipid molecule and thus mimics the effects of lipid dehydration. To a first approximation, the thermodynamic consequences of such solvation-sharing are thus proportional to the deviation of the bulk water activity coefficient from the ideal value of unity.

In addition to this, certain polar solutes, such as many carbohydrates and ions, can interact directly with the phospholipid headgroups. This may lower the phospholipid affinity for water and, moreover, give rise to intermolecular forces that stabilize the more

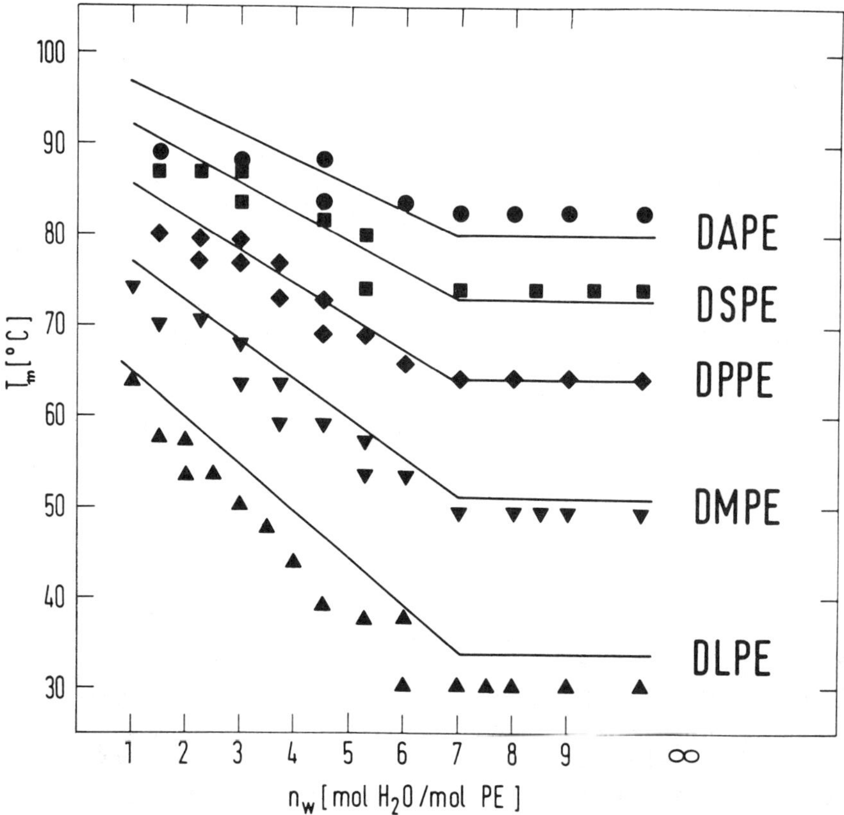

Figure 15 Chain melting phase transition of phosphatidylethanolamines at various degrees of hydration is strongly sensitive to the chain length. Lines are only drawn to guide the eye. Reproduced with permission from [126].

ordered phases. Such solutes, therefore, typically increase the majority of phospholipid phase transition temperatures.

The primary effect of bound monovalent ions is to screen the net membrane surface charges (see [77,129–131] and Chap. 4 for more details). Initially it was thought that this was the only origin of the headgroup ionization effects on lipid polymorphism [77]. Later, however, it was shown that the vicinity of ions to the water binding sites in the phospholipid headgroup region can lower the affinity of the polar lipid headgroups for water, and that this effect is responsible for the major part of the ionization-induced T_m shifts [129,123,120,131]. The only exceptions to this are strongly hydrophilic ions, such as certain halides (I, ClO_4, IO_4) and some organic ions ($N(CH_3)_4^+$) [90]). Divalent and multivalent ions [132–135], which occasionally form multiligand complexes with phospholipid headgroups, also dehydrate lipid headgroups and, moreover, can form ionic bridges between the neighbouring lipid molecules. By doing so, they contribute directly to the interlipid cohesion, increase the stability of ordered lipid phases, and increase the temperature of the chain melting and other order-disorder phase transitions.

Protons [102,77,25,136,137,129,101,138,139], lithium [101,140], and most divalent and polyvalent ions complexed with lipid headgroups [132–135,141,142] are particularly efficient in this respect. These ions replace or repel some of the bound water molecules and cause (partial) membrane condensation [101,140,143]; isothermal fluid-to-gel [144], gel-to-crystal [101,140,134], or even lamellar-to-nonlamellar [101] phase transitions may result from this.

In contrast to this, strongly hydrophilic substances with some affinity for interfaces, such as certain halide and organic ions or monohydric alcohols, promote phospholipid hydration, favor the formation of interdigitated phases [145–147], and hinder nonbilayer fluid phase formation [148].

Sugars and polyhydric alcohols in the bulk solution tend to dehydrate phospholipid bilayers. At low concentrations, this occurs indirectly, via the osmotic mechanism [123]. However, such solutes may also interact directly with the proton donors and acceptors on the phospholipid headgroups and thus partly replace the lipid-bound solvent molecules [149]. The resulting water replacement is often a slow process, owing to the slow diffusion of these solutes across lipid bilayers [89,149]. This sugar binding is particularly important for the cryoprotection of lipid bilayers in membranes of living organisms [93,150–152].

To assess the relative significance of simple electrostatic versus solvation effects, individual chain melting phase transition shifts originating from the different phospholipid headgroup modifications can be evaluated and compared [123,120]. For (zwitterionic) dimyristoylglycerophospholipids the following values were found, for example: $\Delta T_{m,bond} \approx 1.5 \pm 1\ K \leq \Delta T_{m,el} \approx 5.5 \pm 0.5\ K \leq \Delta T^{H}{}_{m,h} \approx 7 \pm 1\ K \ll \Delta T^{PO}4_{m,h} \approx 13 \pm 3\ K$. The first figure gives the significance of direct ordered phase stabilization by interlipid hydrogen bonds. The electrostatic shift $\Delta T_{m,el}$ for the singly ionized state is quite small; the further shift induced by ionization to a doubly-charged state, $\Delta T_{m,el}(++,--) - \Delta T_{m,el}(+,-) \leq 3\ K$, is even smaller. The largest contributions to the total shift come from the two hydrational terms.

G. Lipid Mixtures

The effect of lipid solutes on phospholipid polymorphism and phase transitions has been reviewed by a number of authors [153,154,74,9]. The effects will depend strongly on where the solutes partition, which will in turn depend on the extent to which they are amphiphilic or nonpolar. Amphiphilic solutes can affect both the chain and the headgroup regions, whereas nonpolar molecules such as alkanes will exert their effects primarily in the chain region. In addition to lipid solutes, proteins can also strongly modify lipid polymorphism [155,156].

1. Phospholipid Mixtures

Biological membranes contain complex mixtures of lipids, and so it is important to characterize the polymorphic phase behavior of such mixtures, in well-defined model systems. The asymmetric lipid composition across the bilayer of most biomembranes gives rise to an additional complexity.

Phase diagrams of hydrated binary phospholipid mixtures can exhibit behavior ranging from eutectic and peritectic types to ideal mixing [65,66,13]. Which type of behavior is observed depends upon the immiscibility of the lipids in the various phases; deviations from ideality will increase with increasing difference in chain length and/or headgroup type. Monotectic behavior (fluid phase immiscibility) has been little reported as yet for phospholipid mixtures, but it can occur for mixtures of phospholipids with amphiphilic solutes such as fatty acids [36].

One consequence of forming phospholipid mixtures is that regions of two-phase coexistence can become more extensive in the phase diagrams, and also regions of three-phase coexistence become possible. Furthermore, phases may be observed for mixtures that do not form for either of the pure lipid components.

Different phospholipids exhibit different interactions between bilayers (or other aggregate geometries) due to their specific chemical structure. The hydration interaction between bilayers can be disproportionately modified by forming phospholipid mixtures [157].

The formation and structure of nonlamellar phases is modified for phospholipid mixtures as a consequence of alterations both to the tendency for monolayer curvature and on packing stresses in the system [158]. Generally speaking, for an "H_{II}-forming" system, mixing in a "bilayer-forming" lipid has the effect of tending to stabilize the lamellar phase. Thus for unsaturated phosphatidylethanolamine (PE) systems, addition of phosphatidylcholine (PC) has the effect of stabilizing the bilayer structure (see [9] and references therein). In many cases, isotropic ^{31}P-NMR signals have been seen for certain composition ranges, and it is probable that these are associated with the appearance of cubic phases, which have been observed for various unsaturated PE/PC mixtures containing in the region of 5–50 mol % PC [159,160]. A cubic phase can also be formed by equimolar mixtures of PC with cardiolipin in the presence of low levels of calcium [161].

2. Phospholipid-Amphiphile Mixtures

Most amphiphiles are likely to have dramatic effects on phospholipid polymorphism; only a few common solute systems will be described here.

Incorporation into PC bilayers of long chain fatty alcohols, acids or amines, all of which have the capacity for hydrogen-bonding to the PC phosphate group, has the effect of raising the gel-fluid transition temperature [162–166]. It has been shown that this effect also occurs with phosphatidylglycerol, but not with PE [167].

On the other hand, short chain alcohols such as methanol or ethanol initially lower, then increase the chain melting transition temperature of phosphatidylcholines [145], this effect being due to the induction of an interdigitated $L_{\beta I}$ phase [146,168]. The effect on nonlamellar transitions tends to be to stabilize the L_α phase for short chain alcohols, but for longer chains to tend to promote H_{II} phase formation [169].

Saturated fatty acid-phosphatidylcholine mixtures form stoichiometric 2/1 mol/mol complexes, with sharp calorimetric transitions in water at temperatures some 20°C greater than that of the pure phosphatidylcholine component [165]. The pseudo-binary phase diagrams (in excess water) for a number of DPPC-fatty acid systems have been reported [170]. However, it was subsequently found that the fluid phase for the stoichiometric 2/1 mixtures was not lamellar but an H_{II} phase [171–174]. For the shorter chain length C_{12} and C_{14} mixtures, inverse cubic phases are formed [175,36]. Unsaturated fatty acids, which are fusogenic agents, are known to induce the formation of H_{II} phase on addition to phospholipid or certain cell membranes [176,177].

Incorporation of the fusogen monoolein into bilayers tends to promote formation of nonlamellar phases such as H_{II} [177,178] and bicontinuous cubic phases [179]. This behavior is not entirely unexpected, since pure monoglycerides themselves have a strong tendency to form inverse phases such as H_{II} and bicontinuous cubic phases, in the excess water region [179,180,41].

The effects of diacylglycerols on lipid membranes is of particular interest because they are involved in the activation of protein kinase C [181] and of the intracellular phospholipases A_2 and C [182]. These effects may be related to the fact that diglycerides are potent promoters of nonlamellar phases in various phospholipid systems, such as PE, PC, and phosphatidylserine [183,184,182]. Upon incorporation of as little as 1 mol % diolein or dilinolenin, the L_α–H_{II} transition temperature of fully hydrated palmitoyl-oleoyl PE is lowered by approximately 9°C [185]. When the amount of diacylglycerol incorporated into PC is very large (approximately 70 mol %), the H_{II} phase is transformed into a cubic phase [184], which has been identified as having crystallographic space group Fd3m (Q^{227}) [51]. This Fd3m phase is the first well-established example of a cubic phase whose structure is based on a (complex) packing of discontinuous inverse micelles [186,52].

The effect of cholesterol tends to be to broaden the chain melting transition of phospholipid bilayers, disordering the gel phase but ordering the fluid lamellar phase. The effect on the phase behavior of those phospholipids having a tendency for forming inverted phases is usually to promote H_{II} phase formation. This effect is consistent with its small headgroup size, weak hydrophilicity, and capacity for bonding to phospholipid headgroups, thereby reducing headgroup hydration [187,188].

The L_α-H_{II} transition of most PEs is reduced by incorporation of either cholesterol [189] or various cholesterol esters [190], although the effect can reverse at high concentrations [191]. In addition to the effects on PE systems, it has been reported that incorporation of equimolar cholesterol can also induce H_{II} phase formation in long chain polyunsaturated PCs [192]. For phospholipid mixtures, addition of cholesterol can promote H_{II} phase formation, stabilize the lamellar phase, or induce the appearance of isotropic, probably cubic phases.

Local anesthetics and drugs have a complex variety of effects on lipid polymorphism. The mechanisms probably span all of the effects discussed at the beginning of this section. Thus dibucaine and chlorpromazine were found to induce an L_α-H_{II} phase transition in cardiolipin hydrated at close to neutral pH, probably by charge neutralization [193]. For phosphatidic acid systems, chlorpromazine was effective in inducing H_{II} phase formation at pH 6 but not at pH 4 [194]. On the other hand, dibucaine was found to reverse the calcium-induced L_α-H_{II} transition in PE-PS mixtures [195], and various such charged anesthetics stabilize the bilayer phase of PE systems [169].

3. Phospholipid-Nonpolar Solutes

Although a vast literature exists on ternary phase equilibria in amphiphile-water-nonpolar solute systems [64], the effect of nonpolar solutes on the polymorphism of hydrated phospholipids [153] has been studied only relatively recently [102,158,196–198]. The effect of nonpolar solutes such as alkanes depends strongly on their chain length. Short chain alkanes tend to partition into the center of the bilayer and depress the chain melting transition; conversely, long chain ($>C_{12}$) alkanes tend to align with the phospholipid chains and increase the transition temperature.

Nonpolar solutes can both increase the tendency for a negative mean curvature of the lipid monolayer and, by partitioning into the interstices between the water-lipid cylinders, can also relieve stress in the H_{II} phase caused by the necessity that all of the hydrophobic volume be filled. Thus in general alkanes promote the formation of inverse nonlamellar phases such as H_{II}. For example, incorporation of approximately 12 mol % dodecane (C_{12} *n*-alkane) was found to lower the L_α-H_{II} transition temperature of fully hydrated egg PE by 18°C [169]. For such pure PE systems with a strong tendency to form H_{II} phases, the effects are relatively small. However, for mixtures of PE with added PC, which has the effect of strongly stabilizing the lamellar phase, dramatic effects are observed. For example, addition of 5% dodecane to a 3/1 DOPE/DOPC mixture reduced the L_α-H_{II} transition temperature by 55°C [100].

Alkanes can even induce cubic and H_{II} phases in pure PC systems, when the amount of incorporated alkane is of the order of 8 wt % or greater, depending on temperature, water content, and chain length of the alkane [196–198].

H. Phase Metastability

Lyotropic phases of phospholipids are not necessarily true equilibrium phases. In certain situations they revert spontaneously upon incubation to more ordered, usually lamellar phases. The time scale for this metastability can range from seconds to months or longer. It is thus of paramount importance in any studies of model membrane systems (e.g., by spectroscopy) to control carefully (for example, by x-ray diffraction) which phase the system is actually in during the experiment. Failure to do this may lead to entirely fallacious interpretations of the data (there are a number of publications where this has clearly happened). Metastability is actually a rather general phenomenon in lyotropic systems and is in fact well known in surfactants. There, gel phases may transform to hydrated "coagels," or suspensions of microcrystals, which hydrate, either to gel or to fluid phases, on heating above a critical temperature known as the Krafft point.

Incubation of phosphatidylcholines at low temperatures leads to a conversion of the $L_{\beta'}$ gel phase to so-called subgel phases [82,79,199,200,83,201]. For dipalmitoylphosphatidylcholine (DPPC), this occurs with incubation below approximately 8°C. On subsequent heating, the subgel has an endothermic transition ("subtransition") to the normal ($L_{\beta'}$) gel phase. The structure of the subgel phase appears to have a crystalline chain packing, but it is not fully clear whether the bilayers remain weakly coupled (2-D-crystalline lamellar phase) or are strongly coupled (3-D lamellar crystal). This behavior is not just limited to saturated systems; dioleoyl phosphatidylcholine appears to form a subgel phase below approximately −17°C [202].

For more weakly hydrated phospholipids such as PE, the tendency for metastability is stronger [19,203–208], and the subgel phases tend to have transition temperatures to the fluid phase that are higher than those of the gel phase. This implies that not only the gel but also the fluid lamellar phase may be metastable over certain temperature ranges. For example, dilauroylphosphatidylethanolamine (DLPE), which has a gel-fluid transition temperature of 30.5°C, is metastable below 43°C [19]. In this case the subgel phases are true 3-D lamellar crystals, which are anhydrous (or very nearly so). A scheme of the pattern of metastability observed in L-DLPE is shown in Figure 16a, and that of DPPC is shown in Figure 16b for comparison. The more stable crystalline form $L_{c'}$ (layer spacing 45.5 Å) is stable below 43°C. Another, less stable, crystalline form L_c (layer spacing 37.8Å) can also be adopted on incubation of the L_β gel phase (but forms less readily than $L_{c'}$). This L_c phase has a strong endothermic transition to L_α at 35°C, but with a lower enthalpy than that of the more stable $L_{c'}$ form at 43°C. Incubation of the fluid bilayer L_α phase between 30.5 and 43°C leads to spontaneous reversion (within hours or days) to the most stable $L_{c'}$ crystalline form. For racemic DL-DPPE, the dehydrated crystalline form is particularly stable, having a chain melting transition in water (on the initial heating scan) at 82°C [209].

With increasing chain length, the time required for metastable reversion becomes slower, until by C_{20} (diarachinoyl), the gel phase appears to be the stable phase.

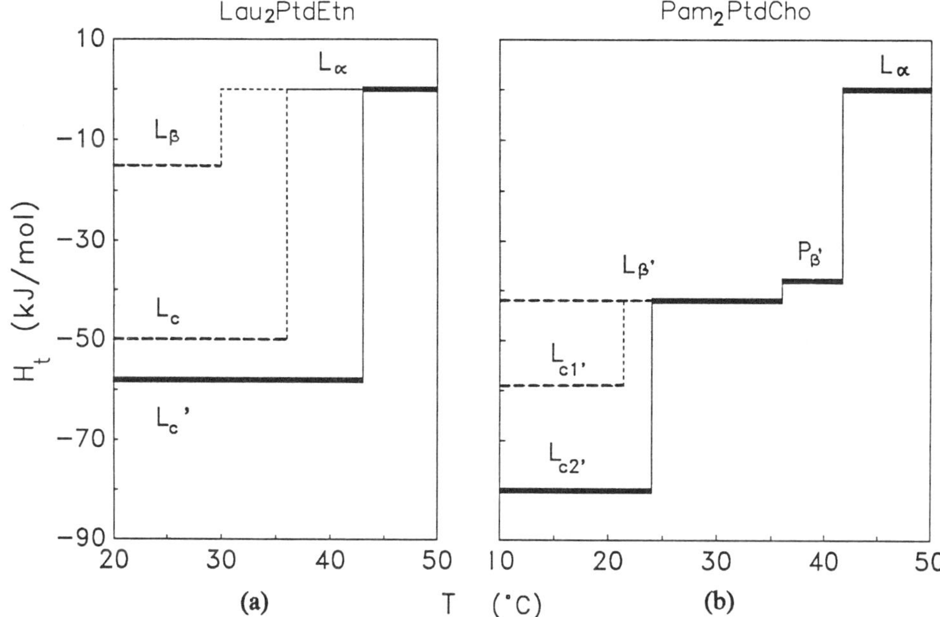

Figure 16 Metastability of: (a) L-dilauroylphosphatidylethanolamine (from Ref. 19) and (b) dipalmitoylphosphatidylcholine (from Ref. 75).

Concomitantly, the stability of the crystalline forms relative to that of the gel phase decreases (i.e., the chain melting temperature rises less rapidly with chain length than that of the gel) so that the two transitions eventually merge.

Similar chain length effects also occur with phosphatidylcholines, although in this case the subgel transition is below the gel-fluid transition temperature for all chain lengths greater than C_{12} [210]. Hydrostatic pressure slows the rate of subgel formation in dipalmitoylphosphatidylcholine, implying that the rate-limiting step is the dehydration of the headgroups rather than the ordering of the chains [211].

Metastability is also observed with headgroup-methylated phosphatidylethanolamines [78], with ether-linked (dialkyl) phosphatidylethanolamines [28] and with phosphatidylglycerol [212]. In all of these cases, the Krafft point of the crystalline form is below the gel-fluid transition temperature, and thus they show endothermic transitions to the gel phase upon heating. For the charged lipid phosphatidylglycerol, the subgel structure has been shown to consist of a lamellar stack of 2-D crystalline bilayers [23].

Stoichiometric fatty acid-phosphatidylcholine (2:1) mixtures, which exhibit transitions direct from the gel phase to nonlamellar inverse phases such as H_{II} or cubic [171,175,36] are also metastable. For example, palmitic acid-DPPC (2:1), which undergoes an L_β-H_{II} transition at 62°C, reverts upon incubation at 4°C to a subgel phase, which has a transition on heating at 38°C to the gel phase [172,173].

Metastability is also not restricted to phospholipids; both diacyl and dialkyl glycolipids form subgel phases, either directly on cooling the gel phase or upon incubation [213–215].

I. Effects of Chirality

Molecular chirality is of paramount importance in biology. Since phospholipids are normally chiral, monolayers or bilayers might be expected to exhibit chiral discrimination/stereoselective recognition [216,217]. However, very little evidence for this has in fact been found. Chiral discrimination has been observed in monolayers of *N*-acylaniline [218], suggesting that the lack of effects in phospholipids is due to the relative inaccessibility of the chiral center of the latter [216].

Two-dimensional chiral crystals have in fact been found to form from monolayers of chiral dipalmitoyl phosphatidylcholine [219]. Furthermore, it has been observed that bilayer dispersions of chiral long chain phosphatidylcholines containing polymerizable diacetylenic groups in the chains roll up into helical structures (tubules) of a specific handedness [220]. When the racemic mixture is used, phase separation of the two enantiomers occurs, leading to the formation of mixtures of left- and right-handed helical tubules. It should be noted that the initial thermal behavior of racemic dipalmitoyl phosphatidylethanolamine has been found to be strikingly different from that of the chiral L enantiomer [209].

Although all cubic phases discovered so far in lipid systems are centrosymmetric, chiral ones could also in principle form, and this might be expected when the lipid contains a chiral center. However, phospholipids appear to form only nonchiral cubic phases, whether they are chiral or racemic [36]. It has been found, however, that the chirality of a dialkyl glycolipid does affect the transitions to nonlamellar phases [221]. The only known example so far of a chiral lyotropic cubic phase is that of a lipid-protein mixture at low water content [17].

J. Transition Kinetics

For reviews of phospholipid phase transition kinetics by diffraction, see [222,223].

1. Chain Melting Transition

It is nearly certain that the phospholipid chain melting phase transition involves the nucleation of fluid lipid domains in those regions containing the chains in which the rotational isomers were originally created [224–227]. The transition kinetics is thus dominated by the domain-growth. This is the reason why the relaxation rate of the phase transition goes through a minimum at the transition midpoint. Consequently, the role of cluster growth in a phase transition can be tested experimentally by plotting the logarithm of the measured transition rate as a function of the degree of completion of a given phase transition; if the transition kinetics is dominated by cluster nucleation and growth, such a plot is linear.

Comparison of measured and calculated chain melting relaxation rates suggests that phospholipid chain melting is a first-order phase transition, albeit one that is already approaching second-order. Being a first-order process, it must be also highly cooperative and restricted to a relatively narrow temperature range. The transition cooperativity can be analyzed in terms of cluster models [228], but normally the more phenomenological van't Hoff thermodynamic relation is used [224].

From the measured phase transition width and transition enthalpy the effective number of molecules that simultaneously participate in phospholipid chain melting, the so-called cooperative unit, is deduced simply by taking the ratio of the van't Hoff and the calorimetrically measured transition enthalpies. (This should not be confused with the cluster size, however; at the center of the transition, the latter is much greater.)

In contrast to the small effect of aggregate size on the cooperative unit, the kinetic parameters of bilayer chain melting are strongly influenced by the vesicle size, at least for phosphatidylcholines [229].

The measured relaxation times for the chain melting of phosphatidylcholine, phosphatidic acid, or phosphatidylserine bilayers exhibit a complex, multicomponent kinetic behavior, typical rates ranging from a few hundred microseconds to <2 s [230,231]. The longest values are typically found with large multilamellar systems. It is possible, therefore, that the shorter relaxation times, which are observed consistently for all systems, refer to the intrinsic molecular relaxation processes, whilst the longer ones are due to kinetic trapping of the system. One possible source of such trapping is from the pools of interlamellar water.

2. Transitions Between Ordered Phases

The lipid pretransition, by and large, occurs on the time scale of (tens of) minutes but may require several hours for completion [232,233]. During the first, rapid stage of a pretransition, the ripple periodicity changes; at the later stage, the remaining imperfections are eliminated [234]. The low speed of bilayer undulation is probably due to the slowness of the solvent and solute exchange across the relatively tightly packed gel-phase bilayers, rather than being a consequence of the slow moleculr relaxation processes [90]. The fact that segmental phospholipid rotation rates, even below the pretransition temperature, are on the order of milliseconds or faster [235,85] (see also Chap. 13) supports this suggestion.

The situation is even more dramatic with lipid subtransitions, which may take days or weeks for completion [82–84] (see also the section on metastability). However, the lower the water content in a system, the more rapid is the chain crystalization.

3. Nonlamellar Transitions

The possible role of nonlamellar structures (such as inverse micelles) as intermediates in the L_α-H_{II} transition has been discussed by Cullis, De Kruijff, Verkleij and coworkers [70–72]. They suggested that the transition proceeds via formation of lipidic particles, or intermembrane inverted micelles.

Siegel has developed a model for lamellar to nonlamellar transitions, which attempts to analyze the role of intermediate structures in the mechanisms of the transitions and to obtain information on the kinetics [236,237,96]. The model takes the primary initial event in the transition to be the formation of "inverted micellar intermediates" (IMI) between pairs of apposed bilayers. After formation, so long as their density is sufficiently high, and the rate of reversion is sufficiently slow, the IMI can either fuse into "rod micellar intermediates" or form "line defects," which then assemble into the H_{II} phase.

Time-resolved diffraction has been used to study the kinetics of the L_α-H_{II} phase transition of phosphatidylethanolamines [238,239,222,223] and the L_β-H_{II} transition of phosphatidylcholine-fatty acid mixtures [173]. The conclusions that emerge from such studies are

1. The transitions are relatively fast, requiring of the order of 1–10 sec for completion, irrespective of whether the lamellar phase is fluid or ordered.
2. The transitions appear to be two-state processes, with no evidence for intermediate structures or phases being observed, at least at the sensitivity of these experiments.

3. The behavior on heating and cooling is fully reversible.
4. The diffraction lines of the different phases remain sharp throughout the transitions, indicating a high degree of long range order within the coexisting phase domains.

By using an IR laser heating method, the T-jump experiment can be carried out with 2 ms time resolution [223]. Using this technique, it has been possible to observe two distinct steps in the L_α-H_{II} transition. First, there is a rapid (<20 ms) thinning of the lamellar spacing. Then the H_{II} phase begins to appear, annealing to the final undistorted structure only after some seconds.

It should be noted that the rate of the L_α-H_{II} transition is probably in general limited by the speed at which water reequilibrates between the two phases, rather than being controlled by the time required for the rearrangement of the lipid molecules at the transition [238].

Little kinetic data is as yet available on phospholipid cubic phase transitions, although time-resolved diffraction studies have been carried out on monoglycerides [240]. In certain phospholipid systems, cubic phases appear to be metastable. A cubic phase in fully hydrated DOPE was very difficult to induce to form yet strongly metastable once formed [241]. Similarly, a sample of fully hydrated *N*-methyl-DOPE, which has an L_α-H_{II} transition at 65–75°C, did not return to L_α upon very slow cooling but instead transformed to two coexisting cubic phases [242].

ACKNOWLEDGMENTS

This work was supported in part by grants GR/C/95428, GR/F/20780, and GR/F/44052 from the Science and Engineering Research Council (U.K.) to J.M.S.

REFERENCES

1. V. Luzzati, X-ray diffraction studies of lipid-water systems, *Biological Membranes*, Vol. 1 (D. Chapman, ed.), Academic Press, London, 1968, pp. 71–123
2. G. G. Shipley, Recent X-ray diffraction studies of biological membranes and membrane components, in *Biological Membranes*, Vol. 2 (D. Chapman and D. F. H. Wallach, eds.), Academic Press, London, 1973. pp. 1–89.
3. J. Charvolin and A. Tardieu, Lyotropic liquid crystals: Structure and molecular motions. *Solid State Phys. Suppl. 14*:209–257 (1978).
4. G. J. T. Tiddy, Surfactant-water liquid crystal phases, *Phys. Rep. 57*:1–46 (1980).
5. D. M. Small, *Handbook of Lipid Research*, Vol. 4, Plenum Press, New York 1986.
6. K. Larsson, Cubic lipid-water phases: Structures and biomembrane aspects, *J. Phys. Chem. 93*:7304–7314 (1989).
7. G. Lindblom and L. Rilfors, Cubic phases and isotropic structures formed by membrane lipids—Possible biological relevance, *Biochim. Biophys. Acta 988*:221–256 (1989).
8. K. Fontell, Cubic phases in surfactant and surfactant-like lipid systems, *Colloid Polym. Sci. 268*:264–285 (1990).
9. J. M. Seddon, Structure of the inverted hexagonal (H_{II}) phase, and non-lamellar transitions of lipids, *Biochim. Biophys. Acta 1031*:1–69 (1990).
10. M. W. Tate, E. F. Eikenberry, D. C. Turner, E. Shyamsunder, and S. M. Gruner, Nonbilayer phases of membrane lipids, *Chem. Phys. Lipids 57*:147–164 (1991).
11. *Chem. Phys. Lipids (Special Issue) 57* (1990).
12. E. Dubois-Violette and B. Pansu, eds., *J. Physique (Colloques)*, International Workshop on Geometry and Interfaces, 1990.

13. G. Cevc and D. Marsh, *Phospholipid Bilayers: Physical Principles and Models*, Wiley, New York, 1987.
14. J. Meunier, D. Langevin, and N. Boccara, eds., *Physics of Amphiphilic Layers,* Springer-Verlag, Berlin, 1987.
15. R. Lipowsky, The conformation of membranes, *Nature 349*:475–481 (1991).
16. J. Prost and F. Rondolez, Structures in colloidal physical chemistry, *Nature (Supplement) 350*:11–23 (1991).
17. P. Mariani, V. Luzzati, and H. Delacroix, Cubic phases of lipid-containing systems. Structure analysis and biological implications, *J. Mol. Biol. 204*:165–189 (1988).
18. T. Gulik-Krzywicki, L. P. Aggerbeck, and K. Larsson, The use of freeze-fracture and freeze-etching electron microscopy for phase analysis and structure determination of lipid systems, In *Surfactants in Solution*, Vol. 1 (K. Mittal and B. Lindman, eds.), Plenum Press, New York, pp. 237–257.
19. J. M. Seddon, K. Harlos, and D. Marsh, Metastability and polymorphism in the gel and fluid bilayer phases of dilauroyl phosphatidylethanolamine, *J. Biol. Chem. 258*:3580–3854 (1983).
20. M. Elder, P. Hitchcock, R. Mason, and G. G. Shipley, A refinement analysis of the crystallography of the phospholipid, 1,2-dilauroyl-DL-phosphatidylethanolamine, and some remarks on lipid-lipid and lipid-protein interactions, *Proc. R. Soc. Lond. A 354*:157–170 (1977).
21. I. Pascher and S. Sundell, Membrane lipids: Preferred conformational states and their interplay. The crystal structure of dilauroylphosphatidyl-*N,N*-dimethylethanolamine, *Biochim. Biophys. Acta 855*:68–78 (1986).
22. R. H. Pearson and I. Pascher, The molecular structure of lecithin dihydrate, *Nature 281*:499–501 (1979).
23. A. E. Blaurock and T. J. McIntosh, Structure of the crystalline bilayer in the subgel phase of phosphatidylglycerol, *Biochemistry 25*:299–305 (1986).
24. A. Tardieu, V. Luzzati, and F. C. Reman, Structure and polymorphism of the hydrocarbon chains of lipids: A study of lecithin-water phases, *J. Mol. Biol. 75*:711–733 (1973).
25. A. Watts, K. Harlos, W. Maschke, and D. Marsh, Control of the structure and fluidity of phosphatidylglycerol bilayers by pH titration, *Biochim. Biophys. Acta 510*:63–74 (1978).
26. J. Stümpel, K. Harlos, and H. Eibl, Charge-induced pretransition in phosphatidylethanolamine multilayers: The occurrence of ripple structures, *Biochim. Biophys. Acta 599*:464–472 (1980).
27. K. Harlos, J. Stümpel, and H. Eibl, Influence of pH on phosphatidic acid multilayers: A rippled structure at high pH values, *Biochim. Biophys. Acta 555*:409–416 (1979).
28. J. M. Seddon, G. Cevc, R. D. Kaye, and D. Marsh, X-ray diffraction study of the polymorphism of hydrated diacyl and dialkyl phosphatidylethanolamines, *Biochemistry 23*:2634–2644 (1984).
29 J. L. Ranck, T. Keira, and V. Luzzati, A novel packing of the hydrocarbon chains in lipids: The low temperature phases of dipalmitoyl phosphatidylglycerol, *Biochim. Biophys. Acta 488*:432–441 (1977).
30. J. C. W. Shepherd and G. Büldt, *Biochim. Biophys. Acta 558*:41–47 (1979).
31. J. Doucet, A. M. Levelut, and M. Lambert, X-ray study of single domains of 1,2-dipalmitoyl-*sn*-phosphatidylcholine with less than 5% water, *Acta Cryst. B39*:724–731 (1983).
32. C. R. Safinya, D. Roux, G. S. Smith, S. K. Sinha, P. Dimon, N. Clark, and A. M. Bellocq, Steric interactions in a model multimembrane system: A synchroton X-ray study, *Phys. Rev. Lett. 57*:2718–2721 (1986).
33. P. Kékicheff and B. Cabane, Between cylinders and bilayers: Structures and intermediate mesophases of the SDS/water system, *J. Physique 48*:1571–1583 (1987).

34. T. Hahn, ed., *International Tables for Crystallography* Volume A, D. Reidel, Dordrecht, 1983.
35. V. Luzzati and P. A. Spegt, Polymorphism of lipids, *Nature 215*:701–704 (1967).
36. J. M. Seddon, J. L. Hogan, N. A. Warrender, and E. Pebay-Peyroula, Structural studies of phospholipid cubic phases, *Progr. Colloid Polym. Sci. 81*:189–197 (1990).
37. A. Tardieu, Ph.D. thesis, Université de Paris-Sud, 1972.
38. W. Longley and T. J. McIntosh, A bicontinuous tetrahedral structure in a liquid-crystalline lipid, *Nature 303*:612–614 (1983).
39. L. E. Scriven, Equilibrium bicontinuous structure, *Nature 263*:123–125 (1976).
40. L. E. Scriven, Equilibrium bicontinuous structures, in *Micellization, Solubilization, and Microemulsions*, Vol. 2 (K. L. Mittal, ed.), Plenum Press, New York, 1977, pp. 877–893.
41. S. T. Hyde, S. Andersson, B. Ericsson, and K. Larsson, A cubic structure consisting of a lipid bilayer forming an infinite periodic minimal surface of the gyroid type in the glycerolmonooleat-water system, *Z. Krystallogr. 168*:213–219 (1984).
42. J. Charvolin, Crystals of interfaces: The cubic phases of amphiphile/water systems, *J. Physique Colloq. 46*(C3):173–190 (1985).
43. J. Charvolin and J. F. Sadoc, Periodic systems of frustrated fluid films and bicontinuous cubic structures in liquid crystals, *J. Physique 48*:1559–1569 (1987).
44. D. M. Anderson, H. T. Davis, L. E. Scriven, and J. C. C. Nitsche, Periodic surfaces of prescribed mean curvature, *Adv. Chem. Phys. 77*:337–396 (1990).
45. A. Tardieu and V. Luzzati, Polymorphism of lipids. A novel cubic phase: A cage-like network of rods with enclosed spherical micelles, *Biochim. Biophys. Acta 219*:11–17 (1970).
46. K. Fontell, K. K. Fox, and E. Hansson, On the structure of the cubic phase I_1 in some lipid-water systems, *Molec. Cryst. Liquid. Cryst. Letters 1*:9–17 (1985).
47. J. Charvolin and J. F. Sadoc, Periodic systems of frustrated fluid films and micellar cubic structures in liquid crystals, *J. Physique 49*:521–526 (1988).
48. R. Vargas, P. Mariani, A. Gulik, and V. Luzzati, The cubic phases of lipid-containing systems. The structure of phase Q^{223} (space group Pm3n): An X-ray scattering study, *J. Mol. Biol.*, 1992 (in press).
49. P.-O. Eriksson, G. Lindblom, and G. Arvidson, NMR studies of 1-palmitoyl-lysophosphatidylcholine in a cubic liquid crystal with a novel structure, *J. Phys. Chem. 89*:1050–1053 (1985).
50. P.-O. Eriksson, G. Lindblom, and G. Arvidson, NMR studies of micellar aggregates in 1-acyl-*sn*-glycerophosphocholine systems. The formation of a cubic liquid crystalline phase, *J. Phys. Chem. 91*:846–853 (1987).
51. J. M. Seddon, An inverse face-centred cubic phase formed by diacylglycerol-phosphatidylcholine mixtures, *Biochemistry 29*:7997–8002 (1990).
52. V. Luzatti, R. Vargas, A. Gulik, P. Mariani, J. M. Seddon, and E. Rivas, Lipid polymorphism: A correction. The structure of the cubic phase of extinction symbol Fd–consists of two types of disjointed reverse micelles embedded in a 3D hydrocarbon matrix, *Biochemistry 31*:279–285 (1992).
53. P. L. Luisi, Enzymes hosted in reverse micelles in hydrocarbon solution, *Angew. Chem. Int. Ed. Engl. 24*:439–528 (1985).
54. K. Martinek, N. L. Klyachko, A. V. Kabanov, Yu. L. Khmelnitsky, and A. V. Levashov, Micellar enzymology: Its relation to membranology, *Biochim. Biophys. Acta 981*:161–172 (1989).
55. P. Walde, A. M. Giuliani, C. A. Boicelli and P. L. Luisi, Phospholipid-based reverse micelles, *Chem. Phys. Lipids 53*:265–288 (1990).
56. R. Scartazzini and P. L. Luisi, Organogels from lecithins, *J. Phys. Chem. 92*:829–833 (1988).
57. P. Schurtenberger, L. J. Magid, J. Penfold, and R. Heenan, Shear aligned lecithin reverse micelles: A small-angle neutron scattering study of the anomalous water-induced micellar growth, *Langmuir 6*:1800–1803 (1990).

58. P. G. de Gennes, and C. Taupin, Microemulsions and the flexibility of oil/water interfaces, *J. Phys. Chem. 86*:2294–2304 (1982).
59. P. Barois, S. Hyde, B. Ninham, and T. Dowling, Observation of two phases within the cubic phase region of a ternary surfactant, *Langmuir 6*:1136–1140 (1990).
60. S. Radiman, C. Toprakacioglu, L. Dai, A. R. Faruqi, R. P. Hjelm, Jr., and A. de Vallera, Structural features of the cubic phase of a ternary surfactant system, *J. Physique (Colloq.) 51*(C7)375–381 (1990).
61. G. Porte, J. Appel, P. Bassereau, and J. Marignan, L_α to L_3: A topology driven transition in phases of infinite fluid membranes, *J. Phys. France 50*:1355–1347 (1989).
62. M. E. Cates and D. Roux, Random bilayer phases of dilute surfactant solutions, *J. Phys. Condens. Matter 2*:SA339–SA346 (1990).
63. P. A. Winsor, Binary and multicomponent solutions of amphiphilic compounds. Solubilization and the formation, structure, and theoretical significance of liquid crystalline solutions, *Chem. Rev. 68*:1–40 (1968).
64. P. Ekwall, Composition, properties and structures of liquid crystalline phases in systems of amphiphilic compounds, *Adv. Liq. Cryst. 1*:1–142 (1975).
65. A. G. Lee, Lipid phase transitions and phase diagrams. II. Mixtures involving lipids, *Biochim. Biophys. Acta 472*:285–344 (1977).
66. A. G. Lee, Calculation of phase diagrams for non-ideal mixtures of lipids, and a possible non-random distribution of lipids in lipid mixtures in the liquid crystalline phase, *Biochim. Biophys. Acta 507*:433–444 (1978).
67. D. Marsh, *Handbook of Lipid Bilayers*, CRC Press, Boca Raton, Fla., 1987.
68. M. Caffrey, D., Moynihan, and J. Hogan, A database of lipid phase transition temperatures and enthalpy changes, *Chem. Phys. Lipids 57*:275–291 (1991).
69. J. R. Silvius, Lipid phase transitions, in *Lipid-Protein Interactions*, (P. C. Jost, and O. H. Griffith, eds.), Wiley Interscience, New York, 1982, Vol. 2, pp. 239–281.
70. P. R. Cullis, M. J. Hope, B. de Kruijff, A. J. Verkleij, and C. P. S. Tilcock, Structural properties and functional roles of phospholipids in biological membranes, in *Phospholipids and Cellular Recognition* (J. F. Kuo, ed.), CRC Press, Boca Raton, Fla., 1985, pp. 1–59.
71. P. R. Cullis, M. J. Hope, and C. P. S. Tilcock, Lipid polymorphism and the roles of lipids in membranes, *Chem. Phys. Lipids 40*:127–144 (1986).
72. B. de Kruijff, P. R. Cullis, A. J. Verkleij, M. J. Hope, C. J. A. van Echteld, and T. F. Taraschi, Lipid polymorphism and membrane function, in *The Enzymes of Biological Membranes*, 2nd edition (A. N. Martinosi, ed.), Plenum Press, New York, 1985. pp. 131–204.
73. C. P. S. Tilcock, Lipid polymorphism, *Chem. Phys. Lipids, 40*:109–125 (1986).
74. S. M. Gruner, Stability of lyotropic phases with curved interfaces. *J. Phys. Chem. 93*:7562–7570 (1989).
75. G. Cevc, Isothermal lipid phase transitions, *Chem. Phys. Lipids 57*:293–307 (1991).
76. D. Marsh, General features of phospholipid phase transitions, *Chem. Phys. Lipids 57*:109–120 (1991).
77. H. Träuble, M. Teubner, P. Wolley, and H. Eibl, Electrostatic interactions at charged lipid membranes. 1. Effects of pH and univalent cations on membrane structure, *Biophys. Chem. 4*:319–346, (1976).
78. S. Mulakutla and G. G. Shipley, Structure and thermotropic properties of phosphatidylethanolamine and its *N*-methyl derivatives, *Biochemistry 23*:2514–2519 (1984).
79. M. J. Ruocco and G. G. Shipley, Characterization of the sub-transition of hydrated dipalmitoylphosphatidylcholine bilayers. Kinetic, hydration and structural study, *Biochim. Biophys. Acta 691*:309–320 (1982).
80. K. Harlos, Pretransitions in the hydrocarbon chains of phosphatidylethanolamines. A wide angle x-ray diffraction study, *Biochim. Biophys. Acta 511*:348–353 (1978).
81. E. F. Gudgin, D. G. Cameron, and H. H. Mantsch, Dependence of acyl chain packing of phospholipids on the head group and acyl chain length, *Biochemistry 20*:4496–4500 (1981).

82. S. C. Chen, J. M. Sturtevant, and B. J. Gaffney, Scanning calorimetric evidence for a third phase transition in phosphatidylcholine bilayers, *Proc. Natl. Acad. Sci. USA 77*:5060–5063 (1980).
83. C. P. Yang, M. C. Wiener, and J. F. Nagle, New phases of DPPC/water mixtures, *Biochim. Biophys. Acta 945*:101–104 (1988).
84. H. H. Füldner, Characterization of a third phase transition in multilamellar dipalmitoyllecithin liposomes, *Biochemistry 20*:5707–5710 (1981).
85. L. W. Trahms, D. Klabe, and E. Boroske, H-NMR study of the three low temperature phases of DPPC-water systems, *Biophys. J. 42*:285–293 (1983).
86. J. Feuer, B. Stamatoff, H. J. Guggenheim, G. Tellez, and T. Yamane, Amplitude of rippling in the P(beta) phase of dipalmitoylphosphatidylcholine bilayers, *Biophys. J. 38*:217–226 (1982).
87. A. V. Parsegian, Dimensions of the "intermediate phase" of dipalmitoylphosphatidylcholine, *Biophys. J. 44*:413–415 (1983).
88. J. A. N. Zasadzinski, J. Schnier, V. Gurley, V. Elings, and P. K. Hansma, Scanning tunneling microscopy of freeze-fracture replicas of biomembranes, *Science 239*:1013–1015 (1988).
89. G. Cevc, Polymorphism of bilayer membranes in the ordered phase and the molecular origin of lipid pretransition and rippled lamellae, *Biochim. Biophys. Acta 1062*:59–69 (1991).
90. S. L. Neidleman, Effects of temperature on lipid unsaturation, *Biotechn. Gen. Engineering Rev. 5*:245–268 (1987).
91. P. L. Steponkus, M. Uemura, R. A. Balsamo, T. Arvinte, and D. V. Lynch, Transformation of the cryobehavior of rye protoplasts by modification of the plasma membrane lipid composition, *Proc. Natl. Acad. Sci. USA 85*:9026–9030 (1989).
92. J. H. Crowe, L. M. Crowe, and D. Chapman, Preservation of membranes in anhydrobiotic organisms: The role of trehalose, *Science 223*:701–703 (1984).
93. G. Bryant and J. Wolfe, Can hydration forces induce lateral phase separations in lamellar phases? *Eur. Biophys. J. 16*:369–374 (1989).
94. G. Ramsay, R. Prabhu, and E. Freire, Direct measurment of the energetics of association between myelin basic protein and phosphatidylserine vesicles, *Biochemistry 25*:2265–2270 (1986).
95. E. Bernard, J.-F. Faucon, and J. Dufourcq, Phase separations induced by melittin in negatively-charged phospholipid bilayers as detected by fluorescence polarization and differential scanning calorimetry, *Biochim. Biophys. Acta 688*:152–162 (1982).
96. D. P. Siegel, Inverted micellar intermediates and the transitions between lamellar, cubic, and inverted hexagonal amphiphile phases. III. Isotropic and inverted cubic state formation via intermediates in transitions between L_α and H_{II} phases, *Chem. Phys. Lipids 42*:279–301 (1986).
97. R. Blumenthal, Membrane fusion, in *Curr. Topics in Membr. and Transp. 29*:203–255 (1985).
98. G. Cevc, Molecular theory of lamellar-to-nonlamellar phase transitions in lipid membranes and creation of non-bilayer structures, submitted for publication.
99. S. M. Gruner, Intrinsic curvature hypothesis for biomembrane lipid composition: A role for nonbilayer lipids, *Proc. Natl. Acad. Sci. USA 82*:3665–3669 (1985).
100. G. L. Kirk and S. M. Gruner, Lyotropic effects of alkanes and headgroup composition on the L (alpha)-H_{II} lipid liquid crystal phase transition: Hydrocarbon packing versus intrinsic curvature, *J. Physique 46*:761–769 (1985).
101. G. Cevc, J. M. Seddon, and D. Marsh, Thermodynamic and structural properties of phosphatidylserine bilayermembranes in the presence of lithium ions and protons, *Biochim. Biophys. Acta 814*:141–150 (1985).
102. G. Cevc, How membrane chain melting properties are controlled by the polar surface of the lipid bilayers, *Biochemistry 26*:6305–6310 (1987).

103. G. Cevc, How membrane chain-melting phase transition temperature is affected by the lipid chain-asymmetry and degree of unsaturation: An effective chain-length model, *Biochemistry 30:*7186–7193 (1991).
104. P. G. Barton and F. D. Gunstone, Hydrocarbon chain packing and molecular motion in phospholipid bilayers formed from unsaturated lecithins, *J. Biol. Chem. 250*:4470–4476 (1975).
105. J. R. Silvius, M. Lyons, P. L. Yeagle, and T. J. O'Leary, *Biochemistry 24:* 5388–5395 (1985).
106. D.V. Lynch and P. L. Steponkus, Lyotropic phase behaviour of unsaturated phosphatidylcholine species: Relevance to the mechanism of plasma membrane destabilization and freezing injury, *Biochim. Biophys. Acta 984*:267–272 (1989).
107. J. Mattai, P. K. Sripada, and G. G. Shipley, Mixed-chain phosphatidylcholine bilayers: Structure and properties, *Biochemistry 26*:3287–3297 (1987).
108. Z. Wang, H. Lin, and C. Huang, Differential scanning calorimetry study of a homologous series of fully hydrated saturated mixed chain C(X):C(X+6) phosphatidylcholines, *Biochemistry 29*:7072–7076 (1990).
109. H. Lin, Z. Wang, and C. Huang, The influence of acyl chain-length asymmetry on the phase transition parameters of phosphatidylcholine dispersions, *Biochim. Biophys. Acta 1067*:17–28 (1991).
110. S. Ali, H.-N. Lin, R. Bittman, and C.-H. Huang, Binary mixtures of saturated and unsaturated mixed-chain phosphatidylcholines. A differential scanning calorimetry study, *Biochemistry 28*:522–528 (1989).
111. E. S. Rowe, Induction of lateral phase separations in binary lipid mixtures by alcohol, *Biochemistry 26*:46–51 (1987).
112. C. Huang, Empirical estimation of the gel to liquid crystalline phase transition temperatures for fully hydrated saturated phosphatidylcholines, *Biochemistry 30*:26–30 (1991).
113. P. Laggner, K. Lohner, G. Degovics, K. Mueller, and A. Schuster, Structure and thermodynamics of the dihexadecylphosphatidylcholine-water system, *Chem. Phys. Lipids 44*:31–60 (1987).
114. K. Lohner, A. Schuster, G. Degovics, K. Mueller, and P. Laggner, Thermal phase behaviour and structure of hydrated mixtures between dipalmitoyl- and dihexadecylphosphatidylcholine, *Chem. Phys. Lipids 44*:61–70 (1987).
115. J. T. Kim, J. Mattai, and G. G. Shipley, Bilayer interactions of ether- and ester-linked phospholipids: Dihexadecyl and dipalmitoylphosphatidylcholines, *Biochemistry 26*:6599–6603 (1987).
116. N. S. Haas, P. K. Sripada, and G. G. Shipley, Effect of chain linkage on the structure of phosphatidylcholine bilayers, *Biophys. J. 57*:117–124 (1990).
117. J. T. Kim, J. Mattai, and G. G. Shipley, Gel phase polymorphism in ether-linked dihexadecylphosphatidylcholine bilayers, *Biochemistry 26*:6592–6598 (1987).
118. J. A. Veiro, P. Nambi, and E. S. Rowe, Effect of alcohols on the phase transitions of dihexadecylphosphatidylcholine, *Biochim. Biophys. Acta 943*:108–111.
119. J. M. Seddon, G. Cevc, and D. Marsh, Calorimetric studies of the gel-fluid (L_β-L_α) and lamellar-inverted hexagonal (L_α-H_{II}) phase transitions in dialkyl- and diacylphosphatidylethanolamines, *Biochemistry 22*:1280–1289 (1983).
120. G. Cevc, Colloidal and phase behaviour of biomacromolecules interdepend and are regulated by the supramolecular surface polarity. Examples with lipid bilayer membranes, *J. de Physique 50*:1117–1134 (1989).
121. T. I. Lotta, I. S. Salonen, J. A. Virtanen, K. K. Eklund, and P. K. J. Kinnunen, Fourier transform infrared study of fully hydrated dimyristoylphosphatidylglycerol. Effects of Na^+ on the *sn*-1 and *sn*-3 head stereoisomers, *Biochemistry 27*:8158–8169 (1988).
122. I. S. Salonen, K. K. Eklund, J. A. Virtanen, and P. K. J. Kinnunen, Comparison of the effects of NaCl on the thermotropic behaviour of *sn*-1' and *sn*-3' stereoisomers of 1,2-

dimyristoyl-*sn*-glycero-3-phosphatidylglycerol, *Biochim. Biophys. Acta 982*:205–215 (1989).

123. G. Cevc, Effects of lipid headgroups and (nonelectrolyte) solution on the structural and phase properties of bilayer membranes, *Ber. Bunsen Ges. 92*:953–961 (1988).
124. D. Bach, I. Bursuker, H. Eibl, and I. R. Miller, Differential scanning calorimetry of dipalmitoyl phosphatidylcholine analogues and their interaction products with basic polypeptides, *Biochim. Biophys. Acta 514*:310–319 (1978).
125. G. Cevc and J. M. Seddon, Structural and dynamic consequences of amphiphile hydration, in *Surfactants in Solution* (K. L. Mittal, ed.), Vol. 4, Plenum Press, New York, 1986, pp. 243–252.
126. G. Cevc and D. Marsh, Hydration of noncharged lipid bilayer membranes. Theory and experiments with phosphatidylethanolamines, *Biophys. J. 47*:21–31 (1985).
127. G. Cevc, Lipid hydration, in *Water and Biological Molecules* (E. Westhof, ed.), Macmillan, New York, 1992 (in press).
128. M. Kodama, M. Kuwabara, and S. Seki, Successive phase-transition phenomena and phase diagram of the phosphatidylcholine-water system as revealed by differential scanning calorimetry, *Biochim. Biophys. Acta 689*:567–570 (1982).
129. G. Cevc, A. Watts, and D. Marsh, Titration of the phase transition of phosphatidylserine bilayer membranes. Effects of pH, surface electrostatics, ion binding, head-group hydration, *Biochemistry 20*:4955–4965 (1981).
130. G. Cevc and D. Marsh, Properties of the electrical double layer near the interface between a charged bilayer membrane and electrolyte solution, *J. Phys. Chem. 87*:376–381 (1983).
131. G. Cevc, Membrane electrostatics, *Biochim. Biophys. Acta. Reviews on Membranes, 1031*:311–382 (1990).
132. K. Arnold, W. Gründer, R. Göldner, and A. Hofmann, P-NMR-Specktroskopischer und kalorimetrischer Nachweis des Einflusses von Pr^{3+}-Ionen auf den thermischen Phasenübergang von Phospholipid-Wasser-Systemen, *Z. Phys. Chemie* (Leipzig) *256*:522–528 (1975).
133. K. Arnold and A. Hofmann, Wechselwirkung von La^{3+} mit Phosphatidycholin-Modellmenbranen und Einfluss auf den thermischen Phasenübergang, *Studia Biophys. 59*:139–147 (1976).
134. H. Hauser and G. G. Shipley, Interactions of divalent cations with phosphatidyserine bilayer membranes, *Biochemistry 23*:34–41 (1984).
135. M. Deleers, J.-P. Servais, and E. Wuelfert, Micromolar concentrations of Al^{3+} induce phase separation, aggregation and dye release in phosphatidyserine-containing lipid vesicles, *Biochim. Biophys. Acta 813*:195–200 (1985).
136. H. Eibl and A. Blume, The influence of charge on phosphatidic acid bilayer membranes, *Biochim. Biophys. Acta 553*:476–488 (1979).
137. A. Blume and H. Eibl, The influence of charge on bilayer membranes: Calorimetric investigations of phosphatidic acid bilayers, *Biochim. Biophys. Acta 558*:13–21 (1979).
138. J. M. Boggs, Lipid intermolecular hydrogen bonding: Influence on structural organization and membrane function, *Biochim. Biophys. Acta 906*:353–404 (1987).
139. Y. Kaminoh, F. Kano, J.-S. Chiou, H. Kamaya, S. H. Lin, and I. Ueda, Effect of surface ionization of dimyristoylphosphatidic acid vesicle membranes on the main phase-transition enthalpy and temperature, *Biochim. Biophys. Acta 943*:522–530 (1988).
140. H. Hauser and G. G. Shipley, Crystallization of phosphatidylserine bilayers induced by lithium, *J. Biol. Chem. 256*:11377–11380 (1981).
141. A. A. Cools and L. H. M. Janssen, The influence of Ca^{2+} on the turbidity of DPPC-DMPA vesicles within the temperature range of the phase transition. *Physiol. Chem. Phys. Med. NMR 18*:171–179 (1986).

142. H. L. Casal, H. H. Mantsch, and H. Hauser, Infrared and ^{31}P-NMR studies of the interaction of Mg^{2+} with phosphatidylserines: Effect of hydrocarbon chain unsaturation, *Biochim. Biophys. Acta 982*:228–236 (1989).
143. L. J. Lis, V. A. Parsegian, and R. P. Rand, Binding of divalent cations to dipalmitoylphosphatidylcholine bilayers and its effect on bilayer interaction, *Biochemistry 20*:1761–1770 (1981).
144. D. M. Haverstick, and M. Glaser, Visualization of Ca^{2+}-induced phospholipid domains, *Proc. Natl. Acad. Sci. USA 84*:4475–4479 (1987).
145. E. S. Rowe, Lipid chain length and temperature dependence of ethanol-phosphatidylcholine interactions, *Biochemistry 22*:3299–3305 (1983).
146. A. Simon and T. J. McIntosh, Interdigitated hydrocarbon chain packing causes the biphasic transition behavior in lipid/alcohol suspensions, *Biochim. Biophys. Acta 773*:169–172 (1984).
147. G. Cevc, J. Käs, S. Kirchner, L. Löbbecke, and N. Nagel, Interdigitated phases in the symmetric-chain lipid bilayer membranes, Int. Workshop on "Structure and Conformation of Amphiphilic Membranes," Jülich, 16–18 Sept. 1991, O2.
148. P. W. Sanderson, L. J. Lis, P. J. Quinn, and W. P. Williams, The Hoffmeister effect in relation to membrane phase stability, *Biochim. Biophys. Acta 1067*:43–50 (1991).
149. L. M. Crowe and J. H. Crowe, Solution effects on the thermotropic phase transition of unilamellar liposomes, *Biochim. Biophys. Acta 1064*:267–274 (1991).
150. L. M. Crowe, C. Womersley, J. H. Crowe, D. Reid, L. Appel, and A. Rudolph, Prevention of fusion and leakage in freeze-dried liposomes by carbohydrates, *Biochim. Biophys. Acta 861*:131–140 (1986).
151. J. H. Crowe, B. J. Spargo, and L. M. Crowe, Preservation of dry liposomes does not require retention of residual water, *Proc. Natl. Acad. Sci. USA 84*:1537–1540 (1987).
152. J. H. Crowe and L. M. Crowe, Factors affecting the stability of dry liposomes, *Biochim. Biophys. Acta 939*:327–334 (1988).
153. K. Lohner, Effects of small organic molecules on phospholipid phase transitions, *Chem. Phys. Lipids 57*:341–362 (1991).
154. G. Lindblom, L. Rilfors, I. Brentel, P.-O. Eriksson, G. Wikander, G. Arvidson, and Å Wieslander, Effect of hydrophobic, amphiphilic and peptide molecules on phase structure and dynamics of membrane lipids, *Chem. Scripta 27B*:221–227 (1987).
155. B. de Kruijff, P. R. Cullis, A. J. Verkleij, M. J. Hope, C. J. A. van Echteld, T. F. Taraschi, P. van Hoogevest, J. A. Killian, A. Rietveld, A. and A. T. M. van der Steen, Modulation of lipid polymorphism by lipid-protein interactions, *Progress in Protein-Lipid Interactions* (A. Watts and J. J. H. H. M. de Pont, eds.), Elsevier, Amsterdam, 1985, pp. 89–142.
156. H. Tournois and B. de Kruijff, Polymorphic phospholipid phase transitions as tools to understand peptide-lipid interactions, *Chem. Phys. Lipids 57*:327–340 (1991).
157. R. P. Rand, N. Fuller, V. A. Parsegian, and D. C. Rau, Variation in hydration forces between neutral phospholipid bilayers: Evidence for hydration attraction, *Biochemistry 27*:7711–7722 (1988).
158. M. W. Tate, and S. M. Gruner, Lipid polymorphism of mixtures of dioleoylphosphatidylethanolamine and saturated and monosaturated phosphatidylcholines of various chain lengths, *Biochemistry 26*:231–236 (1987).
159. L. T. Boni and S. W. Hui, Polymorphic phase behaviour of dilinoleoylphosphatidylethanolamine and palmitoyloleoylphosphatidylcholine mixtures: Structural changes between hexagonal, cubic and bilayer phases. *Biochim. Biophys. Acta 731*:177–185 (1983).
160. P.O. Eriksson, L. Rilfors, G. Lindblom, and G. Arvidson, Multicomponent spectra from ^{31}P NMR studies of the phase equilibria in the system dioleoylphosphatidylcholine-dioleoylphosphatidylethanolamine-water, *Chem. Phys. Lipids 37*:357–371 (1985).

161. B. de Kruijff, A. J. Verkleij, J. Leunissen-Bijvelt, C. J. A. van Echteld, J. Hille, and H. Rijnbout, Further aspects of the Ca^{2+}-dependent polymorphism of bovine heart cardiolipin, *Biochim. Biophys. Acta 693*:1–12 (1982).
162. F. K. Hui, and P. G. Barton, Mesomorphic phase behaviour of some phospholipids with aliphatic alcohols and other non-ionic substances, *Biochim. Biophys. Acta 296*:510–517 (1973).
163. A. W. Eliasz, D. Chapman, D. F. Ewing, Phospholipid phase transitions: Effects of *n*-alcohols, *n*-monocarboxylic acids, phenylalkyl alcohols and quaternary ammonium compounds, *Biochim. Biophys. Acta 448*:220–230 (1976).
164. A. G. Lee, Interactions between anaesthetics and lipid mixtures. Normal alcohols, *Biochemistry 15*:2448–2454 (1976).
165. S. Mabrey and J. Sturtevant, Incorporation of saturated fatty acids into phosphatidylcholine bilayers, *Biochim. Biophys. Acta 486*:444–450 (1977).
166. M. J. Pringle, K. W. Miller, Differential effects on phospholipid phase transitions produced by structurally-related long-chain alcohols, *Biochemistry 18*:3314–3320 (1979).
167. J. M. Boggs, G. Rangaraj, and K. M. Koshy, Effect of hydrogen-bonding and non-hydrogen bonding long chain compounds on the phase transition temperatures of phospholipids, *Chem. Phys. Lipids 40*:23–34 (1986).
168. P. Nambi, E. S. Rowe, and T. J. McIntosh, *Biochemistry 27*:9175–9182 (1988).
169. A. J. Hornby and P. R. Cullis, Influence of local and neutral anaesthetics on the polymorphic phase preferences of egg yolk phosphatidylethanolamine, *Biochim. Biophys. Acta 647*:285–292 (1981).
170. S. E. Schullery, T. A. Seder, D. A. Weinstein, D. A. Bryant, Differential thermal analysis of dipalmitoylphosphatidylcholine-fatty acid mixtures, *Biochemistry 20*:6818–6824 (1981).
171. D. Marsh and J. M. Seddon, Gel-inverted hexagonal (L_α-H_{II}) phase transitions in phosphatidylethanolamines and fatty acid-phosphatidylcholine mixtures, demonstrated by ^{31}P NMR spectroscopy and X-ray diffraction, *Biochim. Biophys. Acta 690*:117–123 (1982).
172. R. D. Koynova, A. I. Boyanov and B. G. Tenchov, Gel-state metastability and nature of the azeotropic points in mixtures of saturated phosphatidylcholines and fatty acids, *Biochim. Biophys. Acta 903*:186–196 (1987).
173. R. D. Koynova, B. G. Tenchov, P. J. Quinn, P. Laggner, Structure and phase behaviour of hydrated mixtures of L-dipalmitoylphosphatidylcholine and palmitic acid. Correlations between structural rearrangements, specific volume changes and endothermic events, *Chem. Phys. Lipids 48*:205–214 (1988).
174. G. Cevc, J. M. Seddon, R. Hartung and W. Eggert, Phosphatidylcholine-fatty acid membranes. I. Effects of protonation, salt concentration, temperature, and chain-length on the colloidal and phase properties of mixed vesicles, bilayers, and non-lamellar structures, *Biochim. Biophys. Acta 940*:219–240 (1988).
175. T. Heimburg, N. J. P. Ryba, U. Würz, D. Marsh, Phase transition from a gel to a fluid phase of cubic symmetry in dimyristoylphosphatidylcholine-myristic acid (1:2, mol/mol) bilayers, *Biochim. Biophys. Acta* 1025:77–81 (1990).
176. P. R. Cullis and M. J. Hope, Effects of fusogenic agent on membrane structure of erythrocyte ghosts and the mechanism of membrane fusion, *Nature 271*:672–674 (1978).
177. M. J. Hope and P. R. Cullis, The role of non-bilayer lipid structures in the fusion of human erythrocytes induced by lipid fusogens, *Biochim. Biophys. Acta 640*:82–90 (1981).
178. C. P. S. Tilcock and D. Fisher, Interactions of glycerol monooleate and dimethylsulphoxide with phospholipids: A differential scanning calorimetry and ^{31}P-NMR study, *Biochim. Biophys. Acta 685*:340–346 (1982).
179. H. Gutman, G. Arvidson, K. Fontell, and G. Lindblom, ^{31}P and ^{2}H NMR studies of phase equilibria in the three component system: monoolein-phosphatidylcholine-water in *Surfactants in Solution* (K. L. Mittal and B. Lindman, eds.), Plenum Press, New York, Vol. 1, 1984, pp. 143–152.

180. E. S. Lutton, Phase behaviour of aqueous systems of monoglycerides, *J. Am. Oil Chem. Soc. 42*:1068–1070 (1965).
181. Y. Nishizuka, *Nature 308*:693–698 (1984).
182. R. M. C. Dawson, R. F. Irvine, J. Bray, and P. J. Quinn, Long-chain unsaturated diacylglycerols cause a perturbation in the structure of phospholipid bilayers rendering them susceptible to phospholipase attack, *Biochem. Biophys. Res. Comm. 125*:836–842 (1984).
183. S. Das and R. P. Rand, Diacylglycerol causes major structural transitions in phospholipid bilayer membranes, *Biochem. Biophys. Res. Comm. 124*:491–496 (1984).
184. S. Das, R. P. Rand, Modification by diacylglycerol of the structure and interaction of various phospholipid bilayer membranes, *Biochemistry 25*:2882–2889 (1986).
185. R. M. Epand, Diacylglycerols, lysolecithin, or hydrocarbons markedly alter the bilayer to hexagonal phase transition temperature of phosphatidylethanolamines, *Biochemistry 24*:7092–7095 (1985).
186. J. M. Seddon, E. A. Bartle, and J. Mingins, Inverse cubic liquid-crystalline phases of phospholipids and related lyotropic systems, *Journal of Physics: Condensed Matter 2*:SA285–SA290 (1990).
187. J. Gallay and B. de Kruijff, Correlation between molecular shape and hexagonal H_{II} phase promoting ability of sterols, *FEBS Lett. 143*:133–136 (1982).
188. S. A. Simon, T. J. McIntosh, and R. Latorre, *Science 216*:65–68 (1982).
189. P. C. Noordam, C. J. A. van Echteld, B. de Kruijff, A. Verkleij, and J. de Gier, Barrier characteristics of membrane model systems containing unsaturated phosphatidylethanolamines, *Chem. Phys. Lipids 27*:221–232 (1980).
190. C. P. S. Tilcock, M. J. Hope, and P. R. Cullis, Influence of cholesterol esters of varying unsaturation on the polymorphic phase preferences of egg phosphatidylethanolamine, *Chem. Phys. Lipids 35*:363–370 (1984).
191. R. M. Epand and R. Bottega, Modulation of the phase transition behaviour of phosphatidylethanolamine by cholesterol and oxysterols, *Biochemistry 26*:1820–1825 (1987).
192. C. J. Dekker, W. S. M. van Kessel, J. P. G. Klom, J. Pieters, and B. de Kruijff, Synthesis and polymorphic phase behaviour of polyunsaturated phosphatidylcholines and phosphatidylethanolamines, *Chem. Phys. Lipids 33*:93–106 (1983).
193. P. R. Cullis, A. J. Verkleij, and P. H. J. T. Ververgaert, Polymorphic phase behaviour of cardiolipin as detected by ^{31}P NMR and freeze-fracture techniques. Effects of calcium, dibucaine and chlorpromazine, *Biochim. Biophys. Acta 513*:11–20 (1978).
194. A. J. Verkleij, R. de Maagd, J. Leunissen-Bijvelt, and B. de Kruijff, Divalent cations and chlorpromazine can induce non-bilayer structures in phosphatidic acid-containing model membranes, *Biochim. Biophys. Acta 684*:255–262 (1982).
195. P. R. Cullis and A. J. Verkleij, Modulation of membrane structure by Ca^{2+} and dibucaine as detected by ^{31}P NMR, *Biochim. Biophys. Acta 552*:546–551 (1979).
196. G. Lindblom, M. Sjölund and L. Rilfors, Effect of *n*-alkanes and peptides on the phase equilibria in phosphatidylcholine-water systems. *Liquid Crystals 3*:783–790 (1988).
197. M. Sjölund, G. Lindblom, L. Rilfors, and G. Arvidson, Hydrophobic molecules in lecithin-water systems. 1. Formation of reversed hexagonal phases at high and low water contents, *Biophys. J. 52*:145–153 (1987).
198. M. Sjölund, L. Rilfors, and G. Lindblom, Reversed hexagonal phase formation in lecithin-alkane-water systems with different acyl chain unsaturation and alkane length, *Biochemistry 28*:1323–1329 (1989).
199. J. F. Nagle and D. A. Wilkinson, Dilatometric study of the subtransition in dipalmitoylphosphatidylcholine, *Biochemistry 21*:3817–3821 (1982).
200. R. N.A. H. Lewis, N. Mak, and R. N. McElhaney, A differential scanning calorimetric study of the thermotropic phase behaviour of model membranes composed of phosphatidylcholines containing linear saturated fatty acyl chains, *Biochemistry 26*:6118–6126 (1987).

201. B. Tenchov, On the reversibility of the phase transitions in lipid-water systems. *Chem. Phys. Lipids 57*:165–177 (1991).
202. R. N. A. H. Lewis, B. D. Sykes, and R. N. McElhaney, Thermotropic phase behaviour of model membranes composed of phosphatidylcholines containing cis-monounsaturated acyl chain homologues of oleic acid. Differential scanning calorimetric and ^{31}P spectroscopic studies, *Biochemistry 27*:880–887 (1988).
203. H. Chang and R. M. Epand, The existence of a highly ordered phase in fully hydrated dilauroylphosphatidylethanolamine, *Biochim. Biophys. Acta 728*:319–324 (1983).
204. H. H. Mantsch, S. C. Hsi, K. W. Butler, and D. G. Cameron, Studies on the thermotropic behaviour of aqueous phsophatidylethanolamines, *Biochim. Biophys. Acta 728*:325–330 (1983).
205. B.Z. Chowdry, G. Lipka, A. W. Dalziel and J. M. Sturtevant, Multicomponent phase transitions of diacyl phosphatidylethanolamines, *Biophys. J. 45*:901–904 (1984).
206. D. A. Wilkinson and J. F. Nagle, Metastability in the phase behaviour of dimyristoylphosphatidylethanolamine bilayers, *Biochemistry 23*:1538–1541 (1984).
207. H. Xu, F. A. Stephenson, H. Lin, and C. Huang, Phase metastability and supercooled metastable state of diundecanoylphosphatidylethanolamine bilayers, *Biochim. Biophys. Acta 943*:63–75 (1988).
208. R. Koynova and H.-J. Hinz, Metastable behaviour of saturated phosphatidylethanolamines: A densitometric study, *Chem. Phys. Lipids 54*:67–72 (1990).
209. B.G. Tenchov, A. I. Boyanov, and R. D. Koynova, Lyotropic polymorphism of racemic dipalmitoyphosphatidylethanolamine. A differential scanning calorimetry study, *Biochemistry 23*:3553–3558 (1984).
210. L. Finegold and M.A. Singer, The metastability of saturated phosphatidylcholines depends on the acyl chain length, *Biochim. Biophys. Acta 855*:417–420 (1986).
211. W.-G. Wu, P. L.-G. Chong, and C.-H. Huang, Pressure effect on the rate of crystalline phase formation of L-dipalmitoylphosphatidylcholines in multilamellar dispersions, *Biophys. J. 47*:237–242 (1985).
212. D. A. Wilkinson and T. J. McIntosh, A subtransition in a phospholipid with a net charge, dipalmitoylphosphatidylglycerol, *Biochemistry 25*:295–298 (1986).
213. R. A. Reed and G. G. Shipley, *Biochim. Biophys. Acta 896*:153–164 (1987).
214. H. Kuttenreich, H.-J. Hinz, M. Inczedy-Marcsek, R. Koynova, B. Tenchov, and P. Laggner, Polymorphism of synthetic 1,2-*O*-dialkyl-3-*O*-β-D-galactosyl-*sn*-glycerols of different alkyl chain lengths, *Chem. Phys. Lipids 47*:245–260 (1990).
215. D. A. Mannock, R. N. A. H. Lewis, A. Sen, and R. N. McElhaney, The physical properties of glycosyldiacylglycerols. Calorimetric studies of a homologous series of 1,2-di-*O*-di-3-*O*-(β-D-glucopyranosyl)-*sn*-glycerols, *Biochemistry 27*:6852–6859 (1988).
216. E. M. Arnett, J. M. Gold, N. Harvey, E. A. Johnson, and L. G. Whitesell, Stereoselective recognition in phospholipid monolayers, in *Biotechnological Applications of Lipid Microstructures* (B. P. Gaber, J. M. Schnur, and D. Chapman, eds.), Plenum Press, New York, 1988, pp. 21–36.
217. D. Andelman, Chiral discrimination and phase transitions in Langmuir monolayers, *J. Am. Chem. Soc. 111*:6536–6544 (1989).
218. O. Bouloussa and M. Dupeyrat, Chiral discrimination in *N*-tetradecanoylaniline and *N*-tetradecanoylaniline/ditetradecanoylphosphatidylcholine monolayers, *Biochim. Biophys. Acta*:938:395–402 (1988).
219. R. M. Weis, and H. M. McConnell, Two-dimensional chiral crystals of phospholipid, *Nature 310*:47–49 (1984).
220. A. Singh, T. G. Burke, J. M. Calvert, J. H. Georger, B. Herendeen, R. R. Price, P. E. Schoen and P. Yager, Lateral phase separation based on chirality in a polymerizable lipid and its influence on formation of tubular microstructures, *Chem. Phys. Lipids 47*:135–148 (1988).

221. D. A. Mannock, R. N. A. H. Lewis, R. N. McElhaney, M. Akiyama, H. Yamada, D. C. Turner and S. M. Gruner, The effect of the chirality of the glycerol backbone on the bilayer and nonbilayer phase transitions in the diastereomers of di-dodecyl-D-glucopyranosyl glycerol, *Biophys. J.* (in press).
222. M. Caffrey, The study of lipid phase transition kinetics by time-resolved X-ray diffraction, *Ann. Rev. Biophys. Biophys. Chem. 18*:159–186 (1989).
223. P. Laggner and M. Kriechbaum, Phospholipid phase transitions: Kinetics and structural mechanisms, *Chem. Phys. Lipids 57*:121–145 (1991).
224. B. Grünewald, On the phase transition kinetics of phospholipid bilayers. Relaxation experiments with detection of fluorescence anisotropy, *Biochim. Biophys. Acta 687*:71–78 (1982).
225. U. Strehlow and F. Jähnig, Electrostatic interactions at charged lipid membranes. Kinetics of the electrostatically triggered phase transition. *Biochim. Biophys. Acta 641*:301–310 (1981).
226. T. Y. Tsong, Kinetics of the crystalline-liquid crystalline phase transition of dimyristoyl L-(alpha)-lecithin bilayers, *Proc. Natl. Acad. Sci. USA* 71:2684–2688 (1974).
227. T. Y. Tsong and M. I. Kanehisa, Relaxation phenomena in aqueous dispersions of synthetic lecithins, *Biochemistry 16*:2674–2680 (1977).
228. M. I. Kanehisa and T. Y. Tsong, Cluster model of lipid phase transitions with application to passive permeation of molecules and structure relaxations in lipid bilayers, *J. Am. Chem. Soc. 100*:424–432 (1978).
229. B. Grünewald, S. Stankowski, and A. Blume, Curvature influence on the cooperativity and the phase transition enthalpy of lecithin vesicles, *FEBS Letts. 102*:227–229 (1979).
230. B. Grünewald, A. Blume, and A. Watanabe, Kinetic investigations on the phase transition of phospholipid bilayers, *Biochim. Biophys. Acta 597*:41–52 (1980).
231. M. Caffrey and D. H. Bilderback, Kinetics of the main phase transition of hydrated lecithin monitored by real-time x-ray diffraction, *Biophys. J. 45*:627–631 (1984).
232. B. R. Lentz, R. L. Biltonen, and E. Freire, Fluorescence and calorimetric studies of phase transitions in phophatidylcholine multilayers: Kinetics of the pretransition, *Biochemistry 17*:4475–4482 (1978).
233. M. Akiyama, Y. Terayama, and N. Matsushima, Kinetics of pretransition in multilamellar dimyristoylphosphatidylcholine vesicle. X-ray diffraction study, *Biochim. Biophys. Acta 687*:337–339 (1982).
234. K. Tsuchida, K. Ohki, T. Sekiya, Y. Nozawa, and I. Hatta, Dynamics of appearance and disappearance of the ripple structure in multilamellar liposomes of dipalmitoylphosphatidylcholine, *Biochim. Biophys. Acta 898*:53–58 (1987).
235. E. Boroske and L. Trahms, A ^{1}H and ^{13}C NMR study of motional changes of dipalmitoyl lecithin associated with the pretransition, *Biophys. J. 42*:275–283 (1983).
236. D. P. Siegel, Inverted micellar intermediates and the transitions between lamellar, cubic and inverted hexagonal lipid phases. I. Mechanism of the L_α-H_{II} phase transitions, *Biophys. J. 49*:1155–1170 (1986).
237. D. P. Siegel, Inverted micellar intermediates and the transitions between lamellar, cubic and inverted hexagonal lipid phases. II. Implications for membrane-membrane interactions and membrane fusion, *Biophys. J. 49*:1171–1183 (1986).
238. M. Caffrey, Kinetics and mechanism of the lamellar gel-lamellar liquid-crystal and lamellar/inverted hexagonal phase transition in phosphatidylethanolamine: A real-time X-ray diffraction study using synchrotron radiation, *Biochemistry 24*:4826–4844 (1985).
239. P. Laggner, K. Lohner, and K. Müller, X-ray cinematography of phospholipid phase transformations with synchrotron radiation, *Mol. Cryst. Liq. Cryst. 151*:373–388 (1987).
240. M. Caffrey, Kinetics and mechanism of transitions involving the lamellar, cubic, inverted hexagonal, and fluid isotropic phases of hydrated monoglycerides monitored by time-resolved X-ray diffraction, *Biochemistry 26*:6349–6363 (1987).

241. E. Shyamsunder, S. M. Gruner, M. W. Tate, D. C. Turner, P. T. C. So, and C. P. S. Tilcock, Observation of inverted cubic phase in hydrated dioleoylphosphatidylethanolamine membranes, *Biochemistry 27*:2332–2336 (1988).
242. S. M. Gruner, M. W. Tate, G. L. Kirk, P. T. C. So, D. C. Turner, D. T. Keane, C. P. S. Tilcock, and P. R. Cullis, X-ray diffraction study of the polymorphic behaviour of *N*-methylated dioleoylphosphatidylethanolamine, *Biochemistry 27*:2853–286 (1988).
243. H. Hauser, I. Pascher, I. H. Pearson, and S. Sundell, Preferred conformation and molecular packing of phosphatidylethanolamine and phosphatidylcholine, *Biochim. Biophys. Acta 650*:21–51 (1981).

13
Dynamic Properties

Alfred Blume *University of Kaiserslautern, Kaiserslautern, Germany*

I. INTRODUCTION

The fluid mosaic membrane model is the currently accepted model for the structure of a biological membrane (Singer and Nicholson, 1972; Capaldi and Green, 1972; for a review see Gennis, 1989). As this name implies, the membrane is a dynamic structure, and the "fluidity" is an essential feature of the whole system, as it is relevant to the proper functioning of the whole membrane and particularly to the functioning of the proteins bound to or embedded in the membrane. In the fluid mosaic model, the proteins are thought to be freely diffusing in the plane of the bilayer built up by the membrane lipids. Consequently, the question of how fast these proteins and the surrounding lipids are diffusing is of considerable interest and of great importance for the understanding of the functions of biological membranes.

Studying membrane dynamics requires the application of sophisticated physicochemical techniques, on preferably simple model systems. For this reason many investigations have been performed using model membranes of synthetic phospholipids. Protein dynamics can also be studied in these systems after reconstituting selected membrane proteins into the lipid model bilayers.

In lipid membranes a wide range of motional modes are present. The characteristic time constants for these motions range from 10^{-14} s for molecular vibrations to hours or days for the so-called lipid flip-flop, i.e., the transbilayer exchange of lipid molecules (Kornberg and McConnell, 1971). No single technique is suited to study time constants for molecular motions over such an enormous time range. Consequently a whole range of different methods, particularly spectroscopic techniques, have been applied to study membrane dynamics (for comprehensive reviews see Grell, 1981).

The spectroscopic techniques have different frequency ranges in which they are sensitive to molecular motions. In the past, this has led to considerable confusion, as the terms "fast" and "slow" are not adequate to describe the reorientational motions as they are related to the particular time scale of the spectroscopic method. These terms should therefore be avoided or only used referring to the time scale of the technique. Using one particular method will possibly give a time averaged picture of the dynamic properties, while using other methods the motion will be slow on their particular time scale, thus giving a snapshot, i.e., a static picture of all possible orientations and conformations of the molecules in the system. We will now discuss the various motional modes present in lipid bilayers.

II. MOTIONAL MODES

Due to the anisotropic nature of the phospholipid bilayer with the polar head groups exposed to athe surrounding water and the apolar fatty acyl chains being located in the interior of the bilayer, the reorientational motions are not isotropic. The reorientations about different molecular axes occur on vastly different time scales. Bilayers can be viewed as two-dimensional fluidlike systems when they are in the liquid crystalline state. At lower temperatures, they form more solid gellike lamellae. In principle, the following reorientational motions are possible, the extent and correlation times of these motions depending on the bilayer state: (a) intramolecular reorientational motions, i.e., torsional oscillations around single bonds, isomerizations between different conformations (such as trans-gauche isomerizations in the fatty acid chains), and free or only marginally hindered rotations of certain groups or segments (such as methyl groups or the whole head group); (b) anisotropic rotation around the long molecular axis (Brownian rotational diffusion or jump diffusion); (c) anisotropic rotation around an axis perpendicular to the long axis. A 180° rotation leads to a flip-flop of whole molecules from one side of the bilayer to the other. This type of motion is very slow (Kornberg and McConnell, 1971; Albert and Yeagle, 1982; Schroit and Zwaal, 1991). However, in the liquid crystalline phase of lipids fast wobbling motions within a restricted angular range determined by the neighboring chains (wobbling in a cone) are possible (Petersen and Chan, 1977; Kinosita et al., 1977; Pace and Chan, 1982a,b); (d) Brownian translational diffusion or jump diffusion of the lipid molecules in the bilayer of the bilayers (Pace and Chan, 1982a,b); (e) collective processes, such as surface undulations of the bilayers. These motions lead to relatively slow director fluctuations (Pace and Chan, 1982a,b).

Figure 1 illustrates these different molecular motions. In Fig. 2 the characteristic correlation times of these reorientational modes in relation to the time ranges where the particular methods are sensitive to these motions are shown. Table 1 (see page 463) summarizes the motional correlation times of phospholipids in their different phase states. In the following we will describe suitable spectroscopic techniques and the type of information on the various motional modes that can be obtained with these different methods.

III. METHODS

A. FT-IR and Raman Spectroscopy

FT-IR and Raman spectroscopy are suitable for the determination of equilibrium conformational characteristics of lipid molecules in their different phases (for reviews see Chapman, 1965; Fringeli and Günthard, 1981; Lord and Mendelsohn, 1981; Amey and Chapman, 1984; Verma and Wallach, 1984; Mendelsohn and Mantsch, 1986; Mantsch et al., 1986). Vibrational spectroscopy is a very fast method. It gives an instantaneous picture of the conformational equilibrium populations because the dynamics of the conformational transitions are much slower.

Several vibrational bands can be used as diagnostic tools. Figure 3 shows an FT-IR spectrum of a film of DMPC monohydrate specifically deuterated at the 4 position of both acyl chains (Hübner, 1988). The most prominent feature in the vibration spectrum of all

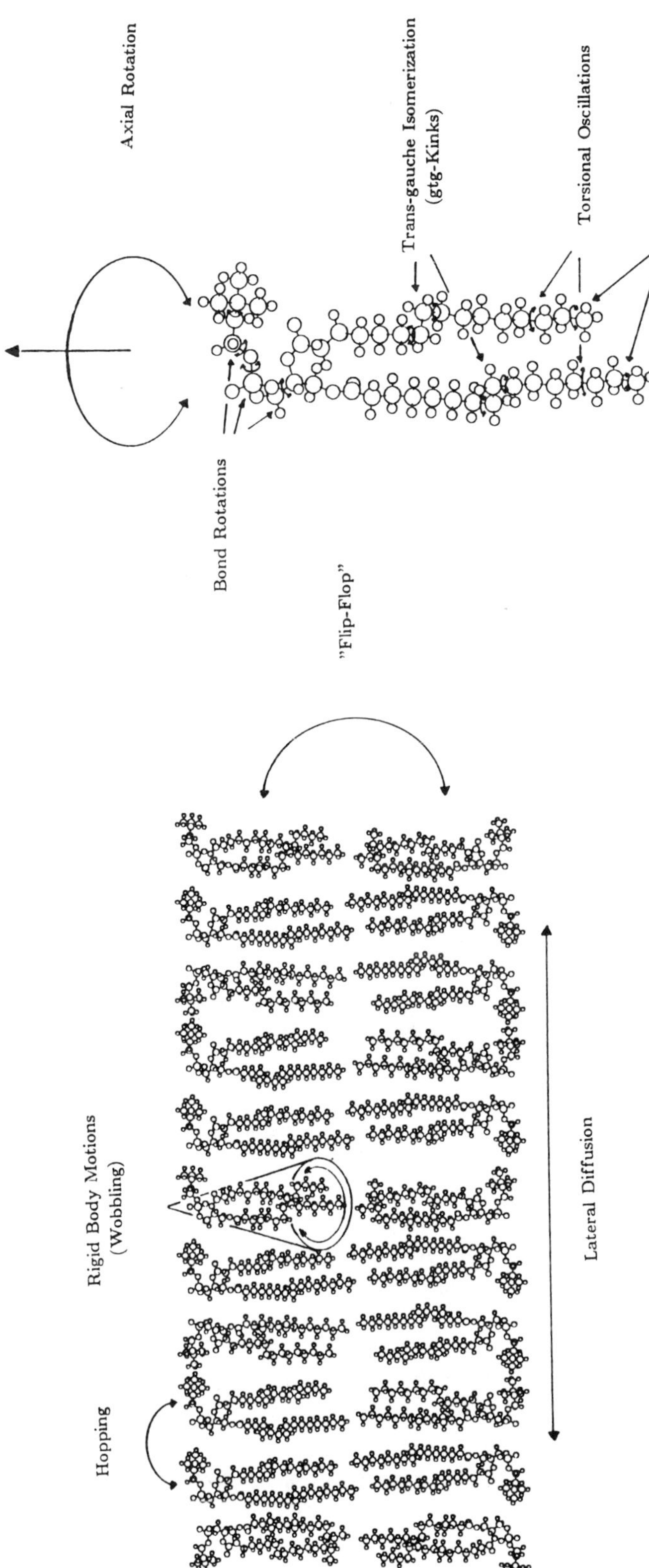

Figure 1 Schematic illustration of the various motional modes observed for lipid molecules in bilayers.

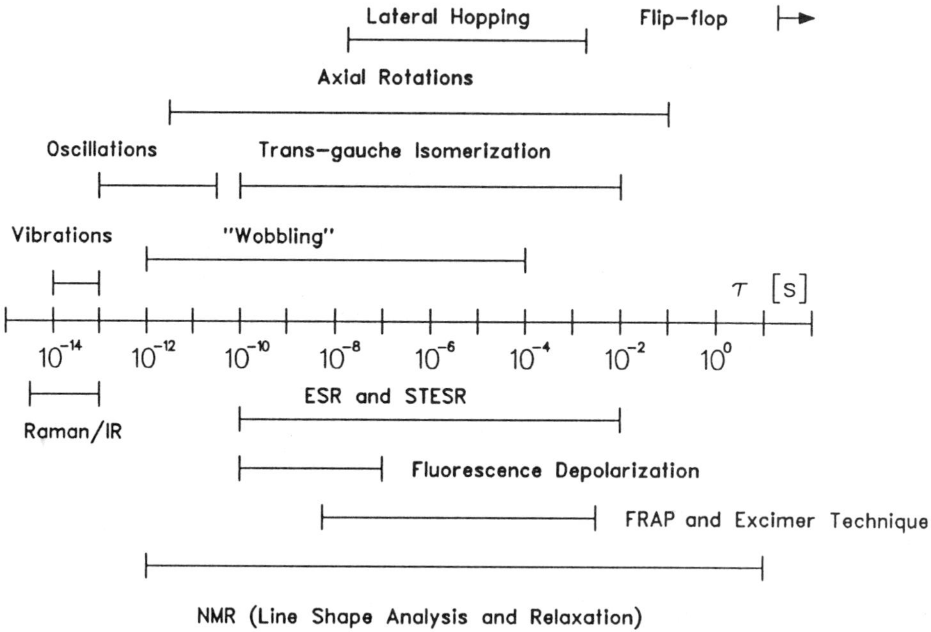

Figure 2 Characteristic ranges for the correlation times for various molecular motions occurring in lipid membranes (top) in comparison with the time ranges in which the spectroscopic techniques are sensitive to reorientational motions (bottom). The time ranges are only approximate.

lipid/water systems are the bands in the region between 2800–3000 cm^{-1} orginating from the symmetric and antisymmetric stretching vibrations of CH_3 and CH_2 groups. The frequencies and bandwidths of these vibrations depend on the phase state of the lipids. These band parameters have been used extensively to study the phase transitions in numerous lipid bilayer systems (Chapman, 1965, 1973; Fringeli et al., 1976; Cameron et al., 1979, 1980; Cortijo and Chapman, 1981; Cortijo et al., 1982; Cameron and Mantsch, 1982; Casal and Mantsch, 1984).

Other diagnostic bands are the CH_2 scissoring modes at ~ 1470 cm^{-1}, the CH_2 wagging mode at 1320 cm^{-1}, the CH_2 rocking modes around 720 cm^{-1}, and in Raman spectroscopy the C—C stretching modes between 1000 and 1200 cm^{-1}. In the IR spectra the C—C stretching region is dominated by vibrations of the phosphate and carbonyl ester groups. The CH_2 wagging modes in IR and the C—C stretching modes in Raman spectroscopy are suitable to determine the number of trans and gauche bonds in the lipid chains (Casal and McElhaney, 1990; Gaber and Peticolas, 1977; Yellin and Levin, 1977a,b; Pink et al., 1980).

The interfacial region and the particularly lipid/water interactions can be monitored by recording the phosphate vibrational bands and also the C=O stretching vibration as a function of temperature. These bands are sensitive to hydrogen bonding, the band frequencies being lowered with increased hydrogen bonding (Fringeli and Günthard, 1976; Mantsch et al., 1981; Blume et al., 1988a). As Fig. 3 shows, there is also the

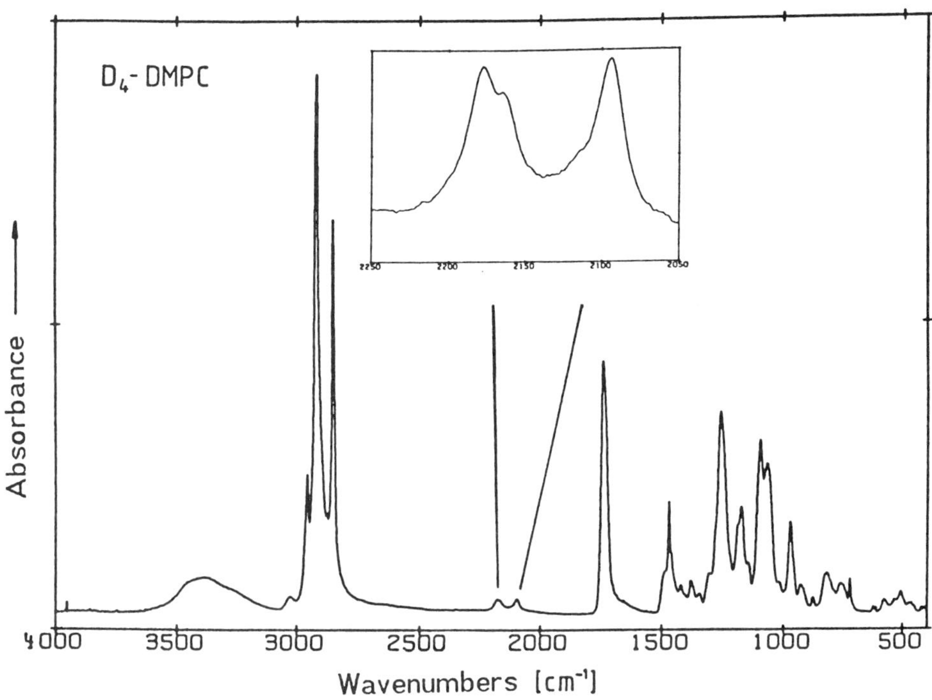

Figure 3 FT-IR spectrum of a lipid film of 1,2(4,4'-d_2)DMPC on a KBr plate. Inset shows the symmetric and antisymmetric DC_2 stretching bands. The characteristic diagnostic bands for the phase state are the CH_2 stretching vibrations at 2800–3000 cm^{-1}, the carbonyl band at ~1740 cm^{-1}, and the CH_2 scissoring modes at 1470 cm^{-1} (see text). (From Hübner, 1988.)

possibility of using specifically deuterated lipids. With these compounds the conformations of particular segments in the acyl chains can be determined from the temperature dependence of the CD_2 stretching vibrational bands between 2000 and 2200 cm^{-1} (Cameron et al., 1981a,b; Hübner, 1988; Blume and Hübner, 1988; Blume et al., 1988b). More quantitative information about the presence of gauche conformers at particular positions along the chains has recently been obtained by an analysis of the CD_2 rocking modes between 600 and 650 cm^{-1} (Mendelsohn et al., 1989; Davies et al., 1990).

B. Fluorescence Spectroscopy

As lipid molecules do not have strongly fluorescing groups, a fluorescent label has to be bound covalently to the lipid molecule. Alternatively, a fluorescent probe can be introduced into the lipid bilayer. As the fluorescent lipid as well as the fluorescent probe have a different molecular shape, and therefore different intermolecular interactions, the probe concentration has to be low so as not to disturb the packing of lipids, particularly in the ordered phases. Figure 4 shows some fluorescent probes and lipids bearing fluorescent groups used in membrane research. With fluorescence probes, the correlation times of mainly two motional modes can be determined, namely the wobbling motion and the lateral diffusion. In addition, information about the "fluidity" and/or the ordering of the molecules can be obtained.

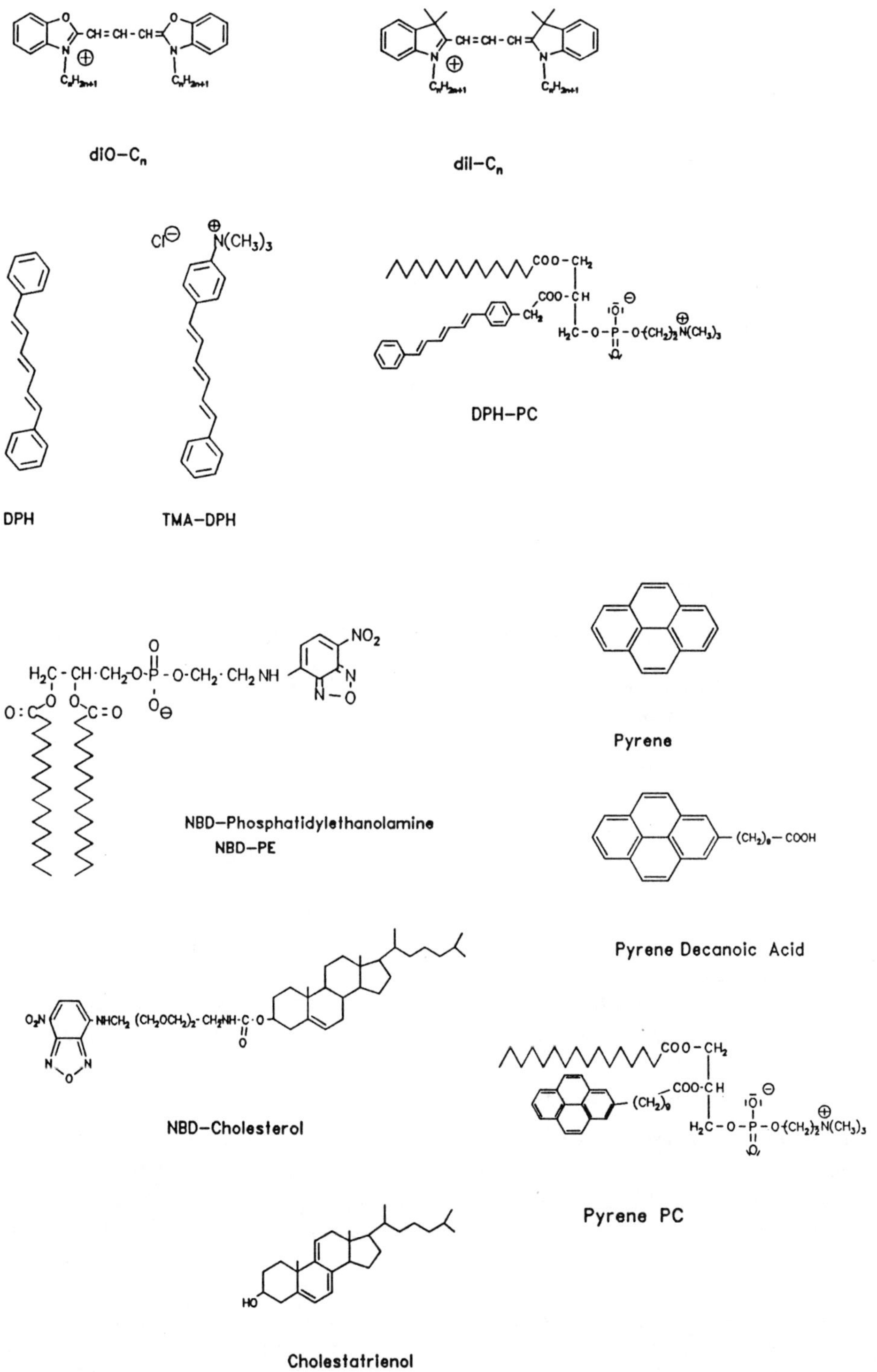

dIO–C_n

diI–C_n

DPH

TMA–DPH

DPH–PC

NBD–Phosphatidylethanolamine
NBD–PE

Pyrene

Pyrene Decanoic Acid

NBD–Cholesterol

Pyrene PC

Cholestatrienol

Figure 4 Chemical structure of fluorescent probes used for the determination of lateral diffusion coefficients.

1. Fluidity and Order of Lipid Membranes

Static or time resolved fluorescence anisotropy measurements can be used to determine molecular motions in membranes. This technique has been particularly useful for determining qualitatively the "fluidity" or "viscosity" of the liquid crystalline membranes (Kinosita et al., 1977; Heyn, 1979; Jähnig, 1979; Heyn et al., 1981; Wolber and Hudson, 1981; Jähnig et al., 1982; Hare, 1983; Vogel and Jähnig, 1985; Best et al., 1987; for reviews see Hoffmann and Restall, 1984; Kinosita et al., 1984; Heyn, 1989). In this method the fluorescence anisotropy of a probe incorporated into the lipid bilayer at low concentration is determined as a function of time and temperature. The fluorescence anisotropy

$$r(t) = \frac{I_{\|}(t) - I_{\perp}(t)}{I_{\|}(t) + 2I_{\perp}(t)} \tag{1}$$

should decay with time acording to the simple exponential function

$$r(t) = r_{\infty} + (r_0 - r_{\infty}) \exp\left(\frac{-t}{\tau_c}\right) \tag{2}$$

τ_c being the rotational correlation time of the fluorescent probe and r_0 and r_{∞} being the values of the anisotropy r at time zero and for $t \rightarrow \infty$. However, in most cases it is necessaery to use a sum of two exponentials and an additional constant term to describe the time dependence of the anisotropy (Rehorek et al., 1985).

From the time dependence of $r(t)$ the steady anisotropy $\bar{r}$ can be determined by integration. The motion of the fluorescent probe is not isotropic but restricted due to the ordering potential of the surrounding lipid chains. The anisotropy, therefore, does not decay to zero but to a finite value r_{∞}. The square root of the ratio r_{∞}/r_0 is related to the order parameter S of the probe (Heyn, 1979; Jähnig, 1979):

$$S = \left\langle \frac{1}{2} (3 \cos^2 \vartheta - 1) \right\rangle = \sqrt{\frac{r_{\infty}}{r_0}} \tag{3}$$

Alternatively, the wobbling angle ϑ_c can be determined when the wobbling in a cone model is used (Kinosita et al., 1977):

$$\frac{r_{\infty}}{r_0} = \left[\frac{\cos \vartheta_c (1 + \cos \vartheta_c)}{2}\right]^2 \tag{4}$$

The time dependent fluorescence anisotropy measurements are restricted to a relatively narrow time interval, because of the limited lifetime of fluorescent probes like 1,6-diphenyl-1,3,5-hexatriene (DPH) or 1-[4-(trimethylamino)phenyl]-6-phenyl-1,3,5-hexatriene (TMA-DPA) of ~ 10^{-8}–10^{-9} s (Best et al., 1987), which is somewhat longer than the rotational correlation times in the liquid crystalline phase of typical phospholipid bilayers.

Some information about the "fluidity" of liquid crystalline bilayers can also be obtained from measurements of the steady state anisotropy $\bar{r}$, which depends on both the value of r_∞ and the difference between the motional correlation time τ_c and the fluorescence life time τ_f (Jähnig, 1979; Heyn, 1979):

$$\bar{r} = r_\infty + (r_0 - r_\infty)\frac{\tau_c}{\tau_c + \tau_f} \tag{5}$$

This equation holds for cases where only one correlation time is observed, which is not always the case (Rehorek et al., 1985; Best et al., 1987; Heyn, 1989). Analysis of this equation shows that $\bar{r}$ is equal to r_∞ when $\tau_c \ll \tau_f$, i.e., when the motional correlation time is very short compared to the fluorescence lifetime. r_∞ can then be determined from the steady state anisotropy, and with $r_0 = 0.4$ the order parameter S can be calculated. When the correlation time is much longer than the fluorescence lifetime, steady state measurements give only r_0, as no depolarization occurs during the lifetime of the fluorescence. In the intermediate time regime, when $\tau_c \approx \tau_f$, $\bar{r}$ depends on r_∞ and τ_c, so that time dependent measurements of r are necessary to determine S and τ_c (Jähnig, 1979; Heyn, 1979).

2. Lateral Diffusion of Lipids

For determination of lateral diffusion coefficients of lipids and proteins in membranes the FRAP (*f*luorescence *r*ecovery *a*fter *p*hotobleaching) technique has been widely employed (for reviews see Yguarabide and Foster, 1981; Hoffman and Restall, 1984; Vaz et al., 1982a,b; Wade, 1985; Jovin and Vaz, 1989). A fluorescent probe like NBD-PE (*N*-(4-nitrobenzo-2-oxa-1,3-diazolyl)phosphatidylethanolamine) is incorporated homogeneously into the membrane. An intense laser beam is then directed toward the surface of the membrane, and the fluorescent molecules in a small spot are irreversibly bleached. After the bleaching pulse, the fluorescence increase in the bleached area is recorded as a function of time. The fluorescence recovery increases due to diffusion of unbleached molecules into the previously bleached area. Figure 5 shows a schematic diagram of this experiment (Vaz et al., 1982b).

The time constant of the fluorescence increase is related to the diffusion coefficient D_L of the fluorescent probe, i.e.,

$$D_L = \frac{\omega^2 \gamma_D}{4\ t_{1/2}} \tag{6}$$

where $\gamma_D = t_{1/2}/\tau_D$ depends on the beam profile and the so-called bleaching parameter, $t_{1/2}$ being the half time of fractional fluorescence recovery, $\tau_D = \omega^2/4D_L$ the characteristic diffusional recovery time, ω the radius of the focussed laser beam, and D_L the translational diffusion coefficient (Axelrod et al., 1976). Several modifications of this method are being used (Hoffmann and Restall, 1984; Jovin and Vaz, 1989). The FRAP technique can also be used for monomolecular lipid films at the air-water interface (Peters and Beck, 1983; Beck and Peters, 1985; Beck, 1987).

Another fluorescence tecnhique for measurements of the lateral diffusion coefficients is the excimer technique. This method relies on probes or lipids with a pyrene moiety (Galla and Sackmann, 1974; Galla et al., 1979a) which can form excimers. The rate of

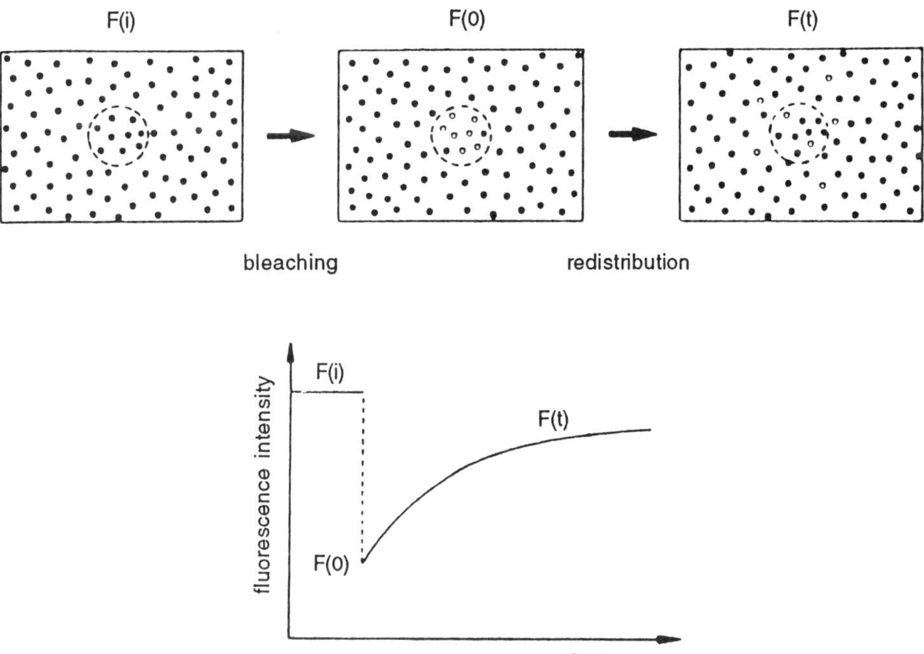

Figure 5 Sketch of a fluorescence recovery after photobleaching (FRAP) experiment. The fluorescence intensity *F(i)* of the chromophores in a small area of the membrane is measured by illumination with the measuring beam. The experiment is started by irreversibly bleaching a fraction of the fluorophores in this area by an intense laser pulse. The recovery of the fluorescence in this area is due to diffusion of bleached molecules out of and unbleached molecules into this area giving rise to the time dependence of *F(t)* shown at the bottom. (From Vaz et al., 1982b.)

excimer formation is related to the translational mobility of the probe or lipid molecules in the bilayer. The dynamical parameters determined by these two methods do not necessarily coincide. The reason for this is that the excimer method measures diffusion over short distances, i.e., a local diffusibility, while using the FRAP technique diffusion is measured over distances of several μm (Jovin and Vaz, 1989). The lateral diffusion coefficients of phosphatidylcholines as determined by different experimental methods are summarized in Table 2.

Table 1 Approximate Correlation Times τ_c for Various Motional Modes in the Different Phases of Phosphatidylcholines and Phosphatidylethanolamines

Phase	Chain isomerization by rotation around single bonds	Long axis rotation	Wobbling of chain axis	Order director fluctuations
L_c ($L_{c'}$)	$< 10^{-3}$	$< 10^{-3}$	—	—
L_β ($L_{\beta'}$)	10^{-9}–10^{-5}	10^{-6}–10^{-5}	10^{-5}–10^{-3}	—
L_α	10^{-12}–10^{-9}	10^{-9}–10^{-8}	10^{-9}–10^{-7}	$\approx 10^{3}$

Table 2 Lateral Diffusion Coefficients in Liquid Crystalline Bilayers of DMPC and DPPC

Lipid	Method	Probe	T/°C	D_L/cm^2·s^{-1} ×10^8	References
DMPC	FRAP	diO-C_{18}	30	5.5	Wu et al. (1977)
	FRAP	NPD-PE	30	8.0	Wu et al. (1977)
	FRAP	diI-C_{18}	30	1.6	Fahey and Webb (1978)
	FRAP	NBD-PE	26	5.0	Rubenstein et al. (1979)
	Excimer	Pyrene	30	12.8	Galla et al. (1979a,b)
	Excimer	Pyrene-PC	30	6.5	Galla et al. (1979a,b)
	NMR	none	31	3.5	Kuo and Wade (1979)
	FRAP	NBD-PE	24	6.3	Derzko and Jacobson (1980)
	FRAP	NBD-Lyso-PE	24	4.5	Derzko and Jacobson (1980)
	FRAP	NBD-C_{12}	24	7.2	Derzko and Jacobson (1980)
	FRAP	diO-C_6	24	8.2	Derzko and Jacobson (1980)
	FRAP	diO-C_{18}	24	6.7	Derzko and Jacobson (1980)
	FRAP	diI-C_{12}	24	5.2	Derzko and Jacobson (1980)
	FRAP	diI-C_{18}	24	8.4	Derzko and Jacobson (1980)
	FRAP	NBD-PE	26	1.8	Alecio et al. (1982)
	FRAP	NBD-Chol	26	1.6	Alecio et al. (1982)
	FRAP	NBD-PE	30	4.0	Tamm and McConnell (1985)
	ESR	16-PCSL	30	12	King et al. (1987)
	Excimer	Pyrene-PC	35	29	Sassaroli et al. (1990)
	Excimer	Pyrene-PC	30	22	Vauhkonen et al. (1990)
DPPC	ESR	Androstane-SL > T_m		~1	Träuble and Sackmann (1972)
	FRAP	diO-C_{18}	45	7.0	Wu et al. (1977)
	FRAP	diI-C_{18}	45	1.0	Fahey and Webb (1978)
	NMR-$T_{1\rho}$	none	45	38	Fisher and James (1978)
	NMR	none	42	5.0	Kuo and Wade (1979)

Table 2 (Continued)

Lipid	Method	Probe	T/°C	D_L/cm^2·s^{-1} ×10^8	References
	Excimer	Pyrene	50	40	Galla et al. (1979a,b)
	Excimer	Pyrene-PC	50	16	Galla et al. (1979a,b)
	FRAP	NBD-PE	45	5.0	Tamm and McConnell (1985)
SUV	Excimer	Pyrene	45	210	Daems et al. (1985)
LUV	Excimer	Pyrene	45	88	Daems et al. (1985)
	Excimer	Pyrene-PC	42	35	Vauhkonen et al. (1990)

C. Electron Spin Resonance (ESR)

ESR spectroscopy has been widely used to study the "fluidity" of lipid bilayers and biological membranes (for reviews see Schreier et al., 1978; Marsh, 1981, 1983; Marsh and Watts, 1982). Its advantage is the great sensitivity. Spectra can be easily recorded as a

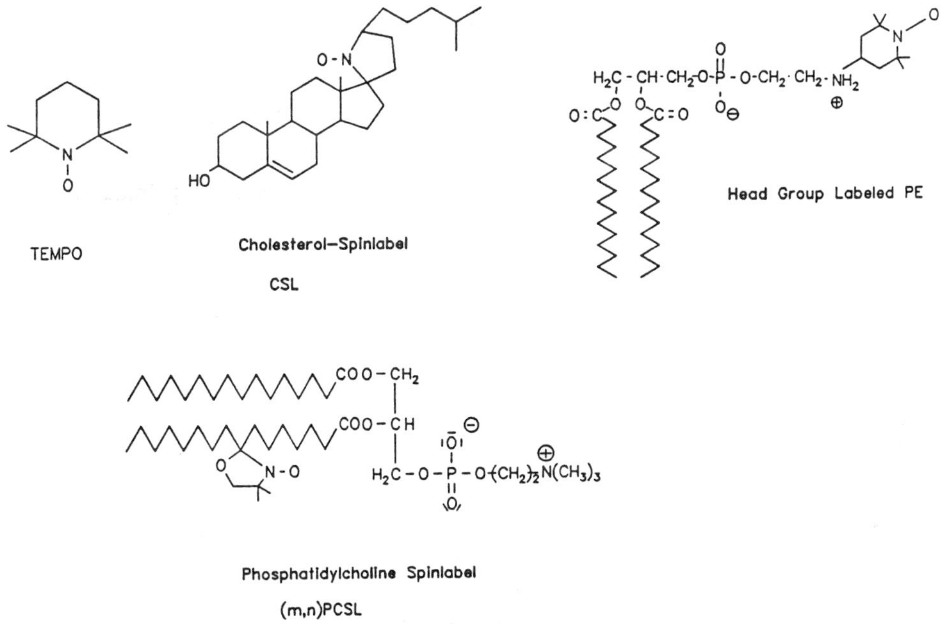

TEMPO

Cholesterol–Spinlabel
CSL

Head Group Labeled PE

Phosphatidylcholine Spinlabel
(m,n)PCSL

Figure 6 ESR spin probes used to study dynamics and lateral diffusion in membranes.

function of temperature. ESR spectra of nitroxide radicals are particularly sensitive to motions with correlation times between 10^{-10} to 10^{-8} s. The time range can be extended by using saturation transfer ESR up to correlation times of 10^{-3} s (Hyde and Dalton, 1972; Robinson and Dalton, 1980; Koole et al., 1981; for a review see Thomas, 1985).

The major disadvantage of ESR is the fact that a spin probe has to be incorporated into the lipid bilayer. Generally, a nitroxide group is covalently bound to a lipid molecule at the particular position of interest, i.e., either to the fatty acyl chains or to the head group. Figure 6 shows some common lipid spin probes used in ESR measurements with membrane systems. As the concentrations of the spin probes are very low, the overall bilayer properties are not altered. However, in the case of complex mixed systems, the spin probe can be distributed inhomogeneously. For instance, in regions of the phase diagram where phase separation is observed, spin probes accumulate in the liquid crystalline phase.

D. Nuclear Magnetic Resonance (NMR)

By far the most versatile method for measuring lipid dynamics is NMR spectroscopy (for reviews see Seelig, 1977, 1978; Bocian and Chan, 1978; Spiess, 1978; Seelig and Seelig, 1980; Chan et al., 1981; Griffin, 1981; Jacobs and Oldfield, 1981; Smith and Jarrell, 1983; Davis, 1983; Smith 1984; Smith and Oldfield, 1984; Spiess, 1985; Davis, 1986, 1989; Trahms, 1985; Seelig and MacDonald, 1987; Blume, 1988).

The principal advantages of NMR over other spectroscopic techiques are

1. No potentially perturbing probes are needed.
2. Different nuclei ^{1}H, ^{2}H, ^{13}C, ^{31}P, etc. giving complementary information can be studied.
3. Motional correlation times for different segments in the molecule can be determined as well as motions of the whole molecule.
4. The range of motional correlation times detectable is very broad, going from 10^{-12} to 10^{-1} s.

The disadvantages are

1. Sensitivity in NMR is much lower than in ESR, necessitating more material and longer accumulation times.
2. The cost of instrumentation is high.
3. Conventional high resolution NMR spectrometers can only be used for a limited number of experiments using small vesicles above the transition temperature.

From multilamellar liposomes no high resolution spectra can be obtained, except when magic-angle spinning techniques are applied. In lipid bilayers the motions of the lipid molecules are anisotropic. Consequently, angular dependent interactions like dipolar interactions or chemical shift anisotropy are not completely averaged, thus giving rise to very broad lines. In small vesicles, the tumbling of the vesicles in combination with the lipid diffusion around the surface are fast enough to lead to isotropic averaging of the NMR lines. With multilamellar liposomes, magic-angle spinning (MAS) can be used to average all angular dependent interactions to obtain high resolution spectra. However, the detailed information about the anisotropy and rates of motions contained in the shapes of the broad lines is now lost. However, ^{1}H- and ^{13}C-NMR-MAS techniques are useful for structure elucidation and conformation determinations in liquid crystalline and for ^{13}C

also gel phase bilayers (Haberkorn et al., 1978; Oldfield et al., 1987; Forbes et al., 1988; Bruzik et al., 1990a,b).

The initial disadvantage of having to deal with broad lines caused by anisotropic interactions can be turned into an advantage using solid-state NMR techniques. The only partially averaged orientation-dependent interactions give rise to line shapes that can be analyzed using appropriate motional models and calculation techniques to yield information on the mechanism and the rates of molecular reorientations (Spiess, 1978, 1985; Griffin, 1981; Wittebort et al., 1982; Meier et al., 1983, 1986, 1987; Wittebort et al., 1987; Mayer et al., 1988). In combination with the use of specifically ^{2}H or ^{13}C labeled molecules, measurements of the longitudinal and transverse relaxation times T_1 and T_2, and the use of macroscopically aligned samples (Jarrell et al., 1988; Hübner, 1988; Speyer et al., 1989; Mayer et al., 1990; Hübner and Blume, 1990), a fairly complete picture of the intramolecular and intermolecular dynamics of lipids can be obtained. In addition, 2D techniques can be used, which give detailed information either on the chemical structure (Forbes et al., 1988; Lee and Griffin, 1989) or on the molecular reorientational mechanisms (Schmidt et al., 1987; Auger and Jarrell, 1990). These methods have great potential. However, most experiments of the latter type are very time consuming, so that 2D techniques cannot routinely be used at the present time.

In the following the results on lipid dynamics obtained for various lipid systems using different methods will be discussed.

IV. ROTATIONAL DIFFUSION AND INTRAMOLECULAR REORIENTATIONS

As the prime example we will discuss the motional behavior of phosphatidylcholines with saturated fatty acyl chains, because for these phospholipids by far the largest body of experimental data has been accumulated over the years. The motional behavior of the phosphatidylcholines is representative for all other lipids. The influences of chain length, different types of fatty acids, and linkage with the glycerol backbone will be described; then we will turn to other phospholipids and glycolipids and the motional modes in lipid mixtures.

A. Phosphatidylcholines

The dipalmitoyl-phosphatidylcholine/water system has been studied very intensively (see Cevc and Marsh, 1987). The principal motions occurring in thes bilayers are very similar for other phospholipid systems, though quantitative differences are observed when the chemical nature of the head groups, the chemical structure of the acyl chains, or the linkage of the alkyl chains is altered.

The DPPC/water phase diagram (see Fig. 7) is well known and is discussed in some detail in Chap. 12 (Janiak et al., 1979; Small, 1986). In the following, we will discuss only the dynamical aspects of the various lamellar phases.

1. Dynamics in the Liquid Crystalline Phase

Based on the results of IR, ^{1}H-NMR, and ^{13}C-NMR spectroscopy, it was recognized very early that in the liquid crystalline phase DPPC molecules are free to rotate and diffuse in the plane of the lipid lamellae, and that extensive trans-gauche isomerization around single bonds of the fatty acyl chains is occurring (Chapman, 1965, 1973; Levine et al., 1972; Lichtenberg et al., 1975; Bloom et al., 1978). The first more quantitative data,

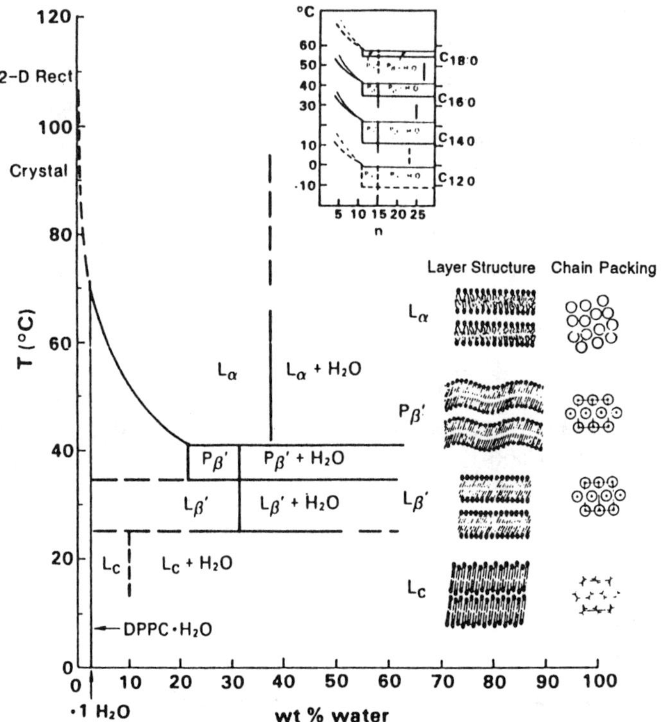

Figure 7 Partial phase diagram of the DPPC/water system with a sketch of the hydrocarbon chain arrangement in the four different lipid/water phases. (From Small, 1986.)

however, came from ^{2}H-NMR measurements of phosphatidylcholines, specifically deuterated in the head groups and in the acyl chains (Seelig and Seelig, 1974, 1975; Seelig and Niederberger, 1974a,b; Seelig and Gally, 1976; Seelig et al., 1977; Seelig and Browning, 1978; Brown et al., 1979).

(a) Order parameters. The ^{2}H-NMR spectra of specifically labeled phospholipids indicate that in the liquid crystalline phase the molecules perform fast anisotropic motions leading to axially symmetric powder patterns of reduced width (Griffin, 1981). The correlation times of these motions are fast on the ^{2}H-NMR time scale, i.e., they are substantially shorter than 10^{-6} s. The spectral line shapes are therefore sharp powder patterns and contain no direct information on the type of motion. These fast limit powder patterns have been commonly interpreted in terms of the order parameter concept (Saupe, 1964; Seelig, 1977). If a cartesian coordinate system x, y, z is fixed to a particular group in a molecule, for instance to a CD_2 group, the order parameters S_{ii} are defined by

$$S_{ii} = \left\langle \frac{1}{2} (3 \cos^2 \vartheta_i - 1) \right\rangle \qquad i = x, y, z \tag{7}$$

These order parameters describe the fluctuations of the x, y, and z axes with respect to a preferred axis, the director axis D, ϑ_i being the angle between the particular coordinate axis and the director D. The brackets $<$ $>$ denote the time average for $\cos^2 \vartheta$ over all

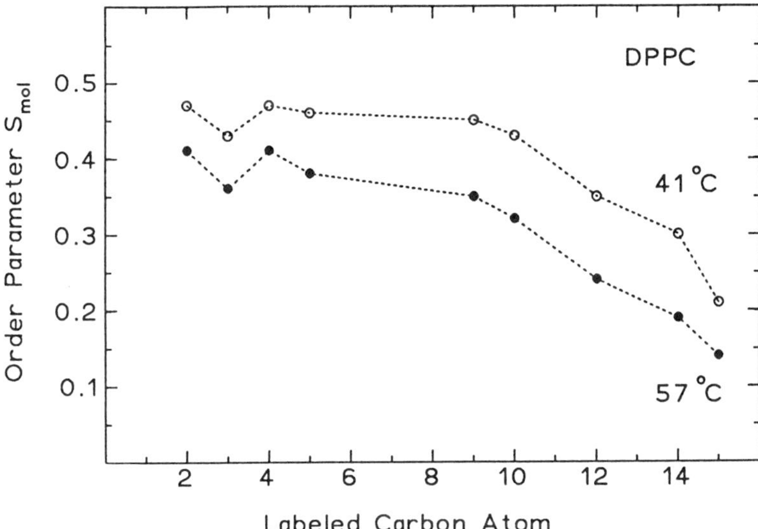

Figure 8 Order parameters S_{mol} determined from ^{2}H-NMR experiments for the *sn*-1 chain of DPPC in the liquid crystalline phase at two different temperatures. (From Seelig and Seelig, 1974.)

orientations. From the orthogonality of the direction cosines it follows that $\Sigma S_{ii} = 0$, so that only two of the order parameters are independent. If the electric field gradient tensor is axially symmetric (this is the case for an aliphatic C—D bond), only one order parameter, commonly called S_{CD}, is independent. The observed quadrupole splittings $\overline{\Delta\nu_Q}$ in the fast limit powder patterns are then related to S_{CD} (Seelig, 1977)

$$\overline{\Delta\nu_Q} = \frac{3}{4}\left(\frac{e^2qQ}{h}\right)S_{CD} \tag{8}$$

with e^2qQ/h being the quadrupole coupling constant (~ 169 kHz for an aliphatic C—D bond). The order parameter S_{mol} is also frequently used. It is related to S_{CD} by $S_{mol} = -2S_{CD}$. S_{mol} is the order parameter of the principal axis, which is perpendicular to the plane spanned by the CD_2 group. The relation between S_{moL} and S_{CD} only holds when there is axial symmetry about the ordering axis, which is usually the long axis of the molecule (Seelig, 1977). Seelig and Seelig (1974) determined order parameters for DPPC in the liquid crystalline phase as a function of label position in the *sn*-1 chain. The characteristic dependence of the order parameter on the position of the CD_2 group is shown in Fig. 8

The order parameters decrease when the lipid sample is heated, indicating that the extent of the angular fluctuations increases upon bilayer expansion (Seelig and Seelig, 1974; Oldfield et al., 1978a,b; Davis, 1979). The temperature dependence is less pronounced for positions close to the glycerol backbone and also relatively small for the highly mobile chain ends (see Fig. 9). The simplest model to describe these effects is that the molecules perform fast long axis rotation and that the probability of trans-gauche isomerization increases toward the chains ends.

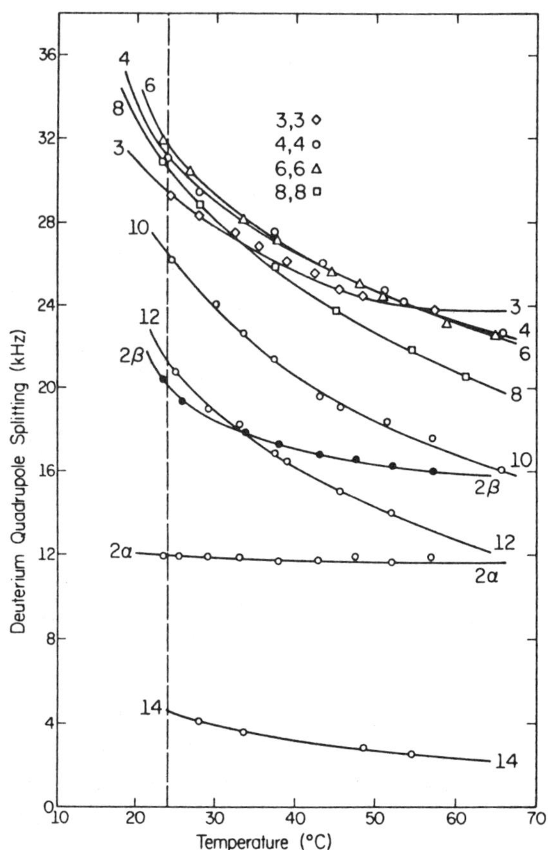

Figure 9 Temperature dependence of the quadrupole splittings for DMPC labeled at various positions in the *sn*-2 chain. (From Oldfield et al., 1978b.)

For the 2 position of the *sn*-2 chain, the order parameters for the two C—D bonds are different (Seelig and Seelig, 1975), showing that these two C—D bonds are conformationally inequivalent. The reason for this is the particular conformation of the *sn*-2 chain, which makes a characteristic bend (McAlister et al., 1973; Pearson and Pascher, 1979). This conformation is also responsible for the slight differences in order parameters between CD_2 groups in the *sn*-1 or *sn*-2 chain (Davis, 1979, 1983).

(b) Trans-gauche isomerization. The correlation times for trans-gauche isomerization can only be determined from relaxation time measurements (Brown et al., 1979; Peters and Kimmich, 1979; Rommel et al., 1988; Mayer et al., 1988) and a comparison with model calculations. In ESR experiments, trans-gauche isomerization of chain labeled phosphatidylcholines is fast on the ESR time scale. Thus the correlation times must be shorter than $2 \cdot 10^{-10}$ s (Moser et al., 1989). From ^{2}H-NMR experiments with specifically labeled DMPC, it was concluded that the correlation time τ_J for trans-gauche isomerization ranges from 10^{-12} to 10^{-10} s. τ_J depends on the label position decreasing towards the chain end (Meier et al., 1986; Mayer et al., 1988). This is in agreement with τ_J values of $\sim 10^{-10}$ s determined from proton spin relaxation dispersion measurements (Rommel et al., 1988).

(c) Molecular rotation. In the liquid crystalline phase, lipid rotation is also in the fast limit on the ^{2}H-NMR time scale. It is thus difficult to obtain reliable data for rotational correlation times $\tau_{\|}$ solely from line shape analysis. ^{1}H-NMR spin lattice relaxation times suggested correlation times $\tau_{\|}$ of $\sim 2 \cdot 10^{-8}$ s (Petersen and Chan, 1977). Based on ESR spectra of chain labeled DMPC, Lange et al. (1985) reported values between 10^{-9} and $7 \cdot 10^{-9}$ s above the T_m of DMPC. This was in relatively good agreement with $\tau_{\|}$ values deduced from ^{2}H-NMR (Meier et al., 1983, 1986). Newer data, including results from ^{2}H-NMR T_{1Z} measurements, confirm these values (Mayer et al., 1988; Moser et al., 1989). ^{1}H-NMR relaxation dispersion measurements gave values of $\sim 5 \cdot 10^{-9}$ for $\tau_{\|}$ (Rommel et al., 1988). T_{1Z} measurements on oriented DMPC bilayers seemed to indicate shorter correlation times $\tau_{\|}$. For this system a value of $5 \cdot 10^{-10}$ s was reported for a temperature of 35°C, compared to $\sim 5 \cdot 10^{-9}$ s in multilamellar systems at the same temperature (Mayer et al., 1990). Whether these differences can be attributed to different mobilities in liposomes compared to oriented systems remains an open question. It is also possible that the differences are caused by the particular model used for the stimulations of the line shapes and the relaxation times.

(d) Wobbling motions and collective fluctuations. Additional motional averaging can occur by a wobbling motion of the whole molecule or of individual chains. Such rigid body motion decreases the order parameter and is one of the major motional modes observed for rigid fluorescent probes or spin probes. The correlation time $\tau_{\perp}$ for this motion is in the ns regime and thus in the fast limit of the ^{2}H-NMR time scale.

Assuming this wobbling motion to be also existent for lipid molecules, the order parameter S_{mol} of chain labeled lipids has to be viewed as a product of an order parameter S_{zz} describing the wobbling motion of the whole molecule and a segmental order parameter $S_{z'z'}$ describing isomerization of the particular chain segment (Petersen and Chan, 1977; Meier et al., 1983, 1986). These two motions occur with different rates, but they are both in the fast limit on the ^{2}H-NMR time scale. The separation of the observed order parameter into these two different contributions is therefore somewhat ambiguous. A reliable determination of S_{zz} and $S_{z'z'}$ can only be accomplished by using lipids labeled at different positions and by performing additional experiments, for instance, determination of the relaxation times T_{1Z} and T_{2E} and/or spectroscopy on oriented bilayer systems (Mayer et al., 1988, 1990).

With this approach it was shown that there is indeed an order parameter gradient of $S_{z'z'}$ along the chains as shown before. However, the segmental order parameters are much higher than previously estimated, because athe order parameter S_{zz} is much less than one. S_{zz} decreases further with increasing temperature, indicating an increase of the wobbling motions of the molecules. The temperature dependence of S_{zz} is larger than for the segmental order parameter $S_{z'z'}$ (see Fig. 10).

The wobbling motions responsible for the low S_{zz} values can be isolated or collective. If they occur as collective modes, they are usually called director fluctuations. Considerable effort has been put into experiments to decide whether these collective modes contribute to spin lattice relaxation in the MHz range or not (Brown et al., 1979; Brown, 1982; Pope et al., 1982; Brown et al., 1983, 1990; Williams et al., 1985; Rommel et al., 1988; Mayer et al., 1988, 1990; Watnick et al., 1990). Over a broad frequency range, the proton T_1 relaxation times were shown to be in in qualitative agreement with rotational and torsional motions and trans-gauche isomerization in the chains (Kimmich and Voigt, 1979; Kimmich et al., 1983). Rommel et al. performed relaxation dispersion studies over a frequency range from 100 Hz to 300 MHz. From the frequency dependence of T_1 it was

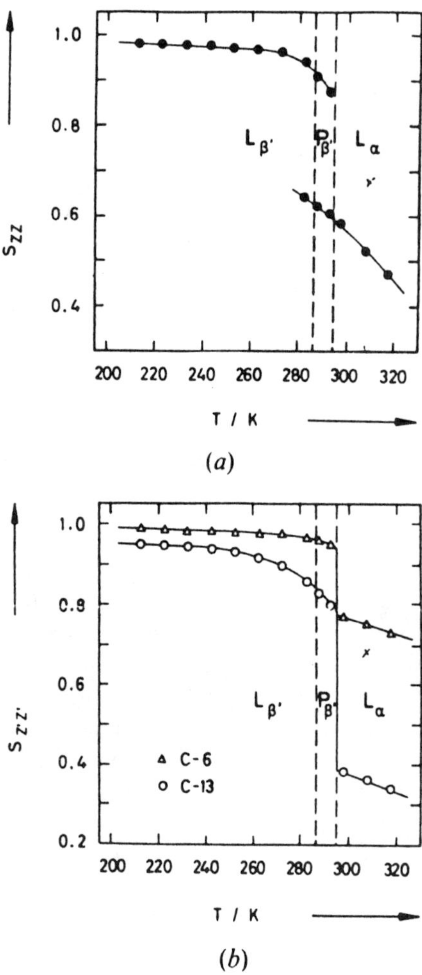

Figure 10 Rigid body order parameter S_{zz} and segmental order parameter $S_{z'z'}$ for DMPC as a function of temperature. $S_{z'z'}$ was determined for the (△) C_6 and (o) the C_{13} position in the chains. (From Mayer et al., 1988.)

concluded, however, that collective motions occur in the 1–10 kHz regime (Rommel et al., 1988). Because of their long correlation times, these collective motions do not affect the T_{1Z} times in th MHz regime and thus do not affect the line shapes. Similar results were obtained by Watnick et al. (1990) measuring T_2 relaxation times in deuterated DMPC.

Because the details of motions contributing to the observed relaxation times for ^1H as well as ^2H are not well understood, several experiments have been performed to measure T_{1Z} in oriented lipid systems (Jarrell et al., 1988; Hübner, 1988; Blume et al., 1988b; Bonmatin et al., 1988; Auger et al., 1990a,b; Mayer et al., 1990). Previous measurements of the spin lattice relaxation times on powder samples revealed no anisotropy, presumably due to the additional motional averaging caused by fast lateral diffusion of lipid molecules around the curved surfaces of the liposomes (Brown and Davis, 1981; Williams et al., 1985). In lipid/cholesterol membranes lateral diffusion is reduced (see below), so that this averaging effect should be eliminated. Indeed, in lipid/cholesterol dispersions T_{1Z} was clearly anisotropic (Siminovitch et al., 1985, 1988). In oriented pure lipid bilayers, however, T_{1Z} is almost orientation independent. When the only relevant motions where trans-gauche isomerization and fast long axis rotation, orientation dependent T_{1Z} values should be observed (Jarrell et al., 1988; Hübner, 1988, Hübner and Blume, 1990; Mayer et al., 1990).

It was suggested that fast wobbling motions ($S_{zz} < 1$) are responsible for the missing T_{1Z} anisotropy. The experiments performed by Mayer et al. (1990) seem to support this notion. In oriented samples of DMPC/40 mol % cholesterol the segmental order parameter $S_{z'z'}$ is increased to almost unity and the rigid body order parameter S_{zz} increases to ~0.68. This increase in S_{zz} then leads to an angular dependent T_{1Z} as observed in the oriented sample.

The interpretations of the ^{2}H-NMR spectra with respect to the two motional modes contributing to the overall order parameter S_{mol} are not unambiguous. In cerebroside dispersions a clear angular dependent T_{1Z} is observed, the sign and value of the anisotropy depending on the position of the CD_2 group in the fatty acyl chains (Speyer et al., 1989). The ^{2}H-NMR spectra of cerebrosides could be simulated without including a wobbling motion. T_{1Z} anisotropy was also found in DPPC labeled at the 3 position in the glycerol backbone and the 12 position of the *sn*-2 chain, but with reversed sign (Speyer et al., 1989). The missing anisotropy found for DMPC and DPPC could thus be fortuitous and only due to the fact that these compounds were labeled at positions close to the 4 position, where indeed T_{1Z} anisotropy is very low. The change in T_{1Z} anisotropy occurring after addition of cholesterol is interpreted by Speyer et al. by relative changes in the rates of rotation and trans-gauche isomerization (see below).

In fluorescence depolarization experiments with probes such as DPH or TMA-DPH, the wobbling motion is responsible for the time dependent decrease of the fluorescence anisotropy and for the reduced steady state anisotropy. The order parameter for this rigid body motion can be determined from the ratio r_∞/r_0 (see Eq. 3). The order parameters determined for molecules with charged groups like TMA-DPH, which are located close to the bilayer water interface and thus mostly reflect the rigid body motions, are very close to those determined by NMR line shape simulations. Best et al. (1987) reported a value of 0.58 for TMPA-DPH in DMPC above T_m. This is almost identical to S_{zz} in DMPC dispersions reported by Mayer et al. (1988) but somewhat lower than the value determined for oriented DMPC bilayers, which is 0.68 (Mayer et al., 1990). For fluorescent probes it was shown that the orientational distribution function has a half width of ~ 20–25° for $S_{zz} = 0.6$ (Best et al., 1987).

The correlation times $\tau_\perp$ for the rigid body motions of DPH and TMA-DPH as measured by fluorescence anisotropy are at least one order of magnitude shorter, namely close to 10^{-9} s (Rehorek et al., 1983; Best et al., 1987), than those determined for DMPC dispersions by ^{2}H-NMR, 10^{-7}–10^{-8} s (Mayer et al., 1988; Moser et al., 1989). ESR measurements with labeled DMPC gave similar results as ^{2}H-NMR (Moser et at., 1989). Possible reasons for these differences are uncertainties of the absolute values in the ^{2}H-NMR line shape and relaxation time simulations, size effects, as probes like TMA-DPA and certainly DPH are smaller than lipid molecules, and perturbations of the packing of the phospholipid chains.

The relative contributions of rigid body order and segmental order to S_{mol} can also be determined by FT-IR spectroscopy, analyzing the CD_2 rocking bands between 600 and 650 cm^{-1} (Mendelsohn et al., 1989; Davies et al., 1990). The low gauche probabilities found by FT-IR seem only to be consistent with the ^{2}H-NMR quadruple splittings when an additional wobbling motion as averaging process is assumed.

In the past, there has been considerable discussion about the differences in order parameters observed with spin labeled lipids and with ^{2}H labeled lipids (Seelig and Niederberger, 1974a,b; Smith et al., 1977; Taylor and Smith, 1983). Moser et al. (1989) have resolved these discrepancies using a simulation model for the ESR spectra that incorporated chain isomerization as well as restricted whole body reorientational motions (wobbling). It was shown that the wobbling motions of the lipids are in the slow-motion regime on the ESR time scale, the correlation times being almost identical in ESR and NMR experiments. However, significant differences were obtained for the segmental motions, showing that the introduction of the nitroxide ring in the spin labeled phosphatidylcholines leads to local distortions. The segmental order parameters for the spin labeled lipids are considerably lower than those for ^{2}H labeled lipids, but the ordered parameter gradient along the chains is preserved (Moser et al., 1989).

(e) Head group motions. Head group labeled phosphatidylcholines display small quadruple splittings between 2 and 10 kHz indicating that the head group can adopt numerous

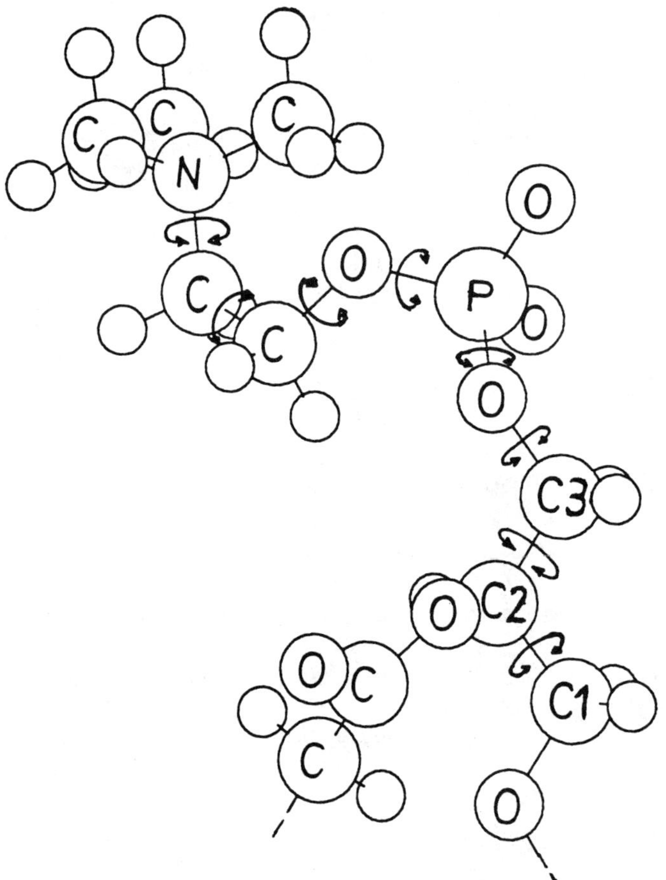

Figure 11 Chemical structure of the PC head group with glycerol backbone and the possible reorientational motions.

different conformations by rotations around single bonds. The average position of the P-N dipole is parallel to the bilayer surface (Seelig and Gally, 1976; Seelig et al., 1977; Yeagle et al., 1976, 1977; Brown and Seelig, 1978; Seelig and Browning, 1978; Griffin et al., 1978; Büldt et al., 1979; Zaccai et al., 1979; Akutsu and Seelig, 1981; Skarjune and Oldfield, 1979b).

Figure 11 shows the structure of the phosphatidylcholine head group and the axes around which various rotations can occur. The quadrupole splittings of head group labeled phosphatidylcholines are sensitive to the surface charge of the membrane, so they can be used as sensors for membrane potentials (Seelig et al., 1987). The correlation times for head group motions have been determined by dielectric relaxation time measurements (Shepherd and Büldt, 1978, 1979) and ^{31}P-NMR spin-lattice relaxation time measurements (Milburn and Jeffrey, 1987, 1989). The correlation times were found to be around $3.0 \cdot 10^{-9}$ s. This is in agreement with ^{31}P line shape simulations (Campbell et al., 1979) for DPPC at 35°C in the P_{β} phase, from which a rotational correlation time of $3.3 \cdot 10^{-9}$ s was calculated. As the motion of the head group does not change dramatically at the main transition, this is in reasonable agreement with the previous value. The rotational correlation times for head group motions are thus very similar to those determined for the overall axial rotation of the whole molecule (see above). They probably describe the relative motion of the head group with respect to the glycerol backbone (Milburn and Jeffrey, 1987, 1989).

The head group motion involves fast rotation of the phosphate group, presumably around the C_2—C_3 bond of the glycerol and the P—O bond next to the glycerol (Kohler and Klein, 1977; Seelig and Gally, 1976; Seelig et al., 1977; Gally et al., 1981; Akutsu and Seelig, 1981; Borle and Seelig, 1983a; Milburn and Jeffrey, 1987, 1989). Other workers (Trahms and Klabe, 1985; Trahms, 1985; Strenk et al., 1985; Braach-Maksvytis and Cornell, 1988) have excluded a rotation around the C_2—C_3 bond as two different quadrupole splittings are observed for deuterons at the C_3 of the glycerol (Gally et al., 1981; Strenk et al., 1985). However, fast threefold rotational jumps with unequal populations of the three sites preserve the conformational inequality of the two deuterons at C_3 and could be consistent with the observation of two splittings. In addition, a slower wobbling motion for head group tilt is suggested by ^{31}P-NMR data (Milburn and Jeffrey, 1987, 1989).

(f) Motions in the glycerol backbone. The reorientational motions in the glycerol backbone, supposedly the most rigid part of the lipid molecule, are complicated and not yet completely clear. For phosphatidylcholines and other phospholipids labeled at the glycerol moiety it was found that deuterons in the 1 and 3 positions of the glycerol give different splittings indicating motional and/or conformational inequivalence (Gally et al., 1975, 1981; Strenk et al., 1985). Strenk et al. (1985) and Braach-Maksvytis and Cornell (1988) suggested a relatively rigid conformation of the glycerol backbone in the liquid crystalline phase with no rotation around the C_1—C_2 and C_2—C_3 bonds but a tilting of the most ordered axis with respect to the bilayer normal. However, isomerization around the glycerol C_1—C_2 and C_2—C_3 bond with different populations for the trans and gauche conformers could also account for the observed effects. This type of motion has been suggested before (Gally et al., 1981) and has definitely been observed for short chain phosphatidylcholines forming micelles (Hauser et al., 1980).

The temperature dependence of the quadrupole splittings of glycerol labeled phosphatidylcholines is much smaller than for chain labeled lipids (Strenk et al., 1985). Different explanations could account for this finding. Either the order parameter for the wobbling motion is temperature independent, or, if there is no wobbling, trans-gauche isomerization in the glycerol backbone is temperature independent, or both effects fortuitously cancel. Recent experiments on glycoglycerolipids labeled at the C_3 position and phospholipid analogs labeled at the C_2 and C_3 positions of the glycerol backbone have revealed that in these lipids internal reorientational motions with correlation times of $\sim 10^{-9}$ s or shorter are even present in the gel phase (Auger et al., 1990b), suggesting that they are also present in the liquid crystalline phase. In the gel phase of glycolipids, motion around the C_2—C_3 bond were simulated on the basis of T_{1Z} data using a three-site jump model with populations of 0.46, 0.34, and 0.20 for the three sites with exchange rates of $\sim 5–6 \cdot 10^8$ rad/s (Auger et al., 1990b). These populations agree surprisingly well with those reported for the g^+, antiplanar, and g^- conformers (0.48, 0.37, 0.15) in the micelle forming phospholipid dihexanoyl-phosphatidylcholine (Hauser et al., 1980). For the phospholipid analog labeled at the C_2 position, a fast torsional oscillation of $\pm 22°$ could best describe the line shapes in the gel state. In the L_α phase, the same type of motion or even conformational jumps around the glycerol C_1—C_2 bond should occur. For dihexanoyl-phosphatidylcholine the equilibrium populations for g^+, g^- and antiplanar conformation are 0.33, 0.03, and 0.06. This shows that there is considerable flexibility aroung the C_1—C_2 bond of the glycerol (Hauser et al., 1980). Newer higher resolution NMR studies of lamellar phospholipids show that there is rapid interconversion between two preferred conformations around the glycerol C_1—C_2 bond (Hauser et al., 1988). Obviously, the high flexibility around the glycerol bonds is due to the low potential energy barrier between conformers (McAlister et al., 1973).

The orientations of the *sn*-1 and *sn*-2 ester carbonyls have been determined in aligned multilayers of egg yolk phosphatidylcholine (Braach-Maksvytis and Cornell, 1988) and found to be consistent with the x-ray crystal structure of DMPC dihydrate (Pearson and Pascher, 1979). The *sn*-1 carbonyl plane adopts an orientation close to the bilayer normal with the C=O bond perpendicular to the bilayer normal, while the *sn*-2 carbonyl group is oriented in such a way that the principal axis of the shielding tensor is at the magic angle. This is also consistent with earlier ^{13}C-NMR results (Cornell, 1980, 1981; Wittebort et al., 1981, 1982).

(g) Summary: dynamics in the liquid crystalline phase. In liquid crystalline bilayers, phosphatidylcholines undergo **bond vibratins and torsional oscillations**, with very short correlation times between 10^{-14} and 10^{-9}s; **free or slightly hindered rotations and jumps around single bonds**, in the head group, the glycerol backbone and the fatty acyl chains with correlation times between 10^{-12} and 10^{-9}s. The fraction of gauche conformers increases towards the chain ends with a concomitant decrease in the correlation time τ_J for trans-gauche isomerization. The fraction of gauche conformers is $\sim$0.2 for CH_2 groups close to the glycerol backbone and $\sim$0.45 at the chain ends, the segmental order parameter $S_{z'z'}$ varying from 0.76 to 0.35. Fast isomerizations in the glycerol backbone, for instance, rotational jumps around the C_2—C_3 and probably also around the C_1—C_2 bond of the glycerol, seem to give rise to unequal populations of the g^+, g^-, and antiplanar conformations; **long axis rotation of the whole molecule**, with correlation times between 10^{-9} and $2 \cdot 10^{-8}$ s; **wobbling motions of the whole molecule**, with correlation times between 10^{-9} and 10^{-7}s. The angular excursions of the molecules can be described by the rigid body order parameter S_{zz}, which can be determined from ^{2}H-NMR or ESR line shape simulations, or time-dependent fluorescence anisotropy measurements with fluorescent probes like DPH, TMA-DPH. All three methods yield a value of $\sim$0.6 for S_{zz} for temperatures just above T_m. Correlation times $\tau_\perp$ determined by fluorescence anisotropy measurements are at least one order of magnitude shorter than those calculated from ESR

and ^{2}H-NMR line shapes; **collective motions of molecules**; these are order director fluctuations with relatively long correlation times of about 10^{-3} s.

In addition to the reorientational modes mentioned above, fast lateral diffusion of the molecules in the plane of the bilayer is observed. Lateral diffusion can contribute to T_1 and T_2 relaxation as the molecular orientation of the molecule change with correlation times $\tau_c > 10^{-5}$ s, when it diffuses along curved surfaces (Cornell and Pope, 1980; Bloom and Sternin, 1987). An additional motional mode is the very slow flip-flop of lipid molecules from one side of the bilayer to the other (Kornberg and McConnell, 1971; Schroit and Zwaal, 1991).

2. Dynamics in the Gel Phases

When phosphatidylcholines are cooled below their main phase transition temperature, dramatic changes in mobility are seen because of the decrease in molecular volume and the concomitant ordering of the acyl chains. DPPC converts to a $P_{\beta(\beta')}$ phase at a temperature of 41.5°C, while for DMPC this transition occurs at 24.5°C. In the $P_{\beta(\beta')}$ phase, the surface is distorted by a periodic ripple (Janiak et al., 1979). These changes in mobility are immediately evident from changes in line width and frequency of the CH_2 or CD_2 vibrations in the FT-IR spectra (Mendelsohn et al., 1981; Cortijo and Chapman, 1981; Mendelsohn and Mantsch, 1986; Cameron et al., 1981a,b; Hübner, 1988; Blume et al., 1988b). Figure 12 shows as an example the frequency for the symmetric CD_2 stretching vibration for 1,2(4,4'-$^2H_2$)DPPC as a function of temperature (Hübner, 1988).

In the P_β phase the number of gauche conformers is drastically reduced, but the molecules are still rotationally disordered and packed in a hexagonal lattice. Also some reorientation fluctuations seem still to be possible, as indicated by the relatively broad bands in the IR spectra. From ESR measurements with biradical spin probes it was deduced that the moelcular long axes show a distribution of tilt angles between 0° and 16° with an average tilt of 0° with respect to the local bilayer surface (Meier et al., 1982). Lowering the temperature in the P_β phase leads to an increased ordering of the molecules and a further decrease in half width of the CH_2 and CD_2 vibrational bands.

(a) P_β phase. According to NMR as well as ESR spectroscopic results, the P_β phase is microscopically heterogeneous, the exchange rate between the two populations varying with temperature (Wittebort et al., 1981, 1982; Blume et al., 1982b; Meier et al., 1983, 1986). Figure 13 shows ^{13}C and ^{2}H-NMR spectra of DPPC as a function of temperature (Blume et al., 1982b). These exchange processes presumably arise from lateral diffusion of the molecules perpendicular to the bilayer ripples. The ^{13}C-NMR spectra of 2(1-^{13}C)DPPC and related phosphatidylcholines in the P_β phase can be simulated by assuming fast axial diffusion on the ^{13}C-NMR time scale resulting in an axially symmetric ^{13}C tensor for one component and a narrow line for the other component, and inclusion of a temperature dependent two-site exchange process (Wittebort et al., 1982). The motional correlation times for long axis rotation are in the fast limit for ^{13}C and close to the fast limit on the ^{2}H-NMR time scale, i.e., $\sim 10^{-6}$ s.

Axial rotation, however, is not the only motional mode present in the gel phase. Additional trans-gauche isomerization is still possible and reduces the widths of the ^{2}H-NMR powder patterns (Davis, 1979; Meier et al., 1983, 1986; Mayer et al., 1988). The ESR spectra of spin labeled lipids show that chain isomerization is in the slow motional regime and the long axis rotation in the rigid limit of the ESR time scale (Lange et al., 1985; Moser et al., 1989). This is in agreement with the ^{2}H-NMR line shape simulations. The correlation times τ_J for trans-gauche isomerization vary between 10^{-9} s

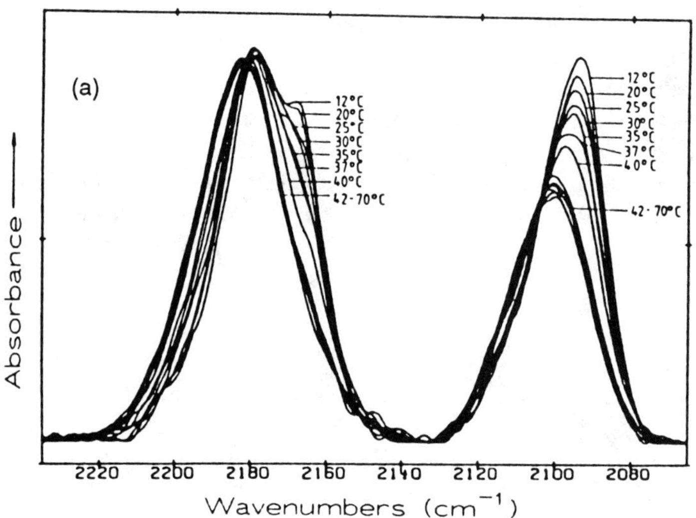

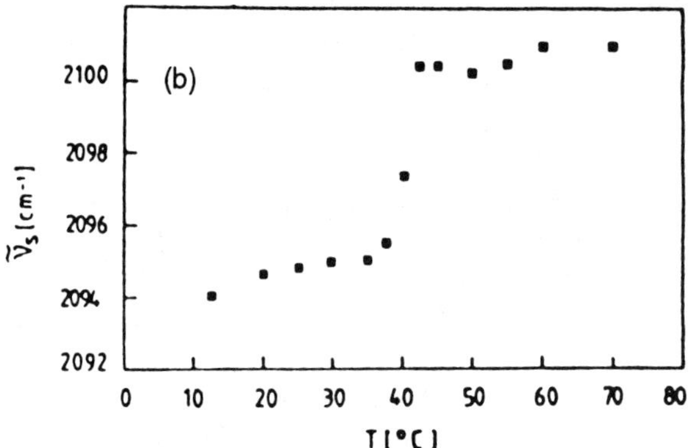

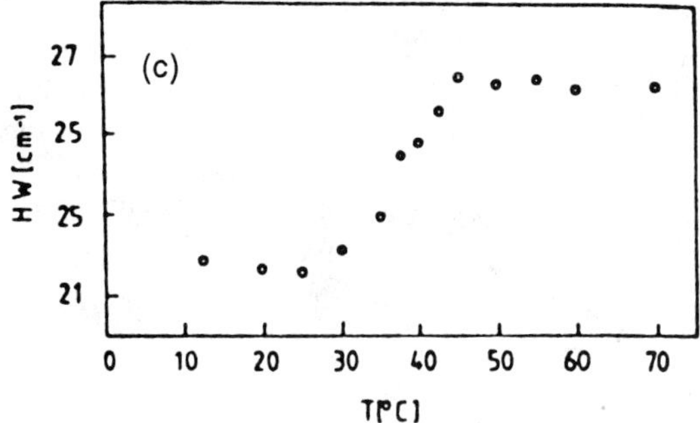

Figure 12 (a) Temperature dependence of the symmetric and antisymmetric CD_2 stretching vibrations for 1,2(4,4'-d_2)DPPC as a function of temperature; (b) frequency $\bar{\nu}_s$ for the symmetric stretching vibration as a function of temperature; (c) half width at half height for the symmetric stretching vibration as a function of temperature. (From Hübner, 1988.)

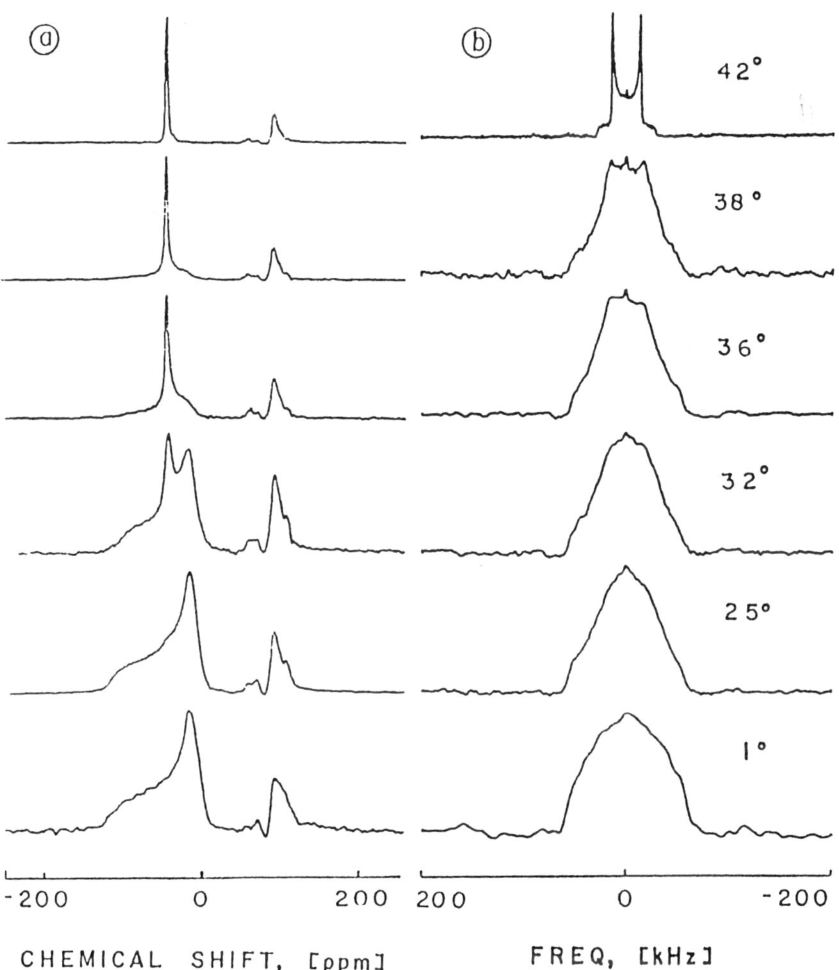

Figure 13 NMR spectra for (a) 2(1-^{13}C)DPPC and (b) 2(4,4'-d_2)DPPC as a function of temperature. (From Blume et al., 1982a.)

for the C_2 and 10^{-11}s for the C_{13} chain segment (Mayer et al., 1988; Moser et al., 1989). Trans-gauche isomerization in the chains of spin labeled DMPC is somewhat slower, presumably caused by a distorting effect of nitroxide groups (Moser et al., 1989).

(b)$L_{\beta'}$ phase. In the $L_{\beta'}$ phase the chains are tilted at an angle of ~30° with respect to the bilayer normal (Tardieu et al., 1973; Janiak et al., 1979). The chains are now packed in a distorted hexagonal lattice. They are still rotationally disordered as indicated by the absence of correlation field splittings of the CH_2 bending (FT-IR and Raman) and CH_2 rocking modes (FT-IR) (Wong et al., 1982, 1988; Wong and Mantsch, 1985). A considerable reduction in mobility is apparent from moment analysis in ^{1}H-NMR (MacKay, 1981; Trahms et al., 1983) and ^{2}H-NMR spectra (Davis, 1979). The line shapes of the ^{2}H- and ^{13}C-NMR spectra are indicative of reorientational motions around the molecular long axis, $\tau_{\|}$ approaching 10^{-5} s just above 0°C. The correlation times for trans-gauche isomerization are still below 10^{-7}s (Mayer et al., 1988). The probability for

trans-gauche isomerization increases towards the chain end and is in the range between 1 and 5% as determined from ^{2}H-NMR (Meier et al., 1983, 1986; Mayer et al., 1988, 1990) and recent FT-IR experiments (Mendelsohn et al., 1989). An alternative explanation for the observed increase in gauche populations and isomerization rates are torsional oscillations of the acyl chains (Cameron et al., 1981a,b). The reorientational modes observed in the glycerol backbone of phospholipid analogs show that torsional oscillations are very probable (Auger et al., 1990b).

Axial motion of lipids in the gel phase can occur via Brownian diffusion around the molecular long axis or an axial reorientation process via large angle rotational jumps, including a small amount of fast trans-gauche isomerization to account for the reduced width of the spectra (Meier et al., 1983, 1986; Wittebort et al., 1987). The latter process seems to be more likely (Bonmatin et al., 1990).

In the gel phase the glycerol backbone does not seem to be completely rigid. Trans-gauche isomerization around the C_2—C_3 bond, and perhaps also around the C_1—C_2 bond, of the glycerol is still possible (Auger and Jarrell, 1990; Auger et al., 1990a,b). The motion around the C_2—C_3 bond was modelled as a three-site exchange with unequal populations between the three sites and a short correlation time of $\sim 7 \cdot 10^{-10}$ s, and the motion around the C_1—C_2 bond as a fast torsional oscillation (see discussion above).

In the P_β phase the phosphocholine head groups are still relatively mobile, a sudden change occurring upon transition into the $L_{\beta'}$ phase. This abrupt change can be detected in ^{2}H- (Ruocco et al., 1985b) and ^{31}P-NMR spectra (Trahms and Klabe, 1985; Frye et al., 1985). The sharp ^{2}H powder pattern is replaced by a broad featureless line of ~20 kHz half width, indicating that the head group can still adopt a large number of different conformations. The reorientational motions approach the intermediate regime on the ^{2}H-NMR time scale (Ruocco et al., 1985b). At 35°C the ^{31}P line shape could be simulated using a rotational diffusion coefficient $R_{\|}$ of $5 \cdot 10^7$ s^{-1} with additional slow wobbling motions, $R_{\|}$ decreasing to $5 \cdot 10^5$ s^{-1} at 25°C in the L_β phase (Campbell et al., 1979). Dielectric relaxation measurements (Shepherd and Büldt, 1979) show that the relaxation times for the phosphatidylcholine head group motions change from $4 \cdot 10^{-9}$ s above to $1.5 \cdot 10^{-8}$ s below the pretransition.

(c) L_c phase. After sufficient annealing times at 0°C some phospholipids convert to an L_c or $L_{c'}$ phase in which most molecular motions are frozen out. FT-IR and Raman spectra of these lipid phases show correlation field splittings of the of the CH_2 scissoring modes indicative of highly ordered chains with the fatty acids packed in a parallel or perpendicular orthorhombic lattice (Wong et al., 1982; Wong and Mantsch, 1985). The ^{2}H-NMR spectra of lipids in these phases are essentially rigid limit powder patterns, and the ^{13}C-NMR spectra show the full asymmetric tensors (Meier et al., 1986; Wittebort et al., 1981, 1982). Also, the head group motions are completely suppressed when the lipid converts into the L_c phase (Füldner, 1981; Trahms et al., 1983; Trahms and Klabe, 1985).

(d) Summary: dynamics in the gel phases. The intra- and intermolecular motions in the gel phase of phosphatidylcholines can be summarized as follows: **bond vibrations and torsional oscillations** have similar correlation times as in the L_α phase (10^{-13}–10^{-10} s). Concerted torsional oscillations of C—C bonds in the chains lead to additional motional averaging around the chain axis, the torsion angles increasing towards the chain ends. **Hindered rotations and trans-gauche isomerization around single bonds** occur mainly in the head group, but also in the chains. The correlation times τ_J are position-dependent decreasing towards the chain ends. τ_J has a value of $\sim 5 \cdot 10^{-10}$s for positions in the middle of the chain. The activation energy for τ_J is relatively low, namely 8–15 kJ/mol (Mayer et al., 1988). The head groups are still highly mobile in the P_β as well as in the $L_{\beta'}$

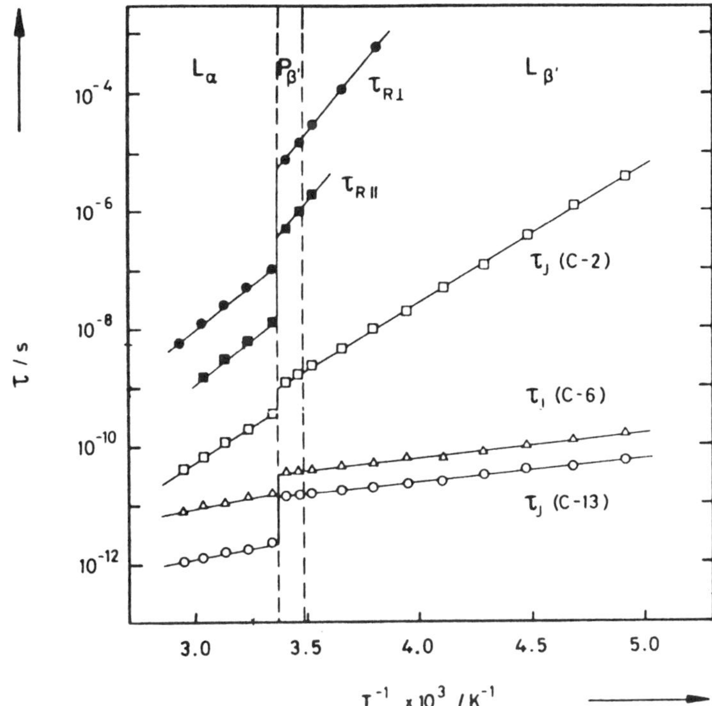

Figure 14 Arrhenius plot of the correlation times for lipid rotation ($\tau_{\|}$), wobbling motion ($\tau_{\perp}$), and trans-gauche isomerization (τ_j) at the C_2 (□), the C_6 (△), and the C_{13} (○) segment of the chains of DMPC. (From Mayer et al., 1988.)

phase, with correlation times for head group reorientations between 10^{-9} and 10^{-6} s. In the L_c phase and at temperatures below –20°C the head group motions are effectively frozen out. **Long axis rotation** occurs with correlation times of $\sim 10^{-6}$ s just below the chain melting transition. The rotational correlation times $\tau_{\|}$ slow down to 10^{-5}s at temperatures around 0°C. **Fluctuations of whole molecules** with small angle amplitudes are possible in the P_β and the $L_{\beta'}$ phase at higher temperatures, the corresponding correlation times ranging from 10^{-5} to 10^{-3} s. The correlation times for all reorientational motions, with the exception of methyl group rotation, become very long when the temperatures are lowered to –100°C or below.

A complete temperature dependence of the motional and orientational parameters for the lipid DMPC labeled with ^2H at different chain positions as reported by Mayer et al. (1988) is shown in Fig. 14. Table 1 shows a summary of the correlation times observed for various reorientational modes in bilayers of phosphatidylcholines and phosphatidylethanolamines (see page 463 for Table 1).

3. Chain Length Dependence

The chain length dependence of various lipid phase transitions is discussed in Chap. 12. For phosphatidylcholines in the L_α phase a slight increase in the ^2H quadrupole splittings is observed for phosphatidylcholines labeled at the same chain position and compared at

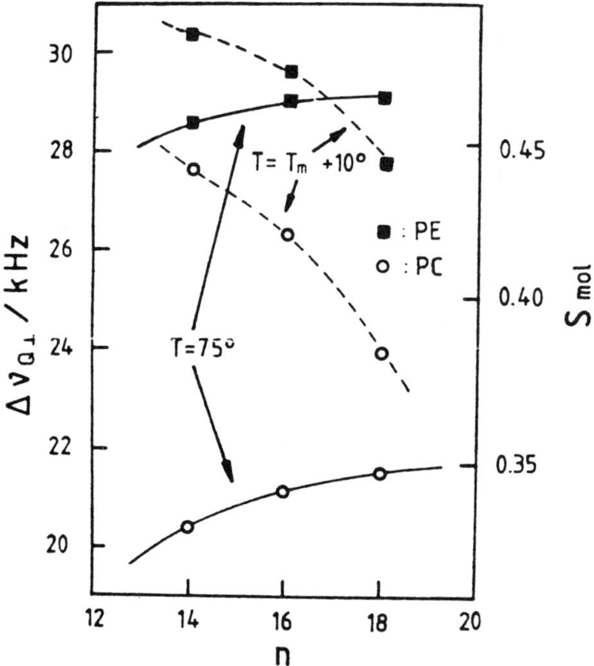

Figure 15 Dependence of the quadrupole splitting and order parameters of PCs and PEs labeled at the C_4 position of the fatty acyl chains as a function of chain length of the lipid n. Data were taken at constant temperature or on a reduced temperature scale. (From Hübner and Blume, 1987.)

the same temperature. On a reduced temperature scale, however, phosphatidylcholines with shorter chains now display larger quadrupole splittings (Hübner and Blume, 1987) (see Fig. 15). Comparisons made on reduced temperature are thus questionable (Hübner, 1988; Blume et al., 1988b).

Chain ordering and motional behavior in the $L_{\beta'}$ phase is only slightly dependent on lipid chain length, as judged from the width of the powder patterns at temperatures just below the pretransition. Differences in spectral intensities at 1°C indicate that rotational mobility decreases with increasing chain length (Hübner, 1988, Blume et al., 1988b).

4. Phosphatidylcholines with Mixed Chains

Lewis et al. (1984) have reported ^{13}C- and ^{2}H-NMR spectra of the mixed chain phosphatidylcholines MPPC and MSPC. These lipids convert directly from a quasi-crystalline L_c phase to the P_β phase and then to the L_α phase, as the $L_{\beta'}$ phase is not stable in these phosphatidylcholines. After sufficient incubation times at temperatures close to 0°C, long axis rotation of the molecules is essentially stopped, as concluded form ^{31}P-NMR as well as the ^{13}C-NMR spectra of C=O labeled lipids. However, the ^{2}H-NMR spectra show that other reorientational motions, presumably trans-gauche isomerization, persist down to relatively low temperatures.

5. Phosphatidylcholines with Alkyl Chains

Dihexadecyl-phosphatidylcholine (DHPC) forms an interdigitated gel phase (Ruocco et al., 1985a). In the liquid crystalline phase, the motional behavior is similar to DPPC. In the interdigitated gel phase, rotational motion of the molecules persists down to very low temperatures below 0°C as indicated by the fast limit axially symmetric ^{2}H-NMR powder patterns (Ruocco et al., 1985a,b). The narrow line shapes of head group labeled DHPC suggest greater motional freedom for the DHPC head group. This is a consequence of an increase of the available surface area by roughly a factor of two upon interdigitation.

6. Influence of Fatty Acyl Chain Structure

Introduction of unsaturated fatty acids lowers the phase transition temperature, a cis double bond normally having a greater effect than a trans double bond (see Chap. 12) (Berde et al., 1980). The order parameter obtained for specifically deuterated 1-palmitoyl-2-oleoylphosphatidylcholine as a function of position of the CD_2 group shows a characteristic dip at the position of the double bond (Seelig and Seelig, 1977; Seelig and Waespe-Sarcevic, 1978). This is a consequence of the particular average orientation of the C-D vector at the double bond position with respect to the director. As rotation around the double bond is not possible, a slight reduction in the average rate of motion at the position of the double bond is probable. This was indicated by the T_1 profile measured for POPC (Perly et al., 1985) and membranes of *Acholeplasma laidlawii* enriched with oleic acid (Rance et al., 1980; Davis, 1986). Incorporation of ^{2}H labeled dioleoyl-phosphatidylcholine into liquid crystalline bilayers of other phosphatidylcholines with an increasing number of double bonds showed no particular change in order parameters or in relaxation times (Yeagle and Frye, 1987). In bilayers of phosphatidylcholines with diunsaturated fatty acids like 1-palmitoyl-2-isolinoleoyl-phosphatidylcholine, slow intramolecular conformational changes might be present consisting of two-site jumps between low energy conformations. These results suggest that polyunsaturated fatty acyl chains are relatively ordered and less flexible than saturated and monounsaturated chains (Baenziger et al., 1988, 1991).

Phosphatidylcholines with cyclopropane fatty acids (Dufourc et al., 1983; Jarrell et al., 1983), branched chain fatty acids (Lindsey et al., 1979; Silvius et al., 1985; Blume and Bäuerle, 1989; Stewart et al., 1990), and ω-cyclohexyl fatty acis (Mantsch et al, 1985; Hübner, 1988; Hübner and Blume, 1988; Blume et al., 1988b) all show fast limit powder patterns in the L_α phase, their reorientational motions thus being very similar to those of phosphatidylcholines with straight chain fatty acids. Introduction of branched chain fatty acids or ω-cyclohexyl fatty acids leads to slight increases in order parameters (Hübner, 1988, Hübner and Blume, 1988; Blume et al., 1988b; Blume and Bäuerle, 1989), presumably caused by a reduction in the wobbling motion of the molecules and/or a reduced degree of trans-gauche isomerization. Another interesting case are phospholipids with phytanyl chains, which have four methyl side groups attached to each chain. These lipids do not show any phase transition in the temperature range between 0 and 100°C (Lindsey et al., 1979). ^{2}H- and ^{31}P-NMR measurements by Stewart et al. (1990) showed that the branched chains in these lipids are more disordered than the acyl or alkyl chains of normal lipids, but that the rates of motion are much slower, the correlation time for segmental reorientation being $\sim 2 \cdot 10^{-9}$ s at a temperature of 8°C.

Mobility in the ordered phases of phosphatidylcholines depends on the type of gel phase. Phosphatidylcholines with iso-branched fatty acids form different L_c phases depending on whether the number of carbons is even or odd. The L_c phase of 17isoPC converts first to an L_β phase and then to an L_α phase (Church et al., 1986). ^{2}H-NMR spectra of 17isoPC labeled at the 12 position of both chains show that the lipids undergo considerable reorientational motions in the L_c phase. The L_β phase of 17isoPC is microscopically heterogeneous as two-component line shapes are observed. 18isoPC converts directly into an $L_{c'}$ phase with tilted chains, when cooled from the L_α phase. In this phase, all motions are almost completely suppressed. ^{2}H-NMR spectra of these two lipids are shown in Fig. 16 (Blume and Bäuerle, unpublished).

Phosphatidylcholines with ω-cyclohexyl fatty acids show similar odd-even effects (Lewis and McElhaney, 1985; Blume et al., 1987). NMR and FT-IR spectra indicate that phosphatidylcholines with odd-numbered ω-cyclohexyl fatty acids form very tightly packed gel phases, while in the gel phase of the corresponding even-numbered analogs some intramolecular motions are retained (Mantsch et al., 1985; Blume et al., 1988b). The line shapes are very similar to those of glycolipids in the gel phase, where simulations using a crankshaft motion of the chains could satisfactorily reproduce the spectra (Huang et al., 1980).

B. Other Lipid Classes

1. Phosphatidylethanolamines

The dynamics of dipalmitoyl-phosphatidylethanolamine (DPPE) has been studied in considerable detail by ^{13}C- and ^{2}H-NMR spectroscopy (Rice et al., 1981; Blume et al., 1982a). DPPE shows a phase transition from a metastable L_β to an L_α phase at a temperature of 64°C (McIntosh, 1980; Cevc and Marsh, 1987). In the liquid crystalline phase, sharp axially symmetric powder patterns indicate fast intramolecular reorientational motions on the ^{2}H-NMR time scale. The correlation times for these motions are probably similar to those of phosphatidylcholines, the spin lattice relaxation time T_{1Z} for DPPE at 70°C being 50 ms and thus only slightly longer than for DPPC at the same temperature (37 ms) (Hübner, 1988).

The order parameters S_{mol} determined from the quadrupole splittings of PEs are definitely larger than the corresponding values observed for phosphatidylcholines (see Fig. 15) (Blume et al., 1982a; Marsh et al., 1983; Hübner and Blume, 1987). For saturated as well as unsaturated phosphatidylethanolamines the differences are small on the reduced temperature scale but become much larger on the absolute temperature scale (Lafleur et al; 1990a,b). This shows that phosphatidylethanolamines, presumably due to their smaller head groups and stronger head group interactions via hydrogen bonds, have less motional freedom in the liquid crystalline phase. Whether these higher order parameters are a consequence of a decrease in trans-gauche isomerization and/or in the wobbling motion of the molecules is still unclear.

Head group labeled DPPE shows a relatively small quadrupole splitting in the L_α phase similar to those found for phosphatidylcholines, indicating high mobility of the head group (Seelig and Gally, 1976; Brown and Seelig, 1978).

Dynamics in the L_β phase can be simulated with a rotational jump model allowing for trans-gauche isomerization in the form of a six-site or nine-site jump model. In the gel phase, the correlation times for the rotational jump motions are between $5 \cdot 10^{-7}$ and $7 \cdot 10^{-8}$ s and thus very similar to those of phosphatidylcholines.

The number of gauche conformers in phosphatidylethanolamines increases towards the chain ends from ~2% at the 4 position to 7% at the 12 position. The correlation times for chain isomerization are longer than for chain rotation, in contrast to the findings for phosphatidylcholines, as checked by simulations of spectra obtained from oriented DMPE bilayers (Hübner and Blume, 1990).

In the gel state, the phosphatidylethanoalmine head group can still adopt a large number of different conformations and orientations, thus yielding a featureless broad

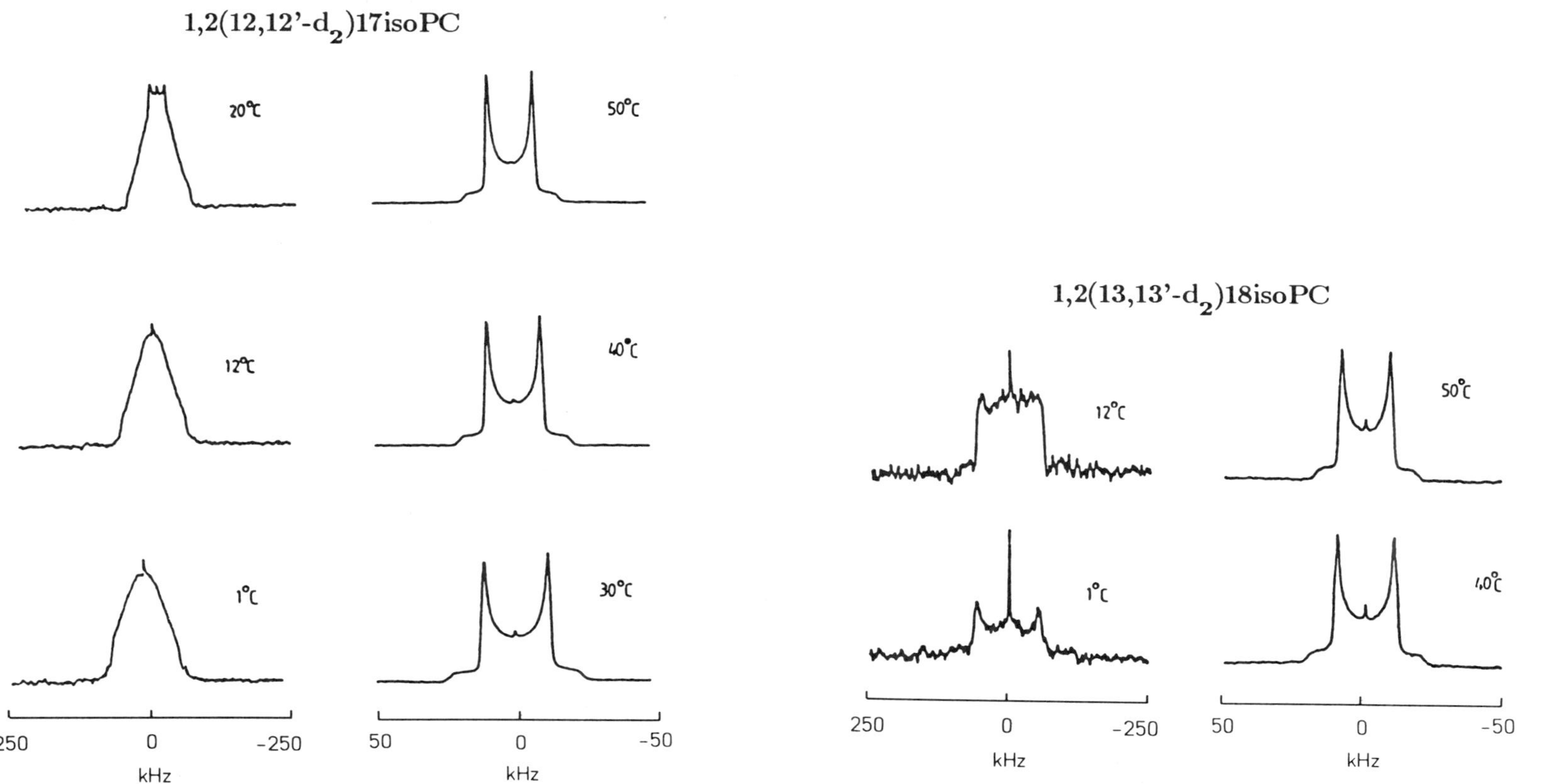

Figure 16 ^{2}H-NMR spectra of 1,2(12,12'-d_2)17isoPC and 1,2(13,13'd_2)18isoPC as a function of temperature. (From Blume and Bäuerle, unpublished.)

^{2}H-NMR line. At lower temperature the correlation times for the reorientational motion of the head group approach the intermediate-time regime (Blume et al., 1982a). Such motional behavior of the phosphatidylethanolamine head group is again qualitatively similar to that observed for phosphatidylcholines (Ruocco et al., 1985a).

^{2}H-NMR spectra of DPPE labeled at the 2 position of the glycerol backbone indicate that torsional oscillations or trans-gauche isomerization occur around the C_1—C_2 bond of the glycerol, as in the case of DPPC (Blume et al., 1982a).

2. Phosphatidylserines, Phosphatidylglycerols, and Cardiolipin

The dynamic behavior of phosphatidylserines with saturated and unsaturated fatty acyl chains labeled at several positions in the chains and in the head group has been studied by ^{2}H-NMR spectroscopy by Browning and Seelig (1980). The motions and the ordering of phosphatidylserine molecules are very similar to those of other phospholipids, except that head group motions are more restricted.

In the H_{II} phase, which is induced by protonation, the quadrupole splitting of chain deuterated DOPS is reduced by a factor of two due to additional fast averaging around the H_{II} phase cylinders (DeKroon et al., 1990). Likewise, the ^{31}P axially symmetric line shape is reversed and narrowed. The ^{2}H spin lattice relaxation times for chain labeled DOPS are hardly affected by the transition into the H_{II} phase. Head group labeled DOPS shows a characteristic T_1 minimum in the H_{II} phase at 45°C, from which a motional correlation time for head group reorientation of $3.45 \cdot 10^{-9}$ s was determined. In DOPS bilayers, the corresponding value is $\sim 0.24 \cdot 10^{-9}$ s at 45°C. Phosphatidylserine head group motions are thus very sensitive to the degree of ionization and the changes in intermolecular hydrogen bonding (De Kroon et al., 1990).

Head group motions of phosphatidylglycerol are similar to those of the zwitterionic lipids like phosphatidylcholine and phosphatidylethanolamine as concluded from ^{31}P and ^{2}H-NMR measurements (Cullis and de Kruijff, 1976; Wohlgemuth et al., 1980; Gally et al., 1981; Borle and Seelig, 1983a,b; Sixl and Watts, 1983).

Specific glycerol deuteration in cardiolipin was achieved using an *E. coli* mutant defective in biosynthesis and degradation of glycerol (Allegrini et al., 1984). Both glycerol backbones of cardiolipin were found to be oriented perpendicular to the bilayer surface. T_1 relaxation time measurements showed that the glycerol residues in the head group are considerably less flexible than the polar groups of other phospholipids due to the bridging structure.

3. Glycolipids

With the exception of phosphatidylinositol, glycolipids are not phospholipids. However, it is instructive to see how molecular motions are affected by head groups with sugar residues. Glycolipids have head groups with numerous hydroxyl groups capable of intermolecular hydrogen bonding. Their hydration in the ordered phase is therefore quite low. Due to this hydrogen bond network, head group motions are quite slow, as observed for instance in palmitoylgalactosylceramide by ^{2}H-NMR. The number of head orientations is small, the head group pointing normally straight up from the bilayer surface (Skarjune and Oldfield, 1979a, 1982). Chain motions in liquid crystalline cerebroside bilayers seem to be identical to those observed in phospholipids (Huang et al., 1980).

In cerebrosides, intermolecular hydrogen bonding can also occur in the backbone region via amide N—H bonds to neighbouring hydroxyl or amide C═O groups of the sphingosine base. Probably as a consequence, the motions in the low temperature gel phase are distinctly different from those of phospholipids at a comparable temperature

(Huang et al., 1980). Rigid-limit powder patterns are observed, for example, for *N*-palmitoyl-galactosyl-cerebroside (NPGS) labeled with ^{13}C at the C═O bond of the amide linkage up to a temperature close to T_m (80°C). This indicates that rotational jump motions around the molecular axis are only possible with correlation times longer than 10^{-2} s. ^2H-NMR spectra of cerebrosides show that residual intramolecular reorientations are still possible. These have been modeled as crankshaft motions in the acyl chains with correlation times of $\sim 10^{-7}$ s at temperatures around 50°C (Huang et al., 1980; Siminovitch et al., 1985).

In the liquid crystalline phase of NPGS, all motions are in the fast limit on the ^2H-NMR time scale. From spin-lattice relaxation times T_{1Z} correlation times τ_c of $9 \cdot 10^{-11}$ s and $5 \cdot 10^{-12}$ s were determined for the 3 and for the 12 position, respectively (Speyer et al., 1989).

Renou et al. (1989) and Carrier et al. (1989) have studied the dynamics and orientation of diglycerides with lactose and gentiobiose as head groups. In both cases, the lipids were specifically labeled with ^2H in the disaccharide head group. Two anomers with α as well as β linkage of the glucose ring of the gentiobiosyl head group to the glycerol were investigated (α- and β-DTDGL). Two conformers exist, for the second sugar ring in the head group, that are in slow exchange on the ^2H-NMR time scale (Carrier et al., 1989). The flexibility around the 1→6 linkage of the sugar rings is low. As T_{1Z} decreases with increasing temperature, the motions dominating T_{1Z} relaxation are in the regime of long correlation times ($(\omega\tau)^2 \gg 1$). The gentiobiosyl head group thus displays much slower reorientational motions than head groups of phospholipids or of other glycolipids like ditetradecyl-glucopyranosul-glycerol (α-DTGL and β-DTGL). For the latter glycolipids it was found that the correlation time for head group reorientation around thc glucose—*O*—glycerol glycosidic bond is approximately 10^{-5}s, the β anomer showing larger angular fluctuations than the α anomer (Jarrell et al., 1986, 1987a,b).

Ditetradecyl-glucopyranosyl-glycerol labeled at the 3 position of the glycerol backbone ((3,3'-d_2)β-DTGL) exhibits fast intramolecular motions in the glycerol backbone consisting of asymmetric trans-gauche isomerization around the C_2—C_3 bond of the glycerol (Auger et al., 1990b). The ^2H-NMR spectra could be simulated using a simple three-site jump model for the intramolecular motion with a correlation time of $6.7 \cdot 10^{-10}$ s at 25°C (Auger et al., 1990b). The authors suggested that this type of motion might be common to all types of glycero lipids, i.e., also for phosphatidylcholines and PEs, as discussed above.

Using 2D-^2H-NMR techniques (Schmidt et al., 1987, Auger et al. 1990a,b) found that in β-DTGL, as well as in NPGS, axial rotation in the gel phase is very slow and best described by large angle threefold jumps with correlation times between 10 and 100 ms at room temperature (see Fig. 17). The rotation rate increases to 10^6 rad s^{-1} at temperatures close to the phase transition.

C. Effects of Ion Binding

NMR spectroscopy can be used to study the effects of ions on the intra- and intermolecular dynamics of the lipid molecules (Brown and Seelig, 1977; Hutton et al., 1977). Depending on the type of nucleus observed, the effects of ion binding are detected either by changes in the chemical shift anisotropy (^{13}C or ^{31}P) or by changes in the quadrupole splittings (^2H and ^{14}N) (Akutsu and Seelig, 1981; Tilcock and Cullis, 1981; Rothgeb and Oldfield, 1981; Altenbach and Seelig, 1984; Tilcock et al., 1984; Siminovitch et al.,

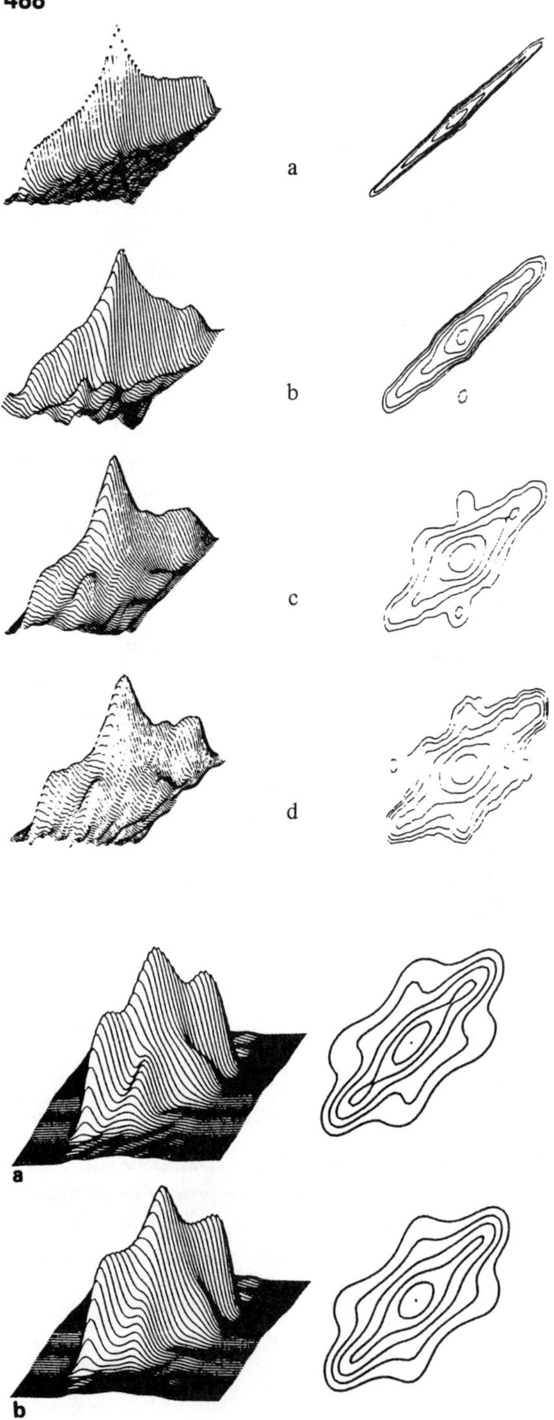

Figure 17 Top four rows: experimental 2D absorption mode exchange spectra and contour plots for a dispersion of (3,3'-d_2)β-DTGL at different temperatures. (a) 25°C, t_{mix} = 1 ms; (b) 35°C, t_{mix} = 0.5 ms; (c) 35°C, t_{mix} = 2 ms; (d) 40°C, t_{mix} = 2 ms. Bottom two rows: 2D spectral simulations using a fast three-site jump model for internal motions and (a) a slow threefold and (b) a slow twelvefold rotational jump for axial rotation. (From Auger and Jarrell, 1990.)

1984; Borle and Seelig, 1985; MacDonald and Seelig, 1987a,b, 1988; Huang et al., 1988; Roux and Bloom, 1990). Cation binding seems to change the average conformation of the phosphatidylcholine head group in such a way that the N^+ end of the P—N dipole is now pointing more to the water phase and not parallel to the bilayer surface as in pure phosphatidylcholine bilayers (Seelig et al., 1987; Scherer and Seelig, 1987, 1989). Ion binding does not change the dynamics of the head groups in the liquid crystalline phase to a measureable extent. As only one quadrupole splitting is observed in systems with added salt, ion exchange between different head groups is fast on the ^{2}H-NMR time scale. For instance, the exchange rate for Ca^{2+} is faster than 10^6–10^7 s^{-1} (MacDonald and Seelig, 1987b).

Addition of divalent or trivalent cations, particularly to negatively charged phospholipids, can lead to drastic changes in lipid phase behavior (see Chap. 12). For lipids that do not convert into the H_{II} phase upon ion binding but stay in the bilayer phase, changes in NMR line shapes are caused by changes in conformation and mobility. One example of a drastic change occuring in lipid dynamics is the binding of UO_{2+} to DPPC (Huang et al., 1988). The uranyl ion binds strongly to the phosphatidylcholine head groups with a stoichiometry of 4:1. UO_{2+} binding does not seem to affect lipid dynamics in the liquid crystalline phase, but dynamics in the gel phase is drastically changed. In pure DPPC bilayers in the $L_{\beta'}$ phase, the molecules rotate rapidly on the ^{2}H-NMR time scale around their long axes. Binding of uranyl ions eliminates this fast axial diffusion, as is evident from the appearance of asymmetric chemical shift spectra for ^{31}P and ^{13}C of C═O labeled DPPC. ^{2}H-NMR spectra of chain labeled DPPC indicate that trans-gauche isomerization is now the dominant motional mode. This motion can be described by two or three site jumps on a tetrahedral lattice with rates of $\sim 10^7$ s^{-1}, very similar to the chain isomerization model described for motions of gel phase glycolipids (Huang et al., 1980, 1988).

D. Lipid Mixtures

1. Lipid-Cholesterol Mixtures

Cholesterol is found in many eukaryotic membranes at high concentrations up to 50 mol %. The properties and function of cholesterol in membranes have been investigated by numerous workers using a variety of physicochemical techniques (for reviews see Demel and de Kurijff, 1976; Yeagle, 1985; Bloom and Mouritsen, 1988; Finean, 1990). Cholesterol at high concentrations has a so-called dual effect: it condenses and orders the liquid crystalline phase and disorders the gel phase leading to some kind of intermediate state.

The condensing effect of cholesterol is immediately apparent from ^{2}H-NMR spectra of chain labeled phospholipids, as the quadrupole splittings become larger with increasing cholesterol content. This effect has been observed in a variety of lipid systems and also in natural membranes (Stockton and Smith, 1976; Stockton et al., 1976; Haberkorn et al., 1977; Oldfield et al., 1978a,b; Jacobs and Oldfield, 1979; Rice et al., 1979; Blume and Griffin, 1982; Ghosh and Seelig, 1982; Rance et al., 1982; Killian et al., 1986; Malthaner et al., 1987; Siminovitch et al., 1987, 1988; Speyer et al, 1989; Vist and Davis, 1990).

(a) DPPC-cholesterol. Cholesterol dynamics. The dynamics of cholesterol molecules in phospholipid bilayers is much clearer than the thermodynamic properties of the systems, as spectroscopic techniques like ESR, NMR, or fluorescence spectroscopy can be used to quantify the molecular motions. ^{2}H-NMR spectra of specifically deuterated cholesterol incorporated into membranes show that the cholesterol molecules, on the average, are

oriented perpendicular to the membrane surface. On the ^{2}H-NMR time scale the molecules rotate fast around their long axis (Oldfield et al., 1978a,b; Taylor and Smith, 1981; Taylor et al., 1981, 1982; Dufourc et al.a, 1984a,b; Dufourc and Smith, 1986; Bonmatin et al., 1988, 1990). A variety of different spectroscopic investigations yielded correlation times for this long axis rotation between 10^{-10} and $4 \cdot 10^{-9}$ s (Yeagle, 1981; Polnaszcek et al., 1981; Siminovitch et al., 1988; Shin and Freed, 1989a; Korstanje et al., 1989; Korstanje et al., 1990).

Whether axial rotation of cholesterol is faster than the rotation of the surrounding phospholipid molecules is not clear at present. For DMPC the correlation time for long axis rotation$\tau_{\parallel}$ was reported to be $\sim 10^{-8}$ s at 25°C just above T_m (Mayer et al., 1988) and thus longer than the value for cholesterol, $3 \cdot 10^{-9}$ s. In NPGS/cholesterol bilayers at $\sim$70°C, the diffusion constant $D_{\parallel}$ of the NPGS molecule is $\sim 8 \cdot 10^{10}$ s^{-1}, corresponding to a value of $\tau_{\parallel}$ of $\sim 2 \cdot 10^{-12}$ s and thus shorter than $\tau_{\parallel}$ for the cholesterol molecule, $4.2 \cdot 10^{-11}$ s (Siminovitch et al., 1988).

The correlation time $\tau_{\perp}$ for the restricted wobbling motion may be somewhat longer by approximately one order of magnitude. For a cholestane spin label in POPC with 20 mol % cholesterol, the value for the rotational rate $R_{\perp}$ is $0.35 \cdot 10^7$ s^{-1}, corresponding to a correlation time $\tau_{\perp}$ of $5 \cdot 10^{-9}$ s (Shin and Freed, 1989a). ^{2}H-NMR measurements give smaller values with $\tau_{\perp} < 10^{-9}$ s (Dufourc and Smith, 1986). With a fluorescent cholesterol analog in bilayers of egg phosphatidylcholine with 15 mol % cholesterol at 25°C, $\tau_{\perp}$ was determined from time resolved fluorescent measurements to $0.9 \cdot 10^{-9}$ s (Yeagle et al., 1990). These differences in values for $\tau_{\perp}$ could well be due to the use of different bilayer systems and different labeled molecules.

(b) DPPC cholesterol. Phospholipid dynamics. The effects of cholesterol on the motional rates of phospholipid molecules seem to be model- and/or method-dependent. For lipid long axis rotation, Siminovitch et al. (1985) and Speyer et al. (189) report a *decrease* in the rotational correlation time $\tau_{\parallel}$ by approximately one order of magnitude in NPGS/cholesterol and DPPE/cholesterol bilayers. ^{2}H-NMR experiments reported by Mayer et al. (1990) give a value of $5 \cdot 10^{-10}$ s, which *increases* to $1.8 \cdot 10^{-8}$ s upon addition of cholesterol. However, these differences could be due to different simulation models used by the two groups. In oriented DMPC bilayers, the correlation time $\tau_{\perp}$ for the wobbling motion *increases* from $5 \cdot 10^{-9}$ to $8.8 \cdot 10^{-7}$ s when 40 mol % cholesterol is added. For unoriented DMPC bilayers at 35°C, a value for $\tau_{\perp}$ of $5 \cdot 10^{-8}$ s and for $\tau_{\parallel}$ of $5 \cdot 10^{-9}$ s was reported (Mayer et al., 1988).

The correlation times τ_J for trans-gauche isomerization in the lipid chains are also dependent on cholesterol content. For example, addition of 40 mol % cholesterol to DMPC increases τ_J from $1 \cdot 10^{-11}$ to $4 \cdot 10^{-11}$ s (Mayer et al., 1990). Addition of 40 mol % cholesterol to NPGS bilayers increases τ_J by about one order of magnitude to $\sim 10^{-10}$ s (Speyer et al., 1989).

(c) DPPE-cholesterol. The DPPE/cholesterol system has been studied in detail using ^{13}C-NMR ad ^{2}H-NMR techniques and specifically ^{13}C=O and ^{2}H labeled DPPE (Wittebort et al., 1982; Blume and Griffin, 1982). ^{13}C spectra of DPPE/cholesterol in the gel phase are two component spectra influenced by exchange. In DPPE/cholesterol, as well as in several phosphatidylcholine/cholesterol mixtures, the spectra can be quantitatively simulated assuming two phospholipid populations. One component is in contact with the cholesterol moleule and gives rise to a relatively sharp line; for the other component an axially symmetric powder pattern is observed. The exchange rates vary with temperature

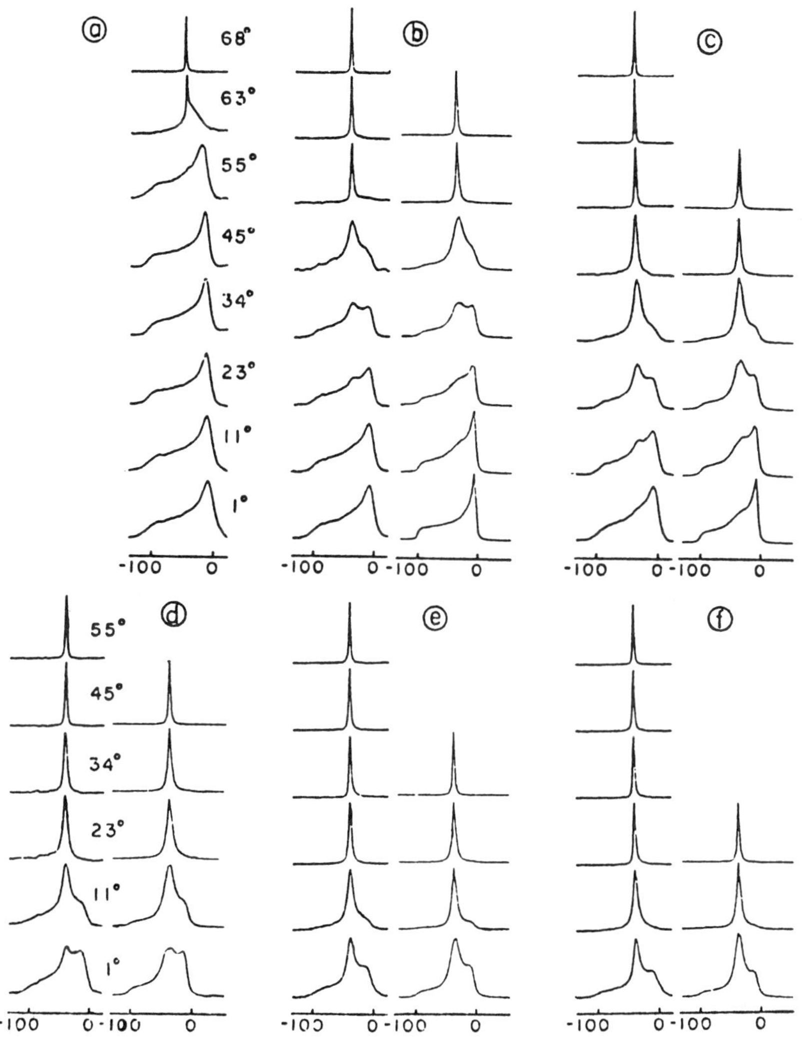

Figure 18 ^{13}C-NMR spectra of 2(1-^{13}C)DPPE/cholesterol mixtures as a function of composition and temperature. On the right line shape simulations using two components in exchange, as described in the text. (a) 0, (b) 10, (c) 20, (d) 30, (e) 40, and (f) 50 mol % cholesterol. (From Blume and Griffin, 1982.)

between 300 and 3000 Hz (see Fig. 18). The ^{2}H-NMR spectra of chain labeled DPPE in bilayers with cholesterol are consistent with the exchange rates determined by ^{13}C-NMR (see Fig. 19) assuming fast axial rotation and reduced trans-gauche isomerization for DPPE molecules in contact with cholesterol, and a line shape identical to unperturbed PE for those lipid molecules farther away from the cholesterol (Blume and Griffin, 1982).

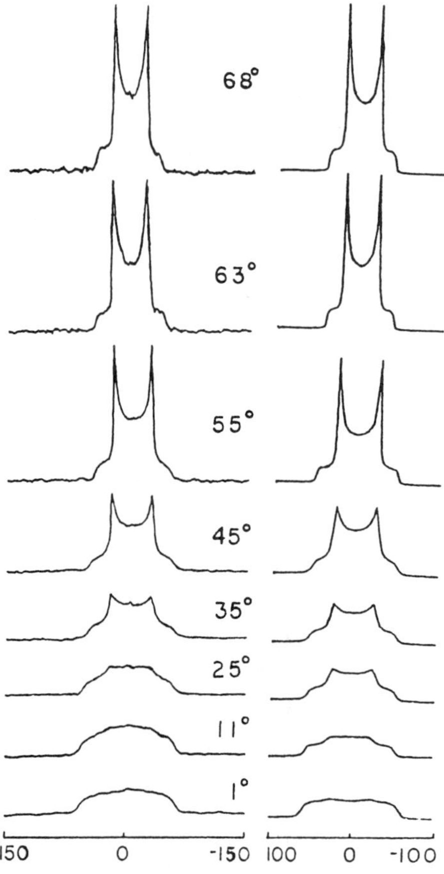

Figure 19 ^{2}H-NMR spectra for 2(4,4'-d_2)DPPE/20mol % cholesterol (left) and line shape simulations using exchange between a fast rotating component with reduced quadrupole splitting and an unperturbed component and exchange rates as determined from the ^{13}C-spectra in Fig. 18. (From Blume and Griffin, 1982.)

The quadrupole splitting of head group labeled DPPE decreases as a function of cholesterol content due to the spacer effect of cholesterol. Simultaneously, the motional correlation time decreases (Brown and Seelig, 1978; Shepherd and Büldt, 1979; Blume and Griffin, 1982).

At high temperature a reduction in trans-gauche isomerization, i.e., an increase in conformational order, is observed, but the molecules still rotate rapidly around their long axes. The extent of reduction of gauche conformers, as calculated from the quadrupole splittings, depends on the model used for the simulations. Inclusion of a wobbling motion suggests that the increased quadrupole splitting $\Delta\nu_Q$ is caused by both a decrease in the wobbling motion and a decrease in the trans-gauche isomerization. Without wobbling motion, the increase in $\Delta\nu_Q$ is only attributed to an increase in trans conformers.

2. Binary Phospholipid Mixtures

Dynamics of lipid molecules in binary lipid mixtures have not been studied in the same detail as pure lipids or lipid-cholesterol mixtures. In binary lipid mixtures in lamellar phases, the change from the liquid crystalline to a gel state is evident from a sudden appearance of a broad component in the ^{2}H-NMR spectra upon decreasing the temperature. On heating, sharp powder patterns, characteristic for liquid crystalline lipid, appear at the onset of melting. Thus ^{2}H-NMR spectroscopy can be used to determine the liquidus and solidus curves of pseudo-binary phase diagrams. DMPC/DPPC, DPPC/DPPE, DSPC/DMPC and DMPC/DSPE mixtures have been studied using ^{13}C-NMR and ^{2}H-NMR methods (Jacobs and Oldfield, 1979; Blume et al., 1982b; Hübner and Blume, 1987; Hübner, 1988). In all cases, two-component spectra were observed in the two-phase region of the phase diagram. This indicates that lipid exchange between the "liquid" and "solid" domains is slow on the ^{2}H-NMR and ^{13}C time scale. In Fig. 20, spectra for a 1:1 mixture of DPPC/DPPE of ^{13}C=O and ^{2}H labeled lipids are shown.

Binary mixtures of DMPC/DSPE (Fig. 21) show a large miscibility gap in the gel phase as two different types of line shapes are observed, a wider one characteristic for DSPE and a somewhat narrower component for DMPC (Hübner, 1988). In the liquid crystalline phase, lateral diffusion is fast on the ^{2}H-NMR time scale, causing quadrupole splittings originating from different environments to be averaged. In case the dependence of $\Delta\nu_Q$ in the liquid crystalline phase on composition is clearly nonlinear, this can be taken as an indication for nonideal mixing and lipid clustering (Blume et al., 1982b; Blume and Hübner, 1987).

In DPPC/DPPE mixtures the motion of the DPPE head groups are somewhat restricted compared to pure DPPE, as the quadrupole splittings increase with increasing DPPC content (Blume et al., 1982b).

Head group interactions in lipid mixtures containing one charged and one zwitterionic lipid have been studied by Sixl and Watts (1982, 1983). The relaxation rates for head group labeled phosphatidylcholine in mixtures with phosphatidylglycerol and phosphatidylserine were not affected. The observed changes in quadrupole splittings are thus mainly due to conformational changes.

V. LATERAL DIFFUSION

A. Pure Lipid Bilayers

The lateral transport of lipids and proteins is of particular interest from a biological point of view, as the reaction rates of membrane bound enzymes are possibly diffusion limited. Therefore lateral diffusion of lipids and proteins in bilayers have been studied in detail by using different techniques.

Methods suitable for the detection of lipid diffusion are optical techniques like the FRAP method or the excimer technique mentioned above. Other methods applied are the ESR (Träuble and Sackmann, 1972; Sackmann and Träuble, 1972; Devaux and McConnell, 1972; Scandella et al., 1972; King et al., 1987; Shin and Freed, 1989a), and the NMR technique (Cullis, 1976; Fisher and James, 1978; Kuo and Wade, 1979; Lindblom and Wennerström, 1977; Lindblom et al., 1981; Silva-Crawford et al., 1980) using line shape analysis in small vesicles, relaxation time measurements, or applying pulsed field gradients.

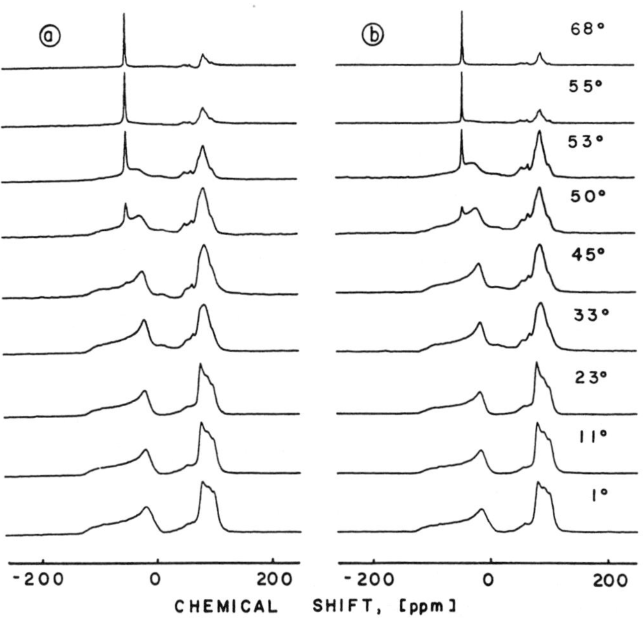

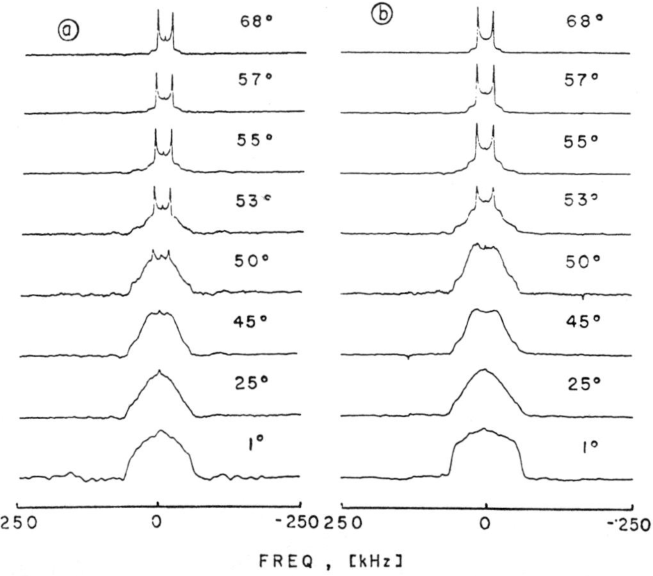

Figure 20 Top: ^{13}C-NMR spectra of a 1:1 mixture of DPPC with DPPE as a function of temperature: (a) 2(1-^{13}C)DPPC-DPPE; (b) DPPC-2(1-^{13}C)DPPE. Bottom: ^{2}H-NMR spectra of a 1:1 mixture of DPPC with DPPE as a function of temperature: (a) 2(4,4'-d_2)DPPC-DPPE, (b) DPPC-2(4,4'-d_2)DPPE. Note that in the phase transition region the spectra consist of superpositions of line shapes characteristic of gel and liquid crystalline like lipid, indicating that exchange between the domains is slow. (From Blume et al., 1982b.)

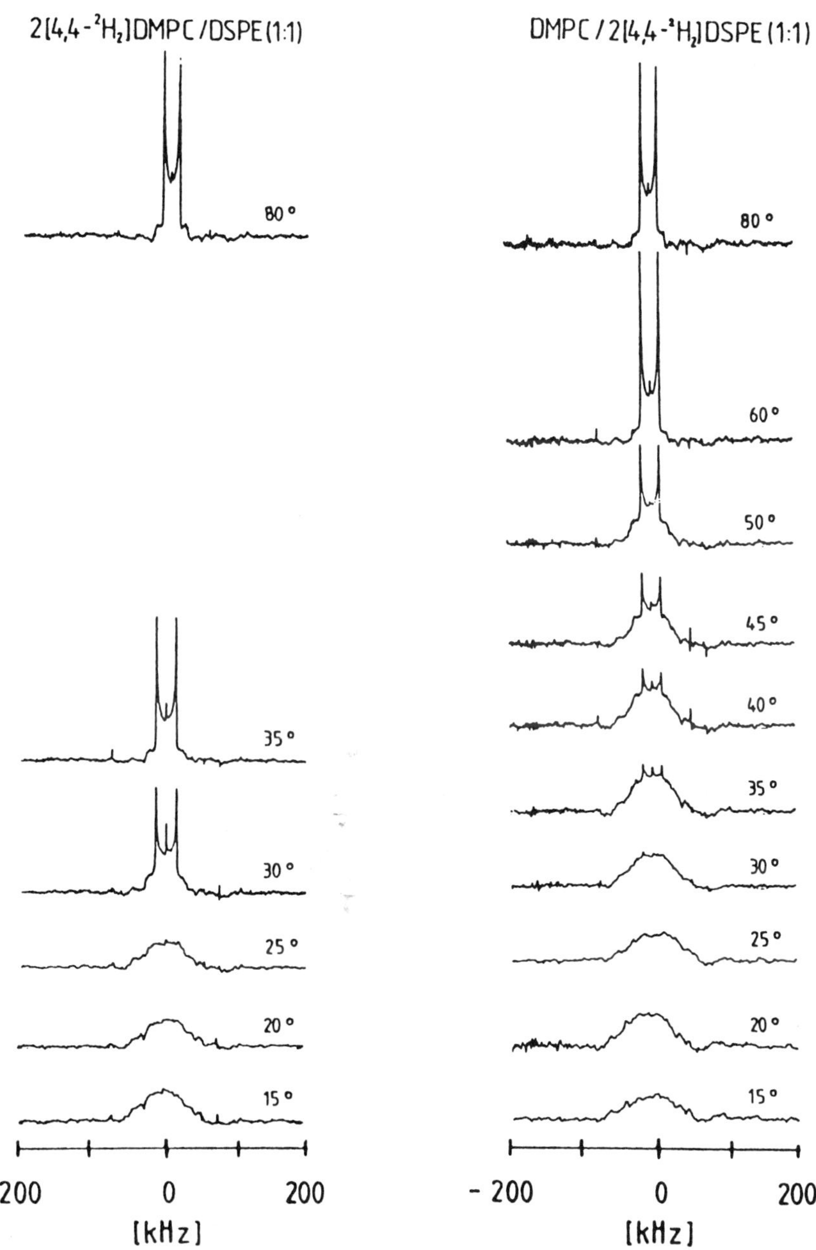

Figure 21 ^{2}H-NMR spectra of a 1:1 mixture of DMPC with DSPE as a function of temperature: left, 2(4,4'-d_2 DMPC)-DSPE; right, DMPC-2(4,4'-d_2 DSPE). (From Hübner and Blume, 1987.)

The first reliable data on lipid lateral diffusion coefficients in the liquid crystalline phase were based on ESR measurements and published by Träuble and Sackmann (1972) and Devaux and McConnell (1972). Träuble and Sackmann used an androstane spin label to determine the diffusion coefficient in liposomes from the exchange broadening of the ESR spectra at different label concentrations and different temperatures. Devaux and McConnell used a head group spin-labeled phosphatidylcholine analog and analyzed the evolution of the ESR spectra of a small spot of highly concentrated spin label surrounded by egg lecithin. Diffusion of the spin label out of this spot leads to a gradual sharpening of the ESR lines due to the dilution of the spin labeled phosphatidylcholine. The values for the diffusion coefficients D_L reported by both groups were $1–2 \cdot 10^{-8}$ cm^2/s. Table 2 shows diffusion data obtained with various techniques on liquid crystalline bilayers of DMPC and DPPC (see page 464 for Table 2).

Small differences in D_L obtained with various methods of the same lipid bilayers can be attributed either to the particular experimental approach or to differences in the determination of D_L. Probe effects originating from differential interactions of the probes with the host lipids may be one source of error. Differences in system preparation and hydration can also lead to variation in D_L. Indeed, the diffusion coefficient is sensitive to the degree of hydration (McCown et al., 1981). Thus the humidity of the bilayers must be controlled very carefully in order to obtain consistent data. Finally, the scale of the experiments is important. The FRAP technique, the ESR dynamic imaging method, and the NMR pulsed field gradient technique all measure diffusion coefficients over distances of several μm, whereas excimer fluorescence measurements and ESR techniques both analyze exchange effects on a local scale. The effects of chemical structure and size of the probe have already been investigated. For fluorescent probes with different chain lengths used in FRAP measurements, Derzko and Jacobson (1980) observed no simple relation between D_L and chain length or chemical structure of a fluorescent probe, the diffusion coefficients varying maximally by a factor of two (see Table 2). With excimer probes like pyrene or pyrene-phosphatidylcholine, the lipid analog always shows the lower diffusion coefficient (Galla et al., 1979b).

Many lipids have similar D_L values in the liquid crystalline phase of phospholipids, such as egg yolk phosphatidylcholine (Fahey et al., 1977; Fahey and Webb, 1978; McCown et al., 1981) or POPC (Shin and Freed, 1989a,b). In POPC at 25°C a cholestane spin label and a phosphatidylcholine spin label give D_L values of $\sim 5.9 \cdot 10^{-8}$ and $3.5 \cdot 10^{-8}$ cm^2/s, respectively. This difference may be due to size effects. Indeed, Vaz et al. (1987) have determined D_L for a bilayer spanning probe in the same system using the FRAP technique to $\sim 3 \cdot 10^{-8}$ cm^2/s at 25°C, a value in very good agreement with the ESR results using a phosphatidylcholine spin label.

While in the liquid crystalline phase the differences between microscopic and macroscopic diffusion are generally small, this is quite different for gel phase lipids. This is because of a potential accumulation of probe molecules in bilayer defects, resulting in diffusion coefficients that are too high.

A representative example for lateral diffusion data of probes in DMPC and DPPC bilayers as a function of temperature is shown in Fig. 22 (Fahey and Webb, 1978; Galla et al., 1979a; Alecio et al., 1982; Tamm and McConnell, 1985; Daems et al., 1985; Sassaroli et al., 1990). D_L values determined by the FRAP technique show an abrupt change at the transition temperature T_m by nearly two orders of magnitude, while diffusion coefficients determined by the excimer technique show no or only minor changes at T_m. NMR pulsed field gradient experiments give lower diffusion coefficients,

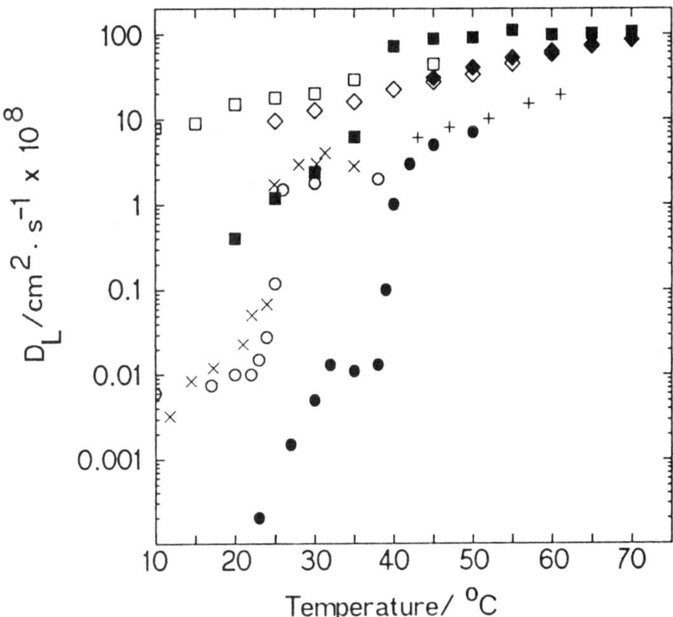

Figure 22 Lateral diffusion coefficients D_L for fluorescent probe diffusion in DMPC and DPPC bilayers as a function of temperature. (○) Fahey and Webb (1978); (+) Kuo and Wade (1979); (◇) and (◆) Galla et al. (1979a); (×) Alecio et al. (1982); (●) Tamm and McConnell (1985); (■) Daems et al. (1985); (□) Sassaroli et al. (1990).

as the technique measures macroscopic diffusion (Kuo and Wade, 1979). With a combination of multiple pulse and multiple pulse gradient techniques it was possible to determine the selfdiffusion coefficients of DPPC in the gel phase at 25°C (Silva-Crawford et al., 1980). The value of D_L of $1.67 \cdot 10^{-10}$ cm^2/s is in good agreement with values obtained by the FRAP technique.

1. Models for Lateral Diffusion

Translational diffusion of lipids in the liquid crystalline phase can be analyzed using different theories. Saffman and Delbrück (1975) have proposed a continuum hydrodynamic model for the diffusion of proteins in lipid bilayers, where the protein was modelled as a cylinder with a height corresponding to the thickness of the membrane. This hydrodynamic model was also tested for lipid diffusion. This model predicts a rather weak dependence of the diffusion coefficient on molecular weight, which was experimentally verified for several proteins (Peters and Cherry, 1982; Vaz et al., 1981; Vaz et al., 1982a,b; Peters, 1985), but it does not account for the diffusion of lipids (Vaz and Hallmann, 1983; Vaz et al., 1987).

Lipid diffusion is more appropriately described with the free volume model developed by Cohen and Turnbull (1959) for diffusion in liquids. It was first applied to lipid diffusion by Galla et al. (1979a,b). In this model, local free volume is assumed to be created by density fluctuations in the bilayer. A neighboring molecule can then jump into this defect or hole, creating a new void at its previous position. This is then filled by another solvent or solute molecule. The free volume v_f is calculated from the difference between the average volume and the van der Waals volume. Statistical mechanical calculations finally yield the relation

$$D_L = D_L^* \cdot \exp\left(\frac{-\gamma v^*}{v_f}\right) \tag{9}$$

with D_L^* being the diffusion coefficient in the void and γ a correction factor for a possible overlap between two voids (0.5–1); v* is a critical value for the void volume into which a molecule has to jump. The diffusion coefficient D_L^* can be calculated from the gas kinetic velocity $<u>$ by $D_L^* = 1/4 \cdot \lambda^* \cdot <u>$, with λ^* being the mean path length.

This free volume model gives satisfactory explanations for the differences in diffusion coefficients of probes with different molecular weights. Also the chain length and pressure dependence of the lipid translational diffusion is adequately described (Galla et al., 1979a,b; Müller and Galla, 1987). Vaz et al. (1985a,b, 1987) extended this model to account for the differences in viscosities of the bilayers and of the surrounding water. Here, the diffusion coefficient D_L^* is related by the Einstein relation $D_L^* = kT/f^*$ to the frictional coefficient f^* in the void. This frictional coefficient results from the interactions of the diffusing molecule with the relatively disordered layer in the middle of the membrane and with the solvent layer at the membrane-water interface. With this approach the temperature dependence of the translational diffusion coefficients of a membrane spanning fluorescent probe incorporated into hydrated or glycerinated bilayes of POPC was successfully simulated (Vaz et al., 1987). A hydrodynamic diffusion model (Hughes et al., 1981) gave considerably worse fits.

For the diffusion of excimeric probes Eisinger et al. (1986) have introduced the so-called "milling crowd" model. In this model diffusion is simulated by a two-dimensional random walk, the lipid molecules exchanging positions with a rate f and a jump length L, being related to the average lipid spacing. The diffusion coefficient is then given by $D_L = 1/4 \cdot f \cdot L^2$. Some data in Fig. 22 for DMPC stem from this model (Sassaroli et al., 1990).

Diffusion in gel phase bilayers probably proceeds via defect diffusion. In the P_{β} and $L_{\beta'}$ phase, the diffusion coefficients D_L vary between 10^{-11} and 10^{-10} cm^2/s (Fahey and Webb, 1978; Smith and McConnell, 1978; Rubenstein et al., 1979; Alecio et al., 1982) (see Fig. 22). In the P_{β} phase of DMPC, diffusion probably proceeds via two pathways, with the diffusion coefficients differing by at least a factor of 100. Diffusion along defect pathways, for example, proceeds with a diffusion coefficient of $\sim 4 \cdot 11^{-11}$ cm^2/s in DMPC at 16°C (Schneider et al., 1982). These defect pathways are probably associated with the ripples in the P_{β} phase, which are ~140 Å apart. Diffusion in highly ordered gel phase bilayers is much slower, characterized by D_L values between 10^{-16} and 10^{-17} cm^2/s (Owicki and McConnell, 1980).

B. Lipid-Cholesterol Mixtures

Quite a few measurements were performed on diffusion in lipid/cholesterol mixtures, addressing the question of the influence of cholesterol on lipid diffusion and the exact nature of the phospholipid/cholesterol phase diagram. Most results were obtained with the FRAP technique in bilayers above the phase transition temperature of the phospholipid component. Rubenstein et al. (1979) have reported D_L values for mixtures of cholesterol with egg phosphatidylcholine and DMPC. Above T_m of the pure phospholipid, cholesterol decreases the diffusion coefficient of fluorescent probes in a nonlinear fashion. Up to ~10 mol % cholesterol, D_L is hardly affected, but at 50 mol % cholesterol content the diffusion coefficient is reduced by a factor of ~4. Shin et al. (1990), however, have reported a continuous decrease in D_L for a cholestane spin label incorporated into POPC/cholesterol and DMPC/POPC/cholesterol bilayers by a factor of ~3.

In contrast to these results obtained by FRAP and ESR measurements, Kuo and Wade (1979) have observed first an increase in D_L in DPPC bilayers up to 10 mol % and then a continuous decrease of the diffusion coefficient up to 50 mol % cholesterol, by using pulsed field gradient NMR techniques. The reason for these differences is not yet clear. Alecio et al. (1982), using the FRAP technique, have investigated the effect of cholesterol on the diffusion of two different fluorescent labels, namely NBD-PE and a fluorescent labeled cholesterol analog, NBD cholesterol. In agreement with the measurements by Wu et al. (1977) and Rubenstein et al. (1979), they found that D_L decreases in DMPC above T_m with increasing cholesterol content, the diffusion coefficients being nearly identical for the two fluorescent probes.

Below T_m of the pure phospholipid the situation is more complicated. In the P_β phase of DMPC, D_L shows a sigmoidal increae with cholesterol content near 20 mol % cholesterol when the temperature is kept constant; (Rubenstein et al., 1979; Alecio et al., 1982) (see Fig. 23). This increase has to be discussed in the context of the phase diagram of phosphatidylcholine/cholesterol bilayers, which is still a matter of intensive debate (see for instance Yeagle, 1985; Ipsen et al., 1987, 1990; Finean, 1990; Scott, 1991; and previous discussion). Owicki and McConnell (1980) have interpreted the results shown in Fig. 23 by also using data from freeze fracture electron microscopy (Copeland and McConnell, 1980). They have suggested that DMPC/cholesterol bilayers below 20 mol %

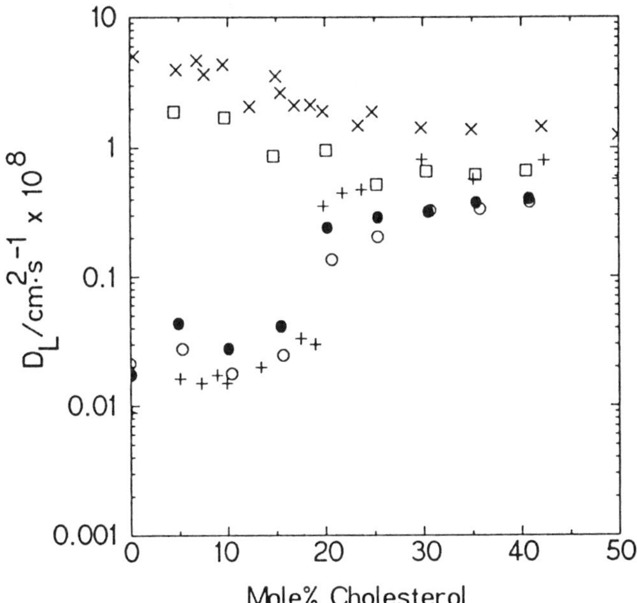

Figure 23 Lateral diffusion coefficients D_L in DMPC bilayers as a function of cholesterol content: (+) NBD-PE (19°C), Rubenstein et al. (1979); (×) NBD-PE (26°C), Rubenstein et al. (1979); (●) NBD-chol (19°C), Alecio et al. (1982); (○) NBD-PE (19°C). Alecio et al. (1982); (□)) NPD-PE (26°C), Alecio et al. (1982).

cholesterol are composed of alternating stripes of "fluid" and "solid" lipids, the "fluid" stripes consisting of a mixture of DMPC with 20 mol % cholesterol separated by nearly "solid" ridges of pure DMPC. The ridge separation increases from 140 Å without cholesterol to ~1000 Å at 18 mol % cholesterol, disappearing completely at higher cholesterol content. Similar microscopic heterogeneities were suggested later by Hatta and Imaizumi (1985) and Mortensen et al. (1988). The cholesterol molecules are probably located in defect stripes. Below the pretransition of pure DMPC, i.e., below 15°C, different secondary ripple structures are observed.

The temperature and concentration dependence of D_L in the ripple phase below 20 mol % cholesterol was explained by Owicki and McConnell (1980) by assuming two different diffusion coefficients, a low one of $\sim 10^{-11}$–10^{-12} cm^2/s for the diffusion in the nearly solid DMPC stripes and a higher one of $\sim 10^{-9}$–10^{-8} cm^2/s for the diffusion in the "fluid" cholesterol-rich stripes, the diffusion parallel and perpendicular to the ridges being anisotropic. In qualitative agreement with this, Alecio et al. (1982) found that the diffusion coefficients for NBD-cholesterol in DMPC/cholesterol bilayers at 16°C between 5 and 20 mol % cholesterol are somewhat higher than for NBD-PE, NPD-PE being preferentially located in the "solid" and NBD-choleserol in the "fluid" stripes.

C. Other Membranes

In general, lipid diffusion in mixed membranes containing only lipids is very similar to that of bilayers of pure lipid membranes (for reviews see Wade, 1985; Jovin and Vaz, 1989). Problems can occur with lipid mixtures in the two phase region of the phase diagram, however, because of differential solubility of the probes. Klausner and Wolf (1980) have investigated the effect of the chain length of fluorescent probes of the diI type on the diffusion coefficient measured in DLPC/DPPC mixtures. Shorter chain probes, which partition preferentially into the liquid crystalline domains, gave considerably higher diffusion coefficients than those with longer chains of 14 or 16 CH_2 groups. When the chain length was further increased, the probes partitioned preferentially into the gel phase. Then no fluorescence recovery was observed, indicating that the fluorescent molecules were essentially immobile and the diffusion coefficients below 10^{-10} cm^2/s.

Vaz et al. (1989) have measured the diffusion coefficient of a probe that partitions preferentially into the liquid crystalline phase as a function of temperature and composition in the binary system DMPC/DSPC. The diffusion data were used to map out the phase boundaries and, more importantly, to analyze whether the domains in the two phase region of the phase diagram were connected or not. Over a broad range of compositions, the line of connectivity is close to the liquidus line in the phase diagram. Here, liquid crystalline domains, comprising ~80% of the whole membrane area, are still disconnected by gel phase domains.

In protein containing membranes the diffusion coefficients of lipids are between 10^{-8} and 10^{-9} cm^2/s and thus somewhat lower than in pure liquid crystalline bilayers. This is due to changes in viscosity and the hindering of free lipid diffusion by the proteins (Saxton, 1987, 1989; Vaz et al., 1987).

VI. CONCLUDING REMARKS

The spectroscopic techniques described in this article, fluorescence spectroscopy, ESR spectroscopy, and NMR spectroscopy, have contributed to the understanding of the order and dynamics in lipid membranes. Fluorecence spectroscopy is particularly well suited for studies of wobbling motions and order in liquid crystalline membranes and has its particular strength examining lipid and protein diffusion. ESR has been the standard tool to investigate lipid/protein interactions. NMR is less sensitive than the former two methods but has its particular strength in its versatility and in the possibility of studying different nuclei. In particular, it is a nonperturbing method, and isotropic labeling makes it possible to gain detailed insight into intramolecular reorientational motion.

The molecular dynamics of lipids in the liquid crystalline phase are characterized by fast lateral diffusion, fast molecular rotations, intramolecular trans-gauche isomerizations in the chains, and wobbling motions of the whole molecule. For some model lipids these motions are now well characterized. The relations between the chemical structure of the lipid (head group and chains) and the order and dynamics in the liquid crystalline phase remain to be clarified.

In ordered gel phases, the mobility of lipids is closely related to the gel phase structure. For studies of gel phases, NMR spectroscopy in combination with low angle x-ray diffraction and FT-IR spectroscopy will lead to further understanding of phospholipid order and dynamics. The residual motions in ordered phase lipids are presently not well understood. In particular, it is still unclear how the rotational motions of a lipid molecule and the trans-gauche isomerization in the chains are related. One extreme case are phosphatidylcholines and phosphatidylethanolamines, which undergo relatively fast rotational motions in the gel phase but have relatively few gauche conformers. The other extreme case are cerebrosides, which show no axial rotations but a high degree of trans-gauche isomerization. The probable reason for these differences is stronger interactions between the glycolipid head groups. How the structure of the fatty acyl chain modulates the influence of head groups remains to be clarified. The use of oriented systems in NMR spectroscopy as well as in ESR measurements, in combination line shape simulations and relaxation time measurements, will lead to further progress in our understanding of lipid dynamics.

ACKNOWLEDGMENTS

I am grateful to W. Hübner who permitted me to use results of his work. I also want to thank J. Tuchtenhagen and W. Ziegler for critically reading the manuscript. Finally, I want to thank the Deutsche Forschungsgemeinschaft and the Fonds der Chemischen Industrie for financial support for this work and the work of my group cited in this paper.

ABBREVIATIONS

DLPC 1,2-dilauroyl-*sn*-glycero-3-phosphorylcholine
DMPC 1,2-dimyristoyl-*sn*-glycero-3-phosphorylcholine
DPPC 1,2-dipalmitoyl-*sn*-glycero-3-phosphorylcholine
DSPC 1,2-distearoyl-*sn*-glycero-3-phosphorylcholine
17isoPC 1,2-diisoheptadecanoyl-*sn*-glycero-3-phosphorylcholine
18isoPC 1,2-diisooctadecanoyl-*sn*-glycero-3-phosphorylcholine
MPPC 1-myristoyl-2-palmitoyl-*sn*-glycero-3-phosphorylcholine
MPSC 1-myristoyl-2-stearoyl-*sn*-glycero-3-phosphorylcholine
POPC 1-palmitoyl-2-oleoyl-*sn*-glycero-3-phosphorylcholine
DHPC 1,2-di-*O*-hexadecyl-rac-glycero-3-phosphorylcholine

DMPE 1,2-dimyristoyl-*sn*-glycero-3-phosphorylethanolamine
DPPE 1,2-dipalmitoyl-*sn*-glycero-3-phosphorylethanolamine
DSPE 1,2-distearoyl-*sn*-glycero-3-phosphorylethanolamine
DOPS 1,2-dioleoyl-*sn*-glycero-3-phosphorylserine
NPGS *N*-palmitoyl-galactosyl-cerebroside
β-DTGL 1,2-di-*O*-tetradecyl-3-*O*-(β-D-glucopryanosyl)-*sn*-gycerol
β-DTDGL 1,2-di-*O*-tetradecyl-3-*O*-(6-*O*-β-D-glucopyranosyl-β-D-glucopyranosyl)-*sn*-glycerol
NBD-PE *N*-(4-nitrobenzo-2-oxa-1,3-diazolyl)-phosphatidylethanolamine
NBD-cholesterol *O*-(4-nitrobenzo-2-oxa-1,3-diazolyl)-cholesterol
DPH 1,6-diphenyl-1,3,5-hexatriene
TMA-DPH 1-[4-(trimethylamine)phenyl]-6-phenyl-1,3,5-hexatriene

REFERENCES

H. Akutsu and J. Seelig, *Biochemistry 20*:7366 (1981).

A. D. Albert, and P. L. Yeagle, *Membr. Biochem. 4*:159 (1982).

M. R. Alecio, D. E. Golan, W. R. Veatch, and R. R. Rando, *Proc. Natl. Acad. Sci. USA 7*: 5171 (1982).

P. R. Allegrini, G. Pluschke, and J. Seelig, *Biochemistry 23*:6452 (1984).

C. Altenbach and J. Seelig, *Biochemistry 23*:3913 (1984).

R. L. Amey and D. Chapman, *Biomembrane Structure and Function* (D. Chapman, ed.), Verlag Chemie, Weinheim, 1984, p. 199.

M. Auger and H. C. Jarrell, *Chem. Phys. Letters 165*:162 (1990).

M. Auger, D. Carrier, I. C. P. Smith, and H. C. Jarrell, *J. Am. Chem. Soc. 112*:1373 (1990a).

M. Auger, M. Van Calsteren, I. C. P. Smith, and H. C. Jarrell, *Biochemistry 29*:5815 (1990b).

D. Axelrod, D. E. Koppel, J. Schlessing, E. Elson, and W. W. Webb, *Biophys. J. 16*:1055 (1976).

J. E. Baenziger, I. C. P. Smith, R. J. Hill, and H. C. Jarrell, *J. Am. Chem. Soc. 110*:8292 (1988).

J. E. Baenziger, H. C. Jarrell, R. J. Hill, and I. C. P. Smith, *Biochemistry 30*:894 (1991).

K. Beck, in *Cytomechanics* (J. Bereiter-Hahn, O. R. Anderson, and W.-E. Reif, eds.), Springer-Verlag, Berlin, Heidelberg, 1987, p. 79.

K. Beck, and R. Peters, in: *Spectroscopy and the Dynamics of Molecular Biological Systems* (P. Bayley and R. E. Dale, eds.), Academic Press, London, 1985, p. 177.

C. B. Berde, H. C. Andersen, and B. S. Hudson, *Biochemistry 19*:4279 (1980).

L. Best, E. John, and F. Jähnig, *Eur. Biophys. J. 15*:87 (1987).

A. Bienvenue, M. Bloom, J. H. Davis, and P. F. Devaux, *J. Biol. Chem. 257*:3032 (1982).

M. Bloom and O. G. Mouritsen, *Can. J. Chem. 66*:706 (1988).

M. Bloom and E. Sternin, *Biochemistry 26*:2101 (1987).

M. Bloom, E. E. Burnell, A. L. McKay, C. P. Nichol, M. I. Valic, and G. Weeks, *Biochemistry 17*:5750 (1978).

A. Blume, in *Physical Properties of Biological Membranes and their Functional Implications* (C. Hidalgo, ed.), Plenum, 1988, p. 21.

A. Blume and H. D. Bäuerle, in *Spectroscopy of Biological Molecules. State of the Art* (A. Bertoluzza, C. Eagnano, and P. Monti, eds.), Società Editrice Esculapio, Bologna, 1989, p. 244.

A. Blume and R. G. Griffin, *Biochemistry 21*:6230 (1982).

A. Blume and W. Hübner, in *Spectroscopy of Biological Molecules—New Advances* (E. D. Schmid, F. W. Schneider, and F. Siebert, eds.), John Wiley, Chichester, 1988. p. 157.

A. Blume, D. M. Rice, R. J. Wittebort, and R. G. Griffin, *Biochemistry 21*:6220 (1982a).

A. Blume, R. J. Wittebort, S. K. Das Gupta, and R. G. Griffin, *Biochemistry 21*:6243 (1982b).

A. Blume, K. Habel, A. Finke, and T. Frey, *Thermochim. Acta 119*:53 (1987).
A. Blume, W. Hübner, and G. Messner, *Biochemistry 27*:8329 (1988a).
A. Blume, W. Hübner, M. Müller, and H. D. Bäuerle, *Ber. Bunsenges. Phys. Chem. 92*:964 (1988b).
D. F. Bocian and S. I. Chan, *Ann. Rev. Phys. Chem. 29*:307 (1978).
J.-M. Bonmatin, I. C. P. Smith, and H. C. Jarrell, *J. Am. Chem. Soc. 110*:8693 (1988).
J.-M. Bonmatin, I. C. P. Smith, H. C. Jarrell, and D. J. Siminovitch, *J. Am. Chem. Soc. 112*:1697 (1990).
F. Borle and J. Seelig, *Biochemistry 22*:5536 (1983a).
F. Borle and J. Seelig, *Biochim. Biophys. Acta 735*:131 (1983b).
F. Borle and M. Seelig, *Chem. Phys. Lipids 36*:263 (1985).
V. L. B. Braach-Maksvytis and B. A. Cornell, *Biophys. J. 53*:839 (1988).
M. F. Brown, *J. Chem. Phys. 77*:1576 (1982).
M. F. Brown, and J. H. Davis, *Chem. Phys. Lett. 79*:431 (1981).
M. F. Brown and J. Seelig, *Nature (London) 269*:721 (1977).
M. F. Brown and J. Seelig, *Biochemistry 17*:381 (1978).
M. F. Brown, J. Seelig, and U. Häberlen, *J. Chem. Phys. 70*:5045 (1979).
M. F. Brown, A. A. Ribeiro, and G. D. Williams, *Proc. Natl. Acad. Sci. USA 80*:4325 (1983).
M. F. Brown, A. Salmon, U. Henriksson, and O. Söderman, *Mol. Phys. 60*:379 (1990).
J. L. Browning and J. Seelig, *Biochemistry 19*:1262 (1980).
K. S. Bruzik, B. Sobon, and G. M. Salamonszyk, *Biochemistry 29*:4017 (1990a).
K. S. Bruzik, B. Sobon, and G. M. Salamonszyk, *Biochim. Biophys. Acta 1023*:143 (1990b).
G. Büldt, H.-U. Gally, J. Seelig, and G. Zaccai, *J. Mol. Biol. 139*:673 (1979).
D. G. Cameron and H. H. Mantsch, *Biophys. J. 38*:175 (1982).
D. G. Cameron, H. L. Casal, and H. H. Mantsch, *J. Biochem. Biophys. Methods 1*:21 (1979).
D. G. Cameron, H. L. Casal, and H. H. Mantsch, *Biochem. 19*:3665 (1980).
D. G. Cameron, H. L. Casal, H. H. Mantsch, Y. Boulanger, and I. C. P. Smith, *Biophys. J. 35*:1 (1981a).
D. G. Cameron, E. F. Gudgin, and H. H. Mantsch, *Biochemistry 20*:4496 (1981b).
R. F. Campbell, E. Meirovitch, and J. H. Freed, *J. Phys. Chem. 83*:525 (1979).
R. A. Capaldi and D. E. Green, *FEBS Lett. 25*:205 (1972).
D. Carrier, J. B. Giziewicz, D. Moir, I. C. P. Smith, and H. C. Jarrell, *Biochim. Biophys. Acta 983*:100 (1989).
H. L. Casal and R. N. McElhaney, *Biochemistry 29*:5423 (1990).
H. L. Casal and H. H. Mantsch, *Biochim. Biophys. Acta 779*:381 (1984).
G. Cevc and D. Marsh, *Phospholipid Bilayers. Physical Principles and Models*, John Wiley, New York, 1987.
S. I. Chan, D. F. Bocian, and N. O. Petersen, in *Membrane Spectroscopy* (E. Grell, ed.), Springer-Verlag, Berlin, 1981, p. 1.
D. Chapman, *The Structure of Lipids,* Methuen, London, 1965.
D. Chapman, in *Biological Membranes* (D. Chapman, and D. F. H. Wallach, eds.), Academic Press, London and New York, Vol. 2, 1973, p. 91.
S. E. Church, D. J. Griffiths, R. N. A. H. Lewis, R. N. McElhaney, and H. H. Wickman, *Biophys. J. 49*:597 (1986).
M. H. Cohen and D. Turnbull, *J. Chem. Phys. 31*:1164 (1959).
B. R. Copeland and H. M. McConnell, *Biochim. Biophys. Acta 599*:95 (1980).
B. A. Cornell, *Chem. Phys. Lett. 72*:462 (1980).
B. A. Cornell, *Chem. Phys. Lipids 28*:69 (1981).
B. A. Cornell, and J. M. Pope, *Chem. Phys. Lipids 27*:151 (1980).
M. Cortijo and D. Chapman, *FEBS Lett. 131*:245 (1981).
M. Cortijo, A. Alonso, J. C. Gomez-Fernandez, and D. Chapman, *J. Mol. Biol. 157*:597 (1982).
P. R. Cullis, *FEBS Lett. 70*:223 (1976).
P. R. Cullis and B. de Kruijff, *Biochim. Biophys. Acta 436*:523 (1976).

D. Daems, M. van den Zegel, N. Boens, and F. C. de Schryver, *Eur. Biophys. J. 12*:97 (1985).
M. A. Davies, H. F. Schuster, J. W. Brauner, and R. Mendelsohn, *Biochemistry 29*:4368 (1990).
J. H. Davis, *Biophys. J. 27*:339 (1979).
J. H. Davis, *Biochim. Biophys. Acta 737*:117 (1983).
J. H. Davis, *Chem. Phys. Lipids 40*:223 (1986).
J. H. Davis, *Adv. Magn. Resonance 13*:195 (1989).
A. I. P. M. De Kroon, J. W. Timmermans, J. A. Killian, and B. de Kruijff, *Chem. Phys. Lipids 54*:33 (1990).
R.A. Demel and B. de Kruijff, *Biochim. Biophys. Acta 457*:109 (1976).
Z. Derzko and K. Jacobson, *Biochemistry 19*:6050 (1980).
P. Devaux and H. M. McConnell, *J. Am. Chem. Soc. 94*:4475 (1972).
E. J. Dufourc and I. C. P. Smith, *Chem. Phys. Lipids 41*:123 (1986).
E. J. Dufourc, I. C. P. Smith, and H. C. Jarrell, *Chem. Phys. Lipids 33*:153 (1983).
E. J. Dufourc, E. J. Parish, S. Chitrakorn, and I. C. P. Smith, *Biochemistry 23*:6062 (1984a).
E. J. Dufourc, I. C. P. Smith, and H. C. Jarrell, *Biochemistry 23*:2300 (1984b).
J. Eisinger, J. Flores, and W. P. Petersen, *Biophys. J. 49*:987 (1986).
P. F. Fahey and W. W. Webb, *Biochemistry 17*:3046 (1978).
P. F. Fahey, D. E. Koppel, L. S. Barak, D. E. Wolf, E. L. Alson, and W. W. Webb, *Science 195*:305 (1977).
J. B. Finean, *Chem. Phys. Lipids 54*:147 (1990).
R. W. Fisher and T. L. James, *Biochemistry 17*:1177 (1978).
J. Forbes, C. Husted, and E. Oldfield, *J. Am. Chem. Soc. 110*:1059 (1988).
U. P. Fringeli and Hs.H. Günthard, *Biochim. Biophys. Acta 450*:101 (1976).
U. P. Fringeli, and Hs.H. Günthard, in *Membrane Spectroscopy* (E. Grell, ed.), Springer-Verlag, Berlin, 1981, p. 270.
U. P. Fringeli, M. Schadt, P. Rihak, and Hs.H. Günthard, *Z. Naturforsch. 31a*:1098 (1976).
J. Frye, A. D. Albert, B. S. Selinsky, and P. L. Yeagle, *Biophys. J. 48*:547 (1985).
H. H. Füldner, *Biochemistry 20*:5707 (1981).
B. P. Gaber and W. L. Peticolas, *Biochim. Biophys. Acta 465*:260 (1977).
H.-J. Galla and E. Sackmann, *Ber. Bunsenges. Phys. Chem. 78*:949 (1974).
H.-J. Galla, W. Hartmann, U. Theilen, and E. Sackmann, *J. Membrane Biol. 48*:215 (1979a).
H.-J. Galla, U. Theilen, and W. Hartmann, *Chem. Phys. Lipids 23*:239 (1979b).
H. U. Gally, W. Niederberger, and J. Seelig, *Biochemistry 14*:3647 (1975).
H. U. Gally, G. Pluschke, P. Overath, and J. Seelig, *Biochemistry 20*:4223 (1981).
R. B. Gennis, *Biomembranes. Molecular Structure and Function*, Springer-Verlag, New York, 1989.
R. Ghosh and J. Seelig, *Biochim. Biophys. Acta 691*:151 (1982).
E. Grell, ed., *Membrane Spectroscopy,* Springer-Verlag, Berlin, 1981.
R. G. Griffin, *Methods Enzym. 72*:108 (1981).
R. G. Griffin, L. Powers, and P. S. Pershan, *Biochemistry 17*:2718 (1978).
R. A. Haberkorn, R. G. Griffin, M. D. Meadows, and E. Oldfield, *J. Am. Chem. Soc. 99*:7353 (1977).
R. A. Haberkorn, J. Herzfeld, and R. G. Griffin, *J. Am. Chem. Soc. 100*:1296 (1978).
J. Hare, *Biophys. J. 42*:205 (1983).
I. Hatta and S. Imaizumi, *Mol. Cryst. Liq. Cryst. 124*:219 (1985).
H. Hauser, W. Guyer, I. Pascher, P. Skrabal, and S. Sundell, *Biochemistry 19*:366 (1980).
H. Hauser, I. Pascher, and S. Sundell, *Biochemistry 27*:9166 (1988).
M. P. Heyn, *FEBS Lett. 108*:359 (1979).
M. P. Heyn, *Methods Enzym. 172*:462 (1989).
M. P. Heyn, A. Blume, M. Rehorek, and N. A. Dencher, *Biochemistry 20*:7109 (1981).
W. Hoffmann and C. J. Restall, in *Biomembrane Structure and Function* (D. Chapman, ed.), Verlag Chemie, Weinheim, 1984, p. 257.

T. H. Huang, R. J. Skarjune, R. J. Wittebort, R. G. Griffin, and E. Oldfield, *J. Am. Chem. Soc.* *102*:7377 (1980).
T. H. Huang, A. Blume, S. K. Das Gupta, and R. G. Griffin, *Biophys. J.* *54*:173 (1988).
W. Hübner, Ph.D. thesis, University of Freiburg, 1988.
W. Hübner and A. Blume, *Ber. Bunsenges. Phys. Chem.* *91*:1127 (1987).
W. Hübner and A. Blume, in *Spectroscopy of Biological Molecules—New Advances* (E. D. Schmid, F. W. Schneider, and F. Siebert, eds.), John Wiley, Chichester, 1988. p. 161.
W. Hübner and A. Blume, *J. Phys. Chem.* *94*:7726 (1990).
B. D. Hughes, B. A. Pailthorpe, and L. R. White, *J. Fluid Mech.* *110*:349 (1981).
W. C. Hutton, P. L. Yeagle, and R. B. Martin, *Chem. Phys. Lipids* *19*:255 (1977).
S. Hyde and L. R. Dalton, *Chem. Phys. Lett.* *16*:568 (1972).
J. H. Ipsen, G. Karlström, O. G. Mouritsen, H. Wennerström, and M. J. Zuckermann, *Biochim. Biophys. Acta* *905*:162 (1987).
J. H. Ipsen, O. G. Mouritsen, and M. Bloom, *Biophys. J.* *57*:405 (1990).
R. Jacobs and E. Oldfield, *Biochemistry* *18*:3280 (1979).
R. Jacobs and E. Oldfield, *Progr. Nucl. Magn. Reson. Spectrosc.* *14*:113 (1981).
F. Jähnig, *Proc. Natl. Acad. Sci. USA* *76*:6361 (1979).
F. Jähnig, H. Vogel, and L. Best, *Biochemistry* *21*:6790 (1982).
M. J. Janiak, D. M. Small, and G. G. Shipley, *J. Biol. Chem.* *254*:6068 (1979).
H. C. Jarrell, A. P. Tulloch, and I. C. P. Smith, *Biochemistry* *22*:5611 (1983).
H. C. Jarrell, J. B. Giziewicz, and I. C. P. Smith, *Biochemistry* *25*:3950 (1986).
H. C. Jarrell, P. A. Jovall, J. B. Giziewicz, L. A. Turner, and I. C. P. Smith *Biochemistry* *26*:1805 (1987a).
H. C. Jarrell, A. J. Wand, J. B. Giziewicz, and I. C. P. Smith, *Biochim. Biophys. Acta* *897*:679 (1987b).
H. C. Jarrell, I. C. P. Smith, P. A. Jovall, H. H. Mantsch, and D. J. Siminovitch, *J. Chem. Phys.* *88*:1260 (1988).
T. M. Jovin, and W. L. C. Vaz, *Methods Enzym.* *172*:471 (1989).
J. A. Killian, F. Borle, B. deKruijff, and J. Seelig, *Biochim, Biophys. Acta* *854*:133 (1986).
R. Kimmich and G. Voigt, *Chem. Phys. Lett.* *65*:181 (1979).
R. Kimmich, G. Schnur, and A. Scheuermann, *Chem. Phys. Lipids* *32*:271 (1983).
M. D. King, J. H. Sachse, and D. Marsh, *J. Magn. Res.* *72*:257 (1987).
K. Kinosita, Jr., S. Kawato, and A. Ikegami, *Biophys. J.* *20*:289 (1977).
K. Kinosita, Jr., S. Kawato, and A. Ikegami, *Adv. Biophys.* *17*:147 (1984).
R. D. Klausner and D. E. Wolf, *Biochemistry* *19*:6199 (1980).
S. J. Kohler and M. P. Klein, *Biochemistry* *16*:519 (1977).
P. Koole, C. Dijkema, G. Casteleijn, and M. A. Hemminga, *FEBS Lett.* *79*:360 (1981).
R. D. Kornberg and H. M. McConnell, *Biochemistry* *10*:1111 (1971).
L. J. Korstanje, E.E. van Faassen, and Y. K. Levine, *Biochim. Biophys. Acta* *980*:225 (1989).
L. J. Korstanje, K.A. Eikelenboom, C. S. van der Reijden, G. van Ginkel, and Y. K. Levine, *Chem. Phys. Lipids* *55*:123 (1990).
A. L. Kuo and C. G. Wade, *Biochemistry* *18*:2300 (1979).
M. Lafleur, P. R. Cullis, and M. Bloom, *Eur. Biophys. J.* *19*:55 (1990a).
M. Lafleur, P. R. Cullis, B. Fine, and M. Bloom, *Biochemistry* *29*:8325 (1990b).
A. Lange, D. Marsh, K.-H. Wassmer, P. Meier, and G. Kothe, *Biochemistry* *24*:4383 (1985).
C. W. B. Lee and R. G. Griffin, *Biophys. J.* *55*:355 (1989).
Y. K. Levine, N. J. M. Birdsall, A. G. Lee, and J. C. Metcalfe, *Biochemistry* *11*:1416 (1972).
B. A. Lewis, S. K. Das Gupta, and R. G. Griffin, *Biochemistry* *23*:1988 (1984).
R. N. A. H. Lewis and R. N. McElhaney, *Biochemistry* *24*:4903 (1985).
P. Lichtenberg, N. P. Petersen, J. Girardet, M. Kainisho, P. A. Kroon, C. H. A. Seiter, G. W. Feigenson, and S. I. Chan, *Biochim. Biophys. Acta* *382*:10 (1975).
G. Lindblom and H. Wennerström, *Biophys. Chem.* *6*:167 (1977).

G. Lindblom, L. B.-A. Johansson, and G. Arvidson, *Biochemistry 20*:2204 (1981).
H. Lindsey, N. O. Petersen, and S. I. Chan, *Biochim. Biophys. Acta 555*:147 (1979).
R. C. Lord and R. Mendelsohn, in *Membrane Spectroscopy* (E. Grell, ed.), Springer-Verlag, Berlin, 1981, p. 377.
J. McAlister, N. Yathindra, and M. Sundaralingam, *Biochemistry 12*:1189 (1973).
J.T. McCown, E. Evans, S. Diehl, and H. C. Wiles, *Biochemistry 20*:3134 (1981).
P. M. MacDonald and J. Seelig, *Biochemistry 26*:6292 (1987a).
P. M. MacDonald and J. Seelig, *Biochemistry 26*:1231 (1987b).
P. M. MacDonald and J. Seelig, *Biochemistry 27*:6769 (1988).
T. J. McIntosh, *Biophys. J. 29*:237 (1980).
A. L. MacKay, *Biophys. J. 35*:301 (1981).
M. Malthaner, A. Hermetter, F. Paltauf, and J. Seelig, *Biochim. Biophys. Acta 900*:191 (1987).
H. H. Mantsch, A. Martin, and D. G. Cameron, *Biochemistry 20*:3138 (1981).
H. H. Mantsch, C. Madec, R. N. A. H. Lewis, and R. N. McElhaney, *Biochemistry 24*:2440 (1985).
H. H. Mantsch, H. L. Casal, and R. N. Jones, in *Spectroscopy of Biological Systems* (R. J. H. Clark and R. E. Hester, eds.), Wiley, New York, 1986, p. 1.
D. Marsh, in *Membrane Spectroscopy* (E. Grell, ed.), Springer-Verlag, Berlin, 1981, p. 51.
D. Marsh, in *Supramolecular Structure and Function* (G. Pifat and J. N. Herck, eds.), Plenum, 1983, p. 127.
D. Marsh and A. Watts, in *Lipid-Protein-Interactions* (P. C. Jost, and H. O. Griffith, eds.), John Wiley, New York, Vol. 2, 1982, p. 53.
D. Marsh, A. Watts, and I. C. P. Smith, *Biochemistry 22*:3023 (1983).
C. Mayer, K. Müller, K. Weisz, and G. Kothe, *Liquid Crystals 3*:797 (1988).
C. Mayer, G. Gröbner, K. Müller, K. Weisz, and G. Kothe, *Chem. Phys. Lett. 165*:155 (1990).
P. Meier, A. Blume, E. Ohmes, F. Neugebauer, and G. Kothe, *Biochemistry 21*:526 (1982).
P. Meier, E. Ohmes, G. Kothe, A. Blume, J. Weidner, and H. Eibl, *J. Phys. Chem. 87*:4904 (1983).
P. Meier, E. Ohmes, and G. Kothe, *J. Chem. Phys. 85*:3592 (1986).
P. Meier, G. Kothe, G. Jonsen, M. Trecoske, and A. Pines, *J. Phys. Chem. 87*:6867 (1987).
R. Mendelsohn, and H. H. Mantsch, in *Progress in Protein-Lipid-Interactions* (A. Watts, and J. J. H. H. M. De Pont, eds.), Elsevier, Amsterdam, Vol. 2, 1986, p. 103.
R. Mendelsohn, R. Dluhy, T. Taraschi, D. G. Cameron, and H. H. Mantsch, *Biochemistry 20*:6699 (1981).
R. Mendelsohn, M. A. Davies, J. W. Brauner, H. Schuster, and R. A. Dluhy, *Biochemistry 28*:8934 (1989).
M. P. Milburn and K. R. Jeffrey, *Biophys. J. 52*:791 (1987).
M. P. Milburn and K. R. Jeffrey, *Biophys. J. 56*:543 (1989).
K. Mortensen, W. Pfeiffer, E. Sackmann, and W. Knoll, *Biochim. Biophys. Acta 945*:221 (1988).
M. Moser, D. Marsh, P. Meier, K.-H. Wassmer, and G. Kothe, *Biophys. J. 55*:111 (1989).
H.-J. Müller and H.-J. Galla, *Eur. Biophys. J. 14*:485 (1987).
E. Oldfield, R. Gilmore, M. Glaser, H. S. Gutowsky, J. C. Hshung, S. Y. Kang, T. E. King, M. Meadows, and D. Rice, *Proc. Natl. Acad. Sci. USA 75*:4657 (1978a).
E. Oldfield, M. Meadows, D. Rice, and R. Jacobs, *Biochemistry 17*:2727 (1978b).
E. Oldfield, J. L. Bowers, and J.F. Forbes, *Biochemistry 26*:6919 (1987).
J. C. Owicki, and H. M. McConnell, *Biophys. J. 30*:383 (1980).
R. J. Pace and S. I. Chan, *J. Chem. Phys. 76*:4228 (1982a).
R. J. Pace and S. I. Chan, *J. Chem. Phys. 76*:4241 (1982b).
R. H. Pearson and I. Pascher, *Nature 281*:499 (1979).
B. Perly, I. C. P. Smith, and H. C. Jarrell, *Biochemistry 24*:4659 (1985).
A. Peters and R. Kimmich, *Z. Naturforsch, 34a*:950 (1979).

R. Peters, in *Structure and Properties of Cell Membranes* (G. Benga, ed.), CRC Press, Boca Raton, Fla., Vol. I, 1985, p. 35.
R. Peters and K. Beck, *Proc. Natl. Acad. Sci. USA 80*:7183 (1983).
R. Peters and R. J. Cherry, *Proc. Natl. Acad. Sci. USA 79*:4317 (1982).
N. O. Petersen and S. I. Chan, *Biochemistry 16*:2657 (1977).
D. A. Pink, T. J. Green, and D. Chapman, *Biochemistry 19*:349 (1980).
C. F. Polnaszek, D. Marsh, and I. C. P. Smith, *J. Magn. Res. 43*:54 (1981).
J. M. Pope, L. Walker, B. A. Cornell, and F. Separovic, *Mol. Cryst. Liq. Cryst. 89*:137 (1982).
M. Rance, K. R. Jeffrey, A. P. Tulloch, K. W. Butler, and I.C. P. Smith, *Biochim. Biophys. Acta 600*:245 (1980).
M. Rance, K. R. Jeffrey, A. P. Tulloch, K. W. Butler, and I. C. P. Smith, *Biochim. Biophys. Acta 688*:191 (1982).
M. Rehorek, N. A. Dencher, and M. P. Heyn, *Biophys. J. 43*:39 (1983).
M. Rehorek, N. A. Dencher, and M. P. Heyn, *Biochemistry 24*:5980 (1985).
J.-P. Renou, J. B. Giziewicz, I. C. P. Smith, and H. C. Jarrell, *Biochemistry 28*:1804 (1989).
D. Rice, M. D. Meadows, A. O. Scheinman, F. M. Goni, J. C. Gomez, M. A. Moscarello, D. Chapman, and E. Oldfield, *Biochemistry 18*:5892 (1979).
D. M. Rice, A. Blume, J. Herzfeld, R. J. Wittebort, T. H. Huang, S. K. Das Gupta, and R. G. Griffin, in *Proceedings of the Second SUNYA Conversation in the Discipline Biomolecular Stereodynamics* (R. H. Sarma, ed.), Adenine Press, New York, 1981, Vol. II, pp. 255.
B. H. Robinson and L. R. Dalton, *J. Chem. Phys. 72*:1312 (1980).
E. Rommel, F. Noack, P. Meier, and G. Kothe, *J. Phys. Chem. 92*:2981 (1988).
T. M. Rothgeb and E. Oldfield, *J. Biol. Chem. 256*:6004 (1981).
M. Roux and M. Bloom, *Biochemistry 29*:7077 (1990).
J. L. R. Rubenstein, B. A. Smith, and H. M. McConnell, *Proc. Natl. Acad. Sci. USA 76*:15 (1979).
M. J. Ruocco, A. Makriyannis, D. J. Siminovitch, and R. G. Griffin, *Biochemistry 24*:4844 (1985a).
M. J. Ruocco, D. J. Siminovitch, and R. G. Griffin, *Biochemistry 24*:2406 (1985b).
E. Sackmann and H. Träuble, *J. Am. Chem. Soc. 94*:4492 (1972).
P. G. Saffman and M. Delbrück, *Proc. Natl. Acad. Sci. USA 72*:3111 (1975).
M. Sassaroli, M. Vauhkonen, D. Perry, and J. Eisinger, *Biophys. J. 57*:281 (1990).
A. Saupe, *Z. Naturforsch., Teil A 19*:161 (1964).
M. J. Saxton, *Biophys. J. 52*:989 (1987).
M. J. Saxton, *Biophys. J. 56*:615 (1989).
C. J. Scandella, P. Devaux, and H. M. McConnell, *Proc. Natl. Acad. Sci. USA 69*:2056 (1972).
P. G. Scherer and J. Seelig, *EMBO J. 6*:2915 (1987).
P. G. Scherer and J. Seelig, *Biochemistry 28*:7720 (1989).
C. Schmidt, B. Blümich, S. Wefing, S. Kaufmann, and H. W. Spiess, *Ber. Bunsenges. Phys. Chem. 91*:1141 (1987).
M. B. Schneider, W. K. Chan, and W. W. Webb, *Biophys. J. 43*:157 (1983).
S. Schreier, C. F. Polnaszek, and I. C. P. Smith, *Biochim. Biophys. Acta 515*:375 (1978).
A. J. Schroit and R. F. A. Zwaal, *Biochim. Biophys. Acta 1071*:313 (1991).
H. L. Scott, *Biophys. J. 59*:445 (1991).
A. Seelig and J. Seelig, *Biochemistry 13*:4389 (1974).
A. Seelig and J. Seelig, *Biochim. Biophys. Acta 406*:1 (1975).
A. Seelig and J. Seelig, *Biochemistry 16*:45 (1977).
J. Seelig, *Quart. Rev. Biophys. 10*:353 (1977).
J. Seelig, *Biochim. Biophys. Acta 515*:105 (1978).
J. Seelig and J. L. Browning, *FEBS Lett. 92*:41 (1978).
J. Seelig and H. U. Gally, *Biochemistry 15*:5199 (1976).
J. Seelig and P. M. MacDonald, *Acc. Chem. Res. 20*:221 (1987).

J. Seelig and W. Niederberger, *Biochemistry 13*:1585 (1974a).
J. Seelig and W. Niederberger, *J. Am. Chem. Soc. 96*:2069 (1974b).
J. Seelig and A. Seelig, *Quart. Rev. Biophys. 13*:19 (1980).
J. Seelig and N. Waespe-Sarcevic, *Biochemistry 17*:3310 (1978).
J. Seelig, H. U. Gally, and R. Wohlgemuth, *Biochim. Biophys. Acta 467*:109 (1977).
J. Seelig, P. M. MacDonald, and P. G. Scherer, *Biochemistry 26*:7535 (1987).
J. C. W. Shepherd and G. Büldt, *Biochim. Biophys. Acta 514*:83 (1978).
J. C. W. Shepherd and G. Büldt, *Biochim. Biophys. Acta 558*:41 (1979).
Y.-K. Shin and J. H. Freed, *Biophys. J. 55*:537 (1989a).
Y.-K. Shin and J. H. Freed, *Biophys. J. 56*:1093 (1989b).
Y.-K. Shin, J. K. Moscicki, and J. H. Freed, *Biophys. J. 57*:445 (1990).
M. Silva-Crawford, B. C. Gerstein, A.-L. Kuo, and C. G. Wade, *J. Am. Chem. Soc. 102*:3728 (1980).
J. R. Silvius, M. Lyons, P. L. Yeagle, and T. J. O'Leary, *Biochemistry 24*:5388 (1985).
D. M. Siminovitch, M. F. Brown, and K. R. Jeffrey, *Biochemistry 24*:2412 (1984).
D. M. Siminovitch, M. J. Ruocco, E. T. Olejniczak, S. K. Das Gupta, and R. G. Griffin, *Chem. Phys. Lett. 119*:251 (1985).
D. M. Siminovitch, M. J. Ruocco, A. Makriyannis, and R. G. Griffin, *Biochim. Biophys. Acta 901*:191 (1987).
D. M. Siminovitch, M. J. Ruocco, E. T. Olejniczak, S. K. Das Gupta, and R. G. Griffin, *Biophys. J. 54*:373 (1988).
S. J. Singer and G. L. Nicholson, *Science 175*:720 (1972).
F. Sixl and A. Watts, *Biochemistry 21*:6446 (1982).
F. Sixl and A. Watts, *Proc. Natl. Acad. Sci. USA 80*:1613 (1983).
R. Skarjune and E. Oldfield, *Biochim. Biophys. Acta 556*:208 (1979a).
R. Skarjune and E. Oldfield, *Biochemistry 18*:5903 (1979b).
R. Skarjune and E. Oldfield, *Biochemistry 21*:3154 (1982).
D. M. Small, in *Handbook of Lipid Research. The Physical Chemistry of Lipids* (D. M. Small, ed.), Plenum, New York, Vol. 4, 1986, p. 475.
B. A. Smith and H. M. McConnell, *Proc. Natl. Acad. Sci. USA 75*:2759 (1978).
I. C. P. Smith, *Biomembranes 12*:133 (1984).
I. C. P. Smith and H. C. Jarrell, *Acc. Chem. Res. 16*:266 (1983).
I. C. P. Smith, G. W. Stockton, A. P. Tulloch, C. F. Polnaszek, and K. G. Johnson, *J. Colloid Interface Sci. 58*:439 (1977).
R. L. Smith and E. Oldfield, *Science 225*:280 (1984).
J. B. Speyer, R. T. Weber, S. K. Das Gupta, and R. G. Griffin, *Biochemistry 28*:9569 (1989).
H. W. Spiess, in *NMR: Basic Principles and Progress* (P. Diehl, E. Fluck, and R. Kosfeld, eds.), Springer-Verlag, Berlin, Vol. 15, 1978, p. 55.
H. W. Spiess, *Adv. Polym. Sci. 66*:21 (1985).
L. C. Stewart, M. Kates, I. H. Ekiel, and I. C. P. Smith, *Chem. Phys. Lipids 54*:115 (1990).
G. W. Stockton and I. C. P. Smith, *Chem. Phys. Lipids 17*:251 (1976).
G. W. Stockton, C. F. Polnaszek, A. P. Tulloch, F. Hasan, and I. C. P. Smith, *Biochemistry 15*:954 (1976).
L. M. Strenk, P. W. Westerman, and J. W. Doane, *Biophys. J. 48*:765 (1985).
L. K. Tamm and H. M. McConnell, *Biophys. J. 47*:105 (1985).
A. Tardieu, V. Luzatti, and F. C. Reman, *J. Mol. Biol. 75*:711 (1973).
M. G. Taylor and I. C. P. Smith, *Biochemistry 20*:5252 (1981).
M. G. Taylor and I. C. P. Smith, *Biochim. Biophys. Acta 733*:256 (1983).
M. G. Taylor, T. Akiyama, and I. C. P. Smith, *Chem. Phys. Lipids 29*:327 (1981).
M. G. Taylor, T. Akiyama, H. Saito, and I. C. P. Smith, *Chem. Phys. Lipids 31*:359 (1982).
D. D. Thomas, in *Enzymes of Biological Membranes* (A. Martonosi, ed.), Plenum, New York, 2nd ed., Vol. 1, 1985, p. 287.

C. P. S. Tilcock and P. R. Cullis, *Biochim. Biophys. Acta 641*:189 (1981).
C. P. S. Tilcock, M. B. Bally, S. B. Farren, P. R. Cullis, and S. M. Gruner, *Biochemistry 23*:2696 (1984).
L. Trahms, in *Structure and Dynamics of Molecular Systems* (R. Daudel et al., eds.), D. Reidel, 1985, p. 203.
L. Trahms and W. D. Klabe, *Mol. Cryst. Liq. Cryst. 123*:333 (1985).
L. Trahms, W. D. Klabe, and E. Boroske, *Biophys. J. 42*:285 (1983).
H. Träuble and E. Sackmann, *J. Am. Chem. Soc. 94*:4499 (1972).
M. Vauhkonen, M. Sassaroli, P. Somerharju, and J. Eisinger, *Biophys. J. 57*:291 (1990).
W. L. C. Vaz and D. Hallmann, *FEBS Lett. 152*:287 (1983).
W. L. C. Vaz, H. G. Kapitza, J. Stümpel, E. Sackmann, and T. M. Jovin, *Biochemistry 20*:1392 (1981).
W. L. C. Vaz, M. Criado, V. M. C. Madeira, G. Schoellmann, and M. Jovin, *Biochemistry 21*:5608 (1982a).
W. L. C. Vaz, Z. I. Derzko, and K. A. Jacobson, *Cell Surface Rev. 8*:83 (1982b).
W. L. C. Vaz, R. M. Clegg, and D. Hallmann, *Biochemistry 24*:781 (1985a).
W. L. C. Vaz, D. Hallmann, A. Gambacorta, and M. De Rosa, *Eur. Biophys. J. 12*:19 (1985b).
W. L. C. Vaz, J. Stümpel, D. Hallmann, A. Gambacorta, and M. De Rosa, *Eur. Biophys. J. 15*:111 (1987).
W. L. C. Vaz, E. C. C. Melo, and T. E. Thompson, *Biophys. J. 56*:869 (1989).
S. P. Verma and D. F. H. Wallach, in *Biomembrane Structure and Function* (D. Chapman, ed.), Verlag Chemie, Weinheim, 1984, p. 167.
M. R. Vist and J. H. Davis, *Biochemistry 29*:451 (1990).
H. Vogel and F. Jähnig, *Proc. Natl. Acad. Sci. USD 82*:2029 (1985).
C. G. Wade, in *Structure and Properties of Cell Membranes* (G. Benga, ed.), CRC Press, Boca Raton, Fla., Vol. 1, 1985, p. 52.
P. I. Watnick, P. Dea, and S. I. Chan, *Proc. Natl. Acad. Sci. USA 87*:2082 (1990).
G. D. Williams, J. M. Beach, S. W. Dodd, and M. F. Brown, *J. Am. Chem. Soc. 107*:6868 (1985).
R. J. Wittebort, C. F. Schmidt, and R. G. Griffin, *Biochemistry 20*:4223 (1981).
R. J. Wittebort, A. Blume, T. H. Huang, S. K. Das Gupta, and R. G. Griffin, *Biochemistry 21*:3487 (1982).
R. J. Wittebort, E. T. Olejniczak, and R. G. Griffin, *J. Chem. Phys. 86*:5411 (1987).
R. Wohlgemuth, N. Waespe-Sarcevic, and J. Seelig, *Biochemistry 19*:3315 (1980).
P. K. Wolber and B. S. Hudson, *Biochemistry 20*:2800 (1981).
P. T. T. Wong and H. H. Mantsch, *J. Chem. Phys. 83*:3268 (1985).
P. T. T. Wong, W. F. Murphy, and H. H. Mantsch, *J. Chem. Phys. 76*:5230 (1982).
P. T. T. Wong, D. J. Siminovitch, and H. H. Mantsch, *Biochim. Biophys. Acta 947*:139 (1988).
E. Wu, K. Jacobson, and D. Papahadjopoulos, *Biochemistry 16*:3936 (1977).
P. L. Yeagle, *Biochim. Biophys. Acta 649*:263 (1981).
P. L. Yeagle, *Biochim. Biophys. Acta 822*:267 (1985).
P. L. Yeagle, and J. Frye, *Biochim. Biophys. Acta 899*:137 (1987).
P. L. Yeagle, W. C. Hutton, C.-H. Huang, and R. B. Martin, *Biochemistry 15*:2121 (1976).
P. L. Yeagle, W. C. Hutton, C.-H. Huang, and R. B. Martin, *Biochemistry 16*:4344 (1977).
P. L. Yeagle, A. D. Albert, K. Boesze-Battaglia, J. Young, and J. Frye, *Biophys. J. 57*:413 (1990).
N. Yellin and I. W. Levin, *Biochim. Biophys. Acta 468*:490 (1977a).
N. Yellin and I. W. Levin, *Biochemistry 16*:642 (1977b).
J. Yguarabide and M. C. Foster, in *Membrane Spectroscopy* (E. Grell, ed.) Springer-Verlag, Berlin, 1981, p. 199.
G. Zaccai, G. Bueldt, A. Seelig, and J. Seelig, *J. Mol. Biol. 134*:693 (1979).

14
Ionization and Ion Binding

Suren A. Tatulian* *Institute of Cytology, Academy of Sciences of the Russian Federation, St. Petersburg, Russia*

I. GENERAL CONSIDERATIONS

To comprehend ion binding to phospholipid molecules or to phospholipid membranes, the behavior of ions in bulk solution and in the vicinity of a membrane-solution interface should be understood. Protons and hydroxyl ions differ distinctly from other ions in this respect by a number of characteristic features. In water, protons usually are associated with an H_2O molecule by a couple of the outer-shell electrons of the oxygen. They thus give rise to a hydroxonium ion, H_3O^+, which then closes its hydration shell by binding three to four water molecules [1]. The H_3O^+ ions accumulated near a protonable group of a phospholipid molecule subsequently may donate their protons to the ligand. This results in an unshared electron-pair formation and in the generation of an excess positive charge at the site of binding.

The stepwise protonation of different phospholipids is depicted in Fig. 1. Hydroxyl ions are formed by the coupling of a pair of electrons between an oxygen and a hydrogen and by accepting an electron in order to complete the octet of the oxygen's L orbital. These ions play a fundamental role in the ionization of phosphatidylcholine molecules, for example, as in this case the supplementary electron comes from the trimethylammonium nitrogen (cf. Fig. 1c). Binding of hydrogen and hydroxyl ions to an ionized lipid, consequently, is often mediated by an electron displacement, i.e., by the formation of new chemical bonds. Such binding thus should be referred to as chemisorption rather than as physical adsorption.

Small or multivalent inorganic ions, with a high charge density, are strongly solvated (hydrated). In polar solvents, such as water, about six or more water molecules are tightly bound to such ions in the primary hydration shell. In the second hydration shell, the orienting effect of the central ion is much weaker. Halide anions, which have larger ionic radii as compared to alkali metal cations of the same periods, are characterized by looser hydration and more flexible primary hydration shell in spite of their ability to form hydrogen bonds with water.

If ion-solvent interactions are stronger than the intermolecular interactions in the solvent, ions are prone to be positively hydrating, or structure-making, or cosmotropic

**Current affiliation*: University of Virginia, Charlottesville, Virginia

(a) Phosphatidic acid

```
   O               O               O
   ↑    +H⁺        ↑     +H⁺       ↑
DAG-O-P-O⁻  ⇒  DAG-O-P-OH  ⇒  DAG-O-P-OH
   |               |               |
   O⁻   6.2        O⁻    1.8       OH
```

(b) Phosphatidylethanolamine

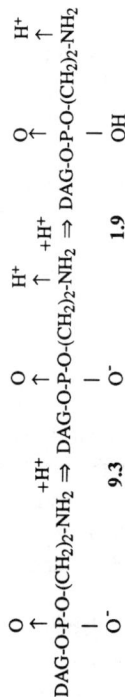

```
   O                               O           H⁺                  O            H⁺
   ↑                   +H⁺         ↑           ↑   +H⁺             ↑            ↑
DAG-O-P-O-(CH2)2-NH2  ⇒  DAG-O-P-O-(CH2)2-NH2  ⇒  DAG-O-P-O-(CH2)2-NH2
   |                               |                               |
   O⁻                  9.3         O⁻              1.9             OH
```

(c) Phosphatidylcholine

```
   O                                 O                                             O
   ↑          δ+          +H⁺        ↑                               +H⁺           ↑
DAG-O-P-O-(CH2)2-N(CH3)3  ⇒  DAG-O-P-O-(CH2)2-N⁺(CH3)3 + H2O  ⇒  DAG-O-P-O-(CH2)2-N⁺(CH3)3
   |                 |               |                              −H2O           |
   O⁻                |    11.0       O⁻                              1.7           OH
                   δ−OH
```

(d) Phosphatidylserine

```
   O         NH2                      O   ⁺H←NH2                     O   ⁺H←NH2                     O   ⁺H←NH2
   ↑          |          +H⁺          ↑        |          +H⁺         ↑        |          +H⁺         ↑        |
DAG-O-P-O-CH2-CH-C=O  ⇒  DAG-O-P-O-CH2-CH-C=O  ⇒  DAG-O-P-O-CH2-CH-C=O  ⇒  DAG-O-P-O-CH2-CH-C=O
   |               |                  |               |              |               |                |               |
   O⁻              O⁻   9.6           O⁻              O⁻   3.0       O⁻              OH   1.8         OH              OH
```

Figure 1 Schematic illustration of the stepwise protonation of different phospholipids. Arrows at P—O⁻ or N—H⁺ define the coordination-covalent bonds and imply that both electrons included in the covalent bonds come from the phosphorus or nitrogen atoms, respectively. Dashed line in N—-OH (phosphatidylcholine) denotes electron transfer bonding, in which an OH group accepts the fifth electron from the nitrogen's L orbital to complete its own L orbital. DAG represents diacylglycerol. Under open arrows the respective intrinsic values of the logarithm of the binding constant (*pK*) are shown; their uncertainty is ±0.2 . . . 0.5 units. The *pK* value of the PtdEtn phosphate group does not include the "abnormally low value" of 0.32 reported by Standish and Pethica (Ref. 31).

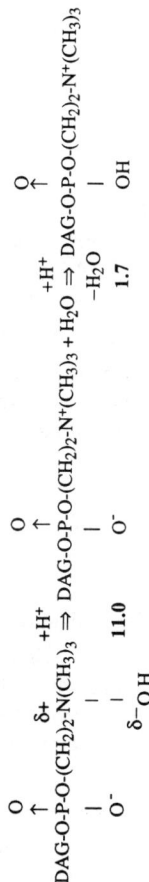

entities. The entropy of water is decreased for such ions, while it is increased near other ion types with a low charge density. The latter ions are thus considered to be negatively hydrating, or structure-breaking, or chaotropic entities [2,3]. The transition between cosmotropes and chaotropes is thought to occur for a hypothetical singly charged ion with an ion radius of 0.11 nm [4]. Sodium ions are weak cosmotropes, and Cl^- ions are weak chaotropes. Sodium chloride, consequently, does not disturb the bulk water structure appreciably [2,3].

When an ion approaches a phospholipid membrane it experiences several forces. The best known of these is the long-range electrostatic, coulombic force. This force is proportional to the product of all involved charges (on ions and phospholipids, for example) and is inversely proportional to the local dielectric constant ϵ. A related force caused by the presence of the water-membrane interface is known under the name of "image force." It arises because the field felt by an ion of charge q is higher in the membrane phase ($\epsilon = \epsilon_m \sim 2$) than in the aqueous phase ($\epsilon = \epsilon_w \sim 80$). This net membrane-ion repulsion is identical, in the first approximation, to the repulsion between a virtual charge, $q(\epsilon_w - \epsilon_m)/(\epsilon_w + \epsilon_m)$, situated in a membrane at the mirror-image position relative to the membrane-water interface, and the ion itself [5,6].

As described in the next chapter, phospholipid polar headgroups in an aqueous medium are typically hydrated (see also Ref. 7). Ion-phospholipid interactions are therefore mediated by dehydration upon binding. As long as several water molecules are intercalated between an ion and its binding site, one should not speak of binding but rather one should use the term "association" (Fig. 2a,d). If one water molecule is shared

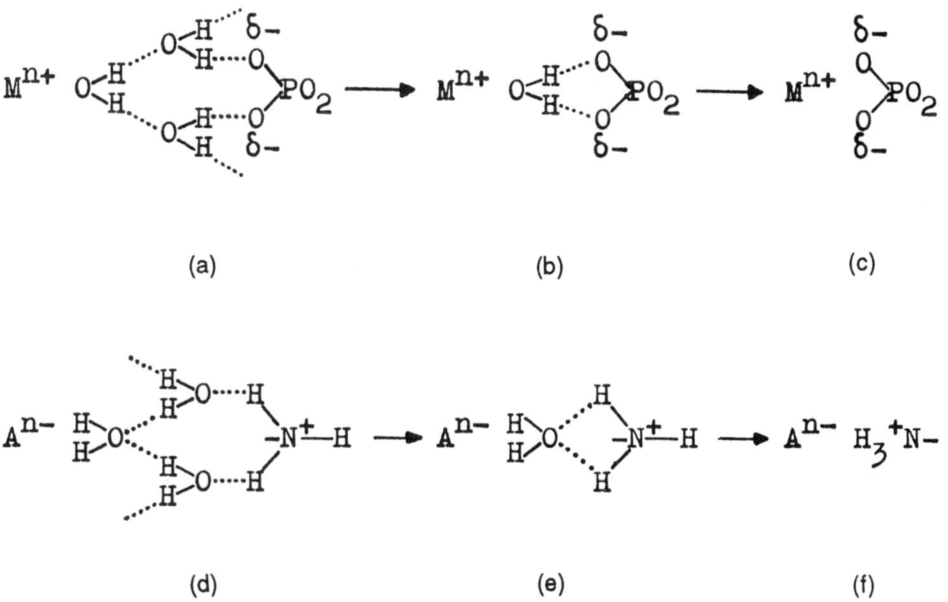

(a) (b) (c)

(d) (e) (f)

Figure 2 A scheme representing the formation of a cation (a,b,c) and an anion (d,e,f) complexes with phosphate- or amino-group, respectively. Hydrated ions interacting with hydrated polar groups via hydrogen bonds (a and d), outer-sphere complex formation with common hydration water molecules (b and e), and inner-sphere complexes with direct quantum-mechanical interactions (c and f) are shown. Dotted lines denote hydrogen bonds. For further interpretation see the text.

between an ion and its ligand, it is customary to speak of outer-sphere complex formation (cf. Figs. 2b and 2e). Complete displacement of the water molecules from the region between an ion and its binding site, finally, corresponds to an inner-sphere complex (cf. Figs. 2c and 2f).

In an inner-sphere complex, the bound ion feels the short-range local atomic field of its binding site, as well as the integral surface potential of a membrane. The ion, as well as membrane constituents, then exert mutual polarization effects on each other. Forces involved in the inner-sphere complex formation thus include ion-dipole, ion-induced dipole, induced dipole-induced dipole, ion-quadrupole forces, etc., in addition to the coulombic interaction. Hydrogen bonding can also participate in inner-sphere complex formation. This is the case, for example, when certain organic ions interact with the ammonium groups of the phosphatidylethanolamine (PtdEtn) or phosphatidylserine (PtdSer) headgroups; hydroxyl groups on the protonated phosphodiester or carboxyl residues also may interact with the halide- or oxy-anions via hydrogen bonds. Under appropriate circumstances the outer-sphere complexes may be stabilized by "through-water" hydrogen bonding as well.

One should only speak about the specificity of binding when ions interact with a ligand stereochemically. This is not to say, however, that the inner-sphere complexes are necessarily more stable than the outer-sphere complexes. Occasionally, even the opposite may be true. Quantum chemical analysis suggests, for example, that the energies gained from a direct or water-shared association of sodioum ions with $H_2PO_4^-$ ions are similar. This makes the competition between inner- and outer-sphere complexes in such a situation probable [8]. Combination of electrophoretic and NMR data, moreover, shows that Co^{2+} cations may form an inner-sphere complex with the phospate group of phospholipids, as well as with carboxyl groups of PtdSer, even though the prevailing fraction of the bound Co^{2+} ions is involved in the outer-sphere complex [9,10].

Organic ions may bind to (phospho)lipid membranes at the level of headgroups or hydrocarbon chains, by virtue of their amphiphilic nature. The final fate of organic ions interacting with a lipid membrane is thus determined by the combination of long-range coulombic forces, image-charge forces, ion-dipole interactions, "neutral" (nonelectrostatic) forces thought to include the dispersion and other van der Waals interactions, and diverse "steric force" components [6,11]. The sign of the dipolar contribution is positive (repulsive) for cations and negative (attractive) for anions. Dispersion forces are always attractive and image forces are always repulsive. Changing relative proportions of these forces in the final ion-phospholipid interaction may result in the creation of potential minima near the membrane-solution interface, deeper for the anions than for the cations, and separated by a permeability barrier in the middle of the membrane (Fig. 3) [11].

II. THEORY

Ion binding to a phospholipid (membrane) is most frequently described by the Gouy-Chapman-Stern theory [12] (for reviews see [13–15]). The surface charge density of a membrane in contact with an electrolyte solution within the framework of this theory is found to be

$$\sigma_{el} = sgn(\psi_{el})\left[2\epsilon\epsilon_0 kT \sum_i c_i(\exp\left(\frac{-Z_i e_0 \psi_{el}}{kT}\right) - 1)\right]^{1/2} \tag{1}$$

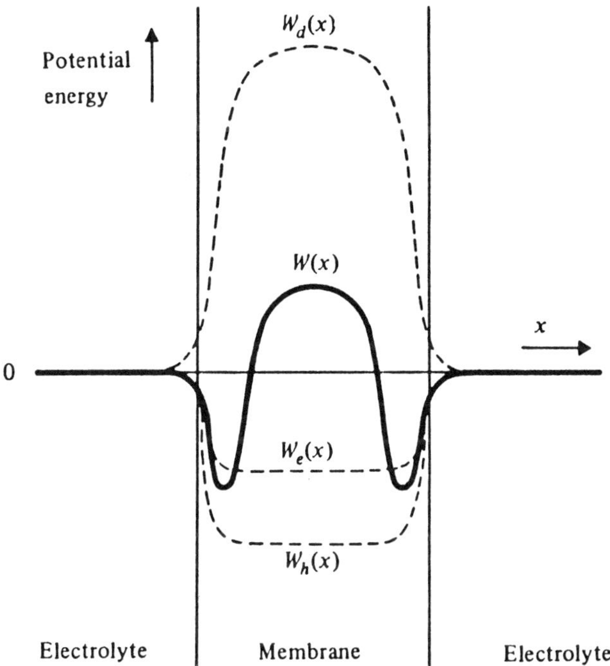

Figure 3 Total potential energy $w(x)$ of a hydrophobic anion in a lipid bilayer membrane as a sum of three main contributions: w_d (image charge potential), w_h (hydrophobic interaction), and w_e (electrostatic interaction due to the interfacial dipole potential). w_e is negative for anions and positive for cations. Consequently, the two interfacial potential minima are deeper for the anions than for the cations. (Reprinted from Ref. 6 with permission from Cambridge University Press.)

where σ_{el} and ψ_{el} are the averaged surface charge density and electrostatic surface potential, respectively, ϵ and ϵ_0 are the local dielectric constant and the permittivity of free space, respectively, c_i and Z_i give the concentration and the valence (or charge) number (including sign) of the i-th ion species; e_0 is the elementary charge, and k and T have their usual meanings. For a symmetrical electrolyte, with $Z_+ = -Z_- \equiv Z$, Eq. 1 takes the form

$$\sigma_{el} = (8\epsilon\epsilon_0 kTc)^{1/2} \sinh\left(\frac{Ze_0\psi_{el}}{2kT}\right) \tag{2}$$

Ion distribution perpendicular to the membrane-solution interface may be calculated from the Boltzmann law:

$$c_i(x) = c_i(x = \infty) \exp\left(\frac{-w_i(x)}{kT}\right) \tag{3}$$

$w_i(x)$ is the mean-force potential, that is, the work required to transfer an ion of the i-th kind from the bulk aqueous phase to the plane x [15]. Very frequently, this potential is assumed to be proportional to the membrane surface potential: $w_i(x) \equiv Z_i e_0 \psi_{el}(x)$.

Gouy-Chapman theory predicts that counterions will accumulate near the charged membrane surface in order to compensate the opposite net charge on the phospholipid molecules. This also ensures overall system electroneutrality. Coions, on the contrary, are depleted from the resulting ionic double layer.

When ions of a certain type bind to a lipid membrane, the resulting change in the surface charge density must be taken into account. In the simplest approximation, this is achieved by combining the Gouy-Chapman result with a Langmuir adsorption isotherm

$$\sigma_1 = \frac{NZe_0c(x = \infty)K \exp\left(\dfrac{-Ze_0\psi_{el}(x_{\text{ads}})}{kT}\right)}{1 + c(x = \infty)\text{K} \exp\left(\dfrac{-Ze_0\psi_{el}(x_{\text{ads}})}{kT}\right)} \tag{4}$$

where

$$K = \frac{\exp(-\Phi/kT)}{55.5} \tag{5}$$

is the binding constant with dimensions of reciprocal molarity. Φ is the intrinsic binding energy, and N is the fixed number of binding sites per unit area. (55.5 is the molar concentration of pure water.)

Ion binding to a phospholipid molecule does not necessarily lead to complete ion dehydration. Therefore no strict distinction between the inner and outer (Helmholtz) binding planes is always possible. In practice, these two planes are often regarded as a common layer that coincides with the slipping plane at which the zeta potential is measured [16].

From Eq. 4, the binding constant is seen to be identical with the reciprocal surface concentration of the binding ions for which $\sigma_{el} = Ze_0N/2$, corresponding to a 50% saturation of the surface binding sites. The intrinsic binding constant K thus differs from the apparent binding constant K_a, as the latter is identical to the reciprocal *bulk* ion concentration for which every second ion binding site is occupied.

III. PROTONATION

Protonation or deprotonation of the phospholipid headgroups changes the chemical nature and net charge of the lipid bilayers. It also may affect crucially the molecular conformation and hydrogen bonding capabilities of the individual lipid molecules and may modify the colloidal properties of the lipid aggregates, phospholipid phase transitions, lipid bilayer interactions with protein molecules, etc. [15,17–21]. Profound understanding of phospholipid protonation, therefore, is of paramount importance for phospholipid studies under essentially all experimental conditions.

Proton binding can be described in terms of the logarithm of the proton binding constant, K_{H}:$pK \equiv \log K_{\text{H}}$. Obviously, one should distinguish between the intrinsic and the apparent proton binding constants and also not mix the corresponding pK_{H} and $pK_{a\text{H}}$ values; this is discussed briefly in the previous section.

In Table 1 the (logarithmic) dissociation constant values of different phospholipids and selected model compounds are summarized [17–52]. It is important to realize that all apparent pK_{aH} values are extremely sensitive to the detailed experimental conditions under which these data were measured. For the same lipid, consequently, widely different apparent dissociation constant values may be found. The intrinsic pK value, however, is a fundamental property of the investigated system and thus is the real quantity of interest [49]. (The difficulty with this latter quantity is that it is often derived by rather *ad hoc* procedures from the apparent dissociation constant values and thus is no more reliable or unique than the pK_{aH} values.)

In the simplest approximation, the apparent and the intrinsic proton binding constants are related as

$$pK_{aH} = pK_{H} - 0.434\frac{e_0\psi_{el}}{kT} = pK_{H} + \Delta pK_{el} \tag{6}$$

which in combination with Equation 2, gives

$$pK_{aH} = pK_{H} - \frac{0.868}{Z}\ln[f + (f^2 + 1)^{1/2}] \tag{7}$$

where

$f = \sigma_{el}/(8\epsilon\epsilon_0 kTc)^{1/2}$.

As long as the intrinsic pK is really a constant, any increase in the negative surface potential will increase the apparent pK_{aH} value; the opposite will be observed for an increasing positive surface potential (see Eq. 6). This is simply a consequence of the electrostatic effect of the surface potential on the interfacial proton concentration.

Increasing the molar PA/PtdCho ratio in mixed lipid vesicles from 1:3 to 1:0.5, for example, has been measured to shift the apparent value of pK_2 of the negatively charged phosphatidic acid component (PA) from 8.15 to 8.65 [27]. Similar shifts were also observed upon decreasing the bulk salt concentration from 5 mM KCl to essentially zero in pure water [27].

Figure 4 gives another example for this. It shows that a tenfold increase in the ionic strength results in similar shifts of the apparent pK values of the carboxyl groups of PtdSer or of the phosphate groups of PA. This is in qualitative agreement with the predictions of Eq. 7 for the highly charged surfaces ($f^2 \gg 1$).

In addition to the coulombic dissociation constant shift, introduced in Eq. 6, a nontrivial hydration-dependent shift is also observed for essentially all phospholipids [53]:

$$pK_{H} = pK_{aH} + \Delta pK_{el} + \Delta pK_{p}$$

This shift is due to the cost of proton charge transfer from the bulk, with a high dielectric constant value, to the interface, with a much smaller value of ϵ.

The deviation of the straight line shown in Fig. 4a from experimental data obtained with highly concentrated ion solutions could be because the high potential approximation underlying Eq. 6 in such a case is not applicable any longer [15]. But it could also be due to the effects of surface polarity, which were unambiguously shown to be rather strong for phospholipid membranes [53]. In fact, the relative permittivity (dielectric constant) of water near a charged phospholipid bilayer has been measured directly to be ~30 at <0.1 M NaCl or CsCl [18,54] similar to the value observed for the ionic surfactant micelles [55].

Table 1 pK Values of Ionizable Groups of Phospholipids and Some Related Compounds

Compound	Chemical group	pK	Methods, conditions	Refs.
		1. Apparent pK		
PA	phosphate	$pK_1 = 3.4–3.8$	Pot., H_2O, 25°C	27
		$pK_2 = 8.0–8.7$		
		$pK_1 = 2.7–3.1$	0.1 M NaCl or KCl	
		$pK_2 = 7.9–8.1$	otherwise the same	
		$pK_1 \leq 3.5$	monolayer	30
		$pK_2 = 6.5–9.5$	air/145 mM NaCl	
Myr_2PA or		$pK_1 = 3.0–4.0$	ESR, DSC, Fl., H_2O [38]	33,37,38[a]
Pam_2PA		$pK_2 = 8.0–9.0$	or 1–100 mM salt [33,37]	
PtdGro		2.9–3.5	ESR, DSC, 0.1 M NaCl	17[a],37
PtdIns		≤3.5	see above	30
PtdInsP_2		$pK_1 = 6.4$	Pot., 20°C,	29
		$pK_2 = 8.4$	0.1 M Pr_4*N*-iodide	
PtdSer	acidic[b]	3.0–5.5	see above	29,30
	carboxyl	4.3–5.2	EP, monolayer, 20–25°C	35
	phosphate	≤1.2	Pot., 25°C	27
	amine	9.2–9.9	see above	29,30
Myr_2Gro*P*Ser	carboxyl	4.4–5.5	ESR, DSC, 0.1 M NaCl	18[a],37
	phosphate	≤2.6		18[a]
	amine	9.8–11.5		18[a],37
PtdEtn	phosphate	≤3.5	monolayer, ^{2}H-NMR	30,50
	amine	9.6–11.2	ESR, DSC, NMR	18[a],46,50
PtdCho	phosphate	≤3.5	see above	30
Gro*P*InsP_2		$pK_1 = 5.70$	see above	29
		$pK_2 = 8.05$		
*P*Ser	acidic[b]	5.80		
	amine	9.90		
*P*Etn	phosphate	5.77		
	amine	10.26		

InsP_3	phosphate	$pK_1 = 6.38$		
		$pK_2 = 8.45$		
		2. Intrinsic pK		
PA	phosphate	$pK_2 = 5.60$	ESR, ^{31}P-NMR, 22°C	41
Myr_2PA		$pK_1 = 1.6–2.26$	ESR, light scattering	43,45,48
		$pK_2 = 7.0$	see [c]	43
		$pK_2 = 6.2$ (11°C)	DSC, pH titration	52
		$pK_2 = 5.4$ (54°C)		
Pam_2PA*		$pK_1 = 4.1$	Pot., 99%	26
		$pK_2 = 10.0$	2-ethoxyethanol	
Myr_2MePA		1.75[a]–1.9	Fl., see [c]	36,43
PtdGro		1.1[a]	ESR. 0.1 M KCl	17
Lau_2Gro*P*Gro		1.2–2.5	monolayer, 20°C	39
Pam_2Gro*P*Gro		1.6–1.7	see above, see [c]	39,43
Ste_2Gro*P*Gro		2.1–2.2	force measurement, 25°C	47
PtdIns		2.0	potentiodynamic, 20°C	44
PtdInsP_2	monoester	$pK_1 = 6.0$	^{31}P-NMR, 0.1 M KCl	51
	phosphate	$pK_2 = 6.7$		
PtdSer	acidic[b]	2.1–2.2	see above	44
	carboxyl	2.6–4.0	EPR, Pot., monolayer,	32,41
	amine	9.7–9.9	25°C	42,49
	carboxyl*	4.6	Pot., 99%	26
	amine*	10.3	2-ethoxyethanol	
Pam_2Gro*P*Ser	carboxyl	2.7[d]–3.3	EP, Pot., 25°C, see [c]	34,43
	amine	9.0	see [c]	43
PamOleGro*P*Ser	carboxyl*	4.4	Pot., 1:1 (v/v)	28
	amine*	9.8	tetrahydrofuran/water	

Table 1 (Continued)

Compound	Chemical group	pK	Methods, conditions	Refs.
PtdEtn	phosphate	1.4–2.5	EPR, Pot., monolayer,	32,44,49
	amine	9.0–9.6	potentiodynamic	
	phosphate*	1.1	Pot., 98% ethanol,	22
	amine*	8.9	25°C	
$Pam_2GroPEtn$	phosphate	0.32	monolayer, 23°C	31
$Myr_2GroPEtn$*	amine	9.1	Pot., 99% 2-ethoxyethanol	26
PtdCho	phosphate	1.4–2.0	Potentiodynamic,	32,44
	Me_3N	10.5–11.5	monolayer, 25°C	
	phosphate*	1.1	Pot., 98% ethanol, 25°C	22
Gro*P**	phosphate	pK_1 = 1.3–1.4	Pot., ESR, 22–25°C	23,25
		pK_2 = 6.1–6.7		
		pK_1 = 4.0	see above	26
		pK_2 = 9.2		
Etn*	amine	8.1–9.2	see above	22,26
DL-Ser*	carboxyl	3.80	see above	26
	amine	9.80		

Phospholipids were arranged in lamellar structure (vesicles or planar bilayers) unless noted by asterisks, indicating that the compound was dissolved in organic solvent or in water. Phospholipids with unidentified acyl chains were originated from egg yolk (PtdCho, PtdEtn) or bovine brain (PtdSer, PtdIns, $InsP_3$). Phosphatidic acid was obtained by enzymatic (phospholipase D) treatment of egg PtdCho. Occasionally spinal cord PtdSer (Ref. 35), egg PtdSer (Ref. 49), or *E. coli* PtdEtn (Ref. 44) were used. Experiments were performed at room temperature unless otherwise specified.

[a]Parameters deduced from pH-dependence of thermotropic phase transition of the lipids.

[b]Not distinguished between phosphate and carboxyl groups.

[c]Data deduced from theoretical treatment of experimental results on pH-dependence of lipid phase transition temperature taken from Refs. 34,36,37.

[d]For $Pam_2GroPSer$ in the fluid state (75°C) pK_H of the carboxyl group was ~2.0.

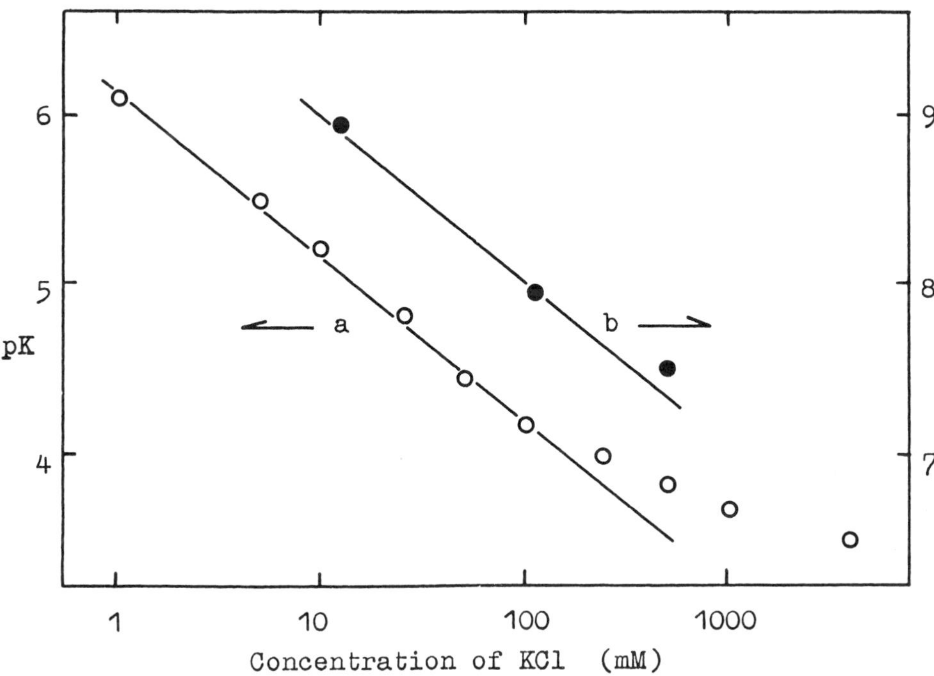

Figure 4 (a) Apparent pK of the phosphatidylserine carboxyl group in the mixed PtdSer/PtdCho (9:1) vesicles as measured by ESR, and (b) pK_2 of the phosphatidic acid vesicles deduced from ^{31}P-NMR, as a function of the bulk KCl concentration. Straight lines were drawn according to Eq. 7 using a "high-potential" approximation ($f^2 \gg 1$), for $pK_H = 2.62$ (a) and 5.60 (b). (Reproduced from Ref. 41 with permission from Elsevier Science Publishers.

Träuble, Eibl and coworkers [19,36,38,56] have shown that with increasing pH the chain-melting phase transition of different phosphatidic acid suspensions was first shifted to a higher temperature, which then reached a maximum, and finally decreased, smoothly at first and then quite sharply, to a very low transition temperature value, as illustrated in Fig. 5.

The effects of changing bulk proton concentration on the phospholipid phase behavior were interpreted in terms of the lipid protonation and electrostatic effects. At low pH, the phosphatidic acid molecules are fully protonated and electroneutral. When pH approaches the first pK value, some PA molecules acquire negative charge. This gives rise to the intermolecular acid-anion complexation, owing to the strong hydrogen bonding between the protonated P-OH and deprotonated P-O$^-$ groups [19–21,57].

Further changes in the ionization and in the hydrogen bonding capacity of the phosphatidic acid suspensions as a function of the bulk pH value are shown in Fig. 6. The observed pH-dependence of the phase transition temperature becomes clear if one takes into account that the bilayer membrane structure is markedly stabilized by intermolecular hydrogen bonds [19–21] and destabilized by electrostatic repulsion between ionized phospholipid molecules [15,17–19,36,37,56]. The highest transition temperature will be deleted at pH = pK_{aH}, corresponding to the maximum H bonding/ionization ratio [21,38,56]. More detailed analysis of these phenomena is given in Ref. [58].

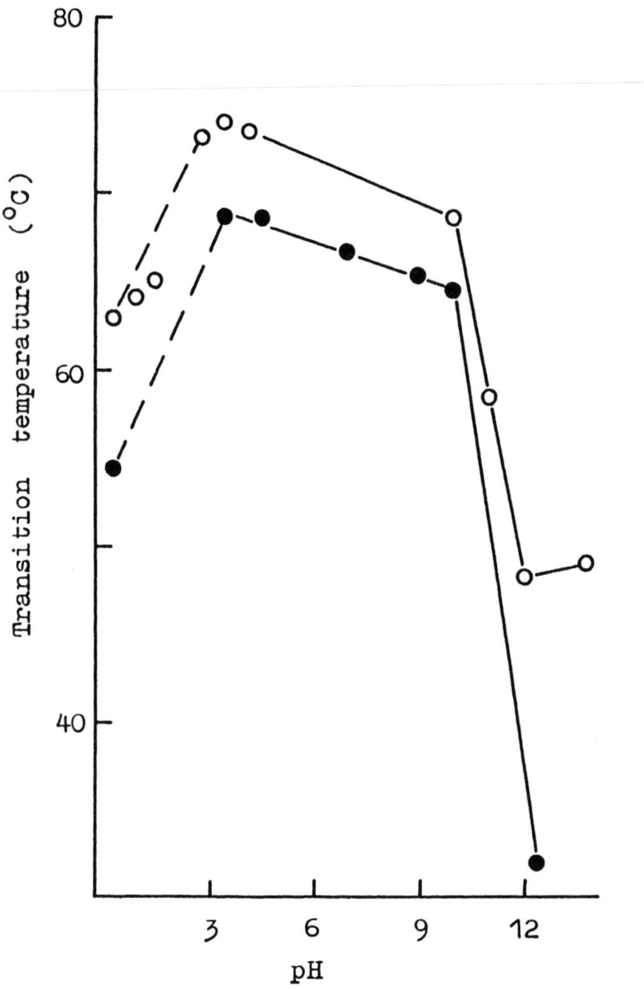

Figure 5 Dependence on pH of the gel-to-liquid-crystal phase transition temperature of ditetradecyl-PA (filled circles) and dihexadecyl-PA (open circles). The dashed lines indicate regions where two transitions were sometimes observed. (Adapted from Ref. 19 with permission from Academic Press.)

The *pK* values of phospholipid molecules dispersed in water depend not only on the electrostatic surface potential but also on all other interactions between the particular chemical residues near a protonable group and the proton. Thus the *pK* value of the carboxyl groups on the PtdSer heads is lower than that of the fatty acids embedded in a lipid bilayer. This is in part owing to the coulombic interactions between the positively charged PtdSer amino-groups and the approaching protons, i.e., shielding the negative charges of carboxyl groups by the vicinal amino groups [20]. The protonation of one monoester phosphate group of PtdInsP_2 markedly reduced the intrinsic association constant of protons with the adjacent monoesters phosphate group [51]. Further examples of this phenomenon may be found in Table 1.

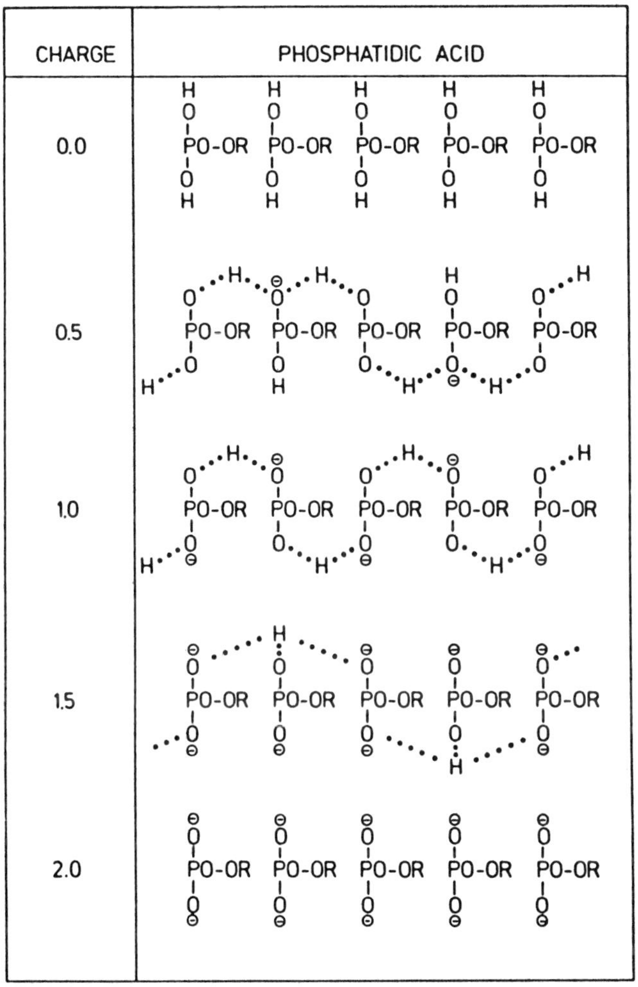

CHARGE	PHOSPHATIDIC ACID
0.0	PO-OR PO-OR PO-OR PO-OR PO-OR
0.5	PO-OR PO-OR PO-OR PO-OR PO-OR
1.0	PO-OR PO-OR PO-OR PO-OR PO-OR
1.5	PO-OR PO-OR PO-OR PO-OR PO-OR
2.0	PO-OR PO-OR PO-OR PO-OR PO-OR

Figure 6 Hydrogen bonding between the phosphatidic acid headgroups (dotted lines) in various ionization states. (Reprinted from Ref. 19 with permission from Academic Press.)

The measured, apparent, and the intrinsic *pK* values differ less in concentrated electrolytes than in the highly diluted salt solutions. There are at least two reasons for this: high salt concentration decreases the bilayer surface potential; but it may also interfere with surface hydration and thus diminish the difference between the surface and the bulk dielectric constant values [54]. Under high ionic strength conditions also the competition between the inorganic cations and protons for the ionizable groups may become important. Indeed, desorption of the bound cations with a decreasing pH or, alternatively, a release of protons with an increasing bulk concentration of the di- or trivalent cations have been detected [20,27,29,32,33,35,36,42,45,51,57–65]. An example of the competition between protons and the Mn^{2+} cations for the phosphate residues of the Pam_2PA headgroups is given in Fig. 7.

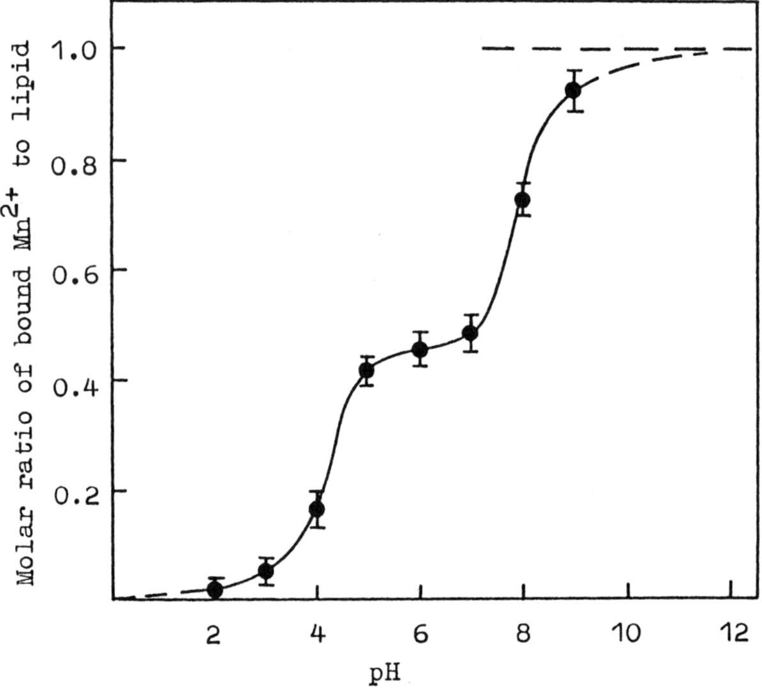

Figure 7 pH-dependence of the bound Mn^{2+}/Pam_2PA molar ratio at 60°C, as derived from the ESR spectra of Mn^{2+} in aqueous dispersions of Pam_2PA. (Adapted from Ref. 33 with permission from Elsevier Science Publishers.)

An increase in the bulk pH value from 8 to 10 caused the apparent binding constant of Mg^{2+} and Myr_2PA to get significantly higher: 1.6×10^4 and 7×10^4 M^{-1} respectively [65]. The intrinsic constants of Ca^{2+} binding to PA bilayers were determined as 40 and 1000 M^{-1} at pH 6.5 and 10, respectively [51]. Likewise, the constant of La^{3+} binding to the PtdSer headgroups changes from 10^5 M^{-1} at pH 7.4 to 4×10^3 M^{-1} at pH 4.5 [63]. Participation of the PtdSer amino-groups in the chelation of La^{3+} is also paralleled by the release of one proton from each PtdSer molecule with an equilibrium constant of $K = [\text{PtdSer}^- \, La^{3+}]/[\text{PtdSer}^{2-} \, La^{3+}][H^+] = 5 \times 10^5$ M^{-1} [63]. It has also been claimed that the pK value of PtdSer amino-group is diminished from ~10 in an indifferent salt solution to 5.7 in the presence of La^{3+} ions [63].

IV. ION BINDING

A. Inorganic Cations

Phospholipid-ion interactions, in the first place, are governed by electrostatic forces, as mentioned already. Let us take pure phosphatidic acid bilayers as an example. At low pH, when all PA molecules are fully protonated, only a few (if any) divalent cations bind to the lipid bilayer. Increasing bulk pH value results in (partial) headgroup deprotonation and ionization. The ion-lipid association ratio, Mn^{2+}/PA, simultaneously increases with

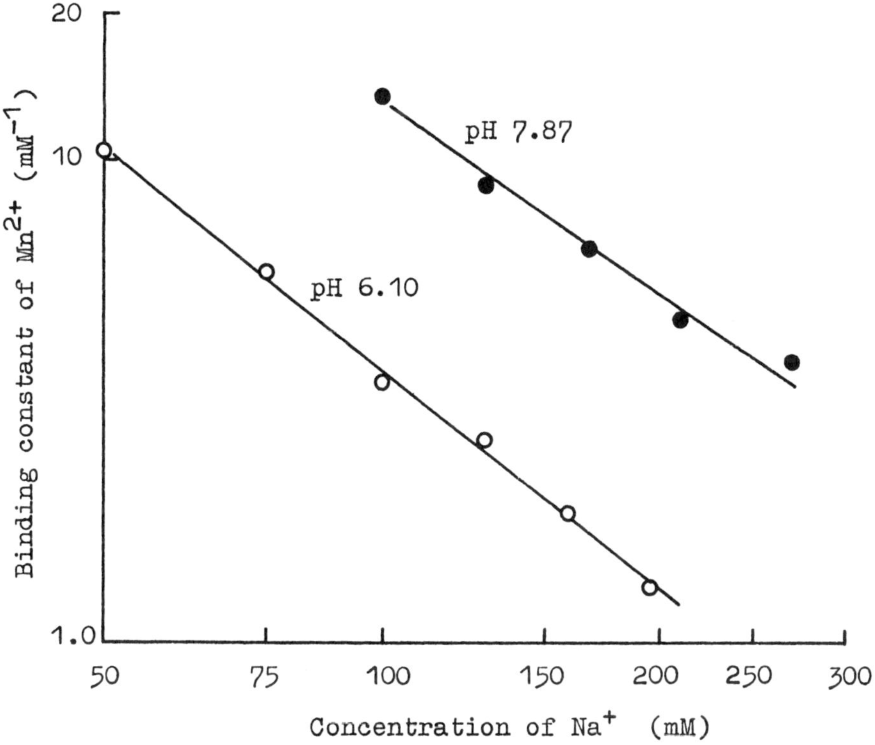

Figure 8 Apparent constant of Mn^{2+} binding to the vesicles composed of bovine brain PtdSer (sodium salt) as function of the bulk NaCl concentration at pH 7.87 (upper curve) and pH 6.10 (lower curve) as determined by EPR (Reproduced from Ref. 59 with permission from Springer-Verlag.)

increasing pH to reach first a stoichiometric ratio of 0.5 for the singly charged and 1.0 for the doubly charged lipids (cf. Fig. 7 and Ref. 33). Introduction of the negative charges on an initially electroneutral PtdCho membrane (by the admixture of negatively charged PtdGro [66] or via binding the ClO_4^- ions to the bilayer surface [67]) significantly increased the binding of calcium or terbium ions to such surface-negative membranes. On the contrary, the affinity of the anionic PtdSer bilayers for Li^+ or Ca^{2+} was suppressed as the membrane surface charge density decreased upon lipid chain unsaturation [68] or incorporation of zwitterionic phospholipids [69].

Alteration of the membrane surface potential by the variation of the ionic strength of the bathing medium effects ion binding as one would expect on the basis of Gouy-Chapman double layer theory [59,66,69–74]. The bulk ionic strength also affects the intrinsic binding constants of cations for both the neutral and the anionic phospholipids [9,47,67,71,75,76]. This clearly cannot be explained in simple electrostatic terms. Lis et al. [76], for example, have interpreted their x-ray data by inferring that the increased NaCl concentration reduces both the apparent and the intrinsic binding constants of the Ca^{2+} ions interacting with the $Pam_2GroPCho$ bilayers in the presence of 30 mM $CaCl_2$. Such a conclusion is supported partly by the results of certain electrophoretic measurements [75]. The intrinsic binding constant of Tb^{3+} interacting with the $Myr_2GroPCho$ bilayers [67]

Table 2 Intrinsic Binding Constants K and Lipid/Ion Stoichiometry Coefficients n Characterizing the Association of Inorganic Cations with Phospholipids and Some Related Compounds

Ion	Compound	K (M^{-1})	n	Methods, conditions	Refs.
Li^+	PA,Ptd(Cho,Gro,Ser)	0.02–1.1	1	EP, monolayer, 20–25°C	15,39,86
Na^+	PtdCho, PtdGro, di-PtdGro, PA	0.0–0.9	1	EP, ^{2}H-NMR, monolayer, ESR, 25–58°C, pH 7.0–7.4	9,39,47 48,74,99
	di-PtdGro	3.3	1	EP, 22°C, pH 7.2–7.4	102
	PtdSer	0.6–1.0	1	20–25°C	see [a]
K^+	PtdCho	0.0–0.1	1	EP, 20°C	15,91,94
	PA,Ptd(Gro,Ser,Ins)	0.1–2.0	1	EP, monolayer, 25°C	39,51,75,86,102
Rb^+,Cs^+	PtdCho	0.0			15
	PA,Ptd(Gro,Ser,Ins)	0.0–0.1	1	EP, monolayer, 25°C	15,39,86
NH_4^+	PtdSer	0.17	1	EP, 25°C	86
Mg^{2+}	PtdCho	1.0–2.5	1	EP (25°C), x-ray (5°C)	75,82,90
		24.4*	1	gel filtration, methanol	78
		30	15	EP, 25°C	72
	$Pam_2GroPCho$ (s)	20	1	interbilayer force	95
	$Lau_2GroPCho$ (f)	10	1	measurement, 16–22°C	
	$Myr_2GroPCho$ (s)	8.6	1		
	$Pam_2GroPEtn$ (s)	4.0	1		
	Ptd(Gro,Ser,Ins)	2–10	1–5	EP, Eq. dial., AAS, monolayer (0.65 nm^2/PtdSer)	9,42,75 83,98,100,103
	$PtdInsP_2$	100	1	EP., 0.1 M NaCl, 25°C	51
	PA	3	0.8	Eq. dial., AAS, 0.1M NaCl, pH 8	103
	Myr_2PA	0.33	?	Fl.	65
Ca^{2+}	PtdCho	1–5	1	EP, EPR, ^{31}P-NMR, 25°C	75,81,82
		28.6*	1	see above	78
		200	?	potentiodynamic	101

	$Pam_2GroPCho$	19–21	1	NMR (59°C), x-ray (5°C)	70,90
		≈100[b] (s)	≈10[c]	x-ray, 25°C	76
		120(s)	1	see above	95
		190(f)	13	EP, pH 7.4	72
		440(s)	14		
	$Myr_2GroPCho$	256(f)	15	see above	72,73
		392(s)	16		
		46(s)	1	see above (16°C)	95
	$Lau_2GroPCho$	15(f)	1	see above (22°C)	95
	PamOleGro*P*Cho	7–14(f)	2	^{2}H-NMR, 25–40°C	71
	PtdEtn	2–3	1	EP, 25°C	75
		6.4×10^5	1[d]	monolayer (1.5 nm^2/PtdEtn)	32
	$Pam_2GroPEtn$	12(s)	1	see above	95
	PtdGro	8–17	1	EP, ^{2}H-NMR. 25°C	9,74
		4	5	Eq. dial., AAS, 0.1 M NaCl, pH 8	103
	$Lau_2GroPGro$	0.03	>1	monolayer, 20°C	39
	Myr_2 or $Ste_2GroPGro$	40–100	1	bilayer interactions, 22°C	47
	di-PtdGro	15.5	1	^{2}H-NMR	98
	PtdSer	3–36	1–2.5	EP, monolayer, ^{31}P-NMR, Eq. dial., AAS, x-ray	42,75,83 89,98,103
	PtdIns	10	1	EP., 0.1 M NaCl, 25°C	51
	$PtdInsP_2$	500	1		
	PA	6–200	1	EPR, Fl., Eq. dial., AAS	79,97,103
	Myr_2MePA	0.5	2	Pot., 5–75°C	87
Sr^{2+}	PtdCho	0.36	1	EP, 25°C	82
		16	17	EP, 25°C	72
	PtdGro, PtdSer	5–14	1	EP, 25°C	9,75,98
Ba^{2+}	PtdCho	0.28	1	EP, 25°C	82
		10	18	EP, 25°C	72

Table 2 (Continued)

Ion	Compound	K (M^{-1})	n	Methods, conditions	Refs.
	PtdGro, PtdSer	5.5–20	1	EP, 25°C	9,75,98
Mn^{2+}	PtdCho	3.3	1	EP, 25°C	82
	$Pam_2GroPCho$	8–18(f)	1	ESR	93
	PtdGro	11.5	1–2	EP, Eq. dial., AAS 25°C	9,103
	$Pam_2GroPGro$	30–100	2	ESR	45
	PtdSer	7–25	1	EP, Eq. dial., AAS, 25°C	75,103
	PA	1.5	1.4	Eq. dial., AAS, pH 8	103
Ni^{2+}	PtdCho	0.83	1	EP, 25°C	82
	PtdGro	6.5	1	EP, 25°C	9
	PtdSer	14–28	1	EP, NMR, 20–25°C	61,75
	Gro*P*Cho	0.2*	1	^{31}P-NMR, 20°C	84
UO_2^{2+}	PtdSer	$(1–3) \times 10^3$	1	EP, 25°C	75
VO^{2+}	$Pam_2GroPGro$	10^4	2	ESR	45
	Pam_2PA	2×10^3	1		
Fe^{3+}	PtdCho	110[e]	300	EP, pH 2.5	77
		7.5[e]	300	EP, pH 4.5	
La^{3+}	$Pam_2GroPCho$	120	?	NMR, pH ≈ 5, 1M NaCl	70
	PtdSer	10^5	1	EP, 25°C, pH 7.4	63
		4×10^3	1	pH 4.5, otherwise the same	
		1–5[f]	1–3[f]	^{31}P-NMR, x-ray, DSC,	89
		≈150[g]	2–3[g]	^{140}La radiotracer, 20°C	
Ce^{3+}	PtdCho*	2500	1	gel filtration, dissolved	78
	Gro*P*Cho*	2500	1	in methanol	
Pr^{3+}	*P*Cho*	5.2	1	^{1}H-NMR, 28°C, pH 5.7	80
	Gro*P*Cho*	7.8	1	pH 6.1, otherwise the same	
	Me_2P	81[h]	1	^{1}H- and ^{31}P-NMR, 52°C	88

	$Pam_2GroPCho$	2 or 300[i]	2		
Eu^{3+}	$GroP$*	78	1	^1H-NMR, 28°C, pH 2.1	80
	$GroPCho$*	2.9	1	^{31}P-NMR, 30°C, 1 M NaCl	85
	PtdCho	700	2	see above	
		900–2500	2	EP, centrifugation, 20°C	92
Tb^{3+}	PtdCho	500–700	3	^{31}P-NMR	81
	$Myr_2GroPCho$	$(2–20) \times 10^4$	2	Eq. dial., 30°C	80

By (s) and (f) are designated the solid and fluid states of lipid bilayers. For asterisks and unidentified phospholipid acyl chains see the footnote to Table 1.

[a]Data deduced from electrophoresis (Refs. 63,75,86,96,98,102), ^{31}P-NMR, DSC, and x-ray diffraction (Ref. 89), EPR and potentiometric titration (Ref. 49), monolayer technique (0.65 nm^2 per PtdSer, Ref. 42), and equilibrium dialysis (Ref. 83).

[b]Intrinsic Ca^{2+} binding constant to $Pam_2GroPCho$ leaflets decreased with both increasing NaCl concentration and reducing interbilayer separation. Data presented correspond to ≈7 nm intermembrane spacing.

[c]Surface density of adsorbed Ca^{2+} decreased as $Pam_2GroPCho$ bilayers were pushed together. The lipid/Ca^{2+} ratio presented is the apparent limiting value for large (>10 nm) bilayer separations in 30 mM $CaCl_2$.

[d]These lipid/ion ratios were achieved only at pH > 8.5–9.0.

[e]Increased binding of Fe^{3+} to PtdCho with decreasing pH was explained by formation of poorly adsorbing hydroxides at high pH.

[f]$0.0043\ mM \le [La^{3+}] \le 0.0275\ mM$.

[g]$0.016\ mM \le [La^{3+}] \le 0.61\ mM$.

[h]Pr^{3+}/Me_2P 1:1 binding constant was 51 and 70 M^{-1} at 28 and 44°C, respectively, whereas the 1:2 binding constant was 349 M^{-1} at 52°C.

[i]In the absence of Pr^{3+}, low affinity sites (K = 2 M^{-1}) were present. With increasing Pr^{3+} concentration, high affinity sites (K = 3000 M^{-1}) appeared.

and that of Ca^{2+} binding to the $Myr_2GroPGro$ vesicles [47] have also been reported to decrease with increasing ionic strength.

A representative survey of the tentative phospholipid/inorganic-ion binding ratios and of the corresponding binding constants [9,15,32,35,42,45,47–49,61,63,65,70–103] is given in Table 2.

Binding of the divalent and trivalent cations to the phospholipid molecules is sensitive to the concentration of the supporting electrolyte, such as NaCl [9,47,67,71,75,76]. It is also affected by the interbilayer separation [69,76], and vice versa [104], the bulk pH-value [51,63,65], temperature [88,105], and phase transition effects [72,73,91,93,105]. Binding cooperativity may also play a decisive role [67,88,89]; the drastic increase in the lanthanide binding to $Pam_2GroPCho$ [88] or PtdSer [89] bilayers upon increasing the bulk cation concentration is indicative of this.

The cohesion or aggregation of phospholipid membrane may affect ion binding as well. The affinity of Ca^{2+} for the PtdSer vesicles, for example, appears to increase significantly if vesicle aggregation is induced by the addition of a 0.6–0.8 mM $CaCl_2$ [69]. In contrast, the binding of Ca^{2+} to the $Pam_2GroPCho$ bilayers is reduced upon decreasing the intermembrane separation [76].

Increased Mn^{2+} binding to the PtdSer, PtdGro, or di-PtdGro headgroups [105] or Pr^{3+} binding to the Me_2P molecules at high temperatures [88] are suggestive of the positive enthalpy change which, in turn, could be indicative of the inner-sphere complex formation between the di- or trivalent cations and the phosphate groups (see also Refs. 9 and 10).

Phospholipid affinity for cations seems to follow the sequence: lanthanides > transition metals > alkaline earths > alkali metals (see, for example, Refs. [35,59,63, 70,87,99,106] and Table 2). This documents, once again, how significant electrostatic interactions are in the process of ion-membrane binding. Such forces are not the only important factor in this respect, however. Ion-phospholipid interactions are also mediated by the local atomic fields of the phospholipid polar residues. This may be the reason why the di- or trivalent cations have a high affinity for the zwitterionic phospholipids [67,70–72,76,81,85,88,90,92,93,95].

Combined ^1H-, ^{13}C-, and ^{31}P-NMR studies strongly imply that the inorganic cations interact predominantly with the phosphodiester groups of the phospholipid headgroups [9,62,64,68,80,84,85,107–109]. This conclusion is also supported by IR spectroscopy [68,109–112], neutron diffraction [113], and ESR data [45]. The data, deduced from IR spectroscopy, showed that ion-phospholipid interactions may cause (partial) dehydration and conformation changes of the lipid polar headgroups [110–114].

Ion-phospholipid interactions may result in partial dehydration of the binding ion, as well. Luminescent lanthanides bound to the zwitterionic phospholipid headgroups tend to displace one or two water molecules from each cation hydration shell, for example. Such dehydration effect is even stronger upon cation binding to the acidic phospholipids, where up to eight water molecules are expelled from the interface once cation-phospholipid association has taken place [115,116].

Association between cations and the phospholipid molecules is a dynamic process. Individual complexes normally have lifetimes in the microsecond time-range. The residence time of Co^{2+} associated with GroPCho, for example, was measured to be $\sim 0.3\ \mu s$ [84]. Lifetimes of the Ca^{2+} cations bound to a PamOleGroPCho bilayer or to the mixed PtdGro/PtdCho vesicles appear to be somewhat longer, however, between 1 and 10

μs [71,74,99]. The upper residence-time limit for the metal cations bound to the Pam_2GroPCho bilayers is around 100 μs [70].

Cation affinities for the small molecules, which have been used as models of phospholipid headgroups, are appreciably lower than those reported for the phospholipid bilayer membranes (Table 2 and Refs. 82, 84, 85, 112, 116). For instance, Eu^{3+} binds to the GroPCho in a solution with a binding constant of 2.9 M^{-1} while in the case of PtdCho bilayers the K value is 700–2500 M^{-1} [85, 92]. Dipropionyl-GroPCho, which forms neither bilayers nor micelles, has an extremely low affinity for Tb^{3+}, even under conditions when this ion binds strongly to the Myr_2GroPCho [116]. Likewise, IR spectroscopy shows that Ca^{2+} does not associate with the glycerophosphoserine molecules but binds strongly to the PtdSer bilayers. This suggests that for tight binding of Ca^{2+} to occur either a charged interface or else chelation by appropriately spaced ligands is required [112]. (Interestingly, Ce^{3+} in methanol solution binds equally strongly to the isolated GroPCho headgroups and to whole PtdCho molecules [78], showing that this ion may interact with just one molecule at a time.)

The stoichiometry of monovalent cation binding to acidic lipids is believed to be 1:1 [74,75,86,99,102]. Several lines of evidence suggest that divalent cations can bind to the negatively charged phospholipids, such as PtdSer, PtdGro, PA, with a 1:1 binding stoichiometry as well; the combination of the 1:1 and 2:1 stoichiometries seems most likely,* however [35,45,60,69,75,98,103,116–118]. A 2:1 lipid-to-ion stoichometry has been proposed to be characteristic of the PtdSer-lanthanide complexes [85,88,89,92].

The effective binding stoichiometry of Ca^{2+} and cardiolipin (di-PtdGro) was estimated to be 0.65:1 [106]. The transition of di-PtdGro aggregates from a lamellar into an inverted hexagonal phase (H_{II}), was accompanied with a shift of this ratio from 0.35:1 to 1:1, however [118].

Increasing divalent cation concentration may reverse the sign of the zeta potential for the PtdSer or PtdGro vesicles. The intrinsic binding constant for 1:1 association therefore may be identical to the inverse value of the bulk ion concentration at which the surface potential becomes zero [9,75,98]

With the PtdCho bilayers, the situation appears to be more complicated. Based on NMR data pertaining to the binding of divalent and trivalent cations to the PtdCho bilayers, lipid-to-ion binding stoichiometries were concluded to vary from 1:1 to 3:1 [62,70,71,74,80,81,85,88,99,108,109]. These conclusions are somewhat tentative, however.

Values of K_{intr} = 120 M^{-1} [95] and 200 M^{-1} [101] were reported for the adsorption of Ca^{2+} to the Pam_2GroPCho and egg PtdCho bilayers, respectively. The binding of Eu^{3+}, Nd^{3+}, and UO_2^{2+} to the outer surface of the Pam_2GroPCho vesicles was saturated at a lipid-to-ion molar ratio between 8:1 and 33:1 [119]. Results of the ^{1}H- and ^{13}C-NMR measurements demonstrate that lanthanide binds to egg PtdCho bilayers with lipid-to-ion stoichiometric ratios between 14:1 and 17:1 [120,121]. Binding of Pr^{3+} to the Pam_2GroPCho bilayers was described using a lipid-to-ion ratio of 7:1 and a binding constant of 435 M^{-1} [88]. Several PtdCho molecules also seem to be involved in UO_2^{2+} binding to the lipid bilayers [122].

*There are unpublished results suggesting that in the case of the binding of alkaline-earth metal cations to PtdSer bilayers approximately 80% of the lipid molecules are involved in 2:1 complex formation and the 20% of them form 1:1 complexes (S.A. Tatulian, unpublished data).

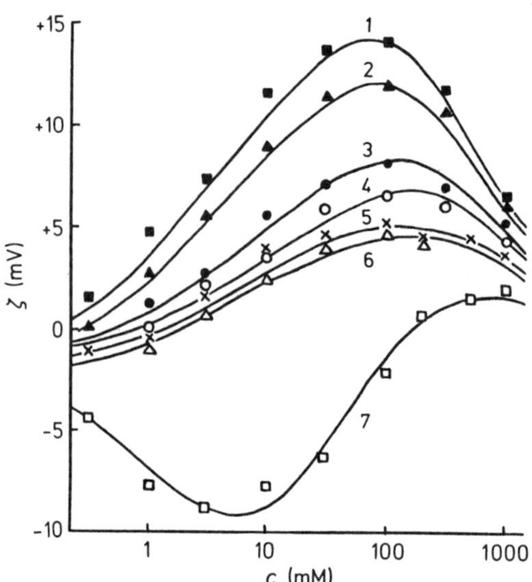

Figure 9 Dependence of the zeta potential of egg PtdCho liposomes on the concentration of $CaCl_2$, $MgCl_2$, $SrCl_2$, $BaCl_2$ $BaBr_2$, $Ba(NO_3)_2$, and $Ba(ClO_4)_2$ (curves 1–7, respectively) in the presence of 5 mM tris-HCl (pH 7.4) at 25°C. Theoretical curves were calculated within the framework of the Gouy-Chapman-Stern theory with parameters given in Tables 2 and 3. (Reprinted from Ref. 72 with permission from Springer-Verlag.)

The relatively low stoichiometry of association between the PtdCho molecules and the divalent and trivalent ions may result from ion binding to the vacant points (structural defects) in the (trigonal) lattice of the PtdCho headgroups [72]. An alternative explanation would be that such ions bind simultaneously to a cluster of adjacent phospholipid molecules. The lipid-to-ion binding ratios of 10–20 speak in favor of the former interpretation, however.

Petersheim et al. [109] have shown, combining fluorescence, NMR, and IR data, that each Ca^{2+} ion is coordinated by the phosphodiester group of PtdCho headgroup in a monodentate manner, i.e., only one oxygen is involved in ion binding. The same is true for the Tb^{3+} ions at low concentrations, but the binding of the latter ion becomes bidentate at higher concentrations.

B. Inorganic Anions

Binding of inorganic anions to phospholipid membranes is less well explored than the association between lipids and cations. In spite of this, there is convincing evidence that the binding of anions to lipid bilayers is as strongly affected by electrostatic forces as lipid-cation association. Binding of the anions Cl^-, Br^-, NO_3^-, and SCN^- to positively charged diacyldimethylammonium membranes, for example, is severalfold stronger than the corresponding ion binding to zwitterionic phospholipid aggregates [96,123].

Anion size also is not an unimportant factor in the process of binding. An increase in the anion radius from 0.181 nm (Cl^-) to 0.236 nm (ClO_4^-) magnifies the binding constant for anion-PtdCho association by more than two orders of magnitude. This is in part due to the transfer of the local excess charges from the anion to the phospholipid headgroups

Table 3 Intrinsic Binding Constants *K* and Lipid/Ion Stoichiometry Coefficients *n* Characterizing Binding of Inorganic Anions to Phospholipid Bilayer Membranes

Lipid	Ion	K (M^{-1})	n	Methods, conditions	Refs.
Egg PtdCho	Cl^-	0.06–1.7	see [a]	^{31}P- or ^{1}H-NMR, EP,	72,81,85
	Br^-	2.0		25–30°C, 0–1 M NaCl	124
	NO_3^-	2.8–8.0			
	SCN^-	23.6			
	ClO_4^-	70.0			
PamOleGro*P*Cho	SCN^-	12.6	7.5	^{2}H- and ^{31}P-NMR, 25°C	123
	SCN^-	80	1	^{2}H-NMR, pH 7.4, 25°C	127
	I^-	32	1		
	ClO_4^-	115	1		
Myr_2Gro*P*Cho	Cl^-,SO_4^{2-}	<1.0		EP, pH 7.2, 25°C	91
	NO_3^-	2.0	13–26[b]		
	Br^-	3.6	9–17[b]		
	SCN^-	10.0	4–9[b]		
	I^-	38.0	12–24[b]		
	ClO_4^-	227	23–48[b]		

[a]The PtdCho/anion ratios were assumed to be 1 [124] and 2 [81,85] or determined as 15–19 [72].

[b]The density of binding sites approximately doubled at gel-to-liquid-crystal phase transition of Myr_2Gro*P*Cho.

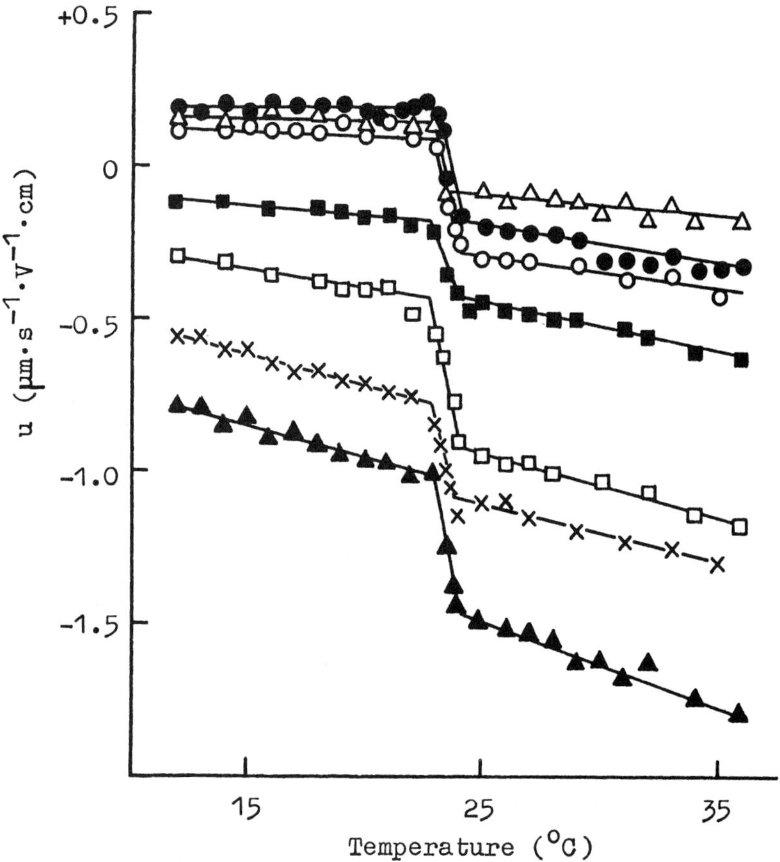

Figure 10 Temperature-dependence of the electrophoretic mobility of $Myr_2GroPCho$ liposomes in a 10 mM solution of K_2SO_4 (open triangles), KCl (filled circles), KNO_3 (open circles), KBr (filled squares), KSCN (open squares), KI (crosses), and $KClO_4$ (filled triangles) buffered to pH 7.4 with 5 mM tris-HCl. (Reproduced from Ref. 91 with permission from Elsevier Science Publishers.)

and vice versa; this phenomenon may also be a consequence of the different ionic polarizabilities in the two cases.

This can be checked by increasing the ion-charge density while keeping the anion radius nearly constant. (To achieve this one can replace ClO_4^- ions with SO_4^{2-}.) Upon such system manipulation, indeed, the anion-PtdCho binding is impaired dramatically (see Table 3 and Ref. 91).

Based on published anion-binding data, the following affinity series for phosphatidylcholine interacting with different anions can be written:

$$ClO_4^- > I^- \geqslant SCN^- > NO_3^- \geqslant Br^- > Cl^- > SO_4^{2-}$$

[72,80,85,91,124–127] (see also Fig. 10.) It is remarkable that the Gibbs free energy of anion hydration increases in the same row [128]. Similar selectivity series have been reported for the PtdEtn bilayers as well [126].

The hydration dependence of anion-phospholipid interaction is not fully explicable by the variable strength of anion-phospholipid association with changing anion radius. If this were so, one would expect the difference between the strengths of association of the Cl^- or ClO_4^- ions and PtdCho to amount to a factor of two rather than two orders of magnitude. Alterations in the binding-site hydration caused by anion adsorption and the differential hydrogen bonding capabilities of the participating groups can provide a partial answer to this problem. Perchlorate, for instance, is a negatively hydrating ion [2] and will thus tend to dehydrate the binding site by its structure-breaking effect, which should enhance the binding constant. The same applies to other chaotropic anions.

The strength of anion binding to phospholipid membranes decreases with the increasing net negative surface charge density of membranes. Halide anions, for example, do not interact significantly with PtdSer bilayers [129].

Data summarized in Table 3 provide support to the recent model of the specificity of ion-lipid binding [130]. This model suggests that the interfacial ion concentration near a polar but electrically neutral surface should increase with ion radius, owing to a partial, hydration-dependent depletion of ions from such an interfacial region. Based on similar considerations, the ion concentration near charged lipid bilayers is postulated to decrease with increasing ion radius. The latter conclusion is directly supported by the electrophoretic data of Eisenberg et al. [86], which pertain to the binding of alkali cations to the PtdSer bilayers.

Proton and deuterium NMR studies, moreover, indicate that some inorganic anions at least may interact specifically with the trimethylammonium residues of the PtdCho headgroups [123–125, 131]. In such studies, the stoichiometry of the PtdCho/anion binding is frequently assumed to be 1:1 or 2:1 [81,85,124,127]; but occasionally, much lower densities of the anion-binding sites are established by both ^{2}H-NMR [123] and electrophoresis data [72,91]. The binding of SCN^- to Myr_2GroPCho was studied by electrophoresis and described by using a binding constant of 10 M^{-1} in combination with the effective lipid-ion binding ratio of 4:1 for the gel and of 9:1 for the fluid phase [91]. In agreement with the results of this study, a binding constant value of 12.6 M^{-1} and a lipid-to-ion binding ratio of 7.5:1 were deduced from the ^{2}H-NMR data on the adsorption of SCN^- to the PamOleGroPCho bilayers in fluid phase [123].

C. Organic Ions

Adsorption of organic ions to phospholipid layers is regulated by many and diverse factors. One of these is the long-range electrostatic force, which causes ion accumulation near the lipid layer surface. The presence of anionic phospholipids PA, PtdGro, or PtdSer, in the zwitterionic PtdCho bilayers, consequently, promotes the adsorption of the organic cations propranolol, tetracaine, lidocaine, procaine [132], gentamicin [133], TPA^+ [134], melittin [135]; but inhibits the binding of hydrophobic anions, such as TPB^-, to similar membranes [134]. Likewise, the insertion of the hexadecyltrimethylammonium surfactant molecules into phospholipid bilayers may interfere effectively with the binding of cationic amphiphiles to such bilayers [132].

Similar effects may arise from the electric field stemming from the interfacially adsorbed ions. Increasing bulk DPA^- or TPP^+ concentrations, at some point, starts to suppress further binding of such ions to the initially uncharged PtdCho bilayers due to the electrostatic saturation effect [136,137]. Highly concentrated NaCl, KCl, or LiCl solutions can partly reverse this trend, by screening the interionic repulsion. The binding of organic anions such as DPA^- [136], TPB^- [134], or ANS^- [138] to the negatively charged [136,137] as well as zwitterionic (PtdCho) bilayers [135,137] in such salt solutions is thus relatively strong. But the adsorption of TPA^+ to the PtdSer/PtdCho membranes is suppressed with increasing ionic strength [134]. These effects are explicable by the reduction of the membrane negative surface potential with increasing salt concentration, which facilitates the adsorption of anions and hinders the binding of cations.

Lanthanides, which have a rather high affinity for the PtdCho bilayers (see Table 2), seem to exert their effects through screening as well as binding mechanisms. Even less than 1 mM of La^{3+} can thus enhance the binding of TPB^- to the PtdCho membranes; this effect is much more pronounced than the effect of a 0.2 M Na^+ or Li^+ [134]. On the other hand, Eu^{3+} may compete quite efficiently with cationic drugs for the common binding sites on PtdCho membranes [92].

Conformational changes or modified molecular packing induced by the bound ions may also play some role. Cationic and anionic amphiphiles appear to change the orientation of lipid headgroups [135,139–141], possibly by electrostatically repelling or attracting the N terminus of the PtdCho headgroup dipole, respectively. Furthermore, 0.5 mM TPB^- adsorbed to the Pam_2GroPCho bilayers lowers the chain melting phase transition by 7°, whilst 0.5 mM TPP^+ has almost no effect [139]. This once again indicates the preferential binding of organic anions to PtdCho membranes.

One of the crucial factors that determine how strongly a hydrophobic ion will bind to the phospholipid membrane is the so-called dipole potential, created by the lipid polar residues. This potential is identical to the potential difference between the membrane-water interface and the membrane interior and may amount to several hundreds of millivolts (positive inside). It is chiefly a result of the dipolar moments on the β chain carbonyl groups of the lipid chains and on the interfacially bound water molecules [11,137].

Bilayer dipole potential favors the binding of hydrophobic anions and hampers the binding of organic cations to phospholipid membranes. Thus the binding of TPP^+ interacting with PamOleGroPCho bilayers is quite weak, corresponding to $K \sim 21\ M^{-1}$ [142]. The binding constants of the multivalent organic cations gentamicin and spermine for negatively charged membranes are shown to be as low as 10–100 M^{-1} [133]. On the other hand, for the binding of organic anions TNP^- [91], ANS^- [138], or TNS^- [143] to PtdCho bilayers the binding constants were reported to be in the range between 5×10^3 and $2 \times 10^4\ M^{-1}$. The probability that a TPB^- anion will partition into a PtdCho membrane is almost four orders of magnitude greater than for a structurally similar cation, TPP^+ [137]. This is indicative of the importance of the dipole potential in the process of ion binding.

Apart from the quantitative distinctions, the processes of organic anion and cation binding to phospholipid bilayers differ thermodynamically; both are characterized by a negative Gibbs free energy change, but the binding of TPP^+ to PtdCho membranes is entropy driven and has a positive enthalpy (decreased order and loss of heat), while the association between the TPB^- ions and PtdCho membranes is both entropy and enthalpy driven (decreased order and gain of heat) [137].

Hydrophobic anions are translocated through the lipid bilayers at much higher rates than organic cations [6,11,137]. Cholesterol in lipid bilayers inhibits such membrane conductance for the cationic amphiphiles but it facilitates, although less extensively, the transbilayer transport of the organic anions [6,144,145]. These phenomena have been explained by cholesterol-induced elevation of the membrane dipole potential and concomitant decrease in bilayer fluidity [144,145]. Interestingly, the insertion of phloretin

Table 4 Intrinsic Binding Constants K or Partition Coefficients β and Lipid/Ion Binding Ratios n Characterizing the Binding of Organic Ions to Phospholipid Bilayers

Lipid	Ion	K or β	n	Methods, conditions	Refs.
		K (M^{-1})			
PtdCho	propranolol$^+$	65–600	1	ESR, EP, ^{31}P-NMR	92,100
	alprenolol$^+$	45–190	1	centrifugation,	150
	metaprolol$^+$	16–33	1	0.1 M NaCl, 20–25°C,	
	tetracaine$^+$	32–135	1	pH 4–6	
	procaine$^+$	0.8–10	1		
	trifluoperazine$^+$	10000	1	EP, 25°C	100
	trichloroacetate$^-$	1.0	1	EP, ^{31}P- and ^{1}H-NMR	151
	ANS$^-$	23000	4	Fl., pH 7.4	138
	TNS$^-$	5000	1	EP, 25°C	143
PamOleGro*P*Cho	TPP$^+$	21	8.3	NMR, 25°C, 0.1 M NaCl	142
	melittin^{5+}	2100	>5	NMR, 25°C, 0.1 M NaCl	135
Myr_2Gro*P*Cho	TNP$^-$	4500	20(s)	EP, pH 7.4, 10 mM KCl	91
		5000	10(f)		
Pam_2Gro*P*Cho	chlorpromazine$^+$	28570	1	Binding constants	149
	dibucaine$^+$	4176	1	deduced from the effect	
	diphenylhydantoin$^+$	3330	1	of drugs on Pam_2Gro*P*Cho	
	benztropine$^+$	2000	1	vesicle phase	
	propranolol$^+$	1250	1	transition,	
	tetracaine$^+$	1000	1	0.1 M NaCl, pH 7.2	
	procaine$^+$	143	1		
	pentobarbitone$^+$	90	1		
	lidocaine$^+$	77	1		
	phenobarbitone$^+$	14	1		
	ANS$^-$	6000[a]	40(s)	Fl.	147
		5000[a]	15(f)		
	BTB$^-$	30000[a]	13(s)		
		18000[a]	5(f)		

Table 4 (Continued)

Lipid	Ion	K or β	n	Methods, conditions	Refs.
$Pam_2GroPCho$	PCP^-	5000	11(s)	EP, 25°C, 30 mM KCl, pH 10	153
		45000	6(f)	42°C,	
	DPA^-	250000	12(s)	25°C,	
		740000	7(f)	42°C	
	TPB^-	310000	10(s)	25°C	
		590000	6(f)	42°C	
PtdCho/PtdEtn	$Me_3Dod\mathit{N}^+$	≈5000		EP, 25°C, pH 8	94
$Pam_2Gro\mathit{P}Etn$	chlorpromazine$^+$	3570	1	see above	149
	dibucaine$^+$	167	1		
PtdSer	$Me_4\mathit{N}^+$	0.05–0.2	1	EP, monolayer, 25°C	86,102
	$ET_4\mathit{N}^+$	0.03	1	EP, 25°C	86
	Tris$^+$	1.1	1		
	choline$^+$	0.11	1	EP, 22°C, pH 7.2–7.4	102
$Myr_2Gro\mathit{P}Ser$	gentamicin^{5+} [b]	100	1	EP, 25°C, 0.1 M NaCl	133
Ptd(Ins,Ser)	gentamicin^{5+} [b]	10	1	see above	133
	spermine^{4+} [b]	10–14	2	see above	51,133
PtdIns	tetracaine$^+$	100	1	EP, 25°C	100
	trifluoperazine$^+$	10000	1		
$PtdInsP_2$	gentamicin^{5+} [b]	500	2	EP, 25°C, 0.1 M NaCl, pH 7.4	51
	spermine^{4+} [b]	500	2		
PamOleGroPSer	diLys^{2+}	10	1	EP, 25°C, 0.1 M KCl, pH 7.0	154
or	triLys^{3+}	100	2		
PamOleGroPGro	tetraLys^{4+}	1000	3		
	pentaLys^{5+}	10000	4		
PtdGro/PtdCho	melittin^{5+}	45000	>5	see above	135

		β (μm)[c]			
PtdCho	TPP^+	0.042	70	Eq. dial.	137
	TPB^-	200–400	43	EPR, 20°C, 0.1 M NaCl	
$Ole_2GroPCho$	ANS^-	10.4		EP, conductance, 20°C	152
	TPB^-	200		current or voltage	6
$Lin_2GroPCho$	DPA^-	150		jump or current noise,	
$Lnn_2GroPCho$	DPA^-	30		0.1 M NaCl, 25°C	
$\Delta Pam_2GroPCho$	DPA^-	180			
$\Delta_4Pam_2GroPCho$	DPA^-	370			
$\Delta Ach_2GroPCho$	DPA^-	460			
$\Delta Beh_2GroPCho$	DPA^-	500–600			6,148
PtdEtn	TPB^-	280		conductance, 1 M NaCl	146
$Ole_2GroPEtn$	DPA^-	110		see above	6
$\Delta_4Pam_2GroPEtn$	DPA^-	55			
PtdSer	TPB^-	2.0			

Values of *n* were not determined when omitted.

By (s) and (f) are designated the solid and fluid states of the bilayers.

[a]Apparent binding constants.

[b]Both gentamicin and spermine have about 3.5 positive charges per molecule at pH 7.4.

[c]Partition coefficients were determined as the surface concentration of adsorbed ion devided by its bulk aqueous concentration. They refer to planar bilayers containing organic solvent (mostly *n*-decane) except for TPP^+ binding to PtdCho vesicles (Ref. 137).

into PtdEtn bilayers reduced the dipole potential and thereby suppressed TPB^- adsorption and promoted the binding of the K^+ complex of proline-valinomycin [146].

More information on organic ion binding to phospholipid bilayers can be found in Table 4 and in Refs. [6,11,92,94,133,137–154]. It should be stressed that relative contributions from the hydrophobic, dipolar, steric, and coulombic interactions as well as from membrane defects may not be the same in all cases. The binding of the cationic peptide melittin to small, sonicated PtdGro-containing vesicles, for example, is 20-fold stronger than to large, unsonicated PtdCho liposomes. This can be ascribed to the tighter lipid packing in the latter case exerting steric hindrance to the peptide penetration between the individual lipid molecules [135]. The dipole potential may play a certain role in these phenomena, however.

The aminoglycoside gentamicin adsorbs about one order of magnitude more strongly to the solid Myr_2GroPSer vesicles than to the fluid PtdSer bilayers [133]. This implies that in this case the coulombic interactions between the multication and the negatively charged membrane surface, which are stronger in the gel phase, are more decisive than the dipole potential or steric effects. The number of ANS^- or BTB^- binding sites per unit area of Pam_2GroPCho surface increases 2.6 times at the lipid transition from the gel to liquid-crystalline phase [147]. This is possibly because the steric constraints are weaker in the latter case. The importance of sterically mediated hydrophobic interactions is also supported by a 20-fold increase of the binding of the neutral fluorescent dye *N*-phenylnaphthylamine to the Myr_2MePA bilayers upon hydrocarbon chain melting [155].

V. TEMPERATURE EFFECTS

The thermotropic lipid phase transitions may affect the lipid ionization state and the adsorption of ions markedly. Phospholipid chain-melting results in a lateral expansion of the lipid bilayers, which for charged systems is also associated with the decrease in the net surface charge density. For negatively charged membranes, this leads to lowering of the interfacial proton concentration and decreases the apparent pK value of the anionic phospholipids [156].

MacDonald et al. [34] have directly detected a decrease in the apparent pK of the Pam_2GroPSer carboxyl groups at the gel-to-fluid phase transition. They concluded that $\approx 30\%$ of the observed pK shift was due to the lowering of surface charge density upon bilayer fluidization (larger area per molecule) while the remainder can be attributed to a diminished intrinsic pK value.

Bilayer fluidization not only decreases the affinity of the phospholipid molecules for protons but also interferes with divalent cation binding to lipid bilayers [72,73,75,93,105,156]. As a result of this, H^+ and/or Ca^{2+} ions are released from the charged Myr_2MePA bilayers at the lipid chain-melting phase transition [156]. As shown in Figs. 11 and 12, such a transition liberates some of the surface-bound cations and thus may increase the bulk proton and ion activities in highly concentrated lipid suspensions. The binding of Mn^{2+} to anionic Pam_2GroPGro bilayers was also shown to drastically decrease at the solid-to-fluid phase transition of the lipid [105].

Changes in ion binding to zwitterionic phospholipids during the phase transition are equally striking but more difficult to explain [72,73,91,93]. The phosphorus chemical shift anisotropy measured for the Pam_2GroPCho molecules suspended in a 0.35 M $CaCl_2$ solution, for example, becomes appreciably smaller as lipids undergo a transition to the fluid-lamellar phase [70]. The gel-to-liquid-crystalline phase transition of Pam_2GroPCho is accompanied with a two-fold decrease in Ca^{2+} binding constant to bilayers [72,73].

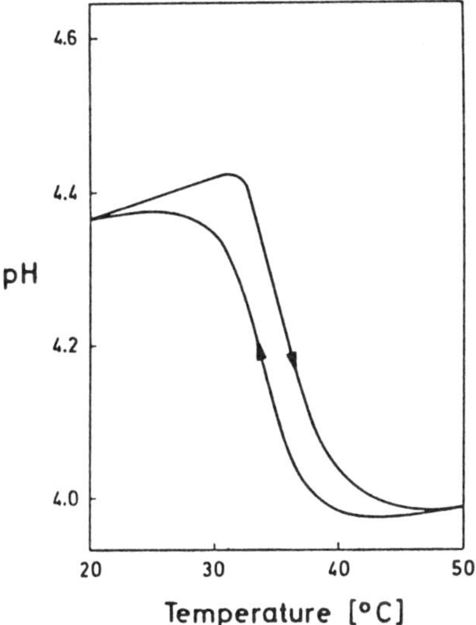

Figure 11 Release of the protons from the surface of the Myr_2MePA vesicles upon lipid chain fluidization, as measured by a pH-electrode. Samples contained 5 mM Myr_2MePA and 0.2 M CsCl. (Reprinted from Ref. 156 with permission from Plenum Press.)

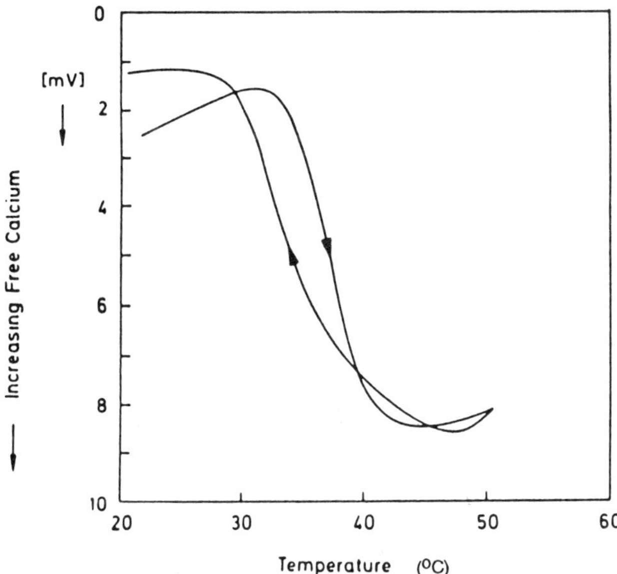

Figure 12 Phase-transition induced release of the Ca^{2+} ions from the surface of Myr_2MePA vesicles, as measured by a Ca^{2+}-sensitive electrode. Each sample contained 2 mM Myr_2MePA, 0.2 M NaCl, and 0.2 mM $CaCl_2$. The amount of released Ca^{2+} upon lipid melting was 0.1 mM. (Reprinted from Ref. 156 with permission from Plenum Press.)

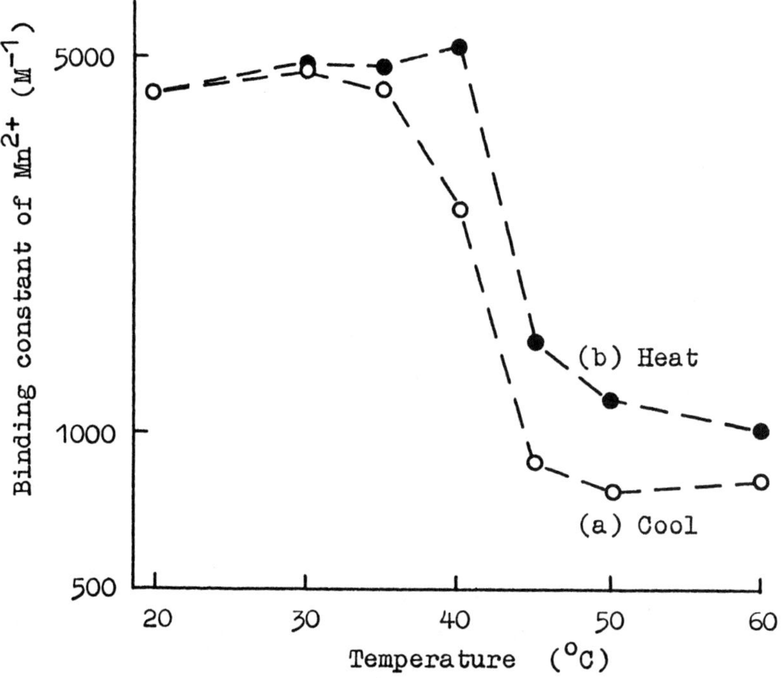

Figure 13 Apparent constant of Mn^{2+} binding to the $Pam_2GroPGro$ vesicles as determined by EPR spectroscopy as a function of the decreasing (a) and increasing (b) temperature. Samples contained 0.14 mM $MnCl_2$, 3 mg/ml $Pam_2GroPGro$, 0.25 M NaCl, and 15 mM HEPES buffer (pH 7.3). (Reproduced from Ref. 105 with permission from Elsevier Science Publishers.)

The apparent constant of Mn^{2+} binding to $Pam_2GroPCho$ also gets smaller upon hydrocarbon chain melting [93].

None of these findings can be accounted for by simple electrostatic reasoning. After all, $Pam_2GroPCho$ bilayers have no net charges that could get diluted upon the lipid chain fluidization. All fluidization-induced changes in ion binding, consequently, must be due either to the altered intrinsic binding constants or to a manifestation of the modified lipid packing and hydration properties. The latter possibility is quite probable. Circumstantial evidence for this comes from the microelectrophoretic data for $Myr_2GroPCho$ and $Pam_2GroPCho$ with bound Ca^{2+} ions. These data show that the ion-induced positive surface potential decreases sharply at the gel-to-fluid phase transition [72,73], as one would expect for the hydration-dependent contribution to this potential. Figure 14 illustrates this in some detail.

Marra and Israelachvili [95] have also reported direct measurements of the intrinsic constants for Ca^{2+} and Mg^{2+} binding to PtdCho bilayers. The binding constants were 5 ± 3-fold lower in the fluid than in the gel phase. The lowering of the PtdCho affinity for the divalent cations has been explained by the lateral expansion of the headgroup lattice upon hydrocarbon chain fluidization and by the subsequent weakening of the ion-dipole interactions between the bound cation and the PtdCho polar groups [72,73].

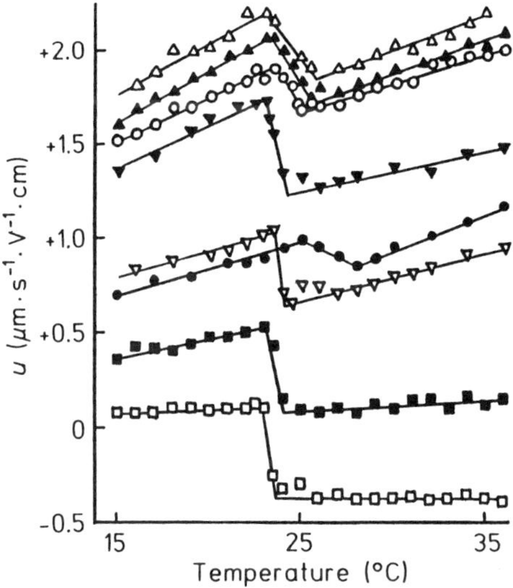

Figure 14 Temperature dependence of the electrophoretic mobility of $Myr_2GroPCho$ liposomes in an electrolyte solution containing 2 mM tris-HCl (pH 7.2). $CaCl_2$ concentrations were (mM): 0.1 (open squares), 0.3 (filled squares), 1.0 (inversed open triangles), 4.0 (inversed filled triangles), 20 (open triangles), 50 (filled triangles), 100 (open circles), and 500 (filled circles). (Reprinted from Ref. 72 with permission from Springer-Verlag.)

As shown in Fig. 10, the negative surface potential of $Myr_2GroPCho$ liposomes, induced by selective adsorption of a series of inorganic anions, abruptly increases at the solid-to-fluid phase transition of the lipid [91]. These data were interpreted in terms of an increase in the surface density of anion binding sites at the lipid hydrocarbon chain melting [91]. This interpretation accords with the data deduced from fluorescence spectroscopy, indicating a 2.6-fold increase in the density of organic anion binding sites of $Pam_2GroPCho$ membranes at their transition from the gel to liquid crystalline phase [147]. Smejtek and Wang [153] also detected an increase in the adsorption of organic anions to $Pam_2GroPCho$ bilayers at the lipid transition to the fluid phase and ascribed this effect to augmented affinities of the anions for liquid-crystalline membranes.

Finally, despite the well substantiated interpretations of the alterations of PtdCho membrane surface potential at thermotropic phase transitions, the question remains, however, whether the melting of lipid acyl chains may stimulate additional adsorption of anions or promote the desorption of cations. It may be that the relative concentrations of the cations and anions in the interfacial region, indeed, are changing upon lipid phase transitions, the net outcome of such changes being determined by the relative binding constant values in either phase.

VI. CONCLUSIONS

Data characterizing ion-phospholipid interactions are numerous but not always conclusive. This is especially true for the absolute values of the ion binding constants. The main sources of experimental ambiguity are the difficulties associated with the direct measurements of bound-ion concentrations. Ion-selective electrodes, for example, can monitor ion concentration in solutions [69,157]. But they are nearly useless for studying interfaces and surfaces. Counterions immobilized in the diffuse part of the double layer, consequently, could be ascribed erroneously to the bound fraction of ions if one relied solely on electrode measurements. This would lead to an overestimation of ion binding affinity. The same is true for the monolayer-radiotracer techniques [32], which are based on the detection of short-range β radiation emerging from $^{45}Ca^{2+}$ nuclei, for example. Such radiation may stem from the accumulated, rather than necessarily bound, ions near the lipid-electrolyte interface.

Electrophoresis seems to be a useful method for the qualitative study of electrostatic properties of the phospholipid-electrolyte interface. But it is difficult to convert the quantity measured by this technique, the vesicle mobility in an external field, into a reliable value for the "surface potential," which could then be used to determine the intrinsic binding constants of ions [9,16,72,73,75,82,91,98,102,122]. Electrophoretic measurements also provide little or no information on the chemical nature of the ion binding site; nor do they help us to distinguish between inner- and outer-sphere complexes.

The most detailed data on ion-phospholipid complex formation to date can be obtained from NMR measurements [9,62,64,66,68,70,71,74,84,85,89,99,107] or IR spectroscopy [68,109–112]. Direct visualization of the lipid-bound ions has been achieved with the neutron diffraction technique [113].

By measuring the fluorescence of membrane-bound ANS^- molecules and its variation with the bulk calcium concentration, Träuble [147] has determined the apparent constant of Ca^{2+} binding to the fluid $Pam_2GroPCho$ bilayers to be 10^3 M^{-1}. From NMR data the same quantity, in the absence of organic anions, was deduced to be $\sim$ 20 M^{-1} [70] (see also Refs. 80,108). The corresponding intrinsic binding constant was evaluated by electrophoresis as 190 M^{-1} [72]. These, seemingly divergent data can be brought into better congruence if the effects of the negative surface potential induced by the adsorbed ANS^- anions [147] and the Ca^{2+}-dependent positive surface potentials [70] are taken into account. But even with allowance being made for such effects, a substantial spread of values remains that illustrates the difficulty of drawing firm conclusions at the molecular level from the experimental data of different sources or experiments.

The reason for obtaining conflicting results is sometimes also the possibility of variable interpretations of the same experimental data. For example, electrophoresis data pertaining to divalent cation binding to egg PtdCho liposomes can be described by using binding constants of the order of 1 M^{-1} and assuming 1:1 binding stoichiometry [82]. But in a similar electrophoresis study the number of lipid molecules per binding site was concluded to be $\sim$ 17:1 based on the extremum of the electrophoretic mobility at high divalent cation concentrations. Using the latter stoichiometric binding ratio the intrinsic binding constant value was calculated to be 10–40 M^{-1} [72]. These considerations indicate that reliable data describing the binding of ions to phospholipid molecules can only be gained by a combination of different methods, taking into account surface potential effects and using experimentally evaluated stoichiometry coefficients.

ABBREVIATIONS

1. *Lipids* (see the first chapter of this volume): (a) polar head-groups: P = phosphoric residue, PA = phosphatidic acid, Ptd = phosphatidyl, Cho = choline, Etn = athanolamine, Gro = glycero(l), Ins = inositol, Ser = serine, MePA = methyl phosphatidic acid;

(b) acyl chains (in parentheses are shown carbon atoms:double bonds per chain): Lau = lauroyl (12:0), Myr = myristoyl (14:0), Pam = palmitoyl (16:0), ΔPam = palmitoleoyl (16:1), Δ_4Pam = phytanoyl (16:4), Ste = stearoyl (18:0), Ole = oleoyl (18:1), Lin = linoleoyl (18:2), Lnn = linolenoyl (18:3), ΔAch = icosenoyl (20:1), ΔBeh = docosenoyl (22:1)

2. *Other compounds*: Me = methyl, Et = ethyl, Pr = propyl, Dod = dodecyl, N = amine or ammonium, CTAB = cetyltrimethylammonium bromide, STS = dosium tetradecyl sulfate, DPA = dipicrylamine, BTB = bromothymol blue, ANS = 1-anilino-8-naphthalene sulfonate, TPA = tetraphenylarsonium, TPB = tetraphenylboron, TPP = tetraphenylphosphonium, TNP = trinitrophenol, TNS = 2,6-toluidinylnaphthalene sulfonate, Tris = Tris(oxymethyl)aminomethan, PCP = pentachlorophenol, Lys = lysine

3. *Methods*: Pot. = potentiometric titration, EP = electrophoresis, Fl. = fluorescence, DSC = differential scanning calorimetry, ESR = electron spin resonance, EPR = electron paramagnetic resonance, IR = infrared spectroscopy, NMR = nuclear magnetic resonance AAS = atomic absorption spectroscopy, Eq. dial. = equilibrium dialysis.

REFERENCES

1. B. E. Conway and M. Salomon, Classical H/D isotope effect in proton conductance and the proton solvation energy, in *Chemical Physics of Ionic Solutions*, Wiley, New York, 1966, pp. 541–556.
2. D. E. Irish, Vibrational spectral studies of electrolyte solutions and fused salts, in *Ionic Interactions* (S. Petrucci, ed.), Academic Press, New York, London, Vol. 2, 1971, pp. 187–258.
3. K. D. Collins and M. W. Washabaugh, The Hofmeister effects and the behaviour of water at interfaces, *Quart. Rev. Biophys. 18*:323–422 (1985).
4. O. Ya. Samoilov, Residence times of ionic hydration, in *Water and Aqueous Solutions. Structure, Thermodynamics, and Transport Processes* (R. A. Horne, ed.), Wiley, New York, 1972, pp. 597–612.
5. A. Warshel and S. T. Russell, Calculations of electrostatic interactions in biological systems and in solutions, *Quart. Rev. Biophys. 17*:283–422 (1984).
6. P. Läuger, R. Benz, G. Stark, E. Bamberg, P. C. Jordan, A. Fahr, and W. Brock, Relaxation studies of ion transport systems in lipid bilayer membranes, *Quart. Rev. Biophys. 14*:513–598 (1981).
7. R. P. Rand and V. A. Parsegian, Hydration forces between phospholipid bilayers, *Biochim. Biophys. Acta 988*:351–378 (1989).
8. A. Pullman, B. Pullman, and H. Berthod, An SCF ab initio investigation of the "through-water" interaction of the phosphate anion with the Na^+ cation, *Theoret. Chim. Acta 47*:175–192 (1978).
9. A. Lau, A. McLaughlin, and S. McLaughlin, The adsorption of divalent cations to phosphatidylglycerol bilayer membranes, *Biochim. Biophys. Acta 645*:279–292 (1981).
10. A. C. McLaughlin, Nuclear magnetic resonance studies of the adsorption of divalent cations of phospholipid bilayer membranes, in *Membranes and Transport* (A. N. Martonosi, ed.), Plenum Press, New York, Vol. 1, 1982, pp. 57–61.
11. R. F. Flewelling and W. L. Hubbell, The membrane dipole potential in a total membrane potential model. Applications of hydrophobic ion interactions with membranes, *Biophys. J. 49*:541–552 (1986).
12. D. C. Grahame, The electrical double layer and the theory of electrocapillarity, *Chem. Rev. 41*:441–501 (1947).
13. S. McLaughlin, Electrostatic potentials at membrane-solution interfaces, *Curr. Top. Membr. Transp. 9*:71–144 (1977).

14. S. McLaughlin, The electrostatic properties of membranes, *Ann. Rev. Biophys. Biophys. Chem. 18*:113–136 (1989).
15. G. Cevc, Membrane electrostatics, *Biochim. Biophys. Acta 1031*:311–382 (1990).
16. R. J. Hunter, in *Zeta Potential in Colloid Science. Principles and Applications* (R. H. Ottewill and R. L. Rowell, eds.), Academic Press, London, 1981, p. 236.
17. A. Watts, K. Harlos, W. Maschke, and D. Marsh, Control of the structure and fluidity of phosphatidylglycerol bilayers by pH titration, *Biochim. Biophys. Acta 510*:63–74 (1978).
18. G. Cevc, A. Watts, and D. Marsh, Titration of the phase transition of phosphatidylserine bilayer membranes. Effects of pH, surface electrostatics, ion binding, head-group hydration, *Biochemistry 20*:4955–4965 (1981).
19. H. Eibl, The effect of the proton and of monovalent cations on membrane fluidity, in *Membrane Fluidity in Biology* (R. C. Aloia, ed.), Academic Press, New York, Vol. 2, 1983, pp. 217–236.
20. J. M. Boggs, Lipid intermolecular hydrogen bonding: influence on structural organization and membrane function, *Biochim. Biophys. Acta 906*:353–404 (1987).
21. J.-F. Tocanne, and J. Teissié, Ionization of phospholipids and phospholipid-supported interfacial lateral diffusion of protons in membrane model systems, *Biochim. Biophys. Acta 1031*:111–142 (1990).
22. T. H. Jukes, The electrometric titration of lecithin and cephalin, *J. Biol. Chem. 107*:783–787 (1934).
23. H. B. Bull, The chemistry of the lipids, *Cold Spring Harbor Symp. Quant. Biol. 8*:63–71 (1940).
24. J. M. Diamond and E. M. Wright, Biological membranes: the physical basis of ion and nonelectrolyte selectivity, *Ann. Rev. Physiol. 31*:581–646 (1969).
25. J. H. Ashby, E. M. Crook, and S. P. Datta, Thermodynamic quantities for the dissociation equilibria of biologically important compounds, *Biochem. J. 56*:198–207 (1954).
26. J. E. Garvin and M. L. Karnovsky, The titration of some phospholipids and related compounds in a nonaqueous medium, *J. Biol. Chem. 221*:211–222 (1956).
27. M. B. Abramson, R. Katzman, C. E. Wilson, and H. P. Gregor, Ionic properties of aqueous dispersions of phosphatidic acid, *J. Biol. Chem. 239*:4066–4072 (1964).
28. G. H. De Haas, H. van Zutphen, P. P. M. Bonsen, and L. L. M. van Deenen, Synthesis of mixed-acid phosphatidylserine containing unsaturated fatty acids, *Rec. Trav. Chim. 83*:99–116 (1964).
29. H. S. Hendrickson and J. G. Fullington, Stabilities of metal complexes of phospholipids: Ca(II), Mg(II), and Ni(II) complexes of phosphatidylserine and triphosphoinositide, *Biochemistry 4*:1599–1605 (1965).
30. D. Papahadjopoulos, Surface properties of acidic phospholipids: interaction of monolayers and hydrated liquid crystals with uni- and bi-valent metal ions, *Biochim. Biophys. Acta 163*:240–254 (1968).
31. M. M. Standish and B. A. Pethica, Surface pressure and surface potential study of a synthetic phospholipid at the air/water interface, *Trans. Faraday Soc. 64*:1113–1122 (1968).
32. T. Seimiya and S. Ohki, Ionic structure of phospholipid membranes and binding of calcium ions, *Biochim. Biophys. Acta 298*:546–561 (1973).
33. H.-J. Galla and E. Sackmann, Chemically induced lipid phase separation in model membranes containing charged lipids: a spin label study, *Biochim. Biophys. Acta 401*:509–529 (1975).
34. R. C. MacDonald, S. A. Simon, and E. Baer, Ionic influences on the phase transition of dipalmitoylphosphatidylserine, *Biochemistry 15*:885–891 (1976).
35. H. Hauser, A. Darke, and M. C. Phillips, Ion-binding to phospholipids. Interaction of calcium with phosphatidylserine, *Eur. J. Biochem. 62*:335–344 (1976).
36. H. Träuble, M. Teubner, P. Woolley, and H. Eibl, Electrostatic interactions at charged lipid membranes. I. Effects of pH and univalent cations on membrane structure, *Biophys. Chem. 4*:319–342 (1976).

37. P. W. M. Van Dijck, B. de Kruijff, A. J. Verkleij, L. L. M. van Deenen, and J. de Gier, Comparative studies on the effects of pH and Ca^{2+} on bilayers of various negatively charged phospholipids and their mixtures with phosphatidylcholine, *Biochim. Biophys. Acta 512*:84–96 (1978).
38. H. Eibl and A. Blume, The influence of charge on phosphatidic acid bilayer membranes, *Biochim. Biophys. Acta 553*:476–488 (1979).
39. K. Toko and K. Yamafuji, Influence of monovalent and divalent cations on the surface area of phosphatidylglycerol monolayers, *Chem. Phys. Lipids 26*:79–99 (1980).
40. M. Ptak, M. Egret-Charlier, A. Sanson, and O. Bouloussa, A NMR study of the ionization of fatty acids, fatty amines and *N*-acylamino acids incorporated in phosphatidylcholine vesicles, *Biochim. Biophys. Acta 600*:387–397 (1980).
41. S. Tokutomi, K. Ohki, S.-I. Ohnishi, Proton-induced phase separation in phosphatidylserine/phosphatidylcholine membranes, *Biochim. Biophys. Acta 596*:192–200 (1980).
42. S. Ohki and R. Kurland, Surface potential of phosphatidylserine monolayers. II. Divalent and monovalent ion binding, *Biochim. Biophys. Acta 645*:170–176 (1981).
43. B. R. Copeland and H. C. Andersen, A theory of effects of protons and divalent cations on phase equilibria in charged bilayer membranes: comparison with experiment, *Biochemistry 21*:2811–2820 (1982).
44. N. S. Matinyan, I. A. Ershler, I. G. Abidor, Proton equilibrium on the bilayer lipid membrane surfaces (in Russian), *Biol. Membrany 1*:254–267 (1984).
45. M. Bozsik, C. Helm, L. Laxhuber, and H. Möhwald, Vanadyl binding to phospholipid membranes, *J. Colloid Interface Sci. 107*:514–524 (1985).
46. S. Akoka, C. Tellier, and S. Poignant, Molecular order, dynamics, and ionization state of phosphatidylethanolamine bilayers as studied by ^{15}N-NMR, *Biochemistry 25*:6972–6977 (1986).
47. J. Marra, Direct measurement of the interaction between phosphatidylglycerol bilayers in aqueous electrolyte solutions, *Biophys. J. 50*:815–825 (1986).
48. C. A. Helm, L. Laxhuber, M. Lösche, and H. Möhwald, Electrostatic interactions in phospholipid membranes. I. Influence of monovalent ions, *Colloid & Polymer Sci., 264*:46–55 (1986).
49. F. C. Tsui, D. M. Ojcius, and W. L. Hubbell, The intrinsic pK_a values for phosphatidylserine and phosphatidylethanolamine in phosphatidylcholine host bilayers, *Biophys. J. 49*:459–468 (1986).
50. A. Watts and T. W. Poile, Direct determination by ^{2}H-NMR of the ionization state of phospholipids and of a local anaesthetic at the membrane surface, *Biochim. Biophys. Acta 861*:368–372 (1986).
51. M. Toner, G. Vaio, A. McLaughlin, and S. McLaughlin, Adsorption of cations to phosphatidylinositol 4,5-bisphosphate, *Biochemistry 27*:7435–7443 (1988).
52. A. Blume, and J. Tuchtenhagen, Thermodynamics of ion binding to phosphatidic acid bilayers. Titration calorimetry of the heat of dissociation of DMPA, Biochemistry 31:4636–4642 (1992).
53. G. Cevc, J. M. Seddon, R. Hartung, and W. Eggert, Properties of phosphatidylcholine-fatty acid bilayer membranes I. Effects of protonation, salt, temperature, and chain length on the colloidal and phase behaviour, *Biochim. Biophys. Acta 940*:219–240 (1988).
54. G. Cevc and D. Marsh, Properties of the electrical double layer near the interface between a charged bilayer membrane and electrolyte solution. Experiment vs. theory. *J. Phys. Chem. 87*:376–379 (1983).
55. M. S. Fernández and P. Fromherz, Lipoid pH indicators as probes of electrical potential and polarity in micelles, *J. Phys. Chem. 81*:1755–1761 (1977).
56. H. Eibl and P. Woolley, Electrostatic interactions at charged lipid membranes. Hydrogen bonds in lipid membrane surfaces, *Biophys Chem. 10*:261–271 (1979).
57. T. H. Haines, Anionic lipid headgroups as a proton conducting pathway along the surface of membranes. A hypothesis, *Proc. Natl. Acad. Sci. USA 80*:160–164 (1983).

58. G. Cevc, How membrane chain melting properties are controlled by the polar surface of the lipid bilayers, *Biochemistry 26*:6305–6310 (1987).
59. J. S. Puskin and M. T. Coene, Na^+ and H^+ dependent Mn^{2+} binding to phosphatidylserine vesicles as a test of the Gouy-Chapman-Stern theory, *J. Membrane Biol. 52*:69–74 (1980).
60. D. L. Holwerda, P. D. Ellis, and R. E. Wuthier, Carbon-13 and phosphorus-31 NMR studies on interaction of calcium with phosphatidylserine, *Biochemistry 20*:418–428 (1981).
61. M. J. Liao and J. H. Prestegard, Structural properties of a Ca^{2+} -phosphatidic acid complex. Small angle X-ray scattering and calorimetric results, *Biochim. Biophys. Acta 645*:149–156 (1981).
62. A. C. McLaughlin, Phosphorus-31 and carbon-13 NMR studies of divalent cation binding to phosphatidylserine membranes: use of cobalt as a paramagnetic probe, *Biochemistry 21*:4879–4885 (1982).
63. J. Bentz, D. Alford, J. Cohen, and N. Düzgünes, La^{3+}-induced fusion of phosphatidylserine liposomes. Close approach, intermembrane intermediates, and the electrostatic surface potential, *Biophys. J. 53*:593–607 (1988).
64. M. Petersheim and J. Sun, On the coordination of La^{3+} by phosphatidylserine, *Biophys. J. 55*:631–636 (1989).
65. H. Träuble and H. Eibl, Electrostatic effects on lipid phase transitions: membrane structure and ionic environment, *Proc. Natl. Acad. Sci. USA 71*:214–219 (1974).
66. F. Borle and J. Seelig, Ca^{2+} binding to phosphatidylglycerol bilayers as studied by DSC and ^{2}H- and ^{31}P-NMR, *Chem. Phys. Lipids 36*:263–283 (1985).
67. J. Conti, H. N. Halladay, and M. Petersheim, An ionotropic phase transition in phosphatidylcholine: cation and anion cooperativity, *Biochim. Biophys. Acta 902*:53–64 (1987).
68. H. L. Casal, H. H. Mantsch, F. Paltauf, and H. Hauser, Infrared and ^{31}P-NMR studies of the effect of Li^+ and Ca^{2+} on phosphatidylserines, *Biochim. Biophys. Acta 919*:275–286 (1987).
69. R. Ekerdt and D. Papahadjopoulos, Intermembrane contact affects calcium binding to phospholipid vesicles, *Proc. Natl. Acad. Sci. USA 79*:2273–2277 (1982).
70. H. Akutsu and J. Seelig, Interaction of metal ions with phosphatidylcholine bilayer membranes, *Biochemistry 20*:7366–7373 (1981).
71. C. Altenbach and J. Seelig, Ca^{2+} binding to phosphatidylcholine bilayers as studied by deuterium magnetic resonance. Evidence for the formation of a Ca^{2+} complex with two phospholipid molecules, *Biochemistry 23*:3913–3920 (1984).
72. S. A. Tatulian, Binding of alkaline-earth metal cations and some anions to phosphatidylcholine liposomes, *Eur. J. Biochem. 170*:413–420 (1987).
73. S. A. Tatulian, Electrophoretic determination of Ca^{2+} adsorption to zwitterionic lipid membranes: structural defects as ion-binding sites (in Russian), *Biol. Membrany 2*:383–394 (1985).
74. P. M. Macdonald and J. Seelig, Calcium binding to mixed phosphatidylglycerol-phosphatidylcholine bilayers as studied by deuterium nuclear magnetic resonance, *Biochemistry 26*:1231–1240 (1987).
75. S. McLaughlin, N. Mulrine, T. Gresalfi, G. Vaio, A. McLaughlin, Adsorption of divalent cations to bilayer membranes containing phosphatidylserine, *J. Gen. Physiol. 77*:445–473 (1981).
76. L. J. Lis, V. A. Parsegian, and R. P. Rand, Binding of divalent cations to dipalmitoylphosphatidylcholine bilayers and its effect on bilayer interaction, *Biochemistry 20*:1761–1770 (1981).
77. R. C. MacDonald and T. E. Thompson, Properties of lipid bilayer membranes separating two aqueous phases: the effects of Fe^{3+} on electrical properties, *J. Membrane Biol. 7*:54–87 (1972).
78. R. L. Misiorowski and M. A. Wells, Competition between cations and water for binding to phosphatidylcholine in organic solvents, *Biochemistry 12*:967–975 (1973).
79. D. H. Haynes, 1-Anilino-8—naphthalenesulfonate: a fluorescent indicator of ion binding and electrostatic potential on the membrane surface, *J. Membrane Biol. 17*:341–366 (1974).

80. H. Hauser, C. C. Hinckley, J. Krebs, B. A. Levine, M. C. Phillips, and R. J. P. Williams, The interaction of ions with phosphatidylcholine bilayers, *Biochim. Biophys. Acta 468*:364–377 (1977).
81. H. Grasdalen, L. E. G. Eriksson, J. Westman, A. Ehrenberg, Surface potential effects on metal ion binding to phosphatidylcholine membranes. ^{31}P-NMR study of lanthanide and calcium ion binding to egg-yolk lecithin vesicles, *Biochim, Biophys. Acta 469*:151–162 (1977).
82. A. McLaughlin, C. Grathwohl, and S. McLaughlin, The adsorption of divalent cations to phosphatidylcholine bilayer membranes, *Biochim. Biophys. Acta 513*:338–357 (1978).
83. S. Nir, C. Newton, and D. Papahadjopoulos, Binding of cations to phosphatidylserine vesicles, *Bioelectrochem. Bioenerg. 5*:116–133 (1978).
84. A. C. McLaughlin, C. Grathwohl, and R. E. Richards, The interaction of cobalt with glycerophosphoryl choline and phosphatidyl choline bilayer membranes, *J. Magn. Res. 31*:283–293 (1978).
85. J. Westman and L. E. G. Eriksson, The interaction of various lanthanide ions and some anions with phosphatidylcholine vesicle membranes. A ^{31}P NMR study of the surface potential effects, *Biochim. Biophys. Acta 557*:62–78 (1979).
86. M. Eisenberg, T. Gresalfi, T. Riccio, and S. McLaughlin, Adsorption of monovalent cations to bilayer membranes containing negative phospholipids, *Biochemistry 18*:5213–5223 (1979).
87. P. Woolley and M. Teubner, Electrostatic interactions at charged lipid membranes. Calcium binding, *Biophys. Chem. 10*:335–350 (1979).
88. A. Chrzeszczyk, A. Wishnia, and C. S. Springer, Jr., Evidence for cooperative effects in the binding of polyvalent metal ions to pure phosphatidylcholine bilayer vesicle surfaces, *Biochim. Biophys. Acta 648*:28–48 (1981).
89. M. M. Hammoudah, S. Nir, J. Bentz, E. Mayhew, T. P. Stewart, S. W. Hui, and R. J. Kurland, Interaction of La^{3+} with phosphatidylserine vesicles. Binding, phase transition, leakage, ^{31}P-NMR and fusion, *Biochim. Biophys. Acta 645*:102–114 (1981).
90. H. Ohshima, Y. Inoko, and T. Mitsui, Hamaker constant and binding constants of Ca^{2+} and Mg^{2+} in dipalmitoyl phosphatidylcholine/water systems, *J. Colloid Interface Sci. 86*:57–72 (1982).
91. S. A. Tatulian, Effect of lipid phase transition on the binding of anions to dimyristoylphosphatidylcholine liposomes, *Biochim. Biophys. Acta 736*:189–195 (1983).
92. J. Westman, L. E. G. Eriksson, and A. Ehrenberg, Interaction of the cationic form of amphiphilic drugs with phosphatidylcholine model membranes. Competition with lanthanide ions, *Biophys. Chem. 19*:57–68 (1984).
93. M. Papánková and D. Chorvát, ESR study of the interaction of manganous ions with zwitterionic phospholipids and their mixtures with cholesterol, *Biochim. Biophys. Acta 778*:17–24 (1984).
94. A. Hattenbach, J. Gündel, D. Haroske, E. Müller, The interaction of a charged amphiphile with lipid model membranes. I. A test of the Gouy-Chapman-Stern theory, *Stud. Biophys., 105*:39–47 (1985).
95. J. Marra and J. Israelachvili, Direct measurements of forces between phosphatidylcholine and phosphatidylethanolamine bilayers in aqueous electrolyte solutions, *Biochemistry 24*:4608–4618 (1985).
96. A. P. Winiski, A. C. McLaughlin, R. V. McDaniel, M. Eisenberg, S. McLaughlin, An experimental test of the discreteness-of-charge effect in positive and negative lipid bilayers, *Biochemistry 25*:8206–8214 (1986).
97. S. A. Sundberg and W. L. Hubbell, Investigation of surface potential asymmetry in phospholipid vesicles by a spin label relaxation method, *Biophys. J. 49*:553–562 (1986).
98. S. A. Tatulian, Electrophoretic determination of ion binding to phosphatidylcholine and phosphatidylserine liposomes, in *Water and Ions in Biological Systems*. (Läuger, P., Packer, L., and Vasilescu, V., eds.) Birkhäuser Verlag, Berlin, Germany (1988) pp. 99–110.

99. P. M. Macdonald and J. Seelig, Calcium binding to mixed cardiolipin-phosphatidylcholine bilayers as studied by deuterium nuclear magnetic resonance. *Biochemistry 26*:6292–6298 (1987).
100. S. McLaughlin and M. Whitaker, Cations that alter surface potentials of lipid bilayers increase the calcium requirement of exocytosis in sea urchin eggs, *J. Physiol 396*:189–204 (1988).
101. N. S. Matinyan, G. B. Melikyan, V. B. Arakelyan, S. L. Kocharov, N. V. Prokazova, Ts. M. Avakian, Interaction of ganglioside-containing planar bilayer with serotonin and inorganic cations, *Biochim. Biophys. Acta 984*:313–318 (1989).
102. Yu. A. Ermakov, The determination of binding site density and associations constants for monovalent cation adsorption onto liposomes made from mixtures of zwitterionic and charged lipids, *Biochim. Biophys. Acta 1023*:91–97 (1990).
103. F. Bellemare and R. Lesage, Mg^{2+}, Ca^{2+}, and Mn^{2+} bound on anionic phospholipids resist desalting dialysis: Evaluation of binding parameters using Stern adsorption isotherm, *J. Colloid Interface Sci. 147*:462–473 (1991).
104. L. J. Lis, W. T. Lis, V. A. Parsegian, and R. P. Rand, Adsorption of divalent cations to a variety of phosphatidylcholine bilayes, *Biochemistry 20*:1771–1777 (1981).
105. J. S. Puskin and T. Martin, Divalent cation binding to phospholipid vesicles. Dependence on temperature and lipid fluidity, *Biochim. Biophys. Acta. 552*:53–65 (1979).
106. P. M. Sokolove, J. M. Brenza, A. E. Shamoo, Ca^{2+}-cardiolipin interaction in a model system. Selectivity and apparent high affinity, *Biochim. Biophys. Acta 732*:41–47 (1983).
107. P. W. Nolden and T. Ackermann, A high-resolution NMR study (^{1}H, ^{13}C, ^{31}P) of the interaction of paramagnetic ions with phospholipids in aqueous dispersions, *Biophys. Chem. 4*:297–304 (1976).
108. H. Hauser, M. C. Phillips, B. A. Levine, and R. J. P. Williams, Ion-binding to phospholipids. Interaction of calcium and lanthanide ions with phosphatidylcholine, *Eur. J. Biochem. 58*:133–144 (1975).
109. M. Petersheim, H. N. Halladay, and J. Blodnieks, Tb^{3+} and Ca^{2+}-binding to phosphatidylcholine. A study comparing data from optical, NMR, and infrared spectroscopies, *Biophys. J. 56*:551–557 (1989).
110. R. A. Dluhy, D. G. Cameron, H. H. Mantsch, and R. Mendelsohn, Fourier transform infrared spectroscopic studies of the effect of calcium ions on phosphatidylserine, *Biochemistry 22*:6318–6325 (1983).
111. H. L. Casal, H. H. Mantsch, and H. Hauser, Infrared studies of fully hydrated saturated phosphatidylserine bilayers. Effect of Li^{+} and Ca^{2+}, *Biochemistry 26*:4408–4416 (1987).
112. H. L. Casal, A. Martin, H. H. Mantsch, F. Paltauf, H. Hauser, Infrared studies of fully hydrated unsaturated phosphatidylserine bilayers. Effect of Li^{+} and Ca^{2+}, *Biochemistry 26*:7395–7401 (1987).
113. L. Herbette, C. Napolitano, R. V. McDaniel, Direct determination of the calcium profile structure for dipalmitoyllecithin multilayers using neutron diffraction, *Biophys. J. 46*:677–685 (1984).
114. H. Hauser, M. C. Phillips, and M. D. Barratt, Differences in the interaction of inorganic and organic (hydrophobic) cations with phosphatidylserine membranes, *Biochim. Biophys. Acta 413*:341–353 (1975).
115. T. P. Herrmann, A. R. Jayaweera, and A. E. Shamoo, Interaction of europeum(III) with phospholipid vesicles as monitored by laser-excited europeum(III) luminescence, *Biochemistry 25*:5834–5838 (1986).
116. H. N. Halladay, and M. Petersheim, Optical properties of Tb^{3+}-phospholipid complexes and their relation to structure, *Biochemistry 27*:2120–2126 (1988).
117. G. W. Feigenson, On the nature of calcium ion binding between phosphatidylserine lamellae, *Biochemistry 25*:5819–5825 (1986).

118. B. De Kruijff, A. J. Verkleij, J. Leunissen-Bijvelt, C. J. A. Van Echteld, J. Hille, and H. Rijnbout, Further aspects of the Ca^{2+}-dependent polymorphism of bovine heart cardiolipin, *Biochim. Biophys. Acta 693*:1–12 (1982).
119. Y. K. Levine, A. G. Lee, N. J. M. Birdsall, J. C. Metcalfe, and J. D. Robinson, The interaction of paramagnetic ions and spin labels with lecithin bilayers, *Biochim. Biophys. Acta 291*:592–607 (1973).
120. C.-H. Huang, J. P. Sipe, S. T. Chow, and R. B. Martin, Differential interaction of cholesterol with phosphatidylcholine on the inner and outer surfaces of lipid bilayer vesicles, *Proc. Natl. Acad. Sci. USA 71*:359–362 (1974).
121. B. Sears, W. C. Hutton, and T. E. Thompson, Effects of paramagnetic shift reagents on the ^{13}C-NMR spectra of egg phosphatidylcholine enriched with ^{13}C in the *N*-methyl carbons, *Biochemistry 15*:1635–1639 (1976).
122. L. Pasquale, A. Winiski, C. Oliva, G. Vaio, and S. McLaughlin, An experimental test of new theoretical models for the electrokinetic properties of biological membranes. The effect of UO_2^{++} and tetracaine on the electrophoretic mobility of bilayer membranes and human erythrocytes, *J. Gen. Physiol. 88*:697–718 (1986).
123. P. M. Macdonald and J. Seelig, Anion binding to neutral and positively charged lipid membranes, *Biochemistry 27*:6769–6775 (1988).
124. L. I. Barsukov, V. I. Volkova, Yu. E. Shapiro, A. V. Viktorov, V. F. Bystrov, and L. D. Bergelson, Selective adsorption of anions by phospholipids (in Russian), *Bioorg. Khimiya 3*:1355–1361 (1977).
125. G. L. Jendrasiak, Halide interaction with phospholipids: proton magnetic resonance studies, *Chem Phys. Lipids 9*:133–146 (1972).
126. S. McLaughlin, A. Bruder, S. Chen, and C. Moser, Chaotropic anions and the surface potential of bilayer membranes, *Biochim. Biophys. Acta 394*:304–313 (1975).
127. J. R. Rydall, and P. M. Macdonald, Investigation of anion binding to neutral lipid membranes using ^2H NMR, *Biochemistry 31*:1092–1099 (1992).
128. J. D. Lamb, J. J. Christensen, S. R. Izatt, K. Bedke, M. S. Astin, and R. M. Izatt, Effects of salt concentration and anion on the rate of carrier-facilitated transport of metal cations through bulk liquid membranes containing crown ethers, *J. Am. Chem. Soc. 102*:3399–3403 (1980).
129. M. E. Loosley-Millman, R. P. Rand, and V. A. Parsegian, Effects of monovalent ion binding and screening on measured electrostatic forces between charged phospholipid bilayers, *Biophys. J. 40*:221–232 (1982).
130. G. Cevc, The molecular mechanism of interaction between monovalent ions and polar surfaces, such as lipid bilayer membranes, *Chem. Phys. Lett. 170*:283–288 (1990).
131. Yu. E. Shapiro, A. V. Viktorov, L. I. Barsukov, L. D. Bergelson, Binding sites of ferrocyanide ions on the surface of lecithin liposomes (in Russian), *Biofizika 23*:727–728 (1978).
132. P. Schlieper and R. Steiner, The effect of different surface chemical groups on drug binding to liposomes, *Chem. Phys. Lipids 34*:81–92 (1983).
133. L. Chung, G. Kaloyanides, R. McDaniel, A. McLaughlin, S. McLaughlin, Interaction of gentamicin and spermine with bilayer membranes containing negatively charged pahospholipids, *Biochemistry 24*:442–452 (1985).
134. B. A. Levine, J. Sackett, and R. J. P. Williams, The binding of organic ions to phospholipid bilayers, *Biochim. Biophys. Acta 550*:201–211 (1979).
135. G. Beschiaschvili and J. Seelig, Melittin binding to mixed phosphatidylglycerol/phosphatidylcholine membranes, *Biochemistry 29*:52–58 (1990).
136. H.-A. Kolb and P. Läuger, Electrical noise from lipid bilayer membranes in the presence of hydrophobic ions, *J. Membrane Biol. 37*:321–345 (1977).
137. R. F. Flewelling and W. L. Hubbell, Hydrophobic ion interactions with membranes. Thermodynamic analysis of tetraphenylphosphonium binding to vesicles, *Biophys. J. 49*:531–540 (1986).

138. R. Gibrat, C. Romieu, and C. Grignon, A procedure for estimating the surface potential of charged or neutral membranes with 8-anilino-1-naphthalenesulfonate probe. Adequacy of the Gouy-Chapman model, *Biochim. Biophys. Acta 736*:196–202 (1983).
139. J. Tanaka, K. Akabori, and Y. Toyoshima, Effect of the membrane boundary potential on the phase transition of dipalmitoylphosphatidylcholine bilayer liposomes, *Mol. Cryst. Liq. Cryst. 74*:287–297 (1981).
140. P. G. Scherer and J. Seelig, Electric charge effects on phospholipid headgroups. Phosphatidylcholine in mixtures with cationic and anionic amphiphiles, *Biochemistry 28*:7720–7728 (1989).
141. H. Akutsu, and T. Nagamori, Conformational analysis of the polar head group in phosphatidylcholine bilayers: A structural change induced by cations, *Biochemistry* 30:4510–4516 (1991).
142. C. Altenbach and J. Seelig, Binding of the lipophilic cation tetraphenylphosphonium to phosphatidylcholine membranes, *Biochim. Biophys. Acta 818*:410–415 (1985).
143. S. McLaughlin and H. Harary, The hydrophobic adsorption of charged molecules to bilayer membranes: a test to the applicability of the Stern equation, *Biochemistry 15*:1941–1948 (1976).
144. A.D. Pickar and R. Benz, Transport of oppositely charged lipophilic probe ions in lipid bilayer membranes having various structures, *J. Membrane Biol. 44*:353–376 (1978).
145. G. Szabo, Dual mechanism for the action of cholesterol on membrane permeability, *Nature 252*:47–49 (1974).
146. E. Melnik, R. Latorre, J. E. Hall, and D. C. Tosteson, Phloretin-induced changes in ion transport across lipid bilayer membranes, *J. Gen. Physiol. 69*:243–257 (1977).
147. H. Träuble, Phasenumwandlungen in Lipiden: Mögliche Schaltprozesse in biologischen Membranen, *Naturwissenschaften 58*:277–284 (1971).
148. L. J. Bruner, The interaction of hydrophobic ions with lipid bilayer membranes. *J. Membrane Biol. 22*:125–141 (1975).
149. A. G. Lee, Effects of charged drugs on the phase transition temperatures of phospholipid bilayers, *Biochim. Biophys. Acta 514*:95–104 (1978).
150. L. E. G. Eriksson and J. Westman, Interaction of some charged amphiphilic drugs with phosphatidylcholine vesicles. A spin label study of surface potential effects, *Biophys. Chem. 13*:253–264 (1981).
151. N. Oku and R. C. MacDonald, Solubilization of phospholipids by chaotropic ion solutions, *J. Biol. Chem. 258*:8733–8738 (1983).
152. Yu. A. Ermakov, V. V. Cherny, S. A. Tatulian, and V. S. Sokolov, Boundary potentials on lipid membranes in the presence of 1-anilino-8-naphthalene sulfonate ions (in Russian), *Biofizika 28*:1010–1013 (1983).
153. P. Smejtek and S. Wang, Adsorption to dipalmitoylphosphatidylcholine membranes in gel and fluid state: Pentachlorophenolate, dipicrylamine, and tetraphenylborate, *Biophys. J. 58*:1285–1294 (1990).
154. J. Kim, M. Mosior, L. A. Chung, H. Wu, and S. McLaughlin, Binding of peptides with basic residues to membranes containing acidic phospholipids, *Biophys. J. 60*:135–148 (1991).
155. P. Woolley, The binding of a neutral aromatic molecule to a negatively charged lipid membrane. I. Thermodynamics and mode of binding, *Biophys. Chem. 10*:289–303 (1979).
156. H. Träuble, Membrane electrostatics, in *Structure of Biological Membranes* (S. Abrahamsson and I. Pascher, eds.), Plenum Press, New York, London, pp. 509–550 (1977).
157. S. J. Rehfeld, L. D. Hansen, E. A. Lewis, and D. J. Eatough, Alkaline earth cation binding to large and small bilayer phosphatidylserine vesicles. A calorimetric and potentiometric study, *Biochim. Biophys. Acta 691*:1–12 (1982).

15
Phosphoilpid Hydration

Thomas J. McIntosh and Alan D. Magid *Duke University Medical Center, Durham, North Carolina*

I. INTRODUCTION

Phospholipids are major components of cell and organelle membranes, blood lipoproteins, and lung surfactant. The interaction of phospholipids with water is critical to the formation, maintenance, and function of each of these important biological complexes. This chapter reviews some of the extensive literature on the hydration properties of various phospholipids. In Sec. II we begin with a brief description of the assembly of lipid aggregates in water and some of the factors responsible for the formation of the various hydrated phospholipid phases. Section III presents data on the amount of water imbibed by phospholipids as obtained from a variety of biophysical techniques. We emphasize the lamellar or bilayer phase, which is the structural backbone of biological membranes. In this section, the hydration properties of the most common membrane phospholipids, including zwitterionic and charged phospholipids, are compared, and the location of water in phospholipid bilayers and the structural effects of dehydration are detailed. Section IV presents an analysis of phospholipid hydration in terms of inter- and intrabilayer forces and the energetics of hydration of phospholipid bilayers. In particular, in this last section we discuss why the various classes of phospholipids have their specific hydration properties as described in Sec. III.

II. AGGREGATION OF PHOSPHOLIPIDS IN WATER

A. Classification

Polar lipids can be classified empirically according to their bulk behavior in water. Small (1986) distinguishes four groupings. Class I lipids (insoluble, nonswelling amphiphiles), exemplified by cholesterol and waxes, do not imbibe water at all. Class II lipids are insoluble (very low critical micelle concentration) but swell in water to form ordered liquid crystalline or mesomorphic phases ("mesophases"). To this class belong many important components of cellular membranes, such as long-chain (more than 8 carbons/chain) phosphatidylcholine, phosphatidylethanolamine, phosphatidylinositol, phosphatidylserine, and sphingomyelin. Class IIIA lipids are soluble, forming micelles above the critical micelle concentration but forming lyotropic liquid crystals at low water content. Class IIIA includes lysolecithins, diacyl phospholipids with short chains, and the salts of long-chain fatty acids. Class IIIB includes the soluble amphiphiles such as bile salts and saponins that form micelles but not smectic liquid crystals. The phospholipid members of classes II and IIIA concern us here.

B. Solubility of Phospholipids

The solubility or critical micelle concentration (cmc) depends on the free energy gained when an isolated amphiphile in solution enters an aggregate (Tanford, 1980). The cmc of diacyl phospholipids in water is, in general, quite low and depends on both the chain length and the head group type. For example, the measured cmcs of dipalmitoylphosphatidylcholine (16 carbons per chain) and didecanoylphosphatidylcholine (10 carbons per chain) are 4.6×10^{-10} M (Smith and Tanford, 1972) and 5×10^{-6} M (Reynolds et al., 1977), respectively. Gershfeld (1989) extended the thermodynamic analysis of aggregate formation to include the electrical work of aggregating a charged lipid such as phosphatidylglycerol. For a given chain length, the solubility of charged phospholipids is higher, as the solubility of dimyristoylphosphatidylglycerol (14 carbons per chain) in water is about 4×10^{-5} M (Gershfeld, 1989).

The cmc of a single-chain phospholipid is higher than that of a diacyl phospholipid with the same head group and the same chain length (Tanford, 1980). For example, the cmc of palmitoyl lysophosphatidylcholine is 7×10^{-6} M (Haberland and Reynolds, 1975).

C. Formation of Lipid Aggregates in Water

At concentrations above their cmc, phospholipids form a variety of ordered structures in water. In this section we briefly outline some of the factors that determine the phase properties of hydrated phospholipids. For a thorough recent review of this topic see Seddon (1990) and also Chap. 12 in this volume.

1. Hydrophobic Energy

Many of the current ideas about the self-organizing properties of polar lipids in an aqueous environment stem from G. S. Hartley's (1936) concept about the dual nature (amphipathy) of such molecules. They possess a hydrophobic moiety that has an antipathy toward water and a polar, hydrophilic moiety that has sympathy toward water. For amphipathic molecules the term amphiphile is more commonly used today. For phospholipids, the hydrophobic moiety generally consists of acyl chains, and the hydrophilic moiety contains a phosphoric-acid derivative. It is the low solubility of the acyl chains in water, combined with the strong hydrogen bonding between the water molecules, that furnishes the "attractive" force that holds together polar lipids into supramolecular complexes (the "hydrophobic bond").

The origin of the hydrophobic effect, first recognized by Hartley (1936), comes from the self-attraction of water rather than the self-attraction of hydrocarbon. The interfacial free energy of attraction of water for itself is what drives the elimination of water-hydrocarbon contact. The attraction of hydrocarbon for itself is about as strong as its attraction for water. In molecular terms, water is a highly structured liquid, probably with an ice-like tetrahedral hydrogen-bonded organization. *Any* solute will disrupt these bonds. However, new H bonds will form to a polar solute but cannot form to a nonpolar solute such as hydrocarbons. These broken bonds reform as a partly ordered cage of water molecules around hydrophobic solutes. When the hydrocarbon chains of an amphiphile make contact, their area of contact with water is reduced, thus reducing free energy (Tanford, 1980).

2. Amphiphile Shape and Aggregate Form

In the presence of water, phospholipids spontaneously organize into characteristic supramolecular structures, such as spherical, disklike, or cylindrical micelles, or into bilayers and vesicles. What forces determine the form that these ordered structures will take and which lipids will form them? These questions were first pursued in detail by Tartar (1955), who considered the case of micelle formation by paraffin chain salts. His analysis provides principles for phospholipids also. He adopted the view of Hartley that self-aggregation was driven thermodynamically by the hydrophobic effect. Tartar assumed that hydrocarbon chains were completely confined to the micelle interior (where it was assumed there was no water present) and were in a "liquid" or disordered state. The polar head groups array on the surface of the aggregate, held there by strong bonding to the water. One dimension of the aggregate cannot exceed the fully extended length of the hydrocarbon chain. This constraint leads to a sphere or oblate spheroid as the simplest shape that minimizes the surface area/volume ratio.

Tanford (1972, 1980) extended and refined Tartar's analysis in several ways. He incorporated the finding that one or more of the methylene groups of the amphiphile alkyl chains does not enter the hydrophobic core. Importantly, he pointed out that the value of area a per lipid molecule is lower for cylindrical micelles than for ellipsoidal micelles and lower for bilayers than any micellar form. He also noted that bilayers can accommodate a limitless number of chains without requiring a change in area/molecule. Bilayers often form closed fluid-filled vesicles; Tanford calculated that the molecular area in the inner and outer lamellae will differ only slightly from that of an infinite planar bilayer.

Another important insight concerns amphiphiles with two alkyl chains such as phospholipids. The head group area will be much the same whether one or two alkyl chains are anchored to the head group, but the area/chain will be half that of single-chain amphiphiles. Only a bilayer organization, which provides the smallest area/chain, will permit the optimal molecular area to be realized. This is why many two-chain amphiphiles form bilayers. Israelachvili et al. (1980) pointed out that single-chain amphiphiles *could* form bilayers as well but entropy favors the smaller aggregates, namely micelles. Notwithstanding this, bilayers from fatty acid/fatty soap mixtures have been observed.

Israelachvili et al. (1980) further developed Tanford's concept of two opposing interfacial forces, one tending to expand and the other to contract the interfacial area. The attractive (hydrophobic) force is given by γa, where γ is the interfacial free energy per unit area (Israelachvili, 1985). The repulsive forces include steric repulsion between lipid hydrocarbon chains and steric and electrostatic forces between head groups. (See also Nagle, 1980, for a detailed discussion of these interactions.) These repulsive forces are represented by Israelachvili (1980) as C/a, where C is a constant. Thus the interfacial energy per amphiphile molecule is given approximately as $\mu^{\circ}_{N} = \gamma a + C/a$. The a at which the free energy reaches a minimum defines an "optimal surface area" a_o at the water-hydrocarbon interface, where $a_o = \sqrt{C/\gamma}$. Recent x-ray difffaction experiments with phospholipids solvated with nonaqueous polar solvents have given direct evidence for the sensitivity of a to the surface tension of the solvent (McIntosh et al., 1989b).

According to Israelachvili et al. (1980), the geometrical packing properties of lipids will be set by a_o, the volume v of the hydrocarbon chains, and the "critical" chain length l_c. The critical length sets a limit on how far the hydrocarbon chain can extend; for liquid crystalline lipids it is expected to be less than the fully extended length of the chain (Israelachvili et al., 1980). These quantities define a dimensionless "critical packing parameter" $v/a_o l_c$ that determines the shape of assembly that an amphiphile will form. These predicted structures are those in which *each* lipid molecule will have a minimum $\mu^o{}_N$, i.e., when $a = a_o$. However, a smaller structure with a higher $\mu^o{}_N$ may be thermodynamically favored due to entropy. Israelachvili et al. (1980) argue that if $v/a_o l_c$ is $< 1/3$ (corresponding to cone-shaped molecules), spherical micelles will form; if $v/a_o l_c$ lies between 1/3 and 1/2 (corresponding to truncated cones), ellipsoidal or cylindrical micelles will form; if $v/a_o l_c$ lies between 1/2 and 1 (corresponding to cylindrical molecules), bilayers or closed vesicles will form; if $v/a_o l_c > 1$ (corresponding to relatively small head group areas), inverted micelles or an inverted hexagonal phase will form. Most diacyl phospholipids are approximately cylindrical in shape, so they form bilayers. However, some phospholipids, such as unsaturated phosphatidylethanolamines, tend to form hexagonal phases, presumably since they have relatively small head groups compared to the excluded volume of their hydrocarbon chain region.

Gruner (1989) has noted that the term $v/a_o l_c$ is not simply a measure of molecular shape but also reflects factors such as head group charge and hydrogen bonding. In addition, Seddon (1990) has pointed out that although the Tanford and Israelachvili approach is quite useful, it does not predict the formation of the cubic phase, a mesophase observed in the phase diagrams of many polar lipids (Lindbolm and Rilfors, 1989), and it neglects the "geometry dependence" of the hydrocarbon chains to the total free energy of the system. That is, according to Seddon (see Seddon, 1990, and references therein), the principal factor that drives the transition from bilayer to nonbilayer phases is the tendency for one or both monolayers of the bilayer to curl. This curling can arise from an imbalance in lateral pressures between the head group region and the hydrocarbon region of each monolayer of the bilayer. That is, if the lateral pressure in the head group region differs from that in the hydrocarbon chain region, the monolayer has a tendency to curl. In particular, Gruner (1989) explained transitions between mesomorphic phases with curved interfaces in terms of a "competition between the elastic energy of bending the interfaces and energies resulting from the constraints of interfacial separation." The bending energy of a thin lipid film depends on its "spontaneous curvature" (Helfrich, 1973), which results if the in-plane forces are balanced for an area that is larger at the head group than at the lipid tail of the molecule (Gruner, 1989). Gruner (1985) and Hui (1987) have advanced the view that the presence in cellular membranes of non-bilayer-forming lipids adjusts the membrane's intrinsic curvature so as to facilitate membrane functions (e.g., membrane fusion, exocytosis) that are topologically forbidden with intact bilayers.

III. UPTAKE OF WATER BY PHOSPHOLIPIDS

The hydration of phospholipids has been studied by a number of methods. The amount of water absorbed by phospholipids from the vapor phase has been directly measured gravimetrically (Elworthy, 1961; Jendrasiak and Hasty, 1974; Lundberg, 1974), whereas in lipid-water dispersions the amount of water imbibed by phospholipid multilayers can be inferred from x-ray diffraction measurements of the lamellar repeat period as a function of water content (Reiss-Husson, 1967; Luzzati et al., 1968; Tardieu et al., 1973; Lundberg,

1974; Franks, 1976; LeNeveu et al., 1977; Janiak et al., 1979; McIntosh and Simon, 1986a). In these studies of dispersions of electrically neutral phospholipids (see below and Fig. 1), it is assumed that the repeat period reaches its limiting value when the lipid has imbibed all the water that it can, and any additional water forms an excess fluid phase. This formation of an excess water phase can be confirmed directly with light microscopy (Bourges et al., 1967). The location of water in phospholipid multilayers has been obtained by neutron diffraction (Zaccai et al., 1975; Worcester and Franks, 1976; Büldt et al., 1978; Büldt et al., 1979) and by the combination of x-ray diffraction and specific capacitance data (Simon et al., 1982). The modification caused by hydration (and dehydration) on the thermal and structural properties of phospholipid bilayers has been investigated by nuclear magnetic resonance (Finer and Darke, 1974; Taylor et al., 1977), electron spin resonance (Griffith et al., 1974; Korstanje et al., 1989), infrared spectroscopy (Finer and Darke, 1974; Taylor et al., 1977), differential scanning calorimetry (Janiak et al., 1979; Kodama et al., 1982; Ter-Minassian-Saraga and Madelmont, 1982; Cevc and Marsh, 1985), dielectric dispersion measurements (Nimtz et al., 1985), as well as by x-ray and neutron diffraction. From the x-ray diffraction experiments, changes in bilayer structure have been analyzed in two different ways. In one method, it is assumed that lipid and water form separate layers, and by measuring the relative volumes of water and lipid in the specimen, one can divide the lamellar repeat period into partial lipid and water thicknesses (Luzzati, 1968), as illustrated in Fig. 1. By this procedure, the area *a*

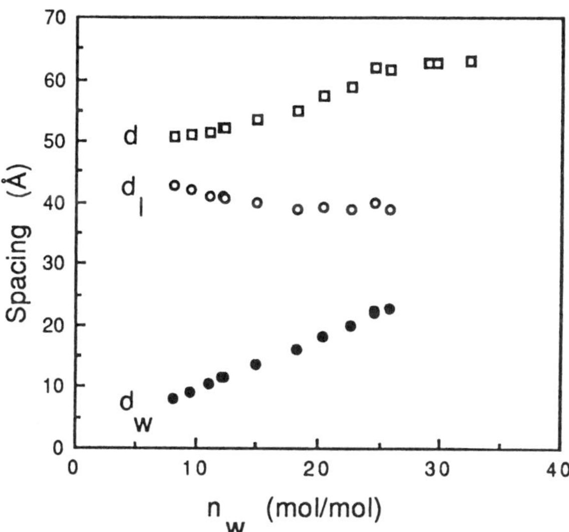

Figure 1 A plot of lamellar repeat period (d), partial lipid thickness (d_l), and partial water thickness (d_w) versus number of water molecules per lipid molecule (n_w) for E-PtdCho bilayers. Data are taken from McIntosh et al. (1989). Note that the lamellar repeat period increases with increasing water content from about 8 to 24 water molecules per lipid molecule. For water contents above about 24 waters per lipid molecule, the repeat period remains nearly constant as an excess water phase forms. The partial fluid thickness increases with increasing water, whereas the partial lipid thickness decreases slightly (about 2 Å) as the water content is increased from 8 to 24 waters per lipid molecule.

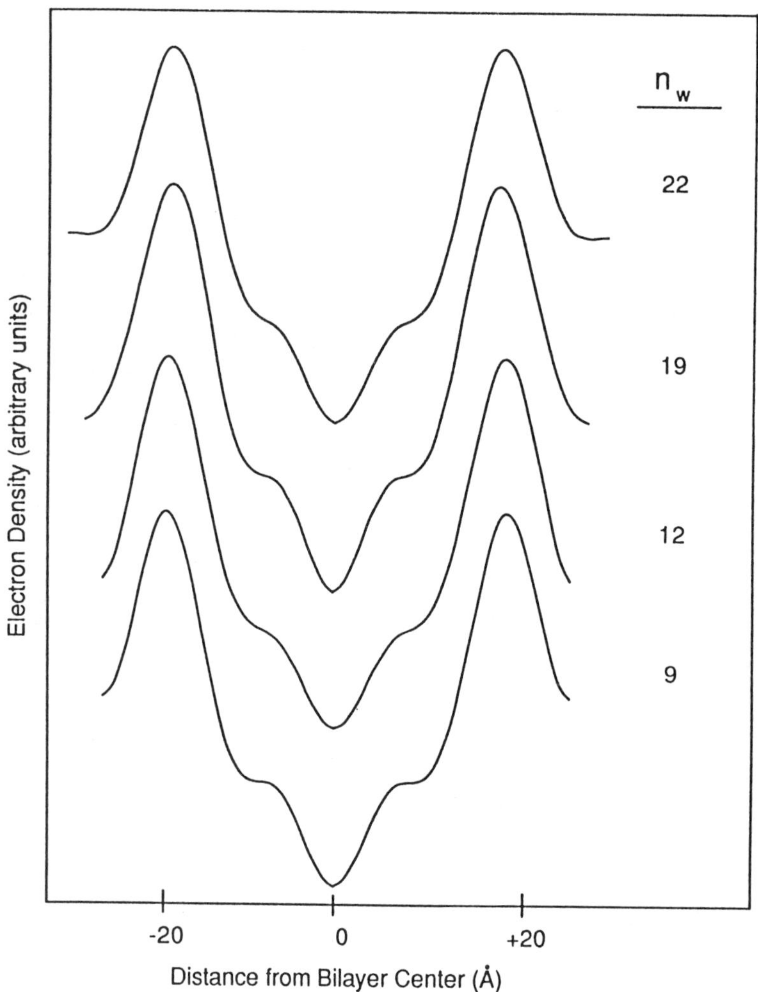

Figure 2 Electron density profiles of E-PtdCho bilayers under applied osmotic stress. The numbers on the right hand side of the figure indicate the approximate number of water molecules per lipid molecule, as determined from the data in McIntosh et al. (1987). The profiles have been offset in the vertical direction. For each profile, the high density peaks centered at about ± 19 Å correspond to the electron dense phospholipid head groups, and the low density trough in the geometric center of the bilayer corresponds to the localization of the lipid terminal methyl groups. The medium density regions between the terminal methyl trough and the head group peaks correspond to the lipid methylene chains, and the medium density regions at the outer edges of each profile correspond to the fluid spaces between adjacent bilayers. Note how the fluid spaces between adjacent bilayers decrease in width with decreasing water content, but the width of the bilayer (as measured by the distance between head group peaks across the bilayer) stays almost constant for this range of water contents.

per lipid molecule can also be obtained (see Table 1), since the area per molecule is inversely proportional to the partial lipid thickness (Luzzati, 1968). In the second method (Franks, 1976; Janiak et al., 1979; McIntosh and Simon, 1986a,b), the thickness of the lipid bilayer can be estimated directly from electron density profiles (see below and Fig. 2).

Here we summarize the results of a number of these experiments. Section A presents data on the amount of water taken up by various phospholipids (see Table 1), Sec. B summarizes recent work on the location of water in lamellar phases, and Sec. C describes some of the effects that may arise because of lipid dehydration. Analysis of these results is deferred to Sec. IV.

A. Amount of Water in Phospholipid Mesophases

1. Lamellar (Bilayer) Phases

Phospholipids are a major component of the bilayer core of most cell membranes. The phospholipids in these membranes, under physiological conditions, fall in two major classes—zwitterionic and charged. At low pH, some of the charged phospholipids may become electrically neutral.

(a) Zwitterionic phospholipids. The two most common zwitterionic membrane phospholipids are phosphatidylcholine (PtdCho) and phosphatidylethanolamine (PtdEtn). Thus the structure and properties (including hydration) of phosphatidylcholine and phosphatidylethanolamine bilayers have been analyzed extensively.

Egg-yolk phosphatidylcholine (E-PtdCho) has been thoroughly studied, since it is naturally occurring, readily available, and is in the physiologically relevant liquid crystalline bilayer (L_α) phase over a broad temperature range, including room temperature. For E-PtdCho bilayers, the adsorption isotherms of water vapor by Elworthy (1961), Jendrasiak and Hasty (1974), and Lundberg (1974) are in general agreement at low relative humidities but differ at high relative humidities. That is, it was found that the amount of water adsorbed by E-PtdCho from the vapor phase increases monotonically from 0 water molecules per E-PtdCho molecule at 0% humidity to between 14 (Jendrasiak and Hasty, 1974) and 20 (Elworthy, 1961) waters per lipid molecule at 100% relative humidity. The basis for these observed differences at the high relative humidities is unknown, although it should be noted that it is very difficult to control accurately relative humitidies near 100%, in part due to temperature gradients in the system. Moreover, as discussed in Sec. IV.C, the hydration energies for the last few waters of hydration are quite small compared to thermal energy. For example, as noted by Rand and Parsegian (1989), a relative humidity of 99.9% corresponds to an osmotic pressure of 10^5 N/m^2, which is enough applied pressure to remove about one-third of the water from E-PtdCho multilayers.

X-ray diffraction studies indicate that E-Ptd Cho imbibes considerably more water when directly mixed with bulk water than it does from the vapor phase. As shown in Table 1, estimated values for the maximum number n_w of water molecules per lipid molecule imbibed by E-PtdCho before an excess phase forms range from $n_w = 24$ (McIntosh et al., 1989b) (see Fig. 1) to $n_w = 34$ (LeNeveu et al., 1977; Small, 1967). There are several possible reasons for this large range of estimates for the maximum n_w. One reason, as mentioned above and in Sec. IV.C, is that the energies required for imbibing the last few water molecules are small compared to thermal energy. This means that, in a plot of lamellar repeat period versus water content, the break where an excess water phase forms is gradual rather than sharp. Thus by this technique it is difficult to determine precisely the

Table 1 Maximum Water Imbibed by Bilayers of Zwitterionic Phospholipids

Lipid	Phase	n_w	a (Å^2)	Technique	Reference
Phosphatidylcholine E-PtdCho	L_α	25	67	XRD	Reiss-Husson (1967)
Phosphatidylcholine E-PtdCho	L_α	24	64	XRD	McIntosh et al. (1989)
Phosphatidylcholine E-PtdCho	L_α	≤27	64	XRD	Klose et al. (1988)
Phosphatidylcholine E-PtdCho	L_α	≥23	—	NMR	Finer and Darke (1974)
Phosphatidylcholine E-PtdCho	L_α	34	72	XRD	Small (1967)
Phosphatidylcholine E-PtdCho	L_α	35	75	XRD	Parsegian et al.(1979)
Lau_2-PtdCho	L_α	25	62	XRD	Janiak et al. (1979)
Lau_2-PtdCho	L_α	30	65	XRD	Lis et al. (1982)
Lau_2-PtdCho	P_β'	15	—	XRD	Janiak et al. (1979)
Lau_2-PtdCho	L_β'	15	—	XRD	Janiak et al. (1979)
Pam_2-PtdCho	L_α	27	—	XRD	Janiak et al. (1979)
Pam_2-PtdCho	L_α	27	—	XRD	Albon (1983)
Pam_2-PtdCho	L_α	25	(64)	XRD	Ruocco and Shipley (1982a,b)
Pam_2-PtdCho	L_α	23	(57)	XRD	Inoko and Mitsui (1978)
Pam_2-PtdCho	L_α	39	71	XRD	Lis et al. (1982
Pam_2-PtdCho	P_β'	15	—	XRD	Janiak et al. (1979)
Pam_2-PtdCho	P_β'	14	—	DSC	Kodama et al. (1982)

Pam_2-PtdCho	L_{β}'	15	—	XRD	Janiak et al. (1979)
Pam_2-PtdCho	L_{β}'	14	—	XRD	Albon (1983)
Pam_2-PtdCho	L_{β}'	13	—	DSC	Ter-Minassian-Saraga et al. (1982)
Pam_2-PtdCho	L_{β}'	8	—	DSC	Kodama et al. (1982)
Pam_2-PtdCho	L_{β}'	19	(52)	XRD	Ruocco and Shipley (1982a,b)
Pam_2-PtdCho	L_{β}'	17	52	XRD	Lis et al. (1982)
Pam_2-PtdCho	L_{β}'	11	46	XRD	Weiner et al. (1989)
Pam_2-PtdCho	L_{β}'	15	(50)	XRD	Inoko and Mitsui (1978)
Pam_2-PtdCho	L_{β}'	14	49	NB & XRD	Nagle and Wiener (1988)
Pam_2-PtdCho	Lc	11	—	XRD	Ruocco and Shipley (1982a,b)
Pam_2-PtdCho	Lc	9	46	NB & XRD	Nagle and Wiener (1988)
Hex_2-PtdCho	L_{α}	22	61	XRD	Kim et al. (1987)
Hex_2-PtdCho	$L_{\beta I}$	24	76	XRD	Kim et al. (1987)
Ole_2-PtdCho	L_{α}	30	70	XRD	Gruner et al. (1988)
Ole_2-PtdCho	L_{α}	44	82	XRD	Lis et al. (1982)
Ste,Ole-PtdCho	L_{α}	28	64	XRD	Rand et al. (1988)
Phosphatidylethanolamine E-PtdEtn	L_{α}	11	—	NMR	Finer and Darke (1974)
Phosphatidylethanolamine E-PtdEtn	L_{α}	26	75	XRD	Lis et al. (1982)
Phosphatidylethanolamine E-PtdEtn	L_{α}	26	72	XRD	Rand et al. (1988)
Lau_2-PtdEtn	L_{α}	10	51	XRD	McIntosh & Simon (1986a,b,c)
Lau_2-PtdEtn	L_{α}	9	51	NB & XRD	Nagle and Wiener (1988)
Lau_2-PtdEtn	L_{β}	7	41	DTD & XRD	McIntosh et al. (1986)

Table 1 (Continued)

Lipid	Phase	n_w	a (Å^2)	Technique	Reference
Lau_2-PtdEtn	L_β	6	41	NB & XRD	Nagle and Wiener (1988)
Lau_2-PtdEtn, Myr_2-PtdEtn	L_β	7	—	DSC	Cevc and Marsh (1985)
Pam_2-PtdEtn, Ste_2-PtdEtn	L_β	7	—	DSC	Cevc and Marsh (1985)
Pam_2-PtdEtn, Ste_2-PtdEtn	L_α	11	—	DSC & XRD	Cevc and Marsh (1985)
Pam,OLE-PtdEtn	L_α	11	57	XRD	Rand et al. (1988)
Ole_2-PtdEtn	L_α	18	65	XRD	Gruner et al. (1988)
Didocedcyl-PtdEtn	L_α	12	55	XRD	Seddon et al. (1984)
Didocedcyl-PtdEtn	L_β	6	42	XRD	Seddon et al. (1984)
Diararchioyl-PtdEtn	L_α	9	58	XRD	Seddon et al. (1984)
Diararchioyl-PtdEtn	L_β'	9	47	XRD	Seddon et al. (1984)

This table shows the number n_w of water molecules per lipid molecule and the area a per lipid molecule as measured by several techniques (XRD = x-ray diffraction; NMR = nuclear magnetic resonance; DTD = differential thermal dilatometry; DSC = differential scanning calorimetry; NB = neutral buoyancy measurements of absolute specific volumes). See text for abbreviations for lipids and lipid phases. Values of a in parentheses were calculated using values of n_w and lamellar repeat period given in each reference and lipid specific volumes from Inoko and Mitsui (1978).

water content where an excess water phase forms. A second possible reason has been discussed in detail recently by Klose et al. (1988), who argue that the morphology and method of sample preparation strongly influence the determination of the maximum water content of the lipid due to the formation of defects in and between the multilayers. These defects can contain additional water molecules that are not incorporated into the ordered lamellar phase and thus are not detected by measurements of lamellar repeat period versus water content. Klose et al. (1988) conclude that the real swelling capacity of E-PtdCho is less than or equal to 27 water molecules per lipid molecule. These x-ray diffraction results can be compared to the deuterium NMR results of Finer and Darke (1974), who found that, for E-PtdCho dispersions, bulk water was present when there were 23 or more water molecules per lipid molecule (see Table 1).

Uptake of water by synthetic saturated phosphatiylcholines has also been studied extensively. These systems have the advantages that the chemical composition of the sample is well defined and sharp thermal phase transitions exist, so that the hydration properties of both gel and liquid crystalline phases can be studied by adsorption isotherms and x-ray diffraction methods. Moreover, differential scanning calorimetry can also be used to estimate the number of water molecules imbibed by the phospholipid by the analysis of the thermal transition as a function of water content (Kodama et al., 1982; Ter-Minassian-Saraga and Madelmont, 1982) (see Table 1).

Adsorption isotherms show that gel phase dipalmitoylphosphatidylcholine (Pam_2-PtdCho) takes up considerably less water from the vapor phase than does liquid crystalline E-PtdCho. Estimates for the water adsorbed at 100% relative humdity range from 9 (Jendrasiak and Hasty, 1974) to 11 (Elworthy, 1962) water molecules per lipid molecule.

In terms of dispersions of saturated phosphatidylcholines, one of the most thorough calorimetric and x-ray diffraction investigations was performed by Janiak et al. (1979). They found that dimyristoylphosphatidylcholine (Myr_2-PtdCho) in the liquid-crystalline (L_α) phase imbibes a maximum of about 25 water molecules per lipid molecule before an excess water phase forms, whereas Myr_2-PtdCho in either the tilted gel ($L_{\beta'}$) phase or the rippled gel ($P_{\beta'}$) phase swells to the extent of only about 15 water molecules per lipid. As shown in Table 1, similar values of n_w have been obtained for other saturated phosphatidylcholines, including Pam_2-PtdCho (Inoko and Mitsui, 1978; Janiak et al., 1979; Albon, 1983). One recent study by Wiener et al. (1989) used low-angle and wide-angle x-ray diffraction data, along with lipid volumes obtained from absolute specific volume measurements, and found that gel phase Pam_2-PtdCho imbibes about 11 water molecules per lipid molecule (see Table 1). This study concluded that previous analyses had over-estimated the amount of water imbibed by gel phase Pam_2-PtdCho because of imperfections in the multilayers, such as those that occur in the centers of liposomes or where liposomes impinge upon one another. Wiener et al. (1989) argue that such regions will likely have a larger water-to-lipid ratio compared to a perfectly oriented system and would therefore make the number of water molcules imbibed appear larger in measurements of repeat period versus water content. This argument is similar to that used by Klose et al. (1988) to explain discrepancies in measurements of water content for E-PtdCho (see above).

Another lipid that has been extensively studied is dihexadecylphosphatidylcholine (Hex_2-PtdCho), the ether-linked analog to Pam_2-PtdCho (Kim et al., 1987). Although the limiting values of water are similar for Pam_2-PtdCho and Hex_2-PtdCho in the liquid crystalline phase (Kim et al., 1987), the gel phase of Hex_2-PtdCho is interesting because the structure of the bilayer changes dramatically as a function of water content. That is, at

20°C and for water contents ranging from about 4 to 14 water molecules per lipid, Hex_2-PtdCho exhibits a normal gel ($L_{\beta'}$) bilayer structure with two hydrocarbon chains per polar group at the interface. However, for water contents from about 17 to 24 water molecules per lipid, Hex_2-PtdCho forms a chain-interdigitated gel phase ($L_{\beta I}$) where hydrocarbon chains from apposing monolayers interpenetrate up to the hydrocarbon/water interface so that there are four chains per polar group at the interface (Kim et al., 1987; Laggner et al., 1987). An excess water phase forms for water contents greater than 24 water molecules per lipid for the interdigitated gel ($L_{\beta I}$) phase of Hex_2-PtdCho, which is considerably more water than is imbibed by the normal gel ($L_{\beta'}$) phase of Pam_2-PtdCho (see Table 1).

Water uptake by synthetic diacyl phosphatidylcholines with mixed acyl chain lengths has also been studied. Of particular interest are lipids with one chain much longer than the other. For instance, it has been shown that a phosphatidylcholine with 18 carbon atoms in one chain and 10 carbons in the other forms a usual gel phase (L_{β}) bilayer at low water contents, but at high water contents converts to an unusual partially interdigitated phase where the long chain spans the entire width of the hydrocarbon region and the short chain abuts with the short chain from the apposing monolayer (McIntosh et al., 1984). Thus this partially interdigitated phase has three hydrocarbon chains per polar group at the interface (Hui et al., 1984; McIntosh et al., 1984). Other mixed-chain phosphatidylcholines also form this unusual partially interdigitated phase (Mattai et al., 1987a), which imbibes about 25 water molecules before an excess phase forms (McIntosh et al., 1984; Mattai et al., 1987a).

Another unusual bilayer structure formed by synthetic phosphatidylcholines is the "subgel" (L_c) phase, characterized several years ago by calorimetry (Chen et al., 1980) and x-ray diffraction (Ruocco and Shipley, 1982a,b). This phase has a crystalline packing of the lipid hydrocarbon chains and imbibes about 10 water molecules per lipid molecule (Ruocco and Shipley, 1982b; Nagle and Wiener, 1988), which is less water than is imbibed by the normal gel ($L_{\beta'}$) phase or the liquid crystalline (L_{α}) phase (see Table 1).

Divalent cations can modify the hydration properties of phosphatidylcholine bilayers. The fluid separation between apposing phosphatidylcholine bilayers has been shown to depend on the concentration and particular divalent cation present in the aqueous solution (Lis et al., 1981a,b). For example, the partial fluid thickness between Pam_2-PtdCho bilayers increases from about 20 Å in water to over 90 Å in 1 mM $CaCl_2$ (Lis et al., 1981b).

The hydration properties of both natural and synthetic phosphatidylethanolamine (PE) have also been systematically studied, although not to the extent of the phosphatidylcholines. One significant finding is that phosphatidylethanolamines, in general, imbibe considerably less water than do the phosphatidylcholines. From the saturated vapor phase, liquid crystalline egg-yolk phosphatidylethanolamine (E-PtdEtn) adsorbs about 10 water molecules per lipid molecule (Jendrasiak and Hasty, 1974) in comparison to the 14 (Jendrasiak and Hasty, 1974) to 20 (Elworthy, 1961) water molecules adsorbed by E-PtdCho under similar experimental conditions, and synthetic gel phase dipalmitoylphosphatidylethanolamine (Pam_2-PtdEtn) adsorbs about 4 water molecules per lipid at 100% relative humidity compared to 11 water molecules adsorbed by Pam_2-PtdCho under the same conditions (Elworthy, 1962).

Compared to phosphatidylcholines, phosphatidylethanolamines also imbibe less water from the bulk aqueous phase. For example, the NMR results of Finer and Darke (1974) indicate that E-PtdEtn takes up a maximum of about 11 water molecules per lipid in comparison to the 23 water molecules imbibed by E-PtdCho, and the lamellar x-ray repeat periods determined by Gruner et al. (1988) reach limiting values at about 18 and 30 water molecules per lipid for dioleoylPE and dioleoyl-PtdCho, respectively (see Table 1). Comparisons of the hydration of gel and liquid crystalline phosphatidylethanolamines have been made by two different types of analysis of x-ray diffraction data. By the standard gravimetric method of measuring repeat period versus water content (see above), Seddon et al. (1984) found that didodecyl-PtdEtn imbibes about 6 and 12 water molecules per lipid in the gel and liquid crystalline phases, respectively, and diarachinoyl-PtdEtn imbibes about 9 water molecules per lipid in both the gel and liquid crystalline phases (see Table 1). Other saturated phosphatidylethanolamines with chain lengths between C_{12} and C_{18} also bind approximately 7 and 11 water molecules in the gel and fluid bilayers, respectively, as concluded from an extensive calorimetric investigation (Cevc and Marsh, 1985). By combining information from wide-angle x-ray diffraction, electron density profiles, and published dilatometry data (Wilkinson and Nagle, 1981), McIntosh and Simon (1986a,b) found that dilauroylphosphatidylethanolamine (Lau_2-PtdEtn) imbibes about 7 and 10 water molecules per lipid molecule in the gel and liquid crystalline phases, respectively. In a similar manner, Nagle and Wiener (1988) used absolute volumes of multilamellar lipid dispersions and diffraction data to calculate that Lau_2-PtdEtn imbibes about 6 and 9 water molecules per lipid molecule in the gel and liquid crystalline phases, respectively.

(b) Charged phospholipids. Several membrane phospholipids carry a net negative charge at physiological conditions, namely phosphatidylinositol (PtdIns), phosphatidylglycerol (PtdGro), and phosphatidylserine (PtdSer). Adsorption isotherms indicate that bovine PtdSer takes up only about 6 water molecules per lipid molecule from the vapor phase at 100% relative humidity (Jendrasiak and Hasty, 1974). However, x-ray diffraction experiments (Atkinson et al., 1974; Cowley et al., 1978; Sugiura, 1981; Hauser et al., 1982; Loosley-Millman et al., 1982; McDaniel, 1988) show that multilayers containing PtdSer, PtdGro, or PtdIns swell indefinitely in bulk water, and NMR measurements (Finer and Darke, 1974) indicate that sodium PtdSer multilayers can "trap" over 120 water molecules per lipid.

The presence of monovalent and/or divalent cations in the fluid phase changes the hydration properties of the charged phospholipids. For most monovalent cations, such as Na^+, K^+, or Cs^+, the fluid spaces between adjacent charged PtdSer or PtdGro bilayers decrease with increasing salt concentration, due to screening of the charge (Loosley-Millman et al., 1982; Hauser and Shipley, 1983). Addition of up to 1 M of NaCl, KCl, or CsCl produces little or no change in the bilayer structure (Hauser and Shipley, 1983). In contrast, the addition of 1 M LiCl produces almost total dehydration and a crystallization of the lipid hydrocarbon chains for multilayers composed of a series of synthetic diacylphosphatidylserines (Hauser and Shipley, 1983; Cevc et al., 1985). (This effect was not observed with PtdGro bilayers (Loosely-Millman et al., 1982), even though recent electrophoretic mobility and ionizing electrode measurements indicate that the electrostatic potential is the same adjacent to PtdSer, PtdGro, or PtdIns bilayers (Langner et al., 1990).)

The addition of divalent cations, moreover, has a dehydrating effect on PtdSer multilayers. The most extensively studied divalent cation is Ca^{2+}. Calcium binds to the phosphate group of PtdSer (Hauser et al., 1977), liberates water from between bilayers and from the lipid polar groups (Hauser et al., 1977), crystallizes the lipid hydrocarbon chains (Hauser et al., 1977; Portis et al., 1979; Hauser and Shipley, 1984), and raises the gel to the liquid crystalline melting temperature of dipalmitoylphosphatidylserine (Pam_2-PtdSer) by over 100°C (Hauser and Shipley, 1984). Calcium ion binding to PtdSer bilayers gives rise to a phase with the composition $Ca(PtdSer)_2$ (Feigenson, 1985). Mg^{2+} also dehydrates Pam_2-PtdSer and causes crystallization of the lipid hydrocarbon chains (Hauser and Shipley, 1984). However, Mg^{2+} raises the melting temperature of Pam_2-PtdSer by about 40°C, or not nearly as much as Ca^{2+} does (Hauser and Shipley, 1984). Sr^{2+} and Ba^{2+} compress the lamellar lipid lattice by dehydrating the PtdSer bilayers, but they do not crystallize the lipid hydrocarbon chains and have relatively modest effects on the melting temperature, raising it about 20°C (Hauser and Shipley, 1984).

2. Non-Lamellar Phases

Several phospholipids, including lysophosphatidylcholines and certain phosphatidylethanolamines, form cubic, hexagonal, and/or micellar phases, depending on temperature, degree of hydrocarbon chain unsaturation, and water content.

First, let us consider the phase behavior and hydration properties of lysophosphatidylcholine. Above its hydrocarbon chain melting temperature of about 35°C, egg-yolk lysophosphatidylcholine is in an hexagonal phase for water contents below about 55% and converts to a micellar phase at higher water contents (Reiss-Husson, 1967). The phase behavior as a function of hydration for synthetic lysophosphatidylcholines is quite complex and has been investigated in depth by Arvidson et al. (1985). For instance, for one synthetic lysophosphatidylcholine, 1-palmitoyl-*sn*-glycero-3-phosphocholine, hydrated crystals are observed at low water contents (below about 10% water by weight), a lamellar phase is observed at water contents between about 10 and 20%, an hexagonal phase forms between 20 and 52% water, a cubic phase is observed between 54 and 61% water, and a micellar phase is observed for water contents greater than about 63%. This study emphasizes the subtle interplay that exists between aggregate structure and amphiphile concentration.

Although, as described above, several synthetic and naturally occurring phosphatidylethanolamines form hydrated lamellar phases at room temperature (Lis et al., 1982; McIntosh and Simon, 1986a,c; Rand and Parsegian, 1989) many of these phases are metastable (Seddon et al., 1983); moreover highly unsaturated phosphatidylethanolamines can form inverted hexagonal and other nonlamellar phases for a broad range of water contents (Seddon et al, 1984). The analysis of Israelachvili et al. (1980) presented in Sec. II.C.2 above would predict that the hexagonal phase would be expected for highly unsaturated PtdEtn since the double bonds increase the effective volume of the hydrocarbon chains. Reiss-Husson (1967) has measured the structural parameters for the hexagonal phase of egg-PtdEtn for water contents ranging from 5 to 20%, and Gruner et al. (1986) have measured the work to remove water from the hexagonal phase of dioleoylphosphatidylethanolamine.

B. Location of Water in Lamellar Phases

The depth to which water penetrates into the lipid bilayer is critical to the organization and properties of both phospholipid and biological membranes (Simon and McIntosh, 1986). For example, the partition coefficient of nonelectrolytes into the bilayer is closely related to the depth of water penetration (Diamond and Katz, 1974). The processes of molecular absorption to the membrane and transport across the membrane both involve the displacement of interfacial water (Dix et al., 1978). In addition, the manner in which extrinsic and intrinsic membrane proteins associate with the bilayer depends on the location of the hydrocarbon/water interface (Tanford, 1980; Engelman and Steitz, 1981; Cevc et al., 1990).

Extensive work determining the depth of water penetration into phospholipid bilayers has used neutron diffraction techniques. Since deuterium has a much larger neutron scattering length than hydrogen, neutron diffraction experiments on lipid bilayers formed in various ratios of D_2O and H_2O can be used to localize water in lipid multilayers (Schoenborn, 1975). Neutron diffraction experiments on multilayers containing phosphatidylcholine (Zaccai et al., 1975; Worcester and Franks, 1976; Büldt et al., 1979), phosphatidylethanolamine (Simon and McIntosh, 1986), and phosphatidylinositol (McDaniel and McIntosh, 1989) have obtained water distributions centered between adjacent bilayers and overlap the head group peaks in the neutron scattering profile of the bilayer. These results imply that water penetrates into the bilayer head group region, but that there are not appreciable quantities of water in the hydrocarbon core. That is, although water can diffuse across the bilayer, the neutron diffraction results indicate that there are not large aggregates of water in the hydrocarbon region.

This conclusion has been supported by studies that combine x-ray diffraction and specific capacitance data. Simon and colleagues (Simon et al., 1982; Simon and McIntosh, 1986) showed that water penetrates to the deeper carbonyl group in PtdEtn bilayers. An excellent match between the structural and capacitance data was obtained when a dielectric constant of 2.1 was chosen for the bilayer hydrocarbon core. It was also shown that the incorporation of cholesterol in these bilayers decreases the depth of water penetration into the bilayer (Simon et al., 1982; Simon and McIntosh, 1986; McIntosh et al., 1989).

By combining x-ray diffraction and dilatometry data, McIntosh and Simon (1986a,b) were able to calculate the number of water molecules located in the interbilayer space and in the head group region for Lau_2-PtdEtn bilayers. They found that there are about 7 and 10 waters per Lau_2-PtdEtn molecule in the gel and liquid crystalline phases, respectively, with about half of these water molecules located between adjacent bilayers and the other half in the head group region. Using a similar procedure, Nagle and Wiener (1988) calculated 2 and 5 waters per lipid molecule in the head group region of Lau_2-PtdEtn in the gel and liquid crystalline phases, respectively, and about 4 waters per lipid molecule in the fluid spaces between Lau_2-PtdEtn bilayers in both phases. By using x-ray diffraction data and calculated densities for the lipid hydrocarbon chain and head group regions of the bilayer, Small (1967) calculated that fully hydrated E-PtdCho contains about 12 and 22 water molecules per lipid molecule in the head group region of the bilayer and in the water space between lipid bilayers, respectively. Although these numbers for E-PtdCho bilayers may be somewhat high (see the discussion above), they do show that the fluid space between E-PtdCho bilayers contains considerably more water than does the fluid space between Lau_2-PtdEtn bilayers.

The location of the first two waters of hydration has been determined from x-ray structure analysis of single crystals of Myr_2-PtdCho dihydrate (Pearson and Pascher, 1979). These two water molecules are located in the lipid head group region, near the phosphate oxygens.

C. Effects of Dehydration on the Structural and Thermal Properties of Lamellar Phases

Phospholipid bilayers must become partially dehydrated (the waters between adjacent bilayers must be removed) as a necessary step in the fusion of phospholipid membranes. A number of studies have determined the structure of the bilayer (the bilayer width and/or area per lipid molecule) and the gel to liquid crystalline phase transition temperature as a function of hydration. The pioneering work of Small (1967) indicated that the partial lipid thickness of E-PtdCho bilayers *increases* upon partial dehydration. Subsequent studies with E-PtdCho (Reiss-Husson, 1967; Torbet and Wilkins, 1976; LeNeveu et al., 1977; Klose et al., 1988; McIntosh et al., 1989b; McIntosh and Simon, 1986c) and with synthetic phosphatidylcholines (Tardieu et al., 1973; Torbet and Wilkins, 1976; Janiak et al., 1979; McIntosh et al., 1984; McIntosh and Simon, 1986c; McIntosh et al., 1987; Mattai et al., 1987b) have also found that the bilayer thickness increases (and area per lipid molecule decreases) upon bilayer dehydration. Two of these studies (Small, 1967; LeNeveu et al., 1977) have found rather large increases ($\approx$ 20%) in partial lipid thickness (and corresponding decreases in area per lipid molecule) upon partial dehydration, whereas other studies (Reiss-Husson, 1967; Janiak et al., 1979, McIntosh and Simon, 1986c; Klose et al., 1988; McIntosh et al., 1989b) have found much smaller changes. For example, on going from excess water to 10 waters per E-PtdCho molecule, LeNeveu et al. (1977) found that the lipid partial thickness increased from about 35 Å to about 41 Å and that the area a per lipid molecule decreased from about 75 Å^2 to about 64 Å^2, whereas McIntosh et al. (1989b) found that on going from excess water to 10 waters per E-PtdCho molecule the partial lipid thickness increased only slightly, from about 40 Å to about 41 Å (see Fig. 1) and a decreased from about 64 Å^2 to about 62 Å^2. These discrepancies in the changes in partial lipid thickness and area per molecule are most likely due to differences in measured values of n_w (see Table 1 and above). As discussed above and by Klose et al. (1988), gravimetric methods for determination of partial lipid thickness can overestimate the amount of water imbibed by the multilayers, which results in an overestimation in the increase in bilayer thickness upon dehydration. However, estimates of changes in bilayer thickness based on electron density maps (Janiak et al., 1979; Mattai et al., 1987b; McIntosh and Simon, 1986c; McIntosh et al., 1987) do not have this problem (see Fig. 2). Structural analyses by Shipley and coworkers (Janiak et al., 1979; Mattai et al., 1987b), Klose et al. (1988), and by us (McIntosh and Simon, 1986c; McIntosh et al., 1987), all indicate that, in the first approximation, there are two stages in lipid dehydration. First, there are only modest changes in bilayer thickness and area per molecule when the water between apposing bilayers is removed, that is from 25 to 10 water molecules per lipid molecule in the case of egg phosphatidylcholine (see Figs. 1 and 2). However, in the second stage, upon further dehydration, when water is removed from the head group region of the bilayer (10 to 2 water molecules per lipid for E-PtdCho), there are relatively large decreases in lipid area per lipid molecule (tighter lipid packing) and increases in bilayer thickness (McIntosh et al., 1987).

Differential scanning calorimetry results are in accord with this two stage model for structural changes upon dehydration. For example, Janiak et al. (1979) showed that for Pam$_2$-PtdCho the thermal transition temperature between the gel and liquid crystalline phases is nearly constant for water contents ranging from excess water to about 10 water molecules per lipid molecule. However, the phase transition temperature rapidly increases as the water content is decreased below 10 waters per Pam$_2$-PtdCho molecule. Thus there is little change in transition temperature as water is removed from between adjacent bilayers (excess water to 10 water molecules per lipid), but there is a large change in the transition temperature as water is removed from the lipid head group region. This is consistent with the idea that there are only modest changes in bilayer structure upon removal of water from between apposing bilayers, but larger decreases in area per lipid molecule occur upon removal of water from the lipid head group region. That is, the waters of hydration located in the head group region of the bilayer occupy an appreciable volume (about 30 Å^3 per water molecule), and their removal decreases the area per lipid molecule and squeezes the hydrocarbon chains closer together, causing an increase in the melting temperature.

Similar arguments can be made for bilayers of phosphatidylethanolamine. For example, McIntosh and Simon (1986a,b) and Nagle and Weiner (1988) calculated that there are about 5 water molecules per lipid molecule in the head group region of Lau$_2$-PtdEtn, and Cevc and Marsh (1985) found that the phase transition temperature of Lau$_2$-PtdEtn was approximately constant as the water content was reduced from excess water to about 6 water molecules per lipid molecule. Further dehydration, from 6 to 1 water molecules per Lau$_2$-PtdEtn molecule, caused the phase transition temperature to increase by over 30°C (Cevc and Marsh, 1985).

The lateral diffusion of fluorescent lipid probes has also been found to be function of water content. For example, McCown et al. (1981) found that there was an eightfold decrease in the diffusion coefficient as multilamellar E-PtdCho bilayers were dehydrated. This decrease in lateral diffusion upon dehydration was due to both the decrease in area per molecule and the increase in proximity of lipid head groups from apposing bilayers.

IV. ANALYSIS OF PHOSPHOLIPID HYDRATION

The amount of water taken up by a given phospholipid depends on interactions between the lipid molecules, both perpendicular to the plane of the bilayer (interbilayer forces) and in the plane of the bilayer (intrabilayer forces). That is, the volume available to water molecules in a multibilayer system will depend both on the distance between adjacent bilayers, which is determined primarily by interbilayer interactions, and on the area per lipid molecule, which is determined primarily by intrabilayer interactions.

A. Interbilayer Forces

In excess water, the equilibrium distance between adjacent bilayers in a multilamellar array is governed by the balance among several repulsive and attractive pressures (forces per unit area). At least four repulsive interactions have been shown to operate between bilayer surfaces. These are the electrostatic, undulation, hydration (solvation), and steric pressures. Attractive pressures include the relatively long-range van der Waals pressure and short-range bonds between the molecules in apposing bilayers, such as hydrogen bonds or bridges formed by divalent salts. Pressure versus distance relationships have been obtained for phospholipid bilayers by two techniques. One method uses a surface force apparatus developed by Israelachvili and colleagues (Israelachvili, 1985; Israelachvili and Marra, 1986; Israelachvili and McGuiggan, 1988) and the second method involves x-ray diffraction measurements of multilamellar liposomes subjected to "osmotic stress" (LeNeveu et al., 1977; Parsegian et al., 1979). The two methods give results that are in general agreement (Horn et al., 1988), although the "osmotic stress" technique has the advantage that the pressure-distance curve can be obtained to smaller (<5 Å) interbilayer distances.

1. Repulsive Interactions

(a) Hydration repulsion. The hydration pressure is thought to be the dominant repulsive pressure for bilayer separations of about 5 to 20 Å (LeNeveu et al., 1977; Parsegian et al., 1979; Marra and Israelachvili, 1985; McIntosh and Simon, 1986c). Several theoretical treatments have been presented to explain the range and magnitude of the hydration

pressure (Marcelja and Radic, 1976; Gruen and Marcelja, 1983; Jonsson and Wennerstrom, 1983; Cevc and Marsh, 1985; Belaya et al., 1986; Graham et al., 1986; Cevc and Marsh, 1988; Dzhavakhidze et al., 1988; Attard and Batchelor, 1988). Although most of these theories do not consider the specific interactions of the bilayer surface with solvent molecules, it is generally agreed that hydration repulsion arises from the binding, polarization, and reorganization of water molecules near the bilayer surface.

Hydration repulsion has been measured for a number of bilayer systems, including both gel and liquid crystalline phospholipids (Parsegian et al., 1979; Lis et al., 1982; McIntosh and Simon, 1986c; Simon et al., 1988), zwitterionic phospholipids (Lis et al., 1982; McIntosh et al., 1989a; McIntosh and Simon, 1986c; Simon et al., 1988), charged phospholipids (Cowley et al., 1978; McIntosh et al., 1990), and uncharged lipids (McIntosh et al., 1989c). For all of these bilayer systems, it has been found that the hydration pressure has the functional form $P_h = P_o \cdot \exp(-d_f/\lambda)$, where the decay length λ depends on the packing density of interlamellar solvent molecules (McIntosh et al., 1989b) and is 1 to 2 Å for water. Recently it has been observed that the magnitude of the hydration pressure P_o is proportional to the Volta potential as measured for lipid monolayers in equilibrium with bilayers (Simon and McIntosh, 1989). For a recent review of the extensive experimental observations regarding hydration pressures between phospholipid membranes see Rand and Parsegian (1989).

(b) Steric repulsion. When apposing E-PtdCho bilayers are squeezed together by applied osmotic pressures, a distinct upward break appears in the pressure-distance relation at a distance between bilayer surfaces of about 5 Å (McIntosh et al., 1987; McIntosh et al., 1989a). This upward break has been attributed to steric repulsion between the mobile lipid head groups that extend 2 to 3 Å into the fluid space between bilayers. The pressure-distance data have been used to separate steric pressure from hydration pressure, as well as to quantitate the range and magnitude of the steric interaction (McIntosh et al., 1987). The steric pressure between phospholipid bilayers has a much larger magnitude and shorter range than does the hydration pressure. For E-PtdCho bilayers the steric pressure can be fitted to an exponential decay with increasing bilayer separation with a decay length of about 0.6 Å (McIntosh et al., 1987). An appreciable fraction of the measured steric energy can be ascribed to a decrease in configurational entropy due to restricted head group motion as adjacent bilayers come together (McIntosh et al., 1987). In addition, it has recently been found that the incorporation of cholesterol into E-PtdCho bilayers decreases the magnitude of the steric pressure by an amount that depends on the cholesterol concentration in the bilayer (McIntosh et al., 1989a). These data indicate that the magnitude of the steric pressure depends on the volume fraction of phospholipid head groups at the bilayer/water interface (McIntosh et al., 1989a).

(c) Electrostatic repulsion. Multilayers formed from phospholipids with a net charge, such as PtdIns, PtdGro, or PtdSer, swell indefinitely in aqueous monovalent electrolyte solutions (< 1 M salt concentration), primarily because of electrostatic repulsion between apposing bilayers (Cowley et al., 1978; Loosely-Millman et al., 1982; Marra, 1986b). For bilayer separations of greater than about 30 Å, the repulsion is dominated by electrostatic pressure, which is well described by Gouy-Chapman double-layer theory (Cowley et al., 1978; McIntosh et al., 1990). The electrostatic pressure between charged bilayers decays exponentially with increasing fluid separation (Israelachvili, 1985). The decay length (approximately equal to the Debye length) of this repulsive pressure depends on the concentration of electrolyte in solution (Loosley-Millman et al., 1982). For example, the decay length has been observed to be about 10 Å for PtdGro bilayers in 0.1 M NaCl

(Loosley-Millman et al., 1982), which makes electrostatic repulsion considerably longer ranged than hydration repulsion (where the decay length is 1 to 2 Å). Bilayers formed from zwitterionic lipids, such as phosphatidylcholine, swell in aqueous solutions containing various divalent cations, such as Ca^{2+} (Lis et al., 1981a,b). This observation has been explained in terms of the Ca^{2+} binding to the phosphate moiety, imparting a net positive charge to the bilayer surface, and therefore adding an electrostatic repulsive pressure (Lis et al., 1981b).

(d) Fluctuation pressure. A fourth repulsive pressure that has been observed between lipid bilayers is the fluctuation pressure, which is due to thermally induced undulations or fluctuations in the bilayer surface (Harbich and Helfrich, 1984; Evans and Parsegian, 1986; Servuss and Helfrich, 1987; Evans and Needham, 1987). Theoretical analyses predict that the fluctuation pressure has the functional form $P_f = 2(kt)^2/Bd_{eff}^3$, where k is Boltzmann's constant, T is temperature, B is the bilayer bending modulus or curvature elastic modulus, and d_{eff} is an effective fluid spacing between bilayers, which is in general larger than d_f of the hydration pressure (Harbich and Helfrich, 1984; Servuss and Helfrich, 1984; Servuss and Helfrich, 1987). Thus since the fluctuation pressure is inversely proportional to B, it would be expected to be larger for liquid crystalline bilayers than for gel phase bilayers. Note also that since P_f is inversely proportional to d_{eff}^3, its influence should extend to longer bilayer separations than the hydration pressure. In agreement with this theoretical analysis, experiments have shown that for stiff gel phase bilayers the undulation pressure is negligible compared to the hydration pressure, whereas for fluid bilayer the undulation pressure dominates at interbilayer spacings greater than 10 to 20 Å (McIntosh et al., 1989c). Thus thermally induced undulations can strongly influence the hydration properties of liquid crystalline bilayers. For example, liquid crystalline monoglyceride bilayers imbibe considerably more water than do gel phase monoglyceride bilayers (McIntosh et al., 1989c). The fluctuation pressure is undoubtedly one reason that liquid crystalline bilayers imbibe more water than do gel phase bilayers (Table 1). (Other reasons, as discussed below, include a smaller van der Waals attractive pressure and a larger area per lipid molecule in liquid crystalline compared to gel phase bilayers.) In addition, the presence of thermal undulations can, at least in part, explain why lipids in multilamellar liposomes imbibe more water from bulk aqueous phase than oriented multilayers adsorb through the vapor phase (see above, Sec. III.A.1.a). That is, bilayers in liposomes presumably contain thermally induced undulations in their surfaces, whereas bilayers oriented on rigid substrates in relative humidity atmospheres do not (Rand and Parsegian, 1989). The presence of this additional repulsive fluctuation pressure would cause liposomes to imbibe more water from bulk phase than oriented multilayers do from the vapor phase, even at 100% relative humidity (Rand and Parsegian, 1989).

2. Attractive Interactions

(a) Van der Waals pressure. The predominant long-range attractive pressure is the van der Waals pressure, which can be approximated by $P_v \approx -H/6\pi d_f^3$, where H is the Hamaker constant (Parsegian and Ninham, 1973). For several types of lipid bilayers, H has been predicted or measured to be on the order of 1×10^{-21} to 6×10^{-21} J (Marra, 1986a; Marra and Israelachvili, 1985; Requena et al., 1977). Since P_v is inversely proportional to d_f^3, it is in general longer-ranged than the hydration pressure. Additional, nonstandard van der Waals attraction may arise from correlations between the charge- and hydration-distribution patterns at opposing interacting interfaces (Cevc and Marsh, 1988; Kornysher and Leikiu, 1989).

For zwitterionic lipids, such as phosphatidylcholine, the amount of water imbibed by the multilayers depends on the equilibrium separation between bilayers—the separation where the repulsive interactions (steric, hydration, and fluctuation pressures) are balanced by the attractive van der Waals pressure. Liquid crystalline phosphatidylcholine bilayers have wider fluid spaces than gel phase bilayers, since for liquid crystalline bilayers the fluctuation pressure is greater and the van der Waals pressure is, in general, expected to be smaller (McDaniel et al., 1983). For charged bilayers such as PtdIns, PtdSer, or PtdGro, the added electrostatic pressure overwhelms the van der Waals pressure and the multilayers swell indefinitely in water.

(b) Intermolecular bonding. For most phospholipid systems, such as bilayers containing zwiterionic PtdCho or charged PItdIns, PtdSer, or Ptd Gro, the distance between bilayer surfaces and the amount of water imbibed from the bulk aqueous phase can be explained in terms of the repulsive and attractive interactions as described above. However, there are at least two cases where the distance between apposing bilayers cannot readily be explained solely in terms of these interactions. These two cases are (1) PtdSer bilayes in the presence of either Li^+ or several different divalent cations and (2) PtdEtn bilayers. In both of these cases the number of water molecules between apposing bilayers is small, so that the fluid space between apposing bilayers is quite narrow (McIntosh and Simon, 1986a). In terms of the nonspecific pressures listed above, such narrow fluid spaces could possibly be explained by either small interbilayer repulsive pressures or a large van der Waals attractive pressure. However, there is no reason to believe that the repulsive pressures are unusually small or that the Hamaker constant is unusually large for either of these systems (Simon and McIntosh, 1989). Rather, for the case of the PtdSer, it has been argued that the dehydration by Li^+ and Ca^{2+} is caused by the specific binding of the hydrophilic cation to the PtdSer head group (Hauser et al., 1977; Newton et al., 1978; Hauser and Shipley, 1983; Cevc et al., 1985), with divalent cations having an intermolecular bridging capability (Portis et al., 1979; Cevc et al., 1985).

The reason for the extremely narrow fluid spacing in phosphatidylethanolamines is, at present, not completely understood, although several possible explanations have been proposed. Seddon et al. (1984) and McIntosh and Simon (1986a) have argued that the tendency for the phosphatidylethanolamines to remain dehydrated might be a consequence of strong hydrogen bond formation through water molecules and electrostatic interactions between the amine and phosphate groups of the PtdEtn molecules, both in the plane of the bilayer and between adjacent bilayers. Rand et al. (1988) have argued for the presence of hydration attractive pressure arising from interbilayer hydrogen-bonded water bridges between apposing bilayer surfaces. Cevc (1989) has proposed that the small hydrophylicity of PtdEtn is a consequence of subcritical polarity caused by the intermolecular hydrogen bonds. Since the close approach of adjacent bilayers is a necessary first step in the fusion of lipid membranes (Duzguneş, et al., 1985), the small equilibrium fluid space between adjacent PtdEtn bilayers may be important in understanding the observed increased fusion rates of vesicles containing phosphatidylethanolamine compared to vesicles containing phosphatidylcholine (Duzguneş et al., 1981; Duzguneş et al., 1985).

B. Intrabilayer Forces

Several of the same forces described above, including electrostatic repulsion, hydration repulsion, steric repulsion, and van der Waals attraction, also act in the plane of the bilayer. In addition, as described in Sec. II.C.1, interfacial tension (Israelachvili, 1985) plays an important role in determining the area per lipid molecule. The balance among thesc intrabilayer interactions is important in terms of the amount of water imbibed by phospholipid bilayers because the total volume available for water is proportional to the area per lipid molecule. That is, the volume of the interlamellar space is equal to the interbilayer separation times the area per molecule. In addition, the larger the area per molecule, the more water can be incorporated into the head group region of the bilayer. For example, consider the case of dilauroyl phosphatidylethanolamine. In the liquid crystalline phase, where the area per molecule is about 51 Å^2 (Table 1), this lipid imbibes about 10 waters per lipid molecule, about 5 in the head group region of the bilayer and 5 in the fluid space between bilayers. However, in the gel phase, where the area per molecule is about 41 Å^2, the lipid imbibes only about 7 waters per lipid molecule, with about 3 in the head group region and 4 in the fluid space between bilayers (McIntosh and Simon, 1986a,b).

The addition of Ca^{2+} to PtdSer bilayers tends to dehydrate the bilayer and crystallize the lipid hydrocarbon chains (see Sec. III). This must be due, at least in part, to Ca^{2+} binding between adjacent PtdSer head groups in the plane of the membrane, thereby decreasing the area per lipid molecule and removing water from the head group region.

As another example of the importance of area per lipid molecule in terms of phospholipid hydration, consider that the interdigitated gel ($L_{\beta I}$) phase of Hex$_2$-PtdCho imbibes considerably more water than the normal gel ($L_{\beta'}$) phase of Pam$_2$-PtdCho (see Sec. III and Table 1). This is undoubtedly because the area per molecule of the interdigitated phase is about twice as great as that of the normal gel phase (Ranck et al., 1977; McIntosh et al., 1983). Thus at a given interbilayer fluid separation there is more volume available for water in the interdigitated phase than in the normal gel phase.

C. Energetics of Phospholipid Hydration

The total work to remove the water of hydration has been calculated for several phospholipid bilayer systems. For example, consider the case of E-PtdCho. Figure 3 is a plot of log P versus the number of water molecules per E-PtdCho molecule, where P is the osmotic pressure applied to the multilamellar system. The circles represent data from the adsorption isotherm of Jendrasiak and Hasty (1974), and the square represent data from our osmotic stress experiments (McIntosh and Simon, 1986c). Note that all of the data points can be fitted quite closely with a single straight line, indicating that the total repulsive pressure, and thus the work to dehydrate the E-PtdCho bilayers, decreases exponentially with increasing water content. Using the data of Fig. 3 we calculated that it takes about 10 kJ/mol to dehydrate E-PtdCho from 23 to 2 waters per lipid molecule (McIntosh et al., 1987). Using a related approach, Cevc and Marsh (1988) obtained a very similar result, about 11 kJ/mol. They also calculated the energy to remove each water of hydration. Much more energy is required to remove the inner waters of hydration than the outer waters of hydration. For example, they found that it takes only 0.14 kJ/mol to remove the 12th water of hydration, but 3.53 kJ/mol to remove the 2^{d} water of hydration, and 4.6 kJ/mol to remove the first water of hydration from E-PtdCho (Cevc and Marsh, 1988). Thus the energy necessary to remove the outer waters of hydration is small compared to thermal energy (2.4 kJ/mol at 20°C), whereas the energy to remove the inner water molecules is significantly larger than thermal energy. As noted above, the first and second waters of hydration have been shown by x-ray crystallography to be located in the head group region of the bilayer, near the phosphate oxygen (Pearson and Pascher, 1979), whereas the outer waters of hydration ($n_w > 10$) are located in the fluid space

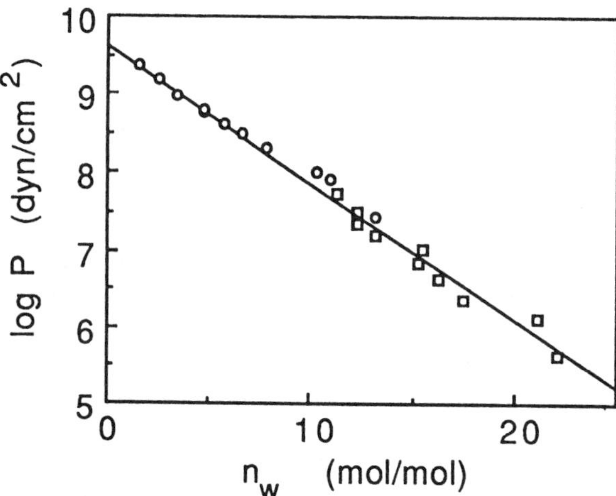

Figure 3 Plot of the logarithm of applied pressure (log P) versus the number of water molecules per E-PtdCho molecule (n_w). This figure has been redrawn from McIntosh et al. (1987). The open circles are taken from the adsorption isotherms of Jendrasiak and Hasty (1974), and the open squares are taken from the phase diagram and osmotic stress data of McIntosh and Simon (1986c).

between adjacent bilayers (McIntosh et al., 1987). Cevc and Marsh (1988) have also calculated the differential hydration energies for other lipids, including $Pam_2PtdCho$, E-PtdEtn, and Lau_2-PtdEtn. For each of these lipids the hydration energies of the inner water molecules are considerably higher than those of the outer water molecules.

V. SUMMARY

In this chapter we have presented some of the extensive data on the amount of water imbibed by phospholipid bilayers (Table 1), the location of water with respect to the bilayer, and the effects of dehydration on bilayer structure. The hydration properties of phospholipid bilayers depend on both the lipid head group type and the organization of the hydrocarbon chains (gel, liquid crystalline, or interdigitated). For example, phosphatidylcholines in the liquid crystalline phase imbibe about 25 waters per lipid molecule, with about 10 of these waters located in the lipid head group region and the remaining waters located in the fluid space between adjacent bilayers. The inner waters are tightly bound to the lipid head group, as it takes over 4 kJ/mol to remove the inner water of hydration, whereas the outer waters are much more weakly bound, as it requires a small fraction of thermal energy (about 0.1 kJ/mol) to remove an outer water molecule. Phosphatidylcholine bilayers in the gel phase imbibe about one-half this amount of water. Phosphatidylethanolamine bilayers imbibe considerably less water than do phosphatidylcholine bilayers, whereas charged lipids, such as phosphatidylinositol or phosphatidylserine, imbibe water without limit. The amount of water taken up to each phospholipid can be explained in terms of nonspecific inter- and intrabilayer pressures acting on the system, including hydration, steric, undulation, and electrostatic repulsive pressures, van der Waals attractive pressure, and intermolecular bonding between headgroups from apposing bilayers.

ACKNOWLEDGMENTS

We thank our colleague Dr. Sidney A. Simon for many helpful comments and suggestions. This work was supported by a grant from the National Institutes of Health (GM 27278).

REFERENCES

N. Albon, *J. Chem. Phys.* 78:4676–4686 (1983).
G. Arvidson, I. Brentel, A. Khan, G. Lindblom, and K. Fontell, *Eur. J. Biochem. 152*:753–759 (1985).
D. Atkinson, H. Hauser, G. G. Shipley, and J. M. Stubbs, *Biochim. Biophys. Acta 339*:10–29 (1974).
P. Attard and M. T. Batchelor, *Chem. Phys. Lett. 149*:206–211 (1988).
M. L. Belaya, M. V. Feigel'man, and V. G. Levadny, *Chem. Phys. Letts. 126*:361–364 (1986).
M. Bourges, D. M. Small, and D. G. Orvichian, *Biochem. Biophys. Acta 137*:157–167 (1967).
G. Buldt, H. U. Gally, A. Seelig, J. Seelig, and G. Zaccai, *Nature 271*:182–184 (1978).
G. Buldt, H. U. Gally, J. Seelig, and G. Zaccai, *J. Mol. Biol. 134*:673–691 (1979).
G. Cevc and D. Marsh, *Biophys. J. 47*:21–32 (1985).
G. Cevc and D. Marsh, in *Phospholipid Bilayers: Physical Principles and Models.* John Wiley, New York, 1988.
G. Cevc, *J. de Phys. 50*:1117–1134 (1989).
G. Cevc, J. M. Seddon, and D. Marsh, *Biochim. Biophys. Acta 813*:343–346 (1985).
G. Cevc, L. Strohmaier, J. Berkholz, and G. Blume, *Stud. Biophys. 138*:57–70 (1990).
S. C. Chen, J. E. Sturtevant, and B. J. Gaffney, *Proc. Natl. Acad. Sci. U.S.A. 77*:5060–5063 (1980).
A. C. Cowley, N. L. Fuller, R. P. Rand, and V. A. Parsegian, *Biochemistry 17*:3163–3138 (1978).
J. M. Diamond and Y. Katz, *J. Membrane Biology 17*:101 (1974).
J. A. Dix, P. Kivelson, and J. M. Diamond, *J. Membrane Biol. 40*:315 (1978).
N. Duzguneş, T. Wilshut, R. Fraley, and D. Papahadjopoulos, *Biochim. Biophys. Acta 642*:182–195 (1981).
N. Duzguneş, R. M. Staubinger, P. A. Baldwin, D. S. Friend, and D. Paphadjopoulos, *Biochemistry 24*:3091–3098 (1985).
P. G. Dzhavakhidze, A. A. Kornyshev, and V. G. Levadny, *Il Nuovo Cimento 10D*:627–654 (1988).
P. H. Elworthy, *J. Chem. Soc.* 5385–5389 (1961).
P. H. Elworthy, *J. Chem. Soc.* 4897–4900 (1962).
D. M. Engelman, and T. A. Steitz, *Cell 23*:411 (1981).
E. Evans, and D. Needham, *J. Phys. Chem. 91*:4219–4228 (1987).
E. A. Evans, and V. A. Parsegian, *Proc. Nat. Acad Sci. U.S.A. 83*:7132–7136 (1986).
G. W. Feigenson, *Biochemistry 25*:5819–5825 (1986).
E. G. Finer, and A. Darke, *Chem. Phys. Lipids 12*:1–16 (1974).
N. P. Franks, *J. Mol. Biol. 100*:345–358 (1976).
N. L. Gershfeld, *Biochemistry 28*:4229–4232 (1989).
I. S. Graham, A. Georgallas, and M. J. Zuckermann, *J. Chem. Phys. 85*:6010–6021 (1986).
O. H. Griffith, P. J. Dehlinger, and S. P. Van, *J. Membr. Biol. 15*:159–192 (1974).
D. W. R. Gruen, and S. Marcelja, *J. Chem. Soc. Faraday TRans. II 79*:225–242 (1983).
S. M. Gruner, *Proc. Nat. Acad. Sci. U.S.A. 82*:3665–3669 (1985).
S. M. Gruner, J. Phys. Chem. 93:7562–7570 (1989).
S. M. Gruner, V. A. Parsegian, and R. P. Rand, *Faraday Discuss. Chem. Soc.* 81:29–37 (1986).
S. M. Gruner, M. W. Tate, G. L. Kirk, P. T. C. So, D. C. Turner, D. T. Keane, C. P. S. Tilcock, and P. R. Cullis, *Biochemistry 27*:2853–2866 (1988).
M. E. Haberland and J. A. Reynolds, *J. Biol. Chem. 250*:6636–6639 (1975).

W. Harbich and W. Helfrich, *Chem. Phys. Lipids 36*:39–63 (1984).
G. S. Hartley, In: *Aqueous Solutions of Paraffin-Chain Salts*, Hermann, Paris, 1936.
H. Hauser and G. G. Shipley, *Biochemistry 22*:2171–2178 (1983).
H. Hauser and G. G. Shipley, *Biochemistry 23*:34–41 (1984).
H. Hauser, E. G. Finer, and A. Darke, *Biochem. Biophys. Res. Commun. 76*:267–274 (1977).
H. Hauser, F. Paltauf, and G. G. Shipley, *Biochemistry 21*:1061–1067 (1982).
W. Helfrich, *Z. Naturforsch. 28C*:693–703 (1973).
R. G. Horn, J. N. Israelachvili, J. Marra, V. P. Parsegian, and R. P. Rand, *Biophys. J. 54*:1185–1186 (1988).
S. W. Hui, *Comments Mol. Cell. Biophys. 4*:233–248 (1987).
S. W. Hui, J. T. Mason, and C. Huang, *Biochemistry 23*:5570–5577 (1984).
Y. Inoko and T. Mitsui, *J. Physical Soc. Jpn 44*:1918–1924 (1978).
J. N. Israelachvili, *Intermolecular and Surface Forces*, Academic Press, London, 1985.
J. N. Israelachvili and P. M. McGuiggan, *Science 241*:795–800 (1988).
J. N. Israelachvili and J. Marra, *Methods Enzymol. 127*:353 (1986).
J. N. Israelachvili, S. Marcelja, and R. G. Horn, *Quart. Rev. Biophysics 13*:121–200 (1980).
M. J. Janiak, D. M. Small, and G. G. Shipley, *J. Biol. Chem. 254*:6068–6078 (1979).
G. L. Jendrasiak, and J. H. Hasty, *Biochim. Biophys. Acta 337*:79–91 (1974).
B. Jonsson and H. Wennerstrom, *J. Chem. Soc. Faraday Trans. II 79*:19–35 (1983).
J. T. Kim, J. Mattai, and G. G. Shipley, *Biochemistry 26*:6592–6598 (1987).
G. Klose, B. Konig, H. W. Meyer, G. Schulze, and G. Degovics, *Chem. Phys. Lipids 47*:225–234 (1988).
M. Kodama, M. Kuwabara, and S. Saki, *Biochim. Biophys. Acta 689*:567–570 (1982).
A. A. Koraysherv and S. Leikiu, *Phys. Rev. A 40*:G431–6437 (1989).
L. J. Korstanje, E. E. Van Faassen, and Y. K. Levine, *Biochim. Biophys. Acta 980*:225–233 (1989).
P. Laggner, K. Lohner, G. Degovics, K. Muller and A. Schuster, *Chem. Phys. Lipids 44*:31–60 (1987).
M. Langner, D. Cafiso, S. Marcelja, and S. McLaughlin, *Biophys. J. 57*:335–350 (1990).
D. M. LeNeveu, R. P. Rand, V. A. Parsegian, and D. Gingell, *Biophys. J. 18*:209–230 (1977).
G. Lindblom and L. Rilfors, *Biochim. Biophys. Acta 988*:221–256 (1989).
L. J. Lis, W. T. Lis, V. A. Parsegian, and R. P. Rand, *Biochemistry 20*:1771–1777 (1981a).
L. J. Lis, V. A. Parsegian, and R. P. Rand, *Biochemistry 20*:1761–1770 (1981b).
L. J. Lis, M. McAlister, N. Fuller, R. P. Rand, and V. A. Parsegian, *Biophys. J. 37*:657–666 (1982).
M. E. Loosley-Millman, R. P. Rand, and V. A. Parsegian, *Biophys. J. 40*:221–232 (1982).
B. Lundberg, *Acta Chem. Scand. B 28*:673–676 (1974).
V. Luzzati, in *Biological Membranes* ed., Academic Press, New York, 1968. pp. 71–123.
V. Luzzati, T., Gulik-Kryzwicki, and A. Tardieu, *Nature 218*:1031–1034 (1968).
S. Marcelja and N. Radic, *Chem. Phys. Lett.* 42:129–130 (1976).
J. Marra, *J. Colloid Interface Sci. 109*:11–20 (1986a).
J. Marra, *Biophysical J. 50*:815–825 (1986b).
J. Marra and J. Israelachvili, *Biochemistry 24*:4608–4618 (1985).
J. Mattai, N. M. Witzke, R. Bittman, and G. G. Shipley, *Biochemistry 26*:623–633 (1987a).
J. Mattai, N. M. Witzke, R. Bittman, and G. G. Shipley, *Biochemistry 26*:623–633 (1987b).
J. T. McCown, E.Evans, S. Diehl, and H. C. Wiles, *Biochemistry 20*:3134–3138 (1981).
R. V. McDaniel, *Biochim. Biophys Acta 940*:158–164 (1988).
R. V. McDaniel, T. J. McIntosh, and S. A. Simon, *Biochim. Biophys Acta 731*:97–108 (1983).
R. V. McDaniel and T. J. McIntosh, *Biochim. Biophys. ACta 983*:241–246 (1989).
T. J. McIntosh and S. A. Simon, *Biochemistry 25*:4948–4952 (1986a).
T. J. McIntosh and S. A. Simon, *Biochemistry 25*:8474 (1986b).
T. J. McIntosh and S. A. Simon, *Biochemistry 25*:4058–4066 (1986c).
T. J. McIntosh, R. V. McDaniel, and S. A. Simon, *Biochim. Biophys. Acta 731*:109–114 (1983).

T. J. McIntosh, S. A. Simon, J. C. Ellington, and N. A. Porter, *Biochemistry 23*:4038–4044 (1984).
T. J. McIntosh, A. D. Magid, and S. A. Simon, *Biochemistry 26*:7325–7332 (1987).
T. J. McIntosh, A. D. Magid, and S. A. Simon, *Biochemistry 28*:17–25 1989a).
T. J. McIntosh, A. D. Magid, and S. A. Simon, *Biochemistry 28*:7904–7912 1989b).
T. J. McIntosh, A. D. Magid, and S. A. Simon, *Biophys. J. 55*:897–904 (1989c).
T. J. McIntosh, S. A. Simon, and J. F. Dilger, in *Water Transport in Biological Membranes* (G. Benga, ed.), CRC Press, Boca Raton, Fla., 1989, pp. 1–15.
T. J. McIntosh, A. D. Magid, and S. A. Simon, *Biophys. J.*, *57*:1187–1197 (1990).
J. F. Nagle, *Ann. Rev. Phys. Chem. 31*:157–195 (1980).
J. F. Nagle and M. C. Wiener, *Biochim. Biophys. Acta 942*:1–10 (1988).
C. Newton, W. Pangborn, S. Nir, and D. Papahadjopoulos, *Biochim. Biophys. Acta 506*:281–287 (1978).
G. Nimtz, A. Endeers, and B. Binggeli, *Ber. Bungenses. Phys. Chem. 89*:842–845 (1985).
V. A. Parsegian and B. W. Ninham, *J. Theor. Biol. 38*:101–109 (1973).
V. A. Parsegian, N. Fuller, and R. P. Rand, *Proc. Natl. Acad. Sci. U.S.A. 76*:2750–2754 (1979).
R. H. Pearson and I. Pascher, *Nature 281*:499–501 (1979).
A. Portis, C. Newton, W. Pangborn, and D. Papahadjopoulos, *Biochemistry 18*:780–790 (1979).
J. L. Ranck, T. Keira, and V. Luzzati, *Biochim. Biophys. Acta 488*:432–441 (1977).
R. P. Rand and V. A. Parsegian, *Biochim. Biophys. Acta 988*:351–376 (1989).
R. P. Rand, N. Fuller, V. A. Parsegian, and D. C. Rau, *Biochemistry 27*:7711–7722 (1988).
F. Reiss-Husson, *J. Mol. Biol. 25*:363–382 (1967).
J. Requena, D. E. Brooks, and D. Haydon, *J. Colloid Interfac. Sci* 58:26–35 (1977).
J. A. Reynolds, C. Tanford, and W. L. Stone, *Proc. Nat. Acad. Sci. U.S.A. 74*:3796–3799 (1977).
M. J. Ruocco and G. G. Shipley, *Biochim. Biophys. Acta 684*:59–66 (1982a).
M. J. Ruocco, and G. G. Shipley, *Biochim. Biophys. Acta 691*:309–320 (1982b).
B. P. Schoenborn, in *Neutron Scattering for the Analysis of Biological Structures*, Brookhaven National Laboratory, Upton, New York, 1975, pp. I10–I17.
J. M. Seddon, *Biochim. Biophys. Acta 1031*:1–69 (1990).
J. M. Seddon, K. Harlos, and D. Marsh, *J. Biol. Chem. 258*:3850–3854 (1983).
J. M. Seddon, G. Cevc, R. D. Kaye, and D. Marsh, *Biochemistry 23*:2634–2644 (1984).
R. M. Servuss and W. Helfrich, in *Physics of Complex and Supermolecular Fluids*, Wiley, New York, 1987, pp. 85–100.
S. A. Simon and T. J. McIntosh, *Methods in Enzymology 127*:511–521 (1986).
S. A. Simon and T. J. McIntosh, *Proc. Nat. Acad. U.S.A. 86*:9263–9267 (1989).
S. A. Simon, T. J. McIntosh, and R. Latorre, *Science 216*:65–67 (1982).
S. A. Simon, T. J. McIntosh, and A. D. Magid, *J. Colloid Interface Sci. 126*:74–83 (1988).
D. M. Small, *J. Lipid Res. 8*:551–557 (1967).
D. M. Small, *Handbook of Lipid Research 4. The Physical Chemistry of Lipids*, pp. 89–96 (1986).
R. Smith and C. Tanford, *J. Mol. Biol. 67*:75–82 (1972).
Y. Sugiura, *Biochim. Biophys. Acta 641*:148–159 (1981).
C. Tanford, *J. Phys. Chem. 76*:3020–3024 (1972).
C. Tanford, *The Hydrophobic Effect*, 2d ed., John Wiley, New York, 1980.
A. Tardieu, V. Luzzati, and F. C. Reman, *J. Mol. Biol. 75*:711–733 (1973).
H. V. Tartar, *J. Phys. Chem. 59*:1195–1199 (1955).
R. P. Taylor, C.-H. Huang, A. V. Broccoli, and L. Leake, *Arch. Biochem. Biophys. 183*:83–89 (1977).
L. Ter-Minassian-Saraga, and G. Madelmont, *J. Colloid Interface Sci. 85*:375–388 (1982).
J. Torbet and M. H. F. Wilkins, *J. Mol. Biol. 62*:447–458 (1976).
M. C. Wiener, R. M. Suter, and J. F. Nagle, *Biophys. J. 55*:315–325 (1989).
D. A. Wilkinson, and J. F. Nagle, *Biochemistry 20*:187–192 (1981).
D. L. Worcester and N. P. Franks, *J. Mol. Biol. 100*:359–378 (1976).
G. Zaccai, J. K. Blasie, B. P. Schoenborn, *Proc. Natl. Acad. Sci. U.S.A. 72*:376–380 (1975).

16
Phospholipid Monolayers

Helmuth Möhwald *Johannes-Gutenberg-University of Mainz, Mainz, Germany*

I. INTRODUCTION

Why study phospholipid monolayers? A monolayer, being half of a membrane, is a well-defined planar system convenient for the study of intermolecular interactions between lipids and also between lipids and proteins. In addition to this, they illustrate many interesting aspects of physics in two dimensions and have some technological relevance.

A phospholipid monolayer shares many properties with other insoluble monolayers [1]. With modern experimental techniques, it can be studied directly without the need for much speculation. Perhaps the best explored are monolayers of glycerophospholipids with saturated aliphatic tails; therefore this chapter will concentrate mostly on these.

The main text contains a description of experimental and theoretical techniques. Their application to phospholipids as well as their limitations, and future tasks and occasions, are discussed next. In the following chapter, present knowledge on phases, phase transitions, and phospholipid structure are introduced, dealing mainly with length scales between molecular and macroscopic dimensions. Results obtained with phospholipid monolayers will be correlated with the corresponding data on other surfactant films. Some general physical principles are derived, and extrapolations to other phospholipids and more complex systems, not yet as extensively studied, are suggested.

II. EXPERIMENTAL METHODS

A. Thermodynamic Measurements

The easiest way to characterize a surfactant monolayer is to measure the lateral pressure π as a function of the molecular area A. The former is given by the difference of the surface pressures in the absence (π_o) and presence (π_1) of surfactant at the air/water interface [2]

$$\pi = \pi_o - \pi_1 \tag{1}$$

and is nominally identical to the derivative of the surface free energy with respect to the intrinsic variable A.

From the pressure/area (π, A) isotherms, the isothermal compressibility χ

$$\chi = -\frac{1}{A}\left(\frac{\partial A}{\partial \pi}\right)_T \tag{2}$$

is obtained, the isobaric thermal expansivity being

$$\lambda = \frac{1}{A}\left(\frac{\partial A}{\partial T}\right)_\pi \tag{3}$$

Usually π versus A is measured, but there are also situations in which special characteristics of the system are shown more clearly in some other detection mode. In any case, a change in the experimental slope corresponds to a phase transition.

As an example, Fig. 1 shows the π, A isotherms of the phospholipid Pam_2Ptd Cho at different temperatures [2,3]. Above 18.6°C, one observes a distinct change in the slope at the pressure π_c and area A_c. On compression, the slope becomes nearly horizontal, suggesting a first order phase transition [4]. Yet there is a finite isotherm slope, and the most probable explanation for this is impurities. Thus the number of components is increased, and the Gibbs phase rule allows phase coexistence over an extended pressure range.

The influence of impurities has been tested in different ways.

1. By extensive purification, the isotherm slope was reduced to virtually zero [5].
2. The above small slope could be measured even for commercially available Myr_2 Ptd Eth, and the slope increase on adding impurities could be interpreted within a simple model [6]. From this one could also estimate that a residual impurity content of 0.2 mol % would suffice to account for the isotherm slope measured.
3. The dependence of slope on impurity content is also suggested theoretically [7].

Compressing the monolayer beyond the nearly horizontal section of the isotherm causes the slope to increase gradually; a discontinuous change, moreover, is detected at molecular area A_s and pressure π_s [2,8]. This is indicative of a second order phase transition.

At lower temperature (<18.6°C), the system illustrated in Fig. 1 does not display a transition at π_c, which is often called the main transition. The transition at π_s is retained, however.

One thus realizes that π_s does not depend on temperature [2]. For charged lipids it was observed that π_s depends on the type of divalent counter ion present in the subphase [8]. However, the transition pressure is independent of the bulk ion concentration, provided that the latter exceeds a limiting value.

For a two-dimensional system the Clausius Clapeyron equation reads [2]

$$S_1 - S_2 = \frac{d\pi_t}{dT} \cdot (A_1 - A_2) \tag{4}$$

where S_i and A_i are the molar entropies and molecular areas of the coexisting phases, respectively. The entropy change is related to the transition enthalpy Q_t according to

$$Q_t = (S_1\text{-}S_2).T \tag{5}$$

Q_t and also (S_1–S_2) often decrease linearly on approaching the critical temperature T_c. T_c can, therefore, be determined from an extrapolation of Q_t towards $Q = 0$ [9].

To determine the changes in entropy and latent heat corresponding to the main lipid phase transition, one has to measure the transition pressure $\pi_t = \pi_c$ as a function of temperature. This value, as well as the molecular area $A_1 = A_c$ of the LE phase, can be derived rather accurately from the abrupt change in the isotherm slope. A problem, however, is the accurate determination of A_2, the molecular area characteristic for the ordered phase, since the termination of the phase transition on compression is not clearly observed in the pressure area isotherm.

Consider a typical example of the main transition of Pam$_2$ Ptd Chol at 20°C. According to the isotherm published by Albrecht et al. [2], the critical molecular area is $A_c = 82$ Å^2. The same authors defined A_2 as the molecular area at which the isotherm deviates from linearity, thus obtaining $A_2 = 62$ Å^2. But other authors derived A_2 from the extrapolation of the very steep part of the isotherm and obtained $A_2 = 50$ Å^2. This shows that the determination of $A_1 - A_2$ and hence of Q_t may differ by 50% depending on the way of deriving A_2. The relative error becomes even larger on approaching the critical temperature T_c. Yet since Q is reduced on approaching T_c, the latter temperature can still be determined with reasonable accuracy.

Consequently, all that one can state at present is that Q_t is of the same order of magnitude as in the case of bilayers [2], but that a more detailed comparison is inadequate at least as long as Q_t is determined from the isotherm data alone.

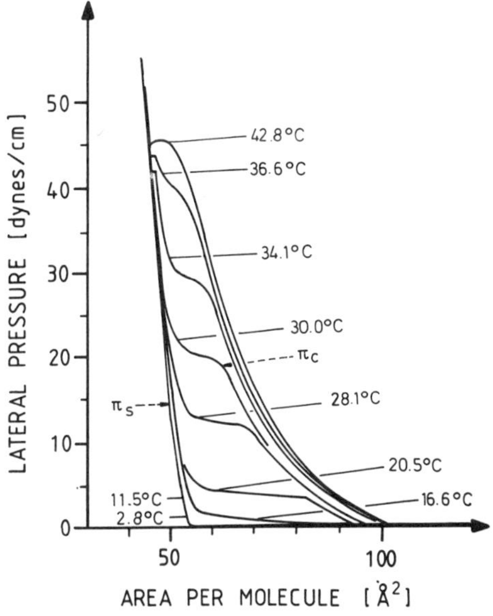

Figure 1 Surface pressure π as a function of molecular area A for a monolayer of Pam$_2$ Ptd Cho at different temperatures. Indicated are also the pressures (π_c, π_s) corresponding to breaks in the isotherm slope. (Data from Ref. 2.)

B. Optical Techniques

1. Fluorescence Microscopy

A more direct picture of the phase diagram of phospholipid monolayers is gained from the fluorescence microscopic studies of the air/water interface. For this purpose a fluorescent dye probe is incorporated into a monolayer and the lateral dye distribution is determined from the analysis of the fluorescence micrographs. A contrast in images is obtained due to the different dye solubility, the fluorescence quantum yield, or the molecular density of the coexisting phases [10–12].

As an example, Fig. 2 presents a series of textures observed while a monolayer is compressed above the pressure π_c [13]. The dark areas can be ascribed to a more ordered phase with low dye solubility. As expected, the area fraction of this phase increases on compression. A quantitative analysis of these images will illuminate two problem areas: (1) What can be learnt about the phase transitions and about the nature of coexisting phases from fluorescence microscopy? (2) Can one understand peculiar domain shapes and superlattices?

As for (1), Fig. 3 gives the number of domains, and the corresponding dark area fraction ϕ, as a function of molecular area [13]. Apparently, on increasing π above π_c, a certain number of domains is formed. On compression, this number remains constant, but the average molecular area increases. The measured area fraction ϕ can then be described by a lever rule, and from the extrapolations to $\phi = 0$ and to $\phi = 1$ one finally obtains the molecular areas of coexisting phases. For $\phi = 0$ one obtains, as expected, a molecular area near A_c, the value characteristic of the LE phase. Extrapolation to $\phi = 1$, however, does not yield a molecular area similar to A_s. Rather than that, a value of 48 ± 2 Å^2 is obtained, which is distinctly larger than A_s. This proves that the phase coexisting with the LE phase is less dense than the pure gel phase.

Fluorescence microscopy clearly shows the coexistence of two phases; it thus demonstrates that the corresponding phase transition is of the first order. This conclusion has been challenged, since the technique depends on the incorporation of dyes as impurities. However, the coexistence of phases meanwhile has been confirmed also by probeless techniques such as electron microscopy [14], surface plasmon microscopy with transferred monolayers [15], and fluorescence microscopy with water soluble dyes adsorbing from the subphase [16], and Brewster angle microscopy [124].

As for (2), the number N of domains has been found to be a nonequilibrium system property [17]. It depends on the nucleation process and cannot be predicted quantitatively. However, its qualitative change with the variation of a nucleation parameter is as expected. Increasing the compression speed, and thus the pressure deviation from π_c during the nucleation period, increases the value of N [17]. The free energy required for the creation of a critical nucleus for growth of ordered phases can be reduced by increasing temperature or surface charge density [8,18] or by adding surface active impurities like dyes [6], cholesterol [19,20], or proteins [21]. This reduces line tension, the energy per interface length between the coexisting phases, and thus increases N.

The domain shape may, but need not, be an equilibrium property, depending on the actual system. In the majority of cases, domains grow far from equilibrium with a rough interface between individual phases. This can be understood as a diffusion limited aggregation process that may lead to fractal structures [6]. For one specific case, this process was analyzed within the framework of the constitutional supercooling model, where domain growth is limited by diffusion of an impurity from the phase boundary [6].

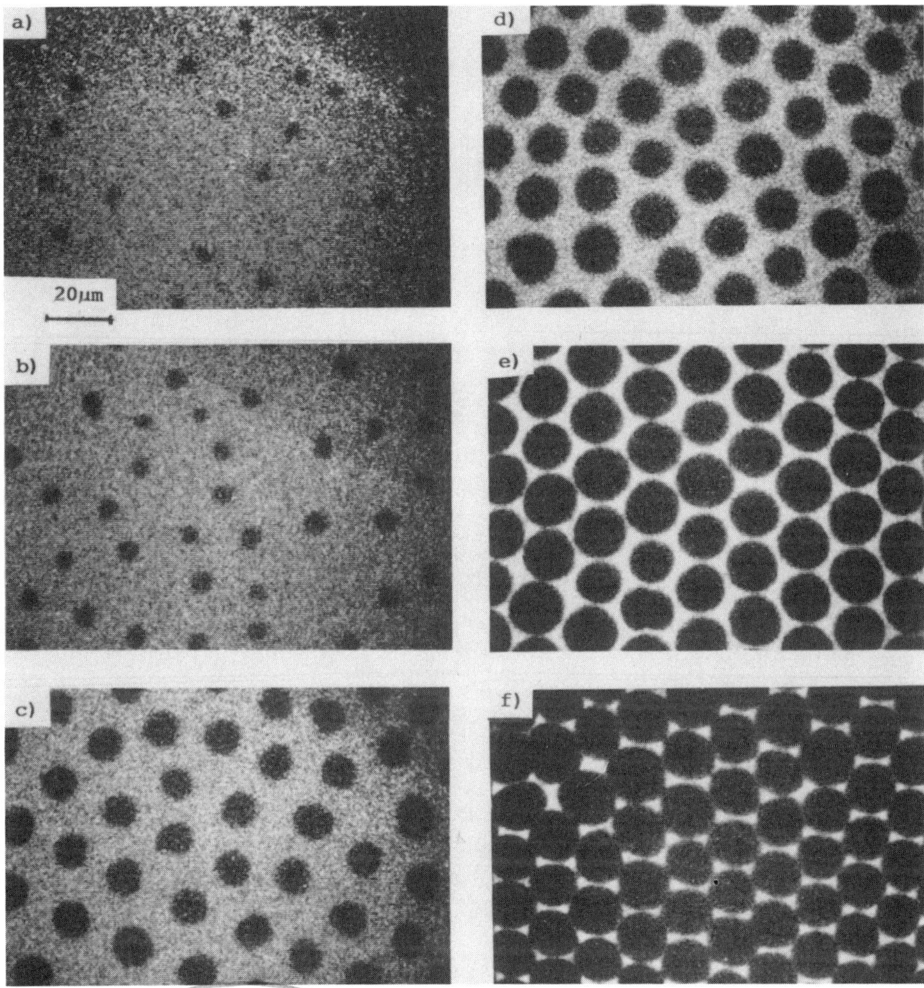

Figure 2 Fluorescence micrograph of a Myr_2 Ptd A monolayer on increasing the lateral pressure (above π_c) from (a) to (f). The film contains 0.25 mol % of a lipidic dye probe. The images were observed at rather complicated subphase ionic conditions [13] but are typical for pH = 5 and the absence of divalent ions and presence of monovalent ions.

The impurity impedes growth and is removed faster for a rough boundary, which therefore is favored. Following this growth period, line tension tends to smoothen the boundary. This often leads to regular shapes like discs, lamellae, or spirals, which will be discussed later (Fig. 4). In other cases, however, smoothing is not terminated even after several hours; then structures like clover leaves or coffee beans may appear.

2. Fluorescence Spectroscopy

Measurements of fluorescence recovery after photobleaching (FRAP) can be used to determine the local diffusion coefficients [22]. In such measurements, a circular area, or a stripe pattern, is photobleached and the recovery of fluorescence intensity due to the

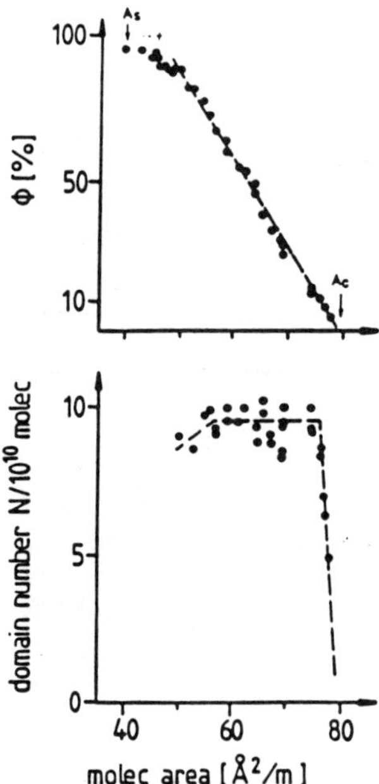

Figure 3 Domain number N (bottom) and condensed phase area fraction ϕ as a function of molecular area for a Myr_2 Ptd A monolayer. Indicated also are the molecular areas corresponding to the phase transitions at π_c and at π_s. (A_c, A_s)

diffusion of dye probes out of nonbleached areas is detected. This is done either in terms of the change of total emission from the photobleached spot [23] or in terms of change in the contrast by comparing previously irradiated and shaded areas [24]. The diffusion coefficient for phospholipid monolayers in the LE phase was shown by such methods to be larger than 10^{-8} cm^2/sec and below 10^{-10} cm^2/sec in the LC phase, in qualitative agreement with data for bilayer membranes [11,25–28].

For studies of diffusion mechanisms, monolayers have been proven to be extremely useful, since they allow a well-defined variation of the molecular area and compressibility. In the diffusion model applicable to the motion of small molecules in a membrane, the diffusion coefficient D is related to the free area $A_f = A - A_o$ per molecule according to [11,26,29]

$$\ln D = \ln a - \frac{b}{A_f} \tag{6}$$

A_o being the minimum molecular area, which is easily determined in monolayer experiments. Figure 5 gives the result of a measurement in which the molecular area A has been varied while maintaining the monolayer in the LE phase. Excellent agreement between theory and experiment proves the adequacy of the diffusion model.

Measurements in the coexistence range have also been analyzed using percolation models [26]. Diffusion under such conditions basically occurs within the LE phase, LC phase domains acting merely as obstacles. At high enough obstacle density, the connections between the large LE phase areas become very small; diffusion is then limited by

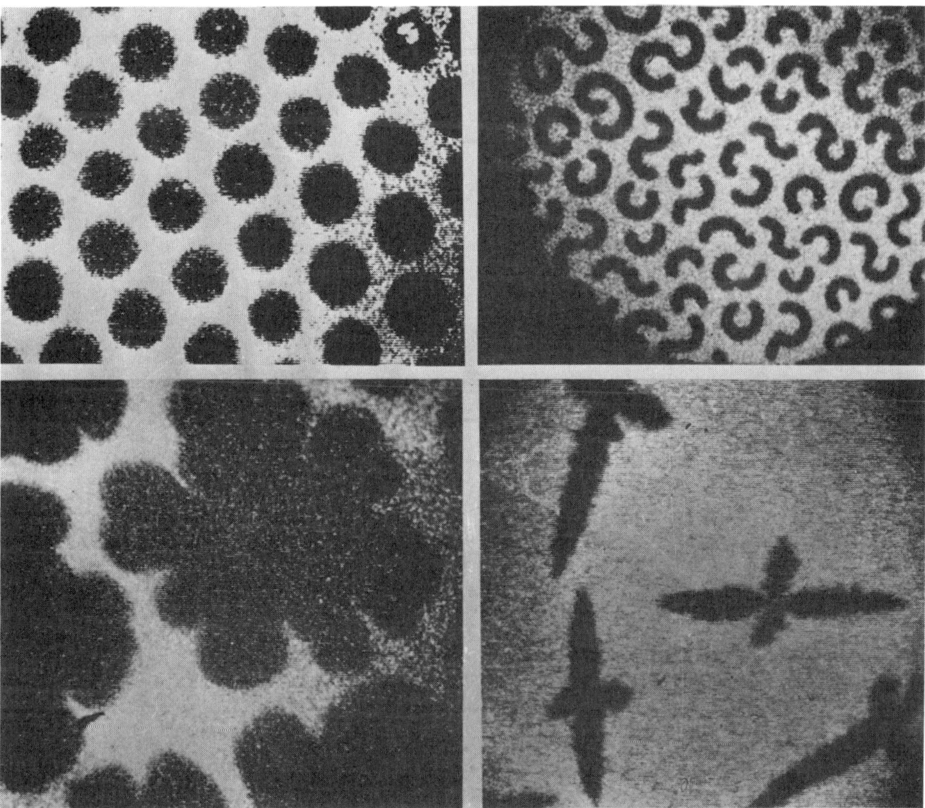

Figure 4 Fluorescence micrographs of various lipid monolayers in the LE/LC phase coexistence range. Upper Left: Myr_2 Ptd A under conditions of Fig. 2. Upper right: Myr_2 Ptd A containing 1 mol % cholesterol at pH 11. Lower left: Myr_2 Ptd A at ph 5, no other ions added. Lower right: A diacetylenic lipid in the unpolymerized state.

passages through these interconnects. The effective diffusion coefficient, therefore, depends on the area fraction of the LE phase.

Incorporating dimerizing dyes or different dyes suitable for the energy transfer measurements provides a means for the detection of local concentration changes at length scales below fluorescence microscopic resolution (~100–1000 Å). By studying fluorescence quenching and energy transfer from porphyrin to phthalocyanin molecules, it has been shown that phase separation occurs in monolayers of Myr_2 Ptd Eth with domains of the ordered phase being too small to be microscopically detected [30].

3. Ellipsometry

Recently it has been become possible to study monolayer microstructure by ellipsometry with films on water. With this technique one basically measures an ellipsometric angle and then tries to develop models for the extraction of structural monolayer parameters, such as film thickness, refractive index, and its anisotropy [31–34]. The results have been shown to be reasonably realistic, at least for one class of lipids (phosphatidylcholines), all along the isotherm.

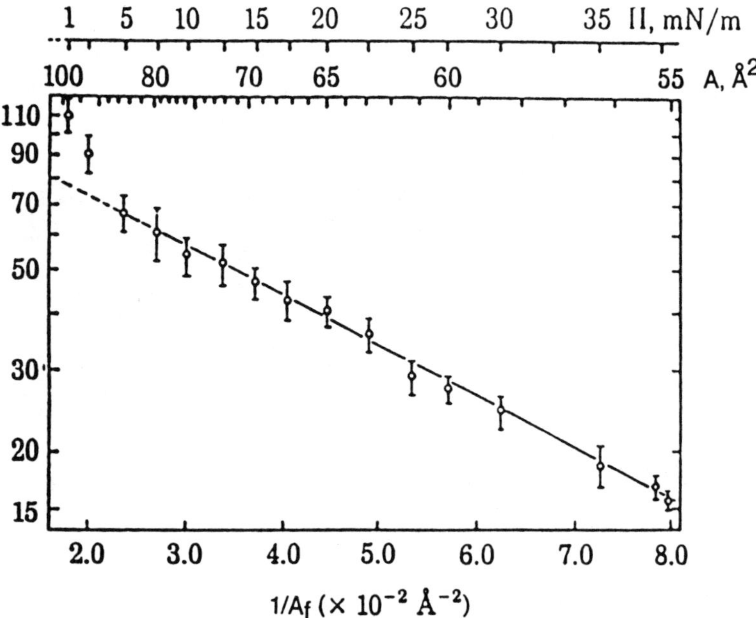

Figure 5 Translational diffusion coefficient D of a dye probe in Lau_2 Ptd Cho monolayers as a function of free area A_f defined in the text. T = 21–22°C. Also given are the corresponding molecular areas A and the surface pressure π. (From Ref. 11.)

C. X-Ray Studies

Use of synchrotons as brilliant x-ray sources has enabled x-ray studies with monolayers at the air/water interface [35,36]. In a typical experiment, a film balance is placed at the sample stage of a diffractometer [37]. The x-ray beam is bent to hit the surface at an angle below that corresponding to total reflection (~0.1°). In this mode, the beam penetrates by only 50 Å into solution and, if the surface is laterally periodical at the nm level, is diffracted. Measuring the diffracted intensity as a function of the in-plane diffraction angle provides information on the lattice structure.

Specular reflection as a function of incidence angle yields information on the electron density distribution ρ along the surface normal z. The measured reflectivity R, divided by the Fresnel reflectivity R_F, calculated for an ideal interface, is related to $\rho(z)$ according to [38]

$$\frac{R}{R_F} = \left[\int \frac{d\rho}{dz} \exp(iQ_z)\, dz\right]^2 \cdot \frac{1}{\rho_w^2} \tag{7}$$

where $Q_z = 4\pi \sin \alpha/\lambda$ is the wave vector transfer in the direction of z (λ = wavelength, α = incidence angle with respect to the surface), and ρ_w is the electron density of water.

X-ray reflectivity measurements can be applied also to nonperiodic structures, but the analysis depends on the choice of suitable models. Fortunately, it is possible to use very simple models with only a few independent and adjustable parameters in order to get information that is at least partly model independent. Experiments can be described with a "two-box model," for example, by dividing the surface into a slice of length l_t and density ρ_t, corresponding to the tail region on a slice (l_h, ρ_h) containing the head groups; the density step between all slices, moreover, is assumed to be smeared by a Gaussian of width σ [38,39].

Figure 6 gives a series of diffraction peaks as a function of lateral pressure. These peaks can be grouped in two regions divided by the pressure π_s. For $\pi < \pi_s$ the lines are weak and broad, whereas for $\pi > \pi_s$ they are much narrower and stronger. These results have been confirmed with other phospholipids and yield the following picture.

The lattice constant, calculated from the peak maximum (~4.2 Å), corresponds to the (1, 0) spacing of the hexagonal lattice (d_{10}) found in electron diffraction studies with monolayers on solid support [40]. d_{10} varies continuously upon going through π_s, whereas the compressibility.

$$\chi = -\frac{2}{d_{10}} \frac{\partial(d_{10})}{\partial\pi} \tag{8}$$

decreases by a factor of two upon increasing π above π_s. The calculated χ value is by about a factor of two smaller than the value of χ derived from isotherms (Eq. (2)).

Although the hexagonal symmetry of the electron diffraction pattern and the observation of merely one x-ray diffraction peak suggest a hexagonal arrangement of the aliphatic tails, in many cases this is not exactly valid. In fact, orthorhombic lattices have been observed frequently for various low temperature phases of phospholipid bilayers [41–43], alkane crystals [44,45] and fatty acid monolayers [46,70]. A break of the hexagonal symmetry, indeed, is expected whenever the aliphatic tails are tilted with respect to thc surface normal. Information on the tilt can be derived from the analysis of Bragg reflections along the surface normal (Bragg rod) [46,48]. When lipid chains modeled as cylindrical rods are uniformly tilted towards a lattice plane, the maximum of the diffraction intensity is moved out of the surface plane. The corresponding wave vector Q_z^{max} is related to the tilt angle t and to the azimuth ψ between the tail projection on the monolayer surface and normal to the surface according to

$$Q_z^{max} = \tan t \,.\, \cos \Psi \,.\, Q_x^{max} \tag{9}$$

As an example, Fig. 7 gives the diffraction intensity along the surface normal for the only diffraction spot measured for Myr_2 Ptd Eth monolayers at different lateral pressures [48]. The intensity has a maximum for $Q_z > 0$ at low pressures, which shifts towards $Q_z = 0$ with increasing pressure. This is diagnostic of the existence of a uniform tilt of the aliphatic tails and of the reduction of this tilt on compression.

Figure 8 reproduces a series of x-ray reflection measurements for Myr_2 Ptd Eth monolayers [49]; the corresponding points in the isotherm are given in the insert. One clearly observes a minimum that shifts to lower Q with increasing pressure. The minimum can be understood as a destructive interference between the wave reflected at the hydrocarbon/air interface with the wave reflected from the centers of the lipid head groups [49]. Hence for the minimum value of Q_z one obtains

$$Q_{min^{-1}} = \frac{1}{2\pi}\left(l_t + \frac{1}{2} l_h\right) \tag{10}$$

and the shift in Q_{min} is due to an increase in monolayer thickness on compression.

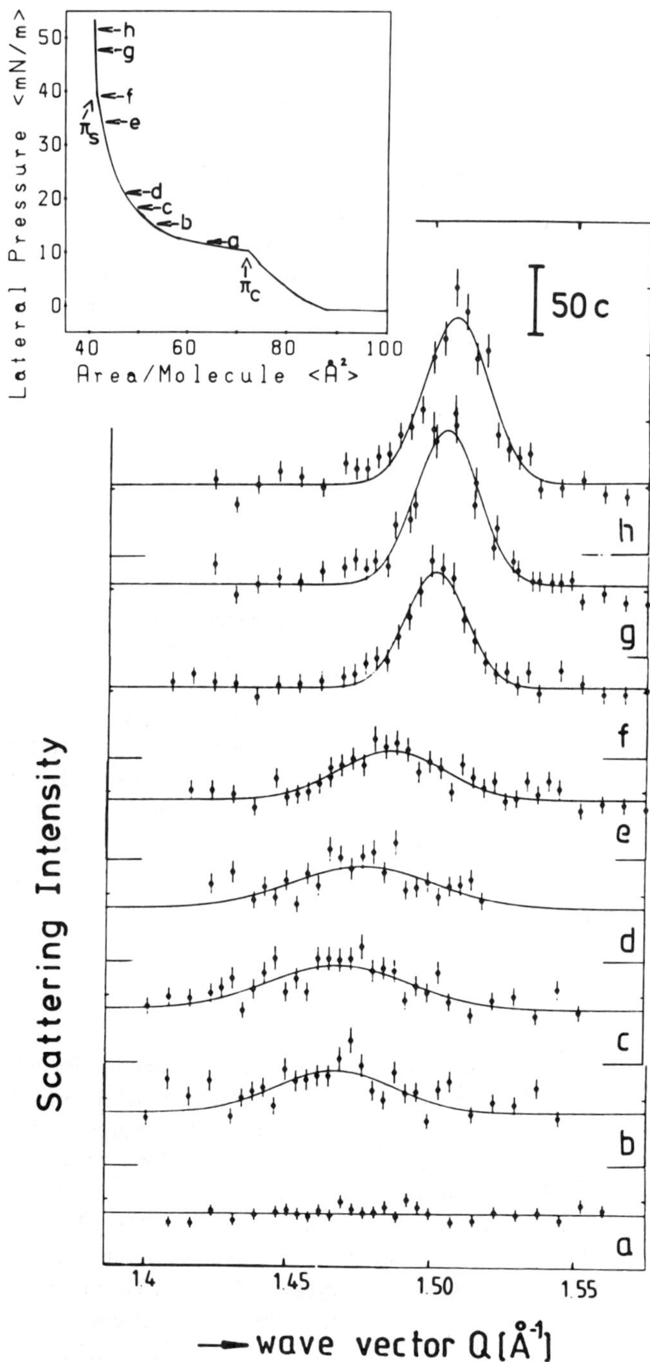

Figure 6 X-ray intensity as a function of in-plane wave vector transfer Q_x for a Myr_2 Ptd A monolayer at various surface pressures indicated by arrows in the isotherms (insert). pH 5.5; $T \approx 20°C$.

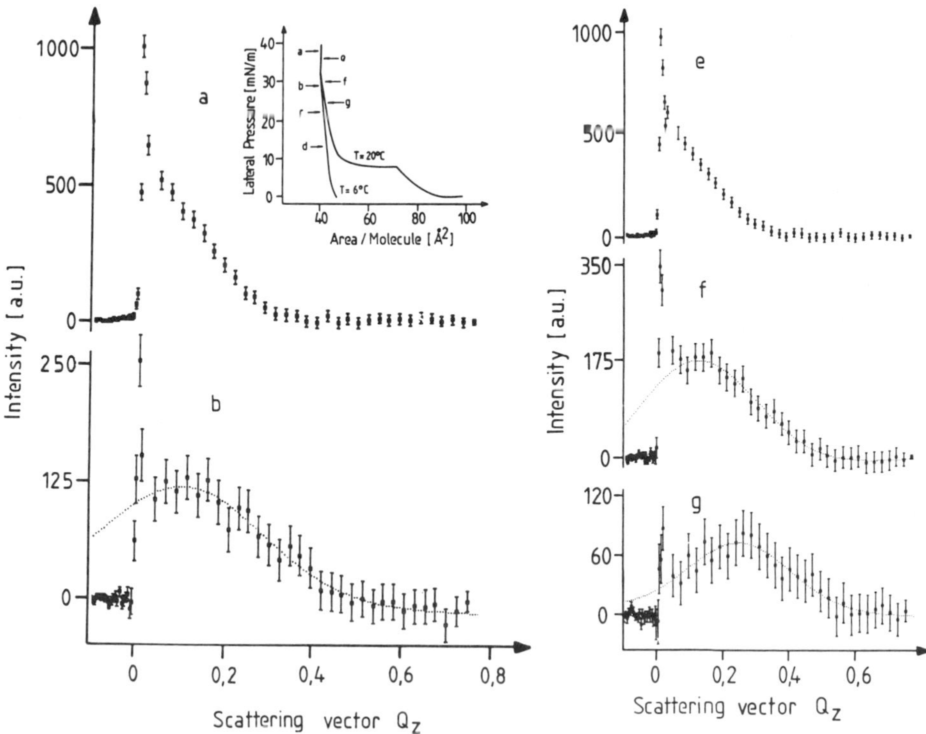

Figure 7 X-ray scattering intensity as a function of wave vector Q_z normal to the surface for an in-plane diffraction angle corresponding to a diffraction maximum. The corresponding pressures and temperatures of the Myr_2 Ptd Eth monolayer are given in the isotherm (insert).

The less pronounced depth of the minimum in curve c of Fig. 8 is because the underlying reflectivity results from the contributions of two phases with different density profiles, such roughness broadening all extrema. In other cases, the increased depth is qualitatively due to an increase in contrast between the maximum head group density and that of the adjacent moieties. The latter is because the contributions to ρ_h result from a large value from the phosphatidylethanolamine and a smaller value of hydration water, and the latter is squeezed out on compression.

D. Neutron Scattering

The vastly different scattering cross-section for hydrogen and deuterium has made neutron scattering a very powerful technique in membrane research [50]. Water can be made invisible to neutrons by index matching with suitable H_2O/D_2O mixtures, and different parts of a membrane can be studied preferentially by selective deuteration. However, compared to x-ray scattering, the gain resulting from high contrast is largely compensated by the loss due to the much lower light flux. Therefore neutron scattering still has to prove its unique features for monolayer research.

It has previously been shown that the neutron technique can yield valuable structural information on surfactant monolayers [51]; recently, monolayers of phosphatidylcholines

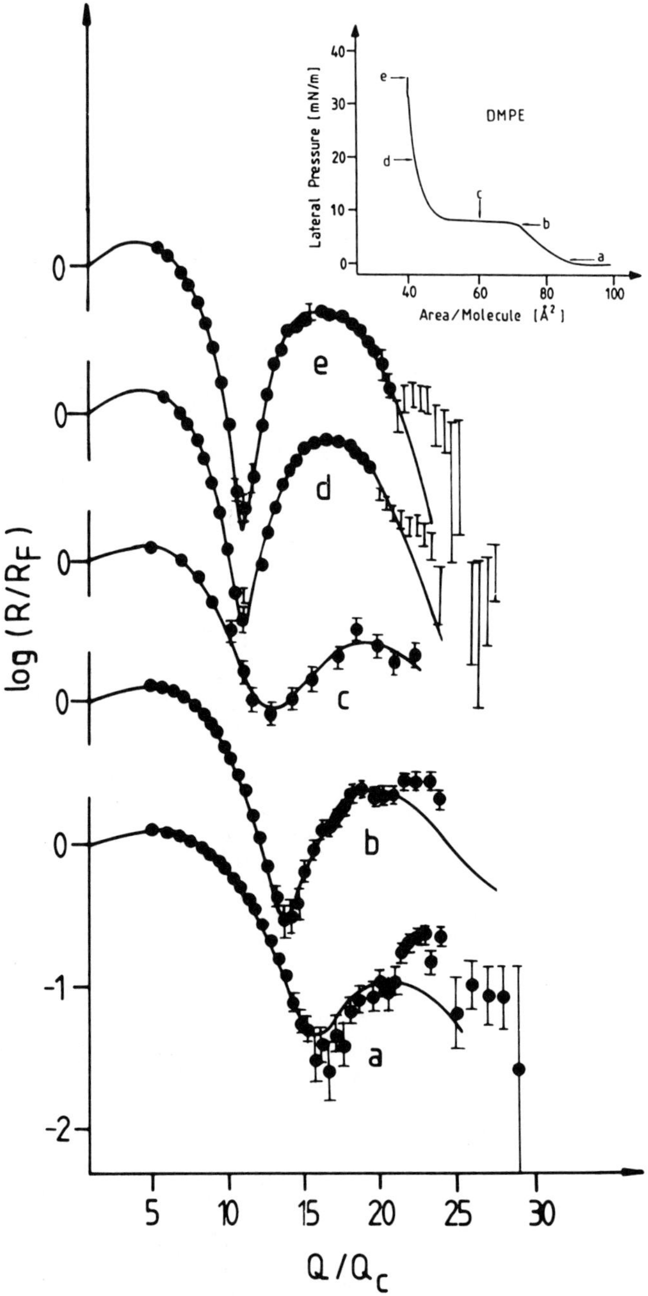

Figure 8 X-ray reflectivity normalized to the Fresnel reflectivity as a function of incidence angle given in units of the critical angle Q_c for a Myr_2 Ptd Eth monolayer at surface pressures indicated in the isotherm insert.

and -glycerols [52,53] have been studied in the same manner. Data analysis is based on various slab models, as in the case of x-ray reflection. To date, however, model parameters determined from neutron scattering data are less accurate than those derived from x-ray studies.

E. Infrared Techniques

Powerful lasers, detectors, and computers have now made Raman and Fourier transform (FT) infrared spectroscopy sensitive enough for studying monolayers on solid and water surfaces. Detailed studies with phospholipids have been performed by FT analysis of IR reflection from a water surface covered with DPPC monolayers in different phases [54–56]. In analogy to studies with bilayer membranes, shifts of frequency and intensity of the CH_2 stretching and bending vibrations have been detected on going from the LE to the LC phase. This supports the view that the corresponding lipid transition involves ordering of aliphatic tails. Changes of vibrations in the head group region have also been reported, but these are more difficult to interpret. These vibrations have the advantage that they respond to changes in orientation and local ordering but do not depend on long-range order. Presumably these changes occur on going from LC to S, so that IR spectroscopy will be a very sensitive tool for their characterization.

F. Scanning Tunneling and Atomic Force Microscopy

Both scanning tunneling and atomic force microscopy are applicable to biological samples in water and therefore have potentially a very great impact on biophysical studies of lipid microstructures [57,58]. Especially atomic force microscopy (AFM) carries many promises, being a technique that does not depend on a conducting substrate or film [59–61]. To date, the sensitivity of the force measurements relies on strong interactions, and therefore the scanning needle distorts the surface to a large extent. This technique, consequently, is applicable only on monolayers stabilized either by a covalent attachment to the substrate or by polymerization. Available data can be interpreted as an indication of the periodic head group arrangement [61]. Such an arrangement has been postulated before and should also show up as a superlattice in diffraction studies, which was never observed. It is possible that AFM, not depending on the long-range order, is more sensitive to the fine head group order. At present, however, while AFM is still in its infancy, one has to be cautious not to overinterpret experimental results.

G. Electron Optical Studies

With rare exceptions [62,63], electron optical studies of lipids have to be performed in vacuo and not on monolayers in situ. Despite this, the development of reliable methods for monolayer transfer has made these studies a very valuable tool for structural and phase studies.

Initial investigations were limited by the small monolayer contrast and thus involved staining methods or Pt shadowing and C evaporation, to obtain replicas. These studies elucidated surface roughness and thus the film stability and phase coexistence criteria [40]. A big step in the development of electron microscopic studies of lipid monolayers came through the invention of the charge decoration technique, however [14,64,65].

This technique is based on the fact that different monolayer phases are charged to a different degree by the primary electron beam. Thus a field develops at the phase boundary that deflects the electron beam. Due to this, the image of the boundary line appears dark, and domains of the different lipid phases can be visualized. This technique can be applied only in the low magnification mode ($\leq \times 5000$), which is sufficient, however, for studying lipid domains with sizes of several μm; domains are thus observed without using any dye probe.

One advantage of this technique is that it can be combined directly with electron diffraction of the selected area. In this way, specific monolayer areas with dimensions in μm can be selected for the local structure studies. In many cases, the correlation of the local crystallographic axes (bond orientational order) extends over more than 10 μm; the diffraction pattern then contains clearly visible spots [14,65]. The analysis of the spot profiles consequently yields information on the bond orientational (azimuthal width) and translational (radial width) order [66]. From a comparison of these profiles with simulations, one concludes that the ordered phases (LC, S) of phospholipids are most probably hexatic or tetratic [66].

III. BASIC FEATURES OF PHOSPHOLIPID MONOLAYERS

A. Thermodynamic Equilibrium?

Strictly speaking, a monolayer is not in thermodynamic equilibrium.

The equilibrium spreading pressure generally is just a few mN/m; for higher lateral pressures, the monolayer would prefer to form a three-dimensional lipid crystal on the water surface. Ordered monolayer phases are, therefore, generally metastable.

Lipid density at the air/water interface is on the order of 10^{-10} mol/cm^2; for a subphase of 1 cm depth this corresponds to a concentration of 10^{-6} M. If the lipid solubility is close to this value, a considerable fraction of the monolayer is dissolved in the subphase in thermodynamic equilibrium. For systems with two saturated aliphatic tails per head group, this is not a problem, but for lysophospholipids and substances with relatively large head groups, such problem may be critical. The question thus arises whether a monolayer can be considered to be a closed system at all.

Under specific conditions, a monolayer may transform spontaneously into a bilayer [67]. This, as well as the collapse into a multilayer film, is another mechanism of monolayer transformation into a more stable three-dimensional phase [68].

In many cases, the lack of thermodynamic equilibrium is not really a practical problem, since the nucleation of more stable phases is normally very slow or even undetectable for days. This means that one may still apply equilibrium thermodynamics, if one is aware of the restraints imposed by a mere local equilibrium. One also has to expect situations where the system is changed because of the induction of a more stable phase. Impurities and local distortions due to the lipid protein interactions may play a role, being nuclei for such processes.

Monolayers are nonergodic and may behave like a glass. Therefore they may be out of equilibrium in terms of long-range translational order, for example, but in equilibrium with respect to other order parameters or variables. It is consequently necessary to specify for each order parameter separately whether or not it is an equilibrium property.

B. Phase Diagrams

A general monolayer phase diagram is given in Fig. 9. At low molecular densities, a gaseous phase prevails. It is not completely disordered, as is the case in three-dimensional

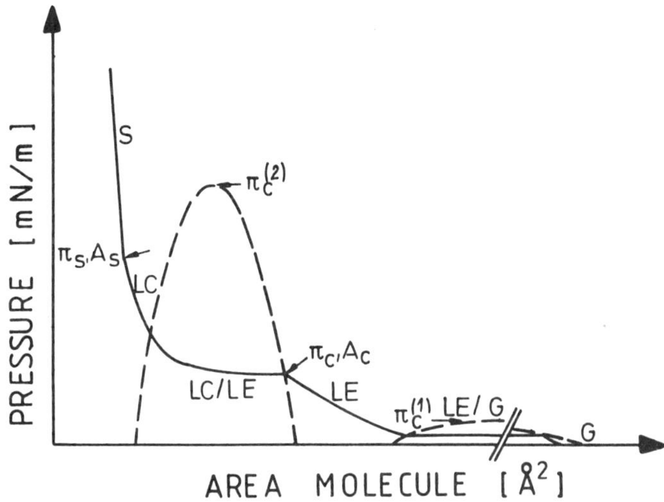

Figure 9 Monolayer phases going along an isotherm. The dashed lines mark coexistence ranges of LE and gaseous (G) phase and of LE and LC phase. Indicated are also the pressures $\pi_c^{(1)}$, $\pi_c^{(2)}$ corresponding to critical points.

systems, molecules still exhibiting a preferential orientation relative to the surface normal. At higher molecular densities, a liquid-expanded (LE) or an even more ordered liquid-condensed (LC) phase is reached through a region of phase coexistence; at high temperatures, one may continuously enter the S phase.

The average molecular area in the LE phase is between 150 and 50 Å^2, depending on the type of lipid molecules and the surface pressure. Its order resembles that of a gaseous phase except in that the lateral compressibility of the LE phase is considerably larger than that of a typical liquid, or of a bilayer membrane, say. In contrast to this, the three-dimensional compressibility χ_t of the hydrophobic membrane moiety, $\chi_t = (1/\rho_t)(\Delta\rho_t/\Delta p)$, is similar to that of simple liquids, being $\chi_t = 5 \cdot 10^{-5}$/atm for Lau_2 Ptd Eth. $\Delta\rho = (\Delta\pi/20$ Å$)$ is here a three-dimensional pressure.

Viscosity and diffusion coefficients of monolayers in the LE phase are similar to those of a fluid. Indirect surface viscosity measurements indicate that there may be yet another fluid phase of the same molecular density [2] with identical positional disorder but a preferential tail arrangement characteristic of the smectic c phases.

Diffusion coefficients in the LC phase are much lower than in the LE phase; the density of kink defects is also low. The aliphatic tails are parallel to each other with a cross-section close to that of alkanes in their rotator phases [44]. The chain tilt may be up to 30°, its value becoming continuously smaller on compression without a change in the volume density ρ_t. Every tail possesses six nearest neighbors, and the projections of tails on the surface form a centered rectangular lattice. In the LC phase, the orientation of local crystallographic axes and the tilt azimuth are correlated over long distances. The positional order extends over merely 10 lattice spacings, however, and is anisotropic, the long correlation range being normal to the tilt direction. At high pressures, the LC phase transforms directly into the S phase. In this S phase, the chain tilt is zero and the positional correlations extend over more than 100 lattice spacings. Head groups in such a phase are well aligned and little hydrated, but their positional and/or orientational order is not precisely known. This phase is characterized by a hexagonal lattice with a compressibility about a factor of three smaller than in the LC phase.

Recent x-ray diffraction experiments with DSPE monolayers have revealed that the centers of the tail projections may form a nonsymmetric triangle pattern with the tail cross section being distinctly below 20 Å^2 [69] and thus not much different from that of fatty acid chains in a monolayer, where the cross-section per chain is 18.6 Å^2 [70]. These structures correspond to the so-called herringbone structures of *n*-alcanes in which the long-axis rotation of chains is hindered [44,45]. In contrast to fatty acids, phospholipids give rise to broad diffraction peaks in the S phase. This is indicative of the interference between the head groups that link the two tails and prevent long-range positional ordering. Nevertheless, in the near future, the richness of phases now detected and characterized for fatty acids, long-chain alcohols, and esters [70–73] is likely to be discovered for phospholipids as well.

C. Phase Transitions

The transition from a gaseous to an LE or LC phase is clearly of first order for temperatures below a certain critical value $T_c^{(1)}$. The corresponding critical pressure $\pi_c^{(1)}$ is below 1 mN/m; monolayers are thus in a liquid phase at pressures above this value. The transitional entropy changes are due to the loss of translational and orientational degrees of freedom, the density being the appropriate order parameter for describing this transition.

For temperatures below a tricritical one, $T_c^{(2)}$, the LE $\Rightarrow$ LC transition is of first order. Its latent heat ΔH is of the same order of magnitude as that of the bilayer chain melting [2,17]. Approaching $T_c^{(2)}$ from below, ΔH approaches zero, often as $(T_c^{(2)}–T)$; from this, $T_c^{(2)}$ can thus be derived by extrapolation. The LE to LC transition is driven by the internal ordering tendency of aliphatic tails, but the three-dimensional density may also change remarkably; the electron density in the hydrophobic moiety may change by more than 10% [49], for example. This is to be compared with a change of at most 3% for the chain-melting phase transition in lipid bilayers [74,75]. Monolayer crystallization (LC phase) also involves a drastic increase in the correlation length of the bond orientational and tilt orientational order, and less so of the translational chain order. Up to now, there is little evidence for LE $\Rightarrow$ LC transitions also involving head group ordering. However, there are indications that in systems with a direct LE $\Rightarrow$ S transition [76] such ordering may occur.

All available data indicate that a LC $\Rightarrow$ S transition in lipid monolayers is of second order. The lattice spacing changes continuously, but the compressibility undergoes a discontinuous change. The lattice symmetry becomes different and the range of positional order correlations increases upon compression. It is as yet unclear whether or not the latter increase is continuous or discontinuous. Transition seems to involve a change in the head group environment, but it is not clear whether this results in an increased ordering.

D. Domain Structure

A unique feature of monolayers is their regular domain structure in the LE/LC phase coexistence range.

It has been pointed out that this can be understood in terms of an interplay between the line tension and long range electrostatic repulsion.

Impurities reduce line tension, by and large, and thus in turn reduce the free energy needed for the creation of a critical nucleus. The force tending to smooth and shorten the domain boundary simultaneously decreases. This can be observed by fluorescence microscopy; with increasing concentration of impurities, the density of domains increases and the domain shapes become carved and do not anneal for hours [6]. Higher compression rates also increase the domain density [17]. Thus the interdomain distance is a nonequilibrium property also influenced by impurities. The domain shape for some systems depends on the growth history, but there are also systems in which a local equilibrium is established after distortion. Thus the width of lamellar domains is uniform and can be reversibly varied, e.g., by temperature cycling. But this width can also be varied via impurity content owing to the influence of the latter on line tension [12].

Considering the gas/LC phase coexistence, one observes a foamlike structure whose dimensions change with time [10,77,78]. In nearly all practical cases, the corresponding transition is not induced by a continuous compression but rather enough material is spread to start a compression in the coexistence range. Domains are thus formed upon solvent evaporation; they increase in density upon compression; their structure and shape is fixed, however. At high domain densities, the interdomain interaction may cause deformation or disruption of domains.

IV. SYSTEM DIVERSITY

Hitherto we have described some basic features of simple phospholipid monolayers. In the following, complex systems will be dealt with, for which far fewer structural data exist and extrapolations therefore must often be made.

A. Saturated Straight Chain Lipids

The influence on latent heat and transition temperature of the chain length variability is reviewed elsewhere in this book.

Monolayers in this sense are qualitatively similar to bilayers: prolongation of aliphatic tails lowers the pressure at which the LE/LC transition occurs for a given temperature and head group. This is also true if head group repulsion is reduced by the addition of ions. Lipids with bulky or strongly hydrated head groups moreover impede the interchain attraction and thus the transition pressure is increased (e.g., compare Myr_2 Ptd Eth and Myr_2 Ptd Cho). The qualitative nature of LE and LC phases, however, remains unchanged with two exceptions. (1) The repulsive forces between the head groups may be so large that an LC phase is not formed at all; (2) the bulky head group may inhibit the formation of the S phase or cause the tilt angle to decrease continuously upon compression. The latter pertains to Pam_2 Ptd Cho (e.g.), which forms a ripple $P_{\beta'}$ phase in bilayer systems. In the corresponding lipid monolayer, the surface is inevitably flat and the tilt angle remains 30° up to the collapse pressure [39].

B. Unsaturated Bonds

It is known that unsaturated bonds tend to create disorder in the hydrophobic region and thus hinder LC phase formation. This again is equivalent to a temperature increase or a reduction of $T_c^{(2)}$.

C. Chain Branching

Chain branching creates orientational disorder with similar consequences as double bonds. A methyl or an ethyl branch near the lipid/air interface increases the transition pressure and reduces the transition enthalpy [79,80]. Similar groups near the head group have only a small influence, however, [9,81] because of the smaller interference with chain packing in the latter case. Short branches in the center of the hydrophobic region or branches of medium length distort the chain packing drastically and this may even prevent formation of the LC phase. If the branch is similar in length to the original tails and connected near the head group, the monolayer behaves as if it had an additional tail per head [82,83]. Subtle structural differences are still under investigation.

D. Different Head Group Features

In most cases, the interactions between lipid heads are repulsive, as has been studied in detail for phosphatidic acid under different ionic conditions [18,84–87]. Attractive interactions may arise, for example, from the most prominent hydrogen bonds between lipid head groups, as in the case of ethanolamine, serine or glycerol termini interacting with the phosphate oxygens. Dipoles connected with the lipid head groups also may attract each other if they are aligned properly [88], for example parallel to the surface.

Recent theoretical calculations have shown that counter ions and water molecules may screen these interactions and give rise to quadrupolar forces [88]. These forces decay over short distances and thus do not diverge with the domain size as is the case for dipolar interactions.

Often, functional groups have been attached in the hydrophilic membrane region, including dyes, specific for protein binding ligands, or polymerizable groups to link the head groups, to mention just a few. The dominant interaction between such modified groups depends on the detailed nature of each group, interesting new features arising from the presence of flexible segments in the head group region. Head group interactions in such systems can be varied by means known for the swelling of three-dimensional polymers, except in that the reduced dimensionality may induce interesting new physical phenomena.

E. Lipid Mixtures

Natural lipids are typical mixtures of phospholipids with different unsaturation and, sometimes, head group patterns. Isotherms, characteristic for such lipids, are typical for the LE phase. Notwithstanding this, phase separations might occur in such systems, as suggested by our knowledge with bilayers on the properties of defined mixtures for the case that one component prefers an ordered phase. It is possible, but difficult to prove, that components and phases in a monolayer are distributed heterogeneously, not lastly because systematic studies on defined mixtures of phospholipids are still scarce. Albrecht et al. have presented a phase diagram for lipids with similar tails but different head groups based on careful isotherm measurements [89]. They have established miscibility in the fluid phases and a miscibility gap in the overall LC phase. Systematic studies with phosphatidylethanolamine mixtures of varying chain length have been performed by the Halle group [90,91]. These show miscibility if the differences in chain lengths are merely $2CH_2$ units in accord with measurements on bilayer vesicles [92]. This is not necessarily so, as there is an important difference between lipid mixing properties in monolayers and in bilayers. In a monolayer the component with the longer tail separating in an LC phase increases its highly energetic interface with the air; in a bilayer, conversely, the corresponding energy is lower due to the solvation of the opposite monolayer.

F. Cholesterol/Lipid Mixtures

Owing to their great importance for the understanding of the functioning of biological membranes, mixtures of cholesterol and phospholipids have been studied in great detail. In most studies, phosphatidylcholines were used, but there are arguments for not generalizing these results. A very interesting feature in the phase diagram of phosphatidylcholine/cholesterol is the existence of fluid/fluid phase separations at low surface pressures [93,94]. Sufficiently far from the critical point, two phases appear to coexist, one with a minimum cholesterol/lipid ratio of about 1:1 and one with a maximum ratio of 1:6. (The latter value has led to specific molecular pictures of the cholesterol arrangement in the lipid matrix, resembling complex formation. The existence of such complexes, indeed, could explain why cholesterol tends to rigidify fluid lipid membranes.)

Recently, the notion has been introduced into theoretical calculations of the phase diagram that a phase cannot be characterized by a single variable [95,96]. One should distinguish between the positional order and the conformational order, the latter concerning the question if the tails are in all-trans configuration or if there is a substantial fraction of kink defects.

Such phase separations have recently gained interest since the domain shapes in fluid coexisting phases can be varied quickly by an external force [97]. Domains at local equilibrium and even thermally activated shape fluctuations can thus be studied in such phases [98].

The incorporation of cholesterol in an ordered lipid phase has been also considered as an "impurity" tending to accumulate at defect points or lines [99]. This could hold for any impurity, but the effects measured with bilayers are very specific for cholesterol. With monolayers, an additional highly specific influence of cholesterol on the domain shape has been observed. Even for cholesterol content below 1 mol %, an elongation of lamellar domains was detected for Pam_2 Ptd Cho and Myr_2 Ptd A at high pH, where the chains are tilted uniformly. This can be quantitatively explained by assuming that cholesterol arranges at the boundaries between LE and LC phases and thus reduces the line tension (the energy per boundary length). The fact that this effect is specific for cholesterol and for lipids with a large tilt suggests that such systems offer suitable steric conditions for the maximal interaction. The influence of cholesterol on the morphology of other phospholipid domains and the corresponding phase diagram is weak.

G. Protein/Lipid Mixtures

Interaction of proteins and other water soluble compounds with the lipid environment has often been studied by measuring the monolayer isotherms [100–103]. Water soluble proteins, like cytochrome c or spectrin, may bind electrostatically to negatively charged membranes [104,105]. This can lead to a reduction of the critical pressure required for the LE/LC phase transition. However, protein/lipid interactions, in general, are rather local; systems with a low protein/lipid ratio ($<$1:1000 mol/mol), consequently, behave as if phase separation occurred, at high protein/lipid ratios the features of pure monolayers being totally lost [106]. To detect lipid/protein interactions, more local probes are therefore needed. Fluorescence microscopy has shown, e.g., that many proteins partition in the LE phase preferentially but could not yield information on local interactions [106,107].

A crucial problem with many studies of protein/lipid interactions is that proteins tend to insert themselves into a membrane, unfold irreversibly, and form clusters. These clusters are seen by fluorescence microscopy and are very difficult to avoid [21]. One solution to this problem is to insert a small fraction ($\sim$1 mol %) of amphiphilic ligands into the monolayer and then study specific interactions. Successful examples of this have been the study of antibody hapten interactions [108–111], glycolipid binding to lectins [21], and lipobiotin steptavidin binding [112]. In the last case the binding constant is exceptionally large, and protein crystallization at the interface occurs [113].

In very elegant experiments, the diffusion of membrane bound antibodies has been investigated by FRAP [114–116]. These results suggest specific transport mechanisms due to partial network formation of the protein.

A great deal of effort has been devoted to studies of interactions of phospholipases with phospholipids by means of surface pressure, potential, and radioactivity measurements [117–122]. The results of these studies suggest that the lytic activity is increased in the phase coexisting range owing to the catalysis of protein action by defect areas at the domain boundaries. Observations by fluorescence microscopy have confirmed this [123].

The latter example has also demonstrated that this protein prefers a specific lipid arrangement for its optimum function, as has also been suggested and confirmed in previous experiments with bilayers for other proteins. Comparable monolayer examples are still missing. With respect to this, one should be aware that a monolayer, being only half of a membrane, is not a good model for studies of integral membrane proteins. But it is suitable for the investigation of the interaction of proteins with lipid head groups or with the adjacent water phase; e.g., electron and proton transport in photosynthesis thus can be explored nicely in monolayers.

V. FUTURE PROSPECTS

In this chapter our understanding of some of the most simple phospholipid systems has been reviewed. Biologically more relevant systems are, of course, not as simple. However, lack of space and partly also of knowledge prevented me from simply listing the wealth of nature. Instead, I wanted to spend more effort on elaborating the physical principles that can be studied by using phospholipid monolayers. This permits me to sketch some future research directions.

Very important questions remain to be solved with regard to ordering in the head group region. New techniques are now emerging that will permit conclusion on the head group arrangement; advanced x-ray and neutron techniques, optical second harmonic generation, FT-IR spectroscopy, and also computer simulations will probably all prove useful for this purpose and also will elucidate certain aspects of the interfacial water structure. These questions are important, since the hydrophilic membrane region mediates interactions with the adjacent water phase and also carries dipoles providing long-range interactions.

Having realized that short- and long-range order has to be described by many parameters, much work will have to be devoted to the understanding of coupling between these parameters. Clearly, answers will be different for different systems, and the question will become important how to control this coupling and how to describe and quantify its intensity. Correlations between various order parameters appear to decay on different length scales, leading to periodic density fluctuations. These may not only be a theoretical curiosity but also enable long distance information transfer.

In this latter aspect, the phospholipid membrane and monolayer has to be considered not only as an electrical but also as a mechanical medium. Techniques are now emerging with which local and global elastic properties can be measured, which surely will renew interest in this field.

Control over the lipid monolayer structure is related to an understanding of domain formation. Interesting aspects of physics in two dimensions (with negligible gravity) with local and global equilibrium properties emerge from this. An essential system parameter in this field is line tension, the energy per length of the boundary between two different phases. It remains to be shown how this line tension is influenced by the environmental parameters and lipid impurities.

In conclusion, I wish to stress that I do not think that these areas can be worked out before devoting more effort to systems more complex than monolayers. Despite this, answering the above questions will contribute much to our understanding of complicated biological systems.

REFERENCES

1. G. L. Gaines, *Insoluble Monolayers at Lipid-Gas Interfaces*, Interscience, New York, 1966.
2. O. Albrecht, H. Gruler, and E. Sackmann, *J. Phys.* (Paris) *39*:301 (1978).
3. D. A. Cadenhead, F. Müller-Landau, and B. M. J. Kellner, In *Ordering in Two Dimensions* (S. K. Sinha, ed.), Elsevier, Amsterdam, 1980, pp. 73 ff.
4. A. Caille, D. Pink, F. de Verteuil, and M. J. Zuckermann, *Can. J. Phys. 58*:581 (1980).
5. N. R. Pallas and B. A. Pethica, *Langmuir 1*:509 (1985).
6. A. Miller and H. Möhwald, *J. Chem. Phys. 86*:4258 (1987).
7. A. Georgallas and D. A. Pink, *J. Coll. Interf. Sci. 89*:107 (1982).
8. M. Lösche and H. Möhwald, *J. Coll. Interf. Sci. 131*:56 (1989).
9. G. Brezesinski, W. Rettig, S. Grunewald, F. Kuschel, and L. Horvath, *Liq. Cryst. 5*:1677 (1989).
10. M. Lösche, E. Sackmann, and H. Möhwald, *Ber. Bunsenges. Phys. Chem. 87*:848 (1983).
11. R. Peters and K. Beck, *Proc. Natl. Acad. Sci. USA 80*:7183 (1983).
12. R. M. Weiss and H. M. McConnell, *Nature 310*:5972 (1984).
13. M. Lösche, H. P. Duwe, and H. Möhwald, *J. Coll. Interf. Sci. 126*:432 (1988).
14. M. Lösche, J. Rabe, A. Fischer, B. U. Rucha, W. Knoll, and H. Möhwald, *Thin Solid Films 117*:269 (1984).
15. W. Hickel, D. Kamp, and W. Knoll, *Nature 339*:186 (1989).
16. S. Kirstein, H. Möhwald, and M. Shimomura, *Chem. Phys. Lett. 154*:303 (1989).
17. C. A. Helm and H. Möhwald, *J. Phys. Chem. 92*:1262 (1988).
18. C. A. Helm, L. Laxhuber, M. Lösche, and H. Möhwald, *Coll. Polym. Sci. 264*:46 (1986).
19. R. M. Weis and H. M. McConnell, *J. Phys. Chem. 89*:4453 (1986).
20. W. M. Heckl, D. A. Cadenhead, and H. Möhwald, *Langmuir 4*:1352 (1988).
21. H. Haas and H. Möhwald, *Thin Solid Films 180*:101 (1989).
22. W. L. C. Vaz, Z. I. Derzko, and K. Jacobson, *Membrane Reconstitution* (G. Poste and G. L. Nicolson, eds.) Elsevier, Amsterdam, 1982, p. 83–136.
23. D. Axelrod, D. E. Koppel, J. Schlessinger, E. Elson, and W. W. Webb, *Biophys. J. 16*:1055 (1976).
24. B.A. Smith, W. R. Clark, and H. M. McConnell, *Proc. Acad. Sci. USA 76*:5641 (1979).
25. D. Zenon and K. Jacobson, *Biochemistry 19*:6050 (1980).

26. K. Beck, in *Cytomechanics* (J. Breiter/Hahn, O. R. Anderson, and W.-E. Reif, eds.), Springer, Berlin, Heidelberg, 1987.
27. M. Seul and H. M. McConnell, *J. Physique 47*:1587 (1986).
28. K. Beck and R. Peters, *Eur. J. Cell. Biol.* Suppl. 2:5 (1983).
29. H.-J. Galla, W. Hartmann, U. Theilen, and E. Sackmann, *J. Membr. Biol. 48*:215 (1979).
30. M. Flörsheimer and H. Möhwald, *Thin Solid Films 159*:115 (1988).
31. D. Den Engelsen and B. De Koning, *J. Chem. Soc. Farad. Trans 170*:1603 (1974).
32. Th. Rasing, H. Hsiung, Y. R. Shen, and M. W. Kim, *Phys. Rev. A37*:2732 (1988).
33. A. F. Antippa, R. M. Leblanc, and D. Ducharme, *J. Opt. Soc. Am. A3*:1974 (1986).
34. D. Ducharme, J.-J. Max, C. Salesse, and R. M. Leblanc, *J. Phys. Chem. 94*:1925 (1990).
35. K. Kjaer, J. Als-Nielsen, C. A. Helm, L. A. Laxhuber, and H. Möhwald, *Phys. Rev. Lett. 58*:2224 (1987).
36. P. Dutta, J. B. Peng, M. Lin, J. B. Ketterson, M. Prakash, P. Geogopoulos, and S. Ehrlich, *Phys. Rev. Lett. 58:*2228 (1987).
37. C. A. Helm, H. Möhwald, K. Kjaer, and J. Als-Nielsen, *Biophys. J. 52*:381 (1987).
38. J. Als-Nielsen, W. Sommers, and P. Blanckenhagen, eds. *Structure and Dynamics of Surfaces,* Vol. 2, Springer, Berlin, Chap. 5, 1986.
39. C. A. Helm, H. Möhwald, K. Kjaer, and J. Als-Nielsen, *Europhys. Lett. 4*:697 (1987).
40. A. Fischer and E. Sackmann, *J. Phys. 45*:269 (1984).
41. M. J. Janiak, D. M. Small, and G. G. Shipley, *Biochim. Biophys. Acta 896*:77 (1976).
42. J. M. Seddon, G. Cevc, R. D. Kaye, and D. Marsh, *Biochemistry 23*:2634 (1984).
43. I. Pascher, S. Sundell, K. Harlos, and H. Eibl, *Biochim. Biophys. Acta 896*:77 (1987).
44. B. Ewen, G. R. Strobl, and D. Richter, *Farad. Disc. 69*:19 (1980).
45. J. Doucet, I. Denicolo, A. F. Craievich, and C. Germain, *J. Chem. Phys. 80*:1647 (1984).
46. J. Als-Nielsen and K. Kjaer, in *Phase Transitions in Soft Condensed Matter* (T. Riste and D. Sherrington, eds.) Plenum, New York, 1989.
47. G. S. Smith, E. B. Sirota, C. R. Safinya, and A. Clark, *Phys. Rev. Lett. 60*:813 (1988).
48. H. Möhwald, R. M. Kenn, D. Degenhardt, K. Kjaer, and J. Als-Nielsen, *Physica 168*:127 (1990).
49. J. Als-Nielsen and H. Möhwald, Synchrotron x-ray scattering studies of Langmuir films, in *Handbook of Synchrotron Radiation* (S. Ebashi, E. Rubenstein, and M. Koch, eds.), Vol. 5, North Holland, 1989.
50. G. Buldt, H. U. Gally, A. Seelig, J. Seelig, and G. Zaccai, *Nature 271*:182 (1978).
51. M. Grundy, R. M. Richardson, S. J. Roser, J. Penfold, and R. C. Ward, *Thin Solid Films 159*:43 (1988).
52. T. M. Bayerl, R. K. Thomas, J. Penfold, A. Rennie, and E. Sackmann, *Biophys. J. 57*:1095 (1990).
53. D. Vaknin, K. Kjaer, J. Als-Nielsen, and M. Lösche, *Biophys. J. 59*:1325 (1991).
54. R. A. Dluhy and D. G. Cornell, *J. Phys. Chem. 89*:3195 (1985).
55. M. L. Mitchell and R. A. Dluhy, *J. Amer. Chem. Soc. 110*:712 (1988).
56. R. D. Hunt, M. M. Mitchell, and R. A. Dluhy, *J. Mol. Struct. 214*:93 (1989).
57. D. P. E. Smith, A. Bryant, C. F. Quate, J. P. Rabe, Ch. Gerber, and J. D. Swalen, *Proc. Natl. Acad. Sci. USA 84*:969 (1987).
58. J. K. H. Hörber, C. A. Lang, T. W. Hänsch, W. M. Heckl, and H. Möhwald, *Chem. Phys. Lett. 145*:151 (1988).
59. W. M. Heckl, K. M. R. Kallury, M. Thompson, Ch. Gerber, H. J. K. Hörber, and G. Binnig, *Langmuir* (1990).
60. G. Binnig, C. F. Quate, and Ch. Gerber, *Phys. Rev. Lett. 12*:930 (1986).
61. A. L. Weisenhorn, M. Egger, F. Ohnesorge, S. A. C. Gould, S.-P. Heyn, H. G. Hansma, R. L. Ginsheimer, H. E. Gaub, and P. K. Hansma, *J. Phys. Chem.* (1991).
62. S. W. Hui, G. G. Hausner, and D. J. Parsons, *J. Phys. E9*:67 (1976).
63. S. W. Hui, and He Neng-Bo, *Biochemistry 22*:1159 (1983).
64. A. Fischer and F. Sackmann, Nature *313*:299 (1985).

65. A. Fischer and E. Sackmann, *J. Colloid Interface Sci. 112*:1 (1986).
66. I. R. Peterson, R. Steitz, H. Krug, and I. Voigt-Martin, *J. Physique 51*:1003 (1990).
67. K. Tajima, N. L. Gershfeld, *Biophys. J. 47*:203, 211 (1985).
68. D. Vollhardt and U. Rettig, *J. Phys. Chem.* submitted for publication (1990).
69. R. M. Kenn, H. Möhwald, and K. Kjaer, and J. Als-Nielsen, to be published.
70. R. M. Kenn, C. Böhm, A. M. Bibo, I. R. Peterson, H. Möhwald, K. Kjaer, and J. Als-Nielsen, *J. Phys. Chem. 95*:2092 (1991).
71. B. Lin, T. M. Bohanon, M. C. Shih, and P. Dutta, *Phys. Rev. Lett.*, in press (1991).
72. S. Ställberg-Stenhagen and E. Stenhagen, *Nature 156*:239 (1945).
73. A. M. Bibo, C. A. Knobler, and I. R. Peterson, *Macromol. Chem., Macromol. Symp., 46*:55. *J. Phys. Chem. 95*:5591 (1991).
74. J. F. Nagle and D. A. Wilkinson, *Biochemistry 20*:187 (1981).
75. G. Schmidt and W. Knoll, *Ber. Bunsenges. Phys. Chem. 89*:36 (1985).
76. M. Flörsheimer and H. Möhwald, *Coll. Surf, 55*:173 (1991).
77. B. Moore, C. M. Knobler, D. Broseta, and F. Rondelez, *J. Chem. Soc. Farad. II. 82*:1753 (1986).
78. A. Miller and H. Möhwald, *Europhys. Lett. 2*:67 (1986).
79. D. K. Rice, D. A. Cadenhead, R. N. A. H. Lewis, and R. N. McElhaney, *Biochemistry 26*:3205 (1987).
80. D. M. Balthasar, D. A. Cadenhead, R. N. A. H. Lewis, and R. N. McElhaney, *Langmuir 4*:180 (1988).
81. G. Förster, G. Brezesinski, *Liq. Cryst. 5*:1659 (1989).
82. P. Nuhn, G. Brezesinski, B. Dobner, G. Förster, M. Gutheil, and H.-D. Dörfler, *Chem. Phys. Lipids 39*:221 (1986).
83. A. Dietrich, H. Möhwald, W. Rettig, and G. Brezesinski, *Langmuir 7*:539 (1991).
84. S. Ohki, *Biochim. Biophys. Acta 689*:1 (1982).
85. M. Lösche, and H. Möhwald, *J. Coll. Interf. Sci. 131*:56 (1989).
86. H. Träuble, U. Teubner, P. Woolley, H.-J. Eibl, *Biophys. Chem. 4*:319 (1976).
87. S. A. McLaughlin, *Current Topics Membr. Transp. 9*:71 (1977).
88. P. Muller and F. Gallet, *J. Phys. Chem., 95*:3257 (1991).
89. O. Albrecht, H. Gruler and E. Sackmann, *J. Coll. Interf. Sci. 79*:319 (1981).
90. W. Rettig, C. Koth, and H. D. Dörfler, *Coll. Polym. Sci. 262*:745 (1984).
91. W. Rettig, H.-D. Dörfler, and C. Koth, *Coll. Polym. Sci. 263*:647 (1985).
92. A. Lee, *Biochim. Biophys. Acta 472*:285 (1977).
93. S. Subramaniam and H. M. McConnell, *J. Phys. Chem. 91*:1715 (1987).
94. C. L. Hirshfeld and M. Seul, *J. Physique 51*:1537 (1990).
95. J. H. Ipsen, G. Karlström, O. G. Mouritsen, H. Wennerström, and M. J. Zuckermann, *Biochim. Biophys. Acta 905*:162 (1987).
96. J. H. Ipsen, O. G. Mouritsen, and M. J. Zuckermann, *Phys. Rev.*, submitted for publication.
97. P. A. Rice and H. M. McConnell, *Proc. Natl. Acad. Sci. USA 86*:6445 (1989).
98. M. Seul and M. J. Sammon, *Phys. Rev. Lett. 64*:1903 (1990).
99. M. K. Jain, in *Membrane Fluidity in Biology*, Academic Press, New York, 1983, Chap. 1.
100. R. J. Davies, G. C. Goodwin, I. G. Lyle, and M. N. Jones, *Coll. Surf. 8*:261 (1984).
101. L. Ter-Minassian-Saraga, *Langmuir 1*:391 (1985).
102. I. I. Panaiotov, L. Ter-Minassian-Saraga, and G. Albrecht, *Langmuir 1*:395 (1985).
103. M. S. Briggs, L. M. Gierasch, A. Zlotnik, J. D. Lear, and W. F. de Grado, *Science 228*:1096 (1985).
104. Z. Kozarac, A. Dhathathreyan, and D. Möbius, *FEBS Lett. 29*:372 (1988).
105. J. Peschke and H. Möhwald, *Coll. Surf. 27*:305 (1987).
106. W. M. Heckl, B. N. Zaba, and H. Möhwald, *Biochim. Biophys. Acta 903*:166 (1987).
107. W. M. Heckl and H. Möhwald, *J. Mol. Electron 3*:67 (1987).
108. M. Piepenstock, H. Möhwald, and M. Lösche, *Makromolek. Chemie*, Macromol. Symp. *46*:301 (1991).

109. H. M. McConnell, T. H. Watts, R. M. Weis, and A. A. Brian, *Biochim. Biophys. Acta 864*:95 (1986).
110. D. G. Hafeman, V. von Tscharner, and H. M. McConnell, *Proc. Nat. Acad. Sci. USA 78*:4552 (1981).
111. R. Blankenburg, P. Meller, H. Ringsdorf, and C. Salesse, *Biochemistry 28*:8214 (1989).
112. S. A. Darst, M. Ahlers, P. H. Meller, E. W. Kubalek, R. Blankenburg, H. O. Ribi, H. Ringsdorf, and R. D. Kornberg, *Biophys. J. 59*:387 (1991).
113. L. Tamm, and H. M. McConnell, *Biophys. J. 47*:105 (1985).
114. S. Subramaniam, M. Seul, and H. M. McConnell, *Proc. Natl. Acad. Sci. USA 83*:1169 (1986).
115. L. K. Tamm, *Biochemistry 27*:1450 (1988).
116. S. Ramsac, H. Moreau, C. Riviere, and R. Verger, *Methods Enzymology*, in press (1990).
117. P. Vainio, J. A. Virtanen, P. K. J. Kinnunen, A. M. Gotto, J. T. Sparrow, F. Pattus, P. Bougis, and R. Verger, *J. Biol. Chem. 258*:5477 (1983).
118. T. Thuren, J. A. Virtanen, P. Vainio, and P. K. J. Kinnunen, *Chem. Phys. Lipids 33*:283 (1983).
119. T. Thuren, P. Vainio, J. A. Virtanen, O. Somerhavju, K. Blomqvist, and P. K. J. Kinnunen, *Biochemistry 23*:5129 (1984).
120. R. Verger and G. H. de Haas, *Ann. Rev. Biophys. Bioengin. 5*:77 (1977).
121. R. A. Demel, K. W. A. Wirtz, H. H. Kamp, W. S. M. Geurts van Kessel, R. F. A. Zwaal, and L. L. M. van Deenen, *Nature 246:*102 (1973).
122. D. W. Grainger, A. Reichert, H. Ringsdorf, and C. Salesse, *FEBS Lett.* (1989).
123. D. Hönig and D. Möbius, *J. Phys. Chem. 95*:4590–4592 (1991).
124. S. Hénon and J. Meunier, *Rev. Sci. Instrum. 62*:936–939 (1991).

17
Phosphotipid Vesicles

Helmut Hauser *ETH Zurich, Zurich, Switzerland*

I. INTRODUCTION

Phospholipid vesicles may be defined as lipid particles surrounded by one or several closed phospholipid bilayers. The space between bilayers and the internal cavity of the vesicle are filled with aqueous solvent. It should be noted that stable or metastable vesicle suspensions require the surrounding phospholipid bilayer to be in the liquid crystalline state. Despite an early start—phospholipids were discovered as early as 1811 [1]—there was not much advance in phospholipid chemistry until the late 1930s. The pioneering discovery of Gorter and Grendel [2] and Danielli and Davson [3] that phospholipid bilayers are fundamental structural elements of biological membranes aroused new interest. Schmitt and his coworkers [4,5] demonstrated for the first time that aqueous dispersions of phospholipids extracted from the myelin sheath give rise to x-ray diffraction patterns very similar to those of intact nerves. The idea of using aqueous phospholipid dispersions as models for biological membranes was born. Bangham and his collaborators showed in the 1960s that "phospholipids spontaneously form closed membrane structures in the presence of water" [6]. They stressed explicitly the usefulness of phospholipid vesicles as models for biological membranes. Most importantly, their work spurred a great deal of research activity devoted to the physicochemical characterization of phospholipids in aqueous dispersions. The final goal of a great number of studies performed by Abrahamsson, Bangham, Chapman, Dawson, de Haas, Luzzati, Saunders, Small, Tanford, Thompson, van Deenen and their coworkers between 1960 and 1980 was to gain insight into the physicochemical properties of cell membranes. As a result of these studies, lipid vesicles (liposomes) have advanced to fundamental tools now widely used in biochemistry, biophysics, and biology.

Phospholipids have been used also commercially for quite some time, e.g., as additives to foodstuffs, emulsifiers, cosmetics, medicinals, lubricants, etc. More recently they have become important as carrier-delivery systems in general and for drug application-targeting in particular. Naturally occuring phospholipids appear to be particularly suited for this purpose, for they are biodegradable and nonantigenic. In the light of the ever-increasing importance of phospholipid vesicles (liposomes), it is not surprising that a flood of original publications, review articles, and monographs on phospholipid vesicles has appeared in the past ten years. In view of the vast expansion of this field, a comprehensive review on phospholipid vesicles is unrealistic. I shall address my discussion to some points that, to my mind, have not been emphasized appropriately or have even been neglected in the past, and to some recent developments. Furthermore, my discussion will be restricted to vesicles prepared from naturally occurring phospholipids and their synthetic analogs. A discussion of synthetic amphiphiles that form bilayer vesicles but do not occur naturally is beyond the scope of this paper. Such amphiphiles may have important applications, because they can be synthesized with reactive groups that undergo interesting polymerization, polycondensation, and other reactions. By this means, specific vesicle properties may be achieved, such as long-term stability or a specific change in bilayer permeability. The reader interested in this research field is referred to a recent, comprehensive review by Ringsdorf et al. [7] and to Chap. 7 in Part I of this handbook.

II. NOMENCLATURE

According to IUPAC's guidelines, the terms lipid vesicle and liposome should be used synonymously. The following terms and abbreviations are now generally accepted. Multilamellar vesicles or liposomes (MLVs) are defined as spherical particles consisting of concentrically arranged, equally spaced bilayers that are separated by layers of water. MLVs give rise to a characteristic x-ray diffraction pattern comprising a series of sharp reflections in the low-angle region and a broad, diffuse reflection in the wide-angle region. The low-angle pattern consists of a series of sharp reflections in the ratio 1:1/2:1/3:1/4, etc., which is due to the lamellar stacking of the bilayers. The single diffuse wide-angle reflection at 1/4.6 $Å^{-1}$ is due to the lateral packing of the hydrocarbon chains in the liquid crystalline state. Unilamellar vesicles are defined as spherical particles consisting of a single, closed bilayer that surrounds an aqueous cavity. The x-ray diffraction pattern of dilute dispersions of unilamellar vesicles is quite different from that of MLVs. Instead of the series of sharp low-angle reflections, a broad scattering envelope is observed, which by computer simulation can be shown to be the one-dimensional Fourier transform of the electron density profile of a single phospholipid bilayer (see below). Unilamellar vesicles are divided, though somewhat arbitrarily, into two classes according to their size; vesicles with diameters larger than about 100 nm are referred to as large unilamellar vesicles (LUV), vesicles under 100 nm as small unilamellar vesicles (SUV). One justification for this definition of LUV and SUV comes from high-resolution NMR: SUV give rise to a fairly good high-resolution NMR spectrum with chemically shifted resonances, whereas LUV give an unresolved, broad-line NMR spectrum. Still, the line of division is arbitrary, and other authors refer to vesicles below 50 nm in diameter as SUVs and above as LUVs, since packing constraints decrease markedly as the vesicle diameter exceeds about 50 nm [8].

III. LYOTROPIC AND THERMOTROPIC MESOMORPHISM

Phospholipids are amphiphilic, surface-active molecules that have a high tendency to form aggregates both in the dry and in the fully hydrated state. The different kinds of aggregates are termed phases. Generally, lipids are much smaller in weight and size than macromolecules such as proteins, polysaccharides, and ribonucleic acids; yet they are sufficiently large to have distinct regions of greatly different polarity: the hydrophobic (apolar) region and the hydrophilic (polar) region. The principle governing the aggregation (association) of lipids is maximization of the interaction energy (or minimization of the free energy). This is apparently accomplished by the association of like regions, i.e., polar groups associate with each other and the same is true for apolar groups. All lipid phases studied to date exhibit this basic principle of lipid packing.

One fundamental property of phospholipids that concerns us here is the thermotropic and lyotropic mesomorphism. Single-crystal structures of phospholipids represent minimum free energy structures (for reviews see Refs. 9 and 10) and are probably the best documentation of the basic principles governing lipid aggregation. In all single-crystal structures of phospholipids, lipid molecules pack as bilayers, or alternatively, but much less frequently, as single-layered structures. Upon heating, lipid crystals do not pass directly to an isotropic liquid but rather form one or several intermediate states or phases. These states have been referred to as mesomorphic or liquid crystalline because they are, in their properties, between the crystalline and the liquid state. The temperature at which the transition from the crystalline to the mesomorphic (liquid crystalline) state (phase) occurs is the transition temperature T_c (analogous to the Krafft temperature of soap-water systems). The phenomenon of (an) additional lipid phase(s) between crystal and isotropic liquid is referred to as thermotropic mesomorphism. Similarly to this phenomenon, lipids do not pass from the crystal to a true solution when interacting with H_2O. Depending on the water content and temperature, different hydrated phases form until, at infinite dilution, an ideal solution of single lipid molecules exists. This behavior is termed lyotropic mesomorphism. Lyotropic phases also exhibit thermotropic mesomorphism, i.e., the particular lipid phase observed depends on both water content and temperature. The phase behavior and polymorphism of lipids has been studied extensively in the past; particular attention has been paid to the mode of lipid packing and dynamics, i.e., the molecular and segmental motion of the liquid crystalline state [11,12]. For a discussion of lipid polymorphism see Chap. 12.

A. The Swelling Properties of Phospholipids in Aqueous Media

It is quite clear that phospholipid vesicles form spontaneously or can be made from a variety of lipids and lipid mixtures. To date, however, most reports have focused on phosphatidylcholine vesicles. In this review, the discussion will be extended to the vesiculation of less frequently occuring phospholipids such as acidic phospholipids.

There are fundamental differences in the swelling behavior of neutral (isoelectric) and negatively charged phospholipids. This is manifest in differences in the resulting equilibrium structures present in excess H_2O. The correlation between the swelling behavior of lipids in water and the equilibrium structure present in excess water has been pointed out before [13] (see also Chap. 15). The fundamental difference in swelling behavior between neutral and charged lipids is depicted in Fig. 1. The swelling in water of pure phosphatidylcholine bilayers and phosphatidylcholine bilayers doped with a few percent of positively or negatively charged lipids is shown in Fig. 1a. Pure phosphatidylcholine bilayers show limiting swelling in H_2O; the lamellar repeat distance d increases with the total water content, reaching a limiting value of $d \sim 6.5$ nm at about 40% H_2O [14–18]. Up to this limit, a single, hydrated lamellar phase exists, and above this limit a two-phase system is present consisting of a fully hydrated, lamellar phase with $d = 6.5$ nm and excess water. It is important to note that the swelling of phsophatidylcholine bilayers is initiated and greatly facilitated at temperatures above the transition temperature T_c.

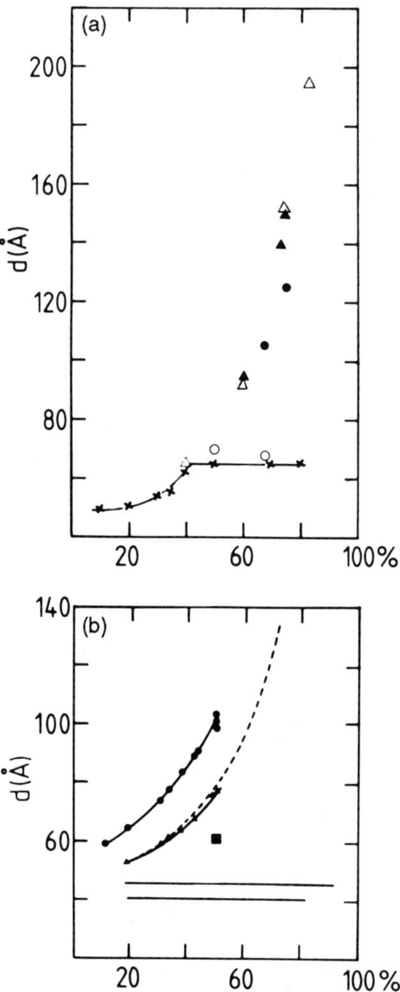

Figure 1 (a) Lamellar repeat distance d (Å) as a function of water content in % for pure egg phosphatidylcholine dispersions (×) and mixed dispersions of egg phosphatidylcholine containing 0.6% cetyltrimethylammonium bromide (CTAB) (○); 0.9% CTAB (●), 1.5% CTAB (△), and 3.5% sodium oleate (▲). (Adapted from Ref. 22.) (b) Lamellar repeat distance d (Å) as a function of water content in % for NH_4^+ dimyristoylphosphatidylserine at 20°C (●) and 50°C (▲). The swelling curves are for phospholipid dispersed in 0.025 M ammonium phosphate buffer pH 6.8. The data point at 50% buffer (■) gives the lamellar repeat d of NH_4^+ dimyristoylphosphatidylserine in buffer containing 0.5 M or more NaCl. For comparison, the swelling curve of the Na^+ salt of ox brain phosphatidylserine in water is included (dotted line). The solid lines at $d = 42$ Å and $d = 45.7$ Å refer to dimyristoylphosphatidylserine and dipalmitoylphosphatidylserine, respectively, both in the protonated form. The suspending medium is 0.01 M HCl. (From Ref. 13.)

Essentially all neutral and isoelectric lipids, e.g., phosphatidylcholines, phosphatidylethanolamines, sphingomyelins, monoacylglycerols, glycoglycerolipids such as mono- and diglycosyl diacylglycerols, and glycolipids such as cerebrosides exibit this type of swelling behavior [14–19].

The swelling behavior typical for an acidic phospholipid is depicted in Fig. 1b. The lamellar repeat distance *d* of NH_4^+ dimyristoylphosphatidylserine, for example, increases continuously with water both in the crystalline and in the liquid crystalline state [20]. The swelling of NH_4^+ dimyristoylphosphatidylserine at 50°C in the liquid crystalline state is within experimental error identical to that of the Na^+ salt of ox brain phosphatidylserine at 20°C [21]. The latter phospholipid is liquid crystalline at 20°C, and continuous swelling is observed up to d = 13.0 nm at ~70% H_2O. At water contents >70%, the x-ray diffraction pattern becomes broad and diffuse as a result of the multilayer stacking disorder [18]. Because of this, excessive swelling of phospholipids with water is not readily detected by x-ray diffraction at water contents exceeding 70–80%.

The presence of electrolyte inhibits or reduces the swelling of lipid bilayers; at room temperature and 50% H_2O containing ~0.5 M NaCl, the lamellar repeat distance *d* of NH_4^+ dimyristoylphosphatidylserine is decreased to d = 6.2 nm, as shown in Fig. 1b [13]. This indicates that swelling in water is affected by electrostatic phenomena depending on the surface charge density of the lipid bilayer and the electrolyte present. Screening of the lipid surface charges by counter ions prevents or reduces swelling of acidic lipid bilayers. For instance, at pH 2 phosphatidylserines are isoelectric, and under these conditions no swelling is observed [13]. Aqueous dispersions of dimyristoyl- and dipalmitoylphosphatidylserine at pH 2 and room temperature show no swelling with water. As shown in Fig. 1b, the lamellar repeat distance *d* of both lipids is invariant over the total water range studied, yielding *d* values of 4.20 ± 0.05 nm and 4.57 ± 0.03 nm, respectively [13]. These values are identical to *d* values obtained for crystalline dimyristoyl- and dipalmitoylphosphatidylserine in the protonated form [20].

Continuous swelling of acidic phospholipids and its salt dependence (Fig. 1) was first reported by Schmitt and his coworkers [4,5]. Figure 1 shows that such behavior can be induced with partly charged bilayers. By doping phosphatidylcholine bilayers with small quantities of amphiphiles, either positively or negatively charged, continuous swelling is induced [22,23] (Fig. 1). Partly charged bilayers thus behave essentially as bilayers of pure acidic phospholipids. The critical surface charge density in 0.1 M electrolytes seems to be 1–2 $\mu C/cm^2$. For comparison, the surface charge density of bilayers of natural and synthetic phosphatidylserines varies between about 20 to 40 $\mu C/cm^2$. Less than 1/10 of this surface charge density is required to produce continuous swelling, i.e., swelling properties characteristic of acidic phospholipids. It is important to note that lipid mixtures widely used in liposome research and technology consist of neutral (isoelectric) lipids (90%) and a minor acidic component (10%). The surface charge density of such mixtures very likely exceeds the threshold value of 1–2 $\mu C/cm^2$, and the equilibrium structure of such lipids in excess water will certainly be different from that of really neutral phospholipids such as phosphatidycholines.

The swelling properties of phospholipids depicted in Fig. 1 allow us to draw some conclusions concerning the thermodynamically stable structures of phospholipids present in excess water. The thermodynamically stable structure of neutral and isoelectric phospholipids present in excess water at temperatures $T > T_c$ is predicted to be the fully hydrated, swollen multilamellar vesicle (MLV). Evidence from x-ray diffraction and electron microscopy supports this notion. In contrast, the thermodynamically stable

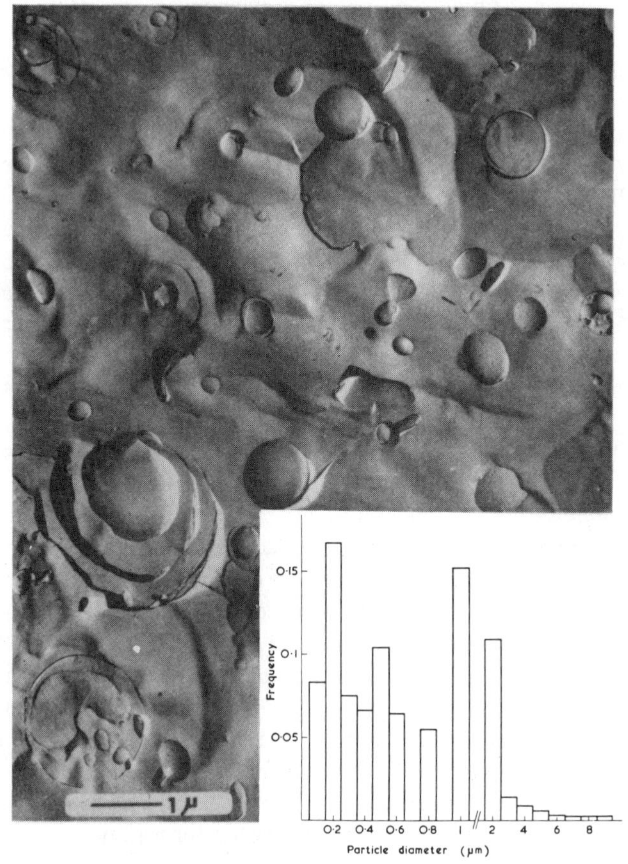

Figure 2 (a) Representative electron micrograph of a freeze-fractured sample of a 5% unsonicated dispersion of sodium ox brain phosphatidylserine in water. Inset: bar histogram derived from electron micrographs similar to that shown. The frequency fi/Σ_i fi is plotted as a function of the particle diameter. (b) Electron micrograph of a thin sectioned sample of a 1% unsonicated, aqueous dispersion of egg phosphatidylcholine containing 10% egg phosphatidic acid. The bar is 200 nm.

structure of acidic phospholipids and lipid mixtures with a surface charge density exceeding 1–2 $\mu C/cm^2$ should be the large unilamellar vesicle (LUV). Freeze-fracture electron microscopy lends support to this view. As an example, an electron micrograph of a freeze-fractured preparation of a 1% ox brain phosphatidylserine dispersion in water is shown. A wide size distribution of LUVs is observed ranging from 0.1 to several micrometers, with 95% of the vesicles having diameters between 0.1 and 2 μm (Fig. 2a). Cross-sections of vesicles reveal that smaller vesicles are frequently enclosed in larger ones. A smaller vesicle entrapped in the cavity of a large vesicle can be made visible by using thin sectioning methods (Fig. 2b). Furthermore, as mentioned already, the sharp low-angle diffraction pattern of acidic phospholipids becomes diffuse at water contents exceeding about 70%, and eventually the low-angle diffraction pattern characteristic of ordered, multilamellar vesicles is replaced by a broad, diffuse scattering peak (Fig. 3). This scattering peak can be shown to be the one-dimensional Fourier transform of the electron density profile of a single phospholipid bilayer [21]. Therefore, small-angle x-ray

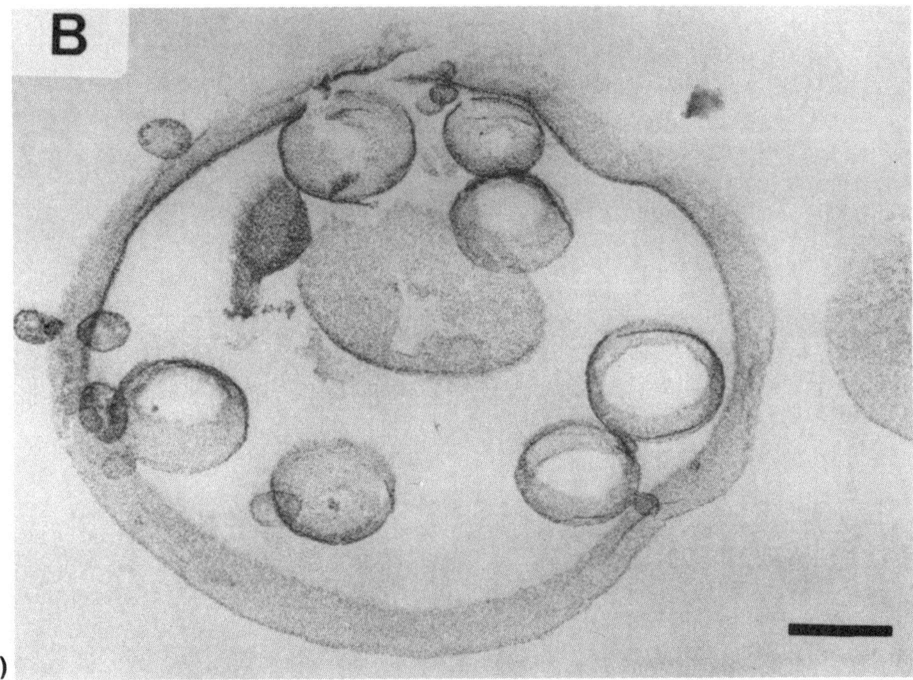

(b)

scattering also provides good evidence for LUV being the dominant particle of acidic phospholipid dispersions in excess H_2O.

The above discussion on the swelling of lipids in H_2O and the resulting thermodynamically stable structures in excess H_2O is summarized schematically in Fig. 4.

IV. PREPARATION OF PHOSPHOLIPID VESICLES

A prerequisite for the preparation of well-defined phospholipid vesicles is the use of pure, chemically well-defined lipids. Frequently, commercial sources of phospholipids are used. These lipids can vary widely in purity from firm to firm and even from batch to batch within the same firm. It is therefore imperative to monitor the purity of lipids, preferably by several independent methods, e.g., C, H, N microanalysis, TLC, HPLC, NMR, etc. (see Chap. 23). Although a few firms provide phospholipids of high quality that may be used without further purification, usually purification of commercial lipids is required. Standard procedures for the purification of naturally occurring phospholipids are now available comprising preparative TLC, HPLC, silicagel chromatography, ion exchange chromatography, etc. [24].

A. Preparation of MLVs and SUVs

In the following section, more general aspects of the preparation of phospholipid vesicles are discussed. The reader interested in details is referred to the book by R. New [25]. Neutral (isoelectric) phospholipids were available in pure form and in relatively large quantities earlier than the less frequently occurring negatively charged phospholipids. Therefore MLVs made from the former phospholipids were first prepared, characterized, and studied in some depth [6,26,27].

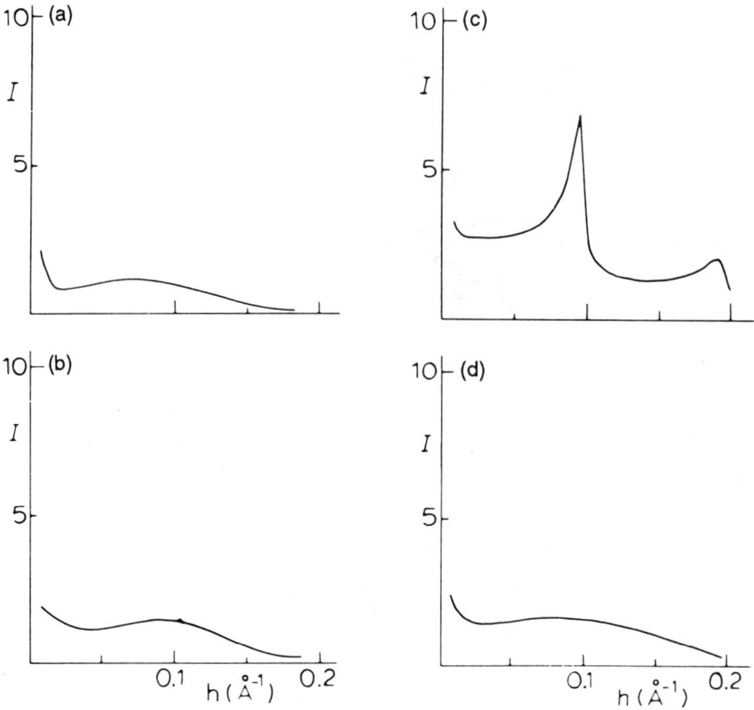

Figure 3 X-ray scattering curves from dilute aqueous dispersions of Na^+ ox brain phosphatidylserine and egg phosphatidylcholine. (a) 2.5% unsonicated phosphatidylserine; (b) 5% sonicated phosphatidylserine; (c) 5% unsonicated phosphatidylcholine; (d) 5% sonicated phosphatidylcholine. The dispersion in (d) was fractionated by gel filtration on Sepharose 4B. The peak fractions containing small unilamellar vesicles of diameter of 20–30 nm were pooled and used for x-ray scattering. I is the experimental scattering intensity; $h = 4\pi \sin \Theta/\lambda$, where 2Θ is the scattering angle and λ the wavelength. (Adapted from Atkinson et al. [21].)

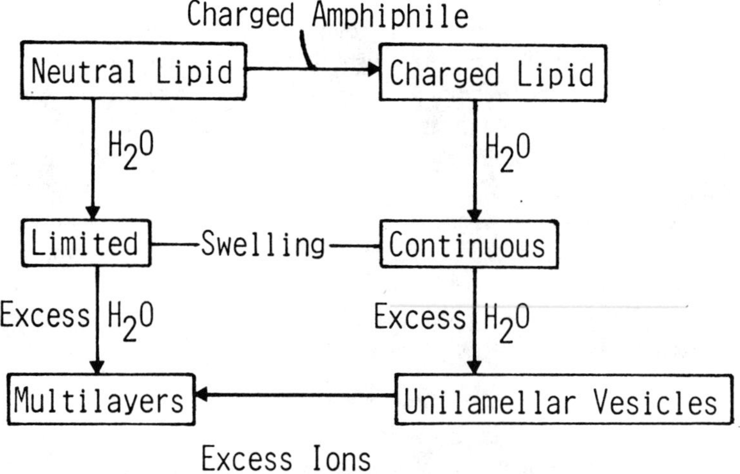

Figure 4 Scheme summarizing the swelling properties of neutral (zwitterionic) lipids and charged lipids. We note that any lipid mixture with a surface charge density exceeding 1–2 $\mu C/cm^2$ will behave like a charged lipid.

The preparation of MLVs is quick and easy: neutral (isoelectric) phospholipids are deposited from organic solvents as thin films on the glass wall of a round-bottom flask, usually by rotary evaporation under reduced pressure. The resulting film is thoroughly dried under vacuo. X-ray diffraction indicates that these films consist of stacks of crystalline or liquid crystalline bilayers depending on the residual water content and the temperature. MLVs are produced from such films by adding aqueous media and by some kind of mechanical agitation (hand-shaking, whirling-mixing, etc.). The addition of the aqueous solvent and the dispersion of the lipid are carried out at temperatures $T > T_c$. The properties of the resulting MLVs, such as size and number of lamellae, depend sensitively on the experimental conditions, not only on the conditions of dispersing the lipid film but also on those of depositing this film. This explains the variation in properties observed with MLV dispersions prepared under assumedly identical conditions as to lipid composition and concentration, pH, ionic strength, and ionic content of the aqueous medium. Unfortunately, the parameters determining the size and number of lamellae of MLV dispersions have not been studied sytematically. The resulting MLVs are heterogeneous in size, usually varying from 0.1 to 10 μm.

The preparation of LUVs is analogous to that of MLVs, except that negatively charged phospholipids or lipid mixtures are used. The resulting LUVs are also heterogeneous in size, larger vesicles frequently entrapping smaller ones, as shown in Fig. 2. The term multivesicular vesicle has been proposed for vesicles within a vesicle [28] in order to distinguish this kind of vesicle from ordinary MLVs.

The method described for preparing MLVs and LUVs is among the simplest and also the quickest ones. Upon the addition of water to the dry lipid film, vesicles form spontaneously, suggesting that the resulting vesicles represent equilibrium structures. MLVs and LUVs differ in their trapped volume (encapsulation capacity) expressed as liter per mole of lipid. The trapped volume of MLVs is relatively low (1–4 L/mol lipid), while that of LUVs is severalfold larger than that of MLVs of comparable size; values as high as 30 L/mol lipid have been reported for LUVs.

B. Methods of Homogenization

The major drawback of MLVs and LUVs prepared as described above is their heterogeneity in size. This is a general difficulty inherent in most methods of vesicle preparation. Homogenization and a concomitant reduction of the number of bilayer lamellae surrounding each vesicle is achieved by exposing phospholipid dispersions to shear forces.

There are several methods based on this principle; in any case, the phospholipid suspension is energized, and as a result of an energy input the bilayers undergo repeated disruption-resealing cycles. By this, unilamellar and oligolamellar vesicles of a more homogeneous size distribution are produced. External energy is supplied to phospholipid suspensions by means of ultrasonic irradiation [29], French press extrusion at generally very high pressures [30,31], and repeated extrusion at low or medium pressures through membrane filters of defined pore size [32]. Ultrasonication using a probe produces mainly SUV with a hydrodynamic diameter between 18 and 80 nm and a minor proportion ($\sim$10%) of MLVs and partly fragmented MLVs. The resulting vesicle size and size distribution depend on the time and input power of ultrasonication, on phospholipid composition and concentration, and on several other experimental parameters (sample volume, geometry of the probe, temperature, etc.) discussed in more detail in Ref. 25. Many of these parameters are difficult to standardize, which accounts for the poor reproducibility of the method. A more homogeneous dispersion containing only SUVs is obtained by centrifuging the phospholipid dispersion after ultrasonication at 100,000 g for 30 min. Titanium particles released from the probe tip and residual MLVs are sedimented and removed from the disperson by this procedure. Prolonged high-speed centrifugation may be used to improve the homogeneity of the SUV dispersion [33]. Because of the high power of ultrasonic irradiation used in probe sonication, SUVs of a minimum size (or minimum radius of curvature) are produced. Qualitatively the minimum radius is determined by packing constraints that prevent sealing of very small bilayer fragments and the formation of SUV smaller than the minimum-size vesicles. High-power sonication suffers from the drawback that it can easily lead to lipid degradation, however [34]. Both oxidative and hydrolytic degradation can be induced by ultrasonication. Another limitation is the small encapsulation capacity of 0.2–1.5 L mol^{-1} lipid. Bath sonication is milder than probe sonication and hence runs less risk of lipid degradation. However, long sonication times (of several hours) are required, and even so the minimum size limit of probe sonicated vesicles is usually not reached. The resulting vesicles are more heterogeneous in size than probe-sonicated ones, and reproducibility is not improved.

In the French press the phospholipid dispersion is extruded through a small orifice at high pressures (5×10^3–40×10^3 p.s.i. = 350 bar to 2.8 kbar = 3.5×10^7 Pa to 2.8×10^5 kPa) yielding uni- and oligolamellar vesicles of diameters of 30–80 nm depending on pressure [30,31]. The average size of the resulting vesicles is inversely proportional to pressure, and the homogeneity of the vesicles is inversely proportional to the flow rate through the orifice [25]. Repeated extrusion improves the homogeneity. For instance, four consecutive extrusions of MLVs made of egg phosphatidylcholine at 138 MPa produce a dispersion of SUV with more than 90% of the vesicles in the size range 30–55 nm [30]. Even working with great care, a small proportion of 5–10% of MLVs is usually not or only partly disrupted by this treatment. This minor vesicle population can be readily removed by centrifugation as described above.

Repeated extrusion of MLVs or LUVs through one or two stacked filters of defined pore size at low (<0.7 MPa) or intermediate pressure (3.5–7 MPa) is even gentler than French-pressing. Nucleopore polycarbonate membrane filters with straight-sided pore holes of exact diameter are widely used now for this purpose. If MLVs are extruded repeatedly, the vesicles change gradually from multilamellar to unilamellar. Five to ten extrusions produce unilamellar vesicles the average size of which is usually somewhat smaller than the nominal pore size of the membrane filter used [8], for pores not smaller than 0.1 μm. Using the appropriate filters and number of extrusions, fairly homogeneous populations of unilamellar vesicles of a size range of 40 to 150 nm can be produced. The major advantages of the method are (1) the repeated extrusion process is rapid, on the order of 10 min; (2) it can be used to process both MLVs and LUVs up to concentrations as high as ~0.5 M, ensuring extremely high trapping (encapsulation)* efficiencies of up to ~50%; (3) the process is mild, and the risk of phospholipid degradation is relatively small. It should be borne in mind that all the methods of homogenization involve repeated cycles of breaking and resealing of phospholipid bilayers. As a result, the vesicle content is exchanged with the external (suspending) medium. If, for instance, LUV loaded with Na^+ or sucrose are subjected to one of the treatments discussed above, the vesicle content is lost, or more precisely, it is equilibrated with the total aqueous medium.

*Encapsulation or entrapment efficiency is defined as the percentage of a water-soluble compound that becomes entrapped.

C. pH-Induced Vesiculation

It was reported that transient exposure of phosphatidic acid dispersions forming smectic (lamellar) phases to high pH (pH 10–12) followed by neutralization of the dispersion induces the formation of SUVs with a diameter of 20–60 nm [35]. This method, originally referred to as the pH jump or pH adjustment method, can be extended to mixed phospholipid dispersions consisting of phosphatidylcholine and phosphatidic acid. It produces mixtures of SUVs and LUVs depending on the phosphatidic acid content. Details of this method are given in Refs. 36 and 37. By using ^{31}P NMR and infrared spectroscopy, the formation of SUVs was shown to be driven by a pH gradient of some 4 ± 1 pH-units imposed on the phosphatidic acid bilayer. The direction of this gradient is such that the external pH is higher than the internal pH of the resulting SUV [38,39]. This method of SUV formation is grouped together with other methods of homogenization because as with all methods of homogenization the phospholipid dispersion is energized, i.e., there is an input of external energy.

A schematic illustration of the process of pH-induced vesiculation is shown in Fig. 5. On the basis of ^{31}P NMR and infrared spectroscopy the following mechanism was postulated [39]: upon exposure of a multilamellar phosphatidic acid vesicle or a stack of phosphatidic acid bilayers deposited on the glass wall of a round-bottom flask to pH 10–12, the outer monolayer of phosphatidic acid molecules becomes fully ionized (cf. the scheme in Fig. 5a), bearing two negative charges per lipid molecule. In contrast, phosphatidic acid molecules located on the inner monolayer of the outermost bilayer and of bilayers in the interior of the multilamellar particles are at nearly neutral pH, hence being in the monoanionic form. Thus a pH gradient is produced across the outermost phosphatidic acid bilayer that causes an electrostatic repulsion between fully ionized molecules in the outer monolayer of the first bilayer and between partially ionized phosphatidic acid molecules present in the inner monolayer. The former repulsion is greater than the latter, however; this causes the outer monolayer to be more expanded than the inner monolayer, which gives rise to a molecular packing gradient (cf. Fig. 5b). As a result of the differential electrostatic repulsion and the difference in molecular packing between the outer and inner monolayers, SUVs start to bud off from the outermost bilayer. The budding process exposes the next bilayer to the high pH of the external medium, a pH gradient is generated across this bilayer, and the budding process spreads from the exterior to the interior of the multilamellar particles (Fig. 5b).

The method of pH-induced formation of SUV was developed with egg phosphatidic acid, which has a chain melting transition temperature close to room temperature (T_c = 21.9°C). Raising the external pH above the second pK probably induces an isothermal phase transition in the outer monolayer of the bilayer but not in the inner monolayer. The mechanism of differential chain melting of the bilayer would enhance the molecular packing gradient between the outer and inner monolayer of a bilayer.

The mechanism of the pH gradient driven formation of SUVs is not restricted to phosphatidic acid bilayers. Its general nature has been recognized: in principle, pH-induced formation of SUVs can be produced in planar bilayers that are constituted of any molecule with one or several ionizable groups. For instance, the pH-induced formation of SUVs was also observed with phosphatidylserine bilayers exposed to pH 11 [39].

D. Methods Involving Organic Solvents

There are several methods of vesicle preparation that start off from stable or metastable solutions (dispersions) of lipids in organic solvents. These include solvent injection and

(a) Phosphatidic acid:

$$\mathrm{R{-}O{-}\overset{\overset{O}{\|}}{\underset{\underset{O(-)}{|}}{P}}{-}OH} \quad \underset{}{\overset{pK\,=\,8.6}{\rightleftharpoons}} \quad \mathrm{R{-}O{-}\overset{\overset{O}{\|}}{\underset{\underset{O(-)}{|}}{P}}{-}O(-)}$$

I   II

Phosphatidylserine:

$$\mathrm{R{-}O{-}\overset{\overset{O}{\|}}{\underset{\underset{O(-)}{|}}{P}}{-}O{\cdot}CH_2{\cdot}\underset{\underset{\underset{(+)}{NH_3}}{|}}{CH}{\cdot}COOH} \quad \overset{pK\,=\,4.5}{\rightleftharpoons} \quad \mathrm{R{-}O{-}\overset{\overset{O}{\|}}{\underset{\underset{O(-)}{|}}{P}}{-}O{\cdot}CH_2{\cdot}\underset{\underset{\underset{(+)}{NH_3}}{|}}{CH}{\cdot}COO(-)}$$

III   IV

R = diacyl- or dialkylglycerol; apparent p*K* values are given.

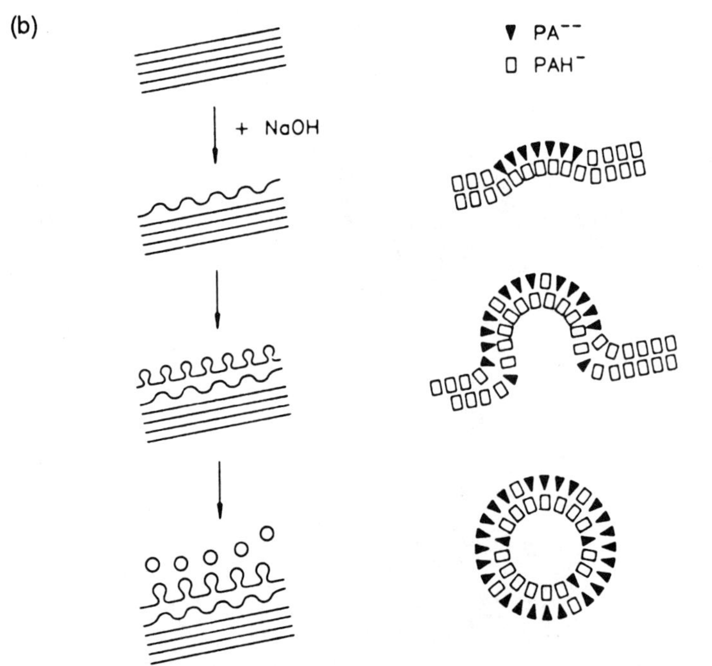

Figure 5 (a) Scheme showing the second dissociation of the phosphate group of phosphatidic acid and the dissociation of the –COOH group of phosphatidylserine. (b) Schematic diagram showing the formation of SUV by pH jump. In this case a pH gradient generated across a phosphatidic acid bilayer is the driving force. PAH^- and PA^{2-} represent monoionic and diionic phosphatidate molecules (structures I and II in (a)).

evaporation, reverse phase evaporation, double emulsion and other less well known methods. Phospholipids dissolved in an organic solvent preferentially form inverted micelles rather than monomeric solutions. It was shown that phosphatidylcholine forms inverted micelles in most organic solvents widely used in lipid chemistry [40]. These inverted micelles are capable of taking up and solubilizing large quantities of water [40], and their hydrodynamic radius has been shown to be directly proportional to the water content of the system [41]. It is therefore reasonable to assume that inverted micelles with their associated water (also referred to as water-in-oil microemulsions) are the starting phase of all these methods of vesicle preparation. The crucial step common to these methods is the perturbation and destabilization of the stable or metastable starting phase of inverted micelles. This step is carried out in a carefully controlled manner and usually involves the removal of the organic solvent by evaporation. Alternatively, in the solvent injection method introduced by Batzri and Korn [42], the starting phase in ethanol is very effectively diluted by injecting the ethanol solution in a huge excess of water. Phospholipids spontaneously rearrange into SUVs with an average diameter of 25 nm [42,43]. Some larger particles, either LUVs or MLVs, are also present, and the population of SUVs generated by ethanol injection appears to be more heterogeneous in size than SUVs produced by ultrasonic irradiation [43]. The advantages of the solvent injection method are the extreme simplicity and small risk of phospholipid degradation. The major drawbacks are the low entrapment capacity and efficiency. These are caused by the limitations pertaining to the total volume of the organic solvent that can be injected into the aqueous medium (less than ~10%).

With the ether infusion technique [44], phospholipids are dissolved in diethylether or diethylether/methanol mixtures, and the lipid solution is injected into the aqueous phase held at a temperature (55–65°C) at which ether vaporizes rapidly. In this way the lipid phase in organic solvent is destabilized and the phospholipids rearrange into LUVs of a size range of 0.1–0.5 μm and with an entrapment capacity of 10–15 L/mol phospholipid. In the reverse phase evaporation [45] and double emulsion methods [25], essentially water-in-oil microemulsions are destabilized by removal of solvent, usually by evaporation [25]. The properties of the resulting vesicles depend on a number of parameters, the lipid-to-aqueous volume ratio being critical. Starting from egg phosphatidylcholine and using the original protocol of Szoka and Papahadjopoulos [45,46], the method yields a heterogeneous population of LUV with a mean diameter of 0.18 μm (size range 0.1–0.26 μm); starting from egg phosphatidylcholine/cholesterol (1:1 mole ratio), the mean diameter of the LUV population is 0.47 μm (size range 0.17–0.8 μm). By increasing the lipid-to-aqueous volume ratio to 300 mg/mL of aqueous phase, and otherwise using the same protocol [45,46], it is possible to produce MLVs with a large aqueous core as described in detail by Pidgeon et al. [47]. The main advantages of the reverse phase evaporation method are that a wide range of lipids and lipid mixtures can be employed and LUVs with a very high entrapment efficiency of 20–70% are formed [25].

Major shortcomings of the methods involving lipid solution/dispersion in organic solvents are (1) the limited solubility of phospholipids and (2) the solvent retention of phospholipid bilayers/aggregates. The limited solubility of phospholipids in diethylether, petroleum ether, and hydrocarbons can be increased by using solvent mixtures, e.g., mixtures of ether and $CHCl_3$ (1:1 by vol.). Organic solvents are tenaciously retained by phospholipid bilayers and lipid aggregates. The removal of the last traces of organic solvent from lipid bilayers is tedious if not impossible. It involves time-consuming procedures such as gel filtration, exhaustive dialysis, or combinations of the two.

E. Detergent Removal

The technique of detergent removal was originally introduced by Kagawa and Racker [48] to reconstitute integral membrane proteins in phospholipid vesicles. These authors added sufficient detergent to solubilize the protein into mixed protein-detergent micelles. Similarly, MLVs of egg phosphatidylcholine are solubilized into mixed phospholipid-detergent micelles. Many different detergents have been employed for this purpose, the most widely used ones being bile salts, 1-0-*n*-alkyl-β-D-glucosides and Triton X-100 [25]. About 2.5% (0.06 M) sodium cholate is sufficient to solubilize MLVs of egg phosphatidylcholine at 10–20 mg lipid/mL, and similar quantities of Na^+cholate are used to solubilize integral membrane proteins at 5 mg protein/mL. To reconstitute proteins into a lipid bilayer, the micellar protein dispersion is mixed with the phospholipid/Na^+cholate mixed micelles to give a final range of lipid/protein weight ratios of ~ 1 to 10. Upon removal of the detergent from this mixture by dialysis [48] or any other suitable method [49], unilamellar phospholipid vesicles are obtained with the integral membrane protein being assembled into the phospholipid bilayer [49]. If no protein is added at the stage of the mixed micellar dispersion, unilamellar phospholipid vesicles are obtained instead. Using specially designed dialysis chambers, the dialysis procedure can be improved considerably. It is now possible to remove the detergent not only efficiently but also continuously and in a controlled and reproducible way. Weder et al. have developed dialysis equipment that allows the removal of detergent at a constant rate and temperature [50,51].

Alternative methods of detergent removal include gel filtration on different materials such as Sephadex and Sepharose [43,49], specific binding of detergent to resins [52], and also simple dilution of the dispersion so that the detergent concentration drops below the critical micellar concentration. A modification of the detergent removal by gel filtration is the centrifugation of mixed lipid-detergent dispersions through a small Sephadex column. This combination of gel filtration and centrifugation is quick, but it has the disadvantage of being limited to rather small volumes of lipid dispersion. Detergent can be efficiently and rapidly removed by the selective binding to a suitable resin. For instance, Triton X-100 was shown to bind to Biobeads SM-2, which can be used to prepare unilamellar phospholipid vesicles from mixed detergent-phospholipid micelles [52]. Resins such as Biobeads SM-2 or Amberlite XAD-2 may also bind phospholipids and proteins, particularly if the adsorbants are used as columns. This problem can be avoided by adding the adsorbant to the dialysis buffer outside the dialysis membrane. In this case, the direct contact of the phospholipid and protein molecules with the adsorbant is avoided and the rate of detergent removal is increased [25,53].

The analogy between the detergent removal technique and the methods involving lipid solutions/dispersions in organic solvents should be clear. Both methodologies start from stable micellar dispersions, the former from ordinary micelles, the latter from inverted micelles. In both cases the initial stable system is destabilized, by detergent removal in the former case and by the removal of the organic solvent in the latter. In the course of this destabilization lipid aggregates rearrange themselves according to the prevailing polarity of the environment, and eventually vesicles consisting of closed lipid bilayers emerge.

The detergent removal technique is an efficient method for the reconstitution or incorporation of integral membrane proteins into the lipid bilayers of unilamellar lipid vesicles. It is widely used in this field, and often it is the only way of assembling functional lipoprotein systems. The method has some remarkable advantages over other methods which, however, are balanced or even outweighed by serious shortcomings. The main advantage is that the resulting vesicle population is almost monodisperse. This can be demonstrated by gel filtration on Sepharose 4B (Fig. 6), analytical ultracentrifugation (Fig. 7), and dynamic light scattering. In Fig. 6a–c, the gel filtration patterns of egg

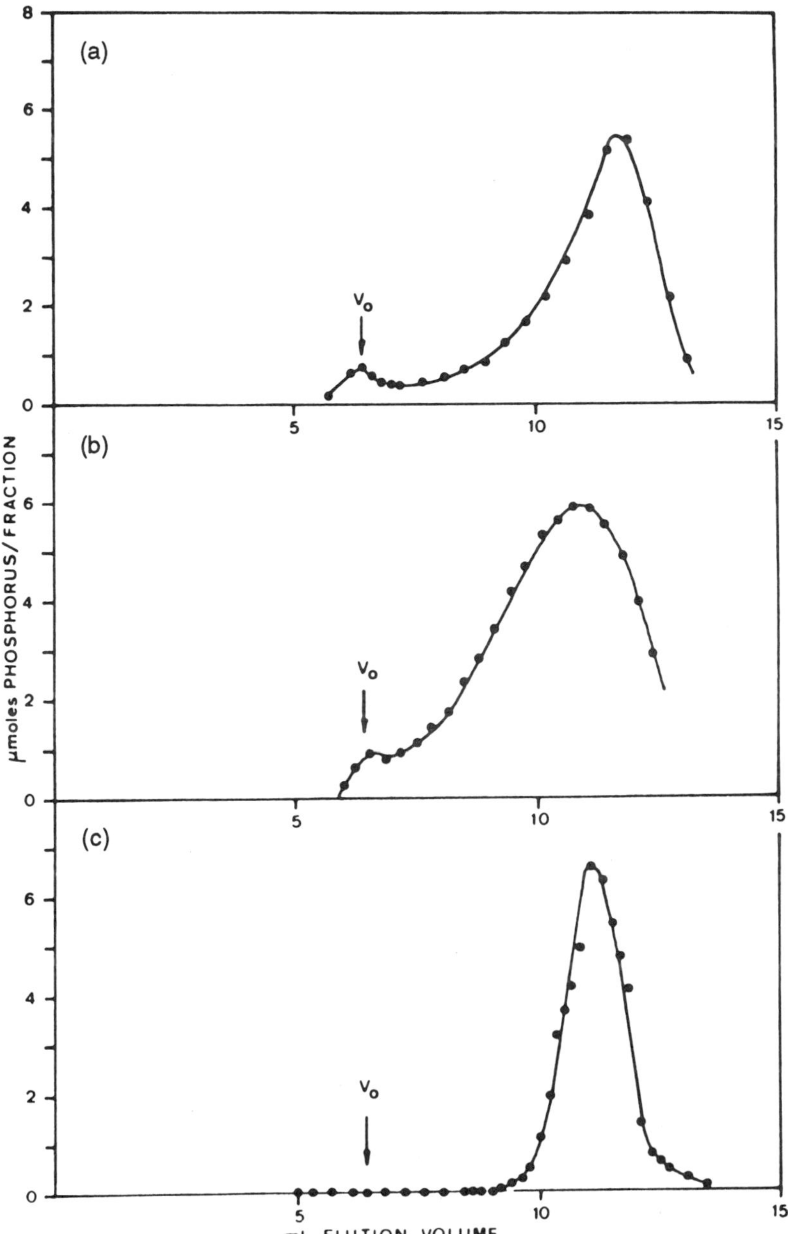

Figure 6 Gel filtration on calibrated Sepharose 4B. SUVs of egg phosphatidylcholine produced by sonication (a), by ethanol injection (b), and by detergent removal (c). The SUVs were dispersed in 0.1 NaCl, 0.01 M Tris. HCl pH 7.3, 0.02% NaN_3 to a final concentration of ~10 mg/mL (= 0.013 M). The SUV preparation in (c) was made from mixed micelles of sodium cholate/egg phosphatidylcholine (cholate conc. ~0.3 M, mole ratio bile salt/phospholipid ~2). Cholate was removed from this micellar dispersion by gel filtration on Sephadex G-50 (for details see Refs. 43, 49), and the resulting SUVs were analyzed by gel filtration on Sepharose 4B.

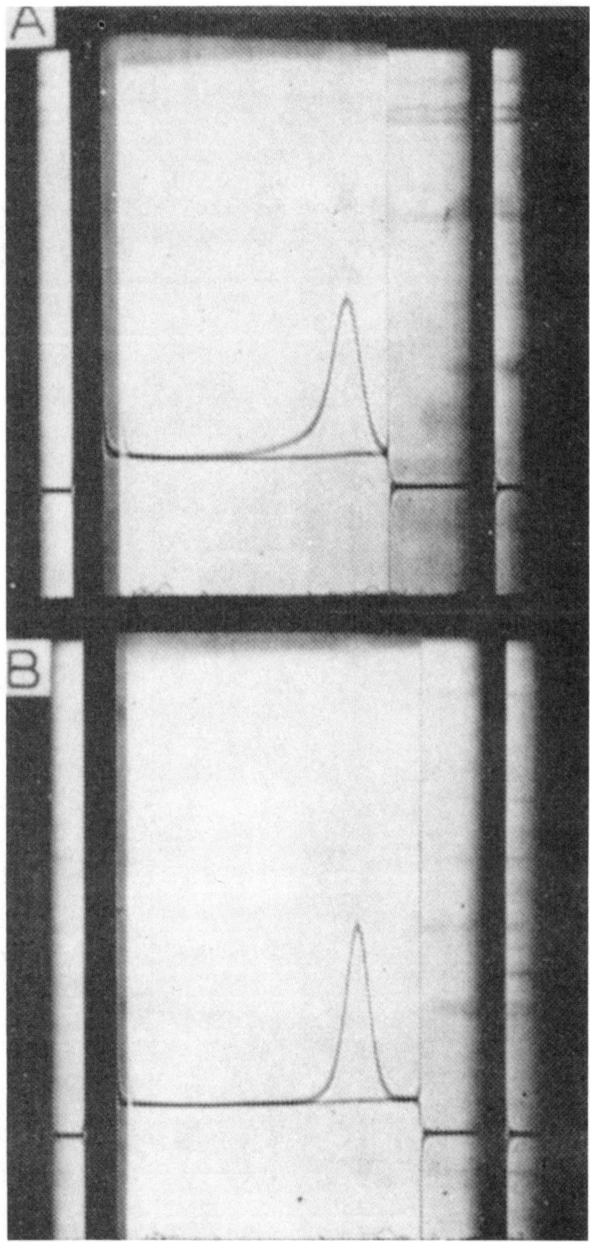

Figure 7 Ultracentrifugal Schlieren pattern of SUVs of egg phosphatidylcholine produced by sonication (A) and detergent removal (B). The SUVs were dispersed in 0.1 M NaCl, 0.01 M Tris. HCl pH 7.3, 0.02 NaN_3 at 20 mg lipid/mL (= 0.027 M). (For experimental details see Ref. 43.)

phosphatidylcholine SUVs are shown produced by ultrasonication (a), ethanol injection (b), and detergent removal (c). As seen in Fig. 6c, the detergent-removal vesicles elute as a narrow, symmetric peak, which indicates that the SUV population is homogeneous in size. Ultrasonication and the ethanol injection method yield less homogeneous SUV populations, as is evident from the comparison of the gel filtration patterns; in both cases broad peaks are obtained, and in contrast to SUVs produced by detergent removal, a small proportion of large particles (LUVs and/or MLVs) is present (Fig. 6a–c). The average size of SUVs produced by the three methods (Fig. 6a–c) is in the range of 20–30 nm, slight intermethod variability being observed. Further evidence for the monodispersity of SUVs produced by detergent removal is provided by analytical ultracentrifugation. The Schlieren pattern of these vesicles is relatively narrow and symmetric compared to that of SUV produced by sonication (Figs. 7A and B). The Schlieren peak of the latter dispersion is asymmetric due to the presence of larger particles with significantly greater sedimentation velocities than the bulk of the vesicles (Fig. 7A).

Further evidence that SUVs produced by detergent removal or simple dilution are practically monodisperse comes from dynamic light scattering (Fig. 8). Mixed micellar dispersions of sodium glycocholate/egg phosphatidylcholine were diluted, and the hydrodynamic radius R_H and polydispersity index V of the resulting particles (micelles or SUVs) were determined by dynamic light scattering [54]. With increasing dilution, both

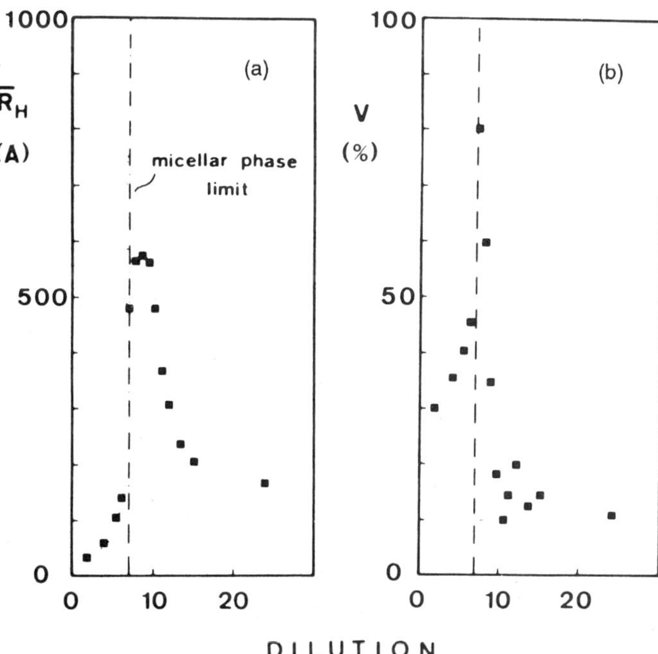

Figure 8 Mean hydrodynamic radius R_H in Å units (a) and polydispersity index V in % (b) of mixed dispersions consisting of sodium glycocholate and egg phosphatidylcholine (0.75 mole ratio, total lipid concentration 50 mg/mL) as a function of dilution. R_H values were determined by dynamic light scattering. The dilution is expressed as the factor n, i.e., the lipid concentration after dilution is $50/n$ (mg/mL). The dashed line represents the phase boundary of the micellar phase. (From Ref. 54.)

R_H and V increase and pass through a maximum that almost coincides with the micellar phase boundary indicated by the dashed line in Fig. 8a. Dilution of micellar dispersions beyond this boundary yields almost monodisperse SUVs. Vesicle size and dispersity both decrease on the right hand side of the phase boundary. The hydrodynamic radius of the resulting SUVs decreases monotonically with increasing dilution and approaches asymptotically a limiting value of about 15 nm (Fig. 8a). This value is in very good agreement with the average radius of egg phosphatidylcholine SUV produced by detergent removal on Sephadex G-50. Simultaneously, the polydispersity index decreases by ~80% to a limiting value of ~10% at 20-fold dilution (Fig. 8b).

The average size of the homogeneous vesicle population can be varied within limits: both SUV and LUV can be produced with average vesicle sizes ranging from 25 nm to about 200 nm [54,56]. The size and size distribution of these vesicles depend on a number of parameters including the nature of the detergent and lipid used, the lipid-to-detergent ratio, the rate of detergent removal, and experimental conditions such as temperature, pH, ionic strength, and composition of the medium. Unfortunately, our understanding of the effect of these parameters on the size and size distribution of phospholipid vesicles is only qualitative or at best semiquantitative. Most studies concerned with the correlations between the vesicle size and different experimental parameters have suffered from a serious limitation: the important question whether or not the resulting vesicles represent equilibrium structures has not been tackled and hence nothing is known about their thermodynamic stability.

The obvious advantage of monodispersity in terms of well defined vesicle properties (e.g., interfacial area, vesicle volume, etc.) achieved by the detergent removal technqiue is balanced by several disadvantages: (1) with most preparations it is unknown whether or not the resulting vesicles are thermodynamically stable; (2) the encapsulation capacity and efficiency are usually low, in the range 0.2–7 L mol^{-1} and 0.1–15%, respectively; (3) the most serious drawback, however, is the retardation of detergent. Residual detergent affects bilayer properties such as bilayer fluidity (microviscosity) and permeability.

Gel filtration on Sephadex G-50 of mixed cholate-egg phosphatidylcholine micelles generates SUV with a residual cholate level of a few mol % [49]. Less efficient removal of detergent by gel filtration was reported by Allen [53], and this may lead to increased average vesicle sizes. To prevent this, the resultant SUV preparation can be subjected to a second gel filtration or 12 h dialysis; this lowers the residual cholate level to 0.2–0.5 mol % [49].

It is thus clear that the removal of residual detergent is problematic and cumbersome. Current procedures therefore employ detergents such as bile salts and *n*-alkylglucosides that can be removed relatively rapidly and completely due to their high critical micellar concentration. Even though detergent-removal vesicles resemble vesicles produced by other techniques in most physicochemical properties, they still may differ in some properties and thus should be used with caution. For instance, LUVs produced from mixed micelles of egg phosphatidylcholine and 1-*n*-octyl-β-D-glucoside by detegent removal exhibit Na^+, Rb^+, and Cl^- permeabilities comparable with permeability data for sonicated egg phosphatidylcholine vesicles [55]. In contrast, egg phospatidylcholine SUV produced from mixed phospholipid/cholate micelles by cholate removal differ in their surface-chemical properties from SUV produced by sonication; residual cholate confers a negative surface potential, as can be shown by ^{31}P NMR in the presence of lanthanides (H. Hauser, unpublished observation).

V. CHARACTERIZATION OF PHOSPHOLIPID VESICLES

A. Size Analysis

The knowledge of the average vesicle size and size distribution (polydispersity) is essential in biochemical and biophysical studies as well as in commercial applications of phospholipid vesicles. In the following section, a brief assessment of different methods currently employed in the size characterization of these vesicles is given. These methods comprise gel filtration on calibrated columns, electron microscopy, laser-based dynamic light scattering, and analytical ultracentrifugation.

1. Gel Filtration

Gel filtration or gel exclusion chromatography is an inexpensive, convenient method for the fractionation of phospholipid vesicles according to their size and for routine size analyses. For a quantitative analysis of vesicle suspensions, each column has to be calibrated with appropriate standards, such as proteins, virus particles, and monodisperse polystyrene latex beads of known size [57,58]. The method of Ackers has been the method of choice for converting the elution volume V_e to an average hydrodynamic radius R_H according to [59]:

$$R_H = a_0 + b_0 \operatorname{erf}^{-1} (1 - K_D) \tag{1}$$

where a_0 and b_0 are constants characteristic of the column material, and $K_D = (V_e - V_0)/(V_t - V_0)$; V_e is the elution volume; V_0 and $+V_t$ are the column void and total volume, respectively. Gel filtration on calibrated columns provides not only an average hydrodynamic radius but also a semiquantitative measure of the size distribution. The upper limit of the size analysis is determined by the exclusion limit of the column material. Sephacryl S-1000 has an exclusion limit of 200 to 250 nm; this is 3–4 times larger than that of Sepharose 4B. It is the column material with the largest exclusion limit presently available. The usefulness and applicability of Sephacryl S-1000 to the fractionation of polydisperse phospholipid vesicles and the routine size analysis of these vesicles were first pointed out by Tanford and coworkers [66].

Vesicles smaller than the exclusion limit of the column material are fractionated, and from their elution volume V_e the average hydrodynamic radius R_H of the vesicle population is derived from Eq. (1). All vesicles with diameters exceeding the exclusion limit of the column are no longer fractionated but eluted at the column void volume V_0. The choice of column material is therefore important: it determines not only the size range of vesicles that can be analyzed but also the size range of maximum resolution. Gel filtration is very well suited for the fractionation and size analysis of SUVs and LUVs up to diameters of ~250 nm and for the separation of MLVs from SUVs. One obvious disadvantage of this method is its size limitation: vesicles larger than ~250 nm cannot be analyzed with column materials presently available. Vesicles of diameter >1 μm usually fail to enter the gel and accumulate on top of the column. Large vesicles trapped on top of the column may either clog the column or cause serious disturbances in gel filtration. Some of the disturbances arise from chemical degradation of the trapped lipids. Therefore samples for gel filtration have to be freed of large vesicles and particles. Another disadvantage is the nonspecific absorption of lipids and proteins to the column material. In order to minimize absorptive effects the column has to be presaturated with the compound(s) to be chromatographed [58].

2. Electron Microscopy

Electron microscopy has become a popular method for the size analysis of phospholipid vesicles. Its major advantage is that individual vesicles can be viewed directly. It yields a good estimate of size range and the size distribution of phospholipid vesicles, provided that the following conditions are fulfilled. (1) In order to produce visible images of phospholipid vesicles by electron microscopy, the aqueous dispersion of lipid vesicles has to be subjected to cryofixation or chemical fixation and staining. Discussing this methodology in detail is beyond the scope of this review. It suffices to say that these procedures may produce artefacts. For instance, the method of negative staining, though quick and extremely simple, is troubled by its inherent probability of generating artificial lipid structures, particles, phases. This has been clearly demonstrated for negatively charged phosphatidylserine dispersions [21,60]. Details derived from x-ray diffraction were found to be in good agreement with structural data derived from electron microscopy of freeze-fractured phosphatidylserine dispersions. However, by comparison of the results obtained by x-ray diffraction with electron microscopy data of negatively stained samples, it was clear that negative staining of phosphatidylserine dispersions can lead to structural artefacts. The general consensus evolving now is that cryofixation methods, such as freeze-fracturing, freeze-etching, and various modifications thereof, e.g., spray-freezing, freeze-drying, etc., preserve the native structure, whereas all kinds of staining procedures have a great potential of inducing structural changes and in turn artefacts. Therefore results obtained by electron microscopy of negatively or positively stained samples have to be treated cautiously and should be counterchecked by independent methods.

A number of different cryofixation procedures are in use, and also different methods exist for the preparation of surface replicas and shadowing. Some of the details are described in Refs. 25, 36, and 61–63. These procedures yield images that may differ in detail. For instance, the image analysis of samples prepared by ordinary freeze-fracture methods is hampered because the position of the fracture plane is unknown. If the angle of shadowing is 45° with respect to the fracture plane, only vesicles fractured equatorially yield circular images that are shadowed to 50%; the boundary between the light and the dark (shadowed) zone then dissects the circular image. This is important, as only these vesicles give the correct vesicle diameter and should be considered in the size analysis. A procedure for deriving the true vesicle size from freeze-fracture electron microscopy has been reported recently. It is based on the assumption that unilamellar vesicles are fractured randomly [65]. In freeze-dried samples [62] that are not subjected to any fracturing processes, whole vesicles are imaged, and the problem just discussed does not appear [66].

Perhaps even more powerful is the technique of cryoelectron microscopy, which uses specimens embedded in ultrathin layers of ice (~100 nm). This method recently introduced requires no shadowing (staining) and has a resolution of ~3 nm [64,70].

(2) Even if artefacts during the sample preparation are avoided, a good estimate of the average size distribution requires unbiased sampling: electron micrographs have to be taken randomly, and the size analysis has to be based on a representative fraction of vesicles. The latter requirement necessitates the analysis of a sufficiently large number of images. A respectable size analysis should be based on no less than 400 to 500 vesicles. This number determines the significance and accuracy of the analysis, for instance, measuring 400 images of vesicles ($n = 400$) yields an average vesicle size with an accuracy of $(100/\sqrt{n})\% = 5\%$.

(3) The size analysis by electron microscopy is based on the assumption that phospholipid vesicles are spherical. This is by and large correct. Some deviations from strict spherical geometry can be tolerated, and methods of approximation for the size analysis of ellipsoidal particles have been worked out [63].

From the above discussion it is clear that a reliable size analysis by electron microscopy is tedious and time-consuming. It requires costly instrumentation and an experienced operator. This clearly limits the use of electron microscopy as a routine method for sizing lipid vesicles.

3. Dynamic Light Scattering

In contrast to electron microscopy, laser light scattering is quick and easy to perform. The only stringent requirement on the lipid dispersion to be analyzed is the removal of dust particles. Because of the dependence of the scattering intensity on the square of the molecular or particle weight, light scattering is particularly sensitive to the presence of large particles such as dust. Removal of dust particles is achieved either by subjecting the sample in the scattering cell to centrifugation or by using a closed loop filtration system that filters the lipid dispersion continuously into the sample cell. The particle size analysis by light scattering should be carried out with dilute lipid dispersions, usually at lipid concentrations below 1 mg/mL in order to eliminate multiple scattering effects. Modern equipment for the size analysis of phospholipid vesicles is commercially available from several manufacturers (for a thorough discussion of the instrumentation consult Ref. 67).

Light scattering is applied to advantage to unimodal vesicle populations. Dynamic light scattering then can be performed at a single scattering angle Θ (usually $\Theta = 90°$), and the method yields a value for the average hydrodynamic radius R_H in a few minutes. Therefore for homogeneous lipid vesicles light scattering is the most convenient method of size analysis and suitable for routine use. If, however, no *a priori* knowledge of the size distribution is available, the size analysis by light scattering is less straightforward. In this case the intensity autocorrelation function should be determined for different values of both the scattering angle Θ and the correlator sample time [68]. A cumulant analysis of the experimentally obtained autocorrelation function yields a z average hydrodynamic radius $R_H(z)$ and an estimate of the polydispersity. Due to the M_r^2 dependence of the scattering intensity, even the presence of a very small quantity of large particles would produce a significant increase in the average vesicle size. Therefore dynamic light scattering applied to polydisperse phospholipid dispersions may give misleading results, or at least results inconsistent with other methods. However, methods of analysis have advanced significantly in recent years. The data analysis can be greatly improved by the use of the inverse Laplace transformation of the autocorrelation function as discussed in more detail in Refs. 68 and 69. Using this treatment, it is now possible to derive a reliable and accurate estimate of the size distribution of phospholipid vesicles.

Static light scattering merits some comments. It can provide useful additional information pertaining to vesicle size and shape and size distribution. Static light scattering yields a weight average molecular (particle) weight and the z average mean square radius of gyration $\langle R_g^2 \rangle_z$. Since macromolecules and particles of different shape have different R_g values for a given hydrodynamic radius R_H, the radius of gyration contains shape information. For instance, for a multilamellar vesicle that to a first approximation may be treated as a solid sphere, we have $R_g = (3/5)^{1/2}R_H$, while for LUVs with the bilayer thickness $d << R_H$, $R_g \approx R_H$. Hence a combination of static and dynamic light scattering measurements may yield information as to the lamellarity of the lipid vesicles. The information on vesicle size, shape, and size distribution derived from static light scattering is very precise for large vesicles with a diameter exceeding about 200 nm.

In the following paragraph a few examples of the size analysis of phospholipid dispersions are given, and results obtained with different methods are compared. Egg phosphatidylcholine SUVs were prepared by cholate dialysis of egg phosphatidylcholine/cholate mixed micelles (mole ratio = 0.5, total lipid concentration 39 mM) in 0.02 M Tris buffer pH 7.4, 70 mM NaCl. For comparison, egg phosphatidylcholine SUVs were also prepared by multiple extrusion (10 times) of MLVs dispersed in the same buffer through two stacked membrane filters with a mean pore size of 50 nm [68]. A standard cumulant analysis of the dynamic light scattering data yield an average hydrodynamic radius of $R_H = 26.4$ nm and $R_H = 34.1$ nm for the two SUV preparations, respectively. The variances V as a measure of the polydispersity of the two SUV preparations are 5% and 11%, respectively. Thus dynamic light scattering of the two vesicle dispersions indicates that SUVs prepared by the detergent removal technique are practically mono-disperse, with a mean radius of 26 nm, while the SUV preparation extruded ten times through a 50 nm pore size filter is more inhomogeneous in size. Furthermore, the mean hydrodynamic diameter of 68 nm of the latter vesicles exceeds the mean pore size of the filter used. Figure 9 shows the size distribution of these two SUV preparations obtained by inverse Laplace transformation of the intensity autocorrelation function. Vesicles prepared by detergent removal are characterized by a narrow, almost symmetrical size distribution (Fig. 9a), while SUVs made by extrusion yield a broader, asymmetric size distribution (Fig. 9b) tailing off at the side of large vesicles. The maxima in the size distribution are consistent with the average hydrodynamic radii.

For comparison, freeze-fractured samples of the extruded vesicles were prepared, and the bar histogram of Fig. 9c was derived from electron micrographs of freeze-fractured preparations. The number distribution represented by the bar histogram (Fig. 9c) is shifted to smaller sizes compared to the intensity distribution derived from light scattering using the Laplace inversion (Fig. 9c). The maximum in the number size distribution is between 20 and 25 nm, which is significantly smaller than the average hydrodynamic radius $R_H = 34.1$ nm derived from dynamic light scattering. The reason for the discrepancy is that dynamic light scattering yields the z average of the hydrodynamic radius, whereas electron microscopy gives the number distribution. The number distribution can be calculated from the intensity distribution by taking into account the weighting of the scattered intensity [68]. As can be seen form Fig. 9c, the number distribution derived from light scattering is quite narrow, with a peak at 23 nm. This is in very good agreement with the bar histogram derived from electron micrscopy. Figure 9c illustrates convincingly how strong the affect of a very small number of large vesicles can be on the intensity distribution. The contribution of a small number of large particles is apparently amplified in the dynamic light scattering experiment due to the weighting of the scattered intensity. After correction for this effect, the average vesicle diameter is 46 nm, slightly smaller than the nominal pore size of 50 nm of the membrane filter used for extrusion. When comparing size parameters derived from different methods of analysis, it should be borne in mind that different methods yield different averages for the size parameters such as hydrodynamic radius, vesicle volume, vesicle weight, etc.

Almost all standard preparative methods for the preparation of phospholipid vesicles produce polydisperse phospholipid vesicles. It is clear from the examples cited that only a combination of several analytical methods can provide reliable information about the vesicle size distribution in such populations. A polydisperse vesicle population can be fractionated by gel filtration prior to analysis by dynamic light scattering. This gives more reliable results than the direct application of dynamic light scattering, even at the most sophisticated level. It was shown previously [58] that satisfactory results can be obtained with even highly polydisperse phospholipid vesicles by a combination of gel filtration and dynamic light scattering techniques. Heterogeneous vesicle dispersions were first fractionated by gel filtration on appropriate columns (Sepharose 4B or Sephacryl S-1000), and fractions thus obtained were subsequently analyzed by dynamic light scattering. The mean hydrodynamic radius derived from dynamic light scattering of individual column fractions is in good agreement with the values derived from gel filtration and freeze-fracture electron microscopy [58].

4. Analytical Ultracentrifugation

Analytical ultracentrifugation has been used, though much less frequently, for the size analysis of phospholipid vesicles. It should be the method of choice if the vesicle weight rather than the vesicle size is required. The method was first applied to sonicated egg phosphatidylcholine dispersions by Saunders et al. [29]. Subsequently it was employed in the weight analysis of phospholipid dispersions by several groups [43,49,56, 71–73] either as the sedimentation/diffusion (*s*/*D*) or the sedimentation equilibrium method.

The (*s*/*D*) method involves the determination of both the sedimentation and the diffusion coefficient at infinite dilution in two series of separate measurements. The vesicle weight M_r is calculated using the Svedberg equation:

$$M_r = \frac{RTs_o}{D_o(1 - \bar{v}\rho)} \tag{2}$$

where M_r is the anhydrous vesicle weight, s_o and D_0 are the sedimentation and diffusion coefficients, respectively, extrapolated to infinite dilution, $\bar{v}$ is the partial specific volume of the phospholipid, and ρ is the density of the solution. For monodisperse samples the (*s*/*D*) method gives vesicle weights in good agreement with the values derived from the sedimentation equilibrium (cf. results compiled in Table 1). The (*s*/*D*) method is advantageous when applied to bimodal or paucimodal lipid dispersions. This was shown for unilamellar phospholipid vesicles with integral membrane proteins incorporated in the phospholipid bilayer. Such proteoliposomes differing in the number of integral membrane protein molecules per lipid vesicle can be separated in a single sedimentation velocity run if the molecular weight of the integral membrane protein is sufficiently large [49]. The vesicle weight of each vesicle species can then be determined, provided that the corresponding diffusion data are available. Schlieren optics are the preferred optical system to be used in the (*s*/*D*) method; they are particularly useful for bi- and paucimodal dispersions. However, with polydisperse vesicle populations, the (*s*/*D*) method is inferior to the sedimentation equilibrium technique. With polydis-

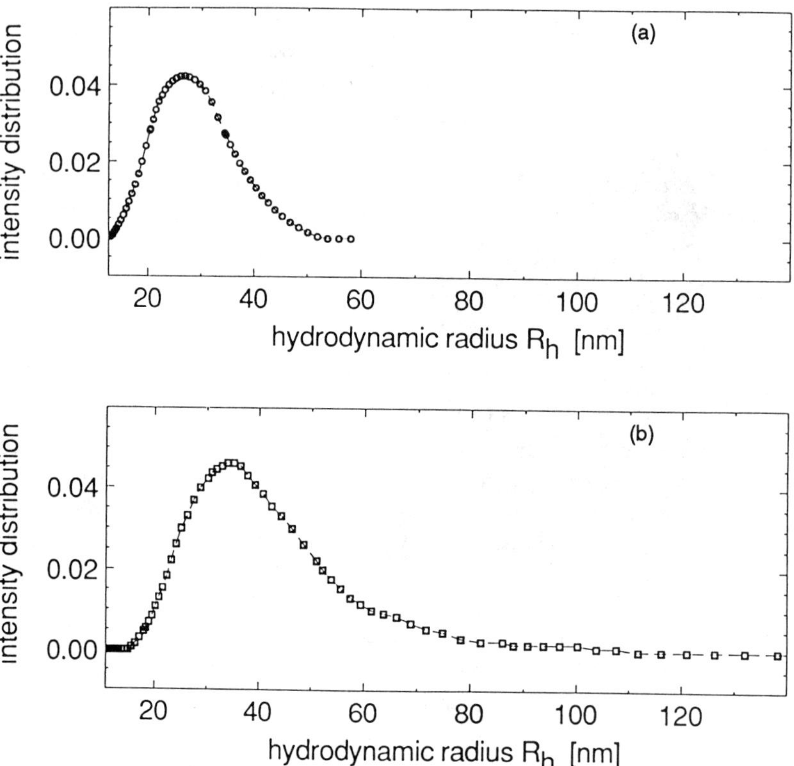

Figure 9 Size distribution of SUV of egg phosphatidylcholine derived from dynamic light scattering. The intensity distributions shown were obtained by inverse Laplace transformation of the intensity autocorrelation functions. (a) Egg phosphatidylcholine SUVs were prepared by cholate removal by dialysis from egg phosphatidylcholine/cholate mixed micelles (mole ratio = 0.5, total lipid concentration 39 mM) dispersed in 0.02 M Tris buffer pH 7.4, 0.07 M NaCl. (b) For comparison, egg phosphatidylcholine was dispersed in the same buffer to a final concentration of 13 mM, and the resulting MLVs were homogenized by multiple extrusion (10×) through two stacked membrane filters of mean pore size 50 nm using a commercially available filtering device (the Extruder ™ from Lipex Biomembranes). (c) The intensity distribution of the sample described under (b) (○) is compared to the number distribution (●) computed from the intensity distribution as described in Ref. 68 and the number distribution (bar histogram) derived from electron microscopy of freeze-fractured samples of the same dispersion.

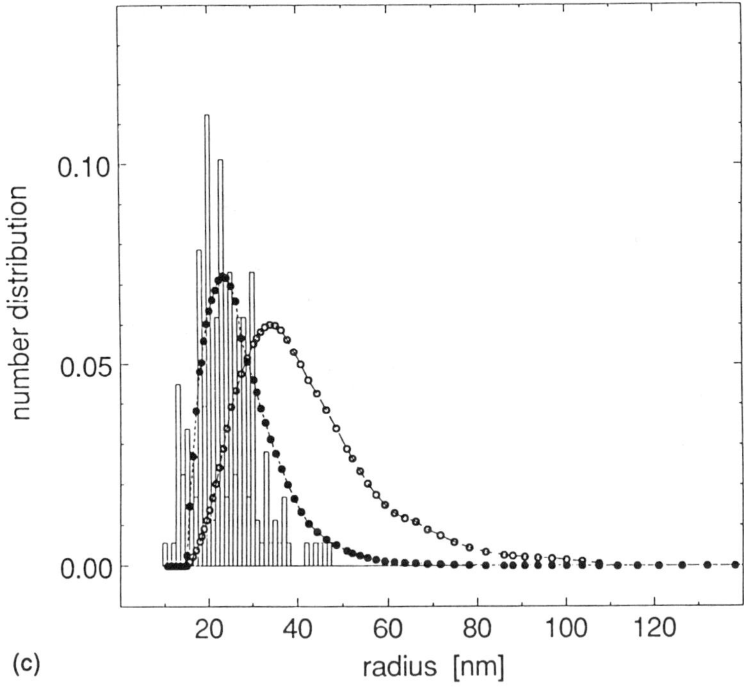

(c)

Table 1 Vesicle Weight Determination by Analytical Ultracentrifugation

Phospholipid[a]	Ultracentrifugal method	Vesicle weight (Da)	Calculated[c] R (nm)
egg PC	(s/D)	2.5×10^6	10.5
DMPC	(s/D)	2.33×10^6	
DOPC	(s/D)	2.72×10^6	
egg PC[b]	sedimentation equilibrium	2.66×10^6	10.8
DMPC	sedimentation equilibrium	2.33×10^6	
DOPC	sedimentation equilibrium	2.66×10^6	
egg PC	sedimentation equilibrium	2.70×10^6	10.8
egg PC	approach to equilibrium	range: $2.1 - 4.5 \times 10^6$	9.8–13.4
		average: 3.2×10^6	11.6

[a]All phospholipids were dispersed in 0.01M Tris buffer pH 7.3 containing 0.1 M NaCl and 0.02% NaN_3 and sonicated in order to produce SUV (experimental details are given in Refs. 49, 72, 73). Abbreviations: PC, phosphatidylcholine; DMPC, 1,2-dimrystoyl-*sn*-phosphatidylcholine; DOPC, 1,2-dioleoyl-sn-phosphatidylcholine.

[b]SUV of egg phosphatidylcholine dispersed in the same buffer were made by the cholate removal method (egg phosphatidylcholine/cholate = 0.35–0.5, mole ratio; final phospholipid concentration = 0.013–0.040 M).

[c]Anhydrous radius R calculated from the anhydrous vesicle weight M_r, the partial specific volume $\bar{v}$ of the phospholipid, and the anhydrous bilayer thickness d using Eq. (4).

perse samples, the (s/D) method yields a mixed average vesicle weight that is difficult to interpret.

The sedimentation equilibrium method has the advantage of a secure theoretical background with only a few assumptions being involved. Sedimentation equilibrium methods are not only simpler experimentally but also more accurate. Furthermore, for polydisperse vesicle populations, sedimentation equilibrium methods have the advantage of yielding a well-defined weight-average vesicle weight and a good assessment of the polydispersity. From a single sedimentation equilibrium run the weight-average vesicle weight can be determined at each point in the cell, and in addition to these "point" averages, the weight and z average vesicle weight averaged over the total vesicle population are obtained. In sedimentation equilibrium, the anhydrous vesicle weight M_r is derived from the measurement of the solute concentration throughout the cell under conditions of thermodynamic equilibrium:

$$M_r(1 - \bar{v}\rho) = \frac{2RT}{\omega^2}\frac{d \ln c}{dr^2} \tag{3}$$

Where ω is the angular velocity in radians/s, $d \ln c/dr^2$ is the slope of the plot of logarithm of lipid concentration ($\ln c$) versus the square of the radial distance (r^2), and other terms have their usual meanings. Since the solute concentration is required as a function of the radial distance r, interference optics and UV absorption are the preferred optical systems. For a homogeneous vesicle population, the plot according to Eq. (3) yields a straight line.

Analytical ultracentrifugation has been used successfully to determine the vesicle weight of egg phosphatidylcholine dispersions produced by sonication [29,43,71,72] and detergent removal [43]. Some representative results are summarized in Table 1. The values for the vesicle weight of sonicated dispersions of different phosphatidylcholines determined by the (s/D) method range between 2.3 and 2.7×10^6 (Table 1). These values agree very well with the results obtained with the sedimentation equilibrium method.

In order to assess the spread in size of sonicated egg phosphatidylcholine dispersions, 12 preparations were analyzed by the approach-to-equilibrium method (or Archibald method). The values for the vesicle weights thus obtained range from 2.1×10^6 to 4.5×10^6 with an average vesicle weight of 3.2×10^6. For comparison, the average vesicle weight of egg phosphatidylcholine SUVs produced by cholate removal was determined by the sedimentation equilibrium method, and a value of 2.66×10^6 was obtained (Table 1). In the last column of Table 1, vesicle radii are listed, calculated from the anhydrous vesicle weight M_r using the equation

$$R = \frac{d}{2} \pm \frac{[(d^2 - 4(d^2/3 - M_r\bar{v}/4\pi Nd)]^{1/2}}{2} \tag{4}$$

where d is the thickness of the anhydrous bilayer and N is the Avogadro constant. R depends on the choice of d. For the calculation of R values included in Table 1, a d value of 4.78 nm [74,75] and a value of $\bar{v} = 0.9839$ cm^3/g [72] typical for egg phosphatidylcholine was used. This d value is based on x-ray diffraction studies of McIntosh and Simon [74], who determined the P-P distance from electron density profiles of egg phosphatidylcholine as 3.78 ± 0.08 nm. Lesslauer et al. [75] have shown the high-density peak in the electron density profile to be close to the center of the phosphatidylcholine head group. From space-filling molecular models, the distance from the center to the outer edge of the phosphatidylcholine head group is taken as 0.5 nm. Therefore the d value of the anhydrous egg phosphatidylcholine bilayer is obtained by adding 1.0 nm to the P-P distance, yielding $d = 4.78$ nm. The anhydrous vesicle radii calculated from Eq. (4) are in satisfactory agreement with experimental values derived from hydrodynamic measurements. For instance, diffusion measurements at 20°C on sonicated egg phosphatidylcholine dispersions yield a value for the diffusion constant $D_0 = 1.80 \pm 0.07 \times 10^{-7}$ cm^2/s, which is consistent with a Stokes radius of $R_H = 12.0 \pm 0.5$ nm [72]. The Stokes radius can be identified with the outer radius of the hydrated phospholipid vesicle; it is obtained from the diffusion coefficient D_0 using the Stokes-Einstein equation

$$R_H = \frac{kT}{6\pi\eta D_o} \tag{5}$$

where η is the viscosity of the suspending medium.

Gel filtration on Sepharose 4B and electron microscopy of the same dispersion of egg phosphatidylcholine SUV give average values for the hydrodynamic radius of 13.1 nm and 12.9 nm, respectively [72]. The hydrodynamic radius of egg phosphatidylcholine SUVs prepared by the cholate removal method was determined by gel filtration and freeze-fracture electron microscopy, and the values thus obtained are 10.5 and 13.5, respectively. As expected, the hydrodynamic radii R_H are greater than the "anhydrous" radii R (Table 1) calculated from Eq. (4).

Despite many advantages, analytical ultracentrifugation has not become a routine method for the weight (size) analysis of phospholipid vesicles. The reason for this is that the equipment required is rather elaborate and relatively expensive compared to the instrumentation needed to perform gel filtration and dynamic light scattering measurements. Moreover, sedimentation equilibrium runs are time-consuming, and even with short sample columns and the overspeeding technique the weight determination of phospholipid vesicles normally takes 1–2 days [73].

B. Lamellarity

A qualitative indication of the lamellarity of phospholipid vesicles is provided by freeze-fracture electron microscopy. Cross-fractures of MLVs reveal the concentric stacking of bilayers. A somewhat better estimate of the lamellarity of phospholipid MLVs is derived from ^{31}P NMR in the presence and absence of shift or broadening reagents. For instance, in the presence of impermeable, paramagnetic broadening reagents such as the transition metal ions Mn^{2+} and Gd^{3+}, the intensity of the ^{31}P NMR resonance is reduced. The reduction in the signal intensity is directly proportional to the fraction of phospholipid present in the outer monolayer of the external bilayer exposed to the broadening reagent. This fraction will be very small for highly multilamellar vesicles, but it approaches unity for LUVs. The approach using paramagnetic probes and also specific labeling methods based on impermeable reagents still yield only an approximate estimate of the lamellarity. This is due to packing defects present in MLVs that make internal lamellae of MLVs accessible to ions and other impermeable reagents.

X-ray diffraction and scattering also yield information concerning the number of lipid bilayers per phospholipid vesicle. As mentioned before, MLVs give rise to clearly defined diffraction (Bragg) maxima at low angles (cf. Chap. 11), while unilamellar vesicles are characterized by a diffuse scattering peak. Oligolamellar vesicles exhibit features in the diffraction pattern from which the number of lipid bilayers can be estimated, at least under appropriate conditions [76]. Even more direct and reliable, but tricky to perform, is the

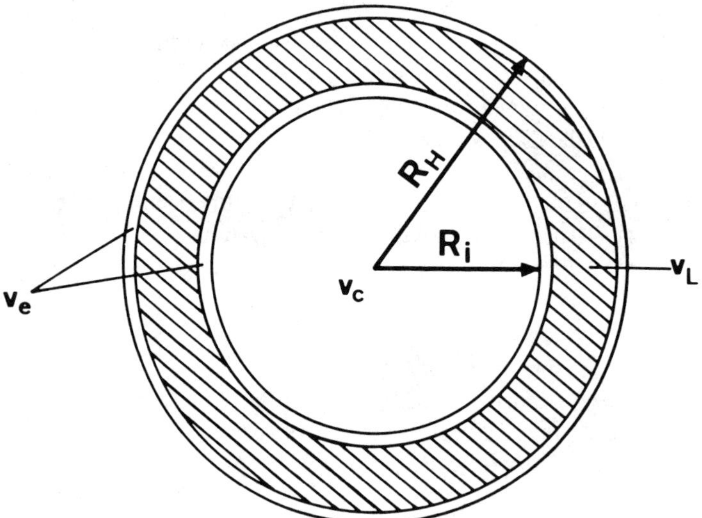

Figure 10 Scale diagram illustrating the various compartments of a phospholipid SUV; v_L is the volume of the anhydrous phospholipid bilayer, v_c is the internal volume, and v_e is the volume of water bound to the phospholipid polar groups.

lamellarity analysis by means of cryoelectron micrscopy using an energy filter for optimal contrast [77].

C. The Relationships Between Vesicle Dimensions Including Those of the Lipid Bilayer

The outer hydrodynamic or Stokes radius R_H of the hydrated vesicle (cf. Fig. 10) is obtained from various hydrodynamic measurements as discussed above. Alternatively, the outer radius R_H can be calculated from the specific volume $\bar{v}_h$ of the hydrated vesicle in mL/g of lipid, obtained from intrinsic viscosity measurements:

$$R_H = \left(\frac{3M_r\bar{v}_h}{4\pi N}\right)^{1/3} \tag{6}$$

$$[\eta] = \frac{\nu\bar{v}_h}{100} \tag{7}$$

where $[\eta]$ is the intrinsic viscosity of the vesicle suspension and ν the Simha viscosity increment. The values of R_H obtained from the measurement of the diffusion coefficient D_0 according to Eq. (5) and from viscosity measurements according to Eqs. (6) and (7) are in good agreement as shown for SUVs of dimyristoyl phosphatidylcholine [78].

The inner radius R_i of the hydrated vesicle is obtained directly from the measurement of the trapped volume v_c (cf. Fig. 10)

$$R_i = \left(\frac{3Mv_c}{4\pi N}\right)^{1/3} \tag{8}$$

For the details of this measurement see Refs. 78 and 79. R_i can also be obtained from R_H if the thickness d_h of the hydrated phospholipid bilayer is known. A value for d_h can be derived experimentally from x-ray diffraction as discussed below.

$$R_i = R_H - d_h \tag{9}$$

The dimensions of the lipid bilayer can be calculated assuming a four-compartment model as depicted in Fig. 10. The specific volume of the hydrated phospholipid vesicle $\bar{v}_h$ can be expressed in this model as

$$\bar{v}_h = v_L + v_e + v_c \tag{10}$$

where v_L is the specific volume of the anhydrous bilayer, v_c is the trapped volume, and v_e is the additional volume of water bound to the outer and inner polar groups of the phospholipid bilayer. All volumes are specific volumes expressed as mL/g of phospholipid. The partial specific volume of the hydrated phospholipid $\bar{v}$ is then

$$\bar{v} = \bar{v}_h - \frac{w_t}{\rho} = v_L + v_c + v_e - \frac{w_t}{\rho} \tag{11}$$

where w_t is the total mass of water associated with the phospholipid in g water per g of lipid and ρ is the density of the suspending medium. The partial specific volume of the phospholipid vesicle can be also expressed in terms of the water of hydration

$$\bar{v} = v_L + \frac{w_b}{\rho'} - \frac{w_b}{\rho} \tag{12}$$

where w_b is the phospholipid hydration expressed as the mass of water bound to the phospholipid polar group in g water/g of phospholipid, and ρ' is the density of bound water. According to Eq. (12), the partial specific volume $\bar{v}$ is not identical to the specific volume v_L of the anhydrous phospholipid. However, if it is assumed to a first approximation that $\rho' \approx \rho$, then $\bar{v} = v_L$.

The specific volume of the hydrated phospholipid bilayer is $v_L + v_e = \bar{v} - v_c + w_t/\rho$; it can be obtained from the measurements of the outer (Stokes) radius R_H and the hydrated bilayer thickness d_h.

$$v_L + v_e = \frac{4\pi}{3} [R_{H^3} - (R_H - d_h)^3] \tag{13}$$

For example, x-ray diffraction measurements performed with unsonicated egg phosphatidylcholine suspensions in water yield a d_h value of 4.56 nm [15]. From the knowledge of $v_L + v_e$ and that of the hydrated volume v_{PL} of one phospholipid molecule, a mean value for the total number n_t, of lipid molecules per vesicle is obtained:

$$n_t = \frac{v_L + v_e}{v_{PL}} \tag{14}$$

For example, the value for the hydrated volume of one egg phosphatidylcholine molecule was determined as $v_{PL} = 1.6368 \text{ nm}^3$ [15]. A mean value for n_t can also be calculated from the mean anhydrous vesicle weight M_r determined by analytical ultracentrifugation:

$$n_t = \frac{M_r}{M_L} \tag{15}$$

where M_L is the molecular weight of the phospholipid.

SUVs produced by exposing the phospholipid suspension to shear forces, such as by sonication and French press treatment, are believed to be under considerable stress due to the vesicle curvature. The relatively small radius of curvature of SUVs causes differences in the packing between the phospholipid polar groups on the outer and inner monolayer of the bilayer. This is manifested in the magnetic nonequivalence of the outer and inner polar groups of phospholipid SUVs observed by ^{1}H NMR. For instance, the polar group resonances of sonicated egg phosphatidylcholine vesicles exhibit a characteristic splitting [43,80–82] that was shown to increase with decreasing vesicle radius [82]. The tighter packing of the glycerophosphorylcholine group in the inner monolayer was suggested to lead to a partial dehydration and to conformational changes in the phospholipid polar group [82]. The coplanar orientation of the phospholipid group present in planar phosphatidylcholine bilayers [9] may change to a space-saving tilted orientation in the inner monolayer. Such a more perpendicular orientation of the inner polar groups gives rise to a net dipole moment along the bilayer normal and a net repulsion between phospholipid molecules in the inner monolayer, thus destabilizing the bilayer.

The splitting of the polar group signals observed in the ^{1}H NMR spectrum of SUVs of phosphatidylcholine can, in principle, be used to determine the molar ratio n_o/n_i where n_o and n_i are the number of phospholipid molecules located in the outer and inner monolayer of the vesicle bilayer, respectively. The splitting of the polar group signals can be artificially enhanced by the addition of paramagnetic shift reagents so that the resonances arising from the outer and inner monolayer are completely separated. In this case the molar ratio n_o/n_i is readily obtained from the integrated areas of the appropriate resonances [83,84]. Knowing the outer (R_H) and inner (R_i) radius of the hydrated phospholipid vesicle, the hydrated phospholipid areas in nm^2 per molecule can be calculated for the outer and inner vesicle surface using the relations

$$n_t = n_o + n_i \tag{16}$$

$$\frac{n_t}{n_i} = 1 + \frac{n_o}{n_i} \tag{17}$$

The hydrated phospholipid areas of the outer and inner vesicle surface, A_o and A_i, respectively, are then

$$A_o = \frac{4\pi R_H^2}{n_o} \tag{18}$$

$$A_i = \frac{4\pi R_i^2}{n_i} \tag{19}$$

D. Thermodynamic Stability of Phospholipid Vesicles

SUVs of phospholipids have been widely used as model membranes in the past, even though the precise relationship between the bilayers of SUVs and those of biological membranes is still unclear. In this context, the question of thermodynamic stability of SUVs is of prime importance. The information available on this question is scarce and often hidden in papers addressed to a different question.

To the best of my knowledge, no systematic study on the thermodynamic stability of phospholipid vesicles has been carried out so far. The information that is available to date may be summarized as follows. Lipid vesicles that form spontaneously on dispersing dry lipid films in aqueous media are considered as thermodynamically stable, at least for all practical purposes. This probably applies to MLVs of phosphatidylcholines and LUVs of acidic phospholipids or lipid mixtures (cf. swelling properties of phospholipids in aqueous media). In contrast, SUVs are usually formed from large vesicles or bilayer sheets by homogenization. The prevailing view is to regard the SUV as a distortion of the planar phospholipid bilayer [87–92]. SUVs usually do not form spontaneously but require the input of external energy for their formation, and they are therefore considered as thermodynamically unstable. There are several lines of evidence indicating that SUVs of phosphatidylcholine produced by sonication or French press treatment of large MLVs are indeed thermodynamically unstable.

(1) Storage of phosphatidylcholine SUVs leads to aggregation and/or fusion. For example, storing a suspension of egg phosphatidylcholine SUVs at room temperature for about one week leads to a significant increase in the mean vesicle size, as is evident from freeze-fracture electron microscopy and analytical ultracentrifugation [72]. The average hydrodynamic radius of the SUV population determined by freeze-fracture electron microscopy was 12.5 nm, and after incubation at room temperature for one week the vesicle radius was increased, ranging from 12.5 nm to about 60 nm. The mean vesicle weight of the original SUV suspension determined by analytical ultracentrifugation was 3.2×10^6 Da, and upon incubation of this suspension at room temperature for one week this value increased to 6.9×10^6 Da [72].

(2) SUVs of egg phosphatidylcholines are known to undergo aggregation and fusion when exposed to stress situations. Such situations are generated by freezing or dehydrating the aqueous suspension of SUVs [85,86]. For example, subjecting a sonicated aqueous suspension of egg phosphatidylcholine SUVs, ranging in diameter from 20–50 nm, to freeze-thaw cycles, freeze-drying-rehydration cycles, or spray-drying-rehydration cycles leads to massive vesicle fusion and the formation of MLVs [85,86].

(3) SUVs of phosphatidylcholines with well-defined hydrocarbon chains and a well-defined crystal-to-liquid-crystal phase transition temperature, e.g., dimyristoyl and dipalmitoyl phosphatidylcholine, aggregate and fuse to MLVs when such suspensions are stored at temperatures $<T_c$ or repeatedly passed through the phase transition temperature [93–95]. All these observations lend support to the notion that large MLVs represent the thermodynamically stable form of aqueous phosphatidylcholine suspensions and SUVs are an energized and hence thermodynamically unstable form.

Even fewer data are available pertaining to the thermodynamic stability of aqueous dispersions of negatively charged phospholipids. As discussed under lyotropic mesomorphism, LUVs and multivesicular vesicles form spontaneously when negatively charged phospholipids or phospholipid mixtures with a surface charge density exceeding 1–2 $\mu C/cm^2$ are dispersed in aqueous media. These particles very likely represent equilibrium structures. What was said for SUVs of isoelectric phospholipids (e.g., phosphatidylcholines) probably holds for negatively charged phospholipids. A recent study on the thermodynamic stability of SUVs of sodium egg phosphatidate carried out in this laboratory shows that the vesicles are unstable when stored at room temperature. SUVs were produced from this negatively charged phospholipid by sonication of aqueous dispersions at neutral pH under conditions that the pH of the suspending medium equals that of the internal cavity of the SUVs. The pH of the small aqueous vesicle cavity can be readily determined by ^{31}P NMR [38]. Egg phosphatidate SUVs thus produced undergo aggregation and fusion when stored at room temperature. The fusion is at least partially leaky as judged by the fluorescence intensity increase measured with SUVs containing carboxy fluorescein at self-quenching concentrations in their internal cavity. Comparing the rate of fusion of these negatively charged SUVs with that of isoelectric phosphatidylcholines, it is clear that the negative surface charge stabilizes the SUV. In contrast to the above experiment, even very small unilamellar vesicles of egg phosphatidate (mean hydrodynamic radius $<$20 nm) can be stabilized by imposing a pH gradient of 2–4 pH units across its phospholipid bilayer. The direction of the pH gradient is such that the external medium is alkaline and the pH of the vesicle cavity is 5–7. Storing such a suspension at 4°C or room temperature for two weeks had no effect on the vesicle size and size distribution as judged from gel filtration and freeze-fracture electron microscopy (B. Lin and H. Hauser, submitted for publication). Experiments with SUVs of negatively charged phospholipids such as phosphatidylserines and phosphatidates indicate that the thermodynamic stability of negatively charged SUVs depends sensitively on experimental conditions such as lipid concentration, pH of the medium, ionic strength and content, temperature and pressure, etc. The correlation between the thermodynamic stability of phospholipid vesicles and these experimental parameters has not been investigated systematically.

Summarizing our present knowledge of phospholipid vesicles, it can be said that the methodology for the preparation of the vesicles has advanced significantly in the past 20 years. This is also the case for the characterization of phospholipid vesicles in terms of their physicochemical properties. However, our knowledge of the thermodynamic stability of phospholipid vesicles is greatly lagging behind. This is a major omission, and it can only be hoped that the important question of thermodynamic stability will be addressed in a more systematic fashion in future studies.

REFERENCES

1. N. L. Vauquelin, *Ann. Mus. Hist. Nat. 18*:212 (1811).
2. E. Gorter and F. Grendel, *J. Exp. Med. 41*:439–443 (1925).
3. J. F. Danielli and H. Davson, *J. Cell. Comp. Physiol. 5*:495–508 (1935).
4. R. S. Bear, K. J. Palmer, and F. O. Schmitt, *J. Cell. Comp. Physiol. 17*:355–368 (1941).
5. K. Palmer and F. O. Schmitt, *J. Cell. Comp. Physiol. 17*:385–394 (1941).
6. A. Bangham, *Prog. Biophys. Mol. Biol. 18*:29–95 (1968).
7. H. Ringsdorf, B. Schlarle, and J. Venzmer, *Angew. Chemie 100*:117–162 (1988).
8. M. J. Hope, M. B. Bally, L. D. Mayer, A. S. Janoff, and P. R. Cullis, *Chem. Phys. Lipids 40*:89–107 (1986).
9. H. Hauser, I. Pascher, R. H. Pearson, and S. Sundell, *Biochim. Biophys. Acta 650*:21–51 (1981).
10. H. Hauser and G. Poupart, in *The Structure of Biological Membranes* (P. Yeagle, ed.), CRC Press, Boca Raton, Fla., Vol. 1, 1991, pp. 3–71.
11. J. L. Browning, in *Liposomes: From Physical Structure to Therapeutic Applications* (C. G. Knight, ed.), Elsevier, Amsterdam, 1981, pp. 189–242.
12. D. Marsh and A. Watts, in *Liposomes: From Physical Structure to Therapeutic Applications* (C. G. Knight, ed.), Elsevier, Amsterdam, 1981, pp. 139–188.
13. H. Hauser, *Biochim. Biophys. Acta 772*:37–50 (1984).
14. D. Chapman, R. M. Williams, B. D. Ladbrooke, *Chem. Phys. Lipids 1*:445–475 (1967).
15. D. M. Small, *J. Lipid Res. 8*:551–557 (1967).
16. G. G. Shipley, in *Biological Membranes* (D. Chapman and D. F. H. Wallach, eds.), Vol. 2, 1973, pp. 1–89.

17. M. J. Janiak, D. M. Small, and G. G. Shipley, *Biochemistry 15*:4575–4580 (1976).
18. M. J. Janiak, D. M. Small, and G. G. Shipley, *J. Biol. Chem. 254*:6068–6078 (1979).
19. G. G. Shipley, L. S. Avecilla, and D. M. Small, *J. Lipid Res. 15*:124–131 (1974).
20. H. Hauser, F. Paltauf, G. G. Shipley, *Biochemistry 21*:1061–1067 (1982).
21. D. Atkinson, H. Hauser, G. G. Shipley, J. M. Stubbs, *Biochim. Biophys. Acta 339*:10–29 (1974).
22. T. Gulik-Krzywicki, A. Tardieu, and V. Luzzati, *Mol. Cryst. Liq. Cryst. 8*:285–291 (1969).
23. H. Hauser, *Chimia 39*:252–264 (1985).
24. M. Kates, in *Techniques of Lipidology, Isolation, Analysis and Identification of Lipids*, Elsevier, Amsterdam, 2nd ed., 1986.
25. R. R. C. New, in *Liposomes A Practical Approach* (R. R. C. New, ed.), Oxford University Press, New York, 1990.
26. A. D. Bangham and R. W. Horne, *J. Mol. Biol. 8*:660–668 (1964).
27. A. D. Bangham, M. M. Standish, and J. C. Watkins, *J. Mol. Biol. 13*:238–252 (1965).
28. H. Talsma, H. Jousma, K. Nicolay, and D. J. A. Crommelin, *Int. J. Pharmaceutics 37*:171–173 (1987).
29. L. Saunders, J. Perrin, and D. B. Gammack, *J. Pharm. Pharmacol. 14*:567–572 (1962).
30. Y. Barenholz, S. Amselem, D. Lichtenberg, *FEBS Lett. 99*:210–214 (1979).
31. R. L. Hamilton, J. Goerke, L. S. S. Guo, M. C. Williams, and R. J. Havel, *J. Lipid Res. 21*:981–992 (1980).
32. F. Olson, T. Hunt, F. C. Szoka, W. J. Vail, D. Papahadjopoulos, *Biochim. Biophys. Acta 557*:9–23 (1979).
33. Y. Barenholz, D. Gibbes, B. J. Litman, J. Goll, T. E. Thompson, and F. D. Carlson, *Biochemistry 16*:2806–2810 (1977).
34. H. Hauser, *Biochem. Biophys. Res. Commun. 45*:1049–1055 (1971).
35. H. Hauser and N. Gains, *Proc. Natl. Acad. Sci. 79*:1683–1687 (1982).
36. H. Hauser, N. Gains, and M. Müller, *Biochemistry 22*:4775–4781 (1983).
37. N. Gains and H. Hauser, in *Liposome Technology* (G. Gregoriadis, ed.), CRC Press Inc., Boca Raton, Fla., Vol. 1, 1984, pp. 67–78.
38. H. Hauser, *Proc. Natl. Acad. Sci. 86*:5351–5355 (1989).
39. H. Hauser, H. H. Mantsch, and H. Casal, *Biochemistry 29*:2321–2329 (1990).
40. P. Walde, A. M. Giuliani, C. A. Boicelli, and P. L. Luisi, *Chem. Phys. Lipids 53*:265–288 (1990).
41. H. Yoshioka, *J. Colloid Interface Sci. 95*:81–86 (1983).
42. S. Batzri and E. D. Korn, *Biochim. Biophys. Acta 298*:1015–1019 (1973).
43. J. Brunner, P. Skrabal, and H. Hauser, *Biochim. Biophys. Acta 455*:322–331 (1976).
44. D. W. Deamer and A. D. Bangham, *Biochim. Biophys. Acta 443*:629–634 (1976).
45. F. Szoka and D. Papahadjopoulos, *Proc. Natl. Acad. Sci. 75*:4194–4198 (1978).
46. F. Szoka and D. Papahadjopoulos, in *Liposomes: From Physical Structure to Therapeutic Applications* (C. G. Knight, ed.), Elsevier, Amsterdam, 1981, pp. 51–82.
47. C. Pidgeon, A. H. Hunt, and K. Dittrich, *Pharm. Res. 3*:23–34 (1986).
48. Y. Kagawa and E. Racker, *J. Biol. Chem. 246*:5477–5487 (1971).
49. J. Brunner, H. Hauser, and G. Semenza, *J. Biol. Chem. 253*:7538–7546 (1978).
50. M. H. W. Milsman, R. A. Schwendener, and H. G. Weder, *Biochim. Biophys. Acta 512*:147–155 (1978).
51. H. G. Weder and O. Zumbuehl, in *Liposome Technology* (G. Gregoriadis, ed.), CRC Press, Boca Raton, Fla., Vol. 1, 1984, pp. 79–107.
52. W. J. Gerritsen, A. J. Verkleij, R. F. Zwaal, and L. L. M. van Deenen, *Eur. J. Biochem. 85*:255–261 (1978).
53. T. M. Allen, in *Liposome Technology* (G. Gregoriadis, ed.), CRC Press, Boca Raton, Fla., Vol. 1, 1984, pp. 109–122.

54. P. Schurtenberger, N. Mazer, and W. Känzig, *J. Phys. Chem. 89*:1042–1049 (1985).
55. L. I. Mimms, G. Zampighi, Y. Nozaki, C. Tanford, and J. A. Reynolds, *Biochemistry 20*:833–840 (1981).
56. O. Zumbuehl and H. G. Weder, *Biochim. Biophys. Acta 640*:252–262 (1981).
57. J. A. Reynolds, Y. Nozaki, and C. Tanford, *Anal. Biochem. 130*:471–474 (1983).
58. P. Schurtenberger and H. Hauser, *Biochim. Biophys. Acta 778*:470–480 (1984).
59. G. K. Ackers, *J. Biol. Chem. 242*:3237–3238 (1967).
60. V. K. Miyamoto and W. Stoeckenius, *J. Membr. Biol. 4*:252–269 (1971).
61. M. Müller, N. Meister, and H. Moor, *Mikroskopie 36*:129–140 (1980).
62. D. Studer, H. Moor, and H. Gross, *J. Cell Biol. 90*:153–157 (1981).
63. P. Guiot and P. Baudhuin, in *Lipsosome Technology* (G. Gregoriadis, ed.), CRC Press, Boca Raton, Fla., Vol. 1, 1984, pp. 163–178.
64. J. Dubochet, M. Adrian, J. J. Chang, J. C. Homo, J. Lepault, A. W. McDowall, and P. Schultz, *Q. Rev. Biophys. 21*:129–228 (1988).
65. F. R. Hallett, J. Watton, and P. Krygsman, *Biophys. J. 59*:357–362 (1991).
66. Y. Nozaki, D. D. Lasic, C. Tanford, and J. A. Reynolds, *Science 217*:366–367 (1982).
67. N. C. Ford, Jr., in *Dynamic Light scattering* (R. Pecora, ed.), Plenum Press, New York, 1985, pp. 7–58.
68. P. Schurtenberger and H. Hauser, in *Liposome Technology* (G. Gregoriadis, ed.), CRC Press, Boca Raton, Fla., Vol. 2, 1992, pp. 253–270.
69. N. Ostrowsky, D. Sornette, P. Parker, and R. Pike, *Opt. Acta 28*:1059–1071 (1981).
70. P. M. Frederik, N. J. Burger, M. C. A. Stuart, and A. J. Verkleij, *Biochim. Biophys. Acta 1062*:133–141 (1991).
71. C. Huang, *Biochemistry 8*:344–352 (1968).
72. H. Hauser and L. Irons, *Hoppe-Seyler's Z. Physiol. Chem. 353*:1579–1590 (1972).
73. M. Spiess, H. Hauser, J. P. Rosenbusch, and G. Semenza, *J. Biol. Chem. 256*:8977–8982 (1981).
74. T. J. McIntosch and S. A. Simon, *Biochemistry 25*:4058–4066 (1986).
75 W. Lesslauer, J. E. Cain, and J. K. Blasie, *Proc. Natl. Acad. Sci. 69*:1499–1503 (1972).
76. H. Jousma, H. Talsma, F. Spies, J. G. H. Joosten, H. E. Junginger, and D. J. A. Crommelin, *Int. J. Pharm. 35*:263–274 (1987).
77. J. Lepault, F. Pattus, and N. Martin, *Biochim. Biophys. Acta 820*:315–318 (1985).
78. A. Watts, D. Marsh, and P. F. Knowles, *Biochemistry 17*:1792–1801 (1978).
79. H. Thurnhofer, B. Kräutler, and H. Hauser, *Biochemistry 28*:2305–2312 (1989).
80. M. P. Sheetz and S. I. Chan, *Biochemistry 11*:4573–4581 (1972).
81. R. J. Kostelnik and S. M. Castellano, *J. Magn. Reson. 9*:291–295 (1973).
82. C. G. Brouilette, J. P. Segrest, T. C. Ng, and J. L. Jones, *Biochemistry 21*:4569–4575 (1982).
83. V. F. Bystrov, N. I. Dubrovina, L. I. Barsukov, L. D. Bergelson, *Chem. Phys. Lipids 6*:343–350 (1971).
84. H. Hauser, M. C. Phillips, B. A. Levine, and R. J. P. Williams, *Nature 261*:390–394 (1976).
85. G. Strauss and H. Hauser, *Proc. Natl. Acad. Sci. 83*:2422–2426 (1986).
86. H. Hauser and G. Strauss, *Biochim. Biophys. Acta 897*:331–334 (1987).
87. W. Helfrich, *Z. Naturforsch. 28c*:693–703 (1973).
88. W. Helfrich, *Phys. Lett. 50a*:115–116 (1974).
89. D. D. Lasic, *Biochim. Biophys. Acta 692*:501–502 (1982).
90. P. Fromherz, *Chem. Phys. Lett. 94*:259–266 (1983).
91. B. A. Cornell, J. Middlehurst, and F. Separovic, *Faraday Discuss. Chem. Soc. 81*:163–178 (1986).
92. D. D. Lasic, *Biochem. J. 256*:1–11 (1988).

93. J. Suurkuusk, B. R. Lentz, Y. Barenholz, R. L. Biltonen, and T. E. Thompson, *Biochemistry* *15*:1393–1401 (1976).
94. H. L. Kantor, S. Mabrey, J. H. Prestegard, and J. M. Sturtevant, *Biochim. Biophys. Acta* *466*:402–410 (1977).
95. P. W. M. Van Dijck, B. de Kruijff, P. A. M. M. Aarts, A. J. Verkleij, and J. de Gier, *Biochim. Biophys. Acta 506*:183–191 (1978).

18
Solute Transport Across Bilayers

Gregor Cevc *Technical University of Munich, Munich, Germany*

I. INTRODUCTION

Each phospholipid bilayer, being essentially an 'amphiphobic' region, is prone to affect the direction or extent of motion of its surrounding molecules dramatically. On the one hand, such a bilayer hinders the transport of hydrophilic molecules through the bilayer interior and sometimes speeds up the motion of such molecules along the bilayer surface. On the other hand, the transport of hydrophobic, fatty substances is often facilitated in the plane of a bilayer; the transport of fat-soluble substances, however, in the direction perpendicular to the bilayer surface is very low.

Diffusion of fatty substances in the lateral direction is one of the topics of Chap. 13 and will therefore not be discussed here. Suffice it to say that this diffusion is by one to two orders of magnitude more rapid in the fluid than in the ordered phase. (The corresponding diffusion constants for phosphatidylcholine are 4×10^{-9} and 10^{-10} $cm^2\ s^{-1}$, respectively [1,2]. The presence of defect lines, moreover, seems to increase the lateral mobility of the lipophilic substances in phospholipid bilayers [3].

The motion of water-soluble molecules along or inside a lipid bilayer in the lateral direction has hardly been studied to date. Consequently, this phenomenon is only incompletely understood. It has been claimed that proton transport in the interfacial region is anomalously high [4]. But the significance of the corresponding experimental data is rather controversial [5]. The following discussion is therefore restricted to material transport through phospholipid bilayers.

The main cause of solute or solvent flow across a phospholipid membrane is the transbilayer concentration (i.e., chemical potential) gradient, the electrostatic potential gradient, or the pressure. The chief transport mechanism, in biological systems at least, is solute diffusion through pores and channels. In principle, however, simple diffusion via hydrocarbon core or carrier facilitated or ionophore facilitated transport are also possible.

II. MODELING TRANSBILAYER SOLUTE FLOW

Material flow across a phospholipid bilayer can be described in a variety of ways. Most frequently, multicomponent Nernst-Planck diffusion equations are applied (see, e.g., Ref. 6). The treatment of diffusion through a phospholipid membrane as an activated process, employing Eyring transition state theory (see, e.g., Ref. 7) or by using nonequilibrium thermodynamics (see, e.g., Ref. 10), is also possible and sensible, however (for a brief introduction see Ref. 11).

In each case, the basic, constitutive relation for transbilayer flow is

$$j_i(x) = \sum_j P_{ij}(x)\Delta c_j(x) \tag{1}$$

where j_i is the flux of the i-th molecular species; $\Delta c_j = c_j(x + dx) - c_j(x)$ is the local concentration gradient, P_{ij} is an element of the permeability matrix, and volume flow has been omitted. For a single molecular species or mutually independent flows, $P_{ij} = 0$, $i \neq j$, and the right hand side of Eq. 1 consists of merely one term. The transmembrane electric current can be deduced directly from Eq. 3 by making use of the definition $I = \Sigma_i(Z_i N_A e_0 j_i)$, $Z_i e_0$ being the charge of the permeating species and N_0 the Avogadro constant.

As a function of the transbilayer diffusivity profile $D_i(x)$ and of the corresponding partition coefficient profile $K_i(x) = \exp(-G_i/kT)$, the flow equation can be written as

$$j_i(x) = - K_i(x)D_i(x)\,\frac{d[c_i(x)/K_i]}{dx} \tag{2}$$

The term in square brackets gives the probability for the permeating species i to be located in the phospholipid bilayer at position x. $G_i(x)$ is the solute free energy at this position. (In the case that an electrical potential difference exists across a bilayer, the electrostatic contribution $Z_i e_0 \psi_{el}(x)/kT$ must also be added to this free energy.)

A. Bilayer Permeability

(a) Permeability of the bilayer core. In order to calculate the transmembrane solute flow, Eqs. 2 and/or 1 must be integrated across the bilayer, which is assumed to extend from $x = 0$ to $x = -d_m$. Comparison of Eqs. 1 and 2 then shows immediately that the diffusive permeability of the entire bilayer interior is given by

$$P_{m,i} = \left[\int_{-d_m}^{0} dx\,\frac{\exp(G_{0,i}(x)/kT)}{D_i(x)}\right]^{-1} \tag{3}$$

This quantity is also identical to the mean of the product $K_i(x)D_i(x)$ divided by the thickness of the membrane core d_m: $P_m = \langle \mathrm{KD}\rangle/d_m$, in analogy with the standard relation $P_m = \mathrm{KD}/d_m$.

Equation 3 clearly shows the limitations and inadequacy of simple versions of Eq. 1, in which all factors are taken to be constant, for the description of transbilayer solute transport. Owing to its complexity, a phospholipid bilayer does not necessarily act as a well-defined uniform permeability barrier. More likely it will change its properties upon changing concentration and/or composition of the solution. Membrane permeability, consequently, is also a function of solute-membrane interactions and the transbilayer gradient. The presence of alcohols, for example, not only results in the transmembrane flow of these molecules but, moreover, may increase the bilayer permeability by changing the effective polarity of the membrane interior [13]. In addition to this, lipid-associated alcohol molecules may induce changes in bilayer packing, such as chain interdigitation [14]. The latter then increases membrane permeability to all kinds of solutes.

For biological membranes, in particular, the permeability factor is often observed to decrease with increasing transbilayer solute concentration gradient [15]. The curve j versus Δc then shows saturation similar to that characteristic of the Michaelis-Menten type of reaction. To minimize the danger of data misinterpretation, it is therefore advisable to measure transport rates for different solute concentrations and then extrapolate the corresponding permeability value to zero solute concentration, $c \rightarrow 0$.

(b) Interfacial resistance. The inverse of the bilayer diffusion permeability—the hydrocarbon resistance—can be represented by a sum of resistance elements $dx/K(x)D(x)$, each arising from a thin membrane section of thickness dx at a distance x from the polar-apolar interface (cf. Eq. 3).

The resistance of the membrane core is not necessarily the sole determinant of the total membrane permeability, however. The latter is also affected by the interfacial permeability barrier [12]. The height of such an interfacial barrier—which may also be a minimum—is proportional to the interfacial partition coefficient value, $K_i(x = \text{interface}) \equiv k_{\text{in},i}(x = \text{interface})/k_{\text{out},i}\,(x = \text{interface})$, where k_{in} and k_{out} are the inward and outward rates of solute transport across the interfacial region. The material flow across an interface with a thickness d_p is thus given by $j_i = \pm d_p[c_{\text{out}}(x)k_{\text{in},i}(x) - c_{\text{in}}(x)k_{\text{out},i}(x)]_{x=\text{interface}}$. In the steady state, this flow must be equal to the transbilayer flow given by Eq. 2.

To calculate the inward rate constant as a function of the interfacial energy $G_i^\ddagger$, one can use the Eyring absolute rate theory. This gives

$$k_{\text{in},i} = \frac{kT}{h} \exp\left(\frac{-G_i^\ddagger}{kT}\right) \tag{4}$$

where h is Planck's constant [7]. The activation enthalpy is directly related to the activation energy $E_i^\ddagger$ by $\Delta H_i^\ddagger - E_i^\ddagger\ ST \sim E_i^\ddagger$.

(c) Total bilayer resistance. The resistance of a typical phospholipid bilayer, therefore, consists of at least two interfacial contributions and a contribution from the hydrocarbon core. These add up as resistances of resistors in a series, suggesting that the total bilayer permeability is equal to

$$P_i = \{P_{m,i}^{-1} + [k_{\text{in},i}(x = 0)^{-1} + k_{\text{in},i}(x = -d_m)^{-1}]d_p{}^{-1}\}^{-1} \tag{5}$$

Solute permeation through a phospholipid bilayer, consequently, can be dominated either by the interfacial resistance or by the resistance of the membrane interior, depending on the relative values of $k_{in}(0)$, $k_{in}(-d_m)$, and P_m. For water-soluble substances, such as inorganic ions, the membrane interior is a permeability barrier; for more lipophilic substances, such as organic ions, the same region acts as an energy well, or even as a trap [8,9] (cf. Fig. 1). In both cases the exchange of material between two aqueous compartments separated by a phospholipid bilayer is hindered.

Many medium size or large organic ions reside preferentially in the interfacial region. When this is the case, the simple three-state picture underlying Eq. 5 must often be modified to account for additional possible energy states of the solute molecules [16].

III. WATER TRANSPORT

The on-off kinetics of water molecules bound at the phospholipid-water interface occurs on the submicrosecond time scale. It is rapid enough for water transport across a

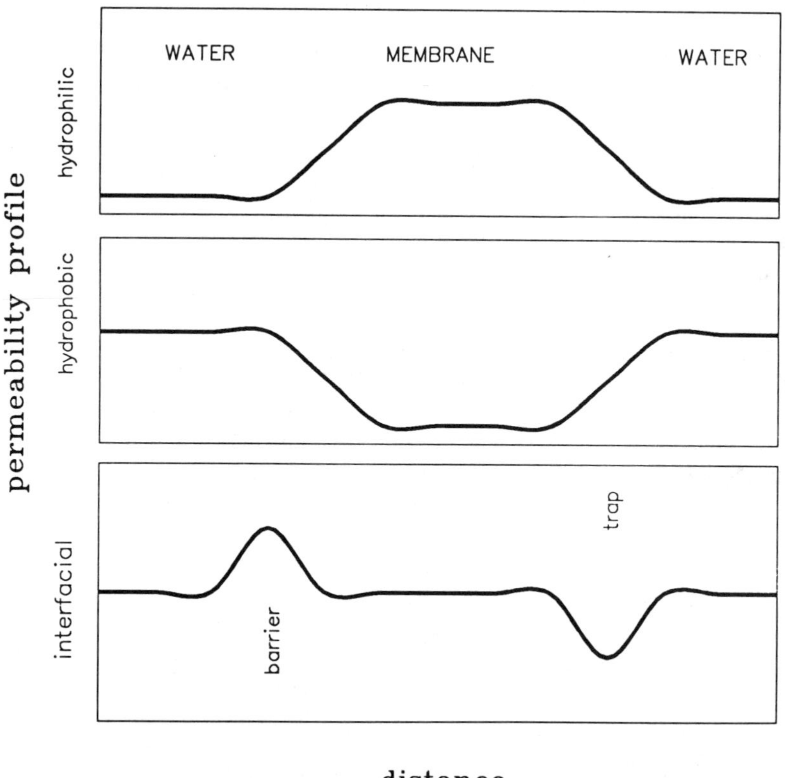

Figure 1 Energy and permeability profile for the hydrophilic (upper) and hydrophobic (middle) solutes moving across a phospholipid bilayer. The lowest panel illustrates the interfacial energy profile for the inorganic (left) and organic ions (right).

phospholipid membrane to be governed nearly entirely [17] by the permeability barrier of the hydrocarbon membrane core. The more ordered and condensed the lipid bilayer is, the lower is its permeability to water molecules. This holds particularly for cholesterol-free lipid membranes below the chain-melting phase transition. For membranes above the chain-melting phase transition, or for cholesterol-containing vesicles in the gel phase, the hydrocarbon membrane interior is a much less effective diffusion barrier. This is seen from model calculations [18] and even more so from a comparison of the self-diffusion coefficients for water across pure ordered-phase phosphatidylcholine membranes, which are on the order of 10^{-7} cm^2s^{-1}, with corresponding values for fluid bilayers. The latter are of the order of 2×10^{-5} cm^2s^{-1}, i.e., comparable to the self-diffusion coefficient of water in the bulk phase [19,20,21].

Water permeability reveals an almost discontinuous break in the temperature curve at the phase transition [22] (cf. Fig. 2), where the permeation increases by a factor of 10–100. Exceptional are the bilayers at the air-water interface, which show a singularly high permeability for water just at the phase transition [23].

For diacylphosphatidylcholine membranes, the activation energy for transmembrane

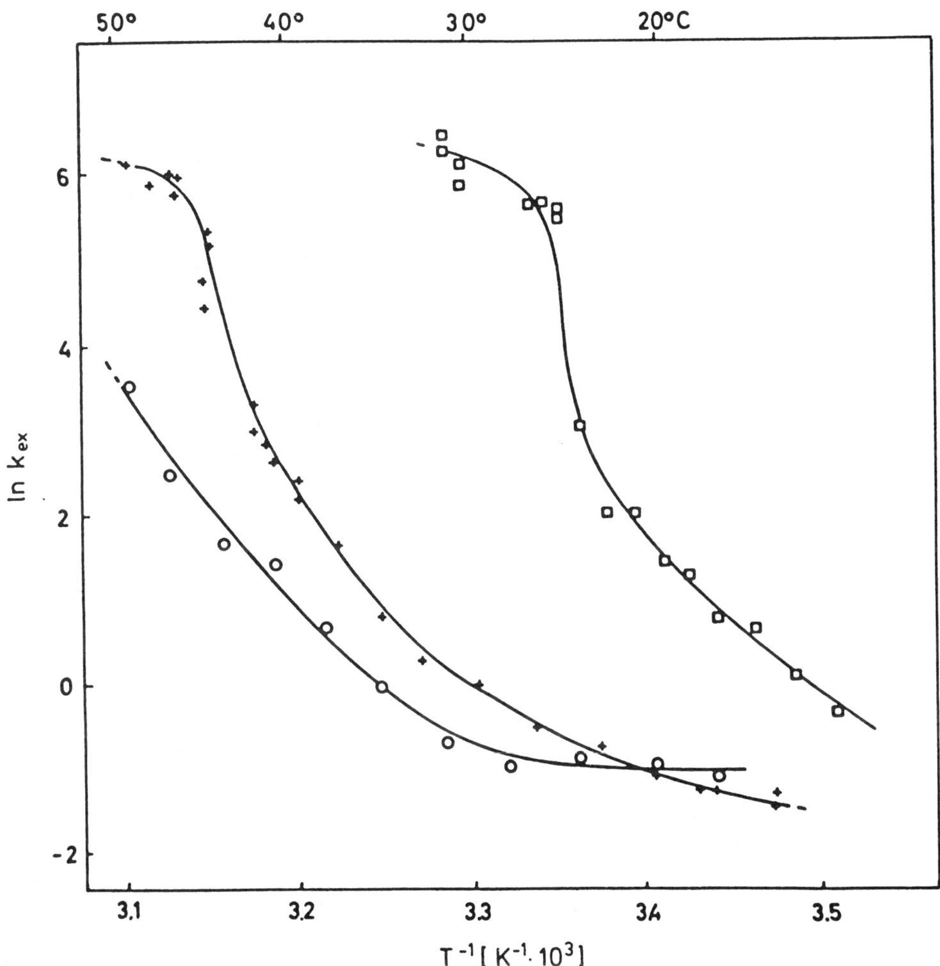

Figure 2 The rate constant for water exchange across various diacylphosphatidylcholine bilayers as a function of (inverse) temperature (□ dimyristoyl-, ◆ dipalmitoyl-, ○ distearoylphosphatidylcholine). k_{ex} is proportional to the water permeability constant. (From Ref. 19.)

water permeation is almost independent of the chain length, being 104.5–117 kJ mol^{-1} and 46 kJ mol^{-1} below and above the chain-melting phase transition temperature, respectively [24]. The corresponding permeability values for water across dipalmitoylphosphatidylcholine bilayers are 4.2×10^{-6} cm s^{-1}, 3.5×10^{-5} cm s^{-1}, and 2.4×10^{-3} at 26.5, 37, and 46°C, respectively. After incorporation of 5–20 mol % cholesterol the activation energy in gel phase bilayers decreases to 92 kJ mol^{-1}.

Incorporation of channels into phospholipid bilayers also may increase the transbilayer water flow appreciably [25,24]. This is owing to the low resistance of the channel's pore, which is normally relatively hydrophilic, to water penetration.

IV. SOLUTE TRANSPORT

To estimate the transmembrane resistance $\langle KD\rangle/d_m$ and thus the permeability rates of different solutes, knowledge of the diffusion and partition profiles is required (cf. Eq. 3). In principle, the former can be determined from the bilayer viscosity and bulk diffusion constant data by making use of the relation $\eta_m(x)D_m(x) = \eta D$, where η and D are the viscosity and diffusion coefficient of water. Some estimates of the average intramembrane diffusion constant are 2–6 $\times$ 10^{-7} (1.8 $\times$ 10^{-5}) cm^2 s^{-1} at $T < T_m$ ($T > T_m$) for water, 10^{-6}–10^{-5} cm^2 s^{-1} (33°C < T < 63°C) for small hydrophilic ions [26], 4 $\times$ 10^{-6} cm^2 s^{-1} for benzene at 22°C [27], and 10^{-8} cm^2 s^{-1} for the tetraphenylborate ion [28]. The bulk membrane viscosity is 1–2 kg m^{-1} s^{-1}, except in the transition region [29].

Experiments suggest that the resistance to solute permeation is inhomogeneously distributed throughout the bilayer, the highest and rate-limiting barrier being often located in the interfacial region. According to the data of Dix and coworkers [26], for small, water soluble molecules the interfacial barrier is 10–10^3 times higher than the remaining transmembrane resistance.

The partition function $K(x)$ corresponds to an intramembrane free energy profile. Three examples of such a profile, including one for hydrophilic (top panel) and one for the hydrophobic (middle panel) solutes, are depicted schematically in Fig. 1. Each partition function therefore depends both on the electrostatic (trans)membrane potential and on the mutual interactions between the permeating molecule, solute, and membrane components.

A. Transport of Nonionic Solutes

For the permeation of noncharged solutes across a phospholipid bilayer only the latter, nonelectrostatic contributions play an important role. In the simplest approximation this can be seen from the average value $\langle K\rangle$ of the hydrocarbon partition function.

For slightly amphiphilic solutes, the value of the partition coefficient is proportional to the length of the apolar side chain. Indeed, the incremental free energy of transfer between aqueous and hydrocarbon compartments has been found for different solutes [30,31,32] to be fairly constant, around 3.3 kJ mol^{-1} (1.3 kT). (For the fatty alcohols the free energy of transfer between water and lipid bilayer as a function of chainlength n_C has been measured to be given by $(-0.9 - 1.11n_C)kT$ [30,31].) Partition coefficients of most other small amphiphilic molecules are also expected to increase with the length of the apolar side chain, within certain limits at least. The proportionality certainly ceases when solute insertion into a bilayer changes the phospholipid properties appreciably.

The permeability of phospholipid bilayers to alcohols with 2–4 carbon atoms per chain, for example, increases nearly linearly with this length [33] (cf. Table 1). The same is true for certain other small water-soluble compounds, such as monocarboxylic acids [35,34] or even cationic short-chain alkylamines [36].[1] But substances with relatively long aliphatic chains (normally with a heyxl residue or longer) are transported *through* the lipid bilayers inefficiently owing to their predominantly hydrophobic character and the resulting inability to diffuse away into the water subphase ($k_{out} \rightarrow 0$).

For various small, uncharged solutes with multiple OH side groups, the partition

[1]This rule is not without exceptions, however: the smallest possible compound of each kind mentioned (i.e., methanol, methylamine, acetic acid) always permeates through the membranes at a slightly higher rate than the corresponding next-higher neighbor.

Table 1 Permeabilities P_m (cm s^{-1}) of Fluid-Phase Phosphatidylcholine Bilayers for Various Organic Solutes

Nonionic					
Alkanols	Methanol	Ethanol	Butanol	Propanol	Pentanol
P_m	3.3×10^{-3}	2×10^{-3}	6×10^{-3}	5.7×10^{-2}	(10^{-2})
Polyalcohols	Glucose $C_6O_6H_{12}$	Fructose $C_6O_6H_{12}$	Sucrose $C_{12}O_{11}H_{22}$	Propantriol (Glycerol=$C_3O_3H_7$)	Ethandiol (Ethyl glycol=$C_2O_2H_5$)
P_m	3×10^{-12}–3×10^{-11}	4×10^{-10}	8×10^{-14}	5×10^{-6}	2.5×10^{-5}
Ionic					
Alkylamines	Methylamine	Ethylamine	Propylamine	Butylamine	Pentylamine
P_m	1×10^{-6}	7×10^{-7}	4.3×10^{-6}	5.2×10^{-5}	4×10^{-4}

(Data for sugars from Refs. 122, 121; for glycerol, Ref. 43; for ethylglycol, Ref. 39. Data for alcohols (in red blood cells) from Ref. 33 and (for lipid vesicles) unpublished; for alkylamines, from Ref. 125. Activation enthalpy for glucose is 145 kJ mol^{-1} and 23 kJ mol^{-1} below and above the chain-melting phase transition temperature [122]. For fructose, $\Delta H^{\ddagger} \sim 90$ kJ mol^{-1} has been reported [121]. For the alkylamines the activation enthalpy and entropy values are nearly independent of the chain length, being approximately 85 kJ mol^{-1} and 150 J/mol K [125].)

coefficient has been found to decrease proportionally with solute capacity for hydrogen bond formation [12,37,38]. This, in the first approximation, is proportional to the number of OH groups. We can estimate the upper limit for the energetic contribution to the partition coefficient from each OH group, or from each COO^- group, on the basis of pyrenyl alkanol, pyrenyl alkyl carboxylate, and pyrenyl alkane transfer data [32] as ≤ 10 kJ mol^{-1} and ≤ 12.5 kJ mol^{-1}, respectively. The incremental change per each hydrogen bond, from other experimental results, appears to be 7.5 kJ mol^{-1}/H-bond [39], in the case of molecular transport across a fluid-phase phosphatidylcholine bilayer. Permeability values calculated from these results by using Eq. 4 are in remarkably good agreement with experimental data for nonionic solutes (cf. Table 1).

Reasonably accurate and simple estimates of the partition coefficients for many small nonionic solutes are possible within the framework of Collander [40] approximation. More refined and modern models, however, e.g., that of Kamlet and coworkers [41], give even superior results. These authors have proposed the following expression for the evaluation of the partition coefficient of nonelectrolytes between the fluid hydrocarbon (octanol) and water [42]:

$$K = \exp\{-2.3[0.2 + 0.0274\bar{V}_{m,i} - 0.92\pi_i^* - 3.49\ \beta_i]\} \tag{6}$$

Because nonelectrolytes are localized predominantly in membrane regions that are only slightly less hydrophilic than octanol [30], the same result can also be used for fluid phospholipid bilayers, however. In Eq. 6, the parameter β_i determines an exoergic term depending on the hydrogen bond donor or acceptor acidity and basicity, the factor π_i^* scales another exoergic term measuring the solute-solvent polarizability, and the cavity term is taken to be proportional to the solute molar volume $\bar{V}_{m,i}$.

Any change in system properties that results in a diminished hydrocarbon packing density, and thus increases the probability of a water-soluble solute partitioning into the phospholipid bilayer, is prone to increase the solute permeation through such a bilayer. The diffusion rate of glycerol into bilayers of distearoyl-, stearoyl-oleoyl-, dioleoyl-, and dilineoylphosphatidylcholine, for example, increases with the number of double bonds in the paraffin chains in the same sequence. Increasing temperature also causes the bilayer permeability for glycerol to increase markedly [43], sometimes by several orders of magnitude [30]. (This presumably reflects differences in the diffusion constant, rather than the partition coefficient. This is concluded from the fact that the entropy and enthalpy of partition increase 10-fold, and the partition coefficient by only a factor of less than 5, upon fluidization of the phosphatidylcholine bilayers.)

In principle, the activation energies for solute permeation can be assigned on the basis of experimental data for individual polar residues of a particular solute. Assigning the number of hydrogen bonds to the different functional groups on the permeating molecules by the criteria of Stein, Cohen [39] has determined an incremental change in the activation energy of 7.5 kJ mol^{-1} per H bond, for example, for fluid-phase phosphatidylcholine bilayers. The extrapolated value for zero H bonds increases from 1 to 17 kJ mol^{-1} upon addition of 30 mol % cholesterol. The corresponding molar entropy change is 9.5 J mol^{-1} K^{-1}.

B. Ion Transport

Ion transport can occur through either pores or channels, with the aid of ionophores or carriers, and in a simple diffusive manner.

1. Diffusive Transport

Simple ion diffusion is the least likely to occur but the most easy to understand of all three above-mentioned processes. Ions going through a phospholipid bilayer must first partition into the interfacial region and then diffuse through the apolar hydrocarbon core. The latter has a characteristic 'polarity profile', which has been directly measured by means of electron spin resonance for small noncharged hydrophilic molecules [44]. Based on these data and the dielectric constant values from other experiments, we conclude [68] that the dielectric constant changes from the bulk value of ~ 78 to a value of ~ 30 in the interfacial region [45] and then to a value of ~ 2 in the membrane core [46]. This variation is quasi-continuous [68] and occurs in two steps at least over a distance of approximately 1 nm in each case.

During the first step—and at latest during the second one—nearly every ion gives up some of its water of hydration. This is, in molecular terms, the chief reason for the low diffusive permeability of ions in phospholipid bilayers and is also part of the reason for the ion selectivity of phospholipid bilayers [47].

In electrostatic terms, the same conclusion can be reached by considering the electrostatic energy cost of transferring an ionic charge from a medium with a high dielectric constant ($\epsilon \sim 80$) into a membrane interior with a low dielectric constant value ($\epsilon_m \sim 2$). From the Born model of ion hydration [11], this energy cost can be estimated [52] as

$$\Delta G_i = \frac{Z_i^2 e_0^2}{4\pi\epsilon_0}\left[\frac{1}{2(r_i + \xi)}\left(\frac{1}{\epsilon} - \frac{1}{\epsilon_m}\right) - \frac{1}{\epsilon_m d_m}\ln\left(\frac{2\epsilon}{\epsilon + \epsilon_m}\right)\right] \tag{7}$$

r_i is the ion radius and $\xi \simeq 0.07$ nm is a correction term [47]. The first term in the square brackets gives the simple Born energy of ion transfer between the bulk electrolyte and the membrane core. The second term allows for finite bilayer thickness effects and according to Eq. 7 becomes important only when $d_m \ll 4$–5 nm. For most phospholipid bilayers this term can thus be omitted. Equations 3 and 7 also suggest that bilayer permeability should decrease nearly exponentially with ion charge. (Divalent ions would thus be transported by a factor of 50 less rapidly than monovalent ions, were it not for their higher affinity for the lipids, the resulting stronger ion-lipid binding, and the subsequent relatively rapid translocation of the complexes across lipid bilayers.) Phospholipid bilayers are essentially refractory to polyions.

In addition to the Born energy, work must be invested to create extra space for the ion in the hydrocarbon region. The simplest way to account for this is to calculate the cavity term, for example from the Uhlig [48] expression

$$\Delta G_{\mathrm{cav},i} = 4\pi r_i^2 \gamma_i \tag{8}$$

where γ_i is the interfacial tension between the solute surface (for hydrated ions: water) and paraffin chains. The energy cost of cavity formation for simple ions is likely to be appreciable if the ion radius r_i is greater than 0.4 nm [49]. For organic ions, however, this term may be negative, giving rise to an attractive energy contribution.

(a) Organic ions. The size dependence of bilayer permeability for complex (e.g., organic) ions is also sensitive to the ion polarizability [50]. From a combination of Eqs. 7 and 8 it follows [51] that for relatively large ions even the total free energy of transfer from water to an organic phase may be negative. (In the case of symmetric

'amino-ions', such a crossover occurs between tetramethyl and tetraethylammonium ions.) This explains why most large ions are imported relatively easily into bilayers but have difficulties in getting out of the lipid phase and back into the aqueous compartments.

Membrane permeability for organic anions and cations is dramatically different [9,53,54,55]. It has been suggested that this is caused by the dipolar membrane potential. It would seem, however, that the significance of this potential, or at least of the dipoles originating from the phospholipid molecules, is exaggerated and that the contribution from the interfacially bound water is equally important. Two things speak in favor of this hypothesis. First, phosphatidylethanolamines and phosphatidylcholines, which have similar dipolar headgroup moments, have entirely different permeation characteristics [56,88]. Second, the rates of transfer of (poly)ions across the bilayers of dialkyl (ether-bound) and diacyl (esterified) phospholipids are similar [58] even though the former lipids carry two additional highly polarizable carbonyl groups with large dipole moments.

(b) Inorganic ions. Bilayer permeability for inorganic and small organic ions decreases with ionic size. It also decreases with increasing bilayer hydration, at least when the rate-limiting step occurs in the interfacial region and no specific ion-lipid association takes place. This conclusion is consistent with the available simple permeability data for alkali ions penetrating through lipid bilayers (see Table 2). These suggest the following permeability series: $Cs^+ > Rb^+ > K^+ > Na^+ \gg Ca^{2+} > Mg^{2+} \gg Al^{3+}$.

Ion binding to the phospholipid molecules modifies this size dependence, however, because lipid-ion binding constants have nearly the opposite sequence (see Chap. 18, Part II). This is because lipid-ion complexes behave as if they were large organic ions (see also the following discussion). Ion-lipid association therefore may increase the bilayer permeability for the associates by several orders of magnitude as compared to the permeability of simple ions. Membrane packing inhomogeneities and the accumulation of ions in the interfacial region, which is caused by membrane electrostatic effects [68], may also promote transbilayer ion transport significantly [67].

Some ions cross the phospholipid bilayers in the form of electroneutral molecular complexes. Membrane transport then becomes relatively insensitive to electrostatic membrane properties, for obvious reasons.

This, on the one hand, provides one possible explanation for the relatively fast permeation of the divalent ions, such as Ag^{2+}, Cd^{2+}, Hg^{2+}, Pb^{2+}, Sn^{2+}, as well as Zn^{2+} [63], which not only have a high affinity for association with substances of opposite charge but also get through lipid bilayers relatively easily in the form of neutral complexes. The same is true for halides [61,62,63]; transbilayer permeation of halides always exceeds the permeation of comparably large cations by at least one order of magnitude (cf. Table 2). In general, anions permeate phosphatidylcholine bilayers in the order $I^- > Br^- > Cl^- > F^+ \gg NO_3^- = SO_4^{2-} > HPO_4^-$ [62,65].

Noncharged ion complexes may contain protons; the rapidly transported HCl is one such case. Quite frequently, however, electroneutral complexes are formed also by a combination of a charged lipid with its counterion.

(c) Temperature dependence. The rate of solute translocation across lipid membranes normally increases with increasing temperature and is particularly strongly affected by the chain-melting phase transition. Some bilayers, for example phosphatidylcholine, exhibit a permeability maximum at the chain-melting phase transition (Refs. 66, 67, 70, 71 and Fig. 3), albeit not for all types of ions. Other lipids, e.g., phosphatidylethanolamine, show no such permeability increase [38]. Small nonelectrolyte molecules also show no critical permeability increase upon chain melting.

Table 2 Permeabilities P_m (cm s^{-1}) and Activation Energies E_a (kJ mol^{-1}) for Ion Transport Across Fluid Phosphatidylcholine Bilayers

Cationic	H^+	Na^+	K^+	Rb^+	Tl^+	Cd^{2+}
P_m	5×10^{-12}–3×10^{-9}	10^{-12}	3×10^{-12}	3.3×10^{-12}	(4×10^{-12})	$3 \times 10^{-12})$
E_a	71–83	70–80	63			25
Anionic	Cl^-	Br^-	J^-	SCN^-	NO_3^-	
P_m	7×10^{-12}	1.8×10^{-11}	3×10^{-11}	(10^{-11})	(5×10^{-11})	
E_a	79					
"Silent"	H_2O	H^+/OH^-	HCl	TlCl	$CdCl_2$	$HgCl_2$
		10^{-9}	1.5×10^{-11}			
P_m	4×10^{-3}	10^{-4}–10^{-3}	7×10^{-8}	1.1×10^{-7}	4×10^{-9}	(10^{-2})

When two values differing by several orders of magnitude are given, the permeation process is biphasic. (Data for protons and H^+/OH^- from Refs. given in the text. Data for alkali ions from Refs. 64, 65, 109, 121; for cadmium from Ref. 124; for halides, Refs. 61, 62, 65, 104; for other anions, Ref. 123.)

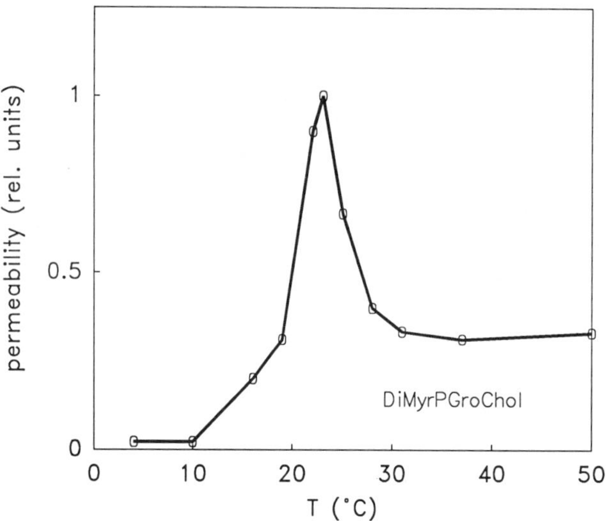

Figure 3 Permeability of dimyristoylphosphatidyl bilayers, with a chain-melting phase transition at 23°C, to monovalent inorganic ions as a function of temperature.

Near the chain-melting phase transition, single ion pores have been reported to occur in pure lipid bilayers (see, e.g., Ref. 72). Such pores may originate in the critical fluctuations in the lateral compressibility of the bilayer close to the phase transition [73,74]. The mismatch in molecular packing between coexisting gel and fluid regions is another possible source of transbilayer pores [71,75]. In addition to this, the increased ion binding in the critical region [76], the increased flip-flop rate upon binding of some divalent, e.g., Cd^{2+} ions [77], and the anomalous increase of the lipid flip-flop rate at the chain-melting phase transition of phosphatidylcholine [78] may all facilitate transbilayer material exchange in the critical region.

(d) Surface potential effects. Membrane potential normally leads to the accumulation of counterions of the opposite sign to that of the membrane charges and the dilution of ions of the same charge, coions, in the interfacial region (see Chap. 13, Part II). Since all common charged phospholipids are anionic, counterions are normally cationic. Ionic layers near the surfaces of charged lipid membranes thus contain an excess of cations, which increases the probability for such ions to get across a charged phospholipid bilayer; conversely, the probability for anion translocation across such a membrane is simultaneously decreased by the negative electrostatic surface potential in the absence of other effects. Simple semiquantitative estimates of the significance of such electrostatic modulation of transbilayer transport is possible within the framework of the Gouy-Chapman approach, which suggests that $k_{\text{in}} = k_{\text{in}}(\psi_{el} = 0) \exp(-Z_i e_0 \psi_{el}/kT)$ (cf. Eq. 5). When membrane charges originate from the interfacially adsorbed ions, however, the situation may be just the opposite, such ions repelling other ions of the same sign from the bilayer surface.

Increasing electrostatic transmembrane potential, moreover, increases the rate of membrane pore formation and thus also indirectly membrane permeability. (This is because such a potential enhances the thermal membrane fluctuations [79], which then act as pore nuclei.) On the other hand, any suppression of membrane fluctuations upon membrane rigidification diminish the probability of pore formation and also the conductivity of transbilayer ion channels. Similar effects are caused by the increased lateral pressure of the chains, which lowers the probability for the internal opening and closing motions in proteins [16,80,81]. Channel conductivity may thus indirectly be regulated via a change in the membrane thickness [82] as well as directly regulated by the membrane surface electrostatics and polarity, and vice versa.

(e) Effects of additives. Addition of cholesterol to the phospholipid molecules in a gel phase increases the bilayer permeability for water; in fluid membranes the effect of cholesterol is just the opposite. Solute transport across lipid bilayers is normally increased by the addition of cholesterol. After the incorporation of cholesterol, moreover, the activation energy for the transmembrane water movement decreases and the corresponding activation energy for various solutes increases significantly.

The permeability pattern of complex artificial lipid bilayers is dominated by short-chain lipids, especially by phosphatidylcholines [88], or by charged lipids. The addition of cholesterol diminishes membrane permeability to ionic solutes, by and large [58].

Protein adsorption to or insertion of proteins into phospholipid bilayers always facilitates the transport of material across a membrane. This is due either to the generation of membrane defects and/or pores, in the former case, or else to the carrier or channel activity of the protein molecules, in the latter situation.

Ion transport through biological membranes seldom proceeds directly through the hydrophobic membrane core. It normally follows the paths along which the energy barrier for the permeation is lowered by the formation of ion-lipid complexes or channels.

2. Transport Through Pores and Channels

When transmembrane pores exist in a bilayer, solutes are normally diverted into such pores, since the permeation resistance of the latter is much lower than the resistance of the bulk hydrocarbon region. Electrostatic calculations for a pore of simple geometry indicates that the permeability barrier is lowered by approximately 80% [52], but this may be an overestimate [83,84].

Pores may arise in phospholipid bilayers spontaneously in the defect-rich bilayer regions. Lipid membranes undergoing phase transitions [72], bilayers under mechanical (ultrasonic [85] or hydrodynamic [86]), osmotic [90], electrical [91], or elastic stress, consequently, are very leaky; so are lipid vesicles with adsorbed proteins [87]. Small sonicated vesicles or lipid vesicles in biological fluids therefore lose their water-soluble contents quite rapidly. Mixed lipid systems also are very poor permeability barriers [88], the shortest-chain or the most polar compound being normally responsible for most of the solute loss.

Far more important and frequent than simple pores are the transbilayer channels, however. Such channels are of paramount importance in the regulation of transmembrane material flow in biological systems (for a series of reviews see Ref. 93, for example.) This is also one of the reasons why such channels have been investigated so often in model lipid membranes. Most frequently the artificial channels in phospholipid bilayers were created by the polyene macrolide antibiotics (cf. amphotericin, filipin, nystatin, etc.) or, and in particular, gramicidins (for reviews see, for example, Refs. 94 and 95).

Ion transport through the majority of artificial channels incorporated into phospholipid bilayers occurs in a single file mode; ions and solute molecules are thus transported in a row. Inside each gramicidin channel, for example, which consists of two superimposed, loosely connected 'barrels', some 6–7 water molecules are entrapped [59]; these fill out a hydrophilic channel core with a radius smaller than 0.2 nm. Charges on the protein [60,96] or lipid [97] molecules normally affect the rate of ion transport through the bilayer either by changing the energy profile for the ion transfer through the channel or by changing the ion concentration near the channel mouth—which affects the probability for ions to enter into the channel.

Channel properties and dimensions are determined largely by the protein structure [98]. This causes the solute selectivity of different channels to be sensitive to the primary protein sequence. But it is also possible to affect the channel properties and conductivity by changing the host lipid matrix and its characteristics. Chain ordering, for example, suppresses entirely the sugar translocation through the lactose permease, when this protein is incorporated into ordered-phase dimyristoylphosphatidylcholine bilayers [81].

3. Carrier Mediated Transport

In the absence of transmembrane channels, solute carriers may be responsible for the rate and efficiency of material translocation across the phospholipid bilayer.

As in the case of trans-channel transport, the solute flow mediated by the ionophore or carrier molecules is controlled by the properties of the latter. The interfacial solute concentration, however, which determines the probability for the solute-carrier association, is also dependent on the membrane surface electrostatics and other solute-membrane interactions. Carrier-mediated solute transport across phospholipid bilayers, consequently, is normally modelled as a multistep kinetic cascade, of which the first step, the interfacial ion-carrier association, is mostly the decisive one.

Halides, for example, have a relatively strong tendency to bind to the phospholipid headgroups, most notably to positively charged ammonium groups. Halides therefore may be transported across bilayers in pairs with phospholipid carriers.

For halide anions, the existence of carrier complexes has been substantiated by the fact that electrically silent ion flow gives rise to ion permeability up to three orders of magnitude larger than for the corresponding counterion diffusion flux. The lipid-carrier idea, furthermore, was supported by the observation that the transbilayer transport rates typically increase with the strength of association of the permeating anion with the phospholipid molecules. The possibility that such association causes membrane defects, which then act as transbilayer pores, cannot be excluded, however.

It thus seems that with halide ions, and possibly with some other ions, the transmembrane flow consists of a fast component with permeability in the order of 10^{-6}–10^{-8} cm s^{-1}, which depends on inter-ion or ion-lipid complex formation, and a slow single-ion diffusion component, characterized by permeability values on the order of $P_m \sim 10^{-11}$ cm s^{-1}, which is sensitive mainly to ion size (see Table 2).

Other ions with strong affinity for phospholipid molecules are also carried, at least partly, across the phospholipid bilayer with the aid of lipidic carriers. Fatty acids and phosphatidic acids are the perhaps best known examples of such lipidic carriers [99], especially when their hydrocarbon chains are polyunsaturated [100]. But it is nearly certain that all charged phospholipids can act as ion carriers [101], provided that their rate of translocation across the bilayer is sufficiently high. In certain cases, carrier transport can also explain the anomalously high permeability of certain nonionic solutes [102].

C. Proton Transport

Transbilayer proton flux is normally by many orders of magnitude higher than the corresponding diffusive flow of other small ions. It is comparable only to the water diffusion rate across lipid bilayers or to ion transport through transbilayer pores or channels. Membrane conductivity to H^+/OH^- ions varies with the method of vesicle preparation, with the level of unsaturation of the membrane phospholipids and in particular, with the magnitude of the transbilayer ΔpH. Such conductivity is particularly high for polyunsaturated phospholipids. It is insensitive, however, to alterations in the membrane dipole field.

All this suggests that H^+/OH^- currents are not rate limited by diffusion over simple electrostatic barriers in the membrane interior [103] and that proton transport across a phospholipid bilayer is not based on a simple proton diffusion through the hydrocarbon membrane core.

Carrier mediated transport of protons bound to phospholipid molecules or complexed with fatty acids in a bilayer could be responsible for this. Alternative explanations are also possible, however.

Part of the transmembrane H^+/OH^--flow has been proposed, for example [104,105,106], but not yet unambiguously proven [107,108], to occur via a network of preassociated water molecules in the bilayer, in close resemblance to proton conductance in ice and water. In fact, the exchange transfer of H^+/OH^- across lipid membranes is biphasic, the fast component being characterized by permeability values 10^{-3} cm s^{-1} $< P < 10^{-6}$ cm s^{-1}, which are very similar to the water permeability values in lipid bilayers.

The slower H^+/OH^--flow has been reliably inferred to depend on the counterion diffusion, giving rise to permeabilities in the order of 10^{-10}–10^{-12} cm s^{-1} [106,109,110]. This latter value is comparable to the permeability measured for other ions.

The proton-wire hypothesis is supported by the observation that the temperature dependence of the transbilayer proton flux, the water partitioning in, and the water penetration through lipid membranes are all very similar: all increase in a steplike manner at the lipid chain-melting phase transition temperature. The large permeability jump for protons, which accompanies hydrocarbon fluidization in lipid bilayers, thus might arise from an increase in the number of intracore water molecules, which takes place without an accompanying increase in activation energy. Certainly, like transmembrane water partitioning (cf. Chap. 13, Part II), the H^+/OH^- exchange permeation never shows the permeability maximum at the phase transition [106].

In spite of this, doubts have been raised as to the true mechanism of transbilayer proton transport based on the observation that the apparent permeability coefficient for proton flux through phosphatidylcholine vesicles is dependent on the direction of flux [111]. These observations were interpreted in terms of a model suggesting that carbonic acid or carbon dioxide together with bicarbonate was an efficient proton carrier across phospholipid bilayers.

V. MEASURING BILAYER PERMEABILITY

Most directly, the rate of molecular translocation across phospholipid bilayers is determined by measuring the loss of material from a lipid vesicle or, alternatively, by measuring the accumulation of the permeating species inside a vesicle. Depending on the type of solute, many different detection methods can be used for this. Most frequently, radioactive tracers (such as $^{109}Cd^{2+}$, $^{204}Tl^+$, $^{17}F^+$, $^{22}Na^+$, $^{42}K^+$, $^{36}Cl^-$, $^{45}Ca^{2+}$, etc.) or NMR-detectable isotopes (such as $^7Li^+$, $^{23}Na^+$, $^{19}F^+$, $^{35}Cl^-$, etc.) are used. Application of shift reagents also has often proven useful for the discrimination of the intra- and extravesicular compartments. Selective or ion-selective electrodes are easy to use, but often they are neither very selective nor very sensitive. Consequently, such electrodes are of limited use for vesicle transport studies.

The time course of solute inflow into lipid vesicles, furthermore, can be determined by measuring the self-quenching of entrapped dye [112] or by observing the solvatochromic effect with many different dyes during the change of composition of the intravesicular solution that results from the transbilayer material exchange.

In any case, it is advantageous to separate the intravesicular and extravesicular material before the beginning of each experiment. Gel chromatography [116] and ion-exchange chromatography are both suitable for this purpose; centrifugation tubes with an builtin low cutoff filters (such as Centrisart Sartorius) are also valuable. (For a detailed description of one specific type of experiment see, for example, Ref. 121.)

Water permeability into phospholipid bilayers is most easily detected by measuring the osmotic swelling of lipid vesicles by a turbidimetric stop-flow technique [21,119]. Alternatively, different fluorescence assays can be used [112,118], which are all based on the solvatochromic effect, i.e., on the changing spectroscopic properties of the dye molecules with the relative H_2O/D_2O ratio.

When interpreting experimental data on solute transport across complex membranes, one should be careful to consider all relevant factors and dangers, however.

System contaminants that can create or catalyze pore formation or else act as solute carriers or channels are the most severe sources of experimental error. This is because the permeability for carrier-solute complexes or solute permeation through membrane pores can result in fluxes many orders of magnitude greater than the spontaneous diffusive transbilayer solute flux. Less than 1% impurity, consequently, may already be fatal for the success of an experiment. Ionic impurities are more dangerous in this respect than nonionic ones; in fact, such impurities may be responsible for the relatively high measured permeability values for divalent ions.

One other possible source of errors is an unstirred layer. This may severely hamper the solute approach to the bilayer surface and contribute an extra resistance term to the total bilayer (inverse) permeability (cf. Eq. 5 and Ref. 120). The precise magnitude of this contribution depends on the thickness of the unstirred layer. Values between 5×10^{-2} and 5×10^{-4} cm s^{-1} have been quoted to pertain to the normal and fast-flow systems, respectively [120]. Unstirred layers thus are important in studies of good membrane permeators but are irrelevant in ion transport studies (cf. Tables 1 and 2).

Yet another danger to the experiment may come from the solute-bilayer association. Interfacially bound ions, for example, will either attract or repel other charged species, depending on the mutual signs of charges involved [68]. It is therefore mandatory to account for this in all calculations of the intrinsic membrane permeability data. Previous sections give a brief account of how this can be done.

This also pertains to channel molecules. Valinomycin, for example, one of the most popular ionophores, not only transports ions across the lipid bilayer but also binds to membranes and thus may influence the membrane charge density and consequently the transmembrane transport in a nontrivial manner [113].

This and other channels may rapidly empty the small lipid vesicles of their initial content. In general, care must be taken to account for the variations in transbilayer concentration gradients during experiments. Sometimes the starting conditions change completely within a few seconds. In other cases, full equilibrium is not attained even after days. Figure 4 illustrates this for a series of representative bilayer permeabilities.

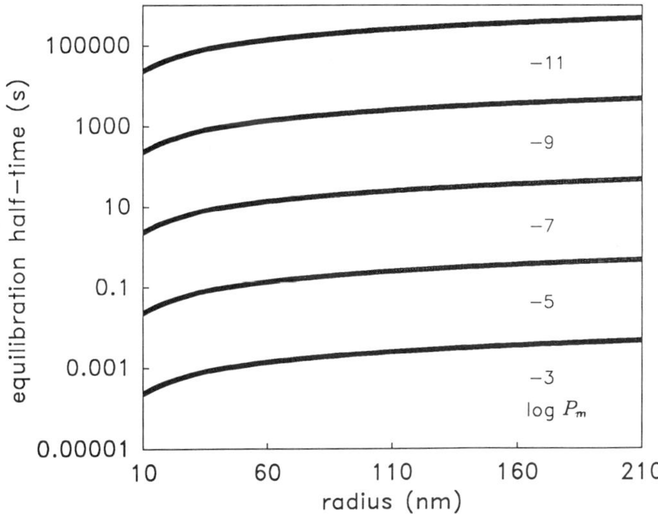

Figure 4 Time required for the unilamellar lipid vesicles of different size r_v to reach 50% of the extravesicular solute concentration as a function of the bilayer permeability P_m.

REFERENCES

1. R. W. Fisher and T. L. James, Lateral diffusion of the phospholipid molecule in dipalmitoylphosphatidylcholine bilayers. An investigation using nuclear spin-lattice relaxation in the rotating frame, *Biochemistry 17*:1177–1183 (1978).
2. W. L. C. Vaz, R. M. Clegg, and D. Hallmann, Translational diffusion of lipids in lipid crystalline phase phosphatidylcholine multibilayers. A comparison of experiment with theory, *Biochemistry 24*:781–786 (1985).
3. M. B. Schneider, W. K. Chan, and W. W. Webb, Fast diffusion along defects and corrugations in phospholipid P(beta) liquid crystals, *Biophys. J.* 43:157–165 (1983).
4. M. Prats, J. F. Tocanne, and J. Teissie, Lateral diffusion of protons along phospholipid monolayers, *J. Memb. Biol. 99*:225–227 (1987).
5. J. Kasianowicz, R. Benz, and S. McLaughlin, How do protons cross the membrane-solution interface? Kinetic studies on bilayer membranes exposed to the protonophore S-13 (5-chloro-3-ter-butyl-2'-chloro-4'-nitrosalicylanilinide, *J. Memb. Biol. 95*:73–89 (1987).
6. D. Rumschitzki and D. Ronis, Microscopic theory of membrane transport. I. Correlation function expressions for the permeability, *J. Chem. Phys. 79*:5628–5638 (1983).
7. B. J. Zwolinsky, H. Eyring, and C. E. Reese, Diffusion and membrane permeability, *J. Phys. Colloid. Chem. 53*:1426–1453 (1949).
8. W. B. Kleijn and L. J. Bruner, Chemical and solvent effects on the interaction of tetraphenylborate with lipid bilayer membranes, *Biochim. Biophys. Acta 769*:33–40 (1984).
9. R. Flewelling and W. Hubbel, The membrane dipole potential in a total membrane model. Applications to hydrophobic ion interactions with membranes, *Biophys. J. 49*:541–552 (1986).
10. O. Kedem and A. Katchalsky, A physical interpretation of the phenomenological coefficients of membrane permeability, *J. Gen. Physiol. 45*:143–179 (1961).
11. G. Cevc and D. Marsh, in *Phospholipid Bilayers*, Wiley-Interscience, New York, 1987.
12. J. M. Diamond and Y. Katz, Interpretation of nonelectrolyte partition coefficients between dimyristoyl lecithin and water, *J. Memb. Biol. 17*:121–154 (1974).

13. F. W. Orme, M. M. Moronne, and R. I. Macey, Modification of the erythrocyte membrane dielectric constant by alcohols, *J. Membr. Biol. 104*:57–68 (1988).
14. A. Simon and T. J. McIntosh, Interdigitated hydrocarbon chain packing causes the biphasic transition behavior in lipid/alcohol suspensions, *Biochim. Biophys. Acta 773*:169–172 (1984).
15. O. Sten-Knudsen, Passive transport processes, in *Membrane Transport in Biology I. Concepts and Models* (G. Giebisch, D. C. Tosteson, and H. H. Ussing, eds.), Springer-Verlag, New York, 1978, pp. 1–113.
16. P. Läuger, Mechanistic properties of ion channels and pumps, *Prog. Clin. Biol. Res. 164*:133–138 (1984).
17. R. Fettiplace, The influence of the lipid on the water permeability of artificial membranes, *Biochim. Biophys. Acta 513*:1–10 (1978).
18. V. B. Arakelian and S. B. Arakelian, Energetic profile of dipole molecules on the border of separation of two phases (in Russian), *Armenian Biol. J. 36*:553–559 (1983).
19. R. Lawaczeck, On the permeability of water molecules across vesicular lipid bilayers, *J. Membrane Biol. 51*:229–261 (1979).
20. H. Engelbert and R. Lawaczeck, The H_2O/D_2O exchange across vesicular lipid bilayers. Lecithins and binary mixtures of lecithins, *Ber. Bunsenges. Phys. Chem. 89*:754–759 (1985).
21. H. Engelbert and R. Lawaczeck, Isotropic light scattering of lipid vesicles. Water permeation and effect of α-tocopherol, *Chem. Phys. Lipids 38*:365–379 (1985).
22. D. W. Deamer and J. Bramhall, Permeability of lipid bilayers to water and ionic solutes, *Chem. Phys. Lipids 40*:167–188 (1986).
23. L. Ginsberg and N. L. Gershfeld, Phospholipid surface bilayers at the air-water interface II. Water permeability of dimyristoylphosphatidylcholine surface bilayers, *Biophys. J. 47*:211–215 (1985).
24. B. A. Böhler, J. de Gier, and L. L. M. van Deenen, The effect of gramicidin A on the temperature dependence of water permeation through liposomal membranes prepared from phosphatidylcholines with different chain lengths, *Biochim. Biophys. Acta 512*:480–488 (1978).
25. P. A. Rosenberg and A. Finkelstein, Water permeability of gramicidin A treated lipid bilayer membranes, *J. Gen. Physiol. 72*:341–350 (1978).
26. J. A. Dix, D. Kivelson, and J. M. Diamond, Molecular motion of small nonelectrolyte molecules in lecithin bilayers, *J. Memb. Biol. 40*:315–342 (1978).
27. J.-L. Rigaud, C. M. Garry-Bobo, and Y. Lange, Diffusion processes in lipid-water lamellar phases, *Biochim. Biophys. Acta 266*:72–84 (1972).
28. O. Sparre and M. Fuchs, Potential energy barriers to ion transport within lipid bilayers. Studies with tetraphenylborate, *Biophys. J. 15*:795–830 (1975).
29. S. Mitaku, A. Ikegami, A. Sakanishi, Ultrasonic studies of lipid bilayer phase transition in synthetic phosphatidylcholine liposomes, *Biophys. Chem. 8*:295–304 (1978).
30. Y. Katz and J. M. Diamond, Thermodynamic constants for nonelectrolyte partition between dimyristoyl lecithin and water, *J. Memb. Biol. 17*:101–120 (1974).
31. M. K. Jain, J. Gleeson, A. Upreti, and G. C. Upreti, *Biochim. Biophys. Acta 509*:1–8 (1978).
32. H. J. Pownall, D. L. Hickson, and L. C. Smith, *J. Am. Chem. Soc. 105*:2440–2445 (1983).
33. J. Brahm, Permeability of human red cells to a homologous series of aliphatic alcohols, *J. Gen. Physiol. 81*:283–304 (1983).
34. A. Walter and J. Gutknecht, Permeability of small non-electrolytes through lipid bilayer membranes, *J. Memb. Biol. 90*:207–217 (1986).
35. A. Walter and J. Gutknecht, Monocarboxylic acid permeation through lipid bilayer membranes, *J. Memb. Biol. 77*:255–264 (1984).
36. R. I. Sha'afi, C. M. Gary-Bobo, A. K. Solomon, Permeability of red cell membranes to small hydrophilic and lipophilic solutes, *J. Gen. Physiol. 58*:238–258 (1971).

37. B. E. Cohen and A. D. Bangham, Diffusion of small non-electrolytes across liposome membranes, *Nature 236*:173–174 (1972).
38. G. J. M. Bresselers, H. L. Goderis, and P. P. Tobback, Measurement of the glucose permeation rate across phospholipid bilayers using small unilamellar vesicles. Effect of membrane composition and temperature, *Biochim. Biophys. Acta 772*:374–382 (1984).
39. E. Cohen, The permeability of liposomes to nonelectrolytes. I. Activation energies for permeation, *J. Memb. Biol. 20*:205–234 (1975).
40. R. Collander, *Acta Physiol. Scand. 13*:363–381 (1947).
41. M. J. Kamlet, J.-L. M. Abbond, M. H. Abraham, and R. W. Taft, Linear solvation energy relationships 23. A comprehensive collection of the solvatochromic parameters pi, alpha, and beta, and some methods for simplifying the generalized solvatochromic equation, *J. Org. Chem. 48*:2877–2887 (1983).
42. R.W. Taft, M. H. Abraham, R. M. Doherty, and M. J. Kamlet, The molecular properties governing solubilities of organic nonelectrolytes in water, *Nature 313*:384–386 (1985).
43. J. DeGier, J. M. Mandersloot, and L. L. M. van Deenen, Liquid composition and permeability of liposomes, *Biochim. Biophys. Acta 150*:666–675 (1968).
44. O. H. Griffith, P. J. Dehlinger, and S. P. Van, Shape of the hydrophobic barrier of phospholipid bilayers (Evidence for water penetration in biological membranes), *J. Memb. Biol. 15*:159–192 (1974).
45. G. Cevc and D. Marsh, Properties of the electrical double layer near the interface between a charged bilayer membrane and electrolyte solution: Experiment vs. theory, *J. Phys. Chem. 87*:376–379 (1983).
46. J. P. Dilger, S. G. A. McLaughlin, T. J. McIntosh, and S. A. Simon, The dielectric constant of phospholipid bilayers and the permeability of membranes to ions. *Science 206*:1196–1198 (1979).
47. G. Cevc, The molecular mechanism of interaction between monovalent ions and polar surfaces, such as lipid bilayer membranes, *Chem. Phys. Letters 170*:283–288 (1990).
48. H. Uhlig, *J. Phys. Chem. 41*:1215–1219 (1937).
49. R. C. MacDonald, Energetics of permeation of a thin lipid membrane by ions. *Biochim. Biophys. Acta 448*:193–198 (1976).
50. R. W. Bradshaw and C. R. Robertson, Effect of ionic polarizability on electrodiffusion in lipid bilayer membranes, *J. Memb. Biol. 25*:93–114 (1975).
51. A. A. Kornyshev and A. G. Volkov, On the evaluation of standard Gibbs energies of ion transfers between two solvents, *J. Electroanal. Chem. 180*:363–381 (1984).
52. A. Parsegian, Energy of an ion crossing a low dielectric membrane: Solutions to four relevant electrostatic problems, *Nature 221*:844–846 (1969).
53. S. B. Hladky and D. A. Haydon, Membrane conductance and surface potential, *Biochim. Biophys. Acta 318*:464–468 (1983).
54. R. F. Flewelling and W. Hubbell, Hydrophobic ion interactions with membranes. Thermodynamic analysis of tetraphenylphosphonium binding to vesicles, *Biophys. J. 49*:531–540 (1986).
55. C. Altenbach and J. Seelig, Binding of the lipophilic cation tetraphenylphosphonium to phosphatidylcholine membranes, *Biochim. Biophys. Acta 818*:410–451 (1985).
56. P. C. Noordam, A. Killian, R. F. M. Oude Elferink, J. de Gier, Comparative study on the properties of saturated phosphatidylethanolamine and phosphatidylcholine bilayers: Barrier characteristics and susceptibility to phospholipase A2 degradation, *Chem. Phys. Lipids 31*:191–204 (1982).
57. M. Singer, Permeability of bilayers composed of mixtures of saturated phospholipids, *Chem. Phys. Lipids 31*:145–159 (1982).
58. R. Bittman, S. Clejan, S. Lund-Katz, and M. C. Phillips, Influence of cholesterol on bilayers of ester- and ether-linked phospholipids. Permeability and ^{13}C-nuclear magnetic resonance measurements, *Biochim. Biophys. Acta 772*:117–126 (1984).

59. P. A. Rosenberg and A. Finkelstein, Interaction of ions and water in gramicidin A channels, *J. Gen. Physiol. 72*:327–340 (1978).
60. F. G. Riddel and S. Arumugam, Surface charge effects upon membrane transport processes: The effects of surface charge on the monensin-mediated transport of lithium ions through phospholipid bilayers studied by ^{7}-Li-NMR spectroscopy, *Biochim. Biophys. Acta 945*:65–72 (1988).
61. Y. Toyoshima and T. E. Thompson, Chloride flux in bilayer membranes: The electrically silent chloride flux in semispherical bilayers, *Biochemistry 14*:1525–1531 (1975).
62. J. Gutknecht, J. S. Graves, and D. C. Tosteson, Electrically silent anion transport through lipid bilayer membranes containing a long-chain secondary amine, *J. Gen. Physiol. 71*:269–284 (1978).
63. J. Gutknecht and A. Walter, Inorganic mercury (Hg^{2+}) transport through lipid bilayer membranes, *Biophys. J. 33*:110a (1981).
64. J. Gutknecht, Cadmium and thallous ion permeabilities through lipid bilayer membranes, *Biochim. Biophys. Acta 735*:185–188 (1983).
65. A. D. Bangham, M. M. Standish, J. C. Watkins, and G. Weissman, Diffusion of univalent ions across the lamellae of swollen phospholipids, *Protoplasma 63* (Suppl.):183–187 (1967).
66. M. C. Blok, E. C. M. Van der Neut-Kok, L. L. M. van Deenen, and J. de Gier, The effect of chain length and lipid phase transitions on the selective permeability properties of liposomes, *Biochim. Biophys. Acta 406*:187–196 (1975).
67. J. Bramhall, Electrostatic forces control the penetration of membranes by charged solutes, *Biochim. Biophys. Acta 778*:393–399 (1984).
68. G. Cevc, Membrane electrostatics, *Biochim. Biophys. Acta. Reviews on Membranes 1031–3*:311–382 (1990).
69. K. Mohr and M. Struve, Differential influence of anionic and cationic charge on the ability of amphiphilic drugs to interact with DPPC liposomes, *Biochem. Pharmacol. 41*:961–965 (1991).
70. K. Jacobson and D. Papahadjopoulos, Phase transitions and phase separations in phospholipid membranes induced by changes in temperature, pH, and concentration of bivalent cations, *Biochemistry 14*:152–161 (1975).
71. D. Marsh, A. Watts, and P. F. Knowles, Evidence for phase boundary lipid. Permeability of tempo-choline into dimyristoylphosphatidylcholine vesicles at the phase transition, *Biochemistry 15*:3570–3578 (1976).
72. V. F. Antonov, V. V. Petrov, A. A. Molnar, D. A. Predvoditelev, and A. S. Ivanov, The appearance of single-ion channels in unmodified lipid bilayer membranes at the phase transition temperature, *Nature 283*:585–586 (1980).
73. J. F. Nagle and H. L. Scott, Jr., Lateral compressibility of lipid mono- and bilayers. Theory of membrane permeability, *Biochim. Biophys. Acta 513*:236–243 (1978).
74. L. Cruzeiro-Hansson and O. G. Mouritsen, Passive ion permeability of lipid membranes modelled via lipid-domain interfacial area, *Biochim. Biophys. Acta 944*:63–72 (1988).
75. M. I. Kanehisa and T. Y. Tsong, Cluster model of lipid phase transitions with application to passive permeation of molecules and structure relaxations in lipid bilayers, *Am. Chem. Soc. 100*:424–432 (1978).
76. H. Träuble, Membrane electrostatics, in *Structure, Function of Biological Membranes* (S. Abrahamsson and I. Pascher, eds.), Plenum Press, New York, 1977, pp. 509–550.
77. B. R. Lentz, S. Madden, and D. R. Alford, Transbilayer redistribution of phosphatidylglycerol in small, unilamellar vesicles induced by specific divalent cations. *Biochemistry 21*:6799–6807 (1982).
78. B. DeKruijff and E. J. J. van Zoelen, Effect of the phase transition on the transbilayer movement of dimyristoyl phosphatidylcholine in unilamellar vesicles, *Biochim. Biophys. Acta 511*:105–115 (1978).

79. J. C. Weawer and R. A. Mintzer, Decreased bilayer stability due to transmembrane potentials, *Physics Lett.* 86A:57–59 (1981).
80. P. Läuger, Internal motions in proteins and gating kinetics of ionic channels, *Biophys. J.* *53*:877–884 (1988).
81. K. Dornmair, P. Overath, and F. Jahnig, Fast measurement of galactoside transport by lactose permease, *J. Biol. Chem.* *264*:342–346 (1989).
82. B. M. Hendry, J. R. Elliot, and D. A. Haydon, Further evidence that membrane thickness influences voltage-gated sodium channels, *Biophys. J.* *47*:841–845 (1985).
83. R. F. Kayser and J. B. Hubbard, Ohmic friction on an ion in a conducting pore, *J. Chem. Phys.* *78*:1935–1937 (1983).
84. W. N. Green and O. S. Andersen, Surface charges and ion channel function, *Ann. Rev. Physiol.* *53*:341–359 (1991).
85. E. Saalman, B. Norden, L. Arvidsson, Y. Hamnerius, P. Hojevik, K. E. Connell, and T. Kurucsev, Effect of 2.45 GHz microwave radiation on permeability of unilamellar liposomes to 5(6)-carboxyfluorescein. Evidence of non-thermal leakage, *Biochim. Biophys. Acta.* *1064*:124–130 (1991).
86. J. Bergholz, Ph.D. thesis, University of Essen, 1992.
87. S. E. Francis, I. G. Lyle, and M. N. Jones, The effect of surface-bound protein on the permeability of proteoliposomes, *Biochim. Biophys. Acta.* *1062*:117–122 (1991).
88. M. Singer, Permeability of bilayers composed of mixtures of saturated phospholipids, *Chem. Phys. Lipids* *31*:145–159 (1982).
89. P. C. Jordan, Electrostatic modeling of ion pores. Energy barriers and electric field profiles, *Biophys. J.* *39*:157–164 (1982).
90. W. Li, T. S. Aurora, T. H. Haines, and H. Z. Cummins, Elasticity of synthetic phospholipid vesicles and submitochondrial particles during osmotic swelling, *Biochemistry* *25*:8220–8229 (1986).
91. J. Teissie and T. Y. Tsong, Electric field induced transient pores in phospholipid bilayer vesicles, *Biochemistry* *20*:1548–1554 (1981).
92. T. H. Haines, W. Li, M. Green, and H. Z. Cummins, The elasticity of uniform unilamellar vesicles of acidic phospholipids during osmotic swelling is dominated by the ionic strength of the media, *Biochemistry* *26*:5439–5447 (1987).
93. F. Bronner, ed., *Current Topics in Membrane Transport. Ion Channels: Molecular and Physiological Aspects*. Academic Press, Orlando, Fla., 1984.
94. J. Bolard, How do the polyene macrolide antibiotics affect the cellular membrane properties? *Biochim. Biophys. Acta* *864*:257–304 (1986).
95. P. Läuger, Dynamics of ion transport systems in membranes, *Phys. Rev.* *67*:1296–1331 (1987).
96. J.-L. Mazet, O. S. Andersen, R. E. Koeppe, Single-channel studies on linear gramicidins with altered amino acid sequences, *Biophys. J.* *45*:263–276 (1984).
97. H. J. Apell, E. Bamberg, and P. Läuger, Effects of surface charge on the conductance of the gramicidin channel, *Biochim. Biophys. Acta* *552*:369–378 (1979).
98. J. Wu, Microscopic model for selective permeation in ion channels, *Biophys. J.* *60*:238–251 (1991).
99. R. Nayer, L. D. Mayer, M. J. Hope, and P. R. Cullis, Phosphatidic acid as a calcium ionophore in large unilamellar vesicle systems, *Biochim. Biophys. Acta* *777*:343–346 (1984).
100. K. Utsumi, K. Nobori, S. Terada, M. Miyahara, and T. Utsumi, Continuous fluorometric assay of Ca^{2+} transport by liposomes with quin 2 entrapped: Effect of phospholipase A2 and unsaturated long-chain fatty acids, *Cell Structure and Function* *10*:339–348 (1985).
101. H. Hauser, D. Oldani, and M. C. Phillips, Mechanism of ion escape from phosphatidylcholine and phosphatidylserine single bilayer vesicles, *Biochemistry* *12*:4507–4517 (1973).

102. D. E. Green, M. Frey, and G. A. Blondin, Phospholipids as the molecular instruments of ion and solute transport in biological membranes, *Proc. Natl. Acad. Sci. USA 77*:257–261 (1980).
103. W. R. Perkins and D. S. Cafiso, An electrical and structural characterization of H+/OH- currents on phospholipid vesicles, *Biochemistry 25*:2270–2276 (1986).
104. D. W. Deamer and J. W. Nichols, Proton-hydroxide permeability of liposomes. *Proc. Natl. Acad. Sci. USA 80*:165–168 (1983).
105. D. W. Deamer, Proton permeation of lipid bilayers, *J. Bioenerg. Biomembr. 19*:457–479 (1987).
106. K. Elamrani and A. Blume, Effect of the lipid phase transition on the kinetics of H+/OH- diffusion across phosphatidic acid bilayers, *Biochim. Biophys. Acta 727*:22–30 (1983).
107. M. P. Conrad and H. L. Strauss, The vibrational spectrum of water in liquid alkanes, *Biophys. J. 48*:117–124 (1985).
108. J. Gutknecht, Proton conductance through phospholipid bilayers, water wires or weak acids? *J. Bioenerg. Biomembr. 19*:427–442 (1987).
109. L. T. Mimms, G. Zampighi, Y. Nozaki, C. Tanford, and J. A. Reynolds, *Biochemistry 20*:833–840 (1981).
110. Y. Nozaki and C. Tanford, Proton and hydroxide ion permeability of phospholipid vesicles, *Proc. Natl. Acad. Sci. USA 78*:4324–4328 (1981).
111. F. A. Norris and G. L. Powell, The apparent permeability coefficient for proton flux through phosphatidylcholine vesicles is dependent on the direction of flux, *Biochim. Biophys. Acta. 1030*:165–171 (1990).
112. P. Y. Chen, D. Pearce, and A. S. Verkman, Membrane water and solute permeability determined quantitatively by self-quenching of an entrapped fluorophore, *Biochemistry 27*:5713–5718 (1988).
113. D. Steverding and B. Kadenbach, Valinomycin binds stoichiometrically to cytochrome C oxidase and changes its structure and function, *Biochem. Biophys. Res. Commun. 160*:1132–1139 (1989).
114. J. G. Mandersloot, W. J. Gerritsen, J. Leunissen-Bijvelt, C. J. A. van Echte, P. C. Noordam, and J. de Gier, Ca^{2+}- induced changes in the barrier properties of cardiolipin/phosphatidylcholine bilayers, *Biochim. Biophys. Acta 640*:106–113 (1981).
115. J. W. Nichols, M. W. Hill, A. D. Bangham, and D. W. Deamer, Measurement of net proton-hydroxyl permeability of large unilamellar liposomes with the fluorescent pH probe, 9-aminoacridine, *Biochim. Biophys. Acta 596*:393–403 (1980).
116. P. Schurtenberger and H. Hauser, Characterization of the size distribution of unilamellar vesicles by gel filtration, quasi-elastic light scattering and electron microscopy, *Biochim. Biophys. Acta 778*:470–480 (1984).
117. A. S. Verkman, R. Takla, B. Sefton, C. Basbaum, and J. H. Widdicombe, Quantitative fluorescence measurement of chloride transport mechanisms in phospholipid vesicles, *Biochemistry 28*:4240–4246 (1989).
118. R. Ye and A. S. Verkman, Simultaneous optical measurement of osmotic and diffusional water permeability in cells and liposomes, *Biochemistry 28*:824–829 (1989).
119. A. Milon, T. Lazrak, A.-M. Albrecht, G. Wolff, G. Ourisson, G. Weill, and Y. Nakatan, Osmotic swelling of unilamellar vesicles by the stopped-flow light scattering method. Influence of vesicle size, solute, temperature, cholesterol and three (alpha),(omega)-dihydroxycarotenoids. *Biochim. Biophys. Acta 859*:1–9 (1986).
120. M. Poznansky, S. Tong, P. C. White, J. M. Milgram, and A. K. Solomon, Non-electrolyte diffusion across lipid bilayer systems, *J. Gen. Physiol. 67*:45–66 (1976).
121. J. Brunner, D. E. Graham, H. Hauser, and G. Semenza, Ion and sugar permeabilities of lecithin bilayers: Comparison of curved and planar bilayers, *J. Memb. Biol. 57*:133–141 (1980).

122. M. E. M. da Cruz, R. Kinne, J. T. Lin, Temperature dependence of D-glucose transport in reconstituted liposomes, *Biochim. Biophys. Acta 732*:691–698 (1983).
123. P. Nicholls and N. Miller, Chloride diffusion from liposomes, *Biochim. Biophys. Acta 356*:184–198 (1974).
124. J.-M. Sequaris and I. R. Miller, Polarographic investigation on the permeability of phospholipid vesicles to cadmium ions, *Bioelec. Bioenerg. 13*:127–141 (1984).
125. Z. Bar-On and H. Degani, Permeability of alkylamines across phosphatidylcholine vesicles as studied by 1H-NMR, *Biochim. Biophys. Acta 813*:207–212 (1985).

19

Intermembrane and Transbilayer Transfer of Phospholipids

J. Wylie Nichols *Emory University School of Medicine, Atlanta, Georgia*

Phospholipids are amphiphilic molecules containing one or two hydrophobic carbon chains and a polar head group. This amphiphilicity is necessary for the formation of stable bilayers. However, the hydrophobic carbon chains and the polar head group, respectively, make transfer into the water phase and translocation across the bilayer highly energetically unfavorable. Although essential to the barrier function, this restricted mobility presents a problem for the organism to distribute phospholipids as necessary to establish the membrane composition and asymmetry required for proper cell function.

Several mechanisms have been described that allow organisms and cells to facilitate the distribution of phospholipids between membranes. Membrane budding and fusion allow phospholipids to be shuttled throughout the cell as intact bilayer vesicles. Phospholipids can equilibrate with other phospholipids in the same leaflet of a bilayer by lateral diffusion. Phospholipids can also be enzymatically hydrolyzed to the more water-soluble products, lysophospholipids and fatty acids, which freely diffuse through the water phase to other membranes where they are recombined to phospholipids. Each of these processes is used to some extent in the regulation and control of membrane composition; they are addressed in other chapters of this book. Two additional mechanisms by which phospholipids are distributed throughout organisms and cells—transfer of intact monomeric phospholipids out of the bilayer, and their translocation across the bilayer—are the topics of this chapter.

Wirtz and Zilversmit in 1968 [1] were the first to demonstrate that cytosolic proteins were capable of facilitating monomeric phospholipid transfer between membranes. Since then numerous other proteins have been discovered with the ability to facilitate the transfer of phospholipids between intra- and extracelluar membranes and lipoproteins. These proteins differ in the head group and acyl chain specificity for the phospholipids they transfer and the mechanisms proposed for their function. They are ubiquitous in animal and plant cells as well as insect and animal plasma, and have been given the generic name of lipid transfer proteins.

The discovery of protein-mediated transbilayer phospholipid transfer (flip-flop) was discovered more recently by Seigneuret and Devaux [2]. They presented evidence that the amino phospholipids phosphatidylethanolamine and phosphatidylserine were transferred preferentially from the outer to the inner leaflet of human red cell membranes by a protein-

and ATP-dependent mechanism. Subsequently, this observation has been confirmed by independent methods [3–5]. Protein-mediated phospholipid flip-flop has also been demonstrated in liver microsomes [6,7]; however, this transfer does not require ATP.

This chapter will present what is known about the spontaneous transfer of phospholipids between membranes and the current state of knowledge about a few intracellular lipid transfer proteins. The reader is referred to a recent review by Quig and Zilversmit [8] for a discussion of phospholipid transfer in the plasma and to a paper by Schroit and Zwaal [8a] for a review of transbilayer movement of phospholipids in red cell and platelet membranes. The spontaneous mechanism of transbilayer movement will be discussed, followed by the current knowledge of membrane proteins that facilitate and even drive phospholipid movement. Rather than present an exhaustive review of intermembrane and transbilayer transfer, this chapter will focus on the methodology developed for their study and efforts to determine their mechanisms of action.

I. TRANSFER OF PHOSPHOLIPIDS OUT OF THE BILAYER

A. Methodology

Implicit in being able to measure the rate of protein-stimulated phospholipid transfer is the ability to measure the spontaneous rate in the absence of protein. Thus the methods for measuring spontaneous and protein-mediated phospholipid transfer are essentially the same. These methods fall into two basic categories. The first group includes methods that depend on physical separation to quantify the amount of phospholipid that has moved from one population of donor membranes to another population of acceptor membranes. The second group includes methods that use spectroscopic or physical techniques (e.g., electron spin resonance, fluorescence, calorimetry) to measure phospholipid movement without physically separating the donor and acceptor membranes.

1. Separation Methods

Stimulation of phospholipid transfer by a cytosolic protein was first observed using a microsome-mitochondria assay [1]. In this assay, the microsome or mitochondrial phospholipids are labeled with radioactive precursors. The radiolabeled donors are mixed with unlabeled acceptors and allowed to equilibrate for a given period of time before the equilibration is terminated by rapid centrifugal separation.

Phospholipids in rat or other small animal mitochondria or microsomes may be radiolabeled by intraperitoneal injection of ^{32}[P] inorganic phosphate or [^{14}C] glycerol prior to subcellular fractionation. This method requires further characterization of the separated donor and acceptor particles by thin-layer chromatography to determine the phospholipid species transferred. To avoid this, specific head-group labeling can also be achieved for phosphatidylcholine [9], phosphatidylinositol [10], phosphatidylserine [11], phosphatidylethanolamine [12], or phosphatidic acid [13].

All the separation techniques require some type of nonexchangeable marker—that is, some readily measurable component in either the donors or the acceptors that is not appreciably transferred during the maximum time course of the experiment. The nonexchangeable marker is necessary to correct for cross-contamination following the separation procedure. Radiolabeled proteins can be used as nonexchangeable markers in the mitochondria-microsome assay.

Several variations on this theme have been successfully employed to measure phospholipid transfer. Transfer can be measured from small unilamellar vesicles to mitochondria using centrifugation to separate the mitochondria away from the vesicles [14,15]. Analogously, transfer can be measured from small unilamellar vesicles to microsomes. In this case, separation is achieved by lowering the pH to 5.1 and pelleting the precipitated microsomes [10,16]. The use of artifical donor vesicles has the advantages of allowing total control of donor phospholipid composition and the incorporation of nonexchangeable radiolabeled lipids such as triacylglycerol or cholesterol esters.

Several assays have been developed using small unilamellar vesicles as donors and acceptors. These methods have the advantage of allowing complete control over the composition of donor as well as acceptor membranes. However, either the donor or the acceptor must be labeled in some fashion to allow for its separation. In one example, negatively charged phospholipids, such as phosphatidic acid or phosphatidylserine, are included in one of the vesicle populations. Donors and acceptors are separated by passing the mixture through a DEAE-cellulose ion exchange column. The negatively charged vesicles stick to the column and the neutral vesicles pass through [17,18]. Free flow electrophoresis can also be used to separate vesicles with different surface charges [19]. Vesicles have been labeled with Forssman antigen from sheep red blood cells and separated from unlabeled vesicles by the addition of anti-Forssman antigen gamma globulin [20]. Vesicles have also been labeled with α-D-mannosyl-1(1→3)-α-D-mannosyl-*sn*-1,2 diglyceride, precipitated with concanavalin A [21], labeled with lactosylceramide, and precipitated with *Ricinus communis* agglutinin [22]. Numerous other separation schemes have been employed to meet the particular needs of a given experiment, and many have been described in the review by Wetterau and Zilversmit [23].

2. Spectroscopic and Other Physical Methods

Spectroscopic and other physical methods can be used to monitor continuously the transfer of phospholipids between membranes without separating donors and acceptors. This has the advantage in most cases of being easier and more accurate, but it has the disadvantage, in the case of ESR or fluorescence, of requiring the attachment of a reporter group that alters the phospholipid structure.

Spontaneous phospholipid transfer between artificial phospholipid vesicles was measured by Martin and MacDonald [24] using shifts in the gel-to-liquid phase transition temperature as dimyristoylphosphatidylcholine vesicles equilibrated with dipalmitoylphosphatidylcholine vesicles. A similar technique was used by Duckwitz-Peterlein et al. [25] to measure phospholipid exchange between two populations of vesicles with different phase transition temperatures prepared from lipid extracts of *Escherichia coli* grown in the presence of different unsaturated fatty acids.

Several techniques have been developed using fluorescent-labeled phospholipids. Roseman and Thompson [26] prepared acyl chain-labeled pyrene phosphatidylcholine and used the concentration dependence of the pyrene eximer formation to monitor its transfer between labeled and unlabeled vesicles with time. This technique has also been used to measure spontaneous and protein-mediated pyrene-labeled phospholipid transfer between lipoproteins [27–29].

Phospholipids labeled with 7-nitro-2,1,4-benzoxadiazol-4-yl (NBD), bound either to the terminal ends of six or twelve carbon fatty acids in the *sn*-2 position of various glycerophospholipids or to the free amino in the head group of phosphatidylethanolamine, have been used to monitor spontaneous and protein-mediated phospholipid transfer [30–32]. Two different assays using NBD-labeled phospholipids have been developed. In the first, donor vesicles were prepared containing sufficient NBD-labeled phospholipid to result in concentration dependent self-quenching. As the NBD-labeled phospholipids equilibrate with unlabeled acceptor vesicles, the increase in fluorescence is directly proportional to the amount of NBD-phospholipid transferred [30]. A similar technique uses long-chained, "nonexchangeable," rhodamine-labeled phosphatidylethanolamine to quench the NBD-phospholipid fluorescence by resonance energy transfer, and transfer is monitored by an increase in NBD fluorescence [31,32]. The latter assay requires probe concentrations as little as 2% of the total phospholipid.

The reduction of self-quenching of 1-acyl-2-parinaroyl-phosphatidylcholine as it transfers from labeled donor vesicles to unlabeled acceptor vesicles provides an additional fluorescent assay for the measurement of phospholipid transfer between membranes [33].

Phospholipids labeled with nitroxide fatty acid have been used to monitor vesicle-vesicle transfer from changes in the electron spin resonance spectrum. Spin-spin interactions in the labeled donor vesicles cause a broadening of the ESR spectrum that narrows as the spin-labeled phospholipid is transferred to the unlabeled acceptor vesicles [34,35].

The continuous and accurate measurements of the transfer rates that can be obtained with these techniques is especially advantageous for kinetic studies of the mechanism of spontaneous and protein-mediated phospholipid transfer. However, a serious disadvantage is the need to estimate the effect of the probe moiety itself on the transfer rates.

B. Spontaneous Phospholipid Transfer

Considering the transfer of phospholipids between two populations of stable membranes, bilayers, or lipid-rich particles that do not fuse, the rate of spontaneous transport is inversely dependent on the number and length of the acyl chains in the phospholipid molecule. That is, the transfer rate depends on the water solubility of the transferring molecule [36,37]. This relationship can be approximated by the rule of thumb that the rate of transfer will decrease approximately by a factor of 3.3 for every unit increase in the number of effective carbons in the acyl chains [38,39]. For molecules with two long carbon chains, the relative contribution to the total hydrophobic effect of the second carbon chain is 0.6 that of the first [40]. Thus the effective number of carbons is equal to the number of methylene groups in the first chain plus the number in the second chain multiplied by 0.6 (see [40] for more details). Unsaturation tends to increase the water solubility and thus the spontaneous transfer rate. Some representative half-times for the spontaneous exchange of radiolabeled phosphatidylcholines between phospholipid vesicles are dimyristoylphosphatidylcholine, $t_{1/2}$ = 2 h; dipalmitoylphosphatidylcholine, $t_{1/2}$ = 83 h; 1-palmitoyl, 2-oleoylphosphatidylcholine, $t_{1/2}$ = 63 h [36]. Changes in the head group structure of the transferring molecule alter the transfer rates modestly compared to the acyl chain dependence [31,32,41].

These relationships imply that the rate limiting step for phospholipid transfer involves its partition into the water phase—that is, that spontaneous phospholipid transfer between cellular membranes, artificial vesicles, and lipoproteins occurs through the water phase and is not dependent on collisional contact between the exchanging particles. Additional support for this mechanism of spontaneous phospholipid transfer has been obtained from kinetic modeling studies of the transfer rates as a function of donor and acceptor particle concentration.

One can propose two kinetically distinct mechanisms to describe the transfer of phospholipids between nonfusing membranes: either the phospholipid molecules transfer between membranes by diffusing through the water separating the membrane or they transfer during transient membrane collisions or associations. Of course transfer could occur by some combination of the two mechanisms.

Both of these models assume an intermediate compartment between the donor and the acceptor particles. In the aqueous diffusion model, the intermediate compartment is the phospholipid that is soluble as monomers or small aggregates in the water phase. The intermediate compartment in the collision dependent model is the transient collision complex. In both models the rate of phospholipid transfer is dependent on the size of the intermediate compartment and will saturate when this compartment is at its maximum.

The aqueous diffusion model is based on a model for micelle association and dissociation developed by Nakagawa [42] and adapted to describe transfer by vesicles by Thilo [43] and by Nichols and Pagano [30,31]. In this model, one assumes that the rate at which an amphiphile escapes from the surface of a membrane is proportional to its concentration on that surface and that the adsorption rate is proportional to the product of the aqueous concentration and the surface area of the membrane (Fig. 1a). The initial rate equations can be written for each step in the transfer of phospholipid from donor (D) to acceptor (A) vesicles. Assuming that the concentration of phospholipid in the water phase reaches steady-state equilibrium quickly in relation to the transfer rate, the initial rate v_0 of transfer can be described by

$$v_0 = \frac{dL_A}{dt} = \frac{k_{-1}k_2L_AL_D}{k_1L_D + k_2L_A} \tag{1}$$

(See Fig. 1a and legend for the definition of rate constants and abbreviations.)

A kinetic description of the collision dependent transfer model was presented by Ferrell et al. in 1985 [37]. In deriving the kinetic model for collision-dependent transfer (Fig. 1b), Ferrell et al. assumed that the surface area of the acceptors that was available for collisional contact was reduced by the number of existing complexes L_{DA}. Thus the initial rate of collision dependent transfer can be predicted by

$$v_0 = \frac{dL_A}{dt} = \frac{k_1k_2L_DL_A}{k_{-1} + k_2 + k_1(\sigma L_D + L_A)} \tag{2}$$

where σ is the ratio of the amount of acceptor surface area that is blocked by a complexed donor vesicle to the total surface area of the donor vesicles. (See Fig. 1b for the definition of rate constants and abbreviations.)

Ferrell et al. were the first to point out that the collision dependent model predicts saturable kinetics when either donors or acceptors are increased while holding the other constant. Prior to this, saturable kinetics of this type were assumed to be limited to aqueous-diffusional transfer. However, understanding the cause of saturation in both cases allows the design of experiments to distinguish easily between the two models. In the collision dependent model, saturation occurs when the formation of the donor-acceptor complex is not rate-limiting—that is, the initial rate is determined solely by the

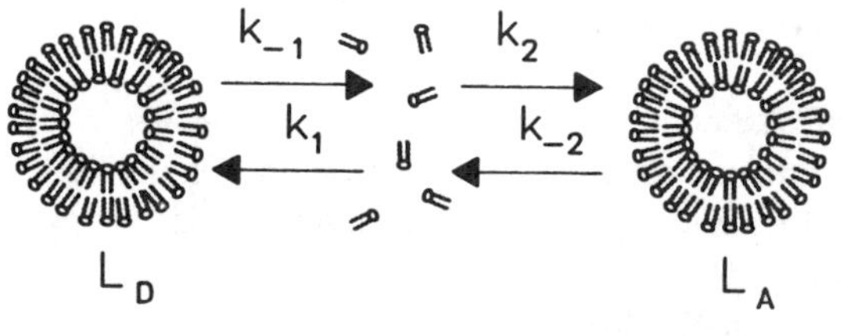

(a) AQUEOUS DIFFUSION TRANSFER

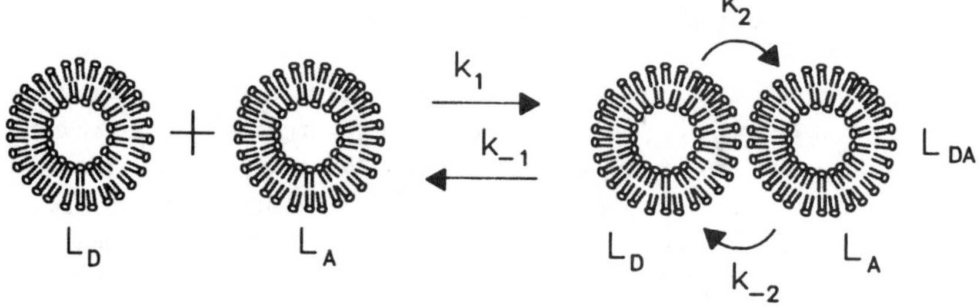

(b) COLLISION-DEPENDENT TRANSFER

Figure 1 Kinetic schemes for spontaneous phospholipid transfer between vesicles. L_D, L_A, and L_{DA} refer to the total bulk solution lipid concentration of the donor vesicles, acceptor vesicles, and donor-acceptor complexes, respectively. (See text for description and discussion.)

rate constant for transfer from the donor-acceptor complex to the acceptors (reaction 2, Fig. 1b). In the case of excess acceptors, the donor-acceptor complex approaches a constant value when all of the donors are complexed with acceptors. On the other hand, in the case of excess donors, saturation occurs when all of the available acceptor sites for donor collision are already bound and no additional donor-acceptor complexes can be formed.

In comparison, transfer by the aqueous diffusion pathway saturates in the presence of excess acceptors because the fraction of donor lipid relative to acceptor lipid in the water phase approaches zero, so that the transfer is limited by the rate of lipid dissociation from the donors into the water phase. In the presence of excess donors, the fraction of donor relative to acceptor phospholipid in water phase approaches 1 and cannot be increased further.

Understanding the above causes for saturation kinetics allows one to distinguish between the two mechanisms when transfer is saturated in the presence of excess acceptors. If collision-mediated transfer were the correct mechanism, one would predict that all of the donor vesicles would be associated with the acceptor vesicles. Thus in an experiment where acceptors (e.g., cells, mitochondria, etc.) could be centrifuged without pelleting unbound donors, one would expect that all of the donors would be bound and spin down with the acceptors. If most of the donors remained in the supernatant, the saturation kinetics cannot result from collision-mediated transfer.

Using a combination of acyl chain dependence, ionic strength dependence, and kinetic modeling studies, most investigatiors have concluded that the predominant mode of phospholipid transfer between vesicles, organelles, cell membranes, and lipoproteins is through the water phase—not requiring contact or collision [18,19,25,26,27,30,31,38]; for a review see [44]. There are several notable exceptions in which second-order kinetics indicating a collision dependent component of transfer have been observed in the presence of excess acceptors [24,45–49]. Jones and Thompson have perhaps explained many of the apparent discrepancies by demonstrating that at low concentrations of phospholipid vesicles (<2 mM) the kinetics of phospholipid transfer are consistent with the aqueous diffusion mechanism, whereas at higher concentrations of vesicles (>2 mM) the kinetics indicate an additional collisional component that acts in parallel with aqueous diffusion to increase the rate of transfer [48]. Jones and Thompson observed that the first-order rate constant (aqueous diffusion) was proportional to the second-order rate constant (collision dependent) when the first-order rate constant was varied by changing the acyl chain length, temperature, and ionic strength [49]. From these observations, they concluded that the second-order process resulted not from partial or transient fusion of the vesicles but instead from collision dependent enhancement of phospholipid dissociation into the water, perhaps by van der Waals interaction between the opposing vesicle and the desorbing lipid molecule.

The observation that high concentrations of vesicles can result in second-order kinetics (i.e., collision dependent transfer) probably explains the discrepancies observed previously for phospholipid transfer between artificial vesicles and may explain the second-order kinetics observed for phospholipid transfer between vesicles and lipoproteins [45]. However, collision dependent transfer of phospholipids between mixed phospholipid-bile salt micelles, which does involve transient fusion complexes, has been observed [50,51]. In this case, the first-order rate constant varies as a function of acyl chain length, as predicted for aqueous diffusion, whereas the second-order rate constant does not.

C. Protein-Mediated Phospholipid Transfer

Since the first demonstration that liver cytosol contains proteins capable of increasing the spontaneous rate of phospholipid transfer between membranes [1], lipid transfer proteins have been found in virtually all mammalian tissues investigated as well as many plant cells, yeast, and prokaryotes [52]. Although the exact nature of their biological function remains unresolved, their abundance and demonstrated in vitro transfer ability strongly suggest a role in the movement of lipids to establish and/or maintain proper lipid composition in the cell membranes. Lipid transfer proteins found in mammalian plasma are likely to play a role in the distribution of lipids among lipoproteins. This chapter will not attempt to catalogue all of the lipid transfer proteins described from the many different sources but instead will focus on three intracellular phospholipid transfer proteins that have been well characterized.

1. Phosphatidylcholine Specific Transfer Protein

Phopsphatidylcholine specific transfer protein (PC-TP) was discovered in response to the attempt to understand how the lipid components of plasma lipoproteins were taken up by liver cells [53]. A protein with phosphatidylcholine transfer activity was discovered that was nondialyzable, heat labile, and protease sensitive [1]. This cytosolic protein has subsequently been purified to homogeneity from several different animal tissues, and the complete amino acid primary structure of the bovine liver PC-TP has been elucidated by proteolytic digestion, chemical cleavage, and Edman degradation [54]. PC-TP contains

213 amino acids and two disulfide bonds, with the *N* terminus blocked by an *N*-acetylmethionine. The calculated molecular weight is 24,680 daltons. An analogous 28,000 dalton protein has been isolated from the rat liver, and its distribution in rat tissues has been tested [55]. The highest levels of PC-TP were measured in the liver and intestinal mucosa, with lower values in the kidney, spleen, and lung. The lowest values were found in the heart and brain.

As its name implies, PC-TP is highly specific for phosphatidylcholine and does not transfer sphingomyelin or glycerolipids with other head groups. Each molecule has a high affinity binding site for phosphatidylcholine [56]. The amino acids forming the lipid binding site have been identified by their reaction with phosphatidylcholine analogs containing photoactivatable carbene precursors [57]. Small perturbations of the phosphorylcholine head group result in significant reduction of transfer efficiency [58]. Addition of two methylenes between the phosphorus and nitrogen atoms completely eliminates transfer, as does removal of only one of the methyls on the quaternary nitrogen. PC-TP transfer efficiency is also dependent on the acyl chain composition of phosphatidylcholine. In general, unsaturated acyl chains larger than fourteen carbons are preferred, as is a liquid crystalline bilayer phase [58–60]. Differences in the fluorescence rotational correlation time measured with a parinaric acid in the *sn*-1 versus the *sn*-2 position of a phosphatidylcholine molecule were used to infer that PC-TP has separate binding sites for the two acyl chains that are located nearly orthogonal to each other [61].

When PC-TP activity is measured between donor and acceptor membranes, both containing phosphatidylcholine, no net transfer of phosphatidylcholine occurs. Only when the acceptor membrane is devoid of phosphatidylcholine can the phosphatidylcholine net transfer be detected, and only then at a much lower rate than when acceptor phosphatidylcholine is available for exchange [32,62]. This evidence, coupled with the demonstration that PC-TP binds phosphatidylcholine with high affinity, led to the hypothesis that PC-TP functions as a shuttle carrier. The general scheme for a shuttle carrier is shown in Fig. 2a. In the case of PC-TP, the dissociation of the unloaded protein (reactions 3 and 4) is very rare, so that after the spontaneous transfer in the absence of protein (reactions 5 and 6) has been subtracted, as is done in all of the PC-TP transfer assays, the general scheme is reduced to only reactions 1 and 2. A mathematical description of this transfer mechanism has been derived [63] and found to predict the kinetics of PC-TP mediated transfer between membranes [35,63–65]. (See Fig. 2a for the definition of rate constants and abbreviations.)

$$v_0 = \frac{k_1 L_1 k_2 L_2 P_{\text{tot}}}{(k_1 L_1 + k_2 L_2)(1 + k_1/k_{-1}[L_1] + k_2/k_{-2}[L_2])}$$

This model assumes that (1) the protein distributes between the membranes quickly relative to the phospholipid transfer rate; (2) the protein always has a bound phospholipid molecule; (3) association of the protein with the membrane always results in transfer. This model predicts the kinetic behavior of PC-TP mediated phosphatidylcholine transfer, but because of the underlying assumptions, the interpretation of the rate constants obtained from the curve fitting is open to question. For example, the van den Besselaar et al. model [63] assumes that each time a protein associates with a membrane an exchange of a bound for an unbound phosphatidylcholine occurs. This is equivalent to the unloading and loading reactions (steps k_{-7}, k_{-8}, k_7, k_8) being very fast relative to the association and dissociation reactions (steps k_1, k_2, k_{-1}, k_{-2}) and the phosphatidylcholine bound state

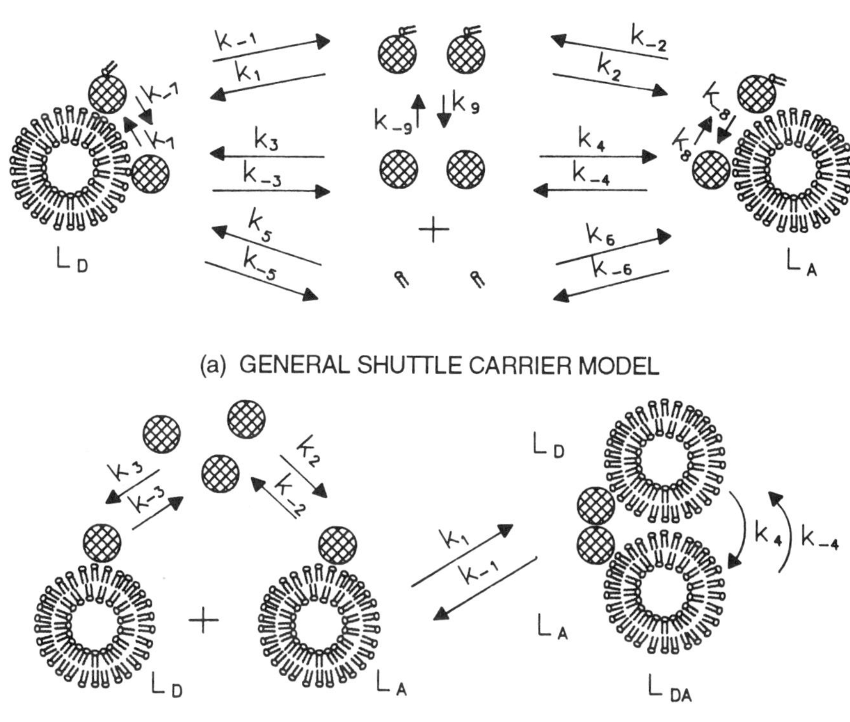

Figure 2 Kinetic schemes for protein-mediated phospholipid transfer. L_D, L_A, L_{DA}, and P_{tot} refer to the total bulk solution lipid concentrations of the donor vesicles, acceptor vesicles, donor-acceptor complexes, and transfer protein, respectively. (See text for description and discussion.)

being highly favored over the unbound. If this is not the case, and there is no evidence to indicate that it is, the calculated association rate constants (k_1 and k_2) would also include the rate constants for loading and unloading. Thus factors thought to alter the rate of protein-membrane association anbd dissociation may actually be affecting only the rate of unloading and loading, and a significant amount of the membrane-associated protein may exist without a bound phosphatidylcholine molecule. This model is also restricted to experiments in which the mole fraction of phosphatidylcholine is equal in donors and acceptors. A model that does not include the assumptions of the van den Besselaar et al. model would probably be too cumbersome to obtain meaningful rate and binding constants. However, it is important to keep in mind the full range of interpretations of the rate constants obtained from the model.

In an attempt to dissociate PC-TP association with the membrane from phosphatidylcholine binding, Runquist and Helmkamp [66] included a nontransferable bis-phosphatidylcholine in acceptor vesicles and found that it had little effect on phosphatidylcholine exchange or net transfer. They concluded that PC-TP did not require a phosphatidylcholine head group for interaction with the membrane and that phosphatidylcholine in the membrane facilitates PC-TP mediated transfer by increasing the rate of unloading and loading of the membrane bound PC-TP (steps k_7, k_{-7}, k_8, k_{-8}).

Although PC-TP has been conclusively demonstrated to transfer phosphatidylcholine between membranes in vitro, its physiological function remains in doubt. Recently, PC-TP has been proposed to regulate the synthesis of phosphatidylcholine via cholinephosphotransferase, perhaps by relief of end product inhibition [67].

2. Phosphatidylinositol Transfer Protein

Cytosolic proteins that facilitate the intermembrane transfer of phosphatidylinositol have been isolated from various tissues of several animals, including bovine brain [68], bovine heart [69], rat liver [70], rat brain [71], human blood platelets [72], and yeast [73]. These phosphatidylinositol transfer proteins (PI-TP) possess structural similarities in molecular weight (35–36 KDa) and pI (4.9–5.3) and a marked preference for phosphatidylinositol. Phosphatidylcholine and phosphatidylglycerol transfer are facilitated, but not as efficiently. Rabbit antibodies to bovine PI-TP react with a 35–36 KDa protein in the cytosolic fraction of mammalian, avian, reptilian, amphibian, and insect tissue [74]. PI-TP activity has been identified in thirteen different tissues in the rat, the highest levels being found in the brain and the lowest in the adipose and skeletal muscle [71]. Two subforms with different pI's have been identified resulting from the noncovalent binding of either phosphatidylinositol or phosphatidylcholine [75].

Recently, Dickeson et al. [74] used immunoscreening techniques to isolate and sequence full-length cDNA clones that encode the rat PI-TP. The proposed amino acid sequence shows no significant similarity to other lipid-binding proteins. The hydrophobic peptide thought to be the phosphatidylcholine binding site in PC-TP [76] was not found in the PI-TP sequence. Studies of the binding site using pyrene-labeled phosphatidylinositol and phosphatidylcholine indiate that separate binding sites exist for the *sn*-1 and *sn*-2 acyl chains [77]. Because of the strong binding of phosphatidylinositol and phosphatidylcholine to the protein, it is thought to function as a shuttle carrier. In all tissues studied, PI-TP prefers to transfer phosphatidylinositol over phosphatidylcholine and does not transfer phosphatidylethanolamine or phosphatidylserine [71]. Because of its specificity for phosphatidylinositol, phosphatidylcholine, and phosphatidylglycerol [78,79], PI-TP can catalyze the net transfer of phosphatidylinositol from one membrane to another in exchange for phosphatidylcholine or phosphatidylglycerol. The kinetic model for phosphatidylinositol heteroexchange for phosphatidylcholine is too complex to be useful in determining specific rate and association constants without additional independent knowledge of the mechanisms involved. However, the homoexchange of phosphatidylcholine by PI-TP has been studied kinetically [80]. Investigators [79–82] have found that a negative surface charge on the membrane increases the apparent association of PI-TP for the membrane (steps k_1/k_{-1} and $k_2/_{-2}$ in Fig. 2a), which results in inhibition of the rate of phosphatidylcholine or phosphatidylinositol transfer. However, from the kinetic modeling studies alone, one cannot rule out the possibility that the negative surface charge has no effect on the association reaction but instead reduces the rate of exchange of the newly inserted lipid for an endogenous lipid in the membrane (steps k_{-7} and k_{-8} in Fig. 2a). However, since the inhibition of PI-TP activity by negatively charged membranes can be reversed by high ionic strength, the most likely interpretation is that relatively nonspecific ionic forces, rather than specific protein-phospholipid head group interactions, contribute to increase the association of PI-TP with membranes.

Although the in vitro activity of PI-TP suggests a physiological role in the distribution of phosphatidylinositol and the maintenance of phosphatodylinositol concentrations in the appropriate membranes for their proper function, this has yet to be demonstrated unequivocally. A promising approach to understanding the physiological role of PI-TP has recently been demonstrated in the yeast *Saccharomyces cerevisiae* [83]. An oligonucleotide based on the amino terminal sequence of the PI-TP from *S. cerevisiae* was used to screen a yeast genomic library for the gene encoding PI-TP. By gene disruption, it was demonstrated that PI-TP was essential for cell viability. Viability could be rescued by the introduction of the PI-TP gene on a plasmid copy. Although this approach has not defined the physiological role of PI-TP, it clearly established that PI-TP is essential for normal cell growth and function.

3. Nonspecific Lipid Transfer Protein

In an effort to identify a rat liver cytosolic protein capable of transferring phosphatidylethanolamine between intracellular organelles, Bloj and Zilversmit [84] identified and purified [85] a 13,500 MW protein which they called nonspecific lipid transfer protein (nsLTP). This name reflects the protein's ability to stimulate the intermembrane transfer of a wide range of lipid molecules including diacylglycerides, gangliosides, glycosphingolipids, and cholesterol [84,86]. nsLTP has subsequently been shown by Trzaskos and Gaylor [87] to be identical to the sterol carrier protein$_2$ (SCP$_2$). SCP$_2$ was initially purified from rat liver cytosol by its ability to activate the microsomal enzymatic conversion of intermediates in the cholesterol synthetic pathway [88]. This activation presumably results from its ability to facilitate the transfer of cholesterol or its intermediates to enzymatic sites in the microsomes [89].

The primary structures for rat [90] and bovine liver [91] nsLTP are very nearly homologous. Rat liver nsLTP has 122 amino acids (one more than bovine), and both contain only one free sulfhydryl group that is necessary for proper function [92–95]. The molecular weight by SDS-PAGE is in the range of 12,400 to 14,800 with a pI of 8.6 [96].

Although radiolabeled unesterified cholesterol has been shown to bind to nsLTP in a one-to-one stoichiometry [89], early attempts failed to demonstrate nsLTP binding of radiolabeled phospholipids. More recently, low affinity binding of fluorescent-labeled phospholipids was demonstrated using spectroscopic detection of binding not requiring physical separation of the nsLTP and donor membranes [92]. Kinetic studies demonstrated that nsLTP binds and releases the phospholipids rapidly (half times < 1 min) [93], which may explain the difficulty in demonstrating nsLTP-phospholipid binding following a lengthy separation procedure. Binding to nsLTP appears to be dependent on the hydrophobicity and not on the molecular structure of the lipid species [92,97]. The amount of phospholipid bound to nsLTP for a given amount of nsLTP and donor phospholipid vesicles decreases as the acyl chain length or hydrophobicity of the phospholipid probe increases [92]. At first thought, this may appear to be the opposite of that expected for a hydrophobic binding site. However, the amount of phospholipid bound to nsLTP in equilibrium with phospholipid vesicles depends on the relative affinity of the phospholipid for nsLTP and the vesicles. If the vesicle bilayer provides a better hydrophobic environment than the binding site on the nsLTP, then the more hydrophobic the phospholipid, the more it will prefer the vesicle to the protein. Thus the amount of lipid associated with the protein will decrease as the hydrophobicity of the lipid increases.

The inhibition of both phospholipid binding to nsLTP [92] and the nsLTP phospholipid transfer activity by thiol reagents [92,94,95] suggests that this binding site is involved in the transfer mechanism. However, to date no definitive evidence has been obtained to distinguish between the three main alternatives for its mechanism of action. nsLTP has been proposed to function as a modified shuttle carrier [93], in which the nsLTP-phospholipid complex forms in the membrane, dissociates from the membrane (reactions k_{-1} and k_{-2} in Fig. 2a) and releases the bound phospholipid to the water (reaction k_{-9}), increasing the total amount of phospholipid available for transfer in the water phase. nsLTP has also been proposed to bind to the membrane and increase the rate of monomeric phospholipid dissociation into the water phase (reaction k_{-5}) [32], thereby increasing the amount available for transfer to another membrane. Finally, nsLTP has been proposed to mediate the formation of transient intermembrane contacts that facilitate the intermembrane transfer of phospholipids (Fig. 2b) [94,95,98,99].

Each of the models is consistent with the observation that the rate of nsLTP stimulated lipid intermembrane transfer is highly correlated with the spontaneous transfer rate [32,98], and each model can account for net transfer of lipids between membranes. All three models are consistent with the observed inability to transfer phospholipids the long distances required for phospholipid transfer between adjacent surface monolayers [98]. Further experiments will be required to determine which if any of the above models best describes the mechanism for nsLTP facilitation of lipid transfer.

At this time, the physiological function of nsLTP is also unknown. Mutated Chinese hamster ovary cells with very low levels of nsLTP appear to grow and function normally with no radical differences in cell phospholipid composition [100]. These mutated CHO cells are lacking in nsLTP and peroxisomes [101], which is consistent with the correlation of low levels of nsLTP with the absence of peroxisomes in Morris hepatoma [102] and liver cells from patients with Zellweger syndrome [103]. Immunoelectron microscopy with antibodies to nsLTP has been used to demonstrate the localization of nsLTP inside the rat liver peroxisomes [104,105]. Immunoblotting experiments of proteins from purified subcellular fractions of rat liver cells with antibodies to nsLTP have also demonstrated that nsLTP is primarily localized in the peroxisomes. However, in addition to the 13.5 kDa nsLTP, another larger 55–60 KDa protein cross-reacts with the anti-nsLTP antibody [104,105]. Similarly, a 58 KDa protein cross-reactive with antibodies to nsLTP co-sediments with peroxisomes from the wild-type CHO cells [101]. In vitro translation studies with mRNA from rat adrenals and liver indicate that these two proteins are synthesized independently [106,107]. Ossendorp et al. [108] have recently used anti-nsLTP antibodies to screen a rat liver cDNA library. A positive 1,851 base pair clone was identified in which the 369 base pairs at the 3' end correspond with the amino acid sequence of nsLTP. Further analysis is required to determine whether the remaining base pairs code for a portion of the 58 kDa protein. The relationship of the 58 KDa protein to nsLTP remains to be determined; however, it is intriguing that rat liver nsLTP has significant homology with the variable domains of the heavy chain of immunoglobulin G [90].

Peroxisomes have been identified as important sites of cholesterol metabolism. They contain the enzymes necessary for cholesterol synthesis [109] and its conversion to bile acids [110]. The association of nsLTP with the peroxisomes suggests that nsLTP's physiological role may be to facilitate the transfer of precursors and end products in the production of cholesterol and bile acids.

Apart from the physiological role of nsLTP, its ability to transfer a wide range of lipids between membranes makes it a valuable experimental tool to manipulate the lipid content of membranes. Bovine liver nsLTP was used to determine the asymmetric distribution of the major phospholipids in rat erythrocytes as well as their rates of transbilayer movements [111]. nsLTP has also been used to alter the membrane phospholipid and cholesterol composition to determine their effects on enzyme activity in mitochondria, microsomes [112], synaptosomes [113], and sarcoplasmic reticulum [114].

Using nsLTP, the effect of membrane composition on enzyme activity may be tested with only minor perturbations of the membrane and protein structure as opposed to other techniques involving lipases and detergents.

II. TRANSBILAYER TRANSFER OF PHOSPHOLIPIDS

A. Spontaneous Transbilayer Transfer

It is now known that there are membrane proteins that either specifically or nonspecifically increase the rate of phospholipid translocation between the two leaflets of a phospholipid bilayer (commonly called flip-flop). Fundamental to the discovery of protein-mediated transbilayer phospholipid transfer was a knowledge of the expected rate of spontaneous transfer [115]. This question was first addressed by Kornberg and McConnell in 1971 [116] using an electron spin-labeled phospholipid. They incorporated the spin-labeled phospholipid into both leaflets of small unilamellar vesicles, and the appearance of the inner leaflet label in the outer leaflet was detected by its accessibility to reduction by the bilayer impermeant molecule ascorbate. The half-time measured for transbilayer movement was 6.5 h. Subsequently, much longer half-times have been measured using different techniques. For example, the transfer of phosphatidylethanolamine from inner to outer leaflets of small unilamellar vesicles was measured by chemically reacting phosphatidylethanolamine initially in the outer leaflet with isethionyl acetamide hydrochloride and detecting the appearance of unlabeled phosphatidylethanolamine on the outer leaflet by reaction with 2,4,6-trinitro-benzenesulfonic acid [117]. The half-timc measured in this experiment was greater than 80 days.

Phosphatidylcholine transfer protein (PC-TP) has been used to measure the transbilayer movement of phosphatidylcholine. PC-TP-mediated exchange of radiolabeled phosphatidylcholine for unlabeled phosphatidylcholine occurs in two phases. The faster phase reflects transfer from the outer leaflet, while the slower correlates with inner to outer leaflet transfer [118,119]. No second phase of transfer was detected, and the lower limit to the half-time for transfer was placed at 4 to 11 days at 37°C. A similarly long half-time for transbilayer phosphatidylcholine transfer was measured using deuterium-labeled phosphatidylcholine. Its rate of appearance in the outer leaflet was measured by the down field shift of the NMR signal by the paramagnetic ion Pr^{3+} [120].

More recently, Homan and Pownall [121] used changes in pyrene eximer fluorescence to measure the intermembrane and transbilayer transfer of pyrene-labeled phospholipids. The short-chain, pyrene-labeled phospholipids transfer rapidly from the outer leaflet to excess unlabeled vesicles, allowing the slower rate of flip-flop from the inner to the outer leaflet to be measured. This technique was used to measure the rate of flip-flop for several different head groups and acyl chain lengths. Increasing the acyl chain length reduced the flip-flop rate by a factor of two for every two methyl carbons. Half-times for flip-flop were more strongly dependent on head group structure with phosphatidylcholine (347 h) < phosphatidylglycerol (69 h) < phosphatidic acid (35 h) < phosphatidylethanolamine (10 h).

In general, the rate of phospholipid flip-flop across pure phospholipid bilayers is a slow process. The transfer of the polar head groups into the apolar bilayer center is highly energetically unfavorable. However, the exact nature of the mechanism of transfer is not known. Studies on the effect of membrane perturbations on the flip-flop rate suggest that membrane discontinuities facilitate phospholipid flip-flop. For example, transbilayer movement of dimyristoyl phosphatidylcholine shows a distinct maximum at the temperature at which the hydrocarbon phase transition occurs [122], and addition of glycophorin to phosphatidylcholine vesicles increases the rate of transbilayer movement by a factor of two [123].

B. Protein-Mediated Transbilayer Transfer

Given the very slow spontaneous transbilayer transfer of phospholipids across pure phospholipid bilayers, there are two cellular membranes where mechanisms for facilitating phospholipid flip-flop would be expected to exist. The endoplasmic reticulum has no asymmetry, although asymmetry is expected, since phospholipids are synthesized and inserted solely into the cytoplasmic leaflet [124]. Maintenance of membrane integrity requires the translocation of phospholipids from the cytoplasmic to the lumenal leaflet [125,126]. On the other hand, the plasma membrane has an asymmetric distribution of phospholipids between its inner and outer leaflets that cannot be explained by asymmetry in other intracellular membranes. With this in mind, several investigators looked for and found specific membrane proteins that have the capacity to facilitate transbilayer phospholipid transfer in these two membranes.

A membrane protein that permits phosphatidylcholine equilibration between the leaflets of rat liver microsomes has been demonstrated by two different techniques. Bishop and Bell [6] measured the uptake of the water soluble *sn*-1,2-dibutyroyl-phosphatidylcholine into rat liver microsomes. They demonstrated that the uptake was saturable and dependent on an intact permeability barrier. Transport was inhibited by the three and five carbon diacyl-phosphatidylcholine analogs but not by *sn*-2,3-dibutyroyl-phosphatidylcholine or phosphorylcholine. This specificity, and the observation that transport is inhibited by proteases and *N*-ethylmaleimide, led the authors to conclude that the transport was mediated by a membrane protein. Subsequently, the same criteria were used to demonstrate protein-mediated transfer of monobutyroyl-PC across microsomal membranes [127].

Transbilayer phosphatidylcholine transfer activity from rat liver microsomes has been reconstituted into liposomes by Backer and Dawidowicz [7]. Vesicles were formed from a mixture of brominated phosphatidylcholine molecules and microsomal membranes. Flip-flop activity was detected from differences in the vesicle density following removal of all of the phosphatidylcholine available for exchange with PC-TP. Vesicles with flip-flop activity were lighter due to the total removal of the heavier brominated phosphatidylcholine. This activity was not sensitive to proteases or *N*-ethylmaleimide. The activity was proposed to be protein-mediated since recombination with only microsomal lipids or with red cell membrane proteins resulted in vesicles with no activity.

Subsequently, the rate of flip-flop of short-chain spin-labeled phospholipids across rat liver microsomes has been determined [128]. The second phase of their biphasic transfer from microsomes to excess bovine serum albumin was used to measure the rate of transbilayer transfer. A half-time of approximately 20 min was measured for a wide range of phospholipids including phosphatidylcholine, phosphatidylserine, phosphatidylethanolamine, sphingomyelin, lysophosphatidylcholine, and lysophosphatidylserine. Treatment with *N*-ethylmaleimide inhibited the rate of flip-flop by a factor of four.

Transbilayer transfer across microsomal membranes does not require energy and does not result in an asymmetric distribution of phospholipids across the bilayer. Further characterization of the protein or proteins involved will require the development of procedures for further isolation and purification.

Several independent techniques have been used to demonstrate the existence of proteins that are capable of tranferring amino phospholipids across the plasma membranes of numerous different cell types [2–5,129–134]. These proteins transfer the amino phospholipids phosphatidylserine and phosphatidylethanolamine from the outer to the inner leaflet and are thought to use the energy from ATP hydrolysis to maintain the preferential location of these phospholipids on the inner leaflet.

Protein-mediated transbilayer phospholipid transfer was first demonstrated in the erythrocyte plasma membrane using spin-labeled phospholipids with a nitroxide radical attached by a short acyl chain in the *sn*-2 position [2]. The membrane impermeant molecule ascorbate was used to reduce the spin-labeled phospholipid in the outer leaflet, allowing measurement of the rate of inward movement. Phosphatidylserine and phosphatidylethanolamine transfer to the inner leaflet with half-times of 5 min and 60 min, respectively, at 37°C, while the half-time for phosphatidylcholine and sphingomyelin transfer are greater than 10 h. Phosphatidylserine and phosphatidylethanolamine compete for the same transporter with K_m's of 5 and 50 μM, respectively. Flip-flop is Mg^{2+} and ATP dependent [2] and is inhibited by *N*-ethylmaleimide, iodoacetamide [2], and high intracellular Ca^{2+} [130,135].

In 1985, Daleke and Huestis [3] inferred inward translocation of PS and PE across erythrocyte membranes from changes in cell shape. Cells induced to form echinocytes by the addition of phosphatidylserine and phosphatidylethanolamine to their outer leaflets reverted to stomatocytes with half-times of less than an hour, whereas shape reversal did not occur following a cell shape change induced by the addition of phosphatidylcholine. Shape reversal was interpreted to result from phospholipid inward translocation. Shape reversal induced by phosphatidylserine and phosphatidylethanolamine is dependent on ATP and Mg^{2+} and is inhibited by sulfhydryl reagents [3].

Additional evidence for an ATP dependent amino phospholipid transfer protein was obtained by incorporating radioactive phospholipids into erythrocytes with nsLTP and following their translocation to the inner leaflet by phospholipase A_2 digestion of the outer leaflet phospholipid [4]. Connor and Schroit [5] used resonance energy transfer between two fluorescent phospholipid probes to determine their inner or outer leaflet location. They confirmed the preference for aminophospholipids and the ATP dependence of the translocation.

Using these techniques and others, the selective translocation of aminophospholipids has been demonstrated in a wide range of mammalian erythrocytes [136–139], in pig lymphocytes [131], and in human platelets [3,129]. Martin and Pagano used fluorescent-labeled phospholipids to demonstrate ATP dependent translocation of phosphatidylserine and phosphatidylethanolamine into cultured Chinese hamster fibroblasts that was not the result of endocytosis or catabolism [132].

Photoactivatable, radioiodinated phosphatidylcholine and phosphatidylserine have been used to label and identify the membrane protein or proteins responsible for amino phospholipids translocation in the red cell [140,141]. Labeling with phosphatidylserine, but not phosphatidylcholine, analogs partially blocks further translocation [141]. However, several different proteins are labeled by this technique, and the identity of the translocase protein remains to be determined.

REFERENCES

1. K. W. A. Wirtz and D. B. Zilversmit, Exchange of phospholipids between liver mitochondria and microsomes in vitro, *J. Biol. Chem. 243*:3596–3602 (1968).
2. M. Seigneuret and P. F. Devaux, ATP-dependent asymmetric distribution of spin-labeled phospholipids in the erythrocyte membrane: Relation to shape changes, *Proc. Natl. Acad. Sci. USA 81*:3751–3755 (1984).
3. D. L. Daleke and W. H. Huestis, Incorporation and translocation of aminophospholipids in human erythrocytes, *Biochemistry 24*:5406–5416 (1985).
4. L. Tilley, S. Cribier, B. Roelofsen, J.A. Op den Kamp, and L. L. M. van Deenen, ATP-dependent translocation of amino phospholipids across the human erythrocyte membrane, *FEBS Lett. 194*:21–27 (1986).
5. J. Connor and A. J. Schroit, Determination of lipid asymmetry in human red cells by resonance energy transfer, *Biochemistry 26*:5099–5105 (1987).
6. W. R. Bishop and R. M. Bell, Assembly of the endoplasmic reticulum phospholipid bilayer: The phosphatidylcholine transporter, *Cell 42*:51–60 (1985).
7. J. M. Backer and E. A. Dawidowicz, Reconstitution of a phospholipid flippase from rat liver microsomes, *Nature 327*:341–343 (1987).
8. D. W. Quig and D. B. Zilversmit, Plasma lipid transfer activities, *Annual Rev. Nutr. 10*:169–193 (1990).

8a. A. J. Schroit and R. F. A. Zwaal, Transbilayer movement of phospholipids in red cell and plasma membranes, *Biochim. Biophys. Acta 1071*:313–329 (1991).

9. K. W. A. Wirtz, H. H. Kamp, and L. L. M. van Deenen, Isolation of a protein from beef liver which specifically stimulates the exchange of phosphatidylcholine, *Biochim. Biophys. Acta 274*:606–617 (1972).
10. G. M. Helmkamp, Jr., M. S. Harvey, K. W. A. Wirtz, and L. L. M. van Deenen, Phospholipid exchange between membranes. Purification of bovine brain proteins that preferentially catalyze the transfer of phosphatidylinositol, *J. Biol. Chem. 249*:6382–6389 (1974).
11. M. M. Butler and W. Thompson, Transfer of phosphatidylserine from liposomes or microsomes to mitochondria. Stimulation by a cell supernatant factor, *Biochim. Biophys. Acta 388*:52–57 (1975).
12. R. H. Lumb, A. D. Kloosterman, K. W. A. Wirtz, and L. L. M. van Deenen, Some properties of phospholipid exchange proteins from rat liver, *Eur. J. Biochim. 69*:15–22 (1976).
13. F. Possmayer, Evidence for a specific phosphatidylinositol transferring protein in rat brain, *Brain Res. 74*:167–174 (1974).
14. L. W. Johnson, and D. B. Zilversmit, Catalytic properties of phospholipid exchange protein from bovine heart, *Biochim. Biophys. Acta 375*:165–175 (1975).
15. R. C. Crain and D. B. Zilversmit, Two nonspecific phospholipid exchange proteins from beef liver. I. Purification and characterization, *Biochemistry 19*:1433–1439 (1980).
16. H. H. Kamp and K. W. A. Wirtz, Phosphatidylcholine exchange protein from beef liver, *Meth. Enzymol. 32(Part B)*:140–146 (1974).
17. J. A. Hellings, H. H. Kamp, K. W. A. Wirtz, and L. L. M. van Deenen, Transfer of phosphatidylcholine between liposomes, *Eur. J. Biochem. 47*:601–605 (1974).
18. L. R. McLean and M. C. Phillips, Mechanism of cholesterol and phosphatidylcholine exchange or transfer between unilamellar vesicles, *Biochemistry 20*:2893–2900 (1981).
19. M. De Cuyper, M. Joniau, and H. Dangreau, Intervesicular phospholipid transfer. A free-flow electrophoresis study, *Biochemistry 22*:415–420 (1983).
20. C. Ehnholm and D. B. Zilversmit, Use of Forssman antigen in the study of phosphatidylcholine exchange between liposomes, *Biochim. Biophys. Acta 274*:652–657 (1972).
21. T. Sasaki and T. Sakagami, A new assay system of phospholipid exchange activities using

concanavalin A in the separation of donor and acceptor liposomes, *Biochim. Biophys. Acta* *512*:461–471 (1978).

22. A. M. Kasper and G. M. Helmkamp, Jr., Protein-catalyzed phospholipid exchange between gel and liquid-crystalline phospholipid vesicles, *Biochemistry 20*:146–151 (1981).
23. J. R. Wetterau and D. B. Zilversmit, Quantitation of lipid transfer activity, *Meth. Biochem. Anal. 30*:199–226 (1984).
24. F. J. Martin and R. C. MacDonald, Phospholipid exchange between bilayer membrane vesicles, *Biochemistry 15*:321–327 (1976).
25. G. Duckwitz-Peterlein, G. Eilenberger, and P. Overath, Phospholipid exchange between bilayer membranes, *Biochim. Biophys. Acta 469*:311–325 (1977).
26. M. A. Roseman and T. E. Thompson, Mechanism of the spontaneous transfer of phospholipids between bilayers, *Biochemistry 19*:439–444 (1980).
27. J. B. Massey, A. M. Gotto, Jr., and H. J. Pownall, Kinetics and mechanism of the spontaneous transfer of fluorescent phosphatidylcholines between apolipoprotein-phospholipid recombinants, *Biochemistry 21*:3630–3636 (1982).
28. J. B. Massey, A. M. Gotto, Jr., and H. J. Pownall, Kinetics and mechanism of the spontaneous transfer of fluorescent phospholipids between apolipoprotein-phospholipid recombinants. Effect of the polar headgroup, *J. Biol. Chem. 257*:5444–5448 (1982).
29. J. B. Massey, D. Hickson-Bick, D. P. Via, A. M. Gotto, Jr., and J. H. Pownall, Fluorescence assay of the specificity of human plasma and bovine liver phospholipid transfer proteins, *Biochim. Biophys. Acta 835*:124–131 (1985).
30. J. W. Nichols and R. E. Pagano, Kinetics of soluble lipid monomer diffusion between vesicles, *Biochemistry 20*:2783–2789 (1981).
31. J. W. Nichols and R. E. Pagano, Use of resonance energy transfer to study the kinetics of amphiphile transfer between vesicles, *Biochemistry 21*:1720–1726 (1982).
32. J. W. Nichols and R. E. Pagano, Resonance energy transfer assay of protein-mediated lipid transfer between vesicles, *J. Biol. Chem. 258*:5368–5371 (1983).
33. P. Somerharju, H. Brockerhoff, and K. W. A. Wirtz, A new fluorimetric method to measure protein-catalyzed phospholipid transfer using 1-acyl-2-parinaroylphosphatidylcholine, *Biochim. Biophys. Acta 649*:521–528 (1981).
34. T. Maeda and S. Ohnishi, Membrane fusion. Transfer of phospholipid molecules between phospholipid bilayer membranes, *Biochem. Biophys. Res Commun. 60*:1509–1516 (1974).
35. K. Machida and S. Ohnishi, A spin-label study of phosphatidylcholine exchange protein. Regulation of the activity by phosphatidylserine and calcium ion, *Biochim. Biophys. Acta 507*:156–164 (1978).
36. L. R. McLean and M. C. Phillips, Kinetics of phosphatidylcholine and lysophosphatidylcholine exchange between unilamellar vesicles, *Biochemistry 23*:4624–4630 (1984).
37. J. E. Ferrell, Jr., K. J. Lee, and W. H. Huestis, Lipid transfer between phosphatidylcholine vesicles and human erythrocytes: Exponential decrease in rate with increasing acyl chain length, *Biochemistry 24*:2857–2864 (1985).
38. J. W. Nichols, Thermodynamics and kinetics of phospholipid monomer-vesicle interaction, *Biochemistry 24*:6390–6398 (1985).
39. H. J. Pownall, D. L. Hickson, and L. C. Smith, Transport of biological lipophiles. Effect of lipophile structure, *J. Am. Chem. Soc. 105*:2440–2445 (1983).
40. C. Tanford, *The Hydrophobic Effect: Formation of Micelles and Biological Membranes*, 2nd ed., Wiley-Interscience, New York, 1980.
41. M. A. Gardam, J. J. Itovitch, and J. R. Silvius, Partitioning of exchangeable fluorescent phospholipids and sphingolipids between different lipid bilayer environments, *Biochemistry 28*:884–893 (1989).
42. T. Nakagawa, Critical examination of published data concerning the rate of micelle dissociation and proposal of a new interpretation, *Coll. Polymer Sci. 252*:56–64 (1974).

43. L. Thilo, Kinetics of phospholipid exchange between bilayer membranes, *Biochim. Biophys. Acta 469*:326–334 (1977).
44. M. C. Phillips, W. J. Johnson, and G. H. Rothblat, Mechansims and consequences of cellular cholesterol exchange and transfer, *Biochim. Biophys. Acta 906*:223–276 (1987).
45. A. Jonas and G. T. Maine, Kinetics and mechanism of phosphatidylcholine and cholesterol exchange between single bilayer vesicles and bovine serum high-density lipoprotein, *Biochemistry 18*:1722–1728 (1979).
46. G. E. Petrie and A. Jonas, Spontaneous phosphatidylcholine exchange between small unilamellar vesciles and lipid apolipoprotein complexes. Effects of particle concentration and compositions, *Biochemistry 23*:720–725 (1984).
47. J. M. Kremer, M. M. Kops-Werkhoven, C. Pathmamanoharan, O. L. Gijzeman, and P. H. Wiersema, Phase diagrams and the kinetics of phospholipid exchange for vesicles of different composition and radius, *Biochim. Biophys. Acta 471*:177–188 (1977).
48. J. D. Jones and T. E. Thompson, Spontaneous phosphatidylcholine transfer by collision between vesicles at high lipid concentration, *Biochemistry 28*:129–134 (1989).
49. J. D. Jones, and T. E. Thompson, Mechanism of spontaneous, concentration-dependent phospholipid transfer between bilayers, *Biochemistry 29*:1593–1600 (1990).
50. J. W. Nichols, Phospholipid transfer between phosphatidylcholine-taurocholate mixed micelles, *Biochemistry 27*:3925–3931 (1988).
51. D. A. Fullington, D. G. Shoemaker, and J. W. Nichols, Characterization of phospholipid transfer between mixed phospholipid-bile salt micelles, *Biochemistry 29*:879–886 (1990).
52. G. M. Helmkamp, Jr., Phospholipid transfer proteins: Mechanism of action, *J. Bioenerg. Biomembr. 18*:71–91 (1986).
53. D. B. Zilversmit, Lipid transfer proteins, *J. Lipid Res. 25*:1563–1569 (1984).
54. R. Akeroyd, P. Moonen, J. Westerman, W. C. Puyk, and K. W. A. Wirtz, The complete primary structure of the phosphatidylcholine-transfer protein from bovine liver. Isolation and characterization of the cyanogen bromide peptides, *Eur. J. Biochem. 114*:385–391 (1981).
55. T. Teerlink, T. P. Van der Krift, M. Post, and K. W. A. Wirtz, Tissue distribution and subcellular localization of phosphatidylcholine transfer protein in rats as determined by radioimmunoassay, *Biochim. Biophys. Acta 713*:61–67 (1982).
56. H. H. Kamp, K. W. A. Wirtz, and L. L. M. van Deenen, Some properties of phosphatidylcholine exchange protein purified from beef liver, *Biochim. Biophys. Acta 318*:313–325 (1973).
57. J. Westerman, K. W. A. Wirtz, T. Berkhout, L. L. M. van Deenen, R. Radhakrishnan, and H. G. Khorana, Identification of the lipid-binding site of phosphatidylcholine-transfer protein with phosphatidylcholine analogs containing photoactivable carbene precursors, *Eur. J. Biochem. 132*:441–449 (1983).
58. H. H. Kamp, K. W. A. Wirtz, P. R. Baer, A. J. Slotboom, A. F. Rosenthal, F. Paltauf, and L. L. M. van Deenen, Specificity of the phosphatidylcholine exchange protein from bovine liver, *Biochemistry 16*:1310–1316 (1977).
59. P. Child, J. J. Myher, F. A. Kuypers, J. A. Op den Kamp, A. Kuksis, and L. L. M. van Deenen, Acyl selectivity in the transfer of molecular species of phosphatidylcholines from human erythrocytes, *Biochim. Biophys. Acta 812*:321–332 (1985).
60. R. Welti and G. M. Helmkamp, Jr., Acyl chain specificity of phosphatidylcholine transfer protein from bovine liver, *J. Biol. Chem. 259*:6937–6941 (1984).
61. P. J. Somerharju, D. van Loon, and K. W. A. Wirtz, Determination of the acyl chain specificity of the bovine liver phosphatidylcholine transfer protein. Application of pyrene-labeled phosphatidylcholine species, *Biochemistry 26*:7193–7199 (1987).
62. K. W. A. Wirtz, P. F. Devaux, and A. Bienvenue, Phosphatidylcholine exchange protein catalyzes the net transfer of phosphatidylcholine to model membranes, *Biochemistry 19*:3395–3399 (1980).
63. A. M. Van den Besselaar, G. M. Helmkamp, Jr., and K. W. A. Wirtz, Kinetic model of the

protein-mediated phosphatidylcholine exchange between single bilayer liposomes, *Biochemistry 14*:1852–1858 (1975).

64. K. W. A. Wirtz, G. Vriend, and J. Westerman, Kinetic analysis of the interaction of the phosphatidylcholine exchange protein with unilamellar vesicles and multilamellar liposomes, *Eur. J. Biochem. 94*:215–221 (1979).
65. E. A. Runquist and G. M. Helmkamp, Jr., Effect of acceptor membrane phosphatidylcholine on the catalytic activity of bovine liver phosphatidylcholine transfer protein, *Biochim. Biophys. Acta 940*:21–32 (1988).
66. E. A. Runquist and G. M. Helmkamp, Jr., Design, synthesis, and characterization of bis-phosphatidylcholine, a mechanistic probe of phosphatidylcholine transfer protein catalytic activity, *Biochim. Biophys. Acta 940*:10–20 (1988).
67. Z. U. Khan and G. M. Helmkamp, Jr., Stimulation of cholinephosphotransferase activity by phosphatidylcholine transfer protein. Regulation of membrane phospholipid synthesis by a cytosolic protein, *J. Biol. Chem. 265*:700–705 (1990).
68. G. M. Helmkamp, Jr., M. S. Harvey, K. W. A. Wirtz, and L. L. M. van Deenen, Phospholipid exchange between membranes. Purification of bovine brain proteins that preferentially catalyze the transfer of phosphatidylinositol, *J. Biol. Chem. 249*:6382–6389 (1974).
69. P. E. DiCorleto, J. B. Warach, and D. B. Zilversmit, Purification and characterization of two phospholipid exchange proteins from bovine heart, *J. Biol. Chem. 254*:7795–7802 (1979).
70. R. H. Lumb, A. D. Kloosterman, K. W. A. Wirtz, and L. M. van Deenen, Some properties of phospholipid exchange proteins from rat liver, *Eur. J. Biochem. 69*:15–22 (1976).
71. S. E. Venuti and G. M. Helmkamp, Jr., Tissue distribution, purification and characterization of rat phosphatidylinositol transfer protein, *Biochim. Biophys. Acta 946*:119–128 (1988).
72. P. Y. George and G. M. Helmkamp, Jr., Purification and characterization of a phosphatidylinositol transfer protein from human platelets, *Biochim. Biophys. Acta 836*:176–184 (1985).
73. G. Daum and F. Paltauf, Phospholipid transfer in yeast, Isolation and partial characterization of a phospholipid transfer protein from yeast cytosol, *Biochim. Biophys. Acta 794*:385–391 (1984).
74. S. K. Dickeson, C. N. Lim, G. T. Schuyler, T. P. Dalton, G. M. Helmkamp, Jr., and L. R. Yarbrough, Isolation and sequence of cDNA clones encoding rat phosphatidylinositol transfer protein, *J. Biol. Chem. 264*:16557–16564 (1989).
75. P. A. Van Paridon, A. J. Visser, and K. W. A. Wirtz, Binding of phospholipids to the phosphatidylinositol transfer protein from bovine brain as studied by steady-state and time-resolved fluorescence spectroscopy, *Biochim. Biophys. Acta 898*:172–180 (1987).
76. D. Van Loon, T. A. Berkhout, R. A. Demel, and K. W. A. Wirtz, The lipid binding site of the phosphatidylcholine transfer protein from bovine liver, *Chem. Phys. Lipids 38*:29–39 (1985).
77. P. A. van Paridon, T. W. Gadella, Jr., P. J. Somerharju, and K. W. A. Wirtz, Properties of the binding sites for the *sn*-1 and *sn*-2 acyl chains on the phosphatidylinositol transfer protein from bovine brain, *Biochemistry 27*:6208–6214 (1988).
78. A. M. Kasper and G. M. Helmkamp, Jr., Intermembrane phospholipid fluxes catalyzed by bovine brain phospholipid exchange protein. *Biochim. Biophys. Acta 664*:22–32 (1981).
79. P. Somerharju, P. Van Paridon, and K. W. A. Wirtz, Phosphatidylinositol transfer protein from bovine brain. Substrate specificity and membrane binding properties, *Biochim. Biophys. Acta 731*:186–195 (1983).
80. P. A. Van Paridon, T. W. Gadella, Jr., and K. W. A. Wirtz, The effect of polyphosphoinositides and phosphatidic acid on the phosphatidylinositol transfer protein from bovine brain: A kinetic study, *Biochim. Biophys. Acta 943*:76–86 (1988).
81. J. Zborowski and R. A. Demel, Transfer properties of the bovine brain phospholipid transfer protein. Effect of charged phospholipids and of phosphatidylcholine fatty acid composition, *Biochim. Biophys. Acta 688*:381–387 (1982).

82. G. M. Helmkamp, Jr., Interaction of bovine brain phospholipid exchange protein with liposomes of different lipid composition, *Biochim. Biophys. Acta 595*:222–234 (1980).
83. J. F. Aitken, G. P. van Heusden, M. Temkin, and W. Dowhan, The gene encoding the phosphatidylinositol transfer protein is essential for cell growth, *J. Biol. Chem. 265*:4711–4717 (1990).
84. B. Bloj and D. B. Zilversmit, Rat liver proteins capable of transferring phosphatidylethanolamine. Purification and transfer activity for other phospholipids and cholesterol, *J. Biol. Chem. 252*:1613–1619 (1977).
85. B. Bloj, M. E. Hughes, D. B. Wilson, and D. B. Zilversmit, Isolation and amino acid analysis of a nonspecific phospholipid transfer protein from rat liver, *FEBS Lett. 96*:87–89 (1978).
86. B. Bloj and D. B. Zilversmit, Accelerated transfer of neutral glycosphingolipids and ganglioside GM1 by a purified lipid transfer protein, *J. Biol. Chem. 256*:5988–5991 (1981).
87. J. M. Trzaskos and J. L. Gaylor, Cytosolic modulators of activities of microsomal enzymes of cholesterol biosynthesis. Purification and characterization of a non-specific lipid-transfer protein, *Biochim. Biophys. Acta 751*:52–65 (1983).
88. B. J. Noland, R. E. Arebalo, E. Hansbury, and T. J. Scallen, Purification and properties of sterol carrier protein$_2$, *J. Biol. Chem. 255*:4282–4289 (1980).
89. R. Chanderbhan, B. J. Noland, T. J. Scallen, and G. V. Vahouny, Sterol carrier protein 2. Delivery of cholesterol from adrenal lipid droplets to mitochondria for pregnenolone synthesis, *J. Biol. Chem. 257*:8928–8934 (1982).
90. A. Pastuszyn, B. J. Noland, J. F. Bazan, R. J. Fletterick, and T. J. Scallen, Primary sequence and structural analysis of sterol carrier protein 2 from rat liver: Homology with immunoglobulins. *J. Biol. Chem. 262*:13219–13227 (1987).
91. J. Westerman and K. W. A. Wirtz, The primary structure of the nonspecific lipid transfer protein (sterol carrier protein$_2$) from bovine liver, *Biochem. Biophys. Res. Commun. 127*:333–338 (1985).
92. J.W. Nichols, Binding of fluorescent-labeled phosphatidylcholine to rat liver nonspecific lipid transfer protein, *J. Biol. Chem. 262*:14172–14177 (1987).
93. J. W. Nichols, Kinetics of fluorescent-labeled phosphatidylcholine transfer between nonspecific lipid transfer protein and phospholipid vesicles, *Biochemistry 27*:1889–1896 (1988).
94. A. Van Amerongen, T. Teerlink, G. P. van Heusden, and K. W. A. Wirtz, The non-specific lipid transfer protein (sterol carrier protein 2) from rat and bovine liver, *Chem. Phys. Lipids 38*:195–204 (1985).
95. F. M. Megli, A. De Lisi, A. van Amerongen, K. W. A. Wirtz, and E. Quagliariello, Nonspecific lipid transfer protein (sterol carrier protein 2) is bound to rat liver mitochondria: Its role in spontaneous intermembrane phospholipid transfer, *Biochim. Biophys. Acta 861*:463–470 (1986).
96. G. V. Vahouny, R. Chanderbhan, A. Kharroubi, B. J. Noland, A. Pastuszyn, and T. J. Scallen, Sterol carrier and lipid transfer proteins, *Adv. Lipid Res. 22*:83–113 (1987).
97. F. Schroeder, P. Butko, G. Nemecz, and T. J. Scallen, Interaction of fluorescent delta 5,7,9(11),22-ergostatetraen-3 beta-ol with sterol carrier protein-2, *J. Biol. chem. 265*:151–157 (1990).
98. A. Van Amerongen, M. van Noort, J. R. van Beckhoven, F. F. Rommerts, J. Orly, and K. W. A. Wirtz, The subcellular distribution of the nonspecific lipid transfer protein (sterol carrier protein 2) in rat liver and adrenal gland, *Biochim. Biophys. Acta 1001*:243–248 (1989).
99. N. Altamura and C. Landriscina, Effect of *N*-ethylmaleimide on rat liver phosphatidylcholine-specific and non-specific transfer protein activities: Its dependence on donor liposome composition, *Int. J. Biochem. 18*:513–517 (1986).

100. R. A. Zoeller and C. R. Raetz, Isolation of animal cell mutants deficient in plasmalogen biosynthesis and peroxisome assembly, *Proc. Natl. Acad. Sci. USA 83*:5170–5174 (1986).
101. G. P. Van Heusden, K. Bos, C. R. Raetz, and K. W. A. Wirtz, Chinese hamster ovary cells deficient in peroxisomes lack the nonspecific lipid transfer protein (sterol carrier protein 2), *J. Biol. Chem. 265*:4105–9110 (1990).
102. T. Teerlink, T. P. Van der Krift, G. P. van Heusden, and K. W. A. Wirtz, Determination of nonspecific lipid transfer protein in rat tissues and Morris hepatomas by enzyme immunoassay, *Biochim. Biophys. Acta 793*:251–259 (1984).
103. A. Van Amerongen, J. B. Helms, T. P. van der Krift, R. B. Schutgens, and K. W. A. Wirtz, Purification of nonspecific lipid transfer proteins (sterol carrier protein 2) from human liver and its deficiency in livers from patients with cerebro-hepato-renal (Zellweger) syndrome, *Biochim. Biophys. Acta 919*:149–155 (1987).
104. G. A. Keller, T. J. Scallen, D. Clarke, P. A. Maher, S. K. Krisans, and S. J. Singer, Subcellular localization of sterol carrier protein-2 in rat hepatocytes: Its primary localization to peroxisomes, *J. Cell Biol. 108*:1353–1361 (1989).
105. M. Tsuneoka, A. Yamamoto, Y. Fujiki, and Y. Tashiro, Nonspecific lipid transfer protein (sterol carrier protein-2) is located in rat liver peroxisomes, *J. Biochem. (Tokyo) 104*:560–564 (1988).
106. Y. Fujiki, M. Tsuneoka, and Y. Tashiro, Biosynthesis of nonspecific lipid transfer proptein (sterol carrier protein 2) on free polyribosomes as a larger precursor in rat liver, *J. Biochem. (Tokyo) 106*:1126–1131 (1989).
107. W. H. Trzeciak, E. R. Simpson, T. J. Scallen, G. V. Vahouny, and M. R. Waterman, Studies on the synthesis of sterol carrier protein-2 in rat adrenocortical cells in monolayer culture. Regulation by ACTH and dibutyryl cyclic 3',5'-AMP, *J. Biol. Chem. 262*:3713–3717 (1987).
108. B. C. Ossendorp, G. P. van Heusden, and K. W. A. Wirtz, The amino acid sequence of rat liver non-specific lipid transfer protein (sterol carrier protein 2) is present in a high molecular weight protein: Evidence from cDNA analysis, *Biochem. Biophys. Res. Commun. 168*:631–636 (1990).
109. S. L. Thompson, R. Burrows, R. J. Laub, and S. K. Krisans, Cholesterol synthesis in rat liver peroxisomes. Conversion of mevalonic acid to cholesterol, *J. Biol. Chem. 262*:17420–17425 (1987).
110. S. K. Krisans, S. L. Thompson, L. A. Pena, E. Kok, and N. B. Javitt, Bile acid synthesis in rat liver peroxisomes: Metabolism of 26-hydroxycholesterol to 3 beta-hydroxy-5-cholenoic acid, *J. Lipid Res. 26*:1324–1332 (1985).
111. R. C. Crain and D. B. Zilversmit, Two nonspecific phospholipid exhcange proteins from beef liver. 2. Use in studying the asymmetry and transbilayer movement of phosphatidylcholine, phosphatidylethanolamine, and sphingomyelin in intact rat erythrocytes, *Biochemsitry 19*:1440–1447 (1980).
112. R. C. Crain, Nonspecific lipid transfer proteins as probes of membrane structure and function, *Lipids 17*:935–943 (1982).
113. P. North and S. Fleischer, Alteration of synaptic membrane cholesterol/phospholipid ratio using a lipid transfer protein. Effect on gamma-aminobutyric acid uptake, *J. Biol. Chem. 258*:1242–1253 (1983).
114. J. Lunardi, P. DeFoor, and S. Fleischer, Modification of phospholipid environment in sarcoplasmic reticulum using nonspecific phospholipid transfer protein, *Meth. Enzymol. 157*:369–377; 693 (1988).
115. T. E. Thompson, Transmembrane compositional asymmetry of lipids in bilayers and biomembranes, in *Molecular Specialization and Symmetry in Membrane Function* (A. K. Solomon and M. Karnovsky, eds.), Harvard University Press, Cambridge, Mass. 1978, pp. 78–98.

116. R. D. Kornberg and H. M. McConnell, Inside outside transitions of phospholipids in vesicle membranes, *Biochemistry* 10:1111–1120 (1971).
117. M. Roseman, B. J. Litman, and T. E. Thompson, Transbilayer exchange of phosphatidylethanolamine for phosphatidylcholine and *N*-acetimidoylphosphatidylethanolamine in single walled bilayer vesicles, *Biochemistry 14*:4826–4830 (1975).
118. L. W. Johnson, M. E. Hughes, and D. B. Zilversmit, Use of phospholipid exchange protein to measure inside outside transposition in phosphatidylcholine liposomes, *Biochim. Biophys. Acta 375*:176–185 (1975).
119. J. E. Rothman and E. A. Dawidowicz, Asymmetric exchange of vesicle phospholipids catalyzed by the phosphatidylcholine exchange protein. Measurement of inside-outside transitions, *Biochemistry* 14:2809–2816 (1975).
120. J. M. Shaw, W. C. Hutton, B. R. Lentz, and T. E. Thompson, Proton nuclear magnetic resonance study of the decay of transbilayer compositional asymmetry generated by a phosphatidylcholine exchange protein. *Biochemistry 16*:4156–4163 (1977).
121. R. Homan and H. J. Pownall, Transbilayer diffusion of phospholipids: Dependence on headgroup structure and acyl chain length, *Biochim. Biophys. Acta 938*:155–166 (1988).
122. B. De Kruijff and E. J. Van Zoelen, Effect of the phase transition on the transbilayer movement of phosphatidylcholine in unilamellar vesicles, *Biochim. Biophys. Acta 511*:105–115 (1978).
123. B. De Kruijff, E. J. van Zoelen, and L. L. M. van Deenen, Glycophorin facilitates the transbilayer movement of phosphatidylcholine in vesicles, *Biochim. Biophys. Acta 509*:537–542
124. W. R. Bishop and R. M. Bell, Assembly of phospholipids into cellular membranes: Biosynthesis, transmembrane movement and intracellular translocation, *Annu. Rev. Cell. Biol. 4*:579–610 (1988).
125. D. B. Zilversmit and M. E. Hughes, Extensive exchange of rat liver microsomal phospholipids, *Biochim. Biophys. Acta* 469:99–110 (1977).
126. A. M. van den Besselaar, B. de Druijff, H. van den Bosch, and L. L. M. van Deenen, Phosphatidylcholine mobility in liver microsomal membranes, *Biochim. Biophys. Acta 510*:242–255 (1978).
127. Y. Kawashima and R. M. Bell, Assembly of the endoplasmic reticulum phospholipid bilayer. Transporters for phosphatidylcholine and metabolites, *J. Biol. Chem. 262*:16495–16502 (1987).
128. A. Herrmann, A. Zachowski, and P. F. Devaux, Protein-mediated phospholipid translocation in the endoplasmic reticulum with a low lipid specificity, *Biochemistry 29*:2023–2027 (1990).
129. A. Sune, P. Bette-Bobillo, A. Bienvenue, P. Fellmann, and P. F. Devaux, Selective outside inside translocation of aminophospholipids in human platelets, *Biochemistry 26*:2972–2978 (1987).
130. A. Zachowski, E. Favre, S. Cribier, P. Herve, and P. F. Devaux, Outside-inside translocation of aminophospholipids in the human erythrocyte membrane is mediated by a specific enzyme, *Biochemistry* 25:2585–2590 (1986). Note also erratum in *Biochemistry 25*:7788
131. A. Zachowski, A. Herrmann, A. Paraf, and P. F. Devaux, Phospholipid outside-inside translocation in lymphocyte plasma membranes is a protein mediated phenomenon, *Biochim. Biophys. Acta 897*:197–200 (1987).
132. O. C. Martin, and R. E. Pagano, Transbilayer movement of fluorescent analogs of phosphatidylserine and phosphatidylethanolamine at the plasma membrane of cultured cells. Evidence for a protein-mediated and ATP-dependent process(es), *J. Biol. Chem. 262*:5890–5898 (1987).
133. M. Vidal, J. Sainte Marie, J. R. Philippot, and A. Bienvenue, Asymmetric distribution of phospholipids in the membrane of vesicles released during in vitro maturation of guinea pig

reticulocytes: evidence precluding a role for "aminophospholipid translocase," *J. Cell. Physiol. 140*:455–462 (1989).

134. R. F. Zwaal, E. M. Bevers, and J. Rosing, Regulation and function of transbilayer movement of phosphatidylserine in activated blood platelets and sickle cells, *Prog. Clin. Biol. Res. 282*:181–192 (1988).

135. M. Bitbol, P. Fellmann, A. Zachowski, and P. F. Devaux, Ion regulation of phosphatidylserine and phosphatidylethanolamine outside-inside translocation and phosphatidylethanolamine outside-inside translocation in human erythrocytes, *Biochim. Biophys. Acta 904*:268–282 (1987).

136. J. Connor and A. J. Schroit, Transbilayer movement of phosphatidylserine in nonhuman erythrocytes: Evidence that the aminophospholipid transporter is a ubiquitous membrane protein, *Biochemistry 28*:9680–9685 (1989).

137. P. F. Devaux, Phospholipid flippases, *FEBS Lett. 234*:8–12 (1988).

138. E. Middelkoop, A. Coppens, M. Llanillo, E. E. Van der Hoek, A. J. Slotboom, B. H. Lubin, J. A. Op den Kamp, L. L. M. van Deenen, B. Roelofsen, Aminophospholipid translocase in the plasma membrane of Friend erythroleukemic cells can induce an asymmetric topology for phosphatidylserine but not for phosphatidylethanolamine, *Biochim. Biophys. Acta 978*:241–248 (1989).

139. G. Morrot, P. Herve, A. Zachowski, P. Fellmann, and P. F. Devaux, Aminophospholipid translocase of human erythrocytes: Phospholipid substrate specificity and effect of cholesterol, *Biochemistry 28*:3456–3462 (1989).

140. A. J. Schroit, J. Madsen, and A. E. Ruoho, Radioiodinated, photoactivatable phosphatidylcholine and phosphatidylserine: Transfer properties and differential photoreactive interaction with human erythrocyte membrane proteins, *Biochemistry 26*:1812–1819 (1987).

141. A. Zachowski, P. Fellmann, P. Herve, and P. F. Devaux, Labeling of human erythrocyte membrane proteins

20

Magnetic Resonance Studies of Phospholipid–Protein Interactions in Bilayers

Anthony Watts *Oxford University, Oxford, England*

I. INTRODUCTION

Much evidence exists to indicate that complete functioning of a biomembrane is controlled by both the protein and the lipid, mainly phospholipid, components. The study of protein-lipid interactions in membranes is usually therefore the identification and description at the molecular level of a detectable interaction between one component and the other and, in the ideal case, a means of interpreting the structural information in such a way as to enable a description of a functional consequence of such structural interactions to be made.

It is true to say that in most biological membranes in viable cells, the composition and activity of the membrane is at least adequate, if not optimized, for the particular function being performed. To elucidate any intermolecular interactions between the various components, it is frequently necessary to perturb or probe the system to be able to describe the nature, origin and type of molecular interactions involved. This is nowhere more true than in magnetic resonance studies of protein-lipid interactions in membranes, where the spectroscopic property to be observed and exploited to report on the interaction, at the molecular level, may have to be introduced into the membrane, or even more drastically the membrane may have to be built around the reporter [1].

Information about the protein-lipid interactions in membranes is necessary if details at the intimate molecular level about the interface between a membrane protein and the bilayer are to be described [1,2,3]. Such details will permit a description of how an integral transmembrane spanning protein is sealed into a membrane. Most ion gradients set up by active transport protein are subsequently used for driving secondary transport processes. If the ion gradients are then lost too quickly by nonspecific leakage that could occur through membrane regions at the protein-lipid interface, then energy (in the form of the potential energy due to the ion gradient) will be lost unnecessarily and not be converted into useful work. Molecular mismatch between the bilayer and the transport species does exist simply as a result of their different architecture and spatial character, and so this interface could be one site of nonspecific leakage.

Another consequence of protein-lipid interactions may be as a result of the mutual dynamic influence of one component on the other. It is possible that a *fluidity window* is required within the bilayer part of a membrane for the proteins to undergo the requisite rates and degrees of molecular motion around the active site for biochemical activity to take place. Some proteins (type II, ion translocating ATPases, for example) undergo significant conformational changes during ion translocation [4]; this gross molecular rearrangement may not be possible in a solid matrix. It is the lipid component of such bilayers that provide this *fluidity window*, and changes in this component can thereby alter the protein activities. Such interactions are likely to occur within the bilayer hydrophobic core, whose chemistry cannot readily or quickly be altered in most situations, although head group interactions with ions and charged species can modulate the bilayer chain motions rather dramatically and quickly [5,6].

Any hydrophobic molecule that partitions into a biomembrane will diffuse rapidly throughout the bilayer, and the site of action for a lipophile that can perturb or modulate the biological activity of a membrane could well be the molecular interface between the protein and the bilayer [7]. Such a region of mismatch could be not only a site for the concentration of lipophiles but also the site of entry into a membrane spanning protein. Perturbation of the protein activity may be a result of direct competition with lipid sensitive sites, which themselves may be responsible for maintaining the 2°, 3°, or 4° structure of the protein in a form ready for activity.

Spectroscopic studies of protein-lipid interactions are usually carried out to identify the structural aspects of the associations of lipids with proteins. This is achieved by exploiting the different motional properties of the lipids in the bulk bilayer from those at the protein-bilayer interface that reveal themselves in the magnetic resonance spectrum and can be analyzed quantitatively, as will be discussed later. Too little effort has, unfortunately, been directed at correlating the structural and functional aspects of the information gained. The methods for performing magnetic resonance studies of protein-lipid interactions are now relatively well established since the pioneering work of Griffith and Jost [8]. The field has progressed to a sophisticated level, and with suitable instrumentation and computational facilities these kinds of studies are now readily performed. However, the major difficulty is frequently in controlling the biochemistry of the systems under study; preventing protein aggregation, resolving reconstitution technology where required, and ensuring that the protein is active and correctly folded under the experimental conditions are all essential prerequisites in such studies.

In this "technical" chapter on the study of protein-lipid interactions, the information that can be gained from such studies will be presented together with the requirements, at the instrumental, chemical, and biochemical level, for carrying out these experiments. The different ways of analyzing the information gained and the ways in which this information can be assessed will be discussed. Pertinent references will be made, but no attempt will be made to produce a comprehensive review (see, for example, Refs. 2, 9–16) or to present a laboratory manual for performing magnetic resonance experiments (see for example, Ref. 17 for spin label ESR and Ref. 18 for solid state NMR).

II. INFORMATION AVAILABLE FROM STUDIES OF PHOSPHOLIPID-PROTEIN INTERACTIONS

A. Proportion of Phospholipids Interacting with a Protein

In a bilayer membrane that contains a heterogeneous distribution of both peripheral and integral proteins, there will be a certain proportion of the phospholipids interacting with the protein component to give the membrane its integrity at both the structural and the functional level. Thus the proportion of phospholipids in the bilayer interacting with protein at any one time is dictated by protein density, protein type, protein size, and aggregation state of the proteins.

1. Influence of Protein Type in a Bilayer

Protein-phospholipid interactions can occur either at the hydrophobic central bilayer core or at the bilayer surface, or at both, depending upon the protein type. The proportion of the lipids at a protein hydrophobic interface at any one time will be determined, for an integral protein, by

The size of that part of the protein in the bilayer
The number of integral proteins in the bilayer

and for a peripheral protein held by electrostatic interactions, with no hydrophobic core penetration, the proportion could be determined by

The number of phospholipids within an area covered by the protein
The number of phospholipids involved in charged amino acids-phospholipid associations

and by a combination of these factors for penetrating peripheral proteins. In addition, it is not yet known how strongly the oligosaccharide portions of a glycosylated protein interact with the membrane phospholipids, although model studies do suggest some change in bilayer properties induced by surface interaction of the oligosaccharides [19].

Although the number of phospholipids and proteins in the bilayer may be given simply by the lipid-protein ratio, as either a weight or a mole ratio, this simple parameter is clearly not the one single determinant of the proportion of phospholipids interacting with protein, especially in membranes with a heterogeneous mix of proteins. A simple relationship between molecular weight and size does not hold, as seen for proteins such as the acetyl choline receptor (M_r ~280,000), which has 20% of its mass in the bilayer, and bacteriorhodopsin (M_r ~ 26,000), which has >90% of its mass in the bilayer. It is for these reasons that the majority of studies of protein-lipid interactions have been carried out on reconstituted bilayer membranes where the protein type, and hence the proportion in the membrane core, is controlled. Only natural membranes that are already of a well-defined composition and of a relatively high degree of predictable homogeneity, can such studies be confidentially performed, an example being the disc membrane from mammalian retina which contains >95% of monomeric rhodopsin when unbleached; ~60% of its mass is in the bilayer as a seven α-helix bundle.

2. Effects of Protein Aggregation

Since the proportion of phospholipids in a membrane interacting with a protein is determined directly by the size of the interfacial region between proteins and bilayer, it is clear that if proteins are aggregated, then potential protein-lipid contacts are occluded by protein-protein contacts, as shown diagramatically in Fig. 1. Thus any studies of protein-lipid interactions need to be viewed with some caution if the protein has a tendency to aggregate, especially if this occurs in reconstituted complexes in an uncontrolled, irreversible way, and where the protein density may need to be rather high to enable a phospholipid-protein interaction to be detected.

In some natural membranes, the integral proteins are naturally aggregated in a well-defined array, such as for bacteriorhodopsin in the purple membrane of *H. halobium* [20]. Thus, although the hydrophobic perimeter of a monomer of rhodopsin and of bacteriorhodopsin are similar, the degree and number of lipids interacting with the monomeric rhodopsin will be greater than for the trimerically arranged bacteriorhodopsin,

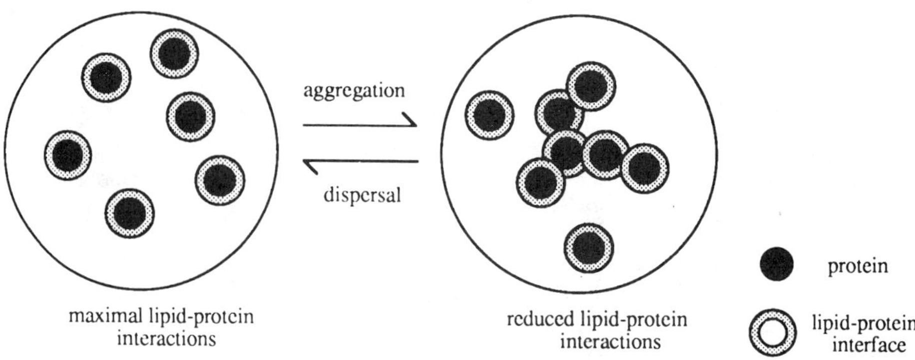

Figure 1 Diagrammatic representation of the way in which protein-lipid contacts in a membrane containing integral membrane proteins will be reduced through induced protein-protein contacts in either an aggregated protein sample or on assembly of a multisubunit complex. Such associations may be induced by temperature, protein density, or chemical means.

where protein-protein contacts reduce lipid accessibility to the protein interface, assuming similar conditions of protein-lipid ratio, which does not of course exist in the natural membranes for these two proteins. The reversible temperature dependent association of band 3 from erythrocytes has been demonstrated by NMR [21], and the control of bacteriorhodopsin array formation by lipids from electron microscopic methods [22].

Therefore, in any system used for the study of lipid-protein interactions, especially reconstituted membranes, some knowledge of the state of protein aggregation, perhaps from freeze fracture electron microscopy or rotational diffusion measurements, is essential.

3. Stoichiometry of Phospholipid-Protein Interactions

When considering the stoichiometry of a protein-lipid interaction, it is usually assumed that it is the number of phospholipids that are at the interface of one protein monomer which is being determined. This interface could be the hydrophobic surface of an integral protein in the hydrophobic core, or the contact area of a peripheral protein on the membrane surface.

At such a hydrophobic interface, there is inevitably a large motional differential between the protein motion and the lipid motion, whether this is of the head group of or the acyl chains. It is this phenomenon that is exploited in determining the number of phospholipids interacting with a protein using methods that are able to distinguish this dynamic difference induced in phospholipid molecular motion, as will be detailed below.

4. Effect of Protein Size and Lateral Diffusion

Since it is the *differential* lateral and rotational motion between a protein and the lipids that is detected and quantitated in spectroscopic studies of protein-lipid interactions, the degree of the protein-lipid intraction is determined to some degree by the size of the protein (in the absence of any restraining mechanism); the larger the protein mass in or on the membrane, the greater the degree of motional restriction of the adjacent lipid and hence the more readily the protein-lipid interaction can be monitored; $D_L \sim \alpha$ (radius of the molecule)$^{-1}$ [23]. Most proteins have been shown to diffuse laterally with values for D_L of 10^{-9} (for small monomeric proteins) to 10^{-11} cm^2 s^{-1} or slower (for larger proteins), depending upon the lipid in the bilayer, temperature, etc., whilst phospholipids diffuse at $\sim 5 \times 10^{-8}$–10^{-7} cm^2 s^{-1} in liquid crystalline bilayers [23].

A single α helix passing through a bilayer membrane ($\sim$23 or 25 amino acids and M_r $\sim$2,500–3,000) has a diameter of about 0.8–1 nm, depending upon side chain extension, which is similar to the long dimension of the cross-section of a diacyl phospholipid ($\sim$0.9–1.0 nm). In the absence of any significant lateral restriction of such a peptide individual helix, the lateral and rotational motion of the peptide will be similar to that for the lipids, and the possibility of detecting the differential motion between the two molecules is small. Nondynamic methods, such as differential scanning calorimetry, do detect the decrease in the cooperativity of the thermotropic phase transitions in bilayers of phospholipids with saturated acyl chains [24]. Also, methods in which the rate of acyl chain motion can be sensitively determined show the presence of small proteins through some slowing down of lipid acyl chain motion.

A monomeric protein with a diameter of about 3.5 nm and relative molecular mass for the hydrophobic part of the protein of $\sim$ 20,000, such as mammalian rhodopsin, has a lateral diffusion coefficient of 3×10^{-8} cm^{2}s^{-1} in a "fluid" membrane and restricts the motion of lipid acyl chains sufficiently that the differential motion of chains at the protein interface and in the bulk bilayer can be detected. For still larger proteins, or protein aggregates, similar "boundary" lipid can be detected, but in this case much slower exchange of lipids onto and off the interface would be anticipated.

The major structural element of the transmembrane part of many integral proteins is the α helix bundle [25, 26], and the disposition and packing of such helices (with M_r $\sim$ 3000/helix) determine rather precisely, in many cases, the degree of protein-lipid interaction. In Fig. 2, the number of lipids that can interact with a number of helices is shown, assuming maximum contact between the hydrophobic part of the amphiphilic helix surface and the lipid chains (that is, with the helices in a row of pairs rather than close packed). A single helix anchors glycophorin in the bilayer [28], and a seven helix bundle a postulated common motif in other receptors [27], is present in rhodopsin and bacteriorhodopsin.

Although there is little difference in the number (n_b) of lipid monomers immediately next to the protein and hence in the first lipid shell at the protein-lipid interface for 2–5 helices, the packing of the helices can alter the value of n_b. At higher helix numbers, as shown in Fig. 3 for up to 12 helices, no monomeric protein has yet been found to contain more than $\sim$13 transmembrane sections from prediction methods, this being band 3 from erythrocytes [29,30]. There are few exceptions to these general observations, and even small single helix proteins, which are self-associated in larger aggregates, do interact with lipid acyl chains in much the same way as for a large protein complex.

The stoichiometry of the phospholipid-protein interaction is thus given by n_b, the number of lipids around the protein hydrophobic interface. This parameter is related to the hydrophobic perimeter of the integral protein and can be used to estimate the hydrophobic diameter d of a protein in the bilayer. Thus

$$d = \frac{(n_b \times 0.96)}{\pi} \text{ nm} \tag{1}$$

where the contact face of a phospholipid is 0.96 nm and the value of n_b is for two halves of the bilayer [9]. This method of estimating protein size has proved successful in the cases for which it is valid (a monomeric, or at least low aggregation number, freely mobile protein) and for which structural information is available from independent techniques and

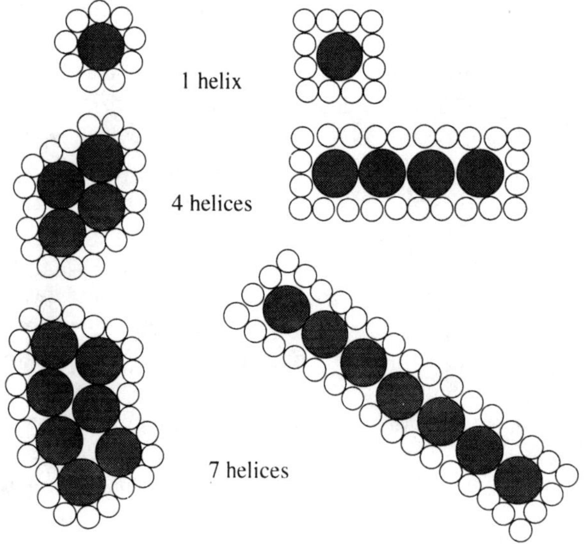

Figure 2 Scale diagram to show the possible packing of transmembrane protein helices with phospholipid acyl chains in a bilayer. Helix packing is shown as either close packed to minimize protein-lipid contact, as with bacteriorhodopsin, or linear to maximize protein-lipid contact, as with phenol-extracted M13 bacteriophage coat protein. Diameter of a lipid acyl chain ~0.48 nm; diameter for a transmembrane α helix ~ 1.0 nm.

can be used to check the method. If the analysis does not hold, then it may be that the protein is associated into aggregates, occluding protein-lipid sites through protein-protein contacts. In principle, then, it should be possible to estimate the degree of protein aggregation if the size of a monomer is known.

B. Protein Function

The study of protein-phospholipid interactions is best described in structural terms that can be related to functional consequences. This comes from the fact that it is the *chemistry* of a biological macromolecule that defines its *structure* and ultimately, through *motion* of specific groups, the *function* [31]. In some cases, site-directed mutagenesis of membrane proteins [32], and in others, alteration of the membrane phospholipid components by genetic means [33], lead to functional changes that confirm this link between chemistry and function, and they provide direct evidence that membrane phospholipids can influence membrane function in a natural membrane.

In studies of phospholipid-protein interactions in membranes using magnetic resonance methods, either isolated membrane fragments are used in which the protein composition is well defined, or proteins are reconstituted into phospholipid bilayers. In either case, the only real criterion that the information being obtained is of direct relevance to the physiological situation is to show that the protein is active. This may be simple in the case of a protein that possesses a chromophore where the absorption spectrum is a sensitive measure of the protein conformation and activity, but it may be less

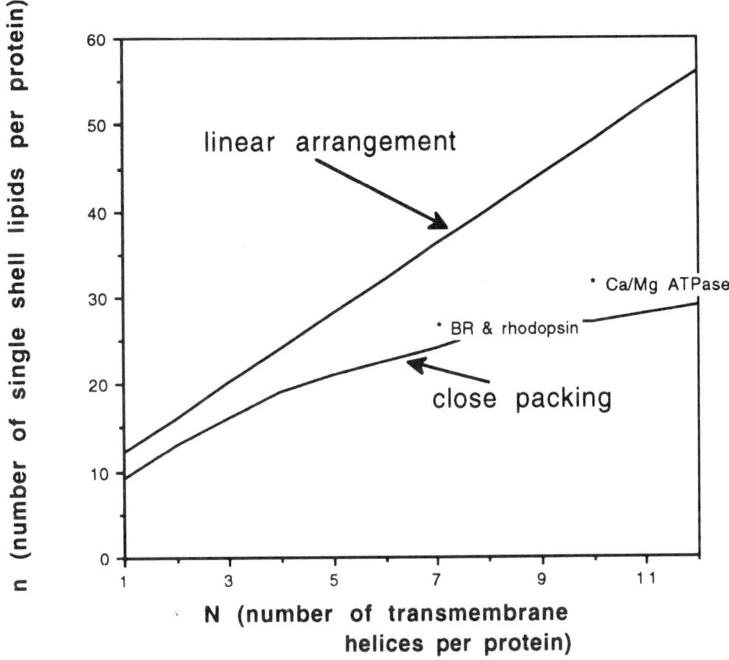

Figure 3 The variation of the number n_b of phospholipids/protein interacting with a monomeric protein composed predominantly of transmembrane helices; dimensions used are the same as in Fig. 2. The lower curve represents the minimum number n_b of lipid acyl chains that interact with one helix bundle (with N helices) because of close packing to the exclusion of lipids. The relationship represented in the upper line is for maximum interaction of the acyl chains with the protein with N helices linearly arranged. The values of n_b determined by spin label ESR methods [13] for rhodopsin (seven helices in a bundle) and the acetyl choline receptor (assuming five subunits of four transmembrane units each) are shown. All integral membrane proteins comprised of N transmembrane α helices will interact with n_b lipid chains, where the value of n_b for any value of N will lie between the two curves.

easy for a protein that has a coupled function, for example an ion translocating ATPase where translocation and ATP hydrolysis need to be coupled to demonstrate full functionality. Less easy again is the demonstration that a protein is in a "natural" folded conformation when it is a structural protein with no measurable biochemical function. Here, efforts need to be made to ensure that the correct structural features exist in the reconstituted protein, such as α-helical content, protease accessibility, etc., which can be compared with the protein in its native environment.

Little biochemically useful information will be gained from studies of phospholipid-protein interactions that involve a denatured or inactive protein.

C. Phospholipid Exchange into and out of the Protein Interface

Phospholipids laterally diffuse through bilayers as a result of acyl chain *trans-gauche* rotational isomerisms, and when they contact the protein interface, their lateral diffusion is reduced. Although protein surfaces are not rigid, indeed they have been described as "squishy" [34], with the amino acid side groups undergoing rapid motions such as ring flipping (τ_c ~ns) and methylene group wobbling (τ_c ~ns), the overall backbone motion is much slower (τ_c ~ms) than lipid chain motion (τ_c ~ns-ps) [35].

For large proteins, lipids exchange onto and off the protein interface about an order of magnitude or more slower than from one lipid site to another [9,13]. For smaller proteins and transmembrane peptides, where the protein and lipid lateral rates are similar, the positional exchange of each component is similar throughout the membrane.

One important situation arises with reversibly aggregated proteins. Here, the exchange of lipids onto and off the large protein aggregate interface, as with any large protein, will be slowed when compared to lipid-lipid exchange. Either the lipids exchange, freely but slowly, from the aggregate interface, or alternatively lipids are trapped in between aggregates and their exchange is limited by the dissociation and then reassociation of the proteins into the aggregate. Freely diffusing lipids in a bilayer, if they are observed to be interacting with the protein aggregate, must be able to penetrate, and exit from, sites in between the protein units. Irreversible aggregation (for longer than, say, seconds, which is a very long time on a magnetic resonance time-scale) clearly can accommodate such lipids in similar interstitial sites, but they will be locked in an aggregate. Thus information about lipid-protein exchange rates can be useful in describing the protein aggregation properties, which themselves may be determined by bilayer properties such as bilayer thickness or protein-protein interactions. Aggregation itself may be determined by lipid-protein interactions involving specific phospholipid types [22].

D. Order of Interfacial Lipid

The surface of integral proteins are not smooth but highly invaginated and rough. The intrinsic order observed in protein-free phospholipid bilayers is therefore reduced when a protein is inserted into a bilayer [9,36]. It is not clear whether the opposite is true in that the order of bilayer lipids is transmitted in any way to the protein side chain dynamics or order, or to the helix-helix arrangement or even subunit interaction; this would be very interesting information to have. It has, however, been shown that bilayer "fluidity" can alter the activity of a membrane protein; rigid bilayers normally reduce or inhibit protein function, and "fluid" bilayers usually permit or enhance protein activity.

To determine the order of interfacial lipid, a spectroscopic method sensitive to motional anisotropy is required, and those lipids interacting with the protein interface need to be examined to see if they are ordered or disordered when compared with the bulk lipids; deuterium NMR and spin label ESR (see later) are able to reveal such information.

E. Selectivity of Phospholipid-Protein Interaction

Molecular selectivity of the protein-phospholipid interface manifests itself in both functional [37] and structural investigations [22]. In the absence of selectivity, the lipids act as solvating species (Fig. 4), maintaining the protein in a suitable form for activity and mobility. Selective interactions are more likely to be determined at the polar-apolar membrane interface, with the type of phospholipid head group being the major determinant of such selectivity, either electrostatically as shown for the Na^+/K^+-ATPase [38] or possibly sterically.

No clear examples yet exist that demonstrate selectively of acyl chains for a protein-lipid interface, although some selective motional changes occur with C22:6 acyl chains on PC when interacting with rhodopsin, when compared with C16 chains [39], as shown by NMR studies.

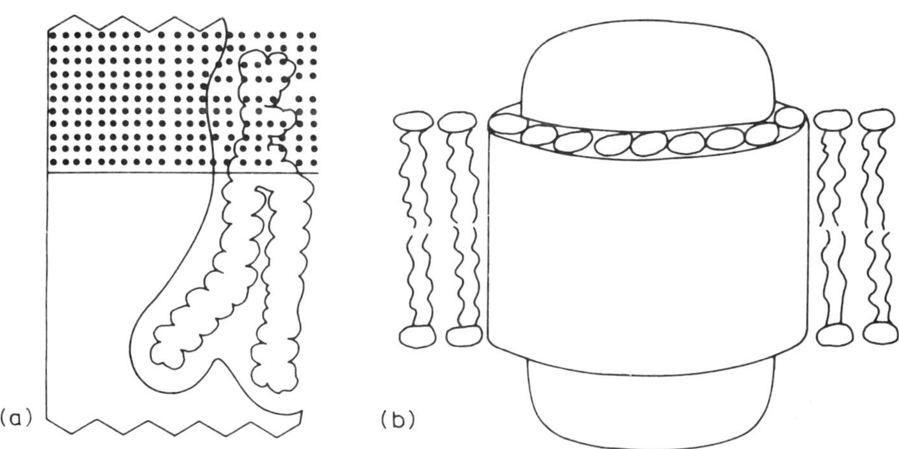

Figure 4 Phospholipids may interact with protein interfaces in selective or nonselective ways. (a) A low specificity site for protein-lipid interaction on a single binding site may occur through associations at the polar-apolar interface of the membrane between polar amino acids and the phospholipid head group. Protein function may be activated by such an interaction. (b) Lipids that occupy multiple binding sites of low chemical specificity around the protein may act as simple solvation molecules, maintaining the protein in allosteric control necessary for activity and preventing aggregation. (Figure from Ref. 37.)

III. EXPERIMENTAL AND ANALYTICAL METHODS

Magnetic resonance methods have their own intrinsic time-scale sensitivity, with, in general, NMR being sensitive to slower molecular motions than spin label ESR (Fig. 5). This arises from the difference in the lifetime of the excited state for each technique, nuclear excitation by a radio frequency in a magnetic field lasting longer ($\sim 10^3$ times longer) before relaxation occurs than for an electron by microwave radiation. However, there are various features of NMR and ESR that themselves are affected by molecular motions and occur in different time domains. Some aspects of the two methods overlap in their sensitivity to molecular motion; for example, spin-lattice relaxation for deuterium NMR and anisotropy averaging for the nitroxide spin label nitrogen-electron hyperfine (coupling) interaction are both sensitive to fast ($\tau_c \sim 10^{-9}$ s) motions characteristic of phospholipid acyl chain motion in bilayers.

The choice of the method to be employed in the study of phospholipid-protein interactions is therefore determined by

The kind of information required (rates or amplitudes of molecular motion or orientation about which the motion occurs)
The time-scale of the motions being investigated
The type and nature of the sample available

The sample preparation and nature may often restrict the choice of magnetic resonance method, and therefore this will be discussed first. If sample type is suitable for both NMR and spin label ESR studies, then in principle, complementary information can be obtained by using both methods, and cases where this has been successfully done will be presented.

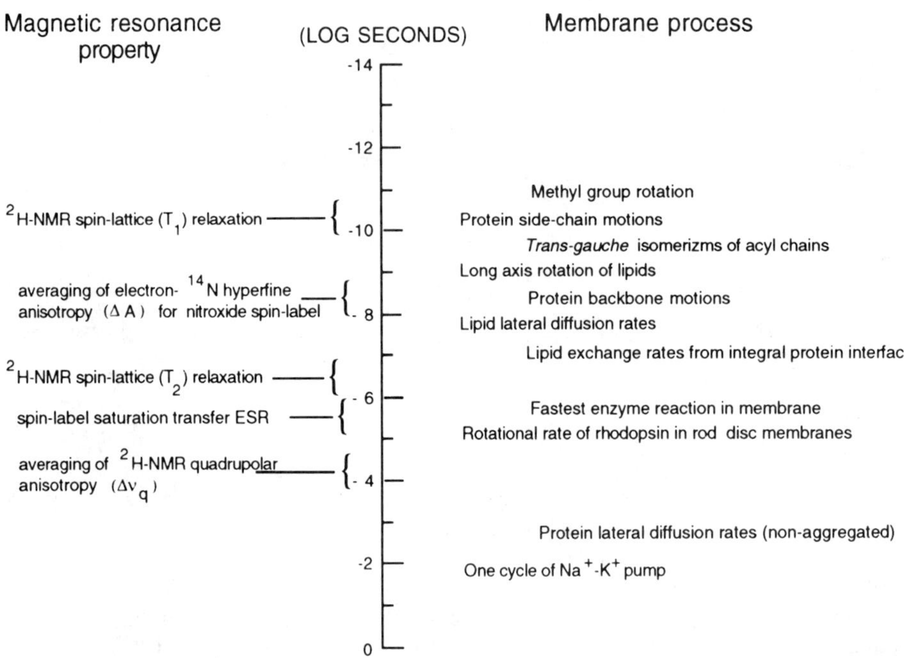

Figure 5 A number of dynamic processes within biological membranes (right) can be accessed by judicious exploitation of an appropriate magnetic resonance property (left). The interaction of lipids with proteins in membranes that can be studied by such methods are those that alter the motional (either rate or amplitude) or conformational properties of the lipid chains or head groups and can thus be resolved by spectral changes that reflect those perturbations. At any one time, a number of motional changes over a range of dynamic time scales may be induced in a lipid upon interaction with a protein (and vice versa), and thus the spectroscopic means for detecting and quantifying that interaction must be chosen rationally. In addition, the different kinds of mutual intermolecular perturbations of lipid and protein upon interaction may be transferred through the membrane by lipid or protein diffusion, and so one type of interaction may appear fast (averaged) on one time scale but slow (discrete) when viewed by another feature of a spectroscopic technique.

A. Biochemical and Chemical Requirements

Both the biochemistry of the proteins and natural membranes, and the chemistry of the phospholipids and the probes used for the magnetic resonance experiments, need consideration when designing an experiment to study phospholipid-protein interations. Usually the major restrictions are on the protein component. Compared with some types of biochemistry and molecular biology, magnetic resonance experiments require rather large amounts of membrane and protein, typically 1–50 mg, with spin label ESR being more sensitive by about 10^3 times when compared with NMR. New refinements and instrumental advances are reducing the amounts of material needed for both NMR and ESR through the use of higher fields and new cavities; thus experimentation is expected to become easier.

Molecular biology methods, in principle, are therefore able to help in these situations where specific proteins are required in reasonable amounts. If proteins not supplied by nature in large amounts can be cloned and expressed in bacterial or viral systems, then the prospect for studying phospholipid-protein interactions in a much wider range of systems and with more diverse protein types becomes a real one.

Bilayer membranes suitable for the study of phospholipid-protein interactions by magnetic resonance are in two basic forms:

Natural membranes that have been isolated and purified with respect to type
Bilayer complexes of phospholipids containing one or more proteins made by reconstitution methods

1. Natural Membranes

Natural membranes may contain predominantly one integral or peripheral protein type, or, as is usually the case, a range of proteins. Only in the case of a biomembrane that contains one dominant protein can the information derived be interpreted as due to phospholipid interactions with a specific protein. Classic examples of biomembranes enriched in one protein are the disc membrane from mammalian retina, which contains > 95% rhodopsin; membranes from the electric organ of the electric fish *Torpedo*, which are rich (~50%) in the nicotinic acetyl choline receptor; the thylakoid membrane from chloroplast, which contains only the light harvesting complex; and the purple membrane from *H. halobium*, which is 100% bacteriorhodopsin. Some membranes can be enriched in a protein during preparation; for example, the kidney medulla can be partially purified using SDS to give a membrane enriched in the Na^+/K^+-ATPase, and erythrocytes can be purified in such a way that the major (~80%) protein is band 3, the anion transporter.

Specific detail is therefore derived from biomembranes containing one protein type, but general information about phospholipid interactions can be obtained from studies of less well-defined systems. For example, the plasma membrane of *E. coli* and erythrocytes have been studied for general changes induced by changing the phospholipid acyl chain composition genetically [32] and the cholesterol content by artificial enrichment methods, respectively. For these kind of studies, therefore, the information is averaged over all the protein in the membranes, and some difficulties may be encountered in producing reproducible experimental results from membranes made at different times, especially if the membranes are derived from cells that grew under different dietary or growth conditions.

2. Reconstituted Complexes

With reconstituted membranes, the protein content, both in type and in concentration, can most readily, but not always predictably, be controlled. In addition, the phospholipid type for the membrane can be determined rather precisely, even if the lipid-protein ratio may not be under complete control.

The methods for reconstitution are not straightforward, however. Firstly, assuming the protein can be isolated satisfactorily and purified to a high degree, a suitable detergent is required in which the protein does not aggregate over an extended time, at least as long as it takes to effect reconstitution with the solubilized phospholipid that will form the complex. Protein aggregation in detergents is a difficult process to control, and some monitor of protein-phospholipid mixing is required. In many reconstitutions, both protein and phospholipid are lost, the lipid through the dialysis bag walls or on the column, depending upon the method used, and protein through aggregation. Some separation of the required complex from the unincorporated lipid or aggregated protein can be effected by density gradient centrifugation (~5–45% sucrose) so that the material with the same density, and, it is anticipated, with the same phospholipid-protein ratio, can be extracted with excess or unincorporated lipid floating and aggregated protein sedimenting. Chemical determinations of the phospholipid-protein ratio of a complex after preparation give only an average value, and the only method that can give an indication of protein distribution within vesicular structures is electron microscopy, ideally using the freeze fracture method, since negative stain methods are unable to show protein distribution within the bilayer of a complex. Negative stain methods do, however, give some indication of sample size and homogeneity, parameters that are important in NMR experiments (Sec. III.B.4).

B. Phospholipid Requirements

Having discussed the protein component of the membrane in which phospholipid-protein interactions are studied, the phospholipid component has equal importance, but it is often not limiting in its production. Some chemistry or molecular biology is needed, but this should not limit the design of an experiment. The phospholipid requirements are different for spin label ESR and NMR studies, and so these will be discussed separately.

1. Spin Labelled Phospholipids for ESR Studies

In the spin label ESR approach to the study of phospholipid-protein interactions, a nitroxide reporter group is attached to a phospholipid. This phospholipid is then inserted by some means (see Sec. III.B.2) into the bilayer membrane on which it is to report, at a probe concentration of 100–200 endogenous lipids/spin label.

To date, it has not been possible to synthesize the whole range of phospholipids found in natural membranes for membrane studies, and so some lipid types are not available for studies of the selectivity of phospholipid-protein interactions [9, 40]. For those types where synthesis is possible, the nitroxide is usually attached to the *sn*-2 chain of a phospholipid, and the information gained in the study is about the hydrophobic interactions of the bilayer with the protein hydrophobic interface; the molecule is a probe, and so it is anticipated that it is immaterial whether the *sn*-1 or the *sn*-2 chain of dioleoyl lipids carries the nitroxide. Some typical phospholipid spin labels are shown in Fig. 6.

If the nitroxide is close to the terminal end of the acyl chain, it is able to report on the lipid motions at the center of the bilayer hydrophobic core. Acyl chains in the protein-free part of the bilayer are large in amplitude and rate, as shown by the well characterized mobility profile [41], but the protein restricts the acyl chain rate of motion probably to the same degree throughout the bilayer. In the spin label method, it is this differential motion of the acyl chains in the bulk phase and at the protein interface that is exploited, hence the requirement for the probe to be close to the end of the chain for a maximum spectral difference between the spectra of labels in the two motionally distinct environments.

The decision of precisely where to position the nitroxide is determined predominantly by experience. The positions that appear to produce the best spectral resolution of protein-interacting and protein-free spin label lipid are with the nitroxide at the C16 and C14 positions. Initial studies were performed with the C16 label, probably because this label was the one that was commercially available as the fatty acid. However, screening of all positions along the chain in a number of systems showed that the C14 label has been a good choice in many situations [9]. The syntheses for the C16 and C14 labels are equally involved, and it is good practice to try both in a system; usually the C14 label is better at higher measuring temperatures and in more "fluid" bilayers, and the C16 is better at lower temperatures and in more "rigid" membranes, for resolution of the phospholipid-protein interactions.

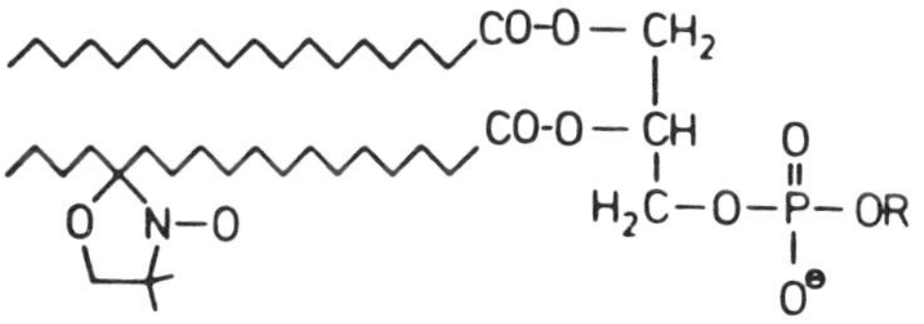

R = H — 14 PASL

R = $(CH_2)_2—\overset{+}{N}H_3$ — 14 PESL

R = $(CH_2)_2—\overset{+}{N}(CH_3)_3$ — 14 PCSL

R = $CH_2—CH(COO^-)—\overset{+}{N}H_3$ — 14 PSSL

R = $CH_2—CHOH—CH_2OH$ — 14 PGSL

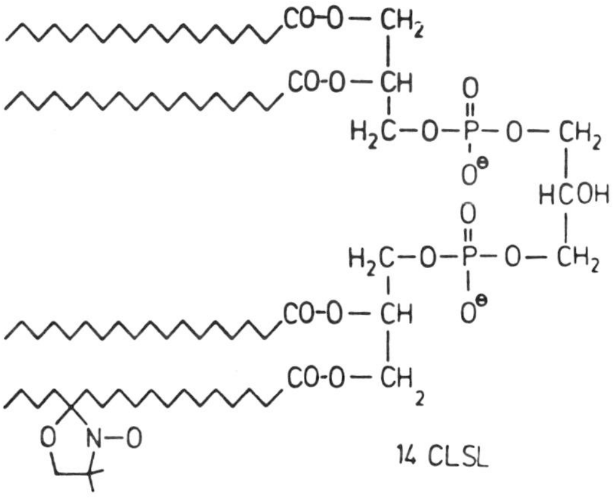

14 CLSL

Figure 6 Formulae of some phospholipid spin labels used for studying phospholipid-protein interactions. The nitroxide reporter group is usually positioned in the *sn*-2 acyl chain of diacyl lipids and close to the end of the chain so that its *amplitude* of motion in the center of the bilayer is large ($S < 0.1$) both in the protein-free part of the membrane and (probably) at the protein interface, due to the relatively unstructured surface of the protein. Spectral resolution is now optimal between the labeled chains, since those away from the protein are undergoing *fast* motion ($\tau_c < 3$ ns, resulting in a narrow ESR spectrum) and those undergoing *slow* motion ($\tau_c > 10$ ns resulting in a broad ESR spectrum) due to inhibition by the protein surface.

Synthetically, the fatty acid spin label is made by established methods [9] and then acylated onto a monoacyl phospholipid. (Fatty acid spin labels can be used effectively in studies of lipid-protein interactions, but the information may not be so relevant in membranes where it is known that fatty acids are not present, since the interations may be dominated by the charge of the fatty acid at nonspecific binding sites on the protein). Acylation to the final product can be straightforward, as with the phosphatidylcholine spin label, where no modification of the head group by the acylation process occurs, but for most other phospholipid spin labels, some modification of the lipid head group can occur during acylation to a monoacyl phospholipid, and therefore further synthesis is required.

One approach to synthesize different phospholipid type spin labels is to make the phosphatidylcholine spin label and then use phospholipase D [9] to exchange the head group in excess alcohol, a method that can be used for making the phosphatidylserine, phosphatidylethanolamine, phosphatidic acid, phosphatidyl glycerol, and phosphatidyl inositol spin labels. Also, cardiolipin spin labels have been produced by enzymatic means from natural analogs. So far, no effort has been invested in making phospholipid spin labels using the established methods for making unlabelled phospholipids [42], that is, making a diacyl glycerol with the nitroxide on one chain and phosphorylating this to the required phospholipid, despite the relatively low yields (1–35%) of the established enzymatic methods.

For a comprehensive study of phospholipid-protein interactions, it is best to have a stock (kept at –20°C as a solid for the major portion but in chloroform, chloroform-methanol, or ethanol at 1–4 mg ml^{-1} for short-term storage and ready for use) of a number of phospholipid type spin labels, each with the nitroxide at the C14 and C16 position, although normally such a portfolio of labels would be built up over an extended period and added to as the need arises.

2. Incorporation of Phospholipid Spin Labels into Membranes

The method of label incorporation into a membrane for studying phospholipid-protein interactions depends upon the membrane type and on the phospholipid spin label type. In every case, some knowledge of the amount of lipid in the sample to be labeled as required. The labeling level is usually one spin label for each 100–200 lipids in the sample; the molecular weight of a phospholipid spin label and an average molecular weight of the sample lipids can both be assumed to be ~800. The absolute level of label incorporation is not crucial within ~20% and this can be determined later if required, although some spin label reduction to a nondetectable, nonparamagnetic species may occur, especially in natural membranes where reducing species may be active.

There are two basic methods of incorporating a spin labeled phospholipid into a bilayer membrane. Both can be used for natural and reconstituted complexes, although certain differences apply for each, and these are given below. The methods are

Addition to the membrane from solvent
Incorporation from a dry spin label film

Addition of a spin labelled phospholipid into a membrane complex is usually effected from ethanol as a solvent in which most phospholipid spin labels are soluble at relatively high concentration (4–5 mg ml^{-1}). Working backwards from an estimate of the spectrometer sensitivity, it can be reasonably assumed that to obtain a suitable ESR signal in most spectrometers with the sample in a flat cell or sealed-off capillary, about 1–10 μg of label are required. Therefore the complex needs to contain typically 0.1–2 mg of membrane phospholipid, and 1–2 μL of the phospholipid spin label stock solution is required. Some alteration in these amounts is possible, and if such a small volume is injected into a much larger volume (> 10 mL) of buffer containing the membrane complex, then the ethanol concentration will be $<10^{-2}$% and far below an ethanol concentration sufficient to perturb the bilayer. Usually spontaneous incorporation of the phospholipid spin label into the membrane occurs, and the membrane can be recovered by centrifugation at moderate speed (10,000 g; 10 min) or even on a bench centrifuge if the membrane is high in protein content. Although the method seems to work well for phosphatidylcholine spin labels and phosphatidyl ethanolamine spin labels, charged phospholipids have a propensity to form small bilayer vesicles in addition to partitioning into the membrane being labeled. This behavior is clearly visible from the resultant ESR spectrum where, underlying the anticipated narrow membrane spectrum, there is an interaction broadened spectrum from unincorporated labels in bilayers made exclusively of the spin label, probably adhering to the surface of the membrane of interest (see below). Removal of such bilayers from the membrane may be effected by one or more of the following procedures

Washing the sample by repeated centrifugations
A suitable density gradient (similar to the last step in isolation or preparation of the membrane)
Some incubation (at elevated or room temperature for some hours) to permit fusion of the bilayers with the membrane being labeled
Reducing the membrane-containing buffer volume

All procedures may be tried to promote better incorporation of the label into the membrane, and only experience with each membrane and label type can tell which is the best method or combination of methods to use. Generally, it appears that

Cholesterol-containing membranes tend to be more difficult to label.
In bilayers in which there is a thermotropic phase transition, labeling is best done above the transition temperature.

There are some phospholipid spin labels that are not soluble in ethanol, notably the phosphatidylserine spin label. Here there is no choice but to use an alternative method to incorporate the label into a membrane complex. One such method is to make a dry film of the label on the inside of a glass test tube, by adding the required amount of label from chloroform to the test tube, removing solvent by blowing with N_2 gas, and then putting under vacuum for an hour or so. The membrane sample is then added directly to the film in a small buffer volume (say 0.5–2 mL), shaken and warmed a little, and the sample concentrated by centrifugation and the ESR spectrum measured. If the signal is significantly interaction broadened, as a result of much of the phosphatidylserine spin label forming bilayer vesicles instead of partitioning into the membrane being labeled, then the procedures above can be used to separate the membrane and the spin label vesicles.

There are two nonsolvent spin labeling methods that should be explored. A lipid transfer protein [43] can be used to transfer a spin labeled phospholipid into a membrane. The proteins themselves have been studied using spin labels [43], and they have been used to incorporate deuterated phospholipids into biological membranes [44], but their use is not established for spin labels. The method is attractive, since some of these proteins have specificity in the type of phospholipid transferred between one membrane and another, and therefore the perturbation of the membrane is not great, being a one-for-one replacement. The level of incorporation is readily controlled by measuring transfer kinetically using a radiolabeled phospholipid marker. One drawback with this method is the long preparation time (some days) for the transfer protein, but the proteins are very stable and can be stored for some while.

Since phospholipid spin labels form bilayer vesicles, it is in principle possible to incubate the membrane to be labeled with a phospholipid spin label vesicle suspension. Some problems may be encountered in separation of vesicles and membrane using the methods detailed above, but the degree of incorporation, and more importantly the contamination in the final sample of spin labeled vesicles, can be readily assessed from the final ESR spectrum.

The morphology of the bilayer phospholipid-protein complex for spin label ESR studies is not important. Either large multibilayer complexes or small single bilayer vesicles, or any intermediate morphology, can be studied. The ESR spectrum is relatively unaffected by such differences, since macroscopic motion is too slow to affect any ESR spin label spectral parameter. If morphological changes do occur, for example with temperature, then spin label ESR methods can detect them [45], but with less sensitivity than, for example, by electron microscopy or ^{31}P NMR [16].

3. Phospholipids for NMR Studies

Some studies of phospholipid-protein interactions have been performed by exploiting the indigenous nuclei of the membrane complexes, but these are few for technical reasons that are now being partially resolved. Clearly, no special labeling requirements for proton or ^{31}P NMR are necessary as far as the nucleus is concerned. Similarly, natural abundance ^{13}C can be studied [46], albeit at low sensitivity [11]. Specific enrichment with ^{13}C is possible but is rarely performed because of the technical difficulties in studying this nucleus, although some instrumental advances now do permit such studies. Here, ^{13}C is incorporated by chemical means into specific parts of the phospholipid using established phospholipid synthetic means. Similarly, replacement of protons by deuterons at specific sites in a phospholipid can be achieved readily using recently improved synthetic means [42]. Any position within a phospholipid can now be labeled, with some precursors, such as fatty acids and polar head group components being readily, although expensively, available.

A relatively unexploited nucleus for membrane studies has been the nitrogen nucleus, both ^{14}N (spin = 1) and ^{15}N (spin = 1/2), which can either be in the phospholipid polar head group [48] or in specific amino acid side chains. With the availability of more ^{15}N labeled precursors for phospholipid synthesis, and genetic and chemical methods that enable incorporation of isotopically labeled amino acids into proteins, this could be a valuable nucleus to study, especially in proteins. Isotopes of oxygen or ^{3}H have not been studied in phospholipid-protein complexes.

In principle, biosynthetic methods can be used for producing labeled phospholipids, but these are not developed sufficiently to yield amounts useful on the preparative scale. Studies on intact membranes into which deuterons have been incorporated into the phospholipids have been reported, but large amounts of membrane are required and the spectral resolution can be low [47].

4. Labeling and Morphology of Membranes for NMR Studies

The method of labeling membranes for NMR studies is dictated exclusively by the nucleus to be exploited. For observation of indigenous nuclei such as ^{1}H and ^{31}P, no labeling is normally required, although potential exists for labeling with another isotope (e.g., ^{2}H) to delete the resonance and magnetic coupling from a natural abundance nucleus from a spectrum.

The low natural abundance of ^{13}C means that labeling is not only desirable to improve the signal-to-noise ratio, but detailed molecular information can be gained if specific parts of the molecular complex can be labeled. For ^{13}C labeled phospholipids, the label need not be in a specific position on the chain, becuase the spectral dispersion for ^{13}C (the frequency span over which the spectrum appears) is rather large (~200 ppm), and assignment of the majority of the resonances is not a great problem, as it can be for ^{1}H NMR, where the spectral dispersion is much less (~ 10 ppm). In addition, dipolar coupling between protons significantly reduces spectral resolution in large membrane complexes.

For observation of natural abundance ^{1}H or ^{13}C nuclei, the resonances due to the protein are often not observed, because either the motions of the nuclei are too slow or they are much lower in concentration compared to those from the phospholipid. Also, for ^{13}C and ^{1}H NMR studies of phospholipid-protein complexes, conventional high resolution NMR approaches are not suitable, because the broadening of the resonances caused by the slow tumbling of the complex in the magnetic field is too slow to average out the magnetic interactions that affect the relaxation and dipolar interations [11]. For this reason, small (diameter < 100 nm) bilayer vesicles are required for such studies, which itself imposes rather difficult constraints on the experimental design, sonication being the only satisfactory way of producing near homogeneously sized small vesicles. An alternative approach is to study large bilayer complexes, such as those produced by dialysis or in a natural membrane sample, and use a rather new method called magic angle sample spinning (MASS) NMR. Little has been reported to date on phospholipid-protein interactions using this approach, although cytochrome *c*-phospholipid interactions have been examined by this method [46,49].

For ^{2}H NMR studies of phospholipid-protein interactions, some means of incorporating the ^{2}H nucleus into the membrane must be found. The objection to using ethanolic injection into a natural or reconstituted membrane as used for spin label ESR studies (see Section III.B.2) is that the amount of ^{2}H phospholipid required to obtain a signal from a complex that can be accommodated in an appropriate NMR tube (5–10 mm o.d.; 0.7–2 mL) is relatively so large (5–100 mg of a phospholipid with one or more deuterons per phospholipid) that the overall change in composition of the complex, and the amount of ethanol-lipid solution required, would perturb the complex to an unacceptable degree. Other ways therefore need to be found. For example, the following can be tried

Biosynthetic incorporation of a deuterated fatty acid or glycerol [50] into a membrane, as accomplished with the outer membrane of *E. coli* using specific auxotrophs, but with inherent low sensitivity and low yields of membrane.

Manufacture of the phospholipid-protein complex itself from the deuterated phospholipid is the most common method for ^{2}H NMR studies, but this then restricts the type of membrane that can be examined to reconstituted complexes with well defined proteins.

A relatively new method is to use a phospholipid transfer protein (a range can be isolated from liver or brain [43], and they are stable to longer-term storage [44]) to replace a natural phospholipid with a deuterated one, possibly with similar acyl chains and to a sufficient level of incorporation for good signal strength.

It is clear that deuterium NMR does have its limitations for the study of phospholipid-protein interactions, but there are advantages in that the sensitivity is not unacceptably low and that assignment of the deuterium resonances observed is unambiguous. In addition, as with ^{31}P-NMR studies, it is the anisotropy of the magnetic interactions that can be exploited to good use. The morphology of the sample is therefore also important, since the rapid ($\tau_c < 10^{-6}$ s) tumbling of small (<200 nm) vesicles does average out the anisotropic contributions to the spectrum and produce less informative single-line NMR spectra. Such spectra can be used for spin-lattice (T_1) relaxation measurements, and the order parameter (S_{CD}) of the lipid segment can, in principle, be determined, since [51]

$$\frac{1}{T_1} = \frac{3\pi^2}{2}\left(\frac{e^2qQ}{h}\right)\left(1 + \frac{1}{2}S_{CD} - \frac{3}{2}S_{CD}{}^2\right)\tau_c \tag{2}$$

where (e^2qQ/h) is the static quadrupole splitting (127 kHz) and τ_c is the rotational correlation time for the vesicle obtained independently from $\tau_c = (4/3)(\pi\eta r^3/kT)$.

Both ^{2}H and ^{31}P NMR experiments are able to yield information about the morphology of the sample, especially if there are changes between bilayer and nonbilayer forms of the complex, changes which may be induced by the protein [16].

C. Instrumental Requirements

The spectrometer requirements for magnetic resonance studies of phospholipid-protein interactions are not unusual for spin label ESR studies, but for NMR a solid state instrument is required, preferably multinuclear.

1. ESR Spectrometer

No attempt will be made to describe ESR spectrometer operation here, and only pertinent aspects to the study of phospholipid-protein interactions will be made. It should however be stated that a standard set of instrumental conditions for use throughout a series of experiments (or even within a research group) is strongly advised, since little, if anything, can be done to ESR spectra once recorded or stored magnetically. This is because in phospholipid-protein interaction studies, direct spectral comparisons are always subsequently required, and such comparisons can only be made if no instrumental distortions of the spectra have occurred between recording.

The majority of studies of phospholipid-protein interactions have been performed with relatively standard ESR spectrometers operating at X-band (9.5 GHz). Thus sensitivity for most spin label experiments on membranes (Sec. III.B.2) is good with 0.1–2 mg of host phospholipid. Spectra should be recorded at good signal-to-noise ratio (better than 100:1 s/n on the recorded spectrum) in a single scan in about 4 min over the full spectral width of 0.10 T (100 gauss) even if data averaging (see below) is to be performed.

The microwave power used should be less than saturating for spin labels ($\sim <40$ m W), since power saturation characteristics may be different for spin labeled phospholipids at the protein interface and those in the bulk bilayer phase. There is clearly a tradeoff with signal strength and saturation, but in general a power of $< \sim 20$ mW is suitable with modulation amplitude of less than half the width (ΔH) of the narrowest spectral line, which may be from an aqueous and therefore unincorporated free spin label with $\Delta H \sim$ 0.15 mT. In the absence of free spin label spectra, a modulation amplitude of ~0.1–0.2 mT is a good starting point, and this should be maintained in any series of experiments, which is particularly important if subseqent computer manipulation of the spectra is to be performed.

In every spin label ESR experiment on any membrane, the whole of the spectrum should be recorded together with a significant amount of base line at high and low field. A good spectral width to use is ± 5 mT around the spectrum center, which will not coincide with the spectral line with (m_I = 0) the maximum amplitude.

Temperature studies are usually carried out in 2–5° steps, and so temperature determinations need be only to ± 0.25–0.5°; ESR spectral changes, even with phospholipid-protein complexes that display a phase transition, are usually not so dramatically changed over a couple of degrees. Most studies are performed over temperature ranges of 273–313 K, and therefore a variable temperature system is required. The most readily constructed system is centered around a nitrogen gas flow through a cold reservoir of acetone or ethanol with dry ice (CO_2), the gas being heated in a transfer line by a heater coil controlled by a servo feedback from a sensor placed in the gas flow through the cavity dewar immediately under, but outside, the detection volume of the cavity itself. Measurement of the sample temperature itself is most accurately done with a separate thermocouple placed immediately above the sample in the cavity, but again with the sensor tip outside the measuring volume of the cavity. If sealed-off capillaries are used, these are best put inside standard 4 mm quartz ESR tubes filled with a light silicon oil for temperature equilibration and the sensing thermocouple tip immersed in the oil, which should fill the ESR tube to a length just in excess of the cavity length, ~3 cm. Both the N_2 gas and the silicon oil should be completely free of water, since water absorbs microwaves (reducing the cavity Q) and the sensitivity will fall and coupling of the cavity may become impossible.

2. ESR Data Acquisition

For studies of phospholipid-protein interactions, the majority of subsequent data manipulations involve spectral subtractions (see Sec. III.D.1). For this reason, the final signal/noise ratio needs to be good, since noise is amplified as signal is subtracted during such manipulations. Therefore much is gained when the spectrometer is equipped with computer averaging facilities so that one spectrum can be recorded and averaged many (N) times before being stored on disc. To adapt a spectrometer to provide this facility is not difficult, and all the triggers required to start and stop spectral accumulation are usually provided on recent spectrometers. However, since the signal/noise ratio improves with $\sqrt{N}$, it is still important to have a relatively good spectrum before extensive accumulations; if accumulation must be carried out over periods of over one hour or more for just one spectrum, then often little extra improvement in the data quality is gained through averaging, and it may be more sensible to examine whether the spin labeling has been efficient or if more sample could or should be used.

ESR spectra can be recorded on chart paper, but then little subsequent analysis is possible. Line heights and positions can be determined, which may be useful for protein-free phospholipid membranes, but in phospholipid-protein complexes the spectra often contain more than one spectral component and simple analysis is not possible because of line overlap between the different components. The overlap between one spectrum and another affects all features, line widths and positions, of each individual spectrum, and only when the different components are resolved from each other can spectral features be measured with confidence. The only satisfactory way to carry out such manipulations therefore is to record the averaged ESR spectra computationally, together with some form of identifier and header if required. Such files are simply series of (x, y) points with between 1 k (which is usually enough) and 4 k data points in the magnetic field y

direction. Although no changes to the spectrum are now possible (as with NMR FIDs), some smoothing can be effected and the spectrum can now be shifted with respect to, subtracted from or added to, another spectrum (see Sec. III.D.1.).

3. NMR Spectrometer

Historically, the study of protein-phospholipid interactions has lagged behind that by ESR spin label methods because of the instrumental requirements and developments that were needed for such work. Not least, sensitivity can be a major problem, especially when studying naturally derived membranes. Indeed, the general study by NMR of macromolecular complexes ($M_r >> 20$ k) places limits on the approaches that can be used; these approaches, called solid-state methods, have been developed to study solid materials and then adapted for the semisolidlike nature of biological material.

Specific instrumental specifications and probes are needed for membrane work, with each nucleus requiring a different specification. In addition, there is considerable conflict between instrumental design for either high-resolution or solid state NMR. High resolution requires good field homogeneity and receiver and digitizer electronics that perform accurately and with minimal signal distortion. Solid state instruments, such as used to study membranelike samples, require [18].

Good quality high-power (up to 1 kW) pulses for simultaneous irradiation of a wide (up to 200 kHz) range of frequencies, or for decoupling strong dipolar interactions
Receivers and amplifiers with very fast recovery (dead") times
Transient recorders (digitizers) capable of operating at high speeds
Where appropriate, specialized probes for performing "line-narrowing" techniques (e.g., for multiple pulse or MASS experiments)

(a) Proton NMR. Proton NMR spectra show that the nuclei are dominated by strong homonuclear dipolar interactions in motionally restricted systems such as membranes (see Sec. III.B.3). Although the nuclei are undergoing fast motion in liquid crystalline bilayers of phospholipids, their motion is incompletely averaged with respect to the field, being of limited amplitude and rate within the proton NMR time scale of ~ 10^{-5} s. Some particular pulse sequences (e.g., WAHUHA) have been developed to average out these dipolar spin interactions, with some success for simple solids. However, mechanical means can be used to facilitate this required averaging, for example, by spinning the sample with respect to the field at the magic angle (the angle of 54.7°, at which all the anisotropic contributions described by $1/2(3\cos^2\Phi - 1) = 0$).

Although such means can help in producing resolved and interpretable spectra, the complexity of such systems and the ubiquitous nature of the protons mean that the spectra are exceptionally difficult to interpret and assign. Therefore unless it is possible to identify particular resonances that are well resolved from the majority of the proton spectrum by some means (ring current shifts, shift reagents, etc.), little detailed molecular information can normally be resolved by proton NMR on protein-phospholipid systems.

(b) ^{31}P NMR. All phospholipid types in solution give rise to high-resolution NMR spectra that have chemical shifts within ~ 5 ppm of each other. This can facilitate their identification (Fig. 7), especially if they are recorded from solubilized complexes for which the type and relative proportions of each type need to be quantified [52].

Phospholipids in macromolecular complexes display a high degree of chemical shift anisotropy (maximum ~ 200 ppm) in their ^{31}P NMR spectra, which, when partially averaged, give spectra with widths of 20–60 ppm for bilayers (Fig. 8). The individual

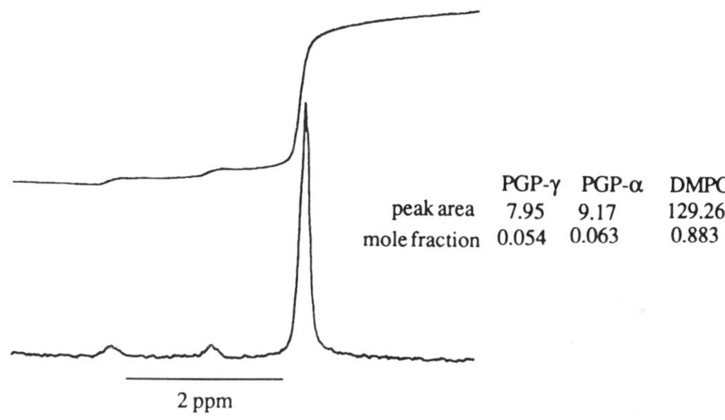

PGP-γ PGP-α DMPC
peak area 7.95 9.17 129.26
mole fraction 0.054 0.063 0.883
2 ppm

Figure 7 High-resolution ^{31}P NMR spectrum of phospholipids (phosphatidylcholine, phosphatidyl glycerol and its phosphorylated derivative) in a solubilized protein-(bacteriorhodopsin)-phospholipid complex to show how lipid types in a mixed phospholipid system can be resolved and quantitated from the integrated spectral intensity of each component. (Adapted from Ref. 52.)

chemical shift differences due to phospholipid type are then small, and each ^{31}P NMR spectrum from bilayers appears rather similar to the others, and the means of identifying the type of phospholipid is lost. Conclusions about the gross morphological state of protein-phospholipid complexes can be made by relatively qualitative comparisons of ^{31}P NMR studies with well-defined systems, such as bilayers, hexagonal H_{II}, or isotropic phases (Fig. 8) [16]. Since it is the spectral shape that is diagnostic in these studies, it is important to record faithful spectra. Because there is a strong magnetic coupling between protons and phosphorus, adequate proton decoupling (some 5–15 W) is essential, combined with an echo-pulse sequence to avoid loss of the initial points in the FID, which would distort the extremes of the relatively wide (some kHz) spectra.

^{31}P NMR spectra can also give useful information about the coexistence of different phospholipid phases that may have been induced by a protein. In this case, limits for the lifetime for any particular phase can be made [53].

The spin lattice (T_1) relaxation times for ^{31}P nuclei in phospholipids are rather long ($\sim$1–3 s), and therefore long relaxation delays are important if intensity measurements are used in quantitations of spectra. In particular, if there is any anisotropy in the relaxation behavior for a phospholipid in a bilayer, then distorted spectra may be recorded if relaxation delays are used that are too short for any one orientational component of the spectrum.

(c) Deuterium NMR. Deuterium NMR has the advantage that the assignments of resonances are generally rather straightforward in that the site at which a deuteron has been introduced chemically is well defined before the phospholipid is used for a membrane experiment. In addition, the single π spin interaction of the electric quadrupole interaction is anisotropic, and hence information about molecular anisotropy is available in the spectrum. The spectral width for a CD bond can maximally be $\sim$ 127 kHz for an oriented system (order parameter = 1), giving a large spectral anisotropy over which spherically averaged spectra from bilayer dispersions can be measured. In practice, motional averaging of the CD bond, whether on a phospholipid head group or on an acyl chain, gives rise to a spectrum that has quadrupole splittings of 1–60 kHz, the quadrupole

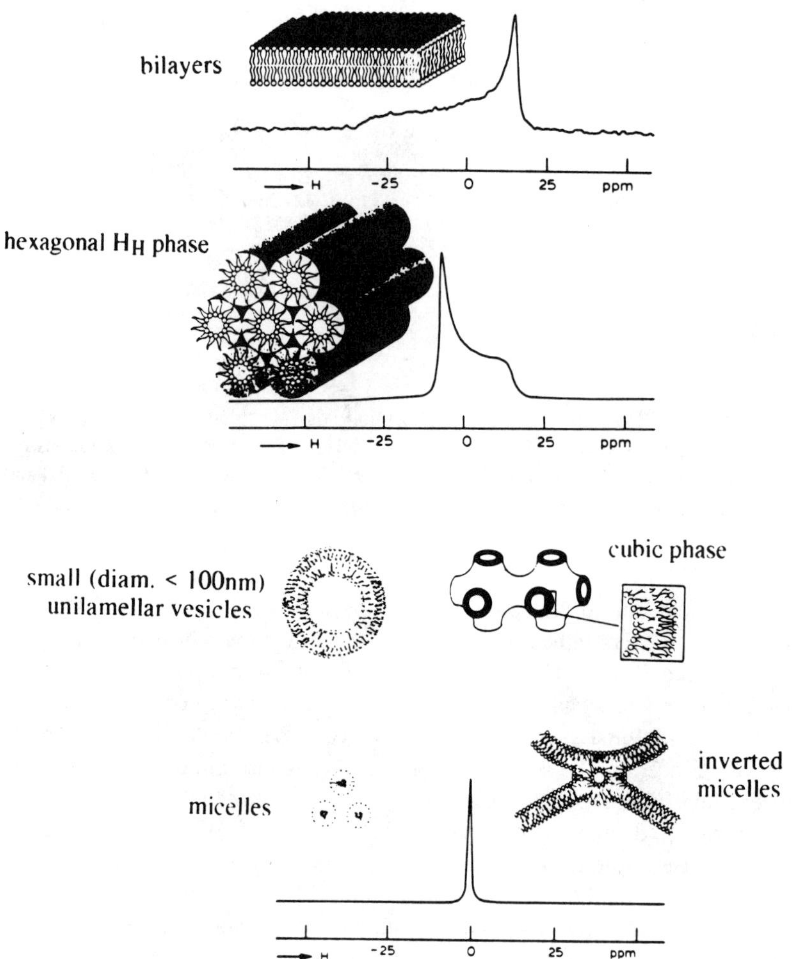

Figure 8 Wide-line, proton decoupled ^{31}P NMR spectrum of macromolecular aqueous dispersions of phospholipids in bilayers, hexagonal, and isotropic phases. The width of the upper two spectra are 6–3 kHz (depending upon spectrometer frequency); compare with the high-resolution spectra in Fig. 7. (From Ref. 16.)

splitting being simply the frequency separation of the two (^2H has two resonant lines; $m_I = 1$) spectral lines from those CD groups with an orientation of 90° with respect to the magnetic field (Fig. 9) [41]. Unlike with spin label ESR, the NMR spectrum from slowly moving deuterons is relatively featureless and cannot be used to give molecular information due to the large degree of inhomogeneous line broadening for slowly moving nuclei.

The rate (τ_c^{-1}) of CD bond motion can be estimated from spin-lattice (T_1) relaxation time measurements (Eq. (2), with T_1's having values in the ms time range and being sensitive to motions in the 10^{-9} s time range (those due to acyl chain isomerizations), and spin-spin (T_2) relaxation time measurements, usually of tens of ms, being sensitive to slower motions with the bilayer (membrane undulations, distortions, etc.) [10,55].

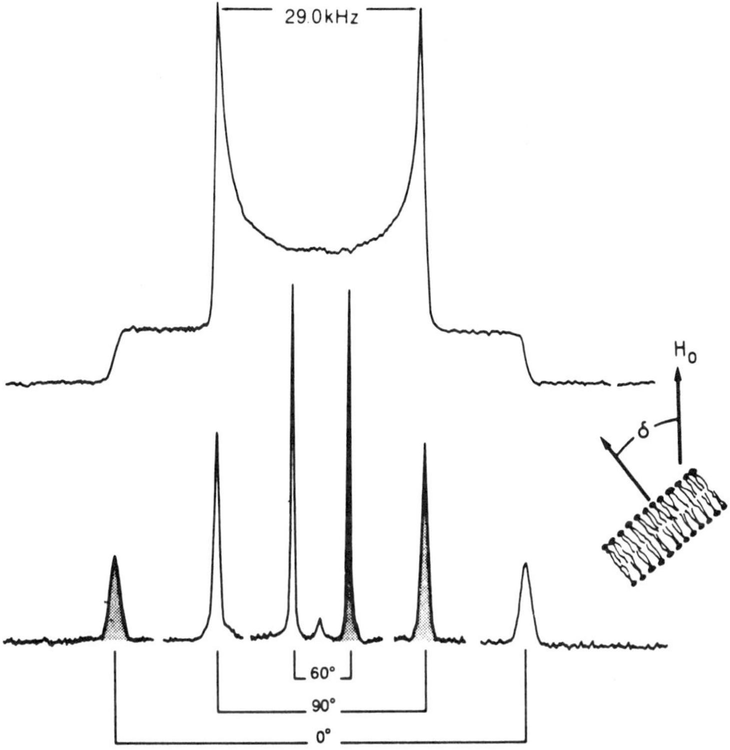

Figure 9 ^2H NMR spherically averaged powder pattern for a deuterated lipid in bilayers. The quadrupole splitting $\Delta \nu_q$ in Hertz in the spherically averaged powder pattern (upper spectrum) is the distance (in frequency units) between the two 90° orientational spectral lines (see lower spectra) and is related to both the motional amplitude of the CD bond and the orientation about which that motion occurs. (From Ref. 15.)

For all these types of experiments with deuterons, it is necessary to irradiate with moderate to high r.f. power (up to 1 kW for wide lines). Sensitivity can be a problem, and high field (6.3 T upwards) is often found to be necessary to obtain good signals from biological membranes. Echo sequences are usually required with broad (>5 kHz) lines so as not to lose the first points in the FID and hence the spectral extremes [18]. However, decoupling is not normally required, since the strength of the quadrupolar interaction is much stronger than any other secondary interaction of the dueteron. Deuterons in a phospholipid head group can give rise to relatively narrow (1–5 kHz) spectral widths and can often be recorded faithfully with low to medium power instruments and even in a single pulse experiment.

(d) ^{13}C NMR. The large chemical shift dispersion for ^{13}C and the possibility of specific labeling permit, in principle, studies of phospholipid-protein interactions. The same problems of spectral broadening as explained for protons apply here also, and unless the drastic operation of sonication or solubilization to allow rapid tumbling of the molecular species with respect to the field can be performed, other methods are needed to obtain a spectrum.

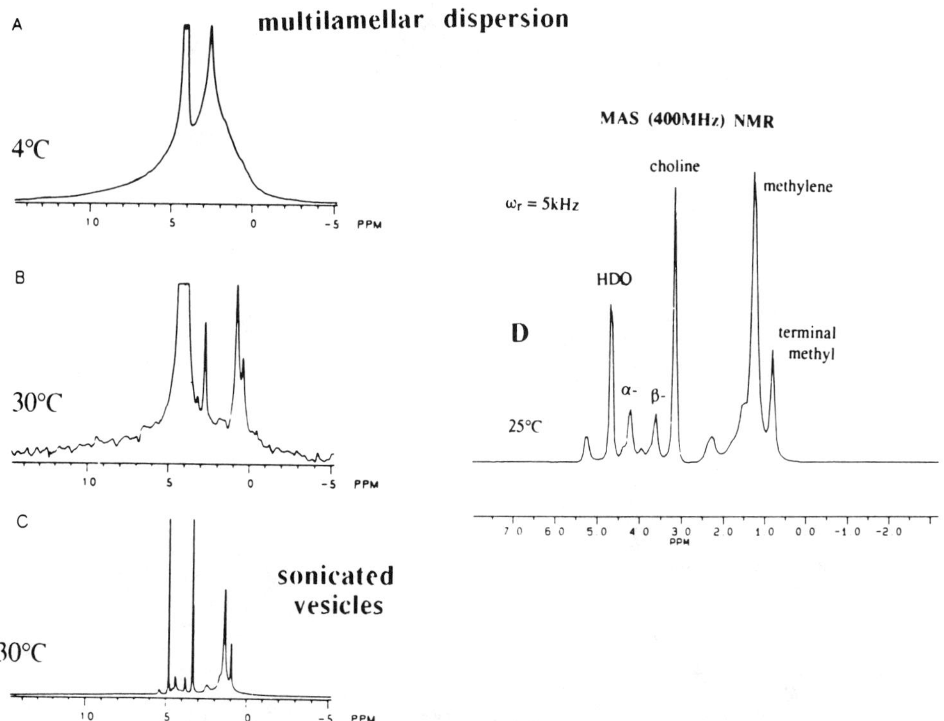

Figure 10 Proton (^1H) NMR spectra of dimyristoyl phosphatidylcholine multilamellar bilayers to show the relative lack of spectral features observed in static samples (left hand spectra, A and B) recorded below and above the liquid crystalline phase transition temperature (23°C) of the bilayers. Much improvement of spectral resolution can be obtained through sonication of the bilayers into small (diameter < 100 nm) single bilayered vesicles (spectrum C). However, to obtain good spectral resolution from large ($\gg 100$ nm) bilayer complexes that are usually formed during protein-lipid reconstitutions, the technique of magic angle sample spinning (MASS) is required (spectrum D).

One solution is to spin the sample itself in the magnet to obtain spectral resolution from an extended phospholipid membrane system, although centrifugation of a pellet may cause a problem, and often low hydration may be necessary; lyophilized samples have been examined. The instrumental requirements are, however, not trivial, and magic angle sample spinning (see Sec. III.C.3.a) is necessary before a high-resolution spectrum suitable for analysis can be recorded (Fig. 10) [53]. This special adaptation of the probe is not complicated, and for critical work, pneumatically controlled spinning rates of some 4–8 kHz are necessary, with a precise and reproducible spinning rate being required in successive experiments where spectra need to be compared or manipulated, for example, by subtraction. Specialist probes and rotors are available, and the method is under continuous development.

D. Data Handling and Spectral Analysis

In magnetic resonance studies of phospholipid-protein interactions, the conventional spectral parameters can be assessed in the same way as in other applications. Such parameters are

Spectral intensity which is proportional to spin concentration
Resonance line position from the chemical shift (NMR) and g value (ESR)
Spin-lattice (T_1) and spin-spin (T_2) relaxation times
Spectral anisotropy where appropriate (^{2}H quadrupolar anisotropy and ^{31}P chemical shift anisotropy in NMR and spin label ^{14}N electron hyperfine coupling anisotropy in ESR) from line separations

Each parameter can be used to give information about independent molecular motions, either rates or amplitudes of motion, as well as about the populations of observed species in each environment. In addition, because of the inherent difference in time scale sensitivity for each method, the parallel parameter deduced from an ESR and NMR spectrum on the same system may give different absolute values for one type of molecular information. For example, a change in the order of acyl chain motion determined in a spin label experiment will most probably not give identical values of chain order (from an order parameter) as in a deuterium NMR experiment, either because of the time scale sensitivity difference or because the nitroxide may be perturbing the local packing of the lipid chains near its site in the bilayer. However, changes and trends in similarly related spectral parameters, for example anisotropy averaging or relaxation times, which are characteristic of similar types of motional behavior, usually change in similar ways when compared in similar systems under similar conditions. Too little information is available at the moment on the same system to make generalizations, but there are very few examples of dissimilar interpretations of phospholipid-protein interactions when examined by both NMR and ESR methods [13].

The different motional sensitivities are given in Fig. 5 for both spin label ESR and NMR methods where applied to phospholipid membranes. This sensitivity arises from the molecular motional rates that are required to change a spectral feature, such as the anisotropy reflected in line positions, or line widths. No attempt will be made here to justify the conclusions or explain how such sensitivity can be extracted for each method, but the reader is referred to some standard magnetic resonance texts [55–57].

1. Spin Label ESR Spectral Analysis

A nitroxide spin label in *isotropic* solution undergoes motion in all orientations, and the ESR spectrum recorded is determined by the rate of molecular motion of the nitroxide. The line shape is narrow for fast motion ($\tau_c < \sim 10^{-9}$ s) (Fig. 11a) but broad for slow motion (Fig. 11b) ($\tau_c > \sim 10^{-7}$ s). The intermediate range shows the time scale sensitivity of spin label ESR, where line shape (reflecting the rate, τ_c^{-1}, at which the ^{14}N electron hyperfine anisotropy is averaged by molecular motion) changes significantly over a certain time range of molecular tumbling ($10^{-7} < \tau_c < 10^{-9}$ s) (Fig. 11a). Estimates for the rate of motion can be made from the line heights and widths [58,59], but some models used to determine τ_c values may assume isotropic motion of the nitroxide, and care must be taken when applying these to bilayers, since the nitroxide motion is then usually anisotropic.

(*a*) Polarity of spin label environment. Line positions do change with the local polarity of the nitroxide environment, but this feature is not used extensively in descriptions of phospholipid-protein interactions. It has, however, been shown that the membrane relative polarity increases due to water penetration into biomembranes if proteins are present when compared with protein-free bilayers (Fig. 12) [9,60].

To assess the relative polarity around the nitroxide from isotropic splitting value a_o it is essential that the hyperfine splitting value really is the isotropic value, which can only

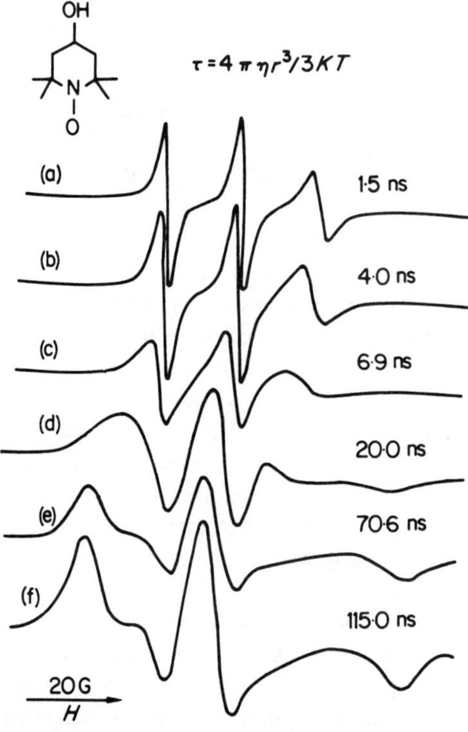

Figure 11a ESR spin label spectra recorded for nitroxides undergoing different rates of isotropic motion to show the time scale for the method. Fast ($\tau_c < \sim 7$ ns) motion gives rise to narrow spectral lines whilst slower ($\tau_c > \sim 70$ ns) motion produces spectra with broader lines (Ref. 17). For motion that is much faster than ~1.5 ns, or much slower than ~70 ns, little further change in the spectral line widths or separations for conventional ESR spectra occurs; saturation transfer can be used to determine slower motional rates ($\mu s < \tau_c <$ ms).

In the intermediate motional range, where line positions and widths are changing significantly, some estimations of nitroxide motional rates can be made from either computer simulation methods or simplified spectral analysis methods. For example, for narrow spectra (1.5 ns $< \tau_c <$ 7 ns),

$$\tau_c = 6.5 \times 10^{-10}\, \Delta H_o \left(\sqrt{\frac{h(0)}{h(-1)}} - 1 \right)$$

where ΔH_o is the width of the central spectral line and $h(0)$ and $h(-1)$ are the heights of the central and low field lines, respectively; the constant is determined from the nitroxide *A* and *g* values.

be determined if the rate of nitroxide motion is fast [61]. If this *is* the case, then a_o values, measured from

$$a_o = 1/3\ (A_{//} + 2A_{\perp})$$

are fairly constant with temperature, where $A_{//}$ and $A_{\perp}$ are measured in magnetic field units (as shown in Fig. 13); $a_o \sim 1.7$ mT and ~ 1.4 mT for a nitroxide in water and a bilayer hydrophobic core or organic solvent, respectively [17]. To carry out such a determination for a protein containing phospholipid bilayer that shows multiple spectral components, it is necessasry first to subtract out broad spectral components from slowly moving labels and analyze only those components (the narrow ones, see later) that fulfill the fast motion criterion.

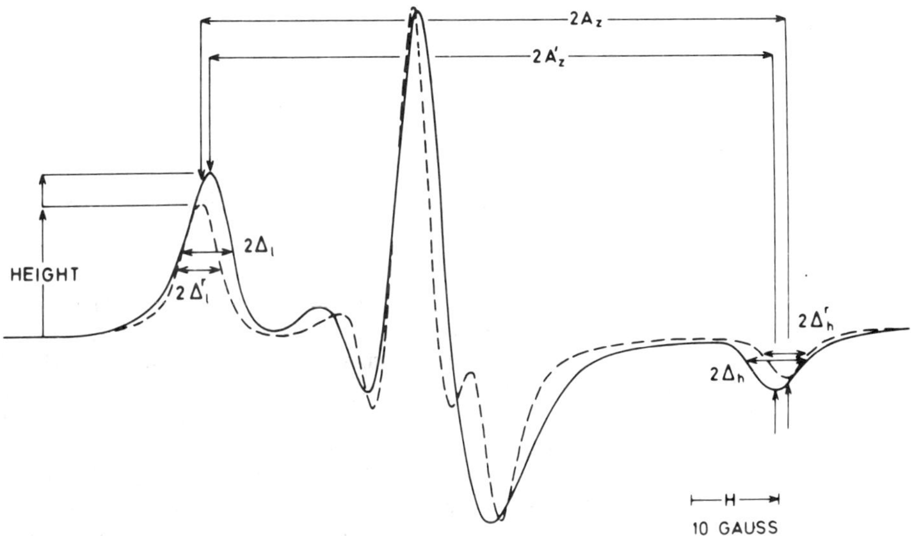

Figure 11b For slower motion of the nitroxide ($\sim$20 ns $< \tau_c < \sim$10 μs), two correlation times τ_c, which should in principle be the same, can be determined, one estimated from the maximum splitting of a spectrum (A'_{zz}) and the other estimated from the line widths (ΔH_m, where m = the high h or low l field line) of the spectral extrema (right hand spectrum) so that (Ref. 62)

$$\tau_r = a'_m \left(\frac{\Delta H_m}{\Delta H^R} - 1 \right)^{b'_m}$$

$$\tau_r = a \left(1 - \frac{A'_{ZZ}}{A^R_{ZZ}} \right)^b$$

in which the R superscript refers to the rigid limit value of the spectral parameter and the a and b terms are calibration constants with values of $b'_m = -1$ and with a'_m = = 2.9×10^{-8}s and 6.57×10^{-8} for $\delta = (2/\surd 3)\Delta H^R_m = 0.3$ and 0.1 mT, respectively. In particular, the values of A^R_{zz} are very sensitive to polarity of the nitroxide (Fig. 12). Suitable values for the rigid limit, peak-peak Lorentzian line widths for phospholipid spin labels are $\Delta H^R_l = 0.239$ mT and $\Delta H^R_h = 0.272$ mT, corresponding to a typical value of $\delta \sim 0.3$ mT (Ref. 59).

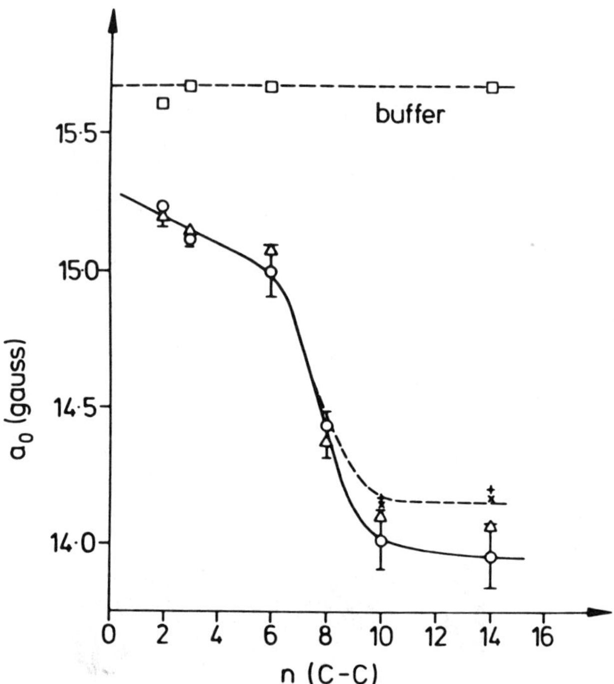

Figure 12 The nitrogen-electron hyperfine coupling for nitroxide spin labels is sensitive to the polarity of the solvent in which the label is dissolved, with a value for the isotropic splitting value a_o varying between 1.39 mT (13.9 gauss) for the nitroxide in organic solvent and up to 1.7 mT in a solvent in which strong hydrogen bonding can occur with the N—O bond. Here, four phosphatidylcholine spin labels with the nitroxide at different positions along the acyl chain (n = carbon number from the ester group of the *sn*-2 chain; see Fig. 6) gave the same value of a_o measure from the ESR spectra when they were monomeric and in buffer (-□-). The same labels, when intercalated into either protein free bilayers (-x-) or protein-containing bilayers (-o-), give a_o values that reflect the less polar environment of the bilayer interior, with the center of the bilayer being much more apolar than the upper half of the bilayer where water probably penetrates. The influence of protein in a bilayer is to make the bilayer interior more polar, either through its own intrinsic polar character or by permitting some water penetration. (From Refs. 9 and 60.)

(*b*) Order parameters. Spin label nitroxide ESR order parameters are readily measured without spectral simulations from the spectra of anisotropically moving phospholipid spin labels in phospholipid bilayers *when the rate of molecular motion is fast* [61]. This is because the averaging of the anisotropy of the nitrogen-electron hyperfine interactions must be complete within the time scale required for a maximum excursion of the label between its limits of motion. This time scale sensitivity for the rate of nitroxide motion is therefore defined, with A values in Hz, as $\tau_c^{-1} > A_{max} - A_{min} = A_{zz} - 1/2(A_{xx} + A_{yy})$, which has a value of $\sim 10^8$ Hz.

Therefore to determine the order parameter for a phospholipid spin label in a protein-containing phospholipid bilayer, it is essential first to deconvolute the spectrum if it is multicomponent (see below), measure $A_{//}$ and $A_{\perp}$ as shown in Fig. 13 for only the component(s) that have fast motion, and then calculate the order parameter S from [17,40,61]

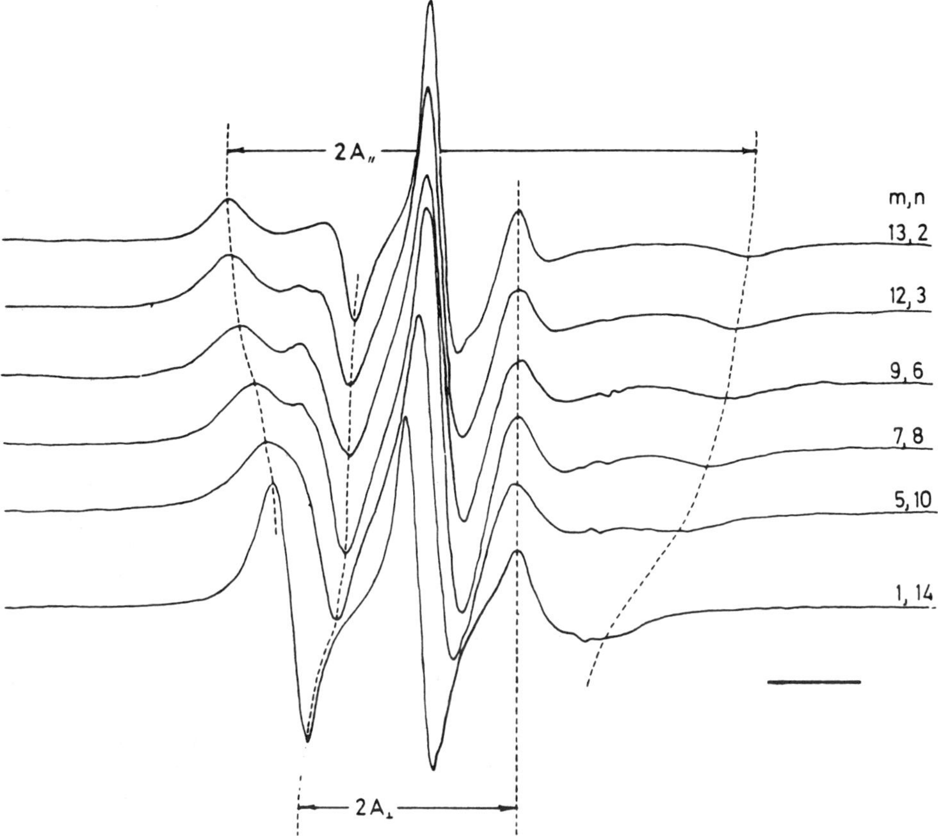

Figure 13 Typical ESR spectra for nitroxide phospholipid spin labels in bilayers. The position of the nitroxide is at successively deeper positions within the bilayer core (increasing value of n), and the spectral shapes reflect the increasing amplitude of motion of the acyl chains. The way in with $A_{//}$ and $A_{\perp}$ are measured is shown. For studies of protein-lipid interactions, normally the label is at $n > 10$ so that the amplitude and rate of label motion is large (i.e., anisotropy small). Any increase in spin label spectral anisotropy caused by motional rate or amplitude changes caused by interaction of the acyl chain with the protein can be readily observed as either a new spectral component (see Fig. 16) or line width changes. Bar = 1 mT.

$$S_{s/l} = \frac{\text{observed anisotropy}}{\text{maximum anisotropy}}$$

$$= \frac{A_{\|} - 2A_{\perp}}{A_{zz} - (1/2)(A_{xx} - A_{yy})} \cdot \frac{a_o'}{a_o}$$

The amplitude of acyl chain motion (Θ in degrees) is then given by $S_{s/1} = (1/2)(3 < \cos^2(\Theta) > - 1)$, and the $\cos^2 \Theta$ term within $< >$ implies fast molecular motion (as defined above, Sec. III.D.1) within the cone of amplitude Θ. The polarity correction term [40] a_o'/a_o accounts for the different polarity of the nitroxide (and hence affecting line positions) in a crystal (giving $A_{xx,yy,zz}$) and in a membrane (giving $A_{//}$ and $A_{\perp}$) [40].

In general, it has been widely reported that $S_{s/1}$ values increase with increasing protein content in a membrane. However, such effects cannot be solely due to changes in the amplitude of acyl chain motion, since line positions are often, especially in natural membranes, seen to change when the lines broaden due to the presence of an integral protein slowing down the rate of spin label motion. Line positions should *not* change with decreasing motional rate as long as the motion is in the fast motional regime where order parameters can be faithfully determined; the only property that should change line position for fast motion of a nitroxide is amplitude (and hence order parameter) of acyl chain motion. Indeed, it has been shown that, as intuitively expected, proteins disorder bilayer lipids when compared with protein-free bilayers (Sec. II.D.) [9,13].

Care must therefore be taken in simply measuring order parameters from spin labels in protein-containing membranes, especially natural membranes where line widths tend to be broad, and interpreting changes in the amplitude of acyl chain motion due to a protein. An S_{apparent} is best quoted, and an explicit statement that S_{app} reflects both *amplitude* and *rate* changes of acyl chains in the phospholipid bilayer core given.

(*c*) Rate of spin label motion. The rate of nitroxide motion is estimated from the spectral line widths. The presence of a protein is always found to decrease the rate of phospholipid acyl chain motion, the degree of motional decrease being determined by both the size of the protein mass interacting with one phospholipid chain; large proteins or aggregates usually cause a very large (by a factor of $10–10^2$ times) decrease in the correlation time for acyl chains in the immediate environment of the protein hydrophobic interface (Sec. II.C). The method for determining motional rates for a phospholipid spin label employed differs depending upon the motional regime in which the spectrum lies (see Fig. 11). For fast ($\tau_c < \sim 10^{-9}$ s) acyl chain motion in bilayers containing a small peptide or protein, and where the spectral lines are relatively narrow, line widths can be justifiably used to estimate the change in the rate of acyl chain motion and these values for rate compared to when the bilayer is protein free. The values obtained are clearly not absolute, but in a *comparative* way some indication of how the rate of nitroxide motion might change in the presence of a protein can be made.

For slow to intermediate ($10^{-6} < \tau_c < \sim 10^{-8}$ s) nitroxide motion, a computer simulation of nitroxide spectra using a stochastic Liouville formulation has been used to identify sensitive spectral parameters (line widths and line positions) (Fig. 11b), which can be employed in estimating empirical correlation times [9,13,62]. With an educated guess about the type of motion of the spin label, one of three motional models (free diffusion, strong jump, and Brownian diffusion) can be used in the analytical procedure, with a strong jump model being found as most appropriate rigid limit parameters of line positions and widths, which can be determined from, for example, the actual spectrum recorded with the sample at very low temperature.

This approach to the description of motional rates for the chains of phospholipid spin labels interacting with an integral protein is useful in describing the trend in motional rates for spin labels at a protein interface when, for example, varying the temperature [9]. Since the major contribution of motional freedom to the chain is probably exchange back into the bulk bilayer away from the protein, increasing temperature might be expected to produce a decrease in τ_r values (that is, faster exchange onto and off the protein interface). Alternatively, for a strongly aggregated protein mass, in which phospholipid exchange is restricted and the acyl chains trapped, a rather small temperature sensitivity of τ_c might be expected. Again, as with other spectral analyses of individual spectral components, only after deconvolution of a multicomponent spectrum can this method normally be used with confidence on one component. However, if the broader spectrum is well resolved from the inner spectral component(s), it may be possible to estimate line positions (for A'_{zz}) and line widths (for ΔH_M) directly, or at least after a minimal subtraction of other narrower spectral components (Sec. III.D.1).

Although saturation transfer ESR (ST-ESR) methods can in principle be used to estimate spin label motions in the $10^{-3} > \tau_c > 10^{-7}$ s time range, this approach is only successful with homogeneous spectra, since spectral overlap from narrower spectral components prevents ready measurement of the three line heights required in the analysis; with this method, spectral density is not proportional to spin concentration.

The ST-ESR method relies on the measurement of three pairs of spectral line height ratios taken at six spectral positions corresponding to the rotational and wobbling motion of the labeled lipid chain in the field anisotropy axis. These three ratios are then related directly to those values for hemoglobin undergoing isotropic motion in solution [63]. To overcome the problem of spectral overlap in mixed spectra, an integral method has been devised in which the ST-ESR spectral integrals are determined and compared with a standard calibration value [64]. Using subtraction methods to arrive at a value for each of the integrals in a multicomponent conventionally recorded ESR spectrum, the value of the integral, and hence the correlation time, of the other component can be found. It is possible that this approach is of general use in phospholipid-protein studies where the nitroxide motional restriction (to give τ_c values $> 10^{-6}$ s) at the interface of a large protein mass is significant and the motional rate of labels away from the protein is fast and the integrals rather small.

(*d*) Spin label ESR spectral subtraction with no exchange. In the study of phospholipid-protein interactions, the most common and informative situation has been the one in which a large ($M_r > \sim 20{,}000$) predominantly integral membrane protein is in a phospholipid bilayer and the interaction is studied by 14C or 16C labeled phospholipid spin label. Here, the added, freely diffusing spin labeled phospholipid is assumed to travel throughout the whole of the complex sampling the bulk bilayer away from the protein as well as the environment immediately next to the protein, the first lipid shell, *boundary* (nothing to do with being bound) or *annular* region. (The description that uses the word "immobilized" is correct inasmuch as the labels are motionally restricted as the ESR spectroscopic time scale defined below, but the spin labels still exchange with the bulk and are not stuck to the protein for any appreciable time, as discussed below).

The most suitable label to use to study boundary lipid is determined by a number of parameters. Ideally, a good resolution between the labels at the boundary and those in the bulk away from the protein interface is required for subsequent analysis, that is, a combination between the uppermost spectrum in Fig. 11a and spectrum in Fig. 11b. Therefore labels undergoing fast, large amplitude motion in the bulk (typically those in the hydrophobic core), but are motionally restricted at the protein interface and thus give rise to a slow motion spectrum, are best resolved into their individual components, as shown in Fig. 14. In this ideal situation, then, a well resolved two component spectrum is recorded, one being from spin labels in the bulk bilayer and one from spin labels motionally restricted in the protein annulus, and its deconvolution into the two respective parts can readily be accomplished by computer methods.

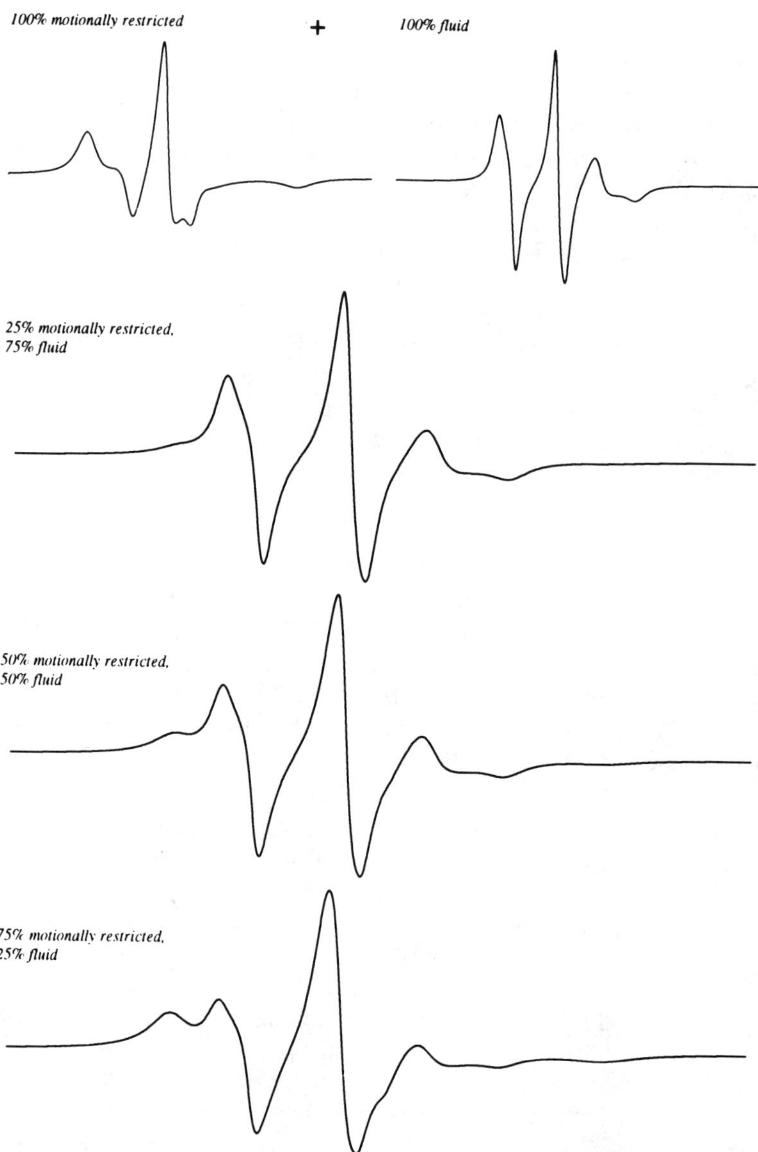

Figure 14 In a bilayer containing a large ($M_r > 20$ K) integral protein in sufficient amount the ESR spin label spectrum for an acyl chain label ($n > \sim 10$) is resolved into two spectral components, one originating from the motionally restricted labels at the protein-lipid interface and the other from labels in the bulk of the bilayer away from the protein-lipid interface. Here, two computer generated spectra have been added at different proportions to show how an ideal situation may produce a two component spectrum with differing proportions of protein (and hence spectral components) in the bilayer. The proportion of each component can be determined by computer subtraction methods.

If it is assumed that initially there is no exchange of the spin labeled phospholipids between the bulk and the annulus *on the spin label ESR time scale*, then each of the two components will be well defined, and subtraction is rather straightforward. Slow exchange, at a rate ν_{ex}, therefore, has a limit in this case that is defined (classically) by the frequency difference between the two spectra that are each characteristic of their own motional environment. For this case, ν_{ex} is given by the difference between the hyperfine splittings of the broad, motionally restricted component ($A_{max}{}^{b}$) and that for the narrower, spectrum ($A_{//}{}^{f}$) from labels in the nonprotein interacting bulk, with the limit of ν_{ex} approximating to

$$\nu_{ex} < \frac{\Delta A}{h} = \frac{A_{max}{}^{b} - A_{//}{}^{f}}{h}$$

where A_{max} is the hyperfine interaction in Teslas for each spectrum (measured as shown in Figs. 11b and 13) and h is Planck's constant; as a rough guide, 0.1 mT = 2.8 MHz, and exchange of the spin labels between two (or more) environments must be slower than ~ 10^8 Hz to give rise to a two (or more) component spectrum.

The exchange limit defined above required to give a resolution of a multicomponent spectrum from a phospholipid spin label in a membrane is about one or two orders of magnitude slower than the lateral diffusion of lipids between a lipid-lipid site ($\nu_{ex} \sim 10^6$ Hz) in a membrane and implies that motional restriction does occur upon collision of a phospholipid with the protein hydrophobic interface. A comprehensive list of the systems, some in natural membranes and some in reconstituted complexes, in which such phospholipid-protein interactions have been studied and the results obtained, are given in recent reviews [2,9–16].

In performing an ESR spectral subtraction, with the assumption that may or may not be true that only *two* motional environments are sensed by the phospholipid spin label in a protein-containing complex, the following practical operations, in summary, are required. If the experimental two component spectrum $M = B + F$, and B = broad spectrum from motionally restricted phospholipid spin labels at the protein interface and F = spectrum from phospholipid spin labels in the bulk, protein-free part of the complex, then with B' = broad experimental or simulated spectrum of a slowly moving phospholipid spin label in bilayers and F' = an experimentally derived or simulated spectrum from a phospholipid spin label in protein-free bilayers, the operations performed iteratively are

Operation I: $M - F' \Rightarrow B$
Operation II: $M - B' \Rightarrow F$

In specific detail, the manipulations for operation I are, for an ideal case,

1. Record the spin label ESR spectrum M from the bilayer complex at a particular phospholipid/protein ratio and temperature T.
2. Record the spin label ESR spectrum F' from protein-free phospholipid bilayers with the same spin label at a range of temperatures, $T \pm \sim 20°$ under identical experimental conditions.
3. If the spectrum from the phospholipid-protein complex is two component, select one of the series of F' spectra that corresponds most closely to F, the narrower, bulk spectrum in M, even though this might be at a different temperature from that used for M.

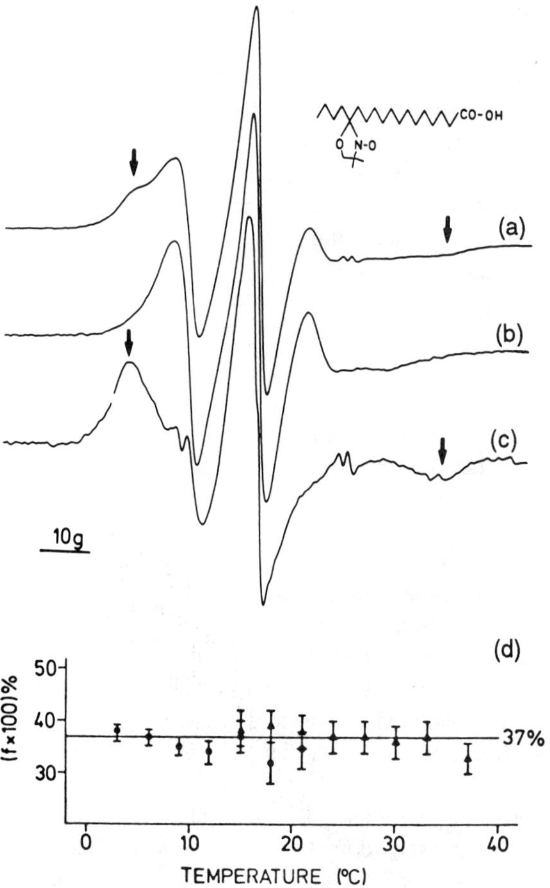

Figure 15 An experimental spectrum (a) from a protein-lipid complex (14-stearic acid spin label in rod outer segment disc membranes; rhodopsin:lipid mole ratio of 1:67; temperature 4°C) that has been resolved into a mobile bulk spectral component (b) and a broad motionally restricted component (c) by operations I and II, respectively, as detailed in the text. The variation of the proportion (f_b) of the motionally restricted component should not change with temperature if the protein aggregation state remains unchanged, as shown for rhodopsin (d). (From Ref. 82).

4. Using digital methods (see Fig. 14), subtract F' from M to leave a residual spectrum B that is broad, has no negative integral contributions (see Fig. 15), and corresponds as closely as possible to the spectrum expected from a motionally restricted spin label (see Fig. 15).
5. Repeat 3 and 4 with different F' spectra until a good end point in the comparison with a motionally restricted spectrum B' is achieved.

For operation II, repeat the above, but this time subtract a generated broad spectrum B' from the membrane spectrum M, thus

1. Generate a spectrum characteristic of a phospholipid spin label motionally restricted in bilayers, perhaps by recording a spectrum of the same spin label at low (<4°C) temperatures, or in another phospholipid bilayer system, which is similar (in line positions and line widths) to the broad spectrum B in the composite membrane

spectrum M. It may be necessary to reduce the scan width of this synthesized spectrum to achieve a good matching in the spectrum, and some indication of spectral parameters will already be available from the process described above. Again, the actual source of this manufactured spectrum is not crucial, as long as it has the correct spectral features and proportions.

2. Subtract spectrum B' from spectrum M until a spectrum F that is characteristic of protein-free bilayers and close in shape to one of the F' series (see above) is obtained.
3. Repeat 1 and 2 until a good spectral form is achieved, using different B' spectra if necessary.

In carrying out both these operations, it is necessary to make valued judgments about the effectiveness of a subtraction that can be described by any or all of the following features.

Symmetry in the low and, especially, high field lines of B
No negative contributions in the spectral integral
Comparison with experimental spectra

In considering operation I, motionally broadened spin label ESR spectra do not contain any information about the order of the spin label when undergoing slow ($\tau_c < 10^{-7}$ s) motion (Fig. 11). Thus the origin of the motionally restricted component B' used for comparison is not critical except that ESR spectral line positions do change with the local polarity of the nitroxide (Sec. III.D.1.a), and so the spectrum from the same phospholipid spin label as used in the protein-containing membrane, but in a protein-free bilayer, is probably the best one to use.

The shapes of the spectra used for the subtraction process itself are critical; they must have a characteristic line shape and separations that, when subtracted from the two component spectrum, will leave an authentic spectral component behind. The combination of the the two line shapes, one narrow and one broad, that make up the phospholipid-protein complex derived spectrum being analysed, are such that spectral subtractions are actually rather critical when the proportions of each component are quantified. Also, by using the negative integral criterion, it is not easy to over subtract. Values for the fraction of total spectral intensity in each component (f_b and f_f for the B and F spectral components, respectively, where $f_b + f_f = 1$) can generally be found to within ± 2–5%, and by approaching a subtraction by both operation I and II, similar values for f should be obtained for both methods, as shown for the case of rhodopsin in retinal rods [82].

There is no reason to expect that the values of f_b (and f_f) will be constant with temperature. If the phospholipid-protein interaction is purely structural, involving no change in the protein aggregation state, then f_b and f_f might well be temperature insensitive for spectra that are only two component with no spectral broadening due to exchange of phospholipids between the bulk and protein interface. However, if the protein changes its aggregation state with temperature, then some discontinuity in the temperature dependence of both f_b as well as the exchange rates (see below) may occur, especially if the process is cooperative.

(e) Spin label ESR spectral subtractions with intermediate exchange. If the exchange rate ν_{ex} of phospholipids between the protein interface and the bulk phase in a phospholipid complex occurs within a specific intermediate exchange range ($10^6 < \nu_{ex} < 10^8$ s), then a spin label will produce an ESR spectrum in which the two spectral lines characteristic of the two environments are broadened through overlap of the components with each other (Fig. 16). This exchange broadening is in the time range for averaging of the two spectra, so that

$$\frac{A_{max}{}^{B} - A_{//}{}^{F}}{h} > \nu_{ex} > \frac{|X^{B}A_{max}{}^{B} - X^{F}A_{//}{}^{F}}{h}$$

where X is the fractional proportion of spin labels in each of the two environments B and F, each giving rise to a spectrum with hyperfine splittings of A_{max} and $A_{//}$ respectively (Figs. 11 and 13) and $X^B + X^F = 1$. Intermediate exchange phenomena usually cause spectral broadening over about one order of magnitude of exchange frequencies, and given the usual range of hyperfine splittings for nitroxides (A_{max} <3.2 mT; $A_{//}$ < 1.6 mT), this defines the range ($10^6 < \nu_{ex}, < 10^8$ s) over which exchange broadening of the spectral anisotropy will be observed.

Spectral broadening due to intermediate exchange of labels between boundary and bulk environment in phospholipid-protein interaction studies, have been observed in a number of systems and is normally the case rather than the exception. Slow to in-

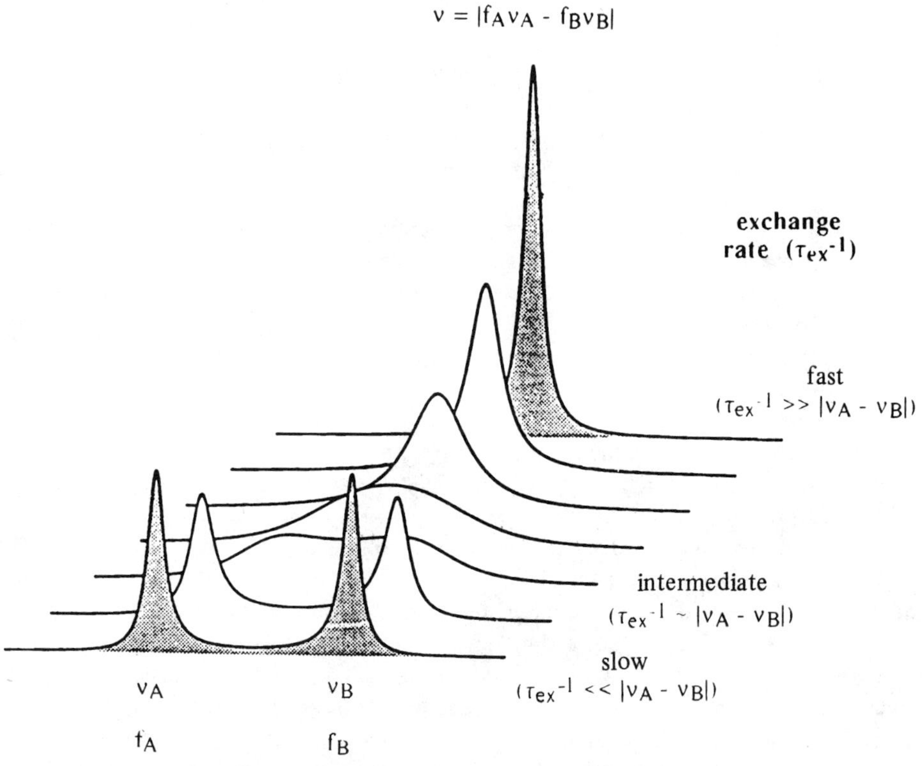

Figure 16 Diagrammatic representation showing the effect of molecular exchange on the spectral line-shape. For a species that rise to a magnetic resonance spectral line at frequencies ν_A and ν_B in environments A and B and in proportions f_A and f_B, if the exchange between A and B is slow, so that $\nu_{ex} = \tau_{ex}{}^{-1} \ll |\nu_A - \nu_B|$, then two spectral lines will be recorded. For fast exchange, where $\nu_{ex} = \tau_{ex}{}^{-1} \gg |\nu_A - \nu_B|$, a single line will be observed with position $\nu = |f_A\nu_A - f_B\nu_B|$. In the intermediate exchange range of $\nu_{ex} \sim \tau_{ex}{}^{-1} \sim |\nu_A - \nu_B|$, the exchange rate of the molecular motion between sites A and B will be given by $\nu_{ex} = |\nu_A - \nu_B|$, where ν_A and ν_B are measured from the line positions (in Hertz).

termediate exchange can only be accurately quantified by spectral simulation, since the integration method for determining the proportion of spin labels in each motionally distinct environment does not apply; the integrated intensity decreases for the intermediate exchange case, and sharp end points cannot be assessed. Exchange broadening also complicates the process of selecting a suitable broad or fluid component for subtraction. Any inconsistency in values of f_b and f_f observed from operation I or II detailed in Sec. III.D.1 is usually due to the phenomenon of exchange broadening.

The way to deal with this complication is to computer simulate the two individual spin label spectra from the protein-free and protein-interacting spin labels as if no exchange occurred, and add the spectra together including a method that enables the exchange overlap to be included through exchange broadening (Fig. 17). The method employs the well-known chemical exchange equations modified from the Bloch equations for magnetization transfer [65]. The choice then has to be made whether to simulate the spectra assuming that the orientation of the nitroxide is preserved on going from the bulk to the interface (model I in Ref. 66) or whether the nitroxide undergoes all orientational excursions in each environment (model II in Ref. 66). For spin labels with the nitroxide at the center of the bilayer, the assumptions in model I are probably applicable, and spectral broadening is not significantly changed when model II is evaluated, with the added advantage of a considerable saving in simulation time for each spectrum with model I.

The practical operations are then

1. Record the experiment, exchange broadened two component membrane spectrum M_{exp}.
2. Use subtraction methods as detailed above (operation I or II) to obtain approximate B_{exp} and F_{exp} spectra and estimate f_b and f_f.
3. Computer simulate each component, B and F, as closely as possible from the line positions to give B_{sim} and F_{sim}.
4. Generate $f_bB_{sim} + f_fF_{sim}$ with no exchange.
5. With the generated M_{sim}, include the exchange parameter to match the spectral overlap observed in M_{exp}.

Although spectral addition is not as critical as subtraction, the end point being less easy to evaluate critically, it is possible to include a mathematically computed error factor that evaluates the difference between the integrals of the simulated (I_{sim}) and experimental (I_{exp}) spectra (for example, using a quadratic minimization function $\sigma^2{}_{err} = (I_{exp-sim})^2/\Sigma\, I^2{}_{exp}$).

The information gained in this process is now both the stoichiometry (from f_b and f_f) and the exchange rate ν_{ex} of phospholipid spin labels from the protein interface to the bulk, which is probably dominated by the leaving process back to the bulk from the motionally restricted environment adjacent to the protein. Whether exchange is included or not in the deconvolution processes, the values of f should be similar, since spectra at different temperatures (usually lower temperatures) are used in the "no exchange" method (Sec. III.D.1.c) to compensate partially for the exchange broadening. The exchange rates are found, in some cases, to be essentially independent of the protein-lipid ratio of the membrane, indicating that the phospholipids in the bulk away from the protein interface are all undergoing similar motion [67].

Not surprisingly, all the situations in which this approach to the study of phospholip-id-protein interactions has been examined has given values for the phospholipid exchange rates in the 10^7 s time range, since this is defined by the spin label technique as described

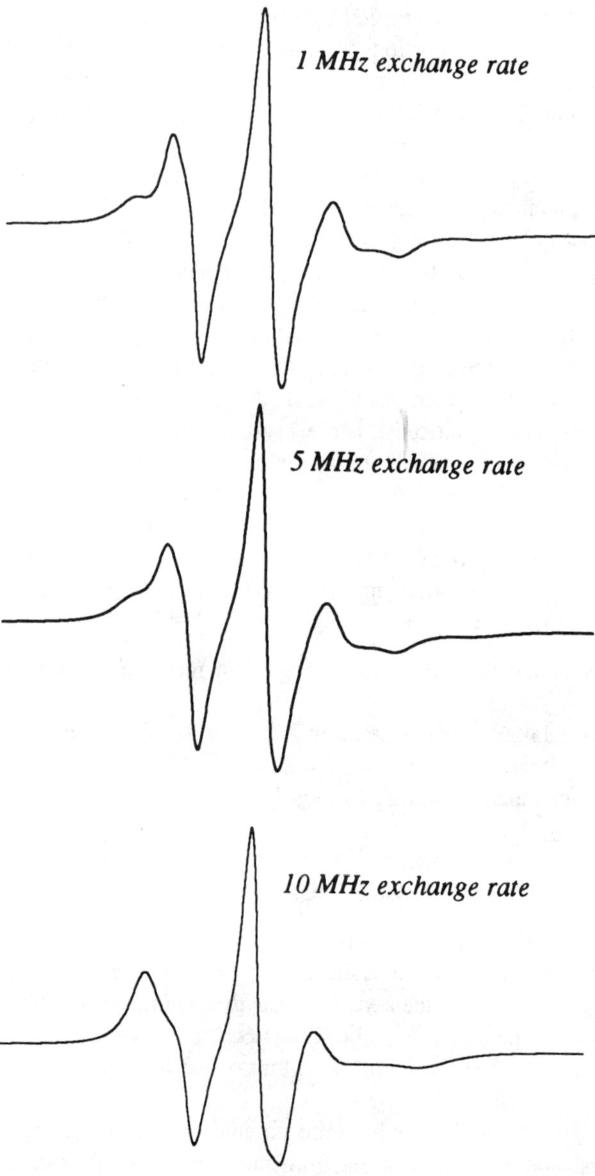

Figure 17 Computer generated nitroxide spin label spectra from Fig. 14 (50% immobile, 50% mobile) in which molecular exchange between the two environments has been simulated at 1 MHz (slow exchange), 5 MHz (intermediate exchange), and 10 MHz (fast exchange). The range over which intermediate exchange occurs, and at which exchange rates can be reliably determined, is usually about an order of magnitude in exchange frequencies.

above. The difficulty arises when trying to interpret spectra that appear single component but are really due to fast exchange of lipids onto and off the protein interface, as might be the case for a small protein that laterally diffuses through the bilayer with a rate that is comparable to that for lipids.

(f) Spin label ESR spectral analysis (fast exchange). As already alluded to, phospholipid spin labels that either are not directly at the interface of a large integral protein, or are in a complex containing a small protein that diffuses at a rate similar to that of phospholipids, do not produce ESR spectra that are two component in nature. In these cases, fast exchange of the phospholipids occurs between the different parts of the membrane in which the acyl chains all have similar fast motion. The information available in these situations is not so readily analyzed in quantitative terms, and certainly it has not been extensively explored so far. The spectral changes that occur are usually confined to line broadening and line position changes, which are indicative of fast exchange ($A_{//}/h$ (in site 1) $> \nu_{ex} > A_{//}/h$ (in site 2)) of the phospholipid spin label between two regions in the bilayer in which the label experiences fast motion (possibly at different rates) or at different amplitudes (giving different values for $A_{//}$) in the two or more environments.

In one case, the retinal disc membrane, which contains ~ 90% of the large 39 kDa integral protein rhodopsin, the motion of phospholipid spin labels not in the annulus but in subsequent shells has been analyzed from line width measurements (Fig. 18). A decrease in motional rate for a phospholipid spin label was observed and compared to the protein-free bilayers, which may have been due to fast-intermediate exchange of the phospholipids between the various protein-free parts of the bilayer [69].

Similarly, the exchange of lipids between subsequent shells away from the annular, motionally restricted lipid in cytochcrome *c* oxidase-phospholipid complexes has been analyzed using the anisotropy parameter $\Delta = 2(A_{//} - A\perp)_{\text{eff}}$. This parameter reflects both line width and line position changes in the spectra and was determined from a series of narrow residual spectra $F_1, F_2, F_3, \ldots$ after subtraction of the broad spectrum B from the composite membrane spectrum $M = B + F$ for complexes at different protein-phospholipid mole ratios n_t where increasing lipid was assumed to produce extra "shells" of lipid, each with spectra $F_2, F_3, F_4, \ldots$ Here, Δ (Fig. 19) appeared to be linear with $1/n_f$, as expected for fast exchange, throughout individual "shells" with n_f lipids. The analysis, and spectra, show that the effect of a large protein is felt by the bilayer lipids some distance away from the protein, and such perturbations may be important in protein-protein communication.

For the case of a small protein diffusing through a phospholipid bilayer, similar behavior in the ESR spectra from phospholipid spin labels might be expected as found in subsequent shells away from large proteins, and similar spectral analysis could be applied. However, information about stoichiometry and exchange rates can only be made if some information about one of the spectral components, in the absence of exchange, is available.

(g) Spectral intersubtractions (slow exchange). One way to prepare samples for determinations of phospholipid selectivity for an integral protein is to produce a reconstituted phospholipid-protein complex containing one major phospholipid type, and separate the complex into a number of aliquots to which a different phospholipid spin-label type is added exogenously by the ethanolic injection method (see Sec. III.B.2). Since the complex now contains identical chemical composition, except for the spin label at 1:100–1:200 label-to-background phospholipid ratio, the spectra can be compared directly. The subtraction procedure used now is an *intersubtraction* method [38,70], in

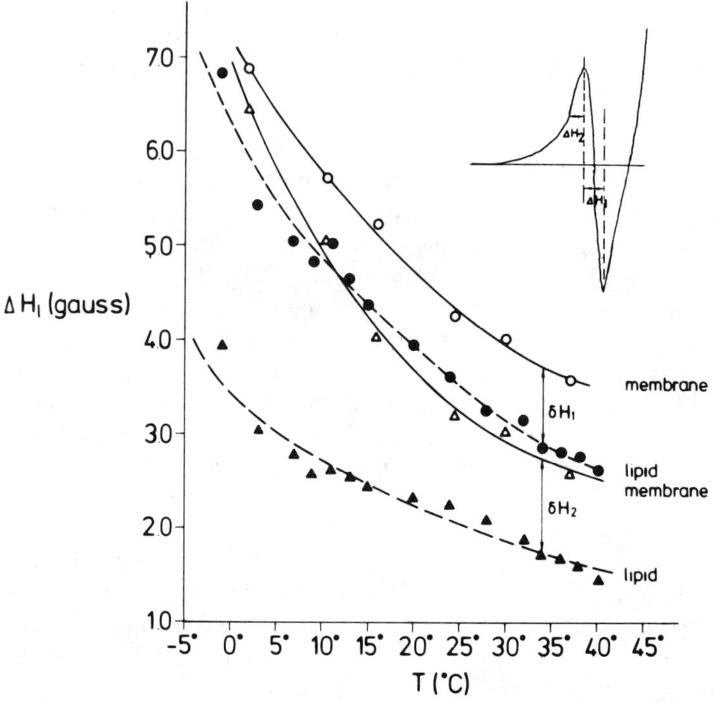

Figure 18 For fast exchange of lipid spin labels, the line widths of the spectra (δH_1 or δH_2) can be used to monitor *relative* changes in exchange rates or diffusion rates, between different environments within the bilayer, for example, lipids within the protein containing bulk phase (not the interfacial region) (open symbols) and protein-free bilayers (filled symbols). The difference between the line widths of 14-stearic acid spin label in two bilayer systems (rod outer segment disc membranes and bilayers made of extracted lipid from the same membranes) imply that the protein (rhodopsin) reduces the diffusion of lipids within the bilayer, as shown by the increased line width for labels in the protein containing bilayers when compared with protein-free bilayers. (From Ref. 69.)

which the two membrane spectra M_1 and M_2 are each composed of the two spectral components B (broad) and F (narrow, fluid-lipid spectrum) as defined above, but in different proportions. With X_1 and $1 - X_1$ for B and F, respectively, in M_1, and X_2 and $1 - X_2$ for B and F, respectively, in M_2, then assuming that B and F are identical in spectral shape when recorded under identical experimental conditions from an aliquot of the same phospholipid-protein complex,

$$M_1 = X_1B + (1 - X_1)F$$
$$M_2 = X_2B + (1 - X_2)F$$

Intersubtraction of M_1 and M_2 (assuming that $X_1B > X_2B$) gives the values of X_1 and X_2, since

$$M_1 - M_2 = \frac{1 - X_1}{1 - X_2} \quad \text{giving the spectrum } B$$

$$M_2 - M_1 = \frac{X_2}{X_1} \quad \text{giving the spectrum } F$$

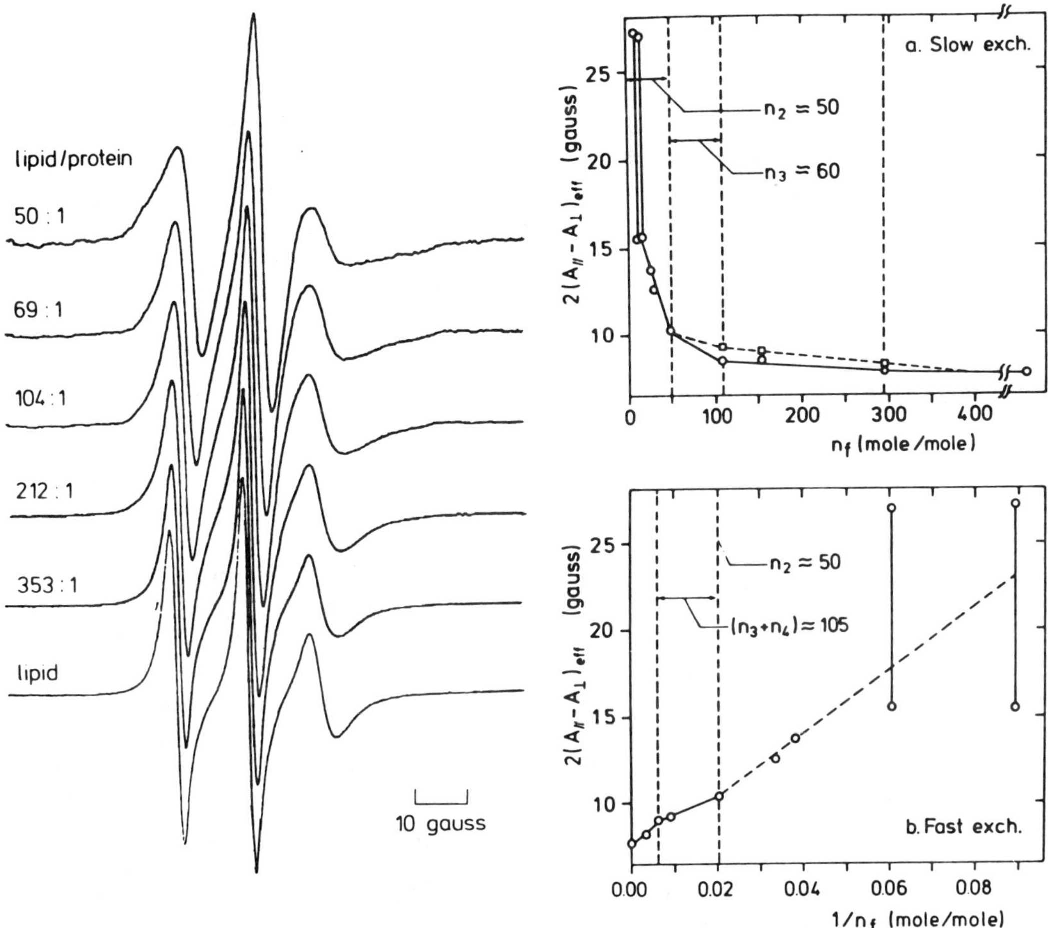

Figure 19 ESR spectra (left hand side) for phospholipid spin labels in the bulk part of phospholipid bilayer complexes containing the large ($M_r \sim 200$ K) integral protein cytochrome *c* oxidase at various protein:lipid mole ratios. These spectra have been generated by subtracting the broad motionally restricted component away from the composite experimental spectrum to reveal the spectra shown, which are from the labels in the bulk part of the bilayer. The values of $A_{//}$ and $A_{\perp}$ (measured as shown in Fig. 13) have been plotted (right hand side) for both slow exchange (a), where $2(A_{//} - A_{\perp})_{eff}$ is assumed to be linear with the protein:lipid mole ratio), and fast exchange (b), where $2(A_{//}–A_{\perp})_{eff}$ is modelled as being linearly proportional to 1/(protein:lipid mole ratio), labels between the subsequent shells of lipid within the bilayers away from the protein-lipid interfacial region, which is ascribed a "first shell". (From Ref. 70.)

The same criteria about evaluation to determine the end pont of an intersubtraction as described above (Sec. III.D.1.d) also applies. The method is particularly useful if selectivity of phospholipids for a protein is being investigated as a function of, for example, pH or salt [38], if the selectivity is thought to be electrostatically mediated, but the method is strictly valid only if

The characteristics of the spin labels in the complex to give M_1 and M_2 are identical in all environments of the complex.

No changes in protein aggregation or morphology occurs in the system on changing external parameters, such as salt or pH.

In principle, the method should be useful for a range of complexes with the same phospholipid and protein and different values of n_t, but because of the changes that occur in the narrow spectrum F due to the lipid further away from the protein being less restricted and giving rise to successively narrower spectra for successively higher values of n_t (the total protein-phospholipid mole ratio), the method is limited, and intersubtractions do not seem to work effectively.

2. NMR Spectral Analysis

NMR studies of phospholipid-protein interactions are scarce in the literature. Problems with amounts of material required and the lack of suitable nuclei for study, without the use of specialist chemistry or microbiology, are significant stumbling blocks. However, some studies have been made using predominantly deuterium NMR methods, and these will be described most fully here.

The slower inherent time scale for NMR experiments (Fig. 5) compared to lipid exchange rates in bilayers means that many of the effects in phospholipids observed from their interaction with proteins are in the fast motional regime for this method, except in one or two cases, where lipid motion is significantly restricted by protein.

(a) Fast phospholipid-protein exchange. Deuterated phospholipids have been used to form phospholipid complexes with large ($M_r > 20{,}000$) integral proteins for study by NMR. The usual spectral change observed is a collapse of the spectral line shape from the normal spherically averaged line shape (Fig. 7) to a single broad line, as shown for rhodopsin (Fig. 20) and band 3 from human erythrocytes. Such spectral broadening is due most probably to the fast exchange of the lipids from a protein-free bilayer to an environment on the protein interface in which the deuterium anisotropy is collapsed. Protein surfaces are not ordered but are highly invaginated (Sec. II.A.1); therefore it is not unreasonable to assume some overall decrease in the order of the phospholipids in such a complex, as shown by spin label ESR methods (see Sec. II.D) in which boundary, motionally restricted phospholipid is observed separately; the two spectroscopic methods are therefore entirely consistent in systems that are themselves directly comparable from the biochemical aspect.

The deuterium NMR spectral parameter most useful is the quadrupole splitting $\Delta\nu_q$, which gives information about the orientation and amplitude of CD bond motion (Sec. III.C.3.c). Assuming fast phospholipid exchange, $\Delta\nu_q$ (observed) should vary with $1/n_t$, where n_t is the phospholipid-protein mole ratio, since [14,71]

$$\Delta\nu_q(\text{observed}) = n_c \cdot \Delta\nu_q(\text{complex}) - n_f \cdot \Delta\nu_q(\text{free}) \tag{3}$$

gives

$$\Delta\nu_q(\text{observed}) = n_c(\Delta\nu_q(\text{complex}) - \Delta\nu_q(\text{free})) \cdot \frac{1}{n_t} + \Delta\nu_q(\text{free}) \tag{4}$$

where $\Delta\nu_q$(complex) and $\Delta\nu_q$(free) are the quadrupole splittings for the deuterated segment when at the protein-phospholipid interface and in protein-free bilayers, respectively, and n_c and n_f are the numbers of phospholipids in the protein complex and free part of the bilayer, respectively. The one case in which this analysis has been carried out has

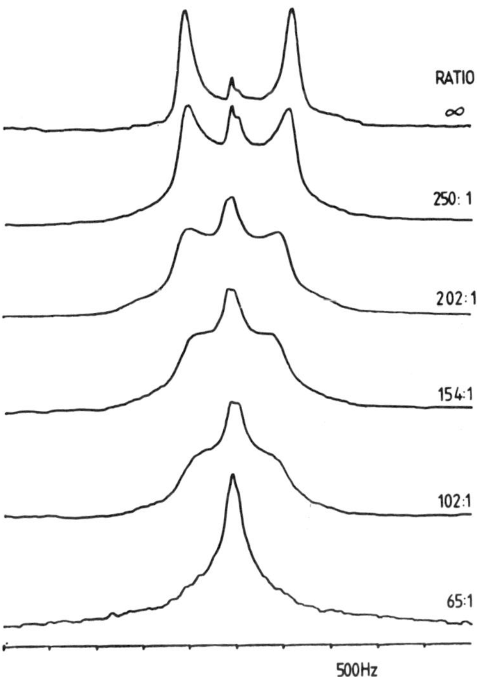

Figure 20 Deuterium NMR spectra for bilayer complexes of rhodopsin and dimyristoyl phosphatidylcholine deuterated at the choline trimethyl group, -$N^+(C^2H_3)_3$. The complexes are in the liquid crystalline state, and the protein is monomeric. Increasing the protein content collapses the well resolved spherically averaged spectrum (see Fig. 7) because the protein disorders the lipid head groups at the bilayer surface when compared to the protein-free bilayers (upper spectrum). The protein and lipid are in fast molecular exchange on the deuterium NMR anisotropy averaging time scale (see Fig. 5); thereby the perturbation of the bilayer by the protein is averaged throughout the whole bilayer. (From Ref. 80.)

confirmed the linearity of $\Delta\nu_q$ with $1/n_t$ for different molar ratios of the peripheral protein, the myelin basic protein, in phosphatidyl glycerol bilayers (Fig. 21) [71].

One difficulty in this method is the lack of information about the spectral parameter for the protein-interacting phospholipid, but by extrapolation it appears that the value of $\Delta\nu_q$ tends to a value of zero for the case studied, implying complete molecular disorder at the site of contact between the protein and phospholipid.

The exchange rate (ν_{ex}) for phospholipids into and out of the protein complex can be estimated from [14,71]

$$(\nu_{ex}) = |(\Delta\nu_q\ (\text{complex}) - \Delta\nu_q(\text{free})| \tag{5}$$

giving at least a fast exchange limit for this interaction.

Relaxation times determined for deuterated phospholipids in protein complexes appear to reflect a decrease in the rate of motions within a bilayer on introduction of a protein. Both the spin-lattice (T_1) and the spin-spin (T_2) relaxation times, which reflect fast (ns) and slower (μs) motions, respectively, seem to be affected, even though segmental motions (from T_1) and bilayer flexing (from T_2) are independent types of motion [10,54]. Although not thoroughly studied, in the measurement of T_1, the

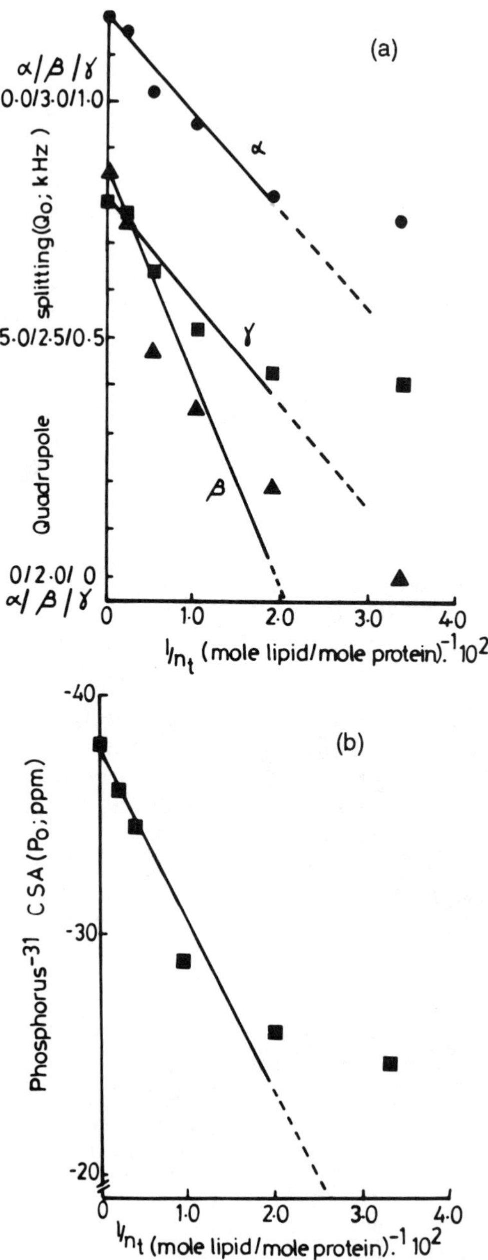

Figure 21 Variation of the observed values of the deuterium quadrupole splitting ($\Delta\nu_q$) (a) and ^{31}P chemical shift anisotropy (b) for bilayers of dimyristoyl phosphatidyl glycerol deuterated at the glycerol group, P-O-C(α)^{2}H$_2$-C(β)^{2}H(OH)-C(γ)^{2}H$_2$(OH), in bilayers as a function of the lipid-to-protein mole ratio ($1/n_t$) for the peripheral myelin basic protein. The close to linear dependence of the four independent experimental parameters at low protein:lipid ratio ($1/n_t$) implies that fast lipid-protein exchange takes place in the complexes, as suggested by Eq. 4. (From Refs. 2 and 71.)

spectral amplitude changes with interpulse time in the inversion recovery experiment and appears to be single component, which is not consistent with ESR spin label studies where the electron-nitrogen hyperfine coupling is sensitive to motions in a similar time range (see above). It has been shown for phospholipid interactions with cytochrome *c* oxidase that T_{2e} varies, given a fast exchange interaction within the T_{2e} time scale, simply as

$$\frac{1}{T_{2e}} = f\left(\frac{1}{T_{2e}}\right)\text{(bulk)} + (1 - f)\left(\frac{1}{T_{2e}}\right)\text{(protein)} + \Omega_{ex} \qquad (6)$$

where Ω_{ex} is an exchange parameter and the bulk value of T_{2e} measured from protein-free bilayers and the (protein) value obtained by extrapolation to bulk-lipid free complexes; Eq. 3 is similar to Eq. 6.

^{31}P NMR methods have been used to study phospholipid-protein interactions, in particular the morphological changes induced in bilayers when proteins are either in or removed from the bilayer [16]. The characteristic bilayer spectrum, which is a spherically averaged powder pattern with a spectral width of ~50ppm due to the two-dimensional motion of the phospholipids in a bilayer (Fig. 9), is a good check on a system to confirm the bilayer nature of a complex for other studies. Since the phospholipids are in fast lateral motion throughout the whole complex on the ^{31}P NMR time scale (defined by the anisotropy averaging of ~ 6 kHz, ≈50ppm, at a field strength of 6.3 T), changes in the ^{31}P chemical shift anisotropy have been used to monitor phospholipid-protein interactions with the myelin basic protein, again using the fast exchange analysis described above (Eq. 3). The interaction of cytochrome *c* with cardiolipin also appears to be averaged by fast motion, probably due to fast up and down protein motion on the bilayer surface, as shown by the considerable changes in the spin-lattice relaxation properties and features of the ^{31}P nucleus due to an interaction with the heme group of the protein [72].

^{13}C NMR has been used to study the interaction of peripheral proteins with small, unilamellar vesicles which are prerequisite for high-resolution spectra for this nucleus in phospholipids. Hence the interaction of cytochrome *c* with phosphatidylcholine and cardiolipin containing bilayers has been explored, with the changes in chemical shift induced by the protein being essentially those averaged out by fast exchange with respect to the frequency of the chemical shift changes [73].

To obtain high-resolution spectra from large extended bilayer complexes of phospholipids containing proteins, new methods of magic angle sample spinning are required. In this approach, the sample is held spinning at the magic angle (54.7°) with respect to the magnetic field to average out the relaxation and dipolar couplings that normally complicate and broaden the spectrum from static samples. The major sources of spectral resonances are the phospholipids, and changes in their positions can be observed in the presence of a protein. Little use of this method has been reported in the study of protein-phospholipid interactions, but cytochorme *c* has been studied in its interaction with cardiolipin bilayers [46,49].

In Fig. 22 is the ^{13}C MASS NMR spectrum of a cytochrome *c* cardiolipin complex in which all the phospholipid resonances can be resolved [74]. The spectral changes induced on protein binding, in particular the reduction in the $\Delta^{10,12}$ resonance intensity, are continuous with protein content, implying fast cytochrome *c* cardiolipin interaction on the time scale (ms) of the spin-lattice relaxation time of the individual groups; no quantitations of protein-lipid stoichiometry have been made here.

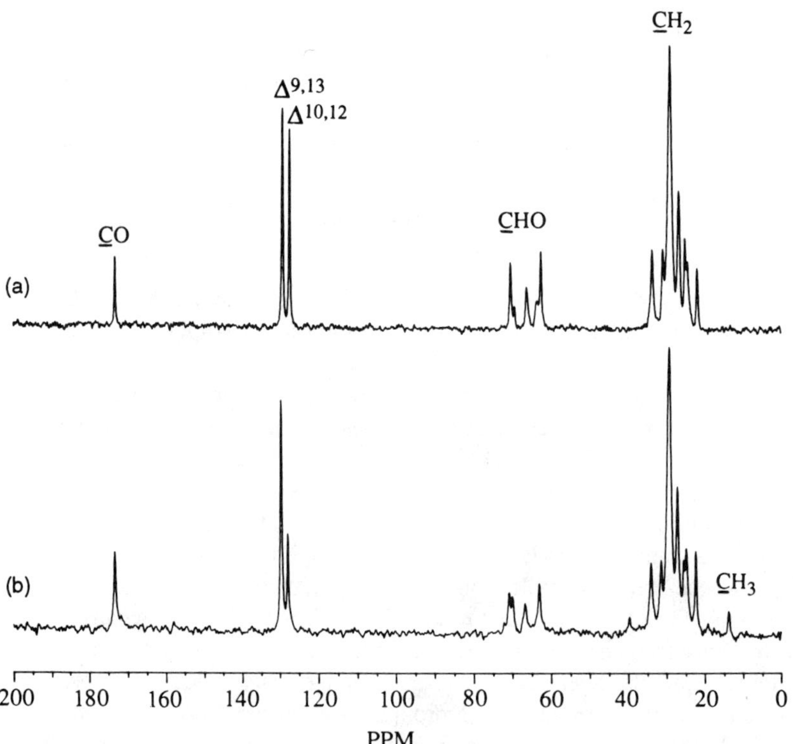

Figure 22 Natural abundance, proton decoupled cross-polarization MASS ^{13}C NMR (161.98 MHz) spectra of hydrated liquid crystalline cardiolipin bilayers without (a) and with (b) bound cytochrome *c* at a 1:15, protein-lipid mole ratio. In the presence of the protein, the intensity of certain of the lipid resonances, under the experimental conditions of cross-polarization employed, are reduced, indicating some selective interaction of the protein with the linoleoyl (*cis*-18:2-$\Delta^{9,13}$) group of the fatty acyl chains, suggesting that the protein permits a greater degree of large amplitude motions within this part of the bilayer. (From Spooner and Watts, Ref. 84.)

(*b*) Slow to intermediate phospholipid-protein exchange. Two-site slow to intermediate exchange is best observed in the NMR spectra in which the anisotropy difference between the spectrum from each environment is large. Slow exchange of phospholipids between a protein complex and the bulk bilayer on the NMR time scale should, therefore, as for the spin label ESR method, produce two spectral components (Fig. 16). The added complication may occur that if one motional environment induces rather slow (on the NMR time scale) motions of the nucleus and thus no spectrum (or at least a much reduced intensity) would be observed for that component, or it could be very broad (this complication does not arise with spin labels since a spectrum is always recorded, regardless of motional rate). Exchange between the motionally distinct environments would then change the spectral features, as shown in Fig. 16.

Slow, two-site exchange ($\nu_{ex} < 10^4$ s^{-1}) between a protein complex site and a protein-free bilayer has been observed for the particular case of the M13 bacteriophage coat protein. The analysis of the deuterium NMR spectra was similar to that used in spin label studies and detailed above (Sec. III.D.1.d). Spectral subtractions (operation I, Sec. III.D.1.d) were performed with a bilayer spectrum from protein-free bilayer used for

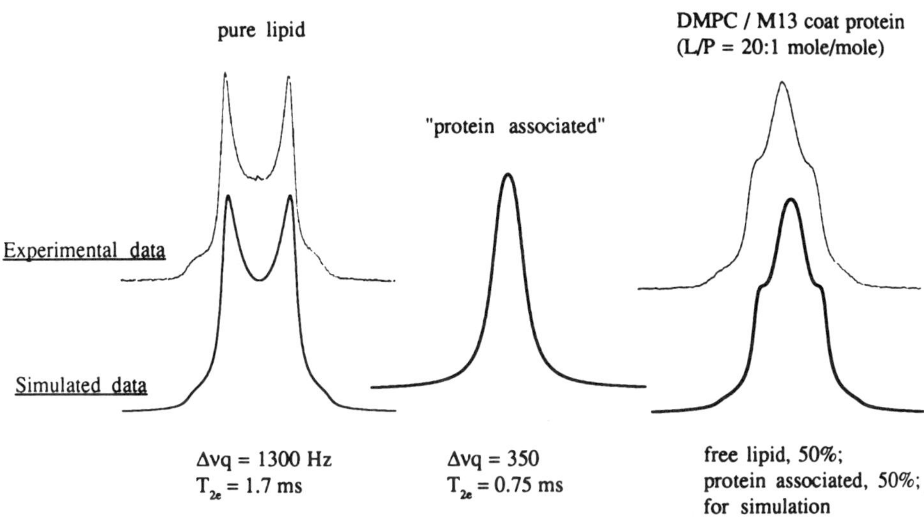

Figure 23 Deuterium NMR spectra of dimyristoyl phosphatidylcholine deuterated at the choline trimethyl group, -$N^+(C^2H_3)_3$, in bilayers with the coat protein from the M13 bacteriophage. Because the protein is highly aggregated, probably into linear aggregates (see Figs. 2 and 3), a certain proportion of lipids is trapped on the deuterium NMR anisotropy averaging time scale, thereby producing a second spectral component in addition to the spectrum from the protein-free bulk part of the bilayer in which the protein is not present. The experimental spectrum can be simulated computationally and added together with a "protein associated" synthesized spectrum, to aid quantitation of the proportion of the two spectral components (right-hand spectra), assuming no protein-lipid exchange in this example. (From Ref. 75.)

subtraction from the membrane spectrum to give the broad, single line spectrum characteristic for deuterated phospholipids in a disordered environment and trapped in protein aggregates (Fig. 23). Both quantitations (to give f_b and f_f) and simulations, which include line broadening due to exchange between the two bilayer environments, were performed, the spectral simulations being performed with a variation of the spin label simulation computer program (Sec. III.D.1.e) [66] in which the anisotropy of the spectrum was included [75] to quantitate the coat protein-phospholipid interaction (Fig. 23).

Although the M13 coat protein is small ($M_r \sim 5240$), with only 50 residues and a single transmembrane pass, the NMR results were interpreted to show that the protein traps phospholipids within aggregates of the protein that might be similarly formed immediately prior to budding of the bacteriophage from the host *E. coli* cell plasma membrane [75, 76].

E. Determining Stoichiometry of Phospholipid-Protein Interactions and Protein Size

Values of f_b and f_f, determined as described above from ESR or NMR experiments, can now be used further to determine the phospholipid-protein stoichiometry. With the chemically determined value for the protein-phospholipid mole ratio n_t of the complex under study, values of n_b and n_f can yield the number n_b of phospholipids in the boundary of the integral protein from

$$n_b = f_b \times n_t \tag{7}$$

$$n_b + n_f = n_t \tag{8}$$

assuming that the reporter lipid samples all parts of the bilayer and that the recorded intensity from the spin-labels in each of the two components in the composite membrane spectrum is proportional to the concentrations of the reporters (spin labels or nuclei in NMR) in each of the two motionally different environments. It is normally assumed that the instrumental settings used in the experiments are such that the intensity of any component is not reduced due to saturation.

In reconstituted complexes where the phospholipid/protein ratio can be manipulated, a titration of n_b with n_t can be performed. Again, if the protein does not aggregate over the range of n_t values of the complexes produced, that is, if the value of n_b has a fixed relationship to the protein size and annular occupancy, then [9,13,59,77]

$$\frac{n_f}{n_b} = \frac{n_t}{N_1 K_r^{av}} - \frac{1}{K_r^{av}} \tag{9}$$

where K_r^{av} is the average relative association constant for all the sites N_1 at the protein interface, which is completely filled by lipid. It is assumed, for the spin label experiment, that the background phospholipid is in large excess over the spin label. By plotting the binding ratio n_f/n_b with the phospholipid/protein mole ratio (Fig. 24), we find that at

$$\frac{n_f}{n_b} = 0, \qquad n_t = \text{the number of phospholipids around the protein}$$

$$\text{at } n_t = 0, \qquad \frac{n_f}{n_b} = \frac{1}{K_r^{av}}$$

If $K_r^{av} = 1$, then the labeled phospholipid, if of the same type as the major phospholipid used for the complex formation, has no selectivity for the protein interface over the unlabeled phospholipid; this is the situation reported in every case studied so far and shows that spin labels do not interact preferentially with the protein when compared with the unlabeled analog. It may be that some sites on a protein can interact specifically with particular phospholipid types, and this formulation is useful in describing this selectivity (see below).

Deviations from the linearity of $1/n_t$ with binding ratio n_f/n_b usually occur at high protein densities (Fig. 24) and indicate that the number of lipids per protein monomer is changing, that is, that protein aggregation is occurring (Sec. II.A.2, Fig. 1). Thus extrapolation of the data may be necessary to obtain the value for n_b and then enable the protein perimeter and size to be determined (Sec. II.A.5 and Eq. 1). The values of n_b measured by ESR spin label methods are given for a range of integral and peripheral proteins in Table 1.

It is often found that there is a close relationship between the value of n_b and the number of phospholipids minimally required to support the protein function, implying that such interactions may be responsible for solvating the protein within the bilayer and maintaining it in a suitable form for biochemical activity [9,37].

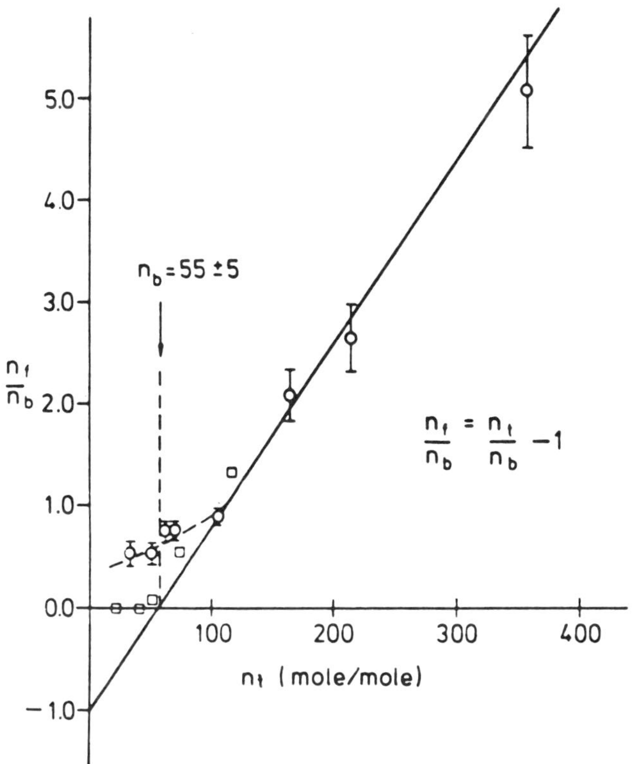

Figure 24 Values of the binding ratio n_f/n_b for lipid spin labels deduced from double integrated intensities of the subtracted ESR spectra (operation I Sec. III.D.1.d) of labels in bilayers of dimyristoyl phosphatidylcholine containing cytochrome *c* oxidase (O) and in complexes of aqueous extracted bovine cytochrome *c* oxidase (□). The linear dependence of n_f/n_b with protein-lipid mole ratio (n_t) for higher protein content (> 100:1) shows that the protein interacts with a fixed number of lipids per protein, a monomer or dimer, as predicted by Eq. 9. The deviation from linearity at higher protein content ($n_t < \sim$ 100:1) is due to the protein becoming aggregated and thus the lipid-protein interface is conserved through protein-protein contacts (see Fig. 1). (From Refs. 9 and 69.)

1. Determining Selectivity of Phospholipid-Protein Interactions

In a spin label experiment, if the value for $K_r \neq 1$ in a titration experiment of n_f/n_b with variable values of n_t (Eq. 9) (Fig. 25), then some selective interactions between the spin labeled phospholipid and the protein probably occur, with $K_r > 1$ implying a greater association for that phospholipid type for the protein interface, and the values of $K_r < 1$ implying that the phospholipid spin label has a repulsion from the interface. In Table 2 are shown the relative selectivities for some proteins studied by ESR methods.

For the case of a selective interaction relative to the major bulk phospholipid used to produce the complex, the relative free energy of association for the selective interaction is given by

$$\Delta G^{\text{rel}} = \Delta G_{\text{L}} - \Delta G_{\text{PL}} = -RT \ln \left[\frac{K_r^{\text{av}}(L)}{K_r^{\text{av}}(PL)} \right] \tag{10}$$

Table 1 Stoichiometries of the Motionally Restricted Lipid Component in Various Lipid-Protein Systems N_1^{exp} and Estimates of the Number of First-Shell Lipids That Can Be Accommodated Around the Integral Proteins N_1^{calc}

Protein/membrane	N_1^{exp} (mol/mol)	N_1^{calc} (mol/mol)
(Na^+, K^+)-ATPase-DOPC	63 ± 3	(~57–72)
(Na^+, K^+)-ATPase shark rectal gland	58 ± 4	(~57–72)
Cytochrome oxidase-DMPC	45 ± 4	40–45
Acetylcholine receptor-DOPC	40 ± 7	43
Ca^{2+}-ATPase-egg PC	22 ± 2	(~23–27)[a]
Sarcoplasmic reticulum/Ca^{2+}-ATPase	24	(~23–27)[a]
Bovine rod outer segment disk/rhodopsin	25 ± 3	24 (±2)
Frog rod outer segment disk/rhodopsin	23 ± 2	(24)
Myelin-proteolipid apoprotein-DMPC	10 ± 2	(~10–12)[b]

DMPC, dimyristoyl phosphatidylcholine; DOPC, dioleoyl phosphatidylcholine;
[a]Calculated assuming a dimer, with monomer radius 20 Å.
[b]Calculated assuming hexamer.
Source: Adapted from Refs. 13 and 77.

where it is assumed that the ΔG^{rel} value is determined with respect to the background phospholipid in the complex (usually phosphatidyl choline); ΔG^{rel} has a value typically 1–5 kJ mol^{-1} for those phospholipids studied to date [9,13,77].

The phospholipid selectivity may be due to specific sites on the protein available for a particular phospholipid type or, more likely, to a stronger binding of the phospholipids at some or all the protein interface sites [37]. It is also possible that a combination of the two effects may contribute to the selectivity of association observed in the spin label experiment.

The effects of selectivity on the exchange rates (see Sec. II.C) have been examined, and it has been shown (from Eq. 9) that

$$\frac{\nu_{ex}{}^{B}}{\nu_{ex}{}^{F}} = \frac{n_t}{N_1 K_r} - \frac{1}{K_r} \tag{11}$$

The on-rate for a phospholipid to the protein interface has been shown reasonably to be diffusion controlled (~5–7 × 10^6 s for a stearic acid and phosphatidylcholine spin label in dimytristoyl phosphatidylcholine bilayers) and the same for both lipid types. It is suggested that it is the off-rate that determines the phospholipid selectivity of the phospholipid-protein interaction [78,79].

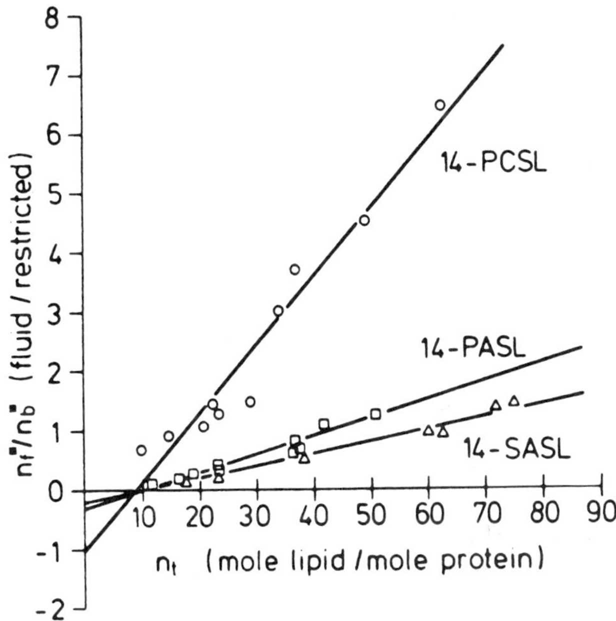

Figure 25 The variation of the ratio of "fluid" to motionally restricted lipids n_f^*/n_b^* as determined from the integrals of the spin-label ESR spectra for the 14-stearic acid (Δ), 14-phosphatidylcholine (○), and 14-phosphatidic acid (□) spin labels in complexes of the myelin proteolipid protein and dimyristoyl phosphatidylcholine as a function of lipid-to-protein mole ratio n_t at 30°C, as suggested by Eq. 9. The intercept of the data on the n_t axis indicates that ten lipids are interacting with the protein at any one time. The intercept on the n_f^*/n_b^* axis reflects any selectivity for the protein by the labels, with an intercept of −1 indicating the $K_r^{av} = 1$ for the phosphatidylcholine spin label, indicating no specific interaction for the protein when compared with the unlabeled lipid; the other two labels show a similar specific interaction with the protein, with the fatty acid-spin label slightly more strongly associated than the phosphatidic acid label. (From Ref. 77.)

Table 2 Order of Selectivity of Spin Labeled Lipids for Association with Integral Membrane Proteins

Protein	Order of lipid selectivity
Myelin proteolipid	SA > PA > CL $\gtrsim$ PS > PG ≈ SM ≈ PC > PE
(Na^+, K^+)-ATPase	CL > PS ≈ SA $\gtrsim$ PA > PG ≈ SM ≈ PC ≈ PE
Cytochrome oxidase	CL > PA ≈ SA > PS ≈ PG ≈ SM ≈ PC ≈ PE
Acetylcholine receptor	SA > PA > PS ≈ PC ≈ PE
Ca^{2+}-ATPase	CL > PS ≈ SA $\gtrsim$ PA $\gtrsim$ PG ≈ SM ≈ PC ≈ PE
Rhodopsin	CL ≈ PA ≈ SA ≈ PS ≈ PG ≈ SM ≈ PC ≈ PE

CL, cardiolipin; PA, phosphatidic acid; PS, phosphatidylserine; PG, phosphatidylglycerol; PC, phosphatidylcholine; PE, phosphatidylethanolamine; SM, sphingomyelin; SA, stearic acid.

Source: From Refs. 13 and 77.

Since it is the polar head group of a phospholipid that defines the electrostatic nature of the molecule, it is not unreasonable to assume that this is the site at which any molecular selectivity may be defined. Such selectivity between a protein and a phospholipid has been shown for the peripheral protein, the myelin basic protein, which binds to bilayer surface through predominantly electrostatic interactions with lysine residues and anionic phospholipids [2,9,13,14]. No interaction was monitored with phosphatidylcholine in similar complexes, implying a rather sensitive and well-defined interaction at the bilayer surface.

IV. CONCLUSIONS AND FUTURE DIRECTIONS

Phospholipid-protein interactions are at an advanced state in that they have been studied now for almost 20 years, with the majority of the work, including attempts at quantifying the interactions rather better, having been performed in the last 14 years. However, problems of controlling the biochemistry of phospholipid-protein complex formation through reconstitution technology still exist, and sometimes we lack the essential assurance required nowadays that the protein under examination really is the form that is functional and that therefore the information gained is of some relevance. The number of proteins accessible to biophysical studies in general is limited by availability, and new genetic methods may solve the problem. Rather little information is available from NMR methods on phospholipid-protein interactions, due in part to the sensitivity problem but also to the severe limitations of conventional high-resolution NMR to the study of large macromolecular complexes.

It would be useful to have more complementary information from both NMR and spin label ESR studies of phospholipid-protein interactions, on the same system and complex, reinforced by other measurements and observations such as calorimetry, microscopy, diffraction, etc. To this end, the stoichiometry n_b of the phospholipid protein interaction was assessed to show that above five phosphatidylcholines were associated on average with each M13 coat protein in the aggregate. On geometric grounds in which the size of the protein and phospholipid were considered, it was suggested that the aggregation state of the protein could not be in a large conglomerate, but rather in a linear (not necessarily straight) array. Similar two component spectra were observed by ^{31}P NMR studies of the same complexes, indicating that the protein-lipid complex was long-lived on both NMR time scales, and probably for longer than 0.3 ms. Spin label ESR experiments on the same complexes also confirmed the entrapment of phospholipids within M13 aggregates, but by this method the stoichiometry was found to be different, with the ESR method identifying ~ 25% more motionally restricted phospholipids, the extra phospholipids being those at the protein-phospholipid interface and not detected by deuterium NMR, being in fast exchange with the bulk on this time scale. There are rather few examples where parallel ESR and NMR experiments have given directly complementary results, and more are needed.

Consistent pictures do not really exist for lipid interactions with any one protein, except perhaps for the well studied cases of rhodopsin and cytochrome *c* oxidase, and even here, conflicting information from NMR and ESR experiments is to be found in the literature [80,81,82,83]. For a systematic study of one protein, one might suggest

1. A successful reconstitution of an active, integral membrane protein, perhaps a transport protein

2. A demonstration of its activity in a variety of phospholipid-type bilayers in terms of kinetic data
3. A spin label ESR study of its phospholipid interactions and selectivity, followed by a parallel NMR (probably deuterium) study giving complementary information
4. Characterization by electron microscopy and thermal denaturation studies to show bilayer disposition and any enhanced stability in folding in response to local phospholipid environment

To give a consistent molecular picture of phospholipid-protein interactions related to protein function, eventually reinforced by a crystal structure and mechanistic description of how phospholipid-protein interactions may control activity.

All these are a long way off and will take much time to accumulate for even one protein, but shreds of relevant information are becoming available to at least make a start on the jigsaw puzzle for one or two proteins.

ACKNOWLEDGMENTS

Some of the work described here was supported by SERC grants nos. GRE/69188, GR/F/54006, GR/F/34084, and GR/F/86925 and by CEC contracts nos. ST002J and ST00368. My thanks to José Areas, David Fraser, Chantal L'Hostis, Saira Malik, Teresa Pinheiro, Sian Renfrey, Jeremy Rowntree, Malkit Sami, Paul Spooner, Anne Ulrich, and Leon van Gorkom for critical comments on this chapter, and to SERC and the EC for financial support for the work described here.

REFERENCES

1. A. Watts, *Nature 294*:512 (1981).
2. A. Watts, *J. Bioenerg. Biomembr. 19*:625–653 (1987).
3. A. Watts, *Current Opinions in Cell Biology* (G. B. Warren and K. Simons, eds.), Current Science, London, Vol. 4, 1989.
4. J. C. Skou and J. G. Nørby, *Na^+–K^+-ATPase. Structure and Kinetics*, Academic Press, New York, 1979.
5. G. Cevc, *Biochemistry 26*:6305 (1987).
6. A. Watts and L. C. M. van Gorkom, in *The Structure of Biological Membranes* (P. Yeagle, ed.), CRC Press, New York, 1991.
7. D. Fraser, S. Louro, L. I. Horvath, K. Miller, and A. Watts, *Biochemistry 29*:2664–2669 (1990).
8. P. C. Jost, O. H. Griffith, R. A. Capaldi, and G. A Vanderkooi, *Proc. Natl. Acad. Sci. USA 70*:4756–4763 (1973).
9. D. Marsh and A. Watts, in *Lipid-Protein Interactions* (P. C. Jost, and O. H. Griffith, eds.), Wiley-Interscience, New York, 1982, Vol. 2, Chap. 2.
10. M. Bloom and I. C. P. Smith, in *Progress in Protein-Lipid Interactions* (A. Watts and J. J. H. H. M. De Pont, eds.), Elsevier, Amsterdam, 1985, Vol. 1, Chap. 2.
11. A. J. Deese and E. A. Dratz, in *Progress in Protein-Lipid Interactions* (A. Watts and J. J. H. H. M. De Pont, eds.), Elsevier, Amsterdam, 1986, Vol. 2, Chap. 2.
12. P. F. Devaux and M. Seigneuret, *Biochim. Biophys. Acta 822*:63–125 (1985).
13. D. Marsh and A. Watts, in *Advances in Membrane Fluidity* (R.A. C. Aloia, ed.), Alan R. Liss, New York, 1987, Vol. 2.
14. A. Watts, in *Biophysics of the Cell Surface* (R. Glaser and D. Gingell, eds.), Springer-Verlag, Berlin-Heidelberg, 1990.

15. J. Seelig, A. Seelig, and L. Tamm, in *Lipid-Protein Interactions*, (P. C. Jost and O. H. Griffith, eds.), Wiley-Interscience, New York, Vol. 1, Chap. 3.
16. B. de Kruijff, P. R. Cullis, A. H. Verkleij, M. J. Hope, C. J. A. van Echteld, T. F. Taraschi, P. Van Hoogevest, J. A. Killian, A. Rietveld, and A. T. M. Van der Steen, in *Progress in Protein-Lipid Interactions* (A. Watts and J. J. H. H. M. De Pont, eds.), Elsevier, Amsterdam, 1985, Vol. 1, Chap. 3.
17. P. C. Jost and O. H. Griffith, in *Spin-labelling. Theory and Application* (L. J. Berliner, ed.), Academic Press, New York, 1976, Chap. 7, pp. 251–272.
18. J. H. Davis, *Biochim. Biophys. Acta 737*:117–171 (1983).
19. R. Glaser, in *Biophysics of the Cell Surface* (R. Glaser and D. Gingell, eds.), Springer-Verlag, Berlin-Heidelberg, 1990, pp. 173–192.
20. D. Oesterhelt and W. Stoekenius, *Nature New Biol. 233*:152–155 (1971).
21. C. E. Dempsey, N. J. P. Ryba, and A. Watts, *Biochemistry 25*:2180–2187 (1986).
22. B. Sternberg, C. L'Hostis, and A. Watts, *Biochim. Biophys. Acta*, in press.
23. R. M. Clegg and W. L. C. Vaz, in *Progress in Protein-Lipid Interactions* (A. Watts and J. J. H. H. M. De Pont, eds.), Elsevier, Amsterdam, 1985, Vol. 1, pp. 173–229.
24. J. R. Silvius, in *Lipid-Protein Interactions* (P. C. Jost, and O. H. Griffith, eds.), John Wiley, 1982, Vol. 2, pp. 239–281.
25. R. Henderson and P. N. T. Unwin, *Nature 257*:28–32 (1975).
26. M. Diesenhofer, *Nature 318*:618–624 (1985).
27. E. Eliopoulous and J. B. C. Findlay, *TIPS 11*:492–500 (1990).
28. J. P. Segrest, in *Mammalian Cell Membranes* (G. A. Jameison, and D. M. Robinson, eds.), Butterworths, London, 1977, Vol. 3, pp. 1–26.
29. R. A. F. Reithmeier, *Ann. NY Acad. Sci. 574*:75 (1989).
30. R. R. Kopito and H. Lodish, *Nature 316*:234–238 (1985).
31. A. Watts, in *Macromolecular Structures and Dynamics* (S. Harding, and A. J. Rowe, eds.), Royal Chemical Society, London, 1988.
32. W. G. Khorana, *J. Biol. Chem. 263*:7439–7442 (1988).
33. C. R. H. Raetz and W. Dowhan, *J. Biol. Chem. 265*:1235–1238 (1990).
34. M. Bloom, *Can. J. Phys. 57*:2227–2230 (1979).
35. J. L. Bowers, R. L. Smith, C. Coretsopoulos, A. C. Kunwar, M. Kreniry, X. Shan, H. S. Gutowsky, and E. Oldfield, in *Progress in Protein-Lipid Interactions* (A. Watts and J. J. H. H. M. De Pont, eds.), Elsevier, Amsterdam, 1986, Vol. 2, Chap. 3.
36. P. C. Jost, O. H. Griffith, and R. A. Capaldi, *Biochim. Biophys. Acta 311*:141–152 (1973).
37. H. Sanderman, in *Progress in Protein-Lipid Interactions* (A. Watts and J. J. H. H. M. De Pont, eds.), Elsevier, Amsterdam, 1986, Vol. 2, Chap. 6.
38. J. K. Brotherus, P. C. Jost, O. H. Griffith, J. F. W. Keana, and L. E. Hokin, *Proc. Nat. Acad. Sci. USA 77*:272–276 (1980).
39. T. S. Weiderman, R. D. Pates, J. M. Beach, A. Salmon, and M. F. Brown, *Biochemistry 27*:6469–6474 (1988).
40. B. J. Gaffney, in *Spin-Labelling. Theory and Application*. (L. J. Berliner, ed.). Academic Press. New York, 1976, Vol. 1, pp. 567–571.
41. J. Seelig and A. Seelig, *Quart. Rev. Biophysics 13*:9–16 (1980).
42. H. Eibl, in *Membrane Fluidity in Biology* (R. C. Aloia, ed.), Academic Press, New York: 1983, pp. 217–236.
43. K. A. W. Wirtz, J. A. F. op den Kamp, and B. Roelofson, in *Progress in protein-Lipid Interactions* (A. Watts, and J. J. H. H. M. de Pont, eds., 1986, Vol. 2, Chap. 7.
44. S. Malik, Protein-protein and protein-lipid interactions of brand 3 in native and model membranes, D. Phil. thesis, Oxford University, 1991.
45. J. M. Seddon, G. Cevc, and D. Marsh, *Biochemistry 22*:1280–1289 (1983).
46. P. J. R. Spooner and A. Watts, *Biochemistry 30*:3871–3879 (1991).
47. P. R. Allegrini, G. Plushke, and J. Seelig, *Biochemistry 2*:6452–6458 (1991).

48. S. Akoka, C. Tellier, and S. Poignant, *Biochemistry 25*:6972–6977 (1986).
49. P. J. R. Spooner and A. Watts, *Biochemistry 30*:3880–3885 (1991).
50. H. U. Gally, G. Pluschke, P. Overath, and J. Seelig, *Biochemistry 20*:1826–1831 (1981).
51. M. F. Brown, J. Seelig, and U. Haberlen, *J. Chem. Phys. 70*:5045–5053 (1979).
52. B. Sternberg, P. Gale, and A. Watts, *Biochim. Biophys. Acta, 980*:117–126 (1989).
53. A. Watts and P. J. R. Spooner, *Chem. Phys. Lipids 57*:195–211 (1991).
54. M. R. Paddy, F. W. Dalquist, J. H. Davis, and M. Bloom, *Biochemistry 20*:3152–3162 (1981).
55. P. Knowles, D. Marsh, and H. W. E. Rattle, *Magnetic Resonance of Biomolecules*, Wiley, London, 1976.
56. A. Abragam, *Principles of Magnetic Resonance*, Oxford University Press, Oxford, 1961.
57. C. P. Poole, Jr., *Electron Spin Resonance. A Comprehensive Treatise on Experimental Technqiues*, Wiley-Interscience, New York, 1967.
58. S. Schreier, C. F. Polnaszek, and I. C. P. Smith, *Biochim. Biophys. Acta* 515:375 (1978).
59. D. Marsh, in *Biological Magnetic Resonance*, Vol. 8, (L. J. Berliner and J. Reuben, eds.), Plenum, 1989.
60. P. Fretten, S. J. Morris, A. Watts, and D. Marsh, *Biochim. Biophys. Acta 598*:247–259 (1980).
61. D. Marsh and A. Watts, in *Liposomes: From Physical Structure to Therapeutic Applications* (C. G. Knight, ed.), Elsevier, Amsterdam. 1981, Chap. 6.
62. J. H. Freed, in *Spin-Labelling, Theory and Application* (L. J. Berliner, ed.), Academic Press, New York, 1970, pp. 53–132.
63. D. D. Thomas, L. R. Dalton, and J. S. Hyde, *J. Chem. Phys. 65*:3006 (1976).
64. L. I. Horváth and D. Marsh, *J. Mag. Res. 54*:363 (1983).
65. H. M. McConnell, *J. Chem. Phys. 28*:430 (1958).
66. J. Davoust and P. F. Devaux, *J. Mag. Res. 48*:475 (1982).
67. N. J. P. Ryba, L. I. Horvath, A. Watts, and D. Marsh, *Biochemistry 26*:3234 (1987).
68. D. Marsh, A. Watts, R. D. Pates, R. Uhl, P. F. Knowles, and M. Esmann, *Biophys. J. 37*:265–274 (1982).
69. P. F. Knowles, A. Watts, and D. Marsh, *Biochemistry 18*:4480–4487 (1979).
70. P. F. Knowles, A. Watts, and D. Marsh, *Biochemistry 20*:5888–5894 (1981).
71. F. Sixl, P. J. Brophy, and A. Watts, *Biochemistry 23*:2032–2039 (1984).
72. T. J. Pinheiro, P. J. R. Spooner, and A. Watts, to be published.
73. L. R. Brown and K. Wüthrich, *Biochim. Biophys. Acta 468*:389–410 (1977).
74. P. J. R. Spooner and A. Watts, *Biochemistry 30*:3880–3885 (1991).
75. L. C. M. van Gorkom, L. I. Horváth, B. Sternberg, M. Hemminga, and A. Watts, *Biochemistry 29*:3828–3834 (1990).
76. C. J. A. M. Wolfs, L. I. Horvath, D. Marsh, A. Watts, and M. Hemminga, *Biochemistry 28*:9995–10001 (1989).
77. D. Marsh, in *Progress in Protein-Lipid Interactions* (A. Watts and J. J. H. H. M. De Pont, eds.), Elsevier, Amsterdam, 1985, Vol. 1, Chap. 4.
78. L. I. Horvath, P. J. Brophy, and D. Marsh, *Biochemistry 27*:46–54 (1988).
79. L. I. Horväth, P. J. Brophy, and D. Marsh, *Biochemistry 27*:52–96 (1988).
80. C. E. Dempsey, N. J. P. Ryba, and A. Watts, *Biochemistry 25*:2180–2187 (1986).
81. G. L. Hoatson, E. Faure, P. Fellman, B. Farren, A. L. MacKay, and M. Bloom, *Biochemistry 25*:3804–3812 (1986).
82. A. Watts, I. D. Volotovski, and D. Marsh, *Biochemistry 18*:5006–5013 (1979).
83. A. Watts, J. Davoust, D. Marsh, and P. F. Devaux, *Biochim. Biophys. Acta* 643:673–676 (1981).
84. P. R. J. Spooner and A. Watts, *Biochemistry 31*:10129 (1992).

III

Biological Aspects

21

Biological Distribution

Mark A. Yorek *Veterans Administration Medical Center, Iowa City, Iowa*

I. INTRODUCTION

Phospholipids are the major lipid component of most cell membranes. The only exceptions are neural tissues, which contain a large portion of cerebrosides and gangliosides, and membranes of plant chloroplasts, which are enriched in diacylglycolipids [1].

The phospholipid content and composition of cellular membranes is highly regulated and varies from one cell type to another. Furthermore, the phospholipid composition of different membranes within a cell also varies. Although the specificity of the content and composition of phospholipids in different cells and organelles and the asymmetry of phospholipid distribution in membranes is well known, the influence this specificity has on cell properties is not understood. The current hypotheses proposed to explain the specificity of phospholipid distribution are that certain enzymes require a specific membrane; lipid order for optimal activity and that cells require a defined membrane order to maintain integrity and functional properties [2,3].

The total phospholipid content and composition in different types of membranes vary considerably. Mammalian membranes generally contain a high proportion (40–80% of dried weight) of phospholipid [4–9]. The lipid content of plant membranes (30–40%) is also high but considerably less than the content of mammalian membranes [10,11]. In microorganisms the membrane lipid content varies to a greater extent than it does in mammalian or plant membranes [12–15]. The lipid composition of membranes from different tissues or cellular organelles also varies (Figs. 1 and 2). This is probably due to the different functional requirements of tissues and organelles and may also be influenced by external factors such as the environment and disease. In this chapter, the biological distribution of phospholipids in bacteria, fungi, plants, invertebrates, fish, and mammals will be discussed. The variability and asymmetry of the phospholipid composition in mammalian membranes will also be reviewed.

The advancements made in the technology for examining phospholipid composition have increased our ability to detect specific changes in cellular membrane structure. Improvements in thin-layer chromatography and high-performance liquid chromatography have increased our ability to separate major phospholipid classes as well as separate phospholipid molecular species [16–46]. Improved techniques in gas liquid chromatogra-

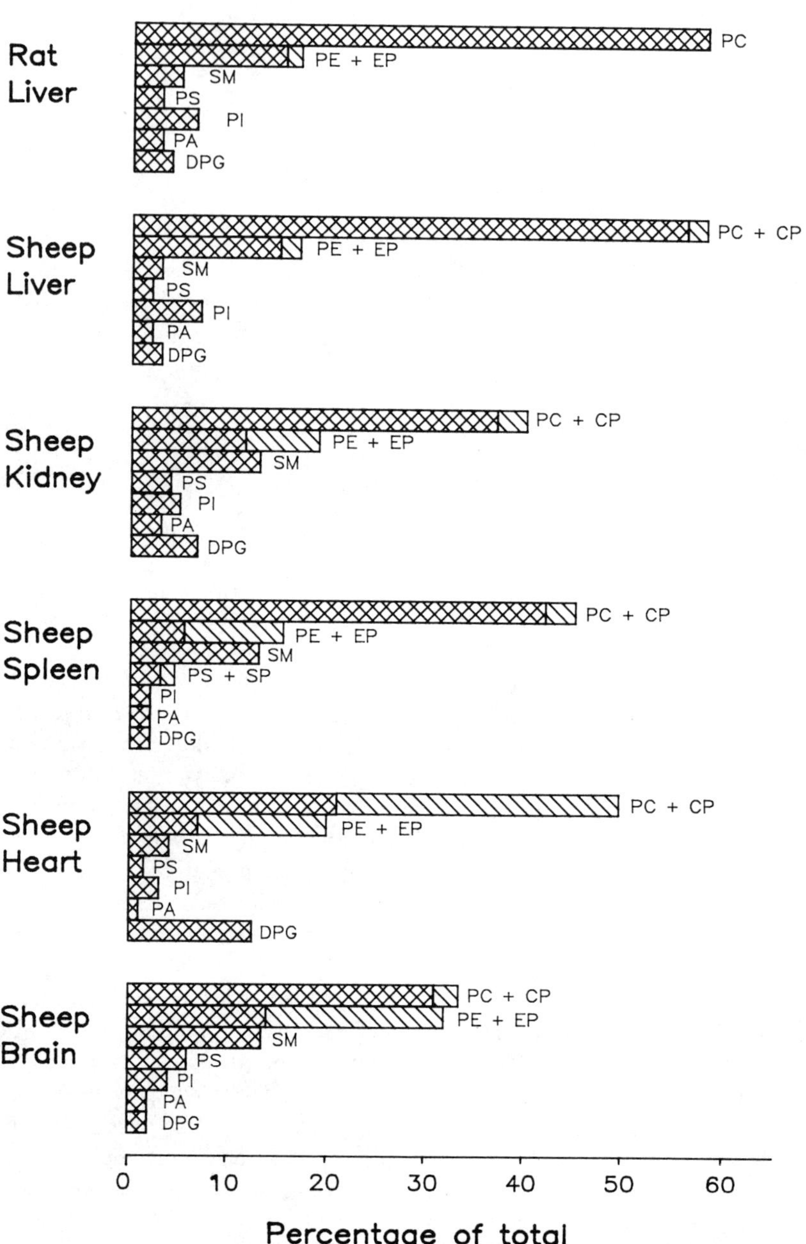

Figure 1 Phospholipid composition of rat and sheep tissues. Values are given as percentages of the total phosphorus in the phospholipid fraction. PC, phosphatidylcholine; CP, choline plasmalogen; PE, phosphatidylethanolamine; EP, ethanolamine plasmalogen; SM, sphingomyelin; PS, phosphatidylserine; SP, serine plasmalogen; PI, phosphatidylinositol; PA, phosphatidic acid; DPG, diphosphatidylglycerol. The plasmalogens are represented by the right hatched area.

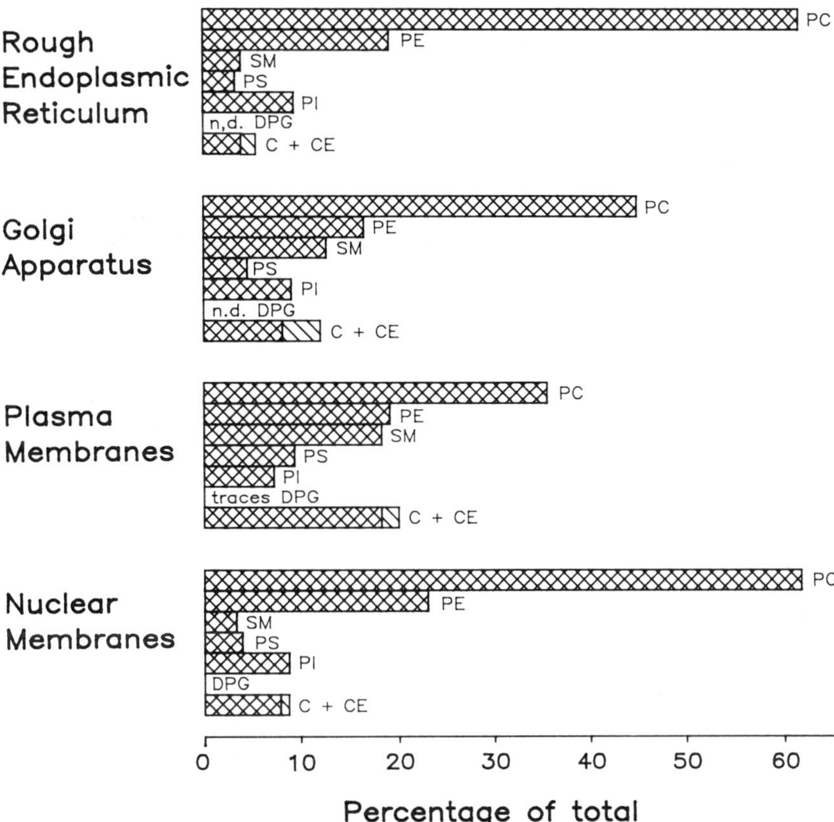

Figure 2 Phospholipid and sterol composition of rough endoplasmic reticulum, Golgi apparatus, plasma membranes, and nuclear membranes from rat liver. Phospholipid values are given as percentages of the total phosphorus of the phospholipid fraction. The abbreviations used for the phospholipids are provided in Fig. 1. C, cholesterol; CE, cholesterol esters (right hatched area). Sterols are presented as percentages of total lipids by weight.

phy and gas chromatography/mass spectroscopy have also increased our ability to detect differences in the composition of phospholipids from different tissues and cellular organelles [24,26–28,47–51]. Furthermore, improved techniques in fractionation of plasma membranes have enhanced the investigation of membrane phospholipid asymmetry [52]. Investigations employing these improved techniques have provided a more detailed picture of the phospholipid composition of cellular membranes.

II. PHOSPHOLIPIDS IN BACTERIA

The phospholipids in *E. coli* are synthesized from CDP-diglyceride, which is formed from phosphatidic acid. *E. coli* membranes contain only three major phospholipid species: phosphatidylethanolamine (accounting for about 75% of the total extractable lipid), phosphatidylglycerol (about 20%) and cardiolipin (about 5%) [53–60]. The phospholipids cover an area equivalent to about 98% of the surface area of a single cell and are located almost exclusively on the inner leaflet of the outer membrane [61,62]. Phosphatidylethanolamine accounts for about 90% of the phospholipid in the inner leaflet of the outer membrane, whereas cytoplasmic membranes are enriched in the other two major species [62–64]. One explanation for the large amount of phosphatidylethanolamine in this membrane is that it forms a stable bilayer with lipopolysaccharide (LPS), which is the major component of the outer leaflet of the outer membrane [62,65,66]. The LPS consists of a hydrophobic region called lipid A and a hydrophilic branched sugar chain. A comparison of the data from various laboratories on the weight ratios of the three major constituents of the outer membrane show that they differ enormously. For *E. coli*, weight ratios for phospholipid to LPS to protein of 1:1:1 to 1:1:6 have been reported [66–67]. Although *E. coli* will tolerate alterations in lipid structure, it seems that the above composition optimizes membrane function and growth of wild-type cells [68–70]. The membranes of *E. coli* also contain a variety of minor phospholipid species (see review [54]). It is plausible that these minor membrane lipid components may have important functional roles in *E. coli*. The major fatty acids in *E. coli* phospholipids are palmitic acid (16:0), palmitoleic acid (16:1), and *cis*-vaccenic acid (18:1). The outer membrane is enriched with saturated fatty acids, whereas the cytoplasmic membranes are enriched in unsaturated fatty acids [56,62,64,67].

III. PHOSPHOLIPIDS IN FUNGI

The total phospholipid content of fungi varies from 1.1 (% mg dry weight) for *Microsporum gypseum* to 7.6 for *Epidermophyton floccosum* [71,72]. Most of the phospholipids in fungi consist of phosphatidylcholine (8–40%), phosphatidylethanolamine (19–30%) and phosphatidylserine (10–28%) [72]. Phosphatidylcholine and lysophosphatidylcholine are generally found on the outer leaflet of the membrane bilayer, whereas phosphatidylethanolamine and phosphatidylserine are located on the inner leaflet [72,73]. The major fatty acids in phosphatidylcholine and phosphatidylethanolamine are palmitic and linoleic. Oleic acid is also present, and the content of oleic acid in phosphatidylcholine is higher than in phosphatidylethanolamine [72].

A microorganism can extensively alter its lipid composition, and changing the lipid order is probably the organism's way of responding to changing environmental conditions. Many laboratories have shown that changing the temperature or modifying the groth medium of fungi can alter the lipid composition [72,74]. Data on the phospholipid composition of some fungi is shown in Table 1 [74].

IV. PHOSPHOLIPIDS IN PLANTS

Phospholipids in plant cells are primarily constituents of lipoprotein membrane structures. Phosphatidylcholine, phosphatidylethanolamine, and phosphatidylglycerol are the major phospholipids in high plants. Plants also contain lesser amounts of phosphatidylinositol and diphosphatidylglycerol. Phosphatidylserine and phosphatidic acid are found in only small amounts. The lipid content in the plant varies with membrane origin. Lipids may account for up to 10% of the dry weight of leaves and up to 40% of the total dry weight of chloroplasts [80]. However, phospholipids account for only about 15% of the total lipids in chloroplasts with rest being galactolipids (40%), sulpholipids (7%), chlorophyll, carotenoids, and quinones (37%). Data in Table 2 show the phospholipid composition from a variety of plant species.

Table 1 Phospholipid Composition of Fungi

	Phospholipid[a] (% of Total)					
Fungal species	PC GroPCho	PE GroPEtn	PI GroPIns	DPG	PS GroPSer	References
Saccharomyces cerevisiae	43	23	21	3	7	75
Endomycopsis selenospora	46	23	17	—	14	76
Saccharomyces cerevisiae	44	24	21	—	11	76
Hansenula anomala	53	23	9	—	15	76
Bullera alba	54	30	8	—	8	76
Trigonopsis variabilis	51	18	15	—	16	76
Candida macedoniensis	41	26	16	—	17	76
Brettanomyces truxellensis	41	21	22	—	16	76
Saccharomyces cerevisiae	39	31	8	11	4	77
Saccharomyces marrianus	42	22	17	—	19	76
Claviceps purpurea	27	19	28	—	13	78
Saccharomyces pombe	45	19	18	—	6	79

[a]PC, phosphatidylcholine; PE, phosphatidylethanolamine; PI, phosphatidylinositol; DPG, diphosphatidylglycerol; PS, phosphatidylserine.

Data presented in Table 3 compare the phospholipid composition of mitochondria and chloroplasts isolated from avocado and cauliflower [80]. The phospholipid content of plant mitochondria is considerably higher than in chloroplasts. The predominant phospholipids in plant mitochondria and chloroplasts are phosphatidylcholine, phosphatidylethanolamine, and phosphatidylinositol. Phosphatidylglycerol is a major phospholipid species in cauliflower chloroplasts.

The fatty acid composition of plant lipids is dependent upon environmental conditions, particularly light and temperature. Many different fatty acids have been detected in plants, but a few common fatty acids account for the majority of the fatty acids in phospholipids. Palmitic, linoleic, and linolenic acids are the major fatty acid constituents of higher plants. Stearic and oleic acids are usually present in smaller amounts, and chain lengths other than 16 or 18 carbons are unusual in other than trace amounts [80,81]. The distribution of fatty acids between the 1 and 2 positions of the glycerol backbone of the phospholipids is similar to that in animal lipids. The saturated fatty acids are located mainly at the 1 position, whereas the 2 position contains predominantly unsaturated fatty acids.

Table 2 Phospholipid Composition of Plant Tissue

Plant	Total phospholipid (μmol/g fresh wt.)	Phospholipid composition (%)[a]					References
		GroPtdCho PC	GroPtdEtn PE	GroPtdGro PG	GroPtdIns PI	DPG	
Leaves							
Corn	1.7	26	14	28	7	14	82
Perennial ryegrass	3.0	44	18	25	7	7	82
Cock's foot	3.4	33	24	33	3	8	82
Lettuce	0.8	38	26	12	7	16	82
Elder (green)	4.4	43	11	39	7	—	83
Tomato	2.2	51	21	20	5	5	82
Squash	3.7	43	27	24	4	2	82
Pumpkin	3.9	36	23	28	13	—	84
Poplar	4.6	40	18	24	11	8	82
Alfalfa	2.8	27	18	24	10	21	82
White clover	3.9	36	22	29	7	7	82
Runner bean	2.7	45	17	22	9	—	85
Rowan	5.5	40	31	18	7	5	82
Rose	3.1	48	18	19	9	6	82
Spinach	2.3	47	17	26	9	1	83
Sugar beet	4.7	47	23	19	11	—	82
Pine	2.3	35	12	24	16	13	82

Leaf stalk							
Sugar beet	14[b]	32	16	8	7	3	86
Root							
Sugar beet	3.5	35	17	6	5	2	86
Parsnip	0.9	38	20	8	13	22	81
Fruit							
Potato	0.7	55	28	2	12	1	87
Apple	0.5	46	31	7	13	2	88
Seed							
Castor bean	1.5[c]	13	83	—	1	—	89
Soybean	1.2[c]	45	15	—	25	4	90
Groundnut	0.7[c]	44	18	—	24	—	90
Pine	1.2[c]	54	11	—	30	—	90
Cotton	—	33	22	—	37	—	91
Sunflower	—	51	23	—	22	—	92
Cocoa	4.0[c]	40	17	—	29	—	93

[a]PC, phosphatidylcholine; PE, phosphatidylethanolamine; PG, phosphatidylglycerol; PI, phosphatidylinositol; DPG, diphosphatidylglycerol.
[b]μmol/g dry weight.
[c]Molar concentration calculated from weight concentration.

Table 3 Lipid Composition of Avocado and Cauliflower Mitochondria and Chloroplasts

	Avocado		Cauliflower	
	Mitochondria	Chloroplasts	Mitochondria	Chloroplasts
Total Lipid[a], %	38	38	33	34
Phospholipid[b]	50	20	54	9
Monogalactosyl diglycerides	2	9	2	19
Digalactosyl diglyceride	3	7	3	7
Sulfolipid	1	4	1	3
Chlorophyll	<.1	8	<.1	34
Neutral lipid	17	ND	19	18
Phospholipid composition[c]				
Phosphatidyl choline	45	34	33	19
Phosphatidyl ethanolamine	21	33	39	13
Phosphatidyl inositol	11	12	10	12
Phosphatidyl serine	4	6	5	9
Phosphatidyl glycerol	5	6	4	44
Diphosphatidyl glycerol	9	9	6	3
Phosphatidic Acid	5	0	3	0

ND = not determined
[a]Total lipid (%) = (lipid/lipid + protein) × 100.
[b]Weight % of total lipid.
[c]Mol %.

V. PHOSPHOLIPIDS IN INVERTEBRATES

A. Mollusca

The lipid composition of the marine invertebrates as well as other marine species is currently a topic of interest because of the role of dietary n-3 polyunsaturated fatty acids in human health and the amelioration of certain cardiovascular diseases. The molluscs and other marine fishery products are excellent sources of n-3 polyunsaturated fatty acids [94–96].

The molluscs constitute one of the more important invertebrate groups in the animal kingdom and are second only to the insects in the number of living species [94]. Molluscs are divided, taxonomically, into seven classes, but lipid data is only available for four: Polyplacaphora, Gastropoda, Bivalvia, and Cephalopoda. Plasmalogens generally account for a large percentage of the total polar lipids in invertebrates, and molluscs contain plasmalogens of phosphatidylethanolamine, phosphatidylcholine, and phosphatidylserine. In the oyster *Crassostrea virginica*, 41%, 28%, 20%, and 11% of the total

Table 4 Seasonal Variation in Distribution of Plasmalogens in Marine Mollusca (% of class)

	Winter			Summer		
Species	Ethanolamine	Choline	Serine	Ethanolamine	Choline	Serine
Gastropoda						
Collisella heroldi	53	30	24	73	21	55
Littorina kurila	69	7	19	70	11	67
Tectonactica janthostoma	52	14	53	70	8	58
Bivalvia						
Crassostrea gigas	55	13	38	81	7	61
Spisula sachalimensis	72	11	21	73	10	46
Mercenaria stimpsoni	19	10	20	69	3	49

phospholipid phosphorus occurs as the diacyl, plasmalogen, sphingolipid, or glycerol ether, respectively [97].

Seasonal changes in the phospholipid composition of marine invertebrates is common and is probably related to the reproductive cycle and is a result of gonadal development or loss of gametes after spawning. Results from studies of the phospholipid seasonal variation of two mollusca species are provided in Table 4 [94].

B. Insects

In most insects, choline phospholipids are the predominant class of phospholipid present [98,99]. However, in two groups of insects, Diptera (flies) and aphids, the ethanolamine phospholipids are the major class of phospholipids. The tissue phospholipid compositions of numerous insects have been reported in a review article by Fast [100]. As is the case in vertebrates, the phospholipid composition of subcellular fractions of a tissue in invertebrates differs from the composition observed when the whole tissue is analyzed. In Diptera, the ratio of ethanolamine and choline in whole tissue is about 1:1, whereas in the mitochondria the ratio is about 2:1 [100].

VI. PHOSPHOLIPIDS IN FISH

The phospholipids from tissues of freshwater and saltwater fish are characteristically rich in polyunsaturated fatty acids [101]. In most cases, docosahexaenoic acid is the major polyunsaturated fatty acid present [101,102]. Generally the phospholipid composition of freshwater and saltwater fish is comparable to that of land animals [103]. Phosphatidylcholine and phosphatidylethanolamine are the predominant phospholipid classes in fish. In lateral line muscle of the goldfish, phosphatidylcholine, phosphatidylethanolamine, and caridolipin accounted for 48%, 34%, and 11% of the total phospholipid [104]. All other phospholipids were present in proportions of less than 3%. The phospholipid content of the goldfish liver accounted for 1.5% of the weight and was composed of 57%

phosphatidylcholine, 22% phosphatidylethanolamine, 7% phosphatidylinositol, 6% sphinogomyelin, and 4% phosphatidylserine [105]. Depending on the diet, the lipid content of livers may vary. The brain of the goldfish contains 9.3% total lipid and 4.4% phospholipid [106]. The phospholipids consisted of 47% phosphatidylcholine, 35% phosphatidylethanolamine, 8% phosphatidylserine, 3% phosphatidylinositol, 2% sphinogomyelin, and 3% phosphatidic acid [105]. The plasmalogens of phosphatidylethanolamine and phosphatidylcholine in goldfish brain account for 60% and 7%, respectively of the phospholipid content of each of these phospholipid classes [107].

The most common molecular species of phosphatidylcholine in trout mitrochondria and microsomal membranes are 16:0/22:6, 16:0/18:1, 16:0/20:3, and 16:0/22:5. In the same membranes, the most abundant molecular species of phosphatidylethanolamine are 18:1/20:4, 14:0/16:0, 18:0, and 18:1/22:6 [108]. The fatty acid content of phosphatidylinositol in trout is enriched in 20:4 and 18:0 [109]. This is similar to the fatty acid content of phosphatidylinositol in terrestrial mammals and may reflect a possible role for the requirement of 20:4 for eicosanoid formation and signal transduction.

Environmental conditions influence the composition of phospholipid in fish. A decrease in temperature has been shown to cause an increase in the polyunsaturated fatty acid content of membrane phospholipids and an increase in ethanolamine phospholipids and a decrease in choline phospholipids and sphinogomyelin [101,110,111].

VII. PHOSPHOLIPIDS IN MAMMALS

A. Introduction

The structure of mammalian biomembranes consists of a lipid bilayer made of phospholipids, which provides the matrix into which the intrinsic proteins are embedded and to which the extrinsic proteins are attached.

Analyses of biomembranes have been conducted for a variety of mammal cells and organelles. The lipids comprising the bilayer matrix are of various types and are present in different amounts. Generally, all mammalian tissues and organelles contain qualitatively the same lipids, except for diphosphatidylglycerol, which predominates in mitochondrial inner membranes and lysosomal membranes [112]. Phosphatidylcholine followed by phosphatidylethanolamine are the major phospholipids of mammalian membranes. Cholesterol is also enriched in animal plasma membranes, lysosomal membranes, and Golgi membranes. Sphingomyelin and gangliosides are also found in the plasma membranes of mammals. Evidence has accumulated over the years to show that membranes are asymmetric. The phospholipid composition of the outer surface of the membrane is different from the inner surface, and the activity and function of the two membrane bilayers are different. Each phospholipid class can also show considerable variation in acyl chain composition. The acyl content of phospholipids is highly specific, and the profile of distribution of the molecular species is highly characteristic of the animal tissue, the organelle, and the metabolic state [112]. In most naturally occurring membrane lipids the acyl chains are even numbered. The C_{16}, C_{18}, and C_{20} fatty acids account for more than 80% of the acyl chains in most membranes. Unsaturated fatty acids with one to six double bonds are present in some membranes. However, 18:1, 18:2, 18:3, and 20:4 account for more than 90% of the unsaturated acyl chains in most membranes. The proportion of saturated to unsaturated fatty acids changes considerably and characteristically from one membrane to the other. Phospholipids generally contain one saturated and one unsaturated acyl chain. In animal membranes, the fatty acid at the *sn*-2 position is usually unsaturated.

The original concept that the phospholipids in the membrane bilayer and the corresponding acyl chains existed in a random order is not correct [113]. Numerous studies have since shown that the phospholipid composition and the particular distribution of fatty acyl residues that occur in a specific biomembrane are present so as to provide the appropriate fluidity and environment for maintaining the optimal metabolic processes required for specific tissues [112,114]. In this section, the phospholipid composition, variability, and asymmetry of mammalian biomembranes will be discussed.

B. Species and Tissue Specificity

The phospholipid composition of membranes can vary considerably with species and source of tissue. This is illustrated by studies of the phospholipid composition of erythroctye ghosts (Fig. 3) and of vertebrate retinas (Table 5) [115]. The total phospholipid content of tissues also varies with species (Table 6) [115,127,128].

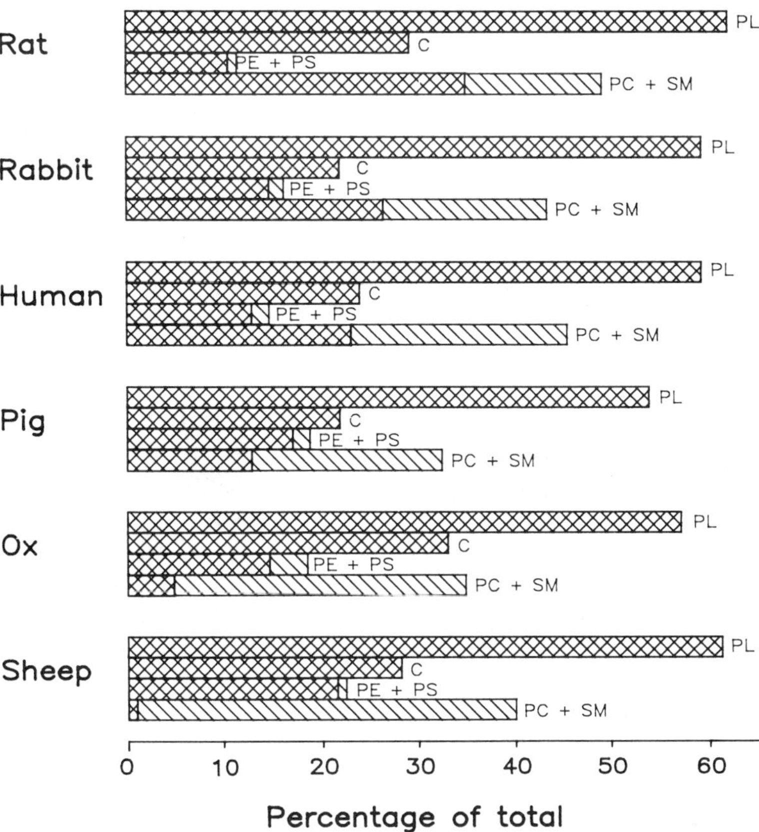

Figure 3 Cholesterol and phospholipid composition of erythrocyte ghosts from several species. Values for each fraction were calculated as the percentage of the total lipids. PL, phospholipids; C, cholesterol; PE, phosphatidylethanolamine; PS (right crosshatching), phosphatidylserine; PC, phosphatidylcholine; SM (right crosshatching), sphingomyelin.

Table 5 Phospholipid Composition of Vertebrate Retinas (mole % of total lipid P)

Phospholipid[a]	Species										
	Bovine	Frog	Toad	Human	Rat	Rabbit	Pig	Sheep	Dog	Duck	Chicken
PC (GroPCho)	46	53	50	48	45	44	47	47	47	42	47
PE (GroPEtn)	33	24	31	32	32	35	32	31	30	35	31
PS (GroPSer)	9	9	10	9	10	7	8	10	9	8	6
PI (GroPIns)	5	2	4	4	4	4	6	5	5	5	5
SM (SphPCho)	4	8	3	4	4	4	4	3	4	7	8
LPC (lyso-GroPCho)	<1	—	<1	<1	—	3	<1	<1	1	—	<1
PA (GroPtd)	<1	<1	<1	—	<1	—	—	—	—	1	—
DPG	3	2	2	—	<1	—	—	—	—	<1	<1
EP	24	13	—	—	14	28	—	—	—	44	33
CP	1	5	—	—	7	—	—	—	—	4	2
Ref	116–118	119,120	121	122	123,124	116	122	122	122	120	118,125,126

[a]PC, phosphatidylcholine; PE, phosphalidylethanolamine; PS, phosphatidylserine; PI, phosphatidylinositol; SM, sphingomyelin; LPC, lysophosphatidylcholine; PA, phosphatidic acid; DPG, diphosphatidylglycerol; EP, ethanolamine plasmalogen; CP, choline plasmalogen.

Table 6 Total Phospholipid Content of Mammalian Tissue

	Tissue[a]							
Species	Brain	Heart	Liver	Kidney	Lung	SM[b]	Spleen	P.nerve[c]
Sheep	2.1	0.8	1.5	1.1	-	-	1.1	1.7
Rat	1.9	0.5	1.3	1.3	0.6	0.4	0.6	-
Rabbit	2.4	0.8	-	-	-	-	-	2.0
Human	2.3	0.7	1.3	0.6	0.1	0.5	0.8	-
Bovine	-	0.5	0.8	0.4	0.5	0.4	0.6	0.4
Lamb	1.1	0.7	0.9	0.6	-	-	-	-
Mouse	2.3	0.6	0.8	0.8	0.7	1.4	0.5	-
Frog	3.0	0.4	0.5	0.4	-	0.1	-	-

[a]Lipid phosphorus mg/g wet weight
[b]Smooth muscle
[c]Peripheral nerve.

Table 7 illustrates the variability of the phospholipid composition from a variety of tissues. For simplicity I have limited the comparison to three mammalian species: human, rat, and guinea pig [129]. The data presented for skeletal muscle was obtained from Swietochowska et al. [130] and from Singh and Swartwout [131]. Information is also available for many other species [127]. It is interesting that even though many differences exist between species, differences of phospholipid composition among various tissues of the same species are greater than the differences between homologous tissues of different species (Fig. 2 and Table 7). The choline phospholipids including sphingomyelin account for more than half of the total phospholipid in most tissues, although the amount of sphingomyelin present may vary considerably. The amount of choline and ethanolamine phospholipids occurring in the plasmalogen form also varies with tissue origin. Plasmalogens account for about half of the choline and ethanolamine phospholipids in heart, and in neural tissues ethanolamine plasmalogen accounts for greater than half of the ethanolamine phospholipids. Only small amounts of choline plasmalogen are present in neural tissue. The inositol phospholipids of mammalian tissue are comprised of a mixture of three different phospholipids, phosphatidylinositol, phosphatidylinositol 4-phosphate, and phosphatidylinositol-4,5-bisphosphate [132]. The latter two phospholipids are generally evenly distributed and account for about 10% of the total inositol phospholipid content [133,134]. Lung tissue also contains measurable amounts of phosphatidylglycerol, and most tissues contain detectable amounts of alkylacyl-glycerophosphocholine [129].

The content of the other components of mammalian membranes, including cholesterol and glycolipids, also varies with species and origin of tissue [135]. In addition, the fatty acid content of the phospholipids is tissue specific. Polyunsaturated fatty acids, especially of the n-3 class, are enriched in neural tissue. Phospholipid classes also demonstrate selectivity for fatty acid content. The inositol phospholipids are enriched in arachidonic acid in the *sn*-2 position. This point will be further discussed when phospholipid molecular species are reviewed in a later section.

Significant differences in the phospholipid composition of tissues exist between fetal and adult mammals. In human liver microsomal membranes, adult tissue contains significantly greater amounts of phosphatidylethanolamine and phosphatidylserine, whereas

Table 7 Phospholipid Composition of Mammalian Tissues

Tissue	Phospholipid composition (% of lipid phosphorus)[a]									
	PC GroPCho	CP	LPC	PE GroPEtn	EP	PI GroPIns	PS GroPSer	SM SphPCho	DPG	PA GroPtd
Brain										
Human	33.9	0.1	—	13.2	20.2	2.6	10.5	13.8	1.9	—
Rat	38.5	0.1	—	15.3	22.4	1.6	10.2	6.9	2.5	0.8
Guinea pig	36.7	0.1	—	17.7	21.1	1.8	8.8	8.8	1.8	0.1
Heart										
Human	22.0	16.9	1.7	14.0	11.9	4.3	3.2	5.5	14.5	—
Rat	39.4	1.6	—	30.4	5.6	2.0	1.7	3.1	15.0	—
Guinea pig	26.5	15.5	0.3	20.7	9.8	2.8	1.7	5.1	14.4	—
Lung										
Human	41.7	1.1	1.1	9.6	10.5	3.8	7.4	15.0	2.7	—
Rat	55.0	0.1	0.6	12.6	10.5	1.6	6.5	11.3	1.2	—
Guinea pig	54.0	0.1	—	13.3	7.7	1.5	3.5	13.0	1.5	—
Liver										
Human	45.2	—	0.7	28.4	1.1	6.5	3.8	6.7	5.5	—
Rat	61.5	—	—	23.5	—	4.9	1.7	4.7	3.6	—
Guinea pig	56.2	—	—	26.7	0.7	4.8	2.0	5.2	2.7	—

Kidney										
Human	32.1	1.3	2.4	17.0	10.0	4.4	5.7	13.5	6.7	—
Rat	38.9	0.1	—	27.4	2.9	3.4	4.2	16.3	6.8	—
Guinea pig	50.8	—	—	18.8	5.2	3.4	3.1	15.1	4.7	—
Testis										
Human	38.8	0.6	—	16.0	11.6	4.7	6.4	11.1	2.4	0.1
Rat	47.7	0.2	—	13.3	9.4	3.2	3.9	9.4	3.7	0.1
Guinea pig	51.6	1.0	—	18.4	7.1	2.3	3.8	10.1	1.3	0.1
Skeletal Muscle										
Human	48.0	—	2.0	26.4	—	9.0	3.3	4.0	5.0	1.0
Rat	52.6	—	3.4	25.6	—	7.9	3.4	4.6	0.6	0.5
Erythrocytes										
Human	31.0	—	0.6	19.2	9.2	1.2	13.4	23.5	—	1.7
Rat	47.0	—	1.3	9.0	12.2	4.5	9.4	13.3	—	—
Guinea pig	48.6	0.3	—	14.8	4.5	2.0	11.3	11.6	—	2.7
Blood Plasma										
Human	76.7	0.2	3.2	1.9	0.4	0.5	0.2	15.2	—	—
Rat	66.7	—	19.0	0.1	—	3.4	—	10.3	—	—
Guinea pig	67.5	—	2.7	8.4	4.0	0.9	0.1	9.1	—	—

[a]PC, phosphatidylcholine; CP, choline plasmalogen; LPC, lysophosphatidylcholine; PE, phosphatidylethanolamine; EP, ethanolamine plasmalogen; PI, phosphatidylinositol; PS, phosphatidylserine; SM, sphingomyelin; DPG, diphosphatidylglycerol; PA, phosphatidic acid.

fetal tissue is enriched in sphingomyelin [136]. The phosphatidylcholine/sphingomyelin and phospholipid/cholesterol molar ratios are also significantly increased in adult tissue, and there is an increase in total phospholipid content. Total phospholipid content in rat brain increases after birth [137]. It doubles after about 10 days and then starts to level off. Wells and Dittmer (Table 8) [138] showed that phospholipids associated with myelination, sphingomyelin, phosphatidylinositol 4,5-bisphosphate, and phosphatidic acid [138] increase rapidly two to threefold, during the period of myelination [138]. Furthermore, ethanolamine plasmalogen, diphosphatidylglycerol, and phosphatidylserine levels also increase following birth. In contrast, phosphatidylcholine, phosphatidylethanolamine, and phosphatidylinositol levels remain fairly constant compared to total phospholipid content during development of the brain. The phospholipid composition of human brain also changes with aging (Table 9). The molar percentage of sphingomyelin and the ethanolamine phospholipids increases from birth to seven months, whereas phosphatidylcholine levels decrease [139].

Phospholipids have other functions besides maintaining membrane integrity. A minor but specific pool of choline phospholipid, 1-alkyl-2-acetyl-*sn*-glycero-3-phosphocholine, is the source for the synthesis of platelet activating factor [140]. Choline and inositol phospholipids have a special function in signal transduction pathways via the production of diacylglycerol (DAG) and inositol trisphosphate (IP_3) [132,141,142]. Diacylglycerol is produced from either phosphatidylcholine or phosphatidylinositol-4,5-bisphosphate following activation of a specific phospholipase C [143]. Choline and inositol phospholipids may also be the source of arachidonic acid for prostaglandin and leukotriene production [144]. The inositol phospholipids are also an important component of a protein membrane anchoring system called PI glycans [145], which may also be part of the signal transduction pathway for insulin action [145]. Phosphatidylcholine has also been proposed to be the source of choline for the synthesis of acetylcholine in neural tissue [146]. Phospholipid composition and fatty acid content of biomembranes have also been shown to regulate specific enzyme activities and membrane transport processes in mammalian cells (for reviews see Refs. 147 and 148). The latter effect may be related to changes in membrane fluidity, which is altered by changing the membrane cholesterol content or polyunsaturated fatty acid composition of membrane phospholipids [147,148]. Therefore

Table 8 Phospholipid Composition of Developing Rat Brain

	Concentration (μmol/g wet wt)				
	Age (days):				
Phospholipid[a]	3	12	24	42	180
PC	14.7	20.4	24.8	25.0	25.0
PE	5.3	8.0	9.4	10.9	10.7
EP	2.2	4.7	11.3	13.5	13.0
PI	1.2	1.6	2.0	2.2	2.2
PS	2.9	4.5	7.0	8.3	8.5
PA	0.1	0.3	0.7	1.0	1.3
SM	0.2	1.0	3.2	3.6	3.7

[a]PC, phosphatidylcholine; PE, phosphatidylethanolamine; EP, ethanolamine plasmalogen; PI, phosphatidylinsoitol; PS, phosphatidylserine; PA, phosphatidic acid; SM, sphingomyelin.
Source: Ref. 138.

Table 9 Phospholipid Composition of Human Cerebrum During Development and Aging

		Molar percentage						
Phospholipid[a]	Tissue[b]	New-born	1 Mo	7 Mo	4 Yr	16 Yr	52 Yr	82 Yr
PC	CC	47	41	39	38	37	37	35
	WM	45	38	35	26	25	25	25
PE	CC	28	35	34	35	36	36	37
	WM	31	35	34	36	35	34	36
PI	CC	3	3	3	2	3	3	4
	WM	4	3	3	2	2	2	3
PS	CC	14	14	14	14	13	14	13
	WM	14	16	16	21	21	20	19
SM	CC	7	7	9	11	11	10	11
	WM	6	8	13	16	17	18	16

[a]PC, choline phospholipids; PE, ethanolamine phospholipids; PI, inositol phospholipids; PS, phosphatidylserine; SM, sphingomyelin.
[b]CC = cerebral cortex; WM = white matter.

the phospholipid components of biomembranes have many specialized functions that are important not only for maintaining membrane integrity but also for regulating the function of mammalian cells.

C. Phospholipid Composition of Cellular Organelles

The characteristic distribution of phospholipids among tissues can be extended to the subcellular level. As illustrated in Figs. 2 and 4, the phospholipid composition of organelles and membranes from rat liver vary considerably [149]. Mitochondria contain less phosphatidylcholine and more phosphatidylethanolamine relative to whole tissue. The diphosphatidylglycerol content is also characteristically high in mitochondria. A recent study examining the phospholipid content of outer and inner membrane mitochondrial fractions indicates that the outer membrane has a higher content of phospholipid per mg of protein than the inner membrane (Table 10) [150]. The inner membrane accounts for a much larger surface area, 25–29% of the total mitochondrial volume, compared to the outer membrane (6–10%) [149]. Therefore the phospholipid content of the inner membrane more closely resembles the whole mitochondrial phospholipid composition. In this study, the phosphatidylcholine and phosphatidylethanolamine content were similar in rat liver mitochondria. The outer membrane fraction of rat liver mitochondria contained a greater amount of phosphatidylcholine and phosphatidylinositol than the inner membrane fraction and a lesser amount of phosphatidylethanolamine. The mitochondria are also enriched in diphosphatidylglycerol (cardiolipin). The cardiolipin content of the inner mitochondrial membrane was increased compared to the outer membrane.

The microsomal fraction (endoplasmic reticulum) of rat liver contains a high concentration of phospholipids. As illustrated in Fig. 4, rough endoplasmic reticulum contains a large amount of phosphatidylcholine [149]. Studies by Glaumann and Dallner [151] have shown that membrane fractions from rough and smooth endoplasmic reticulum contain a similar amount of total phospholipid based upon phospholipid-to-protein ratios and that

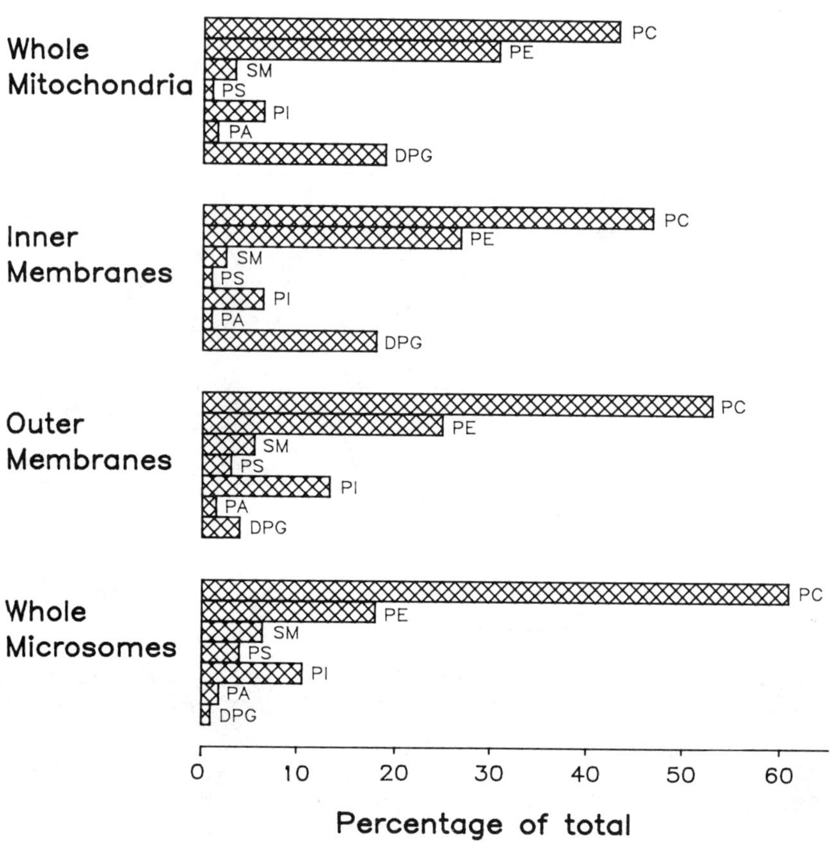

Figure 4 Phospholipid composition of rat liver mitochondria and microsomes. Abbreviations are presented in Fig. 1.

the compositions of phospholipid in the two fractions are similar. Comparing the lipid content of microsomes isolated from human or rat liver shows that microsomes from human liver contain a larger amount of lipid than those from rat liver (Table 11) [152]. The phospholipid composition based upon the percentage of total lipid is similar in human and rat liver microsomes. Rat liver microsomes do contain a greater amount of arachidonic acid than human liver.

The Golgi apparatus of rat liver contains a larger proportion of sphingomyelin than the endoplasmic reticulum, with lower amounts of phosphatidylcholine [149]. The plasma membrane of rat liver compared to other fractions contains a greater proportion of sphingomyelin and phosphatidylserine and a lesser amount of phosphatidylcholine. The cholesterol content of plasma membranes is also higher than in other fractions [149]. The phospholipid composition of nuclear membranes of rat liver has been reported to be similar to that of the endoplasmic reticulum [153].

Although one must always be concerned with cross contamination when separating and characterizing cellular organelles, results thus far suggest that the intracellular membrane system is distinct with regard to the composition of phospholipids. It is tempting to speculate that the specific distribution of phospholipids in intracellular

Table 10 Lipid Composition of Rat Liver Mitochondria

Phospholipid	Mitochondria	Outer membrane	Inner membrane
Total			
(nmol phospholipid/mg protein)	230	1110	450
Fractions[a]			
Phosphatidylcholine	39	48	39
Phosphatidylethanolamine	37	31	43
Cardiolipin	19	9	14
Phosphatidylinositol	2	10	1
Phosphatidylserine	1	1	—

[a]% of total lipid phosphorus.

Table 11 Lipid Composition of Human and Rat Liver Microsomes

Lipid	Human	Rat
Total (μg/mg protein)	1032	610
Phospholipids[a]		
Phosphatidylcholine	62	63
Phosphatidylethanolamine	31	29
Phosphatidylinositol	15	13
Phosphatidylserine	7	5
Sphingomyelin	6	4
Phosphatidic acid	1	1
Other	6	8

[a]% of total phospholipids.

organelles is due to a functional requirement, but additional research will be necessary before a definitive statement describing the reason for phospholipid specificity of membrane organelles can be made.

D. Membrane Asymmetry and Phospholipid Molecular Species

There is considerable evidence that the protein, carbohydrate, and lipid components of biomembranes are asymmetrically distributed. For proteins and carbohydrates, the asymmetry is absolute. However, lipid asymmetry is not absolute; almost every type of lipid is present on both sides of the membrane bilayer, but in different amounts [154]. Not only does lipid asymmetry occur with regard to the arrangement and localization of phospholipid classes, but phospholipid specialization exists according to the fatty acyl content of individual phospholipds.

The erythrocyte membrane is the only mammalian biomembrane that has been shown unequivocally to have an asymmetric lipid arrangement [155]. Unfortunately, most other cells and membranes are less ideally suited for examination of lipid asymmetry. However, it is commonly believed and supported by experimental evidence that lipid asymmetry exists in other mammalian biomembranes [154,156]. The major difficulty in assessing lipid asymmetry in other biomembranes is the problems encountered in achieving the criteria necessary to make accurate evaluations. The current techniques used for assessing biomembrane lipid asymmetry have been reviewed and critically appraised in several articles [155,157,158]. There are certain factors that must be considered for analysis of lipid asymmetry in biomembranes [154]. The membranes must be pure and free of contaminating membranous material from other cells or organelles. All membranes must be present as closed vesicles, and the vesicles must have the same overall sidedness as existed in the intact cell. The examination of isolated membranes from complex cells involves cell disruption, and some organelles such as endoplasmic reticulum and plasma membranes are fragmented and then resealed. The sideness in the resealed vesicles must be established and maintained. The reagents used to analyze lipid asymmetry must be critically evaluated in each system to assure that nonpenetrating probes do not cross the membranes being studied. Both sides of a membrane must be studied to be sure that lipids inaccessible from one side of a membrane are not excluded. Currently this can only be achieved by examining vesicles in a right side out or inside out orientation. In many systems this arrangement is not yet possible. Finally, the purity of the probes being used must be confirmed, or if the analysis being conducted involves radioactive lipids it is necessary that all pools of lipid are labeled to the same specific radioactivity. It is only after these criteria have been met that an accurate assessment of lipid asymmetry of biomembranes can be made.

Membrane asymmetry is achieved by the rapid movement of membrane components from one bilayer to the other. The interesting question that remains to be answered is, What maintains lipid asymmetry in biological membranes? The diffusion of a lipid molecule through the lipid bilayer is termed flip-flop [159], and since it is a diffusion event it is spontaneous and randomizes the phospholipids between the two sides of the bilayer [154]. The transmembrane migration of cholesterol has also been shown to occur by flip-flop mechanism [160]. The rate of flip-flop translocation of phospholipid in the membrane bilayer is very slow due to the high activation energy required to bring the polar group through the hydrocarbon core of the bilayer [161]. Even though flip-flop is extremely slow, it is possible for a rapid transmembrane movement of phospholipid to occur without flip-flop or without the polar head group of the phospholipid passing through the hydrocarbon core of the bilayer. The movement is achieved by specific membrane proteins called phospholipid flippases, first proposed by Bretscher [162]. The existence of phospholipid flippases has been demonstrated in prokaryotic and eukaryotic cells [163]. Several investigators have shown that an ATP dependent protein called aminophospholipid translocase continuously pumps phosphatidylserine and phosphatidylethanolamine from the outer to the inner leaflet in mammalian erythrocytes [164–167]. Other studies have indicated that the same type of protein is present not only in red cells but also in platelets and lymphocytes [163,168,169]. It is possible that similar enzymes are present in other cells and are responsible for creating and maintaining lipid asymmetry [163].

The asymmetry of phospholipid distribution in the membrane has been well established for mammalian erythrocytes [154,155]. The aminophosphatides, phosphatidylserine and phosphatidylethanolamine, occur almost exclusively on the internal leaflet of the membrane bilayer, and the choline phosphatides, phosphatidylcholine, and sphingomyelin, occur in the external leaflet of the bilayer (Table 12) [154,155,163,170,171]. Additional evidence is also available for the asymmetry of phospholipids in other mammalian membranes (Table 12). Membrane lipid asymmetry has also been shown to

Table 12 Phospholipid Asymmetry in Mammalian Membranes

Membrane	Lipids localized[a]			References
	Outside	Inside	Equally distributed	
Erythrocytes	PE,SM	PE,PS		154,155,170,171
Platelets	SM	PC,PE,PS,PI		172
Plasma membrane LM cells	SM	PE	PC	173
Plasma membrane Krebs ascites cells	PC	PS	PE,SM	156

[a]PC, phosphatidylcholine; SM, sphingomyelin; PE, phosphatidylethanolamine; PS, phosphatidylserine; PI, phosphatidylinositol.

exist in cellular organelles. Analysis of intracellular organelles from rat liver has shown that phosphatidylserine and phosphatidylethanolamine are localized on the outer leaflet of the membrane bilayer of microsomes, Golgi membranes, inner mitochondrial membrane, lysosomes, and nuclear membranes [174].

The relevance of lipid asymmetry to membrane function is not known. However, it is likely that specific phospholipid arrangements are necessary for regulating membrane fluidity and the activity of membrane associated enzymes. Membrane phospholipid asymmetry of erythrocytes also influences recognition of erythrocytes by macrophages [175]. Abnormal asymmetric distribution of phospholipids in erythrocyte membranes causes an increase in cell phagocytosis. Since cells lose their lipid asymmetry with time [163,176], maintaining lipid asymmetry is critical to cell survival. If cell death occurs in neuronal cells due to the loss of lipid asymmetry, then maintenance of cellular lipid arrangement could be important for many pathological disorders to the central nervous system [176].

Experimental evidence shows that phospholipid membrane asymmetry occurs with regard to phospholipid classes in human erythrocytes [165–168]. It has also been reported that specific molecular species of phospholipid classes and sphingomyelins exist in human erythrocytes [177–179]. The presence of specific molecular species may be important in regulating erythrocyte cellular properties and cholesterol efflux [180,181]. The major molecular species of human erythrocyte phosphatidylcholine are 16:0/18:1, 16:0/18:2, and 18:0/18:2; phosphatidylethanolamine, 16:0/18:1, 16:0/20:4, and 18:0/20:4; phosphatidylinositol, 16:0/20:4 and 18:0/20:4; and phosphatidylserine, 18:0/20:4, 18:0/22:4, and 18:0/22:6 [178]. The distribution of phospholipids in bovine retina and rat brain has been suggested to be regulated by the fatty acid composition [182–184]. Furthermore, the hydrolysis of the inositol phospholipids by phospholipase C may be specific for a certain 20:4 containing molecular species of phosphatidylinositol 4,5-bisphosphate.

VIII. TISSUE CULTURE SYSTEMS

The use of cultured cell systems has greatly enhanced the understanding of phospholipid metabolism [147,148]. For instance, studies of cultured LM cells have shown that supplementing the growth medium with ethanolamine causes an increase in ethanolamine phosphoglycerides in the plasma membrane [185]. Similar results have been obtained

Table 13 Phospholipid Composition of Cultured Neural Cells

Phospholipid[a]	Phospholipid composition (% of total)		
	Neuroblastoma NIE-115	N/G hybrid NG 108-15	Glioma C-6
PC	50	53	43
PE	17	18	12
EP	6	7	12
LPC	2	1	2
PS	6	3	7
PI	8	8	10
PIP	0.7	0.9	1.5
PIP_2	0.6	0.5	0.7
SM	6	3	9
DPG	5	5	3

[a]PC, phosphatidylcholine; PE, phosphatidylethanolamine; EP, ethanolamine plasmalogen; LPC, lysophosphatidylcholine; PS, phosphatidylserine; PI, phosphatidylinositol; PIP, phosphatidylinositol 4-phosphate; PIP_2, phosphatidylinositol 4,5-bisphosphate; SM, sphingomyelin; DPG, diphosphatidylglycerol.

with studies of human skin fibroblasts, human Y79 retinoblastoma cells, and bovine aorta endothelial cells [186–188]. Furthermore, studies with human Y79 retinoblastoma cells have indicated that extracellular ethanolamine is necessary for the synthesis of ethanolamine plasmalogen and that the availability of extracellular ethanolamine regulates choline and phosphatidylcholine metabolism [188,189].

Cultured neural cells derived from the same original clone and grown in the same medium have a different phospholipid composition (Table 13) [133]. This suggests that

Table 14 Phospholipid Composition of Normal and Neoplastic Tissues

Tissue	Phospholipid composition[a]						
	PC	LPC	PE	PS	PI	DPG	SM
Mouse Liver							
Normal	49	1	27	6	8	4	4
Hepatoma	38	1	29	9	7	4	10
Brain							
Whole tissue	34	—	9	11	7	2	7
Glioblastoma	42	—	10	7	9	2	6
Mouse Epidermis							
Normal	44	—	18	5	<1	2	10
Malignant	44	—	18	5	<1	2	7
Fibroblasts							
Diploid	57	3	12	8	13	—	7
SV-40 transformed	57	2	13	4	10	—	13

[a]PC, phosphatidylcholine; LPC, lysophosphatidylcholine; PE, phosphatidylethanolamine; PS, phosphatidylserine; PI, phosphatidylinositol; DPG, diphosphatidylglycerol; SM, sphingomyelin.

the phospholipid content of mammalian cells is regulated by factors besides the environment.

IX. PHOSPHOLIPID COMPOSITION OF TUMORS

The phospholipid content of neoplasms varies from tumor to tumor [190]. Characteristically the total phospholipid content of tumors is less than the phospholipid content of the corresponding normal tissue. With a few exceptions, this is due to a general decrease in the amount of total lipid in tumor cells [190]. The phospholipid composition of various tumors is similar to the composition of normal tissue. This is illustrated by the data compiled in Table 14 [190]. The major phospholipid in malignant tissues is phosphatidylcholine.

X. CONCLUSION

Both variability and specificity exist in the phospholipid composition of membranes from plant, invertebrate, and vertebrate tissues. Specificity extends to the phospholipid distribution across the inner and outer leaflets of the plasma and cytoplasmic organelle membranes as well as the phospholipid molecular species. Membrane phospholipids are important for maintaining membrane integrity, and the specificity of the phospholipid distribution is likely important in the regulation of membrane properties. Improved techniques have enabled investigators to identify the phospholipid composition of cellular organelles and the phospholipid asymmetry that occurs across membrane bilayers. However, additional questions remain to be answered concerning the regulation of phospholipid specifiity and molecular species and the influence it has on membrane properties.

REFERENCES

1. G. A. Thompson, Jr., Phospholipids, in *The Regulation of Membrane Lipid Metabolism.* CRC Press, Boca Raton, Fla., 1980, p. 175.
2. G. A. Thompson, Jr., A rationale governing the regulation of lipid metabolism, in *The Regulation of Membrane Lipid Metabolism.* CRC Press, Boca Raton, Fla., 1980, p. 1.
3. J. B. Finean, Phospholipids in biological membranes and the study of phospholipid-protein interactions, in *Form and Function of Phospholipids*, (G. B. Ansell, J. N. Hawthorne, R. M. C. Dawson, eds.). Elsevier, New York, 1973, p. 171.
4. Y. Lange, M. H. Swaisgood, B. V. Ramos, and T. L. Steck, Plasma membranes contain half the phospholipid and 90% of the cholesterol and sphingomyelin in cultured human fibroblasts, *J. Biol. Chem. 264*:3786 (1989).
5. A. A. Spector, and M. A. Yorek, Membrane lipid composition and cellular function, *J. Lipid Res. 26:*1015 (1985).
6. A. W. Konings, and A. C. Ruitrok, Role of membrane lipids and membrane fluidity in the thermosensitivity and thermotolerance of mammalian cells, *Radiat. Res. 102*:86 (1985).
7. T. W. Keenan, and D. J. Moore, Phospholipid class and fatty acid composition of Golgi apparatus isolated from rat liver and comparison with other cell functions, *Biochemistry, 9*:19 (1970).
8. N. C. Nielson, S. Fleischer, and D. G. McConnell, Lipid composition of bovine retinal outer segment fragments, *Biochem. Biophys. Acta. 211*:10 (1970).
9. A. A. Spector, S. N. Mathur, T. L. Kaduce, and B. T. Hyman, Lipid nutrition and metabolism of cultured mammalian cells, *Prog. Lipid Res. 19*:155 (1981).

10. J. H. Crowe, L. M. Crowe, and F. A. Hoekstra, Phase transitions and permeability changes in dry membranes during rehydration. *J. Bioenerg. Biomembr. 21*:77 (1989).
11. H. K. Mangold and N. Weber, Biosynthesis and biotransformation of ether lipids, *Lipids 22*:789 (1987).
12. J. A. F. Op den Kamp and L. L. M. Van Deenen, *Structural and Functional Aspects of Lipoproteins in Living Systems* (E. Tria and A. M. Scanu, eds.), Academic Press, London, 1969, p. 227.
13. T. Vanden-Boom and J. E. Cronan, Jr. Genetics and regulation of bacterial lipid metabolism, *Annu. Rev. Microbiol, 43*:317 (1989).
14. A. Chopra and G. K. Khuller, Lipid metabolism in fungi, *Crit. Rev. Microbiol. 11*:209 (1984).
15. A. Chopra, and G. K. Khuller, Lipids of pathogenic fungi, *Prog. Lipid Res. 22*:189 (1983).
16. F. Hullin, H. Y. Kim, and N. Salem, Jr., Analysis of aminophospholipid molecular species by high performance liquid chromatography. *J. Lipid Res. 30*:1963 (1989).
17. C. S. Ramesha, W. C. Pickett, and D. V. Murthy, Sensitive method for the analysis of phospholipid subclasses and molecular species as 1-anthroyl derivatives of their diglycerides, *J. Chromatogr. 491*:37 (1989).
18. J.D. Medh and P. H. Weigel, Separation of phosphatidylinositols and other phospholipids by two-step one-dimensional thin-layer chromatography, *J. Lipid Res. 30*:761 (1989).
19. A. Di-Biase, S. Salvati, and G. S. Crescenzi, Analysis of brain and myelin lipids by high-performance thin layer chromatography and densitometry, *Neurochem. Res. 14*:153 (1989).
20. P. E. Haroldsen and R. C. Murphy, Analysis of phospholipid molecular species in rat lung as dinitrobenzoate diglycerides by electron capture negative chemical ionization mass spectrometry, *Biomed. Environ. Mass. Spectrom. 14*:573 (1987).
21. L. A. Dethloff, L. B. Gilmore, and G. E. Hook, Separation of pulmonary surfactant phospholipids by high-performance liquid chromatography, *J. Chromatogr. 382*:79 (1986).
22. R. D. Paton, A. I. McGillivray, T. F. Speir, M. J. Whittle, C. R. Whitfield, and R. W. Logan, HPLC of phospholipids in biological fluids—application to amniotic fluid for the prediction of fetal lung maturity, *Clin. Chim. Acta, 133*:97 (1983).
23. R. T. Crane, S. C. Goheen, E. C. Larkin, and G. A. Rao, Complexities in lipid quantitation using thin layer chromatography for separation and flame ionization for detection, *Lipids 18*:74 (1983).
24. M. I. Aveldano, M. VanRollins, and L. A. Horrocks, Separation and quantitation of free fatty acids and fatty acid methyl esters by reverse phase high pressure liquid chromatography, *J. Lipid Res. 24*:83 (1983).
25. N. M. Dean and M. A. Beaven, Methods for the analysis of inositol phosphates, *Anal. Biochem. 183*:199 (1989).
26. J. M. Samet, M. Friedman, and D. C. Henke, High-performance liquid chromatography separation of phospholipid classes and arachidonic acid on cyanopropyl columns, *Anal. Biochem. 182*:32 (1989).
27. M. Seewald and H. M. Eichinger, Separation of major phospholipid classes by high-performance liquid chromatography and subsequent analysis of phospholipid-bound fatty acids using gas chromatography, *J. Chromatogr. 469*:271 (1989).
28. T. V. Raglione and R. A. Hartwick, On-line high-performance liquid chromatography-post-column reaction-capillary gas chromatography analysis of lipids in biological samples, *J. Chromatogr. 454*:157 (1988).
29. P. E. Haroldsen and R. C. Murphy, Analysis of phospholipid molecular species in rat lung as dinitrobenzoate diglycerides by electron capture negative chemical ionization mass spectrometry, *Biomed. Environ. Mass. Spectrom 14*:573 (1987).
30. P. Juaneda and G. Rocquelin, Complete separation of phospholipids from human heart combining two HPLC methods, *Lipids 21*:239 (1986).

31. S. J. Ropbins and G. M. Patton, Separation of phospholipid molecular species by high performance liquid chromatography: Potentials for use in metabolic studies, *J. Lipid Res.* *27*:131 (1986).
32. C. A. Mancuso, P. D. Nichols, and D. C. White, A method for the separation and characterization of archaebacterial signature ether lipids, *J. Lipid Res.* *27*:49 (1986).
33. M. H. Creer and R. W. Gross, Separation of isomeric lysophospholipids by reverse phase HPLC, *Lipids* *20*:922 (1985).
34. W. W. Christie, Rapid separation and qualification of lipid classes by high performance liquid chromatography and mass (light-scattering) detection, *J. Lipid Res.* *26*:507 (1985).
35. M. H. Creer, C. Pastor, P. B. Corr, R. W. Gross, and B. E. Sobel, Qualification of choline and ethanolamine phospholipids in rabbit myocardium, *Anal. Biochem.* *144*:65 (1985).
36. O. Itasaka, T. Hori, K. Sasahara, Y. Wakabayashi, F. Takahashi, and H. Rhee, Analysis of phospho- and phosphonosphingololipids by high-performance liquid chromatography, *J. Biochem.* *95*:1671 (1984).
37. H. P. Bissen, and H. W. Kreysel, Analysis of phospholipids in human semen by high-performance liquid chromatography, *J. Chromatogr.* *276*:29 (1983).
38. T. L. Kaduce, K. C. Norton, and A. A. Spector, A rapid, isocratic method for phospholipid separation by high-performance liquid chromatography, *J. Lipid Res.* *24*:1398 (1983).
39. M. L. Blank, E. A. Cress, P. Lee, N. Stephens, C. Piantadosi, and F. Snyder, Quantitative analysis of ether-linked lipids as alkyl and alk-1-enyl-glycerol benzoates by high-performance liquid chromatography, *Anal. Biochem.* *133*:430 (1983).
40. R. E. Barrow, Chemical structure of phospholipids in the lungs and airways of sheep, *Respir. Physiol.* *79*:1 (1990).
41. F. Hullin, H. Y. Kim, and N. Salem, Jr., Analysis of aminophospholipid molecular species by high performance liquid chromatography, *J. Lipid Res.* *30*:1963 (1989).
42. J. J. Myher, A. Kuksis, and S. Pind, Molecular species of glycerophospholipids and sphingomyelins of human plasma: Comparison to red blood cells, *Lipids* *24*:408 (1989).
43. J. J. Myher, A. Kuksis, and S. Pind, Molecular species of glycerophospholipids and sphingomyelins of human erythrocytes: Improved method of analysis *Lipids* *24*:396 (1989).
44. H. Takamura, H. Narita, R. Urade, and M. Kito, Quantitative analysis of polyenoic phospholipid molecular species by high performance liquid chromatography, *Lipids* *21*:356 (1986).
45. M. L. Blank, A. A. Spector, T. L. Kaduce, and F. Snyder, Composition and incorporation of [^{3}H] arachidonic acid into molecular species of phospholipids classes by cultured human endothelial cells, *Biochem. Biophys. Acta* *877*:211 (1986).
46. F. H. Chilton III and R. C. Murphy, Fast atom bombardment analysis of arachidonic acid-containing phosphatidylcholine molecular species, *Biomed. Environ. Mass. Spectrom.* *13*:71 (1986).
47. K. M. Marnela, T. Moilanen, and M. Jutila, Piperidine ester derivatives in mass spectrometric analysis of fatty acids from human serum phospholipids and cholesteryl esters, *Biomed. Environ. Mass Spectrom.* *16*:443 (1988).
48. M.D. Laryea, P. Cieslicki, E. Diekmann, and U. Wendel, Analysis of the fatty acid composition of erythrocyte phospholipids by a base catalysed transesterification method—Prevention of formation of dimethylacetals, *Clin. Chim. Acta* *171*:11 (1988).
49. Y. Nakagawa and K. Waku, Determination of the amounts of free arachidonic acid in resident and activated rabbit alveolar macrophages by fluorometric high-performance liquid chromatography, *Lipids* *20*:482 (1985).
50. S. P. Peters, D. W. McGlashan, Jr., E. S. Schulman, R. P. Schleimer, E. C. Hayes, J. Rokach, N. F. Adkinson, Jr., and L. M. Lichtenstein, Arachidonic acid metabolism in purified human lung mast cells, *J. Immul.* *132*:1972 (1984).
51. M. S. Pessin and D. M. Raben, Molecular species analysis of 1,2-diglycerides stimulated by alpha-thrombin in cultured fibroblasts, *J. Biol. Chem.* *264*:8729 (1989).

52. J. P. Boegheim, Jr., M. Van-Line, J. A. Op-den-Kamp, B. Roelofsen, The sphingomyelin pools in the outer and inner layer of the human erythrocyte membrane are composed of different molecular species, *Biochim. Biophys. Acta. 735*:438 (1983).
53. G. F. Ames, Lipids of *Salmonella typhimuriun* and *Escherichia coli*: Structure and metabolism, *J. Bacteriol. 95*:833 (1968).
54. J.E. Cronan, Jr., and C. O. Rock, Biosynthesis of membrane lipids, *Escherichia coli* and *Salmonella typhimuriun* (F. C. Neidhardt, J. L. Ingraham, K.B. Low, B. Magasanik, M. Schaechter, and H. E. Umbarger, eds.), Vol. 1, Am. Soc. Microbiol., 1987, p. 474.
55. C. R. H. Raetz, Molecular genetics of membrane phospholipid synthesis, *Ann. Rev. Genet. 20*:253 (1986).
56. J. E. cronan and P. R. Vagelos, Metabolism and function of the membrane phospholipids of *Escherichia coli, Biochim. Biophys. Acta 265*:25 (1972).
57. J. E. Cronan, Molecular biology of bacterial membrane lipids, *Ann. Rev. Biochem. 47*:163 (1978).
58. J. E. cronan and E. P. Gelmann, Physical properties of membrane lipids: Biological relevance and regulation, *Bacteriol. Rev. 39*:232 (1975).
59. J. A. F. Op den Kamp, Lipid asymmetry in membranes, *Ann. Rev. Biochem. 48*:47 (1979).
60. C. R. H. Raetz, Enzymology, genetics, and regulation of membrane phospholipid synthesis in *Escherichia coli, Microbiol. Rev. 42*:614 (1978).
61. T. Nakae, Outer membrane permeability of bacteria, *Crit. Rev. Microbiol. 13*:1 (1986).
62. B. Lugtenberg and L. V. Alphen, Molecular architecture and functioning of the outer membrane of *Escherichia coli*, and other gram-negative bacteria, *Biochim. Biophys. Acta 737:51* (1983).
63. M. J. Osborn, J. E. Gandes, E. Parisi, and J. Carson, Mechanism of assembly of the outer membrane of *Salmonella typhinurium, J. Biol. Chem. 247*:3962 (1972).
64. B. Lugtenberg and R. Peters, Distribution of lipids in cytoplasmic and outer membranes of *Escherichia coli. Biochim. Biophys. Acta 441*:38 (1976).
65. H. Nikaido and M. Vaara, Molecular basis of bacteria outer membrane permeability, *Microbiol. Rev. 49*:1 (1985).
66. V. A. Fried and L. I. Rothfield, Interactions between lipopolysaccharide and phosphatidylethanolamine in molecular monolayers, *Biochim. Biophys. Acta 514*:69 (1978).
67. J. Koplow and H. Golfine, Alterations in the outer membrane of the cell envelope of heptose-deficient mutants of *Escherichia coli. J. Bacteriol. 117*:527 (1974).
68. C. Miyazaki, M. Kuroda, A. Ohta, and I. Shibuya, Genetic manipulation of membrane phospholipid composition in *Escherichia coli*: pgsA mutants defective in phosphatidylglycerol synthesis. *Proc. Natl. Acad. Sci. USA 82*:7530 (1985).
69. I. Shibuya, S. Yamagoe, C. Miyazaki, H. Matsuzaki, and A. Ohta, Biosynthesis of novel acidic phospholipid analogs in *Escherichia coli, J. Bacteriol. 161*:473 (1985).
70. T. Vanden-Boom, and J. E. Cronan, Jr., Genetics and regulation of bacterial lipid metabolism. *Ann. Rev. Microbiol. 43*:317 (1989).
71. G. K. Khuller, Lipid composition of *Microsporum gypseum. Experientia 34*:432 (1978).
72. A. Chopra and G. K. Khuller, Lipids of pathogenic fungi, *Prog. Lipid Res 22*:189 (1983).
73. J.E. Rothman and J. Leonard, Membrane asymmetry. *Science 195*:743 (1977).
74. M. K. Wassef, Fungal lipids, *Adv. Lipid Res. 15*:159 (1977).
75. R. Letters, Phospholipids of yeast, *Biochim. Biophys. Acta 116*:489 (1966).
76. G. L. A. Graff, B. Vanderkelen, C. Guening, and J. Humpers, *Soc. Belge Biol.* 1635 (1968).
77. F. Paltauf and G. Schatz, Promitochondrin of anaerobically grown yeast, *Biochemistry 8*:335 (1969).
78. J. A. Anderson, F. Sun, J. K. McDonald, and V. H. Cheldelin, Oxidase activity and lipid composition of respiratory particles from claviceps purpurea, *Arch. Biochem. Biophys. 107*:37 (1964).

79. G. L. White and J. N. Hawthorne, Phosphatidic acid and phosphatidylinositol metabolism in *Schizosaccharomyces, Biochem. J. 117*:203 (1970).
80. H. A. Schwertner and J. B. Biale, Lipid composition of plant mitochondria and of chloroplasts, *J. Lipid Res. 14*:235 (1973).
81. T. Galliard, Phospholipid metabolism in photosynthetic plants, in *Form and Function of Phospholipids*, (G. B. Ansell, J. N. Hawthorne, R. M. C. Dawson, eds.). Elsevier, New York, 1973, p. 253.
82. P. G. Roughan and R. D. Batt, *Phytochemistry 8*:363 (1969).
83. J. F. G. M. Wintermans, Concentration of phosphatides and glycolipids in leaves and chloroplasts, *Biochim. Biophys. Acta 44*:49 (1960).
84. P.G. Roughan, Turnover of the glycerolipids of pumpkin leaves, *Biochem. J. 117*:1 (1970).
85. P. S. Sastry and M. Kates, Lipid components of leaves, *Biochemistry 3*:1271 (1964).
86. U. Beiss, *Landwirtsch. Forsch. 23*:198 (1969).
87. T. Galliard, *Phytochemistry 7*:1907 (1968).
88. T. Galliard, *Phytochemistry 7*:1915 (1968).
89. M. M. Paulose, S. Venkob, and K.T. Achaya, *Indian J. Chem. 4*:529 (1966).
90. H. Wagner and P. Wolff, *Fette. Seifen. Anstrichmittel. 66*:425 (1964).
91. B. Vijayalakshmi, S. Venkob, and K. T. Achaya, *Fette. Seifen. Anstrichmittel. 71*:757 (1969).
92. L. A. Schustanova, A. V. Vmaron, and A. L. Markham, *Khim. Priv. Soedin. 3*:292 (1970).
93. J. G. Parsons, G. Keeney, and S. Patton, Identification and quantitative analysis of phospholipids in cocoa beans, *J. Food Sci. 34*:497 (1969).
94. J. D. Joseph, Lipid composition of marine and estriarine invertebrates, *Prog. Lipid Res. 21*:109 (1982).
95. J. Dyerberg and H. O. Bang, Haemostatic function and platelet polyunsaturated fatty acids in eskimos, *Lancet Sept 1*:433 (1979).
96. J. Dyerberg and K. A. Jorgensen, Marine oils and thrombogenesis, *Prog. Lipid Res. 21*:255 (1982).
97. J. Sampugna, L. Johnson, K. Bachman, and M. Keeney, Lipids of Crassostrea virginica, *Lipids 7*:339 (1972).
98. G. G. Forstner, K. Tanaka, and K. J. Isselbacher, Lipid composition of the isolated rat intestinal microvillus membrane, *Biochem. J. 109*:51 (1968).
99. G. Rouser and S. Fleischer, Phospholipid composition of human, bovine and frog myelin isolated on a large scale from brain and spinal cord, *Lipids 4:239* (1969).
100. P. G. Fast, Insect lipids, *Prog Chem. Fats 11*:179 (1972).
101. R. J. Henderson and D. R. Tocher, The lipid composition and biochemistry of freshwater fish, *Prog. Lipid Res. 26*:281 (1987).
102. R. G. Ackman, and T. Takeuchi, Comparison of fatty acids and lipids of smolting hatchery fed and wild atlantic salmon, *Lipids 21*:117 (1986).
103. D. C. Malins and J.C. Wekell, The lipid biochemistry of marine organisms, *Prog. Chem. Fats 10*:337 (1967).
104. G. Van Den Thillart and G. De Bruin, Influence of environmental temperature on mitochondrial membranes, *Biochim. Biophys. Acta 640*:439 (1981).
105. J. M. Leslie and J.T. Buckley, Phospholipid composition of goldfish liver and brain and temperature-dependence of phosphatidylcholine synthesis, *Comp. Biochem. Physiol. 53B*:335 (1975).
106. B. I. Roots, Phospholipids of goldfish brain, *Comp. Biochem. Physiol. 25*:457 (1968).
107. W. Driedzic, D. P. Selivonchick, and B. I. Roots, Alk-l-enyl ether containing lipids of goldfish, *Comp. Biochem. Physiol. 53B*:311 (1976).
108. J. R. Hazel and E. Zerba, *J. Comp. Physiol. 134*:321 (1986).
109. J. R. Hazel, Influence of thermal acclimation on membrane lipid composition of rainbow trout liver, *Am. J. Physiol. 236*:R91 (1979).

110. D. H. S. Greene and D. P. Selivonchick, Lipid metabolism in fish, *Prog. Lipid Res. 26*:53 (1987).
111. J. R. Hazel, Effects of temperature on the structure and metabolism of cell membranes in fish, *Am. J. Phys. 246*:R460 (1984).
112. D. Chapman, Protein lipid interactions in model and natural biomembranes, in *Membrane Reconstitution* (G. Poste and G. L. Nicholson, eds.). North-Holland, Amsterdam, 1982, pp. 1–41.
113. V. Luzzati, X-ray diffraction studies of lipid-water systems, in *Biological Membranes,* Vol. I (D. Chapman, ed.), Academic Press, London, 1968, pp. 71–123.
114. D. Chapman, P. Byrne, and G. G. Shipley, The physical properties of phospholipids, *Proc. Roy. Soc. Ser. A 290*:115 (1966).
115. S. J. Fliesler and R. E. Anderson, Chemistry and metabolism of lipids in the vertebrate retina, *Prog. Lipid Res. 22*:79 (1983).
116. R. E. Anderson, L. S. Feldman, and G. L. Feldman, Lipids of ocular tissues. *Biochim. Biophys. Acta 202*:367 (1970).
117. R. M. Broekhuyse, Phospholipids in tissues of the eye, *Biochem. Biophys. Acta 152*:307 (1968).
118. R.V. Dorman, H. Dreyfus, L. Freyszll, and L. A. Horrocks, Ether lipid content of phosphoglycerides from the retina and brain of chicken and calf, *Biochem. Biophys. Acta 486*:55 (1977).
119. J. Eichberg, and H. A. Hess, The lipid composition of frog retinal and outer segments, *Experientia 23*:993 (1987).
120. P. F. Urban, H. Dreyfus, N. Neskovic, and P. Mandel, Phospholipid metabolism in light and dark adapted excised retina, *J. Neurochem. 20*:325 (1973).
121. H. E. P. Bazan and N. G. Bazan, Phospholipid composition and [^{14}C]glycerol incorporation into glycerolipids of toad retina and brain, *J. Neurochem, 27*:1051 (1976).
122. R. E. Anderson, Lipids of ocular tissue IV, *Exp. Eye Res. 10*:339 (1970).
123. R. E. Anderson and M. B. Maude, Lipids of ocular tissue, *Arch. Biochem. Biophys. 151*:270 (1972).
124. H. Dreyfus, P. F. Urban, N. Neskovic, and P. Mandel, Distribution des phosphatides et marquaze du phosphore lipidique et du phosphate mineral de retines de rat et de veau, *Biochimie 53*:567 (1971).
125. H. Dreyfus, P. F. Urban, S. Edel-Harth, and P. J. Mandel, Developmental patterns of gangliosides and of phospholipids in chick retina and brain, *J. Neurochem. 25*:245 (1975).
126. D. Johnston and R. A. Hudson, Phospholipids of the cone-rich chicken retina and its photoreceptor outer segment membranes, *Biochim. Biophys. Acta 369*:269 (1974).
127. D. A. White, The phospholipid composition of mammalian tissues, in *Form and Function of Phospholipids* (G.B. Ansell, J. N. Hawthorne, and R. M. C. Dawson, eds.), Elsevier, New York, 1973, p. 441.
128. G. Rouser and G. Kritchervsky, Lipids in the nervous system of different species as a function of age, *Adv. Lipid Res. 10*:261 (1972).
129. A. Diagne, J. Faurel, M. Record, H. Chap, and L. Douste-Blazy, Studies on ether phospholipids, *Biochim. Biophys. Acta 793*:2215 (1984).
130. K. Swietochowska, K. Jaroszewicz, W. Komenda, and T. Januszko, Composition of phospholipids in various rat tissues, *Acta. Phys. Hungar. 62*:145 (1983).
131. E. J. Singh and Jr. Swartwout, The phospholipid composition of human placenta, endometrium and amniotic fluid, *Lipids 7*:26 (1972).
132. R. S. Rana and L. E. Hokin, Role of phosphoinositides in transmembrane signaling, *Phys. Rev. 70*:115 (1990).
133. N. T. Glanville, D. M. Byers, H. W. Cook, M. W. Spence, and F. B. C. Palmer, Differences in the metabolism of inositol and phosphoinositides by cultured cells of neuronal and glial origin, *Biochim. Biophys. Acta 1004*:169 (1989).

134. M. A. Yorek, J. A. Dunlap, and B. H. Ginsberg, Myo-inositol metabolism in 41A3 neuroblastoma cells, *J. Neurochem. 48*:53 (1987).
135. B. L. Slomiany, V. L. N. Murty, Y. H. Liau, and A. Slomiany, Animal glycoglycerolipids, *Prog. Lipid Res. 26*:29 (1987).
136. J. Kapitulnik, E. Weil, R. Rabinowitz, and M. M. Krausz, Fetal and adult human liver differ markedly in the fluidity and lipid composition of their microsomal membranes, *Hepatology 7*:55 (1987).
137. G. Y. Sun, and L. L. Foudin, Phospholipid composition and metabolism in the developing and aging nervous system, *Phospholipids in Nervous Tissue* (J. Eichberg, ed.), Wiley, New York, 1985, p. 79.
138. M. A. Wells and J. C. Dittmer, A comprehensive study of the postnatal changes in the concentration of the lipids of developing rat brain, *Biochemistry 6*:3169 (1967).
139. L. Svennerholm, Distribution and fatty acid composition of phosphoglycerides in normal human brain, *J. Lipid Res. 9*:570 (1968).
140. F. Snyder, T. C. Lee, and M. L. Blank, Platelet activating factor and related ether lipid mediators, *Ann. NY Acad. Sci. 568*:35 (1989).
141. R. H. Michell, Phosphoinositides and inositol phosphates, *Biochem. Soc. Trans. 17*:1 (1988).
142. B. Margolis, A. Zilberstein, C. Franks, S. Felder, S. Kremer, A. Ullrich, S. G. Rhee, K. Skorecki, and J. Schlessinger, Effect of phospholipase-C-c overexpression on PDGF induced second messengers and mitogenesis, *Science 248*:607 (1990).
143. M. J. Berridge and R. F. Irvine, Inositol phosphates and cell signaling, *Nature 341*:197 (1989).
144. A. Nordoy and S. H. Goodnight, Dietary lipids and thrombosis, *Arteriosclerosis 10*:149 (1990).
145. M. G. Low and A. R. Saltiel, Structural and functional roles of glycosylphosphatidylinositol in membranes, *Science 239*:268 (1988).
146. J. K. Blusztain, M. Liscovitch, C. Mauron, U. I. Richardson, and R. J. Wurtman, Phosphatidylcholine as a precursor of choline for acetylcholine synthesis, *J. Neural Trans. 24*:247 (1987).
147. A. A. Spector, S. N. Mathur, T. L. Kaduce, and B. T. Hyman, Lipid nutrition and metabolism of cultured mammalian cells, *Prog. Lipid Res. 19*:155 (1981).
148. A. A. Spector and M. A. Yorek, Membrane lipid composition and cellular function, *J. Lipid Res. 26*:1015 (1985).
149. W. C. McMurray, Phospholipids in subcellular organelles and membranes, *Form and Function of Phospholipids* (G. B. Ansell, J. N. Hawthorne, and R. M. C. Dawson, eds.), Elsevier, Amsterdam, 1973, p. 205.
150. R. Hovius, H. Lambrechts, K. Nicolay, and B. D. Kruijff, Improved methods to isolate and subfractionate rat liver mitochondria, *Biochim. Biophys. Acta 1021*:217 (1990).
151. H. Glaumann and G. Dallner, Lipid composition and turnover or rough and smooth microsomal membranes in rat liver, *J. Lipid Res. 9*:720 (1968).
152. G. Benga, V. I. Pop, M. Ionescu, A. Hodarnau, R. Tilinca, and P. T. Frangopol, Comparison of human and rat liver microsomes by spin label and biochemical analysis, *Biochim. Biophys. Acta 750*:194 (1983).
153. H. Kleinig, Nuclear membranes from mammalian liver, *J. Cell. Biol. 46*:396 (1970).
154. J. E. Rothman and J. Leonard, Membrane asymmetry, *Science 195*:743 (1977).
155. J. A. F. OpdenKamp, Lipid asymmetry in membranes. *Ann. Rev. Biochem. 48*:47 (1979).
156. M. Record, A. E. Tamer, H. Chap, and L. Douste-Blazy, Evidence for a highly asymmetric arrangement of ether-and diacyl-phospholipid subclasses in the plasma membrane of Krebs II ascites cells, *Biochim. Biuophys. Acta 778*:449 (1984).
157. A. H. Etemadi, Membrane asymmetry, *Biochem. Biophys. Acta 604*:423 (1980).
158. L.D. Bergelson and L. I. Barsukov, Topological asymmetry of phospholipids in membranes, *Science 197*:224 (1977).

159. R.D. Kornberg and H. M. McConnell, Inside outside transitions of phospholipids in vesicle membranes, *Biochemistry 10*:1111 (1971).
160. C. J. Kirby and C. Green, Transmembrane migration (flip-flop) of choleserol in erythrocyte membranes, *Biochem. J. 168*:575 (1977).
161. S. J. Singer and G. L. Nicolson, The fluid model of the structure of cell membranes, *Science 175*:720 (1972).
162. M. S. Bretscher, Membrane structure: Some general principles. *Science 181*:622 (1973).
163. P. F. Devaux, Phospholipid flippases, *FEBS. Lett. 234*:8 (1988).
164. M. Seigneuret and P. F. Devaux, ATP-dependent asymmetric distribution of spin-labeled phospholipids in the erythrocyte membrane, *Proc. Natl. Acad. Sci. USA 81*:3751 (1984).
165. A. Zachowski, E. Favre, S. Cribier, P. Herve, and P. F. Devaux, Outside-inside translocation of aminophospholipids in the human erythrocyte membrane is mediated by a specific enzyme, *Biochemistry 25*:2585 (1986).
166. M. Bitbol, P. Fellmann, A. Zachowski, and P. F. Devaux, Ion regulation of phosphatidylserine and phosphatidylethanolamine outside-inside translocation in human erythrocytes, *Biochim. Biophys. Acta 904*:268 (1987).
167. D. L. Daleke and W. H. Huestis, Incorporation and translocation of aminophospholipids in human erythrocytes, *Biochemistry 24*:5406 (1985).
168. A. Sune, P. Bette-Bobillo, A. Bienvenue, P. Fellmann, and P. F. Devaux, Selective outside-inside translocation of aminophospholipids in human platelets, *Biochemistry 26*:2972 (1987).
169. A. Zachowski, A. Herrmann, A. Paraf, and P. F. Devaux, Phospholipid outside-inside translocation in lymphocyte plasma membranes is a protein-mediated phenomenon, *Biochim. Biophys. Acta 897*:197 (1987).
170. W. Renooij, L. M. G. Van Golde, R. F. A. Zwaal, and L. L. M. Van Deenen, Topological asymmetry of phospholipid metabolism in rat erythrocyte membranes, *Eur. J. Biochem. 61*:53 (1976).
171. B. Bloj and D. B. Zilversmit, Asymmetry and transposition rates of phosphatidylcholine in rat erythrocyte ghosts, *Biochemistry 15*:1277 (1976).
172. H. J. Chap, R.F. A. Zwaal, and L. L. M. Van Deenen, Action of highly purified phospholipases of blood platelets, *Biochim. Biophys. Acta 467*:146 (1977).
173. A. Sandra and R. E. Pagano, Phospholipid asymmetry in LM cell plasma membrane derivatives, *Biochemistry 17*:332 (1978).
174. O. S. Nilsson and G. Dallner, Transverse asymmetry of phospholipids in subcellular membranes of rat liver, *Biochim. Biophys. Acta 464*:453 (1977).
175. L. McEvoy, P. Williamson, and R. A. Schlegel, Membrane phospholipid asymmetry as a determinant of erythrocyte recognition by macrophages, *Proc. Natl. Acad. Sci. USA 83*:3311 (1986).
176. F. Schroeder, Role of membrane lipid asymmetry in aging, *Neurobiol. Aging 5*:323 (1984).
177. F. Hullin, H. Y. Kim, and N. Salem, Jr., Analysis of aminophospholipid molecular species by high performance liquid chromatography, *J. Lipid Res. 30*:1963 (1989).
178. J. J. Myher, A. Kuksis, and S. Pind, Molecular species of glycerophospholipids and sphingomyelins of human erythrocytes, *Lipids 24*:396 (1989).
179. J. J. Myher, A. Kuksis, and S. Pind, Molecular species of glycerophospholipds and sphingomyelins of human plasma, *Lipids 24*:408 (1989).
180. F. A. Kuypers, D. Chiu, W. Mohandas, B. Roelofsen, J. A. F. OpdenKamp, and B. Lubin, The molecular species composition of phosphatidylcholine affects cellular properties in normal and sickle erythrocytes, *Blood 70*:1111 (1987).
181. P. Child, J. A. F. OpdenKamp, B. Roelofsen, and L. L. M. Van Deenen, Molecular species composition of membrane phosphatidylcholine influences the rate of cholesterol efflux from human erythrocytes and vesicles of erythrocyte lipid, *Biochim. Biophys. Acta 814*:237 (1985).

182. L. Freysz and H. Van Den Bosch, Distribution and biosynthesis of disaturated phosphatidylcholine in rat brain, *Cellular and Mol. Biol. 27*:377 (1981).
183. M. I. Aveldano and N. G. Bazan, Molecular species of phosphatidylcholine, ethanolamine, serine and inositol in microsomal and photoreceptor membranes of bovine retina, *J. Lipid Res. 24*:620 (1983).
184. M. I. Aveldano, S. J. Pasquare de Garcia, and N. G. Bazan, Biosynthesis of molecular species of inositol, choline, serine, and ethanolamine glycerophospholipids in the bovine retina, *J. Lipid Res. 24*:628 (1983).
185. V. H. Engelhard, J. D. Esko, D. R. Storm, and M. Glaser, Modification of adenylate cyclase activity in LM cells by manipulation of the membrane phospholipid composition in vivo, *Proc. Natl. Acad. Sci. USA 73*:4482 (1976).
186. B. Malkiewicz-Wasowicz, O. Gamst, and J. H. Stromm, The influence of changes in the phospholipid pattern of intact fibroblasts on the activities of four membrane bound enzymes, *Biochim. Biophys. Acta 482*:358 (1977).
187. B. A. Lipton, E. P. Davidson, B. H. Ginsberg, and M. A. Yorek, Ethanolamine metabolism in cultured bovine aortic endothelial cells, *J. Biol. Chem. 265*:7195 (1990).
188. M.A. Yorek, R. T. Rosario, D.T. Dudley, and A. A. Spector, The utilization of ethanolamine and serine for ethanolamine phosphoglyceride synthesis by human Y79 retinoblastoma cells, *J. Biol. Chem. 260*:2930 (1985).
189. M. H. Yorek, J. A. Dunlap, A. A. Spector, and B. H. Ginsberg, Effect of ethanolamine on choline uptake and incorporation into phosphatidylcholine in human Y79 retinoblastoma cells, *J. Lipid Res. 27*:1205 (1986).
190. L.D. Bergelson, Tumor lipids, *Prog. Chem. Fats Lipids 13*:1 (1972).

22
Phospholipid Metabolism in Animal Cells

Gerrit L. Scherphof *University of Groningen, Groningen, The Netherlands*

I. INTRODUCTION

The metabolic pathways of the phospholipids have been textbook knowledge for decades. The biosynthesis of the major phosphoglycerides, phosphatidylcholine and phosphatidylethanolamine, involving phosphatidic acid and CDP-choline and CDP-ethanolamine as crucial intermediates, was unravelled in the 1950s and 1960s. The biosynthetic routes leading to the formation of the anionic phosphoglycerides such as phosphatidylserine, phosphatidylinositol, and diphosphatidylglycerol or cardiolipin were resolved in the 1960s and 1970s. The biosynthesis of sphingomyelin, the only naturally occurring nonglyceride phospholipid in mammalian tissues, has been the subject of debate until a few years ago, but it can now also be considered as established. Also the catabolic pathways along which the various phospholipids are degraded have been known for a long time. The demonstration of the involvement of the phospholipases A_1 and A_2, hydrolyzing the acyl ester bonds at the 1 and 2 positions, respectively, in the initiating deacylation steps of the phosphoglycerides, but also in the remodelling of their acyl group makeup, dates back to the late 1950s. Other phospholipid-specific hydrolases, such as the phospholipases C and D, responsible for the hydrolytic removal of the phosphocholine and the choline group of phosphatidylcholine, respectively, were shown to exist in bacteria and plants but went for a very long time unnoticed in cells of animal origin. It was only a few years ago that the occurrence of these enzymes in animal cells has been known and their important role in the production of a variety of physiologically active metabolites, involved in processes such as membrane signal transduction, has been recognized. Degradation of sphingomyelin was shown a long time ago to start with the hydrolytic removal of the choline moiety by a sphingomyelin-specific phospholipase C, called sphingomyelinase, an enzyme first discovered in bacteria. Again, the significance of sphingomyelin as a precursor of physiologically active metabolites has been recognized for only a few years.

Another aspect of phospholipid metabolism that is relatively new concerns its regulation. In the past decade several studies have shed light on the mechanisms of regulation of phospholipid metabolism.

Some recent developments concerning the role of phospholipid metabolism in important physiological processes as well as some aspects of its regulation will be discussed in this chapter.

II. METABOLISM OF PHOSPHATIDYLCHOLINE AND PHOSPHATIDYLETHANOLAMINE

A. Biosynthesis

The biosynthesis of phosphatidylcholine involves the transfer of the phosphocholine group from CDP-choline to a diacylglycerol moiety formed by the dephosphorylation of phosphatidic acid, which in turn originates from the two-step acylation of glycerol-3-phosphate (Fig. 1). This biosynthetic route has been elucidated by the work of several groups in the 1950s and early 1960s (Wittenberg and Kornberg, 1953; Kennedy and Weiss, 1956; Weiss et al., 1958).

An alternative route along which phosphatidylcholine can be synthesized involves the stepwise methylation of phosphatidylethanolamine (Bremer and Greenberg, 1961). The relative importance of this pathway varies with the tissue studied; in liver, some 20–40% of the total phosphatidylcholine synthesis is accounted for by phosphatidylethanolamine methylation (Sundler and Akesson, 1975). Current knowledge concerning this pathway has been elaborately reviewed by Vance and Ridgway (1988). Since the purification by Ridgway and Vance of the phosphatidylethanolamine methyl-

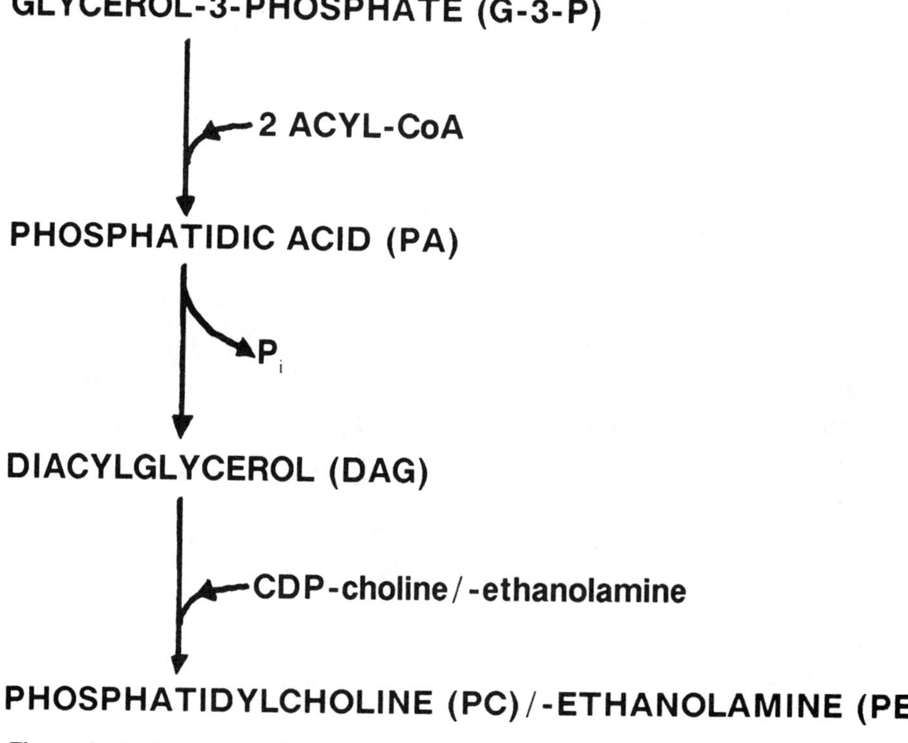

Figure 1 Main pathway of phosphatidylcholine and phosphatidylethanolamine biosynthesis.

transferase from rat liver microsomes (Ridgway and Vance, 1987) it would seem established that the three consecutive methylation steps, using *S*-adenosylmethionine as a methyl donor, are catalyzed by one and the same enzyme, a 18.3 kDa single chain polypeptide.

The formation of phosphatidylethanolamine may also proceed along different pathways. The first, and in most animal tissues probably the most important, is the CDP-ethanolamine pathway, analogous to the formation of phosphatidylcholine, with glycerolphosphate, phosphatidic acid, and diacylglycerol as intermediates (Bell and Coleman, 1980; Tijburg et al., 1989b, Fig. 1). Alternatively, phosphatidylethanolamine may arise from the decarboxylation of phosphatidylserine, catalyzed by the mitochondrial enzyme phosphatidylserine decarboxylase (Bjerve, 1973; Borkenhagen et al., 1961).

B. Regulation of Biosynthesis

The regulation of phospholipid metabolism has been extensively reviewed in recent years by Vance (1989), Tijburg et al. (1989a), Kent (1990), and Kent et al. (1991). The reader is referred to these reviews for a more detailed account of the various aspects of phospholipid metabolism in general and phosphatidylcholine and -ethanolamine metabolism in particular.

The CDP-choline pathway of phosphatidylcholine biosynthesis is mainly regulated by modulation of two enzymes, choline kinase, regulating the amount of phosphocholine, and CTP : phosphocholine cytidylyltransferase, catalyzing the transfer of the phosphocholine group to the nuclcotide. Choline kinase activity is believed to be regulated by modulation of the amounts of enzyme in the cell (Vigo et al., 1981; Paddon et al., 1982; Vigo and Vance, 1981). The occurrence of isoenzymes of the kinase (Cornell, 1989) may explain that not in all cases is an increase in choline kinase activity accompanied by an increase in phosphatidylcholine synthesis (Ishidate et al., 1985b; Tadokoro et al., 1985). The importance of choline kinase in the regulation of phosphatidylcholine synthesis may depend on the prevailing conditions in the particular cell or tissue type under investigation, as was pointed out by Kent (1990). The reactions catalyzed by choline kinase and cytidylyltransferase are both far from equilibrium and, as such, represent candidates qualified to play a regulatory role in phosphatidylcholine biosynthesis.

Cytidylyltransferase activity is modulated by activation or inactivation of preexisting enzyme (Kent et al., 1991). The enzyme is activated by a variety of phospholipids but not by phosphatidylcholine itself. A shortage of the latter may thus activate the enzyme, as has been shown with isolated cells by artificially creating such a shortage by means of phospholipase C treatment (Sleight and Kent, 1980) or choline depletion (Sleight and Kent, 1983; Jamil et al., 1990).

Cytidylyltransferase is in most cells distributed between the cytosolic fraction and the endoplasmic reticulum and, possibly, the Golgi system (Higgins and Fieldsend, 1984; Jelsema and Morré, 1978; Vance and Vance, 1988). In this respect is resembles the rate-limiting enzyme in hepatic triacylglycerol synthesis, phosphatidate phosphatase (E.C. 3.1.3.4). The distribution of this enzyme between soluble and particulate fractions of the cell is also believed to play a crucial role in the regulation of triacylglycerol synthesis (Brindley, 1984). The interconversion of the two different forms of the cytidylyltransferase is facilitated by a number of effector molecules, including fatty acids (Pelech et al., 1984a; Feldman et al., 1985), diacylglycerol, and anionic phospholipids (Feldman et al., 1985; Weinhold et al., 1986; Feldman and Weinhold, 1987; Cornell and

Vance, 1987). Interestingly, aminolipids such as sphingosine antagonize the stimulating effect of these lipids (Sohall and Cornell, 1990). On the other hand, the conversion of the enzyme from a soluble to a membrane-bound form was shown to be induced by reversible phosphorylation (Pelech and Vance, 1982; Vance and Pelech, 1984; Pelech and Vance, 1984; Radika and Possmayer, 1985: Sanghera and Vance, 1989) induced by a cAMP-mediated mechanism. Recent observations by Hatch et al. (1991) on the inhibition of phosphatidylcholine synthesis in rat hepatocytes by a protein phosphatase inhibitor, okadaic acid, lend further support to the involvement of a phosphorylation/dephosphorylation cycle in the conversion of cytidylyltransferase from an active (membrane-bound) to an inactive (cytosolic) form. In addition, evidence has been provided that the transferase may be controlled by a protein kinase C dependent mechanism (Wertz and Mueller, 1978; Guy and Murray, 1982; Hill et al., 1984; Pelech et al., 1984b; Kolesnick, 1987; Nishizuka, 1986), which may explain the stimulating effect of diacylglycerol on cytidylyltransferase activity.

The enzyme from rat liver was purified some years ago by Weinhold et al. (Weinhold et al., 1986), and the cDNA of the enzyme from yeast was isolated by Tsukagoshi et al. (1987). Recently, the cDNA of the rat-liver enzyme was cloned by PCR and expressed in COS cells; from its derived amino acid sequence the authors proposed a mechanism for the interaction of the enzyme with membranes involving a 58-residue amphipathic helix (Kalmar et al., 1990).

Studies on phosphatidylcholine synthesis in the human transformed hepatocytic cell line Hep G2 led Feldman and his associates to postulate the existence of two forms of the transferase in the cytosolic fraction of the cell; an active, lipid-containing, H-form and an inactive, lipid-free, L-form (Weinhold et al., 1989). H- and L-forms are interconvertible in that the H-form is unstable and can be dissociated to the L-form by briefly incubating the cytosolic fraction at 37°C, while the L-form can be reactivated to the H-form with lipid substituents. Recently, this group proposed an interesting hypothesis on the significance of the occurrence of these two enzyme forms (Weinhold et al., 1991). Elaborating on the recent proposal by George et al. (1989) that the individual steps of phosphatidylcholine synthesis in the cytidine pathway are functionally linked, possibly involving cytoskeletal elements, they speculated "that the H-form of the cytidylyltransferase may be part of a multi-enzyme complex that is oriented along cytoskeletal elements such that the cytidylyltransferase complex is associated with both the membrane and the cytoskeleton."

Although the CTP pathway of phosphatidylethanolamine synthesis is very closely similar to that of phosphatidylcholine synthesis, it is clear that at least one of the specific enzymes involved, i.e., ethanolamine-phosphate cytidylyltransferase, is distinctly different from its choline counterpart. The enzyme has been purified from the cytosolic fraction of rat liver (Sundler, 1975), and both its physical and its functional characteristics were shown to be significantly different from those of the choline enzyme. Interestingly, ethanolamine-specific enzyme does not show the typical ambiquitous properties of the choline-phosphate cytidylyltransferase. It has been found to be located exclusively in the cytosolic fraction of a variety of tissues (Sundler, 1975; Schneider and Vance, 1978). Also the lack of control exerted by glucagon over the phosphoethanolamine cytidylyltransferase, as observed by Tijburg et al. (1989c), indicating that this enzyme is not regulated by a phosphorylation/dephosphorylation cycle, reflects the fundamental difference between phosphocholine- and phosphoethanolamine transferase activities.

The other enzyme possibly specific to phosphatidylethanolamine biosynthesis is ethanolamine kinase. There has been considerable debate in the literature about the existence of two different kinases for choline and ethanolamine. The explanation for the long-lasting controversy has probably been provided by Nakazawa and his coworkers (Ishidate et al., 1984; Ishidate et al., 1985a) who reported the complete copurification of the two enzyme activities from kidney tissue. The enzymes were shown to be immunologically indistinguishable, and it was suggested that the two kinases may represent different activities residing on the same polypeptide.

In line with the distinctly different properties of the ethanolamine- and the choline-phosphate cytidylyltransferases, the regulation of phosphatidylethanolamine biosynthesis, although considerably less well investigated than that of phosphatidylcholine synthesis, differs significantly from the latter. More than a decade ago Van Golde and his coworkers observed a remarkable difference in the response to fasting and refeeding protocols between phosphatidylcholine and phosphatidylethanolamine biosynthesis in rat liver (Groener et al., 1979). Work from the same group showed that the synthesis of the two phospholipids is independently controlled by hormones (Haagsman et al., 1984; Tijburg et al., 1987).

C. Phosphatidylserine

The biosynthesis of phosphatidylserine in animal cells proceeds through a Ca^{2+}-dependent base exchange reaction on the endoplasmic reticulum with phosphatidylcholine or phosphatidylethanolamine (Porcellati et al., 1971; Kanfer, 1972; Dennis and Kennedy, 1972; Kennedy, 1986; Baranska, 1982; Bjerve, 1973). The purified enzyme, phosphatidylserine synthase, catalyzing exchange of serine with phosphatidylethanolamine, has been obtained from brain microsomes (Suzuki and Kanfer, 1985). Phosphatidylserine thus formed is transferred from the site of synthesis, the endoplasmic reticulum, to the mitochondria in two steps (Voelker, 1985; 1989), an ATP-requiring step, possibly involving vesicular transport following budding, and an ATP-independent step, which is likely to represent the formation of collision complexes between donor vesicles and the mitochondrion. On the inner mitochondrial membrane, phosphatidylserine thus translocated is readily decarboxylated (Dennis and Kennedy, 1972; Van Golde et al., 1974; Butler and Morell, 1983). Under conditions of restricted ethanolamine supply, this route may be the predominant pathway for phosphatidylethanolamine biosynthesis, prevailing over the CDP-ethanolamine pathway (Miller and Kent, 1986; Voelker, 1984; Bradbury, 1984). Newly synthesized phosphatidylserine does not readily diffuse from the site of production (Corazzi et al., 1987). The base exchange reaction is strongly inhibited by the product, phosphatidylethanolamine. This phenomenon has been proposed to represent a regulatory factor in phosphatidylserine biosynthesis (Corazzi et al., 1991). Alternative pathways for phosphatidylserine formation have been proposed (Infante, 1984; Baranska, 1988), but they are believed to play a minor role, probably only of importance for the synthesis of selected molecular species (Corazzi et al., 1991).

In yeast, phosphatidylserine is synthesized in a different way. CDP-diacylglycerol, formed from phosphatidic acid and CTP (see below under cardiolipin), exchanges the CMP moiety for a serine in a reaction catalyzed by a phosphatidylserine synthase associated with both the mitochondrial and the microsomal fractions of the cell. In a recent review, Kent et al. (1991) discussed the regulation of this enzyme. It was suggested that the observed activation of the enzyme by phosphatidic acid and its inhibition by diacylglycerol are involved in the selection between the methylation and the CDP-choline pathways of phosphatidylcholine synthesis.

Lysophosphatidylserine, a metabolic product of phosphatidylserine arising from phospholipase A2 action on the parent compound, has been shown to possess potent pharmacological properties, partly mediated by its ability to induce histamine release from mast cells (Bigon et al., 1979; Mietto et al., 1984; Kolster et al., 1987; Mietto et al., 1987). Phospholipase A-mediated lysophosphatidylserine formation has also been implicated to play a role in the regulation of macrophage activation (Gilbreath et al. 1985, 1986, 1989).

D. Phosphatidylglycerol and Cardiolipin

Cardiolipin, or diphosphatidylglycerol, in eukaryotic cells is exclusively found in the mitochondria. In addition, it is a component of bacterial membranes. It plays an important role in the activity of some mitochondrial enzymes and transporter proteins, such as cytochrome oxidase (Fry and Green, 1980; Vik et al., 1981), creatine phosphokinase (Muller et al., 1985; Schlame and Augustin, 1985), the ADP/ATP exchange protein (Beyer and Klingenberg, 1985), and the phosphate carrier (Mende et al., 1982). The pathway along which it is synthesized was first reported for bacteria (Fig. 2). Stanacev et al. (1967) initially presented evidence that this pathway involves the transfer of a phosphatidyl group from CDP-diacylglycerol to phosphatidylglycerol:

$$\text{PG} + \text{CDP-DG} \rightarrow \text{DPG} + \text{CMP}$$

The phosphatidylglycerol arises by dephosphorylation of phosphatidylglycerolphosphate, which in turn is formed by transfer of a phosphatidyl moiety to glycerol-3-phosphate:

$$\text{G-3-P} + \text{CDP-DG} \rightarrow \text{PGP} + \text{CMP}$$

$$\text{PGP} + \text{H2O} \rightarrow \text{PG} + \text{Pi}$$

Later it was shown that the formation of cardiolipin in bacteria is actually achieved by the condensation of two molecules of phosphatidylglycerol with the release of a glycerol (Hostetler et al., 1972; Hirschberg and Kennedy, 1972):

$$2\text{PG} \rightarrow \text{DPG} + \text{glycerol}$$

In eukaryotic cells, the formation of PGP and PG were shown to occur in mitochondria by several authors (Kiyasu et al., 1963; Stanacev et al., 1968; Possmayer et al., 1968).

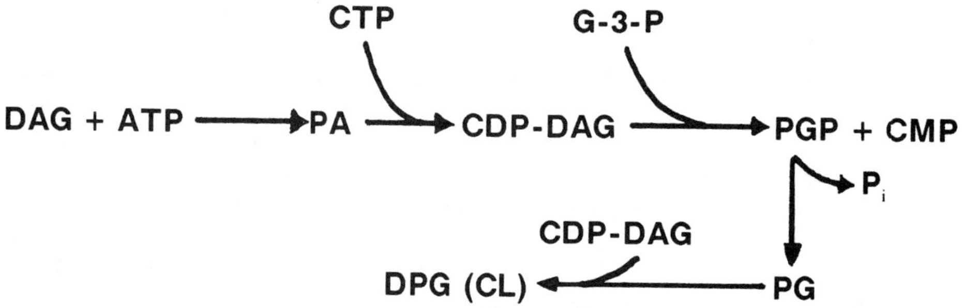

Figure 2 Biosynthesis of diphosphatidylglycerol (cardiolipin). DAG, diacylglycerol; PA, phosphatidic acid; PG(P), phosphatidylglycerol(phosphate); CT(D,M)P, cytidine-tri(di,mono)phosphate; DPG, diphosphatidylglycerol; CL, cardiolipin.

Cardiolipin formation by isolated mitochondria, however, was initially not found. Hostetler et al. (1971) were the first to observe cardiolipin synthesis in rat liver mitochondria; they showed that it proceeds by the transfer of a phosphatidyl moiety from CDP-diacylglycerol to PG, i.e., along the initially proposed pathway in *E. coli* (Stanacev et al., 1967). In a subsequent study Hostetler and Van den Bosch (1972) demonstrated that the formation of cardiolipin from PG exclusively occurs in the inner mitochondrial membrane. PG synthesis was also reported to be localized predominantly in the inner membrane, although a small but significant activity was found in the outer membrane fraction as well (Hostetler and Van den Bosch, 1972).

Little is known about the regulation of the enzymes responsible for cardiolipin synthesis in eukaryotic cells. Cardiolipin synthase, the enzyme catalyzing the final step in cardiolipin synthesis, was found not to be repressed by inositol in *Saccharomyces cerevisiae,* in contrast to several other phospholipid synthesizing enzymes (Tamai and Greenberg, 1990). Recently, thyroid hormone was reported to exert a regulatory effect on cardiolipin synthase in rat liver mitochondria (Hostetler, 1991). This observation may explain the earlier reported increase in cardiolipin content of liver mitochondria as a result of triiodothyronine treatment (Ruggiero et al., 1984; Paradies and Ruggiero, 1990).

III. CATABOLISM OF PHOSPHOGLYCERIDES

A. Phospholipases of the A Type

The main route of catabolic breakdown of the choline and ethanolamine phosphoglycerides involves their stepwise deacylation catalyzed by phospholipases A_1 and A_2. They are designated according to their positional specificity (Fig. 3), the A_1 type catalyzing the hydrolytic removal of the acyl moiety at the *sn*-1 position and the A_2 type removing the acyl group at the *sn*-2 position of the glyceride molecule. Both enzymatic activities have been demonstrated in a large variety of cells and tissues and are probably ubiquitous (Van den Bosch, 1990). The phospholipases A_2 in venoms of several animal species as well as the digestive enzyme produced by the pancreas are by far the most thoroughly investigated enzymes of this type (Verheij et al., 1981; Dennis, 1983). Much less is known about the phospholipases A that are active intracellularly, because they occur only in relatively small amounts and they have proven to be difficult to isolate and purify due to their instability and the low amounts in which they occur. Phospholipases A_2 have attracted a great deal of attention over the past decades because of their potential role in the production of the common precursor molecule of eicosanoids like thromboxanes, leukotrienes, and prostaglandins, i.e., arachidonic acid, which is almost exclusively found at the 2 position of phosphoglycerides and also is a major acyl constituent of plasmalogens (Chilton et al., 1984). These studies have revealed that the phospholipases A_2 described display an overwhelming heterogeneity with respect to substrate specificity, Ca^{2+} dependence, intracellular distribution, and the way in which they are associated with membranes (Van den Bosch, 1990).

Ca^{2+}-dependent phospholipases A_2 have been purified, in some cases to homogeneity, from a variety of animal tissues or secretion fluids (see Van den Bosch, 1990). With a few exceptions the molecular weight of these enzymes was found to be in the same order of magnitude as those of the pancreas and bee and snake venoms, i.e., 13–18 kDa. In rat liver, also in this respect the most intensely investigated tissue, phospholipase A_2 activity

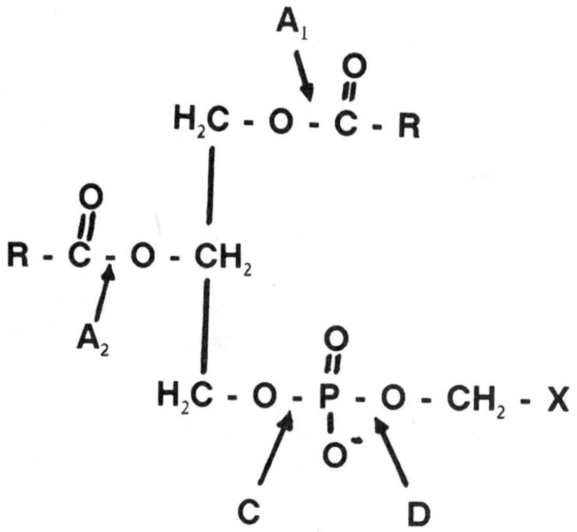

Figure 3 The sites of action of phospholipases A_1, A_2, C, and D on a phosphoglyceride molecule.

was encountered in most subcellular fractions, including mitochondria (Scherphof and Van Deenen, 1965); microsomes (Newkirk and Waite, 1973); plasma membrane (Victoria et al., 1971); Golgi membranes (Van Golde et al., 1971); and lysosomes (Franson et al., 1971; Rahman et al., 1970). The mitochondrial enzyme, whose outer membrane localization was established by Scherphof et al. (1966), has been thoroughly studied by Van den Bosch and his associates (de Winter et al., 1982; de Winter et al., 1984; De Winter et al., 1987; de Jong et al., 1987; Lenting et al., 1987). In a recent study (Aarsman et al., 1989) Van den Bosch and coworkers reported on the subcellular localization of rat liver phospholipase A_2 using monoclonal antibodies against the purified mitochondrial enzyme and concluded that more than 95% of the Ca^{2+}-dependent phospholipase A_2 activities in rat liver homogenates reacted with the monoclonal antibody against the mitochondrial enzyme. This observation was taken to indicate that the phospholipases in the nonmitochondrial fractions are structurally closely related if not identical to the mitochondrial enzyme. Quantitatively, the mitochondria were shown to contain by far the largest enzyme concentration. Only the soluble fraction was found to contain a substantial amount of phospholipase activity not reacting with the antibody against the mitochondrial enzyme. The amino acid sequence of the mitochondrial enzyme revealed that this phospholipase belongs to the so-called group II (Heinrikson et al., 1977) phospholipases A_2, which have in common a lack of a cysteine-11. It shares this property with some snake venom enzymes and a number of intracellular phospholipases.

In recent years a number of Ca^{2+}-independent phospholipases have been described as well, in addition to the Ca^{2+}-independent lysosomal enzyme known already (Waite and Sisson, 1971; Waite et al., 1981). Both neutral and alkaline pH enzymes were reported present in a variety of mammalian tissues, cells, and body fluids. It is tempting to speculate that particularly this class of phospholipase A_2 activities is involved in the release of arachidonic acid under the conditions of submicromolar free Ca^{2+} concentrations prevailing in resting cells. Also the relatively rapid turnover of acyl groups at the 2 position of phosphoglycerides, which is likely to play a role in the remodelling of newly synthesized phospholipids (Lands, 1958), may have to be attributed to these Ca^{2+}-independent enzymes (Van den Bosch, 1990).

It has been suggested that there may be phospholipase A_2 activities with specificity or preference for *sn*-2-arachidonoyl phospholipid species. This was for instance based upon the observation that polymorphonuclear leukocytes, when depleted of arachidonic acid, display a strongly diminished rate of platelet activating factor (PAF) synthesis. As it is known that the precursor of PAF is a 1-alkyl,2-acylphosphatidylcholine with arachidonate as the major acyl moiety, specificity for this substrate would seem indicated (Ramesha and Pickett, 1986). An alternative explanation for specific arachidonate release involves the existence of membrane domains enriched in arachidonate-containing phospholipid species and the presence of the phospholipase A_2 especially in such domains. In a recent study, Schalkwijk et al. (1990) investigated the acyl-chain selectivity of phospholipase A_2 from polymorphonuclear leukocytes and platelets. For leukocytes and rat platelets they found no indication of any substrate specificity either with pure substrates or with mixed bilayers and at either high or low substrate concentrations. With the human platelet enzyme, however, their results were highly suggestive of arachidonate selectivity when using a single phosphatidylcholine species as a substrate. When substrates were presented in mixed bilayers, however, the observed selectivity was nearly completely lost. These observations were explained on the basis of the well-known differences in membrane-penetrating capacity of phospholipase A_2 species (Demel et al., 1975). The authors conclude that the phospholipases that they investigated will not display any species preference as long as the substrate is presented in a mixed bilayer form, as in a biological membrane, although they do not deny the possibility that other phospholipases A_2 with such specificity may exist. In that connection they refer to an observation by Wijkander and Sundler (1989) indicating the presence of such an enzyme in mouse peritoneal macrophages. The alternative, i.e., the presence of phospholipase A_2 in arachidonate-enriched membrane domains remains, however, a conceivable explanation of preferential arachidonate release.

Regulation of phospholipase A_2 activity is obviously of vital importance for the cell. If uncontrolled, this activity will lead to gross destruction of cellular membrane structures. On the one hand, the products of phospholipase action, i.e., fatty acids and lysophospholipids, which by themselves may have deleterious effects on membrane structure, can be removed by a variety of reactions. Fatty acids can be oxidized or esterified; lysophospholipids can be further deacylated by lysophospholipase activity or reacylated to form a diacylphospholipid again (Lands, 1958). On the other hand, phospholipase activity as such may be kept under control. Regulation of phospholipase A_2 activity is still poorly understood, and most if not all knowledge about it refers to the Ca^{2+}-dependent species only. The topic has been reviewed in recent years by a number of specialists (Roberts and Dennis, 1989); Chang et al., 1987; Van den Bosch, 1990). Recently, Vance and coworkers (Tijburg et al., 1991a) concluded that in rat hepatocytes the phosphatidylcholine concentration is an important regulating factor in the overall rate of phosphatidylcholine degradation. The receptor-mediated activation of platelet phospholipases by means of a serine protease has been proposed by Hanahan and coworkers (Sugatani et al., 1987), suggesting the existence of zymogen forms of these enzymes, in analogy to the well-known pancreatic phospholipase A_2 (De Haas et al., 1968). On the other hand, the involvement of both activator and inhibitor molecules has been implicated in several studies. A mellitin-related protein from smooth muscle cells was found selectively to stimulate phospholipase A_2 activity from these cells (Clark et al., 1987). Receptor-mediated activation of phospholipases by G-proteins has been reported by several groups (Burch et al., 1986; Nakashima et al. 1987; Fuse and Tai, 1987). To date, no evidence is available, however, that these proteins directly interact with the phospholipases. Other mediator proteins may be involved, particularly since it is not firmly established whether phospholipases and G-proteins are present in the same membrane.

Calmodulin is another protein whose involvement in phospholipase A_2 activation has been implicated, as most phospholipases A are strictly Ca^{2+} dependent. This is largely based on the often observed inhibitory effect of calmodulin antagonists such as trifluoperazine on phospholipase activity. Because of the structural relationship of trifluoperazine with local anesthetics, this compound exerts a number of additional effects on membranes, which may explain its phospholipase inhibiting potency (Dijkstra et al., 1985). The Ca^{2+} requirement of partially purified phospholipases from different sources was shown not to be mediated by calmodulin (De Winter et al., 1984; Whitnall and Brown, 1982; Teramoto et al., 1983; Ballou and Cheung, 1983).

Another class of proteins that have attracted a great deal of attention as potential modulators of phospholipase activities is represented by the lipocortins. These proteins are synthesized by a variety of cells in response to a glucocorticosteroid stimulus and inhibit prostaglandin production (Flower et al., 1984). A number of different phospholipases of the A, C, and D class were subsequently demonstrated to be inhibited by lipocortin, thus questioning the specific role of these proteins in phospholipase A_2 modulation (Hirata, 1981). Subsequent experiments, including the cloning and expression of the human protein, have revealed that the lipocortins display substantial homology with a number of other Ca^{2+} and phospholipid-binding proteins. Also, the earlier reported elimination of phospholipase inhibiting activity as a result of phosphorylation of lipocortin was put in different perspective when it was found that phosphorylation of Tyr-21 in lipocortin I substantially reduced the Ca^{2+} concentration required for binding to phosphatidylserine vesicles. This and a number of other reports have cast serious doubts on the significance of the lipocortins as physiological phospholipase A_2 modulating proteins.

An elegant and illustrative example of an alternative way in which the consequences of increased phospholipase A activity can be controlled was presented recently by Reinhold et al. (1989). These authors demonstrated that stimulation of human neutrophils by the Ca^{2+}-ionophore A23187, inducing the formation of 1-*O*-alkyl-2-acetyl-*sn*-glycero-3-phosphocholine or platelet activating factor (PAF) as well as that of arachidonic acid and a number of its oxygenated metabolites, simultaneously activates the deacylation/reacylation cycle of phosphatidylcholines (Lands, 1958) in these cells. The authors argued that ionophore-induced stimulation of phospholipase A2 activity caused a substantial increase in both 1-alkyl and 1-acyl lysophospholipids, of which only the 1-alkyl species were needed for the synthesis of PAF. The alkyl lysolipids serve as substrate for acetyltransferase to form PAF, and the excess of potentially lethal acyl lysolipids is efficiently removed by an increase in acyltransferase activity. Interestingly, the reesterification of the acyl lysophospholipids in the stimulated cells was selective for nonarachidonate acyl groups.

B. Phospholipases C and D

In addition to arachidonic acid as a precursor of biologically active eicosanoids and lysoalkylphosphatidylcholine as a precursor of platelet activating factor, as discussed in the previous section, hydrolytic breakdown of phosphoglycerides may produce other physiologically active products as well. Among them, diacylglycerols are well known as activators of protein kinase C, a common mediator in signal transduction events. Phosphoinositides such as phosphatidylinositol (PI) and its phosphorylated derivatives (PIP and PIP2), which in most cells are minor phospholipids, have been known for many years now as precursors of the diacylglycerol moiety through the action of an inositide-specific phospholipase C. In addition to the diacylglycerol, as a protein kinase C activator, the water-soluble reaction products, particularly the inositol-4,5-diphosphate (IP2) and inositol-1,4,5-triphosphate (IP3), are also well documented second messengers, involved in mobilization of intracellular Ca^{2+} stores. Phospholipase C mediated phosphoinositide breakdown has been demonstrated to be a major signaling pathway in the action of a variety of stimuli, including a variety of hormones, growth factors, neurotransmitters, and secretagogues (see Berridge (1984) and Hokin (1985) for reviews).

More recently, however, attention has been focussed on more common phosphoglycerides as important diacylglycerol precursors. In particular, the ubiquitous and highly abundant phosphatidylcholine has emerged as a significant potential source of diacylglycerols. On the basis of significant differences in the fatty acid composition of the diacylglycerols formed by various cell types in response to a variety of stimuli and that of the inositol phospholipids in those cells it was concluded that the latter were unlikely to be the only precursors of the diacylglycerols produced (Bocckino et al., 1985; Bocckino et al., 1987; Pickford et al., 1987; Rosoff et al., 1988; Grove and Schimmel, 1982; Ragab-Thomas et al., 1987). In addition, and more convincingly, it was calculated that the changes in the amounts of phosphoinositides were insufficient to account for the quantities of diacylglycerol formed in response to the relevant stimuli (Augert et al., 1989a; Bocckino et al., 1985; Bocckino et al., 1987; Truett et al., 1988; Wright et al., 1988; Polverino and Barritt, 1988). The role of phosphatidylcholine as a precursor for diacylglycerols was recently reviewed by Loeffelholz (1989).

The most obvious pathway along which the diacylglycerols can be formed from phosphatidylcholine is through the action of a phospholipase C, in analogy to the formation of diacylglycerols from the inositol phosphatides. PI-specific phospholipases C have been isolated from different sources (Tompkins and Moscarello (1991) and references therein). Over the past few years, convincing evidence has also been presented in favor of the existence of phospholipase C activities in animal cells, capable of hydrolyzing phosphatidylcholine (Augert et al., 1989b; Besterman et al., 1986; Irving and Exton, 1987; Muir and Murray, 1987; Rosoff et al., 1988; Martin and Michaelis, 1988; Grillone et al., 1988; Agwu et al., 1989; Daniel et al., 1986; Takuwa et al., 1987; Ragab-Thomas et al., 1987; Glatz et al., 1987; Schrey et al., 1987). On the other hand, several recent reports provide conclusive evidence that diacylglycerol formation from phosphatidylcholine may be catalyzed by phospholipase D activities (Pai et al., 1988; Martin and Michaelis, 1988; Agwu et al., 1989; Liscovitch et al., 1987; Cabot et al., 1988a); Hurst et al., 1990; Metz and Dunlop, 1990; Huang and Cabot, 1990a); the initially formed phosphatidic acid may subsequently yield diacylglycerol when it is hydrolyzed by phosphatidic acid phosphohydrolase (Huang and Cabot, 1990b; Cabot et al., 1988b;

Agwu et al., 1989; Martinson et al., 1989). The presence of phospholipase D activity in cells is often conveniently demonstrated by the formation of phosphatidylethanol when the cells are stimulated in the presence of ethanol, due to the transphosphatidylation properties of the enzyme. Not only the secondarily formed diacylglycerol, however, but also the phosphatidic acid itself nowadays is considered to fulfill a role as a signaling messenger in the cell. Functions proposed for this lipid include inhibition of adenylate cyclase, stimulation of Ca^{2+} influx, stimulation of DNA synthesis, and inhibition of a GTPase-activating protein (McCormick, 1989; Exton, 1988; Murayama and Ui, 1987; Moolenaar et al., 1986; Yu et al., 1988). In several cells, stimulus-induced diacylglycerol is formed in a biphasic manner; the initial phase is believed to derive from phospholipase-C-induced degradation of phosphoinositides while the second, delayed and more sustained, phase is supposed to be the result of phosphatidylcholine breakdown by phospholipase C or D (Anthes et al., 1991; Van Blitterswijk et al., 1991b). An interesting finding is that within one cell the phospholipase C and the phospholipase D/phosphatidic acid phosphohydrolase pathways of diacylglycerol formation from phosphatidylcholine are operational to different extents depending on the stimulus given, as was reported by Van Blitterswijk and coworkers for bradykinin- or PMA-stimulated human fibroblasts (Van Blitterswijk et al., 1991a,b).

IV. METABOLISM OF ETHER PHOSPHOLIPIDS

A substantial proportion of the phosphatidylcholine and phosphatidylethanolamine fraction of cells, particularly in electrically active organs and tissues, belongs to the so-called ether phospholipids. In these phospholipids the 1 position of the glycerol moiety is linked to a fatty alcohol and thus forms an ether linkage. Most often the fatty alcohol is a 1,2-*cis*-monoenoic long-chain alkyl group; the phospholipid containing such a vinyl ether linkage is called plasmalogen. Particularly the phosphatidylethanolamine fraction of cells often has a high plasmalogen content. The biosynthesis of ether phospholipids differs from that of the diacyl phospholipids in that dihydroxyacetone-phosphate (DHAP) replaces the glycero-3-phosphate as the receptor of the alkyl group at the 1 position (Fig. 4). The resulting 1-alkyl-DHAP is reduced to 1-alkyl-G3P, which in turn is acylated to form 1-alkyl,2-acyl-G3P. Dephosphorylation by phosphatidate phosphatase then produces 1-alkyl,2-acyl-glycerol, which can be converted in the usual way to the phosphatidylcholine or -ethanolamine analog by means of the corresponding cytidylyltransferase. Characteristically, the first three enzymes involved in the synthesis of ether phospholipids, i.e., alkyl-DHAP synthase (E.C. 2.5.1.26), alkyl-DHAP reductase (E.C. 1.1.1.101), and alkyl-DHAP acyltransferase (E.C. 2.3.1.42), are mainly or exclusively located in the peroxisomes (Jones and Hajra, 1977; Jones and Hajra, 1980; Hajra and Bishop, 1982; Hardeman and Van den Bosch, 1988; Hardeman and Van den Bosch, 1989; Hardeman and Van den Bosch, 1991). The study of ether phospholipid metabolism became of particular relevance when it was shown that in a number of peroxisomal disorders, such as the cerebro-hepato-renal syndrome of Zellweger, Refsum's disease, and Rhizomelic chondrodysplasia punctata, ether lipid synthesis is impaired, due to a deficiency or insufficiency of alkyl-DHAP synthase and DHAP-acyltransferase activities (Heijmans et al., 1983; Schutgens et al., 1984).

In addition, the ether phospholipids, because of their high arachidonic acid content at the 2 position, may serve as an important source of eicosanoids (Bachelet et al., 1986; Chilton and Connel, 1988). The other product, 1-alkyl lysophosphatidylcholine, may in

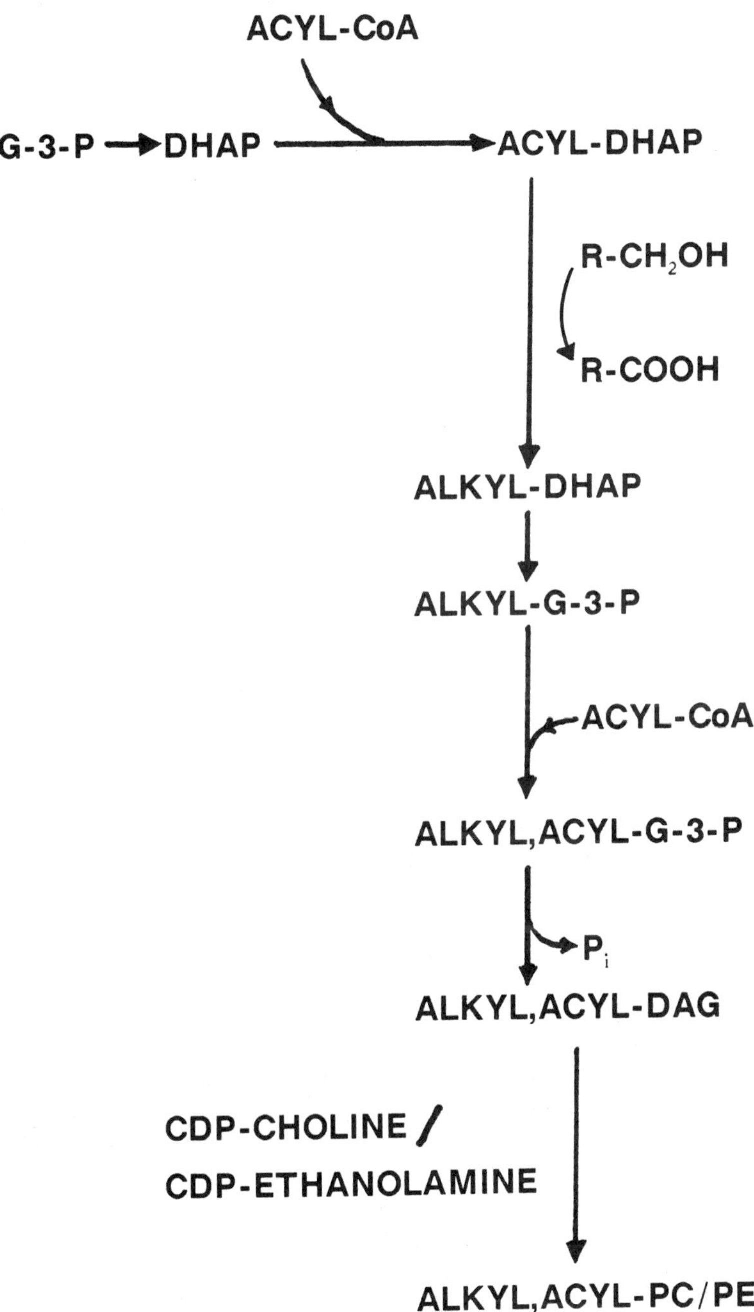

Figure 4 Biosynthetic pathway leading to formation of etherphosphoglycerides. The alkyl group is always attached at the 1 position. In plasmalogens, the alkyl is replaced by an alkenyl group. G-3-P, glycerol-3-phosphate; DHAP, dihydroxyacetone-phosphate.

turn serve as the precursor of 1-alkyl,2-acetyl phosphatidylcholine, also known as platelet activating factor (PAF) (Hanahan, 1986).

V. SPHINGOMYELIN

A. Biosynthesis

Although several biochemistry textbooks still have it that sphingomyelin is synthesized in analogy to the main route of phosphatidylcholine formation, i.e., by means of transfer of a phosphorylcholine moiety from CDP-choline to a free hydroxyl group on the receptor molecule, in this case ceramide (*N*-acylsphingosine) (Stryer, 1988; Martin et al., 1985), it has become clear since the early 1980s that the donor molecule of the phosphorylcholine is phosphatidylcholine (PC):

$$\text{PC} + \text{Cer} \rightarrow \text{SPM} + \text{DAG}$$

The enzyme catalyzing this reaction is called sphingomyelin synthase (ceramide:phosphatidylcholine phosphocholine transferase) (Kishimoto, 1983; Marggraf and Kanfer, 1984; Spence et al., 1983; Eppler et al., 1987); it is mainly localized in the Golgi apparatus, most likely in the cis and medial cisternae (Futerman et al., 1990) and to a smaller extent in the plasma membrane (Lipsky and Pagano, 1985; Van Meer et al., 1987).

The ceramide moiety, which also serves as an acceptor of glycosides in the formation of the glycosphingolipids, is formed from palmitoyl-CoA and L-Serine (Fig. 5). The virtually irreversible condensation of these two molecules, catalyzed by serine palmitoyl transferase, is possibly a point of regulation in the synthesis of sphingolipids (Mandon et al., 1991; Van Echten et al., 1990; Hanada et al., 1990). The 3-ketodihydrosphingosine thus formed is subsequently reduced to dihydrosphingosine, which is *N*-acylated by acyl-CoA to form dihydroceramide; this, upon desaturation, yields ceramide, containing a 4-trans double bond. Alternatively, the dihydrosphingosine may be desaturated to sphingosine before acylation completes the formation of the ceramide molecule (see Madison et al., 1990; Hanada et al., 1990).

Another, minor, pathway for sphingomyelin synthesis involves the transfer of phosphoethanolamine to ceramide from phosphatidylethanolamine, followed by *N*-methylation of the amino group. It was first shown to occur in microsomes and plasma membranes from rat liver and in microsomal and synaptic membranes from rat brain (Malgat et al., 1986) and seems to play a role particularly in tissues with a high phosphatidylethanolamine content (Koval and Pagano, 1991). The active center of the enzyme is in part situated on the external side of synaptic plasma membranes, while another part is embedded in the membrane interior (Maurice and Malgat, 1990).

Sphingomyelin metabolism has received a remarkable increase in interest over the past few years by virtue of its newly discovered role as a precursor of biologically active molecules. One of those molecules is the diacylglycerol arising during the synthesis of sphingomyelin from phosphatidylcholine and ceramide, an alternative way of producing this protein kinase C activator from phosphatidylcholine. However, also enzymatic hydrolysis of sphingomyelin may produce biological signal molecules. The first step in sphingomyelin degradation is the hydrolytic cleavage of the phosphodiester bond by a phospholipase C type enzyme, called sphingomyelinase, yielding phosphocholine and ceramide. There are at least two of such enzyme activities in mammalian tissues. One is a

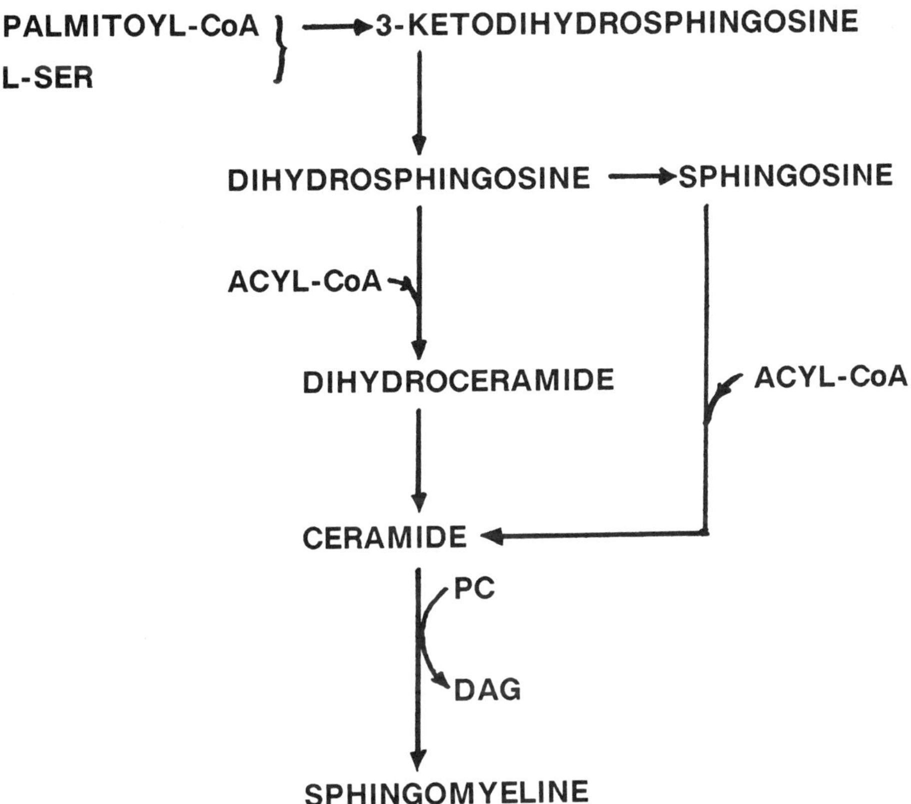

Figure 5 Biosynthesis of sphingomyelin from palmitoyl-CoA and L-serine. Two possible routes from dihydrosphingosine are indicated, differing in the order of acylation and desaturation.

lysosomal enzyme with an acid pH optimum and another, with maximal activity at neutral pH, is probably mainly localized in the plasma membrane (Hostetler and Yazaki, 1979; Bartolf and Franson, 1986; Das et al., 1984). The existence of several other neutral sphingomyelinases has been reported over the years. Koval and Pagano, in a recent review (1991), concluded that "it is difficult to assess how many different mammalian sphingomyelinases exist, given the broad range of tissues and assay systems tested." The lysosomal enzyme has attracted a great deal of interest since it was discovered, almost three decades ago, that it is this enzyme that is lacking in Niemann-Pick sphingolipidosis (Brady et al., 1966).

A relatively recent development in the study of sphingomyelin catabolism relates to the discovery of a family of low molecular weight glycoproteins, which are able to activate the lysosomal sphingomyelinase and glycolipid degrading lysosomal hydrolases. These proteins, after having received a number of different names, are now called saposins, sphingolipid activator proteins. They are all derived from a common precursor that is processed by proteolytic cleavage (O'Brien et al., 1988; Fujibayashi and Wenger, 1986). Interestingly, it may be a defect in one of these regulating peptides that is the cause of Niemann-Pick disease, type C. In vitro assays show that the cells of these patients have quite normal acid sphingomyelinase activities and yet accumulate large quantities of sphingomyelin in their lysosomes. A defect in one of the saposins may prevent the sphingomyelinase from becoming active under in vivo conditions and cause the deleterious accumulation of sphingomyelin associated with this disease (see Koval and Pagano, 1991). Abnormalities in saposin distribution have also been implicated in other sphingolipid storage diseases (Morimoto et al., 1990). In all fairness, it should be noted, however, that in Niemann-Pick disease, type C also a defect in intracellular cholesterol transport mechanisms may be involved (Liscum and Faust, 1987; Pentchev et al., 1986; Slotte et al., 1989). The close association of sphingomyelin and cholesterol, due to the hydrogen-bonding capacity of the former, has been well documented (Barenholz and Thompson, 1980; Demel et al., 1977; Yeagle, 1987). For example, sphingomyelin content of the plasma membrane has been shown to have a profound effect on the steady-state distribution of cellular cholesterol (Slotte et al., 1990), and the sphingomyelin : cholesterol ratio in plasma membranes even has been proposed to play a regulatory role in the activity of HMG-CoA reductase, which catalyzes the rate-limiting step in cholesterol synthesis (Gupta and Rudney, 1991).

Recently it was discovered that vitamin D3 is a potent stimulator of sphingomyelinase (Okazaki et al., 1989). The resulting hydrolysis to ceramide and phosphocholine is followed by resynthesis of sphingomyelin, involving phosphatidylcholine as a donor of the phosphocholine moiety. Not only does this "sphingomyelin cycle" lead to the net formation of diacylglycerol with all its implications involving protein kinase C stimulation, the jump in sphingomyelin turnover as such may also evoke a cellular response. This was shown by Kim et al. (1991), who provided evidence that the vitamin D3-induced increase in sphingomyelin turnover caused a specific *monocytic* differentiation of HL-60 human leukemia cells. Their results indicated that this process was not dependent on the formation of diacylglycerol.

Also of recent date is the recognition of sphingomyelin as the precursor of the biological effector molecule sphingosine (Hannun and Bell, 1989; Nishizuka, 1988; Merrill and Jones, 1990). Sphingosine is produced by deacylation of the ceramide formed from sphingomyelin by sphingomyelinase:

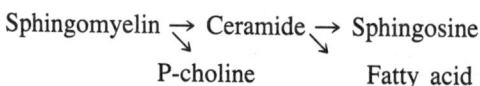

Sphingomyelin → Ceramide → Sphingosine
↘ P-choline ↘ Fatty acid

The deacylation is effectuated by a hydrolase known as ceramidase.

Sphingosine can be considered an antagonist of diacylglycerol in that it was shown to be a potent inhibitor of protein kinase C in a variety of different cells (Hannun et al., 1986; Wilson et al., 1986; Merrill et al., 1986; see Merrill, 1991 for a recent review on regulatory properties of sphingosine and other sphingolipids). With respect to the mechanism of action it has been proposed that the positive charge of the sphingosine may counteract the strong stimulatory action that acidic phospholipids such as phosphatidylserine exert on protein kinase C (Hannun et al., 1986; Merrill et al., 1989). In addition to numerous protein kinase C related cellular events such as platelet aggregation, cell differentiation, phorbol ester- and insulin-induced changes in hexose matabolism, tumorigenesis, and neutrophil stimulation, sphingosine may affect other intracellular signalling systems as well (see Merrill, 1991).

In two recent publications the formation of phosphorylated derivatives of both ceramide and sphingosine was reported. Ceramide-1-phosphate was identified in HL-60 cells treated with exogenous sphingomyelinase, and evidence was provided that it was formed by phosphorylation of the ceramide produced (Dressler and Kolesnick, 1990). Sphingosine-1-phosphate was shown to be formed in Swiss 3T3 cells from exogenously added sphingosine and proposed to be "a component of the intracellular second messenger system involved in calcium release and the regulation of cell growth induced by sphingosine" (Zhang et al., 1991).

An intriguing link between sphingomyelin and phosphoglyceride metabolism was recently proposed by Zhang et al. (1990). These authors observed that mitogenic concentrations of sphingosine in Swiss 3T3 cells induce an early increase in the concentration of cytosolic phosphatidic acid, which is a potent mitogen for these cells. Phosphatidic acid was shown before to play a crucial role in signal transduction and cellular proliferation (Exton, 1990; Moolenaar et al., 1986; Yu et al., 1988; Murayama and Ui, 1987). The results of Zhang and coworkers favor a phospholipase D catalyzed formation of phosphatidic acid over a diacylglycerolkinase-mediated mechanism. Evidence of sphingosine-stimulated phospholipase D activity by unphysiologically high sphingosine concentrations has been reported (Kanoh et al., 1990; Kiss and Anderson, 1990; Lavie and Liscovitch, 1990; references 40–42 in Zhang et al., 1990). Lavie et al. (1990) reported that sphingosine may also act to inhibit phosphatidic acid phosphohydrolase, thus maintaining high phosphatidic acid levels in the cell. Mullmann et al. (1991) provided evidence that sphingosine inhibits phosphatidic acid phosphatase in human neutrophils through a protein kinase C independent mechanism.

ACKNOWLEDGMENTS

I am extremely grateful to my good friend, Gregor Cevc, for his unsurpassed patience, and to my secretary, Rinske Kuperus, for her help in finalizing the manuscript.

REFERENCES

A. J. Aarsman, J. G. N. De Jong, E. Arnoldussen, F. W. Neys, P. D. Van Wassenaar, and H. Van den Bosch, *J. Biol. Chem. 264*:10008–10014 (1989).

D. E. Agwu, L. C. McPhail, M. C. Chabot, L. W. Daniel, R. L. Wykle, and C. E. McCall, *J. Biol. Chem. 264*:1405–1413 (1989).

J. C. Anthes, J. Krasovsky, R. W. Egan, M. I. Siegel, and M. M. Billah, *Arch. Biochem. Biophys. 287*:53–59 (1991).

G. Arthur and L. Page, *Biochem. J. 273*:121–125 (1991).

G. Augert, P. F. Blackmore, and J. H. Exton, *J. Biol. Chem. 264*:2574–2580 (1989a).

G. Augert, B. Bocckino, P. F. Blackmore, and J. H. Exton, *J. Biol. Chem. 264*:21689–21698 (1989b).

M. Bachelet, J. Masliah, B. B. Vargaftig, G. Bereziat, and O. Colard, *Biochim. Biophys. Acta 878*:177–183 (1986).

L. R. Ballou and W. Y. Cheung, *Proc. Natl. Acad. Sci. USA 80*:5203–5207 (1983).

J. Baranska, *Adv. Lipid Res. 19*:163–184 (1982).

J. Baranska, *FEBS Lett. 228*:175–178 (1988).

Y. Barenholz and E. Thompson, *Biochim. Biophys. Acta 604*:129–158 (1980).

M. Bartolf and R. C. Franson, *J. Lipid Res. 27*:57–63 (1986).

J. J. Batenburg, J. N. Den Breejen, M. J. H. Geelen, C. Bijleveld, and L. M. G. Van Golde, in *Basic Research on Lung Surfactant* (P. von Wichert and B. Müller, eds.), Karger Basel, 1990, pp. 96–103.

R. M. Bell and R. A. Coleman, *Ann. Rev. Biochem. 49*:459–487 (1980).

M. J. Berridge, *Biochem. J. 220*:345–360 (1984).

J. M. Besterman, V. Duronio, and P. Cuatrecasas, *Proc. Natl. Acad. Sci. USA 83*:6785–6789 (1986).

K. Beyer and M. Klingenberg, *Biochemistry 24*:3821–3826 (1985).

E. Bigon, E. Boarato, A. Bruni, A. Leon, and G. Toffano, *Br. Jr. Pharmac. 67*:611–616 (1979).

M. M. Billah and J. C. Anthes, *Biochem. J. 269*:281–291 (1990).

K. S. Bjerve, *Biochim. Biophys. Acta 296*:549–562 (1993).

S. B. Bocckino, P. F. Blackmore, and J. H. Exton, *J. Biol. Chem. 260*:14201–14207 (1985).

S. B. Bocckino, P. F. Blackmore, P. B. Wilson, and J. H. Exton, *J. Biol. Chem. 262*:15309–15315 (1987).

L. F. Borkenhagen, E. P. Kennedy, and L. Fielding, *J. Biol. Chem. 236*:PC28–PC30 (1961).

K. Bradbury, *J. Neurochem. 43*:382–387 (1984).

R. O. Brady, J. N. Kanfer, M. B. Mock, and D. S. Frederickson, *Proc. Natl. Acad. Sci. USA 55*:366–369 (1966).

J. Bremer and D. M. Greenberg, *Biochim. Biophys. Acta 46*:205–216 (1961).

D. H. Brindley, *Progr. Lipid Res. 23*:115–133 (1984).

R. M. Burch, A. Luini, and J. Axelrod, *Proc. Natl. Acad. Sci. USA 83*:7201–7205 (1986).

M. Butler and P. Morell, *J. Neurochem. 43*:382–387 (1983).

M. C. Cabot, C. J. Welsh, Z.-C. Zhang, H.-T. Cao, H. Chabbott, and M. Lebowitz, *Biochim. Biophys. Acta 959*:46–57 (1988a).

M. C. Cabot, C. J. Welsh, H.-T. Hao, and H. Chabbott, *FEBS Lett. 238*:153–157 (1988b).

R.-M. Catalioto, G. Ailhaud, and R. Negrel, *Biochem. Biophys. Res. Commun. 173*:840–848 (1990.

J. Chang, J. H. Musser, and H. McGregor, *Biochem. Pharmacol. 36*:2429–2435 (1987).

F. H. Chilton and T. R. Connel, *J. Biol. Chem. 263*:5260–5265 (1988).

F. H. Chilton, J. M. Ellis, S. C. Olson, and R. L. Wykle, *J. Biol. Chem. 259*:12014–12019 (1984).

M. A. Clark, T. M. Conway, R. G. L. Shorr, and S. T. Crooke, *J. Biol. Chem. 262*:4402–4406 (1987).

L. Corazzi, J. Zborowski, R. Roberti, L. Binaglia, and G. Arienti, *Bull. Mol. Biol. Med. 12*:19–31 (1987).

L. Corazzi, R. Pistolesi, and G. Arienti, *J. Neurochem. 56*:207–212 (1991).

R. Cornell, in *Phospholipid Metabolism* (E. D. Vance, ed.), CRC Press, Boca Raton, Fla., 1989, pp. 47–67.

R. B. Cornell and D. E. Vance, *Biochim. Biophys. Acta 919*:37–48 (1987).

L. W. Daniel, M. Waite, and R. L. Wykle, *J. Biol. Chem. 261*:9128–9132 (1986).

D. V. M. Das, H. W. Cook, and M. W. Spence, *Biochim. Biophys. Acta 777*:339–342 (1984).

G. H. De Haas, N. M. Postema, W. Nieuwenhuizen, and L. L. M. Van Deenen, *Biochim. Biophys. Acta 159*:118–129 (1968).

J. G. N. De Jong, H. Amesz, A. J. Aarsman, H. B. M. Lenting, and H. Van den Bosch, *Eur. J. Biochem. 164*:129–135 (1987).

R. A. Demel, W. S. M. Geurts van Kessel, R. F. A. Zwaal, B. Roelofsen, and L. L. M. Van Deenen, *Biochim. Biophys. Acta 406*:97–107 (1975).

R. A. Demel, J. W. C. M. Jansen, P. W. M. Van Dijck, and L. L. M. Van Deenen, *Biochim. Biophys. Acta 465*:1–10 (1977).

E. A. Dennis, in *The Enzymes* (P. D. Boyer, ed.), Academic Press, New York, 1983, Vol. 16, pp. 307–353.

E. A. Dennis and E. P. Kennedy, *J. Lipid Res. 13*:263–267 (1972).

J. M. De Winter, G. M. Vianen, and H. Van den Bosch, *Biochim. Biophys. Acta 712*:332–341 (1982).

J. M. De Winter, J. Korpancova, and H. Van den Bosch, *Arch. Biochem. Biophys. 234*:243–252 (1984).

J. M. De Winter, H. B. M. Lenting, F. W. Neys, and H. Van den Bosch, *Biochim. Biophys. Acta 917*:169–177 (1987).

J. Dijkstra, W. J. M. Van Galen, and G. L. Scherphof, *Biochim. Biphys. Acta 845*:34–42 (1985).
K. A. Dressler, and R. N. Kolesnick, *J. Biol. Chem. 265*:14917–14921 (1990).
J. Duan and M. P. Moffat, *Naunyn-Schmiedeberg's Arch. Pharmacol. 342*:342–348 (1990).
C. M. Eppler, B. Malewicz, H. M. Jenkin, and W. J. Baumann, *Lipids 22*:351–357 (1987).
J. H. Exton, *FASEB J. 2*:2670–2676 (1988).
J. H. Exton, *J. Biol. Chem. 265*:1–4 (1990).
D. A. Feldman and P. A. Weinhold, *J. Biol. Chem. 262*:9075–9081 (1987).
D. A. Feldman, M. E. Rounsifer, and P. A. Weinhold, *Biochim. Biophys. Acta 833*:429–437 (1985).
G. J. Fisher, P. A. H enderson, J. J. Voorhees, and J. J. Baldassare, *J. Cell. Physiol. 146*:309–317 (1991).
R. J. Flower, J. N. Wood, and I. Parente, *Adv. Inflammation Res. 7*:61–69 (1984).
R. Franson, M. Waite, and M. La Via, *Biochemistry 10*:1942–1946 (1971).
M. Fry and D. E. Green, *Biochem. Biophys. Res. Commun. 93*:1238–1246 (1980).
S. Fujibayashi and D. A. Wenger, *Biochim. Biophys. Acta 875*:554–562 (1980).
I. Fuse and H.-H. Tai, *Biochem. Biphys. Res. Commun. 146*:659–667 (1987).
A. Futerman, B. Steiger, A. L. Hubbard, and R. E. Pagano, *J. Biol. Chem. 265*:8650–8657 (1990).
T. P. George, S. C. Morash, H. W. Cook, D. M. Byers, F. B. St. C. Palmer, and M. W. Spence, *Biochim. Biophys. Acta 1004*:283–291 (1989).
T. P. George, H. W. Cook, D. M. Byers, F. B. St. C. Palmer, and M. W. Spence, *J. Biol. Chem. 266*:12419–12423 (1991).
M. J. Gilbreath, C. A. Nacy, D. L. Hoover, C. R. Alving, G. M. Swartz, and M. S. Meltzer, *J. Immunol. 134*:3420–3425 (1985).
M. J. Gilbreath, D. L. Hoover, C. R. Alving, G. M. Swartz, and M. S. Meltzer, *J. Immunol. 137*:1681–1687. (1986).
M. J. Gilbreath, W. E. Fogler, G. M. Swartz, C. R. Alving, and M. S. Meltzer, *Int. J. Immunopharmacol. 11*:103–110 (1989).
J. A. Glatz, J. G. Muir, and A. W. Murray, *Carcinogenesis 8*:194–1945 (1987).
L. R. Grillone, M. A. Clark, R. W. Godfrey, F. Stassen, and S. T. Crooke, *J. Biol. Chem. 263*:2658–2663 (1988).
J. E. M. Groener, W. Klein, and L. M. G. Van Golde, *Arch. Biochem. Biophys. 198*:287–295 (1979).
R. I. Grove and S. D. Schimmel, *Biochim. Biophys. Acta 711*:272–280 (1982).
A. K. Gupta and H. Rudney, *J. Lipid Res. 32*:125–136 (1991).
G. R. Guy and A. W. Murray, *Cancer Res. 42*:1980–1985 (1982).
H. P. Haagsman, J. M. Van den Heuvel, L. M. G. Van Golde, and M. J. H. Geelen, *J. Biol. Chem. 259*:11273–11278 (1984).
K. A. Haines, and G. Weissmann, in *Advances in Prostaglandin, Thromboxane, and Leukotriene Research* (B. Samuelsson et al., eds.), Raven Press, New York, 1990, pp. 545–551.
A. K. Hajra and J. E. Bishop, *Ann. NY Acad. Sci. 386*:170–182 (1982).
K. Hanada, M. Nishijima, and Y. Akamatsu, *J. Biol. Chem. 265*:22137–22142 (1990).
D. J. Hanahan, *Ann. Rev. Biochem. 55*:483–509 (1986).
Y. A. Hannun and R. M. Bell, *Science 243*:500–507 (1989).
Y. A. Hannun, C. R. Loomis, A. Merrill, and R. M. Bell, *J. Biol. Chem. 261*:12604–12609 (1986).
D. Hardeman and H. Van den Bosch, *Biochim. Biophys. Acta 963*:1–9 (1988).
D. Hardeman and H. Van den Bosch, *Biochim. Biophys. Acta 1006*:1–8 (1989).
D. Hardeman and H. Van den Bosch, *Biochim. Biophys. Acta 1081*:285–292 (1991).
G. M. Hatch and P. C. Choy, *Biochem. J. 268*:47–54 (1990).
G. M. Hatch, Y. Tsukitani, and D. E. Vance, *Biochim. Biophys. Acta 1081*:25–32 (1991).
H. S. A. Heijmans, R. B. H. Schutgens, R. Tan, H. Van den Bosch, and P. Borst, *Nature 306*:69–70 (1983).

R. L. Heinrikson, E. T. Krueger, and P. S. Keim, *J. Biol. Chem. 252*:4913–4921 (1977).
J. A. Higgins and J. K. Fieldsend, *J. Lipid Res. 28*:268–278 (1984).
S. A. Hill, W. C. McMurray, and B. D. Sanwal, *Can. J. Biochem. Cell Biol. 62*:369–374 (1984).
F. Hirata, *J. Biol. Chem. 256*:7730–7733 (1981).
C. B. Hirschberg and E. P. Kennedy, *Proc. Natl. Acad. Sci. USA 69*:648–651 (1972).
J. M. Hoffman, M. L. Standaert, G. P. Nair, and R. V. Farese, *Biochemistry 30*:3315–3322 (1991).
L. E. Hokin, *Ann. Res. Biochem. 54*:205–225 (1985).
K. Y. Hostetler, *Biochim. Biophys. Acta 1086*:139–140 (1991).
K. Y. Hostetler and H. Van den Bosch, *Biochim. Biophys. Acta 260*:380–386 (1972).
K. Y. Hostetler and P. J. Yazaki, *J. Lipid Res. 20*:456–463 (1979).
K. Y. Hostetler, H. Van den Bosch, and L. L. M. Van Deenen, *Biochim. Biophys. Acta 239*:113–119 (1971).
K. Y. Hostetler, H. Van den Bosch, and L. L. M. Van Deenen, *Biochim. Biophys. Acta 260*:507–513 (1972).
K. Y. Hostetler, J. M. Galesloot, P. Boer, and H. Van den Bosch, *Biochim. Biophys. Acta 380*:382–389 (1975).
M. Houweling, L. B. M. Tijburg, H. Jamil, D. E. Vance, C. B., Nyathi, W. J. Vaartjes, and L. M. G. Van Golde, *Biochem. J. 278*:347–351 (1991).
C. Huang, and M. C. Cabot, *J. Biol. Chem. 265*:14858–14863 (1990a).
C. Huang and M. C. Cabot, *J. Biol. Chem. 265*:17468–17473 (1990b).
K. M. Hurst, B. P. Hughes, and G. J. Barritt, *Biochem. J. 272*:749–753 (1990).
F. I. Ikegwuonu, J. K. Pai, and G. C. Mueller, *Carginogenesis 11*:1927–1935 (1990).
J. P. Infante, *FEBS Lett. 170*:1–14 (1984).
H. R. Irving and J. H. Exton, *J. Biol. Chem. 262*:3440–3443 (1987).
K. Ishidate, K. Nakagomi, and Y. Nakazawa, *J. Biol. Chem. 259*:14706–14710 (1984).
K. Ishidate, K. Iida, K. Tadokoro, and Y. Nakazawa, *Biochim. Biophys. Acta 833*:1–8 (1985a).
K. Ishidate, K. Furusawa, and Y. Nakazawa, *Biochim. Biophys. Acta 836*:119–124 (1985b).
H. Jamil, Z. Yao, and D. E. Vance, *J. Biol. Chem. 265*:4332–4339 (1990).
C. L. Jelsema and D. J. Morré, *J. Biol. Chem. 253*:7960–7971 (1978).
C. L. Jones and A. K. Hajra, *Biochem. Biophys. Res. Commun. 76*:1138–1143 (1977).
C. L. Jones and A. K. Hajra, *J. Biol. Chem. 255*:8289–8295 (1980).
G. A. Jones and C. Kent, *Arch. Biochem. Biophys. 288*:331–336 (1991).
G. B. Kalmar, R. J. Kay, A. Lachance, R. Aebersold, and R. B. Cornell, *Proc. Natl. Acad. Sci. USA 87*:6029–6033 (1990).
J. N. Kanfer, *J. Lipid Res. 13*:468–476 (1972).
H. Kanoh, K. Yamada, and F. Sakane, *Trends Biochem. Sci. 15*:47–50 (1990).
E. P. Kennedy, in *Lipids and Membranes, Past, Present and Future* (J. A. F. Op den Kamp, B. Roelofsen, and K. W. A. Wirtz, eds.), Elsevier, New York, 1986, pp. 171–206.
E. P. Kennedy and S. B. Weiss, *J. Biol. Chem. 222*:193–215 (1956).
D. A. Kennerly, *J. Immunol. 144*:3912–3919 (1990).
C. Kent, *Prog. Lipid Res. 29*:87–105 (1990).
C. Kent, G. M. Carman, M. W. Spence, and W. Dowhan, *FASEB J. 5*:2258–2266 (1991).
M.-Y. Kim, C. Linardic, L. Obeid, and Y. Hannun, *J. Biol. Chem. 266*:484–489 (1991).
Y. Kishimoto, in *The Enzymes*, Vol. XVI (P. E. Boyer, ed.), Academic Press, New York, 1983, pp. 358–407.
Z. Kiss and W. B. Anderson, *J. Biol. Chem. 265*:7345–7350 (1990).
J. Y. Kiyasu, R. A. Pieringer, H. Paulus, and E. P. Kennedy, *J. Biol. Chem. 238*:2293–2298 (1963).
R. N. Kolesnick, *J. Biol. Chem. 262*:14525–14530 (1987).
L. Kolster, C. Jensen, A. Bruno, L. Mietto, G. Toffano, and S. Norn, *Biochim. Biophys. Acta 929*:196–202 (1987).
M. Koval and R. E. Pagano, *Biochim. Biophys. Acta 1082*:113–125 (1991).

T. Kurayama and M. Ui, *J. Biol. Chem. 260*:7226–7233 (1985).
W. E. Lands, *J. Biol. Chem. 231*:883–888 (1958).
Y. Lavie and M. Liscovitch, *J. Biol. Chem. 265*:3868–3872 (1990).
Y. Lavie, O. Piterman, and M. Liscovitch, *FEBS Lett. 277*:7–10 (1990).
M.-H. Lee and R. M. Bell, *Biochemistry 30*:1041–1049 (1991).
H. B. M. Lenting, F. W. Neys, and H. Van den Bosch, *Biochim. Biophys. Acta 917*:178–185 (1987).
N. G. Lipsky and R. E. Pagano, *J. Cell Biol. 100*:27–34 (1985).
M. Liscovitch, J. K. Blustain, A. Freese, and R. J. Wurtman, *Biochem. J. 241*:81–86 (1987).
L. Liscum and J. R. Faust, *J. Biol. Chem. 262*:17002–17008 (1987).
K. Loeffelholz, *Biochem. Pharmacol. 38*:1543–1549 (1989).
F. McCormick, *Cell 56*:5–8 (1989).
E. E. MacNulty, R. Plevin, and M. J. O. Wakelam, *Biochem. J. 272*:761–766 (1990).
K. C. Madison, D. C. Swartzendruber, P. W. Wertz, and D. T. Downing, *J. Invest. Dermatol. 95*:657–664 (1990).
M. Malgat, A. Maurice, and J. Baraud, *J. Lipid Res. 27*:251–260 (1986).
E. C. Mandon, G. Van Echten, R. Birk, R. R. Schmidt, and K. Sandhoff, *Eur. J. Biochem. 198*:667–674 (1991).
W. D. Marggraf and J. N. Kanfer, *Biochim. Biophys. Acta 793*:346–353 (1984).
T. W. Martin and K. C. Michaelis, *Biochem. Biophys. Res. Commun. 157*:1271–1279 (1988).
D. W. Martin, P. A. Mayes, V. W. Rodwell, and D. K. Granner, *Harpers Review of Biochemistry*, 20th ed., Lange, Medical, Los Altos, Calif., (1985).
E. A. Martinson, D. Goldstein, and J. H. Brown, *J. Biol. Chem. 264*:14748–14754 (1989).
A. Maurice and M. Malgat, *Neurochem. Int. 17*:83–91 (1990).
P. Mende, V. J. Kolbe, B. Kadenbach, I. Stipani, and F. Palmieri, *Eur. J. Biochem. 128*:91–95 (1982).
A. H. Merrill, Jr., *J. Bioenerg. Biomembr. 23*:83–104 (1991).
A. H. Merrill, Jr., and D. D. Jones, *Biochim. Biophys. Acta 1044*:1–12 (1990).
A. H. Merrill, A. M. Sereni, V. Stevens, Y. A. Hannun, R. M. Bell, and J. M. Kinkade, *J. Biol. Chem. 261*:12610–12615 (1986).
A. H. Merrill, Jr., E. Wang, R. E. Mullins, W. C. L. Jamison, S. Nimkar, and D. C. Liotta, *Anal. Biochem. 171*:373–381 (1988).
A. H. Merrill, S. Nimkar, D. Menaldino, Y. A. Hannun, C. Loomis, R. M. Bell, S. R. Tyagi, J. D. Lambeth, V. L. Stevens, R. Hunter, and D. C. Liotta, *Biochemistry 28*:3138–3145 (1989).
S. A. Metz and M. Dunlop, *Arch. Biochem. Biophys. 283*:417–428 (1990).
S. A. Metz and M. Dunlop, *Biochem. Pharmacol. 41*:R1–R4 (1991).
L. Mietto, E. Boarato, G. Toffano, E. Bigon and A. Bruni, *Agents and Action 14*:606–612 (1984).
L. Mietto, E. Boarato, G. Toffano, and A. Bruni, *Biochim. Biophys. Acta 930*:145–153 (1987).
M. A. Miller and C. Kent, *J. Biol. Chem. 261*:9753–9761 (1986).
W. H. Moolenaar, W. Kruijer, B. C. Tilly, I. Verlaan, A. J. Bierman, and S. W. De Laat, *Nature 323*:171–173 (1986).
S. Morimoto, Y. Yamamoto, J. S. O'Brien, and Y. Kishimoto, *Proc. Natl. Acad. Sci. USA 87*:3493–3497 (1990).
J. G. Muir and A. W. Murray, *J. Cell Physiol. 130*:382–391 (1987).
M. Muller, R. Moser, D. Cheneval, and E. Carafoli, *J. Biol. Chem. 260*:3839–3843 (1985).
T. J. Mullmann, M. I. Siegel, R. W. Egan, and M. M. Billah, *J. Biol. Chem. 266*:2013–2016 (1991).
T. Murayama and M. Ui, *J. Biol. Chem. 262*:12463–12467 (1987).
Y. Nakagawa and K. Waku, *Prog. Lipid Res. 28*:205–243 (1989).
S. Nakashima, H. Hattori, L. Shirato, A. Takenaka, and Y. Nozawa, *Biochem. Biophys. Res. Commun. 148*:971–979 (1987).
H. Nazih, D. Devred, F. Martin-Nizard, J. C. Fruchart, and C. Delbart, *Thromb. Res. 59*:913–920 (1990).

J. D. Newkirk and M. Waite, *Biochim. Biophys. Acta 298*:562–576 (1973).
Y. Nishizuka, *Science 233*:305–312 (1986).
Y. Nishizuka, *Nature 334*:661–665 (1988).
U. S. O'Brien, K. A. Kretz, N. Dewji, D. A. Wenger, F. Esch, and A. L. Fluharty, *Science 241*:1098–1101 (1988).
T. Okazaki, R. M. Bell, and Y. A. Hannun, *J. Biol. Chem. 264*:19076–19080 (1989).
H. B. Paddon, C. Vigo, and D. E. Vance, *Biochim. Biophys. Acta 710*:112–115 (1982).
J.-K. Pai, M. Siegel, R. W. Egan, and M. M. Billah, *J. Biol. Chem. 263*:12472–12477 (1988).
G. Paradies and F. M. Ruggiero, *Arch. Biochem. Biophys. 284*:332–337 (1990).
S. L. Pelech and D. E. Vance, *J. Biol. Chem. 257*:14198–14202 (1982).
S. L. Pelech and D. E. Vance, *Biochim. Biophys. Acta 779*:217–251 (1984).
S. L. Pelech, H. W. Cook, H. B. Paddon, and D. E. Vance, *Biochim. Biophys. Acta 795*:433–440 (1984a).
S. L. Pelech, H. B. Paddon, and D. E. Vance, *Biochim. Biophys. Acta 795*:447–451 (1984b).
P. G. Pentchev, H. S. Kruth, M. E. Comly, J. D. Butler, M. T. Vanier, D. A. Wenger, and S. Patel, *J. Biol. Chem. 261*:16775–16780 (1986).
M. Peters-Golden, R. W. McNish, R. Hyzy, C. Shelly, and G. B. Toews, *J. Immunol. 144*:263–270 (1990).
L. B. Pickford, A. J. Polverino, and G. J. Barritt, *Biochem. J. 245*:211–216 (1987).
A. J. Polverino and G. J. Barritt, *Biochim. Biophys Acta 970*:75–82 (1988).
G. Porcellati, G. Arienti, M. G. Pirotta, and D. Giorgini, *J. Neurochem. 18*:1395–1417 (1971).
F. Possmayer, G. Balakrishnan, and K. P. Strickland, *Biochim. Biophys. Acta 164*:79–87 (1968).
S. W. Rabkin, *J. Mol. Cell. Cardiol. 22*:965–974 (1990).
K. Radika and F. Possmayer, *Biochem. J. 232*:833–840 (1985).
J. M.-F. Ragab-Thomas, F. Hullin, H. Chap, and L. Douste-Blazy, *Biochim. Biophys. Acta 917*:388–397 (1987).
Rahman et al., *Biochem. Biophys. Res. Commun. 38*:670–677 (1970).
C. S. Ramesha and W. C. Pickett, *J. Biol. Chem. 261*:7592–7595 (1986).
S. L. Reinhold, G. A. Zimmerman, S. M. Prescott, and T. M. McIntyre, *J. Biol. Chem. 264*:21652–21659 (1989).
N. D. Ridgway and D. E. Vance, *J. Biol. Chem. 262*:17231–17239 (1987).
M. F. Roberts and E. A. Dennis, in *Phosphatidylcholine Metabolism* (D. E. Vance, ed.), CRC Press, Boca Raton, Fla., 1989, pp. 121–142.
P. M. Rosoff, N. Savage, and C. A. Dinarello, *Cell 54*:73–81 (1988).
F. M. Ruggiero, C. Landriscina, G. V. Guoni, and E. Quagliariello, *Lipids 19*:171–178 (1984).
J. S. Sanghera and D. E. Vance, *J. Biol. Chem. 264*:1215–1223 (1989).
C. G. Schalkwijk, F. Märki, and H. Van den Bosch, *Biochim. Biophys. Acta 1044*:139–146 (1990.
G. L. Scherphof and L. L. M. Van Deenen, *Biochim. Biophys. Acta 98*:204–206 (1985).
G. L. Scherphof, M. Waite, and L. L. M. Van Deenen, *Biochim. Biophys. Acta 125*:406–409 (1966).
M. Schlame and W. Augustin, *Biomed. Biochim. Acta 44*:1083–1088 (1985).
G. Schmitz, M. Beuck, G. Fischer, G. Nowicka, and H. Robenek, *J. Lipid Res. 31*:1741–1752 (1990a).
G. Schmitz, H. Fischer, M. Beuck, K.=P. Hoecker, and H. Robenek, *Arteriosclerosis 10*:1010–1019 (1990b).
W. J. Schneider and D. E. Vance, *Eur. J. Biochem. 85*:181–187 (1978).
M. P. Schrey, A. M. Read, and P. J. Steer, *Biochem. J. 246*:705–713 (1987).
R. B. H. Schutgens, G. J. Romeijn, R. J. A. Wanders, H. Van den Bosch, G. Schrakamp, and H. S. A. Heijmans, *Biochem. Biophys. Res. Commun. 120*:179–184 (1984).
R. Sleight and C. Kent, *J. Biol. Chem. 255*:10644–10650 (1980).
R. Sleight and C. Kent, *J. Biol. Chem. 258*:836–839 (1983).
P. J. Slotte, C. Hedstrom, and E. L. Bierman, *Biochim. Biophys. Acta 1005*:303–309 (1989).

J. P. Slotte, A.-S. Harmala, C. Jansson, and M. I. Porn, *Biochim. Biophys. Acta 1030*:251–257 (1990).

P. Sohal and R. Cornell, *J. Biol. Chem. 265*:11746–11750 (1990).

M. W. Spence, J. T. R. Clarke, and H. W. Cook, *J. Biol. Chem. 258*:8595–8600 (1983).

N. Z. Stanacev, Y. Y. Chang, and E. P. Kennedy, *J. Biol. Chem. 242*:3018–3019 (1967).

N. Z. Stanacev, D. C. Isaac, and K. B. Brookes, *Biochim. Biophys. Acta 152*:806–808 (1968).

L. Stryer, *Biochemistry*, 3rd ed., Freeman, New York, 1988.

J. Sugatani, M. Miwa, and D. J. Hanahan, *J. Biol. Chem.* 262:5740–5747 (1987).

R. Sundler, *J. Biol. Chem. 250*:8585–8590 (1975).

R. Sundler and B. Akesson, *J. Biol. Chem. 250*:3359–3367 (1975).

T. T. Suzuki and J. N. Kanfer, *J. Biol. Chem. 260*:1394–1399 (1985).

K. Tadokoro, K. Ishidate, and Y. Nakazawa, *Biochim. Biophys. Acta 835*:501–513 (1985).

N. Takuwa, Y. Takuwa, and H. Rasmussen, *Biochem. J. 243*:647–653 (1987).

K. T. Tamai and M. L. Greenberg, *Biochim. Biophys. Acta 1046*:214–222 (1990).

T. Teramoto, H. Tojo, T. Yamano, and M. Okamoto, *J. Biochem. 93*:1353–1360 (1983).

L. B. M. Tijburg and D. E. Vance, *Biochim. Biophys. Acta 1085*:178–183 (1991c).

L. B. M. Tijburg, E. A. J. M. Schuurmans, M. J. H. Geelen, and L. M. G. Van Golde, *Biochim. Biophys. Acta 919*:49–57 (1987).

L. B. M. Tijburg, M. J. H. Geelen, and L. M. G. Van Golde, *Biochim. Biophys. Acta 1004*:1–19 (1989a).

L. B. M. Tijburg, M. J. H. Geelen, and L. M. G. Van Golde, *Biochem. Biophys. Res. Commun. 160*:1275–1280 (1989b).

L. B. M. Tijburg, M. Houweling, M. J. H. Geelen, and L. M. G. Van Golde, *Biochem. J. 257*:645–650 (1989c).

L. B. M. Tijburg, T. Nishimaki-Mogami, and D. E. Vance, *Biochim. Biophys. Acta 1085*:167–177 (1991a).

L. B. M. Tijburg, R. W. Samborski, and D. E. Vance, *Biochim. Biophys. Acta 1085*:184–190 (1991b).

T. A. Tompkins and M. A. Moscarello, *J. Biol. Chem. 266*:4228–4236 (1991).

A. P. Truett, M. W. Verghese, S. B. Dillon, and R. Snyderman, *Proc. Natl. Acad. Sci. USA 85*:1549–1553 (1988).

Y. Tsukagoshi, J. Nikawa, and S. Yamashita, *Eur. J. Biochem. 169*:477–485 (1987).

W. J. Van Blitterswijk, H. Hilkmann, J. De Widt, and R. L. Van der Bend, *J. Biol. Chem. 266*:10337–10343 (1991a).

W. J. Van Blitterswijk, H. Hilkmann, J. De Widt, and R. L. Van dern Bend, *J. Biol. Chem. 266*:10344–10350 (1991b).

D. E. Vance, in *Phosphatidylcholine Metabolism* (D. E. Vance, ed.), CRC Press, Boca Raton, Fla., 1989, pp. 225–239.

D. E. Vance and S. L. Pelech, *Trends Biochem. Sci. 9*:17–20 (1984).

D. E. Vance and N. D. Ridgeway, *Prog. Lipid Res. 27*:61–79 (1988).

J. E. Vance and D. E. Vance, *J. Biol. Chem. 263*:5898–5909 (1988).

H. Van den Bosch, in *Cell Activation and Signal Initiation: Receptor and Phospholipase Control of Inositol Phosphate, PAF, and Eicosanoid Production*, Alan R. Liss, New York, 1989, pp. 317–321.

H. Van den Bosch, in *Comprehensive Medicinal Chemistry*, Vol. 2 (P. G. Sammes, ed.), Pergamon Press, Oxford, 1990, pp. 515–530.

H. Van den Bosch, A. J. Aarsman, and A. J. Verkleij, in *Leukotrienes and Prostanoids in Health and Disease, New Trends Lipid Mediatores Res.* (U. Zor, Z. Naor, and A. Danon, eds.), Karger, Basel, 1989, pp. 257–261.

H. Van den Bosch, A. J. Aarsman, R. H. N. Van Schaik, C. B. Schalkwijk, F. W. Neijs, and A. Sturk, *Biochem. Soc. Trans. 18*:781–786 (1990).

G. Van Echten, R. Birk, G. Brenner-Weiss, R. R. Schmidt, and K. Sandhoff, *J. Biol. Chem. 265*:9333–9339 (1990).
L. M. G. Van Golde, B. Fleischer, and S. Fleischer, *Biochim. Biophys. Acta 249*:318–330 (1971).
L. M. G. Van Golde, J. Raben, J. J. Batenberg, B. Flesicher, F. Zambrano, and S. Fleischer, *Biochim. Biophys. Acta 360*:179–192 (1974).
G. Van Meer, E. H. K. Stetzer, W. Wijnaendts-Van Resandt, and K. Simons, *J. Cell Biol. 105*:1623–1635 (1987).
H. M. Verheij, A. J. Slotboom, and G. H. De Haas, *Rev. Physiol. Biochem. Pharmacol. 91*:91–203 (1981).
E. J. Victoria, L. M. G. Van Golde, K. Y. Hostetler, G. L. Scherphof, and L. L. M. Van Deenen, *Biochim. Biophys. Acta 239*:443–457 (1971).
C. Vigo and D. E. Vance, *Biochem. J. 200*:321–326 (1981).
C. Vigo, H. B. Paddon, F. C. Millard, P. H. Pritchard, and D. E. Vance, *Biochim. Biophys. Acta 665*:546–550 (1981).
S. B. Vik, G. Georgevich, and R. A. Capaldi, *Proc. Natl. Acad. Sci. USA 78*:1456–1460 (1981).
D. R. Voelker, *Proc. Natl. Acad. Sci. USA 81*:2669–2673 (1984).
D. R. Voelker, *J. Biol. Chem. 260*:14671–14676 (1985).
D. R. Voelker, *J. Biol. Chem. 264*:8019–8025 (1989).
M. Waite and P. Sissons, *Biochemistry 10*:2377–2383 (1971).
M. Waite, R. H. Rao, H. Griffin, R. Franson, C. Miller, P. Sisson, and J. Frye, *Methods Enzymol. 71*:674–689 (1981).
M. Waite, L. King, T. Thornburg, G. Osthoff, and T. Y. Thuren, *J. Biol. Chem. 265*:21720–21726 (1990).
P. A. Weinhold, M. E. Rounsifer, and D. A. Feldman, *J. Biol. Chem. 261*:5104–5110 (1986).
P. Weinhold, M. E. Rounsifer, L. Charles, and D. A. Feldman, *Biochim. Biophys. Acta 1006*:299–310 (1989).
P. A. Weinhold, L. Charles, M. E. Rounsifer, and D. A. Feldman, *J. Biol. Chem. 266*:6093–6100 (1991).
S. B. Weiss, S. W. Smith, and E. P. Kennedy, *J. Biol. Chem. 224*:53–64 (1958).
C. J. Welsh and K. Schmeichel, *Anal. Biochem. 192*:281–292 (1991).
C. J. Welsh, K. Schmeichel, H. Cao, and H. Chabbott, *Lipids 25*:675–684 (1990).
P. W. Wertz and G. C. Mueller, *Cancer Res. 38*:2900–2904 (1978).
R. E. Whatley, G. A. Zimmerman, T. M. McIntyre, and S. M. Prescott, *Prog. Lipid Res. 29*:45–63 (1990).
M. T. Whitnall and T. J. Brown, *Biochem. Biophys. Res. Commun. 106*:1049–1056 (1982).
J. Wijkander and R. Sundler, *FEBS Lett. 244*:51–56 (1989).
E. Wilson, M. C. Olcott, R. M. Bell, Jr., A. H. Merrill, Jr., and J. D. Lambeth, *J. Biol. Chem. 261*:12616–12623 (1986).
J. Wittenberg and A. Kornberg, *J. Biol. Chem. 202*:431–444 (1953).
T. M. Wright, L. A. Rangan, H. S. Shin, and D. M. Raben, *J. Biol. Chem. 263*:9374–9380 (1988).
T. M. Wright, H. S. Shin, and D. M. Raben, *Biochem. J. 267*:501–507 (1990).
Z. Xu, D. M. Byers, F. B. St. C. Palmer, W. Spence, and H. W. Cook, *J. Biol. Chem. 266*:2143–2150 (1991).
P. L. Yeagle, *Membranes of Cells*, Academic Press, Orlando, Fla., 1987, pp. 1–292.
Y.-Y. Yeh, *Pediat. Res. 30*:55–61 (1991).
C.-L. Yu, M.-H. Tsai, and D. W. Stacey, *Cell 52*:63–71 (1988).
C.-L. Yu, F.-S. Wu, and D. W. Stacey, *Science 243*:522–526 (1989).
H. Zhang, N. N. Desai, J. M. Murphey, and S. Spiegel, *J. Biol. Chem. 265*:21309–21316 (1990).
H. Zhang, N. N. Desai, A. Olivera, T. Seki, G. Brooker, and S. Spiegel, *J. Cell Biol. 114*:155–167 (1991).

23
Toxicity and Systemic Effects of Phospholipids

Theresa M. Allen *University of Alberta, Edmonton, Alberta, Canada*

I. INTRODUCTION

Phospholipids are frequently used in applications involving cultured cells, live animals, and, increasingly, humans. Therefore it is important to have an understanding of the ways in which phospholipids, and other forms of lipid, interact with living cells and organisms. Adverse affects associated with in vivo administration of lipids, as liposomes or lipid emulsions, are discussed in this chapter. Because an overview of the biodistribution and pharmacokinetics of phospholipids is necessary to any understanding of adverse effects associated with in vivo administration of phospholipids, these topics are also discussed.

II. THE MONONUCLEAR PHAGOCYTE SYSTEM

The principal site of clearance of particulate matter, including phospholipid vesicles (liposomes), from the circulation in vivo is the mononuclear phagocyte system (MPS), also referred to by the older term reticuloendothelial system. The MPS consists of circulating monocytes and of fixed monocytes of liver (Kupffer cells), spleen, lymph nodes, and bone marrow. A primary function of the MPS is the removal of foreign particulate matter from circulation. Other functions include defence against microorganisms, parasites, and neoplastic cells and involvement in host responses to endotoxin, hemorrhagic shock, drug response, and response to circulating immune complexes [1,2]. Impairment of MPS function, which may occur as a result of a variety of causes, including repeated administration of large doses of liposomes in vivo, puts the host organism at increased risk for infections, systemic endotoxemia, and increased spread and growth of neoplastic disease [3–13].

A. Measurement of MPS Function

Because the MPS functions to clear all manner of circulating foreign particulate matter from circulation, MPS activity is frequently measured by "colloid clearance" methods. A variety of particles can be used to measure the rate of clearance, and these have been reviewed, along with a critique of the various types of particles, by Saba [1]. The rate at which test particles (e.g., colloidal carbon) are removed from circulation is assumed to be proportional to the phagocytic capacity of the MPS. However, accurate measurement of phagocytic activity of the MPS depends on the dose of test particles being below saturating doses for the MPS, and below doses that deplete blood opsonins, which are postulated to be necessary for phagocytosis to occur. The rate of hepatic blood flow must also be constant [1].

The phagocytic index K is determined according to the expression

$$K = \frac{\log C_1 - \log C_2}{T_2 - T_1}$$

where C_1 and C_2 represent the blood colloid concentration at sampling times T_1 and T_2, respectively. For colloidal carbon, blood colloid concentration is proportional to the absorbance at 620 nm of a solution of 0.1 mL blood in 2.9 mL water [14,15].

B. Effects of Single Injections of Phospholipids

The MPS is the principal in vivo site of uptake of foreign particulate matter and therefore is the tissue most likely to sustain damage following administration of phospholipid vesicles in vivo. The reason why naturally occurring phospholipids are treated as foreign material, following in vivo administration, is not clearly understood but is postulated to be due to the opsonization of "naked" phospholipid surfaces by plasma proteins leading to recognition, binding, and uptake by the MPS (see Ref. 16 and references therein). The extent of damage to the MPS is related to the amount and nature of the material ingested by MPS cells. The rate and extent of clearance of single doses of liposomes from the circulation is a function of liposome composition, dose, and size as well as of the route of liposome administration. As with other particulate matter, clearance of phospholipid vesicles following intravenous injection (the most common route of injection), occurs principally into the Kupffer cells of liver and the fixed macrophages of spleen. Uptake of single injections of liposomes by liver and spleen occurs at similar rates in several different species, including mouse [17,18], rat [19,20], monkey [21], and human [22–23].

1. Effect of Liposome Composition

Liposome formulations commonly used in the past have included phospholipids such as phosphatidylcholine (PtdChol), phosphatidylserine (PtdSer), phosphatidylglycerol (PtdGro), and sphingomyelin (Sph), often in combination with each other and in combination with cholesterol (Chol). These compositions are referred to in this chapter as "nonstealth" formulations.[1] For nonstealth liposomes, removal from circulation is rapid for all but the smallest liposomes. The spleen has a higher intrinsic activity for colloid uptake; liposome uptake by spleen macrophages, calculated on the basis of unit tissue weight, is thus several-fold higher than uptake by liver Kupffer cells. However, since the liver is a much larger organ, receiving approximately 30% of the blood flow in each pass, the total uptake of liposomes by liver (L) is on the order of tenfold higher than that of spleen (S) i.e., L/S ratio = 10 [15,18]. For nonstealth liposome compositions, of diameters larger than 80–90 nm, uptake by liver and spleen macrophages can be as much as 80–90% of the injected dose within minutes to hours of injection, depending on size (see below) and composition.

[1] Stealth liposomes are liposomes that substantially avoid uptake by the MPS and have extended circulation half-lives in vivo. Nonstealth liposomes are therefore liposomes that are rapidly removed from circulation by uptake into the MPS.

For newer liposome formulations (see Sec. IV for more detail) that avoid uptake by the MPS (termed stealth liposomes), uptake of liposomes by the liver and spleen is considerably reduced, with total uptake 2 days after injection amounting to only 10–20% of the injected dose [16,24–26].

Liposomes containing negatively charged phospholipids are removed from circulation more rapidly than neutral or positively charged liposomes [20]. Liposomes containing the negatively charged phospholipid phosphatidylserine (PtdSer) have particularly avid uptake by macrophages, and a specific recognition mechanism for this phospholipid may be operating [27–31]. Liposomes containing PtdSer experience increased rates of uptake into cells of the MPS at PtdSer concentrations as low as 3 mol % of total lipid [31]. Uptake is postulated to be mediated by a scavenger receptor [27] having an affinity for acidic phospholipids. PtdSer appears to have a particularly high affinity for this putative receptor and is postulated to play a role in the removal of damaged erythrocytes and other cells from the circulation in response to the loss of the asymmetric distribution of phospholipids across the membrane bilayer [28–31]. PtdSer is normally confined to the interior leaflet of the erythrocyte and other membranes [32].

Liposomes composed of phospholipids with gel to liquid crystalline phase transitions (T_c) above body temperature, e.g., sphingomyelin and distearoylphosphatidylcholine (rigid liposomes), are removed from circulation more slowly than liposomes containing lower phase transition lipids (fluid liposomes) [18,33–36], but the total uptake into the MPS of rigid liposomes is not substantially less, over the long term, than for fluid liposomes. Cholesterol also acts to rigidify bilayers when incorporated into fluid liposomes and also slows MPS uptake [37–40]. Photopolymerized PtdCho liposomes, which would be expected also to have a rigid membrane, were however taken up by macrophages in vitro more avidly than those composed of nonpolymerized PtdCho [41]. Liposomes containing membrane rigidifying agents appear to act synergistically with monosiaylganglioside G_{M1}, a stealth component, to reduce dramatically the MPS uptake of liposomes [23]. The mechanism for the reduced rate of MPS uptake of rigid liposomes appears to be related to decreased opsonization of these liposomes by plasma proteins. There appears to be a relationship between liposomal stability in the presence of plasma, i.e., membrane permeability, and their rate of removal from circulation [18,35,38,42]. However, liposome size and surface charge appear to override membrane permeability in determining rates of liposome clearance [40].

Phospholipid vesicles or phospholipid emulsions of all compositions, given as single injections, appear to have remarkably little acute toxicity. The exception to this general rule is the intracerebral administration of large doses of liposomes containing stearylamine or dicetylphosphate (synthetic positively charged and negatively charged lipids, respectively) [43]. The presence of stearylamine in liposomes also results in increased toxicity for cells in culture [44] and lowers the ED_{50} for a single i.v. injection of multilamellar vesicles (MLV) (egg PtdCho:Chol, 1:1) from 7.2 g/kg to 1.1 g/kg [45]. Little or no information exists in the literature on the toxic effects associated with the administration of other lipids that do not occur naturally in vivo. Examples of such lipids include polymerized liposomes [46], triethoxycholesterol [47], a phospholipase-resistant dialkyl analog of PtdCho [48], and a diether analog of PtdCho [49]. Nonionic surfactant vesicles (niosomes) have also been used as drug carriers, and these molecules may be degraded by esterases in vivo to give triglyceride and palmitic acid [50]. It is likely that lipid derivatives, if they are resistant to metabolism, will result in accumulation of material intracellularly in phagocytic cells, leading to the formation of granulomas in the liver [51] or at sites of subcutaneous injections, particularly after multiple injections (see Sec. II-C).

2. Effect of Size

The rate of clearance of particulate matter is proportional to particle diameter [1]. Smaller liposomes are removed from circulation more slowly than larger liposomes [18,20,40,52–54]. Larger liposomes taken up in the liver are found almost exclusively in Kupffer cells [55]. In addition to their uptake by macrophages, small liposomes, having diameters below approximately 90 nm, are capable of crossing fenestrated capillaries, which have an average pore diameter of around 100 nm [56] and can be taken up into liver parenchymal cells [57,58] in amounts dependent on liposome composition [59]. We have also observed increased uptake of small (0.1 μm) stealth liposomes into kidney [16], another tissue with fenestrated endothelia, raising the possibility that parenchymal and kidney cells could, in addition to the MPS, be potential sites for adverse effects of small liposomes and/or their contents.

In analyzing liposome uptake data it is important to keep in mind that there are complex relationships between weight of lipid ingested, particle numbers, and particle diameters [52,54,60–62]. The total weight of lipid ingested by macrophages increases with increasing particle diameter for unilamellar liposomes (LUV). However, since the number of particles per unit volume decreases in proportion to the cube of the diameter, larger particles are present in much lower numbers than smaller particles and are taken up in much lower numbers at a constant lipid weight. For MLV, which contain additional internal phospholipid lamellae when their diameter is above approximately 0.1 μm, the weight of lipid ingested by macrophages for a particle of a given diameter is greater than for LUV. When extruded to diameters of 0.1 μm of lower, MLV are essentially unilamellar [63]. If MPS impairment is related to amount of lipid ingested (e.g., if the lipids themselves are toxic), then large MLV would cause more impairment than similar particle numbers of LUV of the same diameter. If the phospholipids were nontoxic, but the liposomes contained a cytotoxic drug, than LUV would be more toxic to macrophages than MLV, at the same particle diameter, because LUV contain higher trapped volumes of material. If the availability of binding sites is related to MPS impairment, than smaller liposomes (larger particle numbers) would have greater adverse effects at the same lipid weight. To date no complete studies, taking into account all of these factors, have been done.

3. Pharmacokinetics and Dose

Nonstealth liposomes have very different pharmacokinetics than do stealth liposomes, and both differ dramatically in the manner in which their pharmacokinetics responds to increasing liposome dose. Nonstealth liposomes show saturable (Michaelis-Menton) pharmacokinetics, while stealth liposomes show nonsaturable (first-order) pharmacokinetics (see Sec. IV. B-8).

Early experiments on nonstealth liposomes (e.g., MLV of PtdCho:Chol) showed them to be rapidly removed from circulation following i.v. administration in a multiphasic removal process, with an initial rapid phase of removal followed by one or more slow phases of removal [17,19–20,64]. Initial half-lives for removal of MLV from circulation are on the order of 5 to 15 min. Small unilamellar vesicles (SUV) of similar compositions are cleared with initial half-lives approximately three- to four-fold longer than those seen for MLV [18,20]. Liposomes with more rigid bilayers are also removed from circulation in a multiphasic removal process, but with slower initial half-lives [18,33–36]. By contrast, stealth liposomes containing either monosialoganglioside (GM_1) or a polyethylene glycol derivative of distearoylphosphatidylethanolamine (PEG-Ste_2PtdEtn) as the stealth component (see Sec. IV) are removed from circulation following intravenous administration in a primarily monophasic (log-linear) process with initial half-lives of 12 to 24 h, depending on liposome size [16,23].

As the dosage of a single injection of nonstealth phospholipid vesicles increases, the processes contributing to rapid uptake of liposomes, i.e., recognition, binding, and internalization, become saturated, and the circulatory half-lives of liposomes increase as the slower phases of liposome removal become dominant [54,62,65–68]. Processes that contribute to the slower phases of liposome removal from circulation likely include uptake into other tissues and turnover of binding sites, lipid metabolism, etc. As the injected dose exceeds the capacity of the liver to take up liposomes by rapid process, more liposomes remain in circulation, and spillover into spleen results [18,66]. Saturation of splenic uptake of liposomes results in increased bone marrow accumlation [69]. Depression of liver uptake in mice has been observed at doses of 5–10 μmol lipid/kg body weight [70], although we see evidence of saturation of the rapid phase of uptake at doses as low as 2 μmol/kg of PtdCho : Chol, 2 : 1 (0.1 μm) (Allen and Hansen [116]). Recovery of the MPS after single large (saturating) doses of liposomes has been measured by looking at tissue distributions of second liposome doses given at varying times after injection of the first dose. Recovery of the MPS following doses at high as 1.1 g lipid/kg body weight (MLV) is reported to occur within 24 h [66]. We have found, however, that high doses (10 μmol/mouse) of smaller liposomes (PtdCho : Chol, 2 : 1, 0.1 μm) had approximately 25% of the in vivo liposomes remaining in circulation 24 h after injection, and therefore complete recovery of the MPS, which cannot occur until all circulating particles are cleared, may be delayed beyond 48 h after injection of high doses of small liposomes [116].

The complex interactions between liposome size, dose, and liposomal surface area on liposome disposition and pharmacokinetics in vivo have been examined in detail by Hunt and colleagues [54,62]. The relationship between liposomal size and dose has been studied for liposome uptake by liver parenchymal cells by Chow et al. [71]. For SUV of Sph : Chol, 2 : 1 or Ste_2PtdCho : Chol, 2 : 1 uptake by hepatic parenchymal cells was markedly enhanced by increasing the injected dose [71].

The pharmacokinetics of drugs is dramatically altered upon association with liposomes. In experiments with doxorubicin, comparing free drug to liposome-entrapped drug, the area under the concentration versus time curve (AUC) was increased by more than one order of magnitude when the drug was encapsulated in PtdGro-containing liposomes and more than two orders of magnitude when encapsulated in a stealth formulation of liposomes containing hydrogenated PtdIns [72]. In addition, liposome encapsulation substantially reduces the cardiotoxicity of doxorubicin, but encapsulation of doxorubicin and other cytotoxic drugs, particularly in nonstealth liposomes, appears to significantly increase the toxicity of preparations to the MPS by delivering large amounts of cytotoxic drug to the MPS cells [73]. Although single large doses of nonstealth liposomes saturate the MPS, there appears to be little or no adverse effect associated with this transient blockade, and the MPS appears to recover completely from this blockade within 24 to 72 h after injection. There may, however, be a cause for concern when the liposomes contain cytotoxic drugs. In this case, even low doses of liposomes (10 mg/kg in the case of doxorubicin-containing liposomes) will impair the MPS [73], although it is likely that recovery after a single insult will be rapid due to recruitment of new MPS cells from bone marrow with differentiation in situ to Kupffer cells [74].

4. Effect of Route of Administration

The most common route of administration of liposomes has been i.v., but many other routes of administration have been used. Liposomes have been administered intraperitoneally [34,77,78], subcutaneously [79,80], intracerebroventrically [81,82] intraarticularly [83], intragastically [84], and intraduodenally [85], intraocularly [86,87], interstitially [88,89], topically [90], by the intramuscular route [91], and via the respiratory tract [92–95].

Routes of administration other than the i.v. route reduce the rate of uptake of liposomes by liver and spleen, and they reduce the total amount of liposomes and their contents localizing in liver and spleen, often substantially, depending on the rate at which liposomes are released from the site of injection into blood. Release of intact liposomes into blood appears to take place readily within 2 to 6 h after intraperitoneal (i.p.) administration but very slowly following subcutaneous administration [16]. The use of other sites of administration would be expected to reduce the amount of MPS impairment that occurs, although studies on MPS impairment associated with other routes of administration are absent from the literature. Many of the other routes of administration result in increased localization of liposomes in regional lymph nodes, but to date there is no clear picture of any adverse effects that accompany this increased localization in lymph nodes.

5. Single Doses, Summary

The reason for the lack of noticeable toxicity associated with single, often very large, doses of phospholipids appears to be the remarkable capacity of the MPS for regeneration of binding sites and metabolism of ingested material [34,98,99] and to recruitment of new MPS cells from bone marrow [75,98], proliferation of Kupffer cells in situ [99] and possibly derivation of Kupffer cells from blood monocytes [100]. Phospholipid drug carriers containing cytotoxic drugs, composed of cytotoxic lipids, or carrying nonmetabolizable drugs or lipids, will have a greater chance of causing adverse effects to the MPS, but it would appear that this versatile system has the capacity to deal effectively with most large single doses of phospholipids.

C. Multiple Administrations of Liposomes

The literature on the effects of multiple doses of phospholipids as vesicles or emulsions is small by comparison to the large volume of literature on single administrations of liposomes. Depression of phagocytic index, i.e., suppression of MPS function, is the most commonly described effect of repeated i.v. administrations of liposomes.

One of the first published studies on the effects of multiple liposome injections reported that injections of even low doses of MLV containing dicetylphosphate, stearylamine, or pure PtdSer were toxic when injected i.v. into mice [10]. These lipids were eliminated by the researchers in their subsequent study on mulitple injections of liposome-encapsulated immunomodulators in dogs and mice. Dogs receiving up to 600 mg lipid × 3 weekly injections (MLVs of PtdCho:PtdSer:egg lecithin, 4.94:4.95:0.1 with or without lymphokines) showed slight elevations in serum alkaline phosphatase and glutamine oxaloacetic transaminase levels and occasionally elevated serum bilirubin levels [101]. No histologic evidence of liver damage was found in the animals. Mice receiving the same liposome formulation at doses of 100–1000 mg/kg × 6 over a two-week period had no gross tissue or organ abnormalities, but some animals had microgranulomas in the lung that were not correlated with dose [101].

Because the major site of liposome uptake is the MPS, it is likely that the MPS would be the major site of impairment following multiple injections of phospholipids. Some studies have addressed this problem. Poste [69] has reported that frequent saturation of the phagocytic capacity of the MPS by repeated liposome injections (MLV, PtdSer:PtdCho:Chol, 3:3:4, 150 mg/kg in mice, one injection every three days) resulted in significant reduction in liposome uptake by liver after 5 injections. Increased accumulation of liposomes in spleen and then in bone marrow resulted as liver uptake became saturated. Uptake into spleen and bone marrow became saturated after 10 injections, and small increases in localization of liposomes in kidney, lung, and carcass were then observed [69].

A detailed study of chronic liposome administration in mice has been reported by Allen et al. [15,51]. Multiple i.v. injections (3 per week × 10 injections, 200–400 mg/kg) of liposomes of different sizes (MLV, SUV) and compositions (PtdCho:Chol, 2:1; Ste_2PtdCho:Chol, 2:1; shingomyelin:Chol, 2:1 and sphingomyelin:PtdCho, 4:1) were examined. Phagocytic indices, liver and spleen size and histology, blood chemistry, and the ability of multiple liposome predoses to alter the tissue distribution of subsequent doses of radiolabeled liposomes were determined. Stringent precautions were taken to assure sterility of the phospholipid preparations, freedom from endotoxin contamination, and freedom from lipid peroxidation, all of which can substantially alter the results of these kinds of experiments [15]. We observed that MLV regardless of composition caused greater MPS impairment than small unilamellar vesicles (SUV). Liposomes with fluid bilayers caused less MPS impairment than liposomes with rigid bilayers. Liposomes with rigid bilayers also resulted in pronounced splenomegaly. Oxidized liposomes caused slightly greater MPS impairment than nonoxidized liposomes, and liposomes containing lipid A (from endotoxin) were taken up by the MPS more avidly than liposomes free of lipid A [102]. Finally, liposomes containing the antileishmanial drug, megalumine antimoniate, a cytotoxic drug, caused more MPS impairment than drug-free liposomes [102].

No alterations in liver enzymes were observed [15], but histological studies showed that mice receiving liposomes with rigid bilayers had granulomatous inflammation of the liver, the appearance of which was correlated with depression of MPS function. Splenomegaly and granulomatous inflammation of the liver has been reported in mice for multiple injections of sphingomyelin:Chol, 1:1 liposomes (444 mg/kg/injection) [103]. MPS function was rapidly restored following termination of administation of all compositions of liposomes, and restoration of MPS function was accompanied by disappearance of liver granuloma and resolution of splenomegaly [51]. Multiple doses of PtdCho:Chol, 2:1 (80 mg/kg × 10) liposomes resulted in no histological changes in liver during the time course of the injections, but two weeks following the last injections a large percentage of the mice showed granulomatous inflammation of the liver that had not resolved by nine weeks post-injection [51]. Mice receiving lower liposome doses (20 mg/kg) showed no adverse effects of any kind during or after several weeks of biweekly injections of liposomes [15].

Phospholipid-triglyceride emulsions such as Intralipid® have been widely used for parenteral nutrition in malnourished patients over long periods of time with only minor alterations in MPS function. Multiple i.v. injections of Intralipid®, given at phospholipid doses that resulted in MPS impairment when given as liposomes, did not result in any alteration in MPS function [104]. Intralipid® was retained by liver two to three-fold less than an equivalent dose of liposomes and was retained by spleen ten-fold less than liposomes [105]. It would appear that Intralipid® has a different metabolic pathway in vivo than liposomes [104].

A factor often not taken into account in discussions of possible toxic effects mediated by phospholipids is the possible consequences associated with lipid contaminants, e.g., endotoxins, or breakdown products, e.g., lipid peroxides. The studies quoted above point out that MPS function can be altered in the presence of these substances.

The impairment of MPS function following multiple liposome injections can be interpreted in terms of the various factors involved in the regeneration of normal MPS activity following challenges with particulate material. The ability of Kupffer and other MPS cells to remove particulate matter from circulation over a period of time is a result of two main processes: on the one hand, the rate at which the MPS becomes saturated, i.e., recognition, binding, and uptake sites are occupied and become unavailable for further liposome uptake; and on the other hand, the rate of regeneration of the MPS. Depression of MPS activity could be due either to saturation of binding sites and uptake mechanisms with liposomes or to depletion of plasma opsonins by liposomes. Depletion of plasma opsonins has been rejected as playing a role in depression of RE activity by Abra and Hunt [54] and by Ellens et al. [105]. Recovery of MPS function is undoubtedly a result of a number of processes, some of which are rapid, taking place over a period of a few hours following liposome ingestion, e.g., lipid metabolism [34,96–97], and some of which are slow, taking several days, e.g., those processes involved in Kupffer cell turnover, which takes approximately 21 days [74]. Significant MPS impairment will result if phospholipids are administered at rates or doses that exceed the regenerative capacity of the MPS. Higher doses of liposomes must be given less frequently than lower liposome doses, if MPS impairment is to be avoided.

III. CONSEQUENCES OF MPS IMPAIRMENT

A number of studies have examined the effect of disease and treatment of disease of MPS function. This literature and the literature on the possible consequences of MPS impairment have been covered in detail in a recent review article [13]. There is no doubt that the progression of some diseases is sensitive to alteration in MPS function, but the complex interactions between particulate carriers, associated drug, the MPS, and the disease process make it difficult to make generalizations. It will be necessary to study each liposomal formulation, in combination with each drug, in order to find out which dosage level and which dosage schedule result in minimal MPS impairment.

IV. STRATEGIES FOR AVOIDING MPS IMPAIRMENT

One can, at least in theory, avoid MPS impairment when using liposomal formulations that normally localize in the MPS by keeping the liposome dose below MPS saturating levels, or by using dosage frequencies that allow the MPS to recover completely between doses. In practice, depending on the drug dose and schedule desired, this may prove to be very difficult.

A. Erythrocytes, Lipid Emulsions, and Lipoproteins

One way to avoid possible problems associated with the localization of phospholipid particulate carriers in the MPS is to design carriers that are not recognized as foreign by the MPS and that therefore circulate for extended periods of time. Some strategies incude the use of erythrocytes, lipid emulsions, and lipoproteins as carriers. Another strategy is to modify liposome composition to reduce MPS uptake and increase their circulation half-lives.

Homologous erythrocytes can be used as drug carriers, but the drug loading process often results in damage to the erythrocytes. Damaged erythrocytes are rapidly removed from circulation, primarily by uptake into liver [106–108], likely due to the exposure of PtdSer on the exterior membrane of damaged erythrocytes (see II.B.1). Minimally damaged erythrocytes, in which drugs are entrapped by gentle loading techniques, have circulation half-lives of days [107–110], and in this case MPS damage will be unlikely. Both hydrophilic and hydrophobic drugs could be used in erythrocyte carriers.

Phospholipid-triglyceride emulsions or lipoproteins can be used as drug carriers for hydrophobic drugs. Emulsions such as Intralipid® have been widely used for parenteral nutrition in malnourished patients, who can receive a liter or more of these emulsions daily for long periods of time with only minor alterations in MPS function. Experiments in mice (see II.C) confirm that lipid emulsions cause substantially less MPS impairment than do liposomes.

B. Stealth Liposomes

Stealth liposomes are liposomes in which the lipid composition and liposome surface properties have been altered to result in significant reductions in their recognition and uptake by the cells of the MP system. Stealth liposomes have considerably increased circulation half-lives as compared to conventional liposomes, when measurements are made by the same techniques, and their long circulation times lead to increased uptake of the liposomes by nonreticuloendothelial tissues even in the absence of targeting molecules [16,24–26].

1. Surface Hydrophilicity

From the experiments that we and others have done to date, it appears that the most important condition for achieving liposomes with stealth properties is the presence of functional groups on the liposome surface that increase surface hydrophilicity [16]. Coating of colloidal particles with hydrophilic coatings has also been shown to decrease uptake by liver and by peritoneal macrophages [111]. Increased surface hydrophilicity is postulated to reduce opsonization of the liposomes by plasma proteins, and we have preliminary data in support of this hypothesis [16,122]. Adding a surface glycocalyx to the liposomes, for example by incorporating monosialoganglioside GM_1 into the bilayer, likely works by this mechanism. A distearoylphosphatidylethanolamine derivative of polyethylene glycol (PEG-Ste_2PtdEtn), which has recently been synthesized by C. Redeman of Liposome Technology Inc., has been found to be slightly more effective than GM_1 in decreasing MP system uptake of liposomes (Table 1), again most likely through the mechanism of increasing surface hydrophilicity.

2. Surface Charge

Also important in avoiding MPS uptake is the absence of negative charges directly exposed at the liposome surface. Negative charge can increase surface hydrophilicity, but unprotected surface negative charges, such as those found in phosphatidylglycerol (PtdGro), phosphatidic acid, phosphatidylserine (PtdSer), or di- and trisialogangliosides, may function as binding sites for blood opsonins and/or may be involved in direct binding of liposomes to macrophages [27,28,31], decreasing liposome blood levels (Table 1). Negative charge, shielded from surface exposure by the presence of carbohydrates, as in

Table 1 Comparison of Blood Levels of Liposomes at 2 h after Intravenous Injection

Liposome composition (molar ratio)	Blood levels, % of in vivo cpm	% of in vivo
PtdCho : Chol (2 : 1)	12.0 ± 4.4	71.1 ± 2.4
PtdCho : Chol : GM_2 (2 : 1 : 0.2)	59.4 ± 5.1	86.8 ± 12.2
PtdCho : Chol : PEG-Ste_2PtdEtn (2 : 1 : 0.2)	76.7 ± 3.0	93.7 ± 2.4
Ste_2PtdCho : Chol (2 : 1)	12.0 ± 3.5	30.4 ± 1.2
Ste_2PtdCho : Chol : GM_1 (2 : 1 : 0.2)	76.6 ± 9.5	90.3 ± 5.4
Ste_2PtdCho : Chol : PEG-Ste_2PtdEtn (2 : 1 : 0.2)	80.1 ± 2.0	91.5 ± 2.6
Ste_2PtdCho : Chol : phosphatidic acid (2 : 1 : 0.2)	14.4 ± 1.0	83.9 ± 3.7
Ste_2PtdCho : Chol : PtdSer (2 : 1 : 0.2)	1.1 ± 0.1	91.4 ± 1.2
Ste_2PtdCho : Chol : PtdIns (2 : 1 : 0.2)	49.2 ± 9.3	68.5 ± 1.7
Sphingomyelin : PtdCho : Chol : GM_1	82.3 ± 8.1	87.4 ± 3.3

Liposomes were extruded through 0.1 μm Nuclepore filters and administered to ICR mice at a concentration of 0.5 μm/mouse. Blood levels are expressed as a percentage of intact liposomes remaining in vivo at the time of sampling. The last column gives a measure of the % of injected liposomes remaining intact in the body at 2 h post-injection. Liposomes were labelled with ^{125}I-tyraminylinulin, which is rapidly eliminated from the body upon release from liposomes [24]. Mean ± S.D. (n = 3).

the case of monosialoganglioside GM_1, is postulated to increase surface hydrophilicity, without providing a binding site for blood opsonins. Monosialoganglioside GM_1 has been one of the most successful molecules found to date for conveying stealth properties to liposomes [16,24–26] (Table 1), but liposomes containing phosphatidylinositol (PtdIns) also have increased circulation times [16,26] (Table 1). This latter molecule has a negative charge protected behind only one neutral sugar, which may account for its not being as effective as GM_1 in increasing liposome circulation half-lives (Table 1).

3. Bilayer Rigidity

Increasing the rigidity of the liposomal bilayer in the presence of stealth components can further increase the circulation half-lives of liposomes [16,24–26]. The mechanism is again probably related to the inability of plasma opsonins to insert effectively into more tightly packed bilayers [16,42]. Increased bilayer rigidity can be achieved either through the use of lipid with high phase transitions, e.g., distearoylphosphatidylcholine (Ste_2PtdCho) in combination with stealth components in formulating the liposomes, or through the use of lipids that participate in hydrogen-bonding in the head group region, such as sphingomyelin (Table 1). The effect of bilayer rigidity is, however, minimal when the stealth component is PEG-Ste_2PtdCho (Table 1), possibly due to the large size of the PEG head group (1900 daltons).

4. Concentration of Stealth Components

One begins to see an effect of GM_1 in decreasing MPS uptake of liposomes at a ratio of approximately 5 mol % of phospholipids. Optimum concentrations of GM_1 are between 7 and 15 mol % of GM_1, and in many applications we use a concentration of 10 mol %. At concentrations above 15 mol % a detergent-like effect of GM_1 begins to appear, as evidenced by increased leakage of liposome contents [24]. For PEG-Ste_2PtdEtn, concentrations between 3 and 15 mol % are effective at reducing MP recognition of liposomes. At higher concentrations of this molecule, a surfactant effect of PEG becomes apparent, and increased foaming of the liposome preparations occurs, making them difficult to work with.

5. Phase Separation

At the concentration of GM_1 used in liposomes to achieve long circulation half-lives, GM_1 is monodispersed in the membranes [113,114] and incorporates completely into the liposome bilayer with a symmetric distribution across the bilayer [25]. At the same concentrations, di- and trisialogangliosides appear to be phase separating into domains within the bilayer [114,115], which may in part account for the inability of these gangliosides to impart good stealth properties to liposomes. Similar data regarding the distribution of PEG-Ste_2PtdCho in liposomes is not yet available.

6. Size of Stealth Liposomes

As the size of stealth liposomes decreases, the uptake by the MP system also decreases [16,25]. Stealth liposomes extruded through 0.05 μm Nuclepore filters, and having an average diameter of approximately 90 nm or less, experience increased uptake by liver relative to larger liposomes, possibly because of the ability of the smaller liposomes to pass through liver sinusoids and be taken up by liver hepatocytes [16,25]. Even relatively large stealth liposomes (200–400 nm diameter) are capable of remaining in the circulation for extended periods of time [16].

7. Tissue Distribution of Stealth Liposomes

A comparison of the tissue distribution of two different formulations of stealth liposomes, as compared to liposomes of a conventional formulation (PtdCho:Chol, 2:1), is given in Fig. 1a–d. Blood levels of liposomes containing either GM_1 or PEG-Ste_2PtdEtn were significantly higher than blood levels of liposomes of a similar size (0.1 μm) and dose (0.1 μmol/mouse) but lacking the stealth component. Blood levels of stealth liposomes declined in a log-linear fashion, and the half-life in circulation of the two stealth formulations was close to 24 h, compared to only 90 min for the conventional formulation (Fig. 1a). Liver and spleen uptakes over 48 h were each three-fold less for the stealth liposomes (Figs. 1b and 1c), and uptake into the remaining carcass tissues was four-fold greater (Fig. 1d), as compared to PtdCho:Chol liposomes. Thus significant alterations in the tissue distributions of liposomes and associated drugs are possible with simple surface alterations of liposomes. The stealth liposomes that localize in carcass are widely distributed following either i.v. or i.p. administration [16].

8. Pharmacokinetics and Dose

As the dosage of stealth liposomes increases from 0.1 to 100 μmol/mouse (Sph:PtdCho:Chol:GM_1 of Sph:PtdCho:Chol:PEG-Ste_2PtdCho, 1:1:1:0.1, 0.1 μm), the rate of their removal from blood remains log-linear and independent of dose. In other

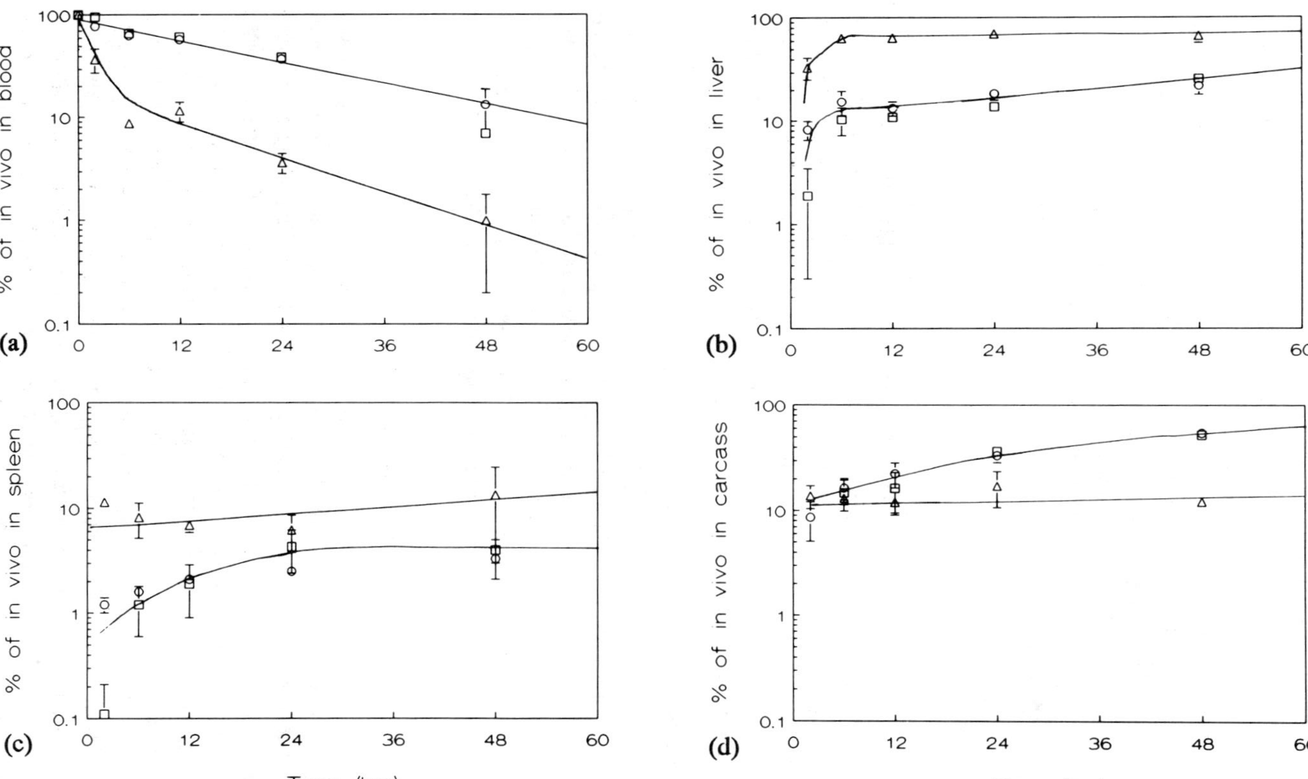

Figure 1 Percentage of liposome-entrapped ^{125}I-tyraminylinulin remaining in ICR mice associated with various tissues at various times post-injection. Mice were injected intravenously with 0.1 μmol (0.2 ml) of liposomes (0.1 μm extruded) composed of (Δ) PtdCho:Chol, 2:1, (O) Sphingomyelin: PtdCho:Chol:PEG-Ste$_2$PtdEtn, 1:1:1:0.2, and (□) Sphingomyelin:PtdCho:Chol:GM$_1$, 1:1:1:0.2. (a) Blood, (b) liver, (c) spleen, (d) remaining carcass. Tissues and remaining carcass were corrected for blood content using correction factors reported in Ref. 25.

words, stealth liposomes demonstrate first-order dose-independent pharmacokinetics and do not appear to saturate MPS uptake over a 100-fold dose range (Allen and Hansen, Ref. 116).

V. CONCLUSIONS

In the past, therapeutic applications of phospholipid and other drug carrier systems have been limited because of the pronounced tendency of these carriers to localize in the cells of the MPS, with the possibility of adverse effects accompanying this localization. The development of lipid carrier systems that have significantly reduced MPS uptake presents many possibilities for improvement of previous applications and for the development of new therapeutic applications.

REFERENCES

1. T. M. Saba, *Arch. Intern. Med. 126*:1031 (1970).
2. B. M. Altura, *Adv. Microcirc. 9*:252 (1980).
3. J. P. Nolan, *Gastroenterology 69*:1346 (1975).
4. M. Grun, and H. Liehr, in *Kupffer Cells and Other Liver Sinusoidal Cells* (E. Wisse and D. L. Knook, eds.), Elsevier, Amsterdam, 1977, p. 437.
5. M. H. Levy, and E. F. Wheelock, *Adv. Cancer Res. 20*:131 (1974).
6. R. Keller, *J. Natl. Cancer Inst. 57*:1355 (1976).
7. D. Glaves, *Int. J. Cancer 26*:115 (1980).
8. T. G. Antikatzides, and T. M. Saba, *J. Reticuloendothelial Soc. 22*:1 (1977).
9. N. R. DiLuzio, in *Kupffer Cells and Other Liver Sinusoidal Cells* (E. Wisse and D. L. Knook, eds.), Elsevier, Amsterdam, 1977, p. 397.
10. A. Kimura, R. L. Sherwood, and E. Goldstein, *J. Reticuloendothelial Soc. 34*:1 (1983).
11. N. R. Adham, M. K. Song, and G. C. Haberfelde, *Gastroenterology 84*:461 (1983).
12. R. L. Souhami, N. Parker, and J. W. B. Bradfielf, in *Kupffer Cells and Other Liver Sinusoidal Cells* (E. Wisse and D. L. Knook, eds.), Elsevier, Amsterdam, 1977, p. 481.
13. T. M. Allen, *Adv. Drug Delivery Rev. 2*:55 (1988).
14. K. Donald, and R. J. Tennent, *J. Pathol. 117*:235 (1975).
15. T. M. Allen, L. Murray, S. MacKeigan, and M. Shah, *J. Pharm. Exp. Therap. 229*:267 (1984).
16. T. M. Allen, C. Hansen, and J. Rutledge, *Biochim. Biophys. Acta 981*:27 (1989).
17. M. M. Jonah, E. A. Cerny, and Y. E. Rahman, *Biochim. Biophys. Acta 401*:336 (1975).
18. T. M. Allen, and J. M. Everest, *J. Pharmacol. Exp. Therap. 226*:539 (1983).
19. G. Gregoriadis, and D. Neerunjun, *Eur. J. Biochem. 47*:179 (1974).
20. R. L. Juliano, and D. Stamp, *Biochem. Biophys. Res. Commun. 63*:651 (1975).
21. H. K. Kimelberg, T. F. Tracy, S. M. Biddlecome, and R. S. Bourke, *Cancer Res. 36*:2949 (1976).
22. V. J. Richardson, B. E. Ryman, R. F. Jewkes, K. Jeyasingh, M. N. H. Tattersall, E. S. Newlands, and S. B. Kaye, *Br. J. Cancer 40*:35 (1979).
23. G. Lopez-Berestein, L. Kasi, M. G. Rosenblum, T. Haynie, J. Monroe, H. Glenn, R. Metha, G. M. Mavligit, and E. M. Hersh, *Cancer Res. 44*:375 (1984).
24. T. M. Allen, and A. Chonn, *FEBS Lett. 223*:42 (1987).
25. T. M. Allen, in *Liposomes in the Therapy of Infectious Diseases and Cancer. UCLA Symposium on Molecular and Cellular Biology* (G. Lopez-Berestein, and I. Fidler, eds.), Vol. 89, Alan R. Liss, New York, 1988, p. 405.
26. A. Gabizon, and D. Papahadjopoulos, *Proc. Natl. Acad. Sci. USA 85*:6949 (1988).
27. K. Nishikawa, H. Aria, and K. Inoue, *J. Biol. Chem. 265*:5226 (1990).

28. Y. Tanaka, and A. J. Schroit, *J. Biol. Chem. 258*:11335 (1983).
29. A. J. Schroit, Y. Tanaka, J. Madsen, and I. J. Fidler, *Biol. Cell 51*:227 (1984).
30. R. A. Schlegel, T. W. Prendergast, and P. Williamson, *J. Cell Physiol. 123*:215 (1985).
31. T. M. Allen, P. Williamson, and R. A. Schelgel, *Proc. Natl. Acad. Sci. USA 85*:8067 (1988).
32. J. A. F. Op den Kamp, *Annu. Rev. Biochem. 48*:47 (1979).
33. K. J. Hwang, K. S. Luk, and P. L. Beaumier, *Proc. Natl. Acad, Sci. USA 77*:4030 (1980).
34. H. Ellens, H. Morselt, and G. Scherphof, *Biochim. Biophys. Acta 674*:10 (1981).
35. J. Senior, and G. Gregoriadis, *Life Sciences 30*:2123 (1982).
36. R. T. Proffitt, L. E. Williams, C. A. Presant, G. W. Tin, J. A. Vliana, R. C. Gamble, and J. D. Baldeschwieler, *Science 220*:502 (1983).
37. G. M. Hwang, and M. R. Mauk, *Proc. Natl. Acad. Sci. 74*:4991 (1977).
38. G. Gregoriadis, and J. Senior, *FEBS Lett. 119*:43 (1980).
39. H. M. Patel, N. S. Tuzel, and B. E. Ryman, *Biochim. Biophys. Acta 761*:142 (1983).
40. J. Senior, J. C. W. Crawley, and G. Gregoriadis, *Biochim. Biophys. Acta 839*:1 (1985).
41. R. L. Juliano, M. J. Hsu, and S. L. Regen, *Biochim. Biophys. Acta 812*:42 (1985).
42. T. M. Allen, *Biochim. Biophys. Acta 640*:385 (1981).
43. D. H. Adams, G. Joyce, V. J. Richardson, B. C. Ryman, and H. M. Wisniewski, *J. Neurol. Sci. 31*:173 (1977).
44. T. M. Allen, L. McAllister, S. Mausolf, and E. Gyorffy, *Biochim. Biophys. Acta 643*:1 (1981).
45. F. Olsen, E. Mayhew, D. Maslow, Y. Rustum, and F. Szoka, *Eur. J. Cancer Clin. Oncol. 18*:167 (1982).
46. S. L. Regen, A. Singh, G. Oehme, and M. Singh, *Biochem Biophys. Res. Commun. 101*:131 (1981).
47. K. R. Patel, M. P. Li, and J. D. Baldeschwieler, *Biochim. Biophys. Acta 797*:20 (1984).
48. D. S. Deshmukh, W. D. Bear, H. M. Wisniewskik, and H. Brockerhoff, *Biochem. Biophys. Res. Commun. 82*:328 (1978).
49. V. Stein, G. Halperin, D. Haratz, and O. Stein, *Biochim. Biophys. Acta 922*:191 (1987).
50. C. A. Hunter, T. F. Dolan, G. H. Coombs, and A. J. Baillie, *J. Pharm. Pharmacol. 40*:161 (1988).
51. T. M. Allen, and E. A. Smuckler, *Res. Commun. Chem. Path. Pharm. 50*:257 (1985).
52. R. A. Schwendener, P. A. Lagocki, and Y. E. Rahman, *Biochim. Biophys. Acta 772*:93 (1984).
53. R. L. Magin, J. M. Hunter, M. R. Niesman, and G. A. Bark, *Cancer Drug Del. 3*:233 (1986).
54. R. M. Abra, and C. A. Hunt, *Biochim. Biophys. Acta 666*:493 (1981).
55. R. Roerdink, J. Dijkstra, G. Hartman, B. Bolscher, and G. Scherphof, *Biochim. Biophys. Acta 677*:79 (1981).
56. E. Wisse, R. DeZanger, and R. Jacobs, *Sinusoidal Liver Cells* (D. L. Knook, and E. Wisse, eds.), Elsevier, Amsterdam, 1982, p. 61.
57. G. Poste, C, Bucana, A. Raz, P. Bugelski, R. Kirsh, and I. J. Fidler, *Cancer Res. 42*:1412 (1982).
58. F. Roerdink, J. Regts, B. V. Leeuwen, and G. Scherphof, *Biochim. Biophys. Acta 770*:195 (1984).
59. J. J. Spanjer, M. Van Galen, R. H. Roerdink, J. Regts, and G. L. Scherphof, *Biochim. Biophys. Acta 770*:195 (1986).
60. M. J. Hsu, and R. L. Juliano, *Biochim. Biophys. Acta 720*:411 (1982).
61. M. K. Pratten, and J. B. Lloyd, *Biochim. Biophys. Acta 881*:307 (1986).
62. R. M. Abra, and C. A. Hunt, *Res. Commun. Chem. Path. Pharmacol. 36*:17 (1982).
63. L. D. Mayer, M. L. Hope, and P. R. Cullis, *Biochim. Biophys. Acta 858*:161 (1986).
64. G. Gregoriadis, and B. E. Ryman, *Eur. J. Biochem. 24*:485 (1972).

65. Y. J. Kao, and R. L. Juliano, *Biochim. Biophys. Acta 677*:453 (1987).
66. R. M. Abra, M. E. Bosworth, and C. A. Hunt, *Res. Commun. Chem. Pathol. Pharmacol. 29*:349 (1980).
67. M. E. Bosworth, and C. A. Hunt, *J. Pharm. Sci. 71*:100 (1981).
68. H. Ellens, H. W. M. Morselt, B. H. J. Dontje, D. Kalicharan, C. E. Hulstaert, and G. L. Scherphof, *Cancer Res. 43*:2927 (1983).
69. G. Poste, *Biol. Cell 47*:19 (1983).
70. M. R. Mauk, and R. E. Gamble, *Proc. Natl. Acad. Sci. USA 76*:765 (1979).
71. D. D. Chow, H. E. Essien, M. M. Padki, and K. J. Hwang, *J. Pharmacol. Exp. Therap. 248*:506 (1989).
72. A. Gabizon, R. Shiota, and D. Papahadjopoulos, *J. Natl. Cancer Inst. 81*:1484.
73. M. A. Bally, R. Nayar, D. Masin, M. J. Hope, P. R. Cullis, and L. D. Mayer, *Biochim. Biophys. Acta 1023*:133 (1990).
74. R. W. Crofton, M. M. C. Diesselhoff-den-Dulk, and R. Van Furth, *J. Exp. Med. 148*:1 (1978).
75. K. Nakatsu, and D. A. Cameron, *Can. J. Physiol. Pharmacol. 57*:756 (1979).
76. R. J. Parker, K. D. Hartman, and S. M. Sieber, *Cancer Res. 41*:1311 (1981).
77. P. Rosa, and F. Clementi, *Pharmacology 26*:211 (1983).
78. G. Delgado, R. K. Potkul, J. A. Treat, G. S. Lewandowski, J. F. Barter, D. F. Forst, and A. Rahman, *Am. J. Obs. Gynecol. 160*:812 (1989).
79. H. M. Patel, K. M. Boodle, and R. Vaughn-Jones, *Biochim. Biophys. Acta 801*:76 (1984).
80. J. Khato, A. A. Del Campo, and S. M. Siever, *Pharmacology 26*:230 (1983).
81. H. K. Kimelberg, T. F. Tracy, Jr., R. E. Watson, D. Kung, R. F. Reiss, and R. S. Bourke, *Cancer Res. 38*:706 (1978).
82. F. Hong, and E. Mayhew, *Cancer Res. 49*:5097 (1989).
83. J. T. Dingle, J. L. Gordon, B. L. Hazleman, C. G. Knight, D. P. Page-Thomas, N. C. Phillips, I. H. Shaw, F. J. T. Fildes, J. E. Oliver, G. Jones, E. H. Turner, and J. S. Lowe, *Nature 271*:372 (1978).
84. D. S. Deshmukh, W. D. Bear, and H. Brockerhoff, *Life Sci. 28*:239 (1981).
85. H. M. Patel, R. W. Stevenson, J. A. Parsons, and B. E. Ryman, *Biochim. Biophys. Acta 716*:188 (1982).
86. D. S. Deshmukh, K. Kristensson, H. M. Wisniewskik, and H. Brockerhoff, *Neurochem. Res. 6*:143 (1981).
87. P. N. Shek, and R. F. Barber, *Biochim. Biophys. Acta 902*:229 (1987).
88. M. P. Osborne, V. J. Richardson, K. Jeyasingh, B. E. Ryman, *Int. J. Nuclear Med. Biol. 6*:75 (1979).
89. V. I. Kaledin, N. A. Matienko, V. P. Nikolin, V. Gruntenko, and V. G. Budker, *J. Nat. Cancer Inst. 66*:881 (1981).
90. A. Gesztes, and M. Mezei, *Anesth. Analg. 6*:1079 (1988).
91. R. Naeff, V. Pliska, and H. G. Weder, *J. Microencapsul. 1*:95 (1990).
92. H. N. McCullough, and R. L. Juliano, *J. Nat. Cancer Inst. 63*:727 (1979).
93. S. G. Wollfrey, I. W. Callaway, G. Taylor, and A. Smith, *J. Controlled Release Soc. 5*:203 (1988).
94. R. M. Fielding, *Proc. West. Pharmacol. Soc. 32*:103 (1989).
95. D. Meisner, J. Pringle, and M. Mezei, *J. Microencapsul. 6*:379 (1989).
96. R. Roerdink, J. Regts, B. Van Leeuwen, and G. Scherphof, *Biochim. Biophys. Acta 770*:195 (1984).
97. J. Dijkstra, M. Van Galen, D. Regts, and G. Scherphof, *Eur. J. Biochem. 148*:391 (1985).
98. R. P. Gale, R. S. Sparkes, and D. W. Golde, *Science 201*:937 (1978).
99. E. Wisse, *Blood Cells 6*:91 (1980).
100. R. Van Furth, *Blood Cells 6*:87 (1980).
101. I. R. Hart, W. E. Fogler, G. Poste, and I. J. Fidler, *Cancer Immunol. Immunother. 10*:157 (1981).

102. T. M. Allen, L. Murray, C. R. Alving, and J. Moe, *Can. J. Physiol. Pharmacol. 65*:185 (1982).
103. E. A. H. Weereratne, G. Gregoriadis, and J. Crow, *Br. J. Exp. Path. 64*:670 (1983).
104. T. M. Allen, and L. Murray, *Can. J. Physiol. Pharmacol. 64*:1006 (1986).
105. H. Ellens, E. Mayhew, and Y. M. Rustum, *Biochim. Biophys. Acta 714*:479 (1982).
106. D. A. Tyrell, and B. E. Ryman, *Biochem. Soc. Trans. 4*:677 (1976).
107. E. Pitt, C. M. Johnson, and D. A. Lewis, *Biochem. Pharmacol. 32*:3359 (1983).
108. G. M. Ihler, *Pharmac. Therap. 20*:151 (1983).
109. K. Kinosita, Jr. and T. V. Tsong, *Nature 272*:258 (1978).
110. H. O. Alpar, and D. A. Lewis, *Biochem. Pharmacol. 34*:257 (1985).
111. L. Illum, I. M. Hunneyball, and S. S. Davis, *Int. J. Pharmaceutics 29*:53 (1986).
112. T. M. Allen, J. L. Ryan, and D. Papahadjopoulos, *Biochim. Biophys. Acta 818*:205 (1985).
113. T. E. Thompson, M. Allietta, R. E. Brown, M. L. Johnson, and T. W. Tilock, *Biochim. Biophys. Acta 817*:229 (1985).
114. H. Kojima, K. Hanada-Yoshikawa, A. Katagiri, and Y. Tamai, *J. Biochem. 103*:126 (1988).
115. M. Myers, C. Wortman, and E. Friere, *Biochemistry 23*:1442 (1984).
116. T. M. Allen and C. Hansen, *Biochim. Biophys. Acta 1068*:133–141 (1991).

24

Immunologic Properties and Activities of Phospholipids

William E. Fogler *Laboratory of Experimental Immunology, Biological Response Modifiers Program, DCT, NCI–FCRDC, Frederick, Maryland*

Modern techniques in lipid biochemistry and immunology have definitively demonstrated the antigenic nature of phospholipids. Both antibody- and cell-mediated recognition of membrane phospholipids have now been reported to occur. Moreover, it is clear that phospholipids and their metabolic products can regulate the function of the immune system. The purpose of this review is to summarize selected areas of representative research involved in determining the immunologic principles of phospholipids.

I. ANTIPHOSPHOLIPID ANTIBODIES

The occurrence of antibodies that bind phospholipids is now well documented. Anticholinephospholipid, antiaminophospholipid, and anticardiolipin antibodies have been described both clinically and experimentally and have been implicated in the pathogenesis of disease as well as in the generation of protective immunity [reviewed in Refs. 1–3].

Several types of experimental evidence have demonstrated that phosphatidylcholine can serve as an antigen. First, antibodies to phosphocholine are readily observed after injection of mice with pristane [4]. Second, murine polyclonal or monoclonal antibodies reacting with bromelin-treated erythrocytes can be obtained by a variety of techniques, and such antibodies are inhibited by phosphatidylcholine and recognize the trimethylammonium group of phosphocholine [5–9]. Third, myeloma B lymphocytes that bind specifically to mouse erythrocyte membranes also bind to liposomes containing phosphatidylcholine [10]. Fourth, antibodies induced by liposomes containing phosphatidylcholine and lipid A can react in a solid phase immunoassay with purified phosphatidylcholine alone [11]. Finally, antibodies to synthetic phosphatidylcholine or to closely related synthetic analoges of phosphatidylcholine have the ability to distinguish between the synthetic phosphatidylcholine analog and synthetic phosphatidylcholine in liposomes [11]. In each of the above immunologic reactions, considerable phospholipid specificity can be demonstrated.

It is not surprising that the essential requirement for specificities of most antibodies to phospholipids is that the antibody recognize the polar head group of the phospholipid. Although the lipid portion of sphingomyelin apparently can be immunogenic [12] and may contribute to the specificities of certain antibodies to sphingomyelin [13,14], the common characteristic of most antibodies to sphingomyelin or lecithin is recognition of the phosphocholine head group by the antibody. Given that the immunodominant feature of phosphocholine resides in either trimethylammonium or phosphate, notable idiotypic differences are found between antibodies to phosphocholine and phosphatidylcholine.

A. Antibodies to Phosphocholine

Antibodies to phosphocholine, the head group of phosphatidylcholine, occur spontaneously in mice treated with pristane and have been influential in the conceptual development of the antigen combining site of antibodies [4–16]. Phosphorylcholine is the immunodominant antigen found in the cell walls of a variety of pathogenic and nonpathogenic microorganisms [17–19], and antibodies that bind phosphorylcholine play an important protective role against pneumococcal bacteria infection in rodents and man [20,21].

In Balb/C mice, antiphosphorylcholine antibodies have been shown to be associated with different idiotypes [22]. The T15/H8 idiotype is dominant in normal BALB/c sera and in the antiphosphorycholine antibodies in mice immunized with C-polysaccharide or phosphorylcholine bound to keyhole limpet hemocyanin [23]. These antibodies belong to the group of germline encoded immunoglobulins of the murine S107/TEPC(T15) V1 gene [24]. Antibody binding is inhibited by phosphocholine and a peptide derived from the CDR2/FR3 region of the heavy chain of murine T15 [25]. It is of interest that the murine monoclonal antibody to the T15 epitope of phosphocholine reacts poorly with phosphatidylcholine in liposomes but demonstrates avid recognition for phosphatidylcholine in lipid emulsions [27] or lipid monolayers [28].

Recently, it has been shown that antiphosphorylcholine antibodies isolated from human sera express the murine T15 idiotype indicating a conserved antigen binding site between mouse and man [26]. The finding of a conserved interspecies idiotype that is associated with an antibacterial response suggests that the T15 idiotype may be maintained in evolution because it is fundamental to survival.

B. Antibodies to Phosphatidylcholine

Antibodies to phosphatidylcholine can be found in normal mouse sera, and titers can be increased by injection of mice with lipopolysaccharide. Identification of these antiphosphatidylcholine antibodies came about through investigations on a "cryptic" antigen made accessible on erythrocytes following treatment with bromelin [reviewed in ref. 3]. Idiotypic analysis of antibodies reactive with bromelin-treated erythrocytes (antiphosphatidylcholine antibodies) has shown a restricted heterogeneity across different strains of mice [29,30]. Indeed, *N*-terminal variable region sequences from heavy and light chains of six NZB-derived or CBA-derived antibodies reactive with bromelin-treated erythrocytes demonstrated almost complete homology [29,31]. However, little if any idiotypic relatedness is exhibited between antibodies to bromelin-treated erythrocytes and two antibodies that bind to phosphocholine [29]. It is evident therefore that the antibodies belong to a highly restricted family of related antibodies with highly conserved gene sequences, and the possibility exists that this represents encoding by unique grem-line genes [29,31].

Natural antibodies to phosphatidylcholine (and phosphatidylserine) have also been demonstrated to occur in humans [reviewed in Ref. 1]. Whether these antibodies share idiotypic determinants with murine antiphosphatidylcholine antibodies, as described above for murine and human antiphosphocholine antibodies [26], is not known. However, distinct similarities have been described for these naturally occurring murine and human antibodies and antibodies induced experimentally against phospholipids presented in liposomes (see below).

C. Antibodies to Liposomes Containing Phosphatidylcholine

Prior to 1979, phospholipids had not been considered antigenic, since repetitive administration of synthetic phospholipid vesicles into experimental animals failed to elicit either a humoral or a cell-mediated response. In 1979, Schuster et al. reported that following the parenteral administration of liposomes containing lipid A (the lipid portion of gram negative bacterial lipopolysaccharide) as an adjuvant into rabbits, polyclonal IgM antibodies were induced that fixed complement and mediated immune lysis of the "immunizing" liposome that lacked lipid A [32]. Antibodies to liposomes were not elicited by liposomes lacking lipid A. A secondary immunization with the liposome preparation yielded an IgG antiliposome antibodies, indicating that both a primary and a secondary immune response against phospholipid was achievable in the host. The liposomes used for immunization contained dipalmitoylphosphatidylcholine, cholesterol, dicetyl phosphate, and lipid A. Adsorption of the antisera with liposomes containing dimyristoylphosphatidylcholine (or sphingomyelin), cholesterol, and dicetyl phosphate caused efficient removal of antibodies. Adsorption was less efficient when liposomes lacking individual components were used, but some adsorption of antibodies did occur with liposomes composed of dimyristoylphosphatidylcholine [32]. The discovery that inclusion of lipid A as an adjuvant into liposomes could result in immune responses to the bulk lipids of the liposomes has provided a novel means to study the immunogenicity of membrane-associated phosphatidylcholine. These techniques have also been applied to elicit an antibody response to other phospholipids [reviewed in Ref. 33].

These data have suggested that antibodies to liposomes recognize the entire liposomal surface as an antigen and that individual liposome constituents are recognized by complementary "subsites" in the antigen combining region [3]. One of the most prominent subsites recognizes phosphate and is inhibited by soluble phosphorylated molecules, including phosphocholine, inositol hexaphosphate, and adenosine triphosphate [33,34]. It is of interest that antibodies that bind to bromelin-treated erythrocytes react with soluble trimethylammonium compounds and also with phosphoserine [29]. Based on this latter reactivity, the possibility exists that the antigen combining site includes a phosphate-binding subsite as described above. It is of further interest that despite the abilities of antiliposome antibodies to recognize phosphocholine, the antibodies lack idiotypes that have been reported to be characteristically associated with murine antiphosphocholine antibodies, including T15 [35]. In addition, strains of immunodeficient mice that lack the ability to produce an immune response to phosphocholine readily produce phosphocholine inhibitable antibodies to liposomes after injection of liposomes containing lipid A [35].

It is not known whether antibodies induced by liposomes containing lipid A share idiotypic determinants with antibodies that react with bromelin-treated erythrocytes. However, in this context it is interesting to note that although antibodies to liposomes did not react with newly obtained nonadherent macrophages, they did react with adherent macrophages, and this reactivity was inhibited by pretreatment of cells with phospholipase C [36]. It is reasonable to conclude, however, that configuration, orientation, charge, packing density, and fluidities of phospholipids in membranes, and complementary conformational specificities of antibodies, are the major factors that form the basis for the specificities of antibodies to the phosphilipids [3].

II. ANTIBODY-INDEPENDENT IMMUNOLOGIC RECOGNITION OF PHOSPHOLIPIDS

It is well recognized that antigen-bound antibody can facilitate the cellular recognition of infectious agents and neoplastic cells. The term "opsonic antibody" was first used to describe the inherent property of immunoglobulin to enhance the phagocytic uptake of antibody-coated microorganisms [37]. This is now known to result from the high affinity binding of the Fc portion of the antibody molecule with specific cell surface receptors on leukocytes [38]. Yet in the absence of any apparent antibody intermediate, cells of the host immune system maintain an ability to recognize and appropriately (homeostasis and immunity) or inappropriately (autoimmune disorders) destroy infectious agents and altered cells. The prominence given to proteins and glycoproteins as a basis for immunologic recognition is firmly established in the literature. Recent investigations on the immunobiology of mononuclear phagocytes (macrophages) have now suggested that phospholipids, in particular phosphatidylserine, may serve as a molecular basis for the discrimination and recognition of "self" and "nonself."

In both invertebrate and vertebrate animals the primary function of macrophages is to discriminate between self and nonself [reviewed in Refs. 39–41]. The macrophage participates in a number of homeostatic mechanisms that include the recognition, phagocytosis, and catabolism of effete cells, e.g., senescent red blood cells, damaged cells, and cellular debris. In this regard, the macrophage directly controls the recycling of extracellular and intracellular iron stores and the metabolism of lipids and cholesterol. When host homeostatic mechanisms are stressed, the macrophage can participate in complex interactions that involve cellular and humoral aspects of the inflammatory and immunological response. Macrophages that line the body cavities provide a first line of defense against microbial and parasitic infections. Just as essentially, the macrophage regulates both the afferent and efferent arms of the immune system in the surveillance against foreign invaders and cancer.

Several reports have demonstrated that phosphatidylserine can be recognized by mononuclear phagocytes and that such recognition may represent a basis for the discrimination of self and nonself. The argument for such a recognition system can be summarized in the following ways: inclusion of negatively charged phospholipids, including phosphatidylserine, within synthetic phospholipid vesicles greatly enhances their binding to and phagocytosis by macrophages [42–45]; red blood cells that express phosphatidylserine in the outer membrane leaflet avidly bind to and are cleared by macrophages in vivo [46–48]; and the extent to which macrophages discriminate between normal cells and tumor cells can be influenced by cellular expression of phosphatidylserine [49–50].

A. Phagocytic Uptake of Liposomes by Macrophages

Studies in many laboratories have demonstrated the utility of synthetic phospholipid vesicles as carrier vehicles for drug delivery [reviewed in Refs. 51–53]. Liposomes provide a unique carrier system for the in vivo delivery of biologically active materials to phagocytic cells. Following i.v. administration, the majority of liposomes are taken up by reticuloendothelial cells in the liver, spleen, lymph nodes, and bone marrow, and by circulating phagocytic cells [54–56]. However, in order for liposomes to deliver compounds to cells of the reticuloendothelial system they must avidly bind to and become phagocytosed by the cells.

The size and lipid composition of liposomes influence the extent of cellular interaction, since these parameters determine the kinetics of liposome uptake by macrophages in vitro and the capillary arrest and organ retention of liposomes in vivo. Extensive experimentation revealed that the inclusion of negatively charged phospholipids within the lipid bilayer of liposomes greatly enhances their binding to and phagocytosis by blood monocytes or tissue macrophages [42,44]. Specifically, negatively charged liposomes containing phosphatidylserine are phagocytosed up to 10-fold faster than are liposomes of the same size bearing a net positive charge (stearylamine) or those composed of phosphatidylcholine alone. The inclusion of 30 mol % phosphatidylserine with phosphatidylcholine was sufficient to increase significantly liposome phagocytosis by macrophages. Similar to phosphatidylserine, inclusion of dicetyl phosphate or phosphatidylglycerol in liposome bilayers leads to enhancement in their internalization by macrophages [43,45]. Collectively, these results have suggested that macrophages might be able to recognize certain classes of phospholipids.

B. Destruction of Senescent Red Cells

Senescent red cells are sequestered in the spleen [57], and their destruction presumably occurs because of a subtle abnormality detected by the splenic macrophages. While considerable data exist describing the mode of destruction of abnormal red cells, relatively little is known about normal red cell aging and the mode of destruction of senescent red cells. A variety of changes have been described in the red cell as it ages, although their significance, if any, in the final destruction of the cell is unknown. It appears that as red cells age there is a decrease in cell size accompanied by an increase in density and mean corpuscular hemoglobin concentration [58–60]. Reticulocytes are rich in membrane lipids, and this excess lipid is lost during reticulocyte maturation [61]. Data concerning a continued loss of membrane lipid during the life span of mature red cells are difficult to interpret because of technical limitations in the separation of cells by age [58,62]; however, it appears that little or no lipid is lost from red cells during the final half of their life span [63]. It has also been proposed that the IgG coating of red cells increases as they age [64], and that this immunoglobulin may provide a recognition signal to splenic macrophages. Older red cells also have a decrease in sialic acid content and, as a result, in zeta potential [65,66]. However, whereas the enzymatic depletion of sialic acid in red cells is associated with a shortened survival [67], red cells in patients with acquired polyagglutinability due to a decreased amount of red cell sialic acid do not [68], and the consequence of a decrease in sialic acid for senescent red cells is not totally resolved.

While little or no phospholipid is lost from red cells during the final half of their life span, there appears to be an alteration in phospholipid asymmetry. Ample evidence supports the general view that both halves of the lipid bilayer of most, if not all, eukaryotic cells have distinctly different lipid compositions. The outer leaflet is generally rich in the cholinephospholipids whereas the aminophospholipids preferentially occupy the inner leaflet. In human red cells, 82% of the sphingomyelin and 76% of the phosphatidylcholine is located in the outer leaflet, whereas 80% of the phosphatidylethanolamine and virtually all of the phosphatidylserine is in the inner leaflet [69–71]. Red cells from other mammalian species exhibit the same general pattern of an outer and an inner membrane leaflet rich in cholinc- and aminophospholipids, respectively [72–74].

The externalization of phosphatidylserine has been reported for aged red cells [75], malaria-infected red cells [76,77], sickle cells [78], and red cells from patients with chronic myeloid leukemia [79] and diabetes mellitus [80]. Moreover, the exposure of phosphatidylserine in the outer leaflet of red cells seems to play a role with subsequent marcrophage interaction. This was first shown by the rapid clearance of red cells containing fluorescent long-chain phosphatidylserine but not phosphatidylcholine analogs from the peripheral circulation of mice [46]. Analysis of the livers and spleens of these animals revealed that the cleared, fluorescent red cells underwent phagocytosis and accumulated in splenic macrophages and Kupffer cells. These observations are reminiscent of the characteristic erythrophagocytosis observed in sickle cell disease [81–83] and suggest that phosphatidylserine might be a determinant for cell-cell recognition. Indeed, malaria-infected red cells [84] and red cells from diabetes mellitus patients [89] display elevated levels of adherence to macrophages. Although the mechanism by which phosphatidylserine is recognized by phagocytes is not known, these findings suggest that phosphatidylserine exposed on the surface of cells may serve as a signal for their recognition by phagocytes and their subsequent elimination from the circulation [56].

C. Macrophage-Mediated Recognition and Destruction of Tumor Cells

Macrophages also play an important role in host defense against cancer and infections [reviewed in Ref. 87]. Normal, noncytotoxic blood monocytes or tissue macrophages can be activated to become tumoricidal or microbicidal subsequent to their interaction with lymphokines, bacterial products, or both [87]. Although tumor cells are heterogeneous with regard to many characteristics, they seem to share susceptibility to destruction by activated macrophages by a process that requires cell-cell contact. Activated macrophages can lyse tumor cells that are resistant to other immune effector cells such as T cells or natural killer cells or to chemotherapeutic drugs [88]. Moreover, activated macrophages can discriminate between tumorigenic cells, which they lyse, and nontumorigenic cells, which they do not, even under cocultivation conditions [89].

Studies in many tumor systems have indicated that macrophage-mediated tumor cell lysis is independent of such tumor cell characteristics as surface receptors, histocompatibility antigens, tumor antigens, cell cycle traverse, expression of endogenous viruses, and metastatic potential [87]. The exact mechanism that regulated macrophage discrimination between normal and tumorigenic cells is not known, but the broad spectrum of tumor cells susceptible to macrophage-mediated lysis suggests that a uniform surface moiety could be involved in target cell recognition.

Recent experimental data suggest that phosphatidylserine may represent a macrophage recognition signal on tumor cells, being similar to the above discussion on senescent and pathologic red blood cells. Utsugi et al. determined whether the presence of phosphatidylserine in the outer membrane leaflet of human tumor cells correlated with their recognition by activated human monocytes [50]. Three tumorigenic cell lines and a normal human epidermal keratinocyte line were incubated with macrophages activated to the tumoricidal state. Activated human monocytes bound to and lysed all tumorigenic targets, while the nontumorigenic normal cell line was neither bound nor killed. A semiquantitative analysis for phosphatidylserine in the outer leaflet of the cells revealed that the tumor cells expressed approximately three- to seven-fold more phosphatidylserine than did the normal cells. Moreover, normal cells containing phosphatidylserine analogs, were bound to activated human monocytes. These data led the authors to suggest a role for phosphatidylserine in monocyte recognition of tumor cells.

III. IMMUNOMODULATORY ROLE OF PHOSPHOLIPIDS

The considerable advances in the field of immunology have not been without inconsequential input from lipid biochemists. As further subsets of immunologic cells are defined, each with individual functions, and as additional interleukins and cytokines are uncovered with distinct immunologic properties, it is comforting that central to this apparent complex network the metabolic pathways of certain phospholipids are intimately involved in the generation of an immunologic response. In this regard, phospholipids can be envisioned as immunomodulatory agents capable of promoting or suppressing both nonspecific (macrophage-mediated) and specific (lymphocyte-mediated) humoral and cellular mechanisms of immunity.

A. Modulatory Effects on Macrophage Function

As previously discussed above, studies in many laboratories have demonstrated that liposomes provide a unique carrier system for the in vivo delivery of biologically active materials to phagocytic cells [51–56]. While these studies have demonstrated that the macrophage activating potential of various agents can be enhanced following their encapsulation within synthetic phospholipid vesicles, it is now clear that the phospholipid composition of the liposomes can influence macrophage activity. Moreover, it is also apparent that liposomes composed of identical phospholipid compositions can lead to either an enhancement or an inhibition of cellular processes involved in macrophage activation. In this context, it is important to point out that the term "macrophage activation" is an operational one and is used by different investigators to denote separate, and even unrelated, changes in phenotype of the same or disparate macrophage population(s).

Following cellular interaction with specific lymphokines, macrophages acquire the capacity to kill intra- and extracellular targets. Lymphokines encapsulated within liposomes composed of phosphatidylcholine and phosphatidylserine (7:3 mol ratio) are highly effective in rodent and human macrophages [reviewed in Ref. 87]. However, liposomes of identical phospholipid composition completely inhibit lymphokine-induced macrophage microbicidal activity (intracellular lytic mechanism) against *L. major* and *L. donovani* [90,91]. Subsequent analyses indicated that liposomes composed of phosphatidylcholine and either phosphatidylethanolamine, phosphatidic acid, or cardiolipin at a 7:3 mol ratio, but not phosphatidylcholine alone or phosphatidylcholine and phosphatidylinositol, caused inhibition of lymphokine-induced macrophage microbicidal activity [45]. Moreover, inhibition was also observed with liposomes containing the lyso forms of phosphatidylserine or phosphatidylethanolamine, and the extent of inhibition was directly related to the degree of fatty acid saturation [45].

These results suggest that the inhibition of lymphokine-activated macrophage microbicidal activity by liposomes was not due to phosphatidylserine directly but rather to the intracellular interconversion and metabolism of phosphatidylserine. From a metabolic standpoint, chemical conversions of the phospholipids could have occurred after ingestion by macrophages. The biosynthetic pathways for phosphatidylserine and phosphatidylethanolamine are closely interrelated, and phosphatidylethanolamine can arise through the decarboxylation of phosphatidylserine, while serine can be incorporated into phospholipids in animal tissue by a direct exchange reaction [92]. In addition, phosphatidylserine (and lysophosphatidylserine) may be generated from the conversion of lysophosphatidylethanolamine [93].

The actions of phospholipase A_2 might also play a role in the modulation of macrophage function by phospholipids. The site of attack for phospholipase A_2 is the fatty acyl ester group on the *sn*-2 carbon in the glyceryl backbone of the phospholipid. The metabolic products of phospholipase A_2 consist of lysophospholipid and free fatty acid. It has been reported that increased levels of phospholipase are associated with activated macrophages [94], and it is well known that phospholipase A_2 activity is strongly influenced by the degree of unsaturation of the *sn*-2 fatty acid [95]. The above observations on specificities relating to phospholipid head groups for inhibition of macrophage microbicidal activity are also compatible with corresponding phospholipase A_2 specificities. Phospholipase A_2 can be influenced by the type of head group of a neutral phospholipid [95–97] and by the charge characteristics of the phospholipid [98,99].

B. Activation and Regulation of T Cell Responses

The T lymphocyte is responsible for the development of a specific cell-mediated immunologic response. The specificity of the T cell response is governed by the T cell antigen receptor, which in turn regulates both a phosphatidylinositol [100] and/or a tyrosine kinase [101] signal transduction pathway. While the relationship between these two signal transduction pathways and their resulting effects on T cell rsponses is unclear, the phosphatidylinostiol pathway clearly has an immunomodulatory role.

Stimulation of T cells with specific antigens or interleukins, via specific cellular receptors, or mitogenic lectins, is accompanied by the hydrolysis of phosphoinositol-4,5 biphosphate by phospholipase C, which results in the formation of inositol triphosphate and diacylglycerol [reviewed in Refs. 102, 103]. Inositol triphosphate mobilizes Ca^{2+} from intracellular stores, while diacylglycerol promotes the translocation of protein kinase C [103]. Protein kinase C, activated in the presence of Ca^{2+}, diacylglycerol, and phospholipid, phosphorylates cellular substrates such as the IL-2 receptor.

Products of membrane phospholipid metabolism can impact significantly on transmembrane signal transduction in T cells. Changes in the content of phospholipid fatty acids, particularly arachidonate, have been shown to modify transmembrane transport processes [104], membrane enzymatic activity [105], and the cytotoxic activity of T cells [106]. Moreover, arachidonate stimulates T cell mitrogenesis and promotes the activation of protein kinase C. Arachidonate comprises approximately 25% of all fatty acids within T cell membranes, being primarily esterified to phosphoglycerides. Normal, resting T cells incorporate traced amounts of radiolabeled arachidonate into membrane phospholipids in a highly predictable fashion, with phosphatidylcholine as the major recipient and lesser amounts incorporated into phosphatidylinositol, phosphatidylethanolamine, and phosphatidylserine [107].

A secondary response to ligand-receptor binding of T cells is the release of arachidonate from and its redistrubition within membrane phospholipids [108,109]. Coincident to this redistribution of arachidonate is the generation of lysophosphatidylcholine. Lysophosphatidylcholine, formed by phospholipase A_2 cleavage of phosphatidylcholine, has been shown to augment directly the activity of protein kinase C [110] and has been implicated in T cell mitogenesis [111]. T cells from autoimmune MRL-*lpr*/*lpr* mice do not exhibit this pattern of phosphatidylcholine and arachidonate metabolism and exhibit defective lymphokine production and responsiveness upon lectin stimulation [112]. Instead, these aberrant T lymphocytes demonstrate a preference for forming arachidonoyl phosphatidylinositol over -phosphatidylcholine [112].

The contribution of enhanced constitutive turnover of phosphatidylinositol to the derangements in transmembrane signal transduction that have been reported in T cells from MRL-*lpr/lpr* mice remains to be determined. If, however, increased metabolism of phosphatidylinositol favors the production of lysophosphatidylinositol over lysophosphatidylcholine, signal transduction may thus be compromised. Another potential consequence of increased phosphatidylinositol/phospholipase A_2-like activity is that of competition with phosphatidylinositol specific phospholipase C for substrate [113]. If phosphatidylinositol is preferentially metabolized by phospholipse A_2, as opposed to phospholipase C, diminished production of inositol phosphates and diacylglycerol might occur. As discussed above, these latter second messengers are required to activate protein kinase C and IL-2 synthesis fully, and receptor expression could conceivably be jeopardized if phosphatidylinositol availability to phospholipase C is limited [112].

The importance of the phosphatidylinositol signal transduction pathway for T cell activation has also been deduced from studies with HIV-infected T cells. Infection with HIV causes impaired cellular immune functions. Preceding development of clinical disease, impaired lymphocyte function has been described in HIV-infected subjects as manifested by a decreased ability to respond to antigen and mitogenic stimulation [114, 115]. It has been shown that mitogen activation of lymphocytes from AIDS patients results in a normal increase in intracellular free Ca^{2+} but impaired production of IL-2 and proliferation [116]. Recently, it has been shown that HIV inhibits the turnover of phosphatidylinositol upon mitogen stimulation of normal lymphocytes, leading to a decreased expression of IL-2 and impaired proliferation [117]. This effect was independent of the initial increase in intracellular free Ca^{2+}, suggesting that enough inositol triphosphate was generated to create the initial Ca^{2+} flux [117]. However, the inhibition of phosphatidylinositol turnover was responsible for the decreased generation of diacylglycerol [117]. Reduced availability of diacylglycerol presumably would interfere with protein kinase C activation and a decreased expression of IL-2 and impaired proliferation.

IV. CONCLUDING REMARKS

A conceptual basis now exists implicating the direct involvement of membrane phospholipids in the immunologic discrimination of self and nonself. Indeed, an inappropriate response to phospholipid has been observed to occur in the generation of autoimmune disease. That the immune response to phospholipid may be highly conserved throughout phylogeny, certainly as determined for murine and human antiphosphocholine antibodies [26], would indicate that this response could represent a primitive recognition system. This remains to be determined, but in this regard it is of interest to note that in the absence of antibody, insect amebocytes (phagocytic cells) bind liposomes composed of phosphatidylserine more avidly than those composed of phosphatidylcholine [87], a finding similar to that observed for mammalian macrophages (see Sec. II.). Moreover, the intimate involvement of phospholipids in the immune response is also illustrated by the increasing attention given to these biochemical moieties as immunomodulatory agents.

Worksheet: Protocol for the Production of Antibodies Against Liposomes and Phosphilipids

Polycolonal or monoclonal antibodies against liposomal phospholipids can be raised by injecting liposomes containing lipid A (the lipid portion of gram negative bacterial lipopolysaccharide) as adjuvant into rabbits or mice. The following details the materials and methods employed to immunize rabbits for the production of such reagents.

1. Source of Lipids

Phospholipids and cholesterol can be obtained from any commercial source. It should be emphasized that the purity of any potential immunogen will determine the subsequent specificity of the immunological response. Chloroform solution of lipids (1.1 pg phosphate/nmol lipid A) from *Ealmonolla Minnesota* R595 can be obtained from List Biological Labs., Campbell, Calif. Alternatively, lipid A can be produced by acetic acid hydrolysis of lipopolysaccharide, followed by treatment of the lipid A precipitate with EDTA to make it chloroform-soluble (see Alving, C. R., *Chem. Phys. Lipids 40*:303 (1986).

2. Preparation of Liposomes

Complete details of liposome preparation have been reviewed in the given reference. Briefly, multilamellar liposomes are prepared from a mixture of an appropriate phospholipid, cholesterol, and lipid A in molar ratios of 100:75:2. All references to molar amounts of lipid A refer to lipid A phosphate. The lipids in chloroform are dried together in a pear-shaped flask on a rotary evaporator, followed by 1 h under high vacuum in a dessicator. The dried lipids are then supplemented with a suitable (e.g., phosphate) buffer unit pH ~ 7.2 and left to swell for at least 1 hour. Subsequently, a suspension of multilamellar vesicles is created by vigorous sleaking. The final concentration of liposomal phospholipid is 10 mM.

3. Immunization Schemes

Male New Zealand white rabbits (2.5 to 3.0 kg) are immunized with the desired liposome preparation. The immunizing liposomes (10 μmol phospholipid/mL) can be administered by various routes and include i.v. (1.0 mL), i.m. (0.E mL), and XX (0.5 mL). Sera containing antiphospholipid antibodies can be obtained on weekly intervals, typically beginning two weeks past immunization. Preimmunization serum is necessary as a control, since naturally occurring antiphospholipid antibodies can be detected.

Following a single administration of the immunizing liposomes, the antiphospholipid antibodies will be of the IgM isotype. At approximately monthly intervals, animals can receive boosting immunizations both to increase the titer of antiphospholipid antibodies and to obtain IgG antibodies.

REFERENCES

1. C. R. Alving, Antibodies to liposomes, phospholipids, and cholesterol: Implications for autoimmunity, atherosclerosis, and aging, *Horizons in Membrane Biotechnology* (C. Nicolau and D. Chapman, eds.), Wiley-Liss, New York, 1990, p. 41.
2. E. N. Harris, Antiphospholipid antibodies, *Br. J. Haemat. 74*:1 (1990).
3. C. R. Alving, Antibodies to lipids and lipid membranes: Reactions with phosphatidylcholine, cholesterol, liposomes, and bromelin-treated erythrocytes, *Phospholipid-Binding Anti-*

bodies (E. N. Harris, T. Exner, G. R. V. Hughes, and R. A. Asherson, eds.), CRC Press, Boca Raton, Fla., 1991, p. 73.

4. M. Potter, and R. Lieberman, Common individual antigenic determinants in five of eight BALB/c IgA myeloma proteins that bind phosphoryl choline, *J. Exp. Med. 132*:737 (1970).
5. J. M. Pages, and A. E. Bussard, Establishment and characterization of a permanent murine hybridoma secreting monoclonal autoantibodies, *Cell. Immunol. 41*:188 (1978).
6. D. H. De Heer, J. M. Pages, and A. E. Bussard, Specificity of anti-erythrocyte autoantibodies secreted by a NZB-derived hybridoma and NZB peritoneal cells, *Cell. Immunol. 49*:135 (1980).
7. D. Serban, J. M. Pages, A. E. Bussard, and I. P. Witz, The participation of trimethylammonium in the mouse erythrocyte epitope recognized by monoclonal autoantibodies, *Immunol. Lett. 3*:315 (1981).
8. J. Pages, P. Poncet, D. Serban, I. Witz, and A. E. Bussard, Relationship between choline derivatives and mouse erythrocyte membrane antigens revealed by mouse monoclonal antibodies. I. Anticholine activity of anti-mouse erythrocyte monoclonal antibodies, *Immunol. Lett. 5*:167 (1982).
9. K. O. Cox, and S. J. Hardy, Autoantibodies against mouse bromelain-modified RBC are specifically inhibited by a common membrane phospholipid, phosphatridylcholine, *Immunology 55*:263 (1985).
10. T. J. Mercolino, L. W. Arnold, and G. Haughton, Phosphatidylcholine is recognized by a series of Ly-1 + murine B cell lymphomas specific for erythrocyte membranes, *J. Exp. Med. 163*:155 (1986).
11. N. M. Wassef, G. M. Swartz, Jr., C. R. Alving, and M. Kates, Antibodies to liposomal phosphatidylcholine and phosphatidylsulfocholine, *Biochem. Cell Biol. 68*:54 (1990).
12. T. Taketomi, and T. Yamakawa, Antigenic properties of a synthetic protein complex with glycolipids and related substances, *Lipids 1*:31 (1966).
13. T. Taketomi, and T. Yamakawa, Immunochemical studies of lipids. II. Antigenic properties of synthetic sphingosylphosphorylcholine-protein conjugate, *Jpn. J. Exp. Med. 37*:423 (1967).
14. R. Arnon, and D. Teitelbaum, Lipid-specific antibodies elicited with synthetic lipid conjugates, *Chem. Phys. Lipids 13*:352 (1974).
15. D. M. Segal, E. A. Padlan, G. H. Cohen, S. Rudikoff, M. Potter, and D. R. Davies, The three dimensional structure of a phosphorylcholine-binding mouse immunoglobulin Fab and the nature of the antigen binding site, *Proc. Natl. Acad. Sci. USA 71*:4298 (1974).
16. E. A. Padlan, D. R. Davies, S. Rudikoff, and M. Potter, Structural basis for the specificity of phosphorylcholine-binding immunoglobulins, *Immunochemistry 13*:945 (1976).
17. M. Potter, and M. A. Leon, Three IgA myeloma immunoglobulins from the BALB/c mouse: Precipitation with pneumococcal C polysaccharide, *Science 162*:369 (1968).
18. C. A. Crandall, and R. B. Crandall, Macroglobulin antibody response to Ascaris suum infection. Detection of a second macroglobulin component in the serum of infected mice, *J. Parasitol. 23*:112 (1969).
19. A. R. Brown, and C. A. Crandall, Phosphorylcholine idiotype related to TEPC 15 in mice infected with Ascaris suum, *J. Immunol. 116*:1105 (1976).
20. D. E. Briles, M. Hanh, K. Schroer, J. Davie, P. Baker, J. Kearney, and R. Barletta, Antiphosphocholine antibodies found in normal mouse serum are protective against intravenous infection with type 3 Streptococcus pneumoniae, *J. Exp. Med. 153*:694 (1981).
21. M. Gray, H. C. Dillman, Jr., and D. E. Briles, Epidemological studies of Streptococcus pneumoniae in infants: Development of antibody to phosphocholine, *J. Clin. Microbiol. 18*:1102 (1983).
22. R. Lieberman, M. Potter, E. B. Mushinski, W. Humphrey, Jr., and S. Rudikoff, Genetics of a new IgV_H (T15 idiotype) marker in the mouse regulating natural antibody to phosphorylcholine, *J. Exp. Med. 139*:983 (1974).

23. W. Lee, H. Cosenza, and H. Kohler, Clonal restriction of the immune response to phosphorylcholine, *Nature 247*:55 (1974).
24. C.-Y. Kang, H.-L. Cheng, S. Rudikoff, and H. Kohler, Idiotypic self-binding of dominant germline idiotype (T15): Autobody activity is affected by antibody valency, *J. Exp. Med. 165*:1332 (1987).
25. C.-Y. Kang, T. K. Brunck, T. Kieber-Emmons, J. E. Blalock, and H. Kohler, Inhibition of self-binding antibodies (autobodies) by a V_H-derived peptide, *Science 240*:1034 (1988).
26. R. Halpern, S. V. Kaveri, and H. Kohler, Human anti-phosphorylcholine antibodies share idiotopes and are self-binding, *J. Clin. Invest. 88*:476 (1991).
27. B. Niedieck, U. Kuck, and H. Gardemin, On the immune precipitation of phosphorylcholine lipids with TEPC 15 mouse myeloma protein and anti-lecithin sera from guinea pigs, *Immunochemistry 15*:471 (1978).
28. M. A. Urbaneja, G. D. Fidelio, J. A. Lucy, and D. Chapman, The interaction of anti-lipid antibody (TEPC 15) with a model biomembrane system (monolayer), *Biochim. Biophys. Acta 898*:253 (1987).
29. P. Poncet, H. P. Kocher, J. Pages, J.-C. Jaton, and A. E. Bussard, Monoclonal antibodies against mouse red blood cells: A family of structurally restricted molecules, *Mol. Immunol. 22*:541 (1980).
30. S. Kawaguchi, M. D. Copper, and J. F. Kearney, Mouse monoclonal antibodies against bromelain-treated mouse erythrocytes: Reactivity with erythrocytes of various species of animals and idiotypes, *Cell. Immunol. 102*:241 (1986).
31. L. Reininger, P. Ollier, P. Poncet, A. Kaushik, and J.-C. Jaton, Novel V genes encode virtually identical variable regions of six murine monoclonal anti-bromelain-treated red blood cell autoantibodies, *J. Immunol. 138*:316 (1987).
32. B. G. Schuster, M. Neidig, B. M. Alving, and C. R. Alving, Production of antibodies against phosphocholine, phosphatidylcholine, sphingomyelin, and lipid A by injection of liposomes containing lipid A, *J. Immunol. 122*:900 (1979).
33. C. R. Alving, Antibodies to liposomes, phospholipids, and phosphate esters, *Chem. Phys. Lipids 40*:303 (1979).
34. B. Banerji, J. A. Lyon, and C. R. Alving, Membrane lipid composition modulates the binding specificity of a monoclonal antibody against liposomes, *Biochim. Biophys. Acta 689*:319 (1982).
35. B. Banerji, J. J. Kenny, I. Scher, and C. R. Alving, Antibodies against liposomes in normal and immune-deficient mice, *J. Immunol. 128*:1603 (1982).
36. W. E. Fogler, G. M. Swartz, Jr., and C. R. Alving, Antibodies to phospholipids and liposomes: Binding of antibodies to cell, *Biochim. Biophys. Acta 903*:265 (1987).
37. T. P. Stossel, The mechanism of phagocytosis, *J. Reticuloendothel. Soc. 19*:237 (1976).
38. M. W. Fanger, L. Shen, R. F. Graziano, and P. M. Guyre, Cytotoxicity mediated by human Fc receptors for IgG, *Immunol. Today 10*:92 (1989).
39. Z. Cohn, The activation of mononuclear phagocytes: Fact, fancy, future, *J. Immunol. 121*:813 (1978).
40. M. L. Karnovsky, and J. K. Lazdins, Biochemical criteria for activated macrophages, *J. Immunol. 121*:809 (1978).
41. G. H. Heppner, and A. M. Fulton, eds., *Macrophages and Cancer*, CRC Press, Boca Raton, Fla., 1988.
42. A. Raz, C. Bucana, W. E. Fogler, G. Poste, and I. J. Fidler, Biochemical, morphological and ultrastructural studies on the uptake of liposomes by murine macrophages, *Cancer Res. 41*:487 (1981).
43. K. Mehta, G. Lopez-Berestein, E. M. Hersh, and R. L. Juliano, Uptake of liposomes and liposome-encapsulated muramyl dipeptide by human peripheral blood monocytes, *J. Reticuloendothel. Soc. 32*:155 (1982).

44. A. J. Schroit, and I. J. Fidler, Effects of liposome structure and lipid composition on the activation of the tumoricidal properties of macrophages by liposomes containing muramyl dipeptide, *Cancer Res. 42*:161 (1982).
45. M. J. Gibreath, W. E. Fogler, G. M. Swartz, Jr., C. R. Alving, and M. S. Meltzer, Inhibition of interferon gamma-induced macrophage microbicidal activity against *Leishmania major* by liposomes: Inhibition is dependent upon composition of phospholipid headgroups and fatty acids, *Int. J. Immunopharmac. 11*:103 (1989).
46. A. J. Schroit, J. W. Madsen, and Y. Tanaka, In vivo recognition and clearance of red blood cells containing phosphatidylserine in their plasma membranes, *J. Biol. Chem. 260*:5131 (1985).
47. R. S. Schwartz, Y. Tanaka, I. J. Fidler, D. Chiu, B. Lubin, and A. J. Schroit, Increased adherence of sickled and phosphatidylserine enriched human erythrocytes to cultured human peripheral blood monocytes, *J. Clin. Invest. 75*:1965 (1985).
48. Y. Tanaka, and A. J. Schroit, Insertion of fluorescent phosphatidylserine into the plasma membrane of red blood cells: Recognition by autologous macrophages, *J. Biol. Chem. 258*:11335 (1983).
49. J. Connor, C. Bucana, I. J. Fidler, and A. J. Schroit, Differentiation-dependent expression of phosphatidylserine in mammalian plasma membrane: Quantitative assessment of outer leaflet lipid by prothrombinase complex formation, *Proc. Natl. Acad. Sci. USA 86*:3184 (1989).
50. T. Utsugi, A. J. Schroit, J. Connor, C. D. Bucana, and I. J. Fidler, Elevated expression of phosphatidylserine in the outer membrane leaflet of human tumor cells and recognition by activated human blood monocytes, *Cancer Res. 51*:3062 (1991).
51. C. R. Alving, Delivery of liposome-encapsulated drugs to macrophages, *Pharmacol. Ther. 22*:407 (1983).
52. G. Gregoriadis, and A. C. Allison, eds., *Liposomes in Biological Systems,* Wiley-Interscience, New York, 1974.
53. C. Nicolau, and A. Paraf, eds., *Liposomes, Drugs, and Immunocompetent Cell Functions*, Academic Press, London, 1981.
54. G. Gregoriadis, D. E. Neorungen, and R. Hunt, Fate of liposome-encapsulated agents injected into normal and tumor-bearing rodents, *Life Sci. 21*:357 (1977).
55. I. J. Fidler, A. Raz, W. E. Fogler, R. Kirsh, P. Bugelski, and G. Poste, The design of liposomes to improve delivery of macrophage-augmenting agents to alveolar macrophages, *Cancer Res. 40*:4460 (1980).
56. G. Poste, C. Bucana, A. Raz, P. Bugelski, R. Kirsh, and I. J. Fidler, Analysis of the fate of systemically administered liposomes and implications for their use in drug delivery, *Cancer Res. 42*:1412 (1982).
57. C. A. Finch, Iron metabolism: The pathophysiology of iron storage, *Blood 5*:983 (1958).
58. D. Danon, and Y. Marikowsky, Determination of density distribution of red cell populations, *J. Lab. Clin. Med. 64*:668 (1964).
59. S. Piomelli, G. Lurinsky, and L. R. Wassermann, The mechanism of red cell aging. I. Relationship between cell age and specific gravity evaluated by ultracentrifugation in a discontinuous density gradient, *J. Lab. Clin. Med. 69*:659 (1967).
60. A. M. Ganzoni, R. Oakes, and R. S. Hellman, Red cell aging in vivo, *J. Clin. Invest. 50*:1373 (1971).
61. S. J. Shattil, and R. A. Cooper, Maturation of macroreticulocyte membranes in vivo, *J. Lab. Clin. Med. 79*:215 (1972).
62. M. P. Westerman, L. E. Pierce, and W. N. Jensen, Erythrocyte lipids: A comparison of normal young and normal old populations, *J. Clin. Med. 62*:394 (1963).
63. C. C. Winterbourn, and R. D. Batt, Lipid composition of human red cells of different ages, *Biochim. Biophys. Acta 202*:1 (1970).
64. M. B. Kay, Mechanism of removal of senescent cells by human macrophages in situ, *Proc. Natl. Acad. Sci. USA. 72*:3521 1975).

65. H. Walter, and F. Selby, Counter-current distribution of red blood cells of slightly different ages, *Biochim. Biophys. Acta 112*:146 (1966).
66. D. Aminoff, J. Anderson, L. Dabich, and W. D. Gathmann, Sialic acid content of erythrocytes in normal individuals and patients with certain hematologic disorders, *Am. J. Hematol. 9*:381 (1980).
67. J. R. Durocher, R. C. Payne, and M. E. Conrad, Role of sialic acid in erythrocyte survival, *Blood 45*:1 (1975).
68. P. Lalezari, and H. Al-Mondhiry, Sialic acid deficiency of human red blood cells associated with red cell, leukocyte and platelet polyagglutinability, *Br. J. Haematol. 25*:399 (1973).
69. A. J. Verkleij, R. F. A. Zwaal, B. Roelofsen, P. Comfurius, D. Kastelijn, and L. L. M. van Deenen, The asymmetric distribution of phospholipids in the human red cell membrane, *Biochim. Biophys. Acta 323*:178 (1973).
70. E. S. Gordesky, and G. V. Marinetti, The arrangement of phospholipids in the human erythrocyte, *Biochim. Biophys. Res. Commun. 50*:1027 (1973).
71. R. F. A. Zwaal, B. Roelofsen, P. Comfurius, and L. L. M. van Deenen, Organization of phospholipids in human red cell membranes as detected by the action of various purified phospholipases, *Biochim. Biophys. Acta 406*:83 (1975).
72. R. F. A. Zwaal, R. Flickiger, S. Moser, and P. Zahler, Lecithinase activities at the external surface of ruminant erythrocyte membranes, *Biochim. Biophys. Acta 373*:416 (1974).
73. W. Renooij, L. M. G. van Golde, R. F. A. Zwaal, and L. L. M. van Deenen, Topological asymmetry of phospholipid metabolism in rat erythrocyte membranes: Evidence for flip flop of lecithin, *Eur. J. Biochem. 61*:53 (1976).
74. H. Chap, R. F. A. Zwaal, and L. L. M. van Deenen, Action of highly purified phospholipases on blood platelets: Evidence for an asymmetric distribution of phospholipids in the surface membrane, *Biochim. Biophys. Acta 467*:146 (1977).
75. A. Herrmann, and P. F. Devauz, Alteration of the aminophospholipid translocase activity during in vivo and artificial aging of human erythrocytes, *Biochim. Biophys. Acta 1027*:41 (1990).
76. P. Joshi, G. P. Dutta, and C. M. Gupta, An intracellular simian malarial parasite (Plasmodium knowlesi) induces stage-dependent alterations in membrane phospholipid organization of its host erythrocyte, *Biochem. J. 246*:103 (1987).
77. C. M. Gupta, A. Alam, P. N. Mathur, and G. P. Dutta, A new look at nonparasitized red cells of malaria-infected monkeys, *Nature 299*:259 (1982).
78. D. Chiu, B. Lubin, and S. B. Shohet, Erythrocyte membrane lipid reorganization during the sickling process, *Br. J. Haematol. 41*:223 (1979).
79. A. Kumar, and C. M. Gupta, Red cell membrane abnormalities in chronic myelogenous leukemia, *Nature 303*:632 (1983).
80. R. K. Wali, S. Jaffe, D. Kumar, and V. K. Kalra, Alterations in organization of phospholipids in erythrocytes as factor in adherence to endothelial cells in diabetes mellitus, *Diabetes 37*:104 (1988).
81. T. W. Bauer, W. Moore, and G. M. Hutchins, The liver in sickle cell disease: A clinicopatholigic study of 70 patients, *Am. J. Med. 69*:833 (1980).
82. R. P. Hebbel, and W. J. Miller, Phagocytosis of sickle erythrocytes: Immunologic and oxidative determinants of hemolytic anemia, *Blood 64*:733 (1984).
83. R. P. Hebbel, Beyond hemoglobin polymerization: The red blood cell membrane and sickle disease pathophysiology, *Blood 77*:214 (1991).
84. M. Hommel, Cytoadherence of malaria-infected erythrocytes, *Blood Cells 16*:605 (1990).
85. A. Zachowski, C. T. Craescu, F. Galacteros, and P. F. Devauz, Abnormality of phospholipid transverse diffusion in sickle erythrocytes, *J. Clin. Invest, 75*:1713 (1985).
86. A. J. Schroit, and R. F. A. Zwaal, Transbilayer movement of phospholipids in red cell and platelet membranes, *Biochem. Biophys. Acta 1071*:313 (1991).
87. I. J. Fidler, Macrophages and metastasis—a biologial approach to cancer therapy: Presidential address, *Cancer Res. 45*:4714 (1985).

88. I. J. Fidler, and A. J. Schroit, Recognition and destruction of neoplastic cells by activated macrophages: Discrimination of altered self, *Biochim. Biophys. Acta 948*:151 (1988).
89. I. J. Fidler, and E. S. Kleinerman, Lymphokine-activated human blood monocytes destroy tumor cells but not normal cells under cocultivation conditions, *J. Clin. Oncol. 2*:937 (1984).
90. M. J. Gilbreath, D. L. Hoover, C. R. Alving, G. M. Swartz, Jr., and M. S. Meltzer, Inhibition of lymphokine-induced macrophage microbicidal activity against leishmania major by liposomes: Characterization of the physiochemical requirements for liposome inhibition, *J. Immunol. 137*:1681 (1986).
91. M. J. Gilbreath, G. M. Swartz, Jr., C. R. Alving, C. A. Nacy, D. L. Hoover, and M. S. Meltzer, Differential inhibition of macrophage microbicidal activity by liposomes, *Infect. Immun. 47*:567 (1985).
92. E. A. Dennis, and E. P. Kennedy, Enzymatic synthesis and decarboxylation of phosphatidylserine in Tetrahymena pyriformis, *J. Lipid Res. 11*:394 (1970).
93. O. Kuge, M. Nishijima, and Y. Akamatsu, Isolation of a somatic-cell mutant defective in phosphatidylserine biosynthesis, *Proc. Natl. Acad. Sci. USA. 82*:1926 (1985).
94. P. Vadas, S. Wasi, H. Z. Movat, and J. B. Hay, Extracellular phospholipase A_2 mediates inflammatory hyperaemis, *Nature 293*:583 (1981).
95. A. J. Slotboom, H. M. Verheij, and G. H. De Hass, On the mechanism of phospholipase A, in *Phospholipids* (J. N. Hawthorne and G. B. Ansell, eds.), Elsevier, Amsterdam, 1982, p. 359.
96. M. F. Roberts, M. Adamich, R. J. Robson, and E. A. Dennis, Phospholipid activation of cobra venom phospholipase A_2. I. Lipid-lipid or lipid-enzyme interaction, *Biochemistry 18*:3301 (1979).
97. M. Adamich, and E. A. Dennis, Specificity reversal in phospholipase A_2 hydrolysis of lipid mixtures, *Biochim. Biophys. Res. Commun. 80*:424 (1978).
98. A. D. Bangham, and R. M. C. Dawson, Control of lecithinase activity by the electrophoretic charge on its substrate surface, *Nature 182*:1292 (1958).
99. G. H. de Haas, N. M. Postema, W. Nieuwenhuizen, and L. L. M. van Deenen, Purification and properties of phospholipase A from porcine pancreas, *Biochim. Biophys. Acta 159*:103 (1968).
100. J. B. Imboden, and J. D. Stobo, Transmembrane signaling by the T cell antigen receptor: Perurbation of the T3-antigen receptor complex generates inositol phosphates and releases calcium ions from intracellular stores, *J. Exp. Med. 161*:446 (1985).
101. L. E. Samuelson, M. D. Patel, A. M. Weissman, J. B. Harford, and R. D. Klausner, Antigen activation of murine T cells induces tyrosine phosphorylation of a polypeptide associated with the T cell antigen receptor, *Cell 46*:1083 (1986).
102. M. J. Berridge, and R. F. Irvine, Inositol triphosphate, a novel second messenger in cellular signal transduction, *Nature 312*:315 (1984).
103. G. Moller, Activation antigens and signal transduction in lymphocyte activation, *Immunol. Rev. 95*:5 (1987).
104. P. Overath, P. H. U. Shairer, and W. Stoffel, Correlation of in vivo and in vitro phase transitions of membrane lipids in Escherichia coli, *Proc. Natl. Acad. Sci. USA. 67*:606 (1970).
105. M. Szamel, and K. Resch, Modulation of enzyme activities in isolated lymphocyte plasma membranes by modification of phospholipid fatty acids, *J. Biol. Chem. 256*:11618 (1987).
106. R. Gill, and W. Clark, Membrane structure-function relationship in cell-mediated cytolysis, *J. Immunol. 125*:689 (1980).
107. J. Trotter, I. Flesch, B. Schmidt, and E. Ferber, Acyltransferase-catalyzed cleavage of arachidonic acid from phospholipids and transfer to lysophosphatides in lymphocytes and macrophages, *J. Biol. Chem. 257*:1816 (1982).
108. C. W. Parker, J. P. Kelly, S. F. Falkenhein, and M. G. Huber, Release of arachidonic acid from human lymphocytes in response to mitogenic lectins, *J. Exp. Med. 149*:1487 (1979).

109. E. Feber, G. G. DePasquale, and K. Resch, Phospholipid metabolism of stimulated lymphocytes. Composition of phospholipid fatty acids, *Biochim. Biophys. Acta 393*:364 (1975).
110. K. Oishi, R. L. Raynor, P. A. Charp, and J. F. Kuo, Regulation of protein kinase C by lysophospholipids, *J. Biol. Chem. 263*:6865 (1988).
111. J. W. Hadden, E. M. Hadden, M. K. Hddox, and N. D. Goldberg, Guanosine 3' : 5'-cyclic monophosphate: A possible intracellular mediator of mitogenic influences in lymphocytes, *Proc. Natl. Acad. Sci. USA. 69*:3024 (1972).
112. M. Tomita-Yamaguchi, J. F. Babich, R. C. Baker, and T. J. Santoro, Incorporation, distribution, and turnover of arachidonic acid within membrane phospholipids of B220+ T cells from autoimmune-prone MRL-lpr/lpr mice, *J. Exp. Med. 171*:787 (1990).
113. S. G. Rhee, P.-G. Suh, S.-H. Ryo, and S. Y. Lee, Studies of inositol phospholipid-specific phospholipase C, *Science 244*:546 (1989).
114. J. W. M. Gold, C. S. Weikel, J. Godbold, C. Garcia, C. S. Urmacher, S. Cunningham-Rundles, B. Koziner, M. Pollack, R. Gallo, M. G. Sarngadharan, and D. Armstrong, Unexplained persistent lymphadenopathy in homosexual men and the acquired immunodeficiency syndrome, *Medicine 64*:203 (1985).
115. B. Hofmann, B. O. Lindhardt, J. Gerstoft, C. S. Petersen, P. Platz, L. P. Ryder, N. Odum, E. Dickmeiss, P. B. Nielsen, S. Ullman, and A. Svejgaard, Lymphocyte tranformation response to pokeweed mitogen as a predictive marker for development of AIDS and AIDS related symptoms in homosexual men with HIV antibodies, *Br. Med. J. 295*:293 (1987).
116. B. Hofmann, J. Moller, E. Langhoff, K. Jakobsen, N. Odum, E. Dickmeiss, L. P. Ryder, O. Thastrup, O. Scharff, B. Foder, P. Platz, L. Ryder, C. Petersen, L. Mathisen, T. Hartvig-Jensen, P. Skinhoj, and A. Svejgaard, Stimulation of AIDS lymphocytes with calcium ionophore (A23187) and Phorbol ester (PMA): Studies of cytoplasmic free Ca^{2+}, IL-2 receptor expression, IL-2 production, and proliferation, *Cell. Immunol. 119*:14 (1989).
117. B. Hofmann, P. Nishanian, R. L. Baldwin, P. Insixiengmay, A. Nel, and J. L. Fahey, HIV inhibits the early steps of lymphocyte activation, including initiation of inositol phospholipid metabolism, *J. Immunol. 145*:3699 (1990).

25
Phosphholipids in Diagnosis

R. Andrew Badley and Paul J. Davis *Unilever Research, Bedford, England*

D. M. Tolley *Unipath Ltd., Bedford, England*

Because of their fundamental and central role and their ubiquitous presence, it is difficult at first to envisage any situations in which phospholipid detection or measurement could be diagnostic. However, the situation is complex, with subtle but important differences between the properties and roles of various types and varieties of phospholipids, allowing the possibility that differential assays, for example, can be of diagnostic value. In addition, the properties that suit them for their membrane-forming role also make them crucial in other situations, such as providing essential surfactant functions in the lungs of the newborn.

In considering phospholipids as tools in diagnosis, it is important to include their ability to form effective solid phases for immunassay of antibodies to polyphosphates, phosphoryl choline, and the like. Again, it is their particular amphiphilic properties that suit them for this purpose, and they fulfil an increasingly useful role embodied in simple commercial assay kits.

Perhaps the most far-reaching combination of their biophysical membrane-forming properties with an indirect diagnostic function is in the field of liposome immunoassays. A wide variety of techniques has been developed for the use of liposomes as signal-generating mechanisms, triggered by some form of immunochemical binding.

These three main areas will be reviewed, to provide an overview of the different aspects of phospholipids in diagnosis, either directly as markers or indirectly as tools.

I. PHOSPHOLIPIDS AS MARKERS

Phospholipids can have a role as diagnostic markers in those diseases or conditions characterized by disruptions to their normal patterns of occurrence, perhaps caused by inborn errors of metabolism, by dietary deficiency, or by the action of infectious organisms, for example. But in practice there are few established clinical diagnostic applications for the detection of phospholipids, analysis of their structure, or assessment of their distribution within the body. There are further potential applications described in the research literature, but these are not likely to expand greatly the list in the forseeable future. In this review we will consider interesting and important phospholipid-related phenomena with either actual or potential diagnostic value.

In general such diagnostic procedures can be (1) measurements of whole phospholipids (types, quantitites, ratios) in body fluids; (2) detection of changes in parts of phospholipids, usually the fatty acid chains; (3) other parameters such as changes in enzymes of phospholipid metabolism.

The most important and well established application for such tests is in connection with lung condition, particularly fetal lung maturity and respiratory distress syndrome, as shown in Table 1, which also gives the literature references for this section. Maintenance of correct lung function depends upon the presence of the so-called pulmonary surfactant, a complex mixture of lipids and proteins. More than 80% of this is phospholipid (predominantly phosphatidyl choline), of which over half normally bears saturated fatty acid chains. Production of surfactant is low in early pregnancy, but it increases from the 26th week and rises rapidly from the 35th week. An active area of diagnosis involves the identification of fetuses and newborns at risk of pulmonary failure arising from a lack of effective surfactant, a defect that is usually treatable. Adults can also suffer from low surfactant levels, either because of inherent lack or as a consequence of disease.

There is a wide choice of analytical approaches, including some enzyme based methods for automated analyzers. The fluids most usefully subjected to analysis range from lung lavage to gastric aspirates and amniotic fluid, since in vivo surfactant is not confined to the lung. The reference method is usually taken to be the ratio of the two choline-containing phospholipids, lecithin (phosphatidyl choline) and sphingomyelin—the L/S ratio. Other important factors include the presence of phosphatidyl glycerol and the saturation of phospholipid fatty acid chains. A very recent report suggests that fluorescence polarization measurement can match the L/S ratio in diagnosing fetal lung immaturity.

Table 1 Phospholipids as Markers in Pulmonary Disease

Condition	Marker	Ref.
RDS	DSPC and surfactant proteins	1
RDS	TLC method for L/S and PG; on gastric aspirate	2
Idiopathic pul. fibrosis	Monitor therapy with phospholipid level in lavage	3
RDS/hyaline membr. dis.	PG and L/S as function of age	4
RDS	Correlation of ventilatory index and L/S	5
FLM	Phospholipid chain saturation by fluorescence polarization	6
Pul. diseases	Bronchial lavage of phospholipds; DSPC, L/S	7
FLM	Autoanalyzer for PC, sphingomyelin correlation	8
RDS (adult) pneumonia	Altered fatty acids in phospholipids	9
FLM	Autoanalyzer for surfactant and albumin	10
FLM	Semiautoanalysis for PG	11
FLM	9 tests compared, L/S and fluorescence polarization best	12

RDS = respiratory distress syndrome
L/S = lecithin/sphingomyelin ratio
DSPC = disaturated phosphatidyl choline
TLC = thin layer chromatography
PG = phosphatidyl glycerol
FLM = fetal lung maturity
PC = phosphatidyl choline

Table 2 Phospholipids as Markers in Nonpulmonary Disease

Condition	Marker	Ref.
Coronary bypass	Serum phospholipid fatty acids/saturation	13
Fetal globoid cell leukodystrophy	Galactosylsphingosine increased	14
Multiple sclerosis	Cerebroside/gangliosides	15
Amniotic fluid	Choline phospholipids by enzyme sensor	16
Krabbe's disease	Galactosylceramidase for detection	17
Infl.-bowel disease	Arachidonate increased	18
Usher's syndrome	Serum phospholipid arachidonate	19
Malaria	Choline kinase into erythrocytes	20
Angina, heart attack	Arachidonate in serum phospholipids	21
Platelet function	Platelet phospholipids	22
T cell function	Arachidonate	23

Table 2 contains examples of phospholipids or their enzymes as diagnostic markers in conditions other than lung disease, together with appropriate references. Saturated and unsaturated fatty acids of phospholipids have been examined in relation to heart disease and inflammatory bowel diseases (including Crohn's disease). In both cases arachidonic acid is significantly raised, implying important links with prostaglandin activity and metabolism.

Usher's syndrome provides a clear example of detectable alterations in plasma phospholipids that relate directly to altered metabolism of phospholipids in affected individuals. The syndrome is an autosomal recessive disorder, leading to blindness through degeneration of rod photoreceptor cells in the retina. These cells contain exceptionally large quantities of docosahexanoic acid (22:6) esterified in phospholipids of the photoreceptor membrane. Arachidonic acid is also present in these membranes, particularly in inositol phospholipids, which are thought to have a functional role in visual cells [24,25].

In Usher's syndrome patients, there is a significant decrease in docosahexanoic and arachidonic acids in plasma phospholipids, even though the fatty acid composition of triglycerides is unchanged. This may reflect abnormal specific carrier mechanisms through the blood interfering with the synthesis of effective photoreceptor membranes [19]. Interestingly, prolonged dietary deprivation also leads to altered retinal function, but it is not known whether similar diagnostic changes in phospholipids can be detected [26,27].

Aspects of phospholipid metabolism can be revealed by detection or analysis of the phospholipids themselves, or by means of the enzymes that work on the phospholpids. Perhaps the clearest example of this is the enzyme choline kinase from Plasmodium (the malarial parasite). This enzyme has been found to be useful as a parasite-specific marker, detectable in infected erythrocytes during isolation procedures [20].

It has been known for some time that various platelet functions are dependent on phospholipid metabolism, particularly the release of arachidonic and other fatty acids, leading to increases in phosphatidyl glycerol. Recently it has been shown that arachidonic acid release and redistribution within T lymphocyte membrane phospholipids is crucial to proper functioning and is involved in the processes of cellular activation [23]. Changes in the content of arachidonic acid result in modified transmembrane transport, membrane enzyme activity, and cytotoxic efficiency of T cells. This redistribution within phospholipids is linked both to ligand-receptor binding and biochemical pathways that generate second messengers. Thus it is suggested that aberrations in the metabolism of arachidonate-containing phospholipids may be associated with, and could potentially contribute to, abnormal T lymphocyte functions, so providing a basis for a new diagnostic approach to the investigation of immune dysfunction.

Although this review is concerned only with phospholipids, it is appropriate to point out that related materials with similar amphiphilic properties also have diagnostic potential. Galactosylceramide is usefully measured in fetal globoid cell leukodystrophy, cerebrosides and gangliosides in multiple sclerosis, and galactosylceramidase in Krabbe's disease. In all of these cases, the lipid markers show correlation to disease states.

However, it is clear that apart from a few specific areas, phospholipids and their enzymes have yet to be fully exploited as diagnostic markers. It is likely that with further improvements in both knowledge and technology there will be increased use of these essential compounds in diagnostic procedures.

II. PHOSPHOLIPIDS AS DIAGNOSTIC TOOLS

A. Phospholipids Used as an Immunochemical Solid Phase

Antibodies that bind to phospholipids are now recognized as important factors in a wide range of important diseases. These antibodies are reviewed fully elsewhere in this book, but it is necessary to consider one particular aspect here. To investigate and detect such antibodies, phospholipids attached to appropriate solid phases must be used. There is now a developing technology in which phospholipids are used as tools to detect the antiphospholipid antibodies that are diagnostic of various diseases, as shown in Table 3, together with appropriate references.

Table 3 Phospholipids Used as Immunological Solid Phases to Detect Antiphospholipid Antibodies in Disease

Disease	Phospholipid used	Ref.
Thrombosis	CL, PI, PS, PC, PA, PE	28,29,30,31
Recurrent abortion	CL, PI, PS, PC	28,32,33,34
Thrombocytopenia	CL, PI, PS, PC	28,35
Systemic lupus erythematosis	CL, PA, PS, PI, PC, PE	33,36,37,38
Rheumatoid arthritis	CL	37,39
Myocardial infarction	CL	40
Infections	CL	40,41
Connective tissue diseases	CL	37
Neurologic disease	CL	42,43,44
Dermatologic disorders	CL	37,45,46
Plastic surgery	CL	47
Filariasis	PC	48

CL = cardiolipin
PI = phosphatidyl inositol
PS = phospatidyl serine
PC = phosphatidyl choline
PA = phosphatidic acid
PE = phosphatidyl ethanolamine

The antibodies first described by Wasserman, diagnostic of syphilis, were not originally recognized as antibodies to phospholipids; in reality these "reaginic" antibodies bind to cardiolipin, although this may not have been the initiating antigen. Evidence has now accumulated to show that the human immune system can generate antibodies to a range of negatively charged phospholipids, especially cardiolipin, phosphatidyl serine, phosphatidyl ethanolamine, phosphatidyl inositol, and phosphatidic acid. Although DNA phosphodiester groups have been thought of as a cross-reactive antigen, polyclonal anticardiolipin antibodies isolated from patient sera do not react with DNA [49]. Thus it is not clear what the natural epitopes for these agents of autoimmune disease really are.

Until a few years ago, antiphospholipid antibodies were detected primarily by agglutination (Venereal Disease Research Laboratory, VDRL, test), by complement fixation (Wasserman) or by the "lupus anticogulant" reaction. All these tests are somewhat nonspecific or too difficult to control for investigation of epitope specificity or, indeed, isotype of bound antibody [50]. Improvements in analytical test methodology have indicated that the phospholipids listed above, when stuck to a plastic surface, can provide an efficient means to capture antiphospholipid antibodies from patients' sera. Generally the abilities of the different phospholipids to capture antibodies are quite similar, and although historically cardiolipin has been the antigen of choice, phosphatidyl serine, being more widely distributed within the body, might be a more sensible choice.

Given the low molecular weights of phospholipids compared to the macromolecular antigens typically used in antibody capture immunoassays, it is perhaps somewhat surprising that useful solid phases can be prepared by the simple expedient of drying down individual, or mixed, phospholipids from organic solvents into microtiter wells. The amphipathic nature of phospholipids may ensure that the epitope-containing head group regions of the molecules face away from the surface, either because the fatty acid chains are preferentially bound to the hydrophobic plastic surface or because they are induced by the addition of water to form a bilayer structure on the plastic surface.

Typically, 1–10 μg of phospholipids seem to form a suitable capture layer for the antibodies. Various prewashing and blocking procedures have been recommended, usually involving detergents and proteins or bovine sera [28,51–53]. Detection of bound antibodies has been accomplished immunochemically with antihuman immunoglobulin antibodies coupled to radiolabels [51] or enzyme labels such as alkaline phosphatase or horse radish peroxidase [28,52,53]. These steps are typical of laboratory based immunoassay systems. The sensitivity and general behavior of the assays [28] suggest that even faster and more convenient (e.g., single step) formats, such as the Unipath Clearview™ immunoassay technology, should be achievable, allowing such assays into the doctor's office or small clinic situation.

Table 3 illustrates the wide range of diseases where antiphospholipid antibodies have been found or suspected. As can be seen, cardiolipin has been implicated most frequently as a reactive antigen, but more recent work suggests that in many cases other phospholipids will also frequently react. Phosphatidyl choline, being a neutral phospholipid, does not react with anticardiolipin antibodies. However, such antibodies can be found in their own right. Much of the data in the table has been acquired only as a consequence of recently developed tests that are more specific and quantitative.

B. Phospholipids Used as Liposomes in Immunoassays

The most varied application of phospholipids in diagnosis has been derived from their use to construct liposomes that can be made to generate some form of detectable signal, as a consequence of antibody-antigen binding. There are several diverse mechanisms by which this can be achieved, as shown in Table 4. There is enormous scope for development of these technologies, and many of the advances in liposome technology made over recent years in pursuit of effective drug targeting (as described elsewhere in this handbook) are of immediate relevance to liposomal immunoassay. Improvements have already been made in the efficiency of marker encapsulation, there are better methods for antibody or antigen attachment, and there have been advances in the ability to formulate membranes with defined properties. In addition, liposome membranes are selectively permeable and can provide the capability for controlled, specific disruption or release of contents. All of these combine to make liposome technology an attractive prospect for those seeking to develop novel immunoassays.

1. Lytic Assays

First, and perhaps most obvious, of these liposome based technologies are the lytic assays, which rely upon the release of some kind of marker from inside a liposome to generate a signal. They are arranged so that the entrapped marker is cryptic (or suppressed) until the membrane is lysed and the contents are released into the surrounding medium, as shown in Fig. 1. The extent of binding between antigen and its antibody is proportional to the extent of liposome lysis, so providing the basis for quantitative assays. There are three basic variants of this approach; complement-mediated lysis, cytolysin-mediated lysis, and divalent cation-mediated lysis (see Table 5).

(a) Complement mediated lysis. Assays that use complement-mediated lysis depend on the ability of complement to lyse liposome membranes to which complement-fixing IgG or IgM antibodies become attached (Fig. 1a). The membrane must carry appropriate antigens, and the surrounding medium must contain antibodies and a source of active complement. The liposome contains a marker that can only be detected when it escapes, or when the external medium can enter. For example, the encapsulated marker may by an enzyme, in which case the substrate would be present in the medium, and contact between the two would be prevented until the membrane is lysed. It would be just as easy (though probably not as sensitive) to put the substrate in the liposomes and keep the enzyme on the outside. Alternatively, the liposomes may contain a flurorescer at a self-quenching

Table 4 Liposome Based Diagnostic Technologies

Type	Assay mode	Procedure	Table
Lytic assays	Complement lysis	Nonseparation	5
	Cytolysin lysis	Nonseparation	5
	Divalent cation lysis	Nonseparation	5
Solid phase assays	Labelled immunoliposome binding	Separation	7
	Immunoliposome destabilization	Nonseparation	7
Agglutination assays	Direct agglutination of immunoliposomes	Nonseparation	8
	Liposome enhanced agglutination procedure (LEAP)	Nonseparation	8
Cellular assays	Cell surface antigen staining	Separation	9
	Leukocyte function tests	N/A	9

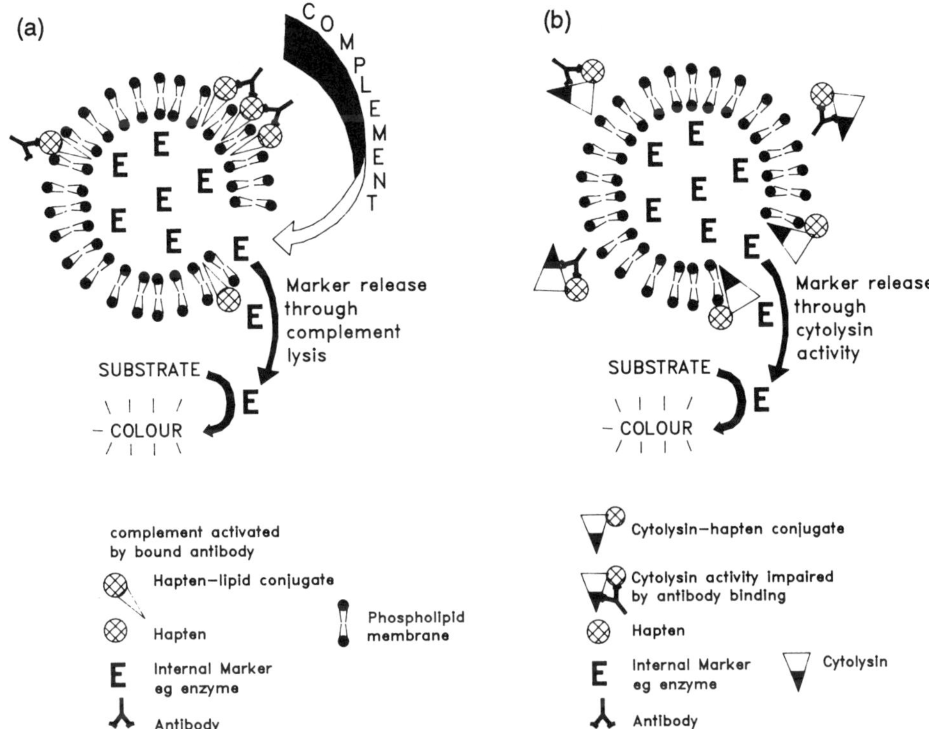

Figure 1 The basis of liposome immunoassay procedures involving lytic mechanisms. In (a), complement is activated by the binding of complement-fixing antibodies to antigens on the liposome surface, but for clarity the details of the complement activation sequence have been omitted. In reality, the activation of the complete sequence would result in many lytic events per liposome. In (b), the cytolysin-hapten conjugate carries the immunological specificity and can only lyse liposomes when it is not bound by antibodies.

concentration; membrane lysis would then allow the contents to be diluted into the surrounding medium, so restoring fluorescence.

Even when all of the necessary components are present, complement is only activated when the antibodies bind to membrane antigens. Thus such systems can be set up to assay *antigens* by competitive inhibition of antibody binding, *antibodies* by direct complement activation or by means of a complement-fixing second antibody, and *antigen* in a double antibody sandwich format, with a complement-fixing second antibody.

Table 6 shows a selection of such complement based assays that have used a variety of entrapped markers and detection systems and that have been applied to a wide range of analytes.

The very first liposome immunoassay of any kind to be described in the literature was for antibodies to lipid A, by Kataoka and colleagues [57]. This could be diagnostic of prior infection with *Salmonella* or related bacteria, but in reality it was more of a model or demonstration system. The entrapped marker was glucose, and its release was detected by means of a complex signal-generating system based on hexokinase and glucose-6 phosphate dehydrogenase. The signal was detected spectrophotometrically.

Table 5 Liposome Immunoassay Procedures Based on Lytic Mechanisms

Technique	Assay principle	Key requirements	Ref.
Complement-mediated lysis	General immunoassay. C-fixing Abs bound to surface ligands activate complement, leading to membrane lysis	Ab or Ag linked to surface membrane; active complement; cryptic internal label.	54
Cytolysin-mediated lysis	Immunoassay for low MW analytes. Lysin/hapten conjugate made nonlytic by binding to free Ab; liposomes used as indicator for free lysin/hapten conj.	Active conjugate of lysin and analyte (hapten); indicator liposomes with a cryptic internal label	55
Divalent cation-mediated lysis	Specialized assay for SLE diagnosis. Divalent cations cause lysis by inducing phase change, unless membrane stabilized by anti-DNA Abs	Liposomes incorporating cardiolipin, with Arsenazo III as internal label	56

C = complement, Ab = antibody, Ag = antigen.

Table 6 Complement Mediated Liposome Immunoassays: Some Examples to Illustrate the Wide Range of Variations Possible

Entrapped marker	Signal	Analyte	Ref.
Glucose	Spectrophotometric	Abs to lipid A	57
Tempocholine chloride	Electron spin resonance	Abs to lipid A	58
Carboxyfluorescein	Fluorescence	Abs to human IgG	59
Alkaline phosphatase	Colorimetry	Thyroxine	54
Horse radish peroxidase	Chemiluminescence	Theophylline	60
Horse radish peroxidase	Electrochemistry	Theophylline	61
Tetra pentyl ammonium ions	Electrochemistry	Abs to gangliosides	62

Ab = antibody.

The technology moved on from this elegant starting point to a much simpler detection mechanism [58]. Tempocholine gives a strong electron spin resonance signal, but at higher concentrations (e.g., 80 mM) the signal is strongly suppressed as a result of exchange broadening intermolecular interactions. Although this is a simple principle and involves no complex reactions, the instrumentation required is formidable and costly.

Carboxyfluorescein, a self-quenching fluorescent dye, has been used widely, but just one example is given in the table. Alkaline phosphatase and peroxidase are two examples of encapsulated enzymes, used as described earlier. The released enzyme activity can be determined in several different ways, including colorimetry. It is worth pointing out that some very interesting work has been done with electrochemical detection of peroxidase activity, with obvious implications for biosensors. A particular advantage of this mode of assay is that electroactive species themselves can be encapsulated, and their release can be directly detected by sensitive ion-specific electrodes, or with differential pulse voltametry

[63]. Complement dependent systems can also be adapted to measure complement activity itself, using appropriately sensitized liposomes and antibodies as the indicator system [64].

Complement-mediated liposomal assays have many virtues that are constantly pursued by the immunodiagnostics industry. They use nonseparation, nonwash procedures, they are quantitative (giving precise numerical readouts), they are sensitive, and they do not need to use hazardous components. So appealing is the basic concept that it seems to have been enthusiastically rediscovered, in various forms, several times since its first description by Kinsky's group. Surprisingly, several companies, including Technicon, Toshiba, and Hitachi, have tried to patent various aspects or embodiments of the technology.

But there are disadvantages, too. Because of the nonseparation procedure, the marker substance comes into contact with the sample, which may contain materials that interfere with the signal (e.g., natural fluorescers, endogenous enzymes, reducing or oxidizing agents). More importantly, the problem of providing a consistent, stable and economical complement source remains one of the main drawbacks of the method and is probably the reason for it not yet being applied commercially.

(b) Cytolysin-mediated lysis. Figure 1b shows the principle of cytolysin-mediated liposome assays, which offer an alternative to those mediated by complement. They are limited, however, to the assay of small molecules or haptens. This technique was developed by Freytag and Litchfield [55]. Unsensitised marker-filled liposomes are used as a general signal-generating reagent, the immunological specificity being built into the cytolysin. A hapten/cytolysin conjugate must be constructed in which the hapten is the substance to be detected and the cytolysin is a molecule, such as mellitin, capable of lysing liposomal membranes, which can be covalently linked to the hapten in question. This conjugate must retain both of the main properties of its two components: it must have membranolytic properties and it must be bindable by an antihapten antibody. The crucial point is that when this conjugate is bound by the antibody, its lytic activity must be impaired.

Thus when a negative control is set up with antibody, cytolysin/hapten conjugate, and liposomes, no lysis takes place, because all of the lytic agent is sequestered by the antibody. When free hapten is introduced into an otherwise identical system, there will be some lysis, because free hapten will compete for the antibody, leaving some of the lytic conjugate free and active.

Obviously, the same considerations for the markers and signal-generating mechanisms apply here as to the complement-mediated assays. Recently, a new variant has been described, in which the encapsulated marker is flavine adenine dinucleotide (FAD) [65]. Apo-glucose oxidase (apo-GOx) is provided in the external medium, together with a microperoxidase/isoluminol chemiluminescent system. Apo-GOx, an inactive version of GOx, can only be recativated if FAD escapes from the liposomes. Thus when lysis results from the action of free conjugate, apo-GOx is reactivated and luminescence occurs, the amount of emitted light being proportional to the extent of lysis. The lower limit of detection for digoxin is claimed to be 0.39 pg/assay tube, some 30–300-fold more sensitive than the alternative tested by the authors. This technique has been named the apoenzyme reactivation immunoassay version of the liposome immunoassay system (ARIS-LIS)!

The advantages with cytolysin-mediated assays are that there is no requirement for complement and no need to conjugate antibodies or antigens to liposomes. As with complement-based systems, it is a nonseparation, nonwash system, making it simple,

Table 7 Liposome Immunoassay Procedures Based on Solid Phase Binding

Technique	Assay principle	Key requirements	Ref.
Labelled immunoliposome binding assay	General immunoassay. Liposomes bind to solid-phase Ab or Ag: extent of binding relates to analyte conc.	Immunoliposomes with Ab or Ag and a detectable internal or membrane label: sensitized solid phase	67
Solid phase binding destabilisation assay	General immunoassay. Special immunoliposomes destabilized by binding to solid phase Ab or Ag: label released on binding	Dioleoyl PE liposomes stabilized by acylated Ab or Ag: cryptic internal label: sensitized surface	68

Ab = antibody, Ag = antigen.

Table 8 Liposome Immunoassay Procedures Based on Agglutination

Technique	Assay principle	Key requirements	Ref.
Liposome agglutination	Immunoassay for agglutinating Ab direct aggultination of Ag-sensitized immunoliposomes	Immunoliposomes (Ag) of large size, with color label	70
Liposome enhanced agglutination procedure	General immunoassay, but especially for serology. Liposomes bearing Ab (or other matched binding partner) complex with other agglutinating indicator particles	Immunoliposomes with antiglob. Ab: agglutinating system to enhance (e.g., latex or RBC)	71

Ab = antibody, Ag = antigen, RBC = red blood cell.

quick, and readily automated. However, it can suffer from serum effects in that serum proteins can inhibit the lytic action of the conjugate either by a direct effect on the lysin or by a stabilization of the liposome membrane through protein adsorption. With exceptionally high sensitivity the problem can be diluted out [65], but dilutions are not always acceptable or convenient. Because of the way it works, the technique will probably always be limited to the analysis of small molecules. It remains to be seen whether this principle will ever be embodied in simple commercial kits.

(c) Divalent cation mediated lysis. Janoff et al. described a neat but very specialized technique for the assay of antibodies that bind cardiolipin, diagnostic of systemic lupus erythematosus in man [56]. Figure 2 explains the principle step by step. The first point to note is that cardiolipin alone will not form liposomes because it reverts to hexagonal phase rather than lamellar phase. Mixed at the rate of 40 mol % with liposome-forming lipids, it will however participate in forming stable vesicles. If these vesicles also contain the dye Arsenazo III, they are conveniently red in color. If magnesium ions are added, the cardiolipin is induced to rearrange and the membrane is disrupted. At the same time, the magnesium ions interact with the Arsenazo III, which turns blue. Normal human serum has no effect on this process, but if serum from an SLE patient is added, the anti-DNA

Table 9 Liposome-Based Procedures for Cell Analysis

Cell property or activity	Assay principle	Ref.
Cell surface markers in flow cytometry (FCM)	Method for immunofluorescent labelling of cells. Liposomes with Ab and fluorescent label bind to cell Ag for analysis by FCM	73,74
Cytolytic or cytotoxic factors and activities	Release of cryptic marker (e.g., spin label or quenched fluor) from target liposome	76–80
Neutrophil and macrophage phagocytosis (resp. burst, and activation)	Changes to spin label lipid hapten in membrane of target liposome on phagocytosis	81–85
Target "killing" by reactive oxygen species from neutrophils	Change in fluorescence of 3,4-dihdroxy mandelic acid in target liposomes in contact with active neutrophil	86

Ab = antibody, Ag = antigen.

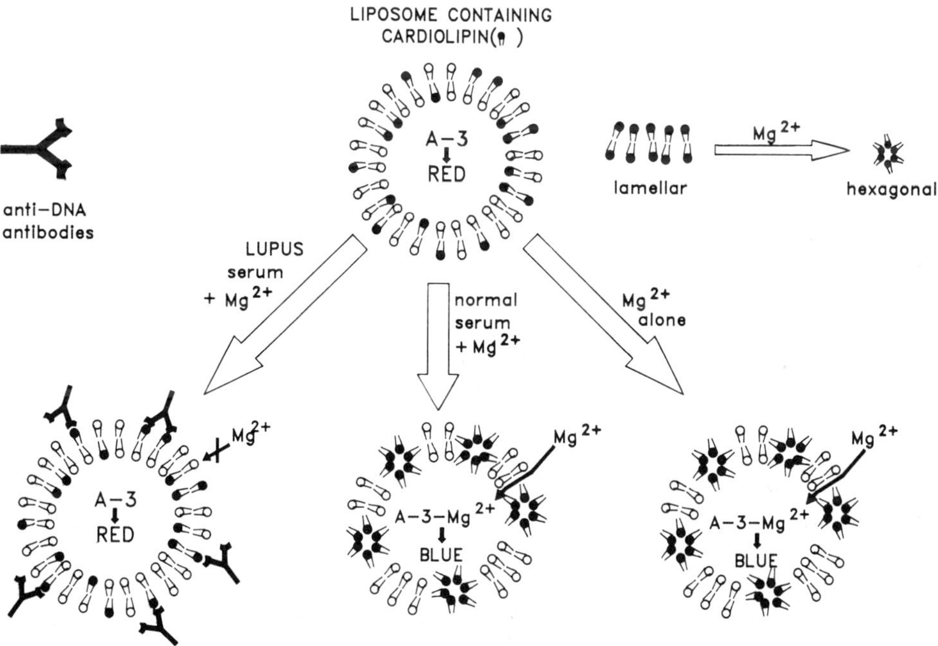

Figure 2 The basis of divalent cation-mediated liposome lysis assays. Magnesium ions fulfil the dual role of disrupting the cardiolipin-containing liposomes and changing the color of the entrapped dye. In the presence of antibodies that bind cardiolipin (e.g., anti-DNA), the liposomes are stabilized and are not disrupted by magnesium ions. (Redrawn from Ref. 56.)

antibodies bind to the cardiolipin and prevent the magnesium ions from inducing the cardiolipin rearrangements. The observed result is that the vesicles stay red.

2. Solid Phase Liposome Binding Assays

Perhaps the simplest form of liposome immunoassay is the solid phase binding assay (Table 7), in which marker-loaded liposomes serve the same function as the tracer label

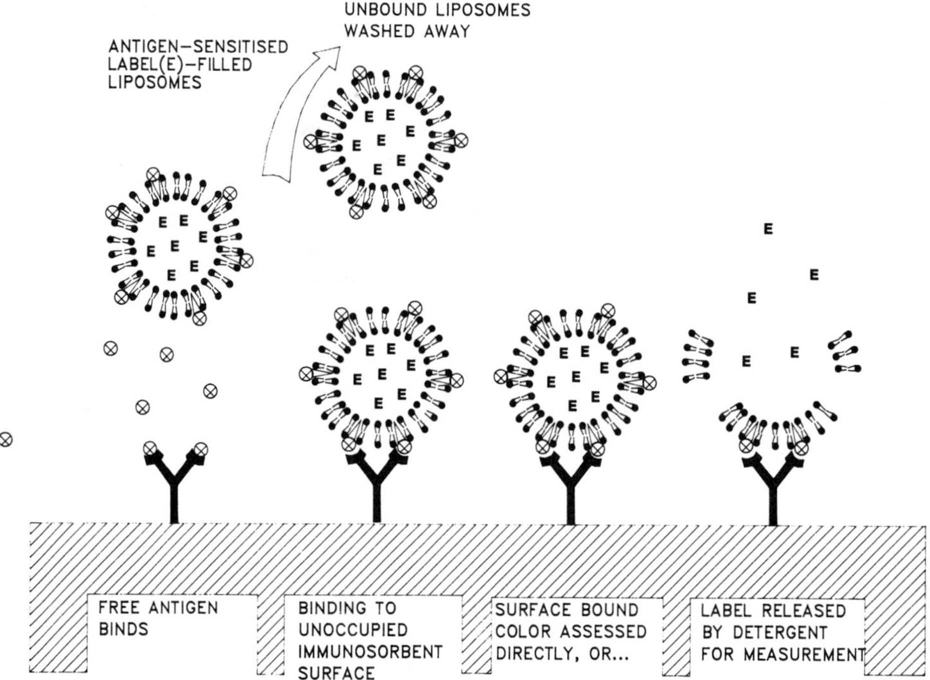

Figure 3 The basis of labelled immunoliposome binding assays. Free antigen binds to solid phase antibodies, and antigen-sensitized liposomes bind to the extent that there are unoccupied antibody binding sites available. Double antibody (two-site) formats are also easily set up.

would have done in radioimmunoassay, immunoradiometric assay, enzyme linked immunosorbent assay, etc. Figure 3 depicts a competitive inhibition liposome immunoassay on an antibody-sensitized solid phase. The label could be an enzyme, a dye, an isotope, etc. It is not absolutely necessary to lyse the bound vesicles, for if enough dye can be bound, the result can be observed in situ. Note that the liposomes could be made to carry a second antibody instead of an antigen, so giving a two site double antibody sandwich assay. This approach was pioneered by Unilever and Becton-Dickinson [66,67].

Becton Dickinson have brought the technology to the market with two products, one qualitative system for use away from the laboratory, and the other a quantitative system for use in the doctor's office or small clinic laboratory. The qualitative version is a self-contained kit based on colored liposomes and a nitrocellulose membrane solid phase, presented in a neat cassette for simple manipulation and reading by eye. It is sold under the name of Directigen™ (BBL Microbiology Systems, Cockeysville, MD 21030, USA). The quantitative version is part of an integrated, instrument-read system for thyroid profiling, therapeutic drug monitoring, etc. and is marketed under the name of IQ™ Immunochemistry System.

Huang and colleagues brought a new sophistication to solid phase liposome binding assays with a technique that in some respects is related to the destabilization assay principle of Janoff for anti-DNA antibodies [68]. Liposomes are made from dioleoyl PE, a lipid that will only make liposomes if it is stabilized by significant levels of other amphiphiles. Antibody molecules bearing acyl chains will insert into the membrane and, at the same time, provide the stabilizing effect. When these antibodies bind to an antigen-bearing solid phase, such as an infected cell [69], they congregate in the area of contact to form a zone of multiple binding. This perturbs the lipid distribution and causes a dramatic destabilization, with leakage of contents. In the original work calcein was the entrapped fluorescer, which was self-quenching until it escaped into the medium.

The technique works with normal solid phases such as polystyrene tubes, etc. and with acylated haptens as the stabilizing amphiphile. It should be noted that this elegant system requires no washing or separation steps and can be used with a wide variety of entrapped labels.

Destabilization assays rely on the lateral mobility of membrane components, and it is this same feature that is exploited in agglutination systems.

3. Agglutination Assays

When antibody molecules are bound by immunoliposomes, aggregates are formed, which can be seen as macroscopic agglutination (Table 8). Liposomes alone can be used in this way, especially if they are large and colored [70], but the best way to use agglutinating liposomes seems to be in enhancing the agglutination of other particles. For example, polystyrene latex agglutination systems are said to be enhanced (in terms of visibility of agglutination and sensitivity to analyte concentration) by the inclusion of antibody-sensitized liposomes that also participate in the immunochemical binding reactions [71]. This has been marketed by Cooper as the LEAP technology (*l*iposome *e*nhanced *a*gglutination *p*rocedure), but opinions are divided as to whether there are real benefits [72].

4. Fluorescent Labels in Flow Cytometry

The most recent application for fluorescent, antibody-sensitized liposomes is that of specific cell labelling in flow cytometry (Table 9). In this technique, cells are caused to flow in a mono-file through a beam of light (from a laser or incandescent source) in such a way that scattered light and fluorescent emmissions are detected with great sensitivity. Up to thousands of cells per second can be analyzed in this way. Cells with particular surface antigens can be analyzed by treating them with fluorescent labelled antibodies and detecting the appropriate emmission wavelength as they pass the detector. Many surface antigens of interest are present at very low density, and these are difficult to analyze with the usual techniques.

However, it has now been found that small unilamellar liposomes filled with carboxyfluorescein and sensitized with appropriate antibodies are especially effective at giving strong emission signals from surface antigens present at low density [73,74]. Such liposomes can give up to a nine-fold increase over the fluorescence derived from the same antibodies directly labelled with fluorescein, without any increase in background staining. Presumably this is a result of efficient binding and a high payload of fluorescent label for each antibody. Obviously, it is important to ensure that the fluorescer concentration is below a self-quenching level. A further advantage of the technique is that a wide variety of fluorescent labels can be used, even those that could not have been linked directly to antibodies by the conventional approaches [74]. This is an interesting area for further developments, which are eagerly awaited.

Many of the applications described above require the preparation of liposomes with antibodies or antigens attached to their outer surface, and an appropriate label entrapped internally or fixed to the membrane. There are many published methods for achieving this, and it is difficult to know which to choose when confronted with the huge variety available. A good survey is given in Chap. 8. Experience in the authors' organization has resulted in a well-defined procedure based on the method of Martin and Heath [75]. The procedure given in Sec. III below is recommended as a reliable means to produce protein-linked, fluorescent labelled liposomes for use in immunoassay systems.

5. Leukocyte Activity Analysis

Finally, there is the area of leukocyte function tests—a developing area of diagnosis and patient monitoring (Table 9). Most analyses of leukocytes are concerned with cell numbers, and proportions of cells with particular surface antigens, but it is often important to measure certain aspects of cell function as well.

T lymphocyte cytolysis of membrane vesicles was studied by Hollander et al. [76] using spin labels to monitor the process. The lytic effects of antibody-mediated cytotoxic cells (ADCC) have been investigated with the aid of marker-filled liposomes as targets. Liposomes immobilized on plastic surfaces, and bearing DNP complexed with appropriate antibodies, could be lysed by human ADCC in an ill-defined process dependent on the presence of free divalent cations [77]. Lysis was indicated by the release of $^{86}Rb^{+}$, which had been encapsulated in the liposomal aqueous compartments, and by the release of tritiated lipid. Other lymphocyte populations could also lyse these target liposomes by a different mechanism, not requiring divalent cations.

DNP-sensitized liposomes were also used in a study of ADCC from mice [78]. These liposomes were filled with a mixture of fluorophore, 1-amino-naphthalene-3,6,8-trisulphonic acid (ANTS) and a quencher, bispyridinium-*p*-xylene (DPX). Whilst entrapped in the liposomes the mixture was quenched but fluoresced brightly when it was released (and thereby diluted) into the external medium. Non-B, non-T lymphocyte fractions from mouse spleen were found to lyse these liposomes (antibody-coated) by a process that involved an initial binding and subsequent lysis, taking up to 8 h. Lewis and McConnell [79] also used liposome lysis, monitored by self-quenching fluorescence, to study antibody dependent cellular immune attack.

Roozemond et al. [80] examined the lytic activity of two different but related cytotoxins, natural killer cytotoxic factor (NKCF) and tumor necrosis factor (TNF). Unilamellar CF-containing liposomes were lysed by NKCF but not by TNF. Depending on the lipid composition, this lysis could be suppressed by the inclusion of cholesterol in the membrane.

Various types of liposomes have been used to investigate aspects of macrophage phagocytic activity, including the respiratory (oxidative) burst [81], the kinetics of antibody dependent binding [82], cell surface receptor and membrane protein changes [83], and surface morphology [84]. The phagocytic activity of neutrophils also has been studied with liposomes. Hafeman et al. [81,85] used liposomes containing spin labelled lipid haptens to examine the process of specific antibody dependent activation of neutrophils, and the triggering of the respiratory burst.

Petty and Francis [86] described a novel liposome based fluorescence method that allowed the direct observation of living cells depositing H_2O_2 and hydroxyl radicals into liposome targets. The authors defined this as target "killing", i.e., "the delivery of reactive oxygen metabolities to the internal volume of a target." The liposomes, made from egg yolk PC (98–99 mol %) and a DNP-modified PE (1–2 mol %), carried 3,4-dihydroxy mandelic acid (DHMA) in their internal aqueous compartments. DHMA is membrane-impermeable and undergoes a free radical dimerization, initiated by hyroxyl radicals, to give a fluorescent compound with spectral properties close to those of fluorescein (though with a low quantum yield). Thus when exposed to reactive oxygen metabolites, these liposomes became fluorescent, so providing a unique tool for analyzing the interaction between effector cells and their targets.

Thus there is an enormous potential for phospholipids to be used as tools in diagnosis. The properties that suit them for their fundamental role in cells also make them ideal for adsorption on solid phases for use in immunoassays, and their ability to form membranes has opened up a wide range of special techniques for immunoassays based on liposomes. Progress in techniques for scale-up, processing, storage, etc. of liposomes will facilitate their wider application and future commercialization. Advances in the basic understanding of phospholipid membranes and the development of "membrane engineering" are sure to provoke yet more creativity and innovation in the use of liposomes in diagnosis.

III. PRODUCTION OF ANTIBODY-LABELLED FLUORESCENT LIPOSOMES

A. Principle of Operation

Antibodies are modified by the insertion of thiol groups, whilst thiol-reactive phospholipids are prepared for inclusion in the liposomes. Thiol-modified liposomes and antibodies are brought together under mildly oxidizing conditions to bring about conjugation. The conjugated products are purified by gel filtration.

B. Literature

F. J. Martin and T. D. Heath, The covalent attachment of proteins to liposomes, in *Liposomes: A Practical Approach* (R. R. New, ed.), Oxford University Press, Oxford, 1990.

C. Materials

Purified antibody solution at about 6 mg/mL; L-alpha phosphatidylcholine (PC), Sigma, P6263; L-alpha phosphatidylethanolamine, distearoyl (PE), Sigma P3531; *N*-succinimidyl pyridyldithiopropionate (SPDP), Sigma, P9398; succinimidyl 4-(*p*-maleimidophenyl) butyrate (SMPB), Pierce, 22315; redistilled triethylamine b.p.89 C (TEA); methanol (dried over molecular sieve type 3A); chloroform (dried over molecular sieve type 3A); silicic acid, Bio-Rad Bio-Sil HA (equilibrated in a glass column with dry chloroform); silica gel 60 TLC plates, Merck 5724; ninhydrin spray reagent, Sigma N0507; acid molybdate spray kit for phosphates, Sigma A 4155; oxygen-free nitrogen, BOC, white spot; dithiothreitol (DTT), BDH 44149; disposable G25 Sephadex columns, Pharmacia, PD10; calcein, Sigma, C0875; phosphate/citrate buffer, pH 5.0 (20 mM citric acid/35 mM Na_2HPO_4/108 mM NaCl/2 mM EDTA, pH adjusted with NaOH soln.); phosphate/citrate buffer, pH 6.7, as above but with pH adjusted to 6.7 with NaOH soln.; acetate buffer, pH 4.5 (0.2 M NaOAc, adjusted to pH 4.5 with acetic acid); phosphate buffered saline (PBS), pH 7.5 (0.1 M Na_2HPO_4/NaH_2PO_4/NaCl). *Note*: all buffers filtered and degassed.

D. Procedure

1. Synthesis of MPB-PE Distearoyl

Dissolve the DSPE in 50 mL of dry chloroform and 25 mL of dry methanol; mix and remove 200 μL for TLC analysis. Add TEA and SMPB, then reflux under nitrogen with stirring for 2 h removing 200 μL samples at 30 min intervals for TLC analysis. Two plates are run in chloroform: methanol: water in the ratio of 65:25:4 and developed with ninhydrin and acid molybdate. Dry down the reaction mixture on a rotary evaporator and resuspend in 10 mL of dry chloroform. Apply the product to the column, followed sequentially by 20 mL aliqots of dry chloroform: methanol in the ratios 20:0, 19:1, 18:2, 17:3, 16:4; collect 5 mL fractions throughout.

Remove 200 μL samples from the fractions and analyze by TLC as above, then pool fractions that are phosphate positive (Rf > 0.52) and ninhydrin negative. Dry down product on a rotary evaporator and resuspend in 10 mL of dry chloroform. Quantitative analysis for amine and phosphorus at this point enables determination of lipid recovery and product purity. Aliquot the product into ampules, seal under argon, and store at –20°C.

2. Synthesis of Antibody-PDP (Ab-PDP)

Desalt the antibody into PBS to give a final concentration of 6 mg/mL. Dissolve SPDP in ethanol at 20 μM/mL (6.24 mg/mL). It is best to use a fresh bottle of SPDP, which has been stored well dessicated. Add SPDP solution dropwise to the antibody solution (stirred), to give a 15:1 molar ratio of SPDP to antibody. This mixture is allowed to react for half an hour at room temperature. Purification of derivatized antibody from reactants is achieved by desalting on a PD10 column equilibrated in acetate buffer, pH 4.5. If this material is to be stored, desalt into PBS preserved with sodium azide (0.05%).

3. Liposome Preparation

Put together in a 25 mL Quickfit tube 32 mg of PC and 4 mg of MPB-PE distearoyl in chloroform. Remove the solvent on the rotary evaporator, forming a film of mixed lipid on the internal surface of the tube. Leave under vacuum for approximately 1 h.

To this lipid film add 5 mL of calcein solution (0.1–1 mM) in phosphate/citrate buffer, pH 5.0. Swirl and gently agitate the tube to resuspend the lipids. Split the suspension into five tubes, seal with "suba seals," and purge with nitrogen. Sonicate the tubes in an ultrasonic water bath, keeping the temperature below 35°C with crushed ice. Sonicate for 40 min or until a clear suspension is obtained in the five tubes. Equilibrate two PD10 columns in phosphate/citrate buffer, pH 6.7, and apply 2.5 mL of suspension to each column. Elute 2 mL of liposome suspension from each column, with the equilibration buffer.

4. Antibody-Liposome Conjugation

Desalt the derivatized antibody into acetate buffer, pH 4.5, if using material stored in PBS. Prepare a 2.5 M (385 mg/mL) solution of DTT in acetate buffer, pH 4.5, and add 10μl to each mL of Ab-PDP (at approximately 5 mg/mL). Leave to react for approximately 1 h and then remove DTT by desalting on a PD10 column equilibrated in phosphate/citrate buffer, pH 6.5. Time procedures 3 and 4 to coincide, so that the liposomes and sulphhydryl-activated antibodies are eluted from the PD10 columns together, and can be mixed in the appropriate proportions (in this case, 3 mL plus 3 mL) for conjugation without delay. Purge this conjugation mixture with nitrogen and leave at room temperature overnight.

5. Purification/Characterization

Purification is ideally achieved by size exclusion chromatography, and optimal separation is achieved on a Sepharose CL-4B column (16 × 400 mm). Initial recoveries are low from a new column, but once the gel is conditioned with lipid, recoveries are acceptable.

Protein incorporation is carried out by means of the Lowry technique, and immunoreactivity is determined by simple assays involving binding to immobilized antigen on nitrocellulose membranes, for example.

E. Recommendations

In step 1 it is essential that all glassware and reagents be dry.

If TLC reveals that the reaction in step 1 has failed to go to completion, then add an extra 30% TEA.

The time required for conjugation in step 4 may be as low as 5 or 6 h, but it is preferable to allow longer, such as overnight.

Other fluorescent labels can be used, such as carboxy fluorescein, Texas red, or sulphorhodamine. With any fluorescent dye, it is essential that optimal concentrations be determined by careful experimentation with different concentrations, because of the problem of self-quenching.

F. Caveats

Severe losses of liposomes can occur on PD10 (Sephadex) or Sepharose columns when they are new. This is because of adsorption onto the gel particles; but the losses are greatly reduced when several batches of liposomes have been passed through. Some workers advocate the use of several batches of crude "sacrificial" liposomes as a conditioning procedure before columns are used on real samples.

The maleimide residue of MBP-PE degrades very quickly in aqueous solution above pH 6.5, so the control of pH throughout the preparation and conjugation steps is very important.

REFERENCES

1. D. B. Wilbur, J. E. Pacign, S. A. Shelley, et al., Prenatal relationship of surfactant lipid and protein constituents in infants with respiratory distress syndrome: A preliminary communication, *Pediatr. Pulmonol. 6*:109 (1989).
2. D. Serrano de la Cruz, E. Santillann, A. Mingo, et al., Improved thin layer chromatographic determination of phospholipids in gastric aspirates from newborns, for assessment of lung maturity, *Clin. Chem. 34*:736–738 (1988).
3. P. C. Robinson, L. C. Walters, T. E. King, et al., Idiopathic pulmonary fibrosis. Abnormalities in bronchoalveolar lavage fluid phospholipids, *Am. Rev. Respir. Dis. 137*:585–591 (1988).
4. J. F. Magny and J. Franconal, Phosphatidylglycerols in tracheal aspirates for diagnosing hyaline membrane disease, *Arch. Dis. Child. 62*:640–641 (1987).
5. M. Hallman, T. A. Merritt, M. Pohjavuori, et al., Effect of surfactant substitution on lung effluent phospholipids in respiratory distress syndrome: Evaluation of surfactant phospholipid turnover, pool size, and the relationship to severity of respiratory failure, *Pediatr. Res. 20*:1228–1235 (1986).
6. J. C. Dohnal, L. J. Bowie, and H. J. Burstein, Degree of unsaturation of the fatty acid chains of phospholipids influences the fluorescence polarisation: Implications of evaluating foetal lung maturity, *Clin. Chem. 32*:425–428 (1986).
7. M. Hallman, P. Arjomaa, J. Tahvanainen, et al., Endobronchial surface active phospholipids in various pulmonary diseases, *Eur. J. Respir. Dis. Suppl. 142*:37–47 (1985).
8. S. H. Teng, A. G. Andrews, and I. Horacek, Rapid enzyme analysis of amniotic fluid phospholipids containing choline: A comparison with the lecithin to sphingomyelin ratio in prenatal assessment of foetal lung maturity, *J. Clin. Pathol. 38*:1304–1308 (1985).

9. R. P. Baughman, E. Stein, J. MacGee, et al., Changes in fatty acids in phospholipids of the bronchoalveolar fluid in bacterial pneumonia and in adult respiratory distress syndrome, *Clin. Chem. 30*:521–523 (1984).
10. J. C. Russell, C. M. Cooper, and C. H. Ketchum, et al., Multicentre evaluation of TDx foetal lung maturity, *Clin. Chem. 35*:1075 (1989).
11. J. C. Phillips and J. F. Chapman, Evaluation of the PG-Numeric test for semi-automated analysis of phosphatidylglycerol (PG) in amniotic fluid, *Clin. Chem. 35*:1072 (1989).
12. H. Crockett, J. A. Knight, E. R. Ashwood, et al., A comparison of tests used to assess foetal maturity, *Clin. Chem. 35*:1069 (1989).
13. J. Marniemi, A. Lehtonen, M. Inberg, et al., Fatty acid composition of serum lipids in patients with a coronary bypass operation, *J. Intern. Med. 225*:343–347 (1989).
14. T. Kobayashi, I. Goto, T. Yamanaka, et al., Infantile and foetal globoid cell leukodystrophy: Analysis of galactosylceramide and galactosylsphingosine, *Ann. Neurol. 24*:517–522 (1988).
15. E. Frich, Immunological studies of the pathogenesis of multiple sclerosis. Cell mediated cytotoxicity by peripheral blood lymphocytes against basic protein of myelin, encephalitogenic peptide, cerebrosides and gangliosides, *Acta Neurol. Scand. 79*:1–11 (1989).
16. L. Campanella, M. Tomassetti, G. DeAngelis, et al., A new assay for choline containing phospholipids in amniotic fluid by an enzyme sensor, *Clin. Chim. Acta 169*:175–182 (1987).
17. M. R. Parvathy, Y. Ben-Yoseph, D. A. Mitchell, et al., Detection of Krabbe disease using tritiated galactosylceramides with medium chain fatty acids, *J. Lab. Clin. Med. 110*:740–746 (1987).
18. S. Pacheco, K. Hillier, and C. Smith, Increased arachidonic acid levels in phospholipids of human colonic mucosa in inflammatory bowel disease, *Clin. Sci. 73*:361–364 (1987).
19. N. G. Bazan, B. L. Scott, T. S. Reddy, et al., Decreased content of docosahexanoate and arachidonate in plasma phospholipids in Usher's Syndrome, *Biochem. Biophys. Res. Commun. 141*:600–604 (1986).
20. M. L. Ancelin, and H. J. Vial, Choline kinase activity in plasmodium-infected erythrocytes: Characterisation and utilisation as a parasite-specific marker in malarial fractionation studies, *Biochim. Biophys. Acta 875*:52–58 (1986).
21. G. Skulladottir, T. Hardarson, N. Sigfusson, et al., Arachidonic acid levels in serum phospholipids of patients with angina pectoris or fatal myocardial infarction, *Acta Med. Scand. 218*:55–58 (1985).
22. R. J. Morin, The role of phospholipids in platelet function, *Ann. Clin. Lab. Sci. 10*:463–473 (1980).
23. M. Tomita-Yamaguchi, J. F. Babich, R. C. Baker, et al., Incorporation, distibution and turnover of arachidonic acid within membrane phospholipids of B220+ T cells from autoimmune-prone MRL-Lpr/Lpr mice, *J. Exp. Med. 171*:787–800 (1990).
24. A. Fein, R. Payne, D. Wesley-Corson, M. J. Berridge, and R. F. Irvine, Photoreceptor excitation and adaption by inositol 1,4,5-triphosphate, *Nature 311*:157–160 (1984).
25. J. E. Brown, L. J. Rubin, A. J. Ghalayini, A. P. Tarver, and R. F. Irvine, Myo-inositol polyphosphate may be a messenger for visual excitation in Limulus photoreceptors, *Nature 311*:160–163 (1984).
26. J. Tinoco, Dietary requirements and functions of *a*-linolenic acid in animals, *Prog. Lipid. Res. 21*:1–45 (1982).
27. T. G. Wheeler, R. M. Benolken, and R. E. Anderson, Visual membranes: Specificity of fatty acid precursors for the electrical response to illumination, *Science 188*:1312–1314 (1975).
28. A. E. Gharavi, E. N. Harris, R. A. Asherson, et al., Anticardiolipin antibodies: Isotype distribution and phospholipid specificity, *Ann. Rheum. Dis. 46*:1–6 (1987).
29. R. A. Asherson and E. N. Harris, Anticardiolipin antibodies in the detection of autoimmune disorders, *Intern. Med. 8*:73–88 (1987).

30. E. N. Harris, A. E. Gharavi, A. Tincani, et al., Affinity purified anticardiolipin and anti-DNA antibodies, *J. Clin. Lab. Immunol. 17*:155–162 (1985).
31. H. L. Staub, E. N. Harris, G. Savidge, et al., Antibody to phosphatidyl ethanolamine in a patient with lupus anticoagulant and thrombosis, *Ann. Rheum. Dis. 48*:166–169 (1989).
32. M. D. Lockshin and A. E. Gharavi, Anticardiolipin antibodies and the lupus anticoagulant. Letter, *Ann. Intern. Med. 107*:431–432 (1987).
33. C. G. Mackworth-Young, S. Loizou, and M. J. Walport, Primary antiphospholipid syndrome: Features of patients with raised anti-cardiolipin antibodies and other disorders, *Ann. Rheum. Dis. 48*:362–367 (1989).
34. M. D. Lockshin, T. Qamar, M. L. Druzin, et al., Antibody to cardiolipin, lupus anticoagulant and foetal death, *J. Rheumatol. 14*:259–262 (1987).
35. E. N. Harris, R. A. Asherson, A. E. Gharavi, et al., Thrombocytopenia in SLE and related autoimmune disorders: Association with anticardiolipin antibody, *Br. J. Haematol. 59*:227–230 (1985).
36. E. N. Harris, G. R. V. Hughes, and A. E. Gharavi, Antiphospholipid antibodies: An elderly statesman dons new garments, *J. Rheumatol. 14(suppl. 13)*:208–213 (1987).
37. R. R. Buchanan, J. R. Wardlaw, A. G. Riglar, et al., Antiphospholipid antibodies in the connective tissue diseases: Their relation to the antiphospholipid syndrome and forme fruste disease, *J. Rheumatol. 16*:757–761 (1989).
38. C. E. Weidemann, D. J. Wallace, J. B. Peter, et al., Studies of IgG, IgM and IgA antiphosphilipid antibody isotypes in systemic lupus erythematosus, *J. Rheumatol. 15*:74–79 (1988).
39. S. Carsons, New Laboratory parameters for the diagnosis of rheumatic disease, *Am. J. Med. 85*:34–38 (1988).
40. K. Mattila, O. Vaarala, T. Palosuo, et al., Serologic response against cardiolipin and enterobacterial common antigen in young patients with acute myocardial infarction, *Clin. Immunol. Immunopathol. 51*:414–418 (1989).
41. M. B. Santiago, W. Cossermelli, M. F. Tuma, et al., Anticardiolipin antibodies in patients with infectious diseases, *Clin. Rheumatol. 8*:23–28 (1989).
42. S. R. Levine and K. M. A. Welch, The spectrum of neurologic disease associated with antiphospholipid antibodies, lupus anticoagulants and anticardiolipin antibodies, *Arch. Neurol. 44*:876–883 (1987).
43. D. P. Briley, B. M. Coull, and S. H. Goodnight, Neurological disease associated with antiphospholipid antibodies, *Ann. Neurol. 25*:221–227 (1989).
44. A. Shuaib, L. Barklay, M. A. Lee, et al., Migraine and antiphospholipid antibodies, *Headache 29*:42–45 (1989).
45. R. D. Southeimer, The anticardiolipin syndrome. A new way to slice an old pie, or a new pie to slice? *Arch. Dermatol. 123*:590–595 (1987).
46. J. J. Grob and J. J. Bonerandi, Thrombotic disease as a marker of the anticardiolipin syndrome. Livedo vasculitis and distal gangrene associated with abnormal serum antiphospholipid activity, *J. Am. Acad. Dermatol. 20*:1063–1069 (1989).
47. S. Alusik, R. Jandova, M. Gebauerova, et al., Anticardiolipin syndrome in plastic surgery of the breast, *Cor. Vasa, 31*:139–144 (1989).
48. R. B. Lal and E. A. Ottesen, Antibodies to phosphocholine bearing antigens in lymphatic filariasis and changes following treatment with diethylcarbamazine, *Clin. Exp. Immunol. 75*:52–57 (1989).
49. E. N. Harris, A. E. Gharavi, and G. R. U. Hughes, Antiphospholipid antibodies, *Clinics in Rheumatic Disease 11*:591–609 (1985).
50. R. D. Catterall, Biological false positive reactions and systemic diseases, in *Ninth Symposium On Advanced Medicine* (G. Walter, ed.), Pitman Medical, London, 1973, pp. 97–111.
51. E. N. Harris, A. E. Gharavi, M. L. Boey, et al., Anticardiolipin antibodies: Detection by radioimmunoassay and association with thrombosis in systemic lupus erythematosus, *Lancet 1983*:1211–1214 (1983).

52. S. Loizou, J. D. McCren, A. C. Rudge, et al., Measurement of anticardiolipin antibodies by an enzyme linked immunosorbent assay (ELISA): Standardisation and quantitation of results, *Clin. Exp. Immunol. 62*:738–745 (1985).
53. D. A. Triplett, J. J. Brandt, K. A. Musgrave, et al., The relationship between lupus anticoagulants and antibodies to phospholipid, *JAMA 259*:550–554 (1988).
54. J. C. Braman, R. J. Broeze, D. W. Bowden, A. Myles, T. R. Fulton, M. Rising, J. Thurston, F. X. Cole, and G. F. Vovis, Enzyme membrane immunoassay (EMIA), *Bio/technology 2*:349–355 (1984).
55. J. W. Freytag and W. J. Litchfield, Liposome mediated immunoassays for small haptens (digoxin) independent of complement, *J. Immunol. Methods 70*:133–140 (1984).
56. A. S. Janoff, S. Carpenter-Green, A. L. Veiner, J. Seibold, G. Weissmann, and M. J. Ostro, Novel liposome composition for a rapid colormetric test for systemic lupus erythematosus, *Clin. Chem. 29*:1587–1592 (1983).
57. T. Kataoka, K. Inoue, C. Galanos, and S. E. Kinsky, Detection and specificity of lipid A antibodies using liposomes sensitized with lipid A and bacterial lipopolysaccharides, *Eur. J. Biochem. 24*:123–127 (1971).
58. K. H. Humphries and H. M. McConnell, Immune lysis of liposomes and erythrocyte ghosts loaded with spin label, *Proc. Natl. Acad. Sci. USA 71*:1691–1694 (1974).
59. Y. Ishimori, M. Notsuki, M. Koyama, and T. Yasuda, European Patent Application, 0 144 084 (1985).
60. J. P. Fox, E. Hedaya, and V. Lippmann, European Patent Application, 0 140 521 (1985).
61. M. Haga, H. Itagaki, S. Sugawara, and T. Okano, Liposome immunosensor for theophylline, *Biochem. Biophys. Res. Commun. 55*:187–192 (1980).
62. K. Shiba, T. Watanabe, Y. Umezawa, S. Fujiwara, and H. Momoi, Liposome immunoelectrode, *Chem. Lett. 2*:155–158 (1980).
63. R. M. Kannuck, J. M. Bellama, and R. A. Durst, Measurement of liposome-released ferrocyanide by a dual function polymer modified electrode, *Anal. Chem. 60*:142–147 (1988).
64. T. Masaki, N. Okada, R. Yasuda, and H. Okada, Assay of complement activity in human serum using large unilamellar liposomes, *J. Immunol. Methods. 123*:19–24 (1989).
65. M. Haga, S. Hoshino, H. Okada, N. Hazemoto, Y. Kato, and Y. Suzuki, An improved chemiluminescence based liposome immunoassay involving apoenzyme, *Chem. Bull. 38*:252 (1990).
66. P. J. Davis, European Patent, 0 014 530 (1981).
67. J. P. O'Connell, V. Piran, and D. B. Wagner, A highly sensitive immunoassay system involving antibody coated tubes and liposome-entrapped dye, *Clin. Chem. 31*:1424–1426 (1985).
68. R. J. Y. Ho and L. Huang, Interactions of antigen-sensitised liposomes with immobilised antibody: A homogeneous solid-phase immunoliposome assay, *J. Immunol. 134*:4035–4040 (1985).
69. R. J. Y. Ho, B. T. Rouse, and L. Huang, Destabilisation of target sensitive immunoliposomes by antigen binding—a rapid assay for virus, *Biochem. Biophys. Res. Commun. 138*:931–937 (1986).
70. V. T. Kung, F. J. Martin, and Y. P. Vollmer, International Patent Application, WO 85/00664 (1985).
71. F. J. Martin, and V. T. Kung, Binding characteristics of antibody-bearing liposomes, *Ann. NY Acad. Sci. 446*:443–456 (1985).
72. L. A. Heath-Fracia and E. G. Estevez, Evaluation of a new latex agglutination test for detection of streptococcal antibodies, *Diagn. Micrbiol. Infect. Dis. 8*:25–30 (1987).
73. A. G. Gray, J. Morgan, D. C. Linch, and E. R. Huehns, Enhanced fluorescence in indirect immunophenotyping by the use of fluorescent liposomes, *J. Immunol. Methods 121*:1–7 (1989).

74. A. Trunch, P. Machy, and P. K. Horan, Antibody-bearing liposomes as multicolor immunofluorescence markers for flow cytometry and imaging, *J. Immunol Methods. 100*:59–71 (1987).
75. F. J. Martin and T. D. Heath, The covalent attachment of proteins to liposomes, in *Liposomes, A Practical Approach* (R. R. New, ed.), Oxford University Press, Oxford, 1990.
76. N. Hollander, S. Q. Mehdi, I. L. Weissman, H. M. McConnell, and J. P. Kriss, Allogenic cytolysis of reconstituted membrane vesicles, *Proc. Natl. Acad. Sci. USA 76*:4024–4025 (1979).
77. K. Ozato, H. K. Ziegler, and C. S. Henney, Liposomes as a model membrane system for immune attack, *J. Immunol. 121*:1383–1388 (1978).
78. B. Geiger and A. D. Schreiber, The use of antibody-coated liposomes as a target cell model for antibody-dependent cell-mediated cytotoxicity, *Clin. Exp. Immunol. 30*:149–154 (1979).
79. J. T. Lewis, and H. M. McConnell, Model lipid bilayer membranes as targets for antibody-dependent, cellular, and complement-mediated immune attack, *Ann. NY Acad. Sci. 308*:124–138 (1978).
80. R. C. Roozemond, D. C. Urli, and B. Bonauidia, Liposomes as targets for NKCF but not for TNF, *Immunobiology 175*:71 (1987).
81. D. G. Hafeman, J. T. Lewis, and H. M. McConnell, Triggering of the macrophage and neutrophil respiratory burst by antibody bound to a spin-label phospholipid hapten in model lipid bilayer membranes, *Biochemistry 19*:5387–5394 (1980).
82. J. T. Lewis, D. G. Hafeman, and H. M. McConnell, Kinetics of antibody-dependent binding of haptenated phospholipid vesicles to a macrophage-related cell line, *Biochemistry 19*:5376–5386 (1980).
83. H. R. Petty, D. G. Hafeman, and H. M. McConnell, Specific antibody-dependent phagocytosis of lipid vesicles by R.A.W.264 macrophages results in the loss of cell surface Fc but not C3b receptor activity, *J. Immunol. 125*:2391–2396 (1980).
84. H. R. Petty and H. M. McConnell, Cytochemical study of liposome and lipid vesicle phagocytosis, *Biochim. Biophys. Acta 735*:77–85 (1983).
85. D. G. Hafeman, J. W. Parce, and H. M. McConnell, Specific antibody-dependent activation of neutrophils by liposomes containing spin-label lipid haptens, *Biochem. Biophys. Res. Commun. 86*:522–528 (1979).
86. H. R. Petty and W. Francis, Novel fluorescence method to visualize antibody-dependent hydrogen peroxide-associated "killing" of liposomes by phagocytes, *Biophys. J. 47*:837–840 (1985).

26

Biological and Biotechnological Applications of Phospholipids

Roger R. C. New* *Cortecs Ltd., London, England*

The very structure of phospholipids determines that one is dealing with a class of molecules with a wide range of contrasting biological roles. Often zwitterionic, with both exceedingly hydrophilic and hydrophobic regions in the same molecule, and a range of different structural options for each of these, phospholipids are a family of compounds that can be used as building bricks to produce entities that display a complexity and diversity on the same scale as proteins composed of amino acids, or complex polysaccharides from single sugars.

The applications to which man has put phospholipids over the years are indeed extremely wide-ranging and make use of physical, chemical, and biological aspects of the behavior of these molecules. In the broad and ill-defined field of biotechnology, phospholipids can be seen to be employed on three distinctly different levels (see Table 1).

On the first level, phospholipids have application by virtue of the intrinsic properties of the individual molecules themselves. They are nutritious, biodegradable, and biocompatible, a source of fatty acids, organic phosphate, and choline. It has been shown that dietary phospholipids are essential for efficient absorption of lipids from the gut lumen and contribute to the phospholipid employed for formation of chylomicrons that export this lipid into the blood stream. The intact molecule has been reported to modulate the activity of lipases in vivo and may have a role in regulating cholesterol levels in blood and tissues. Beneficial effects of dietary supplements on CNS activity have been reported, particularly in organic psychiatric disorders and on memory function in ageing, which are thought to be a result of extra choline being available for incorporation into acetylcholine in the brain.

The second level of complexity displayed by phospholipids is their ability to form supramolecular structures that self-assemble. Phospholipids associate with each other in aqueous media to form micelles or lamellar structures and can interact with lipids and other amphipaths to give mixtures of discrete, immiscible phases. Phospholipids are excellent surfactants and emulsifying agents. Such properties have been exploited in the paint, dye, and food industries for many years (lecithin is a common ingredient of many processed foods under the code name E 322), and all manner of creams and lotions have

**Former affiliation*: Biocompatibles Ltd., Middlesex, England

Table 1 Levels of Complexity in Utilization of Phospholipids in Biotechnology

1. Individual molecule
Nutritional value
2. Closed vesicles
Surfactant properties
Monolayer surface interaction
Compartmentalization
Selective permeability
Targeting
3. Intelligent membranes
Response to external influences
Reversible changes in properties

used natural phospholipids in the cosmetic industry to combine oil and water phases in a stable formulation. The same principle applies in the human body, where an air-water interface in the lung is stabilized by phospholipids, and where phospholipids can be administered in order to correct defects arising from insufficiency of natural lung surfactant.

The sharply defined interface between discontinuous, immiscible phases, stabilized by phospholipids, is a barrier to penetration of molecules from one phase into the other; it can also act as a basis for surface phenomena to take place and can influence the course of chemical interactions by defining an orientation in space and by increasing local concentrations of immobilized reactants. As an example of the participation of phospholipids in surface phenomena may be cited the use of a preparation of PE, PS, and PC (known as "platelet substitute") in stimulating coagulation in the APTT test for clinical detection of clotting deficiencies. Immunological agglutination reactions of antigen-coated liposomes, resonance energy transfer between membrane-bound fluorophores, and modulation of cell-membrane lipids by intermembrane transfer from liposomes are further illustrations of phenomena manifesting as a result of events occurring at a phospholipid interface.

An even more remarkable corollary of the self-association behavior of phospholipids is the observation that they can form bilayer membranes acting as physical barriers that separate compartments of the same aqueous phase from each other geographically. They can form closed, sealed membrane vesicles (liposomes) that are able, in the same aqueous suspension, to maintain different microenvironments (e.g., of pH, solute concentration) for considerable periods of time under conditions that allow considerable scope for variation of properties such as permeability, release rate, and interaction with proteins, cells, and organs by judicious choice of lipid and method of manufacture. Thus observation, coupled with the biocompatible nature of phospholipids alluded to above, has led to an enormous number of potential applications of liposomes in the field of drug targeting, where the distribution and fate of biologically active molecules in vivo can be altered by entrapment within liposomes of appropriate composition.

Phospholipids can self-associate spontaneously to form structures with properties that are predictable. The third level of complexity at which phospholipids may be employed is when the assembly process itself is manipulated, and mechanisms introduced whereby assembly and disassembly of the structures formed may be controlled at will. Liposomes, indeed, are metastable entities, with natural propensities to become more or less permeable, to aggregate, fuse, or disintegrate under appropriate conditions. Depending on the constituent phospholipids, membranes can be constructed that permeabilize in response to heat or light, which fuse upon change of pH, or which disintegrate upon contact with specific ligands.

At the present time, the majority of applications in biotechnology exploit the properties of phospholipids in the first two categories only, although as time progresses, design and use of "intelligent membranes" in such fields as drug targeting, diagnostic testing, or in vitro and in vivo gene delivery will become increasingly more routine. Such applications have been touched upon elsewhere in this book. In the current chapter, emphasis will be placed on use of phospholipids outside the therapeutic arena, and illustrations will be taken from coatings technology, molecular biology, food processing, and biomimetics, in all of which phospholipids are employed to modulate the interaction between manufactured materials and the external environment.

I. INDUSTRIAL APPLICATIONS

A. Paints

Phospholipids have wide application in areas where stabilization of particle dispersions is required. Lecithins were first recognized in the 1930s as valuable components of oil-based paints and more recently are being modified for use in water-based paints, inks, and dyes. They are a cheap and readily-available surfactant and may be incorporated in paints at a level of anywhere between 0.1% and 5% of the total pigment weight, either on their own or in conjunction with other surfactants and stabilizers. The lecithins used are usually crude soya extracts containing a considerable proportion of neutral oils in addition to phospholipids. For applications in water-based paints, the oils may be removed and the phospholipids rendered more water-dispersible by chemical means such as hydroxylation, hydrolysis, or treatment with metal acylates.

A great variety of different benefits have been claimed for use of lecithins in the manufacture of paints and other coating materials. It is added preferably during preparation of the milling paste, where it acts as a lubricant between the solid-solid particle interfaces and reduces the pigment grinding time. All the particles are coated with a layer of phospholipid, which acts to stabilize them upon addition to the solvents, reducing the mixing time and preventing aggregation, settling out, etc. The particles may be smaller, more homogenous, and finely dispersed, resulting in better hiding power and a more stable coating. Furthermore, after storage, the pigments may be redispersed more easily. In paints containing mixtures of pigments of different size and density, differential settling is reduced, so the correct shade of paint is retained even over a long time period. The presence of lecithin also aids in the physical application of paint onto a surface, reducing the effort involved in brushing, and preventing flooding, sagging, or formation of striations in the coating. Because of the content of tocopherols in natural lecithin, together with diacyl phosphatides, which are thought to scavenge radicals, anticorrosive properties have also been claimed for paints containing these materials. For a more detailed review of the use of phospholipids in coating applications see Sipos (1989).

B. Magnetic Recording Media

A special application of the technology described above is the use of phospholipids in coating of recording tape with magnetic particles. As a result of advantages already referred to, such as improved dispersion and homogeneity of particles, coating these particles with surfactants increases the ease with which they are oriented in the initial coating and hence improves the electromagnetic response. Lecithins are of value because they may be employed in a wide range of organic solvents and can absorb to a wide range of pigments such as iron oxide, chromium dioxide, barium hexaferrite, and metallic iron. Comparison of soya lecithin with conventional synthetic surfactants in controlled experiments demonstrated improved performance of tapes prepared with lecithin-coated particles in terms of squareness, switching field distribution, friction, accelerated wear, and audio output (Chagnon and Ferris, 1989).

C. Controlled Microparticle Crystallization

In the applications described above, lecithins have played an important role in the formation of microparticulates by physical means, in their stabilization and in subsequent function. Phospholipids, in the form of bilayer vesicles, can also be employed in preparation of microparticles by chemical means and may facilitate their performance in areas such as heterogeneous catalysis and photosynthesis. While these potential applications are still in the early development stage, and are being explored for a range of surfactant molecules apart from just phospholipids, the basic principles are described here as an illustration of the sort of objectives that may be achieved.

Microcrystals or precipitates can be formed when concentrated solutions of anions and cations are mixed together whose solubility product is lower than the final concentration achieved after mixing. This process can be performed selectively within the internal compartment of a liposome by preparing a liposome suspension with the different ions on the inside and outside of the vesicles. The two ionic components are brought into contact with each other for example by means of an ionophore in the membrane that permits influx of the external ion but not efflux of the entrapped species (see Fig. 1).

Preparation of microparticles in this way permits control of the size of the particle, which is limited by the quantity of ion entrapped in the vesicle, and which in turn is determined by its concentration and the internal volume of the liposome. The nature of the crystals formed may also be controlled, since this will often depend on the rate at which the two components are brought together, which is a function of the concentration of the external ion, the ionophore concentration, and the rate of passage of the ion through the membrane pore. A second factor will be the nature of the nucleation site, which may determine the structure of the crystal formed, and which may be influenced by inclusion of special phospholipids in the liposome membrane, at the requisite surface density.

The crystallization process takes in four stages: encounter between the two crystallizing species, nucleation of crystallization, growth, and finally Ostwald ripening (Fig. 2). In the latter stage, the crystals have grown to such an extent that the local supply of one of the ions is exhausted. One crystal then continues to grow at the expense of the others, by virtue of the presence of ions dissolved at low concentration (below the solubility product) that transfer from one crystal to another. Examples of systems that have been employed in surfactant and/or phospholipid vesicle systems to study these phenomena are $AgNO_3$ + OH^- (Mann and Williams, 1983), Ba^{2+} + PO_4-, and SiO_3^{2-} + H^+ (Perry, Fraser, and Hughes, 1990). An interesting case is the system studied by Mann and coworkers (Mann, Hannington, and Williams, 1986) in which the intravesicular formation of iron oxides was studied, resulting from influx of OH^- into liposomes containing ferrous and ferric ions. In this study, liposomes were used as a model for biological systems for iron storage,

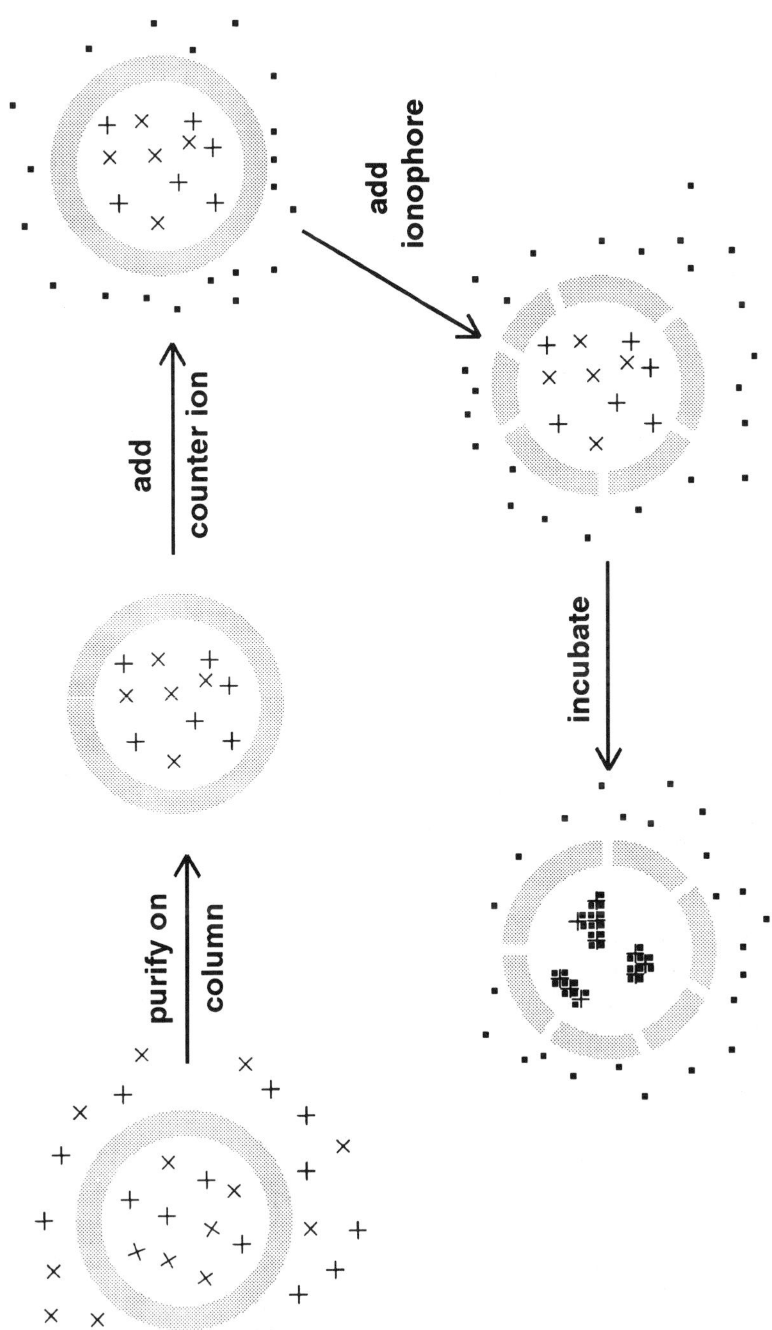

Figure 1 The technique of intravesicular crystallization owes its success to the fact that the membrane is selectively permeable to only one of the species employed. Such behavior may be a natural property of the membrane or may be induced by pH change or addition of ionophore.

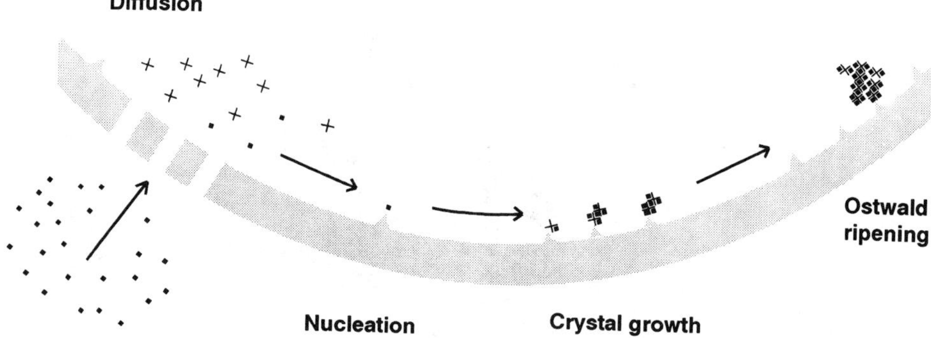

Figure 2 Formation of crystallites within vesicles. Packing irregularities or charged residues on the membrane surface may act as sites for initiation of crystal growth.

structural support, and magnetoreception (as observed in bacteria such as *Aquaspirillum magnetotacticum*). However, their findings could also have application for commercial production of stabilized magnetic microparticles.

Precipitation was brought about by entrapping iron cations within liposomes at pH 2 or pH 4 and then incubating the liposomes at pH 12 and allowing hydroxyl ions to diffuse passively across the membrane. While rapid addition of alkali to ferrous chloride in bulk solution gave mixtures of acicular lepidocrocite and geothite, the same components brought together inside liposomes gave well-defined magnetite crystallites of 30–50 Å diameter. The difference in nature and size between the bulk and intravesicular crystals is attributed to the slower rate of oxidation at low pH. On the other hand, Fe(II)/Fe(III) solutions gave ferrihydrite inside vesicles, while magnetite particles 100–500 Å in diameter were obtained from the same reaction in bulk solution. In contrast to the reactions carried out in bulk solution, no precipitate could be sedimented upon bench centrifugation of vesicle preparations.

Such new techniques for microparticle formation and stabilization may also be of value in production of highly active solid-phase catalysts, whose efficiency is closely related to surface area. Thus particles of colloidal platinum (270 Å in diameter) can be produced inside vesicles by photolysis of K_2PtCl_4 and have been used successfully to demonstrate hydrogenation of double bonds (Kurihara and Fendler, 1983). Apart from physically stabilizing the colloidal particles as an aqueous dispersion (>30 days, cf. 2–3 days for unentrapped particles), the enveloping membrane is able to hold within it redox agents that act as carriers of hydrogen or electrons and can serve as coupling agents between the metal surface and acceptors located in the solution bathing the vesicles (see Fig. 3).

In a similar fashion, coupling agents entrapped at high concentrations within the vesicles can capture energy absorbed by particulate semiconductors upon irradiation with light, before unproductive recombination between electrons and holes has time to occur. Microdispersed semiconductors are efficient light harvesters because of their high extinction coefficients, broad absorption spectra, and high surface area compared with solid state crystalline systems. Semiconductors absorb energy when electrons are promoted from the valence band to the conduction band, and this energy may be used for splitting of water by combining a semiconductor and rare metal in the same vesicle, e.g., rhodium-

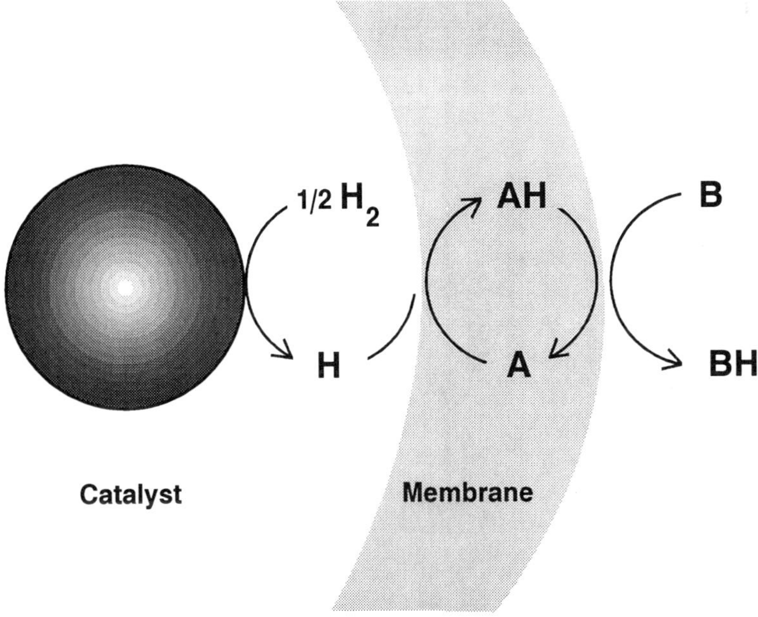

Figure 3 Catalytic hydrogenation by colloidal microparticles stabilized inside liposomes can be coupled to reactions occurring in the external aqueous phase by redox reagents incorporated in the membrane.

coated CdS (Fig. 4). The CdS is produced by exposure of Cd salts to hydrogen sulfide. Using phosphatidyl serine as one component of the membrane, CdS can be induced to form particles embedded in the membrane itself. For an extensive discussion of studies on membrane-stabilized semiconductors the reader is referred to a recent review (Fendler, 1987).

II. MOLECULAR BIOLOGY

For the incorporation of genetic material into the genome of mammalian cells, the methods of choice are electroporation and liposomal delivery, of which the latter is simpler because sophisticated equipment is not required. Several alternative strategies are available when using the liposomal approach (see Fig. 5).

In the first case, lipid vesicles, composed primarily of phospholipids, are constructed that contain the genetic material entrapped in the aqueous core, and the contents of liposome are then delivered to the cell in culture, or even in vivo, by virtue of either nonspecific or targeted uptake, applying the same principles as are employed for liposomes as drug carriers (see elsewhere in this volume). For most efficient incorporation, the use of liposomes that can deliver to the cytoplasm has been recommended. Vesicles containing lipid molecules with readily ionizable head groups have metastable membranes, which cause their contents to be released upon change of pH after ingestion into the primary endosome but before coming into contact with the lysosomal contents (Wang and Huang, 1987). Alternatively, a fusogen such as the F-protein of Sendai virus may be incorporated into the phospholipid membrane, so that the liposomally-entrapped material

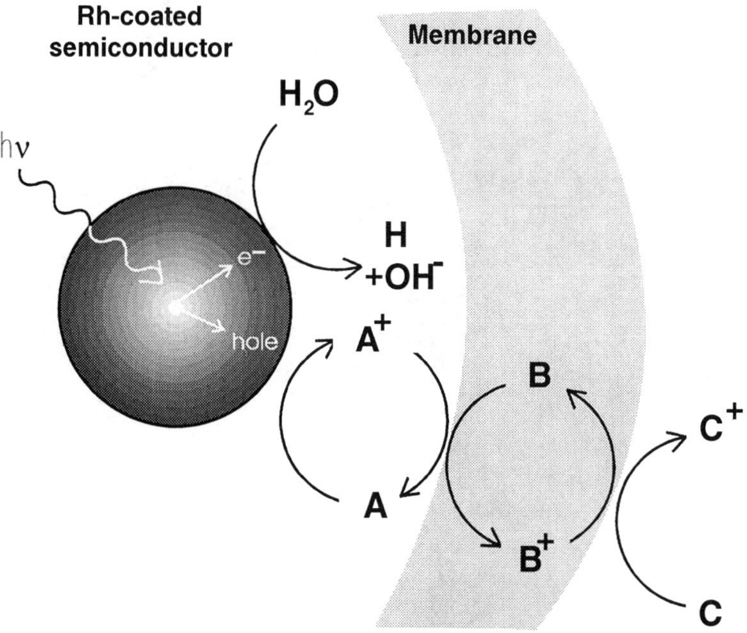

Figure 4 Possible coupling configuration for photolytic reduction of water by vesicle-entrapped semiconductor microparticles.

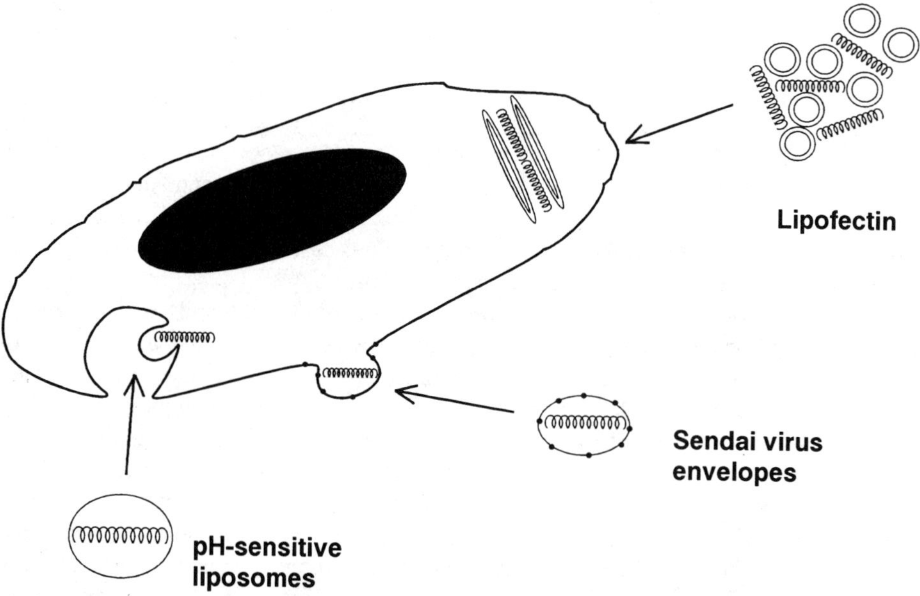

Figure 5 Incorporation of DNA into mammalian cells.

may be "injected" directly through the plasma membrane of the target cell as the two membranes fuse with each other (Loyter et al., 1983).

In the second approach, more widely adopted because of its extreme simplicity, empty cationic liposomes are employed, which, when mixed with DNA or RNA, form a complex in which the nucleic acids are strongly bound to the outer membrane surface of the liposomes (Felgner et al., 1987). The excess positive charge also causes the complex to associate strongly with the cell membrane, and interaction with the cationic lipids causes perturbations of the membrane that facilitate entry of the complex into the cell. Although the active ingredient is a cationic lipid such as DOTMA (2,3 dioleyloxy propyl trimethyl ammonium chloride), phospholipids play an important role in stabilizing the vesicles and the lipid-DNA complexes that they form. The vesicle formulation is currently available commercially under the trade name Lipofectin (Gibco BRL).

III. FOOD TECHNOLOGY

A number of different applications have been identified for microencapsulation of food components using phospholipid liposomes (Kirby, 1990 and 1991). Liposomes can be used to increase the stability of ingredients that in unencapsulated form would be rapidly degraded by contact with other components of the food. They may be used to increase the physical retention of important components during critical stages in the processing of the food; they may be used to control the distribution of ingredients throughout the food, either to be more homogeneously dispersed or to be targeted, actively or passively, to restricted compartments of the food matrix where the activity of these ingredients may be harnessed most efficiently. They may also be used to modulate the activity of these ingredients as a function of time by controlling their rate of release during food processing, either in response to microenvironmental conditions within the food or to some external trigger such as heating, microwave irradiation, or mechanical stimulus. Many of these modes of action are illustrated in work that has been carried out during the development of microencapsulation systems for enzymes and antioxidants, discussed below.

A. Accelerated Cheese Ripening

The production of cheese involves several steps that are common to all types of cheese regardless of their origin or consistency. Milk is first treated in such a way that it coagulates, usually by addition of rennet, a large proportion of which is the enzyme chymosin, which destabilises casein at low pH. Thereafter, the milk solids (curd) are separated from the liquid (whey), either by sedimentation under gravity, or by filtration, and the curd is dried, pressed, and then allowed to mature upon storage. Maturation involves enzymic degradation of the milk proteins, principally B-casein, into hydrolysis products, which are mainly responsible for the flavor of the cheese. The distinctions in taste and structure between different cheeses are due in part to the characteristics of the different bacterial starter cultures, which are introduced in the first step prior to addition of the rennet, and which provide the enzymes responsible for proteolysis. Optimal results are obtained as a result of action by a mixture of different proteases: endopeptidases, which break down proteins to medium chain hydrophobic peptides with a rather bitter taste, and exopeptidases, which convert those peptides to shorter chain acidic moieties with more positive flavor attributes. Carbohydrate metabolism, resulting in formation of acids and aldehydes, and lipolysis and lipoxidation also contribute to the overall background flavor.

At some stage, salt is usually added, either sprinkled as a dry powder together with the milled curd before pressing (e.g., Cheddar, see Table 2) or, as in the case of varieties such as Edam and Gouda, by soaking the pressed curd block in brine.

Maturation of cheese can be a very slow process, since it relies on the action of a small quantity of bacterial enzyme to convert a vast matrix of aggregated protein in semisolid form. In the case of Cheddar cheese, full maturation may take up to one year. This has tremendous implications for the economics of the process, which involves occupancy of storage space for prolonged periods of time under controlled conditions, and which accounts for a considerable proportion of the costs of manufacture. Methods for accelerating the ripening process are actively being sought.

The quantity of bacteria added, which are essentially a form of microencapsulation for the enzymes, cannot easily be augmented, since growth of the bacteria in the milk is pH-limited. Larger numbers of bacteria initially would simply increase the rate at which the pH dropped and halt growth of the culture at an earlier stage. It has been demonstrated, however (Law and Wigmore, 1982), that an extract of bacterial enzymes can be produced that successfully achieves the same flavor characteristics as the original bacteria (lactobacilli or lactococci such as *Bacillus subtilis* or *Streptococcus lactis*) but reduces the maturation time by half, when added in amounts that augment the natural level of enzyme due to intact bacteria. Unfortunately, use of free enzyme in this way results in loss of the majority of the enzyme in the whey, and very little sediments with the curd to go on to participate in the maturation process after pressing. Also, addition of the enzyme at this early stage results in proteolytic attack of the milk proteins before the curd has formed properly, resulting in a very weak-bodied curd and loss of potential cheese components into the whey. This may have implications for production of starter cultures, since the whey is often used as a component of culture medium for the starter bacteria. Use of whey products in other food processes may also be adversely affected. An alternative point of addition of the extra enzyme is after formation of the curd, at the salting stage. This results

Table 2 Steps in the Manufacture of Cheddar Cheese

Initiation	Add starter culture to milk; incubate at 30°C for 2 h. Conversion of lactose to lactic acid induces pH drop.
Coagulation	After 100 min, add rennet to initiate coagulation. When complete, cut solid mass into cubes.
Scalding	Raise temperature to 39°C gradually. Whey exudes from cubes.
Pitching	Drain off whey. Dry curd knits together under own weight.
Chedarring	Mill dried curd into large granules and add salt. Bacteria killed off.
Pressing	Put cheese into molds and press overnight.
Maturation	Store cheese while ripening takes place. Enzymes released from dead bacteria.

in rapid maturation and efficient use of the enzyme but gives rise to a more crumbly texture, since it is not possible to distribute the enzyme homogeneously throughout the protein matrix before pressing, it being restricted to the outside of the milled curd granules, which may be several centimetres in diameter. Some varieties of cheese, which do not include a dry-salting step in their manufacture, cannot have enzyme added in this way.

Microencapsulation of the proteolytic enzymes inside liposomes has been shown to be an approach that overcomes these problems (Law and King, 1982) and that can be developed into a cost-effective process (Kirby, Brooker, and Law, 1987). Using an encapsulation method based upon the dehydration-rehydration method, originally developed for Factor VIII encapsulation (Kirby and Gregoriadis, 1984), high percentage entrapments of enzyme can be achieved within liposomes (~40%) with near complete incorporation of the entrapped protein in the curd, after addition of the liposomes to the milk before onset of coagulation. In this way, the enzyme is evenly distributed throughout the resultant curd, without any degradation in the quality of the protein matrix. Microencapsulated enzyme thus incorporated has been shown to reduce the maturation time by half in the same way as for free enzyme added to the preformed curd at the dry-salting stage.

A critical feature for success of this approach is the stability of the liposomes at different stages throughout the process. It is very important that liposomes remain intact directly after addition to the milk, so that the protein is protected from the enzyme before a strong-bodied curd has had a chance to form. On the other hand, as soon as the curd has been pressed and has started the maturation process, the liposomes need to release their payload as quickly as possible so that the enzymes can begin to act on the protein. Much work has been performed investigating the influence of liposome composition on the stability at these various stages (Kirby and Law, 1987).

The question has been tackled using carboxyfluorescein and horseradish peroxidase as entrapped markers, with assessment of leakage performed by fluorimetry or electron microscopy, respectively. The stability of liposomes in the milk is highly dependent on the presence of cholesterol incorporated in the phospholipid membrane, and instability of cholesterol-free liposomes in milk is thought to be due to interaction between the membrane and fat globules or high density lipoproteins present, similar to the situation observed for the stability profile of liposomes of different composition in blood (Scherphof et al., 1978; Kirby et al., 1980a and 1980b).

The efficacy of liposomes in this system is critically dependent on their stability. Using pure phospholipids, it was found that liposomes containing highest amounts of cholesterol (PC : cholesterol mole ratio 1 : 1 or 1 : 0.5) gave optimal stability in milk, while maturation was retarded with cholesterol-rich liposomes when compared with cholesterol-free liposomes. This finding is interpreted as indicating that cholesterol-rich liposomes are too stable and do not give rapid enough release of the enzyme within the interstices of the protein matrix to enable the maximum possible level of proteolysis to occur. In a separate experiment, in which liposomes with an intermediate level of cholesterol were employed (PC : cholesterol mole ratio 1 : 0.25), ripening was accelerated beyond that seen for either cholesterol-free or cholesterol-rich liposomes, suggesting that an optimal balance may have been struck between lability in milk and stability in the curd matrix.

The effect of a commercial protease (known as Neutrase) inside liposomes composed of high grade egg lecithin alone is shown in Fig. 6, where the liposomes are added to the milk before onset of curd formation. As an index of ripening, β-casein degradation is

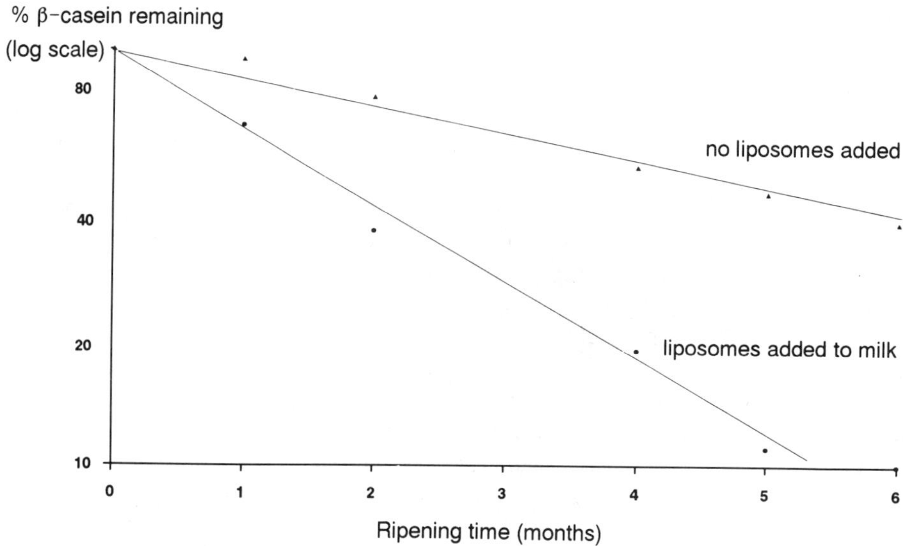

Figure 6 Accelerated degradation of β-casein in ripening cheese treated with liposomal proteinases. (Data redrawn from Kirby et al., 1987.)

used, determined by quantitative analysis after electrophoresis on polyacrylamide gels. Similar results are obtained when looking at TCA-soluble nitrogen products. A marked enhancement in ripening is seen with liposomes, in comparison with control batches of cheese treated normally. Addition of equivalent free protease at the dry salting stage gives similar results to liposomes, but larger amounts of enzyme result in a weak-bodied curd with reduced cheese yields. Addition of larger quantities of liposomes, on the other hand, reduces ripening times still further, without detriment to the quality of the final cheese.

The phospholipids employed for large-scale use in this system were mainly semipurified lecithins, manufactured in bulk from either soya or egg yolk, of a grade suitable for use in the food industry. While these lipids are much cheaper to use, liposomes formed from them were much less stable, and in order to attain sufficient stability to survive in milk, while still breaking down during ripening, higher levels of cholesterol needed to be added. It should be noted that the amount of cholesterol contained in these liposomes makes a negligible contribution to the total cholesterol in the final product.

In conclusion, a simple delivery system has been developed that can markedly accelerate the ripening of cheese, thus reducing production costs, without deterioration of the physical structure of the cheese, as may be seen with use of the enzyme in free form.

B. Reduction of Bacterial Spoilage

An additional problem in the production of cheese (particularly washed curd cheeses such as Gouda, Edam, and Emmental) is that of bacterial spoilage resulting from contamination by bacteria such as *Clostridium tyrobutyricum*, which may contaminate the milk from silage. Apart from the obvious health hazard, the presence of these spore-forming bacteria can result in unpleasant flavor and texture changes as a result of butyric acid fermentation. Routine addition of conventional antibiotics would be costly and highly undesirable in view of the danger of generating resistant strains. The current practice in some countries of adding nitrate has raised health concerns, and the problem of finding alternatives has been approached using natural antibacterial agents such as egg white lysozyme.

Unfortunately, while lysozyme adsorbs to the casein micelles as they sediment, so that good incorporation into the curd is obtained, this binding actually reduces the potency of the enzyme and means that much larger quantities than ideally desirable have to be added in order to maintain sufficient antimicrobial activity in the pressed curd.

Microencapsulation has been used by the same workers as above (Kirby, 1991) in an attempt to overcome this difficulty. In addition to maintaining the potency of the lysozyme by preventing its contact with casein until the surface area of the latter has been much reduced by coagulation, liposome encapsulation targets the enzyme to the same restricted compartment within the curd matrix as that is where bacterial spores are to be found. This is because both entities have a preference for the aqueous phase and are equally excluded from the protein matrix and from the fat globules that make up most of the pressed curd microstructure. Thus both liposomes and bacteria find themselves concentrated in the narrow aqueous space pressed close up against the walls of the interstices in the protein matrix, in between the latter and the fat globules (see Fig. 7). In contrast, free lysozyme would be incorporated uniformly throughout the casein matrix and would therefore be far removed from its desired site of action. Thus in addition to protecting lysozyme from the environment and vice versa, microencapsulation of the enzyme in liposomes alters its distribution within the ripening cheese so as to maximize the efficiency of its action.

For microencapsulated lysozyme to work optimally, it needs to have a slightly different stability profile from that required for ripening applications, so that lysozyme release is delayed until a large part of the casein is already hydrolyzed. From the foregoing discussion one would expect that this could be achieved by employing cholesterol-rich liposomes as encapsulants.

Use of another natural antibiotic, nisin, is currently being developed, to combat contamination in low fat cheeses by *Listeria monocytogenes*. This compound is produced by lactic acid bacteria but requires encapsulation in the early stages of cheese making in order to protect the starter culture from its effects.

C. Encapsulation of Antioxidants

Entrapment of antioxidants inside liposomes may be used as a means either to prevent these compounds from degrading (by premature oxidation) or to enhance their efficacy by linking them with complementary synergistic redox systems and locating them close to the components where protection against oxidation is required (Kirby et al., 1991).

Ascorbic acid (vitamin C) is an important nutritional component of many foodstuffs. Unfortunately it is very readily destroyed by contact with other components of foods, e.g., ascorbate oxidase, often found in plant tissues, amino acids, or iron or copper. Microencapsulated vitamin C is able to remain separated from such undesirable influences, and microencapsulation can help preserve it intact until such time as the food is digested, when the vitamin C may be released and made available for absorption through the gut.

Encapsulated vitamin C itself, however, may also be utilized as an effective antioxidant system for protection of lipids by coupling it with another natural antioxidant, alpha-tocopherol (vitamin E). While ascorbic acid is entrapped in the aqueous compart-

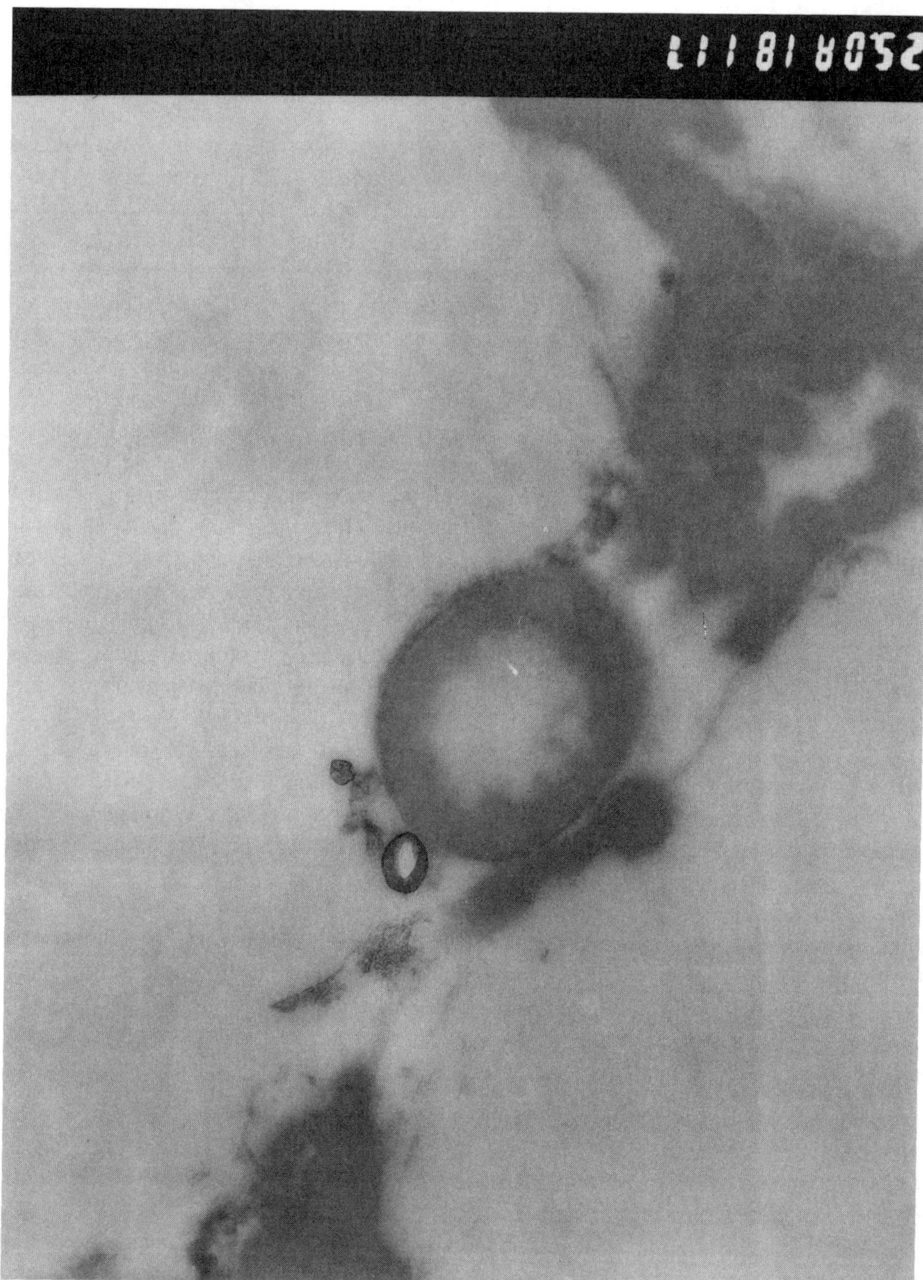

Figure 7 Colocalization of microbes and liposomes in same compartment of ripening cheese. (By kind permission of Dr. C. J. Kirby and B. E. Brooker, Food Research Institute, Shinfield, U.K.)

ment of liposomes, alpha-tocopherol inserts into the membrane, with its polar chromanoxyl head group located at the interface of the aqueous and organic phases.

In lipid-based systems, oxygen and free radical products of oxidation interact preferentially with the double bonds of unsaturated fatty acid chains (or of cholesterol) to give very reactive radical intermediates produced by abstraction of hydrogen. These can quickly lead to formation of toxic peroxides, aldehydes, chain breakage, and continuation of the radical chain reaction, if alpha-tocopherol is not present to act as a radical acceptor (by donating a hydrogen) and delocalizing the extra electron density within the aromatic nucleus. The low-energy tocopherol radical is much more stable than the original fatty acid radical, and its lifetime is sufficiently long that it can remain intact until it meets another tocopherol radical with which it interacts to form a dimer. Thus the danger of further chain initiation and oxidation is averted, and the phospholipid unsaturations are protected. Unfortunately, alpha-tocopherol is consumed in this process, markedly reducing the time span over which its activity is manifested.

One extra feature of liposome behavior makes this system particularly effective for protection from oxidation of fat emulsions composed of high levels of polyunsaturated lipids. Visual examination of such emulsions (e.g., margarine) under the microscope suggest that liposomes have a propensity to cluster round the surface of these emulsion droplets (see Fig. 8), thus bringing the antioxidant system into direct contact with the component requiring protection.

Incorporation of ascorbic acid into the same liposome as alpha-tocopherol permits regeneration by ascorbic acid of the tocopherol radical intermediate before the tocopherol dimer is formed (Kirby et al., 1991). (In the process, ascorbic acid is itself oxidized—initially to dehydroascorbic acid.) In this way, ascorbic acid, leaking slowly across the liposome membrane, can be coupled to oxidation of polyunsaturated lipids via the intermediary of alpha-tocopherol, which is located physically as a bridge between the two species, and which is able to terminate the lipid radical chain reaction at a much earlier stage than can the ascorbic acid on its own. Thus a marked synergistic effect is seen in anti-oxidant activity when the two vitamins are used in concert (see Fig. 7).

In view of the growing trend for replacement of saturated fats by polyunsaturated in

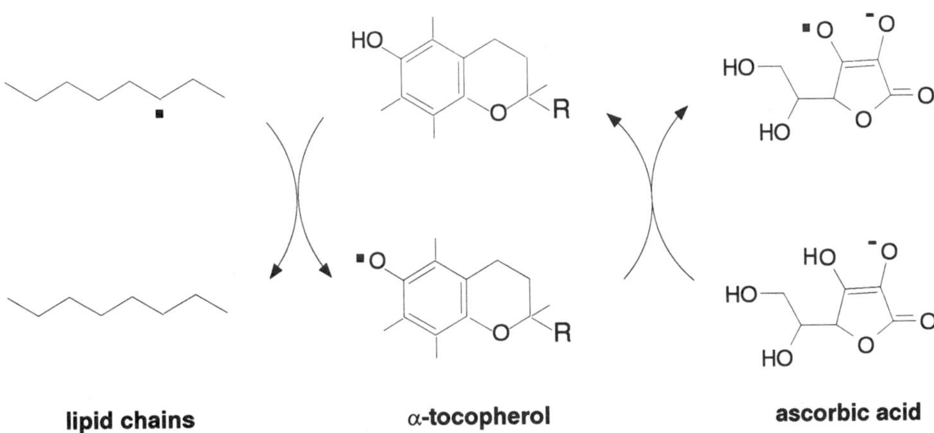

Figure 8 Protection of lipids from radical damage by aqueous phase ascorbate coupled with membrane-bound α-tocopherol.

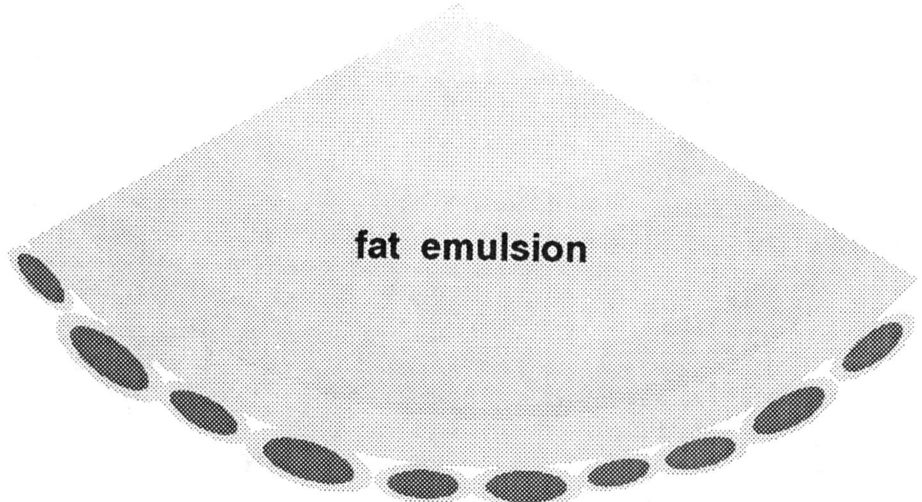

Figure 9 Liposomes containing α-tocopherol and ascorbate provide a protective sheath for emulsion particles against radical damage.

the Western diet, coupled with increasing restrictions on the use of synthetic antioxidants in food, the possibility for use of liposomal vitamin preparations to protect this highly vulnerable class of nutrient from oxidation must assume ever greater importance in the food industry.

IV. BIOCOMPATIBILITY

Although biocompatibility and food technology are completely different applications, these areas have a role in common for phospholipids to play, namely that of acting as a barrier. Instead of separating enzymes from substrate, however, as in cheese ripening, phospholipids improve the biocompatibility of man-made materials by coating their surface so that proteins and cells are protected from contact with chemical structures that can cause undesired binding or activation. The ability to function well in this role is dependent on a property not exploited elsewhere, namely the very low level of interaction between phospholipid (and the phosphoryl choline head group in particular) and proteins.

While charged phospholipids such as PS or PE are known to bind to metal ions, polynucleotides, and proteins, the neutral phosphoryl choline head group has extremely little interaction with such molecules, when in a regular, *close-packed* configuration. Its role in biology as the major component in plasma membranes would suggest that it is specifically designed to display this behavior, in order to permit formation of stable membranes that are not subject to undue perturbation as a result of nonspecific protein binding. In bilayer membranes, the head group has been shown, by NMR, to orient itself partially parallel to the plane of the membrane, with electrostatic interactions taking place between the trimethyl ammonium group of one molecule and the phosphate group of its neighbor. The positive and negatively charged groups have considerable freedom to move, are both bulky, of equal size, and are readily polarizable, so that they can fit together and act to delocalize each other's charge density, leaving no gaps or physical irregularities. Only transient breaches of regular packing due to thermal motion serve to allow ingress of small molecules into the bilayer interior and account for the permeability characteristics of biological membranes. In this way, intermolecular interactions, both between fatty acyl chains and head groups, give rise to a cohesive plane of undulating polarity, with a marked absence of isolated, exposed epitopes capable of forming the basis for molecular recognition and binding.

The zwitterionic nature of the PC head group also leads to its having a very high binding affinity with water. Phosphatidyl choline retains on average four molecules of water even after extensive drying. Surfaces coated with PC are extremely hydrophilic (see Table 3), and it is reasonable to suppose that when confronted with a close-packed PC surface, the only thing that a protein "sees" is an extended wall of fixed water, which rebuffs any attempt at penetration or adherence.

At this stage it is worthwhile digressing on possible differences between phospholipid bilayers in liposomes and those on coated surfaces, which have led some workers to suggest that in plasma, phosphatidyl choline itself binds significant levels of protein. This has also been advanced as the reason why PC liposomes, in the absence of so-called "stealth" coatings, are removed rapidly from the bloodstream.

Coating phospholipids onto a surface from a solution in organic solvent is a much gentler process than that required for manufacture of small liposomes: no sonication or mechanical action is required, no detergents are involved, no osmotic pressure differences are created, and no ions or molecules are entrapped that might interact with the membrane and cause perturbations. The curvature of a liposome membrane is much higher than that for a coated planar surface. In some cases, the phospholipids adsorbed onto a hydrophobic surface may give a monolayer with much lower fluidity than that of a bilayer membrane, because of the more rigid nature of the underlying substrate. The small size of liposomes, their deformability, and the fact that they are moving rapidly in the bloodstream through narrow capillaries at high shear stress suggest that they may be more exposed to conditions that may lead to formation of minor packing defects and irregularities, compared to phospholipids coated onto a planar surface, experiencing boundary layer conditions of flow.

Table 3 Increase in Hydrophilicity of Biomaterial Surfaces when Coated with Phospholipids, Demonstrated by Reduction in Contact Angle

	Contact angle (sessile drop method)	
	Untreated	Treated
DPPC[a]		
Polyethylene	75	15
PVC	77	8
DAPC[b]		
Polyethylene	81	10
PVC	79	13

[a]DPPC, dipalmitoyl phosphatidyl choline.
[b]DAPC, diacetylenyl phosphatidyl choline.
Data kindly provided by M. C. Wiles and G. Wotherspoon.

In such circumstances it is reasonable to expect that intermembrane lipid exchange will take place in vivo much more rapidly, and to a much greater extent, than for a coated surface, so that shortly after injection into the bloodstream the surface of a liposome, which started off composed entirely of phosphatidyl choline, will be contaminated with a small proportion of charged phospholipids. As is well known, charge in a phospholipid membrane strongly influences its ability to interact with proteins, and one may thus expect that upon binding of proteins, uptake by phagocytic cells will quickly follow. In contrast, one may hypothesize that a planar coated surface of phosphatidyl choline in a region of slow flow could remain unchanged indefinitely as a result of absence of conditions that bring it into close contact with lipid-exchanging proteins or membranes.

A. Interaction of "Biomaterials" with Biological Fluids

All materials currently used commercially for manufacture of medical devices suffer from the major drawback that they interact strongly with proteins and/or cells when they come into contact with biological fluids and tissues. Thus although these materials, usually plastics or hydrogel polymers, are termed "biomaterials" because they are an improvement upon glass or steel, they leave much to be desired in terms of their compatibility with biological milieux. Safe introduction of these materials into the bloodstream, even for short-term applications, must invariably be accompanied by administration of anticoagulants to prevent clotting, and catheterization is the single most common route of hospital acquired infection, as a result of the propensity of these materials to bind bacteria and support their growth. In the case of urinary catheters, a further complication is blockage as a result of mineral encrustation, limiting the lifetime of use in a patient, and necessitating frequent changing of the catheter. Implementation of long-term use of implants constructed of synthetic materials, particularly in applications close to the heart, is restricted owing to, amongst other factors, the risk of occlusion as a result of thrombus formation. Even widely accepted applications such as use of hydrogels in contact lenses suffer from problems of protein deposition. Other instances where protein and cell interactions are important are in the functioning of in vivo biosensors, the use of cellulose membranes in dialysis units, and the construction of blood and platelet packs for storage of donor blood prior to transfusion. When body fluids come into contact with conventional synthetic materials, protein adsorption begins in milliseconds. In the case of whole blood or plasma, a complete monolayer of protein will have formed on the surface within ten minutes, of which at least 60% will be fibrinogen. Although fibrinogen is by no means the protein most prevalent in plasma (4 mg/mL, cf. 40 mg/mL for albumin), its structural characteristics and limited solubility predispose it to binding to foreign surfaces. Indeed, one may suppose that this behavior is intentional, since fibrinogen is intimately involved in the process of clot formation, which is an important protective response to wounding by foreign bodies (see Table 4). Fibrinogen interacts with biochemical pathways of clot formation in at least two ways. Firstly, it can bind to platelets that adhere to the surface of the material, and upon activation secrete more fibrinogen, and also factors that cause activation of other platelets. Secondly, bound fibrinogen is converted by thrombin (a product of the clotting cascade that may itself be activated by "biomaterials"—see Fig. 9) to fibrin, which can cross-link to form an insoluble polymer that provides a network to stabilize the platelets already adhering. Thus the combination of bound fibrinogen, activated clotting cascade, and platelets on a biomaterial surface can rapidly give rise to formation of clots and microemboli.

Table 4 Events Occurring on Biomaterials Leading to Abnormalities in Blood Function

Protein deposition
Albumin
Immunoglobulin
Fibrinogen
von Willibrand factor
High molecular weight kininogen
Platelet interactions
Adhesion
Activation
Activation phenomena
Clotting cascade—in particular factor XII
Complement cascade—C3 to C3a and C3b

While there are many other events that can take place at a foreign surface in contact with blood, it is true to say that examination of the interaction of fibrinogen and platelets with a biomaterial surface can give a good indication of the degree of thrombogenicity of a given material. Recently, rapid and sensitive in vitro immunoassay methods have been developed for measurement of fibrinogen binding (Lindon et al., 1985) and activation of platelets (Campbell, New, and Charles, 1992) on foreign surfaces. These techniques have been used to compare the performance of conventional biomaterials with that of biomaterials treated with phosphatidyl cholines.

When performing these tests, biomaterials in the form of thin sheets are cut into small squares approximately 1 cm^2 in surface area and incubated in plasma (for fibrinogen) or fresh whole citrated blood (platelets) in wells of microtiter plates for 10 min to 0.5 h. After washing, the materials are incubated with antibodies directed against the protein of interest (GMP 140 in the case of activated platelets), followed by anti-Ig peroxidase conjugate. After transfer of the washed samples to fresh wells, OPD substrate is added. The color that develops is proportional to the quantity of protein adhering and is measured spectrophotometrically. Platelet adhesion may also be measured quantitatively by luminometric determination of cellular ATP bound to the biomaterial surface.

The type of phospholipid used by Biocompatibles to coat the surface of biomaterials is a diacetylenyl phosphatidyl choline (DAPC; see Chapman, 1979), since this may be cross-linked (after coating) by gamma irradiation to give a polymeric layer that is much more stable in aqueous media than unpolymerized monomeric phosphatidyl choline (either DAPC or fully saturated DPPC). Relative changes in optical density due to fibrinogen adsorption are shown in Fig. 11 for polyethylene coated with DAPC, DPPC, and lyso-PC (palmitoyl derivative), demonstrating that DAPC binds considerably less fibrinogen than either uncoated material or material coated with the other lipids.

It has been shown recently in the author's laboratory that the relationship between protein density and substrate optical density is not linear at high concentrations, and that the reductions in optical density of 70% shown in Fig. 11 correspond to reductions in protein binding of greater than 90%.

The results compare well with activated platelets adhering to the same surfaces after incubation in fresh whole blood (Figs. 10 and 11). Indeed, very good agreement has

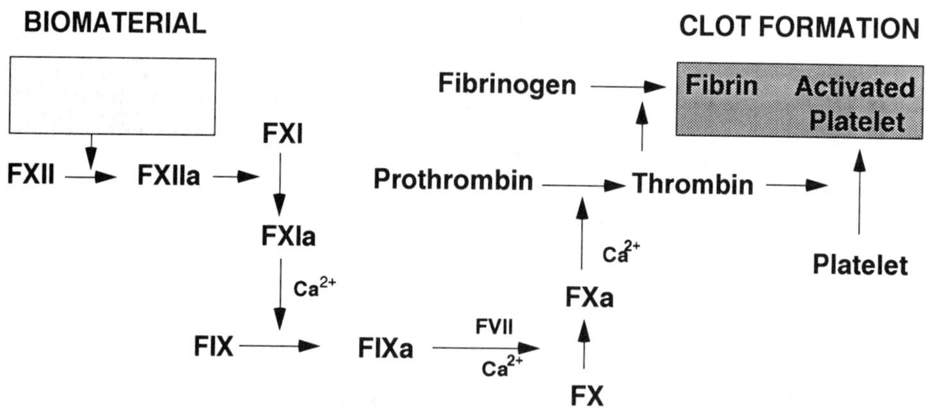

Figure 10 Induction of clot formation by biomaterials via the Intrinsic pathway.

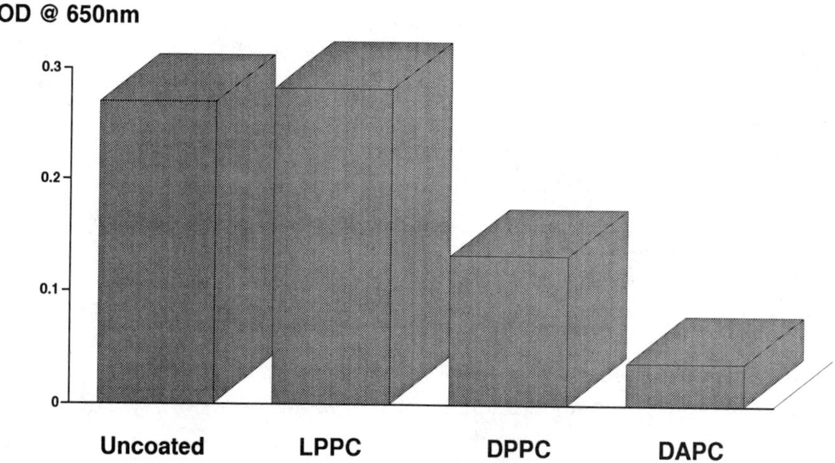

Figure 11 Fibrinogen adhesion on polyethylene coated with phosphatidyl choline.

been seen between the results of platelet and fibrinogen uptake experiments, as is shown in Fig. 12, where surfaces coated with less than the optimum concentration of DAPC are compared in the two tests. As the concentration of PC is increased, progressively less adherence of both fibrinogen and activated platelets is seen.

Phosphatidyl choline not only can be used to coat the surface of biomaterials but also may be incorporated throughout the bulk of the polymer, with the same effect. In this case, egg lecithin was employed and was coextruded in different proportions with PVC to give flat sheets, which were tested as described above. Both fibrinogen and platelet adhesion were dramatically reduced by inclusion of PC at all levels tested (Figs. 13 and 14).

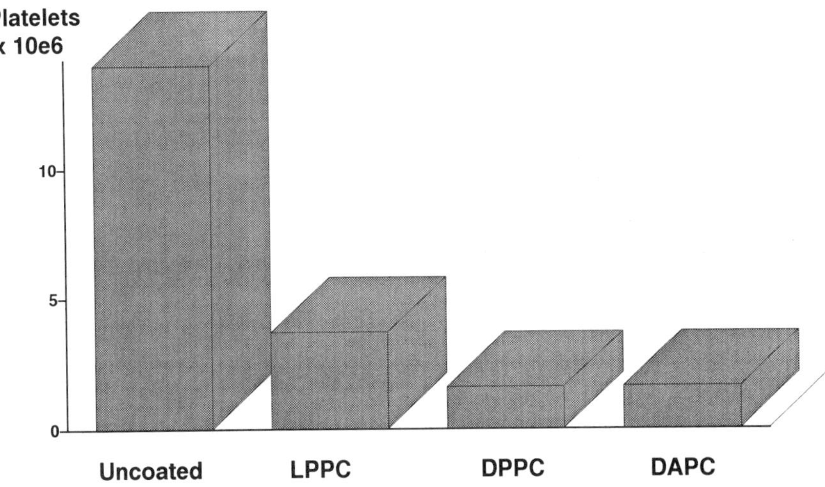

Figure 12 Adhesion of platelets to polyethylene coated with phosphatidyl choline.

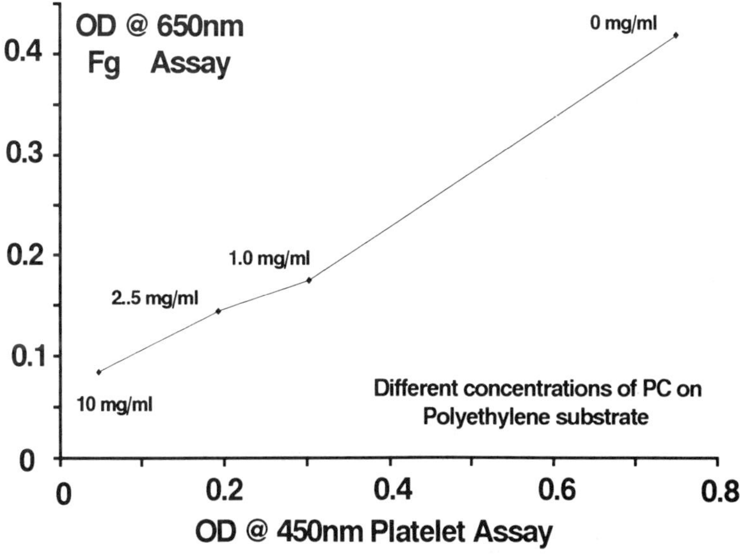

Figure 13 Correlation between fibrinogen adsorption and platelet activation tests.

Recently, recognition of the value of polymerized phospholipid systems in coating of surfaces has led to the development of "pseudo polyphospholipids"—coatable phosphorylcholine copolymers. Here, the essential features of lecithin (polar phosphorylcholine head group and long hydrocarbon chains) are retained, but a polymeric structure is built in at the start. The molecules are constructed with a backbone of polyacrylic acid, to the carboxylate moieties of which are randomly attached phosphorylcholine and hydrocarbon

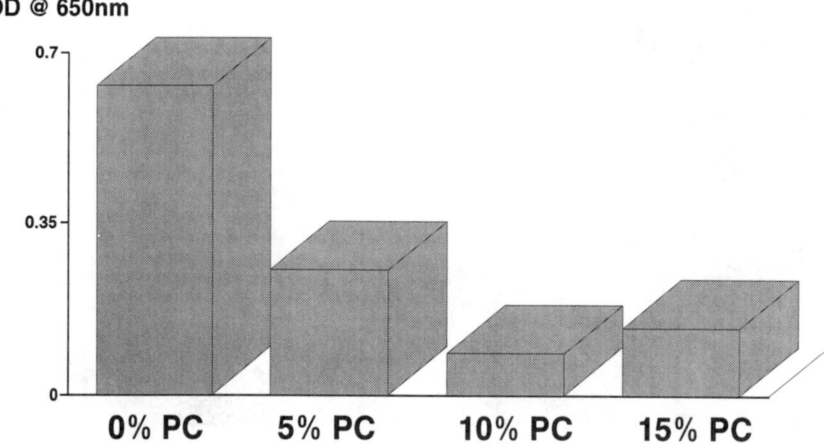

Figure 14 Fibrinogen adsorption to PVC containing phosphatidyl choline.

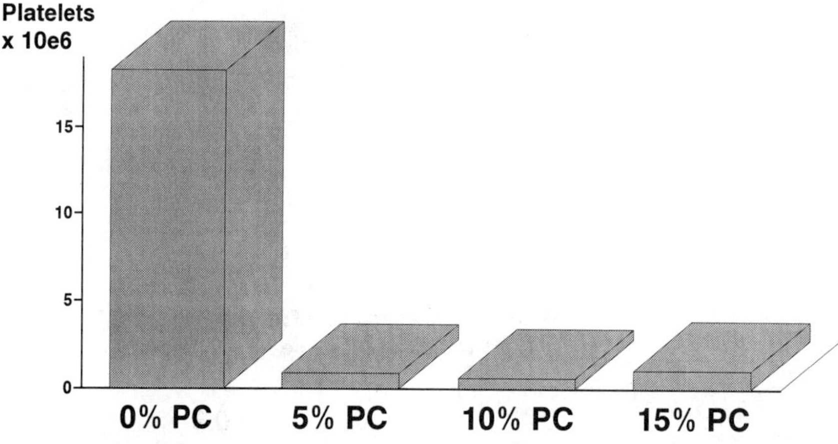

Figure 15 Platelet adhesion to PVC containing phosphatidyl choline.

chains in varying proportions. It is important for the side chains to be of adequate length in order to give satisfactory coating with these compounds. Under the right conditions, results can be obtained for reduction of protein and platelet uptake that are superior to those obtainable for monomeric phosphatidyl cholines (Fig. 15), which raises interesting questions with regard to the packing of the head groups in such structures. Their use to improve on the success already observed with conventional phospholipids in other applications described in this book is also a possibility.

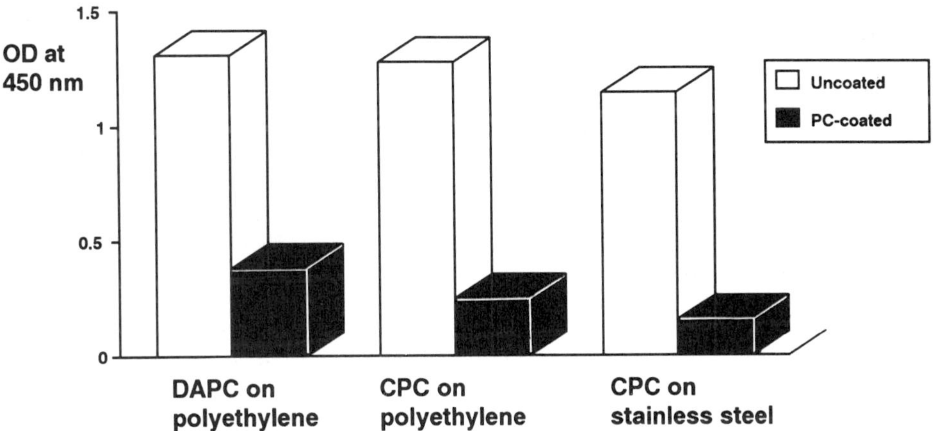

Figure 16 Reduction in fibrinogen bound onto biomaterials pretreated with coatable phosphorylcholine copolymer.

ACKNOWLEDGMENTS

The author gratefully acknowledges Dr. S. A. Charles of Biocompatibles Ltd., United Kingdom, for permission to present data on biomimetic properties of phospholipids shown in this chapter. The author is also very much indebted to Dr. C. J. Kirby for helpful discussions.

REFERENCES

W. Alkhalaf, J. C. Piard, M. El Soda, J. C. Gripon, M. Desmazeaud, and L. Vassal, Liposomes as proteinase carriers for the accelerated ripening of Saint-Paulin type cheese, *J. Food Sci. 53*:1674–1679 (1988).

E. J. Campbell, R. R. C. New, and S. A. Charles, A new enzyme immunoassay for activation of platelets by biomaterials: Reduced activation by phosphorylcholine-coated surfaces, *J. Matl. Res.*, in press.

M. Chagnon and J. Ferris, Phospholipids as surfactants in magnetic recording media, in *Lecithins: Sources, Manufacture and Uses* (B. F. Szuhaj, ed.), American Oil Chemists' Society, 1989, pp. 277–283.

D. Chapman, European Patent, 32622 (1979).

M. El Soda, L. Pannel, and N. Olson, Microencapsulated enzyme systems for the acceleration of cheese ripening, *J. Microencapsulation 6*:319–326 (1989).

P. L. Felgner, T. R. Gadek, M. Holm, R. Roman, H. W. Chan, M. Wenze, J. P. Northrop, G. M. Ringold, and M. Danielson, Lipofection: A highly efficient, lipid-mediated DNA transfection procedure, *Proc. Natl. Acad. Sci. 84*:7413–7417 (1987).

J. H. Fendler, Atomic and molecular clusters in membrane mimetic chemistry, *Chem. Rev. 87*:877–899 (1987).

C. J. Kirby, Delivery systems for enzymes, *Chemistry in Britain* 847–850 (1990).

C. J. Kirby, Microencapsulation and controlled delivery of food ingredients, *Food Sci. Tech. Today 5*(2):74–78 (1991).

C. J. Kirby, B. E. Brooker, and B. A. Law, Accelerated ripening of cheese using liposome-encapsulated enzyme, *Int. J. Food Sci. Tech. 22*:355–375 (1987).

C. J. Kirby, J. Clarke, and G. Gregoriadis, Effect of cholesterol content of small unilamellar liposomes on their stability in vivo and in vitro, *Biochem. J. 186*:591–598 (1980a).

C. J. Kirby, J. Clarke, and G. Gregoriadis, Cholesterol content of small unilamellar liposomes controls phospholipid loss to high density lipoproteins in the presence of serum, *F.E.B.S. Lett. 111*:324–328 (1980b).

C. J. Kirby and G. Gregoriadis, Dehydration-rehydration vesicles: A simple method for high-yield drug entrapment in liposomes, *Biotechnology* 979–984 (November 1984).

C. J. Kirby and B. A. Law, Recent development in cheese flavour technology: Application of enzyme microencapsulation, *Dairy Ind. Int. 52*:19–21 (1987a).

C. J. Kirby and B. A. Law, Developments in the microencapsulation of enzymes in food technology, in *Chemical Aspects of Food Enzymes* (A. T. Andrews, ed.), Royal Society of Chemistry, London, 1987b, pp. 106–119.

C. J. Kirby, C. J. Whittle, N. Rigby, D. T. Coxon, and B. A. Law, Stabilisation of ascorbic acid by microencapsulation in liposomes, *Int. J. Food Sci. Tech. 26*:437–449 (1991).

K. Kurihara and J. H. Fendler, Electron transfer catalysis by surfactant vesicle stabilised colloidal platinum, *J. Am. Chem. Soc. 105*:6152–6153 (1983).

B. Lariviere, M. El Soda, Y. Soucy, G. Trepanier, P. Paquin, and J. C. Vuillemard, Microfluidised liposomes for the acceleration of cheese ripening, *Int. Dairy J. 1*:111–124 (1991).

B. A. Law and J. S. King, Use of liposomes for proteinase addition to cheddar cheese, *J. Dairy Res. 52*:183–188 (1985).

B. A. Law and A. S. Wigmore, Accelerated cheese ripening with food grade proteinases, *J. Dairy Res. 49*:137–146 (1982).

J. N. Lindon, G. McManama, L. Kushner, E. W. Merrill, and E. W. Salzman, Does the conformation of adsorbed fibrinogen dictate platelet interaction with artificial surfaces? *Blood 68*:355 (1985).

A. Loyter, A. Vainstein, M. Graessmann, and A. Graessmann, Fusion-mediated injection of SV40-DNA, *Exp. Cell Res. 143*:415–425 (1983).

S. Mann, J. P. Hannington, and R. J. P. Williams, Phospholipid vesicles as a model system for biomineralisation, *Nature 324*:565–567 (1986).

S. Mann and R. J. P. Williams, Precipitation within unilamellar vesicles. Part 1. Studies of silver (1) oxide formation, *J. Chem. Soc. Dalton Trans.* 311–316 (1983).

C. C. Perry, M. A. Fraser, and N. P. Hughes, Macromolecular assemblages in controlled biomineralization, in *Surface Reactive Peptides and Polymers: Discovery and Commercialisation* (C. S. Sikes and A. S. Wheeler, eds.), ACS Books, Washington, 1990, pp. 316–339.

J. C. Piard, M. El Soda, W. Alkhalaf, M. Rousseau, M. Desmazeaud, L. Vassal, and J. C. Gripon, Acceleration of cheese ripening with liposome-entrapped proteinase, *Biotechnol. Lett. 8*: 241–246 (1986).

G. Scherphof, F. Roerdink, M. Waite, and J. Parks, Disintegration of phospholipid liposomes in plasma as a result of interaction with high density lipoproteins, *Biochim. Biophys. Acta 512*:296–307 (1978).

E. F. Sipos, Industrial coatings applications for lecithin, in *Lecithins: Sources, Manufacture and Uses* (B. F. Szuhaj, ed.), American Oil Chemists' Society, 1989, pp. 261–276.

P. L. Spangler, M. El Soda, M. Johnson, N. F. Olson, C. H. Amundson, and C. G. Hill, Jr., Accelerated ripening of gouda cheese made from ultrifiltrate milk using a liposome-entrapped enzyme and freeze-shocked lactobacilli, *Milchwissenschaft 44*: 199–203 (1989).

C. Y. Wang and L. Huang, pH-sensitive immunoliposomes mediate target-cell-specific delivery and controlled expression of a foreign gene in mouse, *Proc. Natl. Acad. Sci. 84*:7851–7855 (1987).

27
Medical Applications of Phospholipids

Yukihiro Namba *Nippon Fine Chemical Co., Ltd., Hyogo, Japan*

I. INTRODUCTION

It is well known that phospholipids, in many cases, are the main components of the cell membranes in a living body. Some functions of these molecules have been clarified by biochemical studies that have provided basic knowledge for the medical applications of such molecules. Moreover lipids have been investigated from the practical point of view as highly purified phospholipids havc become easily obtainable and now provide suitable starting materials for different kinds of drug carriers. For example, highly purified phospholipids can be used as raw materials for the preparation of lipid vesicles, liposomes, which are the vehicles of choice in some drug delivery systems and are also being explored in the development of artificial lung surfactants. Alkyl lysophosphatidylcholine derivatives have a unique antineoplastic function. In this chapter, developments in liposomal drug delivery and the biological activities of special kinds of phospholipids are described.

II. PHOSPHOLIPIDS FOR USE IN LIPOSOMES

Liposomes are microscopic vesicles composed of one or more phospholipid bilayers separated by an equal number of aqueous interspaces. These capsules can be formed by many procedures. A typical method is to agitate dried lipids warmed to above their transition temperature (the temperature at which lipid changes from the gel into the fluid state in an appropriate aqueous solution). The resulting vesicles can encapsulate water-soluble drugs in their aqueous space and/or lipid-soluble drugs within the hydrocarbon membrane core. Drugs entrapped in liposomes can gain entry into cells by the interaction of liposomes with cells in one of four ways: adsorption, endocytosis, material exchange, and fusion. It seems that these processes are not independent; rather, all are involved to a certain extent. Liposome-entrapped drugs behave differently in a body than free drugs. Most liposomes tend to accumulate in the liver, spleen, lung, bone marrow, and lymph nodes when they are injected intravenously into healthy animals or humans. The lipid vesicles also tend to accumulate at sites of infection and inflammation and, in some cases, in solid tumors. The mechanism of this accumulation has not yet been clarified, but it is possible that its origin is the penetration of lipid vesicles out of the blood capillaries at sites with imperfect microvasculature linkage.

The biodistribution and in vivo behavior of liposomes are affected by three factors:

1. Membrane rigidity. Entrapped water-soluble drugs tend to leak out faster if vesicle membranes do not contain cholesterol, unless lipid bilayers are very rigid.
2. Size of liposomes. Generally speaking, large liposomes are cleared faster from the bloodstream than very small ones, but vesicles of intermediate size may be worse than somewhat larger liposomes.
3. Lipid dose. The half-life of liposomes increases as the lipid dose is increased until saturation is reached.

In addition to these features, liposomes have additional advantages:

1. Phospholipids constituting liposomes are almost nontoxic to the living body because they are the main components of cell membranes.
2. Vesicle size is easily controlled in accordance with the desired target and the route of application.
3. Antigen, antibody, or a carbohydrate moiety can be attached only to the surface of a liposome by the introduction of functional groups onto the liposomal surface. Such modification permits active targeting of the vesicles to the accessible body subsites.
4. Liposomes can entrap most kinds of drugs.
5. Liposomes may prevent the attack of enzymes against drugs that are decomposed by such enzymes in vivo.
6. Liposomes may increase the therapeutic efficiency of drugs by increasing the available amount of a drug in tissues; they may also reduce the drug's side effects.

However, there remain many problems to be solved for the successful use of liposomes in medicine. In this review, the current situations of some liposomal drug preparations are described. The most widely used application of liposome-based therapy is in treatment of systemic fungal infections, especially by using amphotericin B. Liposomes for the treatment of neoplastic disorders are also being used already in human therapy. Antineoplastic agents such as doxorubicin and *cis*-platin, for example, have been loaded into liposomes for use in cancer therapy, and derivatives of immune modulators such as *N*-acetylmuramyl-L-alanine-D-isoglutamine have been attached to lipophilic segments for insertion into lipid bilayers or micelles. Many efforts have been made in order to obtain the targetability of liposomes in drug delivery systems. Recent results provide evidence that liposomes attached monoclonal antibodies against cell-surface-located tumor-associated antigens, and liposomes bearing gangliosides characteristic of hepatocytes can selectively deliver desired drugs to target tumors. Injectable liposomal formulations containing antimicrobial agents and neoplastic agents are undergoing clinical testing and are already on the market in some countries.

A. Therapy for Infections in the Reticuloendothelial System (RES)

Liposomes injected intravenously are trapped by the macrophages in the liver and spleen. It has been shown that antiinfective drugs encapsulated in such liposomes are active against infections caused by facultative intracellular bacteria, parasites such as leishmania, and viruses such as the one causing Rift valley fever. The following drugs have been considered in this respect.

1. Sodium Stibogluconate

Sodium stibogluconate is a typical example of a liposome-delivered drug that is used in leishmaniasis therapy [1]. This drug has a very short half-life owing to its high solubility in water; therefore continuous administration of a free drug solution over 120 days is necessary for good therapeutic efficiency. Such a long dosage time brings about serious side effects, particularly in the heart, liver, and kidney. Moreover it seems not to be able to clean parasites completely from deep tissues such as bone marrow. The use of stibogluconate in a liposomal form, which allows a modification of the drug distribution and excretion pattern, would thus seem to be a promising approach. In fact, Alving et al. [2] have shown that liposomal stibogluconate was 700 times more effective than the free drug in hamster models. Other experiments [1] indicated that the therapeutic efficacy was different in different host animals and sites of infection. In hamster and mouse models, the relative efficacies of the free and liposomal forms of stibogluconate in the liver were similar; they were different, however, in the spleen and bone marrow. The liposomal form was six times more effective than the free form in the case of a mouse application.

Other liposome formulations of different lipid composition, structure, and size have been used for the intravenous administration of antifungal or antiinfectious drugs without inducing toxicity in the tested animals. Clinical experience was obtained with two different liposomal preparations of amphotericin B in the treatment of systemic fungal diseases in cancer patients. Similar efficacy can be achieved with micellar amphotericin B, however.

2. Amphotericin B

Amphotericin B is the first drug of choice for most of the systemic fungal infections, but its acute and chronic toxicity often limit its practical use. Furthermore, this drug is often ineffective in the treatment of fungal infections in neutropenic and immunodeficient patients [3].

Liposomal amphotericin B shows a decreased toxicity and enhanced efficacy in the prophylaxis and treatment of experimental candidiasis in nonneutropenic as well as neutropenic mice. The liposomal form of this, therefore, was used to treat twelve leukemia patients who were also suffering from systemic fungal infections [3]. Patients who had failed to respond to therapy with the drug in the conventional free drug form were treated with liposomal amphotericin B, and most of them improved. Liposomal amphotericin B treatment was nontoxic, even though the patients had severe fever and chills when conventional amphotericin B was given. Lipid vesicles in these experiments have proven to be safe and efficient carriers for intravenous injection of amphotericin B.

The enhanced therapeutic activity of liposomal preparations of amphotericin B can result from several mechanisms. Liposomes injected intravenously are taken up by RES cells in the liver, the spleen, and the target organs of fungal infections. Uptake by blood monocytes and invasion of the endothelial barrier of blood vessels by fungi may also promote the delivery of liposome-entrapped amphotericin B to the site of infection. Large-scale trials of these formulations are in the phase-2 stage in the United States and the European countries, where over 120 people have been treated. Two kinds of combinations of phospholipids are used for amphotericin liposome formulations [4]. One is

DMPC and DMPG (The Liposome Co.) and the other is DSPG and hydrogenated phosphatidylcholine (Vestar, Inc.). The latter product was approved in Ireland in 1990.

3. Other Drugs

Almost the same effects as found with liposomal amphotericin B are also observed in the case of liposomal nystatin [5], a well-known antibiotic. A significant decrease in the toxicity of nystatin was recorded after nystatin had been incorporated into liposomes (maximum tolerated doses: 16 mg/kg for liposomal nystatin and 4 mg/kg for free nystatin) in animal tests with mice. The maximum tolerated dose of free nystatin had no effect on mice infected with *Candida albicans*, whereas liposomal nystatin at an equivalent dose improved the survival of diseased mice. A marked increase in survival was observed when the liposomal form was administered at higher and multiple doses, the total drug amount being up to 80 mg/kg. It was also shown that encapsulation into liposomes could protect erythrocytes from the toxicity of free nystatin [6]. Similar effects were observed for the liposomal form of gentamycin, and clinical trials with the corresponding drug preparation are on the way in the United States.

B. Anticancer Drugs

Liposomal formulations as carriers of anticancer drugs have been investigated for several years. Liposomes can be administered intravenously to carry lipid-soluble drugs, to improve drug stability, to target organs with fenestrated capillaries, to achieve an intravascular slow drug-release system, and to reduce drug levels in certain organs sensitive to toxicity. They also have a significant potential for the local therapy of neoplasia when administered subcutaneously or intraperitoneally.

In general, progress in the development of liposomal anticancer drugs has been slow because of formulation problems. However, in recent years, pharmaceutically acceptable liposomal anticancer drug formulations have been produced by the new drug-loading techniques [7] and by better selection of drugs.

Investigational New Drug Applications of three formulations have now been approved by the Federal Food and Drug Administration (U.S.A). Two contain doxorubicin and one a new lipophilic *cis*-platin analog, *cis*-bisneodecanoate-*trans*-R,R-1,2-diamino-cyclohexane platinum [8].

In preclinical studies, it became clear that these preparations were less toxic than the free drug form and more active in models of liver micrometastases. Clinical phase-1 and -2 studies with these formulations are now in progress. Although liposomal anticancer drugs have not yet been firmly proven to be really successful, the available clinical data indicate that they may improve the therapeutic index or broaden the application range of available anticancer drugs; they may also serve as carriers for the newly synthesized liposome-dependent anticancer drugs [8]. Some representative liposomal anticancer drugs are described in the following.

1. Liposomal Doxorubicin and Daunomycin

The use of the anticancer drugs doxorubicin and daunomycin has clinical limitations due to cardiac toxicity and myelosupressive action; weight loss and resistance after repeated treatment are also disadvantageous. In the case of doxorubicin, chronic administration can result in fatal cardiomyopathy if total lifetime dose exceeds 450–500 mg/m^2 [9]. The frequency of fulminant cardiac failure is low (2.2%) below this limit but rises sharply to 30% at exceeding high dosage [9]. It has been shown in many reports that liposomal doxorubicin can greatly diminish both chronic and acute toxicities of the doxorubicin drug with unaltered or improved antitumor activity [10]. At the same dose level, liposomal doxorubicin showed the same antitumor activity as free doxorubicin in animal survival studies to evaluate the effect of the liposomal drug on the acute toxicity of doxorubicin (Fig. 1) [11]. The LD_{50} of liposomal doxorubicin was found to be 40 mg/kg, 53% higher than that of free doxorubicin, 26 mg/kg (Fig. 2) [11]. Liposomal doxorubicin (25 mg/kg) was also found to be much less toxic than free doxorubicin in the histologic examination of the cardiac sections of mice. Edema, monocytic infiltration, and cell necrosis were observed when the free drug was administered. But in the case of liposomal preparations

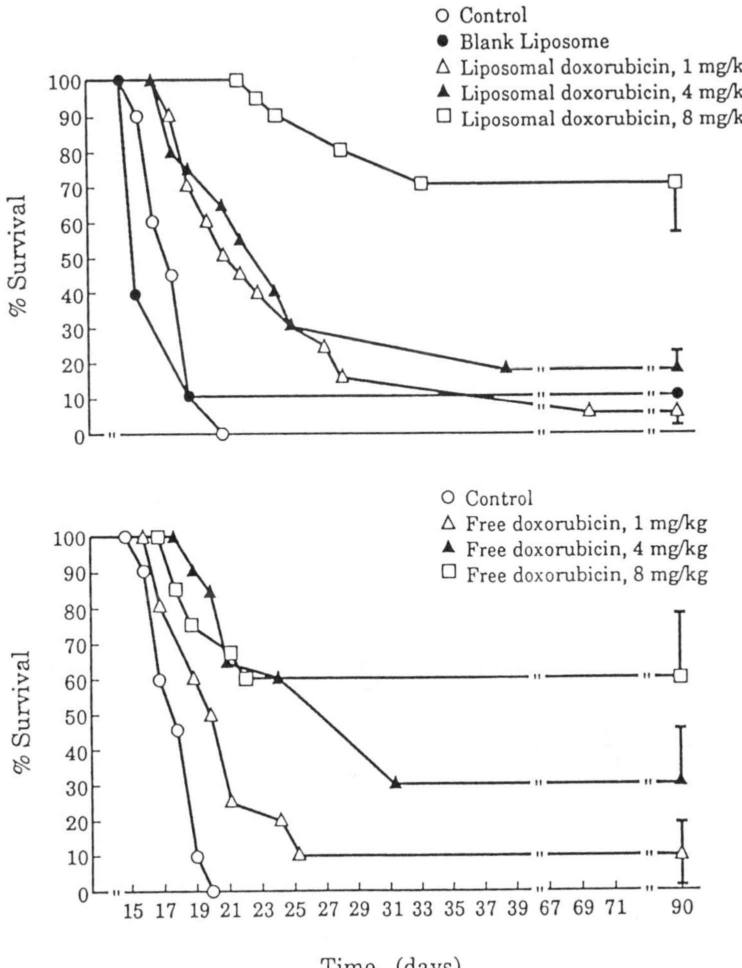

Figure 1 Antitumor activity of free and liposomal doxorubicin against the P815 mastocytoma. DBA/2J mice were inoculated with 1500 P815 cells and treated with i.v. injections of free or liposomal doxorubicin at doses of 1, 4, and 8 mg/kg. Injections of normal saline or blank liposomes were given as controls. On a mg/mg basis, there was no significant difference in survival ($P > 0.05$) between animals treated with free and those treated with liposomal doxorubicin.

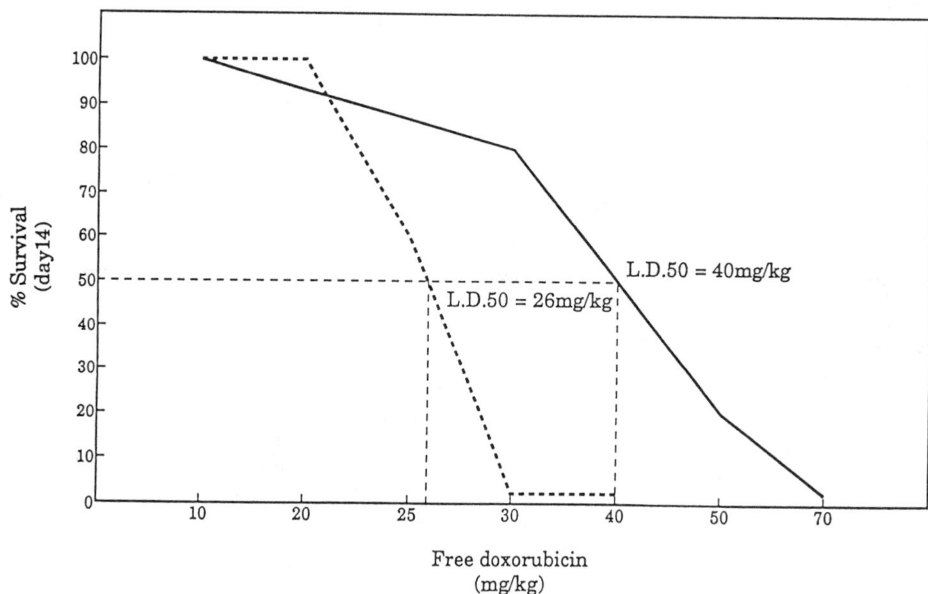

Figure 2 The 14-day survival curve for DBA/2J mice given a single i.v. injection of free doxorubicin (– – – –) and liposomal doxorubicin (——).

only slight cellular edema was observed [11]. In these studies, purified egg phosphatidylcholine and cholesterol were used to construct the liposomes (mean diameter: 200 nm).

The mechanism of liposomal doxorubicin action can be envisaged as follows [12]: the liposomes are trapped in the intracellular spaces of macrophages, where the lipid membrane is gradually degraded by the local enzymes. As membrane degradation proceeds, the doxorubicin molecule leaks out into the blood, providing a small but continuous supply of drug to the tumor. This view is supported by the fact that when free doxorubicin is administered into blood slowly it is less toxic than when administered intermittently at a higher dose without a loss in effectiveness [12].

Daunomycin for the present clinical studies was entrapped into solid, small (50–80 nm) liposomes composed of distearoylphosphatidylcholine (DSPC) and cholesterol. These liposomes have a long half-life, are relatively stable in the blood, and are able to avoid trapping by the RES. As a result, they could accumulate in the tumor by passage through the weak and often porous capillaries feeding the tumors (Fig. 3) [13]. Doxorubicin liposomes are an example of slow drug release; daunomycin liposomes are representative passive targeting.

2. *Cis*-Platin and its Derivatives

Cis-platin is a typical anticancer drug used in chemotherapies. Vomiting, nephrotoxicity, and neurotoxicity are the best known side effects of this drug. Liposomal formulations of *cis*-platin and its derivatives are being examined, which should decrease the toxicity but maintain the drug's activity. Liposomal *cis*-platin projects are proceeding in two ways. One is the encapsulation of ordinary *cis*-platin into liposomes with a high transition temperature, controlled by a combination of dipalmitoylphosphatidylcholine (DPPC) and distearoylphosphatidylcholine [14]. This formulation is used in combination with hyper-

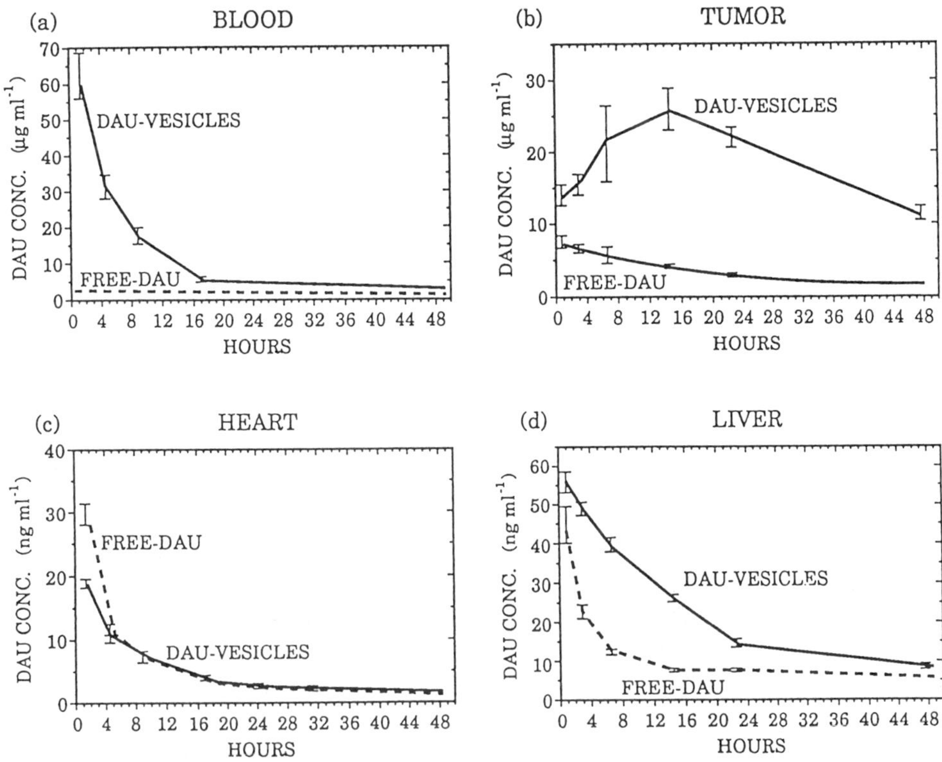

Figure 3 CD2F1 mice bearing six-day-old P-1798 solid tumors received intravenous free or vesicle entrapped daunomycin at a dose of 20 mg/kg, $n = 5$.

thermia. After warming to about 44°C of the cancer site with a high frequency, liposomal *cis*-platin is injected intravenously. The rationale for this therapy is that *cis*-platin should be released from the liposomes at the lipid chain melting phase transition temperature, which is only achieved at the warmed tumor site. The other approach is to use new lipophilic *cis*-platin analogs designed and synthesized specially for the entrapment into liposomes. The structure of a typical such compound, *cis*-bisneodecanoate-*trans*-R,R-1,2-diaminocyclohexane platinum (L-NDDP) is given in Fig. 4 [8]. The features of liposomal *cis*-platin analogs are a lack of cross-resistance in vivo in the L1210 leukemia model and a reduced neophrotoxicity [15]. These *cis*-platin analogs were shown to be encapsulated into liposomes in high yield and to be stable in liposomal formulations. Liposomal L-NDDP was found to be as active as *cis*-platin against L1210/PDD leukemia [16] and more effective than *cis*-platin in inhibiting the growth of microscopic liver metastases of M5076 reticulosarcoma in mice [8,17]. Results of animal experiments indicate that liposomal L-NDDP is initially rapidly cleared from the blood but has a subsequent long elimination phase; it distributes preferentially to the lung, liver, and spleen [8]. The extremely long elimination phase might be due to the slow release of the drug from the tissue into the bloodstream, with significant drug binding to the serum proteins and lipoproteins [18]. These studies show that liposomal L-NDDP can safely be given at therapeutic doses to animals.

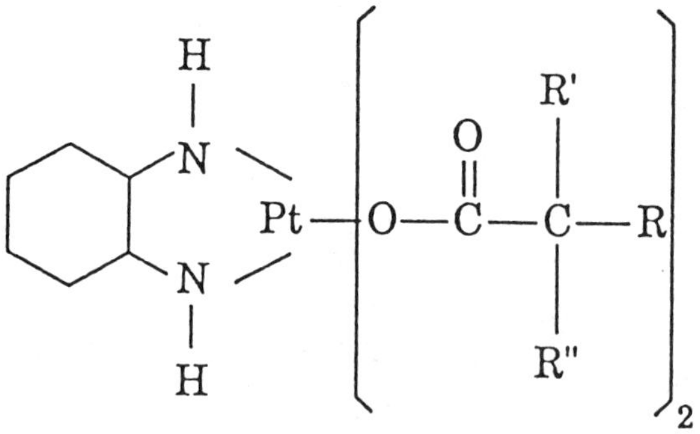

Figure 4 Chemical structure of NDDP. R, R', and R'' can be an aliphatic chain of 2–6 carbons. Combined with the carboxylate group, the neodecanoic moiety has an empirical formula of $C_{10}H_{19}O_2$.

3. Liposomal Cytosine Arabinoside (Ara-C) Derivative

Cytosine arabinoside (Ara-C) is one of the most active drugs for the treatment of acute myelogenous leukemia (AML). Optimal dosage and mode of drug delivery to the tumor cells are, however, still a matter of debate. A liposome-entrapped ara-C derivative (*N*-4-oleyl-cytosine-arabinoside; O-Ara-C) was applied to patients resistant to conventional chemotherapy (Ara-C, daunorubicin) in phase-1 and phase-2 studies [19]. Small size of liposomes (80–100 nm) should ensure that the drug is taken up by the cells with an active pinocytosis, preferentially, which has been known to be the case with certain AML cells. Five hundred milliliters of liposomes containing 420 mg of O-ara-C (corresponding to 200 mg of Ara-C) was continuously infused intravenously over a 3 h period for the first step at 1, 4, and 7 days and for the next step at days 1 and 6. Such treatment ensured a partial remission, thus encouraging further use of liposomal cytosine arabiniside derivatives.

C. Application with Diagnostic Drugs

It has been announced by Vestar Inc. (Los Angeles, Calif.) that one diagnostic tool consisting of small unilamellar liposomes containing In^{111} complex for the imaging of tumors is in the clinical testing phase-3 and -4 in Europe. These liposomes are prepared from synthetic DSPC and are very small (diameter 50–80 nm) and rigid, since they also contain cholesterol (molar ratio of DSPC/cholesterol = 2/1). These characteristics endow liposomes with a long half-life in the bloodstream and result in their accumulation in tumors. Although much effort has been expended to improve the quality of images in the field of magnetic resonance imaging, imaging with free metal complexes is still not satisfactory in the upper abdomen and particularly in the liver [20]. The fast blood and organ clearance rates of such metal complexes (Gd^{3+}, Mn^{2+}, Fe^{3+}) restrict their use to studies of perfusion, capillary integrity, and the extracellular space.

The intrinsic property of liposomes to accumulate in the RES has prompted their use as carriers of contrast agents to such organs. It was shown that manganese chloride entrapped in large unilamellar vesicles can enhance selectively the intensity of the proton signals in the livers of mice [21]. Mn-DTAP(diethylene triamine pentaacetic acid) entrapped in liposomes has increased the manganese target into liver and spleen 2–12 times and has reduced the proton relaxation rate markedly. DTAP stearate was used as the anchoring reagent for the metal complexes in such liposomes, which should prevent the diffusion of complexes from the liposomes. The delivery of gadolinium via such liposomes to liver and spleen reduced the proton relaxation rate even more sharply [22].

From other related studies [23] it has been concluded that Mn-DTAP stearate liposomes (mean diameter 26–36 nm) prepared from soybean phosphatidylcholine and cholesterol are most suitable for the improvement of image quality, having given the strongest signal enhancement among the DTAP metal complexes; this is believed to be due to their rapid accumulation in the liver: 35% of the total applied dose was found in the liver 30 min after application; 80% of it disappeared from the liver rapidly, however (Fig. 5, Table 1) [23]. Gd- and Fe-DTAP stearate liposomes are inferior in these points. Thus Mn-DTAP could prove to be a viable contrast agent for the magnetic resonance imaging of the RES with the aid of lipid vesicles.

In addition to the above-described applications, liposome-encapsulated tumor necrosis factor [24] and liposome-encapsulated macrophage-activating factor [25] are now undergoing clinical studies.

D. Artificial Blood

Demand for transfusion blood is continually increasing owing to the increasing number surgical operations. Adequate supplies of blood of suitable types are not always available, however, because of shortages of blood, blood infections with AIDS or hepatitis viruses,

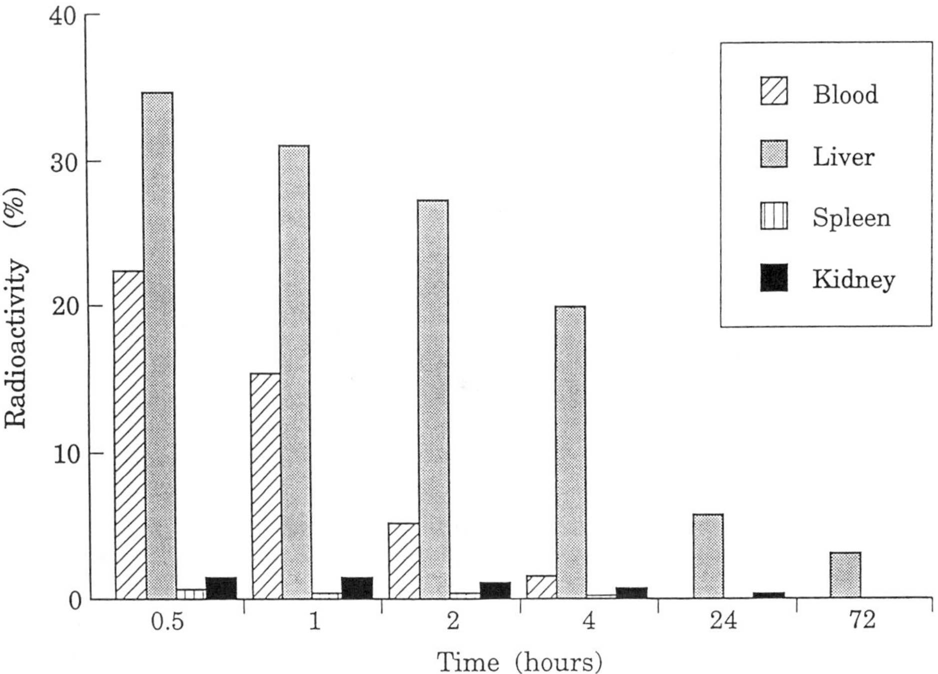

Figure 5 Organ distribution in percent per organ of Mn-DTPA-SA liposomes after intravenous injection (0.03 mM Mn-DTPA-SA/kg b.wt.) into rats. The bars represent the mean of 3 values per time point.

Table 1 Organ Elimination Half-Lives of Gd- and Mn-DTPA-SA Liposomes in Rats[a]

Organ	Gd-DTPA-SA		Mn-DTPA-SA	
	$t_{1/2}(1)^a$ (h)	$t_{1/2}(2)^b$ (h)	$t_{1/2}(1)^a$ (h)	$t_1/2(2)^b$ (h)
Blood	2.07	24.6	0.8	46.2
Liver	—	61[c]	0.25	10.2
Spleen	—	105	0.77	9.2
Kidneys	2.0	580	3.5	89.0
Lung	2.7	580	0.6	13.6
Bones	—	26	—	—
Small intest.	—	—	2.16[d]	—

[a,b]Half-lives of mono- (a) and biphasic (b) elimination curves were calculated by least-squares fitting.
[c]Calculated from 4 to 192 h.
[d]Calculated from 2 to 72 h.

etc. New blood substitutes, especially erythrocyte substitutes, therefore, are required urgently. An artificial red blood cell should satisfy the following criteria, at least:

1. Absence of toxicity and antigenicity
2. Possibility of large-scale production
3. Capability of passing easily through capillaries
4. Oxygen-carrying characteristics similar to those of natural red blood cells
5. Stability and suitable half-life in the bloodstream
6. Compatibility with other blood components

The structures of natural red blood cells and liposomes resemble each other, at least in the first approximation. The use of liposomes with entrapped hemoglobin (Hb) as artificial erythrocytes, therefore, has been extensively investigated over the last year. In this section, some of the characteristics of liposomes containing Hb, called *n*eo *r*ed *c*ells (NRC) are described.

The general properties of NRC are shown in Table 2 [26]. NRC were prepared from lipid mixtures of hydrogenated soy lecithin, cholesterol, myristic acid, and alpha-tocopherol in molar ratios of 7:7:2:0.28, and stroma-free hemoglobin and inositol hexaphosphate[IHP] as an allosteric effector ([IHP]/[Hb] = 1/1). The mean diameter of the NRC was about 200 nm, and the vesicles had an oligomeric lamellar structure [26]. Acute toxicity of NRCs was studied via animal tests (rats, rabbits, and dogs) by exchange transfusion [26]. No evidence of acute toxicity was observed in terms of blood lactate level, pH, body temperature, or number of leukocytes and platelets, except for an increase in the phagocytosis rate by the RES. One of serious problems of NRC liposomes is that liposomes tend to aggregate in serum because of interaction with plasma proteins. This may have lethal effects. When the NRC surfaces are modified with a polyethyleneglycol (PEG) derivative, however, the surface-modified NRC liposomes do not show any hemolitic or other toxicity.

The effect of NRC on dogs suffering from hemorrhagic shock has also been evaluated [27]. After 80% blood exchange with NRCs, the blood gas was analyzed. The hemodynamic parameters were improved to the normal range after an administration of NRCs. Oxygen carried by NRCs was 43.2% of the total body oxygen consumption, when

Table 2 Characteristics of NRC Suspension (Hb = 15 g/dL) Compared with Those of RBC or SFH

Property		NRC	RBC	SFH
Mean particle size	(μm)	0.2	5–8	—
Total lipid conc.	(g/dL)	9.3	0.5	0
Packed volume ratio	(%)	51	45	—
Viscosity[a] at 37°C	(cp)	4.3	3.8	10
Met Hb ratio	(%)	5>	1>	1>
P_{50}[b]	(mmHg)	40	28	11
Hill coefficient	(n)	1.9	2.8	2.7
A-V O_2 delivery	(ml/dL)	7.7	5.6	0.6

[a]Share rate: 192 sec^{-1}. [b]The partial pressure of O_2 at which the Hb is half saturated.[c] The volume of O_2 that can be delivered to tissues by the suspensions was calculated assuming arterial and venous PO_2 value to be 100 and 40 mmHg, respectively.
RBC, Red blood cell; SFH, Stroma-free hemoglobin; NRC, Neo red cell.

normal room air was inhalated. Thus NRC have the potential to provide an artificial blood supply for the treatment of hemorrhagic shock.

III. USE OF PHOSPHOLIPIDS IN LUNG SURFACTANTS

The study of lung surfactants, which began in the 1960s, has been developing rapidly since 1970. Each of approximately three hundred million alveoli contains a substance on its surface that depresses the surface tension of the gaseous phase-liquid phase interface considerably. This substance is called lung surfactant. Lack of this substance or too great a deficiency of it cause the lungs to become gradually weakened, with the alveoli losing both dilatability and contractivity.

Lung surfactant, which comprises phospholipids, fatty acids, protein, cholesterol, and the like, has dipalmitoylphosphatidylcholine (DPPC) as the main active component, this phospholipid occurring at the air-lung interface in a regular arrangement of a monolayer. Exchange between CO_2 in the blood and inspired O_2 takes place smoothly at this interface.

Infant respiratory distress syndrome (IRDS), whose onset involves related factors such as prematurity, asphyxia, pulmonary circulation insufficiency, etc., occurs in situations of deficient or depressed lung surfactant. In recent years, various kinds of artificial lung surfactants with DPPC as the chief component have been developed. Some of these have been confirmed to be effective in animal tests, mainly using rabbits, and are now used clinically. The available surfactant surrogates may contain surfactants extracted from mammals and purified. Alternatively, artificial lung surfactants (bovine lung proteins, DPPC, and fatty acids) are used. Phase surfactants show no particular toxicity on bronchopulmonary dysplasia. Generally, 60–200 mg/kg of surfactant are used prior to the first respiration or in the case of respiratory distress syndrome [28].

Another category is apoprotein-free artificial lung surfactant. One such surfactant comprises only lipoproteins prepared from DPPC and phosphatidylglycerol (PG) [29]; another one contains DPPC, hexadecyl alcohol, and thyloxapol [30]. These surfactants have effects similar to those of protein-containing surfactants and are applicable not only

Table 3 Pulmonary Surfactant Replacements

Type	Derived from	Advantages	Disadvantages
Natural	Human amniotic fluid	High efficacy	Low availability Risk of infection (herpes, AIDS etc.)
Natural	Bovine lung Porcine lung	Cheap, good efficacy	Antigenic
Semisynthetic	Cow/pig lung + phospholipids	Improved efficacy Better yield, inexpensive	Antigenic
Synthetic		Lack biological hazards Long half-life Easy production	Less efficacious
Genetically engineered	Human surfactant protein	High efficacy, pure & safe, reproducible, pathogen-free	Very expensive Early stage of development

Table 4 Competitive Position

Compound	Company	Type	Status
Infasurf	Ony, Inc.	Natural calf lung	Phase-1
SF-RI-1	Boehringer Ingelheim	Natural bovine lung	Preclinical
Curosurf	Chiesi Farmaceutici	Natural porcine lung	Phase-111
Survanta[a]	Abbott Labs	Semisynthetic bovine-based	Approved by FDA in 1990
Surfacten	Tokyo Tanabe		Launched in Japan
Exosurf	Burroughs Wellcome	Synthetic	Approved by FDA in 1990
Alec[b]	Enzymatix (UK) Brittanica Pharmaceutical Amashamu	Synthetic	Approved in UK 1989
	Cal Bio/Genentech	Recombinant human lung	Preclinical

[a]Licenced to Abbott Lab from Tokyo Tanabe.
[b]Alec = Artifical lung-expanding compound (mixture of DPPC and natural PG).

to newborns but also to adults [31]. Tables 3 and 4 show characteristics and developmental status of various lung surfactant preparations.

IV. ANTICANCER ACTIVITY OF LYSOLECITHIN DERIVATIVES

Alkyl lysophosphatidylcholine(ALP) analogs (Fig. 6), which are derivatives of natural lysophosphatidylcholine, have an anticancer function of a new type [32,33,34]. In vitro experiments have revealed that they may be selectively cytotoxic for cancer cells. Effects

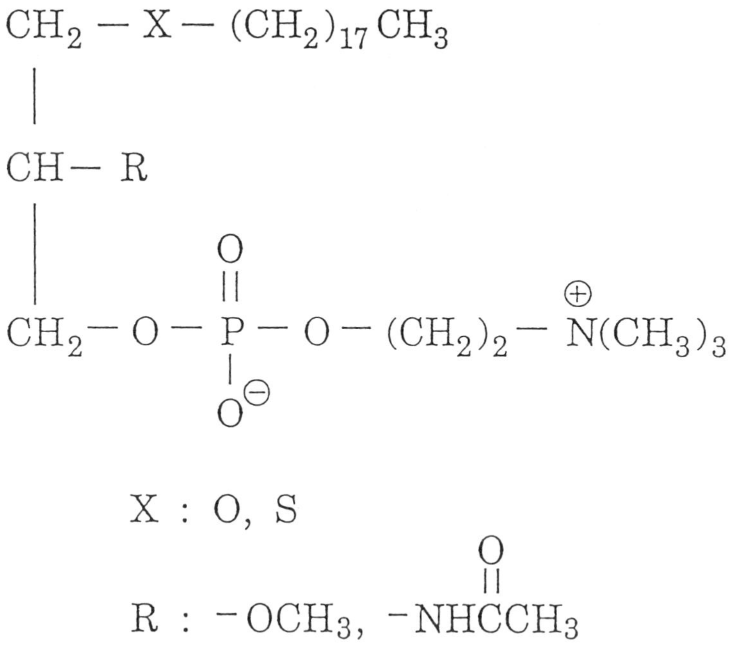

$$\begin{array}{l} CH_2 - X - (CH_2)_{17}CH_3 \\ | \\ CH - R \\ | \\ CH_2 - O - \underset{O^{\ominus}}{\underset{|}{\overset{O}{\overset{\|}{P}}}} - O - (CH_2)_2 - \overset{\oplus}{N}(CH_3)_3 \end{array}$$

$$X : O,\ S$$

$$R : -OCH_3,\ -NH\overset{O}{\overset{\|}{C}}CH_3$$

Figure 6 Structure of alkyl lyso phosphatidylcholine derivative.

of various ALPs were examined in the case of normal human marrow cells and HL-60 leukemic cells; a remarkable difference in the sensitivity to ALP was noted. In the case of ET-18-OMe (1-*O*-octadecyl-2-*O*-methyl-rac-glycero-3-phosphocholine), normal human marrow cell concentration four times higher than that of HL-60 leukemic cells was required in order to obtain an equivalent clonogenicity inhibition after a 24 h incubation with the phospholipid preparation. An even greater difference was noted 6 h after the beginning of incubation. At a 16 μg/mL concentration, the growth of HL-60 colonies was completely inhibited.

1-Thioether (TLP) and 2-acetoamide of phospholipid derivatives have shown similar activities. Cytotoxicity of these substances for leukemic cells was in the order ET-18-OMe > TLP > 2-acetoamide ALP [35]. No great difference was noted between leukemic and nonneoplastic marrow cells [36]. The results of an examination of the antineoplastic activity of ET-18-OMe, according to a human tumor clonogenic assay method [35], have revealed no substantial therapeutic effect after 1 h of incubation at diluted drug suspension ($< 1 \mu$g/mL). A mild dose-response effect was noted at higher concentrations (1–10 μg/mL), however. At this point, 60% of the cells were more or less seriously affected.

Diminution of the colony number was obtained with breast, ovarian and lung cancer cells as well as with mesothelioma. However, a difference in the sensitivity to ET-18-OMe of the cell types ranged between one and five.

TLP was reported to inhibit the ^{3}H-thymidine uptake by 8 kinds of leukemic cells and by 12 kinds of solid cancer cells. Like ET-18-OMe, TLP exerted its effect in a dose and incubation time dependent manner. Examination of the cytotoxicity of TLP toward 19 kinds of marrow cells revealed a remarkably higher cytotoxicity on the leukemic cells than on the normal cells; leukocytoblasts were completely destroyed at drug concentrations >5 μg/mL after 48 h, but control with solid cancers required a concentration of 10–20 μg/mL for a noticeable effect. The ester-bonded lysolecithin did not exert any effect on the tumor cells comparable concentrations, suggesting that the anticancer activity of lysolecithin derivatives requires a stable substituent group at the *sn*-2 position and an alklyl ether group at the *sn*-1 position. The reason for the specificity may be that unlike normal cells, cancer cells lack an alkyl group-decomposing enzyme to decompose ALP absorbed thereby. Accumulation of ALP leads to the inhibition of the turnover of cancer cell membranous phospholipids. Finally, this causes cell membrane destruction and results in the death of the cancer cells.

As a phase-1 test, a trial application of ET-18-*O*-Me is now being performed for the removal of the residual leukemic cells from the bone marrow of patients in a state of remission.

REFERENCES

1. K. C. Carter, and J. O'Grady, *Int. J. Pharm. 53*:129–137 (1989). Direct comparison of sodium stibogluconate treatment in two animal models of human visceral leishmaniasis, mouse and hamster.
2. C. R. Alving, *Proc. Natl. Acad. Sci. USA 75*:2959–2963 (1978). Therapy of leishmaniasis: Superior efficacies of liposome-encapsulated drugs.
3. G. Lopez-Berestein, *J. Infect. Dis. 151*:704–710 (1985). Liposomal amphotericine B for the treatment of systemic fungal infections in patients with cancer.
4. H. J. Clinique, *Infection 16(3)*:141–147 (1988). Liposome as drug delivery system in the treatment of infectious disease. Potential applications and clinical experience.
5. R. T. Metha, R. L. Hopfer, and L. A. Gunner, *Antimicrob. Agents Chemother. 31(12)*:1897–1900 (1987). Formulation, toxicity and antifungal activity in vitro of liposome encapsulated nystatin as therapeutic agent for systemic candidiasis.
6. R. T. Metha, R. L. Hopfer, and T. McQueen, *Antimicrob. Agents Chemother. 31(12)*:1901–1903 (1987). Toxicity and therapeutic effects in mice of liposome-encapsulated nystatin for systemic fungal infections.
7. L. D. Mayer, L. C. L. Tai, M. B. Bally, G. N. Mitilenes, R. S. Ginsberg, and P. R. Cullis, *Biochim. Biophys. Acta 816*:143–151 (1985). Characterization of liposomal systems containing doxorubicin entrapped in responce to pH gradients.
8. R. Perez-Soler, J. Lautersztain, L. C. Stephens, and K. Wright, *Cancer Chemother. Pharmacol. 24*:1–8 (1989). Preclinical toxicity and pharmacology of liposome-entrapped *cis*-bisneodecanoato-*trans*-R,R-1,2-diamino-cyclohexane platinum(II).
9. R. Dorr, and W. Fritz, *Cancer Chemotherapy Handbook*, Elsevier, New York, 1982, p. 388. Drug data sheets, doxorubicin.
10. A. Gabison, D. Goren, Z. Fucks, A. Barenholz, and A. Meshorer, *Cancer Res. 43*:4730–4735 (1983). Enhancement of adriamycin delivery to liver metastatic cells with increased tumoricidal effect using liposomes as drug carriers.
11. J. A. E. Balazsovits, L. D. Mayer, M. B. Bally, P. R. Cullis, M. McDonell, R. S. Ginsberg, and R. E. Falk, *Cancer Chemotherapy Pharmacol. 23*: 81–86 (1989). Analysis of the effect of liposome encapsulation on the vesicant properties, acute and cardiac toxicities, and antitumor efficacy of doxorubicin.
12. M. J. Ostro, *Sci. Amer. 256*:102–111 (1987). Liposomes.
13. E. A. Forssen, *Liposomes as Drug Carriers* (G. Gregoriadis, ed.), Wiley, New York, 1988, p. 355. Chemotherapy with anthracycline-liposomes.
14. K. Iga, N. Hamaguchi, and T. Shimamoto, *Int. J. Pharm. 57*:241–251 (1989). Heat-specific drug release of large unilamellar vesicle as hyperthermia-mediated targeting delivery.

15. J. H. Burchenal, K. Kalaher, T. O'Toole, and J. Chrisholm, *Cancer Res.* *37*:3455–3457 (1977). Lack of cross-resistance to certain platinum coordination compounds in mouse leukemia.
16. R. Perez-Soler, R. K. Abdul, and G. Lopez-Berestein, in *Liposomes as Drug Carriers* (G. Gregoriadis, ed.), Wiley, New York, 1988, p. 401. Development of lipophilic cisplatin analogs encapsulated in liposomes.
17. R. Perez-Soler, R. K. Abdul, and G. Lopez-Berestein, *Cancer Res.* *47*:6462–6466 (1987). Treatment and prophylaxis of experimental liver metastases of M5076 reticulosarcoma with *cis*-bisneodecanoato-*trans*-R,R-1,2-diaminocyclohexaneplatinum(II) encapsulated in multilamellar vesicles.
18. J. Lautersztain, Y. Moreno, R. Perez-Soler, and R. K. Abdul, *Proc. Am. Assoc. Cancer Res.* *29*:4509–4512 (1988). Interaction of liposomal *cis*-bisneodecanoato-*trans*-R,R-1,2-diaminocyclohexane-platinum(II).
19. B. Pestalozzi, C. Sauter, H. Schott, H. Hengatner, and R. Schwendener, *Proc. Ann. Meet. Am. Soc. Clin. Oncol.* *7*:A690 (1988). Acute myelogenious leukemia: Treatment with liposome containing N4-oleyl-cytosine-arabinoside.
20. G. M. Glazzer, *Radiology* *166*:303–312 (1988). MR imaging of the liver, kidneys, and adrenal glands.
21. R. L. Margin, S. M. Wright, M. R. Niesman, H. C. Chan, and H. M. Swartz, *Magnetic Resonance Med.* *3*:440–447 (1986). Liposome delivery of NMR contrast agents for improved tissue imaging.
22. G. Kabalka, E. Buonocore, K. Hubner, T. Moss, N. Norley, and L. Huang, *Radiology* *163*:255–258 (1987). Gadolinium-labeled liposomes: Targeted MR contrast agents for the liver and spleen.
23. R. A. Schwendener, R. Wuthrich, S. Duewell, G. Westera, and G. K. von Schulthess, *Int. J. Pharm.* *49*:249–259 (1989). Small unilamellar liposomes as magnetic resonance contrast agents loaded with paramagnetic Mn-, Gd- and Fe-DTAP-stearate complexes.
24. R. J. Debs, H. J. Fuchs, R. Philips, E. N. Brunette, and N. Duzgunes, *Cancer Res.* *50(2)*:375–380 (1990). Immunomodulatory and toxic effects of free and liposome-encapsulated tumor necrosis factor alpha in rats.
25. F. J. Kos, *Arch. Geschwulstforsch.* *58(6)*:387–393 (1988). Some remarks on application of macrophages in tumor therapy.
26. H. Yosioka, K. Suzuki, Y. Miyauchi, and A. Takahasi, in *9th International Meeting on Artificial Organs, Tokyo, Japan*, 1989, pp. 1–7. Characteristics of neo red cells and their in vivo oxygen transport capacity.
27. A. Usuba, H. Inoue, R. Motoki, K. Suzuki, Y. Miyauchi, and A. Takahasi, in *XVII Congress of the European Society for Artificial Organs, Bologna, Italy*, 1990. Effect of neo red cells on the canine hemorrhagic shock model.
28. M. S. Reynolds, and K. A. Wallander, *Clin. Pharm.* *8(8)*:559–576 (1989). Use of surfactant in the prevention and treatment of neonatal respiratory distress syndrome.
29. A. Liberator, G. Carrera, G. Calciolari, C. Coccia, and B. Coppalini, *Ann. Osp. Maria Vittoria Torino* *30(1)*:28–35 (1987). Substitutive therapy with surfactants in premature infants with respiratory distress syndrome.
30. J. A. Clements, Lung surfactant compositions, *U.S. Patent 4,826,821* (1989).
31. R. J. Mason, *Eur. J. Respir. Dis (Suppl.).* *153*:229–236 (1987). Surfactant in adult respiratory distress syndrome.
32. D. R. Hoffman, L. H. Hoffman, and F. Synder, *Cancer Res.* *46(11)*:5803–5809 (1986). Cytotoxicity and metabolism of alkyl phospholipid analogues in neoplastic cells.
33. D. B. Harmann, and H. A. Heumann, *J. Biol. Chem.* *261(17)*:7742–7747 (1986). Cytotoxic ether phospholipids. Different affinities to lysophosphocholine acytransferases in sensitive and resistant cells.

34. M. R. Berger, P. G. Munder, D. Schmahl, and O. Westphal, *Oncology 41(2)*:109–113 (1984). Influence of the alkyllysophospholipid ET-18-OCH_3.
35. W. E. Berdel, D. D. Von Hoff, C. Unger, H. D. Schick, U. Fink, A. Reichert, H. Eible, and J. Rastetter, *Lipids 21(4)*:301–304 (1986). Ether lipid derivatives: Antineoplastic activity in vitro and the structure-activity relationship.
36. H. D. Schick, W. E. Berdel, M. Fromm, U. Fink, H. Eible, and C. Unger, *Lipids 22(11)*:904–910 (1987). Cytotoxic effects of ether lipids and derivatives on human nonneoplastic bone marrow cells and leukemia cells in vitro.

28
Phospholipids in Disease

C. M. Gupta *Institute of Microbial Technology, Chandigarh, U.T., India*

I. INTRODUCTION

Any disease is primarily characterized by the defective functioning of a given type of cell. This defect is often reflected by some abnormalities in the structure and function of the target cell membranes. As phospholipids form the basic backbone of all animal cell membranes, their abnormal synthesis, structure, or organization invariably leads to some distinct abnormalities in cellular structure and function.

Defects in phospholipid synthesis, besides affecting the absolute amounts of phospholipids per cell, also affect the relative ratios of the various phospholipid species in the membrane, which in turn should lead to changes in the membrane structure and consequently the function. The membrane structure is rendered abnormal also by changes in the phospholipid polar head group and/or fatty acyl chains, which may be caused by some enzymatic defects in the diseased cells. These changes in the phospholipid metabolism generally lead to an altered phospholipid organization within the membrane. Also, changes in the membrane phospholipid organization could be induced by some defects in the membrane protein structure and/or organization or by an altered membrane cholesterol level.

Altered phospholipid metabolism has been reported in a variety of diseases, but except for a few diseases the information available is very scanty. This chapter briefly summarizes most of the relevant information available on this aspect. The chapter is divided into nine further sections, eight of which deal with phospholipid metabolism in eight kinds of diseases: anemias, malaria, cancer, muscular dystrophy, ischemia, diabetes, lung diseases, and liver diseases. The last section attempts to summarize the relevant information available on other diseases.

II. PHOSPHOLIPIDS IN ANEMIAS

Phospholipid levels are decreased in blood plasma and increased in erythrocytes of sickle cell patients, as compared to normal subjects [1]. The sickled erythrocytes seem to be more sensitive to lipid peroxidation than normal cells, since these cells accumulate 2–3 times more malonyldialdehyde (MDA) than the normal erythrocytes, after 24 h of aerobic incubation [2]. Moreover, the irreversibly sickled erythrocyte-enriched fraction is known to contain a novel phospholipid MDA adduct, which is not seen in control cells [2]. Also, this fraction contains less poly phosphoinositides (PtdIns) and more phosphatidate than normal cells, which could be due to activation of poly PtdIns phosphodiesterase in irreversibly sickled cells (ISC) [3]. Further, lipid peroxidation could affect erythrocyte survival in blood circulation, as even a small amount of MDA in in vitro experiments tends to decrease the erythrocyte survival and produces phosphatidylserine (PtdSer) phosphatidylethanolamine (PtdEtn) adduct in both normal and sickle cell membranes [4].

The glycerophospholipid distribution within the sickled erythrocyte membrane is different from that in nonsickled cells [5]. Compared with normal erythrocytes, the outer membrane leaflet of deoxygenated, reversibly sickled cells (RSC) and ISC is enriched in PtdEtn in addition to containing PtdSer. These changes were compensated for by a decrease in phosphatidylcholines (PtdCho) in that layer. The distribution of sphingomyelins (SphPCho) over the two halves of the bilayer remains unaffected by sickling. In contrast to ISC, where the phospholipid organization has been shown to be abnormal under both oxygenated and deoxygenated conditions, reoxygenation of RSC almost completely restores the organization of membrane phospholipids to normal [5]. Further studies on these lines have, however, shown that the PtdSer externalization in sickled cells is limited only to the spiculated regions and that the despiculated sickled cells contain normal transbilayer phospholipid distribution [6]. Contrary to these results, Raval and Allan [7] have observed normal phospholipid distribution in both sickled erythrocytes and spectrin-free spicules derived from these cells. However, altered phospholipid distribution in ISC has also been suggested recently, using spin labeled phospholipid [8].

Unlike the sickled erythrocytes, the total phospholipid content is decreased in erythrocytes of β-thalassemia patients, as compared to normal subjects [9–11]. While the PtdCho and SphPCho contents are increased, PtdEtn and PtdSer concentrations decreased in both β-thalassemia minor and major patients [9,11].

The erythrocyte phospholipid content and distribution are altered in anemias caused by pyruvate kinase or glucose-6-phosphate deficiency [12]. Also lower than normal contents of phospholipid are observed in erythrocyte membranes of patients with nephrogenous anemia; the SphPCho fraction decreased, whereas PtdCho fraction increased [13].

III. PHOSPHOLIPIDS IN CANCERS

In a variety of cancers, the tumorous tissue contains more phospholipid than normal tissue in which the tumor arose [14–17]. However, liver cancer tissues only are characterized by lower phospholipid contents and decreased amounts of stearic and linoleic acids relative to the total fatty acid content [18–23].

Tumorous growth is accompanied by changes in ratios between the various lipids of lysosomes [24]. Phospholipids exhibit the greatest change. The share of PtdCho in the entire lipid fraction is increased and of PtdSer decreased. Differences are observed in the asymmetric arrangement of phospholipids on the membrane surface of intact and tumorous lysosomes [24].

The molecular species of phospholipids in hepatomas are different from those observed with normal liver [25]. The differences are mainly confined to phospholipid fatty acyl chains; there is a marked reduction in degree of unsaturation in hepatomas, as compared to normal liver tissue. Moreover, ethanolamine-containing phospholipid is increased in rat malignant tumors and decreased in livers [26]. Unlike hepatomas, the mammary gland tumor phospholipid contains higher concentrations of polyunsaturated fatty acids as compared to healthy breast tissue [15]. Besides, breast tumor phospholipid has higher proportions of 1,2-diacyl-*sn*-glycero-3-phosphoinositol, 1,2-diacyl-*sn*-glycero-3-phosphoethanolamine, and 1-alkenyl-2-acyl-*sn*-glycero-3-phosphoethanolamine; and lower proportions of 1,2-diacyl-*sn*-glycero-3-phosphocholine than controls [16].

Brain tumors contain substantial levels (0.8–3.4% of total phospholipid) of choline plasmalogen, which is present only in trace amounts in normal brain [27]. In gliomas, a relatively smaller percentage of PtdEtn and a higher percentage of PtdCho are present [28]. In general, brain tumors contain less cholesterol, SphPCho, and serine glycerophospholipid but more choline glycerophospholipid than white matter [29], whereas in lung carcinoma tissues a marked decrease in saturated PtdCho predominantly the dipalmitoyl species, is observed, though these tissues still contain 17–20% of the saturated lysobisphosphatidic acid (PA) and more cardiolipin (CL) and PtdIns than normal lung tissue [30].

Phospholipid analysis in doxorubicin-sensitive and resistant Friend leukemic cells reveals changes in the ratios of PtdCho to PtdEtn and PtdCho to SphPCho [31]. Differences are also seen between the vinblastine-sensitive and resistant leukemic T cells with respect to the head group composition in their plasma membrane phospholipid [32].

Fatty acyl chains in Yoshida ascites hepatoma AH 130 or Ehrlich ascites tumor cell phospholipid are more randomly distributed among the 1 and 2 positions of the glycerol moiety than those observed with normal tissue phospholipid [33]. However, no such disruption of positional specificity has been observed for major phospholipid species of L 1210 leukemic lymphocytes [34], neurosarcoma, and sarcoma 180 [35].

Net phospholipid synthesis is increased in neoplasia [17,36]. In Ehrlich ascites tumor cells, the PtdIns synthesis through pathways other than *de novo* synthesis is increased [37].

Serum phospholipid levels in patients with liver cancers are higher than those observed with normal subjects [33,38]. In animals bearing Yoshida sarcoma, plasmalogen contents of liver PtdCho and PtdEtn are decreased but increased in serum PtdCho and PtdEtn [38], while in patients having renal cell carcinoma, the total serum phospholipid is decreased with decreased percentage of PtdCho in composition [39].

The ratios of phospholipid associated with nonhistone chromosomal protein in chronic lymphocytic leukemic lymphocytes are different from those observed with normal B lymphocytes. The most significant variation concerns the reduction of SphPCho content in leukemic lymphocytes. This phospholipid in vitro affects both DNA stability and transcription [40].

The nature of the phospholipid head group in plasma membranes of tumor cells determines the frequency of tumor metastasis and invasiveness [41].

Inflamed cancer cells release decomposition products of alkyl ether phospholipid and neutral lipids, alkyl lysophospholipid and alkyl glycerols. Administration of alkyl ether analogs of lysophospholipid into mice is able to induce stimulation of macrophages for ingestion with F_c receptor preference [42].

A lipid-like factor of phospholipid nature has been isolated from cell-free media of explants and cell cultures from lipids collected from human carcinoma and malignant melanoma cells [43]. This factor is unique to tumor cells and not present in detectable quantities in normal mammary epithelial cells or skin fibroblasts. It modulates macrophage properties and inhibits the macrophage chemotactic activity, spreading, lipopolysaccharide-induced tumoricidal activities and migration, and also the human normal lymphocyte responses to mitogenic stimulation [43].

IV. PHOSPHOLIPIDS IN DIABETES

Alloxan diabetes in rats is accompanied by a distinct increase in total contents of phospholipids and their fractions, especially acidic phospholipids in whole blood [44,45]. Significant changes occur in individual phospholipids of liver, heart, kidney, and brain [46,47]. A sharp increase in the CL content and a regular reduction of choline-containing lipids, PtdCho and SphPCho, in these organs are characteristics of alloxan diabetes [46]. Besides, acidic phospholipids are absorbed in the brain and neutral phospholipids are discharged, resulting in an increase in the acidic phospholipid levels and a decrease in the levels of neutral phospholipids [45,47]. Also, the levels of monophosphoinositides, phospholipid sulfides, and total phospholipids are higher than those observed with the normal rat brain [47]. Furthermore, the amounts of liver phospholipid, mainly CL and PtdEtn, are increased in the inner and decreased in the outer mitochondrial membranes, whereas in microsomes the amounts of PtdCho and PtdSer are increased in diabetes [48]. Apart from these changes, the cardiac phospholipid levels are decreased. This decrease is restricted only to PtdEtn and lyso PtdCho [49] and is mainly due to impairment of the PtdEtn synthesis and conversion of PtdCho into lyso PtdCho in the hearts of diabetic rats. Treatment of animals with insulin restores the levels of PtdEtn and lyso PtdCho to almost normal [49]. As with rats, increased levels of lyso PtdCho, ethanolamine phospholipids, and acidic phospholipids have been observed also in dog brain arterial and venous blood during alloxan diabetes [50].

The fluidity of RBC membrane is decreased in type I diabetes (insulin-dependent). This change in membrane fluidity is related to a change in the membrane cholesterol to phospholipid ratio and plasma cholesterol content [51]. Besides erythrocytes, lipid changes occur also in platelets during diabetes [52]. Thus the amino phospholipid PtdEtn and PtdSer contents in platelet plasma membrane are increased in poorly controlled diabetes, whereas in well controlled cases only PtdSer has been shown to increase [52]. However, in both groups, the concentration of arachidonic acid in platelet plasma membrane phospholipid is increased [52]. Further, it has been suggested that the spontaneous aggregation of platelets in diabetes may be due to the appearance of anionic phospholipids in the outer half of the plasmalemma [53].

Changes in phospholipid contents and biosynthesis occur in glomeruli of rats 2 and 10 months after induction of diabetes with streptozotocin [54]. The amounts of PtdEtn is decreased and that of SphPCho increased 10 months after inducing diabetes. The main change in the phospholipid fatty acid composition 2 months after inducing diabetes is a decrease in arachidonic acid, while after 10 months there is an increase in linoleic acid and a decrease in arachidonic acid. Acyl-CoA synthetase and acyl transferase activities are increased in diabetic glomeruli with either arachidonic acid or linoleic acid as the substrate 2 months after diabetes. However, after 10 months, acyl-CoA is increased with linoleic acid but not with arachidonic acid as the substrate. Also, acyl transferase activity is decreased with arachidonic acid but not with linoleic acid as the substrate [54].

V. PHOSPHOLIPIDS IN DYSTROPHY

Muscular dystrophy in mice leads to an abnormally high lipid metabolism in nerve and muscle tissues [55]. On the contrary, no differences have been reported in the phospholipid contents of tongue or hind leg muscles of normal and dystrophic mice; though the contents of 18:0, 18:1, and 18:2 fatty acids have been shown to increase in hind leg muscles of dystrophic mice as compared to controls [56]. Also, no significant change is observed in skeletal muscle total phospholipid after dystrophy. However, PtdCho in total phospholipid is decreased by about 9%, and in both PtdCho and PtdEtn fractions a large decrease is observed in 16:0 fatty acid content with an increase in proportions of 18:0 and 18:1 fatty acids [57].

One month after myocardial dystrophy, the amount of PtdSer is increased in mitochondria by 37% and in microsomes by 41%, whereas PtdEtn decreased in all fractions [58]. Moreover, in liver microsomes, total phospholipid, PtdCho, and PtdEtn contents are increased in dystrophic animals as compared to controls [59]. Furthermore, in dystrophic muscle microsomes PA fraction, the incorporation of [^{14}C]glycerophosphate is delayed and decreased [60]. Besides, transverse tubule membrane vesicles and sarcolemma membranes isolated from genetically dystrophic chicken breast muscle display elevated phospholipid levels as compared to controls [61].

Dystrophic cell lines contain a higher relative distribution of individual phospholipids than their normal counterparts [62]. The percentage of acidic phospholipids is increased, with a decrease in PtdCho content of dystrophic cells as compared to normal cells [62].

The dystrophic chicken erythrocyte membrane has an increased content of PtdSer and a decreased concentration of PtdEtn compared to control birds [63]. Also, a distinct polar lipid, adjacent to PtdEtn on thin-layer chromatography plates, has been observed in dystrophic preparation [63]. This lipid could be the unique ethanolamine plasmalogen reported to be present in dystrophic erythrocytes [64].

Phospholipid metabolism appears to be altered in erythrocytes of myotonic dystrophic patients [65]. Perhaps acylation of lyso phospholipid becomes abnormal in this disease [66]. Also, the ratio of plasmalogen form to diacyl form of PtdEtn is decreased in dystrophic erythrocytes as compared to controls [67]. However, the erythrocyte membrane fatty acid composition or metabolism remains essentially unaltered in myotonic dystrophy [68].

PtdCho concentrations are decreased but lyso PtdCho and SphPCho concentrations increased in erythrocytes of patients with Duchenne muscular dystrophy as compared to healthy subjects [69]. Also, the membrane phospholipid organization in dystrophic erythrocytes appears to be abnormal as judged after erythrocyte digestion with phospholipase A_2 [70]. On the contrary, other workers have reported no significant change in membrane phospholipid of dystrophic erythrocytes, but they did observe some changes in erythrocyte membrane phospholipid fatty acid composition during Duchenne muscular dystrophy [71,72]. Besides erythrocytes, the phospholipid contents are normal also in dystrophic platelets, though these cells contain elevated cholesterol levels [73].

Finally, significant increase in the erythrocyte membrane SphPCho is observed in X-chromosomal and autosomal malignant muscular dystrophy [74]. Also, the fatty acyl chain compositions of PtdEtn, SphPCho, and PtdCho are changed in dystrophic erythrocytes as compared to normal controls [74].

VI. PHOSPHOLIPIDS IN ISCHEMIA

Phospholipid metabolism is markedly affected by myocardial ischemia. Membrane phospholipids are degraded in the sarcoplasmic reticulum and mitochondrial fractions of canine heart during acute myocardial ischemia [75,76]. PtdCho, PtdEtn, and CL decreased and lyso PtdCho/lyso PtdEtn contents increased in isoproterenol-induced ischemic heart, as compared to normal controls [77]. It has further been shown that ischemia primarily decreases the membrane PE present at the cytosolic side of the sarcolemmal membranes [78]. Etherphospholipids seem to be more prone to degradation than diacylphospholipids [79].

Plasma and erythrocyte lipids are also altered in patients with ischemic heart disease. Cholesterol levels in plasma and erythrocytes, phospholipid levels in erythrocytes, cholesterol to phospholipid ratio in erythrocytes, and erythrocytes' cholesterol to plasma cholesterol ratios are decreased by about 1.5–2.4, 1.2–2, 2, and 2–3-fold, respectively. The levels of cholesterol and phospholipid in erythrocytes have been positively correlated with the severity of the disease [80]. Besides erythrocytes, phospholipid changes occur also in thrombocytes of patients with heart ischemia and coronary atherosclerosis. In these cells, PtdSer and oleic acid are increased by about 50 and 20% respectively, and linoleic acid and arachidonic acid are decreased by 20–30% [81].

Decreased phospholipid levels in myocardium during ischemia could be caused by several factors. While phospholipid degradation may be caused by ischemia-induced activation of phospholipases [75,82], the reduced PtdCho levels may partly be accounted for by the ischemia-induced inactivation of *de novo* PtdCho synthesis [83]. However, incorporation of long-chain saturated and unsaturated fatty acids into tissue phospholipid has been found to be stimulated by myocardial ischemia [83]. Phospholipid degradation in ischemia may also be caused by free radical mechanisms [84] or ATP depletion [85].

Lyso phospholipids have been implicated in the formation of arrythmias during ischemia [86]. However, other studies have suggested that lyso phospholipids alone may not be responsible for electrophysiological manifestations of ischemia [82,87]. Myocardial injury during ischemia has also been related to the arachidonic acid accumulation [88].

Decreased phospholipid contents have also been reported in brain after cerebral ischemia [75,78,79,82,83,86,89,90]. Brain membrane phospholipids are differentially degraded, with preferential hydrolysis of polyenoic molecular species of individual phospholipid classes, resulting in an increase in free fatty acid concentrations [86]. PtdCho and PtdEtn seem to be the two major phospholipids affected by ischemia [79,82,83,86]. Further, the ether species of PtdEtn are more affected than the diester species [84]. Based on studies of phospholipid metabolites in ischemic rat brain, phospholipid hydrolysis has been proposed to proceed by two pathways involving phospholipases A_1 and A_2, lysophospholipase and glycerophosphorylcholine-diesterase, or phospholipase C and tissue lipases [79]. Further studies have indicated that ethanolamine phosphoglycerides are degraded mainly by phospholipase C, whereas PtdCho are converted to CDP choline and their glycerol derivatives [83].

Apart from heart and brain, phospholipid changes have been recorded in ischemic liver and kidney [78,86,91]. In whole homogenates and post mitochondrial supernatants of ischemic rat livers, PtdCho and PtdEtn are dominantly affected without accumulation of lyso PtdCho or lyso PtdEtn [91].

VII. PHOSPHOLIPIDS IN LIVER DISEASES

The plasma phospholipid patterns in patients with fatty liver remain unaltered, except for an increase in the mean levels of PtdEtn [92]. Likewise, in liver cirrhosis the total plasma phospholipid contents are similar to those of controls, but the mean level of lyso PtdCho is significantly decreased, whereas the PtdEtn level is markedly increased [92–94]. In patients with acute hepatitis, the concentrations of total phospholipids, PtdCho, SphPCho, and PtdEtn are increased, but the mean level of lyso PtdCho significantly decreased [92,95].

The total phospholipids are significantly reduced in both thc plasma and the erythrocytes of patients with liver disease. While plasma SphPCho is decreased, erythrocyte lyso PtdCho is increased and PtdEtn and PtdSer decreased, compared to controls [96]. Moreover, the erythrocyte PtdCho to SphPCho ratio is increased in rats that are chronically fed ethanol [97]. Furthermore, the total phospholipid contents as well as the proportion of arachidonic acid in erythrocytes are significantly decreased in alcoholic liver cirrhotic patients compared to normal controls. The cirrhotic patients display a significantly higher percentage of PtdSer and PtdIns and a lower percentage of PtdEtn than the controls [98]. Besides, the erythrocyte membrane lipid composition in patients with hepatocellular or cholestatic liver disease shows increases of 24% and 58% in the cholesterol to phospholipid and PtdCho to SphPCho ratios, respectively, compared to controls [99].

Erythrocyte and platelet lipid compositions are altered in both parenchymal liver disease and obstructive jaundice. The cholesterol to phospholipid ratio and the proportion of PtdCho in total phospholipids are increased in both erythrocytes and platelets. Also, SphPCho and PtdEtn are decreased in platelets. Further, plasma PtdCho/cholesterol acyl transferase activity shows significant inverse correlation with these erythrocyte membrane lipid changes in parenchymal liver disease but not in obstructive jaundice [100]. In cirrhosis the platelet phospholipids are decreased by about 30% as compared to normal controls, though the percent phospholipid distribution remains nearly normal [101].

VIII. PHOSPHOLIPIDS IN LUNG DISEASES

Respiratory disease syndrome (RDS) in guinea pigs is accompanied by a 35% decrease in total lung phospholipids. These phospholipids are secreted into blood and taken up by the liver [102]. PtdCho is absent in phospholipids of newborns with RDS, whereas it comprises up to 11% of phospholipids in mature newborns. Recovery from RDS is characterized by an increase of PtdIns concentration up to twofold of the initial value in the surviving infants [103].

The content of phsopholipid/mL of bronchoalveolar lavage significantly decreased in idiopathic pulmonary fibrosis (IPF). There is a significant decrease in phosphatidylglycerol (PG) and an increase in PtdIns in IPF but not in sarcoidosis and eosinophilic granuloma. Thus the PG to PtdIns ratio is significantly decreased in IPF. Also, the 1,2-dipalmitoyl-*sn*-glycero-3-phosphocholine (DPPC) is decreased in IPF and sarcoidosis. The decrease in DPPC appears to be the common fcature in interstitial lung disease [104].

The phospholipid fatty acid pattern in bronchial secretion of patients with cystic fibrosis shows an increase of the mole fraction of arachidonic acid in most phospholipids compared to controls. Increase of arachidonic acid in some phospholipid classes is found also in patients with chronic bronchitis or chronically colonized with *Pseudomonas aeruginosa,* but in diphosphatidylglycerol (DPG), lyso PtdEtn, and SphPCho high fractions of arachidonic acid are found exclusively in patients with cystic fibrosis [105]. Contents of various phospholipids like PtdCho, PtdEtn, lyso PtdCho, lyso PtdEtn, PA, and SphPCho are decreased in guinea pig lungs 90 days after intratracheal inoculation of *Candida albicans* [106].

Not all patients with cystic fibrosis show an abnormal in vitro turnover of erythrocyte fatty acids, although they all present abnormal fatty acid patterns for erythrocyte phospholipids, platelet phospholipids, and plasma lipids. Abnormal FA turnover occurs only in incubations where red cells of CF patients are involved and not where red cells of healthy subjects are incubated with serum of CF patients [107]. Also, the PtdIns content is increased in erythrocyte membranes of cystic fibrosis children as compared to normal subjects. Besides, significant changes are observed in phospholipid fatty acid patterns of cystic fibrosis patients' erythrocytes. PtdCho and PtdEtn in particular show strikingly abnormal fatty acid patterns similar to those in the various plasma lipid fractions [108].

Tracheobronchial secretions of patients with cystic fibrosis contain about 30% more lipids than those of normal individuals and exhibit elevated levels of cholesterol, phospholipids, and glycosphingolipids. The phospholipids of cystic fibrosis secretions have higher contents of SphPCho and PtdSer, whereas the normal samples contain more lyso PtdCho [109].

Regulation of arachidonic acid release from membrane phospholipids has been investigated in lymphocytes from patients with cystic fibrosis as well as control patients. No effect of either dexamethasone or fetal calf serum is seen on arachidonic acid release from cystic fibrosis lymphocytes in contrast to the control lymphocytes. The defective regulation of arachidonic acid, resulting in an increased turnover, can explain many of the findings in cystic fibrosis [110].

Phospholipid content is elevated in lung specimens of patients with shock lung (adult respiratory distress syndrome) immediately after death. Also, the DPPC fraction is decreased in total PtdCho, resulting in a shift of palmitic acid to oleic acid ratio in PtdCho from 2.32 for normal lungs to 1.64 for distressed lungs [111]. Since the content of DPPC is closely related to the function of the surfactant system, it has been concluded that the altered surfactant function observed in patients with adult respiratory distress syndrome may be at least in part a result of altered surfactant metabolism [111].

IX. PHOSPHOLIPIDS IN MALARIA

Parasitization of erythrocytes with malarial parasite leads to a significant increase in the absolute amounts of phospholipid per cell [112–118]. Particularly, PtdCho and PtdEtn are considerably increased [112,114–117]. A parallel increase occurs in the ratio of unsaturated to saturated fatty acids in parasitized erythrocyte phospholipids [112,114]. The fatty acids in PtdCho and SphPCho are not as altered as in PtdEtn, PtdSer, PtdIns, and lyso PtdCho [112]. Moreover, the total plasmalogen level is decreased but the PtdIns level increased in infected erythrocytes [112,115–117]. Unlike in infected cells, no phospholipid changes occur in uninfected erythrocytes of malaria-infected animals [113].

About 32% of the total phospholipids present in the erythrocytes infected with the trophozoite stage of the malarial parasite is localized in the erythrocyte membrane, while the remaining 68% is present in the parasite [113]. The parasite contains mainly PtdCho and PtdEtn, which together account for about 87% of the total phospholipid present in the parasite [112,113,115–118]. The remaining about 13% of the total parasite phospholipid consists mainly of PtdIns, which comprises 10% of the parasite phospholipid pool. PtdSer is found predominantly in the erythrocyte membrane of the parasitized cells, while lyso PtdCho is exclusively localized in the erythrocyte membrane of parasitized cells at any stage of parasite development [113]. Besides, no changes have been observed in the plasma membrane phospholipid compositions of the infected erythrocytes [113,119].

Infection of erythrocytes with malarial parasite leads to an impaired phospholipid packing in plasma membranes of the infected cells [120–123]. The various membrane phospholipids are structurally disordered [124] and the transbilayer glycero phospholipid distribution is altered depending on the developmental stage of the parasite inside the host cells [123,125–127]. In contrast to these studies, Van der Schaft et al. [113] have suggested that the membrane phospholipid asymmetry remains unaltered in both uninfected and infected erythrocytes. It has further been indicated that transbilayer phospholipid mobility rather than membrane phospholipid asymmetry is altered after infecting the erythrocytes with malarial parasite [128]. The translocation of PtdEtn and PtdSer from the outer to the inner monolayer is shown to be significantly increased in trophozoite/schizont-infected erythrocytes as compared to the normal uninfected cells [129]. However, no such enhanced PtdEtn and PtdSer uptake has been observed by Haldar [130] in late ring-infected erythrocytes.

X. PHOSPHOLIPIDS IN OTHER DISEASES

Decreased concentrations of some specific phospholipids in erythrocyte membranes have been observed in acute pneumonia (lyso PtdCho and PtdEtn) [131], cerebellar ataxia (PtdSer) [132], and asthma (PtdIns, SphPCho, PtdCho, and PtdEtn) [133], compared to normal controls. However, in patients with chronic renal failure resulting from glomerulonephritis and pyelonephritis, the rise in total erythrocyte membrane phospholipid is confined predominantly to SphPCho and PtdCho, which is often accompanied by a decrease in PtdEtn [134].

Phospholipid changes have been observed in blood plasma or serum of patients with heart disease. Thus during myocardial infarction, the plasma levels of lyso PtdCho [135], total phospholipid, PtdIns, PtdSer, and PA are decreased and that of SphPCho, PtdEtn [136], and phospholipid arachidonate [137] significantly increased, compared to those of normal subjects. However, 2–3 weeks after the infarction, the levels of lyso PtdCho and SphPCho are decreased and that of PtdSer, PtdEtn, PtdIns, CL, and PA increased above the normal levels [136].

Alterations in blood serum phospholipids occur also in several other diseases. For example, distinct increases in total serum phospholipid, PtdCho, SphPCho, and lyso PtdCho have been recorded during X-linked myopathy and Duchenne muscular dystrophy [138], whereas decreased serum levels of total phospholipids and PtdCho have been observed in premature infants (< 3 months) with acute viral pneumonia [139]. Further, the serum phospholipid levels are decreased on day 1 following induction of pancreatitis and subsequently increased to levels threefold greater than control by day 5 [140]. Among the individual phospholipids measured on day 1, lyso PtdCho, SphPCho, and PtdCho are decreased while lyso PtdEtn and CL are increased [140].

Platelets of patients with chronic arterial occlusions contain significantly less PtdEtn (16%) than thrombocytes of normal subject (24%). On the other hand, the plasma PtdEtn levels of these patients have been shown to be significantly elevated (4.8%) in comparison to those of normal subjects (2.7%), possibly as a result of some leakage of PtdEtn from platelets [141]. Moreover, the arachidonate content in platelet phospholipid fraction is increased in patients with chronic heart disease [142].

Serine phosphatides [143,144] and sulfatides [143] are significantly decreased in multiple sclerosis brains. Also, the phospholipids in these brains contain reduced amounts of oleic acid and, except for the lyso compounds, slightly increased quantities of 20:4 and 20:6 fatty acids [144].

Infection of cultured human skin fibroblasts by herpes simplex virus leads to an increased incorporation of labeled phosphate into the host cell membrane SphPCho [145]. Incorporation into other major membrane phospholipids including PtdCho, PtdEtn, PtdIns, and PtdSer remains unaffected. Since SphPCho serves as a precursor to ceramides and since the enhanced synthesis of ceramide-based glycolipids has been shown to occur after herpes simplex virus infection, the observed increase in labeling of SphPCho may reflect mobilization for glycolipid synthesis [145].

PtdEtn metabolism is deranged in cultured fibroblasts from patients with mucolipidosis type IV [146].

Accumulation of SphPCho and bis(monoacylglycero)phosphate has been demonstrated in spleen from types A and B and group C Niemann-Pick disease, whereas only in type A Niemann-Pick brain is the concentration of SphPCho increased. Sphingomyelinase activity is markedly reduced in type A Niemann-Pick brain and spleen, but residual activity of about 12% of the control has been observed in type B Niemann-Pick brain [147].

The amounts of linoleic acid in phospholipid is decreased in hyperthyroidism and increased in hypothyroidism, whereas the opposite is true for arachidonic acid and eicosatrienic acid levels in phospholipid and cholesterol esters. The fatty acids in cholesterol esters and phospholipids are more unsaturated in hyperthyroidism than in hypothyroidism [148].

REFERENCES

1. P. A. Akinyanju, and C. O. Akinyanju, *Ann. Clin. Lab. Sci. 6*:521–524 (1976).
2. S. K. Jain and S. B. Shohet, *Blood 63*:362–367 (1984).
3. P. J. Raval, *Biochim. Biophys. Acta 856*:595–601 (1986).
4. S. B. Shohet and S. K. Jain, *Ann. NY Acad. Sci. 393*:229–236 (1982).
5. B. Lubin, D. Chiu, J. Bastacky, and B. Roelofsen, *J. Clin. Invest. 67*:1643–1649 (1981).
6. P. F. H. Franck, E. M. Bevers, B. Lubin, P. Confurius, D. Chiu, J. A. F. Opden Kamp, R. F. A. Zwaal, L. L. M. Van Deenen, and B. Roelofsen, *J. Clin. Invest. 75*:183–190 (1985).
7. P. J. Raval and D. Allan, *Biochem. J. 223*:555–557 (1984).
8. A. Zachowski, C. T. Craescu, F. Galacteros, and P. F. Devaux, *J. Clin. Invest. 75*:1713–1717 (1985).
9. G. Maggioni, M. Castro, A. Donfrancesco, B. Spano, and O. Giardini, *Acta Haematol. 52*:207–213 (1974).
10. M. Khalil, Y. Aziz, A. Abdel-Malek, N. Shaaban, K. Aref, S. Mahmoud, S. Soliman, and S. Gharib, *Gaz. Egypt Paediatr. Assoc. 23*:273–280 (1975).
11. A. Kalopoutis, E. Diskakis, N. J. Stratakis, and A. Papudemetriou, *Biochem. Med. 23*:1–51 (1980).
12. R. Bleiber, W. Eggert, R. Reichmann, and D. Kunze, *Acta Biol. Med. Ger. 40*:1133–1138 (1981).
13. M. O. Varsanyi-Nagy, L. Takacsi-Nagy, and F. Graf, *Magy. Belory. Arch. 34*:245–249 (1981).

14. E. Hietanen, K. Punnonen, R. Punnonen, and O. Auvinen, *Carcinogenesis 7*:1965–1969 (1986).
15. W. C. Tan, C. Chapman, T. Takatori, and O. S. Privett, *Lipids 10*:70–74 (1975).
16. B. S. Leung and G. Y. Sun, *Proc. Soc. Exp. Biol. Med. 152*:671–676 (1976).
17. W. A. Boggust, *Adv. Tumour Prev. Detect. Charact. (Charact. Hum. Tumours, Proc. Int. Symp.*, 5th, 1973) *1974*:279–289.
18. T. Galeotti, S. Borrello, G. Minotti, and L. Masotti, *Ann. NY Acad. Sci. 488*:468–480 (1986).
19. E. Araki, N. Okazaki, and N. Hattori, *Iqaku no Ayumi 80*:493–498 (1977).
20. T. Kudo, *Kosankinbyo Kenkyu Zasshi 22*:81–97 (1970).
21. H. Filipek-Wender, H. Karon, and L. Torlinski, *Acta Biol. Med. Ger. 29*:823–829 (1972).
22. F. Feo, R. Garcea, R. A. Canuto, and P. Pani, *IRCS Libr. Compend. 1*:2.2.5 (1973).
23. A. Wakui, J. Kikkawa, T. Kudo, K. Kikuchi, and T. Ichikawa, *Tohoku J. Exp. Med. 108*:63–77 (1972).
24. O. S. Dzhishkanani, T. A. Lursmanashvili, M. A. Tsartsidze, and B. A. Lomsadze, *Soobshch Akad. Nauk. Gruz. SSR 92*:441–444 (1978).
25. K. Satouchi, T. Mizuno, Y. Samejima, and K. Saito, *Cancer Res. 44*:1460–1464 (1984).
26. E. E. Tafel'shtein, Yu. A. Vladimirov, and P. I. Pirogova, *Strukt. Funkts. Biol. Membran. 1971*:31–53.
27. D. H. Albert and C. E. Anderson, *Lipids 12*:188–192 (1977).
28. D. Kostić A. Vranešević, S. Vrbaski, I. Nagulic, and L. Rakić, *Acta Med. Iugosl. 30*:369–378 (1976).
29. A. J. Yates, D. K. Thompson, C. P. Boesel, C. Albrightson, and R. W. Hart, *J. Lipid Res. 20*:428–436 (1979).
30. M. Nakamura, O. Tsugutani, and A. Toyoaki, *Lipids 15*:616–623 (1980).
31. H. Tapiero, Z. Mishal, M. Wioland, A. Silber, A. Fourcade, and G. Zwingelstein, *Anticancer Res. 6*:649–652 (1986).
32. K. T. Holmes, R. M. Fox, L. C. Wright, and C. E. Mountford, *Magn. Reson. Cancer Proc. Int. Conf. 1985*:127–128 (1986).
33. R. Takenaka, M. Inoue, T. Hori, and H. Okuyama, *Biochim. Biophys. Acta 754*:28–37 (1983).
34. C. P. Burns, S. P. Wei, D. G. Luttenegger, and A. A. Spence, *Lipids 14*:144–147 (1979).
35. N. Weber, K. Wayss, and M. Volm, *Naturforsch. 36C*:81–83 (1981).
36. D. Pitsin, P. Uzunov, and D. Tsonev, *Acta Med. Bulg. 6*:85–89 (1979).
37. K. Waku, Y. Nakazawa, and W. Mori, *J. Biochem.* (Tokyo) *79*:407–411 (1976).
38. V. Nikolasev, G. Lazar, and I. Karady, *Kiserl. Orvostud. 24*:465–469 (1972).
39. Y. Aso, K. Fujita, A. Tajima, K. Suzuki, and M. Yokoyama, *Acta Urol. Jpn. 27*:1345–1349 (1981).
40. F. A. Manzoli, N. M. Maraldi, L. Cocco, S. Capitani, and A. Facchini, *Cancer Res. 37*:843–849 (1977).
41. A. B. Kier, M. T. Parker, and F. Schroeder, *Biochim. Biophys. Acta 938*:434–446 (1988).
42. N. Yamamoto, and Z. N. Benjamin, *Cancer Res. 47*:2008–2013 (1987).
43. A. A. Hakim, *Cancer Immunol. Immunother. 8*:135–141 (1980).
44. G. S. Vartanyan and K. G. Karagezyan, *Vopr. Med. Khim. 27*:179–181 (1981).
45. K. G. Karagezyan, O. M. Amirkhanyan, and L. T. Amirkhanyan, *Vop. Biokhim. Mozga. 7*:150–157 (1972).
46. Ya. Kh. Turakulov, T. S. Saatov, E. I. Isaev, S. S. Sadykov, and B. R. Zainutdinov, *Probl. Endokrinol. 25*:54–57 (1979).
47. K. G. Karagezyan, G. S. Vartanyan, and M. G. Badalyan, *Byull. Eksp. Biol. Med. 90*:679–681 (1980).
48. K. G. Karagezyan, L. M. Ovsepyan, K. G. Adontis, and L. Z. Mamadzhanyan, *Vopr. Med. Khim. 28*:56–60 (1982).

49. U. P. S. Chauhan and V. N. Singh, *Life Sci. 22*:1771–1776 (1978).
50. G. Kh. Bunyatyan, K. G. Karagezyan, V. B. Egyan, O. M. Amirkhanyan, G. A., Turshyan, L. T. Amirkhanyan, and G. E. Akopyan, *Dokl. Akad. Nauk. Arm. SSR 59*:110–115 (1974).
51. M. Bryszewska, C. Watala, and W. Torzecka, *Br. J. Haematol. 62*:111–116 (1986).
52. A. Kalofoutis and J. Lekakis, *Diabetologia 21*:540–543 (1981).
53. F. Lupu, M. Calb, and A. Fixman, *Thromb. Res. 50*:605–616 (1988).
54. K. Tetsuto, Y. Ishikawa, N. Morisaki, K. Shirai, Y. Saita, and S. Yoshida, *Lipids 22*:704–710 (1987).
55. L. Austin, C. T. Kwok, A. D. Kuffer, and B. Y. Tang, *Adv. Exp. Med. Biol. 72*:367–372 (1976).
56. T. Futo, T. Hitaka, T. Mizutani, H. Okuyama, K. Watanabe, and T. Totsuka, *J. Neurol. Sci. 91*:337–344 (1989).
57. P. H. Pearce and B. A. Kakulas, *Aust. J. Exp. Bio. Med. Sci. 58*:397–408 (1980).
58. E. M. Popova, *Ukr. Biokhim. Zh. 43*:639–644 (1971).
59. H. A. Barakat, D. R. Johnson, and D. S. Kerr, *Proc. Soc. Exp. Biol. Med. 163*:167–170 (1980).
60. D. Kunze, B. Ruestow, and D. Olthoff, *Clin. Chim. Acta 140*:113–124 (1984).
61. G. E. Sumnicht and R. A. Sabbadini, *Arch. Biochem. Biophys. 215*:628–637 (1982).
62. J. L. Rabinowitz and G. Cossu, *Biochim. Biophys. Acta 879*:394–398 (1986).
63. M. Kester and C. A. Privitera, *Biochim. Biophys. Acta 778*:112–120 (1984).
64. M. Kester and C. A. Privitera, *J. Exp. Zool. 230*:159–162 (1984).
65. J. E. Grey, H. J. Gitelman, and A. D. Roses, *J. Clin. Invest. 65*:1478–1482 (1980).
66. A. P. Sherblom, D. J. McAllister, M. V. Macul, and J. L. Howland, *Neurosci. Lett. 20*:115–118 (1980).
67. Y. Antoku, T. Sakai, K. Tsukamoto, I. Goto, H. Iwashita, and Y. Kuroiwa, *J. Neurochem. 44*:1667–1671 (1985).
68. Y. Antoku, T. Sakai, I. Goto, H. Iwashita, and Y. Kuroiwa, *J. Neurol. Sci. 71*:387–393 (1985).
69. A. Kalofoutis, G. Jullien, and V. Spanos, *Clin. Chim. Acta 74*:85–87 (1977).
70. M. I. Hunter, M. S. Lao, and P. J. Devane, *Clin. Chim. Acta 128*:69–74 (1983).
71. P. Beyer, R. Bieth, J. Robert, H. Freysz, and T. Uhl, *Ann. Pediatr. 27*:233–236 (1980).
72. P. Beyer, R. Bieth, J. Robert, H. Freysz, and T. Uhl, *Sem. Hop. 57*:56–59 (1981).
73. D. G. Hassall, R. A. Hutton, and K. R. Bruckdorfer, *Biochem. Soc. Trans. 11*:378–379 (1983).
74. D. Kunze, G. Reichmann, E. Egger, G. Leuschner, and H. Eckhardt, *Clin. Chim. Acta 43*:333–341 (1973).
75. K. Umetsu, M. Mochizuki, T. Yanagishita, and T. Katagiri, *Showa Igakkai Zasshi 48*:205–213 (1988).
76. T. Yanagishita, N. Konno, E. Geshi, and T. Katagiri *Jpn. Circ. J. 51*:41–50 (1987).
77. P. Chatelain, M. Gremel, and R. Brotelle, *Eur. J. Pharmacol. 144*:83–90 (1987).
78. K. Takahashi and K. J. Kako, *Biochem. Med. Metab. Biol. 35*:308–321 (1986).
79. D. A. Ford and R. W. Gross, *Circ. Res. 64*:173–177 (1989).
80. Yu. M. Lopukhin, E. A. Borodin, V. I. Sergienko, E. M. Khalikov, A. I. Zyulyaev, V. A. Lyusov, and A. I. Archakov, *Vopr. Med. Khim. 25*:466–468 (1979).
81. S. S. Vladimirov, T. I. Torkhovskaya, V. P. Zykova, and E. N. Gerasimova, *Kardiologiya 19*:73–80 (1979).
82. N. A. Shaikh and E. Downar, *Dev. Cardiovasc. Med. 49*:298–316 (1985).
83. A. Lochner and M. Devilliers, *J. Mol. Cell. Cardiol. 21*:151–163 (1989).
84. W. B. Weglicki, B. F. Dickens, I. T. Mak, J. H. Kramer, *Life Chem. Rep. 3*:189–198 (1985).
85. M. D. Gunn, A. Sen, A. Chang, J. T. Willerson, L. M. Buja, and K. R. Chien, *Am. J. Physiol. 249*(2):H1188–H1194 (1985).

86. B. E. Slobel and P. B. Corr, *Adv. Cardiol.* *26*:76–85 (1979).
87. N. A. Shaikh and E. Downer, *Circ. Res.* *49*:316–325 (1981).
88. K. P. Burton, L. M. Buja, A. Sen, J. T. Willerson, and K. R. Chien, *Am. J. Pathol.* *124*:238–245 (1986).
89. H. Otani, M. R. Prasad, R. M. Jones, and D. K. Das, *Am. J. Physiol.* *257*(2):H252–H258 (1989).
90. A. Toleikis, A. Tumpickas, and A. Praskevicius, *Vopr. Med. Khim.* *32*:76–80 (1986).
91. K. R. Chien, J. Abrams, A. Serroni, J. T. Martin, and J. L. Farber, *J. Biol. Chem.* *253*:4809–4817 (1978).
92. P. Jipp, K. G. Ravens, and M. Salm, *Klin. Wochenschr.* *52*:759–762 (1974).
93. T. Hashimoto, *Iwate Igaku Zasshi* *38*:55–66 (1986).
94. S. B. Curri, L. Cantoni, P. Andreuzzi, and P. Rochetti *Biochem. Exp. Biol.* *11*:91–104 (1974).
95. V. S. Vasil'ev and S. B. Yushkevich, *Ter. Arkh.* *50*:104–107 (1978).
96. L. Cantoni, S. B. Curri, P. Andreuzzi, and P. Rochetti, *Clin. Chim. Acta* *60*:405–408 (1975).
97. K. Araki, R. Shinozaki, and M. Maezawa, *Tokyo Joshi Ika Daigaku Zasshi* *57*:321–328 (1987).
98. D. M. A. Alvardo, A. F. Attili, A. DeSantis, U. Pieche, and L. Capocaccia, *Biochem. Med.* *28*:157–164 (1982).
99. J. S. Owen, K. R. Bruckdorfer, and N. McIntyre, *Biochem. Soc. Trans.* *7*:1272–1274 (1979).
100. J. S. Owen, R. A. Hutton, M. J. Hope, D. S. Harry, K. R. Bruckdorfer, R. C. Day, N. McIntyre, and J. A. Lucy, *Scand. J. Clin. Lab. Invest. Suppl.* *38*:228–232 (1978).
101. A. Z. Aktulga, M. Gurakar, and O. N. Ulutin, *Acta Hepatogastroenterol.* *24*:411–414 (1977).
102. K. Winsel, B. Lachmann, and H. Iwainsky in *Lung Lipid Metabolism: Mech. Its Regul. Alveolar Surfactant. Proc. Int. Symp. 1976* (G. A. Georgiev, ed.), Ind. BAN: Sofia, Bulgaria, 1978, pp. 83–96.
103. M. Obladen, *Fortschr. Med.* *97*:403–408 (1979).
104. Y. Honda, K. Tsunematsu, A. Suzuki, and T. Akino, *Lung* *166*:293–302 (1988).
105. H. Gilljam, B. Strandvik, A. Ellin, and L. G. Wiman, *Scand. J. Clin. Lab. Invest.* *46*:511–518 (1986).
106. A. K. Jaiswal, *Toxicon* *19*:910–912 (1982).
107. V. Rogiers, I. Dab, Y. Michotte, A. Vercruysse, R. Crokaert, and H.-L. Vis, *Pediatr. Res.* *18*:704–709 (1984).
108. V. Rogiers, R. Crokaert, and H.-L. Vis, *Clin. Chim. Acta* *105*:105–116 (1980).
109. A. Slomiany, V. L. N. Murty, M. Aono, C. E. Snyder, A. Herp, and B. L. Slomiany, *Biochim. Biophys. Acta* *710*:106–111 (1982).
110. J. Carlstedt-Duke, M. Bronnegard, and B. Strandvik, *Proc. Natl. Acad. Sci. USA* *83*:9202–9206 (1986).
111. P. Von Wichert, and F. V. Kohl, *Intensive Care Med.* *3*:27–30 (1977).
112. B. D. Beaumelle and H. J. Vial, *Biochim. Biophys. Acta* *877*:262–270 (1986).
113. P. H. Van der Schaft, B. Beaumelle, H. Vial, B. Roelofsen, J. A. F. Opden Kamp, and L. L. M. Van Deenen, *Biochim. Biophys. Acta* *901*:1–14 (1987).
114. R. Stocker, W. B. Cowden, R. L. Tellam, M. J. Weidemann, and N. H. Hunt, *Lipids* *22*:51–57 (1987).
115. R. C. Rock, J. C. Standefer, R. T. Cook, W. Little, and H. Sprinz, *Comp. Biochem. Physiol.* *38B*:425–437 (1971).
116. R. A. DeZeeuw, J. Wijsbeek, R. C. Rock, and G. McCormick, *Proc. Helminthol. Soc. Wash.* *39*:412–418 (1972).
117. S. McClean, W. C. Purdy, A. Kabat, J. Sampugna, R. A. DeZeeuw, and G. McCormick, *Anal. Chim. Acta* *82*:175–185 (1976).

118. H. J. Vial, J. R. Philippot, and D. F. Wallach, *Mol. Biochem. Parasitol. 13*:53–65 (1984).
119. J. Gruenberg, and I. W. Sherman, *Proc. Natl. Acad. Sci. USA 80*:1087–1091 (1983).
120. R. J. Howard, and W. H. Sawyer, *Parasitology 80*:331–342 (1980).
121. D. R. Allred, C. R. Sterling, and P. D. Morse, *Mol. Biochem. Parasitol. 7*:27–39 (1983).
122. I. W. Sherman, and J. Greenan, *Trans. R. Soc. Trop. Med. Hyg. 78*:641–644 (1984).
123. P. Joshi, G. P. Datta, and C. M. Gupta *Biochem. J. 246*:103–108 (1987).
124. T. F. Taraschi, A. Parashar, M. Hooks, and H. Rubin, *Science 232*:102–104 (1986).
125. R. S. Schwartz, J. A. Olson, C. Raventos-Suarez, M. Yee, R. H. Heath, B. Lubin, and R. L. Nagel, *Blood 69*:401–407 (1987).
126. P. Joshi, and C. M. Gupta, *Br. J. Haemat. 68*:255–259 (1988).
127. C. M. Gupta and G. C. Mishra, *Science 212*:1047–1049 (1981).
128. B. D. Beaumelle, H. J. Vial, and Bienvenue, *J. Cell Physiol. 135*:94–100 (1988).
129. G. N. Moll, H. J. Vial, M. L. Ancelin, J. A. F. Opden Kamp, B. Roelofsen, and L. L. M. Van Deenen, *FEBS Lett. 232*:341–346 (1988).
130. K. Haldar *Dynamics and Biogenesis of Membranes, NATO ASI Series H40*:109–120 (1989).
131. A. V. Volkora, V. I. Krylov, A. V. Shantarina, S. I. Eremeeva, S. I. Borodzich, and A. D. Petrushina, *Khromatogr. Elektroforeticheskie Metody Issled. Biol. Aktiv. Soedin 1976*:57–59.
132. V. Gallai, and C. Firenze, *Eur. Neurol. 22*:340–343 (1983).
133. D. Tsambaos and M. Gustav, *Arch. Dermatol. Res. 266*:177–180 (1979).
134. J. Pielichowski, J. Kwiatkowska, and W. Zatonski, *Arch. Immunol. Ther. Exp. 25*:213–217 (1977).
135. M. P. P. DeNeuman, J. L. Martiarena, and J. Neuman, *Rev. Asoc. Bioquim. Argent. 38*:207–212 (1973).
136. K. Yu. Yuldashev, Yu. M. Fuzailov, and Sh. N. Nasyrov, *Med. Zh. Uzb. 4*:38–42 (1978).
137. G. Skulldottir, T. Hardarson, N. Sigfusson, G. Oddsson, and G. Sigmundur, *Acta Med. Scand. 218*:55–58 (1985).
138. L. P. Grinio and T. I. Turkina, *Vopr. Med. Khim. 27*:647–649 (1981).
139. E. B. Leonova and G. N. Savel'eva, *Vopr. Det. Pul'Monol. 2*:23–26 (1976).
140. T. Wesolowska, S. Wujkowski, K. Chrzanowska-Ansilewska, and J. Gregorczyk, *Diagn. Lab. 15*:173–179 (1979).
141. H. D. Bruhn, P. Jipp, and K. G. Ravens, *Haemostasis 6*:197–202 (1977).
142. J. Aznar, J. Valles, M. T. Santos, and M. A. Fernandez, *Thromb. Res. 18*:505–512 (1980).
143. M. T. Vanier, and L. Svennerholm, *Brain Res. 35*:325–336 (1971).
144. H. Woelk and B. Piero, *Eur. Neurol. 10*:250–260 (1973).
145. W. L. Steinhart, C. M. Nicolet, and J. L. Howland *Intervirology 16*:80–85 (1981).
146. G. Bach and R. J. Desnick, *Enzyme* (Basel) *40*:40–44 (1988).
147. G. T. N. Besley and M. Elleder, *J. Inherited Metab. Dis. 9*:59–71 (1986).
148. V. Felt and P. Husek *Rev. Czech. Med. 23*:109–119 (1977).

Appendix A: Structural Parameters of Phospholipids

John M. Seddon *Imperial College, London, England*

INTRODUCTION

The figures and tables are taken largely from the publications of Pascher and coworkers (the unpublished atomic fractional coordinates for DMPC.$2H_2O$ were kindly supplied by Prof. I. Pascher). Following these authors, the atomic numbering and notation of torsion angles is as introduced by Sundaralingam [1]. The glycerol carbon atom to which the polar head group is attached is designated C(1).

The following abbreviations are used:

DLPE, 2,3-dilauroyl-DL-glycero-1-phosphorylethanolamine acetic acid [2,3,4]
DLPEM_2, 2,3-dilauroyl-DL-glycero-1-phospho-*N*,*N*-dimethylethanolamine [5]
DMPC, 2,3-dimyristoyl-D-glycero-1-phosphorylcholine dihydrate [2,6]
DMPA, monosodium 2,3-dimyristoyl-D-glycero-1-phosphate [7]
DMPG, monosodium 2,3-dimyristoyl-D-glycero-1-phospho-DL-glycerol [8]
PPE, 3-palmitoyl-DL-glycero-1-phosphorylethanolamine [9]
LPPC, 3-lauroyl-propanediol-1-phosphorylcholine [10]

For comparison, in the stereospecific numbering *sn* convention, DLPE is named 1,2-dilauroyl-*rac*-glycero-3-phosphorylethanolamine acetic acid, and DMPC is named 1,2-dimyristoyl-*sn*-glycero-3-phosphorylcholine dihydrate.

Table A1 Crystal Data and Structure Parameters

Lipid	DLPE	$DLPEM_2$	DMPC	DMPA	DMPG	PPE	LPPC
Crystal system	Monoclinic	Triclinic	Monoclinic	Monoclinic	Monoclinic	Monoclinic	Monoclinic
Space group	$P2_1/c$	$P\bar{1}$	$P2_1$	$P2_1$	$P2_1$	$P2_1/a$	$P2_1/c$
Unit cell a (Å)	47.7	5.636	8.72	5.439	10.370	7.66	24.83
b (Å)	7.77	8.200	8.92	7.953	8.482	9.08	9.53
c (Å)	9.95	39.863	55.4	43.98	45.52	37.08	10.94
α (°)		94.56					
β(°)	92.00	90.15	97.40	114.19	95.2	90.2	99.66
γ(°)		101.94					
No. of molecules per unit cell	4	2	4	2	4	4	4

Source: Refs. 4 and 11.

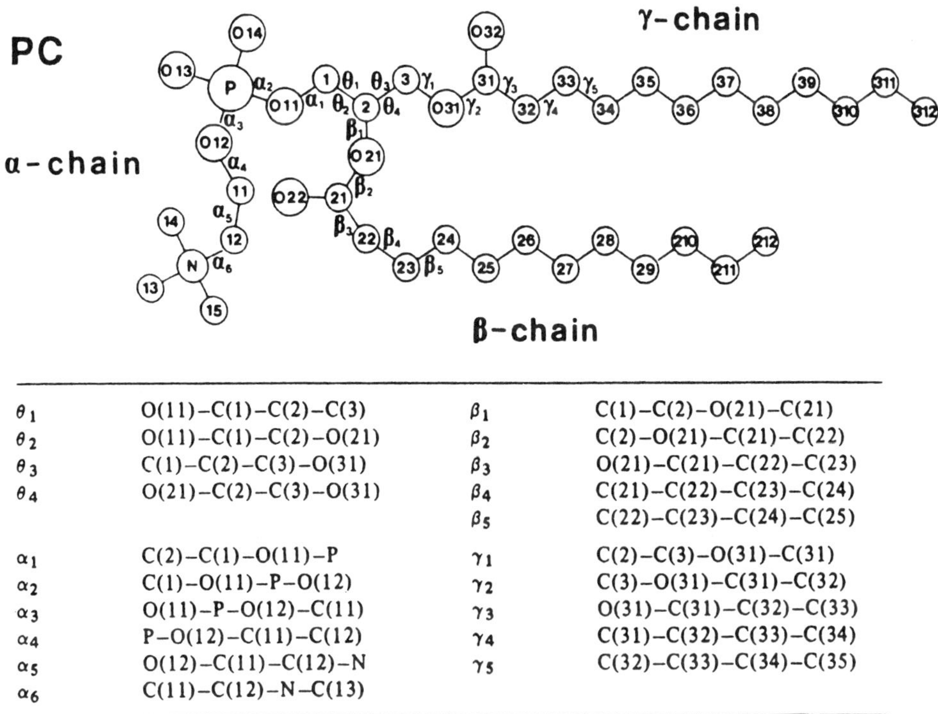

θ_1	O(11)–C(1)–C(2)–C(3)	β_1	C(1)–C(2)–O(21)–C(21)
θ_2	O(11)–C(1)–C(2)–O(21)	β_2	C(2)–O(21)–C(21)–C(22)
θ_3	C(1)–C(2)–C(3)–O(31)	β_3	O(21)–C(21)–C(22)–C(23)
θ_4	O(21)–C(2)–C(3)–O(31)	β_4	C(21)–C(22)–C(23)–C(24)
		β_5	C(22)–C(23)–C(24)–C(25)
α_1	C(2)–C(1)–O(11)–P	γ_1	C(2)–C(3)–O(31)–C(31)
α_2	C(1)–O(11)–P–O(12)	γ_2	C(3)–O(31)–C(31)–C(32)
α_3	O(11)–P–O(12)–C(11)	γ_3	O(31)–C(31)–C(32)–C(33)
α_4	P–O(12)–C(11)–C(12)	γ_4	C(31)–C(32)–C(33)–C(34)
α_5	O(12)–C(11)–C(12)–N	γ_5	C(32)–C(33)–C(34)–C(35)
α_6	C(11)–C(12)–N–C(13)		

DLPE

(*a*) Interatomic distances (Å) and, in parentheses, their standard deviations

P–O(11)	1.62(2)	C(28)–(29)	1.64(5)
P–O(12)	1.63(2)	C(29)–C(210)	1.66(5)
P–O(13)	1.47(2)	C(210)–C(211)	1.52(6)
P–O(14)	1.43(2)	C(211)–C(212)	1.67(7)
N–C(12)	1.51(3)	O(31)–C(31)	1.34(4)
O(11)–C(1)	1.48(4)	O(32–C(31)	1.14(6)
O(12)–C(11)	1.50(3)	C(31)–C(32)	1.62(5)
C(11)–C(12)	1.61(4)	C(32)–C(33)	1.53(5)
C(1)–C(2)	1.57(4)	C(33)–C(34)	1.59(5)
C(2)–C(3)	1.50(4)	C(34)–C(35)	1.46(5)
C(2)–O(21)	1.52(4)	C(35)–C(36)	1.55(5)
C(3)–O(31)	1.51(3)	C(36)–C(37)	1.56(6)
O(21)–C(21)	1.31(4)	C(37)–C(38)	1.56(6)
O(22)–C(21)	1.19(4)	C(38)–C(39)	1.58(7)
C(21)–C(22)	1.61(4)	C(39)–C(310)	1.53(8)
C(22)–C(23)	1.61(4)	C(310)–C(311)	1.58(9)
C(23)–C(24)	1.49(4)	C(311)–C(312)	1.62(12)
C(24)–C(25)	1.62(4)	C(41)–C(42)	1.40(5)
C(25)–C(26)	1.50(4)	C(41)–O(41)	1.24(4)
C(26)–C(27)	1.59(5)	C(41)–O(42)	1.41(3)
C(27)–C(28)	1.62(5)		

Figure A1 Atomic numbering, notation for torsion angles, and bond distances (Å) and angles (°). (From Refs. 2, 4, 5, 7, and 9.)

(*b*) Bond angles (degrees) and their standard deviations

O(12)-P-O(11)	101(1)	C(36)–C(35)-C(34)	110(3)
O(14)-P-O(11)	108(1)	C(38)–C(37)–C(36)	109(3)
O(13)-P-O(12)	102(1)	C(310)–C(39)–C(38)	106(4)
O(13)-P-O(11)	110(1)	C(312)–C(311)–C(310)	100(6)
O(14)-P-O(12)	115(1)	O(42)–C(41)–C(42)	116(3)
O(14)-P-O(13)	119(1)	C(3)–C(2)–C(1)	113(2)
C(11)-O(12)-P	119(2)	O(21)-C(2)–C(3)	106(2)
C(11)–C(12)-N	107(2)	C(21)-O(21)-C(2)	110(2)
C(12)–C(11)-O(12)	109(2)	C(22)-C(21)-O(21)	133(3)
C(1)-O(11)-P	116(1)	C(22)–C(21)–O(22)	121(3)
C(2)–C(1)-O(11)	103(2)	C(24)–C(23)–C(22)	111(3)
O(21)-C(2)-C(1)	108(2)	C(26)–C(25)-C(24)	113(3)
O(31)-C(3)–C(2)	104(2)	C(28)–C(27)–C(26)	105(3)
C(31)-O(31)–C(3)	113(2)	C(210)–C(29)–C(28)	100(3)
C(22)-C(21)-O(21)	106(3)	C(212)–C(211)–C(210)	105(4)
C(23)-C(22)-C(21)	110(2)	C(32)-C(31)-O(31)	101(3)
C(25)-C(24)-C(23)	112(3)	C(33)-C(32)-C(31)	108(3)
C(27)-C(26)-C(25)	104(3)	C(35)–C(34)–C(33)	108(3)
C(29)-C(28)-C(27)	103(3)	C(37)–C(36)–C(35)	113(3)
C(211) C(210)-C(29)	99(3)	C(39)–C(38)–C(37)	106(4)
O(32)-C(31)-O(31)	125(3)	C(311)–C(310)–C(39)	106(5)
C(32)–C(31)-O(32)	132(4)	O(41)-C(41)-C(42)	130(3)
C(34)-C(33)-C(32)	108(3)	O(42)–C(41)–O(41)	114(3)

DLPEM$_2$

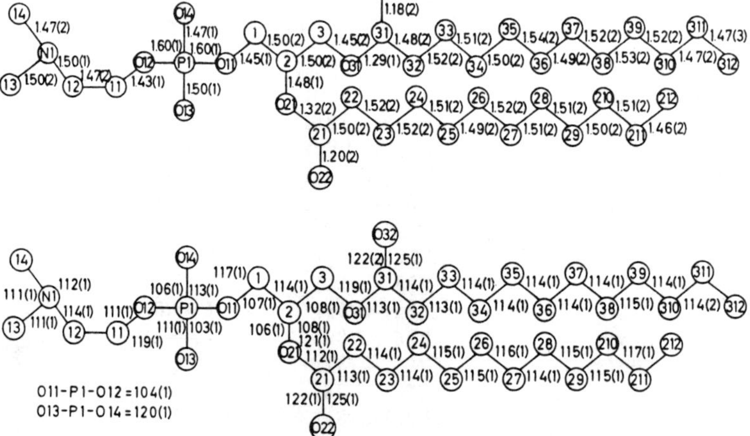

Atom numbering, bond distances (Å) and bond angles (°) of DLPEM$_2$.

Figure A1 Continued

DMPA

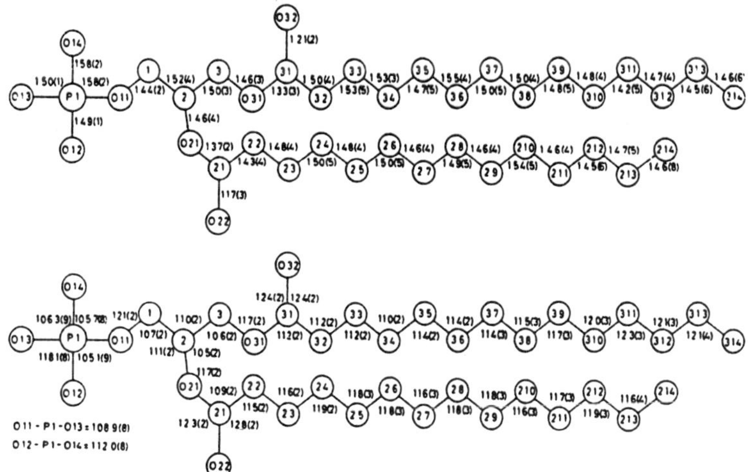

Atom numbering and bond distances (Å) and angles(°) of DMPA.

PPE

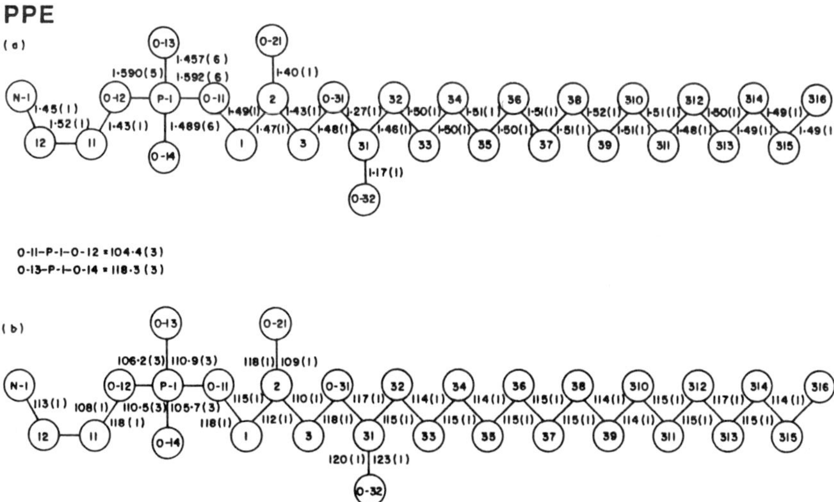

Atom numbering of PPE; (a) bond lengths (Å) and (b) bond angles (deg.).

Table A2 Atomic Fractional Coordinates and Equivalent Isotropic Temperature Factors ($Å^2 \times 10^2$, Except Where Otherwise Stated) for Nonhydrogen Atoms

DLPE

Atom	x/a	y/b	z/c	$10^3 U_{iso}$ ($Å^2$)
P	0.0779(1)	0.9522(11)	−0.1428(11)	34(3)
N	0.0529(4)	1.4790(25)	−0.0285(22)	24(6)
O(11)	0.1083(3)	0.8832(20)	−0.0891(19)	32(5)
O(12)	0.0733(3)	1.1154(21)	−0.0431(18)	38(5)
O(13)	0.0557(3)	0.8341(23)	−0.1016(18)	47(6)
O(14)	0.0799(3)	0.9943(20)	−0.2817(22)	38(6)
C(11)	0.0922(5)	1.2684(34)	−0.0523(29)	34(8)
C(12)	0.0753(5)	1.4249(36)	−0.1239(30)	40(9)
C(1)	0.1104(5)	0.8333(34)	0.0539(31)	32(8)
C(2)	0.1425(5)	0.8583(40)	0.0908(29)	40(9)
C(3)	0.1614(5)	0.7676(35)	−0.0041(31)	45(9)
O(21)	0.1492(3)	1.0483(28)	0.0803(23)	57(6)
O(22)	0.1424(3)	1.0669(26)	0.3063(24)	53(7)
C(21)	0.1485(5)	1.1198(37)	0.1994(39)	24(8)
C(22)	0.1562(5)	1.3187(41)	0.1775(32)	51(9)
C(23)	0.1844(6)	1.3661(41)	0.2622(33)	71(11)
C(24)	0.2089(6)	1.2701(37)	0.2115(32)	54(10)
C(25)	0.2376(6)	1.3104(43)	0.2983(35)	81(11)
C(26)	0.2631(6)	1.2275(42)	0.2416(36)	75(11)
C(27)	0.2884(7)	1.2960(45)	0.3356(38)	93(12)
C(28)	0.3163(6)	1.2264(41)	0.2651(36)	77(11)
C(29)	0.3416(7)	1.3050(49)	0.3620(40)	106(14)
C(210)	0.3690(8)	1.2305(50)	0.2827(42)	114(14)
C(211)	0.3924(9)	1.2971(56)	0.3771(46)	138(16)
C(212)	0.4221(11)	1.2332(66)	0.3096(53)	180(21)
O(31)	0.1904(4)	0.7878(25)	0.0607(22)	70(7)
O(32)	0.2077(6)	0.6648(42)	−0.1182(39)	157(13)
C(31)	0.2108(7)	0.7183(44)	−0.0117(47)	68(12)
C(32)	0.2384(7)	0.7839(46)	0.0730(38)	95(13)
C(33)	0.2642(6)	0.7120(40)	0.0047(34)	69(11)
C(34)	0.2912(7)	0.7508(42)	0.0988(37)	83(12)
C(35)	0.3158(6)	0.7022(40)	0.0246(34)	68(10)
C(36)	0.3429(8)	0.7551(49)	0.1054(42)	112(14)
C(37)	0.3704(9)	0.7113(52)	0.0314(43)	121(15)
C(38)	0.3962(9)	0.7602(55)	0.1239(46)	134(16)
C(39)	0.4230(12)	0.7192(66)	0.0411(55)	176(20)
C(310)	0.4482(11)	0.7609(62)	0.1349(51)	157(19)
C(311)	0.4748(16)	0.6965(93)	0.0594(73)	263(31)
C(312)	0.4992(20)	0.8045(124)	0.1387(90)	351(44)
C(41)	−0.0052(6)	1.1880(48)	−0.1388(39)	61(11)
C(42)	0.0098(5)	1.1446(40)	−0.2520(34)	62(11)
O(41)	−0.0004(3)	1.3005(26)	−0.0530(21)	47(6)
O(42)	−0.0281(3)	1.0811(24)	−0.1109(19)	56(6)

$DLPEM_2$				
Atom	x	y	z	U_{eq}
P(1)	0.0574 (6)	0.2620 (4)	0.0508 (1)	5.2 (0.2)
N(1)	−0.1951 (16)	0.2802 (11)	−0.0525 (2)	5.0 (0.6)
O(11)	−0.0975 (13)	0.3507 (9)	0.0774 (2)	5.1 (0.5)
O(12)	−0.1423 (12)	0.1621 (8)	0.0237 (2)	4.4 (0.4)
O(13)	0.2152 (13)	0.4066 (9)	0.0360 (2)	5.5 (0.5)
O(14)	0.1685 (14)	0.1356 (9)	0.0654 (2)	6.0 (0.5)
O(21)	−0.1918 (15)	0.4446 (11)	0.1444 (2)	7.0 (0.6)
O(22)	−0.3507 (22)	0.6688 (13)	0.1507 (3)	11.8 (0.9)
O(31)	−0.4893 (18)	0.1428 (12)	0.1592 (2)	9.1 (0.7)
O(32)	−0.6569 (28)	0.2147 (22)	0.2050 (3)	20.9 (1.5)
C(11)	−0.2998 (19)	0.2504 (15)	0.0078 (3)	5.6 (0.7)
C(12)	−0.3537 (19)	0.1842 (14)	−0.0274 (3)	5.0 (0.7)
C(13)	0.0637 (19)	0.2656 (15)	−0.0485 (3)	6.4 (0.8)
C(14)	−0.2835 (26)	0.2312 (16)	−0.0873 (3)	7.3 (0.9)
C(1)	−0.2709 (21)	0.2433 (14)	0.0971 (3)	5.3 (0.7)
C(2)	−0.3868 (21)	0.3542 (15)	0.1207 (3)	5.5 (0.7)
C(3)	−0.5861 (23)	0.2607 (19)	0.1409 (3)	7.6 (1.0)
C(21)	−0.1934 (27)	0.5971 (18)	0.1572 (3)	7.6 (1.0)
C(22)	0.0044 (24)	0.6580 (16)	0.1833 (3)	6.8 (0.9)
C(23)	−0.0088 (23)	0.5447 (18)	0.2120 (3)	7.1 (0.9)
C(24)	0.1998 (23)	0.5989 (17)	0.2374 (3)	6.4 (0.8)
C(25)	0.1886 (24)	0.4859 (17)	0.2656 (3)	6.9 (0.9)
C(26)	0.3977 (23)	0.5301 (18)	0.2901 (3)	7.0 (0.9)
C(27)	0.3898 (24)	0.4152 (18)	0.3184 (3)	7.2 (0.9)
C(28)	0.5983 (24)	0.4619 (19)	0.3437 (3)	7.7 (1.0)
C(29)	0.5801 (25)	0.3467 (17)	0.3718 (3)	7.3 (0.9)
C(210)	0.7864 (27)	0.3898 (20)	0.3968 (4)	8.9 (1.1)
C(211)	0.7692 (29)	0.2794 (22)	0.4255 (4)	10.1 (1.3)
C(212)	0.9670 (34)	0.3177 (26)	0.4504 (4)	12.3 (1.5)
C(31)	−0.5244 (26)	0.1391 (18)	0.1911 (3)	8.2 (1.0)
C(32)	−0.3725 (23)	0.0411 (17)	0.2076 (3)	6.9 (0.8)
C(33)	−0.3620 (22)	0.0686 (17)	0.2459 (3)	6.8 (0.8)
C(34)	−0.1882 (22)	−0.0211 (17)	0.2619 (3)	6.4 (0.8)
C(35)	−0.1758 (24)	0.0056 (16)	0.2995 (3)	6.8 (0.9)
C(36)	−0.0020 (26)	−0.0878 (19)	0.3159 (4)	8.3 (1.0)
C(37)	0.0155 (25)	−0.0606 (17)	0.3533 (3)	7.5 (0.9)
C(38)	0.1920 (24)	−0.1500 (17)	0.3692 (3)	7.3 (0.9)
C(39)	0.2063 (26)	−0.1235 (19)	0.4077 (3)	8.1 (1.0)
C(310)	0.3868 (28)	−0.2092 (20)	0.4242 (4)	8.9 (1.1)
C(311)	0.3993 (33)	−0.1831 (24)	0.4612 (4)	11.4 (1.4)
C(312)	0.5774 (38)	−0.2621 (29)	0.4770 (4)	14.8 (1.8)

Table A2 Continued

DMPC.2H_2O							
Molecule A				Molecule B			
Atom	*x*	*y*	*z*	Atom	*x*	*y*	*z*
O 2W	0.6950	–0.1190	0.0290	O 1W	–0.0690	0.0730	0.0762
O 4W	0.5180	–0.3170	0.0501	O 3W	0.2880	0.6480	0.0191
N 1	0.4090	0.1960	0.0884	N 1B	–0.2120	–0.6480	0.0176
P 1	0.6620	–0.2010	0.1125	P 1B	0.1130	–0.3090	0.0573
O 11	0.6920	–0.2400	0.1390	O 11B	0.1640	–0.3460	0.0843
O 12	0.4960	–0.1340	0.1089	O 12B	–0.0530	–0.3830	0.0501
O 13	0.6430	–0.3390	0.0955	O 13B	0.2050	–0.3960	0.0404
O 14	0.7660	–0.0870	0.1089	O 14B	0.0720	–0.1490	0.0527
C 1	0.6110	–0.3670	0.1489	C 1B	0.0930	–0.2620	0.1010
C 2	0.6790	–0.4120	0.1714	C 2B	0.1370	–0.3330	0.1268
C 3	0.7070	–0.2960	0.1906	C 3B	0.0780	–0.2070	0.1467
C 11	0.4870	0.0260	0.1252	C 11B	–0.0890	–0.5490	0.0574
C 12	0.3980	0.1380	0.1127	C 12B	–0.2180	–0.6220	0.0447
C 13	0.5420	0.2590	0.0963	C 13B	–0.2050	–0.5060	0.0030
C 14	0.2900	0.2940	0.0810	C 14B	–0.1000	–0.7580	0.0152
C 15	0.4320	0.1080	0.0685	C 15B	–0.3380	–0.7380	0.0068
O 21	0.5760	–0.5190	0.1832	C 21B	0.0350	–0.4640	0.1278
O 22	0.6240	–0.7220	0.1597	O 22B	0.2140	–0.6170	0.1437
C 21	0.5670	–0.6600	0.1740	C 21B	0.0820	–0.6200	0.1331
C 22	0.4560	–0.7510	0.1883	C 22B	–0.0160	–0.7420	0.1378
C 23	0.5260	–0.8150	0.2090	C 23B	0.0210	–0.8350	0.1610
C 24	0.5990	–0.7450	0.2310	C 24B	0.0170	–0.7510	0.1850
C 25	0.6320	–0.8360	0.2553	C 25B	0.0560	–0.8620	0.2046
C 26	0.7100	–0.7670	0.2756	C 26B	0.0810	–0.7630	0.2311
C 27	0.7260	–0.8580	0.2985	C 27B	0.1400	–0.8810	0.2500
C 28	0.8090	–0.7990	0.3197	C 28B	0.1740	–0.7840	0.2747
C 29	0.8160	–0.8730	0.3440	C 29B	0.2530	–0.8950	0.2955
C210	0.8920	–0.8110	0.3649	C210B	0.2850	–0.7950	0.3186
C211	0.9210	–0.8880	0.3878	C211B	0.3430	–0.9110	0.3390
C212	0.9940	–0.8220	0.4094	C212B	0.3940	–0.8030	0.3636
C213	1.0400	–0.8920	0.4318	C213B	0.4270	–0.9280	0.3860
C214	1.0470	–0.8410	0.4549	C214B	0.5140	–0.8070	0.4082
O 31	0.7900	–0.3720	0.2102	O 31B	0.1110	–0.2870	0.1688
O 32	0.8230	–0.1610	0.2308	O 32B	0.3090	–0.1630	0.1794
C 31	0.8160	–0.2920	0.2317	C 31B	0.2300	–0.2540	0.1834
C 32	0.9010	–0.3690	0.2533	C 32B	0.2250	–0.3160	0.2089
C 33	0.9410	–0.2760	0.2773	C 33B	0.3610	–0.2660	0.2272
C 34	1.0020	–0.3800	0.2980	C 34B	0.3500	–0.3550	0.2545
C 35	1.0360	–0.2890	0.3208	C 35B	0.4890	–0.2920	0.2733
C 36	1.0860	–0.3770	0.3429	C 36B	0.4610	–0.3690	0.2980
C 37	1.1300	–0.3050	0.3682	C 37B	0.5640	–0.3000	0.3179
C 38	1.1620	–0.3980	0.3891	C 38B	0.5780	–0.3940	0.3405
C 39	1.2220	–0.3100	0.4111	C 39B	0.6590	–0.3290	0.3610
C310	1.2630	–0.4210	0.4327	C310B	0.6400	–0.3830	0.3867
C311	1.3170	–0.3320	0.4563	C311B	0.7570	–0.3350	0.4058

DMPC.$2H_2O$

Molecule A				Molecule B			
Atom	x	y	z	Atom	x	y	z
C312	1.3710	–0.3290	0.4797	C312B	0.7500	–0.4060	0.4303
C313	1.3240	–0.3200	0.5031	C313B	0.8440	–0.3100	0.4502
C314	1.4370	–0.3660	0.5198	C314B	0.8630	–0.3840	0.4735

DMPA

Atom	x	y	z	U_{eq}
Na(1)	0.5343(15)	0.2279(12)	0.0159(2)	6.4(0.7)
P(1)	1.0431(10)	0.5008(—)	0.0236(1)	3.2(0.2)
O(11)	1.1408(28)	0.5979(19)	0.0581(4)	7.1(1.5)
O(12)	0.7448(21)	0.4937(20)	0.0111(3)	4.6(1.1)
O(13)	1.1948(27)	0.3378(17)	0.0289(4)	6.3(1.3)
O(14)	1.1366(26)	0.6117(19)	0.0006(3)	7.6(1.4)
O(21)	1.3133(29)	0.7606(21)	0.1209(4)	7.2(1.5)
O(22)	1.7116(28)	0.6622(23)	0.1568(4)	8.7(1.7)
O(31)	1.2691(31)	1.0766(21)	0.0960(4)	6.4(1.5)
O(32)	0.8275(31)	1.1255(24)	0.0682(4)	8.9(1.7)
C(1)	1.4193(38)	0.6433(31)	0.0761(5)	5.9(1.9)
C(2)	1.4262(45)	0.7980(28)	0.0967(6)	6.7(2.2)
C(3)	1.2484(52)	0.9322(33)	0.0746(6)	6.9(2.4)
C(21)	1.4875(46)	0.6979(29)	0.1510(5)	6.1(2.1)
C(22)	1.3534(45)	0.6926(36)	0.1730(6)	8.6(2.4)
C(23)	1.4864(51)	0.5853(40)	0.2027(6)	12.8(2.9)
C(24)	1.3549(48)	0.5783(35)	0.2268(7)	8.7(2.6)
C(25)	1.4761(59)	0.4740(38)	0.2571(7)	10.2(3.0)
C(26)	1.3421(54)	0.4751(43)	0.2809(7)	11.9(3.0)
C(27)	1.4632(63)	0.3738(51)	0.3112(7)	13.0(3.5)
C(28)	1.3254(61)	0.3805(40)	0.3342(7)	12.6(3.4)
C(29)	1.4477(61)	0.2896(59)	0.3657(7)	12.0(3.8)
C(210)	1.3078(61)	0.2987(45)	0.3899(7)	13.0(3.5)
C(211)	1.4344(61)	0.2049(51)	0.4210(7)	12.8(3.4)
C(212)	1.2934(62)	0.2036(48)	0.4427(8)	14.9(3.7)
C(213)	1.4067(95)	0.1068(56)	0.4740(9)	18.7(5.4)
C(214)	1.2561(93)	0.1131(84)	0.4949(11)	25.9(6.4)
C(31)	1.0449(44)	1.1638(34)	0.0896(6)	8.2(2.4)
C(32)	1.0888(62)	1.3082(31)	0.1130(7)	10.4(3.0)
C(33)	0.9184(48)	1.2925(30)	0.1333(6)	7.5(2.3)
C(34)	1.0259(46)	1.1569(27)	0.1601(6)	8.5(2.3)
C(35)	0.8758(52)	1.1609(36)	0.1813(7)	11.8(3.0)
C(36)	0.9870(51)	1.0403(36)	0.2115(7)	11.7(2.9)
C(37)	0.8506(53)	1.0573(42)	0.2348(7)	11.5(3.2)
C(38)	0.9671(56)	0.9490(36)	0.2653(6)	7.2(2.8)
C(39)	0.8307(53)	0.9611(39)	0.2882(7)	13.6(3.2)
C(310)	0.9483(61)	0.8620(41)	0.3194(7)	12.6(3.4)
C(311)	0.8216(58)	0.8620(38)	0.3419(6)	10.5(2.9)

Table A2 Continued

DMPA

Atom	x	y	z	U_{eq}
C(312)	0.9333(58)	0.7728(45)	0.3740(6)	9.6(3.1)
C(313)	0.8033(85)	0.7800(57)	0.3969(9)	20.1(5.3)
C(314)	0.9111(108)	0.6858(70)	0.4281(9)	24.5(6.4)

DMPG

Molecule A

Atom	x	y	z	U_{eq}
Na(1)	0.1593 (9)	0.4944 (14)	0.0075 (3)	4.7 (0.7)
P(1)	0.0790 (7)	0.1754 (10)	0.0391 (2)	4.1 (0.5)
O(11)	0.1192 (20)	0.1510 (25)	0.0731 (4)	5.3 (1.3)
O(12)	−0.0131 (19)	0.3275 (23)	0.0381 (5)	5.0 (1.3)
O(13)	0.0049 (17)	0.0226 (23)	0.0298 (5)	6.0 (1.4)
O(14)	0.1883 (16)	0.2313 (25)	0.0226 (5)	5.6 (1.3)
C(11)	−0.1521 (26)	0.3090 (39)	0.0417 (8)	5.7 (2.2)
C(12)	−0.1908 (24)	0.4541 (41)	0.0598 (7)	4.9 (1.9)
C(13)	−0.3406 (33)	0.4364 (40)	0.0639 (9)	6.8 (2.3)
C(15)	−0.1717 (20)	0.6001 (25)	0.0441 (5)	5.6 (1.4)
O(16)	−0.4172 (18)	0.4488 (28)	0.0379 (5)	5.6 (1.3)
C(1)	0.2181 (37)	0.2613 (46)	0.0866 (12)	10.6 (3.3)
C(2)	0.3033 (51)	0.1729 (64)	0.1109 (10)	12.2 (3.8)
C(3)	0.4403 (30)	0.2402 (44)	0.1151 (10)	8.1 (2.5)
0(21)	0.2324 (22)	0.1610 (35)	0.1382 (6)	8.0 (1.8)
0(22)	0.3546 (25)	−0.0469 (40)	0.1537 (7)	10.5 (2.3)
C(21)	0.2740 (45)	0.0474 (59)	0.1573 (10)	8.3 (3.2)
C(22)	0.1975 (35)	0.0559 (44)	0.1828 (8)	7.0 (2.4)
C(23)	0.2234 (34)	−0.0644 (57)	0.2067 (9)	9.0 (2.8)
C(24)	0.1434 (37)	−0.0569 (65)	0.2330 (9)	10.0 (3.2)
C(25)	0.1866 (41)	−0.1868 (63)	0.2554 (9)	10.1 (3.3)
C(26)	0.1092 (45)	−0.1843 (69)	0.2809 (12)	12.0 (3.9)
C(27)	0.1384 (39)	−0.3100 (65)	0.3035 (8)	9.8 (3.2)
C(28)	0.0656 (36)	−0.3090 (63)	0.3311 (10)	10.0 (3.2)
C(29)	0.0958 (47)	−0.4339 (62)	0.3530 (9)	11.3 (3.6)
C(210)	0.0239 (48)	−0.4350 (69)	0.3791 (10)	12.2 (3.9)
C(211)	0.0636 (49)	−0.5389 (89)	0.4032 (10)	14.1 (4.5)
C(212)	−0.0031 (68)	−0.5506 (105)	0.4301 (15)	18.7 (6.4)
C(213)	0.0224 (67)	−0.6426 (131)	0.4583 (18)	22.0 (8.1)
C(214)	−0.0490 (79)	−0.6283 (118)	0.4843 (15)	22.0 (8.4)
O(31)	0.4140 (25)	0.4033 (32)	0.1258 (6)	9.4 (1.9)
O(32)	0.6302 (27)	0.4079 (48)	0.1391 (10)	13.7 (2.9)
C(31)	0.5301 (64)	0.4729 (66)	0.1389 (10)	11.9 (4.2)
C(32)	0.4805 (48)	0.6121 (52)	0.1520 (12)	10.6 (3.6)
C(33)	0.4899 (55)	0.6050 (55)	0.1852 (13)	12.4 (4.2)
C(34)	0.4372 (38)	0.4922 (54)	0.2000 (9)	7.7 (2.7)
C(35)	0.4441 (48)	0.4752 (65)	0.2308 (13)	12.2 (4.1)
C(36)	0.3899 (47)	0.3542 (61)	0.2480 (10)	11.1 (3.6)

DMPG

Molecule A

Atom	x	y	z	U_{eq}
C(37)	0.4051 (45)	0.3496 (69)	0.2802 (11)	11.4 (3.9)
C(38)	0.3510 (43)	0.2285 (75)	0.2969 (13)	11.4 (4.0)
C(39)	0.3643 (58)	0.2136 (70)	0.3279 (16)	14.5 (5.0)
C(310)	0.3080 (44)	0.1049 (60)	0.3442 (13)	10.3 (3.6)
C(311)	0.3289 (59)	0.0881 (95)	0.3768 (15)	16.2 (5.9)
C(312)	0.2683 (66)	–0.0047 (95)	0.3954 (14)	16.1 (5.8)
C(313)	0.2900 (63)	–0.0258 (89)	0.4308 (21)	18.2 (6.7)
C(314)	0.2200 (90)	–0.1420 (122)	0.4437 (22)	24.2 (9.3)

DMPG

Molecule B

Atom	x	y	z	U_{eq}
Na(2)	0.6580 (9)	–0.3216 (14)	0.0081 (2)	4.2 (0.6)
P(1)	0.5778 (7)	0.0000 (–)	0.0390 (2)	4.2 (0.4)
O(11)	0.6116 (18)	0.0175 (24)	0.0730 (4)	5.1 (1.2)
O(12)	0.4794 (17)	–0.1549 (24)	0.0370 (5)	5.7 (1.4)
O(13)	0.5097 (16)	0.1492 (24)	0.0296 (4)	4.7 (1.2)
O(14)	0.6904 (15)	–0.0608 (24)	0.0239 (4)	5.2 (1.2)
C(11)	0.3476 (27)	–0.1235 (37)	0.0437 (9)	6.1 (2.2)
C(12)	0.3104 (27)	–0.2795 (47)	0.0584 (9)	7.1 (2.4)
C(13)	0.1621 (36)	–0.2783 (38)	0.0641 (9)	7.0 (2.4)
O(15)	0.3314 (19)	–0.4190 (25)	0.0427 (6)	5.7 (1.4)
O(16)	0.0839 (19)	–0.2785 (24)	0.0377 (5)	4.9 (1.2)
C(1)	0.6724 (61)	–0.1066 (55)	0.0918 (9)	12.1 (3.8)
C(2)	0.8015 (51)	–0.0500 (82)	0.1041 (11)	11.1 (4.1)
C(3)	0.8018 (44)	0.0617 (56)	0.1241 (11)	9.3 (3.3)
O(21)	0.8366 (38)	–0.2067 (42)	0.1228 (9)	11.6 (2.8)
O(22)	1.0382 (43)	–0.1420 (63)	0.1235 (9)	15.3 (3.9)
C(21)	0.9583 (89)	–0.2258 (71)	0.1289 (17)	12.9 (5.9)
C(22)	1.0015 (51)	–0.3761 (99)	0.1472 (13)	17.1 (5.8)
C(23)	0.9187 (52)	–0.3694 (59)	0.1722 (11)	13.1 (4.2)
C(24)	0.9712 (41)	–0.5164 (61)	0.1939 (10)	10.1 (3.3)
C(25)	0.9033 (40)	–0.5219 (59)	0.2205 (13)	11.4 (3.7)
C(26)	0.9360 (46)	–0.6473 (61)	0.2440 (11)	10.7 (3.6)
C(27)	0.8649 (46)	–0.6516 (69)	0.2704 (14)	13.7 (4.6)
C(28)	0.9038 (43)	–0.7640 (59)	0.2939 (10)	10.3 (3.3)
C(29)	0.8264 (40)	–0.7625 (59)	0.3179 (11)	10.5 (3.3)
C(210)	0.8629 (50)	–0.8717 (67)	0.3439 (11)	11.6 (4.0)
C(211)	0.7817 (48)	–0.8884 (74)	0.3667 (13)	13.4 (4.5)
C(212)	0.8191 (47)	–0.9873 (69)	0.3940 (11)	12.3 (3.9)
C(213)	0.7429 (63)	–1.0022 (90)	0.4194 (14)	17.0 (5.8)
C(214)	0.7834 (60)	–1.1137 (92)	0.4421 (13)	17.7 (5.9)
O(31)	0.7167 (22)	0.0450 (31)	0.1466 (5)	7.8 (1.7)
O(32)	0.8714 (33)	0.0595 (58)	0.1824 (7)	15.4 (3.3)

Table A2 Continued

DMPG

Molecule B

Atom	x	y	z	U_{eq}
C(31)	0.7516 (42)	0.0430 (48)	0.1733 (10)	8.9 (3.0)
C(32)	0.6726 (36)	0.0292 (56)	0.1979 (8)	7.9 (2.7)
C(33)	0.7179 (31)	−0.0619 (51)	0.2239 (8)	6.9 (2.3)
C(34)	0.6240 (40)	−0.0865 (74)	0.2476 (9)	11.1 (3.7)
C(35)	0.6719 (38)	−0.1818 (61)	0.2740 (12)	10.4 (3.5)
C(36)	0.5914 (41)	−0.2008 (69)	0.2977 (10)	10.3 (3.5)
C(37)	0.6307 (40)	−0.2914 (69)	0.3227 (11)	10.8 (3.6)
C(38)	0.5444 (45)	−0.3108 (85)	0.3465 (11)	13.3 (4.3)
C(39)	0.5922 (37)	−0.4229 (68)	0.3723 (12)	11.2 (3.7)
C(310)	0.5117 (56)	−0.4450 (69)	0.3965 (11)	12.7 (4.3)
C(311)	0.5535 (45)	−0.5510 (78)	0.4210 (13)	12.6 (4.3)
C(312)	0.4786 (59)	−0.5684 (91)	0.4448 (12)	15.3 (5.0)
C(313)	0.5093 (66)	−0.6586 (120)	0.4742 (17)	20.0 (7.1)
C(314)	0.4268 (73)	−0.6595 (173)	0.4942 (14)	25.9 (9.6)

PPE

Atom	x	y	z
P-1	0.2694 (2)	0.5016 (2)	0.0604 (1)
N-1	0.8261 (7)	0.3951 (7)	0.0402 (2)
O-11	0.2643 (7)	0.5869 (6)	0.0979 (1)
O-12	0.4693 (6)	0.4598 (6)	0.0552 (1)
O-13	0.1749 (6)	0.3626 (6)	0.0627 (1)
O-14	0.2136 (7)	0.6108 (6)	0.0327 (1)
O-21	0.1045 (19)	0.3103 (14)	0.1357 (3)
O-31	−0.0749 (10)	0.4608 (9)	0.1860 (2)
O-32	−0.2426 (16)	0.6467 (15)	0.1825 (4)
C-11	0.5885 (10)	0.5770 (10)	0.0476 (3)
C-12	0.7697 (10)	0.5267 (10)	0.0590 (3)
C-1	0.3239 (14)	0.5074 (15)	0.1308 (2)
C-2	0.1841 (16)	0.4348 (13)	0.1511 (3)
C-3	0.0502 (15)	0.5360 (14)	0.1619 (3)
C-31	−0.2153 (17)	0.5281 (14)	0.1938 (4)
C-32	−0.3412 (14)	0.4475 (12)	0.2158 (3)
C-33	−0.4800 (13)	0.5399 (12)	0.2331 (3)
C-34	−0.6120 (14)	0.4533 (12)	0.2540 (3)
C-35	−0.7575 (13)	0.5445 (12)	0.2701 (3)
C-36	−0.8949 (14)	0.4555 (12)	0.2889 (3)
C-37	−1.0398 (14)	0.5446 (12)	0.3055 (3)
C-38	−1.1763 (14)	0.4555 (12)	0.3250 (3)
C-39	−1.3214 (14)	0.5453 (12)	0.3420 (3)
C-310	−1.4551 (15)	0.4548 (12)	0.3618 (3)
C-311	−1.5972 (14)	0.5439 (12)	0.3793 (3)
C-312	−1.7336 (15)	0.4569 (13)	0.3980 (3)
C-313	−1.8732 (16)	0.5452 (15)	0.4160 (3)

PPE			
Atom	x	y	z
C-314	–2.0137 (16)	0.4612 (17)	0.4348 (3)
C-315	–2.1520 (19)	0.5515 (20)	0.4522 (4)
C-316	–2.2872 (22)	0.4643 (29)	0.4717 (5)
H (N-11)	0.750 (11)	0.296 (10)	0.054 (2)
H (N-12)	0.943 (12)	0.356 (10)	0.042 (2)
H (N-13)	0.803 (12)	0.447 (10)	0.020 (2)
H (O-21)	0.133 (12)	0.253 (10)	0.112 (2)

LPPC				
Atom	x	y	z	U
P(1)	0.899 (1)	0.114 (2)	0.377 (2)	21 (4)
N(1)	0.924 (4)	–0.132 (10)	0.738 (8)	120 (30)
O(11)	0.837 (3)	0.186 (8)	0.337 (7)	137 (29)
O(12)	0.902 (3)	0.004 (8)	0.503 (6)	139 (28)
O(13)	0.931 (3)	0.018 (11)	0.283 (7)	205 (38)
O(14)	0.942 (3)	0.248 (8)	0.439 (5)	119 (24)
O(31)	0.701 (3)	0.207 (7)	0.332 (5)	170 (56)
O(32)	0.675 (4)	0.416 (13)	0.382 (9)	103 (24)
C(1)	0.805 (5)	0.078 (14)	0.250 (11)	150 (47)
C(2)	0.756 (6)	0.176 (16)	0.185 (14)	198 (65)
C(3)	0.735 (5)	0.308 (14)	0.251 (11)	170 (56)
C(11)	0.874 (4)	0.062 (11)	0.611 (8)	112 (35)
C(12)	0.930 (3)	0.023 (9)	0.707(7)	60 (25)
C(13)	0.978 (6)	–0.144 (15)	0.869 (13)	188 (57)
C(14)	0.871 (4)	–0.173 (9)	0.775 (8)	81 (31)
C(15)	0.946 (5)	–0.245 (15)	0.637 (11)	166 (52)
C(31)	0.658 (9)	0.271 (26)	0.373 (18)	290 (100)
C(32)	0.630 (5)	0.188 (13)	0.450 (11)	137 (46)
C(33)	0.587 (6)	0.279 (17)	0.494 (13)	208 (69)
C(34)	0.561 (4)	0.193 (11)	0.575 (9)	108 (37)
C(35)	0.512 (5)	0.291 (13)	0.622 (9)	128 (43)
C(36)	0.468 (4)	0.200 (11)	0.682 (8)	90 (32)
C(37)	0.437 (4)	0.301 (11)	0.748 (8)	104 (35)
C(38)	0.392 (4)	0.221 (10)	0.810 (7)	85 (31)
C(39)	0.351 (5)	0.327 (12)	0.858 (9)	124 (42)
C(310)	0.308 (4)	0.228 (11)	0.906 (8)	92 (33)
C(311)	0.283 (6)	0.315 (15)	0.996 (12)	187 (60)
C(312)	0.230 (5)	0.216 (14)	1.024 (10)	156 (50)
O(H_2O)	0.898 (3)	0.048(9)	0.047(7)	184(36)

Source: Refs. 4, 5, and 7–10.

Table A3 Torsion Angles

Lipid	α_1	α_2	α_3	α_r	α_5	α_6	θ_1	θ_2	θ_3	θ_4
DLPE (D)	–154	58	66	106	67	—	–52	65	–172	69
DLPE (L)	154	–58	–66	–106	–67	—	52	–65	172	–69
DLPEM$_2$ (D)	179	65	54	144	–96	—	176	–66	56	–60
DLPEM$_2$ (L)	–179	–65	–54	–144	96	—	–176	66	–56	60
DMPC 1	163	62	68	143	–64	179	58	177	–178	63
DMPC 2	177	–74	–47	–150	54	176	168	–82	166	51
DMPA	153	—	—	—	—	—	–54	62	–178	61
DMPG A	–146	–76	–86	143	180	–66	151	–78	64	–63
DMPG B	116	58	78	–147	–173	67	71	179	45	–58
PPE (D)	–94	–60	–71	157	60	—	–57	70	–173	55
PPE (L)	94	60	71	–157	–60	—	57	–70	173	–55
LPPC 1	162	86	45	129	84	166	28	—	78	—
LPPC 2	–162	–86	–45	–129	–84	–166	–28	—	–78	—

Lipid	β_1	β_2	β_3	β_4	β_5	γ_1	γ_2	γ_3	γ_4	γ_5
DLPE (D)	97	179	–119	65	–178	–178	173	179	–171	–173
DLPE (L)	–97	–179	119	–65	178	178	–173	–179	171	173
DLPEM$_2$ (D)	148	173	–57	176	—	129	–167	166	175	—
DLPEM$_2$ (L)	–148	–173	57	–176	—	–129	167	–166	–175	—
DMPC 1	82	172	–81	45	171	–177	168	–173	178	–180
DMPC 2	120	179	–134	67	180	102	176	180	180	–170
DMPA	87	172	164	179	180	–142	180	–119	73	173
DMPG A	159	178	178	–179	—	164	–170	110	–57	—
DMPG B	157	180	–50	–175	—	122	179	142	174	—
PPE (D)	—	—	—	—	—	–170	176	164	177	—
PPE (L)	—	—	—	—	—	170	–176	–164	–177	—
LPPC 1	—	—	—	—	—	156	176	–175	178	178
LPPC 2	—	—	—	—	—	–156	–176	175	–178	–178

For the racemic crystals DLPE, DLPEM$_2$, and PPE, the D and L enantiomers necessarily have equal and opposite angles.
Source: Refs. 2, 5, and 7–10.

Table A4 Inclination to Layer Plane, and Intra- and Intermolecular Distances

	Inclination towards layer plane (°)				Intramolecular distances (Å)				Intermolecular distances (Å)			
	C(1)-C(2) bond	P-N dipole	$P\bar{O}_2$-N dipole	P<O(13), O(14) plane	O(12)…N	P…N	O(13)…N, O(14)…N	$P\bar{O}_2$…N	P…P	N…N	P<O(13)…, O(14)… N	P<O(13)…, O(14)… O(W)
DLPE	75	15	8	51	2.99	4.42	4.73 5.07	4.74	5.88	6.12	2.74 2.86	— —
DMPC1	61	17	9	53	3.21	4.29	4.05 5.19	4.48	5.37 5.42	6.42 5.62	1: 4.62 2: 4.72 2:4.53	W1: 2.84 W2: 2.61
DMPC2	63	27	18	57	3.18	4.52	4.32 5.34	4.68	—	—	2: 3.91*, 4.23*	W1: 2.80 W2: 2.77 W3: 2.91
LPPC	48	7	−3	68	2.85	4.55	4.95 5.21	4.89	6.05	5.92	3.71 4.23 3.8*, 4.3* 4.6*	2.62 2.62

*Refers to contact distances across the layer interface.
Source: Ref. 2.

DLPE

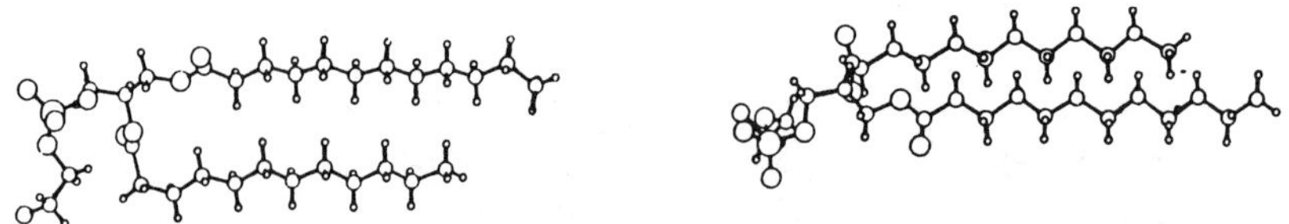

DMPC

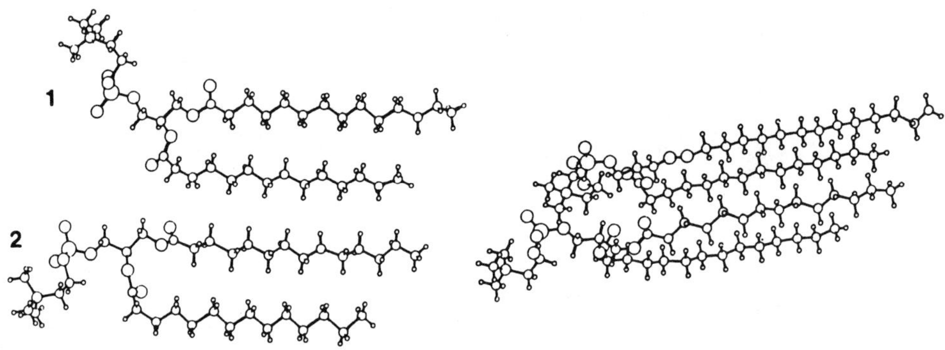

DLPEM$_2$

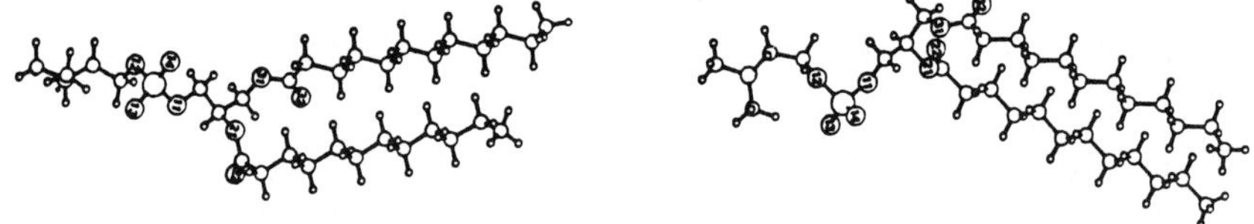

Figure A2 Molecular conformation. (From Refs. 2, 5, 7, 8, and 10.)

DMPA

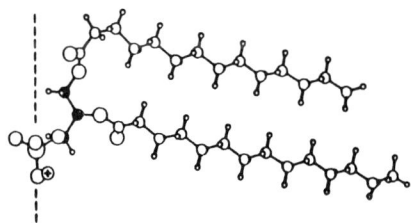

DMPG

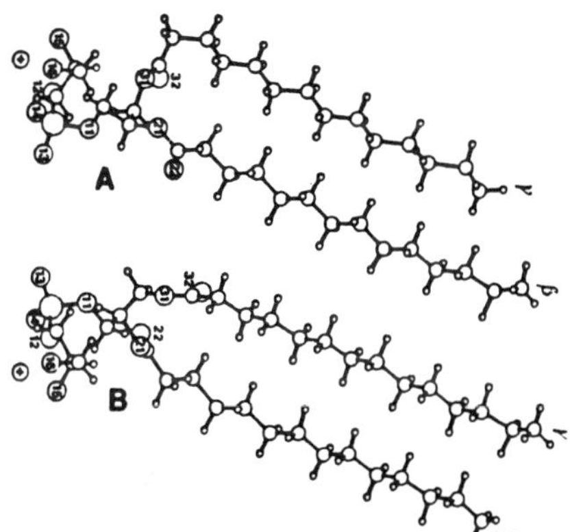

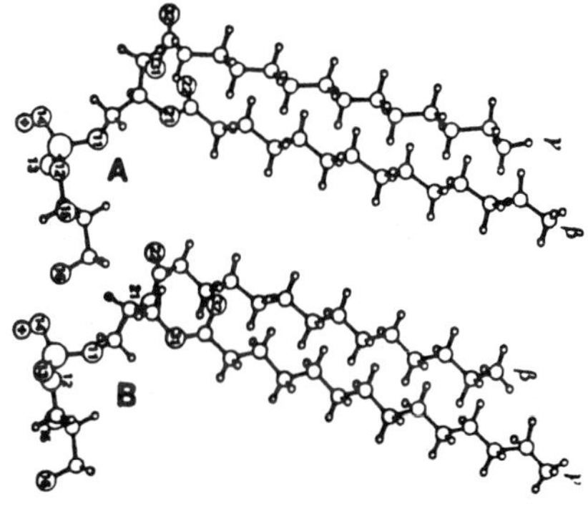

LPPC

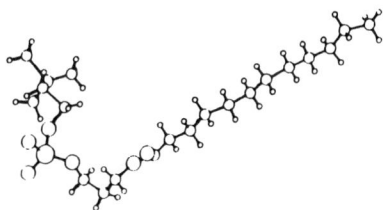

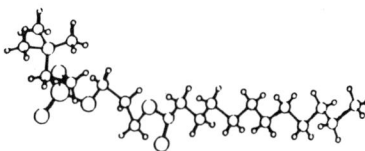

Figure A2 Continued

DLPE

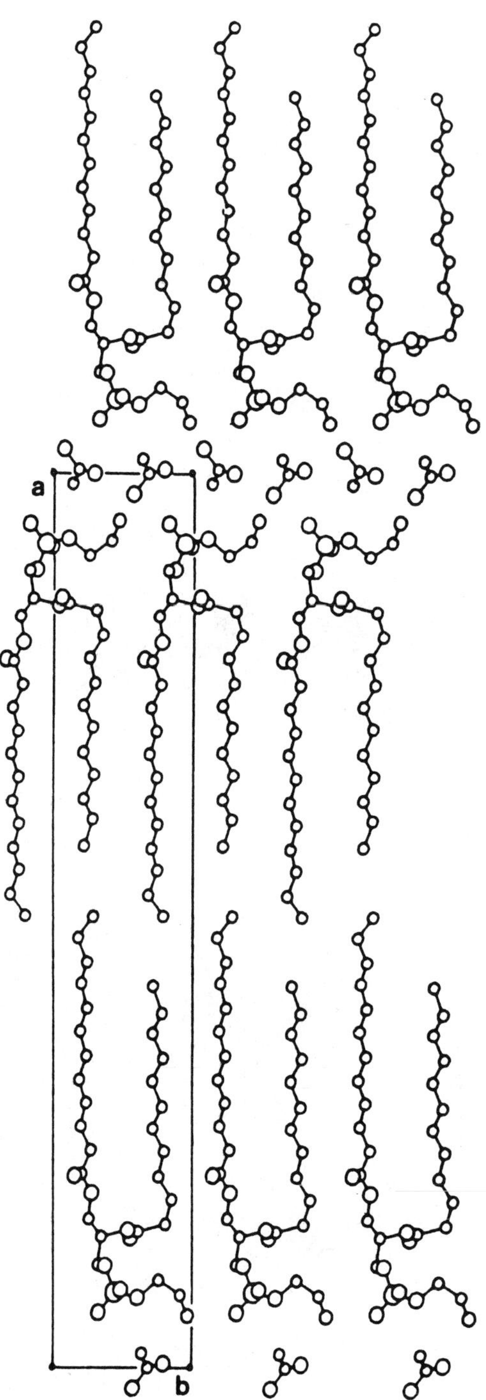

Figure A3 Molecular packing. (From Refs. 2, 5, and 7–10.)

DLPEM$_2$

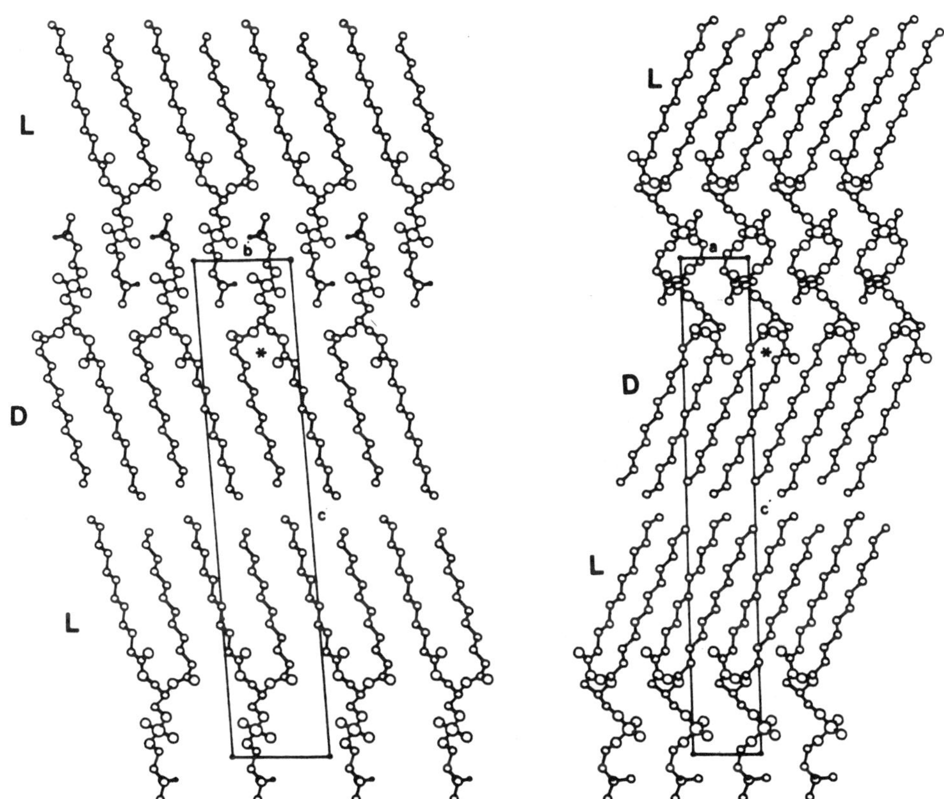

Figure A3 Continued

DMPC

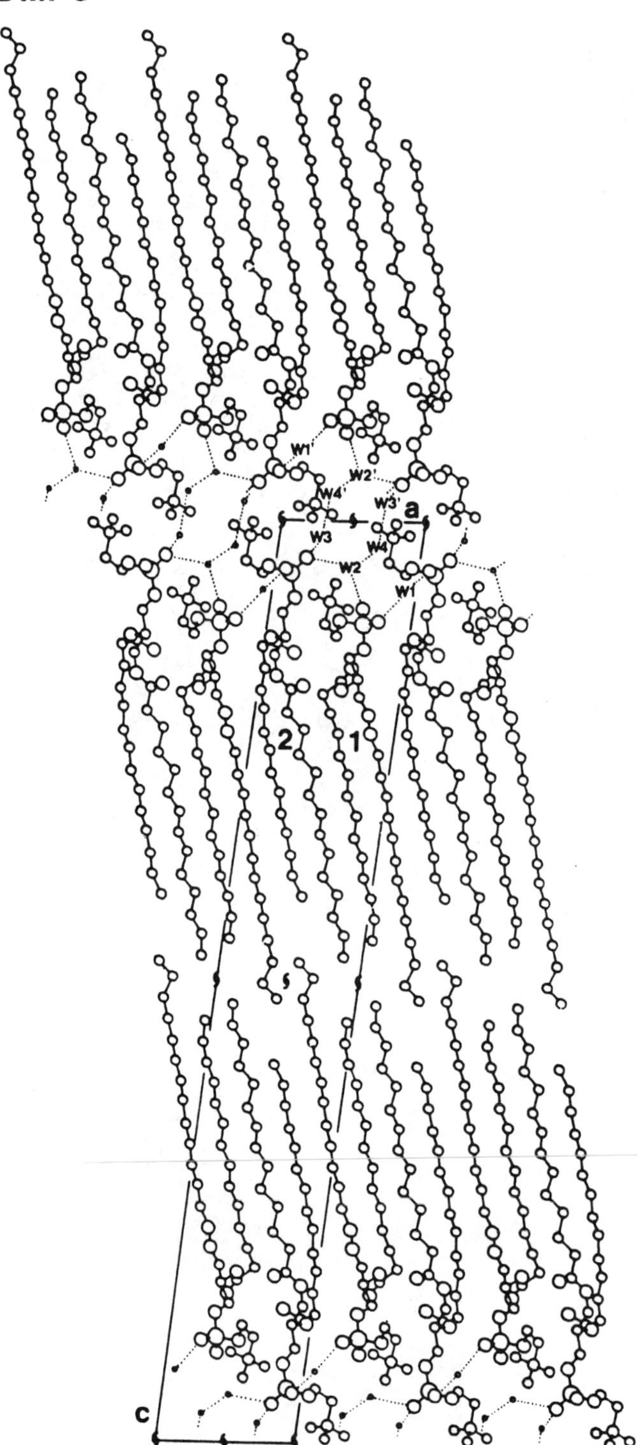

Figure A3 Continued

DMPA

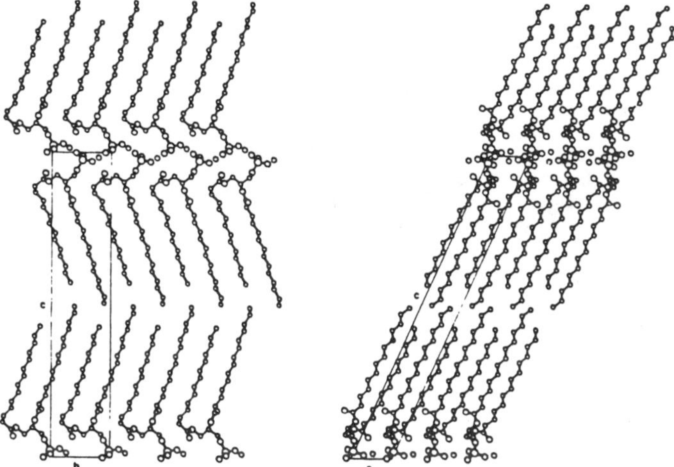

DMPG

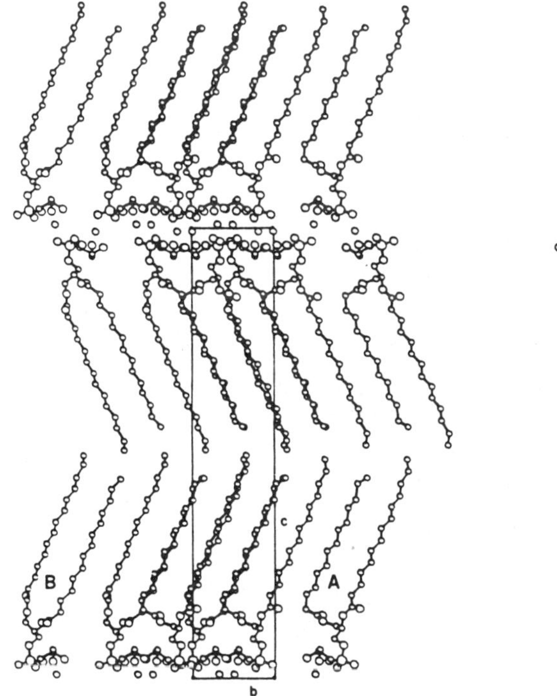

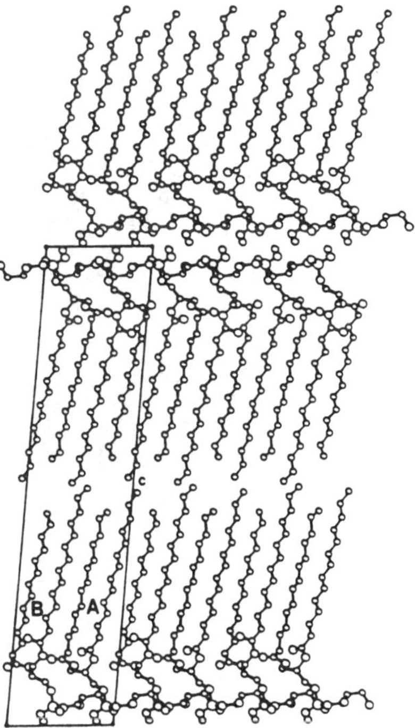

Figure A3 Continued

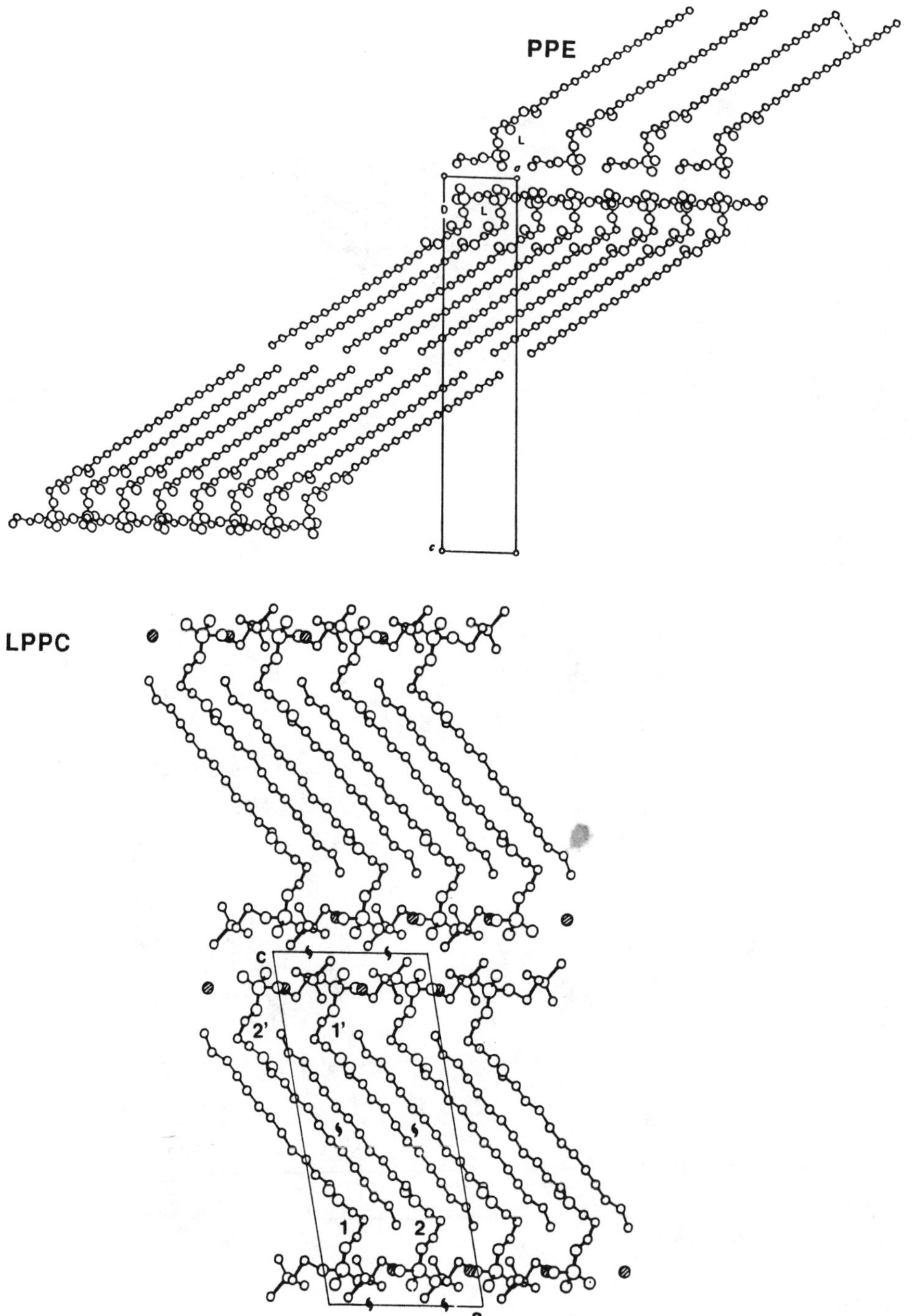

Figure A3 Continued

DLPE

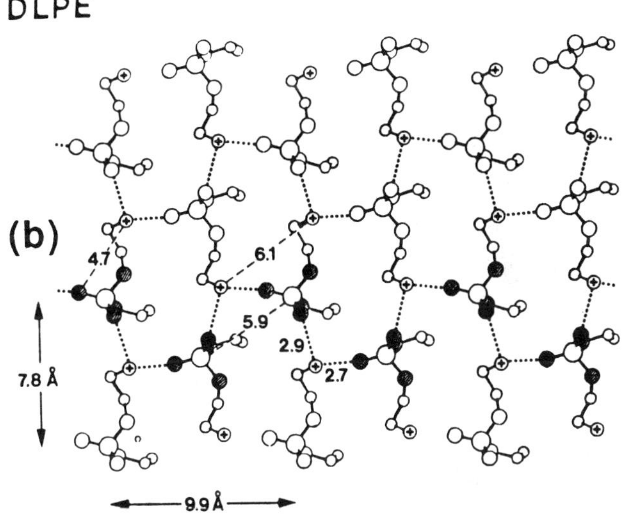

DLPEM2

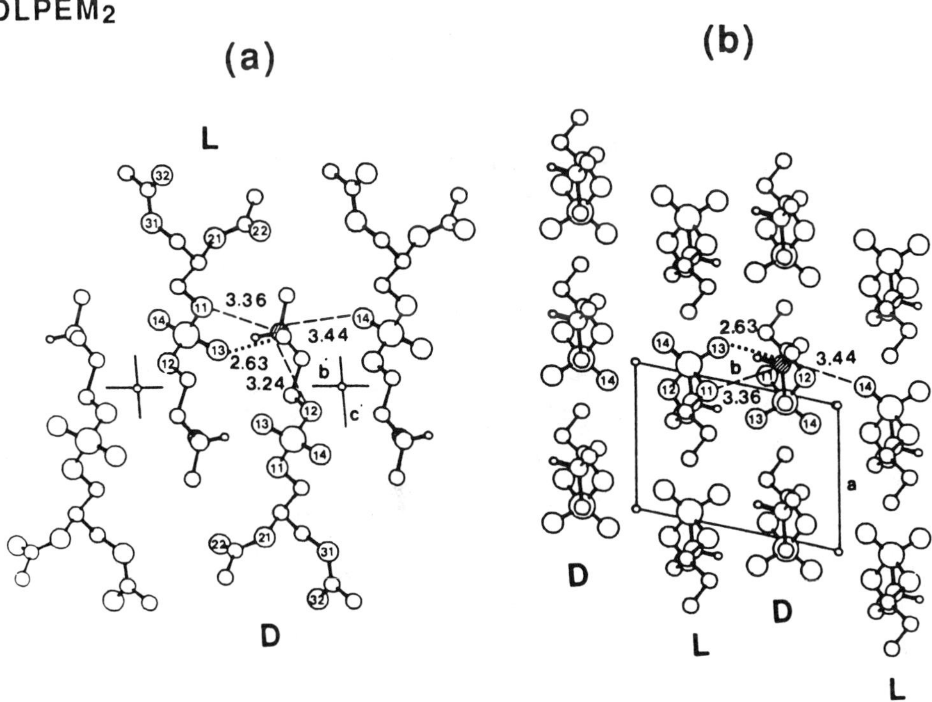

Figure A4 Head group packing and interactions viewed (a) parallel and (b) perpendicular to the plane of the bilayer. (From Refs. 2, 5, and 7–10.)

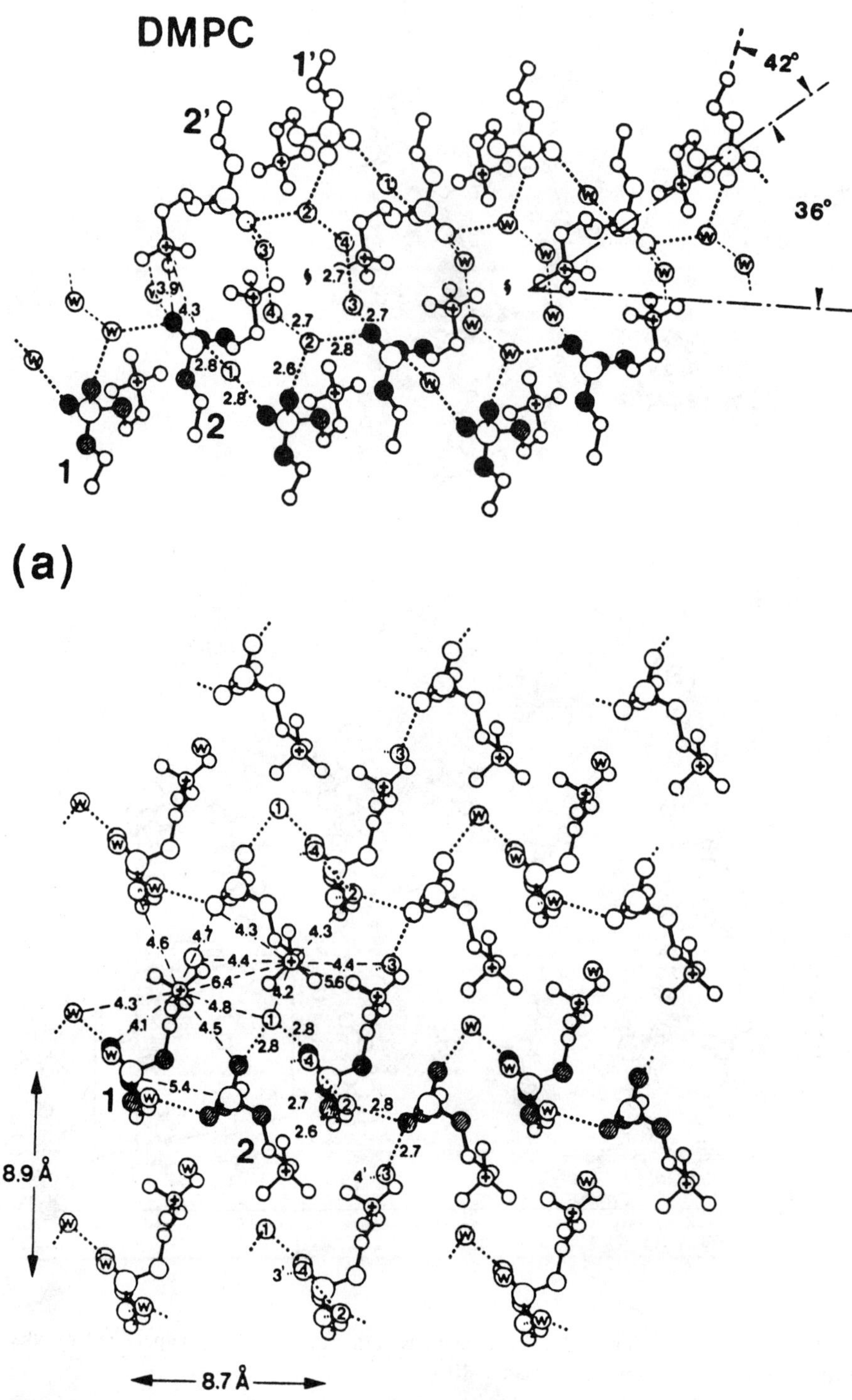

Figure A4 Continued

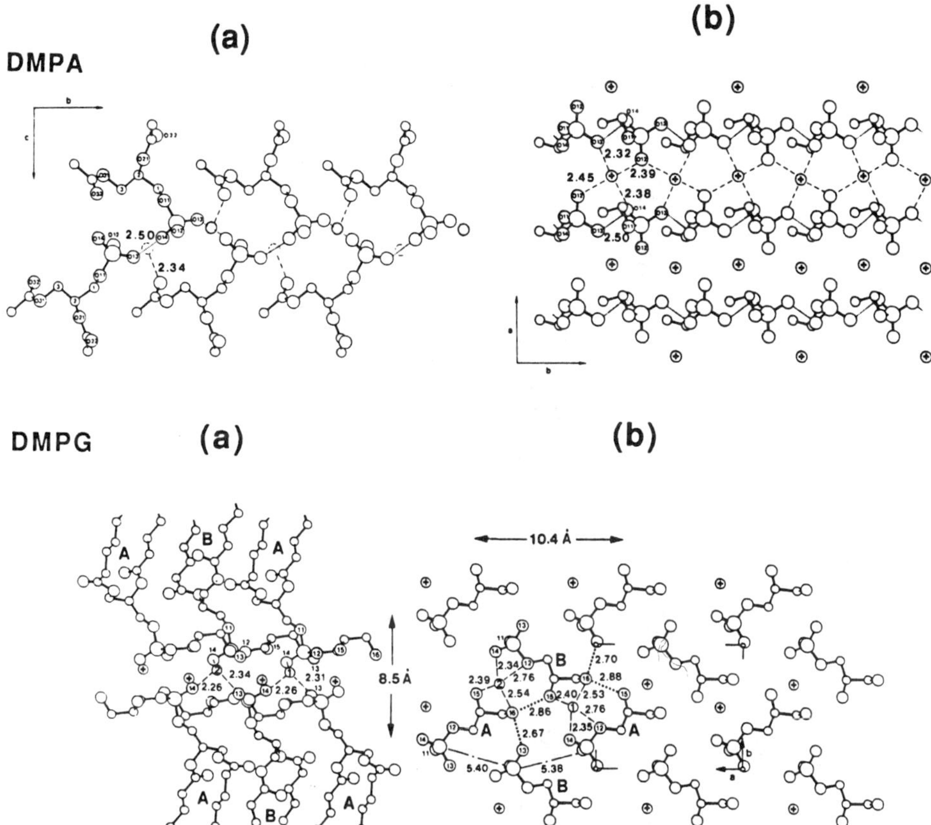

Figure A4 Continued

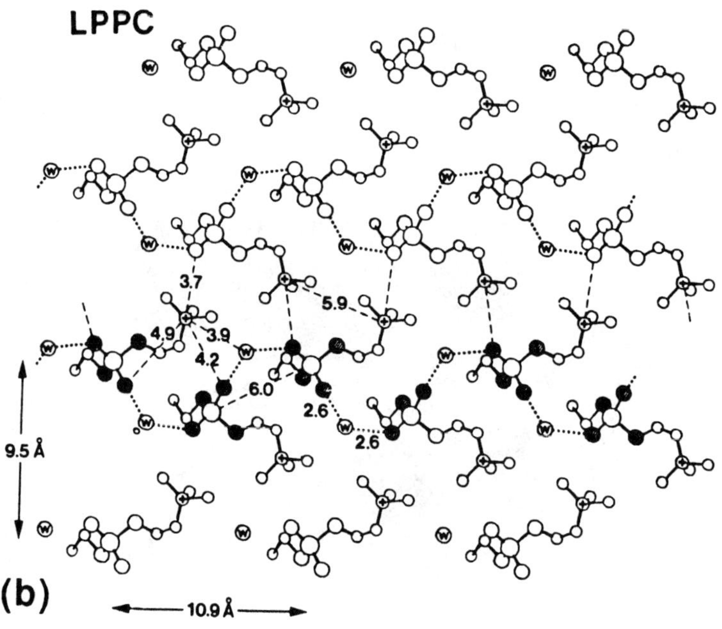

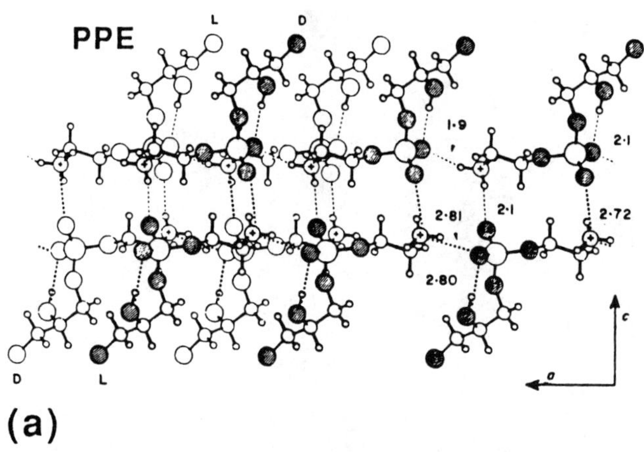

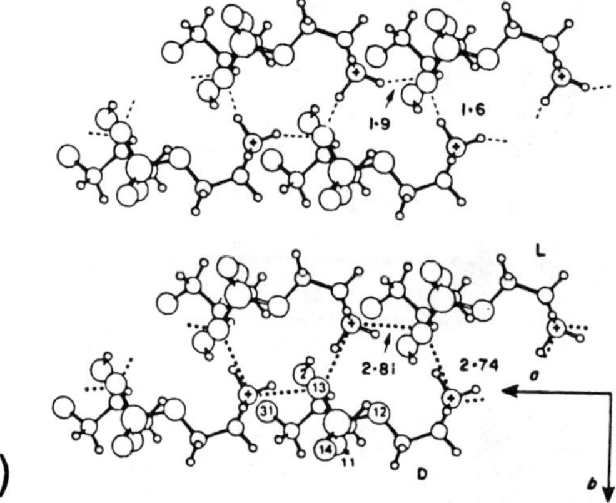

Figure A4 Continued

Table A5 Best Values of Average Structural Features of Gel Phase DPPC at T = 20°C with Estimated Uncertainties

D	63.7 ± 0.3 Å
D_c	17.5 ± 0.5 Å
D_W	13.9 ± 2.6 Å
D_H	7.4 ± 0.5 Å
D_W	11.8 ± 0.8 Å
D_H	8.5 ± 0.4 Å
D_B	49.8 ± 2.3 Å
$X_{H\text{-}H}$	45.0 ± 1.0 Å
A	45.9 ± 2.0 Å^2
$2A_c$	39.8 ± 0.4 Å^2
A'	40.0 ± 3.1 Å^2
V_L	1,144.0 ± 2.0 Å^3
V_X	1,462.0 ± 62.0 Å^3
V_c	804.0 ± 12.0 Å^3
V_H	340.0 ± 10.0 Å^3
V_{CH_2}	25.3 ± 0.2 Å^3
V_{CH_3}	48.8 ± 1.9 Å^3
θ	30.0 ± 3.0°
n_W	10.6 ± 2.0
n'_W	1.7 ± 1.5

D_c: hydrocarbon single layer thickness; D_W: interlamellar water layer thickness; D_H: headgroup layer thickness; D_B: bilayer thickness; $X_{H\text{-}H}$: distance between the centres of two Gaussians representing the headgroups (so-called phosphate-phosphate separation). D: lamellar repeat distance ($D_B + D_W$); A: area per molecule; A_C: chain cross-section area; in general: $V_i = D_i \cdot A_i$; V_L: lipid volume; V_C: chains volume; V_H: headgroup volume; V_{CH2}: volume per methylene group; V_{CH3}: volume per methyl group; V_X: lipid volume obtained by using $X_{H\text{-}H}$ instead of D_B; θ: hydrocarbon tilt angle; n_W: number of waters of hydration; n'_W: number of waters in the headgroup region.

All values with a dash are obtained after the assumption that n'_W water molecules are intercalated in the headgroup region, leaving a region of thickness D'_W of 'pure water' between the lipid layers ($2D_C + 2D'_H + D' = D$).

Source: Ref. 12.

Table A6 Partial Specific Volume and Main Structural Parameters of Diacyl Phosphatidylcholines and Phosphatidylethanolmines at Maximum Hydration (G. Cevc)

1,2-Diacyl-	PC						PE			
	V_L (μL/g)	A_L (nm^2)	$d_b^\$$ (nm)	$d_p^\$$ (nm)	d_w (nm)	V_L (μL/g)	A_L (nm^2)	d_b (nm)	$d_p^\$$ (nm)	$d_w^\$$ (nm)
L_c or L'_c										
C12:0/C12:0							0.39 ± 0.01	4.5 ± 0.1	0.7 ± 0.1	0
C14:0/C14:0				0.7 ± 0.15				5.0 ± 0.1	0.7 ± 0.1	0
C16:0/C16:0	925	< 0.4	5 ± 0.1		1.1 ± 0.2			5.5 ± 0.1	0.7 ± 0.1	0
C20:0/C20:0						(945)	0.39 ± 0.01	6.5	(0.55)	0
($L'_c \rightarrow L'_\beta$[†])										
C12:0/C12:0							*0.02*	*0*	*0*	*0.5*
C14:0/C14:0							*0.02*	*0*	*0*	*0.5*
C16:0/C16:0	*9 ± 0.5*	*> 0.05*	*–0.25*	*0.15*	*1*					
C20:0/C20:0						(40)				
L_β or L'_β										
C12:0/C12:0						898	0.145 ± 0.005	4.5	0.65 ± 0.15	0.5[$]
C14:0/C14:0	937	0.46 ± 0.02	4.2 ± 0.2			930	0.42 ± 0.01	4.7	0.65 ± 0.15	
C16:0/C16:0	955	0.5 ± 0.05	4.5 ± 0.1	0.85 ± 0.15	2.3 ± 0.15	958	0.42 ± 0.01	5.15	0.65 ± 0.15	
C18:0/C18:0	970		4.75 ± 0.15		2 ± 0.2		(0.43)	(5.6)		
C20:0/C20:0			5.1 ± 0.2		2 ± 0.2	985	0.47	5.62	0.5 ± 0.2	1.2
C22:0/C22:0			5.5 ± 0.2		2.15 ± 0.2					
($L'_\beta \rightarrow P'_\beta$)										
C14:0/C14:0	*3 ± 0.75*									
C16:0/C16:0	*3.3 ± .5*	*15 ± 5%*	*–0.15*	*(0.4)*	*0.2*	—*	—	—	—	—
C18:0/C18:0	*2 ± 1*									

$(L'_\beta(P'_\beta) \rightarrow L_\alpha)$										
C12:0/C12:0						*16 ± 2*	*0.11*	*–1*	*0.6*	*0.5*
C14:0/C14:0	*24 ± 3*	*11.5 ± 3 %*	*–0.78*			*19 ± 2*				
C16:0/C16:0	*30 ± 4*		*–0.7*	*0.4 ± 0.05*	*0.6*	*22 ± 2*				
C18:0/C18:0	*36 ± 2*		*–0.65*							
C20:0/C20:0			*–0.7*			*28 ± 2*	*0.11*	*–0.9*		
L_α‡										
C12:0/C12:0						977	0.52 ± 0.025[a]	3.25		1.3
C14:0/C14:0	975	0.62	3.42		2.7 ± 0.1			3.65		1.2
C16:0/C16:0	998	0.65	3.8	1.2	2.9 ± 0.1			(3.95)	1.05 ± 0.15	(1.2)
C18:0/C18:0	1012		4.1		2.8 ± 0.1			(4.4)		(1.1)
C20:0/C20:0			4.4		2.8 ± 0.1	1043	0.580	4.73	1.0	
C22:0/C22:0			4.8		2.9 ± 0.1					
H_α (H_{II})										
C12:0/C12:0[a]	—*	—	—	—	—		0.57	2.03	(0.7)	3.2[b]
C20:0/C20:0	—	—	—	—	—	1043	0.49	3.53	(0.7)	4.35[b]

Partial specific volume (V_L), total bilayer thickness (d_b), head group region thickness (d_p), interbilayer water layer thickness (d_w), and area per molecule (A_L). All values in italic give changes at the corresponding phase transition.

$Water intercalation between the lipid heads will cause d_w to be overestimated and d_p to be underestimated.

†This transition may involve phases with tilted or untilted chains.

‡Normally 10° below or above T_c.

*PE does not form undulated bilayers.

*PC does not form nonbilayer phases.

[a]This value pertains to didodecyl-PE. Ether (dialkyl) lipids typically have approximately 30 μL/g higher density than the corresponding ester (diacyl) phospholipids.

[b]This water layer thickness is not directly comparable with that of the lamellar phase systems, owing to the differences in geometry.

REFERENCES

1. M. Sundaralingam, Molecular structures and conformations of the phospholipids and sphingolipids, *Ann. N. Y. Acad. Sci, USA 195*: 324–355 (1972).
2. H. Hauser, I. Pascher, I. H. Pearson, and S. Sundell, Preferred conformation and molecular packing of phosphatidylethanolamine and phosphatidylcholine, *Biochim. Biophys. Acta 650*:21–51 (1981).
3. P. Hitchcock, R. Mason, K. M. Thomas, and G. G. Shipley, Structural chemistry of 1,2-dilauroyl-DL-phosphatidylethanolamine: Molecular conformation and intermolecular packing of phospholipids, *Proc. Natl. Acad. Sci. USA 71*:3036–3040 (1974).
4. M. Elder, P. Hitchcock, R. Mason, and G. G. Shipley, A refinement analysis of the crystallography of the phospholipid, 1,2-dilauroyl-DL-phosphatidylethanolamine, and some remarks on lipid-lipid and lipid-protein interactions, *Proc. R. Soc. Lond. A 354*:157–170 (1977).
5. I. Pascher, and S. Sundell, Membrane lipids: Preferred conformational states and their interplay. The crystal structure of dilauroylphosphatidyl-*N,N*-dimethylethanolamine, *Biochim. Biophys. Acta 855*:68–78 (1986).
6. R. H. Pearson and I. Pascher, The molecular structure of lecithin dihydrate, Nature 281:499–501 (1979).
7. K. Harlos, H. Eibl, I. Pascher, and S. Sundell, Conformation and packing properties of phosphatidic acid: The crystal structure of monosodium dimyristoylphosphatidate, *Chem. Phys. Lipids 34*:115–126 (1987).
8. I. Pascher, S. Sundell, K. Harlos, and H. Eibl, Conformation and packing properties of membrane lipids: The crystal structure of sodium dimyristoylphosphatidylglycerol, *Biochim. Biophys. Acta 896*:77–88 (1987).
9. I. Pascher, S. Sundell, and H. Hauser, Polar group interaction and molecular packing of membrane lipids: The crystal structure of lysophosphatidylethanolamine, *J. Mol. Biol. 153*:807–824 (1981).
10. H. Hauser, I. Pascher, and S. Sundell, Conformation of phospholipids: Crystal structure of a lysophosphatidylcholine analogue, *J. Mol. Biol. 137*:249–264 (1980).
11. I. Pascher, M. Lundmark, P.-G Nyholm, and S. Sundell, Crystal structures of membrane lipids, *Biochim. Biophys. Acta 1113*:339–373 (1992).
12. M. C. Wiener, R. M. Suter, and J. F. Nagle, Structure of the Fully Hydrated Gel Phase of Dipalmitoylphosphatidylcholine, *Biophys. J. 55*:315–325 (1989).

Appendix B: Thermodynamic Parameters of Phospholipids

Gregor Cevc *Technical University of Munich, Munich, Germany*

Table B1 Chain Melting (Normally Gel to Fluid) Phase Transition Temperature (°C)[a] of Some Common Diacyl Chain Phospholipids in Water or Low Ionic Strength Monovalent Electrolyte as a Function of the Hydrocarbon Chain Length and Site of Attachment

Head group	PC	$PE(CH_3)_2$	$PE(CH_3)$	PE	PS^-	PG^-	PA^-	$PA(CH_3)^-$	CL^-
1,2-Glycero-									
C10:0/C10:0	(–5.5)[b]			2					
C11:0/C11:0	(–1)			17					
C12:0/C12:0	–1		9	30.5	13	–2	32	9	41.5
C13:0/C13:0	14								
C14:0/C14:0	23.5	31	42.5	49.5	36	24	50	32	59
C15:0/C15:0	33.5					33	60		
C16:0/C16:0	41.5	48	58	65	54	41.5	68	48	77
C17:0/C17:0	48			70.5					
C18:0/C18:0	55.5	61	69.5	74	68	54.5	66[b]	57.5	
C19:0/C19:0	61.5			79					
C20:0/C20:0	66	71	78	83			65		
C21:0/C21:0	71								
C22:0/C22:0	75			90					
C23:0/C23:0	78								
C24:0/C24:0	80								
1,3-Glycero-									
C14:0/C14:0	19						42		
C16:0/C16:0	36.5	42.5	42.5/50	43/53					
C18:0/C18:0	53								
2,3-Glycero-									
C14:0/C14:0	24				37		5.5		
C16:0/C16:0	41.5				54				
C18:0/C18:0	53				70				

All values in this and following tables are rounded to within half a digit, typical experimental scatter for the transition temperature being greater than 2°.
[a]Values in brackets are from a lipid crystal into a fluid chain micellar solution.
[b]The relatively low transition temperature of this lipid may be due to the large chain tilt.

Table B2 Representative Chain Melting Phase Transition Enthalpies (kJ mol^{-1}) of Common Diacyl Chain Phospholipids as a Function of the Hydrocarbon Chain Length

Head group	PC	$PE(CH_3)_2$	$PE(CH_3)$	PE	PS	PG	PA	CL
1,2-Glycero-								
C10:0/C10:0				8				
C11:0/C11:0				13.5				
C12:0/C12:0	16			17	13	19	14	36
C13:0/C13:0	18			21.5				
C14:0/C14:0	26	26	24	25	36	28.5	23	52
C15:0/C15:0	32			30				
C16:0/C16:0	36.5	36	35	34.5	37	37	33	74
C17:0/C17:0	37			39				
C18:0/C18:0	46	45	45	44	46	44		
C19:0/C19:0	45			49.5				
C20:0/C20:0	50			55				
C21:0/C21:0	51							
C22:0/C22:0	62			64				
1,3-Glycero-								
C14:0/C14:0	31							
C16:0/C16:0	38.5							
C18:0/C18:0	41.5							
2,3-Glycero-								
C14:0/C14:0	15							
C16:0/C16:0	36							

[a]Best-choice estimates rounded to within half a digit; literature values differ by more than 100%.

Table B3 Chain Melting Phase Transition Temperatures (°C)[a] of Ether Phospholipids in Water or Low Ionic Strength Monovalent Electrolyte as a Function of Hydrocarbon Chain Length and Site of Attachment

Head group	PC	$PE(CH_3)_2$	$PE(CH_3)$	PE	PS^-	PG^-	PA^-	$PA(CH_3)^-$
1,2-Dialkyl-glycero-								
C-O-12:0/C-O-12:0				35	19			
C-O-14:0/C-O-14:0	28.5	36	47	55	41	28	55	37
C-O-16:0/C-O-16:0	43.5	50.5	61.5	68.5	56	46	71	52
C-O-18:0/C-O-18:0				77				
1,3-Dialkyl-glycero-								
C-O-14:0/C-O-14:0	28.5							
C-O-16:0/C-O-16:0	42.5							
C-O-18:0/C-O-18:0	59							
*1-Acyl,2-alkyl-glycero**								
C16:1(3:0)2/C-O-16:0	31							
C16:1(4:0)2/C-O-16:0	29.5							
C16:1(5:0)2/C-O-16:0	26.5							
C16:1(6:0)2/C-O-16:0	25							
C16:1(8:0)2/C-O-16:0	13							
C16:1(10:0)2/C-O-16:0	11.5							
C16:1(12:0)2/C-O-16:0	33.2							
C16:1(14:0)2/C-O-16:0	43							

**sn*-1 chain branched at the 2 position with alkanes of increasing length.

Sources: Most of the data from G. Cevc, J. M. Seddon, and D. Marsh, *Faraday Discus. 81*:179–189 (1986); G. Cevc, *Biochemistry 26*:6305–6310 (1987), G. Cevc, *J. de Physique 50*:1117–1134 (1989) Data for the 1,3-derivatives from T. Kunitake, Y. Okahata, and S.-I. Tawaki, *J. Colloid Interface Sci. 103*:190–197 (1985). Data for the branched, mixed chains from G. Brezesinski, B. Dobner, H. Dorfler, M. Fischer, S. Haas, and P. Nuhn, *Chem. Phys. Lipids 43*:257–264 (1987).

Table B4 Chain Melting Phase Transition Temperatures (°C)[a] of Symmetric Chain Phospholipids in ~pH 7 Low-Salt Buffers or Water as a Function of the Head Group Length

Segment length *n*	1	2	3	4	5	6	7	8	9	10	11	12	16	18
C-O-14:0/C-O-14:0-P-$(Ch_2)_nN_3$		55	45.5	38	28									
C16:0/C16:0-P-$(CH_2)_nN(CH_3)_3$		42	41	42	43.5	43	45	42.5	43.5	42.5	41			
C16:0/C16:0-P-$(CH_2)_2NH_2$-O-$(CH_2)_nH$				31		32.5		18		27		48	58	68
C16:0/C16:0-P-$(CH_2)_nNH_3$		64.5		51		40		40						
C-O-16:0/C-O-16:0-P-$(CH_2)_nNH_3$		68.5	62.5	57.5										
C18:0/C18:0-P-$(CH_2)_nH$	57.5	57	56.5	55										
C18:0/C18:0-P-$(CH_2)(COH)_nH$	56	55	54											
C18:0/C18:0-P-$(CH_2)_nNH_3$		74		63										
C18:0/C18:0-P-$(CH_2)_nN(CH_3)_3$		55		54										

Sources: Data from D. Bach, I. Bursuker, H.-J. Eibl, and I. R. Miller, *Biochim. Biophys. Acta 514*:310–319 (1978); G. Cevc, J. M. Seddon, and D. Marsh, *Faraday Discus. 81*:179–189 (1986); G. Cevc, *J. de Physique 50*:1117–1134 (1989); D. Lafrance, D. Marion, and M. Pezolet, *Biochemistry 29*: 45922–4599 (1990) and unpublished.

Table B5 Chain Melting Phase Transition Temperatures (°C)[a] of Fully Hydrated Phospholipids From Various Biological Sources

	Head group					
Source	PC	PE	$PE_{H_{II}}$	PS^-	PG^-	PA^-
Egg yolk	−10 ± 5	10 ± 2	(30 ± 8)			18
Egg yolk (hydrogenated)	46					
Soybean	−15 ± 5	−5	(30 ± 8)			
Soybean (hydrogenated)	51					
Bovine brain or spinal cord				13 ± 3		
E. coli		30 ± 8	(30 ± 8)		36 ± 2	

[a]Transitions into a nonbilayer phase are given in parentheses.

Table B6 Chain Melting Phase Transition[a] Temperatures (°C) and Enthalpies (kJ mol^{-1}) of Fully Saturated, Asymmetric Chain Phosphatidylcholines as a Function of the Hydrocarbon Chain Length

Chains	T_m	ΔH_m	Chains	T_m	ΔH_m	Chains	T_m	ΔH_m
			C18/C1	26	29			
			C18/C2	18.5	21	C16/C9	3.5	25.5
			C18/C4	14	16.5	C16/C10	5	27
C6/C18	5/–14	7.5	C18/C6	8	15	C18/C10	20.5	37
C8/C18	10	29.5	C18/C8	0.5	26.5	C18/C11	21.5	38.5
C10/C18	12	26.5	C18/C10	18	35	C20/C11	24	47
C12/C18	21	26.5	C18/C12	16	31	C20/C12	34	48.5
C14/C18	38	32	C18/C14	30	22.5	C22/C12	43	55
			C18/C15	38		C22/C13	44.5	59
C16/C18	49	37.5	C18/C16	44.5	35			
			C18/C24	63				
			C18/C26	64				
C8/C20	21.5	51				C20/C12	34	48
C9/C19	19.5	47	C19/C9	13.5	34.5	C19/C13	24	21
C10/C18	11	26	C18/C10	18	35	C18/C14	30.5	22.5
C11/C17	14	20.5	C17/C11	13	29	C17/C15	37.5	31
C12/C16	21.5	24	C16/C12	11.5	19.5			
C13/C15	25.5	25	C15/C13	19	19			
C9/C21	29.5	46	C21/C9	20	33.5			
C10/C20	27	38.5	C20/C10	25	45			
C11/C19	17.5	18.5	C19/C11	28.5	42			

C12/C18	23.5	24	C18/C12	17.5	35			
C13/C17	30.5	29	C17/C13	21	22			
C14/C16	35	34	C16/C14	28	28.5			
C10/C22	37	51						
			C11/C17	12	13	C11/C21	32.5	39
			C12/C18	21	24.5	C12/C20	25.5	23
C13/C21	34	23	C13/C19	31	28.5	C13/C19	31	28.5
C14/C20	39	30	C14/C20	39	31.5	C14/C18	39	31.5
C15/C19	45	36.5	C15/C21	44	46	C15/C17	41.5	42
C16/C18	49	39	C16/C22	50.5	53			
C17/C17	49	40.5	C17/C23	56	58.5			
C18/C16	44.5	33.4	C18/C24	61	65			
C19/C15	39	27	C19/C25	—				
C20/C14	33	19.5	C20/C26	68.5	77.5			
C21/C13	(34)	23						
C-O-12/C20	25	19.5	C20/C-O-12	35	47			
C18/C11:1Δ^{10}	13.5	25.5						

[a]Transition is chiefly from a gel $P_\beta^!$ (or, perhaps, sometimes $L_\beta^!$) phase into a fluid lamellar, L_α phase; notable exceptions are phosphatidylcholines with an effective chain length difference greater than 7 (such as C10/C18, C17/C11, C9/C19, C18/C10, C8/C20, (C19/C9,) C18/C12, C10/C20, C19/C11, C9/C21, C20/C10), which form low temperature structures with interdigitated chains.

Sources: Most of the data from S. Ali, H.-N Lin, R. Bittman, and C.-H. Huang, *Biochemistry 28*:522–528 (1989); C.-H. Huang, *Biochemistry 30*:26–30 (1991); H. Lin, Z. Wang, C. Huang, *Biochemistry 29*:7063–7072 (1990); H. Lin, Z. Wang, C. Huang, *Biochim. Biophys. Acta 1067*:17–28 (1991); J. Shah, P. K. Sripada, and G. G. Shipley, *Biochemistry 29*:4254–4262 (1990); E. N. Serralach, G. H. De Haas, G. G. Shipley, *Biochemistry 23*:713–720 (1984); Z. Wang, H. Lin, and C. Huang, *Biochemistry 29*:7072–7076 (1990); H. Xu and C. Huang, *Biochemistry 26*:1036–1043 (1987).

Table B7 Chain Melting Phase Transition Temperatures (°C) of Common Phospholipids with Mixed Chains and/or Double Bonds Near the Middle of the Chain

With *cis* configuration	T_m PC	T_m PE	With *trans* configuration	T_m PC	T_m PE
C16:0/C16:1cΔ^{9}	–4		C16:0/C16:1tΔ^{9}		24
C16:1cΔ^{9}/C16:1cΔ^{9}	–35.5	–31.5	C16:1tΔ^{9}/C16:1tΔ^{9}	–4	
C16:0/C18:0	48				
C16:0/C18:1cΔ^{9}	–3	25	C16:0/C18:1tΔ^{9}	35	
C16:0/C18:2c$\Delta^{9,12}$	–20				
C18:0/C16:0					
C18:1cΔ^{9}/C16:0	–11	–8			
C17:1cΔ^{9}/C17:1cΔ^{9}	–27.5				
C18:0/C18:cΔ^{6}	31				
C18:0/C18:1cΔ^{9}	6.5	30.5	C18:0/C18:tΔ^{9}	26	
C18:0/C18:2c$\Delta^{9,12}$	–16	–40			
C18:0/C18:3c$\Delta^{9,12,15}$	–13				
C18:0/C18.4c$\Delta^{5,8,11,14}$	–13				
C18:1cΔ^{9}/C18:0	6.5	9			
C18:1cΔ^{9}/C18:1cΔ^{9}	–18	–16	C181tΔ^{9}/C18:1tΔ^{9}	12.5	37.5
C18:1cΔ^{11}/C18:1cΔ^{11}	–19.5		C18:1tΔ^{11}/C18:1tΔ^{11}	13	
C18:2c$\Delta^{5,9}$/C18:2c$\Delta^{5,9}$	<–30				
C18:2c$\Delta^{9,12}$/C18:2c$\Delta^{9,12}$	–53	–15/–25			
C18:3c$\Delta^{9,12,15}$/C18:3c$\Delta^{9,12,15}$	–63	–30			
C18:1cΔ^{9}/C20:0	16				
C19:1cΔ^{10}/C19:1cΔ^{10}	–9				
C20:0/C20:1cΔ^{9}	12				
C20:0/C20:1cΔ^{11}	20.5				
C20:1/cΔ^{11}/C20:1cΔ^{11}	–4.5				
C20:0/C20:2c$\Delta^{11,14}$	0				
C20:0/C20:3c$\Delta^{11,14,17}$	4				
C20:0/C20:4c$\Delta^{5,8,11,14}$	–7				
C21:1cΔ^{12}/C21:1cΔ^{12}	6.5				
C22:1cΔ^{13}/C22:1cΔ^{13}	13		C22:1tΔ^{13}/C22:1tΔ^{13}		41
C22:2c$\Delta^{5,9}$/C22:2c$\Delta^{5,9}$	7				
C23:1cΔ^{14}/C23:1cΔ^{14}	21				
C24:1cΔ^{5}/C24:1cΔ^{5}	59/51				
C24:1cΔ^{9}/C24:1cΔ^{9}	34/25				
C24:1cΔ^{15}/C24:1cΔ^{15}	26/20				
C24:2c$\Delta^{5,9}$/C24:2c$\Delta^{5,9}$	31/27	29			
C26:2c$\Delta^{5,9}$/C26:2c$\Delta^{5,9}$	42	29	C26:2cΔ^{5}*t*Δ^{9}/C26:2cΔ^{5}*t*Δ^{9}	49	
C26:2c$\Delta^{6,9}$/C26:2c$\Delta^{6,9}$	45	46			

Table B8 Effect of the Double-Bond Position on the Chain Melting Phase Transition Temperature (°C) and Enthalpy (kJ/mol) of Phosphatidylcholine

Alkenoyl-alkenoyl	T_m	ΔH_m	Acyl-alkenoyl	T_m
C18:1cΔ^2/C18:1cΔ^2	41	40		
C18:1cΔ^3/C18:1cΔ^3	35	36.3		
C18:1cΔ^4/C18:1cΔ^4	23	34.3		
C18:1cΔ^5/C18:1cΔ^5	11	32.6		
C18:1cΔ^6/C18:1cΔ^6	1	32.6	C18:0/C18:1cΔ^6	31
C18:1cΔ^7/C18:1cΔ^7	−8	31.8		
C18:1cΔ^8/C18:1cΔ^8	−13	13.3		
C18:1cΔ^9/C18:1cΔ^9	−21	32.2	C18:0/C18:1cΔ^9	13
C18:1cΔ^{10}/C18:1cΔ^{10}	−21	31.8		
C18:1cΔ^{11}/C18:1cΔ^{11}	−19	32.6		
C18:1cΔ^{12}/C18:1cΔ^{12}	−8	33	C18:0/C18:1cΔ^{12}	19
C18:1cΔ^{13}/C18:1cΔ^{13}	1	34.3		
C18:1cΔ^{14}/C18:1cΔ^{14}	7	35.9		
C18:1cΔ^{15}/C18:1cΔ^{15}	24	37.2		
C18:1cΔ^{16}/C18:1cΔ^{16}	35	40	C18:0/C18:1cΔ^{16}	44
C18:1cΔ^{17}/C18:1cΔ^{17}	45			

Source: Data from P. G. Barton and F. D. Gunstone, *J. Biol. Chem.* *250*:4470–4476 (1975).

Table B9 Typical Chain Melting Phase Transition Temperatures (°C) and Enthalpies (kJ mol^{-1}) of some ω-Cyclohexyl and Cyclopropane Phospholipids

	PC		PE		PG			PC
Cyclohexyl	T_m	ΔH_m	T_m	ΔH_m	T_m	ΔH_m	Cyclopropane	T_m
C9-ωch/C9-ωch	–7	18.5						
C10-ωch/C10-ωch	–11	5						
C11-ωch/C11-ωch	18.5	49						
C12-ωch/C12-ωch	16	28						
C13-ωch/C13-ωch	33	55	37	26	31.5	56		
C14-ωch/C14-ωch	34.5	32.5	46		33	50		
C15-ωch/C15-ωch	46.5	68	54		44.5	58		
C16-ωch/C16-ωch	48.5	36.5						
C17-ωch/C17-ωch	57.5	(75)					C17cp,tΔ^9/C17cp,tΔ^9	–0.5
							C17cp,cΔ^9/C17cp,cΔ^9	–20
C18-ωch/C18-ωch	60.5	(46)						
C19ωch/C19-ωch			35			35	C19cp,tΔ^9/C19cp,tΔ^9	16.5
							C19cp,cΔ^9/C19cp,cΔ^9	–0.5
							C19cp,tΔ^{11}/C19cp,tΔ^{11}	14
							C19cp,cΔ^{11}/C19cp,cΔ^{11}	–3.5
C20-ωch/C20ωch			46		46			
C20-ωch/C21-ωch			54		54			

Sources: Most of the PC data stem from R. N. A. H. Lewis and R. N. McElhaney, *Biochemistry 24*:4903–4911 (1985); J. R. Silvius and R. N. McElhaney, *Chem. Phys. Lipids 24*:2287–2296 (1979); J. R. Silvius and R. N. McElhaney, *Chem. Phys. Lipids 25*:125–134 (1979); PG data from A. Blume, K. Habel, A. Finke, T. Frey, *Thermochim. Acta 119*:53–58 (1987); PE data from R. N. A. H. Lewis, D. A. Mannock, R. N. McElhaney, D. C. Turner, and S. M. Gruner, *Biochemistry 28*:541–548 (1989).

Table B10 Typical Chain Melting Phase Transition Temperatures (°C) and Enthalpies (kJ mol^{-1}) of the Branched or Dimethylated Chain Phospholipids[a]

Chains	PC	PE	Chains	PC	PE
Iso-acyl			*±Anteiso-acyl*		
C12:1me^{10}/C12:1me^{10}	–18.5	45(30)			
C13:1me^{11}/C13:1me^{11}	–9.5		C13:1me^{10}/C13:1me^{10}	<60	
C14:1me^{12}/C14:1me^{12}	6.5	–36/–42	C14:1me^{11}/C14:1me^{11}	–33.5	–21.5
C15:1me^{13}/C15:1me^{13}	6.5	–31	C15:1me^{12}/C15:1me^{12}	–15	
C16:1me^{14}/C16:1me^{14}	22		C16:1me^{13}/C16:1me^{13}	–3	
C17:1me^{15}/C17:1me^{15}	27	42.5	C17:1me^{14}/C17:1me^{14}	8	
C18:1me^{16}/C18:1me^{16}	36.5	52	C18:1me^{15}/C18:1me^{15}	18.5	
C19:1me^{17}/C19:1me^{17}		59	C19:1me^{16}/C19:1me^{16}	27	44.5
C20:1me^{18}/C20:1me^{18}		64[b]	C20:1me^{17}/C20:1me^{17}		51.5
Other types					
C15:1me^{2}/C15:1me^{2}	3.5		C19:1me^{2}/C19:1me^{2}	46	
C15:1me^{3}/C15:1me^{3}	–4		C19:1me^{3}/C19:1me^{3}	39	
C15:1me^{4}/C15:1me^{4}	–23		C19:1me^{4}/C19:1me^{4}	29.5	
C15:1me^{11}/C15:1me^{11}	–31		C19:1me^{15}/C19:1me^{15}	19	
C15:2me^{12}/C15:2me^{12}	–31		C19:2me^{16}/C19:2me^{16}	22	38.5
C16:1me^{2}/C16:1me^{2}	–14		C20:1me^{16}/C20:1me^{16}	29.5	
C16:2me^{13}/C16:2me^{13}	–10		C20:2me^{17}/C20:2me^{17}	28	43
C17:1me^{2}/C17:1me^{2}	28		C21:1me^{17}/C21:1me^{17}	36.5	
C17:1me^{3}/C17:1me^{3}	20		C21:2me^{18}/C21:2me^{18}	35.5	
C17:1me^{4}/C17:1me^{4}	7				
C17:1me^{5}/C17:1me^{5}	–3		C22:1me^{18}/C22:1me^{18}	43.5	
C17:1me^{6}/C17:1me^{6}	–21.5				
C17:1;me^{13}/C17:1me^{13}	–0.5		C23:1me^{19}/C23:1me^{19}	48.5	
C17:2me^{14}/C17:2me^{14}	5.5/10				
C18:1me^{14}/C18:1me^{14}	9				
C18:2me^{15}/C18:2me^{15}	11				
C19:2me^{15}/C19:2me^{15}		38.5			
C20:2me^{16}/C20:2me^{16}		43			

[a]First number gives the number of the carbon atoms in the main chain; total number of the side chains is given by the second number; superscript gives the position along the main chain at which the side chains are attached: C18:1me^{14}/C18:1me^{14} consequently denotes a distearoyl-(14-methyl) phospholipid; C18:0i/C18:0i denotes an isobranched distearoylphospholipid (or C18:1me^{16}/C18:1me^{16}); an anteiso phospholipid, C20:0ai/C20:0ai, could be also written as C18:1me^{15}/C18:1me^{15}.
[b]Ethyl *iso*-branched PE has a chain melting phase transition temperature of 24°C.
[c]Ethyl *anteiso*-branched PE has a chain melting phase transition temperature of 24°C.
Source:R. N. A. H. Lewis, D. A. Mannock, R. N. McElhaney, D. C. Turner, and S. M. Gruner, *Biochemistry 28*:541–548 (1989); R. N. A. H. Lewis, R. N. McElhaney, N. A. H. Ruthven, *Biochemistry 24*:4903–4911 (1985); R. N. A. H. Lewis and R. N. McElhaney, *Biochemistry 24*:2431–2439 (1985); R. N. A. H. Lewis, R. N. McElhaney, and N. A. H. Ruthven, *Biochemistry 24*:4903–4911 (1985); R. N. A. H. Lewis, B. Sykes, and R. N. McElhaney, *Biochemistry 26*:4036–4044 (1987); P. Nuhn, G. Brezesinki, B. Dobner, G. Forster, M. Gutheil and M. Dorfler, *Chem. Phys. Lipids 39*:221–236 (1986); J. R. Silvius and R. N. McElhaney, *Chem. Phys. Lipids 25*:125–134 (1979); J. R. Silvius, M. Lyons, P. L. Yeagle, and T. J. O'Leary, *Biochemistry 24*:5388–5395 (1985).

Table B11 'Subtransition' ('Rotator' Transition, $L_c \rightarrow L_\beta$ ($L_c \rightarrow L_\alpha$), all in °C) Temperatures and Enthalpies (kJ mol^{-1}) of Some Diacyl Phospholipids in Water or a Dilute 1:1 Electrolyte as a Function of the Hydrocarbon Chain Length

Head group	T_m(PC)	ΔH_m(PC)	T_m(PE(CH_3))	T_m(PE)	ΔH_m(PE)	T_m(PG)
1,2-Glycero-						
C10:0/C10:0	*–5.5*	76.5		*26.5*[a]	41	
C11:0/C11:0	*–1*	39.5		*36*	53	
C12:0/C12:0	*7*	54	9	*45*	68	
C13:0/C13:0	*11.5*			*53*	76	
C14:0/C14:0			21	*55.5*	75	
C16:0/C16:0	19'[b]			*64*		28.5'
C20:0/C20:0				*82*		
C22:0/C22:0	32					
C24:0/C24:0	80					
1,3-Glycero-						
C16:0/C16:0	25					

[a]Values printed in italic correspond to a transition between the crystalline low-temperature and a fluid-lamellar (L_α) high-temperature phase; PE values stem from R. N. A. H. Lewis and R. N. McElhaney, *Biophys. J.*, in press.

Chain melting transitions of the monounsaturated phosphatidylcholines with more than 18 carbon atoms per chain and a single double bond in the *cis* conformation have also been reported to be of the $L_c \rightarrow L_\alpha$ type (R. N. A. H. Lewis, B. Sykes, and R. N. McElhaney, *Biochemistry 26*:4036–4044 (1987)).

[b]Transitions for which the high temperature phase has been confirmed independently to be of the L'_β type (with tilted chains) are denoted in italics.

Table B12 'Pretransition' Temperature (°C) of Symmetric, Diacyl-Glycero-Phospholipids in Water or a Low-Ionic Strength Monovalent Electrolyte as a Function of the Hydrocarbon Chain Length and Head Group Type

Head group	PC ≡ $PE(CH_3)_3$	$PE(CH_3)_2$	$PE(CH_3)$	PS^{2-}	PG^{-}	PA^{2-}
Symmetric chains						
C12:0/C12:0	(3.5)					
C13:0/C13:0	(1)					
C14:0/C14:0	11	19	—	6	11	
C15:0/C15:0	20					
C16:0/C16:0	34	29	—	20	33	18
C17:0/C17:0	43.5					
C18:0/C18:0	49				48.5	
C19:0/C19:0	57.5					
C20:0/C20:0	63					
C21:0/C21:0	79					
C22:0/C22:0	—					
Asymmetric chains						
C14:0/C16:0	23					
C14:0/C18:0	(25)					
C16:0/C14:0	(11)					
C16:0/C18:0	40					
C18:0/C14:0	19					
C18:0/C16:0	30.5					

Source: Most of the data for symmetric chain PCs from R. N. A. Lewis, N. Mak, and R. N. McElhaney, *Biochemistry 26*:6118–6126 (1987); data for the asymmetric PCs chiefly from S. C. Chen and J. M. Sturtevant, *Biochemistry 20*:713–718 (1981); E. N. Serralach, G. H. Haas, and G. G. Shipley, *Biochemistry 23*:713–720 (1984).

Table B13 Temperature (°C) and Enthalpy (kJ mol^{-1}) of the Lamellar to Nonlamellar Phase Transition (Normally into an Inverted-Hexagonal Phase (H_{II})) of Different Phospholipids in Excess Water or Low Ionic Strength Buffer* as a Function of Hydrocarbon Chain Length and Desaturation

Head group	PE(CH_3) T_m	PE T_m	PE ΔH_m	PS T_m	PA T_m
1,2-Diacyl-glycero-					
C14:0/C14:0					(45)
C16:0/C16:0		120	1.7		(64)
C17:0/C17:0		107.5	2.5		
C18:0/C18:0		100.5	3.3	103	
C19:0/C19:0		98	3.8		
C20:0/C20:0		95.5	5.4		
C22:0/C22:0		90	3.3		
1,2-Dialkyl-glycero-[b]					
C12:0/C12:0		(90[c])			
C14:0/C14:0	90[d]	96			
C16:0/C16:0	86	86		86	63
C18:0/C18:0		81		80	
1,2-(Di)Alkenoyl-glycero-					
C16:1cΔ^9/C16:1cΔ^9	(40)	42.5			
C16:0/C18:1tΔ^9		71.5			
C16:0/C18:1cΔ^9		74			
C18:1tΔ^9/C18:1tΔ^9		63			
C18:1cΔ_9/C18:1cΔ^9	64	10			
C18:2c$\Delta^{9,12}$/C18:2c$\Delta^{9,12}$		–25			
C18:3c$\Delta^{9,12,15}$/C18:3c$\Delta^{9,12,15}$		–30			
C20:4c/C20:4c		<–30			
C22:6c/C22:6c		<–30			

*For the noncharged phospholipids typically pH ~7; for the charged lipids in the fully protonated state.
[a]Best-choice estimates rounded to within half a digit.
[b]In saturated NaCl solution, the transition temperatures of PEs are shifted downwards to 72, 74, 75, and 81°C.
[c]Initially a cubic phase.
[d]Unidentified nonlamellar high-temperature phase.
Source: Most of the data stem from: R. N. A. H. Lewis and R. N. McElhaney, *Biophys. J.*, in press; J. M. Seddon, G. Cevc, and D. Marsh, *Biochemistry 22*:1280–1289 (1983); J. M. Seddon, G. Cevc, R. D. Kaye, and D. Marsh, *Biochemistry 23*:2634–2644 (1984); C. Dekker, G. Van Kessel, J. Klomp, J. Pieters, and B. DeKruijff, *Chem. Phys. Lipids 33*:93–106 (1983).

Table B14 Temperature (°C) of the Transition into an Inverted Hexagonal, Nonbilayer Phase of Branched-, Cyclohexyl-, or Dimethylated-Chain Phosphatidylethanolamines[a]

Chains	T_m(PE)	Chains	T_m(PE)
Iso-acyl		*Anteiso-acyl*	
C17:1me[15]/C17:1me[15]	104		
C18:1me[16]/C18:1me[16]	94		
C19:1me[17]/C19:1me[17]	88	C19:1me[16]/C19:1me[16]	80.5
C20:1me[18]/C20:1me[18]	86.5	C20:1me[17]/C20:1me[17]	75
Other types			
C19:2me[16]/C19:2me[16]	77		
C20:2me[17]/C20:2me[17]	75		
C13-ωch/C13-ωch	106		
C14-ωch/C14-ωch	84		
C15-ωch/C15-ωch	83		

[a]For abbreviations see table B10 in this appendix. *Source:* R. N. A. H. Lewis, D. A. Mannock, R. N. McElhaney, D. C. Turner, S. M. Gruner, *Biochemistry 28*:541–548 (1989).

Table B15 Calorimetrically Determined Heating Endothermic Transition Temperatures of Aqueous Dispersons of 1,2-Di-*n*-acylphosphatidylcholines

	Calorimetrically obsd transition temp[a] (°C)					
PC	$L_c \to L_\alpha$[b]	$L_c \to L_{\beta'}$[b]	$L_c \to P_{\beta'}$[b]	L ′ → $P_{\beta'}$[c]	$L_{\beta'} \to L_\alpha$	$P_{\beta'} \to L_\alpha$[d]
10:0	−5.7					
11:0	−0.8					
12:0	7.0					−2.1
13:0	14.4		11.7	−0.8		13.7
14:0			16.2	14.3		23.9
15:0		22.3		24.8		34.7
16:0		21.2		34.2		41.4
17:0		25.8		43.0		49.8
18:0		28.2		50.7		55.3
19:0		33.0		57.8		61.8
20:0		37.8		63.7		66.4
21:0		28.0		68.7		71.1
22:0		32.1			74.8	

[a]These are the calorimetrically determined transition temperatures. Since many of the processes are kinetically limited, they may not be the true equilibrium transition temperatures for the observed event. [b]In cases where more than one L_c phase was observed the $L_C \to$?? transition temperature refers to that of the most stable L_c phase observed. [c]Pretransition temperatures: these were determined at the slowest rate feasible with the instrument used (18.75°C·h^{-1} for 13:0-PC, 5–6°C·h^{-1} for all other PCs). [d]Gel/liquid-crystalline phase transition temperatures.

Source: R. N. A. Lewis, N. Mak, and R. N. McElhaney, *Biochemistry 26*:6118–6126 (1987).

Table B16 Enthalpy Changes Associated with Thermotropic Phase Transitions of Aqueous Dispersions of 1,2-Di-*n*-acylphosphatidylcholines

PC	Enthalpy change (kcal·mol⁻1)					
	$L_c \rightarrow L_{\alpha}$ [a]	$L_c \rightarrow L_{\beta'}$ [a]	$L_c \rightarrow P_{\beta'}$ [a]	$L_{\beta}' \rightarrow P_{\beta'}$	$L_{\beta'} \rightarrow L_{\alpha}$	$P_{\beta}' \rightarrow L_{\alpha}$
10:0	18.3					
11:0	9.5					
12:0	13.4					1.8
13:0	12.3			0.5		4.4
14:0			6.2	1.1		5.9
15:0		6.3		0.9		6.9
16:0		6.2		1.1		7.7
17:0		6.5		1.1		8.7
18:0		6.7		1.2		9.8[b]
19:0		6.9		1.3		10.7[b]
20:0		4.9		1.4		11.4[b]
21:0		2.9		1.4		12.2[b]
22:0		4.3			14.9	

[a]The formation of stable L_c phases in these PCs is a complex process which is believed to proceed via at least two intermediate L_c phases. Furthermore, the kinetics of formation of those phases are slow, and in the case of the longer chain compounds, the process of formation was not complete after a 2-year period. Thus, for all transitions involving a transformation of an L_c phase, the enthalpy change observed is the maximum that has been obtained under our conditions and must be regarded as the lower limit of the enthalpy change associated with these PC processes. Due to a partial [< 1%] hydrolysis, these values are slightly underestimated.
Source: R. N. A. Lewis, N. Mak, and R. N. McElhaney, *Biochemistry 26*:6118–6126 (1987).

Table B17 Effect of Lipid Head Group Modification on the Chain Melting Temperature (°C) of Lipid Lamellae in a 0.1 mol dm⁻³ Monovalent Salt Solution

Ammonium group methylation						
DHPE	68.5	71.5	DHP[CH_2]$_3$$NH_3$[a]	62.5	DEPE[b]	38.5
DHPE(1CH_3)	61.5	66	DHP[CH_2]$_3$$NH_2$(1$CH_3$)	58.5	DEPE(1CH_3)	32
DHPE(2CH_3)	50.5	60.5	DHP[CH_2]$_3$NH(2CH_3)	49	DEPE(2CH_3)	21
DHPC	43.5	54	DHP[CH_2]$_3$N(3CH_3)	43	DEPC	12
pH =	7	0		7		7

Phosphate alkylation		Zwitterionic headgroup lengthening			
DMPA	50	DTP[CH_2]$_2$$NH_3$[c]	55	DPP[CH_2]$_2$N(3CH_3)	42
DMPA(1CH_3)	32	DTP[CH_2]$_2$$NH_3$	45.5	DPP[CH_2]$_3$N(3CH_3)	41
DMPA[CH_2](1CH_3)	23	DTP[CH_2]$_4$$NH_3$	38	DPP[CH_2]$_4$N(3CH_3)	42
DMPA[CH_2]$_2$(1CH_3)	19	DTP[CH_2]$_5$$NH_3$	28	DPP[CH_2]$_5$N(3CH_3)	43.5
pH =	6		7		7

[a]DHP(CH_2)$_2$$NH_3$ = DHPE and so on. [b]DEPx: 1,2-dielaidoyl-*sn*-glycerophospholipid; DEPC = DEPE(3CH_3).
[c]The transition temperature of DTP[CH_2]$_2$N(3CH_3) = DTPC is 28°C.
Source: G. Cevc, J. M. Seddon, and D. Marsh, *Faraday Discus. 81*:179–189 (1986).

Table B18 Chain Melting Phase Transition Temperature (°C) [a] of Bilayers of Various 1,2-Distearoyl-*sn*-Glycerophospholipids[b] as a Function of the Headgroup Structure, Degree of Methylation, or Protonation as Regulators of the Interfacial Polarity and Hydrophilicity

Lipid[b]	pH ≤ 1	pH = 7	pH = 13	Lipid	pH ≤ 1	pH = 7	pH = 13
DSPA	69[c]	66	59				
DSPA(CH_2)H	68[c]	57.5	57.5				
DSPA(CH_2)$_2$H	67.5[c]	57	57	DSPA(CH_2)COH)H	67.5	56	56
DSPA(CH_2)$_3$H	67[c]	56.5	56.5	DSPA(CH_2)(COH)$_2$H	66	55	55
DSPA(CH_2)$_4$H	66.5[c]	55	55	DSPA(CH_2)(COH)$_3$H	65	54	54
DSPS	76	68	48	DSPS(CH_3)	71	68	53
DSPE ≡				DSPE(CH_3)	72	69	56.5
DSPA(CH_2)$_2$$NH_3$	76	74	56	DSP(CH_2)$_2$N(CH_3)$_3$	64[d]	55	55
DSPA(CH_2)$_4$$NH_3$	70	63	55.5	DSPA(CH_2)$_4$N(CH_3)$_3$	63[d]	54	54

[a]Phase transition temperatures were determined calorimetrically or by measuring the optical density at 300–400 nm of lipid solutions (pH = 7, > 12) or samples contained in flat mica containers (pH = 7, < 1). The values given were obtained from the heating scans, performed at a scanning rate of ≤ 1 K min^{-1}, and are rounded to within 0.5 K.
[b]DS ≡ 1,2-distearoyl-*sn*-glycero; PE ≡ PA(CH_2)$_2$$NH_3$ ≡ phosphorylethanolamine; PE(CH_3) ≡ phosphoryl-*N*-methylethanolamine; PA(CH_2)$_4$$NH_3$ ≡ phosphorylbutanolamine; PE(CH_3)$_3$ ≡ PA(CH_2)$_2$N(CH_3)$_3$ ≡ PC ≡ phosphorylcholine; PA(CH_2)$_4$N(CH_3)$_3$ ≡ phosphoryl-trimethyl-butanolamine; PS ≡ phosphorylserine; PS(CH_3) ≡ phosphorylserine methyl-ester. PA(CH_2)COH)H ≡ phosphorylethyleneglycol; PA(CH_2)(COH)$_2$H ≡ PG ≡ phosphorylglycerol; PA(CH_2)(COH)$_3$H ≡ phosphorylerithriol; PA ≡ phosphoric acid; PA(CH_2)H ≡ phosphoric acid methyl-ester; PA(CH_2)$_2$H ≡ phosphoric acid ethyl-ester; PA(CH_2)$_4$H ≡ phosphoric acid butyl-ester. For similar data pertaining to other lipids see [7,26,55].
[c]The relatively low transition temperature may be due to the tendency of this lipid to form systems with titled hydrocarbon chains at low pH.
[d]It is impossible to achieve complete protonation of PC, event at such low pH, in contrast to PE.
Source: G. Cevc, *J. de Physique 50*:1117–1134 (1989).

Table B19 Effect of Lipid Chain Length, Head Group Protonation, or Methylation on the Chain Melting Phase Transition Temperature (°C) of Lipid Bilayers

	pH				pH		
Lipid[a]	0	8	13	Lipid[a]	0	8	13
DMPE[b]	54	49.5	24	DTPE	59	55	29
DMPE(CH_3)	48	42.5	25	DTPE(CH_3)	53	47	30
DMPE(CH_3)$_2$	42	31	25	DTPE(CH_3)$_2$	46	36	30
DMPC	36	23	23	DTPC	39	28.5	28.5
DPPE	67	63.5	42	DHPE	71.5	68.5	44
DPPE(CH_3)	61.5	58	43	DHPE(CH_3)	66	61.5	45
DPPE(CH_3)$_2$	56	48	43	DHPE(CH_3)$_2$	60.5	50.5	45
DPPC	50	42	42	DHPC	54	43.5	43.5

[a]DM = 1,2-dimyristoyl-*sn*-glycero; DT = 1,2-ditetradecyl-*rac*-glycero; DP = 1,2-dipalmitoyl-*sn*-glycero; DH = 1,2-dihexadecyl-*rac*-glycero; PE ≡ P(CH_2)$_2$$NH_3$ ≡ phosphatidylethanolamine; PE(CH_3) ≡ phosphoryl-*N*-methylethanolamine; PE(CH_3)$_2$ ≡ phosphoryl-*N,N*-dimethylethanolamine; PE(CH_3)$_3$ ≡ P(CH_2)$_2$N(CH_3)$_3$ ≡ PC ≡ phosphatidylcholine. [b]Some related data concerning methylated PE are also found in works by Mulukutla and Shipley (1984), Seddon et al. (1983a), and Vaughan and Keough (1974).
All lipids are nearly completely protonated at pH 0 and maximally deprotonated at pH 13.
Source: G. Cevc, *Biochemistry 26*:6305–6310 (1987).

Table B19 Continued

Effect of Lipid Ionization[a] and Chain Length on the Bilayer Chain Melting Phase Transition Temperature (°C)

Lipid[b]	pH				Lipid[b]	pH			
	0[d]	~3	8	13		0[d]	~3	8	13
DMPG	42		24	24	DTPG	46		28	28
DMPS	52	44	36	(15)	DTPS	56	48	41	19.5
DMPA[c]	45	55	50	28	DTPA	48	59	55	34
$DMPA(CH_3)$	48	47	32	32	$DTPA(CH_3)$	52	51	37	37
DPPG	58		42	42	DHPG	62		46	46
DPPS	68.5	61.5	54	32	DHPS	72	64	56	34
DPPA[c]	62	74	68	45	DHPA	62	76	71	50
$DPPA(CH_3)$	63	62	48	48	$DHPA(CH_3)$	66	65	52	52

[a]Ionization states are at pH 0 for PG, $PA(CH_3)$, PA, and PS^+, at pH 3–4 for $PA(CH_3)$, $PA^{0.5-}$, and PS^-, at pH 8 for PG^-, PS^-, $PA(CH_3)^-$, and PA^-, and at pH 13 for PG^-, PS^{2-}, $PA(CH_3)^-$, and PA^{2-}. [b]DM = 1,2-dimyristoyl-*sn*-glycero; DT = 1,2-ditetradecyl-*rac*-glycero; DP = 1,2-dipalmitoyl-*rac*-glycero; DH = 1,2-dihexadecyl-*rac*-glycero; PG = phosphatidylglycerol; PA = phosphatidic acid; $PA(CH_3)$ = phosphatidic acid methyl ester; PS = phosphatidylserine. [c]For the data concerning PA, see also papers by Eibl and Blume (1979) and Blume and Eibl (1979). [d]Occasionally the transition temperature of the samples that have been stored for longer periods of time or prepared in highly concentrated salt solutions is 3–5 K higher. This indicates the possibility that lipid under such conditions has reverted into another state, which may imply that values given here are characteristic of the bilayers with tilted chains.

Source: G. Cevc, *Biochemistry 26*:6305–6310 (1987).

Table B20 Thermal Volume Expansion Coefficients (μL/g K) of the Diacyl-Phosphatidylcholines and Phosphatidylethanolamines in Excess Water

Phase	Diacyl-PC	Diacyl-PE	Dialkyl-PC	Dialkyl-PE	Long alkanes
L_c or L_c'	0.9 ± 0.2		0.9	0.8 ± 0.1	0.5
L_β or L_β' or $L_{\beta i}$	0.8 ± 0.1	1 ± 0.1	0.8	1	
P_β'	1 ± 0.2	—[a]	2	—	
L_α	<1 ± 0.1[b]	1.2 ± 0.2	1	1	2

[a]Does not exist for this class of lipids.

[b]This expansivity coefficient appears to decrease nonlinearly with temperature; at T_m + 30 degrees it is only 50% of the given value.

Appendix C: Mechanical, Solubility, and Related Parameters of Phospholipids

Gregor Cevc *Technical University of Munich, Munich, Germany*

Table C1 Some Elasto-Mechanical Properties of Lipid Bilayers

Lipid composition (Phase)	K_B (N m^{-2})	K_A[a] (N m^{-1})	K_t (N m^{-2})	K_c (N m)	$\bar{K}_{c,app}$ (N m)
DLPC (L_α)			$(1.9 \pm 1.0) \cdot 10^{7}$		
DMPC (L_β)		0.85 ± 0.14			
DMPC (P_β)		0.065 ± 0.005			
DMPC (L_α)		0.145 ± 0.01		$(1.15 \pm 0.15) \cdot 10^{-19}$[b]	
				$(5.6 \pm 0.6) \cdot 10^{-19}$[b]	
DMPC, 12.5% Chol (15°C)		0.4			
DMPC, 20% Chol (30°C)				$(2.1 \pm 0.25) \cdot 10^{-19}$	
DMPC, 30% Chol (30°C)				$(4.0 \pm 0.8) \cdot 10^{-19}$	
DMPC, 33% Chol (15°C)		0.65			
DMPC, 33% Chol (25°C)		0.56			
DMPC, 40% Chol (35°C)		0.60			
DMPC, 50% Chol (25°C)		0.63			
DAPC (18°C, L_α)		0.135 ± 0.02		$(4.4 \pm 0.5) \cdot 10^{-19}$	
SOPC (18°C, L_α)		0.19 ± 0.01		$(9.0 \pm 0.6) \cdot 10^{-19}$	
SOPC and Chol (15°C)		0.64 ± 0.032		$(24.6 \pm 3.9) \cdot 10^{-19}$	
Egg-yolk PC (L_α)	$(1 \cdots 3) \cdot 10^{9}$	0.14	$1.1 \cdot 10^{8}$	$(2.3 \pm 0.3) \cdot 10^{-19}$	$-(1.9 \pm 0.3) \cdot 10^{-19}$
E. coli lipids		0.24			
DOPS		0.02			
DOPE		0.32			
DMPC with a bola-lipid				$\sim 2 \cdot 10^{-21}$	
DPPC in 1.5 M ethanol (L_β)				$\sim 5 \cdot 10^{-21}$	
PC/sodium cholate 4/1				10^{-22}	
Galactosylglycerol				$(1.5 \cdots 4) \cdot 10^{-20}$	

Lifetime (L_α)	
Local fluctuations ($q \to \infty$)	$\sim 10^{-6}$ s
Correlation time for $q = 4$	0.40 s
Correlation time for $q = 3$	0.68 s
Correlation time for $q = 2$	2.55 s
Long wave-length undulations	≥ 1 s

DMPC: Dimyristoylphosphatidylcholine; DPPC: Dipalmitoylphosphatidylcholine; DAPC: Diarachidoylphosphatidylcholine; SOPC: Stearoyloleoylphosphatidylcholine; DOPS: Dioleoylphosphatidylserine; DOPE: Dioleoylphosphatidylethanolamine; PC: Phosphatidylcholine; Chol: cholesterol. $K_B = V(\partial p/\partial V)_T$, bulk compressibility modulus; $K_A = A(\partial T/\partial A)_T$, area compressibility modulus, where T is an isotropic tension. According to Evans and Rawicz (*Phys. Rev. Lett. 64*: 2094 (1990)), microscopic fluctuations lead to a renormalization of the elastic compressibility modulus. The 'apparent' surface expansion modulus is approximated by: $\bar{K} \simeq K_A(1 + K_A kT/8\pi K_c\bar{\tau}$, where $\bar{\tau}$ is the (time-averaged) membrane tension. $K_t = d_b(\partial\sigma_t/\partial d_b)_{T,V} \simeq K_A/d_b$, isothermal thickness compressibility, where σ_t is normal stress applied to the bilayer surface and d_b is bilayer thickness. High limit stems from electrocompressibility measurements; the lower from the analysis of thermal fluctuations in DLPC (Wack & Webb, *Phys. Rev. 40A*: 1627 (1989)). $K_c = (\partial M/\partial(1/R))_T$ (also sometimes written as B or κ or k_c) bending stiffnes or elastic bending curvature modulus, where M is a bending moment acting on the bilayer edge and R is vesicle radius. The precise value of K_c may increase with decreasing vesicle radius (cf. Helfrich, *J. Physique 47*: 321 (1986). $K_{c,app}$, apparent Gaussian curvature modulus (from Lorenzen et al., *Biophys. J. 50*: 565 (1986)). Data for the correlation times (q: wave-vector) from Duwe et al., *J. Phys. (France) 51*: 945 (1990); estimates from Data for egg-yolk PC from: Servuss et al., *Biochem. Biophys. Acta 436*: 900 (1976); Kwok & Evans *Biophys. J. 21*: 637 (1981); Data for DMPC from: Needham et al., *Biochem. 227*: 4668 (1988); Evans & Kwok, *Biochem. 21*: 4874 (1982); Duwe et al. *op. cit.* Most of the data for area compressibility moduli are taken from the obovecited works of Evans and colleagues, except for those pertaining to DOPS, DOPE and E. coli lipids, which stem from Haines et al., *Biochem. 26*: 5439 (1887) ($K_A \equiv M/d_b$). Data for bola lipids from Duwe et al, *op. cit.*

[a]Apparent membrane rigidity may vary with the size of lipid vesicles and with the solute concentration (Tarlok et al., *Biochem. 25*: 8220 (1985). Theoretically one would expect a logarithmic size dependence (Helfrich, *J. Phys. (France) 47*: 321 (1986)).

[b]Small values are typically obtained in the vesicle-flicker experiments; high values are measured by the pipette aspiration technique; intermediate results $(2.3 \pm 0.3) \times 10^{-19}$ have been deduced from the microscopic study of thermally fluctuating bilayer tubes.

I. CRITICAL CONCENTRATION FOR THE SELF-AGGREGATION OF VARIOUS PHOSPHOLIPIDS AS A FUNCTION OF POLAR HEAD GROUP AND HYDROCARBON CHAIN LENGTH

$$\ln[\mathrm{CMC}] = \Delta G_{tr}^{pol}/RT - 0.4 - 1.7\, n_{CH}^{s} - 1.1(n_{CH}^{l} - n_{CH}^{s}) + 1.2\, n_{unsat}^{0} - 0.4(n_{unsat}^{0} - 1) + 1.6\, n_{unsat}^{n} - 0.5(n_{unsat}^{n} - 1) \quad (1)$$

where n_{CH}^{l} and n_{CH}^{s} are the numbers of aliphatic C-atoms in the longer and shorter chains, respectively, and n_{unsat}^{0} and n_{unsat}^{n} are the numbers of double bonds in the overlapping and nonoverlapping regions of the chains, respectivley. Conversion from mole fraction units to molar concentrations can be made by adding 4.0 to the right-hand side of Eq. 1

Table C2 Free Energies of Monomer-Micelle Transfer ($\Delta G_{tr}^{pol}/RT$) Relative to PC (in the Absence of Salt) for Different *sn*-2 Spin Labelled Phospholipids with Fixed Chain Composition in NaCl at 20°C and pH 7, Except for PA^{2-} and PAH^{-}, Which Represent Phosphatidic Acid at pH 8 and pH 5, Respectively

[NaCl] (M)	PA^{2-}	PA^{-}	PS^{-}	PG^{-}	PC	PE
0.0	—	—	+2.7	+2.2	0.0	–0.7
0.15	+1.9	+0.1	+.07	+0.4	–0.2	–0.9
0.5	+0.7	–0.5	–0.2	–0.3	–0.3	–0.9
1.0	+0.1	–1.1	–0.9	–0.8	–0.8	–1.1
2.0	–1.0	–2.0	–1.7	–1.8	–1.3	–2.0

Table C3 Experimentally Determined CMCs Compared with the Predictions from Eq. (1), for 1-Acyl-2-Acetyl- and 1-Palmitoyl-2-Acyl-*sn*-Glycero-3-Phosphocholines

Chains	[CMC] (mol liter^{-1}) Expt.	[CMC] (mol liter^{-1}) Eqn. (1)
1-Acyl-2-acetyl		
1-(12:0)-2-(2:0)	$(1.1 \pm 0.25) \cdot 10^{-4}$	$1.1 \cdot 10^{-4}$
1-(14:0)-2-(2:0)	$(1.1 \pm 0.12) \cdot 10^{-5}$	$1.2 \cdot 10^{-5}$
1-(16:0)-2-(2:0)	$(1.3 \pm 0.06) \cdot 10^{-6}$	$1.4 \cdot 10^{-6}$
1-(18:0)-2-(2:0)	$(2.2 \pm 0.1) \cdot 10^{-7}$	$1.5 \cdot 10^{-7}$
1-Palmitoyl-2-acyl		
1-(16:0)-2-(2:0)	$(1.3 \pm 0.06) \cdot 10^{-6}$	$1.4 \cdot 10^{-6}$
1-(16:0)-2-(3:0)	$(7.2 \pm 0.3) \cdot 10^{-7}$	$7.5 \cdot 10^{-7}$
1-(16:0)-2-(4:0)	$(4.1 \pm 0.2) \cdot 10^{-7}$	$4.1 \cdot 10^{-7}$
1-(16:0)-2-(6:0)	$(2.2 \pm 0.1) \cdot 10^{-7}$	$1.2 \cdot 10^{-7}$

CMCs determined from W. Kramp, G. Pieroni, R. N. Pinckard, and D. J. Hanahan, *Chem. Phys. Lipids 35*:49–62 (1984).

Source: D. Marsh and M. C. King. *Chem. Phys. Lipids 42*:271–277 (1986).

II. KINETIC DATA FOR THE TRANSFER OF CHARGED PHOSPHOLIPIDS ACROSS AN AQUEOUS SUBPHASE

Table C4 Chemical Structures and Transfer Rates of Anionic Dimyristoylphospholipids at Neutral pH

Part	—X	Half-time
A	—H	156 min
B	$-CH_2-CH_3$	270 min
	$-CH_2-CH_2-CH_3$	379 min
	$-CH_2-CH_2-CH_2-CH_3$	758 min
	$-CH_2-CH_2-CH_2-CH_2-CH_3$	> 24 h
	$-CH_2-CH_2-CH_2-CH_2-CH_2-CH_3$	> 24 h
C	$-CH_2-CH_2OH$	71 min
	$-CH_2-CHOH-CH_2OH$	41 min
	$-CH_2-CHOH-CHOH-CH_2OH$	47 min
D	$-CH_2-CH(NH_3^+)-COO^-$	41 min

$$
\begin{array}{l}
CH_3\text{-}(CH_2)_{12}\text{-}CO\text{-}O\text{-}CH_2 \\
\qquad\qquad\qquad\qquad\;| \\
CH_3\text{-}(CH_2)_{12}\text{-}CO\text{-}O\text{-}CH \qquad O^- \\
\qquad\qquad\qquad\qquad\;| \qquad\quad | \\
\qquad\qquad\qquad\qquad CH_2\text{-}O\text{-}P\text{-}O\text{-}X \\
\qquad\qquad\qquad\qquad\qquad\quad \| \\
\qquad\qquad\qquad\qquad\qquad\quad O
\end{array}
$$

Table C5 Kinetic Data for the Transfer of Negatively Charged Phospholipids Bearing the Same Total Number of Methyl(ene) Groups

t_∞ means the time at which equilibrium is reached.

Number of $-CH_2-(-CH_3)$ groups	Transferable phospholipid	Rate
30	DPPA	No transfer observed
	DMPButanol	$t_{1/2}$ = 758 min
28	$DC_{15:0}PA$	No transfer observed
	DMPEthanol	$t_{1/2}$ = 270 min
26	DMPA	$t_{1/2}$ = 156 min
	$DC_{13:0}PEthanol$	t_∞ < 10 min
	DLPButanol	t_∞ < 10 min

Experiments are performed at 33°C in 5 mM Tes (pH 7.0)/10 mM potassium chloride. The donors consist of DMPC-anionic phospholipid mixtures (molar ratio 9 : 1); the recipient vesicles are built up of DMPC. Donors and acceptors are mixed in equimolar amounts. Lipid concentration equals 2 $mg \cdot ml^{-1}$.

Source: M. Cuyper, and M. Joniau, Biochim. Biophys. Acta 814:374–380 (1985). The measured half-times of transfer may change by up to a factor of 2 as a function of the receptor vesicle composition (M. Cuyper, M. Joniau, J. B. F. N. Engberts, and E. Sudholter, Jr.), *Colloids and Surfaces 10:*313–319 (1984).

Table C6 Kinetic Data for the Phosphatidylcholine Transfer Across an Aqueous Subphase as a Function of the Hydrocarbon Chain Length

Rate coefficients for lipid transfer

Collision: Monomer:	k_2 k_1 (min^{-1})	σ k_{-1}/k_2	$(k_{-1} + k_2)k_1$ (M)[a]	CBC ≡ k_1/k_{-1} (M)[a]	k_{-1} (min^{-1} M^{-1})[a]	k_2 (min^{-1} M^{-1})[a]
C12:0/C12:0	$(5.6 \pm 1.5) \times 10^{-1}$	4.7 ± 3.4	0	3.4×10^{-7}[b]	$(1.5 \pm 0.4) \times 10^{6}$	$(4.1 \pm 2.3) \times 10^{5}$
C13:0/C13:0	$(4.0 \pm 0.7) \times 10^{-2}$	5.7 ± 4.2	0	6.6×10^{-8}[b]	$(6.0 \pm 1.1) \times 10^{1}$	$(1.5 \pm 1.3) \times 10^{4}$
C14:0/C14:0	$(9.3 \pm 3.9) \times 10^{-3}$	15.1 ± 5.6	$(5.8 \pm 2.4) \times 10^{-4}$	1.3×10^{-8}[b]	$(7.1 \pm 3.0) \times 10^{5}$	$(5.5 \pm 4.0) \times 10^{4}$
C15:0/C15:0	$(2.2 \pm 1.5) \times 10^{-4}$	*c*	*c*	2.4×10^{-9}[b]	$(9.2 \pm 6.3) \times 10^{4}$	*c*
C16:0/C16:0	$(0.3 - 2.7) \times 10^{-4}$[d]	*c*	*c*	4.7×10^{-10}[e]	$(0.7 - 5.7) \times 10^{5}$	*c*
Egg lyso-PC	>6	*c*	*c*	7×10^{-6}[f]	$>9 \times 10^{5}$	*c*

[a]Molar units are taken to mean moles per liter of supernatant. [b]Interpolated from critical micelle/bilayer concentrations reported for diheptanoyl-PC, dioctanoyl-PC, didecanoyl-PC, and DPPC [summarized in Tanford (1980)]. [c]Insufficient data. [d]Range calculated as described in the text. Other ranges denote standard deviations. [e]Reported by Smith and Tanford (1972). [f]Reported by Haberland and Reynolds (1975).

In collision models, k_2 represents the rate of material transfer between the surface-aggregated vesicles and k_1, k_{-1} the rates of lipid transport between the surface-attached and free vesicles and vice versa. In the monomer model k_1, k_{-1} represent the material flow between the bulk subphase and a free vesicle and k_2 the rate of transport from the bulk monomer-pool to the acceptor surface. CBC denotes critical bilayer concentration (corresponds to CMC for the single chain amphiphiles). σ is the ratio of total acceptor bilayer to vesicle area in each complex.

Source: J. E. Ferrell, K.-J. Lee, and W. H. Huestis, *Biochemistry 24*:2857–2864 (1985). By the method used to get the data from previous table, the half-time of dimyrystoylphosphatidylcholine (C14:0/C14:0) transfer has been measured to be 105 ± 25 min.

Index